GØRDIAN®

Electrical Costs with RSMeans data

Michael Ouillette, Senior Editor

2019
42nd annual edition

Chief Data Officer
Noam Reininger

Engineering Director
Bob Mewis (1, 3, 4, 5, 11, 12)

Contributing Editors
Brian Adams (21, 22)
Christopher Babbitt
Sam Babbitt
Michelle Curran
Matthew Doheny (8, 9, 10)
John Gomes (13, 41)
Derrick Hale, PE (2, 31, 32, 33, 34, 35, 44, 46)
Michael Henry

Joseph Kelble (14, 23, 25,)
Charles Kibbee
Gerard Lafond, PE
Thomas Lane (6, 7)
Jake MacDonald
Elisa Mello
Michael Ouillette (26, 27, 28, 48)
Gabe Sirota
Matthew Sorrentino
Kevin Souza
David Yazbek

Product Manager
Andrea Sillah

Production Manager
Debbie Panarelli

Production
Jonathan Forgit
Mary Lou Geary
Sharon Larsen
Sheryl Rose

Data Quality Manager
Joseph Ingargiola

Technical Support
Kedar Gaikwad
Todd Klapprodt
John Liu

Cover Design
Blaire Collins

Data Analytics
Tim Duggan
Todd Glowac
Matthew Kelliher-Gibson

Numbers in italics are the divisional responsibilities for each editor. Please contact the designated editor directly with any questions.

RSMeans data from Gordian
Construction Publishers & Consultants
1099 Hingham Street, Suite 201
Rockland, MA 02370
United States of America
1.800.448.8182
RSMeans.com

Printed in the United States of America
ISSN 1540-6318
ISBN 978-1-946872-54-8

| 0039 | $293.99 per copy (in United States)
Price is subject to change without prior notice. |

Related Data and Services

Our engineers recommend the following products and services to complement *Electrical Costs with RSMeans data:*

Annual Cost Data Books

2019 Electrical Change Order Costs with RSMeans data
2019 Building Construction Costs with RSMeans data
2019 Mechanical Costs with RSMeans data

Reference Books

Electrical Estimating Methods
Estimating Building Costs
RSMeans Estimating Handbook
Green Building: Project Planning & Estimating
How to Estimate with RSMeans data
Plan Reading & Material Takeoff
Project Scheduling & Management for Construction

Seminars and In-House Training

Unit Price Estimating
Training for our online estimation solution
Practical Project Management for Construction Professionals
Scheduling with MSProject for Construction Professionals
Mechanical & Electrical Estimating

RSMeans data Online

For access to the latest cost data, an intuitive search, and an easy-to-use estimate builder, take advantage of the time savings available from our online application. To learn more visit: RSMeans.com/2019online.

Enterprise Solutions

Building owners, facility managers, building product manufacturers, and attorneys across the public and private sectors engage with RSMeans data Enterprise to solve unique challenges where trusted construction cost data is critical. To learn more visit: RSMeans.com/Enterprise.

Custom Built Data Sets

Building and Space Models: Quickly plan construction costs across multiple locations based on geography, project size, building system component, product options, and other variables for precise budgeting and cost control.

Predictive Analytics: Accurately plan future builds with custom graphical interactive dashboards, negotiate future costs of tenant build-outs, and identify and compare national account pricing.

Consulting

Building Product Manufacturing Analytics: Validate your claims and assist with new product launches.

Third-Party Legal Resources: Used in cases of construction cost or estimate disputes, construction product failure vs. installation failure, eminent domain, class action construction product liability, and more.

API

For resellers or internal application integration, RSMeans data is offered via API. Deliver Unit, Assembly, and Square Foot Model data within your interface. To learn more about how you can provide your customers with the latest in localized construction cost data visit: RSMeans.com/API.

Table of Contents

Foreword

The Value of RSMeans data from Gordian

Since 1942, RSMeans data has been the industry-standard materials, labor, and equipment cost information database for contractors, facility owners and managers, architects, engineers, and anyone else that requires the latest localized construction cost information. More than 75 years later, the objective remains the same: to provide facility and construction professionals with the most current and comprehensive construction cost database possible.

With the constant influx of new construction methods and materials, in addition to ever-changing labor and material costs, last year's cost data is not reliable for today's designs, estimates, or budgets. Gordian's cost engineers apply real-world construction experience to identify and quantify new building products and methodologies, adjust productivity rates, and adjust costs to local market conditions across the nation. This adds up to more than 22,000 hours in cost research annually. This unparalleled construction cost expertise is why so many facility and construction professionals rely on RSMeans data year over year.

About Gordian

Gordian originated in the spirit of innovation and a strong commitment to helping clients reach and exceed their construction goals. In 1982, Gordian's chairman and founder, Harry H. Mellon, created Job Order Contracting while serving as chief engineer at the Supreme Headquarters Allied Powers Europe. Job Order Contracting is a unique indefinite delivery/indefinite quantity (IDIQ) process, which enables facility owners to complete a substantial number of repair, maintenance, and construction projects with a single, competitively awarded contract. Realizing facility and infrastructure owners across various industries could greatly benefit from the time and cost saving advantages of this innovative construction procurement solution, he established Gordian in 1990.

Continuing the commitment to providing the most relevant and accurate facility and construction data, software, and expertise in the industry, Gordian enhanced the fortitude of its data with the acquisition of RSMeans in 2014. And in an effort to expand its facility management capabilities, Gordian acquired Sightlines, the leading provider of facilities benchmarking data and analysis, in 2015.

Our Offerings

Gordian is the leader in facility and construction cost data, software, and expertise for all phases of the building life cycle. From planning to design, procurement, construction, and operations, Gordian's solutions help clients maximize efficiency, optimize cost savings, and increase building quality with its highly specialized data engineers, software, and unique proprietary data sets.

Our Commitment

At Gordian, we do more than talk about the quality of our data and the usefulness of its application. We stand behind all of our RSMeans data—from historical cost indexes to construction materials and techniques—to craft current costs and predict future trends. If you have any questions about our products or services, please call us toll-free at 800.448.8182 or visit our website at gordian.com.

How the Cost Data Is Built: An Overview

Unit Prices*
All cost data have been divided into 50 divisions according to the MasterFormat® system of classification and numbering.

Assemblies*
The cost data in this section have been organized in an "Assemblies" format. These assemblies are the functional elements of a building and are arranged according to the 7 elements of the UNIFORMAT II classification system. For a complete explanation of a typical "Assembly", see "RSMeans data: Assemblies—How They Work."

*Residential Models**
Model buildings for four classes of construction—economy, average, custom, and luxury—are developed and shown with complete costs per square foot.

*Commercial/Industrial/Institutional Models**
This section contains complete costs for 77 typical model buildings expressed as costs per square foot.

*Green Commercial/Industrial/Institutional Models**
This section contains complete costs for 25 green model buildings expressed as costs per square foot.

*References**
This section includes information on Equipment Rental Costs, Crew Listings, Historical Cost Indexes, City Cost Indexes, Location Factors, Reference Tables, and Change Orders, as well as a listing of abbreviations.

- **Equipment Rental Costs:** Included are the average costs to rent and operate hundreds of pieces of construction equipment.
- **Crew Listings:** This section lists all the crews referenced in the cost data. A crew is composed of more than one trade classification and/or the addition of power equipment to any trade classification. Power equipment is included in the cost of the crew. Costs are shown both with bare labor rates and with the installing contractor's overhead and profit added. For each, the total crew cost per eight-hour day and the composite cost per labor-hour are listed.

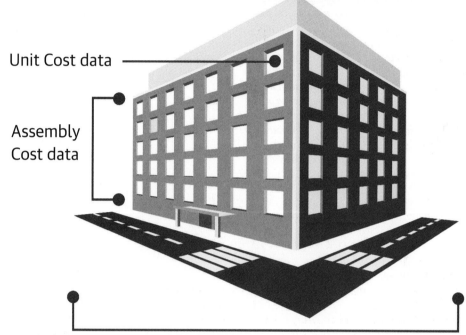

Unit Cost data

Assembly Cost data

Square Foot Models

- **Historical Cost Indexes:** These indexes provide you with data to adjust construction costs over time.
- **City Cost Indexes:** All costs in this data set are U.S. national averages. Costs vary by region. You can adjust for this by CSI Division to over 730 cities in 900+ 3-digit zip codes throughout the U.S. and Canada by using this data.
- **Location Factors:** You can adjust total project costs to over 730 cities in 900+ 3-digit zip codes throughout the U.S. and Canada by using the weighted number, which applies across all divisions.
- **Reference Tables:** At the beginning of selected major classifications in the Unit Prices are reference numbers indicators. These numbers refer you to related information in the Reference Section. In this section, you'll find reference tables, explanations, and estimating information that support how we develop the unit price data, technical data, and estimating procedures.
- **Change Orders:** This section includes information on the factors that influence the pricing of change orders.

- **Abbreviations:** A listing of abbreviations used throughout this information, along with the terms they represent, is included.

Index (printed versions only)
A comprehensive listing of all terms and subjects will help you quickly find what you need when you are not sure where it occurs in MasterFormat®.

Conclusion
This information is designed to be as comprehensive and easy to use as possible.

The Construction Specifications Institute (CSI) and Construction Specifications Canada (CSC) have produced the 2016 edition of MasterFormat®, a system of titles and numbers used extensively to organize construction information.

All unit prices in the RSMeans cost data are now arranged in the 50-division MasterFormat® 2016 system.

---

*   Not all information is available in all data sets

Note: The material prices in RSMeans cost data are "contractor's prices." They are the prices that contractors can expect to pay at the lumberyards, suppliers'/distributors' warehouses, etc. Small orders of specialty items would be higher than the costs shown, while very large orders, such as truckload lots, would be less. The variation would depend on the size, timing, and negotiating power of the contractor. The labor costs are primarily for new construction or major renovation rather than repairs or minor alterations. With reasonable exercise of judgment, the figures can be used for any building work.

Estimating with RSMeans data: Unit Prices

Following these steps will allow you to complete an accurate estimate using RSMeans data Unit Prices.

1. Scope Out the Project

- Think through the project and identify the CSI divisions needed in your estimate.
- Identify the individual work tasks that will need to be covered in your estimate.
- The Unit Price data have been divided into 50 divisions according to CSI MasterFormat® 2016.
- In printed versions, the Unit Price Section Table of Contents on page 1 may also be helpful when scoping out your project.
- Experienced estimators find it helpful to begin with Division 2 and continue through completion. Division 1 can be estimated after the full project scope is known.

2. Quantify

- Determine the number of units required for each work task that you identified.
- Experienced estimators include an allowance for waste in their quantities. (Waste is not included in our Unit Price line items unless otherwise stated.)

3. Price the Quantities

- Use the search tools available to locate individual Unit Price line items for your estimate.
- Reference Numbers indicated within a Unit Price section refer to additional information that you may find useful.
- The crew indicates who is performing the work for that task. Crew codes are expanded in the Crew Listings in the Reference Section to include all trades and equipment that comprise the crew.
- The Daily Output is the amount of work the crew is expected to complete in one day.
- The Labor-Hours value is the amount of time it will take for the crew to install one unit of work.
- The abbreviated Unit designation indicates the unit of measure upon which the crew, productivity, and prices are based.
- Bare Costs are shown for materials, labor, and equipment needed to complete the Unit Price line item. Bare costs do not include waste, project overhead, payroll insurance, payroll taxes, main office overhead, or profit.
- The Total Incl O&P cost is the billing rate or invoice amount of the installing contractor or subcontractor who performs the work for the Unit Price line item.

4. Multiply

- Multiply the total number of units needed for your project by the Total Incl O&P cost for each Unit Price line item.
- Be careful that your take off unit of measure matches the unit of measure in the Unit column.
- The price you calculate is an estimate for a completed item of work.
- Keep scoping individual tasks, determining the number of units required for those tasks, matching each task with individual Unit Price line items, and multiplying quantities by Total Incl O&P costs.
- An estimate completed in this manner is priced as if a subcontractor, or set of subcontractors, is performing the work. The estimate does not yet include Project Overhead or Estimate Summary components such as general contractor markups on subcontracted work, general contractor office overhead and profit, contingency, and location factors.

5. Project Overhead

- Include project overhead items from Division 1-General Requirements.
- These items are needed to make the job run. They are typically, but not always, provided by the general contractor. Items include, but are not limited to, field personnel, insurance, performance bond, permits, testing, temporary utilities, field office and storage facilities, temporary scaffolding and platforms, equipment mobilization and demobilization, temporary roads and sidewalks, winter protection, temporary barricades and fencing, temporary security, temporary signs, field engineering and layout, final cleaning, and commissioning.
- Each item should be quantified and matched to individual Unit Price line items in Division 1, then priced and added to your estimate.
- An alternate method of estimating project overhead costs is to apply a percentage of the total project cost—usually 5% to 15% with an average of 10% (see General Conditions).
- Include other project related expenses in your estimate such as:
  - Rented equipment not itemized in the Crew Listings
  - Rubbish handling throughout the project (see section 02 41 19.19)

6. Estimate Summary

- Include sales tax as required by laws of your state or county.
- Include the general contractor's markup on self-performed work, usually 5% to 15% with an average of 10%.
- Include the general contractor's markup on subcontracted work, usually 5% to 15% with an average of 10%.
- Include the general contractor's main office overhead and profit:
  - RSMeans data provides general guidelines on the general contractor's main office overhead (see section 01 31 13.60 and Reference Number R013113-50).
  - Markups will depend on the size of the general contractor's operations, projected annual revenue, the level of risk, and the level of competition in the local area and for this project in particular.
- Include a contingency, usually 3% to 5%, if appropriate.
- Adjust your estimate to the project's location by using the City Cost Indexes or the Location Factors in the Reference Section:
  - Look at the rules in "How to Use the City Cost Indexes" to see how to apply the Indexes for your location.
  - When the proper Index or Factor has been identified for the project's location, convert it to a multiplier by dividing it by 100, then multiply that multiplier by your estimated total cost. The original estimated total cost will now be adjusted up or down from the national average to a total that is appropriate for your location.

Editors' Note:
We urge you to spend time reading and understanding the supporting material. An accurate estimate requires experience, knowledge, and careful calculation. The more you know about how we at RSMeans developed the data, the more accurate your estimate will be. In addition, it is important to take into consideration the reference material such as Equipment Listings, Crew Listings, City Cost Indexes, Location Factors, and Reference Tables.

How to Use the Cost Data: The Details

What's Behind the Numbers? The Development of Cost Data

RSMeans data engineers continually monitor developments in the construction industry in order to ensure reliable, thorough, and up-to-date cost information. While overall construction costs may vary relative to general economic conditions, price fluctuations within the industry are dependent upon many factors. Individual price variations may, in fact, be opposite to overall economic trends. Therefore, costs are constantly tracked and complete updates are performed yearly. Also, new items are frequently added in response to changes in materials and methods.

Costs in U.S. Dollars

All costs represent U.S. national averages and are given in U.S. dollars. The City Cost Index (CCI) with RSMeans data can be used to adjust costs to a particular location. The CCI for Canada can be used to adjust U.S. national averages to local costs in Canadian dollars. No exchange rate conversion is necessary because it has already been factored in.

G The processes or products identified by the green symbol in our publications have been determined to be environmentally responsible and/or resource-efficient solely by RSMeans data engineering staff. The inclusion of the green symbol does not represent compliance with any specific industry association or standard.

Material Costs

RSMeans data engineers contact manufacturers, dealers, distributors, and contractors all across the U.S. and Canada to determine national average material costs. If you have access to current material costs for your specific location, you may wish to make adjustments to reflect differences from the national average. Included within material costs are fasteners for a normal installation. RSMeans data engineers use manufacturers' recommendations, written specifications, and/or standard construction practices for the sizing and spacing of fasteners. Adjustments to material costs may be required for your specific application or location. The manufacturer's warranty is assumed. Extended warranties are not included in the material costs. **Material costs do not include sales tax.**

Labor Costs

Labor costs are based upon a mathematical average of trade-specific wages in 30 major U.S. cities. The type of wage (union, open shop, or residential) is identified on the inside back cover of printed publications or selected by the estimator when using the electronic products. Markups for the wages can also be found on the inside back cover of printed publications and/or under the labor references found in the electronic products.

- If wage rates in your area vary from those used, or if rate increases are expected within a given year, labor costs should be adjusted accordingly.

Labor costs reflect productivity based on actual working conditions. In addition to actual installation, these figures include time spent during a normal weekday on tasks, such as material receiving and handling, mobilization at the site, site movement, breaks, and cleanup.

Productivity data is developed over an extended period so as not to be influenced by abnormal variations and reflects a typical average.

Equipment Costs

Equipment costs include not only rental but also operating costs for equipment under normal use. The operating costs include parts and labor for routine servicing, such as the repair and replacement of pumps, filters, and worn lines. Normal operating expendables, such as fuel, lubricants, tires, and electricity (where applicable), are also included. Extraordinary operating expendables with highly variable wear patterns, such as diamond bits and blades, are excluded. These costs are included under materials. Equipment rental rates are obtained from industry sources throughout North America—contractors, suppliers, dealers, manufacturers, and distributors.

Rental rates can also be treated as reimbursement costs for contractor-owned equipment. Owned equipment costs include depreciation, loan payments, interest, taxes, insurance, storage, and major repairs.

Equipment costs do not include operators' wages.

Equipment Cost/Day—The cost of equipment required for each crew is included in the Crew Listings in the Reference Section (small tools that are considered essential everyday tools are not listed out separately). The Crew Listings itemize specialized tools and heavy equipment along with labor trades. The daily cost of itemized equipment included in a crew is based on dividing the weekly bare rental rate by 5 (number of working days per week), then adding the hourly operating cost times 8 (the number of hours per day). This Equipment Cost/Day is shown in the last column of the Equipment Rental Costs in the Reference Section.

Mobilization, Demobilization—The cost to move construction equipment from an equipment yard or rental company to the job site and back again is not included in equipment costs. Mobilization (to the site) and demobilization (from the site) costs can be found in the Unit Price Section. If a piece of equipment is already at the job site, it is not appropriate to utilize mobilization or demobilization costs again in an estimate.

Overhead and Profit

Total Cost including O&P for the installing contractor is shown in the last column of the Unit Price and/or Assemblies. This figure is the sum of the bare material cost plus 10% for profit, the bare labor cost plus total overhead and profit, and the bare equipment cost plus 10% for profit. Details for the calculation of overhead and profit on labor are shown on the inside back cover of the printed product and in the Reference Section of the electronic product.

General Conditions

Cost data in this data set are presented in two ways: Bare Costs and Total Cost including O&P (Overhead and Profit). General Conditions, or General Requirements, of the contract should also be added to the Total Cost including O&P when applicable. Costs for General Conditions are listed in Division 1 of the Unit Price Section and in the Reference Section.

General Conditions for the installing contractor may range from 0% to 10% of the Total Cost including O&P. For the general or prime contractor, costs for General Conditions may range from 5% to 15% of the Total Cost including O&P, with a figure of 10% as the most typical allowance. If applicable, the Assemblies and Models sections use costs that include the installing contractor's overhead and profit (O&P).

Factors Affecting Costs

Costs can vary depending upon a number of variables. Here's a listing of some factors that affect costs and points to consider.

Quality—The prices for materials and the workmanship upon which productivity is based represent sound construction work. They are also in line with industry standard and manufacturer specifications and are frequently used by federal, state, and local governments.

Overtime—We have made no allowance for overtime. If you anticipate premium time or work beyond normal working hours, be sure to make an appropriate adjustment to your labor costs.

Productivity—The productivity, daily output, and labor-hour figures for each line item are based on an eight-hour work day in daylight hours in moderate temperatures and up to a 14' working height unless otherwise indicated. For work that extends beyond normal work hours or is performed under adverse conditions, productivity may decrease.

Size of Project—The size, scope of work, and type of construction project will have a significant impact on cost. Economies of scale can reduce costs for large projects. Unit costs can often run higher for small projects.

Location—Material prices are for metropolitan areas. However, in dense urban areas, traffic and site storage limitations may increase costs. Beyond a 20-mile radius of metropolitan areas, extra trucking or transportation charges may also increase the material costs slightly. On the other hand, lower wage rates may be in effect. Be sure to consider both of these factors when preparing an estimate, particularly if the job site is located in a central city or remote rural location. In addition, highly specialized subcontract items may require travel and per-diem expenses for mechanics.

Other Factors—

- season of year
- contractor management
- weather conditions
- local union restrictions
- building code requirements
- availability of:
  - adequate energy
  - skilled labor
  - building materials
- owner's special requirements/restrictions
- safety requirements
- environmental considerations
- access

Unpredictable Factors—General business conditions influence "in-place" costs of all items. Substitute materials and construction methods may have to be employed. These may affect the installed cost and/or life cycle costs. Such factors may be difficult to evaluate and cannot necessarily be predicted on the basis of the job's location in a particular section of the country. Thus, where these factors apply, you may find significant but unavoidable cost variations for which you will have to apply a measure of judgment to your estimate.

Rounding of Costs

In printed publications only, all unit prices in excess of $5.00 have been rounded to make them easier to use and still maintain adequate precision of the results.

How Subcontracted Items Affect Costs

A considerable portion of all large construction jobs is usually subcontracted. In fact, the percentage done by subcontractors is constantly increasing and may run over 90%. Since the workers employed by these companies do nothing else but install their particular products, they soon become experts in that line. As a result, installation by these firms is accomplished so efficiently that the total in-place cost, even with the general contractor's overhead and profit, is no more, and often less, than if the principal contractor had handled the installation. Companies that deal with construction specialties are anxious to have their products perform well and, consequently, the installation will be the best possible.

Contingencies

The allowance for contingencies generally provides for unforeseen construction difficulties. On alterations or repair jobs, 20% is not too much. If drawings are final and only field contingencies are being considered, 2% or 3% is probably sufficient and often nothing needs to be added. Contractually, changes in plans will be covered by extras. The contractor should consider inflationary price trends and possible material shortages during the course of the job. These escalation factors are dependent upon both economic conditions and the anticipated time between the estimate and actual construction. If drawings are not complete or approved, or a budget cost is wanted, it is wise to add 5% to 10%. Contingencies, then, are a matter of judgment.

Important Estimating Considerations

The productivity, or daily output, of each craftsman or crew assumes a well-managed job where tradesmen with the proper tools and equipment, along with the appropriate construction materials, are present. Included are daily set-up and cleanup time, break time, and plan layout time. Unless otherwise indicated, time for material movement on site (for items

that can be transported by hand) of up to 200'
into the building and to the first or second floor
is also included. If material has to be transported
by other means, over greater distances, or to
higher floors, an additional allowance should be
considered by the estimator.

While horizontal movement is typically a
sole function of distances, vertical transport
introduces other variables that can significantly
impact productivity. In an occupied building,
the use of elevators (assuming access, size, and
required protective measures are acceptable)
must be understood at the time of the estimate.
For new construction, hoist wait and cycle
times can easily be 15 minutes and may result in
scheduled access extending beyond the normal
work day. Finally, all vertical transport will
impose strict weight limits likely to preclude the
use of any motorized material handling.

The productivity, or daily output, also assumes
installation that meets manufacturer/designer/
standard specifications. A time allowance for
quality control checks, minor adjustments, and
any task required to ensure proper function
or operation is also included. For items that
require connections to services, time is included
for positioning, leveling, securing the unit,
and making all the necessary connections
(and start up where applicable) to ensure a
complete installation. Estimating of the services
themselves (electrical, plumbing, water, steam,
hydraulics, dust collection, etc.) is separate.

In some cases, the estimator must consider the
use of a crane and an appropriate crew for the
installation of large or heavy items. For those
situations where a crane is not included in the
assigned crew and as part of the line item cost,
then equipment rental costs, mobilization and
demobilization costs, and operator and support
personnel costs must be considered.

Labor-Hours

The labor-hours expressed in this publication are
derived by dividing the total daily labor-hours for
the crew by the daily output. Based on average
installation time and the assumptions listed
above, the labor-hours include: direct labor,
indirect labor, and nonproductive time. A typical
day for a craftsman might include but is not
limited to:

- Direct Work
  - □ Measuring and layout
  - □ Preparing materials
  - □ Actual installation
  - □ Quality assurance/quality control
- Indirect Work
  - □ Reading plans or specifications
  - □ Preparing space
  - □ Receiving materials
  - □ Material movement
  - □ Giving or receiving instruction
  - □ Miscellaneous
- Non-Work
  - □ Chatting
  - □ Personal issues
  - □ Breaks
  - □ Interruptions (i.e., sickness, weather,
    material or equipment shortages, etc.)

If any of the items for a typical day do not apply
to the particular work or project situation,
the estimator should make any necessary
adjustments.

Final Checklist

Estimating can be a straightforward process
provided you remember the basics. Here's
a checklist of some of the steps you should
remember to complete before finalizing your
estimate.

Did you remember to:

- factor in the City Cost Index for your locale?
- take into consideration which items have
  been marked up and by how much?
- mark up the entire estimate sufficiently for
  your purposes?
- read the background information on
  techniques and technical matters that could
  impact your project time span and cost?
- include all components of your project in the
  final estimate?
- double check your figures for accuracy?
- call RSMeans data engineers if you have any
  questions about your estimate or the data
  you've used? Remember, Gordian stands
  behind all of our products, including our
  extensive RSMeans data solutions. If you have
  any questions about your estimate, about
  the costs you've used from our data, or even
  about the technical aspects of the job that
  may affect your estimate, feel free to call the
  Gordian RSMeans editors at 1.800.448.8182.

Included with Purchase
RSMeans Data Online
Free Trial

rsmeans.com/2019freetrial

Unit Price Section

Table of Contents

RSMeans data: Unit Prices— How They Work

All RSMeans data: Unit Prices are organized in the same way.

03 30 Cast-In-Place Concrete

03 30 53 – Miscellaneous Cast-In-Place Concrete

03 30 53.40 Concrete In Place	Crew	Daily Output	Labor-Hours	Unit	Material	2019 Bare Costs Labor	Equipment	Total	Total Incl O&P
0010 **CONCRETE IN PLACE**									
0020 Including forms (4 uses), Grade 60 rebar, concrete (Portland cement									
0050 Type I), placement and finishing unless otherwise indicated									
3540 Equipment pad (3000 psi), 3' x 3' x 6" thick	C-14H	45	1.067	Ea.	46.50	53.50	.59	100.59	133
3550 4' x 4' x 6" thick		30	1.600		72.50	80.50	.88	153.88	201
3560 5' x 5' x 8" thick		18	2.667		132	134	1.47	267.47	350
3570 6' x 6' x 8" thick		14	3.429		181	172	1.89	354.89	460
3580 8' x 8' x 10" thick		8	6		385	300	3.30	688.30	880
3590 10' x 10' x 12" thick		5	9.600		665	485	5.30	1,155.30	1,475
3800 Footings (3000 psi), spread under 1 C.Y.	C-14C	28	4	C.Y.	193	196	.96	389.96	510
3825 1 C.Y. to 5 C.Y.		43	2.605		227	128	.63	355.63	445
3850 Over 5 C.Y.		75	1.493		212	73	.36	285.36	345
3900 Footings, strip (3000 psi), 18" x 9", unreinforced	C-14L	40	2.400		149	115	.67	264.67	340
3920 18" x 9", reinforced	C-14C	35	3.200		174	157	.77	331.77	430

It is important to understand the structure of RSMeans data: Unit Prices so that you can find information easily and use it correctly.

---

① Line Numbers
Line Numbers consist of 12 characters, which identify a unique location in the database for each task. The first 6 or 8 digits conform to the Construction Specifications Institute MasterFormat® 2016. The remainder of the digits are a further breakdown in order to arrange items in understandable groups of similar tasks. Line numbers are consistent across all of our publications, so a line number in any of our products will always refer to the same item of work.

② Descriptions
Descriptions are shown in a hierarchical structure to make them readable. In order to read a complete description, read up through the indents to the top of the section. Include everything that is above and to the left that is not contradicted by information below. For instance, the complete description for line 03 30 53.40 3550 is "Concrete in place, including forms (4 uses), Grade 60 rebar, concrete (Portland cement Type 1), placement and finishing unless otherwise indicated; Equipment pad (3000 psi), 4' × 4' × 6" thick."

③ RSMeans data
When using **RSMeans data**, it is important to read through an entire section to ensure that you use the data that most closely matches your work. Note that sometimes there is additional information shown in the section that may improve your price. There are frequently lines that further describe, add to, or adjust data for specific situations.

④ Reference Information
Gordian's RSMeans engineers have created **reference** information to assist you in your estimate. **If** there is information that applies to a section, it will be indicated at the start of the section. The Reference Section is located in the back of the data set.

⑤ Crews
Crews include labor and/or equipment necessary to accomplish each task. In this case, Crew C-14H is used. Gordian's RSMeans staff selects a crew to represent the workers and equipment that are

typically used for that task. In this case, Crew C-14H consists of one carpenter foreman (outside), two carpenters, one rodman, one laborer, one cement finisher, and one gas engine vibrator. Details of all crews can be found in the Reference Section.

Crews - Standard

Crew No.	Bare Costs		Incl. Subs O&P		Cost Per Labor-Hour	
Crew C-14H	Hr.	Daily	Hr.	Daily	Bare Costs	Incl. O&P
1 Carpenter Foreman (outside)	$53.65	$429.20	$81.15	$649.20	$50.29	$75.79
2 Carpenters	51.65	826.40	78.15	1250.40		
1 Rodman (reinf.)	54.85	438.80	83.00	664.00		
1 Laborer	41.05	328.40	62.10	496.80		
1 Cement Finisher	48.90	391.20	72.20	577.60		
1 Gas Engine Vibrator		26.55		29.20	.55	.61
48 L.H., Daily Totals		$2440.55		$3667.20	$50.84	$76.40

6 Daily Output

The **Daily Output** is the amount of work that the crew can do in a normal 8-hour workday, including mobilization, layout, movement of materials, and cleanup. In this case, crew C-14H can install thirty 4' × 4' × 6" thick concrete pads in a day. Daily output is variable and based on many factors, including the size of the job, location, and environmental conditions. RSMeans data represents work done in daylight (or adequate lighting) and temperate conditions.

7 Labor-Hours

The figure in the **Labor-Hours** column is the amount of labor required to perform one unit of work–in this case the amount of labor required to construct one 4' × 4' equipment pad. This figure is calculated by dividing the number of hours of labor in the crew by the daily output (48 labor-hours divided by 30 pads = 1.6 hours of labor per pad). Multiply 1.6 times 60 to see the value in minutes: 60 × 1.6 = 96

minutes. Note: the labor-hour figure is not dependent on the crew size. A change in crew size will result in a corresponding change in daily output, but the labor-hours per unit of work will not change.

8 Unit of Measure

All RSMeans data: Unit Prices include the typical **Unit of Measure** used for estimating that item. For concrete-in-place the typical unit is cubic yards (C.Y.) or each (Ea.). For installing broadloom carpet it is square yard and for gypsum board it is square foot. The estimator needs to take special care that the unit in the data matches the unit in the take-off. Unit conversions may be found in the Reference Section.

9 Bare Costs

Bare Costs are the costs of materials, labor, and equipment that the installing contractor pays. They represent the cost, in U.S. dollars, for one unit of work. They do not include any markups for profit or labor burden.

10 Bare Total

The **Total column** represents the total bare cost for the installing contractor in U.S. dollars. In this case, the sum of $72.50 for material + $80.50 for labor + $.88 for equipment is $153.88.

11 Total Incl O&P

The **Total Incl O&P column** is the total cost, including overhead and profit, that the installing contractor will charge the customer. This represents the cost of materials plus 10% profit, the cost of labor plus labor burden and 10% profit, and the cost of equipment plus 10% profit. It does not include the general contractor's overhead and profit. Note: See the inside back cover of the printed product or the Reference Section of the electronic product for details on how the labor burden is calculated.

National Average

The RSMeans data in our print publications represent a "national average" cost. This data should be modified to the project location using the **City Cost Indexes** or **Location Factors** tables found in the Reference Section. Use the Location Factors to adjust estimate totals if the project covers multiple trades. Use the City Cost Indexes (CCI) for single trade projects or projects where a more detailed analysis is required. All figures in the two tables are derived from the same research. The last row of data in the CCI—the weighted average—is the same as the numbers reported for each location in the location factor table.

RSMeans data: Unit Prices— How They Work (Continued)

Project Name: Pre-Engineered Steel Building				Architect: As Shown				
Location:	**Anywhere, USA**						**01/01/19**	**STD**
Line Number	**Description**	**Qty**	**Unit**	**Material**	**Labor**	**Equipment**	**SubContract**	**Estimate Total**
03 30 53.40 3940	Strip footing, 12" x 24", reinforced	15	C.Y.	$2,460.00	$1,710.00	$8.40	$0.00	
03 30 53.40 3950	Strip footing, 12" x 36", reinforced	34	C.Y.	$5,372.00	$3,111.00	$15.30	$0.00	
03 11 13.65 3000	Concrete slab edge forms	500	L.F.	$155.00	$1,305.00	$0.00	$0.00	
03 22 11.10 0200	Welded wire fabric reinforcing	150	C.S.F.	$3,000.00	$4,275.00	$0.00	$0.00	
03 31 13.35 0300	Ready mix concrete, 4000 psi for slab on grade	278	C.Y.	$35,584.00	$0.00	$0.00	$0.00	
03 31 13.70 4300	Place, strike off & consolidate concrete slab	278	C.Y.	$0.00	$5,184.70	$133.44	$0.00	
03 35 13.30 0250	Machine float & trowel concrete slab	15,000	S.F.	$0.00	$9,750.00	$300.00	$0.00	
03 15 16.20 0140	Cut control joints in concrete slab	950	L.F.	$47.50	$408.50	$57.00	$0.00	
03 39 23.13 0300	Sprayed concrete curing membrane	150	C.S.F.	$1,890.00	$1,035.00	$0.00	$0.00	
Division 03	**Subtotal**			**$48,508.50**	**$26,779.20**	**$514.14**	**$0.00**	**$75,801.84**
08 36 13.10 2650	Manual 10' x 10' steel sectional overhead door	8	Ea.	$10,600.00	$3,680.00	$0.00	$0.00	
08 36 13.10 2860	Insulation and steel back panel for OH door	800	S.F.	$4,120.00	$0.00	$0.00	$0.00	
Division 08	**Subtotal**			**$14,720.00**	**$3,680.00**	**$0.00**	**$0.00**	**$18,400.00**
13 34 19.50 1100	Pre-Engineered Steel Building, 100' x 150' x 24'	15,000	SF Flr.	$0.00	$0.00	$0.00	$367,500.00	
13 34 19.50 6050	Framing for PESB door opening, 3' x 7'	4	Opng.	$0.00	$0.00	$0.00	$2,240.00	
13 34 19.50 6100	Framing for PESB door opening, 10' x 10'	8	Opng.	$0.00	$0.00	$0.00	$9,200.00	
13 34 19.50 6200	Framing for PESB window opening, 4' x 3'	6	Opng.	$0.00	$0.00	$0.00	$3,300.00	
13 34 19.50 5750	PESB door, 3' x 7', single leaf	4	Opng.	$2,700.00	$716.00	$0.00	$0.00	
13 34 19.50 7750	PESB sliding window, 4' x 3' with screen	6	Opng.	$2,910.00	$612.00	$44.10	$0.00	
13 34 19.50 6550	PESB gutter, eave type, 26 ga., painted	300	L.F.	$2,160.00	$840.00	$0.00	$0.00	
13 34 19.50 8650	PESB roof vent, 12" wide x 10' long	15	Ea.	$562.50	$3,360.00	$0.00	$0.00	
13 34 19.50 6900	PESB insulation, vinyl faced, 4" thick	27,400	S.F.	$12,330.00	$9,864.00	$0.00	$0.00	
Division 13	**Subtotal**			**$20,662.50**	**$15,392.00**	**$44.10**	**$382,240.00**	**$418,338.60**
			Subtotal	$83,891.00	$45,851.20	$558.24	$382,240.00	$512,540.44
Division 01	**General Requirements @ 7%**			**5,872.37**	**3,209.58**	**39.08**	**26,756.80**	
			Estimate Subtotal	$89,763.37	$49,060.78	$597.32	$408,996.80	$512,540.44
			Sales Tax @ 5%	4,488.17		29.87	10,224.92	
			Subtotal A	94,251.54	49,060.78	627.18	419,221.72	
			GC O & P	9,425.15	25,560.67	62.72	41,922.17	
			Subtotal B	103,676.69	74,621.45	689.90	461,143.89	$640,131.94
			Contingency @ 5%					32,006.60
			Subtotal C					$672,138.53
			Bond @ $12/1000 +10% O&P					8,872.23
			Subtotal D					$681,010.76
			Location Adjustment Factor		113.90			94,660.50
			Grand Total					**$775,671.26**

This estimate is based on an interactive spreadsheet. You are free to download it and adjust it to your methodology.
A copy of this spreadsheet is available at **RSMeans.com/2019books.**

4

Sample Estimate

This sample demonstrates the elements of an estimate, including a tally of the RSMeans data lines and a summary of the markups on a contractor's work to arrive at a total cost to the owner. The Location Factor with RSMeans data is added at the bottom of the estimate to adjust the cost of the work to a specific location.

---

❶ Work Performed

The body of the estimate shows the RSMeans data selected, including the line number, a brief description of each item, its take-off unit and quantity, and the bare costs of materials, labor, and equipment. This estimate also includes a column titled "SubContract." This data is taken from the column "Total Incl O&P" and represents the total that a subcontractor would charge a general contractor for the work, including the sub's markup for overhead and profit.

❷ Division 1, General Requirements

This is the first division numerically but the last division estimated. Division 1 includes project-wide needs provided by the general contractor. These requirements vary by project but may include temporary facilities and utilities, security, testing, project cleanup, etc. For small projects a percentage can be used—typically between 5% and 15% of project cost. For large projects the costs may be itemized and priced individually.

❸ Sales Tax

If the work is subject to state or local sales taxes, the amount must be added to the estimate. Sales tax may be added to material costs, equipment costs, and subcontracted work. In this case, sales tax was added in all three categories. It was assumed that approximately half the subcontracted work would be material cost, so the tax was applied to 50% of the subcontract total.

❹ GC O&P

This entry represents the general contractor's markup on material, labor, equipment, and subcontractor costs. Our standard markup on materials, equipment, and subcontracted work is 10%. In this estimate, the markup on the labor performed by the GC's workers uses "Skilled Workers Average" shown in Column F on the table "Installing Contractor's Overhead & Profit," which can be found on the inside back cover of the printed product or in the Reference Section of the electronic product.

❺ Contingency

A factor for contingency may be added to any estimate to represent the cost of unknowns that may occur between the time that the estimate is performed and the time the project is constructed. The amount of the allowance will depend on the stage of design at which the estimate is done and the contractor's assessment of the risk involved. Refer to section 01 21 16.50 for contingency allowances.

❻ Bonds

Bond costs should be added to the estimate. The figures here represent a typical performance bond, ensuring the owner that if the general contractor does not complete the obligations in the construction contract the bonding company will pay the cost for completion of the work.

❼ Location Adjustment

Published prices are based on national average costs. If necessary, adjust the total cost of the project using a location factor from the "Location Factor" table or the "City Cost Index" table. Use location factors if the work is general, covering multiple trades. If the work is by a single trade (e.g., masonry) use the more specific data found in the "City Cost Indexes."

Estimating Tips
01 20 00 Price and Payment Procedures

- Allowances that should be added to estimates to cover contingencies and job conditions that are not included in the national average material and labor costs are shown in Section 01 21.
- When estimating historic preservation projects (depending on the condition of the existing structure and the owner's requirements), a 15–20% contingency or allowance is recommended, regardless of the stage of the drawings.

01 30 00 Administrative Requirements

- Before determining a final cost estimate, it is good practice to review all the items listed in Subdivisions 01 31 and 01 32 to make final adjustments for items that may need customizing to specific job conditions.
- Requirements for initial and periodic submittals can represent a significant cost to the General Requirements of a job. Thoroughly check the submittal specifications when estimating a project to determine any costs that should be included.

01 40 00 Quality Requirements

- All projects will require some degree of quality control. This cost is not included in the unit cost of construction listed in each division. Depending upon the terms of the contract, the various costs of inspection and testing can be the responsibility of either the owner or the contractor. Be sure to include the required costs in your estimate.

01 50 00 Temporary Facilities and Controls

- Barricades, access roads, safety nets, scaffolding, security, and many more requirements for the execution of a safe project are elements of direct cost. These costs can easily be overlooked when preparing an estimate. When looking through the major classifications of this subdivision, determine which items apply to each division in your estimate.
- Construction equipment rental costs can be found in the Reference Section in Section 01 54 33. Operators' wages are not included in equipment rental costs.
- Equipment mobilization and demobilization costs are not included in equipment rental costs and must be considered separately.

- The cost of small tools provided by the installing contractor for his workers is covered in the "Overhead" column on the "Installing Contractor's Overhead and Profit" table that lists labor trades, base rates, and markups. Therefore, it is included in the "Total Incl. O&P" cost of any unit price line item.

01 70 00 Execution and Closeout Requirements

- When preparing an estimate, thoroughly read the specifications to determine the requirements for Contract Closeout. Final cleaning, record documentation, operation and maintenance data, warranties and bonds, and spare parts and maintenance materials can all be elements of cost for the completion of a contract. Do not overlook these in your estimate.

Reference Numbers

Reference numbers are shown at the beginning of some major classifications. These numbers refer to related items in the Reference Section. The reference information may be an estimating procedure, an alternate pricing method, or technical information.

Note: Not all subdivisions listed here necessarily appear. ■

Did you know?

RSMeans data is available through our online application:

- Search for costs by keyword
- Leverage the most up-to-date data
- Build and export estimates

Try it free
rsmeans.com/2019freetrial

---

01 11 Summary of Work

01 11 31 – Professional Consultants

01 11 31.20 Construction Management Fees	Crew	Daily Output	Labor-Hours	Unit	Material	2019 Bare Costs Labor	Equipment	Total	Total Incl O&P
0010 **CONSTRUCTION MANAGEMENT FEES**									
0020 $1,000,000 job, minimum				Project				4.50%	4.50%
0050 Maximum								7.50%	7.50%
0060 For work to $100,000								10%	10%
0300 $50,000,000 job, minimum								2.50%	2.50%
0350 Maximum								4%	4%

01 11 31.30 Engineering Fees

01 11 31.30 Engineering Fees	Crew	Daily Output	Labor-Hours	Unit	Material	2019 Bare Costs Labor	Equipment	Total	Total Incl O&P
0010 **ENGINEERING FEES** R011110-30									
0020 Educational planning consultant, minimum				Project				.50%	.50%
0100 Maximum				"				2.50%	2.50%
0200 Electrical, minimum				Contrct				4.10%	4.10%
0300 Maximum								10.10%	10.10%
0400 Elevator & conveying systems, minimum								2.50%	2.50%
0500 Maximum								5%	5%
0600 Food service & kitchen equipment, minimum								8%	8%
0700 Maximum								12%	12%
1000 Mechanical (plumbing & HVAC), minimum								4.10%	4.10%
1100 Maximum								10.10%	10.10%

01 21 Allowances

01 21 16 – Contingency Allowances

01 21 16.50 Contingencies

01 21 16.50 Contingencies	Crew	Daily Output	Labor-Hours	Unit	Material	2019 Bare Costs Labor	Equipment	Total	Total Incl O&P
0010 **CONTINGENCIES**, Add to estimate									
0020 Conceptual stage				Project				20%	20%
0050 Schematic stage								15%	15%
0100 Preliminary working drawing stage (Design Dev.)								10%	10%
0150 Final working drawing stage								3%	3%

01 21 53 – Factors Allowance

01 21 53.50 Factors

01 21 53.50 Factors	Crew	Daily Output	Labor-Hours	Unit	Material	2019 Bare Costs Labor	Equipment	Total	Total Incl O&P
0010 **FACTORS** Cost adjustments									
0100 Add to construction costs for particular job requirements									
0500 Cut & patch to match existing construction, add, minimum				Costs	2%	3%			
0550 Maximum					5%	9%			
0800 Dust protection, add, minimum					1%	2%			
0850 Maximum					4%	11%			
1100 Equipment usage curtailment, add, minimum					1%	1%			
1150 Maximum					3%	10%			
1400 Material handling & storage limitation, add, minimum					1%	1%			
1450 Maximum					6%	7%			
1700 Protection of existing work, add, minimum					2%	2%			
1750 Maximum					5%	7%			
2000 Shift work requirements, add, minimum						5%			
2050 Maximum						30%			
2300 Temporary shoring and bracing, add, minimum					2%	5%			
2350 Maximum					5%	12%			

01 21 53.60 Security Factors

01 21 53.60 Security Factors	Crew	Daily Output	Labor-Hours	Unit	Material	2019 Bare Costs Labor	Equipment	Total	Total Incl O&P
0010 **SECURITY FACTORS** R012153-60									
0100 Additional costs due to security requirements									
0110 Daily search of personnel, supplies, equipment and vehicles									
0120 Physical search, inventory and doc of assets, at entry				Costs		30%			

01 21 Allowances

01 21 53 – Factors Allowance

01 21 53.60 Security Factors	Crew	Daily Output	Labor-Hours	Unit	Material	2019 Bare Costs Labor	Equipment	Total	Total Incl O&P
0130	At entry and exit				Costs		50%			
0140	Physical search, at entry						6.25%			
0150	At entry and exit						12.50%			
0160	Electronic scan search, at entry						2%			
0170	At entry and exit						4%			
0180	Visual inspection only, at entry						.25%			
0190	At entry and exit						.50%			
0200	ID card or display sticker only, at entry						.12%			
0210	At entry and exit				▼		.25%			
0220	Day 1 as described below, then visual only for up to 5 day job duration									
0230	Physical search, inventory and doc of assets, at entry				Costs		5%			
0240	At entry and exit						10%			
0250	Physical search, at entry						1.25%			
0260	At entry and exit						2.50%			
0270	Electronic scan search, at entry						.42%			
0280	At entry and exit				▼		.83%			
0290	Day 1 as described below, then visual only for 6-10 day job duration									
0300	Physical search, inventory and doc of assets, at entry				Costs		2.50%			
0310	At entry and exit						5%			
0320	Physical search, at entry						.63%			
0330	At entry and exit						1.25%			
0340	Electronic scan search, at entry						.21%			
0350	At entry and exit				▼		.42%			
0360	Day 1 as described below, then visual only for 11-20 day job duration									
0370	Physical search, inventory and doc of assets, at entry				Costs		1.25%			
0380	At entry and exit						2.50%			
0390	Physical search, at entry						.31%			
0400	At entry and exit						.63%			
0410	Electronic scan search, at entry						.10%			
0420	At entry and exit				▼		.21%			
0430	Beyond 20 days, costs are negligible									
0440	Escort required to be with tradesperson during work effort				Costs		6.25%			

01 21 55 – Job Conditions Allowance

01 21 55.50 Job Conditions

		Crew	Daily Output	Labor-Hours	Unit	Material	Labor	Equipment	Total	Total Incl O&P
0010	**JOB CONDITIONS** Modifications to applicable									
0020	cost summaries									
0100	Economic conditions, favorable, deduct				Project				2%	2%
0200	Unfavorable, add								5%	5%
0300	Hoisting conditions, favorable, deduct								2%	2%
0400	Unfavorable, add								5%	5%
0700	Labor availability, surplus, deduct								1%	1%
0800	Shortage, add								10%	10%
0900	Material storage area, available, deduct								1%	1%
1000	Not available, add								2%	2%
1100	Subcontractor availability, surplus, deduct								5%	5%
1200	Shortage, add								12%	12%
1300	Work space, available, deduct								2%	2%
1400	Not available, add				▼				5%	5%

01 21 57 – Overtime Allowance

01 21 57.50 Overtime

		Crew	Daily Output	Labor-Hours	Unit	Material	Labor	Equipment	Total	Total Incl O&P
0010	**OVERTIME** for early completion of projects or where	R012909-90								
0020	labor shortages exist, add to usual labor, up to					Costs		100%			

01 21 Allowances

01 21 63 – Taxes

01 21 63.10 Taxes

		Crew	Daily Output	Labor-Hours	Unit	Material	2019 Bare Costs Labor	2019 Bare Costs Equipment	Total	Total Incl O&P
0010	**TAXES** R012909-80									
0020	Sales tax, State, average				%	5.08%				
0050	Maximum R012909-85					7.50%				
0200	Social Security, on first $118,500 of wages						7.65%			
0300	Unemployment, combined Federal and State, minimum R012909-86						.60%			
0350	Average						9.60%			
0400	Maximum						12%			

01 31 Project Management and Coordination

01 31 13 – Project Coordination

01 31 13.20 Field Personnel

		Crew	Daily Output	Labor-Hours	Unit	Material	2019 Bare Costs Labor	2019 Bare Costs Equipment	Total	Total Incl O&P
0010	**FIELD PERSONNEL**									
0020	Clerk, average				Week	495			495	750
0100	Field engineer, junior engineer					1,175			1,175	1,775
0120	Engineer					1,525			1,525	2,325
0140	Senior engineer					1,725			1,725	2,650
0160	General purpose laborer, average					1,650			1,650	2,475
0180	Project manager, minimum					2,175			2,175	3,300
0200	Average					2,500			2,500	3,800
0220	Maximum					2,850			2,850	4,325
0240	Superintendent, minimum					2,125			2,125	3,225
0260	Average					2,325			2,325	3,525
0280	Maximum					2,650			2,650	4,025
0290	Timekeeper, average					1,350			1,350	2,050

01 31 13.30 Insurance

		Crew	Daily Output	Labor-Hours	Unit	Material	2019 Bare Costs Labor	2019 Bare Costs Equipment	Total	Total Incl O&P
0010	**INSURANCE** R013113-40									
0020	Builders risk, standard, minimum				Job				.24%	.24%
0050	Maximum R013113-50								.64%	.64%
0200	All-risk type, minimum								.25%	.25%
0250	Maximum R013113-60								.62%	.62%
0400	Contractor's equipment floater, minimum				Value				.50%	.50%
0450	Maximum				"				1.50%	1.50%
0600	Public liability, average				Job				2.02%	2.02%
0800	Workers' compensation & employer's liability, average									
0850	by trade, carpentry, general				Payroll	11.97%				
1000	Electrical					4.91%				
1150	Insulation					10.07%				
1450	Plumbing					5.77%				
1550	Sheet metal work (HVAC)					7.56%				

01 31 13.50 General Contractor's Mark-Up

		Crew	Daily Output	Labor-Hours	Unit	Material	2019 Bare Costs Labor	2019 Bare Costs Equipment	Total	Total Incl O&P
0010	**GENERAL CONTRACTOR'S MARK-UP** on Change Orders									
0200	Extra work, by subcontractors, add				%				10%	10%
0250	By General Contractor, add								15%	15%
0400	Omitted work, by subcontractors, deduct all but								5%	5%
0450	By General Contractor, deduct all but								7.50%	7.50%
0600	Overtime work, by subcontractors, add								15%	15%
0650	By General Contractor, add								10%	10%

01 31 13.80 Overhead and Profit

		Crew	Daily Output	Labor-Hours	Unit	Material	2019 Bare Costs Labor	2019 Bare Costs Equipment	Total	Total Incl O&P
0010	**OVERHEAD & PROFIT** Allowance to add to items in this R013113-50									
0020	book that do not include Subs O&P, average				%				25%	

01 31 Project Management and Coordination

01 31 13 – Project Coordination

01 31 13.80 Overhead and Profit

		Crew	Daily Output	Labor-Hours	Unit	Material	2019 Bare Costs Labor	Equipment	Total	Total Incl O&P
0100	Allowance to add to items in this book that	R013113-55									
0110	do include Subs O&P, minimum					%				5%	5%
0150	Average									10%	10%
0200	Maximum									15%	15%
0300	Typical, by size of project, under $100,000									30%	
0350	$500,000 project									25%	
0400	$2,000,000 project									20%	
0450	Over $10,000,000 project									15%	

01 31 13.90 Performance Bond

0010	**PERFORMANCE BOND**	R013113-80									
0020	For buildings, minimum					Job				.60%	.60%
0100	Maximum					"				2.50%	2.50%

01 32 Construction Progress Documentation

01 32 33 – Photographic Documentation

01 32 33.50 Photographs

		Crew	Daily Output	Labor-Hours	Unit	Material	2019 Bare Costs Labor	Equipment	Total	Total Incl O&P
0010	**PHOTOGRAPHS**										
0020	8" x 10", 4 shots, 2 prints ea., std. mounting					Set	545			545	600
0100	Hinged linen mounts						550			550	605
0200	8" x 10", 4 shots, 2 prints each, in color						510			510	560
0300	For I.D. slugs, add to all above						5.10			5.10	5.60
1500	Time lapse equipment, camera and projector, buy					Ea.	2,775			2,775	3,050
1550	Rent per month					"	1,300			1,300	1,425
1700	Cameraman and processing, black & white					Day	1,275			1,275	1,400
1720	Color					"	1,450			1,450	1,600

01 41 Regulatory Requirements

01 41 26 – Permit Requirements

01 41 26.50 Permits

0010	**PERMITS**										
0020	Rule of thumb, most cities, minimum					Job				.50%	.50%
0100	Maximum					"				2%	2%

01 51 Temporary Utilities

01 51 13 – Temporary Electricity

01 51 13.50 Temporary Power Equip (Pro-Rated Per Job)

		Crew	Daily Output	Labor-Hours	Unit	Material	2019 Bare Costs Labor	Equipment	Total	Total Incl O&P
0010	**TEMPORARY POWER EQUIP (PRO-RATED PER JOB)**	R015113-65									
0020	Service, overhead feed, 3 use										
0030	100 Amp		1 Elec	1.25	6.400	Ea.	390	385		775	1,000
0040	200 Amp			1	8		465	480		945	1,225
0050	400 Amp			.75	10.667		825	640		1,465	1,850
0060	600 Amp			.50	16		1,175	960		2,135	2,725
0100	Underground feed, 3 use										
0110	100 Amp		1 Elec	2	4	Ea.	395	240		635	795
0120	200 Amp			1.15	6.957		505	420		925	1,175
0130	400 Amp			1	8		860	480		1,340	1,675
0140	600 Amp			.75	10.667		1,050	640		1,690	2,125

01 51 13.50 Temporary Power Equip (Pro-Rated Per Job)	Crew	Daily Output	Labor-Hours	Unit	Material	2019 Bare Costs Labor	2019 Bare Costs Equipment	Total	Total Incl O&P
0150	800 Amp	1 Elec	.50	16	Ea.	1,525	960		2,485	3,100
0160	1000 Amp		.35	22.857		1,675	1,375		3,050	3,900
0170	1200 Amp		.25	32		1,850	1,925		3,775	4,925
0180	2000 Amp		.20	40		2,250	2,400		4,650	6,050
0200	Transformers, 3 use									
0210	30 kVA	1 Elec	1	8	Ea.	1,900	480		2,380	2,825
0220	45 kVA		.75	10.667		2,300	640		2,940	3,475
0230	75 kVA		.50	16		3,650	960		4,610	5,450
0240	112.5 kVA		.40	20		3,800	1,200		5,000	6,000
0250	Feeder, PVC, CU wire in trench									
0260	60 Amp	1 Elec	96	.083	L.F.	2.61	5		7.61	10.30
0270	100 Amp		85	.094		4.98	5.65		10.63	13.95
0280	200 Amp		59	.136		12.40	8.15		20.55	26
0290	400 Amp		42	.190		32	11.45		43.45	52.50
0300	Feeder, PVC, aluminum wire in trench									
0310	60 Amp	1 Elec	96	.083	L.F.	2.77	5		7.77	10.50
0320	100 Amp		85	.094		3.13	5.65		8.78	11.90
0330	200 Amp		59	.136		7.25	8.15		15.40	20
0340	400 Amp		42	.190		15.60	11.45		27.05	34
0350	Feeder, EMT, CU wire									
0360	60 Amp	1 Elec	90	.089	L.F.	3.26	5.35		8.61	11.55
0370	100 Amp		80	.100		6	6		12	15.55
0380	200 Amp		60	.133		13.20	8		21.20	26.50
0390	400 Amp		35	.229		34	13.75		47.75	58
0400	Feeder, EMT, aluminum wire									
0410	60 Amp	1 Elec	90	.089	L.F.	4.49	5.35		9.84	12.90
0420	100 Amp		80	.100		5.30	6		11.30	14.75
0430	200 Amp		60	.133		9.30	8		17.30	22
0440	400 Amp		35	.229		17.55	13.75		31.30	40
0500	Equipment, 3 use									
0510	Spider box, 50 Amp	1 Elec	8	1	Ea.	730	60		790	895
0520	Lighting cord, 100'		8	1		123	60		183	225
0530	Light stanchion		8	1		78	60		138	176
0540	Temporary cords, 100', 3 use									
0550	Feeder cord, 50 Amp	1 Elec	16	.500	Ea.	279	30		309	350
0560	Feeder cord, 100 Amp		12	.667		1,125	40		1,165	1,275
0570	Tap cord, 50 Amp		12	.667		615	40		655	735
0580	Tap cord, 100 Amp		6	1.333		1,075	80		1,155	1,300
0590	Temporary cords, 50', 3 use									
0600	Feeder cord, 50 Amp	1 Elec	16	.500	Ea.	325	30		355	400
0610	Feeder cord, 100 Amp		12	.667		610	40		650	730
0620	Tap cord, 50 Amp		12	.667		305	40		345	395
0630	Tap cord, 100 Amp		6	1.333		610	80		690	790
0700	Connections									
0710	Compressor or pump									
0720	30 Amp	1 Elec	7	1.143	Ea.	16.95	68.50		85.45	121
0730	60 Amp		5.30	1.509		29	90.50		119.50	167
0740	100 Amp		4	2		75	120		195	262
0750	Tower crane									
0760	60 Amp	1 Elec	4.50	1.778	Ea.	29	107		136	191
0770	100 Amp	"	3	2.667	"	75	160		235	320
0780	Manlift									
0790	Single	1 Elec	3	2.667	Ea.	34	160		194	276

12

01 51 Temporary Utilities

01 51 13 – Temporary Electricity

01 51 13.50 Temporary Power Equip (Pro-Rated Per Job)

		Crew	Daily Output	Labor-Hours	Unit	Material	2019 Bare Costs Labor	2019 Bare Costs Equipment	Total	Total Incl O&P
0800	Double	1 Elec	2	4	Ea.	71.50	240		311.50	440
0810	Welder with disconnect									
0820	50 Amp	1 Elec	5	1.600	Ea.	179	96		275	340
0830	100 Amp		3.80	2.105		345	126		471	565
0840	200 Amp		2.50	3.200		640	192		832	985
0850	400 Amp		1	8		1,350	480		1,830	2,200
0860	Office trailer									
0870	60 Amp	1 Elec	4.50	1.778	Ea.	49	107		156	213
0880	100 Amp		3	2.667		89.50	160		249.50	340
0890	200 Amp		2	4		290	240		530	680

01 51 13.80 Temporary Utilities

		Crew	Daily Output	Labor-Hours	Unit	Material	2019 Bare Costs Labor	2019 Bare Costs Equipment	Total	Total Incl O&P
0010	**TEMPORARY UTILITIES**									
0350	Lighting, lamps, wiring, outlets, 40,000 S.F. building, 8 strings	1 Elec	34	.235	CSF Flr	5.70	14.15		19.85	27.50
0360	16 strings	"	17	.471		11.40	28.50		39.90	54.50
0400	Power for temp lighting only, 6.6 KWH, per month								.92	1.01
0430	11.8 KWH, per month								1.65	1.82
0450	23.6 KWH, per month								3.30	3.63
0600	Power for job duration incl. elevator, etc., minimum								47	51.50
0650	Maximum								110	121
0675	Temporary cooling				Ea.	1,000			1,000	1,100

01 52 Construction Facilities

01 52 13 – Field Offices and Sheds

01 52 13.20 Office and Storage Space

		Crew	Daily Output	Labor-Hours	Unit	Material	2019 Bare Costs Labor	2019 Bare Costs Equipment	Total	Total Incl O&P
0010	**OFFICE AND STORAGE SPACE**									
0020	Office trailer, furnished, no hookups, 20' x 8', buy	2 Skwk	1	16	Ea.	9,050	855		9,905	11,300
0250	Rent per month					201			201	222
0300	32' x 8', buy	2 Skwk	.70	22.857		14,500	1,225		15,725	17,800
0350	Rent per month					254			254	280
0400	50' x 10', buy	2 Skwk	.60	26.667		29,900	1,425		31,325	35,100
0450	Rent per month					365			365	400
0500	50' x 12', buy	2 Skwk	.50	32		25,500	1,700		27,200	30,600
0550	Rent per month					455			455	500
0700	For air conditioning, rent per month, add					51			51	56
0800	For delivery, add per mile				Mile	12.15			12.15	13.35
0890	Delivery each way				Ea.	2,800			2,800	3,075
0900	Bunk house trailer, 8' x 40' duplex dorm with kitchen, no hookups, buy	2 Carp	1	16		88,000	825		88,825	98,000
0910	9 man with kitchen and bath, no hookups, buy		1	16		90,000	825		90,825	100,500
0920	18 man sleeper with bath, no hookups, buy		1	16		97,000	825		97,825	108,000
1000	Portable buildings, prefab, on skids, economy, 8' x 8'		265	.060	S.F.	25.50	3.12		28.62	32.50
1100	Deluxe, 8' x 12'		150	.107	"	28	5.50		33.50	39.50
1200	Storage boxes, 20' x 8', buy	2 Skwk	1.80	8.889	Ea.	3,250	475		3,725	4,300
1250	Rent per month					85			85	93.50
1300	40' x 8', buy	2 Skwk	1.40	11.429		3,800	610		4,410	5,100
1350	Rent per month					111			111	122

01 52 13.40 Field Office Expense

		Crew	Daily Output	Labor-Hours	Unit	Material	2019 Bare Costs Labor	2019 Bare Costs Equipment	Total	Total Incl O&P
0010	**FIELD OFFICE EXPENSE**									
0100	Office equipment rental average				Month	207			207	227
0120	Office supplies, average				"	85			85	93.50
0125	Office trailer rental, see Section 01 52 13.20									

For customer support on your Electrical Costs with RSMeans data, call 800.448.8182.

13

01 52 Construction Facilities

01 52 13 – Field Offices and Sheds

01 52 13.40 Field Office Expense	Crew	Daily Output	Labor-Hours	Unit	Material	2019 Bare Costs Labor	Equipment	Total	Total Incl O&P
0140	Telephone bill; avg. bill/month incl. long dist.				Month	87			87	95.50
0160	Lights & HVAC				"	162			162	179

01 54 Construction Aids

01 54 16 – Temporary Hoists

01 54 16.50 Weekly Forklift Crew

		Crew	Daily Output	Labor-Hours	Unit	Material	Labor	Equipment	Total	Total Incl O&P
0010	**WEEKLY FORKLIFT CREW**									
0100	All-terrain forklift, 45' lift, 35' reach, 9000 lb. capacity	A-3P	.20	40	Week		2,075	2,325	4,400	5,650

01 54 19 – Temporary Cranes

01 54 19.50 Daily Crane Crews

		Crew	Daily Output	Labor-Hours	Unit	Material	Labor	Equipment	Total	Total Incl O&P
0010	**DAILY CRANE CREWS** for small jobs, portal to portal R015433-15									
0100	12-ton truck-mounted hydraulic crane	A-3H	1	8	Day		460	710	1,170	1,475
0200	25-ton	A-3I	1	8			460	785	1,245	1,550
0300	40-ton	A-3J	1	8			460	1,275	1,735	2,100
0400	55-ton	A-3K	1	16			855	1,450	2,305	2,850
0500	80-ton	A-3L	1	16	↓		855	2,200	3,055	3,700
0900	If crane is needed on a Saturday, Sunday or Holiday									
0910	At time-and-a-half, add				Day		50%			
0920	At double time, add				"		100%			

01 54 19.60 Monthly Tower Crane Crew

		Crew	Daily Output	Labor-Hours	Unit	Material	Labor	Equipment	Total	Total Incl O&P
0010	**MONTHLY TOWER CRANE CREW**, excludes concrete footing									
0100	Static tower crane, 130' high, 106' jib, 6200 lb. capacity	A-3N	.05	176	Month		10,100	26,000	36,100	43,800

01 54 23 – Temporary Scaffolding and Platforms

01 54 23.70 Scaffolding

		Crew	Daily Output	Labor-Hours	Unit	Material	Labor	Equipment	Total	Total Incl O&P
0010	**SCAFFOLDING** R015423-10									
0015	Steel tube, regular, no plank, labor only to erect & dismantle									
0090	Building exterior, wall face, 1 to 5 stories, 6'-4" x 5' frames	3 Carp	8	3	C.S.F.		155		155	234
0200	6 to 12 stories	4 Carp	8	4			207		207	315
0301	13 to 20 stories	5 Clab	8	5			205		205	310
0460	Building interior, wall face area, up to 16' high	3 Carp	12	2			103		103	156
0560	16' to 40' high		10	2.400	↓		124		124	188
0800	Building interior floor area, up to 30' high	↓	150	.160	C.C.F.		8.25		8.25	12.50
0900	Over 30' high	4 Carp	160	.200	"		10.35		10.35	15.65
0906	Complete system for face of walls, no plank, material only rent/mo				C.S.F.	33			33	36.50
0908	Interior spaces, no plank, material only rent/mo				C.C.F.	3.80			3.80	4.18
0910	Steel tubular, heavy duty shoring, buy									
0920	Frames 5' high 2' wide				Ea.	92			92	101
0925	5' high 4' wide					97.50			97.50	107
0930	6' high 2' wide					99			99	109
0935	6' high 4' wide				↓	116			116	128
0940	Accessories									
0945	Cross braces				Ea.	17.10			17.10	18.80
0950	U-head, 8" x 8"					18.90			18.90	21
0955	J-head, 4" x 8"					13.75			13.75	15.15
0960	Base plate, 8" x 8"					15.15			15.15	16.65
0965	Leveling jack				↓	36			36	39.50
1000	Steel tubular, regular, buy									
1100	Frames 3' high 5' wide				Ea.	89.50			89.50	98.50
1150	5' high 5' wide					108			108	119
1200	6'-4" high 5' wide				↓	106			106	117

14

01 54 Construction Aids

01 54 23 – Temporary Scaffolding and Platforms

01 54 23.70 Scaffolding

		Crew	Daily Output	Labor-Hours	Unit	Material	2019 Bare Costs Labor	Equipment	Total	Total Incl O&P
1350	7'-6" high 6' wide				Ea.	157			157	173
1500	Accessories, cross braces					18.05			18.05	19.85
1550	Guardrail post					19.90			19.90	22
1600	Guardrail 7' section					8.15			8.15	8.95
1650	Screw jacks & plates					25.50			25.50	28.50
1700	Sidearm brackets					22.50			22.50	25
1750	8" casters					37.50			37.50	41
1800	Plank 2" x 10" x 16'-0"					67			67	74
1900	Stairway section					284			284	310
1910	Stairway starter bar					32.50			32.50	35.50
1920	Stairway inside handrail					53			53	58.50
1930	Stairway outside handrail					84.50			84.50	93
1940	Walk-thru frame guardrail					42			42	46
2000	Steel tubular, regular, rent/mo.									
2100	Frames 3' high 5' wide				Ea.	4.45			4.45	4.90
2150	5' high 5' wide					4.45			4.45	4.90
2200	6'-4" high 5' wide					5.40			5.40	5.95
2250	7'-6" high 6' wide					9.85			9.85	10.85
2500	Accessories, cross braces					.89			.89	.98
2550	Guardrail post					.89			.89	.98
2600	Guardrail 7' section					.89			.89	.98
2650	Screw jacks & plates					1.78			1.78	1.96
2700	Sidearm brackets					1.78			1.78	1.96
2750	8" casters					7.10			7.10	7.80
2800	Outrigger for rolling tower					2.67			2.67	2.94
2850	Plank 2" x 10" x 16'-0"					9.85			9.85	10.85
2900	Stairway section					31.50			31.50	34.50
2940	Walk-thru frame guardrail					2.22			2.22	2.44
3000	Steel tubular, heavy duty shoring, rent/mo.									
3250	5' high 2' & 4' wide				Ea.	8.35			8.35	9.20
3300	6' high 2' & 4' wide					8.35			8.35	9.20
3500	Accessories, cross braces					.89			.89	.98
3600	U-head, 8" x 8"					2.46			2.46	2.71
3650	J-head, 4" x 8"					2.46			2.46	2.71
3700	Base plate, 8" x 8"					.89			.89	.98
3750	Leveling jack					2.46			2.46	2.71
5700	Planks, 2" x 10" x 16'-0", labor only to erect & remove to 50' H	3 Carp	72	.333			17.20		17.20	26
5800	Over 50' high	4 Carp	80	.400			20.50		20.50	31.50

01 54 23.75 Scaffolding Specialties

		Crew	Daily Output	Labor-Hours	Unit	Material	2019 Bare Costs Labor	Equipment	Total	Total Incl O&P
0010	**SCAFFOLDING SPECIALTIES**									
1500	Sidewalk bridge using tubular steel scaffold frames including									
1510	planking (material cost is rent/month)	3 Carp	45	.533	L.F.	8.35	27.50		35.85	50.50
1600	For 2 uses per month, deduct from all above					50%				
1700	For 1 use every 2 months, add to all above					100%				
1900	Catwalks, 20" wide, no guardrails, 7' span, buy				Ea.	151			151	166
2000	10' span, buy					212			212	233
3720	Putlog, standard, 8' span, with hangers, buy					81.50			81.50	90
3750	12' span, buy					101			101	111
3760	Trussed type, 16' span, buy					266			266	293
3790	22' span, buy					279			279	305
3795	Rent per month					32.50			32.50	35.50
3800	Rolling ladders with handrails, 30" wide, buy, 2 step					294			294	325

15

For customer support on your Electrical Costs with RSMeans data, call 800.448.8182.

01 54 Construction Aids

01 54 23 – Temporary Scaffolding and Platforms

01 54 23.75 Scaffolding Specialties	Crew	Daily Output	Labor-Hours	Unit	Material	2019 Bare Costs Labor	Equipment	Total	Total Incl O&P
4000	7 step				Ea.	965			965	1,050
4050	10 step					1,375			1,375	1,525
4100	Rolling towers, buy, 5' wide, 7' long, 10' high					1,325			1,325	1,450
4200	For additional 5' high sections, to buy				▼	253			253	278
4300	Complete incl. wheels, railings, outriggers,									
4350	21' high, to buy				Ea.	2,275			2,275	2,500
4400	Rent/month = 5% of purchase cost				"	199			199	219

01 54 26 – Temporary Swing Staging

01 54 26.50 Swing Staging	Crew	Daily Output	Labor-Hours	Unit	Material	Labor	Equipment	Total	Total Incl O&P
0010	**SWING STAGING**, 500 lb. cap., 2' wide to 24' long, hand operated									
0020	Steel cable type, with 60' cables, buy				Ea.	6,075			6,075	6,675
0030	Rent per month				"	605			605	670
2200	Move swing staging (setup and remove)	E-4	2	16	Move		905	48	953	1,500

01 54 36 – Equipment Mobilization

01 54 36.50 Mobilization	Crew	Daily Output	Labor-Hours	Unit	Material	Labor	Equipment	Total	Total Incl O&P
0010	**MOBILIZATION** (Use line item again for demobilization) R015436-50									
0015	Up to 25 mi. haul dist. (50 mi. RT for mob/demob crew)									
1200	Small equipment, placed in rear of, or towed by pickup truck	A-3A	4	2	Ea.		103	30.50	133.50	189
1300	Equipment hauled on 3-ton capacity towed trailer	A-3Q	2.67	3			155	55.50	210.50	294
1400	20-ton capacity	B-34U	2	8			395	218	613	835
1500	40-ton capacity	B-34N	2	8			410	335	745	985
1600	50-ton capacity	B-34V	1	24			1,250	920	2,170	2,875
1700	Crane, truck-mounted, up to 75 ton (driver only)	1 Eqhv	4	2			115		115	173
1800	Over 75 ton (with chase vehicle)	A-3E	2.50	6.400			335	48.50	383.50	555
2400	Crane, large lattice boom, requiring assembly	B-34W	.50	144	▼		7,175	6,850	14,025	18,400
2500	For each additional 5 miles haul distance, add						10%	10%		
3000	For large pieces of equipment, allow for assembly/knockdown									
3100	For mob/demob of micro-tunneling equip, see Section 33 05 23.19									

01 55 Vehicular Access and Parking

01 55 23 – Temporary Roads

01 55 23.50 Roads and Sidewalks	Crew	Daily Output	Labor-Hours	Unit	Material	Labor	Equipment	Total	Total Incl O&P
0010	**ROADS AND SIDEWALKS** Temporary									
0050	Roads, gravel fill, no surfacing, 4" gravel depth	B-14	715	.067	S.Y.	3.39	2.90	.45	6.74	8.60
0100	8" gravel depth	"	615	.078	"	6.80	3.37	.52	10.69	13.10
1000	Ramp, 3/4" plywood on 2" x 6" joists, 16" OC	2 Carp	300	.053	S.F.	1.65	2.75		4.40	6
1100	On 2" x 10" joists, 16" OC	"	275	.058	"	2.36	3.01		5.37	7.15

01 56 Temporary Barriers and Enclosures

01 56 13 – Temporary Air Barriers

01 56 13.90 Winter Protection

		Crew	Daily Output	Labor-Hours	Unit	Material	2019 Bare Costs Labor	Equipment	Total	Total Incl O&P
0010	**WINTER PROTECTION**									
0100	Framing to close openings	2 Clab	500	.032	S.F.	.46	1.31		1.77	2.49
0200	Tarpaulins hung over scaffolding, 8 uses, not incl. scaffolding		1500	.011		.25	.44		.69	.94
0300	Prefab fiberglass panels, steel frame, 8 uses		1200	.013		2.58	.55		3.13	3.67

01 56 16 – Temporary Dust Barriers

01 56 16.10 Dust Barriers, Temporary

		Crew	Daily Output	Labor-Hours	Unit	Material	2019 Bare Costs Labor	Equipment	Total	Total Incl O&P
0010	**DUST BARRIERS, TEMPORARY**									
0020	Spring loaded telescoping pole & head, to 12', erect and dismantle	1 Clab	240	.033	Ea.		1.37		1.37	2.07
0025	Cost per day (based upon 250 days)				Day	.22			.22	.24
0030	To 21', erect and dismantle	1 Clab	240	.033	Ea.		1.37		1.37	2.07
0035	Cost per day (based upon 250 days)				Day	.68			.68	.74
0040	Accessories, caution tape reel, erect and dismantle	1 Clab	480	.017	Ea.		.68		.68	1.04
0045	Cost per day (based upon 250 days)				Day	.32			.32	.35
0060	Foam rail and connector, erect and dismantle	1 Clab	240	.033	Ea.		1.37		1.37	2.07
0065	Cost per day (based upon 250 days)				Day	.10			.10	.11
0070	Caution tape	1 Clab	384	.021	C.L.F.	2.69	.86		3.55	4.25
0080	Zipper, standard duty		60	.133	Ea.	7.40	5.45		12.85	16.45
0090	Heavy duty		48	.167	"	9.95	6.85		16.80	21.50
0100	Polyethylene sheet, 4 mil		37	.216	Sq.	2.66	8.90		11.56	16.40
0110	6 mil		37	.216	"	3.87	8.90		12.77	17.70
1000	Dust partition, 6 mil polyethylene, 1" x 3" frame	2 Carp	2000	.008	S.F.	.33	.41		.74	.99
1080	2" x 4" frame	"	2000	.008	"	.35	.41		.76	1.02
1085	Negative air machine, 1800 CFM				Ea.	910			910	1,000
1090	Adhesive strip application, 2" width	1 Clab	192	.042	C.L.F.	6.90	1.71		8.61	10.15

01 56 23 – Temporary Barricades

01 56 23.10 Barricades

		Crew	Daily Output	Labor-Hours	Unit	Material	2019 Bare Costs Labor	Equipment	Total	Total Incl O&P
0010	**BARRICADES**									
0020	5' high, 3 rail @ 2" x 8", fixed	2 Carp	20	.800	L.F.	6.40	41.50		47.90	69.50
0150	Movable		30	.533		5.35	27.50		32.85	47.50
1000	Guardrail, wooden, 3' high, 1" x 6" on 2" x 4" posts		200	.080		1.34	4.13		5.47	7.70
1100	2" x 6" on 4" x 4" posts		165	.097		2.66	5		7.66	10.50
1200	Portable metal with base pads, buy					14.50			14.50	15.95
1250	Typical installation, assume 10 reuses	2 Carp	600	.027		2.37	1.38		3.75	4.69
1300	Barricade tape, polyethylene, 7 mil, 3" wide x 300' long roll				Ea.	8.05			8.05	8.90
3000	Detour signs, set up and remove									
3010	Reflective aluminum, MUTCD, 24" x 24", post mounted	1 Clab	20	.400	Ea.	2.66	16.40		19.06	28
4000	Roof edge portable barrier stands and warning flags, 50 uses	1 Rohe	9100	.001	L.F.	.06	.03		.09	.12
4010	100 uses	"	9100	.001	"	.03	.03		.06	.09

01 56 26 – Temporary Fencing

01 56 26.50 Temporary Fencing

		Crew	Daily Output	Labor-Hours	Unit	Material	2019 Bare Costs Labor	Equipment	Total	Total Incl O&P
0010	**TEMPORARY FENCING**									
0020	Chain link, 11 ga., 4' high	2 Clab	400	.040	L.F.	1.79	1.64		3.43	4.45
0100	6' high		300	.053		4.68	2.19		6.87	8.45
0200	Rented chain link, 6' high, to 1000' (up to 12 mo.)		400	.040		3.04	1.64		4.68	5.80
0250	Over 1000' (up to 12 mo.)		300	.053		3.82	2.19		6.01	7.50
0350	Plywood, painted, 2" x 4" frame, 4' high	A-4	135	.178		6.45	8.70		15.15	20
0400	4" x 4" frame, 8' high	"	110	.218		12.35	10.65		23	29.50
0500	Wire mesh on 4" x 4" posts, 4' high	2 Carp	100	.160		10.50	8.25		18.75	24
0550	8' high	"	80	.200		15.80	10.35		26.15	33

01 56 Temporary Barriers and Enclosures

01 56 29 – Temporary Protective Walkways

01 56 29.50 Protection	Crew	Daily Output	Labor-Hours	Unit	Material	2019 Bare Costs Labor	Equipment	Total	Total Incl O&P
0010 **PROTECTION**									
0020 Stair tread, 2" x 12" planks, 1 use	1 Carp	75	.107	Tread	5.45	5.50		10.95	14.35
0100 Exterior plywood, 1/2" thick, 1 use		65	.123		1.90	6.35		8.25	11.70
0200 3/4" thick, 1 use		60	.133		2.81	6.90		9.71	13.50
2200 Sidewalks, 2" x 12" planks, 2 uses		350	.023	S.F.	.91	1.18		2.09	2.79
2300 Exterior plywood, 2 uses, 1/2" thick		750	.011		.32	.55		.87	1.18
2400 5/8" thick		650	.012		.39	.64		1.03	1.39
2500 3/4" thick		600	.013		.47	.69		1.16	1.56

01 58 Project Identification

01 58 13 – Temporary Project Signage

01 58 13.50 Signs

	Crew	Daily Output	Labor-Hours	Unit	Material	2019 Bare Costs Labor	Equipment	Total	Total Incl O&P
0010 **SIGNS**									
0020 High intensity reflectorized, no posts, buy				Ea.	25.50			25.50	28

01 66 Product Storage and Handling Requirements

01 66 19 – Material Handling

01 66 19.10 Material Handling

	Crew	Daily Output	Labor-Hours	Unit	Material	2019 Bare Costs Labor	Equipment	Total	Total Incl O&P
0010 **MATERIAL HANDLING**									
0020 Above 2nd story, via stairs, per C.Y. of material per floor	2 Clab	145	.110	C.Y.		4.53		4.53	6.85
0030 Via elevator, per C.Y. of material		240	.067			2.74		2.74	4.14
0050 Distances greater than 200', per C.Y. of material per each addl 200'		300	.053			2.19		2.19	3.31

01 74 Cleaning and Waste Management

01 74 13 – Progress Cleaning

01 74 13.20 Cleaning Up

	Crew	Daily Output	Labor-Hours	Unit	Material	2019 Bare Costs Labor	Equipment	Total	Total Incl O&P
0010 **CLEANING UP**									
0020 After job completion, allow, minimum				Job				.30%	.30%
0040 Maximum				"				1%	1%
0050 Cleanup of floor area, continuous, per day, during const.	A-5	24	.750	M.S.F.	2.34	31	2.03	35.37	52
0100 Final by GC at end of job	"	11.50	1.565	"	2.48	65	4.24	71.72	106

01 91 Commissioning

01 91 13 – General Commissioning Requirements

01 91 13.50 Building Commissioning

	Crew	Daily Output	Labor-Hours	Unit	Material	2019 Bare Costs Labor	Equipment	Total	Total Incl O&P
0010 **BUILDING COMMISSIONING**									
0100 Systems operation and verification during turnover				%				.25%	.25%
0150 Including all systems subcontractors								.50%	.50%
0200 Systems design assistance, operation, verification and training								.50%	.50%
0250 Including all systems subcontractors								1%	1%

01 93 13 – Facility Maintenance Procedures

01 93 13.16 Electrical Facilities Maintenance	Crew	Daily Output	Labor-Hours	Unit	Material	2019 Bare Costs Labor	Equipment	Total	Total Incl O&P
0010	**ELECTRICAL FACILITIES MAINTENANCE**									
0700	Cathodic protection systems									
0720	Check and adjust reading on rectifier	1 Elec	20	.400	Ea.		24		24	36
0730	Check pipe to soil potential		20	.400			24		24	36
0740	Replace lead connection		4	2			120		120	179
0800	Control device, install		5.70	1.404			84.50		84.50	126
0810	Disassemble, clean and reinstall		7	1.143			68.50		68.50	102
0820	Replace		10.70	.748			45		45	67
0830	Trouble shoot		10	.800			48		48	71.50
0900	Demolition, for electrical demolition see Section 26 05 05.10									
1000	Distribution systems and equipment install or repair a breaker									
1010	In power panels up to 200 amps	1 Elec	7	1.143	Ea.		68.50		68.50	102
1020	Over 200 amps		2	4			240		240	360
1030	Reset breaker or replace fuse		20	.400			24		24	36
1100	Megger test MCC (each stack)		4	2			120		120	179
1110	MCC vacuum and clean (each stack)		5.30	1.509			90.50		90.50	135
2500	Remove and replace or maint., road fixture & lamp		3	2.667		1,325	160		1,485	1,700
2510	Fluorescent fixture		7	1.143		76	68.50		144.50	186
2515	Relamp (fluor.) facility area each tube, spot	1 Clab	24	.333		2.16	13.70		15.86	23
2516	Group		100	.080		2.16	3.28		5.44	7.35
2518	Fluorescent fixture, clean (area)		44	.182		.04	7.45		7.49	11.35
2520	Incandescent fixture	1 Elec	11	.727		58	43.50		101.50	129
2530	Lamp (incandescent or fluorescent)	1 Clab	60	.133		2.16	5.45		7.61	10.70
2535	Replace cord in socket lamp	1 Elec	13	.615		1.75	37		38.75	57
2540	Ballast electronic type for two tubes		8	1		43.50	60		103.50	137
2541	Starter		30	.267		.86	16		16.86	25
2545	Replace other lighting parts		11	.727		20.50	43.50		64	87.50
2550	Switch		11	.727		2.96	43.50		46.46	68.50
2555	Receptacle		11	.727		9.55	43.50		53.05	75.50
2560	Floodlight		4	2		243	120		363	445
2570	Christmas lighting, indoor, per string	1 Clab	16	.500			20.50		20.50	31
2580	Outdoor		13	.615			25.50		25.50	38
2590	Test battery operated emergency lights		40	.200			8.20		8.20	12.40
2600	Repair and replace component in communication system	1 Elec	6	1.333		53.50	80		133.50	178
2700	Repair misc. appliances (incl. clocks, vent fan, blower, etc.)	"	6	1.333			80		80	119
2710	Reset clocks & timers	1 Clab	50	.160			6.55		6.55	9.95
2720	Adjust time delay relays	1 Elec	16	.500			30		30	45
2730	Test specific gravity of lead-acid batteries	1 Clab	80	.100			4.11		4.11	6.20
3000	Motors and generators									
3020	Disassemble, clean and reinstall motor, up to 1/4 HP	1 Elec	4	2	Ea.		120		120	179
3030	Up to 3/4 HP		3	2.667			160		160	239
3040	Up to 10 HP		2	4			240		240	360
3050	Replace part, up to 1/4 HP		6	1.333			80		80	119
3060	Up to 3/4 HP		4	2			120		120	179
3070	Up to 10 HP		3	2.667			160		160	239
3080	Megger test motor windings		5.33	1.501			90		90	134
3082	Motor vibration check		16	.500			30		30	45
3084	Oil motor bearings		25	.320			19.20		19.20	28.50
3086	Run test emergency generator for 30 minutes		11	.727			43.50		43.50	65
3090	Rewind motor, up to 1/4 HP		3	2.667			160		160	239
3100	Up to 3/4 HP		2	4			240		240	360
3110	Up to 10 HP		1.50	5.333			320		320	480
3150	Generator, repair or replace part		4	2			120		120	179

01 93 13.16 Electrical Facilities Maintenance	Crew	Daily Output	Labor-Hours	Unit	Material	2019 Bare Costs Labor	Equipment	Total	Total Incl O&P
3160	Repair DC generator	1 Elec	2	4	Ea.		240		240	360
4000	Stub pole, install or remove		3	2.667			160		160	239
4500	Transformer maintenance up to 15 kVA		2.70	2.963			178		178	265

Estimating Tips
02 30 00 Subsurface Investigation

In preparing estimates on structures involving earthwork or foundations, all information concerning soil characteristics should be obtained. Look particularly for hazardous waste, evidence of prior dumping of debris, and previous stream beds.

02 40 00 Demolition and Structure Moving

The costs shown for selective demolition do not include rubbish handling or disposal. These items should be estimated separately using RSMeans data or other sources.

- Historic preservation often requires that the contractor remove materials from the existing structure, rehab them, and replace them. The estimator must be aware of any related measures and precautions that must be taken when doing selective demolition and cutting and patching. Requirements may include special handling and storage, as well as security.

- In addition to Subdivision 02 41 00, you can find selective demolition items in each division. Example: Roofing demolition is in Division 7.
- Absent of any other specific reference, an approximate demolish-in-place cost can be obtained by halving the new-install labor cost. To remove for reuse, allow the entire new-install labor figure.

02 40 00 Building Deconstruction

This section provides costs for the careful dismantling and recycling of most low-rise building materials.

02 50 00 Containment of Hazardous Waste

This section addresses on-site hazardous waste disposal costs.

02 80 00 Hazardous Material Disposal/Remediation

This subdivision includes information on hazardous waste handling, asbestos remediation, lead remediation, and mold remediation. See reference numbers RO28213-20 and RO28319-60 for further guidance in using these unit price lines.

02 90 00 Monitoring Chemical Sampling, Testing Analysis

This section provides costs for on-site sampling and testing hazardous waste.

Reference Numbers

Reference numbers are shown at the beginning of some major classifications. These numbers refer to related items in the Reference Section. The reference information may be an estimating procedure, an alternate pricing method, or technical information.

Note: Not all subdivisions listed here necessarily appear. ■

Did you know?

RSMeans data is available through our online application:

- Search for costs by keyword
- Leverage the most up-to-date data
- Build and export estimates

Try it free
rsmeans.com/2019freetrial

02 41 Demolition

02 41 13 – Selective Site Demolition

02 41 13.17 Demolish, Remove Pavement and Curb		Crew	Daily Output	Labor-Hours	Unit	Material	2019 Bare Costs Labor	Equipment	Total	Total Incl O&P
0010	**DEMOLISH, REMOVE PAVEMENT AND CURB** R024119-10									
5010	Pavement removal, bituminous roads, up to 3" thick	B-38	690	.058	S.Y.		2.69	1.62	4.31	5.85
5050	4"-6" thick		420	.095			4.42	2.67	7.09	9.60
5100	Bituminous driveways		640	.063			2.90	1.75	4.65	6.30
5200	Concrete to 6" thick, hydraulic hammer, mesh reinforced		255	.157			7.30	4.39	11.69	15.80
5300	Rod reinforced		200	.200			9.30	5.60	14.90	20
5400	Concrete, 7"-24" thick, plain		33	1.212	C.Y.		56	34	90	122
5500	Reinforced		24	1.667	"		77.50	46.50	124	169
5600	With hand held air equipment, bituminous, to 6" thick	B-39	1900	.025	S.F.		1.09	.12	1.21	1.79
5700	Concrete to 6" thick, no reinforcing		1600	.030			1.29	.15	1.44	2.12
5800	Mesh reinforced		1400	.034			1.48	.17	1.65	2.42
5900	Rod reinforced		765	.063			2.71	.31	3.02	4.43

02 41 13.23 Utility Line Removal

		Crew	Daily Output	Labor-Hours	Unit	Material	Labor	Equipment	Total	Total Incl O&P
0010	**UTILITY LINE REMOVAL**									
0015	No hauling, abandon catch basin or manhole	B-6	7	3.429	Ea.		153	45.50	198.50	282
0020	Remove existing catch basin or manhole, masonry		4	6			267	80	347	495
0030	Catch basin or manhole frames and covers, stored		13	1.846			82.50	24.50	107	151
0040	Remove and reset		7	3.429			153	45.50	198.50	282

02 41 13.30 Minor Site Demolition

		Crew	Daily Output	Labor-Hours	Unit	Material	Labor	Equipment	Total	Total Incl O&P
0010	**MINOR SITE DEMOLITION** R024119-10									
4000	Sidewalk removal, bituminous, 2" thick	B-6	350	.069	S.Y.		3.06	.91	3.97	5.60
4010	2-1/2" thick		325	.074			3.29	.99	4.28	6.05
4050	Brick, set in mortar		185	.130			5.80	1.73	7.53	10.60
4100	Concrete, plain, 4"		160	.150			6.70	2	8.70	12.30
4110	Plain, 5"		140	.171			7.65	2.29	9.94	14.05
4120	Plain, 6"		120	.200			8.90	2.67	11.57	16.40
4200	Mesh reinforced, concrete, 4"		150	.160			7.15	2.13	9.28	13.10
4210	5" thick		131	.183			8.15	2.44	10.59	15
4220	6" thick		112	.214			9.55	2.86	12.41	17.55

02 41 19 – Selective Demolition

02 41 19.13 Selective Building Demolition

		Crew	Daily Output	Labor-Hours	Unit	Material	Labor	Equipment	Total	Total Incl O&P
0010	**SELECTIVE BUILDING DEMOLITION**									
0020	Costs related to selective demolition of specific building components									
0025	are included under Common Work Results (XX 05)									
0030	in the component's appropriate division.									

02 41 19.16 Selective Demolition, Cutout

		Crew	Daily Output	Labor-Hours	Unit	Material	Labor	Equipment	Total	Total Incl O&P
0010	**SELECTIVE DEMOLITION, CUTOUT** R024119-10									
0020	Concrete, elev. slab, light reinforcement, under 6 C.F.	B-9	65	.615	C.F.		25.50	3.61	29.11	42.50
0050	Light reinforcing, over 6 C.F.	"	75	.533	"		22	3.13	25.13	37
6000	Walls, interior, not including re-framing,									
6010	openings to 5 S.F.									
6100	Drywall to 5/8" thick	1 Clab	24	.333	Ea.		13.70		13.70	20.50
6200	Paneling to 3/4" thick		20	.400			16.40		16.40	25
6300	Plaster, on gypsum lath		20	.400			16.40		16.40	25
6340	On wire lath		14	.571			23.50		23.50	35.50

02 41 19.19 Selective Demolition

		Crew	Daily Output	Labor-Hours	Unit	Material	Labor	Equipment	Total	Total Incl O&P
0010	**SELECTIVE DEMOLITION**, Rubbish Handling R024119-10									
0020	The following are to be added to the demolition prices									
0600	Dumpster, weekly rental, 1 dump/week, 6 C.Y. capacity (2 tons)				Week	415			415	455
0700	10 C.Y. capacity (3 tons)					480			480	530
0725	20 C.Y. capacity (5 tons) R024119-20					565			565	625

02 41 Demolition

02 41 19 – Selective Demolition

02 41 19.19 Selective Demolition

		Crew	Daily Output	Labor-Hours	Unit	Material	2019 Bare Costs Labor	Equipment	Total	Total Incl O&P
0800	30 C.Y. capacity (7 tons)				Week	730			730	800
0840	40 C.Y. capacity (10 tons)				↓	775			775	850
2000	Load, haul, dump and return, 0'-50' haul, hand carried	2 Clab	24	.667	C.Y.		27.50		27.50	41.50
2005	Wheeled		37	.432			17.75		17.75	27
2040	0'-100' haul, hand carried		16.50	.970			40		40	60
2045	Wheeled		25	.640			26.50		26.50	39.50
2050	Forklift	A-3R	25	.320			16.55	5.75	22.30	31.50
2080	Haul and return, add per each extra 100' haul, hand carried	2 Clab	35.50	.451			18.50		18.50	28
2085	Wheeled		54	.296			12.15		12.15	18.40
2120	For travel in elevators, up to 10 floors, add		140	.114			4.69		4.69	7.10
2130	0'-50' haul, incl. up to 5 riser stairs, hand carried		23	.696			28.50		28.50	43
2135	Wheeled		35	.457			18.75		18.75	28.50
2140	6-10 riser stairs, hand carried		22	.727			30		30	45
2145	Wheeled		34	.471			19.30		19.30	29
2150	11-20 riser stairs, hand carried		20	.800			33		33	49.50
2155	Wheeled		31	.516			21		21	32
2160	21-40 riser stairs, hand carried		16	1			41		41	62
2165	Wheeled		24	.667			27.50		27.50	41.50
2170	0'-100' haul, incl. 5 riser stairs, hand carried		15	1.067			44		44	66
2175	Wheeled		23	.696			28.50		28.50	43
2180	6-10 riser stairs, hand carried		14	1.143			47		47	71
2185	Wheeled		21	.762			31.50		31.50	47.50
2190	11-20 riser stairs, hand carried		12	1.333			54.50		54.50	83
2195	Wheeled		18	.889			36.50		36.50	55
2200	21-40 riser stairs, hand carried		8	2			82		82	124
2205	Wheeled		12	1.333			54.50		54.50	83
2210	Haul and return, add per each extra 100' haul, hand carried		35.50	.451			18.50		18.50	28
2215	Wheeled		54	.296	↓		12.15		12.15	18.40
2220	For each additional flight of stairs, up to 5 risers, add		550	.029	Flight		1.19		1.19	1.81
2225	6-10 risers, add		275	.058			2.39		2.39	3.61
2230	11-20 risers, add		138	.116			4.76		4.76	7.20
2235	21-40 risers, add	↓	69	.232	↓		9.50		9.50	14.40
3000	Loading & trucking, including 2 mile haul, chute loaded	B-16	45	.711	C.Y.		30.50	12.60	43.10	60
3040	Hand loading truck, 50' haul	"	48	.667			28.50	11.80	40.30	56.50
3080	Machine loading truck	B-17	120	.267			12.05	5.50	17.55	24
5000	Haul, per mile, up to 8 C.Y. truck	B-34B	1165	.007			.32	.49	.81	1.03
5100	Over 8 C.Y. truck	"	1550	.005	↓		.24	.37	.61	.77

02 41 19.20 Selective Demolition, Dump Charges

		Crew	Daily Output	Labor-Hours	Unit	Material	Labor	Equipment	Total	Total Incl O&P
0010	**SELECTIVE DEMOLITION, DUMP CHARGES** R024119-10									
0020	Dump charges, typical urban city, tipping fees only									
0100	Building construction materials				Ton	74			74	81
0200	Trees, brush, lumber					63			63	69.50
0300	Rubbish only					63			63	69.50
0500	Reclamation station, usual charge				↓	74			74	81

02 41 19.27 Selective Demolition, Torch Cutting

		Crew	Daily Output	Labor-Hours	Unit	Material	Labor	Equipment	Total	Total Incl O&P
0010	**SELECTIVE DEMOLITION, TORCH CUTTING** R024119-10									
0020	Steel, 1" thick plate	E-25	333	.024	L.F.	.88	1.39	.04	2.31	3.23
0040	1" diameter bar	"	600	.013	Ea.	.15	.77	.02	.94	1.41
1000	Oxygen lance cutting, reinforced concrete walls									
1040	12"-16" thick walls	1 Clab	10	.800	L.F.		33		33	49.50
1080	24" thick walls	"	6	1.333	"		54.50		54.50	83

For customer support on your Electrical Costs with RSMeans data, call 800.448.8182.

23

02 82 Asbestos Remediation

02 82 13 – Asbestos Abatement

02 82 13.47 Asbestos Waste Pkg., Handling, and Disp.	Crew	Daily Output	Labor-Hours	Unit	Material	2019 Bare Costs Labor	Equipment	Total	Total Incl O&P
0010 **ASBESTOS WASTE PACKAGING, HANDLING, AND DISPOSAL**									
0100 Collect and bag bulk material, 3 C.F. bags, by hand	A-9	400	.160	Ea.	.90	9.20		10.10	15.20
0200 Large production vacuum loader	A-12	880	.073		1.01	4.18	.82	6.01	8.45
1000 Double bag and decontaminate	A-9	960	.067		.90	3.83		4.73	6.90
2000 Containerize bagged material in drums, per 3 C.F. drum	"	800	.080		18.50	4.59		23.09	27.50
3000 Cart bags 50' to dumpster	2 Asbe	400	.040			2.29		2.29	3.54
5000 Disposal charges, not including haul, minimum				C.Y.				61	67
5020 Maximum				"				355	395
9000 For type B (supplied air) respirator equipment, add				%				10%	10%

Estimating Tips
General

- Carefully check all the plans and specifications. Concrete often appears on drawings other than structural drawings, including mechanical and electrical drawings for equipment pads. The cost of cutting and patching is often difficult to estimate. See Subdivision 03 81 for Concrete Cutting, Subdivision 02 41 19.16 for Cutout Demolition, Subdivision 03 05 05.10 for Concrete Demolition, and Subdivision 02 41 19.19 for Rubbish Handling (handling, loading, and hauling of debris).

- Always obtain concrete prices from suppliers near the job site. A volume discount can often be negotiated, depending upon competition in the area. Remember to add for waste, particularly for slabs and footings on grade.

03 10 00 Concrete Forming and Accessories

- A primary cost for concrete construction is forming. Most jobs today are constructed with prefabricated forms. The selection of the forms best suited for the job and the total square feet of forms required for efficient concrete forming and placing are key elements in estimating concrete construction. Enough forms must be available for erection to make efficient use of the concrete placing equipment and crew.

- Concrete accessories for forming and placing depend upon the systems used. Study the plans and specifications to ensure that all special accessory requirements have been included in the cost estimate, such as anchor bolts, inserts, and hangers.

- Included within costs for forms-in-place are all necessary bracing and shoring.

03 20 00 Concrete Reinforcing

- Ascertain that the reinforcing steel supplier has included all accessories, cutting, bending, and an allowance for lapping, splicing, and waste. A good rule of thumb is 10% for lapping, splicing, and waste. Also, 10% waste should be allowed for welded wire fabric.

- The unit price items in the subdivisions for Reinforcing In Place, Glass Fiber Reinforcing, and Welded Wire Fabric include the labor to install accessories such as beam and slab bolsters, high chairs, and bar ties and tie wire. The material cost for these accessories is not included; they may be obtained from the Accessories Subdivisions.

03 30 00 Cast-In-Place Concrete

- When estimating structural concrete, pay particular attention to requirements for concrete additives, curing methods, and surface treatments. Special consideration for climate, hot or cold, must be included in your estimate. Be sure to include requirements for concrete placing equipment and concrete finishing.

- For accurate concrete estimating, the estimator must consider each of the following major components individually: forms, reinforcing steel, ready-mix concrete, placement of the concrete, and finishing of the top surface. For faster estimating, Subdivision 03 30 53.40 for Concrete-In-Place can be used; here, various items of concrete work are presented that include the costs of all five major components (unless specifically stated otherwise).

03 40 00 Precast Concrete
03 50 00 Cast Decks and Underlayment

- The cost of hauling precast concrete structural members is often an important factor. For this reason, it is important to get a quote from the nearest supplier. It may become economically feasible to set up precasting beds on the site if the hauling costs are prohibitive.

Reference Numbers

Reference numbers are shown at the beginning of some major classifications. These numbers refer to related items in the Reference Section. The reference information may be an estimating procedure, an alternate pricing method, or technical information.

Note: Not all subdivisions listed here necessarily appear. ■

03 01 Maintenance of Concrete

03 01 30 – Maintenance of Cast-In-Place Concrete

03 01 30.62 Concrete Patching

	03 01 30.62 Concrete Patching	Crew	Daily Output	Labor-Hours	Unit	Material	2019 Bare Costs Labor	2019 Bare Costs Equipment	Total	Total Incl O&P
0010	**CONCRETE PATCHING**									
0100	Floors, 1/4" thick, small areas, regular grout	1 Cefi	170	.047	S.F.	1.54	2.30		3.84	5.10
0150	Epoxy grout	"	100	.080	"	8.75	3.91		12.66	15.40
2000	Walls, including chipping, cleaning and epoxy grout									
2100	1/4" deep	1 Cefi	65	.123	S.F.	8.60	6		14.60	18.35
2150	1/2" deep		50	.160		17.20	7.80		25	30.50
2200	3/4" deep		40	.200		26	9.80		35.80	43

03 11 Concrete Forming

03 11 13 – Structural Cast-In-Place Concrete Forming

03 11 13.40 Forms In Place, Equipment Foundations

		Crew	Daily Output	Labor-Hours	Unit	Material	Labor	Equipment	Total	Total Incl O&P
0010	**FORMS IN PLACE, EQUIPMENT FOUNDATIONS**									
0020	1 use	C-2	160	.300	SFCA	2.99	15.05		18.04	26.50
0050	2 use		190	.253		1.64	12.70		14.34	21
0100	3 use		200	.240		1.20	12.05		13.25	19.55
0150	4 use		205	.234		.98	11.75		12.73	18.85

03 11 13.45 Forms In Place, Footings

		Crew	Daily Output	Labor-Hours	Unit	Material	Labor	Equipment	Total	Total Incl O&P
0010	**FORMS IN PLACE, FOOTINGS**									
0020	Continuous wall, plywood, 1 use	C-1	375	.085	SFCA	7	4.18		11.18	14.05
0050	2 use		440	.073		3.85	3.56		7.41	9.65
0100	3 use		470	.068		2.80	3.34		6.14	8.15
0150	4 use		485	.066		2.28	3.23		5.51	7.40
5000	Spread footings, job-built lumber, 1 use		305	.105		2.31	5.15		7.46	10.35
5050	2 use		371	.086		1.29	4.23		5.52	7.80
5100	3 use		401	.080		.93	3.91		4.84	6.90
5150	4 use		414	.077		.75	3.79		4.54	6.60

03 11 13.65 Forms In Place, Slab On Grade

		Crew	Daily Output	Labor-Hours	Unit	Material	Labor	Equipment	Total	Total Incl O&P
0010	**FORMS IN PLACE, SLAB ON GRADE**									
3000	Edge forms, wood, 4 use, on grade, to 6" high	C-1	600	.053	L.F.	.31	2.61		2.92	4.29
6000	Trench forms in floor, wood, 1 use		160	.200	SFCA	2.05	9.80		11.85	17.10
6050	2 use		175	.183		1.13	8.95		10.08	14.80
6100	3 use		180	.178		.82	8.70		9.52	14.10
6150	4 use		185	.173		.67	8.50		9.17	13.55

03 15 Concrete Accessories

03 15 05 – Concrete Forming Accessories

03 15 05.75 Sleeves and Chases

		Crew	Daily Output	Labor-Hours	Unit	Material	Labor	Equipment	Total	Total Incl O&P
0010	**SLEEVES AND CHASES**									
0100	Plastic, 1 use, 12" long, 2" diameter	1 Carp	100	.080	Ea.	1.89	4.13		6.02	8.35
0150	4" diameter		90	.089		5.30	4.59		9.89	12.80
0200	6" diameter		75	.107		9.30	5.50		14.80	18.60
0250	12" diameter		60	.133		27.50	6.90		34.40	41
5000	Sheet metal, 2" diameter [G]		100	.080		1.51	4.13		5.64	7.90
5100	4" diameter [G]		90	.089		1.89	4.59		6.48	9.05
5150	6" diameter [G]		75	.107		2.03	5.50		7.53	10.60
5200	12" diameter [G]		60	.133		3.97	6.90		10.87	14.75
6000	Steel pipe, 2" diameter [G]		100	.080		3.12	4.13		7.25	9.70
6100	4" diameter [G]		90	.089		10.85	4.59		15.44	18.90
6150	6" diameter [G]		75	.107		35	5.50		40.50	47

03 15 Concrete Accessories

03 15 05 – Concrete Forming Accessories

03 15 05.75 Sleeves and Chases

		Crew	Daily Output	Labor-Hours	Unit	Material	2019 Bare Costs Labor	Equipment	Total	Total Incl O&P
6200	12" diameter	G 1 Carp	60	.133	Ea.	83.50	6.90		90.40	102

03 15 16 – Concrete Construction Joints

03 15 16.20 Control Joints, Saw Cut

		Crew	Daily Output	Labor-Hours	Unit	Material	Labor	Equipment	Total	Total Incl O&P
0010	**CONTROL JOINTS, SAW CUT**									
0100	Sawcut control joints in green concrete									
0120	1" depth	C-27	2000	.008	L.F.	.04	.39	.06	.49	.68
0140	1-1/2" depth		1800	.009		.05	.43	.06	.54	.77
0160	2" depth	↓	1600	.010	↓	.07	.49	.07	.63	.88
0180	Sawcut joint reservoir in cured concrete									
0182	3/8" wide x 3/4" deep, with single saw blade	C-27	1000	.016	L.F.	.05	.78	.11	.94	1.34
0184	1/2" wide x 1" deep, with double saw blades		900	.018		.10	.87	.12	1.09	1.52
0186	3/4" wide x 1-1/2" deep, with double saw blades	↓	800	.020		.21	.98	.14	1.33	1.82
0190	Water blast joint to wash away laitance, 2 passes	C-29	2500	.003			.13	.03	.16	.23
0200	Air blast joint to blow out debris and air dry, 2 passes	C-28	2000	.004	↓		.20	.01	.21	.30
0300	For backer rod, see Section 07 91 23.10									
0342	For joint sealant, see Section 07 92 13.20									

03 15 19 – Cast-In Concrete Anchors

03 15 19.10 Anchor Bolts

		Crew	Daily Output	Labor-Hours	Unit	Material	Labor	Equipment	Total	Total Incl O&P
0010	**ANCHOR BOLTS**									
0015	Made from recycled materials									
0025	Single bolts installed in fresh concrete, no templates									
0030	Hooked w/nut and washer, 1/2" diameter, 8" long	G 1 Carp	132	.061	Ea.	1.48	3.13		4.61	6.35
0040	12" long	G	131	.061		1.64	3.15		4.79	6.55
0050	5/8" diameter, 8" long	G	129	.062		4.10	3.20		7.30	9.35
0060	12" long	G	127	.063		5.05	3.25		8.30	10.45
0070	3/4" diameter, 8" long	G	127	.063		5.05	3.25		8.30	10.45
0080	12" long	G	↓ 125	.064	↓	6.30	3.31		9.61	11.95

03 15 19.45 Machinery Anchors

		Crew	Daily Output	Labor-Hours	Unit	Material	Labor	Equipment	Total	Total Incl O&P
0010	**MACHINERY ANCHORS,** heavy duty, incl. sleeve, floating base nut,									
0020	lower stud & coupling nut, fiber plug, connecting stud, washer & nut.									
0030	For flush mounted embedment in poured concrete heavy equip. pads.									
0200	Stud & bolt, 1/2" diameter	G E-16	40	.400	Ea.	57.50	23	2.39	82.89	103
0300	5/8" diameter	G	35	.457		67	26	2.73	95.73	118
0500	3/4" diameter	G	30	.533		79.50	30.50	3.19	113.19	140
0600	7/8" diameter	G	25	.640		89.50	36.50	3.83	129.83	161
0800	1" diameter	G	20	.800		89.50	45.50	4.78	139.78	176
0900	1-1/4" diameter	G	↓ 15	1.067	↓	128	61	6.40	195.40	245

03 21 Reinforcement Bars

03 21 11 – Plain Steel Reinforcement Bars

03 21 11.60 Reinforcing In Place

		Crew	Daily Output	Labor-Hours	Unit	Material	Labor	Equipment	Total	Total Incl O&P
0010	**REINFORCING IN PLACE,** 50-60 ton lots, A615 Grade 60									
0020	Includes labor, but not material cost, to install accessories									
0030	Made from recycled materials									
0502	Footings, #4 to #7	G 4 Rodm	4200	.008	Lb.	.51	.42		.93	1.20
0552	#8 to #18	G	7200	.004		.51	.24		.75	.94
0602	Slab on grade, #3 to #7	G	↓ 4200	.008	↓	.51	.42		.93	1.20
0900	For other than 50-60 ton lots									
1000	Under 10 ton job, #3 to #7, add					25%	10%			
1010	#8 to #18, add					20%	10%			

03 21 Reinforcement Bars

03 21 11 – Plain Steel Reinforcement Bars

03 21 11.60 Reinforcing In Place	Crew	Daily Output	Labor-Hours	Unit	Material	2019 Bare Costs Labor	Equipment	Total	Total Incl O&P
1050	10-50 ton job, #3 to #7, add					10%			
1060	#8 to #18, add					5%			
1100	60-100 ton job, #3 to #7, deduct					5%			
1110	#8 to #18, deduct					10%			
1150	Over 100 ton job, #3 to #7, deduct					10%			
1160	#8 to #18, deduct					15%			

03 22 Fabric and Grid Reinforcing

03 22 11 – Plain Welded Wire Fabric Reinforcing

03 22 11.10 Plain Welded Wire Fabric

		Crew	Daily Output	Labor-Hours	Unit	Material	Labor	Equipment	Total	Total Incl O&P
0010	**PLAIN WELDED WIRE FABRIC** ASTM A185									
0020	Includes labor, but not material cost, to install accessories									
0050	Sheets									
0100	6 x 6 - W1.4 x W1.4 (10 x 10) 21 lb./C.S.F. [G]	2 Rodm	35	.457	C.S.F.	15.90	25		40.90	55.50

03 22 13 – Galvanized Welded Wire Fabric Reinforcing

03 22 13.10 Galvanized Welded Wire Fabric

		Crew	Daily Output	Labor-Hours	Unit	Material	Labor	Equipment	Total	Total Incl O&P
0010	**GALVANIZED WELDED WIRE FABRIC**									
0100	Add to plain welded wire pricing for galvanized welded wire				Lb.	.24			.24	.27

03 22 16 – Epoxy-Coated Welded Wire Fabric Reinforcing

03 22 16.10 Epoxy-Coated Welded Wire Fabric

		Crew	Daily Output	Labor-Hours	Unit	Material	Labor	Equipment	Total	Total Incl O&P
0010	**EPOXY-COATED WELDED WIRE FABRIC**									
0100	Add to plain welded wire pricing for epoxy-coated welded wire				Lb.	.42			.42	.46

03 30 Cast-In-Place Concrete

03 30 53 – Miscellaneous Cast-In-Place Concrete

03 30 53.40 Concrete In Place

		Crew	Daily Output	Labor-Hours	Unit	Material	Labor	Equipment	Total	Total Incl O&P
0010	**CONCRETE IN PLACE**									
0020	Including forms (4 uses), Grade 60 rebar, concrete (Portland cement									
0050	Type I), placement and finishing unless otherwise indicated									
3540	Equipment pad (3000 psi), 3' x 3' x 6" thick	C-14H	45	1.067	Ea.	46.50	53.50	.59	100.59	133
3550	4' x 4' x 6" thick		30	1.600		72.50	80.50	.88	153.88	201
3560	5' x 5' x 8" thick		18	2.667		132	134	1.47	267.47	350
3570	6' x 6' x 8" thick		14	3.429		181	172	1.89	354.89	460
3580	8' x 8' x 10" thick		8	6		385	300	3.30	688.30	880
3590	10' x 10' x 12" thick		5	9.600		665	485	5.30	1,155.30	1,475
3800	Footings (3000 psi), spread under 1 C.Y.	C-14C	28	4	C.Y.	193	196	.96	389.96	510
3825	1 C.Y. to 5 C.Y.		43	2.605		227	128	.63	355.63	445
3850	Over 5 C.Y.		75	1.493		212	73	.36	285.36	345
3900	Footings, strip (3000 psi), 18" x 9", unreinforced	C-14L	40	2.400		149	115	.67	264.67	340
3920	18" x 9", reinforced	C-14C	35	3.200		174	157	.77	331.77	430
3925	20" x 10", unreinforced	C-14L	45	2.133		146	103	.60	249.60	315
3930	20" x 10", reinforced	C-14C	40	2.800		166	137	.67	303.67	390
3935	24" x 12", unreinforced	C-14L	55	1.745		143	84	.49	227.49	286
3940	24" x 12", reinforced	C-14C	48	2.333		164	114	.56	278.56	355
3945	36" x 12", unreinforced	C-14L	70	1.371		139	66	.38	205.38	253
3950	36" x 12", reinforced	C-14C	60	1.867		158	91.50	.45	249.95	310
4000	Foundation mat (3000 psi), under 10 C.Y.		38.67	2.896		231	142	.70	373.70	470
4050	Over 20 C.Y.		56.40	1.986		205	97.50	.48	302.98	375

03 30 Cast-In-Place Concrete

03 30 53 – Miscellaneous Cast-In-Place Concrete

03 30 53.40 Concrete In Place

		Crew	Daily Output	Labor-Hours	Unit	Material	2019 Bare Costs Labor	2019 Bare Costs Equipment	Total	Total Incl O&P
4650	Slab on grade (3500 psi), not including finish, 4" thick	C-14E	60.75	1.449	C.Y.	147	72	.43	219.43	270
4700	6" thick	"	92	.957	↓	141	47.50	.29	188.79	228
4701	Thickened slab edge (3500 psi), for slab on grade poured									
4702	monolithically with slab; depth is in addition to slab thickness;									
4703	formed vertical outside edge, earthen bottom and inside slope									
4705	8" deep x 8" wide bottom, unreinforced	C-14L	2190	.044	L.F.	3.99	2.11	.01	6.11	7.60
4710	8" x 8", reinforced	C-14C	1670	.067		6.40	3.29	.02	9.71	12.05
4715	12" deep x 12" wide bottom, unreinforced	C-14L	1800	.053		8.20	2.56	.01	10.77	12.90
4720	12" x 12", reinforced	C-14C	1310	.086		12.60	4.19	.02	16.81	20
4725	16" deep x 16" wide bottom, unreinforced	C-14L	1440	.067		13.90	3.20	.02	17.12	20
4730	16" x 16", reinforced	C-14C	1120	.100		19.15	4.90	.02	24.07	28.50
4735	20" deep x 20" wide bottom, unreinforced	C-14L	1150	.083		21	4.01	.02	25.03	29
4740	20" x 20", reinforced	C-14C	920	.122		28	5.95	.03	33.98	39.50
4745	24" deep x 24" wide bottom, unreinforced	C-14L	930	.103		30	4.96	.03	34.99	40.50
4750	24" x 24", reinforced	C-14C	740	.151	↓	38.50	7.40	.04	45.94	53.50

03 31 Structural Concrete

03 31 13 – Heavyweight Structural Concrete

03 31 13.35 Heavyweight Concrete, Ready Mix

		Crew	Daily Output	Labor-Hours	Unit	Material	2019 Bare Costs Labor	2019 Bare Costs Equipment	Total	Total Incl O&P
0010	**HEAVYWEIGHT CONCRETE, READY MIX**, delivered									
0012	Includes local aggregate, sand, Portland cement (Type I) and water									
0015	Excludes all additives and treatments									
0020	2000 psi				C.Y.	110			110	121
0100	2500 psi					113			113	125
0150	3000 psi					124			124	136
0200	3500 psi					125			125	137
0300	4000 psi					128			128	141
1000	For high early strength (Portland cement Type III), add					10%				
1300	For winter concrete (hot water), add					5.35			5.35	5.90
1410	For mid-range water reducer, add					4.01			4.01	4.41
1420	For high-range water reducer/superplasticizer, add					6.25			6.25	6.90
1430	For retarder, add					3.25			3.25	3.58
1440	For non-Chloride accelerator, add					6.40			6.40	7.05
1450	For Chloride accelerator, per 1%, add					3.91			3.91	4.30
1460	For fiber reinforcing, synthetic (1 lb./C.Y.), add					8.05			8.05	8.85
1500	For Saturday delivery, add				↓	8.25			8.25	9.10
1510	For truck holding/waiting time past 1st hour per load, add				Hr.	105			105	115
1520	For short load (less than 4 C.Y.), add per load				Ea.	92			92	101
2000	For all lightweight aggregate, add				C.Y.	45%				

03 31 13.70 Placing Concrete

		Crew	Daily Output	Labor-Hours	Unit	Material	2019 Bare Costs Labor	2019 Bare Costs Equipment	Total	Total Incl O&P
0010	**PLACING CONCRETE**									
0020	Includes labor and equipment to place, level (strike off) and consolidate									
1900	Footings, continuous, shallow, direct chute	C-6	120	.400	C.Y.		17.10	.44	17.54	26
1950	Pumped	C-20	150	.427			18.80	6	24.80	35
2000	With crane and bucket	C-7	90	.800			35.50	11.80	47.30	66.50
2100	Footings, continuous, deep, direct chute	C-6	140	.343			14.65	.38	15.03	22.50
2150	Pumped	C-20	160	.400			17.60	5.60	23.20	32.50
2200	With crane and bucket	C-7	110	.655			29	9.65	38.65	54.50
2400	Footings, spread, under 1 C.Y., direct chute	C-6	55	.873			37.50	.97	38.47	57
2450	Pumped	C-20	65	.985			43.50	13.80	57.30	80.50
2500	With crane and bucket	C-7	45	1.600	↓		71.50	23.50	95	133

03 31 Structural Concrete

03 31 13 – Heavyweight Structural Concrete

03 31 13.70 Placing Concrete

		Crew	Daily Output	Labor-Hours	Unit	Material	2019 Bare Costs Labor	Equipment	Total	Total Incl O&P
2600	Over 5 C.Y., direct chute	C-6	120	.400	C.Y.		17.10	.44	17.54	26
2650	Pumped	C-20	150	.427			18.80	6	24.80	35
2700	With crane and bucket	C-7	100	.720			32	10.60	42.60	60
2900	Foundation mats, over 20 C.Y., direct chute	C-6	350	.137			5.85	.15	6	8.95
2950	Pumped	C-20	400	.160			7.05	2.24	9.29	13.05
3000	With crane and bucket	C-7	300	.240			10.70	3.53	14.23	20

03 35 Concrete Finishing

03 35 13 – High-Tolerance Concrete Floor Finishing

03 35 13.30 Finishing Floors, High Tolerance

		Crew	Daily Output	Labor-Hours	Unit	Material	2019 Bare Costs Labor	Equipment	Total	Total Incl O&P
0010	**FINISHING FLOORS, HIGH TOLERANCE**									
0012	Finishing of fresh concrete flatwork requires that concrete									
0013	first be placed, struck off & consolidated									
0015	Basic finishing for various unspecified flatwork									
0100	Bull float only	C-10	4000	.006	S.F.		.28		.28	.41
0125	Bull float & manual float		2000	.012			.56		.56	.83
0150	Bull float, manual float & broom finish, w/edging & joints		1850	.013			.60		.60	.89
0200	Bull float, manual float & manual steel trowel		1265	.019			.88		.88	1.31
0210	For specified Random Access Floors in ACI Classes 1, 2, 3 and 4 to achieve									
0215	Composite Overall Floor Flatness and Levelness values up to FF35/FL25									
0250	Bull float, machine float & machine trowel (walk-behind)	C-10C	1715	.014	S.F.		.65	.02	.67	.99
0300	Power screed, bull float, machine float & trowel (walk-behind)	C-10D	2400	.010			.46	.05	.51	.75
0350	Power screed, bull float, machine float & trowel (ride-on)	C-10E	4000	.006			.28	.06	.34	.48
0352	For specified Random Access Floors in ACI Classes 5, 6, 7 and 8 to achieve									
0354	Composite Overall Floor Flatness and Levelness values up to FF50/FL50									
0356	Add for two-dimensional restraightening after power float	C-10	6000	.004	S.F.		.19		.19	.28
0358	For specified Random or Defined Access Floors in ACI Class 9 to achieve									
0360	Composite Overall Floor Flatness and Levelness values up to FF100/FL100									
0362	Add for two-dimensional restraightening after bull float & power float	C-10	3000	.008	S.F.		.37		.37	.55
0364	For specified Superflat Defined Access Floors in ACI Class 9 to achieve									
0366	Minimum Floor Flatness and Levelness values of FF100/FL100									
0368	Add for 2-dim'l restraightening after bull float, power float, power trowel	C-10	2000	.012	S.F.		.56		.56	.83

03 54 Cast Underlayment

03 54 16 – Hydraulic Cement Underlayment

03 54 16.50 Cement Underlayment

		Crew	Daily Output	Labor-Hours	Unit	Material	2019 Bare Costs Labor	Equipment	Total	Total Incl O&P
0010	**CEMENT UNDERLAYMENT**									
2510	Underlayment, P.C. based, self-leveling, 4100 psi, pumped, 1/4" thick	C-8	20000	.003	S.F.	1.67	.13	.04	1.84	2.07
2520	1/2" thick		19000	.003		3.34	.13	.04	3.51	3.92
2530	3/4" thick		18000	.003		5	.14	.05	5.19	5.75
2540	1" thick		17000	.003		6.65	.15	.05	6.85	7.60
2550	1-1/2" thick		15000	.004		10	.17	.06	10.23	11.30
2560	Hand placed, 1/2" thick	C-18	450	.020		3.34	.83	.13	4.30	5.05
2610	Topping, P.C. based, self-leveling, 6100 psi, pumped, 1/4" thick	C-8	20000	.003		2.12	.13	.04	2.29	2.58
2620	1/2" thick		19000	.003		4.25	.13	.04	4.42	4.92
2630	3/4" thick		18000	.003		6.35	.14	.05	6.54	7.25
2660	1" thick		17000	.003		8.50	.15	.05	8.70	9.60
2670	1-1/2" thick		15000	.004		12.75	.17	.06	12.98	14.30
2680	Hand placed, 1/2" thick	C-18	450	.020		4.25	.83	.13	5.21	6.05

03 63 Epoxy Grouting

03 63 05 – Grouting of Dowels and Fasteners

03 63 05.10 Epoxy Only

		Crew	Daily Output	Labor-Hours	Unit	Material	2019 Bare Costs Labor	2019 Bare Costs Equipment	Total	Total Incl O&P
0010	**EPOXY ONLY**									
1500	Chemical anchoring, epoxy cartridge, excludes layout, drilling, fastener									
1530	For fastener 3/4" diam. x 6" embedment	2 Skwk	72	.222	Ea.	5.20	11.85		17.05	24
1535	1" diam. x 8" embedment		66	.242		7.75	12.95		20.70	28.50
1540	1-1/4" diam. x 10" embedment		60	.267		15.55	14.25		29.80	38.50
1545	1-3/4" diam. x 12" embedment		54	.296		26	15.80		41.80	52.50
1550	14" embedment		48	.333		31	17.80		48.80	61
1555	2" diam. x 12" embedment		42	.381		41.50	20.50		62	76.50
1560	18" embedment	↓	32	.500	↓	52	26.50		78.50	97.50

03 81 Concrete Cutting

03 81 13 – Flat Concrete Sawing

03 81 13.50 Concrete Floor/Slab Cutting

		Crew	Daily Output	Labor-Hours	Unit	Material	2019 Bare Costs Labor	2019 Bare Costs Equipment	Total	Total Incl O&P
0010	**CONCRETE FLOOR/SLAB CUTTING**									
0050	Includes blade cost, layout and set-up time									
0300	Saw cut concrete slabs, plain, up to 3" deep	B-89	1060	.015	L.F.	.13	.73	.43	1.29	1.71
0320	Each additional inch of depth		3180	.005		.04	.24	.14	.42	.58
0400	Mesh reinforced, up to 3" deep		980	.016		.14	.79	.46	1.39	1.86
0420	Each additional inch of depth		2940	.005		.05	.26	.15	.46	.62
0500	Rod reinforced, up to 3" deep		800	.020		.18	.97	.57	1.72	2.29
0520	Each additional inch of depth	↓	2400	.007	↓	.06	.32	.19	.57	.77

03 81 13.75 Concrete Saw Blades

		Crew	Daily Output	Labor-Hours	Unit	Material	2019 Bare Costs Labor	2019 Bare Costs Equipment	Total	Total Incl O&P
0010	**CONCRETE SAW BLADES**									
3000	Blades for saw cutting, included in cutting line items									
3020	Diamond, 12" diameter				Ea.	221			221	244
3040	18" diameter					425			425	470
3080	24" diameter					665			665	730
3120	30" diameter					965			965	1,075
3160	36" diameter					1,325			1,325	1,450
3200	42" diameter	↓				2,525			2,525	2,775

03 82 Concrete Boring

03 82 13 – Concrete Core Drilling

03 82 13.10 Core Drilling

		Crew	Daily Output	Labor-Hours	Unit	Material	2019 Bare Costs Labor	2019 Bare Costs Equipment	Total	Total Incl O&P
0010	**CORE DRILLING**									
0015	Includes bit cost, layout and set-up time									
0020	Reinforced concrete slab, up to 6" thick									
0100	1" diameter core	B-89A	17	.941	Ea.	.18	44.50	6.60	51.28	75
0150	For each additional inch of slab thickness in same hole, add		1440	.011		.03	.52	.08	.63	.92
0200	2" diameter core		16.50	.970		.29	46	6.80	53.09	77.50
0250	For each additional inch of slab thickness in same hole, add		1080	.015		.05	.70	.10	.85	1.22
0300	3" diameter core		16	1		.39	47	7	54.39	79.50
0350	For each additional inch of slab thickness in same hole, add		720	.022		.07	1.05	.16	1.28	1.83
0500	4" diameter core		15	1.067		.51	50.50	7.50	58.51	85.50
0550	For each additional inch of slab thickness in same hole, add		480	.033		.08	1.57	.23	1.88	2.74
0700	6" diameter core		14	1.143		.78	54	8	62.78	91.50
0750	For each additional inch of slab thickness in same hole, add		360	.044		.13	2.10	.31	2.54	3.67
0900	8" diameter core		13	1.231		1.09	58	8.65	67.74	98.50
0950	For each additional inch of slab thickness in same hole, add	↓	288	.056	↓	.18	2.62	.39	3.19	4.61

For customer support on your Electrical Costs with RSMeans data, call 800.448.8182.

31

03 82 13 – Concrete Core Drilling

03 82 13.10 Core Drilling

		Crew	Daily Output	Labor-Hours	Unit	Material	2019 Bare Costs Labor	Equipment	Total	Total Incl O&P
1100	10" diameter core	B-89A	12	1.333	Ea.	1.34	63	9.35	73.69	107
1150	For each additional inch of slab thickness in same hole, add		240	.067		.22	3.15	.47	3.84	5.55
1300	12" diameter core		11	1.455		1.87	68.50	10.20	80.57	117
1350	For each additional inch of slab thickness in same hole, add		206	.078		.31	3.67	.54	4.52	6.50
1500	14" diameter core		10	1.600		2.14	75.50	11.20	88.84	130
1550	For each additional inch of slab thickness in same hole, add		180	.089		.36	4.20	.62	5.18	7.45
1700	18" diameter core		9	1.778		2.99	84	12.45	99.44	144
1750	For each additional inch of slab thickness in same hole, add		144	.111		.50	5.25	.78	6.53	9.35
1754	24" diameter core		8	2		4.18	94.50	14	112.68	163
1756	For each additional inch of slab thickness in same hole, add		120	.133		.70	6.30	.93	7.93	11.35
1760	For horizontal holes, add to above						20%	20%		
1770	Prestressed hollow core plank, 8" thick									
1780	1" diameter core	B-89A	17.50	.914	Ea.	.24	43	6.40	49.64	73
1790	For each additional inch of plank thickness in same hole, add		3840	.004		.03	.20	.03	.26	.36
1794	2" diameter core		17.25	.928		.39	44	6.50	50.89	74
1796	For each additional inch of plank thickness in same hole, add		2880	.006		.05	.26	.04	.35	.49
1800	3" diameter core		17	.941		.52	44.50	6.60	51.62	75.50
1810	For each additional inch of plank thickness in same hole, add		1920	.008		.07	.39	.06	.52	.73
1820	4" diameter core		16.50	.970		.68	46	6.80	53.48	78
1830	For each additional inch of plank thickness in same hole, add		1280	.013		.08	.59	.09	.76	1.09
1840	6" diameter core		15.50	1.032		1.04	49	7.25	57.29	83
1850	For each additional inch of plank thickness in same hole, add		960	.017		.13	.79	.12	1.04	1.46
1860	8" diameter core		15	1.067		1.45	50.50	7.50	59.45	86.50
1870	For each additional inch of plank thickness in same hole, add		768	.021		.18	.98	.15	1.31	1.85
1880	10" diameter core		14	1.143		1.79	54	8	63.79	93
1890	For each additional inch of plank thickness in same hole, add		640	.025		.22	1.18	.18	1.58	2.23
1900	12" diameter core		13.50	1.185		2.49	56	8.30	66.79	97
1910	For each additional inch of plank thickness in same hole, add		548	.029		.31	1.38	.20	1.89	2.66

03 82 16 – Concrete Drilling

03 82 16.10 Concrete Impact Drilling

		Crew	Daily Output	Labor-Hours	Unit	Material	2019 Bare Costs Labor	Equipment	Total	Total Incl O&P
0010	**CONCRETE IMPACT DRILLING**									
0020	Includes bit cost, layout and set-up time, no anchors									
0050	Up to 4" deep in concrete/brick floors/walls									
0100	Holes, 1/4" diameter	1 Carp	75	.107	Ea.	.07	5.50		5.57	8.45
0150	For each additional inch of depth in same hole, add		430	.019		.02	.96		.98	1.47
0200	3/8" diameter		63	.127		.05	6.55		6.60	9.95
0250	For each additional inch of depth in same hole, add		340	.024		.01	1.22		1.23	1.86
0300	1/2" diameter		50	.160		.06	8.25		8.31	12.55
0350	For each additional inch of depth in same hole, add		250	.032		.01	1.65		1.66	2.52
0400	5/8" diameter		48	.167		.10	8.60		8.70	13.15
0450	For each additional inch of depth in same hole, add		240	.033		.03	1.72		1.75	2.63
0500	3/4" diameter		45	.178		.13	9.20		9.33	14.05
0550	For each additional inch of depth in same hole, add		220	.036		.03	1.88		1.91	2.88
0600	7/8" diameter		43	.186		.18	9.60		9.78	14.75
0650	For each additional inch of depth in same hole, add		210	.038		.05	1.97		2.02	3.03
0700	1" diameter		40	.200		.17	10.35		10.52	15.85
0750	For each additional inch of depth in same hole, add		190	.042		.04	2.18		2.22	3.34
0800	1-1/4" diameter		38	.211		.28	10.85		11.13	16.75
0850	For each additional inch of depth in same hole, add		180	.044		.07	2.30		2.37	3.55
0900	1-1/2" diameter		35	.229		.51	11.80		12.31	18.40
0950	For each additional inch of depth in same hole, add		165	.048		.13	2.50		2.63	3.93
1000	For ceiling installations, add						40%			

Estimating Tips
05 05 00 Common Work Results for Metals

- Nuts, bolts, washers, connection angles, and plates can add a significant amount to both the tonnage of a structural steel job and the estimated cost. As a rule of thumb, add 10% to the total weight to account for these accessories.

- Type 2 steel construction, commonly referred to as "simple construction," consists generally of field-bolted connections with lateral bracing supplied by other elements of the building, such as masonry walls or x-bracing. The estimator should be aware, however, that shop connections may be accomplished by welding or bolting. The method may be particular to the fabrication shop and may have an impact on the estimated cost.

05 10 00 Structural Steel

- Steel items can be obtained from two sources: a fabrication shop or a metals service center. Fabrication shops can fabricate items under more controlled conditions than crews in the field can. They are also more efficient and can produce items more economically. Metal service centers serve as a source of long mill shapes to both fabrication shops and contractors.

- Most line items in this structural steel subdivision, and most items in 05 50 00 Metal Fabrications, are indicated as being shop fabricated. The bare material cost for these shop fabricated items is the "Invoice Cost" from the shop and includes the mill base price of steel plus mill extras, transportation to the shop, shop drawings and detailing where warranted, shop fabrication and handling, sandblasting and a shop coat of primer paint, all necessary structural bolts, and delivery to the job site. The bare labor cost and bare equipment cost for these shop fabricated items are for field installation or erection.

- Line items in Subdivision 05 12 23.40 Lightweight Framing, and other items scattered in Division 5, are indicated as being field fabricated. The bare material cost for these field fabricated items is the "Invoice Cost" from the metals service center and includes the mill base price of steel plus mill extras, transportation to the metals service center, material handling, and delivery of long lengths of mill shapes to the job site. Material costs for structural bolts and welding rods should be added to the estimate. The bare labor cost and bare equipment cost for these items are for both field fabrication and field installation or erection, and include time for cutting, welding, and drilling in the fabricated metal items. Drilling into concrete and fasteners to fasten field fabricated items to other work is not included and should be added to the estimate.

05 20 00 Steel Joist Framing

- In any given project the total weight of open web steel joists is determined by the loads to be supported and the design. However, economies can be realized in minimizing the amount of labor used to place the joists. This is done by maximizing the joist spacing and therefore minimizing the number of joists required to be installed on the job. Certain spacings and locations may be required by the design, but in other cases maximizing the spacing and keeping it as uniform as possible will keep the costs down.

05 30 00 Steel Decking

- The takeoff and estimating of a metal deck involve more than the area of the floor or roof and the type of deck specified or shown on the drawings. Many different sizes and types of openings may exist. Small openings for individual pipes or conduits may be drilled after the floor/roof is installed, but larger openings may require special deck lengths as well as reinforcing or structural support. The estimator should determine who will be supplying this reinforcing. Additionally, some deck terminations are part of the deck package, such as screed angles and pour stops, and others will be part of the steel contract, such as angles attached to structural members and cast-in-place angles and plates. The estimator must ensure that all pieces are accounted for in the complete estimate.

05 50 00 Metal Fabrications

- The most economical steel stairs are those that use common materials, standard details, and most importantly, a uniform and relatively simple method of field assembly. Commonly available A36/A992 channels and plates are very good choices for the main stringers of the stairs, as are angles and tees for the carrier members. Risers and treads are usually made by specialty shops, and it is most economical to use a typical detail in as many places as possible. The stairs should be pre-assembled and shipped directly to the site. The field connections should be simple and straightforward enough to be accomplished efficiently, and with minimum equipment and labor.

Reference Numbers

Reference numbers are shown at the beginning of some major classifications. These numbers refer to related items in the Reference Section. The reference information may be an estimating procedure, an alternate pricing method, or technical information.

Note: Not all subdivisions listed here necessarily appear. ∎

Did you know?

RSMeans data is available through our online application:

- Search for costs by keyword
- Leverage the most up-to-date data
- Build and export estimates

Try it free
rsmeans.com/2019freetrial

05 05 19.10 Chemical Anchors

		Crew	Daily Output	Labor-Hours	Unit	Material	2019 Bare Costs Labor	Equipment	Total	Total Incl O&P
0010	**CHEMICAL ANCHORS**									
0020	Includes layout & drilling									
1430	Chemical anchor, w/rod & epoxy cartridge, 3/4" diameter x 9-1/2" long	B-89A	27	.593	Ea.	10.90	28	4.15	43.05	59
1435	1" diameter x 11-3/4" long		24	.667		18.80	31.50	4.67	54.97	73.50
1440	1-1/4" diameter x 14" long		21	.762		37.50	36	5.35	78.85	102
1445	1-3/4" diameter x 15" long		20	.800		72	38	5.60	115.60	143
1450	18" long		17	.941		86	44.50	6.60	137.10	169
1455	2" diameter x 18" long		16	1		114	47	7	168	204
1460	24" long		15	1.067		148	50.50	7.50	206	248

05 05 19.20 Expansion Anchors

			Crew	Daily Output	Labor-Hours	Unit	Material	2019 Bare Costs Labor	Equipment	Total	Total Incl O&P
0010	**EXPANSION ANCHORS**										
0100	Anchors for concrete, brick or stone, no layout and drilling										
0200	Expansion shields, zinc, 1/4" diameter, 1-5/16" long, single	G	1 Carp	90	.089	Ea.	.45	4.59		5.04	7.45
0300	1-3/8" long, double	G		85	.094		.59	4.86		5.45	8
0400	3/8" diameter, 1-1/2" long, single	G		85	.094		.70	4.86		5.56	8.10
0500	2" long, double	G		80	.100		1.18	5.15		6.33	9.10
0600	1/2" diameter, 2-1/16" long, single	G		80	.100		1.37	5.15		6.52	9.30
0700	2-1/2" long, double	G		75	.107		2.17	5.50		7.67	10.75
0800	5/8" diameter, 2-5/8" long, single	G		75	.107		2.19	5.50		7.69	10.75
0900	2-3/4" long, double	G		70	.114		2.97	5.90		8.87	12.20
1000	3/4" diameter, 2-3/4" long, single	G		70	.114		3.78	5.90		9.68	13.10
1100	3-15/16" long, double	G		65	.123		5.55	6.35		11.90	15.70
2100	Hollow wall anchors for gypsum wall board, plaster or tile										
2300	1/8" diameter, short	G	1 Carp	160	.050	Ea.	.37	2.58		2.95	4.32
2400	Long	G		150	.053		.32	2.75		3.07	4.52
2500	3/16" diameter, short	G		150	.053		.52	2.75		3.27	4.74
2600	Long	G		140	.057		.68	2.95		3.63	5.20
2700	1/4" diameter, short	G		140	.057		.76	2.95		3.71	5.30
2800	Long	G		130	.062		.86	3.18		4.04	5.75
3000	Toggle bolts, bright steel, 1/8" diameter, 2" long	G		85	.094		.26	4.86		5.12	7.65
3100	4" long	G		80	.100		.29	5.15		5.44	8.10
3400	1/4" diameter, 3" long	G		75	.107		.46	5.50		5.96	8.85
3500	6" long	G		70	.114		.59	5.90		6.49	9.60
3600	3/8" diameter, 3" long	G		70	.114		.86	5.90		6.76	9.90
3700	6" long	G		60	.133		1.71	6.90		8.61	12.30
3800	1/2" diameter, 4" long	G		60	.133		1.68	6.90		8.58	12.25
3900	6" long	G		50	.160		2.25	8.25		10.50	15
4000	Nailing anchors										
4100	Nylon nailing anchor, 1/4" diameter, 1" long		1 Carp	3.20	2.500	C	24	129		153	222
4200	1-1/2" long			2.80	2.857		25.50	148		173.50	252
4300	2" long			2.40	3.333		28	172		200	292
4400	Metal nailing anchor, 1/4" diameter, 1" long	G		3.20	2.500		22.50	129		151.50	220
4500	1-1/2" long	G		2.80	2.857		29.50	148		177.50	256
4600	2" long	G		2.40	3.333		36.50	172		208.50	300
5000	Screw anchors for concrete, masonry,										
5100	stone & tile, no layout or drilling included										
5700	Lag screw shields, 1/4" diameter, short	G	1 Carp	90	.089	Ea.	.48	4.59		5.07	7.50
5900	3/8" diameter, short	G		85	.094		.73	4.86		5.59	8.15
6100	1/2" diameter, short	G		80	.100		1.04	5.15		6.19	8.95
6300	5/8" diameter, short	G		70	.114		1.56	5.90		7.46	10.65
6600	Lead, #6 & #8, 3/4" long	G		260	.031		.19	1.59		1.78	2.61
6700	#10 - #14, 1-1/2" long	G		200	.040		.40	2.07		2.47	3.57

05 05 Common Work Results for Metals

05 05 19 – Post-Installed Concrete Anchors

05 05 19.20 Expansion Anchors

		Crew	Daily Output	Labor-Hours	Unit	Material	2019 Bare Costs Labor	Equipment	Total	Total Incl O&P
6800	#16 & #18, 1-1/2" long	G 1 Carp	160	.050	Ea.	.42	2.58		3	4.37
6900	Plastic, #6 & #8, 3/4" long		260	.031		.05	1.59		1.64	2.46
7100	#10 & #12, 1" long		220	.036		.08	1.88		1.96	2.93
8000	Wedge anchors, not including layout or drilling									
8050	Carbon steel, 1/4" diameter, 1-3/4" long	G 1 Carp	150	.053	Ea.	.54	2.75		3.29	4.76
8100	3-1/4" long	G	140	.057		.71	2.95		3.66	5.25
8150	3/8" diameter, 2-1/4" long	G	145	.055		.54	2.85		3.39	4.91
8200	5" long	G	140	.057		.95	2.95		3.90	5.50
8250	1/2" diameter, 2-3/4" long	G	140	.057		1.05	2.95		4	5.65
8300	7" long	G	125	.064		1.80	3.31		5.11	7
8350	5/8" diameter, 3-1/2" long	G	130	.062		1.79	3.18		4.97	6.80
8400	8-1/2" long	G	115	.070		3.81	3.59		7.40	9.65
8450	3/4" diameter, 4-1/4" long	G	115	.070		2.79	3.59		6.38	8.50
8500	10" long	G	95	.084		6.35	4.35		10.70	13.60
8550	1" diameter, 6" long	G	100	.080		9	4.13		13.13	16.15
8575	9" long	G	85	.094		11.70	4.86		16.56	20
8600	12" long	G	75	.107		12.65	5.50		18.15	22.50
8650	1-1/4" diameter, 9" long	G	70	.114		22	5.90		27.90	33
8700	12" long	G	60	.133		28	6.90		34.90	41.50
8750	For type 303 stainless steel, add					350%				
8800	For type 316 stainless steel, add					450%				
8950	Self-drilling concrete screw, hex washer head, 3/16" diam. x 1-3/4" long	G 1 Carp	300	.027	Ea.	.19	1.38		1.57	2.29
8960	2-1/4" long	G	250	.032		.21	1.65		1.86	2.73
8970	Phillips flat head, 3/16" diam. x 1-3/4" long	G	300	.027		.19	1.38		1.57	2.29
8980	2-1/4" long	G	250	.032		.21	1.65		1.86	2.73

05 05 21 – Fastening Methods for Metal

05 05 21.15 Drilling Steel

		Crew	Daily Output	Labor-Hours	Unit	Material	2019 Bare Costs Labor	Equipment	Total	Total Incl O&P
0010	**DRILLING STEEL**									
1910	Drilling & layout for steel, up to 1/4" deep, no anchor									
1920	Holes, 1/4" diameter	1 Sswk	112	.071	Ea.	.09	4		4.09	6.50
1925	For each additional 1/4" depth, add		336	.024		.09	1.33		1.42	2.23
1930	3/8" diameter		104	.077		.09	4.31		4.40	6.95
1935	For each additional 1/4" depth, add		312	.026		.09	1.44		1.53	2.39
1940	1/2" diameter		96	.083		.10	4.67		4.77	7.55
1945	For each additional 1/4" depth, add		288	.028		.10	1.56		1.66	2.59
1950	5/8" diameter		88	.091		.17	5.10		5.27	8.30
1955	For each additional 1/4" depth, add		264	.030		.17	1.70		1.87	2.89
1960	3/4" diameter		80	.100		.19	5.60		5.79	9.15
1965	For each additional 1/4" depth, add		240	.033		.19	1.87		2.06	3.19
1970	7/8" diameter		72	.111		.27	6.20		6.47	10.25
1975	For each additional 1/4" depth, add		216	.037		.27	2.07		2.34	3.61
1980	1" diameter		64	.125		.25	7		7.25	11.40
1985	For each additional 1/4" depth, add		192	.042		.25	2.33		2.58	3.99
1990	For drilling up, add						40%			

05 05 23 – Metal Fastenings

05 05 23.10 Bolts and Hex Nuts

		Crew	Daily Output	Labor-Hours	Unit	Material	2019 Bare Costs Labor	Equipment	Total	Total Incl O&P
0010	**BOLTS & HEX NUTS**, Steel, A307									
0100	1/4" diameter, 1/2" long	G 1 Sswk	140	.057	Ea.	.06	3.20		3.26	5.15
0200	1" long	G	140	.057		.07	3.20		3.27	5.20
0300	2" long	G	130	.062		.10	3.45		3.55	5.60
0400	3" long	G	130	.062		.15	3.45		3.60	5.65
0500	4" long	G	120	.067		.17	3.73		3.90	6.15

For customer support on your Electrical Costs with RSMeans data, call 800.448.8182.

35

05 05 23 – Metal Fastenings

05 05 23.10 Bolts and Hex Nuts

		Crew	Daily Output	Labor-Hours	Unit	Material	2019 Bare Costs Labor	2019 Bare Costs Equipment	Total	Total Incl O&P
0600	3/8" diameter, 1" long	G 1 Sswk	130	.062	Ea.	.14	3.45		3.59	5.65
0700	2" long	G	130	.062		.18	3.45		3.63	5.70
0800	3" long	G	120	.067		.24	3.73		3.97	6.20
0900	4" long	G	120	.067		.30	3.73		4.03	6.30
1000	5" long	G	115	.070		.38	3.90		4.28	6.60
1100	1/2" diameter, 1-1/2" long	G	120	.067		.48	3.73		4.21	6.50
1200	2" long	G	120	.067		.56	3.73		4.29	6.55
1300	4" long	G	115	.070		.92	3.90		4.82	7.20
1400	6" long	G	110	.073		1.30	4.07		5.37	7.95
1500	8" long	G	105	.076		1.73	4.27		6	8.70
1600	5/8" diameter, 1-1/2" long	G	120	.067		.85	3.73		4.58	6.90
1700	2" long	G	120	.067		.94	3.73		4.67	7
1800	4" long	G	115	.070		1.35	3.90		5.25	7.70
1900	6" long	G	110	.073		1.73	4.07		5.80	8.40
2000	8" long	G	105	.076		2.56	4.27		6.83	9.60
2100	10" long	G	100	.080		3.22	4.48		7.70	10.70
2200	3/4" diameter, 2" long	G	120	.067		1.13	3.73		4.86	7.20
2300	4" long	G	110	.073		1.61	4.07		5.68	8.25
2400	6" long	G	105	.076		2.07	4.27		6.34	9.10
2500	8" long	G	95	.084		3.12	4.72		7.84	10.95
2600	10" long	G	85	.094		4.09	5.25		9.34	12.90
2700	12" long	G	80	.100		4.79	5.60		10.39	14.20
2800	1" diameter, 3" long	G	105	.076		2.91	4.27		7.18	10
2900	6" long	G	90	.089		4.24	4.98		9.22	12.60
3000	12" long	G	75	.107		7.60	5.95		13.55	17.90
3100	For galvanized, add					75%				
3200	For stainless, add					350%				

05 05 23.30 Lag Screws

		Crew	Daily Output	Labor-Hours	Unit	Material	2019 Bare Costs Labor	2019 Bare Costs Equipment	Total	Total Incl O&P
0010	**LAG SCREWS**									
0020	Steel, 1/4" diameter, 2" long	G 1 Carp	200	.040	Ea.	.10	2.07		2.17	3.24
0100	3/8" diameter, 3" long	G	150	.053		.28	2.75		3.03	4.48
0200	1/2" diameter, 3" long	G	130	.062		.69	3.18		3.87	5.55
0300	5/8" diameter, 3" long	G	120	.067		1.31	3.44		4.75	6.65

05 05 23.35 Machine Screws

		Crew	Daily Output	Labor-Hours	Unit	Material	2019 Bare Costs Labor	2019 Bare Costs Equipment	Total	Total Incl O&P
0010	**MACHINE SCREWS**									
0020	Steel, round head, #8 x 1" long	G 1 Carp	4.80	1.667	C	4.13	86		90.13	135
0110	#8 x 2" long	G	2.40	3.333		8.40	172		180.40	270
0200	#10 x 1" long	G	4	2		5.05	103		108.05	162
0300	#10 x 2" long	G	2	4		9	207		216	325

05 05 23.50 Powder Actuated Tools and Fasteners

		Crew	Daily Output	Labor-Hours	Unit	Material	2019 Bare Costs Labor	2019 Bare Costs Equipment	Total	Total Incl O&P
0010	**POWDER ACTUATED TOOLS & FASTENERS**									
0020	Stud driver, .22 caliber, single shot				Ea.	162			162	178
0100	.27 caliber, semi automatic, strip				"	460			460	505
0300	Powder load, single shot, .22 cal, power level 2, brown				C	6.50			6.50	7.15
0400	Strip, .27 cal, power level 4, red					9.30			9.30	10.20
0600	Drive pin, .300 x 3/4" long	G 1 Carp	4.80	1.667		4.43	86		90.43	135
0700	.300 x 3" long with washer	G "	4	2		13.50	103		116.50	171

05 05 23.70 Structural Blind Bolts

		Crew	Daily Output	Labor-Hours	Unit	Material	2019 Bare Costs Labor	2019 Bare Costs Equipment	Total	Total Incl O&P
0010	**STRUCTURAL BLIND BOLTS**									
0100	1/4" diameter x 1/4" grip	G 1 Sswk	240	.033	Ea.	1.76	1.87		3.63	4.92
0150	1/2" grip	G	216	.037		1.87	2.07		3.94	5.35
0200	3/8" diameter x 1/2" grip	G	232	.034		3.46	1.93		5.39	6.90

05 05 Common Work Results for Metals

05 05 23 – Metal Fastenings

05 05 23.70 Structural Blind Bolts		Crew	Daily Output	Labor-Hours	Unit	Material	2019 Bare Costs Labor	Equipment	Total	Total Incl O&P
0250	3/4" grip	G	1 Sswk	208	.038	Ea.	3.33	2.15		5.48	7.10
0300	1/2" diameter x 1/2" grip	G		224	.036		6.15	2		8.15	10
0350	3/4" grip	G		200	.040		8.35	2.24		10.59	12.75
0400	5/8" diameter x 3/4" grip	G		216	.037		11.10	2.07		13.17	15.55
0450	1" grip	G		192	.042		15.50	2.33		17.83	21

05 12 Structural Steel Framing

05 12 23 – Structural Steel for Buildings

05 12 23.60 Pipe Support Framing

		Crew	Daily Output	Labor-Hours	Unit	Material	2019 Bare Costs Labor	Equipment	Total	Total Incl O&P
0010	**PIPE SUPPORT FRAMING**									
0020	Under 10#/L.F., shop fabricated	G	E-4	3900	.008	Lb.	1.77	.46	.02	2.25	2.72
0200	10.1 to 15#/L.F.	G		4300	.007		1.74	.42	.02	2.18	2.61
0400	15.1 to 20#/L.F.	G		4800	.007		1.72	.38	.02	2.12	2.51
0600	Over 20#/L.F.	G		5400	.006		1.69	.34	.02	2.05	2.41

05 35 Raceway Decking Assemblies

05 35 13 – Steel Cellular Decking

05 35 13.50 Cellular Decking

		Crew	Daily Output	Labor-Hours	Unit	Material	2019 Bare Costs Labor	Equipment	Total	Total Incl O&P
0010	**CELLULAR DECKING**										
0015	Made from recycled materials										
0200	Cellular units, galv, 1-1/2" deep, Type BC, 20-20 ga., over 15 squares	G	E-4	1460	.022	S.F.	9.95	1.24	.07	11.26	13
0250	18-20 ga.	G		1420	.023		11.45	1.27	.07	12.79	14.70
0300	18-18 ga.	G		1390	.023		11.75	1.30	.07	13.12	15.05
0320	16-18 ga.	G		1360	.024		13.80	1.33	.07	15.20	17.40
0340	16-16 ga.	G		1330	.024		15.40	1.36	.07	16.83	19.15
0400	3" deep, Type NC, galvanized, 20-20 ga.	G		1375	.023		10.95	1.31	.07	12.33	14.25
0500	18-20 ga.	G		1350	.024		13.40	1.34	.07	14.81	16.95
0600	18-18 ga.	G		1290	.025		13.35	1.40	.07	14.82	17
0700	16-18 ga.	G		1230	.026		14.85	1.47	.08	16.40	18.80
0800	16-16 ga.	G		1150	.028		16.20	1.57	.08	17.85	20.50
1000	4-1/2" deep, Type JC, galvanized, 18-20 ga.	G		1100	.029		15.45	1.64	.09	17.18	19.70
1100	18-18 ga.	G		1040	.031		15.35	1.74	.09	17.18	19.75
1200	16-18 ga.	G		980	.033		17.10	1.84	.10	19.04	22
1300	16-16 ga.	G		935	.034		18.60	1.93	.10	20.63	23.50
1900	For multi-story or congested site, add							50%			

Division Notes

	CREW	DAILY OUTPUT	LABOR-HOURS	UNIT	BARE COSTS				TOTAL INCL O&P
					MAT.	LABOR	EQUIP.	TOTAL	

Estimating Tips
06 05 00 Common Work Results for Wood, Plastics, and Composites

- Common to any wood-framed structure are the accessory connector items such as screws, nails, adhesives, hangers, connector plates, straps, angles, and hold-downs. For typical wood-framed buildings, such as residential projects, the aggregate total for these items can be significant, especially in areas where seismic loading is a concern. For floor and wall framing, the material cost is based on 10 to 25 lbs. of accessory connectors per MBF. Hold-downs, hangers, and other connectors should be taken off by the piece.

  Included with material costs are fasteners for a normal installation. Gordian's RSMeans engineers use manufacturers' recommendations, written specifications, and/or standard construction practice for the sizing and spacing of fasteners. Prices for various fasteners are shown for informational purposes only. Adjustments should be made if unusual fastening conditions exist.

06 10 00 Carpentry

- Lumber is a traded commodity and therefore sensitive to supply and demand in the marketplace. Even with "budgetary" estimating of wood-framed projects, it is advisable to call local suppliers for the latest market pricing.

- The common quantity unit for wood-framed projects is "thousand board feet" (MBF). A board foot is a volume of wood—1" x 1' x 1' or 144 cubic inches. Board-foot quantities are generally calculated using nominal material dimensions—dressed sizes are ignored. Board foot per lineal foot of any stick of lumber can be calculated by dividing the nominal cross-sectional area by 12. As an example, 2,000 lineal feet of 2 x 12 equates to 4 MBF by dividing the nominal area, 2 x 12, by 12, which equals 2, and multiplying that by 2,000 to give 4,000 board feet. This simple rule applies to all nominal dimensioned lumber.

- Waste is an issue of concern at the quantity takeoff for any area of construction. Framing lumber is sold in even foot lengths, i.e., 8', 10', 12', 14', 16', and depending on spans, wall heights, and the grade of lumber, waste is inevitable. A rule of thumb for lumber waste is 5–10% depending on material quality and the complexity of the framing.

- Wood in various forms and shapes is used in many projects, even where the main structural framing is steel, concrete, or masonry. Plywood as a back-up partition material and 2x boards used as blocking and cant strips around roof edges are two common examples. The estimator should ensure that the costs of all wood materials are included in the final estimate.

06 20 00 Finish Carpentry

- It is necessary to consider the grade of workmanship when estimating labor costs for erecting millwork and an interior finish. In practice, there are three grades: premium, custom, and economy. The RSMeans daily output for base and case moldings is in the range of 200 to 250 L.F. per carpenter per day. This is appropriate for most average custom-grade projects. For premium projects, an adjustment to productivity of 25–50% should be made, depending on the complexity of the job.

Reference Numbers
Reference numbers are shown at the beginning of some major classifications. These numbers refer to related items in the Reference Section. The reference information may be an estimating procedure, an alternate pricing method, or technical information.

Note: Not all subdivisions listed here necessarily appear. ■

06 16 Sheathing

06 16 36 – Wood Panel Product Sheathing

06 16 36.10 Sheathing		Crew	Daily Output	Labor-Hours	Unit	Material	2019 Bare Costs Labor	Equipment	Total	Total Incl O&P
0010	**SHEATHING**									
0012	Plywood on roofs, CDX									
0030	5/16" thick	2 Carp	1600	.010	S.F.	.59	.52		1.11	1.43
0035	Pneumatic nailed		1952	.008		.59	.42		1.01	1.29
0050	3/8" thick		1525	.010		.61	.54		1.15	1.50
0055	Pneumatic nailed		1860	.009		.61	.44		1.05	1.35
0100	1/2" thick		1400	.011		.63	.59		1.22	1.59
0105	Pneumatic nailed		1708	.009		.63	.48		1.11	1.43
0200	5/8" thick		1300	.012		.78	.64		1.42	1.82
0205	Pneumatic nailed		1586	.010		.78	.52		1.30	1.65
0300	3/4" thick		1200	.013		.94	.69		1.63	2.07
0305	Pneumatic nailed		1464	.011		.94	.56		1.50	1.88
0500	Plywood on walls, with exterior CDX, 3/8" thick		1200	.013		.61	.69		1.30	1.72
0505	Pneumatic nailed		1488	.011		.61	.56		1.17	1.52
0600	1/2" thick		1125	.014		.63	.73		1.36	1.81
0605	Pneumatic nailed		1395	.011		.63	.59		1.22	1.60
0700	5/8" thick		1050	.015		.78	.79		1.57	2.05
0705	Pneumatic nailed		1302	.012		.78	.63		1.41	1.82
0800	3/4" thick		975	.016		.94	.85		1.79	2.31
0805	Pneumatic nailed		1209	.013		.94	.68		1.62	2.06

Estimating Tips

07 10 00 Dampproofing and Waterproofing

- Be sure of the job specifications before pricing this subdivision. The difference in cost between waterproofing and dampproofing can be great. Waterproofing will hold back standing water. Dampproofing prevents the transmission of water vapor. Also included in this section are vapor retarding membranes.

07 20 00 Thermal Protection

- Insulation and fireproofing products are measured by area, thickness, volume, or R-value. Specifications may give only what the specific R-value should be in a certain situation. The estimator may need to choose the type of insulation to meet that R-value.

07 30 00 Steep Slope Roofing
07 40 00 Roofing and Siding Panels

- Many roofing and siding products are bought and sold by the square. One square is equal to an area that measures 100 square feet.

  This simple change in unit of measure could create a large error if the estimator is not observant. Accessories necessary for a complete installation must be figured into any calculations for both material and labor.

07 50 00 Membrane Roofing
07 60 00 Flashing and Sheet Metal
07 70 00 Roofing and Wall Specialties and Accessories

- The items in these subdivisions compose a roofing system. No one component completes the installation, and all must be estimated. Built-up or single-ply membrane roofing systems are made up of many products and installation trades. Wood blocking at roof perimeters or penetrations, parapet coverings, reglets, roof drains, gutters, downspouts, sheet metal flashing, skylights, smoke vents, and roof hatches all need to be considered along with the roofing material. Several different installation trades will need to work together on the roofing system. Inherent difficulties in the scheduling and coordination of various trades must be accounted for when estimating labor costs.

07 90 00 Joint Protection

- To complete the weather-tight shell, the sealants and caulkings must be estimated. Where different materials meet—at expansion joints, at flashing penetrations, and at hundreds of other locations throughout a construction project—caulking and sealants provide another line of defense against water penetration. Often, an entire system is based on the proper location and placement of caulking or sealants. The detailed drawings that are included as part of a set of architectural plans show typical locations for these materials. When caulking or sealants are shown at typical locations, this means the estimator must include them for all the locations where this detail is applicable. Be careful to keep different types of sealants separate, and remember to consider backer rods and primers if necessary.

Reference Numbers

Reference numbers are shown at the beginning of some major classifications. These numbers refer to related items in the Reference Section. The reference information may be an estimating procedure, an alternate pricing method, or technical information.

Note: Not all subdivisions listed here necessarily appear. ■

07 84 13 – Penetration Firestopping

07 84 13.10 Firestopping		Crew	Daily Output	Labor-Hours	Unit	Material	2019 Bare Costs Labor	2019 Bare Costs Equipment	Total	Total Incl O&P
0010	**FIRESTOPPING** R078413-30									
0100	Metallic piping, non insulated									
0110	Through walls, 2" diameter	1 Carp	16	.500	Ea.	3.95	26		29.95	43.50
0120	4" diameter		14	.571		6.90	29.50		36.40	52
0130	6" diameter		12	.667		10.15	34.50		44.65	63
0140	12" diameter		10	.800		20	41.50		61.50	85
0150	Through floors, 2" diameter		32	.250		2.01	12.90		14.91	22
0160	4" diameter		28	.286		3.53	14.75		18.28	26.50
0170	6" diameter		24	.333		5.15	17.20		22.35	31.50
0180	12" diameter		20	.400		10.35	20.50		30.85	43
0190	Metallic piping, insulated									
0200	Through walls, 2" diameter	1 Carp	16	.500	Ea.	6.90	26		32.90	46.50
0210	4" diameter		14	.571		10.15	29.50		39.65	55.50
0220	6" diameter		12	.667		13.60	34.50		48.10	67
0230	12" diameter		10	.800		23	41.50		64.50	88
0240	Through floors, 2" diameter		32	.250		3.53	12.90		16.43	23.50
0250	4" diameter		28	.286		5.15	14.75		19.90	28
0260	6" diameter		24	.333		6.95	17.20		24.15	33.50
0270	12" diameter		20	.400		10.60	20.50		31.10	43
0280	Non metallic piping, non insulated									
0290	Through walls, 2" diameter	1 Carp	12	.667	Ea.	49	34.50		83.50	106
0300	4" diameter		10	.800		97	41.50		138.50	170
0310	6" diameter		8	1		183	51.50		234.50	280
0330	Through floors, 2" diameter		16	.500		32.50	26		58.50	75
0340	4" diameter		6	1.333		61	69		130	171
0350	6" diameter		6	1.333		118	69		187	234
0370	Ductwork, insulated & non insulated, round									
0380	Through walls, 6" diameter	1 Carp	12	.667	Ea.	13.60	34.50		48.10	67
0390	12" diameter		10	.800		23	41.50		64.50	88
0400	18" diameter		8	1		28	51.50		79.50	109
0410	Through floors, 6" diameter		16	.500		6.95	26		32.95	46.50
0420	12" diameter		14	.571		11.85	29.50		41.35	57.50
0430	18" diameter		12	.667		14.40	34.50		48.90	68
0440	Ductwork, insulated & non insulated, rectangular									
0450	With stiffener/closure angle, through walls, 6" x 12"	1 Carp	8	1	Ea.	18.15	51.50		69.65	98
0460	12" x 24"		6	1.333		35.50	69		104.50	143
0470	24" x 48"		4	2		76	103		179	240
0480	With stiffener/closure angle, through floors, 6" x 12"		10	.800		9.05	41.50		50.55	72.50
0490	12" x 24"		8	1		18.15	51.50		69.65	98
0500	24" x 48"		6	1.333		38	69		107	146
0510	Multi trade openings									
0520	Through walls, 6" x 12"	1 Carp	2	4	Ea.	49.50	207		256.50	370
0530	12" x 24"	"	1	8		172	415		587	815
0540	24" x 48"	2 Carp	1	16		590	825		1,415	1,900
0550	48" x 96"	"	.75	21.333		2,200	1,100		3,300	4,100
0560	Through floors, 6" x 12"	1 Carp	2	4		46.50	207		253.50	365
0570	12" x 24"	"	1	8		63	415		478	695
0580	24" x 48"	2 Carp	.75	21.333		151	1,100		1,251	1,850
0590	48" x 96"	"	.50	32		360	1,650		2,010	2,900
0600	Structural penetrations, through walls									
0610	Steel beams, W8 x 10	1 Carp	8	1	Ea.	75	51.50		126.50	161
0620	W12 x 14		6	1.333		109	69		178	224
0630	W21 x 44		5	1.600		147	82.50		229.50	287

For customer support on your Electrical Costs with RSMeans data, call 800.448.8182.

07 84 Firestopping

07 84 13 – Penetration Firestopping

07 84 13.10 Firestopping

07 84 13.10 Firestopping		Crew	Daily Output	Labor-Hours	Unit	Material	2019 Bare Costs Labor	2019 Bare Costs Equipment	Total	Total Incl O&P
0640	W36 x 135	1 Carp	3	2.667	Ea.	241	138		379	475
0650	Bar joists, 18" deep		6	1.333		44.50	69		113.50	153
0660	24" deep		6	1.333		59	69		128	169
0670	36" deep		5	1.600		89	82.50		171.50	223
0680	48" deep		4	2		119	103		222	286
0690	Construction joints, floor slab at exterior wall									
0700	Precast, brick, block or drywall exterior									
0710	2" wide joint	1 Carp	125	.064	L.F.	9.85	3.31		13.16	15.85
0720	4" wide joint	"	75	.107	"	16.35	5.50		21.85	26.50
0730	Metal panel, glass or curtain wall exterior									
0740	2" wide joint	1 Carp	40	.200	L.F.	10.65	10.35		21	27.50
0750	4" wide joint	"	25	.320	"	16.65	16.55		33.20	43.50
0760	Floor slab to drywall partition									
0770	Flat joint	1 Carp	100	.080	L.F.	14.55	4.13		18.68	22.50
0780	Fluted joint		50	.160		16.15	8.25		24.40	30.50
0790	Etched fluted joint		75	.107		20.50	5.50		26	31
0800	Floor slab to concrete/masonry partition									
0810	Flat joint	1 Carp	75	.107	L.F.	14.50	5.50		20	24.50
0820	Fluted joint	"	50	.160	"	16.55	8.25		24.80	30.50
0830	Concrete/CMU wall joints									
0840	1" wide	1 Carp	100	.080	L.F.	20.50	4.13		24.63	29.50
0850	2" wide		75	.107		25.50	5.50		31	36.50
0860	4" wide		50	.160		35.50	8.25		43.75	51.50
0870	Concrete/CMU floor joints									
0880	1" wide	1 Carp	200	.040	L.F.	20.50	2.07		22.57	25.50
0890	2" wide		150	.053		25.50	2.75		28.25	32
0900	4" wide		100	.080		36.50	4.13		40.63	46.50

07 91 Preformed Joint Seals

07 91 13 – Compression Seals

07 91 13.10 Compression Seals

07 91 13.10 Compression Seals		Crew	Daily Output	Labor-Hours	Unit	Material	2019 Bare Costs Labor	2019 Bare Costs Equipment	Total	Total Incl O&P
0010	**COMPRESSION SEALS**									
4900	O-ring type cord, 1/4"	1 Bric	472	.017	L.F.	.41	.87		1.28	1.77
4910	1/2"		440	.018		1.06	.93		1.99	2.59
4920	3/4"		424	.019		2.05	.96		3.01	3.73
4930	1"		408	.020		4.15	1		5.15	6.10
4940	1-1/4"		384	.021		7.90	1.06		8.96	10.30
4950	1-1/2"		368	.022		9.70	1.11		10.81	12.35
4960	1-3/4"		352	.023		13.60	1.16		14.76	16.70
4970	2"		344	.023		22.50	1.19		23.69	26.50

07 91 23 – Backer Rods

07 91 23.10 Backer Rods

07 91 23.10 Backer Rods		Crew	Daily Output	Labor-Hours	Unit	Material	2019 Bare Costs Labor	2019 Bare Costs Equipment	Total	Total Incl O&P
0010	**BACKER RODS**									
0030	Backer rod, polyethylene, 1/4" diameter	1 Bric	4.60	1.739	C.L.F.	2.42	89		91.42	139
0050	1/2" diameter		4.60	1.739		4.09	89		93.09	141
0070	3/4" diameter		4.60	1.739		6.95	89		95.95	144
0090	1" diameter		4.60	1.739		12.45	89		101.45	150

07 91 Preformed Joint Seals

07 91 26 – Joint Fillers

07 91 26.10 Joint Fillers

		Crew	Daily Output	Labor-Hours	Unit	Material	2019 Bare Costs Labor	Equipment	Total	Total Incl O&P
0010	**JOINT FILLERS**									
4360	Butyl rubber filler, 1/4" x 1/4"	1 Bric	290	.028	L.F.	.23	1.41		1.64	2.40
4365	1/2" x 1/2"		250	.032		.91	1.63		2.54	3.49
4370	1/2" x 3/4"		210	.038		1.37	1.95		3.32	4.47
4375	3/4" x 3/4"		230	.035		2.05	1.78		3.83	4.96
4380	1" x 1"	↓	180	.044	↓	2.73	2.27		5	6.45
4390	For coloring, add					12%				
4980	Polyethylene joint backing, 1/4" x 2"	1 Bric	2.08	3.846	C.L.F.	13.85	196		209.85	315
4990	1/4" x 6"		1.28	6.250	"	29.50	320		349.50	520
5600	Silicone, room temp vulcanizing foam seal, 1/4" x 1/2"		1312	.006	L.F.	.46	.31		.77	.99
5610	1/2" x 1/2"		656	.012		.92	.62		1.54	1.97
5620	1/2" x 3/4"		442	.018		1.38	.92		2.30	2.93
5630	3/4" x 3/4"		328	.024		2.08	1.25		3.33	4.19
5640	1/8" x 1"		1312	.006		.46	.31		.77	.99
5650	1/8" x 3"		442	.018		1.38	.92		2.30	2.93
5670	1/4" x 3"		295	.027		2.77	1.38		4.15	5.15
5680	1/4" x 6"		148	.054		5.55	2.76		8.31	10.30
5690	1/2" x 6"		82	.098		11.10	4.98		16.08	19.80
5700	1/2" x 9"		52.50	.152		16.60	7.80		24.40	30
5710	1/2" x 12"	↓	33	.242	↓	22	12.40		34.40	43.50

07 92 Joint Sealants

07 92 13 – Elastomeric Joint Sealants

07 92 13.20 Caulking and Sealant Options

		Crew	Daily Output	Labor-Hours	Unit	Material	2019 Bare Costs Labor	Equipment	Total	Total Incl O&P
0010	**CAULKING AND SEALANT OPTIONS**									
0050	Latex acrylic based, bulk				Gal.	30			30	33
0055	Bulk in place 1/4" x 1/4" bead	1 Bric	300	.027	L.F.	.09	1.36		1.45	2.18
0060	1/4" x 3/8"		294	.027		.16	1.39		1.55	2.29
0065	1/4" x 1/2"		288	.028		.21	1.42		1.63	2.40
0075	3/8" x 3/8"		284	.028		.24	1.44		1.68	2.46
0080	3/8" x 1/2"		280	.029		.32	1.46		1.78	2.58
0085	3/8" x 5/8"		276	.029		.39	1.48		1.87	2.69
0095	3/8" x 3/4"		272	.029		.47	1.50		1.97	2.81
0100	1/2" x 1/2"		275	.029		.42	1.49		1.91	2.73
0105	1/2" x 5/8"		269	.030		.53	1.52		2.05	2.90
0110	1/2" x 3/4"		263	.030		.63	1.55		2.18	3.06
0115	1/2" x 7/8"		256	.031		.74	1.60		2.34	3.25
0120	1/2" x 1"		250	.032		.84	1.63		2.47	3.42
0125	3/4" x 3/4"		244	.033		.95	1.67		2.62	3.60
0130	3/4" x 1"		225	.036		1.26	1.82		3.08	4.16
0135	1" x 1"	↓	200	.040	↓	1.68	2.04		3.72	4.97
0190	Cartridges				Gal.	33			33	36
0200	11 fl. oz. cartridge				Ea.	2.82			2.82	3.10
0600	1/2" x 1/2"	1 Bric	275	.029	L.F.	.46	1.49		1.95	2.78
3200	Polyurethane, 1 or 2 component				Gal.	55.50			55.50	61
3300	Cartridges				"	68			68	75
3500	Bulk, in place, 1/4" x 1/4"	1 Bric	300	.027	L.F.	.18	1.36		1.54	2.28
3655	1/2" x 1/4"		288	.028		.36	1.42		1.78	2.57
3800	3/4" x 3/8"		272	.029		.82	1.50		2.32	3.19
3900	1" x 1/2"	↓	250	.032		1.44	1.63		3.07	4.08

07 92 Joint Sealants

07 92 19 – Acoustical Joint Sealants

07 92 19.10 Acoustical Sealant		Crew	Daily Output	Labor-Hours	Unit	Material	2019 Bare Costs Labor	Equipment	Total	Total Incl O&P
0010	**ACOUSTICAL SEALANT**									
0020	Acoustical sealant, elastomeric, cartridges				Ea.	8.60			8.60	9.45
0025	In place, 1/4" x 1/4"	1 Bric	300	.027	L.F.	.35	1.36		1.71	2.47
0030	1/4" x 1/2"		288	.028		.70	1.42		2.12	2.94
0035	1/2" x 1/2"		275	.029		1.40	1.49		2.89	3.81
0040	1/2" x 3/4"		263	.030		2.11	1.55		3.66	4.69
0045	3/4" x 3/4"		244	.033		3.16	1.67		4.83	6.05
0050	1" x 1"		200	.040		5.60	2.04		7.64	9.30

Division Notes

	CREW	DAILY OUTPUT	LABOR-HOURS	UNIT	BARE COSTS				TOTAL INCL O&P
					MAT.	LABOR	EQUIP.	TOTAL	

Estimating Tips
08 10 00 Doors and Frames

All exterior doors should be addressed for their energy conservation (insulation and seals).

- Most metal doors and frames look alike, but there may be significant differences among them. When estimating these items, be sure to choose the line item that most closely compares to the specification or door schedule requirements regarding:
  - □ type of metal
  - □ metal gauge
  - □ door core material
  - □ fire rating
  - □ finish
- Wood and plastic doors vary considerably in price. The primary determinant is the veneer material. Lauan, birch, and oak are the most common veneers. Other variables include the following:
  - □ hollow or solid core
  - □ fire rating
  - □ flush or raised panel
  - □ finish
- Door pricing includes bore for cylindrical locksets and mortise for hinges.

08 30 00 Specialty Doors and Frames

- There are many varieties of special doors, and they are usually priced per each. Add frames, hardware, or operators required for a complete installation.

08 40 00 Entrances, Storefronts, and Curtain Walls

- Glazed curtain walls consist of the metal tube framing and the glazing material. The cost data in this subdivision is presented for the metal tube framing alone or the composite wall. If your estimate requires a detailed takeoff of the framing, be sure to add the glazing cost and any tints.

08 50 00 Windows

- Steel windows are unglazed and aluminum can be glazed or unglazed. Some metal windows are priced without glass. Refer to 08 80 00 Glazing for glass pricing. The grade C indicates commercial grade windows, usually ASTM C-35.
- All wood windows and vinyl are priced preglazed. The glazing is insulating glass. Add the cost of screens and grills if required and not already included.

08 70 00 Hardware

- Hardware costs add considerably to the cost of a door. The most efficient method to determine the hardware requirements for a project is to review the door and hardware schedule together. One type of door may have different hardware, depending on the door usage.
- Door hinges are priced by the pair, with most doors requiring 1-1/2 pairs per door. The hinge prices do not include installation labor because it is included in door installation.

Hinges are classified according to the frequency of use, base material, and finish.

08 80 00 Glazing

- Different openings require different types of glass. The most common types are:
  - □ float
  - □ tempered
  - □ insulating
  - □ impact-resistant
  - □ ballistic-resistant
- Most exterior windows are glazed with insulating glass. Entrance doors and window walls, where the glass is less than 18" from the floor, are generally glazed with tempered glass. Interior windows and some residential windows are glazed with float glass.
- Coastal communities require the use of impact-resistant glass, dependent on wind speed.
- The insulation or 'u' value is a strong consideration, along with solar heat gain, to determine total energy efficiency.

Reference Numbers

Reference numbers are shown at the beginning of some major classifications. These numbers refer to related items in the Reference Section. The reference information may be an estimating procedure, an alternate pricing method, or technical information.

Note: Not all subdivisions listed here necessarily appear. ■

08 31 13.10 Types of Framed Access Doors	Crew	Daily Output	Labor-Hours	Unit	Material	2019 Bare Costs Labor	Equipment	Total	Total Incl O&P
0010 **TYPES OF FRAMED ACCESS DOORS**									
4000　Recessed door for drywall									
4100　　Metal 12" x 12"	1 Carp	6	1.333	Ea.	74.50	69		143.50	186
4150　　12" x 24"		5.50	1.455		119	75		194	245
4200　　24" x 36"	↓	5	1.600	↓	167	82.50		249.50	310

Estimating Tips

General

- Room Finish Schedule: A complete set of plans should contain a room finish schedule. If one is not available, it would be well worth the time and effort to obtain one.

09 20 00 Plaster and Gypsum Board

- Lath is estimated by the square yard plus a 5% allowance for waste. Furring, channels, and accessories are measured by the linear foot. An extra foot should be allowed for each accessory miter or stop.

- Plaster is also estimated by the square yard. Deductions for openings vary by preference, from zero deduction to 50% of all openings over 2 feet in width. The estimator should allow one extra square foot for each linear foot of horizontal interior or exterior angle located below the ceiling level. Also, double the areas of small radius work.

- Drywall accessories, studs, track, and acoustical caulking are all measured by the linear foot. Drywall taping is figured by the square foot. Gypsum wallboard is estimated by the square foot. No material deductions should be made for door or window openings under 32 S.F.

09 60 00 Flooring

- Tile and terrazzo areas are taken off on a square foot basis. Trim and base materials are measured by the linear foot. Accent tiles are listed per each. Two basic methods of installation are used. Mud set is approximately 30% more expensive than thin set. The cost of grout is included with tile unit price lines unless otherwise noted. In terrazzo work, be sure to include the linear footage of embedded decorative strips, grounds, machine rubbing, and power cleanup.

- Wood flooring is available in strip, parquet, or block configuration. The latter two types are set in adhesives with quantities estimated by the square foot. The laying pattern will influence labor costs and material waste. In addition to the material and labor for laying wood floors, the estimator must make allowances for sanding and finishing these areas, unless the flooring is prefinished.

- Sheet flooring is measured by the square yard. Roll widths vary, so consideration should be given to use the most economical width, as waste must be figured into the total quantity. Consider also the installation methods available—direct glue down or stretched. Direct glue-down installation is assumed with sheet carpet unit price lines unless otherwise noted.

09 70 00 Wall Finishes

- Wall coverings are estimated by the square foot. The area to be covered is measured—length by height of the wall above the baseboards—to calculate the square footage of each wall. This figure is divided by the number of square feet in the single roll which is being used. Deduct, in full, the areas of openings such as doors and windows. Where a pattern match is required allow 25–30% waste.

09 80 00 Acoustic Treatment

- Acoustical systems fall into several categories. The takeoff of these materials should be by the square foot of area with a 5% allowance for waste. Do not forget about scaffolding, if applicable, when estimating these systems.

09 90 00 Painting and Coating

- A major portion of the work in painting involves surface preparation. Be sure to include cleaning, sanding, filling, and masking costs in the estimate.

- Protection of adjacent surfaces is not included in painting costs. When considering the method of paint application, an important factor is the amount of protection and masking required. These must be estimated separately and may be the determining factor in choosing the method of application.

Reference Numbers

Reference numbers are shown at the beginning of some major classifications. These numbers refer to related items in the Reference Section. The reference information may be an estimating procedure, an alternate pricing method, or technical information.

Note: Not all subdivisions listed here necessarily appear. ■

09 22 Supports for Plaster and Gypsum Board

09 22 03 – Fastening Methods for Finishes

09 22 03.20 Drilling Plaster/Drywall	Crew	Daily Output	Labor-Hours	Unit	Material	2019 Bare Costs Labor	Equipment	Total	Total Incl O&P
0010 **DRILLING PLASTER/DRYWALL**									
1100 Drilling & layout for drywall/plaster walls, up to 1" deep, no anchor									
1200 Holes, 1/4" diameter	1 Carp	150	.053	Ea.	.01	2.75		2.76	4.18
1300 3/8" diameter		140	.057		.01	2.95		2.96	4.48
1400 1/2" diameter		130	.062		.01	3.18		3.19	4.82
1500 3/4" diameter		120	.067		.02	3.44		3.46	5.20
1600 1" diameter		110	.073		.02	3.76		3.78	5.70
1700 1-1/4" diameter		100	.080		.04	4.13		4.17	6.30
1800 1-1/2" diameter		90	.089		.06	4.59		4.65	7
1900 For ceiling installations, add						40%			

09 57 Special Function Ceilings

09 57 53 – Security Ceiling Assemblies

09 57 53.10 Ceiling Assem., Security, Radio Freq. Shielding

09 57 53.10 Ceiling Assem., Security, Radio Freq. Shielding	Crew	Daily Output	Labor-Hours	Unit	Material	2019 Bare Costs Labor	Equipment	Total	Total Incl O&P
0010 **CEILING ASSEMBLY, SECURITY, RADIO FREQUENCY SHIELDING**									
0020 Prefabricated, galvanized steel	2 Carp	375	.043	SF Surf	5.35	2.20		7.55	9.25
0050 5 oz., copper ceiling panel		155	.103		4.58	5.35		9.93	13.10
0110 12 oz., copper ceiling panel		140	.114		9.15	5.90		15.05	19
0250 Ceiling hangers	E-1	45	.533	Ea.	37	29.50	2.13	68.63	89
0300 Shielding transition, 11 ga. preformed angles	"	1365	.018	L.F.	37	.97	.07	38.04	42

09 69 Access Flooring

09 69 13 – Rigid-Grid Access Flooring

09 69 13.10 Access Floors

09 69 13.10 Access Floors	Crew	Daily Output	Labor-Hours	Unit	Material	2019 Bare Costs Labor	Equipment	Total	Total Incl O&P
0010 **ACCESS FLOORS**									
0015 Access floor pkg. including panel, pedestal, & stringers 1500 lbs. load									
0100 Package pricing, conc. fill panels, no fin., 6" ht.	4 Carp	750	.043	S.F.	17.20	2.20		19.40	22
0105 12" ht.		750	.043		17.40	2.20		19.60	22.50
0110 18" ht.		750	.043		17.60	2.20		19.80	22.50
0115 24" ht.		750	.043		17.85	2.20		20.05	23
0120 Package pricing, steel panels, no fin., 6" ht.		750	.043		13.45	2.20		15.65	18.15
0125 12" ht.		750	.043		18.30	2.20		20.50	23.50
0130 18" ht.		750	.043		18.45	2.20		20.65	24
0135 24" ht.		750	.043		18.55	2.20		20.75	24
0140 Package pricing, wood core panels, no fin., 6" ht.		750	.043		13.55	2.20		15.75	18.30
0145 12" ht.		750	.043		13.80	2.20		16	18.55
0150 18" ht.		750	.043		13.80	2.20		16	18.55
0155 24" ht.		750	.043		14.05	2.20		16.25	18.85
0160 Pkg. pricing, conc. fill pnls, no stringers, no fin., 6" ht, 1250 lbs. load		700	.046		10.90	2.36		13.26	15.55
0165 12" ht.		700	.046		10.10	2.36		12.46	14.65
0170 18" ht.		700	.046		11.05	2.36		13.41	15.70
0175 24" ht.		700	.046		12	2.36		14.36	16.75
0250 Panels, 2' x 2' conc. fill, no fin.	2 Carp	500	.032		23	1.65		24.65	28
0255 With 1/8" high pressure laminate		500	.032		56.50	1.65		58.15	65
0260 Metal panels with 1/8" high pressure laminate		500	.032		67	1.65		68.65	76.50
0265 Wood core panels with 1/8" high pressure laminate		500	.032		44	1.65		45.65	51
0400 Aluminum panels, no fin.		500	.032		34	1.65		35.65	39.50
0600 For carpet covering, add					9.20			9.20	10.10
0700 For vinyl floor covering, add					9.45			9.45	10.40

09 69 Access Flooring

09 69 13 – Rigid-Grid Access Flooring

09 69 13.10 Access Floors		Crew	Daily Output	Labor-Hours	Unit	Material	2019 Bare Costs Labor	Equipment	Total	Total Incl O&P
0900	For high pressure laminate covering, add				S.F.	7.80			7.80	8.55
0910	For snap on stringer system, add	2 Carp	1000	.016	↓	1.61	.83		2.44	3.02
1000	Machine cutouts after initial installation	1 Carp	50	.160	Ea.	20.50	8.25		28.75	35
1050	Pedestals, 6" to 12"	2 Carp	85	.188		8.80	9.70		18.50	24.50
1100	Air conditioning grilles, 4" x 12"	1 Carp	17	.471		71.50	24.50		96	116
1150	4" x 18"	"	14	.571	↓	98	29.50		127.50	153
1200	Approach ramps, steel	2 Carp	60	.267	S.F.	27	13.75		40.75	51
1300	Aluminum	"	40	.400	"	34.50	20.50		55	69
1500	Handrail, 2 rail, aluminum	1 Carp	15	.533	L.F.	120	27.50		147.50	174

Division Notes

	DAILY	LABOR-		BARE COSTS				TOTAL
	CREW	OUTPUT	HOURS	UNIT	MAT.	LABOR	EQUIP.	TOTAL	INCL O&P

Estimating Tips

General

- The items in this division are usually priced per square foot or each.
- Many items in Division 10 require some type of support system or special anchors that are not usually furnished with the item. The required anchors must be added to the estimate in the appropriate division.
- Some items in Division 10, such as lockers, may require assembly before installation. Verify the amount of assembly required. Assembly can often exceed installation time.

10 20 00 Interior Specialties

- Support angles and blocking are not included in the installation of toilet compartments, shower/dressing compartments, or cubicles. Appropriate line items from Division 5 or 6 may need to be added to support the installations.
- Toilet partitions are priced by the stall. A stall consists of a side wall, pilaster, and door with hardware. Toilet tissue holders and grab bars are extra.
- The required acoustical rating of a folding partition can have a significant impact on costs. Verify the sound transmission coefficient rating of the panel priced against the specification requirements.

- Grab bar installation does not include supplemental blocking or backing to support the required load. When grab bars are installed at an existing facility, provisions must be made to attach the grab bars to a solid structure.

Reference Numbers

Reference numbers are shown at the beginning of some major classifications. These numbers refer to related items in the Reference Section. The reference information may be an estimating procedure, an alternate pricing method, or technical information.

Note: Not all subdivisions listed here necessarily appear. ■

10 28 16.20 Medicine Cabinets	Crew	Daily Output	Labor-Hours	Unit	Material	2019 Bare Costs Labor	Equipment	Total	Total Incl O&P
0010 **MEDICINE CABINETS**									
0020 With mirror, sst frame, 16" x 22", unlighted	1 Carp	14	.571	Ea.	96.50	29.50		126	151
0100 Wood frame		14	.571		136	29.50		165.50	195
0300 Sliding mirror doors, 20" x 16" x 4-3/4", unlighted		7	1.143		126	59		185	229
0400 24" x 19" x 8-1/2", lighted		5	1.600		216	82.50		298.50	365
0600 Triple door, 30" x 32", unlighted, plywood body		7	1.143		355	59		414	480
0700 Steel body		7	1.143		355	59		414	480
0900 Oak door, wood body, beveled mirror, single door		7	1.143		176	59		235	283
1000 Double door		6	1.333		370	69		439	510
1200 Hotel cabinets, stainless, with lower shelf, unlighted		10	.800		217	41.50		258.50	300
1300 Lighted		5	1.600		335	82.50		417.50	495

Estimating Tips
General

- The items in this division are usually priced per square foot or each. Many of these items are purchased by the owner for installation by the contractor. Check the specifications for responsibilities and include time for receiving, storage, installation, and mechanical and electrical hookups in the appropriate divisions.

- Many items in Division 11 require some type of support system that is not usually furnished with the item. Examples of these systems include blocking for the attachment of casework and support angles for ceiling-hung projection screens. The required blocking or supports must be added to the estimate in the appropriate division.

- Some items in Division 11 may require assembly or electrical hookups. Verify the amount of assembly required or the need for a hard electrical connection and add the appropriate costs.

Reference Numbers

Reference numbers are shown at the beginning of some major classifications. These numbers refer to related items in the Reference Section. The reference information may be an estimating procedure, an alternate pricing method, or technical information.

Did you know?

RSMeans data is available through our online application:

- Search for costs by keyword
- Leverage the most up-to-date data
- Build and export estimates

Try it free
rsmeans.com/2019freetrial

11 12 Parking Control Equipment

11 12 13 – Parking Key and Card Control Units

11 12 13.10 Parking Control Units	Crew	Daily Output	Labor-Hours	Unit	Material	2019 Bare Costs Labor	Equipment	Total	Total Incl O&P
0010 **PARKING CONTROL UNITS**									
5100 Card reader	1 Elec	2	4	Ea.	1,425	240		1,665	1,925
5120 Proximity with customer display	2 Elec	1	16		5,975	960		6,935	8,000
6000 Parking control software, basic functionality	1 Elec	.50	16		24,000	960		24,960	27,800
6020 Multi-function	"	.20	40		95,500	2,400		97,900	108,500

11 12 16 – Parking Ticket Dispensers

11 12 16.10 Ticket Dispensers

	Crew	Daily Output	Labor-Hours	Unit	Material	Labor	Equipment	Total	Total Incl O&P
0010 **TICKET DISPENSERS**									
5900 Ticket spitter with time/date stamp, standard	2 Elec	2	8	Ea.	6,725	480		7,205	8,100
5920 Mag stripe encoding	"	2	8	"	18,900	480		19,380	21,500

11 12 26 – Parking Fee Collection Equipment

11 12 26.13 Parking Fee Coin Collection Equipment

	Crew	Daily Output	Labor-Hours	Unit	Material	Labor	Equipment	Total	Total Incl O&P
0010 **PARKING FEE COIN COLLECTION EQUIPMENT**									
5200 Cashier booth, average	B-22	1	30	Ea.	10,600	1,450	195	12,245	14,100
5300 Collector station, pay on foot	2 Elec	.20	80		111,500	4,800		116,300	130,000
5320 Credit card only	"	.50	32		20,600	1,925		22,525	25,600

11 12 26.23 Fee Equipment

	Crew	Daily Output	Labor-Hours	Unit	Material	Labor	Equipment	Total	Total Incl O&P
0010 **FEE EQUIPMENT**									
5600 Fee computer	1 Elec	1.50	5.333	Ea.	12,400	320		12,720	14,100

11 12 33 – Parking Gates

11 12 33.13 Lift Arm Parking Gates

	Crew	Daily Output	Labor-Hours	Unit	Material	Labor	Equipment	Total	Total Incl O&P
0010 **LIFT ARM PARKING GATES**									
5000 Barrier gate with programmable controller	2 Elec	3	5.333	Ea.	3,200	320		3,520	4,000
5020 Industrial		3	5.333		5,250	320		5,570	6,250
5050 Non-programmable, with reader and 12' arm		3	5.333		1,775	320		2,095	2,450
5500 Exit verifier		1	16		19,400	960		20,360	22,800
5700 Full sign, 4" letters	1 Elec	2	4		1,225	240		1,465	1,700
5800 Inductive loop	2 Elec	4	4		209	240		449	590
5950 Vehicle detector, microprocessor based	1 Elec	3	2.667		465	160		625	755

11 21 Retail and Service Equipment

11 21 73 – Commercial Laundry and Dry Cleaning Equipment

11 21 73.13 Dry Cleaning Equipment

	Crew	Daily Output	Labor-Hours	Unit	Material	Labor	Equipment	Total	Total Incl O&P
0010 **DRY CLEANING EQUIPMENT**									
2000 Dry cleaners, electric, 20 lb. capacity, not incl. rough-in	L-1	.20	80	Ea.	34,500	4,925		39,425	45,300
2050 25 lb. capacity		.17	94.118		51,500	5,800		57,300	65,500
2100 30 lb. capacity		.15	107		54,500	6,575		61,075	70,000
2150 60 lb. capacity		.09	178		78,500	11,000		89,500	102,500

11 21 73.19 Finishing Equipment

	Crew	Daily Output	Labor-Hours	Unit	Material	Labor	Equipment	Total	Total Incl O&P
0010 **FINISHING EQUIPMENT**									
3500 Folders, blankets & sheets, minimum	1 Elec	.17	47.059	Ea.	35,300	2,825		38,125	43,000
3700 King size with automatic stacker		.10	80		64,000	4,800		68,800	77,000
3800 For conveyor delivery, add		.45	17.778		16,400	1,075		17,475	19,600

11 21 73.23 Commercial Ironing Equipment

	Crew	Daily Output	Labor-Hours	Unit	Material	Labor	Equipment	Total	Total Incl O&P
0010 **COMMERCIAL IRONING EQUIPMENT**									
4500 Ironers, institutional, 110", single roll	1 Elec	.20	40	Ea.	34,200	2,400		36,600	41,200

11 21 Retail and Service Equipment

11 21 73 – Commercial Laundry and Dry Cleaning Equipment

11 21 73.26 Commercial Washers and Extractors

		Crew	Daily Output	Labor-Hours	Unit	Material	2019 Bare Costs Labor	Equipment	Total	Total Incl O&P
0010	**COMMERCIAL WASHERS AND EXTRACTORS**, not including rough-in									
6000	Combination washer/extractor, 20 lb. capacity	L-6	1.50	8	Ea.	6,625	495		7,120	8,050
6100	30 lb. capacity		.80	15		10,100	930		11,030	12,500
6200	50 lb. capacity		.68	17.647		12,800	1,100		13,900	15,700
6300	75 lb. capacity		.30	40		18,900	2,475		21,375	24,500
6350	125 lb. capacity	↓	.16	75	↓	29,200	4,650		33,850	39,200

11 21 73.33 Coin-Operated Laundry Equipment

		Crew	Daily Output	Labor-Hours	Unit	Material	2019 Bare Costs Labor	Equipment	Total	Total Incl O&P
0010	**COIN-OPERATED LAUNDRY EQUIPMENT**									
5290	Clothes washer									
5300	Commercial, coin operated, average	1 Plum	3	2.667	Ea.	1,325	168		1,493	1,700

11 30 Residential Equipment

11 30 13 – Residential Appliances

11 30 13.15 Cooking Equipment

		Crew	Daily Output	Labor-Hours	Unit	Material	2019 Bare Costs Labor	Equipment	Total	Total Incl O&P
0010	**COOKING EQUIPMENT**									
0020	Cooking range, 30" free standing, 1 oven, minimum	2 Clab	10	1.600	Ea.	470	65.50		535.50	620
0050	Maximum		4	4		2,325	164		2,489	2,825
0700	Free standing, 1 oven, 21" wide range, minimum		10	1.600		475	65.50		540.50	620
0750	21" wide, maximum	↓	4	4		670	164		834	990
0900	Countertop cooktops, 4 burner, standard, minimum	1 Elec	6	1.333		325	80		405	475
0950	Maximum		3	2.667		1,850	160		2,010	2,300
1050	As above, but with grill and griddle attachment, minimum		6	1.333		1,425	80		1,505	1,700
1100	Maximum		3	2.667		3,925	160		4,085	4,575
1200	Induction cooktop, 30" wide		3	2.667		1,525	160		1,685	1,925
1250	Microwave oven, minimum		4	2		96	120		216	285
1300	Maximum		2	4		455	240		695	860

11 30 13.17 Kitchen Cleaning Equipment

		Crew	Daily Output	Labor-Hours	Unit	Material	2019 Bare Costs Labor	Equipment	Total	Total Incl O&P
0010	**KITCHEN CLEANING EQUIPMENT**									
2750	Dishwasher, built-in, 2 cycles, minimum	L-1	4	4	Ea.	305	246		551	705
2800	Maximum		2	8		455	495		950	1,225
2950	4 or more cycles, minimum		4	4		420	246		666	835
2960	Average		4	4		560	246		806	985
3000	Maximum	↓	2	8		1,900	495		2,395	2,825

11 30 13.18 Waste Disposal Equipment

		Crew	Daily Output	Labor-Hours	Unit	Material	2019 Bare Costs Labor	Equipment	Total	Total Incl O&P
0010	**WASTE DISPOSAL EQUIPMENT**									
3300	Garbage disposal, sink type, minimum	L-1	10	1.600	Ea.	109	98.50		207.50	266
3350	Maximum	"	10	1.600	"	208	98.50		306.50	375

11 30 13.19 Kitchen Ventilation Equipment

		Crew	Daily Output	Labor-Hours	Unit	Material	2019 Bare Costs Labor	Equipment	Total	Total Incl O&P
0010	**KITCHEN VENTILATION EQUIPMENT**									
4150	Hood for range, 2 speed, vented, 30" wide, minimum	L-3	5	3.200	Ea.	96	179		275	375
4200	Maximum		3	5.333		965	299		1,264	1,500
4300	42" wide, minimum		5	3.200		155	179		334	440
4330	Custom		5	3.200		1,625	179		1,804	2,075
4350	Maximum	↓	3	5.333		1,975	299		2,274	2,625
4500	For ventless hood, 2 speed, add					17.80			17.80	19.60
4650	For vented 1 speed, deduct from maximum	↓			↓	64			64	70.50

11 30 13.24 Washers

		Crew	Daily Output	Labor-Hours	Unit	Material	2019 Bare Costs Labor	Equipment	Total	Total Incl O&P
0010	**WASHERS**									
5000	Residential, 4 cycle, average	1 Plum	3	2.667	Ea.	980	168		1,148	1,325

For customer support on your Electrical Costs with RSMeans data, call 800.448.8182.

57

11 30 Residential Equipment

11 30 15 – Miscellaneous Residential Appliances

11 30 15.23 Water Heaters

		Crew	Daily Output	Labor-Hours	Unit	Material	2019 Bare Costs Labor	2019 Bare Costs Equipment	Total	Total Incl O&P
0010	**WATER HEATERS**									
6900	Electric, glass lined, 30 gallon, minimum	L-1	5	3.200	Ea.	895	197		1,092	1,275
6950	Maximum		3	5.333		1,250	330		1,580	1,875
7100	80 gallon, minimum		2	8		1,650	495		2,145	2,550
7150	Maximum	↓	1	16	↓	2,300	985		3,285	4,025

11 30 15.43 Air Quality

		Crew	Daily Output	Labor-Hours	Unit	Material	2019 Bare Costs Labor	2019 Bare Costs Equipment	Total	Total Incl O&P
0010	**AIR QUALITY**									
2450	Dehumidifier, portable, automatic, 15 pint	1 Elec	4	2	Ea.	208	120		328	410
2550	40 pint		3.75	2.133		240	128		368	455
3550	Heater, electric, built-in, 1250 watt, ceiling type, minimum		4	2		119	120		239	310
3600	Maximum		3	2.667		185	160		345	440
3700	Wall type, minimum		4	2		199	120		319	400
3750	Maximum		3	2.667		194	160		354	455
3900	1500 watt wall type, with blower		4	2		189	120		309	385
3950	3000 watt	↓	3	2.667		490	160		650	780
4850	Humidifier, portable, 8 gallons/day					164			164	180
5000	15 gallons/day				↓	198			198	218

11 32 Unit Kitchens

11 32 13 – Metal Unit Kitchens

11 32 13.10 Commercial Unit Kitchens

		Crew	Daily Output	Labor-Hours	Unit	Material	2019 Bare Costs Labor	2019 Bare Costs Equipment	Total	Total Incl O&P
0010	**COMMERCIAL UNIT KITCHENS**									
1500	Combination range, refrigerator and sink, 30" wide, minimum	L-1	2	8	Ea.	1,400	495		1,895	2,275
1550	Maximum		1	16		1,150	985		2,135	2,750
1570	60" wide, average		1.40	11.429		1,275	705		1,980	2,450
1590	72" wide, average		1.20	13.333		1,800	820		2,620	3,225
1600	Office model, 48" wide		2	8		1,400	495		1,895	2,250
1620	Refrigerator and sink only	↓	2.40	6.667	↓	2,650	410		3,060	3,550
1640	Combination range, refrigerator, sink, microwave									
1660	Oven and ice maker	L-1	.80	20	Ea.	4,900	1,225		6,125	7,250

11 41 Foodservice Storage Equipment

11 41 13 – Refrigerated Food Storage Cases

11 41 13.20 Refrigerated Food Storage Equipment

		Crew	Daily Output	Labor-Hours	Unit	Material	2019 Bare Costs Labor	2019 Bare Costs Equipment	Total	Total Incl O&P
0010	**REFRIGERATED FOOD STORAGE EQUIPMENT**									
2350	Cooler, reach-in, beverage, 6' long	Q-1	6	2.667	Ea.	3,225	152		3,377	3,775

11 42 Food Preparation Equipment

11 42 10 – Commercial Food Preparation Equipment

11 42 10.10 Choppers, Mixers and Misc. Equipment		Crew	Daily Output	Labor-Hours	Unit	Material	2019 Bare Costs Labor	Equipment	Total	Total Incl O&P
0010	**CHOPPERS, MIXERS AND MISC. EQUIPMENT**									
1850	Coffee urn, twin 6 gallon urns	1 Plum	2	4	Ea.	2,375	253		2,628	2,975
3800	Food mixers, bench type, 20 quarts	L-7	7	4		3,175	199		3,374	3,800
3900	60 quarts	"	5	5.600	↓	11,400	279		11,679	13,000

11 44 Food Cooking Equipment

11 44 13 – Commercial Ranges

11 44 13.10 Cooking Equipment

		Crew	Daily Output	Labor-Hours	Unit	Material	2019 Bare Costs Labor	Equipment	Total	Total Incl O&P
0010	**COOKING EQUIPMENT**									
0020	Bake oven, gas, one section	Q-1	8	2	Ea.	5,850	114		5,964	6,600
1300	Broiler, without oven, standard	"	8	2		3,875	114		3,989	4,450
1550	Infrared	L-7	4	7		7,500	350		7,850	8,775
6350	Kettle, w/steam jacket, tilting, w/positive lock, SS, 20 gallons		7	4		8,500	199		8,699	9,650
6600	60 gallons		6	4.667		20,300	232		20,532	22,800
8850	Steamer, electric 27 KW		7	4		12,100	199		12,299	13,600
9100	Electric, 10 KW or gas 100,000 BTU	↓	5	5.600		7,450	279		7,729	8,625
9150	Toaster, conveyor type, 16-22 slices/minute				↓	1,250			1,250	1,375
9200	For deluxe models of above equipment, add					75%				
9400	Rule of thumb: Equipment cost based									
9410	on kitchen work area									
9420	Office buildings, minimum	L-7	77	.364	S.F.	95.50	18.10		113.60	133
9450	Maximum		58	.483		156	24		180	208
9550	Public eating facilities, minimum		77	.364		125	18.10		143.10	166
9600	Maximum		46	.609		203	30.50		233.50	270
9750	Hospitals, minimum		58	.483		124	24		148	173
9800	Maximum	↓	39	.718	↓	228	36		264	305

11 46 Food Dispensing Equipment

11 46 16 – Service Line Equipment

11 46 16.10 Commercial Food Dispensing Equipment

		Crew	Daily Output	Labor-Hours	Unit	Material	2019 Bare Costs Labor	Equipment	Total	Total Incl O&P
0010	**COMMERCIAL FOOD DISPENSING EQUIPMENT**									
3300	Food warmer, counter, 1.2 KW				Ea.	710			710	785
3550	1.6 KW					2,175			2,175	2,400
5750	Ice dispenser 567 pound	Q-1	6	2.667	↓	5,600	152		5,752	6,375

11 48 Foodservice Cleaning and Disposal Equipment

11 48 13 – Commercial Dishwashers

11 48 13.10 Dishwashers

			Crew	Daily Output	Labor-Hours	Unit	Material	2019 Bare Costs Labor	Equipment	Total	Total Incl O&P
0010	**DISHWASHERS**										
2700	Dishwasher, commercial, rack type										
2720	10 to 12 racks/hour		Q-1	3.20	5	Ea.	3,325	284		3,609	4,075
2730	Energy star rated, 35 to 40 racks/hour	G		1.30	12.308		5,000	700		5,700	6,550
2740	50 to 60 racks/hour	G	↓	1.30	12.308	↓	11,100	700		11,800	13,300

For customer support on your Electrical Costs with RSMeans data, call 800.448.8182.

59

11 52 Audio-Visual Equipment

11 52 16 – Projectors

11 52 16.10 Movie Equipment

		Crew	Daily Output	Labor-Hours	Unit	Material	2019 Bare Costs Labor	Equipment	Total	Total Incl O&P
0010	**MOVIE EQUIPMENT**									
0020	Changeover, minimum				Ea.	510			510	560
0100	Maximum					995			995	1,100
0800	Lamphouses, incl. rectifiers, xenon, 1,000 watt	1 Elec	2	4		7,325	240		7,565	8,425
0900	1,600 watt		2	4		7,825	240		8,065	8,950
1000	2,000 watt		1.50	5.333		8,500	320		8,820	9,825
1100	4,000 watt		1.50	5.333		11,700	320		12,020	13,400
3700	Sound systems, incl. amplifier, mono, minimum		.90	8.889		3,675	535		4,210	4,850
3800	Dolby/Super Sound, maximum		.40	20		19,900	1,200		21,100	23,700
4100	Dual system, 2 channel, front surround, minimum		.70	11.429		5,100	685		5,785	6,625
4200	Dolby/Super Sound, 4 channel, maximum		.40	20		18,800	1,200		20,000	22,500
5300	Speakers, recessed behind screen, minimum		2	4		1,175	240		1,415	1,650
5400	Maximum		1	8		3,400	480		3,880	4,475
7000	For automation, varying sophistication, minimum		1	8	System	2,625	480		3,105	3,625
7100	Maximum	2 Elec	.30	53.333	"	6,125	3,200		9,325	11,500

11 52 16.20 Movie Equipment- Digital

		Crew	Daily Output	Labor-Hours	Unit	Material	2019 Bare Costs Labor	Equipment	Total	Total Incl O&P
0010	**MOVIE EQUIPMENT- DIGITAL**									
1000	Digital 2K projection system, 98" DMD	1 Elec	2	4	Ea.	52,000	240		52,240	58,000
1100	OEM lens		2	4		6,000	240		6,240	6,950
2000	Pedestal with power distribution		2	4		2,300	240		2,540	2,875
3000	Software		2	4		1,975	240		2,215	2,525

11 53 Laboratory Equipment

11 53 19 – Laboratory Sterilizers

11 53 19.13 Sterilizers

		Crew	Daily Output	Labor-Hours	Unit	Material	2019 Bare Costs Labor	Equipment	Total	Total Incl O&P
0010	**STERILIZERS**									
0700	Glassware washer, undercounter, minimum	L-1	1.80	8.889	Ea.	6,875	550		7,425	8,400
0710	Maximum	"	1	16	"	14,900	985		15,885	17,900

11 53 33 – Emergency Safety Appliances

11 53 33.13 Emergency Equipment

		Crew	Daily Output	Labor-Hours	Unit	Material	2019 Bare Costs Labor	Equipment	Total	Total Incl O&P
0010	**EMERGENCY EQUIPMENT**									
1400	Safety equipment, eye wash, hand held				Ea.	425			425	470
1450	Deluge shower				"	840			840	920

11 53 43 – Service Fittings and Accessories

11 53 43.13 Fittings

		Crew	Daily Output	Labor-Hours	Unit	Material	2019 Bare Costs Labor	Equipment	Total	Total Incl O&P
0010	**FITTINGS**									
8000	Alternate pricing method: as percent of lab furniture									
8050	Installation, not incl. plumbing & duct work				% Furn.				22%	22%
8100	Plumbing, final connections, simple system								10%	10%
8110	Moderately complex system								15%	15%
8120	Complex system								20%	20%
8150	Electrical, simple system								10%	10%
8160	Moderately complex system								20%	20%
8170	Complex system								35%	35%

11 61 Broadcast, Theater, and Stage Equipment

11 61 33 – Rigging Systems and Controls

11 61 33.10 Controls		Crew	Daily Output	Labor-Hours	Unit	Material	2019 Bare Costs Labor	Equipment	Total	Total Incl O&P
0010	**CONTROLS**									
0050	Control boards with dimmers and breakers, minimum	1 Elec	1	8	Ea.	19,100	480		19,580	21,700
0100	Average		.50	16		53,000	960		53,960	59,500
0150	Maximum		.20	40		143,500	2,400		145,900	161,000

11 66 Athletic Equipment

11 66 43 – Interior Scoreboards

11 66 43.10 Scoreboards		Crew	Daily Output	Labor-Hours	Unit	Material	2019 Bare Costs Labor	Equipment	Total	Total Incl O&P
0010	**SCOREBOARDS**									
7000	Baseball, minimum	R-3	1.30	15.385	Ea.	3,925	920	100	4,945	5,800
7200	Maximum		.05	400		21,100	23,900	2,600	47,600	62,000
7300	Football, minimum		.86	23.256		5,175	1,400	151	6,726	7,950
7400	Maximum		.20	100		19,400	5,975	650	26,025	31,000
7500	Basketball (one side), minimum		2.07	9.662		2,525	575	62.50	3,162.50	3,700
7600	Maximum		.30	66.667		7,625	3,975	435	12,035	14,800
7700	Hockey-basketball (four sides), minimum		.25	80		9,375	4,775	520	14,670	18,000
7800	Maximum		.15	133		16,400	7,975	865	25,240	30,900

11 71 Medical Sterilizing Equipment

11 71 10 – Medical Sterilizers & Distillers

11 71 10.10 Sterilizers and Distillers		Crew	Daily Output	Labor-Hours	Unit	Material	2019 Bare Costs Labor	Equipment	Total	Total Incl O&P
0010	**STERILIZERS AND DISTILLERS**									
6200	Steam generators, electric 10 kW to 180 kW, freestanding									
6250	Minimum	1 Elec	3	2.667	Ea.	11,400	160		11,560	12,800
6300	Maximum	"	.70	11.429	"	22,900	685		23,585	26,100

11 72 Examination and Treatment Equipment

11 72 53 – Treatment Equipment

11 72 53.13 Medical Treatment Equipment		Crew	Daily Output	Labor-Hours	Unit	Material	2019 Bare Costs Labor	Equipment	Total	Total Incl O&P
0010	**MEDICAL TREATMENT EQUIPMENT**									
6700	Surgical lights, doctor's office, single arm	2 Elec	2	8	Ea.	3,150	480		3,630	4,200
6750	Dual arm	"	1	16	"	5,325	960		6,285	7,275

11 76 Operating Room Equipment

11 76 10 – Operating Room Equipment

11 76 10.10 Surgical Equipment		Crew	Daily Output	Labor-Hours	Unit	Material	2019 Bare Costs Labor	Equipment	Total	Total Incl O&P
0010	**SURGICAL EQUIPMENT**									
6800	Surgical lights, major operating room, dual head, minimum	2 Elec	1	16	Ea.	5,250	960		6,210	7,200
6850	Maximum	"	1	16	"	37,500	960		38,460	42,600

For customer support on your Electrical Costs with RSMeans data, call 800.448.8182.

61

11 78 Mortuary Equipment

11 78 13 – Mortuary Refrigerators

11 78 13.10 Mortuary and Autopsy Equipment	Crew	Daily Output	Labor-Hours	Unit	Material	2019 Bare Costs Labor	Equipment	Total	Total Incl O&P
0010 **MORTUARY AND AUTOPSY EQUIPMENT**									
0015 Autopsy table, standard	1 Plum	1	8	Ea.	10,200	505		10,705	12,000

11 78 16 – Crematorium Equipment

11 78 16.10 Crematory

	Crew	Daily Output	Labor-Hours	Unit	Material	Labor	Equipment	Total	Total Incl O&P
0010 **CREMATORY**									
1500 Crematory, not including building, 1 place	Q-3	.20	160	Ea.	78,500	9,625		88,125	101,000
1750 2 place	"	.10	320	"	112,500	19,200		131,700	152,500

11 82 Facility Solid Waste Handling Equipment

11 82 26 – Facility Waste Compactors

11 82 26.10 Compactors

	Crew	Daily Output	Labor-Hours	Unit	Material	Labor	Equipment	Total	Total Incl O&P
0010 **COMPACTORS**									
0020 Compactors, 115 volt, 250 lbs./hr., chute fed	L-4	1	24	Ea.	12,400	1,175		13,575	15,500
1000 Heavy duty industrial compactor, 0.5 C.Y. capacity		1	24		11,000	1,175		12,175	13,900
1050 1.0 C.Y. capacity		1	24		16,800	1,175		17,975	20,300
1100 3.0 C.Y. capacity		.50	48		28,100	2,325		30,425	34,500
1150 5.0 C.Y. capacity		.50	48		35,400	2,325		37,725	42,600
1200 Combination shredder/compactor (5,000 lbs./hr.)		.50	48		69,000	2,325		71,325	79,000
1400 For handling hazardous waste materials, 55 gallon drum packer, std.					21,400			21,400	23,500
1410 55 gallon drum packer w/HEPA filter					26,400			26,400	29,000
1420 55 gallon drum packer w/charcoal & HEPA filter					35,200			35,200	38,700
1430 All of the above made explosion proof, add					1,550			1,550	1,700
5800 Shredder, industrial, minimum					25,500			25,500	28,100
5850 Maximum					136,500			136,500	150,000
5900 Baler, industrial, minimum					10,200			10,200	11,200
5950 Maximum					595,500			595,500	655,000

Estimating Tips
General

- The items in this division are usually priced per square foot or each. Most of these items are purchased by the owner and installed by the contractor. Do not assume the items in Division 12 will be purchased and installed by the contractor. Check the specifications for responsibilities and include receiving, storage, installation, and mechanical and electrical hookups in the appropriate divisions.

- Some items in this division require some type of support system that is not usually furnished with the item. Examples of these systems include blocking for the attachment of casework and heavy drapery rods. The required blocking must be added to the estimate in the appropriate division.

Reference Numbers

Reference numbers are shown at the beginning of some major classifications. These numbers refer to related items in the Reference Section. The reference information may be an estimating procedure, an alternate pricing method, or technical information.

12 46 Furnishing Accessories

12 46 19 – Clocks

12 46 19.50 Wall Clocks	Crew	Daily Output	Labor-Hours	Unit	Material	2019 Bare Costs Labor	Equipment	Total	Total Incl O&P
0010 **WALL CLOCKS**									
0080 12" diameter, single face	1 Elec	8	1	Ea.	89	60		149	187
0100 Double face	"	6.20	1.290	"	160	77.50		237.50	292

Estimating Tips
General

- The items and systems in this division are usually estimated, purchased, supplied, and installed as a unit by one or more subcontractors. The estimator must ensure that all parties are operating from the same set of specifications and assumptions, and that all necessary items are estimated and will be provided. Many times the complex items and systems are covered, but the more common ones, such as excavation or a crane, are overlooked for the very reason that everyone assumes nobody could miss them. The estimator should be the central focus and be able to ensure that all systems are complete.

- It is important to consider factors such as site conditions, weather, shape and size of building, as well as labor availability as they may impact the overall cost of erecting special structures and systems included in this division.

- Another area where problems can develop in this division is at the interface between systems.

The estimator must ensure, for instance, that anchor bolts, nuts, and washers are estimated and included for the air-supported structures and pre-engineered buildings to be bolted to their foundations. Utility supply is a common area where essential items or pieces of equipment can be missed or overlooked because each subcontractor may feel it is another's responsibility. The estimator should also be aware of certain items which may be supplied as part of a package but installed by others, and ensure that the installing contractor's estimate includes the cost of installation. Conversely, the estimator must also ensure that items are not costed by two different subcontractors, resulting in an inflated overall estimate.

13 30 00 Special Structures

- The foundations and floor slab, as well as rough mechanical and electrical, should be estimated, as this work is required for the assembly and erection of the structure. Generally, as noted in the data set, the pre-engineered building comes as a shell. Pricing is based on the size and structural design parameters stated in the reference section. Additional features, such as windows and doors with their related structural framing, must also be included by the estimator. Here again, the estimator must have a clear understanding of the scope of each portion of the work and all the necessary interfaces.

Reference Numbers

Reference numbers are shown at the beginning of some major classifications. These numbers refer to related items in the Reference Section. The reference information may be an estimating procedure, an alternate pricing method, or technical information.

Note: Not all subdivisions listed here necessarily appear. ■

Did you know?

RSMeans data is available through our online application:

- Search for costs by keyword
- Leverage the most up-to-date data
- Build and export estimates

Try it free
rsmeans.com/2019freetrial

13 11 Swimming Pools

13 11 13 – Below-Grade Swimming Pools

13 11 13.50 Swimming Pools		Crew	Daily Output	Labor-Hours	Unit	Material	2019 Bare Costs Labor	Equipment	Total	Total Incl O&P
0010	**SWIMMING POOLS** Residential in-ground, vinyl lined									
0020	Concrete sides, w/equip, sand bottom	B-52	300	.187	SF Surf	27	8.85	2	37.85	45.50
0100	Metal or polystyrene sides	B-14	410	.117		22.50	5.05	.78	28.33	33.50
0200	Add for vermiculite bottom					1.73			1.73	1.90
0500	Gunite bottom and sides, white plaster finish									
0600	12' x 30' pool	B-52	145	.386	SF Surf	50.50	18.35	4.14	72.99	87.50
0720	16' x 32' pool		155	.361		45.50	17.15	3.88	66.53	80.50
0750	20' x 40' pool		250	.224		40.50	10.65	2.40	53.55	63
0810	Concrete bottom and sides, tile finish									
0820	12' x 30' pool	B-52	80	.700	SF Surf	51	33	7.50	91.50	114
0830	16' x 32' pool		95	.589		42	28	6.35	76.35	95.50
0840	20' x 40' pool		130	.431		33.50	20.50	4.62	58.62	73
1100	Motel, gunite with plaster finish, incl. medium									
1150	capacity filtration & chlorination	B-52	115	.487	SF Surf	62	23	5.25	90.25	109
1200	Municipal, gunite with plaster finish, incl. high									
1250	capacity filtration & chlorination	B-52	100	.560	SF Surf	80.50	26.50	6	113	135
1350	Add for formed gutters				L.F.	118			118	130
1360	Add for stainless steel gutters				"	350			350	385
1700	Filtration and deck equipment only, as % of total				Total				20%	20%
1800	Deck equipment, rule of thumb, 20' x 40' pool				SF Pool				1.18	1.30
1900	5,000 S.F. pool				"				1.73	1.90

13 11 23 – On-Grade Swimming Pools

13 11 23.50 Swimming Pools		Crew	Daily Output	Labor-Hours	Unit	Material	2019 Bare Costs Labor	Equipment	Total	Total Incl O&P
0010	**SWIMMING POOLS** Residential above ground, steel construction									
0100	Round, 15' diam.	B-80A	3	8	Ea.	810	330	82	1,222	1,475
0120	18' diam.		2.50	9.600		910	395	98.50	1,403.50	1,700
0140	21' diam.		2	12		1,025	495	123	1,643	2,000
0160	24' diam.		1.80	13.333		1,125	545	137	1,807	2,225
0180	27' diam.		1.50	16		1,325	655	164	2,144	2,625
0200	30' diam.		1	24		1,450	985	246	2,681	3,350
0220	Oval, 12' x 24'		2.30	10.435		1,500	430	107	2,037	2,425
0240	15' x 30'		1.80	13.333		2,075	545	137	2,757	3,250
0260	18' x 33'		1	24		2,350	985	246	3,581	4,375

13 11 46 – Swimming Pool Accessories

13 11 46.50 Swimming Pool Equipment		Crew	Daily Output	Labor-Hours	Unit	Material	2019 Bare Costs Labor	Equipment	Total	Total Incl O&P
0010	**SWIMMING POOL EQUIPMENT**									
0020	Diving stand, stainless steel, 3 meter	2 Carp	.40	40	Ea.	17,300	2,075		19,375	22,200
2100	Lights, underwater, 12 volt, with transformer, 300 watt	1 Elec	1	8		365	480		845	1,125
2200	110 volt, 500 watt, standard		1	8		320	480		800	1,075
2400	Low water cutoff type		1	8		296	480		776	1,050
2800	Heaters, see Section 23 52 28.10									

13 12 Fountains

13 12 13 – Exterior Fountains

13 12 13.10 Outdoor Fountains

		Crew	Daily Output	Labor-Hours	Unit	Material	2019 Bare Costs Labor	2019 Bare Costs Equipment	Total	Total Incl O&P
0010	**OUTDOOR FOUNTAINS**									
0100	Outdoor fountain, 48" high with bowl and figures	2 Clab	2	8	Ea.	460	330		790	1,000
0200	Commercial, concrete or cast stone, 40-60" H, simple		2	8		845	330		1,175	1,425
0220	Average		2	8		1,625	330		1,955	2,300
0240	Ornate		2	8		3,700	330		4,030	4,575
0260	Metal, 72" high		2	8		1,575	330		1,905	2,225
0280	90" high		2	8		2,125	330		2,455	2,850
0300	120" high		2	8		5,150	330		5,480	6,150
0320	Resin or fiberglass, 40-60" H, wall type		2	8		595	330		925	1,150
0340	Waterfall type		2	8		1,000	330		1,330	1,600

13 12 23 – Interior Fountains

13 12 23.10 Indoor Fountains

		Crew	Daily Output	Labor-Hours	Unit	Material	2019 Bare Costs Labor	2019 Bare Costs Equipment	Total	Total Incl O&P
0010	**INDOOR FOUNTAINS**									
0100	Commercial, floor type, resin or fiberglass, lighted, cascade type	2 Clab	2	8	Ea.	365	330		695	900
0120	Tiered type		2	8		430	330		760	970
0140	Waterfall type		2	8		276	330		606	800

13 21 Controlled Environment Rooms

13 21 13 – Clean Rooms

13 21 13.50 Clean Room Components

		Crew	Daily Output	Labor-Hours	Unit	Material	2019 Bare Costs Labor	2019 Bare Costs Equipment	Total	Total Incl O&P
0010	**CLEAN ROOM COMPONENTS**									
1100	Clean room, soft wall, 12' x 12', Class 100	1 Carp	.18	44.444	Ea.	16,400	2,300		18,700	21,600
1110	Class 1,000		.18	44.444		15,000	2,300		17,300	20,000
1120	Class 10,000		.21	38.095		13,600	1,975		15,575	18,000
1130	Class 100,000		.21	38.095		13,000	1,975		14,975	17,300
2800	Ceiling grid support, slotted channel struts 4'-0" OC, ea. way				S.F.				5.90	6.50
3000	Ceiling panel, vinyl coated foil on mineral substrate									
3020	Sealed, non-perforated				S.F.				1.27	1.40
4000	Ceiling panel seal, silicone sealant, 150 L.F./gal.	1 Carp	150	.053	L.F.	.29	2.75		3.04	4.49
4100	Two sided adhesive tape	"	240	.033	"	.13	1.72		1.85	2.74
4200	Clips, one per panel				Ea.	1.01			1.01	1.11
6000	HEPA filter, 2' x 4', 99.97% eff., 3" dp beveled frame (silicone seal)					365			365	400
6040	6" deep skirted frame (channel seal)					480			480	525
6100	99.99% efficient, 3" deep beveled frame (silicone seal)					450			450	495
6140	6" deep skirted frame (channel seal)					480			480	530
6200	99.999% efficient, 3" deep beveled frame (silicone seal)					470			470	520
6240	6" deep skirted frame (channel seal)					515			515	570
7000	Wall panel systems, including channel strut framing									
7020	Polyester coated aluminum, particle board				S.F.				18.20	18.20
7100	Porcelain coated aluminum, particle board								32	32
7400	Wall panel support, slotted channel struts, to 12' high								16.35	16.35

For customer support on your Electrical Costs with RSMeans data, call 800.448.8182.

67

13 24 Special Activity Rooms

13 24 16 – Saunas

13 24 16.50 Saunas and Heaters

		Crew	Daily Output	Labor-Hours	Unit	Material	2019 Bare Costs Labor	Equipment	Total	Total Incl O&P
0010	**SAUNAS AND HEATERS**									
0020	Prefabricated, incl. heater & controls, 7' high, 6' x 4', C/C	L-7	2.20	12.727	Ea.	5,250	635		5,885	6,725
0050	6' x 4', C/P		2	14		4,350	695		5,045	5,825
0400	6' x 5', C/C		2	14		5,425	695		6,120	7,000
0450	6' x 5', C/P		2	14		5,150	695		5,845	6,700
0600	6' x 6', C/C		1.80	15.556		7,100	775		7,875	8,975
0650	6' x 6', C/P		1.80	15.556		5,450	775		6,225	7,175
0800	6' x 9', C/C		1.60	17.500		7,850	870		8,720	9,975
0850	6' x 9', C/P		1.60	17.500		6,450	870		7,320	8,425
1000	8' x 12', C/C		1.10	25.455		10,700	1,275		11,975	13,700
1050	8' x 12', C/P		1.10	25.455		8,850	1,275		10,125	11,700
1200	8' x 8', C/C		1.40	20		8,600	995		9,595	11,000
1250	8' x 8', C/P		1.40	20		7,675	995		8,670	9,925
1400	8' x 10', C/C		1.20	23.333		9,400	1,150		10,550	12,100
1450	8' x 10', C/P		1.20	23.333		8,325	1,150		9,475	10,900
1600	10' x 12', C/C		1	28		12,600	1,400		14,000	16,000
1650	10' x 12', C/P	▼	1	28		13,600	1,400		15,000	17,000
2500	Heaters only (incl. above), wall mounted, to 200 C.F.					960			960	1,050
2750	To 300 C.F.					1,050			1,050	1,150
3000	Floor standing, to 720 C.F., 10,000 watts, w/controls	1 Elec	3	2.667		2,475	160		2,635	2,975
3250	To 1,000 C.F., 16,000 watts	"	3	2.667	▼	3,600	160		3,760	4,200

13 24 26 – Steam Baths

13 24 26.50 Steam Baths and Components

		Crew	Daily Output	Labor-Hours	Unit	Material	2019 Bare Costs Labor	Equipment	Total	Total Incl O&P
0010	**STEAM BATHS AND COMPONENTS**									
0020	Heater, timer & head, single, to 140 C.F.	1 Plum	1.20	6.667	Ea.	2,300	420		2,720	3,150
0500	To 300 C.F.		1.10	7.273		2,625	460		3,085	3,600
1000	Commercial size, with blow-down assembly, to 800 C.F.		.90	8.889		5,750	560		6,310	7,175
1500	To 2,500 C.F.	▼	.80	10		8,225	630		8,855	9,975
2000	Multiple, motels, apts., 2 baths, w/blow-down assm., 500 C.F.	Q-1	1.30	12.308		6,675	700		7,375	8,400
2500	4 baths	"	.70	22.857	▼	10,700	1,300		12,000	13,800

13 47 Facility Protection

13 47 13 – Cathodic Protection

13 47 13.16 Cathodic Prot. for Underground Storage Tanks

		Crew	Daily Output	Labor-Hours	Unit	Material	2019 Bare Costs Labor	Equipment	Total	Total Incl O&P
0010	**CATHODIC PROTECTION FOR UNDERGROUND STORAGE TANKS**									
1000	Anodes, magnesium type, 9 #	R-15	18.50	2.595	Ea.	36	152	15.30	203.30	284
1010	17 #		13	3.692		76	217	22	315	435
1020	32 #		10	4.800		119	282	28.50	429.50	580
1030	48 #	▼	7.20	6.667		164	390	39.50	593.50	810
1100	Graphite type w/epoxy cap, 3" x 60" (32 #)	R-22	8.40	4.438		124	244		368	500
1110	4" x 80" (68 #)		6	6.213		240	340		580	775
1120	6" x 72" (80 #)		5.20	7.169		1,525	395		1,920	2,275
1130	6" x 36" (45 #)	▼	9.60	3.883		770	213		983	1,175
2000	Rectifiers, silicon type, air cooled, 28 V/10 A	R-19	3.50	5.714		2,150	345		2,495	2,875
2010	20 V/20 A		3.50	5.714		2,200	345		2,545	2,950
2100	Oil immersed, 28 V/10 A		3	6.667		2,850	400		3,250	3,725
2110	20 V/20 A	▼	3	6.667	▼	3,175	400		3,575	4,075
3000	Anode backfill, coke breeze	R-22	3850	.010	Lb.	.29	.53		.82	1.11
4000	Cable, HMWPE, No. 8		2.40	15.533	M.L.F.	590	855		1,445	1,925
4010	No. 6	▼	2.40	15.533	▼	845	855		1,700	2,200

13 47 Facility Protection

13 47 13 – Cathodic Protection

13 47 13.16 Cathodic Prot. for Underground Storage Tanks		Crew	Daily Output	Labor-Hours	Unit	Material	2019 Bare Costs Labor	2019 Bare Costs Equipment	Total	Total Incl O&P
4020	No. 4	R-22	2.40	15.533	M.L.F.	1,275	855		2,130	2,675
4030	No. 2		2.40	15.533		1,975	855		2,830	3,450
4040	No. 1		2.20	16.945		2,700	930		3,630	4,350
4050	No. 1/0		2.20	16.945		3,500	930		4,430	5,250
4060	No. 2/0		2.20	16.945		5,025	930		5,955	6,950
4070	No. 4/0	↓	2	18.640	↓	6,250	1,025		7,275	8,400
5000	Test station, 7 terminal box, flush curb type w/lockable cover	R-19	12	1.667	Ea.	79	100		179	237
5010	Reference cell, 2″ diam. PVC conduit, cplg., plug, set flush	"	4.80	4.167	"	156	251		407	545

13 49 Radiation Protection

13 49 13 – Integrated X-Ray Shielding Assemblies

13 49 13.50 Lead Sheets

		Crew	Daily Output	Labor-Hours	Unit	Material	2019 Bare Costs Labor	2019 Bare Costs Equipment	Total	Total Incl O&P
0010	**LEAD SHEETS**									
0300	Lead sheets, 1/16″ thick	2 Lath	135	.119	S.F.	11.10	5.95		17.05	21
0400	1/8″ thick		120	.133		23	6.70		29.70	35
0500	Lead shielding, 1/4″ thick		135	.119		42	5.95		47.95	55
0550	1/2″ thick	↓	120	.133	↓	79.50	6.70		86.20	97.50
0950	Lead headed nails (average 1 lb. per sheet)				Lb.	8.50			8.50	9.35
1000	Butt joints in 1/8″ lead or thicker, 2″ batten strip x 7′ long	2 Lath	240	.067	Ea.	31.50	3.35		34.85	39.50
1200	X-ray protection, average radiography or fluoroscopy									
1210	room, up to 300 S.F. floor, 1/16″ lead, economy	2 Lath	.25	64	Total	11,300	3,225		14,525	17,100
1500	7′-0″ walls, deluxe	"	.15	107	"	13,600	5,350		18,950	22,800
1600	Deep therapy X-ray room, 250 kV capacity,									
1800	up to 300 S.F. floor, 1/4″ lead, economy	2 Lath	.08	200	Total	31,500	10,100		41,600	49,500
1900	7′-0″ walls, deluxe	"	.06	267	"	38,900	13,400		52,300	62,500

13 49 19 – Lead-Lined Materials

13 49 19.50 Shielding Lead

		Crew	Daily Output	Labor-Hours	Unit	Material	2019 Bare Costs Labor	2019 Bare Costs Equipment	Total	Total Incl O&P
0010	**SHIELDING LEAD**									
0100	Laminated lead in wood doors, 1/16″ thick, no hardware				S.F.	53.50			53.50	59
0200	Lead lined door frame, not incl. hardware,									
0210	1/16″ thick lead, butt prepared for hardware	1 Lath	2.40	3.333	Ea.	880	168		1,048	1,200
0850	Window frame with 1/16″ lead and voice passage, 36″ x 60″	2 Glaz	2	8		4,575	395		4,970	5,625
0870	24″ x 36″ frame		4	4	↓	2,375	199		2,574	2,900
0900	Lead gypsum board, 5/8″ thick with 1/16″ lead		160	.100	S.F.	15.90	4.97		20.87	25
0910	1/8″ lead	↓	140	.114		25	5.65		30.65	36
0930	1/32″ lead	2 Lath	200	.080	↓	8.60	4.02		12.62	15.40

13 49 21 – Lead Glazing

13 49 21.50 Lead Glazing

		Crew	Daily Output	Labor-Hours	Unit	Material	2019 Bare Costs Labor	2019 Bare Costs Equipment	Total	Total Incl O&P
0010	**LEAD GLAZING**									
0600	Lead glass, 1/4″ thick, 2.0 mm LE, 12″ x 16″	2 Glaz	13	1.231	Ea.	395	61		456	525
0700	24″ x 36″		8	2		1,375	99.50		1,474.50	1,675
0800	36″ x 60″	↓	2	8	↓	3,800	395		4,195	4,775
2000	X-ray viewing panels, clear lead plastic									
2010	7 mm thick, 0.3 mm LE, 2.3 lb./S.F.	H-3	139	.115	S.F.	282	5.10		287.10	320
2020	12 mm thick, 0.5 mm LE, 3.9 lb./S.F.		82	.195		390	8.65		398.65	445
2030	18 mm thick, 0.8 mm LE, 5.9 lb./S.F.		54	.296		480	13.10		493.10	545
2040	22 mm thick, 1.0 mm LE, 7.2 lb./S.F.		44	.364		585	16.10		601.10	670
2050	35 mm thick, 1.5 mm LE, 11.5 lb./S.F.		28	.571		895	25.50		920.50	1,025
2060	46 mm thick, 2.0 mm LE, 15.0 lb./S.F.	↓	21	.762	↓	1,175	33.50		1,208.50	1,325
2090	For panels 12 S.F. to 48 S.F., add crating charge				Ea.				50	50

13 49 23 – Integrated RFI/EMI Shielding Assemblies

13 49 23.50 Modular Shielding Partitions	Crew	Daily Output	Labor-Hours	Unit	Material	2019 Bare Costs Labor	Equipment	Total	Total Incl O&P
0010 MODULAR SHIELDING PARTITIONS									
4000 X-ray barriers, modular, panels mounted within framework for									
4002 attaching to floor, wall or ceiling, upper portion is clear lead									
4005 plastic window panels 48"H, lower portion is opaque leaded									
4008 steel panels 36"H, structural supports not incl.									
4010 1-section barrier, 36"W x 84"H overall									
4020 0.5 mm LE panels	H-3	6.40	2.500	Ea.	8,750	111		8,861	9,775
4030 0.8 mm LE panels		6.40	2.500		9,425	111		9,536	10,600
4040 1.0 mm LE panels		5.33	3.002		12,100	133		12,233	13,600
4050 1.5 mm LE panels		5.33	3.002		15,300	133		15,433	17,000
4060 2-section barrier, 72"W x 84"H overall									
4070 0.5 mm LE panels	H-3	4	4	Ea.	12,700	177		12,877	14,200
4080 0.8 mm LE panels		4	4		14,400	177		14,577	16,200
4090 1.0 mm LE panels		3.56	4.494		16,000	199		16,199	17,900
5000 1.5 mm LE panels		3.20	5		25,200	221		25,421	28,000
5010 3-section barrier, 108"W x 84"H overall									
5020 0.5 mm LE panels	H-3	3.20	5	Ea.	19,000	221		19,221	21,200
5030 0.8 mm LE panels		3.20	5		21,700	221		21,921	24,100
5040 1.0 mm LE panels		2.67	5.993		24,000	265		24,265	26,700
5050 1.5 mm LE panels		2.46	6.504		37,700	288		37,988	41,900
7000 X-ray barriers, mobile, mounted within framework w/casters on									
7005 bottom, clear lead plastic window panels on upper portion,									
7010 opaque on lower, 30"W x 75"H overall, incl. framework									
7020 24"H upper w/0.5 mm LE, 48"H lower w/0.8 mm LE	1 Carp	16	.500	Ea.	3,900	26		3,926	4,325
7030 48"W x 75"H overall, incl. framework									
7040 36"H upper w/0.5 mm LE, 36"H lower w/0.8 mm LE	1 Carp	16	.500	Ea.	6,875	26		6,901	7,600
7050 36"H upper w/1.0 mm LE, 36"H lower w/1.5 mm LE	"	16	.500	"	8,150	26		8,176	9,000
7060 72"W x 75"H overall, incl. framework									
7070 36"H upper w/0.5 mm LE, 36"H lower w/0.8 mm LE	1 Carp	16	.500	Ea.	8,150	26		8,176	9,000
7080 36"H upper w/1.0 mm LE, 36"H lower w/1.5 mm LE	"	16	.500	"	10,200	26		10,226	11,200

13 49 33 – Radio Frequency Shielding

13 49 33.50 Shielding, Radio Frequency	Crew	Daily Output	Labor-Hours	Unit	Material	2019 Bare Costs Labor	Equipment	Total	Total Incl O&P
0010 SHIELDING, RADIO FREQUENCY									
0020 Prefabricated, galvanized steel	2 Carp	375	.043	SF Surf	5.35	2.20		7.55	9.25
0040 5 oz., copper floor panel		480	.033		4.58	1.72		6.30	7.65
0050 5 oz., copper wall/ceiling panel		155	.103		4.58	5.35		9.93	13.10
0100 12 oz., copper floor panel		470	.034		9.15	1.76		10.91	12.70
0110 12 oz., copper wall/ceiling panel		140	.114		9.15	5.90		15.05	19
0150 Door, copper/wood laminate, 4' x 7'		1.50	10.667	Ea.	8,550	550		9,100	10,300

70

For customer support on your Electrical Costs with RSMeans data, call 800.448.8182.

Estimating Tips

Pipe for fire protection and all uses is located in Subdivisions 21 11 13 and 22 11 13.

The labor adjustment factors listed in Subdivision 22 01 02.20 also apply to Division 21.

Many, but not all, areas in the U.S. require backflow protection in the fire system. Insurance underwriters may have specific requirements for the type of materials to be installed or design requirements based on the hazard to be protected. Local jurisdictions may have requirements not covered by code. It is advisable to be aware of any special conditions.

For your reference, the following is a list of the most applicable Fire Codes and Standards, which may be purchased from the NFPA, 1 Batterymarch Park, Quincy, MA 02169-7471.

- NFPA 1: Uniform Fire Code
- NFPA 10: Portable Fire Extinguishers
- NFPA 11: Low-, Medium-, and High-Expansion Foam
- NFPA 12: Carbon Dioxide Extinguishing Systems (Also companion 12A)
- NFPA 13: Installation of Sprinkler Systems (Also companion 13D, 13E, and 13R)
- NFPA 14: Installation of Standpipe and Hose Systems
- NFPA 15: Water Spray Fixed Systems for Fire Protection
- NFPA 16: Installation of Foam-Water Sprinkler and Foam-Water Spray Systems
- NFPA 17: Dry Chemical Extinguishing Systems (Also companion 17A)
- NFPA 18: Wetting Agents
- NFPA 20: Installation of Stationary Pumps for Fire Protection
- NFPA 22: Water Tanks for Private Fire Protection
- NFPA 24: Installation of Private Fire Service Mains and their Appurtenances
- NFPA 25: Inspection, Testing and Maintenance of Water-Based Fire Protection

Reference Numbers

Reference numbers are shown at the beginning of some major classifications. These numbers refer to related items in the Reference Section. The reference information may be an estimating procedure, an alternate pricing method, or technical information.

Did you know?

RSMeans data is available through our online application:

- Search for costs by keyword
- Leverage the most up-to-date data
- Build and export estimates

Try it free
rsmeans.com/2019freetrial

21 21 Carbon-Dioxide Fire-Extinguishing Systems

21 21 16 – Carbon-Dioxide Fire-Extinguishing Equipment

21 21 16.50 CO2 Fire Extinguishing System

	Crew	Daily Output	Labor-Hours	Unit	Material	2019 Bare Costs Labor	Equipment	Total	Total Incl O&P
0010 **CO₂ FIRE EXTINGUISHING SYSTEM**									
0100 Control panel, single zone with batteries (2 zones det., 1 suppr.)	1 Elec	1	8	Ea.	970	480		1,450	1,800
0150 Multizone (4) with batteries (8 zones det., 4 suppr.)	"	.50	16		2,925	960		3,885	4,650
1000 Dispersion nozzle, CO₂, 3" x 5"	1 Plum	18	.444		154	28		182	211
2000 Extinguisher, CO₂ system, high pressure, 75 lb. cylinder	Q-1	6	2.667		1,350	152		1,502	1,700
2100 100 lb. cylinder	"	5	3.200		1,750	182		1,932	2,200
3000 Electro/mechanical release	L-1	4	4		985	246		1,231	1,450
3400 Manual pull station	1 Plum	6	1.333		83	84		167	218
4000 Pneumatic damper release	"	8	1	▼	163	63		226	274

21 22 Clean-Agent Fire-Extinguishing Systems

21 22 16 – Clean-Agent Fire-Extinguishing Equipment

21 22 16.50 Clean-Agent Extinguishing Systems

	Crew	Daily Output	Labor-Hours	Unit	Material	2019 Bare Costs Labor	Equipment	Total	Total Incl O&P
0010 **CLEAN-AGENT EXTINGUISHING SYSTEMS**									
0020 FM200 fire extinguishing system									
1100 Dispersion nozzle FM200, 1-1/2"	1 Plum	14	.571	Ea.	191	36		227	264
2400 Extinguisher, FM200 system, filled, with mounting bracket									
2460 26 lb. container	Q-1	8	2	Ea.	2,025	114		2,139	2,400
2480 44 lb. container		7	2.286		2,675	130		2,805	3,150
2500 63 lb. container		6	2.667		3,200	152		3,352	3,750
2520 101 lb. container		5	3.200		5,275	182		5,457	6,075
2540 196 lb. container	▼	4	4		6,700	227		6,927	7,725
6000 FM200 system, simple nozzle layout, with broad dispersion				C.F.	1.79			1.79	1.97
6010 Extinguisher, FM200 system, filled, with mounting bracket									
6020 Complex nozzle layout and/or including underfloor dispersion				C.F.	3.56			3.56	3.92
6100 20,000 C.F. 2 exits, 8' clng					2.13			2.13	2.34
6200 100,000 C.F. 4 exits, 8' clng					1.92			1.92	2.11
6300 250,000 C.F. 6 exits, 8' clng				▼	1.62			1.62	1.78
7010 HFC-227ea fire extinguishing system									
7100 Cylinders with clean-agent									
7110 Does not include pallete jack/fork lift rental fees									
7120 70 lb. cyl, w/35 lb. agent, no solenoid	Q-12	14	1.143	Ea.	2,425	63.50		2,488.50	2,775
7130 70 lb. cyl w/70 lb. agent, no solenoid		10	1.600		3,175	88.50		3,263.50	3,625
7140 70 lb. cyl w/35 lb. agent, w/solenoid		14	1.143		3,550	63.50		3,613.50	4,000
7150 70 lb. cyl w/70 lb. agent, w/solenoid		10	1.600		4,400	88.50		4,488.50	4,950
7220 250 lb. cyl, w/125 lb. agent, no solenoid		8	2		5,600	111		5,711	6,325
7230 250 lb. cyl, w/250 lb. agent, no solenoid		5	3.200		5,600	177		5,777	6,425
7240 250 lb. cyl, w/125 lb. agent, w/solenoid		8	2		7,525	111		7,636	8,450
7250 250 lb. cyl, w/250 lb. agent, w/solenoid		5	3.200		10,800	177		10,977	12,200
7320 560 lb. cyl, w/300 lb. agent, no solenoid		4	4		10,300	221		10,521	11,600
7330 560 lb. cyl, w/560 lb. agent, no solenoid		2.50	6.400		15,400	355		15,755	17,500
7340 560 lb. cyl, w/300 lb. agent, w/solenoid		4	4		13,900	221		14,121	15,600
7350 560 lb. cyl, w/560 lb. agent, w/solenoid	▼	2.50	6.400		21,200	355		21,555	23,800
7420 1,200 lb. cyl, w/600 lb. agent, no solenoid	Q-13	4	8		19,400	470		19,870	22,100
7430 1,200 lb. cyl, w/1,200 lb. agent, no solenoid		3	10.667		31,900	625		32,525	36,000
7440 1,200 lb. cyl, w/600 lb. agent, w/solenoid		4	8		49,400	470		49,870	55,000
7450 1,200 lb. cyl, w/1,200 lb. agent, w/solenoid	▼	3	10.667	▼	76,000	625		76,625	84,500
7500 Accessories									
7510 Dispersion nozzle	1 Spri	16	.500	Ea.	98	31		129	155
7520 Agent release panel	1 Elec	4	2		62	120		182	247
7530 Maintenance switch	"	6	1.333		39.50	80		119.50	163

21 22 Clean-Agent Fire-Extinguishing Systems

21 22 16 – Clean-Agent Fire-Extinguishing Equipment

21 22 16.50 Clean-Agent Extinguishing Systems	Crew	Daily Output	Labor-Hours	Unit	Material	2019 Bare Costs Labor	Equipment	Total	Total Incl O&P
7540	Solenoid valve, 12v dc	1 Spri	8	1	Ea.	234	61.50		295.50	350
7550	12v ac		8	1		510	61.50		571.50	660
7560	12v dc, explosion proof		8	1		395	61.50		456.50	530

Division Notes

		CREW	DAILY OUTPUT	LABOR-HOURS	UNIT	BARE COSTS				TOTAL INCL O&P
						MAT.	LABOR	EQUIP.	TOTAL	

Estimating Tips
22 10 00 Plumbing Piping and Pumps

This subdivision is primarily basic pipe and related materials. The pipe may be used by any of the mechanical disciplines, i.e., plumbing, fire protection, heating, and air conditioning.

Note: CPVC plastic piping approved for fire protection is located in 21 11 13.

- The labor adjustment factors listed in Subdivision 22 01 02.20 apply throughout Divisions 21, 22, and 23. CAUTION: the correct percentage may vary for the same items. For example, the percentage add for the basic pipe installation should be based on the maximum height that the installer must install for that particular section. If the pipe is to be located 14' above the floor but it is suspended on threaded rod from beams, the bottom flange of which is 18' high (4' rods), then the height is actually 18' and the add is 20%. The pipe cover, however, does not have to go above the 14' and so the add should be 10%.

- Most pipe is priced first as straight pipe with a joint (coupling, weld, etc.) every 10' and a hanger usually every 10'. There are exceptions with hanger spacing such as for cast iron pipe (5')

and plastic pipe (3 per 10'). Following each type of pipe there are several lines listing sizes and the amount to be subtracted to delete couplings and hangers. This is for pipe that is to be buried or supported together on trapeze hangers. The reason that the couplings are deleted is that these runs are usually long, and frequently longer lengths of pipe are used. By deleting the couplings, the estimator is expected to look up and add back the correct reduced number of couplings.

- When preparing an estimate, it may be necessary to approximate the fittings. Fittings usually run between 25% and 50% of the cost of the pipe. The lower percentage is for simpler runs, and the higher number is for complex areas, such as mechanical rooms.

- For historic restoration projects, the systems must be as invisible as possible, and pathways must be sought for pipes, conduit, and ductwork. While installations in accessible spaces (such as basements and attics) are relatively straightforward to estimate, labor costs may be more difficult to determine when delivery systems must be concealed.

22 40 00 Plumbing Fixtures

- Plumbing fixture costs usually require two lines: the fixture itself and its "rough-in, supply, and waste."

- In the Assemblies Section (Plumbing D2010) for the desired fixture, the System Components Group at the center of the page shows the fixture on the first line. The rest of the list (fittings, pipe, tubing, etc.) will total up to what we refer to in the Unit Price section as "Rough-in, supply, waste, and vent." Note that for most fixtures we allow a nominal 5' of tubing to reach from the fixture to a main or riser.

- Remember that gas- and oil-fired units need venting.

Reference Numbers

Reference numbers are shown at the beginning of some major classifications. These numbers refer to related items in the Reference Section. The reference information may be an estimating procedure, an alternate pricing method, or technical information.

Note: Not all subdivisions listed here necessarily appear. ∎

22 01 Operation and Maintenance of Plumbing

22 01 02 – Labor Adjustments

		Crew	Daily Output	Labor-Hours	Unit	Material	2019 Bare Costs Labor	2019 Bare Costs Equipment	Total	Total Incl O&P
22 01 02.10 Boilers, General										
0010	**BOILERS, GENERAL,** Prices do not include flue piping, elec. wiring,									
0020	gas or oil piping, boiler base, pad, or tankless unless noted									
0100	Boiler H.P.: 10 KW = 34 lb./steam/hr. = 33,475 BTU/hr.									
0150	To convert SFR to BTU rating: Hot water, 150 x SFR;									
0160	Forced hot water, 180 x SFR; steam, 240 x SFR									
22 01 02.20 Labor Adjustment Factors										
0010	**LABOR ADJUSTMENT FACTORS** (For Div. 21, 22 and 23) R220102-20									
0100	Labor factors: The below are reasonable suggestions, but									
0110	each project must be evaluated for its own peculiarities, and									
0120	the adjustments be increased or decreased depending on the									
0130	severity of the special conditions.									
1000	Add to labor for elevated installation (Above floor level)									
1080	10' to 14.5' high						10%			
1100	15' to 19.5' high						20%			
1120	20' to 24.5' high						25%			
1140	25' to 29.5' high						35%			
1160	30' to 34.5' high						40%			
1180	35' to 39.5' high						50%			
1200	40' and higher						55%			
2000	Add to labor for crawl space									
2100	3' high						40%			
2140	4' high						30%			
3000	Add to labor for multi-story building									
3010	For new construction (No elevator available)									
3100	Add for floors 3 thru 10						5%			
3110	Add for floors 11 thru 15						10%			
3120	Add for floors 16 thru 20						15%			
3130	Add for floors 21 thru 30						20%			
3140	Add for floors 31 and up						30%			
3170	For existing structure (Elevator available)									
3180	Add for work on floor 3 and above						2%			
4000	Add to labor for working in existing occupied buildings									
4100	Hospital						35%			
4140	Office building						25%			
4180	School						20%			
4220	Factory or warehouse						15%			
4260	Multi dwelling						15%			
5000	Add to labor, miscellaneous									
5100	Cramped shaft						35%			
5140	Congested area						15%			
5180	Excessive heat or cold						30%			
9000	Labor factors: The above are reasonable suggestions, but									
9010	each project should be evaluated for its own peculiarities.									
9100	Other factors to be considered are:									
9140	Movement of material and equipment through finished areas									
9180	Equipment room									
9220	Attic space									
9260	No service road									
9300	Poor unloading/storage area									
9340	Congested site area/heavy traffic									

22 05 Common Work Results for Plumbing

22 05 23 – General-Duty Valves for Plumbing Piping

22 05 23.20 Valves, Bronze	Crew	Daily Output	Labor-Hours	Unit	Material	2019 Bare Costs Labor	Equipment	Total	Total Incl O&P
0010 **VALVES, BRONZE**									
8800 Water heater water & gas safety shut off									
8810 Protection against a leaking water heater									
8814 Shut off valve	1 Plum	16	.500	Ea.	193	31.50		224.50	260
8818 Water heater dam		32	.250		32	15.80		47.80	58.50
8822 Gas control wiring harness	↓	32	.250	↓	22	15.80		37.80	48
8830 Whole house flood safety shut off									
8834 Connections									
8838 3/4" NPT	1 Plum	12	.667	Ea.	950	42		992	1,125
8842 1" NPT		11	.727		975	46		1,021	1,150
8846 1-1/4" NPT	↓	10	.800	↓	1,000	50.50		1,050.50	1,175

22 05 29 – Hangers and Supports for Plumbing Piping and Equipment

22 05 29.10 Hangers & Supp. for Plumb'g/HVAC Pipe/Equip.

	Crew	Daily Output	Labor-Hours	Unit	Material	2019 Bare Costs Labor	Equipment	Total	Total Incl O&P
0010 **HANGERS AND SUPPORTS FOR PLUMB'G/HVAC PIPE/EQUIP.**									
0011 TYPE numbers per MSS-SP58									
0050 Brackets									
0060 Beam side or wall, malleable iron, TYPE 34									
0070 3/8" threaded rod size	1 Plum	48	.167	Ea.	4.24	10.55		14.79	20.50
0080 1/2" threaded rod size		48	.167		3.88	10.55		14.43	20
0090 5/8" threaded rod size		48	.167		11.45	10.55		22	28.50
0100 3/4" threaded rod size		48	.167		20.50	10.55		31.05	39
0110 7/8" threaded rod size	↓	48	.167	↓	12.70	10.55		23.25	30
0120 For concrete installation, add						30%			
0150 Wall, welded steel, medium, TYPE 32									
0160 0 size, 12" wide, 18" deep	1 Plum	34	.235	Ea.	180	14.85		194.85	221
0170 1 size, 18" wide, 24" deep		34	.235		216	14.85		230.85	261
0180 2 size, 24" wide, 30" deep	↓	34	.235	↓	285	14.85		299.85	340
0300 Clamps									
0310 C-clamp, for mounting on steel beam flange, w/locknut, TYPE 23									
0320 3/8" threaded rod size	1 Plum	160	.050	Ea.	4.04	3.16		7.20	9.20
0330 1/2" threaded rod size		160	.050		4.88	3.16		8.04	10.10
0340 5/8" threaded rod size		160	.050		5.35	3.16		8.51	10.65
0350 3/4" threaded rod size	↓	160	.050	↓	6.85	3.16		10.01	12.30
0750 Riser or extension pipe, carbon steel, TYPE 8									
0760 3/4" pipe size	1 Plum	48	.167	Ea.	3.38	10.55		13.93	19.50
0770 1" pipe size		47	.170		3.51	10.75		14.26	20
0780 1-1/4" pipe size		46	.174		5.50	11		16.50	22.50
0790 1-1/2" pipe size		45	.178		10.15	11.25		21.40	28
0800 2" pipe size		43	.186		5.60	11.75		17.35	24
0810 2-1/2" pipe size		41	.195		5.50	12.30		17.80	24.50
0820 3" pipe size		40	.200		5.30	12.65		17.95	25
0830 3-1/2" pipe size		39	.205		5.50	12.95		18.45	25.50
0840 4" pipe size		38	.211		9	13.30		22.30	30
0850 5" pipe size		37	.216		11.45	13.65		25.10	33
0860 6" pipe size	↓	36	.222	↓	15.70	14.05		29.75	38.50
1150 Insert, concrete									
1160 Wedge type, carbon steel body, malleable iron nut, galvanized									
1170 1/4" threaded rod size	1 Plum	96	.083	Ea.	8.15	5.25		13.40	16.90
1180 3/8" threaded rod size		96	.083		19.90	5.25		25.15	30
1190 1/2" threaded rod size		96	.083		10.65	5.25		15.90	19.65
1200 5/8" threaded rod size		96	.083		55	5.25		60.25	68.50
1210 3/4" threaded rod size	↓	96	.083	↓	12.10	5.25		17.35	21

For customer support on your Electrical Costs with RSMeans data, call 800.448.8182.

77

22 05 29 – Hangers and Supports for Plumbing Piping and Equipment

22 05 29.10 Hangers & Supp. for Plumb'g/HVAC Pipe/Equip.	Crew	Daily Output	Labor-Hours	Unit	Material	2019 Bare Costs Labor	Equipment	Total	Total Incl O&P
1220	7/8" threaded rod size	1 Plum	96	.083	Ea.	12.45	5.25		17.70	21.50
2650	Rods, carbon steel									
2660	Continuous thread									
2670	1/4" thread size	1 Plum	144	.056	L.F.	2.30	3.51		5.81	7.80
2680	3/8" thread size		144	.056		2.46	3.51		5.97	7.95
2690	1/2" thread size		144	.056		3.87	3.51		7.38	9.50
2700	5/8" thread size		144	.056		5.50	3.51		9.01	11.30
2710	3/4" thread size		144	.056		9.65	3.51		13.16	15.90
2720	7/8" thread size		144	.056		12.15	3.51		15.66	18.60
2725	1/4" thread size, bright finish		144	.056		1.33	3.51		4.84	6.70
2726	1/2" thread size, bright finish		144	.056		4.35	3.51		7.86	10.05
2730	For galvanized, add					40%				
4400	U-bolt, carbon steel									
4410	Standard, with nuts, TYPE 42									
4420	1/2" pipe size	1 Plum	160	.050	Ea.	1.17	3.16		4.33	6.05
4430	3/4" pipe size		158	.051		1.17	3.20		4.37	6.10
4450	1" pipe size		152	.053		1.21	3.32		4.53	6.30
4460	1-1/4" pipe size		148	.054		1.52	3.41		4.93	6.75
4470	1-1/2" pipe size		143	.056		1.62	3.53		5.15	7.10
4480	2" pipe size		139	.058		1.75	3.63		5.38	7.40
4490	2-1/2" pipe size		134	.060		2.86	3.77		6.63	8.80
4500	3" pipe size		128	.063		3.21	3.95		7.16	9.45
4510	3-1/2" pipe size		122	.066		3.46	4.14		7.60	10
4520	4" pipe size		117	.068		3.52	4.32		7.84	10.35
4530	5" pipe size		114	.070		4.08	4.43		8.51	11.15
4540	6" pipe size		111	.072		7.80	4.55		12.35	15.45
4580	For plastic coating on 1/2" thru 6" size, add					150%				
8800	Wire cable support system									
8810	Cable with hook terminal and locking device									
8830	2 mm (.079") diam. cable (100 lb. cap.)									
8840	1 m (3.3') length, with hook	1 Shee	96	.083	Ea.	4.05	5.10		9.15	12.15
8850	2 m (6.6') length, with hook		84	.095		4.59	5.80		10.39	13.85
8860	3 m (9.9') length, with hook		72	.111		3.84	6.75		10.59	14.50
8870	5 m (16.4') length, with hook	Q-9	60	.267		5.05	14.65		19.70	27.50
8880	10 m (32.8') length, with hook	"	30	.533		7.15	29.50		36.65	52.50
8900	3 mm (.118") diam. cable (200 lb. cap.)									
8910	1 m (3.3') length, with hook	1 Shee	96	.083	Ea.	8.20	5.10		13.30	16.75
8920	2 m (6.6') length, with hook		84	.095		9.10	5.80		14.90	18.85
8930	3 m (9.9') length, with hook		72	.111		9.90	6.75		16.65	21
8940	5 m (16.4') length, with hook	Q-9	60	.267		12.35	14.65		27	35.50
8950	10 m (32.8') length, with hook	"	30	.533		16.35	29.50		45.85	62.50
9000	Cable system accessories									
9010	Anchor bolt, 3/8", with nut	1 Shee	140	.057	Ea.	1.10	3.48		4.58	6.50
9020	Air duct corner protector		160	.050		1.26	3.05		4.31	6
9030	Air duct support attachment		140	.057		2.03	3.48		5.51	7.55
9040	Flange clip, hammer-on style									
9044	For flange thickness 3/32"-9/64", 160 lb. cap.	1 Shee	180	.044	Ea.	.38	2.71		3.09	4.53
9048	For flange thickness 1/8"-1/4", 200 lb. cap.		160	.050		.31	3.05		3.36	4.97
9052	For flange thickness 5/16"-1/2", 200 lb. cap.		150	.053		.64	3.25		3.89	5.65
9056	For flange thickness 9/16"-3/4", 200 lb. cap.		140	.057		.86	3.48		4.34	6.25
9060	Wire insulation protection tube		180	.044	L.F.	.46	2.71		3.17	4.62
9070	Wire cutter				Ea.	43			43	47

22 33 Electric Domestic Water Heaters

22 33 13 – Instantaneous Electric Domestic Water Heaters

22 33 13.10 Hot Water Dispensers

22 33 13.10 Hot Water Dispensers		Crew	Daily Output	Labor-Hours	Unit	Material	2019 Bare Costs Labor	Equipment	Total	Total Incl O&P
0010	**HOT WATER DISPENSERS**									
0160	Commercial, 100 cup, 11.3 amp	1 Plum	14	.571	Ea.	470	36		506	570
3180	Household, 60 cup	"	14	.571	"	249	36		285	325

22 33 30 – Residential, Electric Domestic Water Heaters

22 33 30.13 Residential, Small-Capacity Elec. Water Heaters

22 33 30.13			Crew	Daily Output	Labor-Hours	Unit	Material	2019 Bare Costs Labor	Equipment	Total	Total Incl O&P
0010	**RESIDENTIAL, SMALL-CAPACITY ELECTRIC DOMESTIC WATER HEATERS**										
1000	Residential, electric, glass lined tank, 5 yr., 10 gal., single element	R224000-10	1 Plum	2.30	3.478	Ea.	445	220		665	815
1040	20 gallon, single element	R224000-20		2.20	3.636		590	230		820	995
1060	30 gallon, double element			2.20	3.636		990	230		1,220	1,450
1080	40 gallon, double element			2	4		1,075	253		1,328	1,550
1100	52 gallon, double element			2	4		1,200	253		1,453	1,700
1120	66 gallon, double element			1.80	4.444		1,625	281		1,906	2,225
1140	80 gallon, double element			1.60	5		1,850	315		2,165	2,500
1180	120 gallon, double element			1.40	5.714		2,550	360		2,910	3,350

22 33 33 – Light-Commercial Electric Domestic Water Heaters

22 33 33.10 Commercial Electric Water Heaters

22 33 33.10		Crew	Daily Output	Labor-Hours	Unit	Material	2019 Bare Costs Labor	Equipment	Total	Total Incl O&P
0010	**COMMERCIAL ELECTRIC WATER HEATERS**									
4000	Commercial, 100° rise. NOTE: for each size tank, a range of									
4010	heaters between the ones shown is available									
4020	Electric									
4100	5 gal., 3 kW, 12 GPH, 208 volt	1 Plum	2	4	Ea.	4,625	253		4,878	5,450
4120	10 gal., 6 kW, 25 GPH, 208 volt		2	4		5,100	253		5,353	5,975
4130	30 gal., 24 kW, 98 GPH, 208 volt		1.92	4.167		8,275	263		8,538	9,500
4136	40 gal., 36 kW, 148 GPH, 208 volt		1.88	4.255		10,800	269		11,069	12,200
4140	50 gal., 9 kW, 37 GPH, 208 volt		1.80	4.444		7,525	281		7,806	8,700
4160	50 gal., 36 kW, 148 GPH, 208 volt		1.80	4.444		11,600	281		11,881	13,100
4180	80 gal., 12 kW, 49 GPH, 208 volt		1.50	5.333		8,600	335		8,935	9,975
4200	80 gal., 36 kW, 148 GPH, 208 volt		1.50	5.333		12,200	335		12,535	13,900
4220	100 gal., 36 kW, 148 GPH, 208 volt		1.20	6.667		13,300	420		13,720	15,200
4240	120 gal., 36 kW, 148 GPH, 208 volt		1.20	6.667		13,900	420		14,320	15,900
4260	150 gal., 15 kW, 61 GPH, 480 volt		1	8		32,700	505		33,205	36,700
4280	150 gal., 120 kW, 490 GPH, 480 volt		1	8		45,800	505		46,305	51,500
4300	200 gal., 15 kW, 61 GPH, 480 volt	Q-1	1.70	9.412		35,300	535		35,835	39,600
4320	200 gal., 120 kW, 490 GPH, 480 volt		1.70	9.412		45,200	535		45,735	50,500
4340	250 gal., 15 kW, 61 GPH, 480 volt		1.50	10.667		36,400	605		37,005	40,900
4360	250 gal., 150 kW, 615 GPH, 480 volt		1.50	10.667		49,600	605		50,205	55,500
4380	300 gal., 30 kW, 123 GPH, 480 volt		1.30	12.308		40,200	700		40,900	45,300
4400	300 gal., 180 kW, 738 GPH, 480 volt		1.30	12.308		57,000	700		57,700	63,500
4420	350 gal., 30 kW, 123 GPH, 480 volt		1.10	14.545		42,500	825		43,325	48,000
4440	350 gal., 180 kW, 738 GPH, 480 volt		1.10	14.545		60,000	825		60,825	67,500
4460	400 gal., 30 kW, 123 GPH, 480 volt		1	16		48,200	910		49,110	54,500
4480	400 gal., 210 kW, 860 GPH, 480 volt		1	16		69,000	910		69,910	77,500
4500	500 gal., 30 kW, 123 GPH, 480 volt		.80	20		52,500	1,125		53,625	59,000
4520	500 gal., 240 kW, 984 GPH, 480 volt		.80	20		89,500	1,125		90,625	100,000
4540	600 gal., 30 kW, 123 GPH, 480 volt	Q-2	1.20	20		64,500	1,175		65,675	72,500
4560	600 gal., 300 kW, 1,230 GPH, 480 volt		1.20	20		97,500	1,175		98,675	109,500
4580	700 gal., 30 kW, 123 GPH, 480 volt		1	24		67,500	1,425		68,925	76,500
4600	700 gal., 300 kW, 1,230 GPH, 480 volt		1	24		101,500	1,425		102,925	114,000
4620	800 gal., 60 kW, 245 GPH, 480 volt		.90	26.667		73,000	1,575		74,575	83,000
4640	800 gal., 300 kW, 1,230 GPH, 480 volt		.90	26.667		104,000	1,575		105,575	116,500
4660	1,000 gal., 60 kW, 245 GPH, 480 volt		.70	34.286		80,000	2,025		82,025	91,000
4680	1,000 gal., 480 kW, 1,970 GPH, 480 volt		.70	34.286		133,000	2,025		135,025	149,000

22 33 Electric Domestic Water Heaters

22 33 33 – Light-Commercial Electric Domestic Water Heaters

22 33 33.10 Commercial Electric Water Heaters	Crew	Daily Output	Labor-Hours	Unit	Material	2019 Bare Costs Labor	Equipment	Total	Total Incl O&P
4700	1,200 gal., 60 kW, 245 GPH, 480 volt	Q-2	.60	40	Ea.	84,500	2,350		86,850	96,500
4720	1,200 gal., 480 kW, 1,970 GPH, 480 volt		.60	40		137,000	2,350		139,350	154,500
4740	1,500 gal., 60 kW, 245 GPH, 480 volt		.50	48		111,500	2,825		114,325	127,000
4760	1,500 gal., 480 kW, 1,970 GPH, 480 volt		.50	48		165,500	2,825		168,325	186,500
5400	Modulating step control for under 90 kW, 2-5 steps	1 Elec	5.30	1.509		805	90.50		895.50	1,025
5440	For above 90 kW, 1 through 5 steps beyond standard, add		3.20	2.500		238	150		388	485
5460	For above 90 kW, 6 through 10 steps beyond standard, add		2.70	2.963		465	178		643	775
5480	For above 90 kW, 11 through 18 steps beyond standard, add		1.60	5		690	300		990	1,200

22 34 Fuel-Fired Domestic Water Heaters

22 34 30 – Residential Gas Domestic Water Heaters

22 34 30.13 Residential, Atmos, Gas Domestic Wtr Heaters

		Crew	Daily Output	Labor-Hours	Unit	Material	2019 Bare Costs Labor	Equipment	Total	Total Incl O&P
0010	**RESIDENTIAL, ATMOSPHERIC, GAS DOMESTIC WATER HEATERS**									
2000	Gas fired, foam lined tank, 10 yr., vent not incl.									
2040	30 gallon	1 Plum	2	4	Ea.	1,775	253		2,028	2,325
2060	40 gallon		1.90	4.211		1,775	266		2,041	2,350
2080	50 gallon		1.80	4.444		1,925	281		2,206	2,525
2090	60 gallon		1.70	4.706		1,800	297		2,097	2,425
2100	75 gallon		1.50	5.333		2,775	335		3,110	3,550
2120	100 gallon		1.30	6.154		2,950	390		3,340	3,825
2900	Water heater, safety-drain pan, 26" round		20	.400		21	25.50		46.50	61
3000	Tank leak safety, water & gas shut off see 22 05 23.20 8800									

22 34 46 – Oil-Fired Domestic Water Heaters

22 34 46.10 Residential Oil-Fired Water Heaters

		Crew	Daily Output	Labor-Hours	Unit	Material	2019 Bare Costs Labor	Equipment	Total	Total Incl O&P
0010	**RESIDENTIAL OIL-FIRED WATER HEATERS**									
3000	Oil fired, glass lined tank, 5 yr., vent not included, 30 gallon	1 Plum	2	4	Ea.	1,275	253		1,528	1,775
3040	50 gallon		1.80	4.444		1,500	281		1,781	2,075
3060	70 gallon		1.50	5.333		2,275	335		2,610	3,025

Estimating Tips

The labor adjustment factors listed in Subdivision 22 01 02.20 also apply to Division 23.

23 10 00 Facility Fuel Systems

- The prices in this subdivision for above- and below-ground storage tanks do not include foundations or hold-down slabs, unless noted. The estimator should refer to Divisions 3 and 31 for foundation system pricing. In addition to the foundations, required tank accessories, such as tank gauges, leak detection devices, and additional manholes and piping, must be added to the tank prices.

23 50 00 Central Heating Equipment

- When estimating the cost of an HVAC system, check to see who is responsible for providing and installing the temperature control system. It is possible to overlook controls, assuming that they would be included in the electrical estimate.

- When looking up a boiler, be careful on specified capacity. Some manufacturers rate their products on output while others use input.

- Include HVAC insulation for pipe, boiler, and duct (wrap and liner).

- Be careful when looking up mechanical items to get the correct pressure rating and connection type (thread, weld, flange).

23 70 00 Central HVAC Equipment

- Combination heating and cooling units are sized by the air conditioning requirements. (See Reference No. R236000-20 for the preliminary sizing guide.)

- A ton of air conditioning is nominally 400 CFM.

- Rectangular duct is taken off by the linear foot for each size, but its cost is usually estimated by the pound. Remember that SMACNA standards now base duct on internal pressure.

- Prefabricated duct is estimated and purchased like pipe: straight sections and fittings.

- Note that cranes or other lifting equipment are not included on any lines in Division 23. For example, if a crane is required to lift a heavy piece of pipe into place high above a gym floor, or to put a rooftop unit on the roof of a four-story building, etc., it must be added. Due to the potential for extreme variation—from nothing additional required to a major crane or helicopter—we feel that including a nominal amount for "lifting contingency" would be useless and detract from the accuracy of the estimate. When using equipment rental cost data from RSMeans, do not forget to include the cost of the operator(s).

Reference Numbers

Reference numbers are shown at the beginning of some major classifications. These numbers refer to related items in the Reference Section. The reference information may be an estimating procedure, an alternate pricing method, or technical information.

Note: Not all subdivisions listed here necessarily appear. ■

23 05 Common Work Results for HVAC

23 05 02 – HVAC General

23 05 02.10 Air Conditioning, General	Crew	Daily Output	Labor-Hours	Unit	Material	2019 Bare Costs Labor	Equipment	Total	Total Incl O&P
0010 **AIR CONDITIONING, GENERAL** Prices are for standard efficiencies (SEER 13)									
0020 for upgrade to SEER 14 add					10%				

23 09 Instrumentation and Control for HVAC

23 09 23 – Direct-Digital Control System for HVAC

23 09 23.10 Control Components/DDC Systems

	Crew	Daily Output	Labor-Hours	Unit	Material	2019 Bare Costs Labor	Equipment	Total	Total Incl O&P
0010 **CONTROL COMPONENTS/DDC SYSTEMS** (Sub's quote incl. M & L)									
0100 Analog inputs									
0110 Sensors (avg. 50' run in 1/2" EMT)									
0120 Duct temperature				Ea.				415	415
0130 Space temperature								665	665
0140 Duct humidity, +/- 3%								695	695
0150 Space humidity, +/- 2%								1,075	1,075
0160 Duct static pressure								565	565
0170 CFM/transducer								765	765
0180 KW/transducer								1,350	1,350
0182 KWH totalization (not incl. elec. meter pulse xmtr.)								625	625
0190 Space static pressure								1,075	1,075
1000 Analog outputs (avg. 50' run in 1/2" EMT)									
1010 P/I transducer				Ea.				635	635
1020 Analog output, matl. in MUX								305	305
1030 Pneumatic (not incl. control device)								645	645
1040 Electric (not incl. control device)								380	380
2000 Status (alarms)									
2100 Digital inputs (avg. 50' run in 1/2" EMT)									
2110 Freeze				Ea.				435	435
2120 Fire								395	395
2130 Differential pressure (air)								595	595
2140 Differential pressure (water)								975	975
2150 Current sensor								435	435
2160 Duct high temperature thermostat								570	570
2170 Duct smoke detector								705	705
2200 Digital output (avg. 50' run in 1/2" EMT)									
2210 Start/stop				Ea.				340	340
2220 On/off (maintained contact)				"				585	585
3000 Controller MUX panel, incl. function boards									
3100 48 point				Ea.				5,275	5,275
3110 128 point				"				7,225	7,225
3200 DDC controller (avg. 50' run in conduit)									
3210 Mechanical room									
3214 16 point controller (incl. 120 volt/1 phase power supply)				Ea.				3,275	3,275
3229 32 point controller (incl. 120 volt/1 phase power supply)				"				5,425	5,425
3230 Includes software programming and checkout									
3260 Space									
3266 VAV terminal box (incl. space temp. sensor)				Ea.				840	840
3280 Host computer (avg. 50' run in conduit)									
3281 Package complete with PC, keyboard,									
3282 printer, monitor, basic software				Ea.				3,150	3,150
4000 Front end costs									
4100 Computer (P.C.) with software program				Ea.				6,350	6,350
4200 Color graphics software								3,925	3,925

23 09 23 – Direct-Digital Control System for HVAC

23 09 23.10 Control Components/DDC Systems		Crew	Daily Output	Labor-Hours	Unit	Material	2019 Bare Costs Labor	Equipment	Total	Total Incl O&P
4300	Color graphics slides				Ea.				490	490
4350	Additional printer				↓				980	980
4400	Communications trunk cable				L.F.				3.80	3.80
4500	Engineering labor (not incl. dftg.)				Point				94	94
4600	Calibration labor				↓				120	120
4700	Start-up, checkout labor				↓				120	120
4800	Programming labor, as req'd									
5000	Communications bus (data transmission cable)									
5010	#18 twisted shielded pair in 1/2" EMT conduit				C.L.F.				380	380
8000	Applications software									
8050	Basic maintenance manager software (not incl. data base entry)				Ea.				1,950	1,950
8100	Time program				Point				6.85	6.85
8120	Duty cycle								13.65	13.65
8140	Optimum start/stop								41.50	41.50
8160	Demand limiting								20.50	20.50
8180	Enthalpy program				↓				41.50	41.50
8200	Boiler optimization				Ea.				1,225	1,225
8220	Chiller optimization				"				1,625	1,625
8240	Custom applications									
8260	Cost varies with complexity									

23 09 43 – Pneumatic Control System for HVAC

23 09 43.10 Pneumatic Control Systems

		Crew	Daily Output	Labor-Hours	Unit	Material	2019 Bare Costs Labor	Equipment	Total	Total Incl O&P
0010	**PNEUMATIC CONTROL SYSTEMS**									
0011	Including a nominal 50' of tubing. Add control panelboard if req'd.									
0100	Heating and ventilating, split system									
0200	Mixed air control, economizer cycle, panel readout, tubing									
0220	Up to 10 tons	G	Q-19	.68	35.294	Ea.	4,500	2,050		6,550	8,050
0240	For 10 to 20 tons	G		.63	37.915		4,775	2,225		7,000	8,575
0260	For over 20 tons	G		.58	41.096		5,175	2,400		7,575	9,275
0270	Enthalpy cycle, up to 10 tons			.50	48.387		4,925	2,825		7,750	9,650
0280	For 10 to 20 tons			.46	52.174		5,325	3,050		8,375	10,400
0290	For over 20 tons		↓	.42	56.604	↓	5,825	3,300		9,125	11,400
0300	Heating coil, hot water, 3 way valve,									
0320	Freezestat, limit control on discharge, readout		Q-5	.69	23.088	Ea.	3,350	1,325		4,675	5,675
0500	Cooling coil, chilled water, room									
0520	Thermostat, 3 way valve		Q-5	2	8	Ea.	1,500	460		1,960	2,350
0600	Cooling tower, fan cycle, damper control,									
0620	Control system including water readout in/out at panel		Q-19	.67	35.821	Ea.	5,925	2,100		8,025	9,650
1000	Unit ventilator, day/night operation,									
1100	freezestat, ASHRAE, cycle 2		Q-19	.91	26.374	Ea.	3,275	1,550		4,825	5,925
2000	Compensated hot water from boiler, valve control,									
2100	readout and reset at panel, up to 60 GPM		Q-19	.55	43.956	Ea.	6,075	2,575		8,650	10,500
2120	For 120 GPM			.51	47.059		6,575	2,750		9,325	11,400
2140	For 240 GPM			.49	49.180		6,800	2,875		9,675	11,800
3000	Boiler room combustion air, damper to 5 S.F., controls			1.37	17.582		2,975	1,025		4,000	4,775
3500	Fan coil, heating and cooling valves, 4 pipe control system			3	8		1,325	465		1,790	2,150
3600	Heat exchanger system controls		↓	.86	27.907	↓	2,875	1,625		4,500	5,625
3900	Multizone control (one per zone), includes thermostat, damper									
3910	motor and reset of discharge temperature		Q-5	.51	31.373	Ea.	2,975	1,800		4,775	5,950
4000	Pneumatic thermostat, including controlling room radiator valve		"	2.43	6.593		890	380		1,270	1,550
4040	Program energy saving optimizer	G	Q-19	1.21	19.786		7,400	1,150		8,550	9,875
4060	Pump control system		"	3	8	↓	1,375	465		1,840	2,200

For customer support on your Electrical Costs with RSMeans data, call 800.448.8182.

83

23 09 43 – Pneumatic Control System for HVAC

23 09 43.10 Pneumatic Control Systems		Crew	Daily Output	Labor-Hours	Unit	Material	2019 Bare Costs Labor	Equipment	Total	Total Incl O&P
4080	Reheat coil control system, not incl. coil	Q-5	2.43	6.593	Ea.	1,150	380		1,530	1,850
4500	Air supply for pneumatic control system									
4600	Tank mounted duplex compressor, starter, alternator,									
4620	piping, dryer, PRV station and filter									
4630	1/2 HP	Q-19	.68	35.139	Ea.	10,900	2,050		12,950	15,100
4640	3/4 HP		.64	37.383		11,500	2,175		13,675	16,000
4650	1 HP		.61	39.539		12,600	2,300		14,900	17,400
4660	1-1/2 HP		.58	41.739		13,500	2,425		15,925	18,500
4680	3 HP		.55	43.956		18,300	2,575		20,875	24,100
4690	5 HP		.42	57.143		31,600	3,325		34,925	39,800
4800	Main air supply, includes 3/8" copper main and labor	Q-5	1.82	8.791	C.L.F.	365	505		870	1,175
4810	If poly tubing used, deduct					30%				30%
7000	Static pressure control for air handling unit, includes pressure									
7010	sensor, receiver controller, readout and damper motors	Q-19	.64	37.383	Ea.	8,675	2,175		10,850	12,800
7020	If return air fan requires control, add					70%				70%
8600	VAV boxes, incl. thermostat, damper motor, reheat coil & tubing	Q-5	1.46	10.989		1,375	630		2,005	2,475
8610	If no reheat coil, deduct					204				204

23 09 53 – Pneumatic and Electric Control System for HVAC

23 09 53.10 Control Components

		Crew	Daily Output	Labor-Hours	Unit	Material	2019 Bare Costs Labor	Equipment	Total	Total Incl O&P
0010	**CONTROL COMPONENTS**									
0700	Controller, receiver									
0850	Electric, single snap switch	1 Elec	4	2	Ea.	490	120		610	715
0860	Dual snap switches	"	3	2.667	"	655	160		815	960
3590	Sensor, electric operated									
3620	Humidity	1 Elec	8	1	Ea.	269	60		329	385
3650	Pressure		8	1		305	60		365	425
3680	Temperature		10	.800		118	48		166	202
5000	Thermostats									
5200	24 hour, automatic, clock [G]	1 Shee	8	1	Ea.	236	61		297	350
5220	Electric, low voltage, 2 wire	1 Elec	13	.615		62	37		99	123
5230	3 wire	"	10	.800		53	48		101	130
7090	Valves, motor controlled, including actuator									
7100	Electric motor actuated									
7200	Brass, two way, screwed									
7210	1/2" pipe size	L-6	36	.333	Ea.	365	20.50		385.50	430
7220	3/4" pipe size		30	.400		615	25		640	710
7230	1" pipe size		28	.429		730	26.50		756.50	845
7240	1-1/2" pipe size		19	.632		825	39		864	965
7250	2" pipe size		16	.750		1,175	46.50		1,221.50	1,375
7350	Brass, three way, screwed									
7360	1/2" pipe size	L-6	33	.364	Ea.	320	22.50		342.50	385
7370	3/4" pipe size		27	.444		350	27.50		377.50	425
7380	1" pipe size		25.50	.471		405	29		434	490
7384	1-1/4" pipe size		21	.571		475	35.50		510.50	575
7390	1-1/2" pipe size		17	.706		525	44		569	640
7400	2" pipe size		14	.857		765	53.50		818.50	925
7550	Iron body, two way, flanged									
7560	2-1/2" pipe size	L-6	4	3	Ea.	1,150	186		1,336	1,550
7570	3" pipe size		3	4		1,175	248		1,423	1,675
7580	4" pipe size		2	6		2,200	375		2,575	2,975
7850	Iron body, three way, flanged									
7860	2-1/2" pipe size	L-6	3	4	Ea.	1,150	248		1,398	1,650

23 09 Instrumentation and Control for HVAC

23 09 53 – Pneumatic and Electric Control System for HVAC

23 09 53.10 Control Components	Crew	Daily Output	Labor-Hours	Unit	Material	2019 Bare Costs Labor	2019 Bare Costs Equipment	Total	Total Incl O&P
7870	3" pipe size	L-6	2.50	4.800	Ea.	1,325	298		1,623	1,900
7880	4" pipe size	↓	2	6	↓	1,675	375		2,050	2,400

23 34 HVAC Fans

23 34 13 – Axial HVAC Fans

23 34 13.10 Axial Flow HVAC Fans

		Crew	Daily Output	Labor-Hours	Unit	Material	Labor	Equipment	Total	Total Incl O&P
0010	**AXIAL FLOW HVAC FANS**									
0020	Air conditioning and process air handling									
1500	Vaneaxial, low pressure, 2,000 CFM, 1/2 HP	Q-20	3.60	5.556	Ea.	2,450	310		2,760	3,175
1520	4,000 CFM, 1 HP		3.20	6.250		2,575	350		2,925	3,350
1540	8,000 CFM, 2 HP		2.80	7.143		3,650	400		4,050	4,625
1560	16,000 CFM, 5 HP	↓	2.40	8.333	↓	5,100	465		5,565	6,300

23 34 14 – Blower HVAC Fans

23 34 14.10 Blower Type HVAC Fans

		Crew	Daily Output	Labor-Hours	Unit	Material	Labor	Equipment	Total	Total Incl O&P
0010	**BLOWER TYPE HVAC FANS**									
2000	Blowers, direct drive with motor, complete									
2020	1,045 CFM @ 0.5" S.P., 1/5 HP	Q-20	18	1.111	Ea.	345	62		407	475
2040	1,385 CFM @ 0.5" S.P., 1/4 HP		18	1.111		360	62		422	490
2060	1,640 CFM @ 0.5" S.P., 1/3 HP		18	1.111		330	62		392	455
2080	1,760 CFM @ 0.5" S.P., 1/2 HP	↓	18	1.111	↓	340	62		402	470
2090	4 speed									
2100	1,164 to 1,739 CFM @ 0.5" S.P., 1/3 HP	Q-20	16	1.250	Ea.	345	70		415	485
2120	1,467 to 2,218 CFM @ 1.0" S.P., 3/4 HP	"	14	1.429	"	395	80		475	555
2500	Ceiling fan, right angle, extra quiet, 0.10" S.P.									
2520	95 CFM	Q-20	20	1	Ea.	297	56		353	410
2540	210 CFM		19	1.053		350	59		409	475
2560	385 CFM		18	1.111		445	62		507	585
2580	885 CFM		16	1.250		880	70		950	1,075
2600	1,650 CFM		13	1.538		1,225	86		1,311	1,475
2620	2,960 CFM	↓	11	1.818		1,625	102		1,727	1,925
2680	For speed control switch, add	1 Elec	16	.500	↓	162	30		192	223
7500	Utility set, steel construction, pedestal, 1/4" S.P.									
7520	Direct drive, 150 CFM, 1/8 HP	Q-20	6.40	3.125	Ea.	940	175		1,115	1,300
7540	485 CFM, 1/6 HP		5.80	3.448		1,075	193		1,268	1,475
7560	1,950 CFM, 1/2 HP		4.80	4.167		1,400	233		1,633	1,875
7580	2,410 CFM, 3/4 HP		4.40	4.545		2,575	254		2,829	3,200
7600	3,328 CFM, 1-1/2 HP	↓	3	6.667	↓	2,575	375		2,950	3,400
7680	V-belt drive, drive cover, 3 phase									
7700	800 CFM, 1/4 HP	Q-20	6	3.333	Ea.	1,025	186		1,211	1,400
7720	1,300 CFM, 1/3 HP		5	4		1,075	224		1,299	1,525
7740	2,000 CFM, 1 HP		4.60	4.348		1,275	243		1,518	1,775
7760	2,900 CFM, 3/4 HP		4.20	4.762		1,700	266		1,966	2,275
7780	3,600 CFM, 3/4 HP		4	5		2,100	279		2,379	2,725
7800	4,800 CFM, 1 HP		3.50	5.714		2,475	320		2,795	3,200
7820	6,700 CFM, 1-1/2 HP		3	6.667		3,050	375		3,425	3,950
7830	7,500 CFM, 2 HP		2.50	8		4,150	445		4,595	5,250
7840	11,000 CFM, 3 HP		2	10		5,550	560		6,110	6,950
7860	13,000 CFM, 3 HP		1.60	12.500		5,625	700		6,325	7,250
7880	15,000 CFM, 5 HP		1	20		5,825	1,125		6,950	8,125
7900	17,000 CFM, 7-1/2 HP		.80	25		6,250	1,400		7,650	9,000

For customer support on your Electrical Costs with RSMeans data, call 800.448.8182.

85

23 34 HVAC Fans

23 34 14 – Blower HVAC Fans

23 34 14.10 Blower Type HVAC Fans	Crew	Daily Output	Labor-Hours	Unit	Material	2019 Bare Costs Labor	Equipment	Total	Total Incl O&P
7920 20,000 CFM, 7-1/2 HP	Q-20	.80	25	Ea.	7,450	1,400		8,850	10,300

23 34 16 – Centrifugal HVAC Fans

23 34 16.10 Centrifugal Type HVAC Fans

	Crew	Daily Output	Labor-Hours	Unit	Material	2019 Bare Costs Labor	Equipment	Total	Total Incl O&P
0010 **CENTRIFUGAL TYPE HVAC FANS**									
0200 In-line centrifugal, supply/exhaust booster									
0220 aluminum wheel/hub, disconnect switch, 1/4" S.P.									
0240 500 CFM, 10" diameter connection	Q-20	3	6.667	Ea.	1,625	375		2,000	2,350
0260 1,380 CFM, 12" diameter connection		2	10		1,675	560		2,235	2,700
0280 1,520 CFM, 16" diameter connection		2	10		1,700	560		2,260	2,700
0300 2,560 CFM, 18" diameter connection		1	20		1,850	1,125		2,975	3,725
0320 3,480 CFM, 20" diameter connection		.80	25		2,150	1,400		3,550	4,500
0326 5,080 CFM, 20" diameter connection		.75	26.667		2,200	1,500		3,700	4,675
3500 Centrifugal, airfoil, motor and drive, complete									
3520 1,000 CFM, 1/2 HP	Q-20	2.50	8	Ea.	2,150	445		2,595	3,025
3540 2,000 CFM, 1 HP		2	10		2,325	560		2,885	3,400
3560 4,000 CFM, 3 HP		1.80	11.111		2,975	620		3,595	4,225
3580 8,000 CFM, 7-1/2 HP		1.40	14.286		4,300	800		5,100	5,925
3600 12,000 CFM, 10 HP		1	20		5,925	1,125		7,050	8,225
5000 Utility set, centrifugal, V belt drive, motor									
5020 1/4" S.P., 1,200 CFM, 1/4 HP	Q-20	6	3.333	Ea.	1,900	186		2,086	2,375
5040 1,520 CFM, 1/3 HP		5	4		2,475	224		2,699	3,050
5060 1,850 CFM, 1/2 HP		4	5		2,450	279		2,729	3,125
5080 2,180 CFM, 3/4 HP		3	6.667		2,850	375		3,225	3,700
5100 1/2" S.P., 3,600 CFM, 1 HP		2	10		3,050	560		3,610	4,225
5120 4,250 CFM, 1-1/2 HP		1.60	12.500		3,650	700		4,350	5,050
5140 4,800 CFM, 2 HP		1.40	14.286		4,475	800		5,275	6,125
5160 6,920 CFM, 5 HP		1.30	15.385		5,400	860		6,260	7,225
5180 7,700 CFM, 7-1/2 HP		1.20	16.667		6,550	930		7,480	8,600
5200 For explosion proof motor, add					15%				
5500 Fans, industrial exhauster, for air which may contain granular matl.									
5520 1,000 CFM, 1-1/2 HP	Q-20	2.50	8	Ea.	3,150	445		3,595	4,150
5540 2,000 CFM, 3 HP		2	10		3,850	560		4,410	5,075
5560 4,000 CFM, 7-1/2 HP		1.80	11.111		5,400	620		6,020	6,875
5580 8,000 CFM, 15 HP		1.40	14.286		6,950	800		7,750	8,850
5600 12,000 CFM, 30 HP		1	20		10,700	1,125		11,825	13,500
7000 Roof exhauster, centrifugal, aluminum housing, 12" galvanized									
7020 curb, bird screen, back draft damper, 1/4" S.P.									
7100 Direct drive, 320 CFM, 11" sq. damper	Q-20	7	2.857	Ea.	805	160		965	1,125
7120 600 CFM, 11" sq. damper		6	3.333		950	186		1,136	1,325
7140 815 CFM, 13" sq. damper		5	4		1,025	224		1,249	1,475
7160 1,450 CFM, 13" sq. damper		4.20	4.762		1,600	266		1,866	2,150
7180 2,050 CFM, 16" sq. damper		4	5		1,825	279		2,104	2,425
7200 V-belt drive, 1,650 CFM, 12" sq. damper		6	3.333		1,475	186		1,661	1,900
7220 2,750 CFM, 21" sq. damper		5	4		1,750	224		1,974	2,275
7230 3,500 CFM, 21" sq. damper		4.50	4.444		1,975	248		2,223	2,550
7240 4,910 CFM, 23" sq. damper		4	5		2,400	279		2,679	3,075
7260 8,525 CFM, 28" sq. damper		3	6.667		3,200	375		3,575	4,100
7280 13,760 CFM, 35" sq. damper		2	10		4,450	560		5,010	5,750
7300 20,558 CFM, 43" sq. damper		1	20		8,825	1,125		9,950	11,400
7320 For 2 speed winding, add					15%				
7340 For explosion proof motor, add					765			765	840
7360 For belt driven, top discharge, add					15%				

23 34 HVAC Fans

23 34 16 – Centrifugal HVAC Fans

23 34 16.10 Centrifugal Type HVAC Fans		Crew	Daily Output	Labor-Hours	Unit	Material	2019 Bare Costs Labor	Equipment	Total	Total Incl O&P
8500	Wall exhausters, centrifugal, auto damper, 1/8" S.P.									
8520	Direct drive, 610 CFM, 1/20 HP	Q-20	14	1.429	Ea.	420	80		500	580
8540	796 CFM, 1/12 HP		13	1.538		1,025	86		1,111	1,250
8560	822 CFM, 1/6 HP		12	1.667		1,050	93		1,143	1,325
8580	1,320 CFM, 1/4 HP		12	1.667		1,400	93		1,493	1,675
8600	1,756 CFM, 1/4 HP		11	1.818		1,250	102		1,352	1,525
8620	1,983 CFM, 1/4 HP		10	2		1,275	112		1,387	1,575
8640	2,900 CFM, 1/2 HP		9	2.222		1,350	124		1,474	1,700
8660	3,307 CFM, 3/4 HP	↓	8	2.500	↓	1,475	140		1,615	1,825
9500	V-belt drive, 3 phase									
9520	2,800 CFM, 1/4 HP	Q-20	9	2.222	Ea.	1,900	124		2,024	2,275
9540	3,740 CFM, 1/2 HP		8	2.500		1,975	140		2,115	2,375
9560	4,400 CFM, 3/4 HP		7	2.857		2,000	160		2,160	2,450
9580	5,700 CFM, 1-1/2 HP	↓	6	3.333	↓	2,075	186		2,261	2,550

23 34 23 – HVAC Power Ventilators

23 34 23.10 HVAC Power Circulators and Ventilators

			Crew	Daily Output	Labor-Hours	Unit	Material	2019 Bare Costs Labor	Equipment	Total	Total Incl O&P
0010	**HVAC POWER CIRCULATORS AND VENTILATORS**										
3000	Paddle blade air circulator, 3 speed switch										
3020	42", 5,000 CFM high, 3,000 CFM low	G	1 Elec	2.40	3.333	Ea.	116	200		316	425
3040	52", 6,500 CFM high, 4,000 CFM low	G	"	2.20	3.636	"	151	218		369	490
3100	For antique white motor, same cost										
3200	For brass plated motor, same cost										
3300	For light adaptor kit, add	G				Ea.	40			40	44
6000	Propeller exhaust, wall shutter										
6020	Direct drive, one speed, 0.075" S.P.										
6100	653 CFM, 1/30 HP		Q-20	10	2	Ea.	209	112		321	400
6120	1,033 CFM, 1/20 HP			9	2.222		310	124		434	530
6140	1,323 CFM, 1/15 HP			8	2.500		350	140		490	590
6160	2,444 CFM, 1/4 HP		↓	7	2.857	↓	435	160		595	715
6300	V-belt drive, 3 phase										
6320	6,175 CFM, 3/4 HP		Q-20	5	4	Ea.	1,075	224		1,299	1,550
6340	7,500 CFM, 3/4 HP			5	4		1,150	224		1,374	1,625
6360	10,100 CFM, 1 HP			4.50	4.444		1,375	248		1,623	1,875
6380	14,300 CFM, 1-1/2 HP			4	5		1,625	279		1,904	2,225
6400	19,800 CFM, 2 HP			3	6.667		2,175	375		2,550	2,950
6420	26,250 CFM, 3 HP			2.60	7.692		2,300	430		2,730	3,175
6440	38,500 CFM, 5 HP			2.20	9.091		2,825	510		3,335	3,900
6460	46,000 CFM, 7-1/2 HP			2	10		2,900	560		3,460	4,050
6480	51,500 CFM, 10 HP		↓	1.80	11.111	↓	3,000	620		3,620	4,250
6650	Residential, bath exhaust, grille, back draft damper										
6660	50 CFM		Q-20	24	.833	Ea.	53	46.50		99.50	129
6670	110 CFM			22	.909		106	51		157	193
6680	Light combination, squirrel cage, 100 watt, 70 CFM		↓	24	.833	↓	117	46.50		163.50	199
6700	Light/heater combination, ceiling mounted										
6710	70 CFM, 1,450 watt		Q-20	24	.833	Ea.	144	46.50		190.50	229
6800	Heater combination, recessed, 70 CFM			24	.833		75.50	46.50		122	154
6820	With 2 infrared bulbs			23	.870		96	48.50		144.50	180
6900	Kitchen exhaust, grille, complete, 160 CFM			22	.909		107	51		158	195
6910	180 CFM			20	1		97	56		153	192
6920	270 CFM			18	1.111		235	62		297	350
6930	350 CFM		↓	16	1.250		150	70		220	271
6940	Residential roof jacks and wall caps										

23 34 HVAC Fans

23 34 23 – HVAC Power Ventilators

	23 34 23.10 HVAC Power Circulators and Ventilators	Crew	Daily Output	Labor-Hours	Unit	Material	2019 Bare Costs Labor	Equipment	Total	Total Incl O&P
6944	Wall cap with back draft damper									
6946	3" & 4" diam. round duct	1 Shee	11	.727	Ea.	25	44.50		69.50	95
6948	6" diam. round duct	"	11	.727	"	77	44.50		121.50	153
6958	Roof jack with bird screen and back draft damper									
6960	3" & 4" diam. round duct	1 Shee	11	.727	Ea.	20.50	44.50		65	90
6962	3-1/4" x 10" rectangular duct	"	10	.800	"	37	49		86	115
6980	Transition									
6982	3-1/4" x 10" to 6" diam. round	1 Shee	20	.400	Ea.	35	24.50		59.50	75.50
8020	Attic, roof type									
8030	Aluminum dome, damper & curb									
8080	12" diameter, 1,000 CFM (gravity)	1 Elec	10	.800	Ea.	610	48		658	740
8090	16" diameter, 1,500 CFM (gravity)		9	.889		735	53.50		788.50	890
8100	20" diameter, 2,500 CFM (gravity)		8	1		900	60		960	1,075
8110	26" diameter, 4,000 CFM (gravity)		7	1.143		1,100	68.50		1,168.50	1,300
8120	32" diameter, 6,500 CFM (gravity)		6	1.333		1,500	80		1,580	1,775
8130	38" diameter, 8,000 CFM (gravity)		5	1.600		2,225	96		2,321	2,600
8140	50" diameter, 13,000 CFM (gravity)	↓	4	2	↓	3,225	120		3,345	3,725
8160	Plastic, ABS dome									
8180	1,050 CFM	1 Elec	14	.571	Ea.	179	34.50		213.50	248
8200	1,600 CFM	"	12	.667	"	268	40		308	355
8240	Attic, wall type, with shutter, one speed									
8250	12" diameter, 1,000 CFM	1 Elec	14	.571	Ea.	415	34.50		449.50	505
8260	14" diameter, 1,500 CFM		12	.667		450	40		490	550
8270	16" diameter, 2,000 CFM	↓	9	.889	↓	505	53.50		558.50	635
8290	Whole house, wall type, with shutter, one speed									
8300	30" diameter, 4,800 CFM	1 Elec	7	1.143	Ea.	1,075	68.50		1,143.50	1,300
8310	36" diameter, 7,000 CFM		6	1.333		1,175	80		1,255	1,425
8320	42" diameter, 10,000 CFM		5	1.600		1,325	96		1,421	1,600
8330	48" diameter, 16,000 CFM	↓	4	2		1,650	120		1,770	1,975
8340	For two speed, add				↓	99			99	109
8350	Whole house, lay-down type, with shutter, one speed									
8360	30" diameter, 4,500 CFM	1 Elec	8	1	Ea.	1,150	60		1,210	1,375
8370	36" diameter, 6,500 CFM		7	1.143		1,250	68.50		1,318.50	1,475
8380	42" diameter, 9,000 CFM		6	1.333		1,375	80		1,455	1,625
8390	48" diameter, 12,000 CFM	↓	5	1.600		1,550	96		1,646	1,850
8440	For two speed, add					74.50			74.50	82
8450	For 12 hour timer switch, add	1 Elec	32	.250	↓	74.50	15		89.50	105

23 52 Heating Boilers

23 52 13 – Electric Boilers

23 52 13.10 Electric Boilers, ASME

		Crew	Daily Output	Labor-Hours	Unit	Material	2019 Bare Costs Labor	Equipment	Total	Total Incl O&P
0010	**ELECTRIC BOILERS, ASME**, Standard controls and trim									
1000	Steam, 6 KW, 20.5 MBH	Q-19	1.20	20	Ea.	4,300	1,175		5,475	6,475
1040	9 KW, 30.7 MBH		1.20	20		4,275	1,175		5,450	6,450
1060	18 KW, 61.4 MBH		1.20	20		4,400	1,175		5,575	6,600
1080	24 KW, 81.8 MBH		1.10	21.818		5,100	1,275		6,375	7,525
1120	36 KW, 123 MBH		1.10	21.818		5,650	1,275		6,925	8,125
1160	60 KW, 205 MBH		1	24		7,075	1,400		8,475	9,875
1220	112 KW, 382 MBH		.75	32		9,600	1,875		11,475	13,300
1240	148 KW, 505 MBH		.65	36.923		10,200	2,150		12,350	14,400
1260	168 KW, 573 MBH		.60	40	↓	20,500	2,325		22,825	26,000

23 52 13.10 Electric Boilers, ASME		Crew	Daily Output	Labor-Hours	Unit	Material	2019 Bare Costs Labor	Equipment	Total	Total Incl O&P
1280	222 KW, 758 MBH	Q-19	.55	43.636	Ea.	23,900	2,550		26,450	30,100
1300	296 KW, 1,010 MBH		.45	53.333		25,100	3,125		28,225	32,300
1320	300 KW, 1,023 MBH		.40	60		26,400	3,500		29,900	34,400
1340	370 KW, 1,263 MBH		.35	68.571		29,300	4,000		33,300	38,200
1360	444 KW, 1,515 MBH		.30	80		31,400	4,675		36,075	41,600
1380	518 KW, 1,768 MBH	Q-21	.36	88.889		32,700	5,325		38,025	44,000
1400	592 KW, 2,020 MBH		.34	94.118		35,500	5,625		41,125	47,400
1420	666 KW, 2,273 MBH		.32	100		36,900	5,975		42,875	49,600
1460	740 KW, 2,526 MBH		.28	114		38,400	6,825		45,225	52,500
1480	814 KW, 2,778 MBH		.25	128		40,800	7,650		48,450	56,500
1500	962 KW, 3,283 MBH		.22	145		45,000	8,700		53,700	62,500
1520	1,036 KW, 3,536 MBH		.20	160		47,800	9,575		57,375	67,000
1540	1,110 KW, 3,788 MBH		.19	168		49,700	10,100		59,800	69,500
1560	2,070 KW, 7,063 MBH		.18	178		68,500	10,600		79,100	91,000
1580	2,250 KW, 7,677 MBH		.17	188		80,000	11,300		91,300	105,000
1600	2,340 KW, 7,984 MBH		.16	200		86,500	12,000		98,500	113,500
2000	Hot water, 7.5 KW, 25.6 MBH	Q-19	1.30	18.462		5,200	1,075		6,275	7,350
2020	15 KW, 51.2 MBH		1.30	18.462		5,250	1,075		6,325	7,400
2040	30 KW, 102 MBH		1.20	20		5,600	1,175		6,775	7,900
2060	45 KW, 164 MBH		1.20	20		5,700	1,175		6,875	8,025
2070	60 KW, 205 MBH		1.20	20		5,800	1,175		6,975	8,150
2080	75 KW, 256 MBH		1.10	21.818		6,150	1,275		7,425	8,675
2100	90 KW, 307 MBH		1.10	21.818		6,150	1,275		7,425	8,675
2120	105 KW, 358 MBH		1	24		6,650	1,400		8,050	9,400
2140	120 KW, 410 MBH		.90	26.667		6,675	1,550		8,225	9,650
2160	135 KW, 461 MBH		.75	32		7,550	1,875		9,425	11,100
2180	150 KW, 512 MBH		.65	36.923		7,950	2,150		10,100	12,000
2200	165 KW, 563 MBH		.60	40		8,025	2,325		10,350	12,300
2220	296 KW, 1,010 MBH		.55	43.636		15,600	2,550		18,150	21,000
2280	370 KW, 1,263 MBH		.40	60		17,300	3,500		20,800	24,300
2300	444 KW, 1,515 MBH		.35	68.571		20,100	4,000		24,100	28,100
2340	518 KW, 1,768 MBH	Q-21	.44	72.727		22,300	4,350		26,650	31,000
2360	592 KW, 2,020 MBH		.43	74.419		25,300	4,450		29,750	34,500
2400	666 KW, 2,273 MBH		.40	80		26,400	4,775		31,175	36,200
2420	740 KW, 2,526 MBH		.39	82.051		27,500	4,900		32,400	37,700
2440	814 KW, 2,778 MBH		.38	84.211		29,500	5,025		34,525	40,100
2460	888 KW, 3,031 MBH		.37	86.486		32,000	5,175		37,175	43,000
2480	962 KW, 3,283 MBH		.36	88.889		32,800	5,325		38,125	44,000
2500	1,036 KW, 3,536 MBH		.34	94.118		35,200	5,625		40,825	47,200
2520	1,110 KW, 3,788 MBH		.33	96.970		38,500	5,800		44,300	51,000
2540	1,440 KW, 4,915 MBH		.32	100		42,900	5,975		48,875	56,000
2560	1,560 KW, 5,323 MBH		.31	103		46,900	6,175		53,075	61,000
2580	1,680 KW, 5,733 MBH		.30	107		50,000	6,375		56,375	64,500
2600	1,800 KW, 6,143 MBH		.29	110		53,500	6,600		60,100	68,500
2620	1,980 KW, 6,757 MBH		.28	114		58,500	6,825		65,325	74,000
2640	2,100 KW, 7,167 MBH		.27	119		64,500	7,075		71,575	81,500
2660	2,220 KW, 7,576 MBH		.26	123		65,000	7,350		72,350	82,500
2680	2,400 KW, 8,191 MBH		.25	128		68,000	7,650		75,650	86,500
2700	2,610 KW, 8,905 MBH		.24	133		72,500	7,975		80,475	92,000
2720	2,790 KW, 9,519 MBH		.23	139		76,000	8,325		84,325	96,000
2740	2,970 KW, 10,133 MBH		.21	152		77,500	9,100		86,600	99,000
2760	3,150 KW, 10,748 MBH		.19	168		83,000	10,100		93,100	106,000
2780	3,240 KW, 11,055 MBH		.18	178		84,500	10,600		95,100	109,000

23 52 Heating Boilers

23 52 13 – Electric Boilers

23 52 13.10 Electric Boilers, ASME	Crew	Daily Output	Labor-Hours	Unit	Material	2019 Bare Costs Labor	Equipment	Total	Total Incl O&P
2800 3,420 KW, 11,669 MBH	Q-21	.17	188	Ea.	90,500	11,300		101,800	116,500
2820 3,600 KW, 12,283 MBH	↓	.16	200	↓	94,500	12,000		106,500	122,000

23 52 28 – Swimming Pool Boilers

23 52 28.10 Swimming Pool Heaters

	Crew	Daily Output	Labor-Hours	Unit	Material	2019 Bare Costs Labor	Equipment	Total	Total Incl O&P
0010 **SWIMMING POOL HEATERS**, Not including wiring, external									
0020 piping, base or pad									
2000 Electric, 12 KW, 4,800 gallon pool	Q-19	3	8	Ea.	2,125	465		2,590	3,050
2020 15 KW, 7,200 gallon pool		2.80	8.571		2,150	500		2,650	3,100
2040 24 KW, 9,600 gallon pool		2.40	10		2,500	585		3,085	3,625
2060 30 KW, 12,000 gallon pool		2	12		2,550	700		3,250	3,875
2080 36 KW, 14,400 gallon pool		1.60	15		2,950	875		3,825	4,550
2100 57 KW, 24,000 gallon pool	↓	1.20	20	↓	3,700	1,175		4,875	5,800
9000 To select pool heater: 12 BTUH x S.F. pool area									
9010 X temperature differential = required output									
9050 For electric, KW = gallons x 2.5 divided by 1,000									
9100 For family home type pool, double the									
9110 Rated gallon capacity = 1/2°F rise per hour									

23 54 Furnaces

23 54 13 – Electric-Resistance Furnaces

23 54 13.10 Electric Furnaces

	Crew	Daily Output	Labor-Hours	Unit	Material	2019 Bare Costs Labor	Equipment	Total	Total Incl O&P
0010 **ELECTRIC FURNACES**, Hot air, blowers, std. controls									
1000 Electric, UL listed									
1070 10.2 MBH	Q-20	4.40	4.545	Ea.	350	254		604	770
1080 17.1 MBH		4.60	4.348		430	243		673	845
1100 34.1 MBH		4.40	4.545		540	254		794	975
1120 51.6 MBH		4.20	4.762		590	266		856	1,050
1140 68.3 MBH		4	5		610	279		889	1,100
1160 85.3 MBH	↓	3.80	5.263	↓	635	294		929	1,150

23 82 Convection Heating and Cooling Units

23 82 16 – Air Coils

23 82 16.20 Duct Heaters

	Crew	Daily Output	Labor-Hours	Unit	Material	2019 Bare Costs Labor	Equipment	Total	Total Incl O&P
0010 **DUCT HEATERS**, Electric, 480 V, 3 Ph.									
0020 Finned tubular insert, 500°F									
0100 8" wide x 6" high, 4.0 kW	Q-20	16	1.250	Ea.	825	70		895	1,000
0120 12" high, 8.0 kW		15	1.333		1,375	74.50		1,449.50	1,625
0140 18" high, 12.0 kW		14	1.429		1,900	80		1,980	2,225
0160 24" high, 16.0 kW		13	1.538		2,450	86		2,536	2,825
0180 30" high, 20.0 kW		12	1.667		3,000	93		3,093	3,450
0300 12" wide x 6" high, 6.7 kW		15	1.333		875	74.50		949.50	1,075
0320 12" high, 13.3 kW		14	1.429		1,400	80		1,480	1,675
0340 18" high, 20.0 kW		13	1.538		1,975	86		2,061	2,300
0360 24" high, 26.7 kW		12	1.667		2,550	93		2,643	2,950
0380 30" high, 33.3 kW		11	1.818		3,100	102		3,202	3,550
0500 18" wide x 6" high, 13.3 kW		14	1.429		940	80		1,020	1,150
0520 12" high, 26.7 kW		13	1.538		1,625	86		1,711	1,900
0540 18" high, 40.0 kW		12	1.667		2,150	93		2,243	2,500
0560 24" high, 53.3 kW	↓	11	1.818		2,850	102		2,952	3,300

23 82 16 – Air Coils

23 82 16.20 Duct Heaters		Crew	Daily Output	Labor-Hours	Unit	Material	2019 Bare Costs Labor	Equipment	Total	Total Incl O&P
0580	30" high, 66.7 kW	Q-20	10	2	Ea.	3,575	112		3,687	4,100
0700	24" wide x 6" high, 17.8 kW		13	1.538		1,025	86		1,111	1,275
0720	12" high, 35.6 kW		12	1.667		1,750	93		1,843	2,075
0740	18" high, 53.3 kW		11	1.818		2,450	102		2,552	2,850
0760	24" high, 71.1 kW		10	2		3,175	112		3,287	3,675
0780	30" high, 88.9 kW		9	2.222		3,900	124		4,024	4,475
0900	30" wide x 6" high, 22.2 kW		12	1.667		1,100	93		1,193	1,350
0920	12" high, 44.4 kW		11	1.818		1,850	102		1,952	2,175
0940	18" high, 66.7 kW		10	2		2,600	112		2,712	3,050
0960	24" high, 88.9 kW		9	2.222		3,375	124		3,499	3,900
0980	30" high, 111.0 kW		8	2.500		4,125	140		4,265	4,725
1400	Note decreased kW available for									
1410	each duct size at same cost									
1420	See line 5000 for modifications and accessories									
2000	Finned tubular flange with insulated									
2020	terminal box, 500°F									
2100	12" wide x 36" high, 54 kW	Q-20	10	2	Ea.	4,050	112		4,162	4,625
2120	40" high, 60 kW		9	2.222		4,650	124		4,774	5,325
2200	24" wide x 36" high, 118.8 kW		9	2.222		5,125	124		5,249	5,825
2220	40" high, 132 kW		8	2.500		5,125	140		5,265	5,825
2400	36" wide x 8" high, 40 kW		11	1.818		2,050	102		2,152	2,400
2420	16" high, 80 kW		10	2		2,750	112		2,862	3,200
2440	24" high, 120 kW		9	2.222		3,600	124		3,724	4,150
2460	32" high, 160 kW		8	2.500		4,600	140		4,740	5,275
2480	36" high, 180 kW		7	2.857		5,675	160		5,835	6,500
2500	40" high, 200 kW		6	3.333		6,300	186		6,486	7,200
2600	40" wide x 8" high, 45 kW		11	1.818		2,150	102		2,252	2,525
2620	16" high, 90 kW		10	2		2,975	112		3,087	3,425
2640	24" high, 135 kW		9	2.222		3,750	124		3,874	4,325
2660	32" high, 180 kW		8	2.500		5,000	140		5,140	5,700
2680	36" high, 202.5 kW		7	2.857		5,775	160		5,935	6,600
2700	40" high, 225 kW		6	3.333		6,575	186		6,761	7,500
2800	48" wide x 8" high, 54.8 kW		10	2		2,250	112		2,362	2,650
2820	16" high, 109.8 kW		9	2.222		3,125	124		3,249	3,650
2840	24" high, 164.4 kW		8	2.500		3,975	140		4,115	4,575
2860	32" high, 219.2 kW		7	2.857		5,300	160		5,460	6,100
2880	36" high, 246.6 kW		6	3.333		6,100	186		6,286	7,000
2900	40" high, 274 kW		5	4		6,900	224		7,124	7,950
3000	56" wide x 8" high, 64 kW		9	2.222		2,575	124		2,699	3,025
3020	16" high, 128 kW		8	2.500		3,550	140		3,690	4,125
3040	24" high, 192 kW		7	2.857		4,325	160		4,485	5,000
3060	32" high, 256 kW		6	3.333		6,025	186		6,211	6,900
3080	36" high, 288 kW		5	4		6,925	224		7,149	7,950
3100	40" high, 320 kW		4	5		7,650	279		7,929	8,850
3200	64" wide x 8" high, 74 kW		8	2.500		2,650	140		2,790	3,100
3220	16" high, 148 kW		7	2.857		3,650	160		3,810	4,275
3240	24" high, 222 kW		6	3.333		4,650	186		4,836	5,400
3260	32" high, 296 kW		5	4		6,250	224		6,474	7,225
3280	36" high, 333 kW		4	5		7,450	279		7,729	8,625
3300	40" high, 370 kW		3	6.667		8,225	375		8,600	9,625
3800	Note decreased kW available for									
3820	each duct size at same cost									
5000	Duct heater modifications and accessories									

For customer support on your Electrical Costs with RSMeans data, call 800.448.8182.

91

23 82 Convection Heating and Cooling Units

23 82 16 – Air Coils

23 82 16.20 Duct Heaters		Crew	Daily Output	Labor-Hours	Unit	Material	2019 Bare Costs Labor	Equipment	Total	Total Incl O&P
5120	T.C.O. limit auto or manual reset	Q-20	42	.476	Ea.	135	26.50		161.50	189
5140	Thermostat		28	.714		575	40		615	695
5160	Overheat thermocouple (removable)		7	2.857		815	160		975	1,125
5180	Fan interlock relay		18	1.111		197	62		259	310
5200	Air flow switch		20	1		170	56		226	272
5220	Split terminal box cover	▼	100	.200	▼	56	11.20		67.20	78.50
8000	To obtain BTU multiply kW by 3413									

23 83 Radiant Heating Units

23 83 33 – Electric Radiant Heaters

23 83 33.10 Electric Heating

		Crew	Daily Output	Labor-Hours	Unit	Material	2019 Bare Costs Labor	Equipment	Total	Total Incl O&P
0010	**ELECTRIC HEATING**, not incl. conduit or feed wiring									
1100	Rule of thumb: Baseboard units, including control	1 Elec	4.40	1.818	kW	112	109		221	286
1300	Baseboard heaters, 2' long, 350 watt		8	1	Ea.	27.50	60		87.50	120
1400	3' long, 750 watt		8	1		32	60		92	125
1600	4' long, 1,000 watt		6.70	1.194		39	71.50		110.50	150
1800	5' long, 935 watt		5.70	1.404		46.50	84.50		131	177
2000	6' long, 1,500 watt		5	1.600		51.50	96		147.50	200
2200	7' long, 1,310 watt		4.40	1.818		56	109		165	225
2400	8' long, 2,000 watt		4	2		65.50	120		185.50	251
2600	9' long, 1,680 watt		3.60	2.222		86	133		219	294
2800	10' long, 1,875 watt	▼	3.30	2.424	▼	154	146		300	385
2950	Wall heaters with fan, 120 to 277 volt									
3160	Recessed, residential, 750 watt	1 Elec	6	1.333	Ea.	94.50	80		174.50	223
3170	1,000 watt		6	1.333		94.50	80		174.50	223
3180	1,250 watt		5	1.600		102	96		198	255
3190	1,500 watt		4	2		102	120		222	291
3210	2,000 watt		4	2		94.50	120		214.50	283
3230	2,500 watt		3.50	2.286		237	137		374	465
3240	3,000 watt		3	2.667		365	160		525	645
3250	4,000 watt		2.70	2.963		370	178		548	670
3260	Commercial, 750 watt		6	1.333		202	80		282	340
3270	1,000 watt		6	1.333		202	80		282	340
3280	1,250 watt		5	1.600		202	96		298	365
3290	1,500 watt		4	2		202	120		322	400
3300	2,000 watt		4	2		202	120		322	400
3310	2,500 watt		3.50	2.286		202	137		339	430
3320	3,000 watt		3	2.667		425	160		585	710
3330	4,000 watt		2.70	2.963		495	178		673	810
3600	Thermostats, integral		16	.500		28	30		58	75.50
3800	Line voltage, 1 pole		8	1		17.15	60		77.15	108
3810	2 pole	▼	8	1	▼	22	60		82	114
4000	Heat trace system, 400 degree R238313-10									
4020	115 V, 2.5 watts/L.F.	1 Elec	530	.015	L.F.	11.35	.91		12.26	13.80
4030	5 watts/L.F. R238313-20		530	.015		11.35	.91		12.26	13.80
4050	10 watts/L.F.		530	.015		11.35	.91		12.26	13.80
4060	208 V, 5 watts/L.F.		530	.015		11.35	.91		12.26	13.80
4080	480 V, 8 watts/L.F.	▼	530	.015	▼	11.35	.91		12.26	13.80
4200	Heater raceway									
4260	Heat transfer cement									
4280	1 gallon				Ea.	63.50			63.50	70

23 83 Radiant Heating Units

23 83 33 – Electric Radiant Heaters

23 83 33.10 Electric Heating	Crew	Daily Output	Labor-Hours	Unit	Material	2019 Bare Costs Labor	Equipment	Total	Total Incl O&P
4300	5 gallon				Ea.	246			246	271
4320	Cable tie									
4340	3/4" pipe size	1 Elec	470	.017	Ea.	.02	1.02		1.04	1.55
4360	1" pipe size		444	.018		.02	1.08		1.10	1.63
4380	1-1/4" pipe size		400	.020		.02	1.20		1.22	1.81
4400	1-1/2" pipe size		355	.023		.02	1.35		1.37	2.04
4420	2" pipe size		320	.025		.05	1.50		1.55	2.30
4440	3" pipe size		160	.050		.05	3		3.05	4.54
4460	4" pipe size		100	.080		.06	4.80		4.86	7.20
4480	Thermostat NEMA 3R, 22 amp, 0-150 degree, 10' cap.		8	1		201	60		261	310
4500	Thermostat NEMA 4X, 25 amp, 40 degree, 5-1/2' cap.		7	1.143		254	68.50		322.50	380
4520	Thermostat NEMA 4X, 22 amp, 25-325 degree, 10' cap.		7	1.143		650	68.50		718.50	815
4540	Thermostat NEMA 4X, 22 amp, 15-140 degree		6	1.333		575	80		655	750
4580	Thermostat NEMA 4, 7, 9, 22 amp, 25-325 degree, 10' cap.		3.60	2.222		845	133		978	1,125
4600	Thermostat NEMA 4, 7, 9, 22 amp, 15-140 degree		3	2.667		835	160		995	1,150
4720	Fiberglass application tape, 36 yard roll		11	.727		79	43.50		122.50	152
5000	Radiant heating ceiling panels, 2' x 4', 500 watt		16	.500		395	30		425	480
5050	750 watt		16	.500		500	30		530	595
5200	For recessed plaster frame, add		32	.250		134	15		149	170
5300	Infrared quartz heaters, 120 volts, 1,000 watt		6.70	1.194		345	71.50		416.50	485
5350	1,500 watt		5	1.600		345	96		441	525
5400	240 volts, 1,500 watt		5	1.600		395	96		491	580
5450	2,000 watt		4	2		345	120		465	560
5500	3,000 watt		3	2.667		390	160		550	670
5550	4,000 watt		2.60	3.077		390	185		575	705
5570	Modulating control		.80	10		140	600		740	1,050
5600	Unit heaters, heavy duty, with fan & mounting bracket									
5650	Single phase, 208-240-277 volt, 3 kW	1 Elec	6	1.333	Ea.	695	80		775	885
5750	5 kW		5.50	1.455		785	87.50		872.50	995
5800	7 kW		5	1.600		1,300	96		1,396	1,575
5850	10 kW		4	2		1,025	120		1,145	1,300
5950	15 kW		3.80	2.105		1,650	126		1,776	2,000
6000	480 volt, 3 kW		6	1.333		535	80		615	710
6020	4 kW		5.80	1.379		745	83		828	945
6040	5 kW		5.50	1.455		740	87.50		827.50	945
6060	7 kW		5	1.600		1,125	96		1,221	1,375
6080	10 kW		4	2		1,175	120		1,295	1,475
6100	13 kW		3.80	2.105		1,750	126		1,876	2,125
6120	15 kW		3.70	2.162		2,525	130		2,655	2,975
6140	20 kW		3.50	2.286		2,975	137		3,112	3,450
6300	3 phase, 208-240 volt, 5 kW		5.50	1.455		575	87.50		662.50	760
6320	7 kW		5	1.600		935	96		1,031	1,175
6340	10 kW		4	2		875	120		995	1,150
6360	15 kW		3.70	2.162		1,625	130		1,755	1,975
6380	20 kW		3.50	2.286		2,825	137		2,962	3,300
6400	25 kW		3.30	2.424		3,550	146		3,696	4,125
6500	480 volt, 5 kW		5.50	1.455		730	87.50		817.50	930
6520	7 kW		5	1.600		1,100	96		1,196	1,375
6540	10 kW		4	2		975	120		1,095	1,250
6560	13 kW		3.80	2.105		1,700	126		1,826	2,050
6580	15 kW		3.70	2.162		1,425	130		1,555	1,750
6600	20 kW		3.50	2.286		2,425	137		2,562	2,875
6620	25 kW		3.30	2.424		3,325	146		3,471	3,900

23 83 33 – Electric Radiant Heaters

23 83 33.10 Electric Heating		Crew	Daily Output	Labor-Hours	Unit	Material	2019 Bare Costs Labor	Equipment	Total	Total Incl O&P
6630	30 kW	1 Elec	3	2.667	Ea.	3,375	160		3,535	3,950
6640	40 kW		2	4		4,075	240		4,315	4,825
6650	50 kW	↓	1.60	5	↓	6,375	300		6,675	7,475
6800	Vertical discharge heaters, with fan									
6820	Single phase, 208-240-277 volt, 10 kW	1 Elec	4	2	Ea.	985	120		1,105	1,250
6840	15 kW		3.70	2.162		1,700	130		1,830	2,075
6900	3 phase, 208-240 volt, 10 kW		4	2		930	120		1,050	1,200
6920	15 kW		3.70	2.162		1,700	130		1,830	2,050
6940	20 kW		3.50	2.286		3,025	137		3,162	3,525
6960	25 kW		3.30	2.424		3,275	146		3,421	3,825
6980	30 kW		3	2.667		4,150	160		4,310	4,800
7000	40 kW		2	4		4,650	240		4,890	5,475
7020	50 kW		1.60	5		6,625	300		6,925	7,725
7100	480 volt, 10 kW		4	2		1,075	120		1,195	1,350
7120	15 kW		3.70	2.162		1,750	130		1,880	2,125
7140	20 kW		3.50	2.286		2,825	137		2,962	3,300
7160	25 kW		3.30	2.424		3,550	146		3,696	4,150
7180	30 kW		3	2.667		3,925	160		4,085	4,575
7200	40 kW		2	4		4,075	240		4,315	4,825
7410	Sill height convector heaters, 5" high x 2' long, 500 watt		6.70	1.194		335	71.50		406.50	470
7420	3' long, 750 watt		6.50	1.231		395	74		469	545
7430	4' long, 1,000 watt		6.20	1.290		455	77.50		532.50	615
7440	5' long, 1,250 watt		5.50	1.455		515	87.50		602.50	700
7450	6' long, 1,500 watt		4.80	1.667		585	100		685	795
7460	8' long, 2,000 watt		3.60	2.222		795	133		928	1,075
7470	10' long, 2,500 watt	↓	3	2.667	↓	990	160		1,150	1,350
7900	Cabinet convector heaters, 240 volt, three phase,									
7920	3' long, 2,000 watt	1 Elec	5.30	1.509	Ea.	2,300	90.50		2,390.50	2,650
7940	3,000 watt		5.30	1.509		2,400	90.50		2,490.50	2,750
7960	4,000 watt		5.30	1.509		2,450	90.50		2,540.50	2,825
7980	6,000 watt		4.60	1.739		2,725	104		2,829	3,150
8000	8,000 watt		4.60	1.739		2,775	104		2,879	3,225
8020	4' long, 4,000 watt		4.60	1.739		2,500	104		2,604	2,900
8040	6,000 watt		4	2		2,575	120		2,695	3,025
8060	8,000 watt		4	2		2,650	120		2,770	3,100
8080	10,000 watt	↓	4	2	↓	2,700	120		2,820	3,150
8100	Available also in 208 or 277 volt									
8200	Cabinet unit heaters, 120 to 277 volt, 1 pole,									
8220	wall mounted, 2 kW	1 Elec	4.60	1.739	Ea.	2,150	104		2,254	2,525
8230	3 kW		4.60	1.739		2,200	104		2,304	2,575
8240	4 kW		4.40	1.818		2,325	109		2,434	2,725
8250	5 kW		4.40	1.818		2,350	109		2,459	2,775
8260	6 kW		4.20	1.905		2,400	114		2,514	2,825
8270	8 kW		4	2		2,450	120		2,570	2,875
8280	10 kW		3.80	2.105		2,500	126		2,626	2,950
8290	12 kW		3.50	2.286		2,625	137		2,762	3,100
8300	13.5 kW		2.90	2.759		3,175	166		3,341	3,750
8310	16 kW		2.70	2.963		3,200	178		3,378	3,800
8320	20 kW		2.30	3.478		4,000	209		4,209	4,700
8330	24 kW		1.90	4.211		4,200	253		4,453	4,975
8350	Recessed, 2 kW		4.40	1.818		2,175	109		2,284	2,575
8370	3 kW		4.40	1.818		2,275	109		2,384	2,675

23 83 33 – Electric Radiant Heaters

23 83 33.10 Electric Heating	Crew	Daily Output	Labor-Hours	Unit	Material	2019 Bare Costs Labor	Equipment	Total	Total Incl O&P
8380	4 kW	1 Elec	4.20	1.905	Ea.	2,350	114		2,464	2,775
8390	5 kW		4.20	1.905		2,375	114		2,489	2,800
8400	6 kW		4	2		2,425	120		2,545	2,850
8410	8 kW		3.80	2.105		2,550	126		2,676	3,000
8420	10 kW		3.50	2.286		3,200	137		3,337	3,725
8430	12 kW		2.90	2.759		3,200	166		3,366	3,775
8440	13.5 kW		2.70	2.963		3,175	178		3,353	3,750
8450	16 kW		2.30	3.478		3,325	209		3,534	3,950
8460	20 kW		1.90	4.211		3,850	253		4,103	4,625
8470	24 kW		1.60	5		3,825	300		4,125	4,650
8490	Ceiling mounted, 2 kW		3.20	2.500		2,150	150		2,300	2,575
8510	3 kW		3.20	2.500		2,250	150		2,400	2,700
8520	4 kW		3	2.667		2,300	160		2,460	2,775
8530	5 kW		3	2.667		2,375	160		2,535	2,875
8540	6 kW		2.80	2.857		2,400	172		2,572	2,875
8550	8 kW		2.40	3.333		2,525	200		2,725	3,075
8560	10 kW		2.20	3.636		2,650	218		2,868	3,250
8570	12 kW		2	4		2,700	240		2,940	3,325
8580	13.5 kW		1.50	5.333		2,675	320		2,995	3,425
8590	16 kW		1.30	6.154		2,750	370		3,120	3,575
8600	20 kW		.90	8.889		3,800	535		4,335	5,000
8610	24 kW		.60	13.333		3,825	800		4,625	5,400
8630	208 to 480 V, 3 pole									
8650	Wall mounted, 2 kW	1 Elec	4.60	1.739	Ea.	2,275	104		2,379	2,650
8670	3 kW		4.60	1.739		2,375	104		2,479	2,750
8680	4 kW		4.40	1.818		2,425	109		2,534	2,850
8690	5 kW		4.40	1.818		2,500	109		2,609	2,925
8700	6 kW		4.20	1.905		2,525	114		2,639	2,950
8710	8 kW		4	2		2,600	120		2,720	3,050
8720	10 kW		3.80	2.105		2,675	126		2,801	3,125
8730	12 kW		3.50	2.286		2,725	137		2,862	3,200
8740	13.5 kW		2.90	2.759		2,800	166		2,966	3,350
8750	16 kW		2.70	2.963		2,825	178		3,003	3,375
8760	20 kW		2.30	3.478		3,975	209		4,184	4,675
8770	24 kW		1.90	4.211		4,200	253		4,453	5,000
8790	Recessed, 2 kW		4.40	1.818		2,275	109		2,384	2,675
8810	3 kW		4.40	1.818		2,300	109		2,409	2,700
8820	4 kW		4.20	1.905		2,425	114		2,539	2,850
8830	5 kW		4.20	1.905		2,425	114		2,539	2,825
8840	6 kW		4	2		2,450	120		2,570	2,850
8850	8 kW		3.80	2.105		2,525	126		2,651	2,975
8860	10 kW		3.50	2.286		2,550	137		2,687	3,025
8870	12 kW		2.90	2.759		2,850	166		3,016	3,400
8880	13.5 kW		2.70	2.963		2,425	178		2,603	2,925
8890	16 kW		2.30	3.478		4,725	209		4,934	5,500
8900	20 kW		1.90	4.211		4,275	253		4,528	5,075
8920	24 kW		1.60	5		4,350	300		4,650	5,225
8940	Ceiling mount, 2 kW		3.20	2.500		2,400	150		2,550	2,875
8950	3 kW		3.20	2.500		2,150	150		2,300	2,575
8960	4 kW		3	2.667		2,575	160		2,735	3,075
8970	5 kW		3	2.667		2,475	160		2,635	2,975
8980	6 kW		2.80	2.857		2,500	172		2,672	3,000
8990	8 kW		2.40	3.333		2,575	200		2,775	3,150

23 83 Radiant Heating Units

23 83 33 - Electric Radiant Heaters

23 83 33.10 Electric Heating		Crew	Daily Output	Labor-Hours	Unit	Material	2019 Bare Costs Labor	Equipment	Total	Total Incl O&P
9000	10 kW	1 Elec	2.20	3.636	Ea.	2,350	218		2,568	2,925
9020	13.5 kW		1.50	5.333		2,950	320		3,270	3,725
9030	16 kW		1.30	6.154		4,425	370		4,795	5,425
9040	20 kW		.90	8.889		4,275	535		4,810	5,500
9060	24 kW		.60	13.333		4,750	800		5,550	6,425
9230	13.5 kW, 40,956 BTU		2.20	3.636		1,725	218		1,943	2,225
9250	24 kW, 81,912 BTU	↓	2	4	↓	1,750	240		1,990	2,275

Estimating Tips

- When estimating material costs for special systems, it is always prudent to obtain manufacturers' quotations for equipment prices and special installation requirements that may affect the total cost.

- For cost modifications for elevated tray installation, add the percentages to labor according to the height of the installation and only to the quantities exceeding the different height levels, not to the total tray quantities. Refer to subdivision 26 01 02.20 for labor adjustment factors.

- Do not overlook the costs for equipment used in the installation.

Reference Numbers

Reference numbers are shown at the beginning of some major classifications. These numbers refer to related items in the Reference Section. The reference information may be an estimating procedure, an alternate pricing method, or technical information.

Note: Trade Service, in part, has been used as a reference source for some of the material prices used in Division 25.

Did you know?

RSMeans data is available through our online application:

- Search for costs by keyword
- Leverage the most up-to-date data
- Build and export estimates

Try it free
rsmeans.com/2019freetrial

25 05 Common Work Results for Integrated Automation

25 05 28 – Pathways for Integrated Automation

25 05 28.39 Surface Raceways for Integrated Automation	Crew	Daily Output	Labor-Hours	Unit	Material	2019 Bare Costs Labor	Equipment	Total	Total Incl O&P
0010 **SURFACE RACEWAYS FOR INTEGRATED AUTOMATION**									
1000 Din rail, G type, 15 mm, steel, galvanized, 3', for termination blocks	1 Elec	48	.167	Ea.	12.35	10		22.35	28.50
1010 32 mm, steel, galvanized, 3', for termination blocks		48	.167		17.50	10		27.50	34
1020 35 mm, steel, galvanized, 3', for termination blocks		48	.167		14.55	10		24.55	31
1030 Din rail, rail mounted, 35 mm, aluminum, 3', for termination blocks		48	.167		20.50	10		30.50	37.50
1040 35 mm, aluminum, 6', for termination blocks		48	.167		25.50	10		35.50	43.50
2000 Terminal block, Din mounted, 35 mm, medium screw, connector end bracket		48	.167		2.19	10		12.19	17.35
2010 35 mm, large screw, connector end bracket		48	.167		3.43	10		13.43	18.70
2020 Large screw, connector end bracket		48	.167		4.21	10		14.21	19.60
3000 Terminal block, feed through, 35 mm, Din mounted, 600V, 20A		48	.167		1	10		11	16.05
3010 35 mm, Din mounted, 600V, 30A		48	.167		1.29	10		11.29	16.35
3020 Terminal block, fuse holder, LED indicator, Din mounted, 110V, 6A		48	.167		8.55	10		18.55	24.50
3030 Din mounted, 220V, 6A		48	.167		7.85	10		17.85	23.50
3040 Din mounted, 300V, 6A		48	.167		7.65	10		17.65	23.50

25 36 Integrated Automation Instrumentation & Terminal Devices

25 36 19 – Integrated Automation Current Sensors

25 36 19.10 Integrated Automation Current Sensors

	Crew	Daily Output	Labor-Hours	Unit	Material	Labor	Equipment	Total	Total Incl O&P
0010 **INTEGRATED AUTOMATION CURRENT SENSORS**									
1000 Current limiter inrush, 100 to 480 VAC, 10A, power supply	1 Elec	48	.167	Ea.	80.50	10		90.50	103

Estimating Tips
26 05 00 Common Work Results for Electrical

- Conduit should be taken off in three main categories—power distribution, branch power, and branch lighting—so the estimator can concentrate on systems and components, therefore making it easier to ensure all items have been accounted for.

- For cost modifications for elevated conduit installation, add the percentages to labor according to the height of installation and only to the quantities exceeding the different height levels, not to the total conduit quantities. Refer to subdivision 26 01 02.20 for labor adjustment factors.

- Remember that aluminum wiring of equal ampacity is larger in diameter than copper and may require larger conduit.

- If more than three wires at a time are being pulled, deduct percentages from the labor hours of that grouping of wires.

- When taking off grounding systems, identify separately the type and size of wire, and list each unique type of ground connection.

- The estimator should take the weights of materials into consideration when completing a takeoff. Topics to consider include: How will the materials be supported? What methods of support are available? How high will the support structure have to reach? Will the final support structure be able to withstand the total burden? Is the support material included or separate from the fixture, equipment, and material specified?

- Do not overlook the costs for equipment used in the installation. If scaffolding or highlifts are available in the field, contractors may use them in lieu of the proposed ladders and rolling staging.

26 20 00 Low-Voltage Electrical Transmission

- Supports and concrete pads may be shown on drawings for the larger equipment, or the support system may be only a piece of plywood for the back of a panelboard. In either case, they must be included in the costs.

26 40 00 Electrical and Cathodic Protection

- When taking off cathodic protection systems, identify the type and size of cable, and list each unique type of anode connection.

26 50 00 Lighting

- Fixtures should be taken off room by room using the fixture schedule, specifications, and the ceiling plan. For large concentrations of lighting fixtures in the same area, deduct the percentages from labor hours.

Reference Numbers

Reference numbers are shown at the beginning of some major classifications. These numbers refer to related items in the Reference Section. The reference information may be an estimating procedure, an alternate pricing method, or technical information.

Note: Not all subdivisions listed here necessarily appear. ■

26 01 02.20 Labor Adjustment Factors	Crew	Daily Output	Labor-Hours	Unit	Material	2019 Bare Costs Labor	Equipment	Total	Total Incl O&P
0010 **LABOR ADJUSTMENT FACTORS** (For Div. 26, 27 and 28) R260519-90									
0100 Subtract from labor for Economy of Scale for Wire									
0110 4-5 wires						25%			
0120 6-10 wires						30%			
0130 11-15 wires						35%			
0140 over 15 wires						40%			
0150 Labor adjustment factors (For Div. 26, 27, 28 and 48)									
0200 Labor factors: The below are reasonable suggestions, but									
0210 each project must be evaluated for its own peculiarities, and									
0220 the adjustments be increased or decreased depending on the									
0230 severity of the special conditions.									
1000 Add to labor for elevated installation (above floor level)									
1010 10' to 14.5' high						10%			
1020 15' to 19.5' high						20%			
1030 20' to 24.5' high						25%			
1040 25' to 29.5' high						35%			
1050 30' to 34.5' high						40%			
1060 35' to 39.5' high						50%			
1070 40' and higher						55%			
2000 Add to labor for crawl space									
2010 3' high						40%			
2020 4' high						30%			
3000 Add to labor for multi-story building									
3100 For new construction (No elevator available)									
3110 Add for floors 3 thru 10						5%			
3120 Add for floors 11 thru 15						10%			
3130 Add for floors 16 thru 20						15%			
3140 Add for floors 21 thru 30						20%			
3150 Add for floors 31 and up						30%			
3200 For existing structure (Elevator available)									
3210 Add for work on floor 3 and above						2%			
4000 Add to labor for working in existing occupied buildings									
4010 Hospital						35%			
4020 Office building						25%			
4030 School						20%			
4040 Factory or warehouse						15%			
4050 Multi-dwelling						15%			
5000 Add to labor, miscellaneous									
5010 Cramped shaft						35%			
5020 Congested area						15%			
5030 Excessive heat or cold						30%			
6000 Labor factors: the above are reasonable suggestions, but									
6100 each project should be evaluated for its own peculiarities									
6200 Other factors to be considered are:									
6210 Movement of material and equipment through finished areas						10%			
6220 Equipment room min security direct access w/authorization						15%			
6230 Attic space						25%			
6240 No service road						25%			
6250 Poor unloading/storage area, no hydraulic lifts or jacks						20%			
6260 Congested site area/heavy traffic						20%			
7000 Correctional facilities (no compounding division 1 adjustment factors)									
7010 Minimum security w/facilities escort						30%			
7020 Medium security w/facilities and correctional officer escort						40%			

26 01 Operation and Maintenance of Electrical Systems

26 01 02 – Labor Adjustment

26 01 02.20 Labor Adjustment Factors	Crew	Daily Output	Labor-Hours	Unit	Material	2019 Bare Costs Labor	Equipment	Total	Total Incl O&P
7030 Max security w/facilities & correctional officer escort (no inmate contact)						50%			

26 01 40 – Operation and Maintenance of Electrical Protection Systems

26 01 40.51 Electrical Systems Repair and Replacement

0010 **ELECTRICAL SYSTEMS REPAIR AND REPLACEMENT**									
3000 Remove and replace (reinstall), switch cover	1 Elec	60	.133	Ea.		8		8	11.95
3020 Outlet cover	"	60	.133	"		8		8	11.95

26 01 50 – Operation and Maintenance of Lighting

26 01 50.81 Luminaire Replacement

0010 **LUMINAIRE REPLACEMENT**									
0100 Fluorescent fixtures, elevate from work area	1 Elec	96	.083	Ea.	92.50	5		97.50	109
0962 Remove and install new replacement lens, 2' x 2' troffer		16	.500		20.50	30		50.50	68
0964 2' x 4' troffer		16	.500		14.80	30		44.80	61.50
3200 Remove and replace (reinstall), lighting fixture	↓	4	2	↓		120		120	179

26 05 Common Work Results for Electrical

26 05 05 – Selective Demolition for Electrical

26 05 05.10 Electrical Demolition

0010 **ELECTRICAL DEMOLITION** R260105-30									
0020 Electrical demolition, conduit to 10' high, incl. fittings & hangers									
0100 Rigid galvanized steel, 1/2" to 1" diameter	1 Elec	242	.033	L.F.		1.99		1.99	2.96
0120 1-1/4" to 2"	"	200	.040			2.40		2.40	3.58
0140 2-1/2" to 3-1/2"	2 Elec	302	.053			3.18		3.18	4.75
0160 4" to 6"	"	160	.100			6		6	8.95
0170 PVC #40, 1/2" to 1"	1 Elec	410	.020			1.17		1.17	1.75
0172 1-1/4" to 2"		350	.023			1.37		1.37	2.05
0174 2-1/2"	↓	250	.032			1.92		1.92	2.87
0176 3" to 3-1/2"	2 Elec	340	.047			2.83		2.83	4.22
0178 4" to 6"	"	230	.070			4.18		4.18	6.25
0200 Electric metallic tubing (EMT), 1/2" to 1"	1 Elec	394	.020			1.22		1.22	1.82
0220 1-1/4" to 1-1/2"		326	.025			1.47		1.47	2.20
0240 2" to 3"	↓	236	.034			2.04		2.04	3.04
0260 3-1/2" to 4"	2 Elec	310	.052	↓		3.10		3.10	4.62
0270 Armored cable (BX) avg. 50' runs									
0280 #14, 2 wire	1 Elec	690	.012	L.F.		.70		.70	1.04
0290 #14, 3 wire		571	.014			.84		.84	1.26
0300 #12, 2 wire		605	.013			.79		.79	1.18
0310 #12, 3 wire		514	.016			.93		.93	1.39
0320 #10, 2 wire		514	.016			.93		.93	1.39
0330 #10, 3 wire		425	.019			1.13		1.13	1.69
0340 #8, 3 wire	↓	342	.023	↓		1.40		1.40	2.10
0350 Non metallic sheathed cable (Romex)									
0360 #14, 2 wire	1 Elec	720	.011	L.F.		.67		.67	1
0370 #14, 3 wire		657	.012			.73		.73	1.09
0380 #12, 2 wire		629	.013			.76		.76	1.14
0390 #10, 3 wire	↓	450	.018	↓		1.07		1.07	1.59
0400 Wiremold raceway, including fittings & hangers									
0420 No. 3000	1 Elec	250	.032	L.F.		1.92		1.92	2.87
0440 No. 4000		217	.037			2.21		2.21	3.30
0460 No. 6000		166	.048			2.89		2.89	4.32
0462 Plugmold with receptacle	↓	114	.070	↓		4.21		4.21	6.30

For customer support on your Electrical Costs with RSMeans data, call 800.448.8182.

101

26 05 05 – Selective Demolition for Electrical

26 05 05.10 Electrical Demolition		Crew	Daily Output	Labor-Hours	Unit	Material	2019 Bare Costs Labor	Equipment	Total	Total Incl O&P
0465	Telephone/power pole	1 Elec	12	.667	Ea.		40		40	59.50
0470	Non-metallic, straight section	↓	480	.017	L.F.		1		1	1.49
0500	Channels, steel, including fittings & hangers									
0520	3/4" x 1-1/2"	1 Elec	308	.026	L.F.		1.56		1.56	2.33
0540	1-1/2" x 1-1/2"	↓	269	.030			1.79		1.79	2.66
0560	1-1/2" x 1-7/8"	↓	229	.035	↓		2.10		2.10	3.13
0600	Copper bus duct, indoor, 3 phase									
0610	Including hangers & supports									
0620	225 amp	2 Elec	135	.119	L.F.		7.10		7.10	10.60
0640	400 amp		106	.151			9.05		9.05	13.50
0660	600 amp		86	.186			11.15		11.15	16.65
0680	1,000 amp		60	.267			16		16	24
0700	1,600 amp		40	.400			24		24	36
0720	3,000 amp	↓	10	1.600	↓		96		96	143
0800	Plug-in switches, 600 V 3 ph., incl. disconnecting									
0820	wire, conduit terminations, 30 amp	1 Elec	15.50	.516	Ea.		31		31	46.50
0840	60 amp		13.90	.576			34.50		34.50	51.50
0850	100 amp		10.40	.769			46		46	69
0860	200 amp	↓	6.20	1.290			77.50		77.50	116
0880	400 amp	2 Elec	5.40	2.963			178		178	265
0900	600 amp		3.40	4.706			283		283	420
0920	800 amp		2.60	6.154			370		370	550
0940	1,200 amp		2	8			480		480	715
0960	1,600 amp	↓	1.70	9.412	↓		565		565	845
1010	Safety switches, 250 or 600 V, incl. disconnection									
1050	of wire & conduit terminations									
1100	30 amp	1 Elec	12.30	.650	Ea.		39		39	58.50
1120	60 amp		8.80	.909			54.50		54.50	81.50
1140	100 amp		7.30	1.096			66		66	98
1160	200 amp	↓	5	1.600			96		96	143
1180	400 amp	2 Elec	6.80	2.353			141		141	211
1200	600 amp	"	4.60	3.478			209		209	310
1202	Explosion proof, 60 amp	1 Elec	8.80	.909			54.50		54.50	81.50
1204	100 amp		5	1.600			96		96	143
1206	200 amp	↓	3.33	2.402	↓		144		144	215
1210	Panel boards, incl. removal of all breakers,									
1220	conduit terminations & wire connections									
1230	3 wire, 120/240 V, 100A, to 20 circuits	1 Elec	2.60	3.077	Ea.		185		185	276
1240	200 amps, to 42 circuits	2 Elec	2.60	6.154			370		370	550
1241	225 amps, to 42 circuits		2.40	6.667			400		400	595
1250	400 amps, to 42 circuits	↓	2.20	7.273			435		435	650
1260	4 wire, 120/208 V, 125A, to 20 circuits	1 Elec	2.40	3.333			200		200	299
1270	200 amps, to 42 circuits	2 Elec	2.40	6.667			400		400	595
1280	400 amps, to 42 circuits		1.92	8.333			500		500	745
1285	600 amps, to 42 circuits	↓	1.60	10	↓		600		600	895
1300	Transformer, dry type, 1 phase, incl. removal of									
1320	supports, wire & conduit terminations									
1340	1 kVA	1 Elec	7.70	1.039	Ea.		62.50		62.50	93
1360	5 kVA		4.70	1.702			102		102	153
1380	10 kVA	↓	3.60	2.222			133		133	199
1400	37.5 kVA	2 Elec	3	5.333			320		320	480
1420	75 kVA	"	2.50	6.400	↓		385		385	575
1440	3 phase to 600 V, primary									

26 05 05 – Selective Demolition for Electrical

26 05 05.10 Electrical Demolition	Crew	Daily Output	Labor-Hours	Unit	Material	2019 Bare Costs Labor	2019 Bare Costs Equipment	Total	Total Incl O&P
1460	3 kVA	1 Elec	3.87	2.067	Ea.		124		124	185
1480	15 kVA	2 Elec	3.67	4.360			262		262	390
1490	25 kVA		3.51	4.558			274		274	410
1500	30 kVA		3.42	4.678			281		281	420
1510	45 kVA		3.18	5.031			300		300	450
1520	75 kVA		2.69	5.948			355		355	535
1530	112.5 kVA	R-3	2.90	6.897			410	45	455	665
1540	150 kVA		2.70	7.407			440	48	488	715
1550	300 kVA		1.80	11.111			665	72	737	1,075
1560	500 kVA		1.40	14.286			855	92.50	947.50	1,375
1570	750 kVA		1.10	18.182			1,075	118	1,193	1,750
1600	Pull boxes & cabinets, sheet metal, incl. removal									
1620	of supports and conduit terminations									
1640	6" x 6" x 4"	1 Elec	31.10	.257	Ea.		15.45		15.45	23
1660	12" x 12" x 4"		23.30	.343			20.50		20.50	31
1680	24" x 24" x 6"		12.30	.650			39		39	58.50
1700	36" x 36" x 8"		7.70	1.039			62.50		62.50	93
1720	Junction boxes, 4" sq. & oct.		80	.100			6		6	8.95
1740	Handy box		107	.075			4.49		4.49	6.70
1760	Switch box		107	.075			4.49		4.49	6.70
1780	Receptacle & switch plates		257	.031			1.87		1.87	2.79
1790	Receptacles & switches, 15 to 30 amp		135	.059			3.56		3.56	5.30
1800	Wire, THW-THWN-THHN, removed from									
1810	in place conduit, to 10' high									
1830	#14	1 Elec	65	.123	C.L.F.		7.40		7.40	11.05
1840	#12		55	.145			8.75		8.75	13.05
1850	#10		45.50	.176			10.55		10.55	15.75
1860	#8		40.40	.198			11.90		11.90	17.75
1870	#6		32.60	.245			14.75		14.75	22
1880	#4	2 Elec	53	.302			18.15		18.15	27
1890	#3		50	.320			19.20		19.20	28.50
1900	#2		44.60	.359			21.50		21.50	32
1910	1/0		33.20	.482			29		29	43
1920	2/0		29.20	.548			33		33	49
1930	3/0		25	.640			38.50		38.50	57.50
1940	4/0		22	.727			43.50		43.50	65
1950	250 kcmil		20	.800			48		48	71.50
1960	300 kcmil		19	.842			50.50		50.50	75.50
1970	350 kcmil		18	.889			53.50		53.50	79.50
1980	400 kcmil		17	.941			56.50		56.50	84.50
1990	500 kcmil		16.20	.988			59.50		59.50	88.50
2000	Interior fluorescent fixtures, incl. supports									
2010	& whips, to 10' high									
2100	Recessed drop-in 2' x 2', 2 lamp	2 Elec	35	.457	Ea.		27.50		27.50	41
2110	2' x 2', 4 lamp		30	.533			32		32	48
2120	2' x 4', 2 lamp		33	.485			29		29	43.50
2140	2' x 4', 4 lamp		30	.533			32		32	48
2160	4' x 4', 4 lamp		20	.800			48		48	71.50
2180	Surface mount, acrylic lens & hinged frame									
2200	1' x 4', 2 lamp	2 Elec	44	.364	Ea.		22		22	32.50
2210	6" x 4', 2 lamp		44	.364			22		22	32.50
2220	2' x 2', 2 lamp		44	.364			22		22	32.50
2260	2' x 4', 4 lamp		33	.485			29		29	43.50

26 05 05.10 Electrical Demolition	Crew	Daily Output	Labor-Hours	Unit	Material	2019 Bare Costs Labor	Equipment	Total	Total Incl O&P
2280	4' x 4', 4 lamp	2 Elec	23	.696	Ea.		42		42	62.50
2281	4' x 4', 6 lamp	▼	23	.696	▼		42		42	62.50
2300	Strip fixtures, surface mount									
2320	4' long, 1 lamp	2 Elec	53	.302	Ea.		18.15		18.15	27
2340	4' long, 2 lamp		50	.320			19.20		19.20	28.50
2360	8' long, 1 lamp		42	.381			23		23	34
2380	8' long, 2 lamp	▼	40	.400	▼		24		24	36
2400	Pendant mount, industrial, incl. removal									
2410	of chain or rod hangers, to 10' high									
2420	4' long, 2 lamp	2 Elec	35	.457	Ea.		27.50		27.50	41
2421	4' long, 4 lamp		35	.457			27.50		27.50	41
2440	8' long, 2 lamp	▼	27	.593	▼		35.50		35.50	53
2460	Interior incandescent, surface, ceiling									
2470	or wall mount, to 12' high									
2480	Metal cylinder type, 75 Watt	2 Elec	62	.258	Ea.		15.50		15.50	23
2500	150 Watt		62	.258			15.50		15.50	23
2502	300 Watt	▼	53	.302			18.15		18.15	27
2520	Metal halide, high bay									
2540	400 Watt	2 Elec	15	1.067	Ea.		64		64	95.50
2560	1,000 Watt		12	1.333			80		80	119
2580	150 Watt, low bay		20	.800			48		48	71.50
2585	Globe type, ceiling mount		62	.258			15.50		15.50	23
2586	Can type, recessed mount	▼	62	.258	▼		15.50		15.50	23
2600	Exterior fixtures, incandescent, wall mount									
2620	100 Watt	2 Elec	50	.320	Ea.		19.20		19.20	28.50
2640	Quartz, 500 Watt		33	.485			29		29	43.50
2660	1,500 Watt	▼	27	.593	▼		35.50		35.50	53
2680	Wall pack, mercury vapor									
2700	175 Watt	2 Elec	25	.640	Ea.		38.50		38.50	57.50
2720	250 Watt	"	25	.640			38.50		38.50	57.50
7000	Weatherhead/mast, 2"	1 Elec	16	.500			30		30	45
7002	3"		10	.800			48		48	71.50
7004	3-1/2"		8.50	.941			56.50		56.50	84.50
7006	4"		8	1	▼		60		60	89.50
7100	Service entry cable, #6, +#6 neutral		420	.019	L.F.		1.14		1.14	1.71
7102	#4, +#4 neutral		360	.022			1.33		1.33	1.99
7104	#2, +#4 neutral		345	.023			1.39		1.39	2.08
7106	#2, +#2 neutral	▼	330	.024	▼		1.46		1.46	2.17
9900	Add to labor for higher elevated installation									
9905	10' to 14.5' high, add						10%			
9910	15' to 20' high, add						20%			
9920	20' to 25' high, add						25%			
9930	25' to 30' high, add						35%			
9940	30' to 35' high, add						40%			
9950	35' to 40' high, add						50%			
9960	Over 40' high, add						55%			

26 05 05.15 Electrical Demolition, Grounding

		Crew	Daily Output	Labor-Hours	Unit	Material	Labor	Equipment	Total	Total Incl O&P
0010	**ELECTRICAL DEMOLITION, GROUNDING** Addition									
0100	Ground clamp, bronze	1 Elec	64	.125	Ea.		7.50		7.50	11.20
0140	Water pipe ground clamp, bronze, heavy duty		24	.333			20		20	30
0150	Ground rod, 8' to 10'	▼	13	.615	▼		37		37	55
0200	Ground wire, bare armored	2 Elec	900	.018	L.F.		1.07		1.07	1.59

26 05 Common Work Results for Electrical

26 05 05 – Selective Demolition for Electrical

26 05 05.15 Electrical Demolition, Grounding		Crew	Daily Output	Labor-Hours	Unit	Material	2019 Bare Costs Labor	Equipment	Total	Total Incl O&P
0240	Bare copper or aluminum	2 Elec	2800	.006	L.F.		.34		.34	.51

26 05 05.20 Electrical Demolition, Wiring Methods

26 05 05.20	Electrical Demolition, Wiring Methods		Crew	Daily Output	Labor-Hours	Unit	Material	2019 Bare Costs Labor	Equipment	Total	Total Incl O&P
0010	**ELECTRICAL DEMOLITION, WIRING METHODS** Addition										
0100	Armored cable, w/PVC jacket, in cable tray										
0110	#6	R024119-10	1 Elec	9.30	.860	C.L.F.		51.50		51.50	77
0120	#4		2 Elec	16.20	.988			59.50		59.50	88.50
0130	#2			13.80	1.159			69.50		69.50	104
0140	#1			12	1.333			80		80	119
0150	1/0			10.80	1.481			89		89	133
0160	2/0			10.20	1.569			94		94	141
0170	3/0			9.60	1.667			100		100	149
0180	4/0			9	1.778			107		107	159
0190	250 kcmil		3 Elec	10.80	2.222			133		133	199
0210	350 kcmil			9.90	2.424			146		146	217
0230	500 kcmil			9	2.667			160		160	239
0240	750 kcmil			8.10	2.963			178		178	265
1100	Control cable, 600 V or less										
1110	3 wires		1 Elec	24	.333	C.L.F.		20		20	30
1120	5 wires			20	.400			24		24	36
1130	7 wires			16.50	.485			29		29	43.50
1140	9 wires			15	.533			32		32	48
1150	12 wires		2 Elec	26	.615			37		37	55
1160	15 wires			22	.727			43.50		43.50	65
1170	19 wires			19	.842			50.50		50.50	75.50
1180	25 wires			16	1			60		60	89.50
1500	Mineral insulated (MI) cable, 600 V										
1510	#10		1 Elec	4.80	1.667	C.L.F.		100		100	149
1520	#8			4.50	1.778			107		107	159
1530	#6			4.20	1.905			114		114	171
1540	#4		2 Elec	7.20	2.222			133		133	199
1550	#2			6.60	2.424			146		146	217
1560	#1			6.30	2.540			153		153	228
1570	1/0			6	2.667			160		160	239
1580	2/0			5.70	2.807			169		169	252
1590	3/0			5.40	2.963			178		178	265
1600	4/0			4.80	3.333			200		200	299
1610	250 kcmil		3 Elec	7.20	3.333			200		200	299
1620	500 kcmil		"	5.90	4.068			244		244	365
2000	Shielded cable, XLP shielding, to 35 kV										
2010	#4		2 Elec	13.20	1.212	C.L.F.		73		73	109
2020	#1			12	1.333			80		80	119
2030	1/0			11.40	1.404			84.50		84.50	126
2040	2/0			10.80	1.481			89		89	133
2050	4/0			9.60	1.667			100		100	149
2060	250 kcmil		3 Elec	13.50	1.778			107		107	159
2070	350 kcmil			11.70	2.051			123		123	184
2080	500 kcmil			11.20	2.143			129		129	192
2090	750 kcmil			10.80	2.222			133		133	199
3000	Modular flexible wiring										
3010	Cable set		1 Elec	120	.067	Ea.		4		4	5.95
3020	Conversion module			48	.167			10		10	14.95
3030	Switching assembly			96	.083			5		5	7.45

For customer support on your Electrical Costs with RSMeans data, call 800.448.8182.

105

26 05 05.20 Electrical Demolition, Wiring Methods	Crew	Daily Output	Labor-Hours	Unit	Material	2019 Bare Costs Labor	Equipment	Total	Total Incl O&P
3200	Undercarpet									
3210	Power or telephone, flat cable	1 Elec	1320	.006	L.F.		.36		.36	.54
3220	Transition block assemblies w/fitting		75	.107	Ea.		6.40		6.40	9.55
3230	Floor box with fitting		60	.133			8		8	11.95
3300	Data system, cable with connection		50	.160			9.60		9.60	14.35
4000	Cable tray, including fitting & support									
4010	Galvanized steel, 6" wide	2 Elec	310	.052	L.F.		3.10		3.10	4.62
4020	9" wide		295	.054			3.26		3.26	4.86
4030	12" wide		285	.056			3.37		3.37	5.05
4040	18" wide		270	.059			3.56		3.56	5.30
4050	24" wide		260	.062			3.70		3.70	5.50
4060	30" wide		240	.067			4		4	5.95
4070	36" wide		220	.073			4.37		4.37	6.50
4110	Aluminum, 6" wide		420	.038			2.29		2.29	3.41
4120	9" wide		415	.039			2.31		2.31	3.45
4130	12" wide		390	.041			2.46		2.46	3.68
4140	18" wide		370	.043			2.60		2.60	3.87
4150	24" wide		350	.046			2.74		2.74	4.10
4160	30" wide		325	.049			2.96		2.96	4.41
4170	36" wide		300	.053			3.20		3.20	4.78
4310	Cable channel, aluminum, 4" wide straight	1 Elec	240	.033			2		2	2.99
4320	Cable tray fittings, 6" to 9" wide	2 Elec	27	.593	Ea.		35.50		35.50	53
4330	12" to 18" wide	"	19	.842	"		50.50		50.50	75.50
5000	Conduit nipples, with locknuts and bushings									
5020	1/2"	1 Elec	108	.074	Ea.		4.45		4.45	6.65
5040	3/4"		96	.083			5		5	7.45
5060	1"		81	.099			5.95		5.95	8.85
5080	1-1/4"		69	.116			6.95		6.95	10.40
5100	1-1/2"		60	.133			8		8	11.95
5120	2"		54	.148			8.90		8.90	13.25
5140	2-1/2"		45	.178			10.70		10.70	15.95
5160	3"		36	.222			13.35		13.35	19.90
5180	3-1/2"		33	.242			14.55		14.55	21.50
5200	4"		27	.296			17.80		17.80	26.50
5220	5"		21	.381			23		23	34
5240	6"		18	.444			26.50		26.50	40
5500	Electric nonmetallic tubing (ENT), flexible, 1/2" to 1" diameter		690	.012	L.F.		.70		.70	1.04
5510	1-1/4" to 2" diameter		300	.027			1.60		1.60	2.39
5600	Flexible metallic tubing, steel, 3/8" to 3/4" diameter		600	.013			.80		.80	1.19
5610	1" to 1-1/4" diameter		260	.031			1.85		1.85	2.76
5620	1-1/2" to 2" diameter		140	.057			3.43		3.43	5.10
5630	2-1/2" diameter		100	.080			4.80		4.80	7.15
5640	3" to 3-1/2" diameter	2 Elec	150	.107			6.40		6.40	9.55
5650	4" diameter	"	100	.160			9.60		9.60	14.35
5700	Sealtite flexible conduit, 3/8" to 3/4" diameter	1 Elec	420	.019			1.14		1.14	1.71
5710	1" to 1-1/4" diameter		180	.044			2.67		2.67	3.98
5720	1-1/2" to 2" diameter		120	.067			4		4	5.95
5730	2-1/2" diameter		80	.100			6		6	8.95
5740	3" diameter	2 Elec	150	.107			6.40		6.40	9.55
5750	4" diameter	"	100	.160			9.60		9.60	14.35
5800	Wiring duct, plastic, 1-1/2" to 2-1/2" wide	1 Elec	360	.022			1.33		1.33	1.99
5810	3" wide		330	.024			1.46		1.46	2.17
5820	4" wide		300	.027			1.60		1.60	2.39

26 05 Common Work Results for Electrical

26 05 05 – Selective Demolition for Electrical

26 05 05.20 Electrical Demolition, Wiring Methods

		Crew	Daily Output	Labor-Hours	Unit	Material	2019 Bare Costs Labor	2019 Bare Costs Equipment	Total	Total Incl O&P
6000	Floor box and carpet flange	1 Elec	9	.889	Ea.		53.50		53.50	79.50
6300	Wireway, with fittings and supports, to 10' high									
6310	2-1/2" x 2-1/2"	1 Elec	180	.044	L.F.		2.67		2.67	3.98
6320	4" x 4"	"	160	.050			3		3	4.48
6330	6" x 6"	2 Elec	240	.067			4		4	5.95
6340	8" x 8"		160	.100			6		6	8.95
6350	10" x 10"		120	.133			8		8	11.95
6360	12" x 12"		80	.200			12		12	17.90
6400	Cable reel with receptacle, 120 V/208 V		13	1.231	Ea.		74		74	110
6500	Equipment connection, to 10 HP	1 Elec	23	.348			21		21	31
6510	To 15 to 30 HP		18	.444			26.50		26.50	40
6520	To 40 to 60 HP		15	.533			32		32	48
6530	To 75 to 100 HP		10	.800			48		48	71.50
7008	Receptacle, explosion proof, 120 V, to 30 A		16	.500			30		30	45
7020	Dimmer switch, 2,000 W or less		62	.129			7.75		7.75	11.55

26 05 05.25 Electrical Demolition, Electrical Power

		Crew	Daily Output	Labor-Hours	Unit	Material	2019 Bare Costs Labor	2019 Bare Costs Equipment	Total	Total Incl O&P
0010	**ELECTRICAL DEMOLITION, ELECTRICAL POWER** R024119-10									
0100	Meter centers and sockets									
0120	Meter socket, 4 terminal R260105-30	1 Elec	8.40	.952	Ea.		57		57	85.50
0140	Trans-socket, 13 terminal, 400 A	"	3.60	2.222			133		133	199
0160	800 A	2 Elec	4.40	3.636			218		218	325
0200	Meter center, 400 A		5.80	2.759			166		166	247
0210	600 A		4	4			240		240	360
0220	800 A		3.30	4.848			291		291	435
0230	1,200 A		2.80	5.714			345		345	510
0240	1,600 A		2.50	6.400			385		385	575
0300	Base meter devices, 3 meter		3.60	4.444			267		267	400
0310	4 meter		3.30	4.848			291		291	435
0320	5 meter		2.90	5.517			330		330	495
0330	6 meter		2.20	7.273			435		435	650
0340	7 meter		2	8			480		480	715
0350	8 meter		1.90	8.421			505		505	755
0400	Branch meter devices									
0410	Socket w/circuit breaker 200 A, 2 meter	2 Elec	3.30	4.848	Ea.		291		291	435
0420	3 meter		2.90	5.517			330		330	495
0430	4 meter		2.60	6.154			370		370	550
0450	Main circuit breaker, 400 A		5.80	2.759			166		166	247
0460	600 A		4	4			240		240	360
0470	800 A		3.30	4.848			291		291	435
0480	1,200 A		2.80	5.714			345		345	510
0490	1,600 A		2.50	6.400			385		385	575
0500	Main lug terminal box, 800 A		3.40	4.706			283		283	420
0510	1,200 A		2.60	6.154			370		370	550
1000	Motors, 230/460 V, 60 Hz, 3/4 HP	1 Elec	10.70	.748			45		45	67
1010	5 HP		9	.889			53.50		53.50	79.50
1020	10 HP		8	1			60		60	89.50
1030	15 HP		6.40	1.250			75		75	112
1040	20 HP	2 Elec	10.40	1.538			92.50		92.50	138
1050	50 HP		9.60	1.667			100		100	149
1060	75 HP		5.60	2.857			172		172	256
1070	100 HP	3 Elec	5.40	4.444			267		267	400
1080	150 HP		3.60	6.667			400		400	595

For customer support on your Electrical Costs with RSMeans data, call 800.448.8182.

107

26 05 05.25 Electrical Demolition, Electrical Power	Crew	Daily Output	Labor-Hours	Unit	Material	2019 Bare Costs Labor	Equipment	Total	Total Incl O&P
1090	200 HP	3 Elec	3	8	Ea.		480		480	715
1200	Variable frequency drive, 460 V, for 5 HP motor size	1 Elec	3.20	2.500			150		150	224
1210	10 HP motor size	"	2.70	2.963			178		178	265
1220	20 HP motor size	2 Elec	3.60	4.444			267		267	400
1230	50 HP motor size	"	2.10	7.619			460		460	685
1240	75 HP motor size	R-3	2.20	9.091			545	59	604	875
1250	100 HP motor size		2	10			595	65	660	960
1260	150 HP motor size		2	10			595	65	660	960
1270	200 HP motor size	↓	1.70	11.765	↓		705	76.50	781.50	1,125
2000	Generator set w/accessories, 3 phase 4 wire, 277/480 V									
2010	7.5 kW	3 Elec	2	12	Ea.		720		720	1,075
2020	20 kW		1.70	14.118			850		850	1,275
2040	30 kW		1.33	18.045			1,075		1,075	1,625
2060	50 kW		1	24			1,450		1,450	2,150
2080	100 kW		.75	32			1,925		1,925	2,875
2100	150 kW		.63	38.095			2,300		2,300	3,425
2120	250 kW		.59	40.678			2,450		2,450	3,650
2140	400 kW		.50	48			2,875		2,875	4,300
2160	500 kW	↓	.44	54.545			3,275		3,275	4,875
2180	750 kW	4 Elec	.56	57.143			3,425		3,425	5,125
2200	1,000 kW	"	.46	69.565	↓		4,175		4,175	6,225
3000	Uninterruptible power supply system (UPS)									
3010	Single phase, 120 V, 1 kVA	1 Elec	3.20	2.500	Ea.		150		150	224
3020	2 kVA	2 Elec	3.60	4.444			267		267	400
3030	5 kVA	3 Elec	2.50	9.600			575		575	860
3040	10 kVA	↓	2.30	10.435			625		625	935
3050	15 kVA	↓	1.80	13.333	↓		800		800	1,200
4000	Transformer, incl. support, wire & conduit termination									
4010	Buck-boost, single phase, 120/240 V, 0.1 kVA	1 Elec	25	.320	Ea.		19.20		19.20	28.50
4020	0.5 kVA		12.50	.640			38.50		38.50	57.50
4030	1 kVA		6.30	1.270			76.50		76.50	114
4040	5 kVA		3.80	2.105			126		126	189
4800	5 kV or 15 kV primary, 277/480 V second, 112.5 kVA	R-3	3.50	5.714			340	37	377	550
4810	150 kVA		2.70	7.407			440	48	488	715
4820	225 kVA		2.30	8.696			520	56.50	576.50	835
4830	500 kVA		1.45	13.793			825	89.50	914.50	1,325
4840	750 kVA		1.33	15.038			900	97.50	997.50	1,450
4850	1,000 kVA		1.25	16			955	104	1,059	1,550
4860	2,000 kVA		1.05	19.048			1,150	124	1,274	1,825
4870	3,000 kVA	↓	.75	26.667			1,600	173	1,773	2,575
6010	Isolation panel, 3 kVA	1 Elec	2.30	3.478			209		209	310
6020	5 kVA		2.20	3.636			218		218	325
6030	7.5 kVA		2.10	3.810			229		229	340
6040	10 kVA		1.80	4.444			267		267	400
6050	15 kVA	↓	1.40	5.714	↓		345		345	510
7000	Power filters & conditioners									
7100	Automatic voltage regulator	2 Elec	6	2.667	Ea.		160		160	239
7200	Capacitor, 1 kVAR	1 Elec	8.60	.930			56		56	83.50
7210	5 kVAR		5.80	1.379			83		83	124
7220	10 kVAR		4.80	1.667			100		100	149
7230	15 kVAR		4.20	1.905			114		114	171
7240	20 kVAR		3.50	2.286			137		137	205
7250	30 kVAR		3.40	2.353	↓		141		141	211

For customer support on your Electrical Costs with RSMeans data, call 800.448.8182.

26 05 05 – Selective Demolition for Electrical

26 05 05.25 Electrical Demolition, Electrical Power	Crew	Daily Output	Labor-Hours	Unit	Material	2019 Bare Costs Labor	Equipment	Total	Total Incl O&P
7260	50 kVAR	1 Elec	3.20	2.500	Ea.		150		150	224
7400	Computer isolator transformer									
7410	Single phase, 120/240 V, 0.5 kVAR	1 Elec	12.80	.625	Ea.		37.50		37.50	56
7420	1 kVAR		8.50	.941			56.50		56.50	84.50
7430	2.5 kVAR		6.40	1.250			75		75	112
7440	5 kVAR		3.70	2.162			130		130	194
7500	Computer regulator transformer									
7510	Single phase, 240 V, 0.5 kVAR	1 Elec	8.50	.941	Ea.		56.50		56.50	84.50
7520	1 kVAR		6.40	1.250			75		75	112
7530	2 kVAR		3.20	2.500			150		150	224
7540	Single phase, plug-in unit 120 V, 0.5 kVAR		26	.308			18.50		18.50	27.50
7550	1 kVAR		17	.471			28.50		28.50	42
7600	Power conditioner transformer									
7610	Single phase 115 V - 240 V, 3 kVA	2 Elec	5.10	3.137	Ea.		188		188	281
7620	5 kVA	"	3.70	4.324			260		260	385
7630	7.5 kVA	3 Elec	4.80	5			300		300	450
7640	10 kVA	"	4.30	5.581			335		335	500
7700	Transient suppressor/voltage regulator									
7710	Single phase 115 or 220 V, 1 kVA	1 Elec	8.50	.941	Ea.		56.50		56.50	84.50
7720	2 kVA		7.30	1.096			66		66	98
7730	4 kVA		6.80	1.176			70.50		70.50	105
7800	Transient voltage suppressor transformer									
7810	Single phase, 115 or 220 V, 3.6 kVA	1 Elec	12.80	.625	Ea.		37.50		37.50	56
7820	7.2 kVA		11.50	.696			42		42	62.50
7830	14.4 kVA		10.20	.784			47		47	70.50
7840	Single phase, plug-in, 120 V, 1.8 kVA		26	.308			18.50		18.50	27.50
8000	Power measurement & control									
8010	Switchboard instruments, 3 phase 4 wire, indicating unit	1 Elec	25	.320	Ea.		19.20		19.20	28.50
8020	Recording unit		12	.667			40		40	59.50
8100	3 current transformers, 3 phase 4 wire, 5 to 800 A		6.40	1.250			75		75	112
8110	1,000 to 1,500 A		4.20	1.905			114		114	171
8120	2,000 to 4,000 A		3.20	2.500			150		150	224

26 05 05.30 Electrical Demolition, Transmission and Distribution

		Crew	Daily Output	Labor-Hours	Unit	Material	2019 Bare Costs Labor	Equipment	Total	Total Incl O&P
0010	**ELECTRICAL DEMOLITION, TRANSMISSION & DISTRIBUTION**									
0100	Load interrupter switch, 600 A, NEMA 1, 4.8 kV	R-3	1.33	15.038	Ea.		900	97.50	997.50	1,450
0120	13.8 kV	R024119-10	"	1.27	15.748			940	102	1,042	1,500
0200	Lightning arrester, 4.8 kV		1 Elec	9	.889			53.50		53.50	79.50
0220	13.8 kV	R260105-30		6.67	1.199			72		72	107
0300	Alarm or option items			3.33	2.402			144		144	215

26 05 05.35 Electrical Demolition, L.V. Distribution

		Crew	Daily Output	Labor-Hours	Unit	Material	2019 Bare Costs Labor	Equipment	Total	Total Incl O&P
0010	**ELECTRICAL DEMOLITION, L.V. DISTRIBUTION** Addition	R024119-10									
0100	Circuit breakers in enclosure										
0120	Enclosed (NEMA 1), 600 V, 3 pole, 30 A	R260105-30	1 Elec	12.30	.650	Ea.		39		39	58.50
0140	60 A			8.80	.909			54.50		54.50	81.50
0160	100 A			7.30	1.096			66		66	98
0180	225 A			5	1.600			96		96	143
0200	400 A		2 Elec	6.80	2.353			141		141	211
0220	600 A			4.60	3.478			209		209	310
0240	800 A			3.60	4.444			267		267	400
0260	1,200 A			3.10	5.161			310		310	460
0280	1,600 A			2.80	5.714			345		345	510
0300	2,000 A			2.50	6.400			385		385	575

For customer support on your Electrical Costs with RSMeans data, call 800.448.8182.

109

26 05 05.35 Electrical Demolition, L.V. Distribution		Crew	Daily Output	Labor-Hours	Unit	Material	2019 Bare Costs Labor	2019 Bare Costs Equipment	Total	Total Incl O&P
0400	Enclosed (NEMA 7), 600 V, 3 pole, 50 A	1 Elec	8.80	.909	Ea.		54.50		54.50	81.50
0410	100 A		5.80	1.379			83		83	124
0420	150 A		3.80	2.105			126		126	189
0430	250 A	2 Elec	6.20	2.581			155		155	231
0440	400 A	"	4.60	3.478			209		209	310
0460	Manual motor starter, NEMA 1	1 Elec	24	.333			20		20	30
0480	NEMA 4 or NEMA 7		15	.533			32		32	48
0500	Time switches, single pole single throw		15.40	.519			31		31	46.50
0520	Photo cell		30	.267			16		16	24
0540	Load management device, 4 loads		7.70	1.039			62.50		62.50	93
0550	8 loads		3.80	2.105			126		126	189
0560	Master light control panel	2 Elec	2	8			480		480	715
0600	Transfer switches, enclosed, 30 A	1 Elec	12	.667			40		40	59.50
0610	60 A		9.50	.842			50.50		50.50	75.50
0620	100 A		6.50	1.231			74		74	110
0630	150 A	2 Elec	12	1.333			80		80	119
0640	260 A		10	1.600			96		96	143
0660	400 A		8	2			120		120	179
0670	600 A		5	3.200			192		192	287
0680	800 A		4	4			240		240	360
0690	1,200 A		3.50	4.571			275		275	410
0700	1,600 A		3	5.333			320		320	480
0710	2,000 A		2.50	6.400			385		385	575
0800	Air terminal with base and connector	1 Elec	24	.333			20		20	30
1000	Enclosed controller, NEMA 1, 30 A		13.80	.580			35		35	52
1020	60 A		11.50	.696			42		42	62.50
1040	100 A		9.60	.833			50		50	74.50
1060	150 A		7.70	1.039			62.50		62.50	93
1080	200 A		5.40	1.481			89		89	133
1100	400 A	2 Elec	6.90	2.319			139		139	208
1120	600 A		4.60	3.478			209		209	310
1140	800 A		3.80	4.211			253		253	375
1160	1,200 A		3.10	5.161			310		310	460
1200	Control station, NEMA 1	1 Elec	30	.267			16		16	24
1220	NEMA 7		23	.348			21		21	31
1230	Control switches, push button		69	.116			6.95		6.95	10.40
1240	Indicating light unit		120	.067			4		4	5.95
1250	Relay		15	.533			32		32	48
2000	Motor control center components									
2020	Starter, NEMA 1, size 1	1 Elec	9	.889	Ea.		53.50		53.50	79.50
2040	Size 2	2 Elec	13.30	1.203			72		72	108
2060	Size 3		6.70	2.388			143		143	214
2080	Size 4		5.30	3.019			181		181	270
2100	Size 5		3.30	4.848			291		291	435
2120	NEMA 7, size 1	1 Elec	8.70	.920			55		55	82.50
2140	Size 2	2 Elec	12.70	1.260			75.50		75.50	113
2160	Size 3		6.30	2.540			153		153	228
2180	Size 4		5	3.200			192		192	287
2200	Size 5		3.20	5			300		300	450
2300	Fuse, light contactor, NEMA 1, 30 A	1 Elec	9	.889			53.50		53.50	79.50
2310	60 A		6.70	1.194			71.50		71.50	107
2320	100 A		3.30	2.424			146		146	217
2330	200 A		2.70	2.963			178		178	265

26 05 05.35 Electrical Demolition, L.V. Distribution	Crew	Daily Output	Labor-Hours	Unit	Material	2019 Bare Costs Labor	2019 Bare Costs Equipment	Total	Total Incl O&P
2350	Motor control center, incoming section	2 Elec	4	4	Ea.		240		240	360
2352	Structure per section	"	4.20	3.810			229		229	340
2400	Starter & structure, 10 HP	1 Elec	9	.889			53.50		53.50	79.50
2410	25 HP	2 Elec	13.30	1.203			72		72	108
2420	50 HP		6.70	2.388			143		143	214
2430	75 HP		5.30	3.019			181		181	270
2440	100 HP		4.70	3.404			204		204	305
2450	200 HP		3.30	4.848			291		291	435
2460	400 HP		2.70	5.926			355		355	530
2500	Motor starter & control									
2510	Motor starter, NEMA 1, 5 HP	1 Elec	8.90	.899	Ea.		54		54	80.50
2520	10 HP	"	6.20	1.290			77.50		77.50	116
2530	25 HP	2 Elec	8.40	1.905			114		114	171
2540	50 HP		6.90	2.319			139		139	208
2550	100 HP		4.60	3.478			209		209	310
2560	200 HP		3.50	4.571			275		275	410
2570	400 HP		3.10	5.161			310		310	460
2610	Motor starter, NEMA 7, 5 HP	1 Elec	6.20	1.290			77.50		77.50	116
2620	10 HP	"	4.20	1.905			114		114	171
2630	25 HP	2 Elec	6.90	2.319			139		139	208
2640	50 HP		4.60	3.478			209		209	310
2650	100 HP		3.50	4.571			275		275	410
2660	200 HP		1.90	8.421			505		505	755
2710	Combination control unit, NEMA 1, 5 HP	1 Elec	6.90	1.159			69.50		69.50	104
2720	10 HP	"	5	1.600			96		96	143
2730	25 HP	2 Elec	7.70	2.078			125		125	186
2740	50 HP		5.10	3.137			188		188	281
2750	100 HP		3.10	5.161			310		310	460
2810	NEMA 7, 5 HP	1 Elec	5	1.600			96		96	143
2820	10 HP	"	3.90	2.051			123		123	184
2830	25 HP	2 Elec	5.10	3.137			188		188	281
2840	50 HP		3.10	5.161			310		310	460
2850	100 HP		2.30	6.957			420		420	625
2860	200 HP		1.50	10.667			640		640	955
3000	Panelboard or load center circuit breaker									
3010	Bolt-on or plug in, 15 A to 50 A	1 Elec	20	.400	Ea.		24		24	36
3020	60 A to 70 A		16	.500			30		30	45
3030	Bolt-on, 80 A to 100 A		14	.571			34.50		34.50	51
3040	Up to 250 A		7.30	1.096			66		66	98
3050	Motor operated, 30 A		13	.615			37		37	55
3060	60 A		10	.800			48		48	71.50
3070	100 A		8	1			60		60	89.50
3200	Switchboard circuit breaker									
3210	15 A to 60 A	1 Elec	19	.421	Ea.		25.50		25.50	37.50
3220	70 A to 100 A		14	.571			34.50		34.50	51
3230	125 A to 400 A		10	.800			48		48	71.50
3240	450 A to 600 A		5.30	1.509			90.50		90.50	135
3250	700 A to 800 A		4.30	1.860			112		112	167
3260	1,000 A		3.30	2.424			146		146	217
3270	1,200 A		2.70	2.963			178		178	265
3500	Switchboard, incoming section, 400 A	2 Elec	3.70	4.324			260		260	385
3510	600 A		3.30	4.848			291		291	435
3520	800 A		2.90	5.517			330		330	495

26 05 05.35 Electrical Demolition, L.V. Distribution

		Crew	Daily Output	Labor-Hours	Unit	Material	2019 Bare Costs Labor	Equipment	Total	Total Incl O&P
3530	1,200 A	2 Elec	2.40	6.667	Ea.		400		400	595
3540	1,600 A		2.20	7.273			435		435	650
3550	2,000 A		2.10	7.619			460		460	685
3560	3,000 A		1.90	8.421			505		505	755
3570	4,000 A		1.70	9.412			565		565	845
3610	Distribution section, 600 A		4	4			240		240	360
3620	800 A		3.60	4.444			267		267	400
3630	1,200 A		3.10	5.161			310		310	460
3640	1,600 A		2.90	5.517			330		330	495
3650	2,000 A		2.70	5.926			355		355	530
3710	Transition section, 600 A		3.80	4.211			253		253	375
3720	800 A		3.30	4.848			291		291	435
3730	1,200 A		2.70	5.926			355		355	530
3740	1,600 A		2.40	6.667			400		400	595
3750	2,000 A		2.20	7.273			435		435	650
3760	2,500 A		2.10	7.619			460		460	685
3770	3,000 A		1.90	8.421			505		505	755
4000	Bus duct, aluminum or copper, 30 A	1 Elec	200	.040	L.F.		2.40		2.40	3.58
4020	60 A		160	.050			3		3	4.48
4040	100 A		140	.057			3.43		3.43	5.10
5000	Feedrail, trolley busway, up to 60 A		160	.050			3		3	4.48
5020	100 A		120	.067			4		4	5.95
5040	Busway, 50 A		200	.040			2.40		2.40	3.58
6000	Fuse, 30 A		133	.060	Ea.		3.61		3.61	5.40
6010	60 A		133	.060			3.61		3.61	5.40
6020	100 A		105	.076			4.58		4.58	6.85
6030	200 A		95	.084			5.05		5.05	7.55
6040	400 A		80	.100			6		6	8.95
6050	600 A		53	.151			9.05		9.05	13.50
6060	601 A to 1,200 A		42	.190			11.45		11.45	17.05
6070	1,500 A to 1,600 A		34	.235			14.15		14.15	21
6080	1,800 A to 2,500 A		26	.308			18.50		18.50	27.50
6090	4,000 A		21	.381			23		23	34
6100	4,500 A to 5,000 A		18	.444			26.50		26.50	40
6120	6,000 A		15	.533			32		32	48
6200	Fuse plug or fustat		100	.080			4.80		4.80	7.15

26 05 05.50 Electrical Demolition, Lighting

		Crew	Daily Output	Labor-Hours	Unit	Material	2019 Bare Costs Labor	Equipment	Total	Total Incl O&P
0010	**ELECTRICAL DEMOLITION, LIGHTING** Addition R024119-10									
0100	Fixture hanger, flexible, 1/2" diameter	1 Elec	36	.222	Ea.		13.35		13.35	19.90
0120	3/4" diameter R260105-30	"	30	.267	"		16		16	24
3000	Light pole, anchor base, excl. concrete bases									
3010	Metal light pole, 10'	2 Elec	24	.667	Ea.		40		40	59.50
3020	16'	"	18	.889			53.50		53.50	79.50
3030	20'	R-3	8.70	2.299			137	14.90	151.90	221
3035	25'		8.60	2.326			139	15.10	154.10	225
3038	30'		6	3.333			199	21.50	220.50	320
3039	35'		6	3.333			199	21.50	220.50	320
3040	40'		6	3.333			199	21.50	220.50	320
3100	Wood light pole, 10'	2 Elec	36	.444			26.50		26.50	40
3120	20'	"	24	.667			40		40	59.50
3140	Bollard light, 42"	1 Elec	9	.889			53.50		53.50	79.50
3160	Walkway luminaire	"	8.10	.988			59.50		59.50	88.50

26 05 Common Work Results for Electrical

26 05 05 – Selective Demolition for Electrical

26 05 05.50 Electrical Demolition, Lighting

26 05 05.50 Electrical Demolition, Lighting	Crew	Daily Output	Labor-Hours	Unit	Material	2019 Bare Costs Labor	2019 Bare Costs Equipment	Total	Total Incl O&P
4000	Explosion proof									
4010	Metal halide, 175 W	1 Elec	8.70	.920	Ea.		55		55	82.50
4020	250 W	"	8.10	.988			59.50		59.50	88.50
4030	400 W	2 Elec	14.40	1.111			66.50		66.50	99.50
4050	High pressure sodium, 70 W	1 Elec	9	.889			53.50		53.50	79.50
4060	100 W		9	.889			53.50		53.50	79.50
4070	150 W		8.10	.988			59.50		59.50	88.50
4100	Incandescent		8.70	.920			55		55	82.50
4200	Fluorescent		8.10	.988			59.50		59.50	88.50
5000	Ballast, fluorescent fixture		24	.333			20		20	30
5040	High intensity discharge fixture		24	.333			20		20	30
5300	Exit and emergency lighting									
5310	Exit light	1 Elec	24	.333	Ea.		20		20	30
5320	Emergency battery pack lighting unit		12	.667			40		40	59.50
5330	Remote lamp only		80	.100			6		6	8.95
5340	Self-contained fluorescent lamp pack		30	.267			16		16	24
5500	Track lighting, 8' section	2 Elec	40	.400			24		24	36
5510	Track lighting fixture	1 Elec	64	.125			7.50		7.50	11.20
5800	Energy saving devices									
5810	Occupancy sensor	1 Elec	21	.381	Ea.		23		23	34
5820	Automatic wall switch		72	.111			6.65		6.65	9.95
5830	Remote power pack		30	.267			16		16	24
5840	Photoelectric control		24	.333			20		20	30
5850	Fixture whip		100	.080			4.80		4.80	7.15
6000	Lamps									
6010	Fluorescent	1 Elec	200	.040	Ea.		2.40		2.40	3.58
6030	High intensity discharge lamp, up to 400 W		68	.118			7.05		7.05	10.55
6040	Up to 1,000 W		45	.178			10.70		10.70	15.95
6050	Quartz		90	.089			5.35		5.35	7.95
6070	Incandescent		360	.022			1.33		1.33	1.99
6080	Exterior, PAR		290	.028			1.66		1.66	2.47
6090	Guards for fluorescent lamp		24	.333			20		20	30

26 05 13 – Medium-Voltage Cables

26 05 13.10 Cable Terminations

		Crew	Daily Output	Labor-Hours	Unit	Material	2019 Bare Costs Labor	2019 Bare Costs Equipment	Total	Total Incl O&P
0010	**CABLE TERMINATIONS**, 5 kV to 35 kV									
0100	Indoor, insulation diameter range 0.64" to 1.08"									
0300	Padmount, 5 kV	1 Elec	8	1	Ea.	80.50	60		140.50	179
0400	10 kV		6.40	1.250		98.50	75		173.50	221
0500	15 kV		6	1.333		126	80		206	257
0600	25 kV		5.60	1.429		181	86		267	325
0700	Insulation diameter range 1.05" to 1.8"									
0800	Padmount, 5 kV	1 Elec	8	1	Ea.	43.50	60		103.50	138
0900	10 kV		6	1.333		138	80		218	271
1000	15 kV		5.60	1.429		149	86		235	292
1100	25 kV		5.30	1.509		207	90.50		297.50	360
1200	Insulation diameter range 1.53" to 2.32"									
1300	Padmount, 5 kV	1 Elec	7.40	1.081	Ea.	141	65		206	252
1400	10 kV		5.60	1.429		159	86		245	305
1500	15 kV		5.30	1.509		175	90.50		265.50	325
1600	25 kV		5	1.600		298	96		394	470
1700	Outdoor systems, #4 stranded to 1/0 stranded									
1800	5 kV	1 Elec	7.40	1.081	Ea.	94.50	65		159.50	201

For customer support on your Electrical Costs with RSMeans data, call 800.448.8182.

113

26 05 13 – Medium-Voltage Cables

26 05 13.10 Cable Terminations		Crew	Daily Output	Labor-Hours	Unit	Material	2019 Bare Costs Labor	Equipment	Total	Total Incl O&P
1900	15 kV	1 Elec	5.30	1.509	Ea.	120	90.50		210.50	267
2000	25 kV		5	1.600		179	96		275	340
2100	35 kV		4.80	1.667		325	100		425	505
2200	#1 solid to 4/0 stranded, 5 kV		6.90	1.159		124	69.50		193.50	241
2300	15 kV		5	1.600		170	96		266	330
2400	25 kV		4.80	1.667		232	100		332	405
2500	35 kV		4.60	1.739		325	104		429	515
2600	2/0 solid to 350 kcmil stranded, 5 kV		6.40	1.250		178	75		253	310
2700	15 kV		4.80	1.667		192	100		292	360
2800	25 kV		4.60	1.739		298	104		402	485
2900	35 kV		4.40	1.818		395	109		504	595
3000	400 kcmil compact to 750 kcmil stranded, 5 kV		6	1.333		170	80		250	305
3100	15 kV		4.60	1.739		232	104		336	410
3200	25 kV		4.40	1.818		340	109		449	540
3300	35 kV		4.20	1.905		345	114		459	545
3400	1,000 kcmil, 5 kV		5.60	1.429		241	86		327	395
3500	15 kV		4.40	1.818		315	109		424	510
3600	25 kV		4.20	1.905		345	114		459	545
3700	35 kV	▼	4	2	▼	500	120		620	730

26 05 13.16 Medium-Voltage, Single Cable

		Crew	Daily Output	Labor-Hours	Unit	Material	2019 Bare Costs Labor	Equipment	Total	Total Incl O&P
0010	**MEDIUM-VOLTAGE, SINGLE CABLE** Splicing & terminations not included									
0040	Copper, XLP shielding, 5 kV, #6	2 Elec	4.40	3.636	C.L.F.	162	218		380	505
0050	#4		4.40	3.636		210	218		428	555
0100	#2		4	4		268	240		508	655
0200	#1		4	4		325	240		565	715
0400	1/0		3.80	4.211		350	253		603	760
0600	2/0		3.60	4.444		430	267		697	875
0800	4/0	▼	3.20	5		575	300		875	1,075
1000	250 kcmil	3 Elec	4.50	5.333		690	320		1,010	1,250
1200	350 kcmil		3.90	6.154		895	370		1,265	1,525
1400	500 kcmil	▼	3.60	6.667		1,150	400		1,550	1,875
1600	15 kV, ungrounded neutral, #1	2 Elec	4	4		345	240		585	740
1800	1/0		3.80	4.211		415	253		668	830
2000	2/0		3.60	4.444		470	267		737	920
2200	4/0	▼	3.20	5		625	300		925	1,150
2400	250 kcmil	3 Elec	4.50	5.333		690	320		1,010	1,250
2600	350 kcmil		3.90	6.154		875	370		1,245	1,525
2800	500 kcmil	▼	3.60	6.667		1,075	400		1,475	1,775
3000	25 kV, grounded neutral, 1/0	2 Elec	3.60	4.444		570	267		837	1,025
3200	2/0		3.40	4.706		630	283		913	1,100
3400	4/0	▼	3	5.333		790	320		1,110	1,350
3600	250 kcmil	3 Elec	4.20	5.714		980	345		1,325	1,575
3800	350 kcmil		3.60	6.667		1,150	400		1,550	1,875
3900	500 kcmil	▼	3.30	7.273		1,350	435		1,785	2,125
4000	35 kV, grounded neutral, 1/0	2 Elec	3.40	4.706		620	283		903	1,100
4200	2/0		3.20	5		710	300		1,010	1,225
4400	4/0	▼	2.80	5.714		895	345		1,240	1,500
4600	250 kcmil	3 Elec	3.90	6.154		1,050	370		1,420	1,700
4800	350 kcmil		3.30	7.273		1,250	435		1,685	2,025
5000	500 kcmil	▼	3	8		1,475	480		1,955	2,350
5050	Aluminum, XLP shielding, 5 kV, #2	2 Elec	5	3.200		222	192		414	530
5070	#1	↓	4.40	3.636	▼	230	218		448	580

114

26 05 Common Work Results for Electrical

26 05 13 – Medium-Voltage Cables

26 05 13.16 Medium-Voltage, Single Cable

		Crew	Daily Output	Labor-Hours	Unit	Material	2019 Bare Costs Labor	Equipment	Total	Total Incl O&P
5090	1/0	2 Elec	4	4	C.L.F.	266	240		506	650
5100	2/0		3.80	4.211		298	253		551	705
5150	4/0		3.60	4.444		355	267		622	790
5200	250 kcmil	3 Elec	4.80	5		425	300		725	920
5220	350 kcmil		4.50	5.333		505	320		825	1,025
5240	500 kcmil		3.90	6.154		645	370		1,015	1,250
5260	750 kcmil		3.60	6.667		855	400		1,255	1,525
5300	15 kV aluminum, XLP, #1	2 Elec	4.40	3.636		285	218		503	640
5320	1/0		4	4		294	240		534	685
5340	2/0		3.80	4.211		320	253		573	725
5360	4/0		3.60	4.444		390	267		657	830
5380	250 kcmil	3 Elec	4.80	5		465	300		765	960
5400	350 kcmil		4.50	5.333		520	320		840	1,050
5420	500 kcmil		3.90	6.154		720	370		1,090	1,350
5440	750 kcmil		3.60	6.667		960	400		1,360	1,650

26 05 19 – Low-Voltage Electrical Power Conductors and Cables

26 05 19.13 Undercarpet Electrical Power Cables

		Crew	Daily Output	Labor-Hours	Unit	Material	2019 Bare Costs Labor	Equipment	Total	Total Incl O&P
0010	**UNDERCARPET ELECTRICAL POWER CABLES** R260519-80									
0020	Power System									
0100	Cable flat, 3 conductor, #12, w/attached bottom shield	1 Elec	982	.008	L.F.	5.35	.49		5.84	6.60
0200	Shield, top, steel		1768	.005	"	4.83	.27		5.10	5.70
0250	Splice, 3 conductor		48	.167	Ea.	14.60	10		24.60	31
0300	Top shield		96	.083		1.35	5		6.35	8.95
0350	Tap		40	.200		18.75	12		30.75	38.50
0400	Insulating patch, splice, tap & end		48	.167		47	10		57	66.50
0450	Fold		230	.035			2.09		2.09	3.12
0500	Top shield, tap & fold		96	.083		1.35	5		6.35	8.95
0700	Transition, block assembly		77	.104		68.50	6.25		74.75	84.50
0750	Receptacle frame & base		32	.250		37.50	15		52.50	63.50
0800	Cover receptacle		120	.067		3.17	4		7.17	9.45
0850	Cover blank		160	.050		3.75	3		6.75	8.60
0860	Receptacle, direct connected, single		25	.320		80.50	19.20		99.70	117
0870	Dual		16	.500		132	30		162	190
0880	Combination high & low, tension		21	.381		97	23		120	141
0900	Box, floor with cover		20	.400		81.50	24		105.50	126
0920	Floor service w/barrier		4	2		230	120		350	430
1000	Wall, surface, with cover		20	.400		53.50	24		77.50	95
1100	Wall, flush, with cover		20	.400		37.50	24		61.50	77
1450	Cable flat, 5 conductor #12, w/attached bottom shield		800	.010	L.F.	8.80	.60		9.40	10.55
1550	Shield, top, steel		1768	.005	"	8.70	.27		8.97	9.95
1600	Splice, 5 conductor		48	.167	Ea.	23.50	10		33.50	41
1650	Top shield		96	.083		1.35	5		6.35	8.95
1700	Tap		48	.167		31.50	10		41.50	49.50
1750	Insulating patch, splice tap & end		83	.096		55	5.80		60.80	69
1800	Transition, block assembly		77	.104		49	6.25		55.25	63.50
1850	Box, wall, flush with cover		20	.400		50	24		74	91
1900	Cable flat, 4 conductor #12		933	.009	L.F.	6.25	.51		6.76	7.60
1950	3 conductor #10		982	.008		5.95	.49		6.44	7.30
1960	4 conductor #10		933	.009		7.80	.51		8.31	9.35
1970	5 conductor #10		884	.009		9.55	.54		10.09	11.30
2500	Telephone System									
2510	Transition fitting wall box, surface	1 Elec	24	.333	Ea.	61.50	20		81.50	97.50

26 05 19.13 Undercarpet Electrical Power Cables	Crew	Daily Output	Labor-Hours	Unit	Material	2019 Bare Costs Labor	2019 Bare Costs Equipment	Total	Total Incl O&P
2520	Flush	1 Elec	24	.333	Ea.	61.50	20		81.50	97.50
2530	Flush, for PC board		24	.333		61.50	20		81.50	97.50
2540	Floor service box		4	2		239	120		359	440
2550	Cover, surface					20.50			20.50	22.50
2560	Flush					20.50			20.50	22.50
2570	Flush for PC board					20.50			20.50	22.50
2700	Floor fitting w/duplex jack & cover	1 Elec	21	.381		46	23		69	84.50
2720	Low profile		53	.151		17.20	9.05		26.25	32.50
2740	Miniature w/duplex jack		53	.151		32.50	9.05		41.55	49.50
2760	25 pair kit		21	.381		49.50	23		72.50	88.50
2780	Low profile		53	.151		17.55	9.05		26.60	33
2800	Call director kit for 5 cable		19	.421		91.50	25.50		117	139
2820	4 pair kit		19	.421		94	25.50		119.50	141
2840	3 pair kit		19	.421		116	25.50		141.50	165
2860	Comb. 25 pair & 3 conductor power		21	.381		91.50	23		114.50	135
2880	5 conductor power		21	.381		90	23		113	133
2900	PC board, 8 per 3 pair		161	.050		68.50	2.98		71.48	79.50
2920	6 per 4 pair		161	.050		68.50	2.98		71.48	79.50
2940	3 pair adapter		161	.050		62	2.98		64.98	73
2950	Plug		77	.104		3.04	6.25		9.29	12.65
2960	Couplers		321	.025		8.55	1.50		10.05	11.65
3000	Bottom shield for 25 pair cable		4420	.002	L.F.	.86	.11		.97	1.11
3020	4 pair		4420	.002		.40	.11		.51	.60
3040	Top shield for 25 pair cable		4420	.002		.86	.11		.97	1.11
3100	Cable assembly, double-end, 50', 25 pair		11.80	.678	Ea.	241	40.50		281.50	325
3110	3 pair		23.60	.339		74.50	20.50		95	112
3120	4 pair		23.60	.339		91.50	20.50		112	132
3140	Bulk 3 pair		1473	.005	L.F.	1.27	.33		1.60	1.89
3160	4 pair		1473	.005	"	1.57	.33		1.90	2.22
3500	Data System									
3520	Cable 25 conductor w/connection 40', 75 ohm	1 Elec	14.50	.552	Ea.	82.50	33		115.50	141
3530	Single lead		22	.364		226	22		248	281
3540	Dual lead		22	.364		241	22		263	298
3560	Shields same for 25 conductor as 25 pair telephone									
3570	Single & dual, none required									
3590	BNC coax connectors, plug	1 Elec	40	.200	Ea.	14.45	12		26.45	34
3600	TNC coax connectors, plug	"	40	.200	"	15.55	12		27.55	35
3700	Cable-bulk									
3710	Single lead	1 Elec	1473	.005	L.F.	3.50	.33		3.83	4.34
3720	Dual lead	"	1473	.005	"	5.05	.33		5.38	6.10
3730	Hand tool crimp				Ea.	525			525	580
3740	Hand tool notch				"	19.55			19.55	21.50
3750	Boxes & floor fitting same as telephone									
3790	Data cable notching, 90°	1 Elec	97	.082	Ea.		4.95		4.95	7.40
3800	180°		60	.133			8		8	11.95
8100	Drill floor		160	.050		3.89	3		6.89	8.75
8200	Marking floor		1600	.005	L.F.		.30		.30	.45
8300	Tape, hold down		6400	.001	"	.17	.08		.25	.30
8350	Tape primer, 500' per can		96	.083	Ea.	54.50	5		59.50	67.50
8400	Tool, splicing				"	168			168	185

26 05 19.20 Armored Cable		Crew	Daily Output	Labor-Hours	Unit	Material	2019 Bare Costs Labor	Equipment	Total	Total Incl O&P
0010	**ARMORED CABLE** R260519-20									
0050	600 volt, copper (BX), #14, 2 conductor, solid	1 Elec	2.40	3.333	C.L.F.	40.50	200		240.50	345
0100	3 conductor, solid		2.20	3.636		63.50	218		281.50	395
0120	4 conductor, solid		2	4		89	240		329	460
0150	#12, 2 conductor, solid		2.30	3.478		41	209		250	355
0200	3 conductor, solid		2	4		70.50	240		310.50	440
0220	4 conductor, solid		1.80	4.444		95.50	267		362.50	505
0250	#10, 2 conductor, solid		2	4		79	240		319	445
0300	3 conductor, solid		1.60	5		108	300		408	570
0320	4 conductor, solid		1.40	5.714		177	345		522	705
0340	#8, 2 conductor, stranded		1.50	5.333		240	320		560	745
0350	3 conductor, stranded		1.30	6.154		240	370		610	815
0370	4 conductor, stranded		1.10	7.273		340	435		775	1,025
0380	#6, 2 conductor, stranded		1.30	6.154		251	370		621	825
0390	#4, 3 conductor, stranded		1.40	5.714		425	345		770	975
0400	3 conductor with PVC jacket, in cable tray, #6		3.10	2.581		535	155		690	815
0450	#4	2 Elec	5.40	2.963		655	178		833	985
0500	#2		4.60	3.478		810	209		1,019	1,200
0550	#1		4	4		1,025	240		1,265	1,475
0600	1/0		3.60	4.444		1,125	267		1,392	1,625
0650	2/0		3.40	4.706		1,350	283		1,633	1,900
0700	3/0		3.20	5		1,725	300		2,025	2,325
0750	4/0		3	5.333		2,100	320		2,420	2,775
0800	250 kcmil	3 Elec	3.60	6.667		2,700	400		3,100	3,575
0850	350 kcmil		3.30	7.273		3,250	435		3,685	4,225
0900	500 kcmil		3	8		4,500	480		4,980	5,675
0910	4 conductor with PVC jacket, in cable tray, #6	1 Elec	2.70	2.963		890	178		1,068	1,250
0920	#4	2 Elec	4.60	3.478		810	209		1,019	1,200
0930	#2		4	4		1,275	240		1,515	1,750
0940	#1		3.60	4.444		1,475	267		1,742	2,000
0950	1/0		3.40	4.706		1,400	283		1,683	1,975
0960	2/0		3.20	5		1,725	300		2,025	2,350
0970	3/0		3	5.333		2,275	320		2,595	2,975
0980	4/0		2.40	6.667		2,650	400		3,050	3,525
0990	250 kcmil	3 Elec	3.30	7.273		3,200	435		3,635	4,175
1000	350 kcmil		3	8		4,275	480		4,755	5,425
1010	500 kcmil		2.70	8.889		5,925	535		6,460	7,300
1050	5 kV, copper, 3 conductor with PVC jacket,									
1060	non-shielded, in cable tray, #4	2 Elec	380	.042	L.F.	7.35	2.53		9.88	11.85
1100	#2		360	.044		9.55	2.67		12.22	14.50
1200	#1		300	.053		12.15	3.20		15.35	18.20
1400	1/0		290	.055		14.05	3.31		17.36	20.50
1600	2/0		260	.062		16.25	3.70		19.95	23.50
2000	4/0		240	.067		21.50	4		25.50	30
2100	250 kcmil	3 Elec	330	.073		29.50	4.37		33.87	39
2150	350 kcmil		315	.076		36.50	4.58		41.08	47
2200	500 kcmil		270	.089		54.50	5.35		59.85	68
2400	15 kV, copper, 3 conductor with PVC jacket galv., steel armored									
2500	grounded neutral, in cable tray, #2	2 Elec	300	.053	L.F.	15.75	3.20		18.95	22
2600	#1		280	.057		16.45	3.43		19.88	23
2800	1/0		260	.062		19.20	3.70		22.90	26.50
2900	2/0		220	.073		25	4.37		29.37	34

For customer support on your Electrical Costs with RSMeans data, call 800.448.8182.

117

26 05 Common Work Results for Electrical

26 05 19 – Low-Voltage Electrical Power Conductors and Cables

26 05 19.20 Armored Cable		Crew	Daily Output	Labor-Hours	Unit	Material	2019 Bare Costs Labor	Equipment	Total	Total Incl O&P
3000	4/0	2 Elec	190	.084	L.F.	28.50	5.05		33.55	38.50
3100	250 kcmil	3 Elec	270	.089		31.50	5.35		36.85	43
3150	350 kcmil		240	.100		37	6		43	50
3200	500 kcmil	↓	210	.114	↓	50	6.85		56.85	65.50
3400	15 kV, copper, 3 conductor with PVC jacket,									
3450	ungrounded neutral, in cable tray, #2	2 Elec	260	.062	L.F.	16.65	3.70		20.35	24
3500	#1		230	.070		18.45	4.18		22.63	27
3600	1/0		200	.080		21	4.80		25.80	30
3700	2/0		190	.084		26	5.05		31.05	36
3800	4/0	↓	160	.100		31.50	6		37.50	43.50
4000	250 kcmil	3 Elec	210	.114		36.50	6.85		43.35	50.50
4050	350 kcmil		195	.123		48	7.40		55.40	64
4100	500 kcmil		180	.133	↓	58.50	8		66.50	76
4200	600 volt, aluminum, 3 conductor in cable tray with PVC jacket									
4300	#2	2 Elec	540	.030	L.F.	4.05	1.78		5.83	7.10
4400	#1		460	.035		4.79	2.09		6.88	8.35
4500	1/0		400	.040		5.95	2.40		8.35	10.15
4600	2/0		360	.044		6.05	2.67		8.72	10.65
4700	3/0		340	.047		7.10	2.83		9.93	12
4800	4/0	↓	320	.050		8.55	3		11.55	13.95
4900	250 kcmil	3 Elec	450	.053		10.30	3.20		13.50	16.10
5000	350 kcmil		360	.067		12.25	4		16.25	19.45
5200	500 kcmil		330	.073		15.25	4.37		19.62	23.50
5300	750 kcmil	↓	285	.084	↓	19.65	5.05		24.70	29
5400	600 volt, aluminum, 4 conductor in cable tray with PVC jacket									
5410	#2	2 Elec	520	.031	L.F.	4.73	1.85		6.58	7.95
5430	#1		440	.036		5.85	2.18		8.03	9.65
5450	1/0		380	.042		6.80	2.53		9.33	11.25
5470	2/0		340	.047		7	2.83		9.83	11.90
5480	3/0		320	.050		8.20	3		11.20	13.50
5500	4/0		300	.053		9.60	3.20		12.80	15.40
5520	250 kcmil	3 Elec	420	.057		10.30	3.43		13.73	16.45
5540	350 kcmil		330	.073		13.35	4.37		17.72	21
5560	500 kcmil		300	.080		16.70	4.80		21.50	25.50
5580	750 kcmil	↓	270	.089		23	5.35		28.35	33.50
5600	5 kV, aluminum, unshielded in cable tray, #2 with PVC jacket	2 Elec	380	.042		6.05	2.53		8.58	10.40
5700	#1 with PVC jacket		360	.044		6.75	2.67		9.42	11.45
5800	1/0 with PVC jacket		300	.053		6.95	3.20		10.15	12.40
6000	2/0 with PVC jacket		290	.055		7.10	3.31		10.41	12.75
6200	3/0 with PVC jacket		260	.062		8.50	3.70		12.20	14.85
6300	4/0 with PVC jacket	↓	240	.067		10	4		14	16.95
6400	250 kcmil with PVC jacket	3 Elec	330	.073		10.95	4.37		15.32	18.55
6500	350 kcmil with PVC jacket		315	.076		12.80	4.58		17.38	21
6600	500 kcmil with PVC jacket		300	.080		15.20	4.80		20	24
6800	750 kcmil with PVC jacket		270	.089		18.70	5.35		24.05	28.50
6900	15 kV, aluminum, shielded-grounded, #2 with PVC jacket	2 Elec	320	.050		13.75	3		16.75	19.65
7000	#1 with PVC jacket		300	.053		14.15	3.20		17.35	20.50
7200	1/0 with PVC jacket		280	.057		15.25	3.43		18.68	22
7300	2/0 with PVC jacket		260	.062		15.50	3.70		19.20	22.50
7400	3/0 with PVC jacket		240	.067		17.40	4		21.40	25
7500	4/0 with PVC jacket	↓	220	.073		17.85	4.37		22.22	26
7600	250 kcmil with PVC jacket	3 Elec	300	.080		19.60	4.80		24.40	28.50
7700	350 kcmil with PVC jacket	↓	270	.089	↓	23	5.35		28.35	33.50

118

26 05 19 – Low-Voltage Electrical Power Conductors and Cables

26 05 19.20 Armored Cable	Crew	Daily Output	Labor-Hours	Unit	Material	2019 Bare Costs Labor	Equipment	Total	Total Incl O&P
7800	500 kcmil with PVC jacket	3 Elec	240	.100	L.F.	27.50	6		33.50	39.50
8000	750 kcmil with PVC jacket	▼	204	.118		32.50	7.05		39.55	46.50
8200	15 kV, aluminum, shielded-ungrounded, #1 with PVC jacket	2 Elec	250	.064		17.25	3.84		21.09	24.50
8300	1/0 with PVC jacket		230	.070		17.85	4.18		22.03	26
8400	2/0 with PVC jacket		210	.076		19.60	4.58		24.18	28.50
8500	3/0 with PVC jacket		200	.080		19.90	4.80		24.70	29
8600	4/0 with PVC jacket	▼	190	.084		21.50	5.05		26.55	31.50
8700	250 kcmil with PVC jacket	3 Elec	270	.089		23	5.35		28.35	33.50
8800	350 kcmil with PVC jacket		240	.100		26.50	6		32.50	38
8900	500 kcmil with PVC jacket		210	.114		32	6.85		38.85	45.50
8950	750 kcmil with PVC jacket	▼	174	.138	▼	40	8.30		48.30	56.50
9010	600 volt, copper (MC) steel clad, #14, 2 wire	R-1A	8.16	1.961	C.L.F.	42	106		148	204
9020	3 wire		7.21	2.219		76	120		196	263
9030	4 wire		6.69	2.392		196	129		325	410
9040	#12, 2 wire		7.21	2.219		56	120		176	241
9050	3 wire		6.69	2.392		69	129		198	269
9060	4 wire		6.04	2.649		114	143		257	340
9070	#10, 2 wire		6.04	2.649		101	143		244	325
9080	3 wire		5.65	2.832		131	153		284	370
9090	4 wire		5.23	3.059		212	165		377	480
9100	#8, 2 wire, stranded		3.94	4.061		155	219		374	500
9110	3 wire, stranded		3.40	4.706		189	254		443	590
9120	4 wire, stranded		2.80	5.714		261	310		571	745
9130	#6, 2 wire, stranded		3.45	4.638		204	251		455	600
9200	600 volt, copper (MC) aluminum clad, #14, 2 wire		8.16	1.961		40	106		146	202
9210	3 wire		7.21	2.219		72.50	120		192.50	259
9220	4 wire		6.69	2.392		87	129		216	289
9230	#12, 2 wire		7.21	2.219		45	120		165	229
9240	3 wire		6.69	2.392		7.40	129		136.40	201
9250	4 wire		6.04	2.649		135	143		278	360
9260	#10, 2 wire		6.04	2.649		86.50	143		229.50	310
9270	3 wire		5.65	2.832		114	153		267	355
9280	4 wire		5.23	3.059		204	165		369	470
9600	Alum (MC) aluminum clad, #6, 3 conductor w/#6 grnd		2.75	5.818		292	315		607	790
9610	4 conductor w/#6 grnd	▼	2.29	6.987		222	380		602	810
9620	#4, 3 conductor w/#6 grnd	R-1B	3.13	7.668		212	400		612	830
9630	4 conductor w/#6 grnd		2.67	8.989		229	470		699	950
9640	#2, 3 conductor w/#4 grnd		2.59	9.266		410	480		890	1,175
9650	4 conductor w/#4 grnd		2.21	10.860		465	565		1,030	1,350
9660	#1, 3 conductor w/#4 grnd		2.15	11.163		315	580		895	1,200
9670	4 conductor w/#4 grnd		1.85	12.973		630	675		1,305	1,700
9680	1/0, 3 conductor w/#4 grnd		2.16	11.111		360	580		940	1,250
9690	4 conductor w/#4 grnd		1.73	13.873		435	720		1,155	1,550
9700	2/0, 3 conductor w/#4 grnd		1.91	12.565		640	655		1,295	1,675
9710	4 conductor w/#4 grnd		1.53	15.686		490	815		1,305	1,750
9720	3/0, 3 conductor w/#4 grnd		1.75	13.714		820	715		1,535	1,975
9730	4 conductor w/#4 grnd		1.40	17.143		605	890		1,495	2,000
9740	4/0, 3 conductor w/#2 grnd		1.54	15.584		915	810		1,725	2,200
9750	4 conductor w/#2 grnd	▼	1.23	19.512		360	1,025		1,385	1,925
9760	250 kcmil, 3 conductor w/#1 grnd	R-1C	1.90	16.842		1,175	910	80.50	2,165.50	2,750
9770	4 conductor w/#1 grnd		1.52	21.053		1,600	1,150	100	2,850	3,575
9780	350 kcmil, 3 conductor w/1/0 grnd		1.76	18.182		1,375	985	86.50	2,446.50	3,100
9790	4 conductor w/1/0 grnd	▼	1.41	22.695	▼	1,150	1,225	108	2,483	3,225

26 05 19 – Low-Voltage Electrical Power Conductors and Cables

26 05 19.20 Armored Cable

		Crew	Daily Output	Labor-Hours	Unit	Material	2019 Bare Costs Labor	2019 Bare Costs Equipment	Total	Total Incl O&P
9800	500 kcmil, 3 conductor w/#1 grnd	R-1C	1.64	19.512	C.L.F.	1,750	1,050	93	2,893	3,600
9810	4 conductor w/2/0 grnd		1.31	24.427		1,575	1,325	117	3,017	3,825
9840	750 kcmil, 3 conductor w/1/0 grnd		1.31	24.427		2,800	1,325	117	4,242	5,175
9850	4 conductor w/3/0 grnd		1.04	30.769		3,150	1,675	147	4,972	6,075

26 05 19.25 Cable Connectors

		Crew	Daily Output	Labor-Hours	Unit	Material	2019 Bare Costs Labor	2019 Bare Costs Equipment	Total	Total Incl O&P
0010	**CABLE CONNECTORS**									
0100	600 volt, nonmetallic, #14-2 wire	1 Elec	160	.050	Ea.	1.53	3		4.53	6.15
0200	#14-3 wire to #12-2 wire		133	.060		1.53	3.61		5.14	7.10
0300	#12-3 wire to #10-2 wire		114	.070		1.53	4.21		5.74	8
0400	#10-3 wire to #14-4 and #12-4 wire		100	.080		1.53	4.80		6.33	8.85
0500	#8-3 wire to #10-4 wire		80	.100		2.90	6		8.90	12.15
0600	#6-3 wire		40	.200		4.05	12		16.05	22.50
0800	SER, 3 #8 insulated + 1 #8 ground		32	.250		3.64	15		18.64	26.50
0900	3 #6 + 1 #6 ground		24	.333		4.55	20		24.55	35
1000	3 #4 + 1 #6 ground		22	.364		4.76	22		26.76	38
1100	3 #2 + 1 #4 ground		20	.400		8.90	24		32.90	46
1200	3 1/0 + 1 #2 ground		18	.444		18.85	26.50		45.35	60.50
1400	3 2/0 + 1 #1 ground		16	.500		26.50	30		56.50	74
1600	3 4/0 + 1 #2/0 ground		14	.571		27.50	34.50		62	81
1800	600 volt, armored, #14-2 wire		80	.100		1.28	6		7.28	10.35
2200	#14-4, #12-3 and #10-2 wire		40	.200		.93	12		12.93	18.90
2400	#12-4, #10-3 and #8-2 wire		32	.250		3.51	15		18.51	26.50
2600	#8-3 and #10-4 wire		26	.308		4.03	18.50		22.53	32
2650	#8-4 wire		22	.364		6.40	22		28.40	39.50
2652	Non-PVC jacket connector, #6-3 wire, #6-4 wire		22	.364		55.50	22		77.50	93.50
2660	1/0-3 wire		11	.727		79	43.50		122.50	152
2670	300 kcmil-3 wire		6	1.333		120	80		200	251
2680	700 kcmil-3 wire		4	2		185	120		305	380
2700	PVC jacket connector, #6-3 wire, #6-4 wire		16	.500		5.45	30		35.45	51
2800	#4-3 wire, #4-4 wire		16	.500		5.45	30		35.45	51
2900	#2-3 wire		12	.667		5.45	40		45.45	65.50
3000	#1-3 wire, #2-4 wire		12	.667		14.05	40		54.05	75
3200	1/0-3 wire		11	.727		14.05	43.50		57.55	80.50
3400	2/0-3 wire, 1/0-4 wire		10	.800		14.05	48		62.05	87
3500	3/0-3 wire, 2/0-4 wire		9	.889		16.50	53.50		70	97.50
3600	4/0-3 wire, 3/0-4 wire		7	1.143		16.50	68.50		85	120
3800	250 kcmil-3 wire, 4/0-4 wire		6	1.333		32	80		112	154
4000	350 kcmil-3 wire, 250 kcmil-4 wire		5	1.600		32	96		128	178
4100	350 kcmil-4 wire		4	2		209	120		329	410
4200	500 kcmil-3 wire		4	2		209	120		329	410
4250	500 kcmil-4 wire, 750 kcmil-3 wire		3.50	2.286		248	137		385	475
4300	750 kcmil-4 wire		3	2.667		248	160		408	510
4400	5 kV, armored, #4		8	1		67.50	60		127.50	164
4600	#2		8	1		67.50	60		127.50	164
4800	#1		8	1		88.50	60		148.50	187
5000	1/0		6.40	1.250		108	75		183	231
5200	2/0		5.30	1.509		108	90.50		198.50	254
5500	4/0		4	2		147	120		267	340
5600	250 kcmil		3.60	2.222		147	133		280	360
5650	350 kcmil		3.20	2.500		171	150		321	415
5700	500 kcmil		2.50	3.200		216	192		408	525
5720	750 kcmil		2.20	3.636		265	218		483	615

26 05 19.25 Cable Connectors

		Crew	Daily Output	Labor-Hours	Unit	Material	2019 Bare Costs Labor	Equipment	Total	Total Incl O&P
5750	1,000 kcmil	1 Elec	2	4	Ea.	320	240		560	710
5800	15 kV, armored, #1		4	2		129	120		249	320
5900	1/0		4	2		162	120		282	355
6000	3/0		3.60	2.222		256	133		389	480
6100	4/0		3.40	2.353		282	141		423	520
6200	250 kcmil		3.20	2.500		315	150		465	570
6300	350 kcmil		2.70	2.963		288	178		466	580
6400	500 kcmil		2	4		445	240		685	845

26 05 19.30 Cable Splicing

		Crew	Daily Output	Labor-Hours	Unit	Material	2019 Bare Costs Labor	Equipment	Total	Total Incl O&P
0010	**CABLE SPLICING** URD or similar, ideal conditions									
0100	#6 stranded to #1 stranded, 5 kV	1 Elec	4	2	Ea.	146	120		266	340
0120	15 kV		3.60	2.222		209	133		342	430
0140	25 kV		3.20	2.500		272	150		422	525
0200	#1 stranded to 4/0 stranded, 5 kV		3.60	2.222		152	133		285	365
0210	15 kV		3.20	2.500		230	150		380	475
0220	25 kV		2.80	2.857		272	172		444	555
0300	4/0 stranded to 500 kcmil stranded, 5 kV		3.30	2.424		152	146		298	385
0310	15 kV		2.90	2.759		375	166		541	655
0320	25 kV		2.50	3.200		435	192		627	760
0400	500 kcmil, 5 kV		3.20	2.500		224	150		374	470
0410	15 kV		2.80	2.857		375	172		547	665
0420	25 kV		2.30	3.478		435	209		644	785
0500	600 kcmil, 5 kV		2.90	2.759		224	166		390	495
0510	15 kV		2.40	3.333		490	200		690	835
0520	25 kV		2	4		510	240		750	920
0600	750 kcmil, 5 kV		2.60	3.077		224	185		409	520
0610	15 kV		2.20	3.636		490	218		708	860
0620	25 kV		1.90	4.211		510	253		763	935
0700	1,000 kcmil, 5 kV		2.30	3.478		224	209		433	555
0710	15 kV		1.90	4.211		530	253		783	960
0720	25 kV		1.60	5		520	300		820	1,025

26 05 19.35 Cable Terminations

		Crew	Daily Output	Labor-Hours	Unit	Material	2019 Bare Costs Labor	Equipment	Total	Total Incl O&P
0010	**CABLE TERMINATIONS**									
0015	Wire connectors, screw type, #22 to #14	1 Elec	260	.031	Ea.	.06	1.85		1.91	2.83
0020	#18 to #12		240	.033		.07	2		2.07	3.07
0035	#16 to #10		230	.035		.27	2.09		2.36	3.42
0040	#14 to #8		210	.038		.30	2.29		2.59	3.74
0045	#12 to #6		180	.044		.45	2.67		3.12	4.48
0050	Terminal lugs, solderless, #16 to #10		50	.160		.33	9.60		9.93	14.70
0100	#8 to #4		30	.267		.66	16		16.66	24.50
0150	#2 to #1		22	.364		.84	22		22.84	33.50
0200	1/0 to 2/0		16	.500		1.56	30		31.56	46.50
0250	3/0		12	.667		4.40	40		44.40	64.50
0300	4/0		11	.727		4.37	43.50		47.87	70
0350	250 kcmil		9	.889		3.23	53.50		56.73	83
0400	350 kcmil		7	1.143		4.24	68.50		72.74	107
0450	500 kcmil		6	1.333		7.35	80		87.35	127
0500	600 kcmil		5.80	1.379		8.15	83		91.15	133
0550	750 kcmil		5.20	1.538		10.80	92.50		103.30	150
0600	Split bolt connectors, tapped, #6		16	.500		3.09	30		33.09	48.50
0650	#4		14	.571		3.57	34.50		38.07	55
0700	#2		12	.667		5.45	40		45.45	65.50

For customer support on your Electrical Costs with RSMeans data, call 800.448.8182.

121

26 05 19.35 Cable Terminations		Crew	Daily Output	Labor-Hours	Unit	Material	2019 Bare Costs Labor	Equipment	Total	Total Incl O&P
0750	#1	1 Elec	11	.727	Ea.	7.40	43.50		50.90	73
0800	1/0		10	.800		7.40	48		55.40	79.50
0850	2/0		9	.889		11.95	53.50		65.45	92.50
0900	3/0		7.20	1.111		24.50	66.50		91	127
1000	4/0		6.40	1.250		32.50	75		107.50	148
1100	250 kcmil		5.70	1.404		32.50	84.50		117	162
1200	300 kcmil		5.30	1.509		31	90.50		121.50	169
1400	350 kcmil		4.60	1.739		31	104		135	190
1500	500 kcmil	▼	4	2	▼	79.50	120		199.50	267
1600	Crimp 1 hole lugs, copper or aluminum, 600 volt									
1620	#14	1 Elec	60	.133	Ea.	.57	8		8.57	12.60
1630	#12		50	.160		1.06	9.60		10.66	15.50
1640	#10		45	.178		1.06	10.70		11.76	17.10
1780	#8		36	.222		1.40	13.35		14.75	21.50
1800	#6		30	.267		1.79	16		17.79	26
2000	#4		27	.296		3.40	17.80		21.20	30
2200	#2		24	.333		2.83	20		22.83	33
2400	#1		20	.400		5.25	24		29.25	42
2500	1/0		17.50	.457		6.20	27.50		33.70	48
2600	2/0		15	.533		7.80	32		39.80	56.50
2800	3/0		12	.667		7.10	40		47.10	67.50
3000	4/0		11	.727		5.85	43.50		49.35	71.50
3200	250 kcmil		9	.889		7.35	53.50		60.85	87.50
3400	300 kcmil		8	1		11.55	60		71.55	102
3500	350 kcmil		7	1.143		12.60	68.50		81.10	116
3600	400 kcmil		6.50	1.231		16.70	74		90.70	128
3800	500 kcmil		6	1.333		19.15	80		99.15	140
4000	600 kcmil		5.80	1.379		27	83		110	154
4200	700 kcmil		5.50	1.455		29.50	87.50		117	163
4400	750 kcmil	▼	5.20	1.538	▼	29.50	92.50		122	171
4500	Crimp 2-way connectors, copper or alum., 600 volt									
4510	#14	1 Elec	60	.133	Ea.	2.49	8		10.49	14.70
4520	#12		50	.160		2.65	9.60		12.25	17.25
4530	#10		45	.178		2.79	10.70		13.49	19
4540	#8		27	.296		2.82	17.80		20.62	29.50
4600	#6		25	.320		4.03	19.20		23.23	33
4800	#4		23	.348		3.93	21		24.93	35.50
5000	#2		20	.400		4.84	24		28.84	41.50
5200	#1		16	.500		5.75	30		35.75	51.50
5400	1/0		13	.615		5.45	37		42.45	61
5420	2/0		12	.667		7.15	40		47.15	67.50
5440	3/0		11	.727		7.30	43.50		50.80	73
5460	4/0		10	.800		7.65	48		55.65	80
5480	250 kcmil		9	.889		15.95	53.50		69.45	97
5500	300 kcmil		8.50	.941		16.85	56.50		73.35	103
5520	350 kcmil		8	1		16.70	60		76.70	108
5540	400 kcmil		7.30	1.096		20.50	66		86.50	121
5560	500 kcmil		6.20	1.290		25.50	77.50		103	144
5580	600 kcmil		5.50	1.455		45	87.50		132.50	180
5600	700 kcmil		4.50	1.778		52.50	107		159.50	217
5620	750 kcmil		4	2		55	120		175	240
7000	Compression equipment adapter, aluminum wire, #6		30	.267		7.70	16		23.70	32.50
7020	#4	▼	27	.296	▼	7.75	17.80		25.55	35

26 05 19 – Low-Voltage Electrical Power Conductors and Cables

26 05 19.35 Cable Terminations

		Crew	Daily Output	Labor-Hours	Unit	Material	2019 Bare Costs Labor	Equipment	Total	Total Incl O&P
7040	#2	1 Elec	24	.333	Ea.	6.45	20		26.45	37
7060	#1		20	.400		8.10	24		32.10	45
7080	1/0		18	.444		13.80	26.50		40.30	55
7100	2/0		15	.533		16.45	32		48.45	66
7140	4/0		11	.727		17.40	43.50		60.90	84
7160	250 kcmil		9	.889		18.05	53.50		71.55	99.50
7180	300 kcmil		8	1		18.55	60		78.55	110
7200	350 kcmil		7	1.143		24	68.50		92.50	129
7220	400 kcmil		6.50	1.231		30.50	74		104.50	144
7240	500 kcmil		6	1.333		34	80		114	157
7260	600 kcmil		5.80	1.379		40	83		123	168
7280	750 kcmil	▼	5.20	1.538		42.50	92.50		135	185
8000	Compression tool, hand					1,475			1,475	1,625
8100	Hydraulic					1,925			1,925	2,125
8500	Hydraulic dies				▼	325			325	360

26 05 19.50 Mineral Insulated Cable

		Crew	Daily Output	Labor-Hours	Unit	Material	2019 Bare Costs Labor	Equipment	Total	Total Incl O&P
0010	**MINERAL INSULATED CABLE** 600 volt									
0100	1 conductor, #12	1 Elec	1.60	5	C.L.F.	360	300		660	845
0200	#10		1.60	5		565	300		865	1,075
0400	#8		1.50	5.333		690	320		1,010	1,250
0500	#6	▼	1.40	5.714		745	345		1,090	1,325
0600	#4	2 Elec	2.40	6.667		1,100	400		1,500	1,800
0800	#2		2.20	7.273		1,500	435		1,935	2,300
0900	#1		2.10	7.619		1,725	460		2,185	2,575
1000	1/0		2	8		2,025	480		2,505	2,950
1100	2/0		1.90	8.421		2,525	505		3,030	3,525
1200	3/0		1.80	8.889		3,025	535		3,560	4,125
1400	4/0	▼	1.60	10		3,500	600		4,100	4,750
1410	250 kcmil	3 Elec	2.40	10		3,950	600		4,550	5,250
1420	350 kcmil		1.95	12.308		4,525	740		5,265	6,075
1430	500 kcmil	▼	1.95	12.308		5,700	740		6,440	7,375
1500	2 conductor, #12	1 Elec	1.40	5.714		970	345		1,315	1,575
1600	#10		1.20	6.667		1,200	400		1,600	1,900
1800	#8		1.10	7.273		1,325	435		1,760	2,125
2000	#6	▼	1.05	7.619		1,875	460		2,335	2,750
2100	#4	2 Elec	2	8		2,650	480		3,130	3,625
2200	3 conductor, #12	1 Elec	1.20	6.667		1,150	400		1,550	1,850
2400	#10		1.10	7.273		1,300	435		1,735	2,075
2600	#8		1.05	7.619		1,725	460		2,185	2,575
2800	#6	▼	1	8		2,275	480		2,755	3,225
3000	#4	2 Elec	1.80	8.889		2,950	535		3,485	4,025
3100	4 conductor, #12	1 Elec	1.20	6.667		1,200	400		1,600	1,925
3200	#10		1.10	7.273		1,450	435		1,885	2,225
3400	#8		1	8		2,000	480		2,480	2,925
3600	#6		.90	8.889		2,700	535		3,235	3,775
3620	7 conductor, #12		1.10	7.273		1,625	435		2,060	2,450
3640	#10		1	8	▼	2,225	480		2,705	3,175
3800	Terminations, 600 volt, 1 conductor, #12		8	1	Ea.	19.30	60		79.30	111
4000	#10		7.60	1.053		19.30	63		82.30	116
4100	#8		7.30	1.096		19.30	66		85.30	119
4200	#6		6.70	1.194		19.30	71.50		90.80	128
4400	#4		6.20	1.290		19.30	77.50		96.80	137

123

26 05 19.50 Mineral Insulated Cable

26 05 19.50 Mineral Insulated Cable		Crew	Daily Output	Labor-Hours	Unit	Material	2019 Bare Costs Labor	Equipment	Total	Total Incl O&P
4600	#2	1 Elec	5.70	1.404	Ea.	29	84.50		113.50	158
4800	#1		5.30	1.509		29	90.50		119.50	167
5000	1/0		5	1.600		29	96		125	175
5100	2/0		4.70	1.702		29	102		131	185
5200	3/0		4.30	1.860		29	112		141	199
5400	4/0		4	2		64.50	120		184.50	250
5410	250 kcmil		4	2		64.50	120		184.50	250
5420	350 kcmil		4	2		107	120		227	296
5430	500 kcmil		4	2		107	120		227	296
5500	2 conductor, #12		6.70	1.194		19.30	71.50		90.80	128
5600	#10		6.40	1.250		29	75		104	144
5800	#8		6.20	1.290		29	77.50		106.50	148
6000	#6		5.70	1.404		29	84.50		113.50	158
6200	#4		5.30	1.509		64.50	90.50		155	206
6400	3 conductor, #12		5.70	1.404		29	84.50		113.50	158
6500	#10		5.50	1.455		29	87.50		116.50	162
6600	#8		5.20	1.538		29	92.50		121.50	170
6800	#6		4.80	1.667		29	100		129	181
7200	#4		4.60	1.739		64.50	104		168.50	227
7400	4 conductor, #12		4.60	1.739		32.50	104		136.50	192
7500	#10		4.40	1.818		32.50	109		141.50	199
7600	#8		4.20	1.905		32.50	114		146.50	207
8400	#6		4	2		67.50	120		187.50	254
8500	7 conductor, #12		3.50	2.286		32.50	137		169.50	241
8600	#10	▼	3	2.667		70	160		230	315
8800	Crimping tool, plier type					74			74	81.50
9000	Stripping tool					280			280	310
9200	Hand vise			▼		83.50			83.50	92

26 05 19.55 Non-Metallic Sheathed Cable

26 05 19.55 Non-Metallic Sheathed Cable		Crew	Daily Output	Labor-Hours	Unit	Material	2019 Bare Costs Labor	Equipment	Total	Total Incl O&P
0010	**NON-METALLIC SHEATHED CABLE** 600 volt									
0100	Copper with ground wire (Romex)									
0150	#14, 2 conductor	1 Elec	2.70	2.963	C.L.F.	18.50	178		196.50	286
0200	3 conductor		2.40	3.333		25	200		225	325
0220	4 conductor		2.20	3.636		39	218		257	370
0250	#12, 2 conductor		2.50	3.200		24.50	192		216.50	315
0300	3 conductor		2.20	3.636		39.50	218		257.50	370
0320	4 conductor		2	4		59	240		299	425
0350	#10, 2 conductor		2.20	3.636		41	218		259	370
0400	3 conductor		1.80	4.444		62	267		329	470
0420	4 conductor		1.60	5		88.50	300		388.50	545
0430	#8, 2 conductor		1.60	5		76.50	300		376.50	535
0450	3 conductor		1.50	5.333		116	320		436	610
0500	#6, 3 conductor	▼	1.40	5.714		195	345		540	725
0520	#4, 3 conductor	2 Elec	2.40	6.667		400	400		800	1,025
0540	#2, 3 conductor	"	2.20	7.273	▼	770	435		1,205	1,500
0550	SE type SER aluminum cable, 3 RHW and									
0600	1 bare neutral, 3 #8 & 1 #8	1 Elec	1.60	5	C.L.F.	55.50	300		355.50	510
0650	3 #6 & 1 #6	"	1.40	5.714		66	345		411	585
0700	3 #4 & 1 #6	2 Elec	2.40	6.667		70	400		470	670
0750	3 #2 & 1 #4		2.20	7.273		122	435		557	785
0800	3 #1/0 & 1 #2		2	8		179	480		659	910
0850	3 #2/0 & 1 #1	▼	1.80	8.889	▼	208	535		743	1,025

26 05 19 – Low-Voltage Electrical Power Conductors and Cables

26 05 19.55 Non-Metallic Sheathed Cable	Crew	Daily Output	Labor-Hours	Unit	Material	2019 Bare Costs Labor	Equipment	Total	Total Incl O&P
0900	3 #4/0 & 1 #2/0	2 Elec	1.60	10	C.L.F.	290	600		890	1,225
1000	URD - triplex underground distribution cable, alum. 2 #4 + #4 neutral		2.80	5.714		75.50	345		420.50	595
1010	2 #2 + #4 neutral		2.65	6.038		111	365		476	660
1020	2 #2 + #2 neutral		2.55	6.275		95	375		470	665
1030	2 1/0 + #2 neutral		2.40	6.667		114	400		514	720
1040	2 1/0 + 1/0 neutral		2.30	6.957		111	420		531	745
1050	2 2/0 + #1 neutral		2.20	7.273		130	435		565	795
1060	2 2/0 + 2/0 neutral		2.10	7.619		145	460		605	845
1070	2 3/0 + 1/0 neutral		1.95	8.205		177	495		672	930
1080	2 3/0 + 3/0 neutral		1.95	8.205		174	495		669	925
1090	2 4/0 + 2/0 neutral		1.85	8.649		182	520		702	975
1100	2 4/0 + 4/0 neutral		1.85	8.649		202	520		722	1,000
1450	UF underground feeder cable, copper with ground, #14, 2 conductor	1 Elec	4	2		40.50	120		160.50	224
1500	#12, 2 conductor		3.50	2.286		35	137		172	244
1550	#10, 2 conductor		3	2.667		54	160		214	299
1600	#14, 3 conductor		3.50	2.286		37.50	137		174.50	247
1650	#12, 3 conductor		3	2.667		53.50	160		213.50	298
1700	#10, 3 conductor		2.50	3.200		85	192		277	380
1710	#8, 3 conductor		2	4		134	240		374	510
1720	#6, 3 conductor		1.80	4.444		222	267		489	645
2400	SEU service entrance cable, copper 2 conductors, #8 + #8 neutral		1.50	5.333		103	320		423	595
2600	#6 + #8 neutral		1.30	6.154		132	370		502	695
2800	#6 + #6 neutral		1.30	6.154		154	370		524	720
3000	#4 + #6 neutral	2 Elec	2.20	7.273		202	435		637	870
3200	#4 + #4 neutral		2.20	7.273		223	435		658	895
3400	#3 + #5 neutral		2.10	7.619		335	460		795	1,050
3600	#3 + #3 neutral		2.10	7.619		269	460		729	980
3800	#2 + #4 neutral		2	8		305	480		785	1,050
4000	#1 + #1 neutral		1.90	8.421		345	505		850	1,125
4200	1/0 + 1/0 neutral		1.80	8.889		940	535		1,475	1,825
4400	2/0 + 2/0 neutral		1.70	9.412		685	565		1,250	1,600
4600	3/0 + 3/0 neutral		1.60	10		850	600		1,450	1,825
4620	4/0 + 4/0 neutral		1.45	11.034		2,075	665		2,740	3,275
4800	Aluminum 2 conductors, #8 + #8 neutral	1 Elec	1.60	5		49	300		349	505
5000	#6 + #6 neutral	"	1.40	5.714		56.50	345		401.50	575
5100	#4 + #6 neutral	2 Elec	2.50	6.400		65	385		450	645
5200	#4 + #4 neutral		2.40	6.667		64	400		464	665
5300	#2 + #4 neutral		2.30	6.957		79	420		499	710
5400	#2 + #2 neutral		2.20	7.273		92	435		527	750
5450	1/0 + #2 neutral		2.10	7.619		122	460		582	820
5500	1/0 + 1/0 neutral		2	8		127	480		607	855
5550	2/0 + #1 neutral		1.90	8.421		135	505		640	905
5600	2/0 + 2/0 neutral		1.80	8.889		143	535		678	950
5800	3/0 + 1/0 neutral		1.70	9.412		217	565		782	1,075
6000	3/0 + 3/0 neutral		1.70	9.412		183	565		748	1,050
6200	4/0 + 2/0 neutral		1.60	10		187	600		787	1,100
6400	4/0 + 4/0 neutral		1.60	10		204	600		804	1,125
6500	Service entrance cap for copper SEU									
6600	100 amp	1 Elec	12	.667	Ea.	7.95	40		47.95	68.50
6700	150 amp		10	.800		10.70	48		58.70	83.50
6800	200 amp		8	1		17.25	60		77.25	108

For customer support on your Electrical Costs with RSMeans data, call 800.448.8182.

125

26 05 19.70 Portable Cord

26 05 19.70 Portable Cord		Crew	Daily Output	Labor-Hours	Unit	Material	2019 Bare Costs Labor	2019 Bare Costs Equipment	Total	Total Incl O&P
0010	**PORTABLE CORD** 600 volt									
0100	Type SO, #18, 2 conductor	1 Elec	980	.008	L.F.	.29	.49		.78	1.05
0110	3 conductor		980	.008		.33	.49		.82	1.09
0120	#16, 2 conductor		840	.010		.38	.57		.95	1.27
0130	3 conductor		840	.010		4	.57		4.57	5.25
0140	4 conductor		840	.010		.57	.57		1.14	1.48
0240	#14, 2 conductor		840	.010		.53	.57		1.10	1.43
0250	3 conductor		840	.010		.72	.57		1.29	1.64
0260	4 conductor		840	.010		1.07	.57		1.64	2.03
0280	#12, 2 conductor		840	.010		.78	.57		1.35	1.71
0290	3 conductor		840	.010		.91	.57		1.48	1.85
0300	4 conductor		840	.010		1.21	.57		1.78	2.18
0320	#10, 2 conductor		765	.010		.85	.63		1.48	1.88
0330	3 conductor		765	.010		1.27	.63		1.90	2.34
0340	4 conductor		765	.010		1.30	.63		1.93	2.37
0360	#8, 2 conductor		555	.014		2.03	.87		2.90	3.52
0370	3 conductor		540	.015		2.43	.89		3.32	4
0380	4 conductor		525	.015		2.14	.92		3.06	3.72
0400	#6, 2 conductor		525	.015		2.96	.92		3.88	4.63
0410	3 conductor		490	.016		3.41	.98		4.39	5.20
0420	4 conductor		415	.019		3.26	1.16		4.42	5.30
0440	#4, 2 conductor	2 Elec	830	.019		8.15	1.16		9.31	10.75
0450	3 conductor		700	.023		5.70	1.37		7.07	8.35
0460	4 conductor		660	.024		6.50	1.46		7.96	9.30
0480	#2, 2 conductor		450	.036		5.20	2.14		7.34	8.90
0490	3 conductor		350	.046		7.15	2.74		9.89	11.95
0500	4 conductor		280	.057		9.05	3.43		12.48	15.05
2000	See 26 27 26.20 for Wiring Devices Elements									

26 05 19.75 Modular Flexible Wiring System

26 05 19.75 Modular Flexible Wiring System		Crew	Daily Output	Labor-Hours	Unit	Material	2019 Bare Costs Labor	2019 Bare Costs Equipment	Total	Total Incl O&P
0010	**MODULAR FLEXIBLE WIRING SYSTEM**									
0020	Commercial system grid ceiling									
0100	Conversion Module	1 Elec	32	.250	Ea.	8	15		23	31.50
0120	Fixture cable, for fixture to fixture, 3 conductor, 15' long		40	.200		34	12		46	55.50
0150	Extender cable, 3 conductor, 15' long		48	.167		31.50	10		41.50	49.50
0200	Switch drop, 1 level, 9' long		32	.250		22	15		37	46.50
0220	2 level, 9' long		32	.250		30	15		45	55.50
0250	Power tee, 9' long		32	.250		27.50	15		42.50	52.50
1020	Industrial system open ceiling									
1100	Converter, interface between hardwiring and modular wiring	1 Elec	32	.250	Ea.	21	15		36	46
1120	Fixture cable, for fixture to fixture, 3 conductor, 21' long		32	.250		112	15		127	146
1125	3 conductor, 25' long		24	.333		147	20		167	192
1130	3 conductor, 31' long		19	.421		139	25.50		164.50	191
1150	Fixture cord drop, 10' long		40	.200		37.50	12		49.50	59.50

26 05 19.90 Wire

26 05 19.90 Wire			Crew	Daily Output	Labor-Hours	Unit	Material	2019 Bare Costs Labor	2019 Bare Costs Equipment	Total	Total Incl O&P
0010	**WIRE**, normal installation conditions in wireway, conduit, cable tray										
0020	600 volt, copper type THW, solid, #14	R260519-90	1 Elec	13	.615	C.L.F.	5.80	37		42.80	61.50
0030	#12			11	.727		9.70	43.50		53.20	75.50
0040	#10			10	.800		15.75	48		63.75	89
0050	Stranded, #14			13	.615		7.60	37		44.60	63.50
0100	#12			11	.727		13.65	43.50		57.15	80
0120	#10	R260519-92		10	.800		21.50	48		69.50	95
0140	#8			8	1		33.50	60		93.50	127

126

26 05 19 – Low-Voltage Electrical Power Conductors and Cables

26 05 19.90 Wire			Crew	Daily Output	Labor-Hours	Unit	Material	2019 Bare Costs Labor	Equipment	Total	Total Incl O&P
0160	#6	R260519-93	1 Elec	6.50	1.231	C.L.F.	42.50	74		116.50	157
0180	#4		2 Elec	10.60	1.509		79.50	90.50		170	223
0200	#3	R260519-94		10	1.600		112	96		208	266
0220	#2			9	1.778		100	107		207	269
0240	#1	R260533-22		8	2		133	120		253	325
0260	1/0			6.60	2.424		212	146		358	450
0280	2/0			5.80	2.759		201	166		367	470
0300	3/0			5	3.200		230	192		422	540
0350	4/0		▼	4.40	3.636		425	218		643	795
0400	250 kcmil		3 Elec	6	4		345	240		585	740
0420	300 kcmil			5.70	4.211		440	253		693	855
0450	350 kcmil			5.40	4.444		455	267		722	900
0480	400 kcmil			5.10	4.706		530	283		813	1,000
0490	500 kcmil			4.80	5		800	300		1,100	1,325
0500	600 kcmil			3.90	6.154		1,225	370		1,595	1,900
0510	750 kcmil			3.30	7.273		1,375	435		1,810	2,150
0520	1,000 kcmil		▼	2.70	8.889		1,975	535		2,510	2,975
0540	600 volt, aluminum type THHN, stranded, #6		1 Elec	8	1		25	60		85	117
0560	#4		2 Elec	13	1.231		30.50	74		104.50	144
0580	#2			10.60	1.509		42.50	90.50		133	182
0600	#1			9	1.778		62	107		169	227
0620	1/0			8	2		74.50	120		194.50	261
0640	2/0			7.20	2.222		88.50	133		221.50	297
0680	3/0			6.60	2.424		111	146		257	340
0700	4/0		▼	6.20	2.581		123	155		278	365
0720	250 kcmil		3 Elec	8.70	2.759		149	166		315	410
0740	300 kcmil			8.10	2.963		206	178		384	490
0760	350 kcmil			7.50	3.200		208	192		400	515
0780	400 kcmil			6.90	3.478		237	209		446	570
0800	500 kcmil			6	4		267	240		507	655
0850	600 kcmil			5.70	4.211		340	253		593	745
0880	700 kcmil			5.10	4.706		415	283		698	875
0900	750 kcmil			4.80	5		410	300		710	900
0910	1,000 kcmil		▼	3.78	6.349		585	380		965	1,200
0920	600 volt, copper type THWN-THHN, solid, #14		1 Elec	13	.615		6.60	37		43.60	62.50
0940	#12			11	.727		9.80	43.50		53.30	76
0960	#10			10	.800		14.65	48		62.65	87.50
1000	Stranded, #14			13	.615		7.55	37		44.55	63.50
1200	#12			11	.727		10.50	43.50		54	76.50
1250	#10			10	.800		16.10	48		64.10	89
1300	#8			8	1		28.50	60		88.50	121
1350	#6		▼	6.50	1.231		48.50	74		122.50	163
1400	#4		2 Elec	10.60	1.509		69	90.50		159.50	211
1450	#3			10	1.600		97	96		193	250
1500	#2			9	1.778		117	107		224	287
1550	#1			8	2		151	120		271	345
1600	1/0			6.60	2.424		195	146		341	430
1650	2/0			5.80	2.759		221	166		387	490
1700	3/0			5	3.200		283	192		475	595
2000	4/0		▼	4.40	3.636		350	218		568	710
2200	250 kcmil		3 Elec	6	4		430	240		670	835
2400	300 kcmil			5.70	4.211		505	253		758	930
2600	350 kcmil		▼	5.40	4.444		590	267		857	1,050

127

26 05 19 – Low-Voltage Electrical Power Conductors and Cables

26 05 19.90 Wire		Crew	Daily Output	Labor-Hours	Unit	Material	2019 Bare Costs Labor	Equipment	Total	Total Incl O&P
2700	400 kcmil	3 Elec	5.10	4.706	C.L.F.	675	283		958	1,175
2800	500 kcmil		4.80	5		840	300		1,140	1,375
2802	600 kcmil		3.90	6.154		755	370		1,125	1,375
2804	750 kcmil		3.30	7.273		1,425	435		1,860	2,225
2805	1,000 kcmil		1.94	12.371		2,225	745		2,970	3,550
2900	600 volt, copper type XHHW, solid, #14	1 Elec	13	.615		11.40	37		48.40	67.50
2920	#12		11	.727		17.95	43.50		61.45	85
2940	#10		10	.800		28	48		76	103
3000	Stranded, #14		13	.615		9.15	37		46.15	65
3020	#12		11	.727		13.75	43.50		57.25	80
3040	#10		10	.800		20.50	48		68.50	94
3060	#8		8	1		30	60		90	123
3080	#6		6.50	1.231		51	74		125	166
3100	#4	2 Elec	10.60	1.509		78.50	90.50		169	221
3110	#3		10	1.600		97	96		193	250
3120	#2		9	1.778		122	107		229	293
3140	#1		8	2		166	120		286	360
3160	1/0		6.60	2.424		205	146		351	440
3180	2/0		5.80	2.759		257	166		423	530
3200	3/0		5	3.200		320	192		512	640
3220	4/0		4.40	3.636		400	218		618	765
3240	250 kcmil	3 Elec	6	4		445	240		685	850
3260	300 kcmil		5.70	4.211		515	253		768	940
3280	350 kcmil		5.40	4.444		595	267		862	1,050
3300	400 kcmil		5.10	4.706		675	283		958	1,175
3320	500 kcmil		4.80	5		835	300		1,135	1,375
3340	600 kcmil		3.90	6.154		905	370		1,275	1,550
3360	750 kcmil		3.30	7.273		1,525	435		1,960	2,325
3380	1,000 kcmil		2.40	10		1,950	600		2,550	3,050
5020	600 volt, aluminum type XHHW, stranded, #6	1 Elec	8	1		25	60		85	118
5040	#4	2 Elec	13	1.231		31	74		105	145
5060	#2		10.60	1.509		43	90.50		133.50	183
5080	#1		9	1.778		62	107		169	228
5100	1/0		8	2		74	120		194	261
5120	2/0		7.20	2.222		91.50	133		224.50	300
5140	3/0		6.60	2.424		109	146		255	335
5160	4/0		6.20	2.581		121	155		276	365
5180	250 kcmil	3 Elec	8.70	2.759		153	166		319	415
5200	300 kcmil		8.10	2.963		202	178		380	490
5220	350 kcmil		7.50	3.200		211	192		403	520
5240	400 kcmil		6.90	3.478		240	209		449	575
5260	500 kcmil		6	4		262	240		502	650
5280	600 kcmil		5.70	4.211		340	253		593	745
5300	700 kcmil		5.40	4.444		380	267		647	820
5320	750 kcmil		5.10	4.706		370	283		653	830
5340	1,000 kcmil		3.60	6.667		575	400		975	1,225
5390	600 volt, copper type XLPE-USE (RHW), solid, #14	1 Elec	12	.667		13.20	40		53.20	74
5400	#12		11	.727		17.05	43.50		60.55	84
5420	#10		10	.800		22.50	48		70.50	96
5440	Stranded, #14		13	.615		15.30	37		52.30	72
5460	#12		11	.727		18.35	43.50		61.85	85
5480	#10		10	.800		29.50	48		77.50	104
5500	#8		8	1		34	60		94	127

26 05 19 – Low-Voltage Electrical Power Conductors and Cables

26 05 19.90 Wire		Crew	Daily Output	Labor-Hours	Unit	Material	2019 Bare Costs Labor	Equipment	Total	Total Incl O&P
5520	#6	1 Elec	6.50	1.231	C.L.F.	57	74		131	173
5540	#4	2 Elec	10.60	1.509		86.50	90.50		177	230
5560	#2		9	1.778		141	107		248	315
5580	#1		8	2		190	120		310	390
5600	1/0		6.60	2.424		237	146		383	480
5620	2/0		5.80	2.759		315	166		481	590
5640	3/0		5	3.200		390	192		582	715
5660	4/0		4.40	3.636		425	218		643	790
5680	250 kcmil	3 Elec	6	4		495	240		735	905
5700	300 kcmil		5.70	4.211		615	253		868	1,050
5720	350 kcmil		5.40	4.444		695	267		962	1,175
5740	400 kcmil		5.10	4.706		795	283		1,078	1,300
5760	500 kcmil		4.80	5		835	300		1,135	1,375
5780	600 kcmil		3.90	6.154		1,475	370		1,845	2,175
5800	750 kcmil		3.30	7.273		1,650	435		2,085	2,450
5820	1,000 kcmil		2.70	8.889		2,175	535		2,710	3,200
5840	600 volt, aluminum type XLPE-USE (RHW), stranded, #6	1 Elec	8	1		31	60		91	124
5860	#4	2 Elec	13	1.231		37	74		111	151
5880	#2		10.60	1.509		52	90.50		142.50	192
5900	#1		9	1.778		74	107		181	240
5920	1/0		8	2		86	120		206	274
5940	2/0		7.20	2.222		101	133		234	310
5960	3/0		6.60	2.424		118	146		264	345
5980	4/0		6.20	2.581		131	155		286	375
6000	250 kcmil	3 Elec	8.70	2.759		178	166		344	445
6020	300 kcmil		8.10	2.963		232	178		410	520
6040	350 kcmil		7.50	3.200		234	192		426	545
6060	400 kcmil		6.90	3.478		288	209		497	625
6080	500 kcmil		6	4		320	240		560	710
6100	600 kcmil		5.70	4.211		405	253		658	820
6110	700 kcmil		5.40	4.444		440	267		707	880
6120	750 kcmil		5.10	4.706		475	283		758	945

26 05 23 – Control-Voltage Electrical Power Cables

26 05 23.10 Control Cable

		Crew	Daily Output	Labor-Hours	Unit	Material	2019 Bare Costs Labor	Equipment	Total	Total Incl O&P
0010	**CONTROL CABLE**									
0020	600 volt, copper, #14 THWN wire with PVC jacket, 2 wires	1 Elec	9	.889	C.L.F.	39	53.50		92.50	122
0030	3 wires		8	1		52	60		112	147
0100	4 wires		7	1.143		53	68.50		121.50	161
0150	5 wires		6.50	1.231		75.50	74		149.50	193
0200	6 wires		6	1.333		102	80		182	231
0300	8 wires		5.30	1.509		137	90.50		227.50	286
0400	10 wires		4.80	1.667		153	100		253	315
0500	12 wires		4.30	1.860		199	112		311	385
0600	14 wires		3.80	2.105		213	126		339	425
0700	16 wires		3.50	2.286		243	137		380	470
0800	18 wires		3.30	2.424		273	146		419	515
0810	19 wires		3.10	2.581		287	155		442	545
0900	20 wires		3	2.667		300	160		460	570
1000	22 wires		2.80	2.857		335	172		507	620

For customer support on your Electrical Costs with RSMeans data, call 800.448.8182.

129

26 05 23.20 Special Wires and Fittings	Crew	Daily Output	Labor-Hours	Unit	Material	2019 Bare Costs Labor	Equipment	Total	Total Incl O&P
0010 **SPECIAL WIRES & FITTINGS**									
0100 Fixture TFFN 600 volt 90°C stranded, #18	1 Elec	13	.615	C.L.F.	9.25	37		46.25	65
0150 #16		13	.615		12.80	37		49.80	69
0500 Thermostat, jacket non-plenum, twisted, #18-2 conductor		8	1		9.85	60		69.85	100
0550 #18-3 conductor		7	1.143		12.10	68.50		80.60	115
0600 #18-4 conductor		6.50	1.231		19.95	74		93.95	132
0650 #18-5 conductor		6	1.333		21.50	80		101.50	143
0700 #18-6 conductor		5.50	1.455		25	87.50		112.50	158
0750 #18-7 conductor		5	1.600		34.50	96		130.50	181
0800 #18-8 conductor		4.80	1.667		32	100		132	185
2460 Tray cable, type TC, copper, #16-2 conductor		9.40	.851		28.50	51		79.50	108
2464 #16-3 conductor		8.40	.952		46.50	57		103.50	137
2468 #16-4 conductor		7.30	1.096		50	66		116	153
2472 #16-5 conductor		6.70	1.194		42	71.50		113.50	153
2476 #16-7 conductor		5.50	1.455		57.50	87.50		145	194
2480 #16-9 conductor		5	1.600		72	96		168	223
2484 #16-12 conductor	2 Elec	8.80	1.818		121	109		230	296
2488 #16-15 conductor		7.20	2.222		125	133		258	335
2492 #16-19 conductor		6.60	2.424		154	146		300	385
2496 #16-25 conductor		6.40	2.500		217	150		367	465
2500 #14-2 conductor	1 Elec	9	.889		31.50	53.50		85	114
2520 #14-3 conductor		8	1		39	60		99	133
2540 #14-4 conductor		7	1.143		52	68.50		120.50	159
2560 #14-5 conductor		6.50	1.231		62	74		136	179
2564 #14-7 conductor		5.30	1.509		90	90.50		180.50	234
2568 #14-9 conductor	2 Elec	9.60	1.667		108	100		208	267
2572 #14-12 conductor		8.60	1.860		144	112		256	325
2576 #14-15 conductor		7.20	2.222		172	133		305	390
2578 #14-19 conductor		6.20	2.581		216	155		371	470
2582 #14-25 conductor		5.30	3.019		285	181		466	585
2590 #12-2 conductor	1 Elec	8.40	.952		48.50	57		105.50	139
2592 #12-3 conductor		7.60	1.053		65.50	63		128.50	167
2594 #12-4 conductor		6.60	1.212		103	73		176	222
2596 #12-5 conductor		6.20	1.290		106	77.50		183.50	232
2598 #12-7 conductor	2 Elec	10.40	1.538		117	92.50		209.50	267
2602 #12-9 conductor		8.80	1.818		214	109		323	400
2604 #12-12 conductor		7.80	2.051		231	123		354	440
2606 #12-15 conductor		6.80	2.353		257	141		398	495
2608 #12-19 conductor		6	2.667		315	160		475	585
2610 #12-25 conductor	3 Elec	7.70	3.117		430	187		617	750
2618 #10-2 conductor	1 Elec	8	1		82.50	60		142.50	181
2622 #10-3 conductor	"	7.30	1.096		83.50	66		149.50	190
2624 #10-4 conductor	2 Elec	12.80	1.250		143	75		218	270
2626 #10-5 conductor		11.80	1.356		135	81.50		216.50	269
2628 #10-7 conductor		9.40	1.702		194	102		296	365
2630 #10-9 conductor		8.40	1.905		225	114		339	420
2632 #10-12 conductor		7.20	2.222		251	133		384	475
2640 300 V, copper braided shield, PVC jacket									
2650 2 conductor #18 stranded	1 Elec	7	1.143	C.L.F.	25.50	68.50		94	130
2660 3 conductor #18 stranded		6	1.333		38.50	80		118.50	162
2670 4 conductor #18 stranded		6	1.333		162	80		242	297
3000 Strain relief grip for cable									

26 05 23 – Control-Voltage Electrical Power Cables

26 05 23.20 Special Wires and Fittings		Crew	Daily Output	Labor-Hours	Unit	Material	2019 Bare Costs Labor	Equipment	Total	Total Incl O&P
3050	Cord, top, #12-3	1 Elec	40	.200	Ea.	9.40	12		21.40	28.50
3060	#12-4		40	.200		9.40	12		21.40	28.50
3070	#12-5		39	.205		11.65	12.30		23.95	31
3100	#10-3		39	.205		11.65	12.30		23.95	31
3110	#10-4		38	.211		11.65	12.65		24.30	31.50
3120	#10-5		38	.211		13.95	12.65		26.60	34
3200	Bottom, #12-3		40	.200		25	12		37	45.50
3210	#12-4		40	.200		25	12		37	45.50
3220	#12-5		39	.205		26	12.30		38.30	47.50
3230	#10-3		39	.205		26	12.30		38.30	47.50
3300	#10-4		38	.211		26	12.65		38.65	48
3310	#10-5		38	.211		32.50	12.65		45.15	55
3400	Cable ties, standard, 4" length		190	.042		.14	2.53		2.67	3.92
3410	7" length		160	.050		.18	3		3.18	4.68
3420	14.5" length		90	.089		.41	5.35		5.76	8.40
3430	Heavy, 14.5" length		80	.100		.61	6		6.61	9.60
3500	Cable gland, nylon, 1/4 NPT thread		60	.133		1.95	8		9.95	14.10
3510	Cable gland, nylon, 3/8 NPT thread		60	.133		2.30	8		10.30	14.50
3520	Cable gland, nylon, 1/2 NPT thread		60	.133		2.89	8		10.89	15.15
3530	Cable gland, nylon, 3/4 NPT thread		60	.133		4.25	8		12.25	16.65
3540	Cable gland, nylon, 1" NPT thread		60	.133		7.10	8		15.10	19.80
3550	Cable gland, nylon, 1-1/4 NPT thread		60	.133		7.90	8		15.90	20.50
3970	U-cable guards									
3980	.75" x 5' U-cable guard, 14 gauge	1 Elec	24	.333	Ea.	11.10	20		31.10	42.50
3990	.75" x 8' U-cable guard, 14 gauge		24	.333		17.65	20		37.65	49.50
4000	1 1/8" x 5' U-cable guard		24	.333		15.65	20		35.65	47.50
4010	1 1/8" x 8' U-cable guard		24	.333		19	20		39	51
4020	2 3/16" x 5' U-cable guard		24	.333		30.50	20		50.50	63.50
4030	2 3/16" x 8' U-cable guard		24	.333		31	20		51	64.50
4040	3 3/16" x 5' U-cable guard		24	.333		37.50	20		57.50	71.50
4050	3 3/16" x 8' U-cable guard		24	.333		47.50	20		67.50	82
4060	4" x 8' U-cable guard		24	.333		41	20		61	75
4070	2" x 9' flanged-cable guard		20	.400		85.50	24		109.50	130
4080	3" x 9' flanged-cable guard		20	.400		104	24		128	151
4090	3.5" x 9' flanged-cable guard		20	.400		141	24		165	192
5000	2" x 5' flanged-cable guard, extension		20	.400		65.50	24		89.50	108
5010	3" x 5' flanged-cable guard, extension		20	.400		97.50	24		121.50	143
5020	.05" x 8' U-cable guard, black plastic		26	.308		5.45	18.50		23.95	33.50
5030	.75" x 8' U-cable guard, black plastic		26	.308		7.15	18.50		25.65	35.50
5040	1" x 8' U-cable guard, black plastic		26	.308		11.50	18.50		30	40

26 05 26 – Grounding and Bonding for Electrical Systems

26 05 26.80 Grounding

0010	GROUNDING	R260526-80								
0030	Rod, copper clad, 8' long, 1/2" diameter	1 Elec	5.50	1.455	Ea.	19.75	87.50		107.25	152
0040	5/8" diameter		5.50	1.455		22	87.50		109.50	155
0050	3/4" diameter		5.30	1.509		35.50	90.50		126	174
0080	10' long, 1/2" diameter		4.80	1.667		22.50	100		122.50	174
0090	5/8" diameter		4.60	1.739		23.50	104		127.50	182
0100	3/4" diameter		4.40	1.818		49.50	109		158.50	217
0130	15' long, 3/4" diameter		4	2		49.50	120		169.50	234
0150	Coupling, bronze, 1/2" diameter					5.20			5.20	5.70
0160	5/8" diameter					5.40			5.40	5.95

26 05 26 – Grounding and Bonding for Electrical Systems

26 05 26.80 Grounding		Crew	Daily Output	Labor-Hours	Unit	Material	2019 Bare Costs Labor	Equipment	Total	Total Incl O&P
0170	3/4" diameter				Ea.	15.35			15.35	16.85
0190	Drive studs, 1/2" diameter					11.35			11.35	12.50
0210	5/8" diameter					18.65			18.65	20.50
0220	3/4" diameter					19.90			19.90	22
0230	Clamp, bronze, 1/2" diameter	1 Elec	32	.250		5.40	15		20.40	28.50
0240	5/8" diameter		32	.250		4.96	15		19.96	28
0250	3/4" diameter		32	.250		5.60	15		20.60	28.50
0260	Wire ground bare armored, #8-1 conductor		2	4	C.L.F.	69	240		309	435
0270	#6-1 conductor		1.80	4.444		84	267		351	495
0280	#4-1 conductor		1.60	5		136	300		436	600
0320	Bare copper wire, #14 solid		14	.571		6.20	34.50		40.70	58
0330	#12		13	.615		14	37		51	70.50
0340	#10		12	.667		19.25	40		59.25	80.50
0350	#8		11	.727		24.50	43.50		68	91.50
0360	#6		10	.800		38.50	48		86.50	114
0370	#4		8	1		97.50	60		157.50	197
0380	#2		5	1.600		129	96		225	285
0390	Bare copper wire, stranded, #8		11	.727		43	43.50		86.50	113
0400	#6		10	.800		41	48		89	117
0450	#4	2 Elec	16	1		86.50	60		146.50	185
0600	#2		10	1.600		89	96		185	241
0650	#1		9	1.778		128	107		235	299
0700	1/0		8	2		148	120		268	340
0750	2/0		7.20	2.222		262	133		395	485
0800	3/0		6.60	2.424		315	146		461	560
1000	4/0		5.70	2.807		440	169		609	735
1200	250 kcmil	3 Elec	7.20	3.333		445	200		645	790
1210	300 kcmil		6.60	3.636		510	218		728	885
1220	350 kcmil		6	4		465	240		705	870
1230	400 kcmil		5.70	4.211		560	253		813	990
1240	500 kcmil		5.10	4.706		715	283		998	1,200
1260	750 kcmil		3.60	6.667		1,175	400		1,575	1,900
1270	1,000 kcmil		3	8		1,550	480		2,030	2,425
1360	Bare aluminum, stranded, #6	1 Elec	9	.889		14.45	53.50		67.95	95.50
1370	#4	2 Elec	16	1		29	60		89	122
1380	#2		13	1.231		39.50	74		113.50	154
1390	#1		10.60	1.509		42	90.50		132.50	181
1400	1/0		9	1.778		52.50	107		159.50	217
1410	2/0		8	2		66.50	120		186.50	252
1420	3/0		7.20	2.222		89.50	133		222.50	298
1430	4/0		6.60	2.424		95	146		241	320
1440	250 kcmil	3 Elec	9.30	2.581		118	155		273	360
1450	300 kcmil		8.70	2.759		136	166		302	395
1460	400 kcmil		7.50	3.200		178	192		370	485
1470	500 kcmil		6.90	3.478		212	209		421	545
1480	600 kcmil		6	4		239	240		479	620
1490	700 kcmil		5.70	4.211		262	253		515	665
1500	750 kcmil		5.10	4.706		274	283		557	720
1510	1,000 kcmil		4.80	5		390	300		690	875
1800	Water pipe ground clamps, heavy duty									
2000	Bronze, 1/2" to 1" diameter	1 Elec	8	1	Ea.	32	60		92	125
2100	1-1/4" to 2" diameter		8	1		29	60		89	122
2200	2-1/2" to 3" diameter		6	1.333		69.50	80		149.50	196

26 05 26.80 Grounding		Crew	Daily Output	Labor-Hours	Unit	Material	2019 Bare Costs Labor	Equipment	Total	Total Incl O&P
2400	Grounding, exothermic welding									
2500	Exothermic welding reusable mold, cable to cable, parallel, vertical	1 Elec	8	1	Ea.	315	60		375	435
2505	Splice single		8	1		296	60		356	415
2510	Splice single to double		8	1		315	60		375	435
2520	Cable to cable, termination, Tee		8	1		305	60		365	430
2530	Cable to rod, termination, Tee		8	1		282	60		342	400
2535	Cable to rebar over, termination, Tee		8	1		93.50	60		153.50	193
2540	Cable to rod, termination, 90 Deg		8	1		300	60		360	420
2550	Cable to verticle flat steel		8	1		315	60		375	435
2555	Cable to 45 Deg		8	1		315	60		375	435
2730	Exothermic weld, 4/0 wire to 1" ground rod		7	1.143		9.40	68.50		77.90	112
2740	4/0 wire to building steel		7	1.143		9.40	68.50		77.90	112
2750	4/0 wire to motor frame		7	1.143		9.40	68.50		77.90	112
2760	4/0 wire to 4/0 wire		7	1.143		9.40	68.50		77.90	112
2770	4/0 wire to #4 wire		7	1.143		9.40	68.50		77.90	112
2780	4/0 wire to #8 wire		7	1.143		9.40	68.50		77.90	112
2790	Mold, reusable, for above					140			140	154
2800	Brazed connections, #6 wire	1 Elec	12	.667		16.15	40		56.15	77.50
3000	#2 wire		10	.800		21.50	48		69.50	95.50
3100	3/0 wire		8	1		32	60		92	125
3200	4/0 wire		7	1.143		36.50	68.50		105	143
3400	250 kcmil wire		5	1.600		42.50	96		138.50	190
3600	500 kcmil wire		4	2		52.50	120		172.50	237
3700	Insulated ground wire, copper #14		13	.615	C.L.F.	7.55	37		44.55	63.50
3710	#12		11	.727		10.50	43.50		54	76.50
3720	#10		10	.800		16.10	48		64.10	89
3730	#8		8	1		28.50	60		88.50	121
3740	#6		6.50	1.231		48.50	74		122.50	163
3750	#4	2 Elec	10.60	1.509		69	90.50		159.50	211
3770	#2		9	1.778		117	107		224	287
3780	#1		8	2		151	120		271	345
3790	1/0		6.60	2.424		195	146		341	430
3800	2/0		5.80	2.759		221	166		387	490
3810	3/0		5	3.200		283	192		475	595
3820	4/0		4.40	3.636		350	218		568	710
3830	250 kcmil	3 Elec	6	4		430	240		670	835
3840	300 kcmil		5.70	4.211		505	253		758	930
3850	350 kcmil		5.40	4.444		590	267		857	1,050
3860	400 kcmil		5.10	4.706		675	283		958	1,175
3870	500 kcmil		4.80	5		840	300		1,140	1,375
3880	600 kcmil		3.90	6.154		1,225	370		1,595	1,900
3890	750 kcmil		3.30	7.273		1,375	435		1,810	2,150
3900	1,000 kcmil		2.70	8.889		1,975	535		2,510	2,975
3960	Insulated ground wire, aluminum, #6	1 Elec	8	1		25	60		85	117
3970	#4	2 Elec	13	1.231		30.50	74		104.50	144
3980	#2		10.60	1.509		42.50	90.50		133	182
3990	#1		9	1.778		62	107		169	227
4000	1/0		8	2		74.50	120		194.50	261
4010	2/0		7.20	2.222		88.50	133		221.50	297
4020	3/0		6.60	2.424		111	146		257	340
4030	4/0		6.20	2.581		123	155		278	365
4040	250 kcmil	3 Elec	8.70	2.759		149	166		315	410
4050	300 kcmil		8.10	2.963		206	178		384	490

26 05 26 – Grounding and Bonding for Electrical Systems

26 05 26.80 Grounding		Crew	Daily Output	Labor-Hours	Unit	Material	2019 Bare Costs Labor	Equipment	Total	Total Incl O&P
4060	350 kcmil	3 Elec	7.50	3.200	C.L.F.	208	192		400	515
4070	400 kcmil		6.90	3.478		237	209		446	570
4080	500 kcmil		6	4		267	240		507	655
4090	600 kcmil		5.70	4.211		340	253		593	745
4100	700 kcmil		5.10	4.706		415	283		698	875
4110	750 kcmil		4.80	5		410	300		710	900
5000	Copper electrolytic ground rod system									
5010	Includes augering hole, mixing bentonite clay,									
5020	Installing rod, and terminating ground wire									
5100	Straight vertical type, 2" diam.									
5120	8.5' long, clamp connection	1 Elec	2.67	2.996	Ea.	720	180		900	1,075
5130	With exothermic weld connection		1.95	4.103		730	246		976	1,175
5140	10' long		2.35	3.404		850	204		1,054	1,250
5150	With exothermic weld connection		1.78	4.494		860	270		1,130	1,350
5160	12' long		2.16	3.704		1,125	222		1,347	1,575
5170	With exothermic weld connection		1.67	4.790		1,075	288		1,363	1,625
5180	20' long		1.74	4.598		1,400	276		1,676	1,925
5190	With exothermic weld connection		1.40	5.714		1,400	345		1,745	2,025
5195	40' long with exothermic weld connection	2 Elec	2	8		3,050	480		3,530	4,075
5200	L-shaped, 2" diam.									
5220	4' vert. x 10' horiz., clamp connection	1 Elec	5.33	1.501	Ea.	1,325	90		1,415	1,575
5230	With exothermic weld connection	"	3.08	2.597	"	1,325	156		1,481	1,700
5300	Protective box at grade level, with breather slots									
5320	Round 12" long, fiberlyte	1 Elec	32	.250	Ea.	70.50	15		85.50	100
5330	Concrete	"	16	.500		110	30		140	166
5400	Bentonite clay, 50# bag, 1 per 10' of rod					50			50	55
5500	Equipotential earthing bar	1 Elec	2	4		139	240		379	515
7000	Exothermic welding kit, multi vertical	"	8	1		680	60		740	835

26 05 29 – Hangers and Supports for Electrical Systems

26 05 29.20 Hangers

		Crew	Daily Output	Labor-Hours	Unit	Material	Labor	Equipment	Total	Total Incl O&P
0010	**HANGERS** R260533-21									
0015	See section 22 05 29.10 for additional items									
0030	Conduit supports									
0050	Strap w/2 holes, rigid steel conduit									
0100	1/2" diameter	1 Elec	470	.017	Ea.	.09	1.02		1.11	1.63
0150	3/4" diameter		440	.018		.10	1.09		1.19	1.74
0200	1" diameter		400	.020		.14	1.20		1.34	1.94
0300	1-1/4" diameter		355	.023		.23	1.35		1.58	2.27
0350	1-1/2" diameter		320	.025		.30	1.50		1.80	2.57
0400	2" diameter		266	.030		.40	1.81		2.21	3.14
0500	2-1/2" diameter		160	.050		.59	3		3.59	5.15
0550	3" diameter		133	.060		.86	3.61		4.47	6.35
0600	3-1/2" diameter		100	.080		1.87	4.80		6.67	9.20
0650	4" diameter		80	.100		.88	6		6.88	9.90
0700	EMT, 1/2" diameter		470	.017		.16	1.02		1.18	1.71
0800	3/4" diameter		440	.018		.15	1.09		1.24	1.80
0850	1" diameter		400	.020		.26	1.20		1.46	2.08
0900	1-1/4" diameter		355	.023		.52	1.35		1.87	2.59
0950	1-1/2" diameter		320	.025		.37	1.50		1.87	2.65
1000	2" diameter		266	.030		.70	1.81		2.51	3.47
1100	2-1/2" diameter		160	.050		1.17	3		4.17	5.75
1150	3" diameter		133	.060		1.49	3.61		5.10	7.05

26 05 29.20 Hangers		Crew	Daily Output	Labor-Hours	Unit	Material	2019 Bare Costs Labor	Equipment	Total	Total Incl O&P
1200	3-1/2" diameter	1 Elec	100	.080	Ea.	2.91	4.80		7.71	10.35
1250	4" diameter		80	.100		2.39	6		8.39	11.60
1400	Hanger, with bolt, 1/2" diameter		200	.040		.34	2.40		2.74	3.95
1450	3/4" diameter		190	.042		.36	2.53		2.89	4.17
1500	1" diameter		176	.045		.46	2.73		3.19	4.58
1550	1-1/4" diameter		160	.050		.57	3		3.57	5.10
1600	1-1/2" diameter		140	.057		.76	3.43		4.19	5.95
1650	2" diameter		130	.062		1.25	3.70		4.95	6.90
1700	2-1/2" diameter		100	.080		1.35	4.80		6.15	8.65
1750	3" diameter		64	.125		1.92	7.50		9.42	13.30
1800	3-1/2" diameter		50	.160		1.65	9.60		11.25	16.15
1850	4" diameter		40	.200		2.55	12		14.55	20.50
1900	Riser clamps, conduit, 1/2" diameter		40	.200		7.20	12		19.20	26
1950	3/4" diameter		36	.222		7.40	13.35		20.75	28
2000	1" diameter		30	.267		7.70	16		23.70	32.50
2100	1-1/4" diameter		27	.296		9.60	17.80		27.40	37
2150	1-1/2" diameter		27	.296		9.90	17.80		27.70	37.50
2200	2" diameter		20	.400		10.75	24		34.75	48
2250	2-1/2" diameter		20	.400		12.05	24		36.05	49.50
2300	3" diameter		18	.444		12.75	26.50		39.25	54
2350	3-1/2" diameter		18	.444		10.70	26.50		37.20	52
2400	4" diameter		14	.571		12.45	34.50		46.95	64.50
2500	Threaded rod, painted, 1/4" diameter		260	.031	L.F.	1.42	1.85		3.27	4.32
2600	3/8" diameter		200	.040		4.96	2.40		7.36	9.05
2700	1/2" diameter		140	.057		5.15	3.43		8.58	10.80
2800	5/8" diameter		100	.080		7.05	4.80		11.85	14.90
2900	3/4" diameter		60	.133		6.35	8		14.35	18.90
2940	Couplings painted, 1/4" diameter				C	276			276	305
2960	3/8" diameter					445			445	490
2970	1/2" diameter					555			555	610
2980	5/8" diameter					750			750	825
2990	3/4" diameter					955			955	1,050
3000	Nuts, galvanized, 1/4" diameter					10.30			10.30	11.30
3050	3/8" diameter					16.85			16.85	18.50
3100	1/2" diameter					42.50			42.50	46.50
3150	5/8" diameter					155			155	171
3200	3/4" diameter					128			128	140
3250	Washers, galvanized, 1/4" diameter					13.80			13.80	15.20
3300	3/8" diameter					15.85			15.85	17.45
3350	1/2" diameter					38			38	42
3400	5/8" diameter					82			82	90.50
3450	3/4" diameter					144			144	158
3500	Lock washers, galvanized, 1/4" diameter					9.10			9.10	10
3550	3/8" diameter					15.25			15.25	16.75
3600	1/2" diameter					20.50			20.50	23
3650	5/8" diameter					42			42	46.50
3700	3/4" diameter					79			79	87
3705	Metal channel slotted, 304 stainless steel, aluminum									
3710	304 stainless steel, 1' long, 12GA, 1-5/8" x 1-5/8"	1 Elec	24	.333	L.F.	19.90	20		39.90	52
3715	304 stainless steel, 1'6" long, 12GA, 1-5/8" x 1-5/8"		24	.333		26.50	20		46.50	59.50
3720	304 stainless steel, 2' long, 12GA, 1-5/8" x 1-5/8"		24	.333		37.50	20		57.50	71.50
3725	304 stainless steel, 3' long, 12GA, 1-5/8" x 1-5/8"		24	.333		53	20		73	88.50
3730	304 stainless steel, 4' long, 12GA, 1-5/8" x 1-5/8"		24	.333		60	20		80	96

26 05 29 – Hangers and Supports for Electrical Systems

26 05 29.20 Hangers		Crew	Daily Output	Labor-Hours	Unit	Material	2019 Bare Costs Labor	Equipment	Total	Total Incl O&P
3735	304 stainless steel, 5' long, 12GA, 1-5/8" x 1-5/8"	1 Elec	24	.333	L.F.	90.50	20		110.50	130
3740	304 stainless steel, 10' long, 12GA, 1-5/8" x 1-5/8"		24	.333		124	20		144	167
3745	Aluminum, 1' length, 12GA, 1-5/8" x 1-5/8"		24	.333		5.45	20		25.45	36
3750	Aluminum, 1'6" length, 12GA, 1-5/8" x 1-5/8"		24	.333		7.85	20		27.85	38.50
3755	Aluminum, 2' length, 12GA, 1-5/8" x 1-5/8"		24	.333		10.35	20		30.35	41.50
3760	Aluminum, 3' length, 12GA, 1-5/8" x 1-5/8"		24	.333		16.30	20		36.30	48
3765	Aluminum, 4' length, 12GA, 1-5/8" x 1-5/8"		24	.333		23	20		43	55.50
3770	Aluminum, 5' length, 12GA, 1-5/8" x 1-5/8"		24	.333		24.50	20		44.50	57
3775	Aluminum, 10' length, 12GA, 1-5/8" x 1-5/8"		24	.333		45.50	20		65.50	80
3800	Channels, steel, 3/4" x 1-1/2", 14 ga.		80	.100		2.63	6		8.63	11.85
3900	1-1/2" x 1-1/2", 12 ga.		70	.114		3.43	6.85		10.28	14
4000	1-7/8" x 1-1/2"		60	.133		17.60	8		25.60	31.50
4100	3" x 1-1/2"		50	.160		17.95	9.60		27.55	34
4110	1-5/8" x 13/16", 16 ga.		80	.100		6.40	6		12.40	16
4114	Trapeze channel support, 12" wide, steel, 12 ga.		8.80	.909	Ea.	23.50	54.50		78	108
4116	18" wide, steel		8.30	.964	"	25.50	58		83.50	115
4120	1-5/8" x 1-5/8", 14 ga.		70	.114	L.F.	6.20	6.85		13.05	17.05
4130	1-5/8" x 7/8", 12 ga.		70	.114		5.20	6.85		12.05	15.95
4140	1-5/8" x 1-3/8", 12 ga.		60	.133		7.20	8		15.20	19.90
4150	1-5/8" x 1-5/8", 12 ga.		50	.160		6.30	9.60		15.90	21.50
4160	Flat plate fitting, 2 hole, 3-1/2", material only				Ea.	3.02			3.02	3.32
4170	3 hole, 1-7/16" x 4-1/8"					3.84			3.84	4.22
4200	Spring nuts, long, 1/4"	1 Elec	120	.067		.63	4		4.63	6.65
4250	3/8"		100	.080		.71	4.80		5.51	7.95
4300	1/2"		80	.100		.79	6		6.79	9.80
4350	Spring nuts, short, 1/4"		120	.067		1.28	4		5.28	7.35
4400	3/8"		100	.080		1.58	4.80		6.38	8.90
4450	1/2"		80	.100		1.66	6		7.66	10.80
4500	Closure strip		200	.040	L.F.	3.01	2.40		5.41	6.90
4550	End cap		60	.133	Ea.	1.05	8		9.05	13.10
4600	End connector 3/4" conduit		40	.200		7.10	12		19.10	25.50
4650	Junction box, 1 channel		16	.500		32.50	30		62.50	81
4700	2 channel		14	.571		52.50	34.50		87	109
4750	3 channel		12	.667		72	40		112	139
4800	4 channel		10	.800		77.50	48		125.50	157
4850	Splice plate		40	.200		9.20	12		21.20	28
4900	Continuous concrete insert, 1-1/2" deep, 1' long		16	.500		19.30	30		49.30	66.50
4950	2' long		14	.571		24	34.50		58.50	77.50
5000	3' long		12	.667		23.50	40		63.50	85.50
5050	4' long		10	.800		33.50	48		81.50	109
5100	6' long		8	1		50.50	60		110.50	145
5150	3/4" deep, 1' long		16	.500		17.55	30		47.55	64.50
5200	2' long		14	.571		22	34.50		56.50	75
5250	3' long		12	.667		22.50	40		62.50	84.50
5300	4' long		10	.800		30	48		78	105
5350	6' long		8	1		45	60		105	139
5400	90° angle fitting 2-1/8" x 2-1/8"		60	.133		3.66	8		11.66	16
5450	Supports, suspension rod type, small		60	.133		56.50	8		64.50	74
5500	Large		40	.200		65.50	12		77.50	90
5550	Beam clamp, small		60	.133		28	8		36	42.50
5600	Large		40	.200		25	12		37	45.50
5650	U-support, small		60	.133		4.55	8		12.55	16.95
5700	Large		40	.200		9.90	12		21.90	29

26 05 Common Work Results for Electrical

26 05 29 – Hangers and Supports for Electrical Systems

26 05 29.20 Hangers		Crew	Daily Output	Labor-Hours	Unit	Material	2019 Bare Costs Labor	Equipment	Total	Total Incl O&P
5750	Concrete insert, cast, for up to 1/2" threaded rod	1 Elec	16	.500	Ea.	11.30	30		41.30	57.50
5800	Beam clamp, 1/4" clamp, for 1/4" threaded drop rod		32	.250		3.02	15		18.02	26
5900	3/8" clamp, for 3/8" threaded drop rod		32	.250		3.21	15		18.21	26
6000	Strap, rigid conduit, 1/2" diameter		540	.015		1.11	.89		2	2.55
6050	3/4" diameter		440	.018		1.17	1.09		2.26	2.92
6100	1" diameter		420	.019		1.39	1.14		2.53	3.24
6150	1-1/4" diameter		400	.020		1.61	1.20		2.81	3.56
6200	1-1/2" diameter		400	.020		1.89	1.20		3.09	3.87
6250	2" diameter		267	.030		2.16	1.80		3.96	5.05
6300	2-1/2" diameter		267	.030		2.23	1.80		4.03	5.15
6350	3" diameter		160	.050		2.78	3		5.78	7.55
6400	3-1/2" diameter		133	.060		3.08	3.61		6.69	8.80
6450	4" diameter		100	.080		3.58	4.80		8.38	11.10
6500	5" diameter		80	.100		8.30	6		14.30	18.10
6550	6" diameter		60	.133		13.55	8		21.55	27
6600	EMT, 1/2" diameter		540	.015		.87	.89		1.76	2.29
6650	3/4" diameter		440	.018		.91	1.09		2	2.63
6700	1" diameter		420	.019		.96	1.14		2.10	2.77
6750	1-1/4" diameter		400	.020		1.21	1.20		2.41	3.12
6800	1-1/2" diameter		400	.020		1.55	1.20		2.75	3.50
6850	2" diameter		267	.030		1.47	1.80		3.27	4.30
6900	2-1/2" diameter		267	.030		1.83	1.80		3.63	4.69
6950	3" diameter		160	.050		2.26	3		5.26	6.95
6970	3-1/2" diameter		133	.060		3.05	3.61		6.66	8.75
6990	4" diameter		100	.080		3.01	4.80		7.81	10.45
7000	Clip, 1 hole for rigid conduit, 1/2" diameter		500	.016		.06	.96		1.02	1.50
7050	3/4" diameter		470	.017		.68	1.02		1.70	2.28
7100	1" diameter		440	.018		1.66	1.09		2.75	3.46
7150	1-1/4" diameter		400	.020		1.60	1.20		2.80	3.55
7200	1-1/2" diameter		355	.023		1.86	1.35		3.21	4.07
7250	2" diameter		320	.025		1.65	1.50		3.15	4.06
7300	2-1/2" diameter		266	.030		4.52	1.81		6.33	7.65
7350	3" diameter		160	.050		4.38	3		7.38	9.30
7400	3-1/2" diameter		133	.060		6.30	3.61		9.91	12.30
7450	4" diameter		100	.080		13.30	4.80		18.10	22
7500	5" diameter		80	.100		97.50	6		103.50	116
7550	6" diameter		60	.133		175	8		183	205
7600	Hammer on purlin clip, 1/4" hole, 1/16" flange		60	.133		.57	8		8.57	12.60
7610	1/4" hole threaded,1/4" flange, rod hanger		60	.133		1.29	8		9.29	13.35
7620	3/8" hole threaded,1/4" flange, rod hanger		60	.133		1.31	8		9.31	13.40
7630	Z purlin clip, 1/4" & 3/8" flange threaded rod hanger		60	.133		1.07	8		9.07	13.15
7640	1/2" rigid, 3/4" conduit, 1" flange		60	.133		1.35	8		9.35	13.45
7650	1" conduit, 1" flange		60	.133		1.53	8		9.53	13.65
7660	1-1/4" conduit, 1" flange		60	.133		1.81	8		9.81	13.95
7670	Push-in type conduit clip 3/4" EMT, with 9/32" hole		60	.133		.44	8		8.44	12.45
7680	3/4" EMT, 1/2" rigid		60	.133		1.17	8		9.17	13.25
7690	Push-in type tandem conduit clip 3/4" - 1" EMT, 1/2" - 3/4" rigid		60	.133		1.21	8		9.21	13.30
7820	Conduit hangers, with bolt & 12" rod, 1/2" diameter		150	.053		5.30	3.20		8.50	10.65
7830	3/4" diameter		145	.055		5.30	3.31		8.61	10.80
7840	1" diameter		135	.059		5.40	3.56		8.96	11.25
7850	1-1/4" diameter		120	.067		5.55	4		9.55	12.05
7860	1-1/2" diameter		110	.073		5.70	4.37		10.07	12.80
7870	2" diameter		100	.080		6.40	4.80		11.20	14.20

26 05 29.20 Hangers		Crew	Daily Output	Labor-Hours	Unit	Material	2019 Bare Costs Labor	Equipment	Total	Total Incl O&P
7880	2-1/2" diameter	1 Elec	80	.100	Ea.	6.50	6		12.50	16.10
7890	3" diameter		60	.133		7.10	8		15.10	19.75
7900	3-1/2" diameter		45	.178		6.80	10.70		17.50	23.50
7910	4" diameter		35	.229		7.70	13.75		21.45	29
7920	5" diameter		30	.267		9.60	16		25.60	34.50
7930	6" diameter		25	.320		11.95	19.20		31.15	41.50
7950	Jay clamp, 1/2" diameter		32	.250		6.05	15		21.05	29
7960	3/4" diameter		32	.250		6.05	15		21.05	29
7970	1" diameter		32	.250		4.59	15		19.59	27.50
7980	1-1/4" diameter		30	.267		8.90	16		24.90	34
7990	1-1/2" diameter		30	.267		8	16		24	33
8000	2" diameter		30	.267		12.35	16		28.35	37.50
8010	2-1/2" diameter		28	.286		19.05	17.15		36.20	46.50
8020	3" diameter		28	.286		19.05	17.15		36.20	46.50
8030	3-1/2" diameter		25	.320		42	19.20		61.20	74.50
8040	4" diameter		25	.320		42	19.20		61.20	74.50
8050	5" diameter		20	.400		97	24		121	143
8060	6" diameter		16	.500		185	30		215	249
8070	Channels, 3/4" x 1-1/2" w/12" rods for 1/2" to 1" conduit		30	.267		8.90	16		24.90	34
8080	1-1/2" x 1-1/2" w/12" rods for 1-1/4" to 2" conduit		28	.286		9.25	17.15		26.40	35.50
8090	1-1/2" x 1-1/2" w/12" rods for 2-1/2" to 4" conduit		26	.308		11.35	18.50		29.85	40
8100	1-1/2" x 1-7/8" w/12" rods for 5" to 6" conduit		24	.333		41.50	20		61.50	75.50
8110	Beam clamp, conduit, plastic coated steel, 1/2" diam.		30	.267		29	16		45	55.50
8120	3/4" diameter		30	.267		32	16		48	59.50
8130	1" diameter		30	.267		28	16		44	54.50
8140	1-1/4" diameter		28	.286		41	17.15		58.15	71
8150	1-1/2" diameter		28	.286		50	17.15		67.15	80.50
8160	2" diameter		28	.286		59.50	17.15		76.65	91
8170	2-1/2" diameter		26	.308		52.50	18.50		71	85.50
8180	3" diameter		26	.308		81	18.50		99.50	117
8190	3-1/2" diameter		23	.348		71	21		92	109
8200	4" diameter		23	.348		79.50	21		100.50	119
8210	5" diameter		18	.444		201	26.50		227.50	261
8220	Channels, plastic coated									
8250	3/4" x 1-1/2", w/12" rods for 1/2" to 1" conduit	1 Elec	28	.286	Ea.	35	17.15		52.15	64
8260	1-1/2" x 1-1/2", w/12" rods for 1-1/4" to 2" conduit		26	.308		39.50	18.50		58	71
8270	1-1/2" x 1-1/2", w/12" rods for 2-1/2" to 3-1/2" conduit		24	.333		42.50	20		62.50	77
8280	1-1/2" x 1-7/8", w/12" rods for 4" to 5" conduit		22	.364		71	22		93	111
8290	1-1/2" x 1-7/8", w/12" rods for 6" conduit		20	.400		78	24		102	122
8320	Conduit hangers, plastic coated steel, with bolt & 12" rod, 1/2" diam.		140	.057		20	3.43		23.43	27
8330	3/4" diameter		135	.059		21	3.56		24.56	28.50
8340	1" diameter		125	.064		21	3.84		24.84	29.50
8350	1-1/4" diameter		110	.073		22.50	4.37		26.87	31
8360	1-1/2" diameter		100	.080		25	4.80		29.80	34.50
8370	2" diameter		90	.089		27.50	5.35		32.85	38.50
8380	2-1/2" diameter		70	.114		33	6.85		39.85	46.50
8390	3" diameter		50	.160		40	9.60		49.60	58.50
8400	3-1/2" diameter		35	.229		40.50	13.75		54.25	65
8410	4" diameter		25	.320		59	19.20		78.20	93.50
8420	5" diameter		20	.400		64.50	24		88.50	107
9000	Parallel type, conduit beam clamp, 1/2"		32	.250		9.55	15		24.55	33
9010	3/4"		32	.250		5.70	15		20.70	29
9020	1"		32	.250		6.50	15		21.50	29.50

26 05 29.20 Hangers

		Crew	Daily Output	Labor-Hours	Unit	Material	2019 Bare Costs Labor	2019 Bare Costs Equipment	Total	Total Incl O&P
9030	1-1/4"	1 Elec	30	.267	Ea.	13.05	16		29.05	38.50
9040	1-1/2"		30	.267		14.80	16		30.80	40.50
9050	2"		30	.267		13.50	16		29.50	39
9060	2-1/2"		28	.286		20.50	17.15		37.65	48
9070	3"		28	.286		21.50	17.15		38.65	49
9090	4"		25	.320		14.95	19.20		34.15	45
9110	Right angle, conduit beam clamp, 1/2"		32	.250		2.49	15		17.49	25
9120	3/4"		32	.250		2.69	15		17.69	25.50
9130	1"		32	.250		2.43	15		17.43	25
9140	1-1/4"		30	.267		3.62	16		19.62	28
9150	1-1/2"		30	.267		3.75	16		19.75	28
9160	2"		30	.267		5.50	16		21.50	30
9170	2-1/2"		28	.286		6.75	17.15		23.90	33
9180	3"		28	.286		7.40	17.15		24.55	33.50
9190	3-1/2"		25	.320		9.50	19.20		28.70	39
9200	4"		25	.320		9.75	19.20		28.95	39.50
9230	Adjustable, conduit hanger, 1/2"		32	.250		7.75	15		22.75	31
9240	3/4"		32	.250		5.70	15		20.70	29
9250	1"		32	.250		7.55	15		22.55	31
9260	1-1/4"		30	.267		8.10	16		24.10	33
9270	1-1/2"		30	.267		9.45	16		25.45	34.50
9280	2"		30	.267		4.10	16		20.10	28.50
9290	2-1/2"		28	.286		21	17.15		38.15	48.50
9300	3"		28	.286		25.50	17.15		42.65	53.50
9310	3-1/2"		25	.320		26	19.20		45.20	57
9320	4"		25	.320		29.50	19.20		48.70	61
9330	5"		20	.400		38	24		62	77.50
9340	6"		16	.500		48	30		78	98
9350	Combination conduit hanger, 3/8"		32	.250		13.90	15		28.90	38
9360	Adjustable flange, 3/8"		32	.250		16.25	15		31.25	40.50

26 05 29.30 Fittings and Channel Support

		Crew	Daily Output	Labor-Hours	Unit	Material	2019 Bare Costs Labor	2019 Bare Costs Equipment	Total	Total Incl O&P
0010	**FITTINGS & CHANNEL SUPPORT**									
0020	Rooftop channel support									
0200	2-7/8" L x 1-5/8" W, 12 ga. pre galv., dbl. base	1 Elec	46	.174	Ea.	3.87	10.45		14.32	19.85
0210	2-7/8" L x 1-5/8" W, 12 ga. hot dip galv., dbl. base		46	.174		2.25	10.45		12.70	18.10
0220	2-7/16" L x 1-5/8" W, 12 ga. pre galv., sngl base		48	.167		1.86	10		11.86	17
0230	2-7/16" L x 1-5/8" W, 12 ga. hot dip galv., sngl base		48	.167		6.60	10		16.60	22
0240	13/16" L x 1-5/8" W, 14 ga. pre galv., sngl base		56	.143		1.58	8.60		10.18	14.55
0250	13/16" L x 1-5/8" W, 14 ga. hot dip galv., sngl base		56	.143		6.45	8.60		15.05	19.90
0260	1-5/8" L x 1-5/8" W, 12 ga. pre galv., sngl base		54	.148		1.71	8.90		10.61	15.15
0270	1-5/8" L x 1-5/8" W, 12 ga. hot dip galv., sngl base		54	.148		1.81	8.90		10.71	15.25
0280	1-5/8" x 28", 12 ga. pre galv., dbl. H block base		42	.190		3.92	11.45		15.37	21.50
0290	1-5/8" x 36", 12 ga. pre galv., dbl. H block base		40	.200		4.15	12		16.15	22.50
0300	1-5/8" x 42", 12 ga. pre galv., dbl. H block base		34	.235		4.66	14.15		18.81	26
0310	1-5/8" x 50", 12 ga. pre galv., dbl. H block base		38	.211		4.97	12.65		17.62	24.50
0320	1-5/8" x 60", 12 ga. pre galv., dbl. H block base		36	.222		5.45	13.35		18.80	26
0330	13/16" x 8" H, threaded rod pre galv., H block base		35	.229		2.63	13.75		16.38	23.50
0340	13/16" x 12" H, threaded rod pre galv., H block base		34.60	.231		3.29	13.90		17.19	24
0350	1-5/8" x 16" H, threaded rod dbl. H block base		34	.235		4.92	14.15		19.07	26.50
0360	1-5/8" x 12" H, pre galv., dbl. H block base		34.60	.231		9.40	13.90		23.30	31
0370	1-5/8" x 24"H, pre galv., dbl. H block base		34	.235		9.95	14.15		24.10	32
0380	3-1/4" x 24"H, pre galv., dbl. H block base		32	.250		22.50	15		37.50	47.50

26 05 29 – Hangers and Supports for Electrical Systems

26 05 29.30 Fittings and Channel Support	Crew	Daily Output	Labor-Hours	Unit	Material	2019 Bare Costs Labor	Equipment	Total	Total Incl O&P
0390 3-1/4" x 36"H, pre galv., dbl. H block base	1 Elec	30	.267	Ea.	24.50	16		40.50	51

26 05 33 – Raceway and Boxes for Electrical Systems

26 05 33.13 Conduit

		Crew	Daily Output	Labor-Hours	Unit	Material	2019 Bare Costs Labor	Equipment	Total	Total Incl O&P
0010	**CONDUIT** To 10' high, includes 2 terminations, 2 elbows, R260533-20									
0020	11 beam clamps, and 11 couplings per 100 L.F.									
0300	Aluminum, 1/2" diameter	1 Elec	100	.080	L.F.	1.74	4.80		6.54	9.05
0500	3/4" diameter R260533-25		90	.089		2.31	5.35		7.66	10.50
0700	1" diameter		80	.100		2.84	6		8.84	12.05
1000	1-1/4" diameter		70	.114		4.07	6.85		10.92	14.70
1030	1-1/2" diameter		65	.123		4.69	7.40		12.09	16.20
1050	2" diameter		60	.133		6.60	8		14.60	19.20
1070	2-1/2" diameter		50	.160		9.25	9.60		18.85	24.50
1100	3" diameter	2 Elec	90	.178		12.25	10.70		22.95	29.50
1130	3-1/2" diameter		80	.200		15.95	12		27.95	35.50
1140	4" diameter		70	.229		17.95	13.75		31.70	40.50
1150	5" diameter		50	.320		37.50	19.20		56.70	70
1160	6" diameter		40	.400		54	24		78	95.50
1161	Field bends, 45° to 90°, 1/2" diameter	1 Elec	53	.151	Ea.		9.05		9.05	13.50
1162	3/4" diameter		47	.170			10.20		10.20	15.25
1163	1" diameter		44	.182			10.90		10.90	16.30
1164	1-1/4" diameter		23	.348			21		21	31
1165	1-1/2" diameter		21	.381			23		23	34
1166	2" diameter		16	.500			30		30	45
1170	Elbows, 1/2" diameter		40	.200		5.85	12		17.85	24.50
1200	3/4" diameter		32	.250		7.60	15		22.60	31
1230	1" diameter		28	.286		10.80	17.15		27.95	37.50
1250	1-1/4" diameter		24	.333		17.50	20		37.50	49.50
1270	1-1/2" diameter		20	.400		23	24		47	61.50
1300	2" diameter		16	.500		33.50	30		63.50	82
1330	2-1/2" diameter		12	.667		55.50	40		95.50	121
1350	3" diameter		8	1		70	60		130	167
1370	3-1/2" diameter		6	1.333		113	80		193	243
1400	4" diameter		5	1.600		145	96		241	305
1410	5" diameter		4	2		425	120		545	645
1420	6" diameter		2.50	3.200		555	192		747	895
1430	Couplings, 1/2" diameter		320	.025		1.91	1.50		3.41	4.34
1450	3/4" diameter		192	.042		2.89	2.50		5.39	6.90
1470	1" diameter		160	.050		3.76	3		6.76	8.60
1500	1-1/4" diameter		120	.067		4.55	4		8.55	10.95
1530	1-1/2" diameter		96	.083		5.25	5		10.25	13.25
1550	2" diameter		80	.100		7.65	6		13.65	17.40
1570	2-1/2" diameter		69	.116		14.95	6.95		21.90	27
1600	3" diameter		64	.125		22	7.50		29.50	35.50
1630	3-1/2" diameter		56	.143		27	8.60		35.60	43
1650	4" diameter		53	.151		28.50	9.05		37.55	45
1670	5" diameter		48	.167		101	10		111	126
1690	6" diameter		46	.174		161	10.45		171.45	193
1691	See note on line 26 05 33.13 9995 R260533-30									
1750	Rigid galvanized steel, 1/2" diameter	1 Elec	90	.089	L.F.	2.30	5.35		7.65	10.50
1770	3/4" diameter		80	.100		4.74	6		10.74	14.15
1800	1" diameter		65	.123		7.05	7.40		14.45	18.80
1830	1-1/4" diameter		60	.133		4.82	8		12.82	17.25

26 05 33.13 Conduit		Crew	Daily Output	Labor-Hours	Unit	Material	2019 Bare Costs Labor	2019 Bare Costs Equipment	Total	Total Incl O&P
1850	1-1/2" diameter	1 Elec	55	.145	L.F.	7.75	8.75		16.50	21.50
1870	2" diameter		45	.178		9.95	10.70		20.65	27
1900	2-1/2" diameter	▼	35	.229		12.05	13.75		25.80	34
1930	3" diameter	2 Elec	50	.320		14	19.20		33.20	44
1950	3-1/2" diameter		44	.364		17.55	22		39.55	52
1970	4" diameter		40	.400		20.50	24		44.50	59
1980	5" diameter		30	.533		38.50	32		70.50	90.50
1990	6" diameter	▼	20	.800	▼	56	48		104	133
1991	Field bends, 45° to 90°, 1/2" diameter	1 Elec	44	.182	Ea.		10.90		10.90	16.30
1992	3/4" diameter		40	.200			12		12	17.90
1993	1" diameter		36	.222			13.35		13.35	19.90
1994	1-1/4" diameter		19	.421			25.50		25.50	37.50
1995	1-1/2" diameter		18	.444			26.50		26.50	40
1996	2" diameter		13	.615			37		37	55
2000	Elbows, 1/2" diameter		32	.250		4.35	15		19.35	27.50
2030	3/4" diameter		28	.286		4.35	17.15		21.50	30.50
2050	1" diameter		24	.333		7.20	20		27.20	38
2070	1-1/4" diameter		18	.444		10.05	26.50		36.55	51
2100	1-1/2" diameter		16	.500		16.95	30		46.95	63.50
2130	2" diameter		12	.667		24.50	40		64.50	86.50
2150	2-1/2" diameter		8	1		32	60		92	125
2170	3" diameter		6	1.333		44	80		124	167
2200	3-1/2" diameter		4.20	1.905		69.50	114		183.50	248
2220	4" diameter		4	2		79.50	120		199.50	266
2230	5" diameter		3.50	2.286		213	137		350	440
2240	6" diameter		2	4		380	240		620	780
2250	Couplings, 1/2" diameter		267	.030		1.12	1.80		2.92	3.91
2270	3/4" diameter		160	.050		1.61	3		4.61	6.25
2300	1" diameter		133	.060		2.29	3.61		5.90	7.90
2330	1-1/4" diameter		100	.080		2.58	4.80		7.38	10
2350	1-1/2" diameter		80	.100		4.82	6		10.82	14.25
2370	2" diameter		67	.119		6.30	7.15		13.45	17.65
2400	2-1/2" diameter		57	.140		10.70	8.45		19.15	24.50
2430	3" diameter		53	.151		13.80	9.05		22.85	28.50
2450	3-1/2" diameter		47	.170		17.30	10.20		27.50	34.50
2470	4" diameter		44	.182		25.50	10.90		36.40	44.50
2480	5" diameter		40	.200		41.50	12		53.50	64
2490	6" diameter	▼	38	.211	▼	62.50	12.65		75.15	87.50
2491	See note on line 26 05 33.13 9995 R260533-30									
2500	Steel, intermediate conduit (IMC), 1/2" diameter	1 Elec	100	.080	L.F.	1.69	4.80		6.49	9
2530	3/4" diameter		90	.089		2.06	5.35		7.41	10.20
2550	1" diameter		70	.114		2.58	6.85		9.43	13.10
2570	1-1/4" diameter		65	.123		3.69	7.40		11.09	15.10
2600	1-1/2" diameter		60	.133		4.67	8		12.67	17.10
2630	2" diameter		50	.160		6.25	9.60		15.85	21.50
2650	2-1/2" diameter	▼	40	.200		9.25	12		21.25	28
2670	3" diameter	2 Elec	60	.267		12.15	16		28.15	37.50
2700	3-1/2" diameter		54	.296		17.30	17.80		35.10	45.50
2730	4" diameter	▼	50	.320	▼	11.80	19.20		31	41.50
2731	Field bends, 45° to 90°, 1/2" diameter	1 Elec	44	.182	Ea.		10.90		10.90	16.30
2732	3/4" diameter		40	.200			12		12	17.90
2733	1" diameter		36	.222			13.35		13.35	19.90
2734	1-1/4" diameter	▼	19	.421	▼		25.50		25.50	37.50

26 05 33 – Raceway and Boxes for Electrical Systems

26 05 33.13 Conduit		Crew	Daily Output	Labor-Hours	Unit	Material	2019 Bare Costs Labor	Equipment	Total	Total Incl O&P
2735	1-1/2" diameter	1 Elec	18	.444	Ea.		26.50		26.50	40
2736	2" diameter		13	.615			37		37	55
2750	Elbows, 1/2" diameter		32	.250		6.50	15		21.50	29.50
2770	3/4" diameter		28	.286		11.60	17.15		28.75	38.50
2800	1" diameter		24	.333		13.70	20		33.70	45
2830	1-1/4" diameter		18	.444		17.90	26.50		44.40	59.50
2850	1-1/2" diameter		16	.500		31.50	30		61.50	80
2870	2" diameter		12	.667		60.50	40		100.50	126
2900	2-1/2" diameter		8	1		42	60		102	136
2930	3" diameter		6	1.333		58.50	80		138.50	184
2950	3-1/2" diameter		4.20	1.905		206	114		320	400
2970	4" diameter		4	2		83	120		203	271
3000	Couplings, 1/2" diameter		293	.027		1.12	1.64		2.76	3.68
3030	3/4" diameter		176	.045		1.61	2.73		4.34	5.85
3050	1" diameter		147	.054		2.29	3.27		5.56	7.40
3070	1-1/4" diameter		110	.073		2.58	4.37		6.95	9.35
3100	1-1/2" diameter		88	.091		4.82	5.45		10.27	13.45
3130	2" diameter		73	.110		6.30	6.60		12.90	16.75
3150	2-1/2" diameter		63	.127		10.70	7.65		18.35	23
3170	3" diameter		59	.136		13.80	8.15		21.95	27.50
3200	3-1/2" diameter		52	.154		17.30	9.25		26.55	33
3230	4" diameter		49	.163		25.50	9.80		35.30	42.50
3231	See note on line 26 05 33.13 9995 R260533-30									
4100	Rigid steel, plastic coated, 40 mil thick									
4130	1/2" diameter	1 Elec	80	.100	L.F.	8.50	6		14.50	18.30
4150	3/4" diameter		70	.114		9.70	6.85		16.55	21
4170	1" diameter		55	.145		11.10	8.75		19.85	25.50
4200	1-1/4" diameter		50	.160		15.90	9.60		25.50	32
4230	1-1/2" diameter		45	.178		17.95	10.70		28.65	35.50
4250	2" diameter		35	.229		22.50	13.75		36.25	45.50
4270	2-1/2" diameter		25	.320		31.50	19.20		50.70	63.50
4300	3" diameter	2 Elec	44	.364		40.50	22		62.50	77
4330	3-1/2" diameter		40	.400		52.50	24		76.50	93.50
4350	4" diameter		36	.444		56	26.50		82.50	102
4370	5" diameter		30	.533		120	32		152	180
4400	Elbows, 1/2" diameter	1 Elec	28	.286	Ea.	17.45	17.15		34.60	44.50
4430	3/4" diameter		24	.333		18.20	20		38.20	50
4450	1" diameter		18	.444		21.50	26.50		48	63.50
4470	1-1/4" diameter		16	.500		28.50	30		58.50	76.50
4500	1-1/2" diameter		12	.667		32.50	40		72.50	95.50
4530	2" diameter		8	1		42	60		102	136
4550	2-1/2" diameter		6	1.333		85	80		165	213
4570	3" diameter		4.20	1.905		134	114		248	320
4600	3-1/2" diameter		4	2		189	120		309	385
4630	4" diameter		3.80	2.105		193	126		319	400
4650	5" diameter		3.50	2.286		490	137		627	745
4680	Couplings, 1/2" diameter		213	.038		4.99	2.26		7.25	8.85
4700	3/4" diameter		128	.063		5.80	3.75		9.55	12
4730	1" diameter		107	.075		6.95	4.49		11.44	14.35
4750	1-1/4" diameter		80	.100		9.35	6		15.35	19.20
4770	1-1/2" diameter		64	.125		10.35	7.50		17.85	22.50
4800	2" diameter		53	.151		14.60	9.05		23.65	29.50
4830	2-1/2" diameter		46	.174		34	10.45		44.45	53

26 05 33 – Raceway and Boxes for Electrical Systems

26 05 33.13 Conduit		Crew	Daily Output	Labor-Hours	Unit	Material	2019 Bare Costs Labor	Equipment	Total	Total Incl O&P
4850	3" diameter	1 Elec	43	.186	Ea.	39	11.15		50.15	59
4870	3-1/2" diameter		38	.211		51	12.65		63.65	75.50
4900	4" diameter		36	.222		56.50	13.35		69.85	82.50
4950	5" diameter		32	.250		216	15		231	260
4951	See note on line 26 05 33.13 9995 R260533-30									
5000	Electric metallic tubing (EMT), 1/2" diameter	1 Elec	170	.047	L.F.	.81	2.83		3.64	5.10
5020	3/4" diameter		130	.062		1.18	3.70		4.88	6.80
5040	1" diameter		115	.070		1.90	4.18		6.08	8.35
5060	1-1/4" diameter		100	.080		3.03	4.80		7.83	10.50
5080	1-1/2" diameter		90	.089		3.48	5.35		8.83	11.75
5100	2" diameter		80	.100		3.89	6		9.89	13.25
5120	2-1/2" diameter		60	.133		4.54	8		12.54	16.95
5140	3" diameter	2 Elec	100	.160		5.55	9.60		15.15	20.50
5160	3-1/2" diameter		90	.178		7.15	10.70		17.85	24
5180	4" diameter		80	.200		12.55	12		24.55	31.50
5200	Field bends, 45° to 90°, 1/2" diameter	1 Elec	89	.090	Ea.		5.40		5.40	8.05
5220	3/4" diameter		80	.100			6		6	8.95
5240	1" diameter		73	.110			6.60		6.60	9.80
5260	1-1/4" diameter		38	.211			12.65		12.65	18.85
5280	1-1/2" diameter		36	.222			13.35		13.35	19.90
5300	2" diameter		26	.308			18.50		18.50	27.50
5320	Offsets, 1/2" diameter		65	.123			7.40		7.40	11.05
5340	3/4" diameter		62	.129			7.75		7.75	11.55
5360	1" diameter		53	.151			9.05		9.05	13.50
5380	1-1/4" diameter		30	.267			16		16	24
5400	1-1/2" diameter		28	.286			17.15		17.15	25.50
5420	2" diameter		20	.400			24		24	36
5700	Elbows, 1" diameter		40	.200		3.61	12		15.61	22
5720	1-1/4" diameter		32	.250		4.25	15		19.25	27
5740	1-1/2" diameter		24	.333		4.98	20		24.98	35.50
5760	2" diameter		20	.400		7.20	24		31.20	44
5780	2-1/2" diameter		12	.667		19.60	40		59.60	81
5800	3" diameter		9	.889		27	53.50		80.50	109
5820	3-1/2" diameter		7	1.143		35.50	68.50		104	141
5840	4" diameter		6	1.333		42	80		122	166
6200	Couplings, set screw, steel, 1/2" diameter		235	.034		1.70	2.04		3.74	4.92
6220	3/4" diameter		188	.043		2.40	2.56		4.96	6.45
6240	1" diameter		157	.051		3.84	3.06		6.90	8.80
6260	1-1/4" diameter		118	.068		7.85	4.07		11.92	14.70
6280	1-1/2" diameter		94	.085		1.75	5.10		6.85	9.60
6300	2" diameter		78	.103		13.35	6.15		19.50	24
6320	2-1/2" diameter		67	.119		5.50	7.15		12.65	16.75
6340	3" diameter		59	.136		5.25	8.15		13.40	17.95
6360	3-1/2" diameter		47	.170		7.25	10.20		17.45	23.50
6380	4" diameter		43	.186		53	11.15		64.15	74.50
6381	See note on line 26 05 33.13 9995 R260533-30									
6500	Box connectors, set screw, steel, 1/2" diameter	1 Elec	120	.067	Ea.	.37	4		4.37	6.35
6520	3/4" diameter		110	.073		.60	4.37		4.97	7.15
6540	1" diameter		90	.089		1.15	5.35		6.50	9.20
6560	1-1/4" diameter		70	.114		2.36	6.85		9.21	12.85
6580	1-1/2" diameter		60	.133		2.84	8		10.84	15.05
6600	2" diameter		50	.160		3.87	9.60		13.47	18.60
6620	2-1/2" diameter		36	.222		7.80	13.35		21.15	28.50

26 05 33 – Raceway and Boxes for Electrical Systems

26 05 33.13 Conduit		Crew	Daily Output	Labor-Hours	Unit	Material	2019 Bare Costs Labor	Equipment	Total	Total Incl O&P
6640	3" diameter	1 Elec	27	.296	Ea.	13.95	17.80		31.75	42
6680	3-1/2" diameter		21	.381		10.55	23		33.55	45.50
6700	4" diameter		16	.500		12.60	30		42.60	59
6740	Insulated box connectors, set screw, steel, 1/2" diameter		120	.067		.66	4		4.66	6.70
6760	3/4" diameter		110	.073		1.55	4.37		5.92	8.20
6780	1" diameter		90	.089		2.71	5.35		8.06	10.95
6800	1-1/4" diameter		70	.114		2.59	6.85		9.44	13.10
6820	1-1/2" diameter		60	.133		4.34	8		12.34	16.70
6840	2" diameter		50	.160		6.40	9.60		16	21.50
6860	2-1/2" diameter		36	.222		19.90	13.35		33.25	42
6880	3" diameter		27	.296		31	17.80		48.80	60.50
6900	3-1/2" diameter		21	.381		54	23		77	93.50
6920	4" diameter		16	.500		44	30		74	93.50
7000	EMT to conduit adapters, 1/2" diameter (compression)		70	.114		3.55	6.85		10.40	14.15
7020	3/4" diameter		60	.133		4.85	8		12.85	17.30
7040	1" diameter		50	.160		7.45	9.60		17.05	22.50
7060	1-1/4" diameter		40	.200		17.95	12		29.95	37.50
7080	1-1/2" diameter		30	.267		21.50	16		37.50	48
7100	2" diameter		25	.320		37	19.20		56.20	69
7200	EMT to Greenfield adapters, 1/2" to 3/8" diameter (compression)		90	.089		2.08	5.35		7.43	10.25
7220	1/2" diameter		90	.089		4.46	5.35		9.81	12.85
7240	3/4" diameter		80	.100		5.80	6		11.80	15.35
7260	1" diameter		70	.114		15.20	6.85		22.05	27
7270	1-1/4" diameter		60	.133		17.90	8		25.90	31.50
7280	1-1/2" diameter		50	.160		20.50	9.60		30.10	37
7290	2" diameter		40	.200		30	12		42	51
7400	EMT, LB, LR or LL fittings with covers, 1/2" diameter, set screw		24	.333		4.92	20		24.92	35.50
7420	3/4" diameter		20	.400		6.30	24		30.30	43
7440	1" diameter		16	.500		9	30		39	55
7450	1-1/4" diameter		13	.615		13.50	37		50.50	70
7460	1-1/2" diameter		11	.727		17.65	43.50		61.15	84.50
7470	2" diameter		9	.889		29	53.50		82.50	112
7600	EMT, "T" fittings with covers, 1/2" diameter, set screw		16	.500		5.15	30		35.15	50.50
7620	3/4" diameter		15	.533		6.65	32		38.65	55.50
7640	1" diameter		12	.667		8.15	40		48.15	68.50
7650	1-1/4" diameter		11	.727		13.05	43.50		56.55	79.50
7660	1-1/2" diameter		10	.800		16.15	48		64.15	89.50
7670	2" diameter		8	1		22	60		82	114
8000	EMT, expansion fittings, no jumper, 1/2" diameter		24	.333		96	20		116	136
8020	3/4" diameter		20	.400		80.50	24		104.50	125
8040	1" diameter		16	.500		101	30		131	156
8060	1-1/4" diameter		13	.615		133	37		170	201
8080	1-1/2" diameter		11	.727		181	43.50		224.50	264
8100	2" diameter		9	.889		410	53.50		463.50	530
8110	2-1/2" diameter		7	1.143		365	68.50		433.50	500
8120	3" diameter		6	1.333		540	80		620	715
8140	4" diameter		5	1.600		785	96		881	1,000
8200	Split adapter, 1/2" diameter		110	.073		3.03	4.37		7.40	9.85
8210	3/4" diameter		90	.089		2.98	5.35		8.33	11.25
8220	1" diameter		70	.114		5.20	6.85		12.05	15.95
8230	1-1/4" diameter		60	.133		8.10	8		16.10	21
8240	1-1/2" diameter		50	.160		12.10	9.60		21.70	27.50
8250	2" diameter		36	.222		34	13.35		47.35	57.50

26 05 33.13 Conduit		Crew	Daily Output	Labor-Hours	Unit	Material	2019 Bare Costs Labor	Equipment	Total	Total Incl O&P
8300	1 hole clips, 1/2" diameter	1 Elec	500	.016	Ea.	.42	.96		1.38	1.89
8320	3/4" diameter		470	.017		.23	1.02		1.25	1.78
8340	1" diameter		444	.018		.34	1.08		1.42	1.98
8360	1-1/4" diameter		400	.020		.47	1.20		1.67	2.31
8380	1-1/2" diameter		355	.023		.95	1.35		2.30	3.07
8400	2" diameter		320	.025		1.31	1.50		2.81	3.68
8420	2-1/2" diameter		266	.030		3.60	1.81		5.41	6.65
8440	3" diameter		160	.050		4.12	3		7.12	9
8460	3-1/2" diameter		133	.060		7.70	3.61		11.31	13.90
8480	4" diameter		100	.080		7.30	4.80		12.10	15.15
8500	Clamp back spacers, 1/2" diameter		500	.016		.83	.96		1.79	2.34
8510	3/4" diameter		470	.017		.98	1.02		2	2.61
8520	1" diameter		444	.018		1.65	1.08		2.73	3.43
8530	1-1/4" diameter		400	.020		3.87	1.20		5.07	6.05
8540	1-1/2" diameter		355	.023		7.25	1.35		8.60	9.95
8550	2" diameter		320	.025		5.25	1.50		6.75	8
8560	2-1/2" diameter		266	.030		12	1.81		13.81	15.90
8570	3" diameter		160	.050		33	3		36	41
8580	3-1/2" diameter		133	.060		42	3.61		45.61	51.50
8590	4" diameter		100	.080		97	4.80		101.80	114
8600	Offset connectors, 1/2" diameter		40	.200		3.16	12		15.16	21.50
8610	3/4" diameter		32	.250		3.62	15		18.62	26.50
8620	1" diameter		24	.333		4.96	20		24.96	35.50
8650	90° pulling elbows, female, 1/2" diameter, with gasket		24	.333		8.05	20		28.05	39
8660	3/4" diameter		20	.400		13.30	24		37.30	50.50
8700	Couplings, compression, 1/2" diameter, steel		78	.103		1.48	6.15		7.63	10.85
8710	3/4" diameter		67	.119		1.69	7.15		8.84	12.55
8720	1" diameter		59	.136		2.94	8.15		11.09	15.40
8730	1-1/4" diameter		47	.170		5.90	10.20		16.10	22
8740	1-1/2" diameter		38	.211		8.75	12.65		21.40	28.50
8750	2" diameter		31	.258		11.85	15.50		27.35	36
8760	2-1/2" diameter		24	.333		46.50	20		66.50	81.50
8770	3" diameter		21	.381		27	23		50	64
8780	3-1/2" diameter		19	.421		100	25.50		125.50	148
8790	4" diameter		16	.500		86.50	30		116.50	140
8791	See note on line 26 05 33.13 9995 R260533-30									
8800	Box connectors, compression, 1/2" diam., steel	1 Elec	120	.067	Ea.	2.34	4		6.34	8.50
8810	3/4" diameter		110	.073		3.21	4.37		7.58	10.05
8820	1" diameter		90	.089		4.58	5.35		9.93	13
8830	1-1/4" diameter		70	.114		11.35	6.85		18.20	22.50
8840	1-1/2" diameter		60	.133		14.10	8		22.10	27.50
8850	2" diameter		50	.160		19	9.60		28.60	35.50
8860	2-1/2" diameter		36	.222		48	13.35		61.35	73
8870	3" diameter		27	.296		65	17.80		82.80	98
8880	3-1/2" diameter		21	.381		101	23		124	145
8890	4" diameter		16	.500		102	30		132	157
8900	Box connectors, insulated compression, 1/2" diam., steel		120	.067		1.80	4		5.80	7.95
8910	3/4" diameter		110	.073		2.06	4.37		6.43	8.75
8920	1" diameter		90	.089		3.80	5.35		9.15	12.15
8930	1-1/4" diameter		70	.114		7.95	6.85		14.80	18.95
8940	1-1/2" diameter		60	.133		11.75	8		19.75	25
8950	2" diameter		50	.160		17.05	9.60		26.65	33
8960	2-1/2" diameter		36	.222		65.50	13.35		78.85	92

26 05 33.13 Conduit		Crew	Daily Output	Labor-Hours	Unit	Material	2019 Bare Costs Labor	Equipment	Total	Total Incl O&P
8970	3" diameter	1 Elec	27	.296	Ea.	57.50	17.80		75.30	90
8980	3-1/2" diameter		21	.381		94.50	23		117.50	138
8990	4" diameter		16	.500		86.50	30		116.50	141
9100	PVC, schedule 40, 1/2" diameter		190	.042	L.F.	.93	2.53		3.46	4.79
9110	3/4" diameter		145	.055		1.06	3.31		4.37	6.10
9120	1" diameter		125	.064		1.58	3.84		5.42	7.50
9130	1-1/4" diameter		110	.073		1.77	4.37		6.14	8.45
9140	1-1/2" diameter		100	.080		2.08	4.80		6.88	9.45
9150	2" diameter		90	.089		2.85	5.35		8.20	11.10
9160	2-1/2" diameter		65	.123		3.30	7.40		10.70	14.70
9170	3" diameter	2 Elec	110	.145		4.78	8.75		13.53	18.30
9180	3-1/2" diameter		100	.160		6.40	9.60		16	21.50
9190	4" diameter		90	.178		5.50	10.70		16.20	22
9200	5" diameter		70	.229		7.90	13.75		21.65	29
9210	6" diameter		60	.267		11.85	16		27.85	37
9220	Elbows, 1/2" diameter	1 Elec	50	.160	Ea.	.62	9.60		10.22	15.05
9225	3/4" diameter		42	.190		.68	11.45		12.13	17.80
9230	1" diameter		35	.229		1.30	13.75		15.05	22
9235	1-1/4" diameter		28	.286		2.93	17.15		20.08	28.50
9240	1-1/2" diameter		20	.400		3.01	24		27.01	39.50
9245	2" diameter	R-1A	36.40	.440		3.97	24		27.97	40
9250	2-1/2" diameter		26.70	.599		8.25	32.50		40.75	57.50
9255	3" diameter		22.90	.699		13.15	38		51.15	71
9260	3-1/2" diameter		22.20	.721		17.70	39		56.70	77.50
9265	4" diameter		18.20	.879		23	47.50		70.50	96.50
9270	5" diameter		12.10	1.322		35.50	71.50		107	146
9275	6" diameter		11.10	1.441		36	78		114	156
9312	Couplings, 1/2" diameter	1 Elec	50	.160		.11	9.60		9.71	14.45
9314	3/4" diameter		42	.190		.32	11.45		11.77	17.40
9316	1" diameter		35	.229		.24	13.75		13.99	21
9318	1-1/4" diameter		28	.286		.34	17.15		17.49	26
9320	1-1/2" diameter		20	.400		.40	24		24.40	36.50
9322	2" diameter	R-1A	36.40	.440		1.25	24		25.25	37
9324	2-1/2" diameter		26.70	.599		.91	32.50		33.41	49.50
9326	3" diameter		22.90	.699		1.43	38		39.43	58
9328	3-1/2" diameter		22.20	.721		1.75	39		40.75	60
9330	4" diameter		18.20	.879		2.28	47.50		49.78	73.50
9332	5" diameter		12.10	1.322		5.95	71.50		77.45	114
9334	6" diameter		11.10	1.441		7.70	78		85.70	124
9335	See note on line 26 05 33.13 9995 R260533-30									
9340	Field bends, 45° & 90°, 1/2" diameter	1 Elec	45	.178	Ea.		10.70		10.70	15.95
9350	3/4" diameter		40	.200			12		12	17.90
9360	1" diameter		35	.229			13.75		13.75	20.50
9370	1-1/4" diameter		32	.250			15		15	22.50
9380	1-1/2" diameter		27	.296			17.80		17.80	26.50
9390	2" diameter		20	.400			24		24	36
9400	2-1/2" diameter		16	.500			30		30	45
9410	3" diameter		13	.615			37		37	55
9420	3-1/2" diameter		12	.667			40		40	59.50
9430	4" diameter		10	.800			48		48	71.50
9440	5" diameter		9	.889			53.50		53.50	79.50
9450	6" diameter		8	1			60		60	89.50
9460	PVC adapters, 1/2" diameter		50	.160		.29	9.60		9.89	14.65

26 05 33.13 Conduit		Crew	Daily Output	Labor-Hours	Unit	Material	2019 Bare Costs Labor	Equipment	Total	Total Incl O&P
9470	3/4" diameter	1 Elec	42	.190	Ea.	.48	11.45		11.93	17.60
9480	1" diameter		38	.211		.48	12.65		13.13	19.40
9490	1-1/4" diameter		35	.229		.82	13.75		14.57	21.50
9500	1-1/2" diameter		32	.250		.80	15		15.80	23.50
9510	2" diameter		27	.296		1.53	17.80		19.33	28
9520	2-1/2" diameter		23	.348		2.20	21		23.20	33.50
9530	3" diameter		18	.444		3.11	26.50		29.61	43.50
9540	3-1/2" diameter		13	.615		5.35	37		42.35	61
9550	4" diameter		11	.727		5.15	43.50		48.65	70.50
9560	5" diameter		8	1		10.30	60		70.30	101
9570	6" diameter	↓	6	1.333	↓	19.80	80		99.80	141
9580	PVC-LB, LR or LL fittings & covers									
9590	1/2" diameter	1 Elec	20	.400	Ea.	2.33	24		26.33	38.50
9600	3/4" diameter		16	.500		5.60	30		35.60	51
9610	1" diameter		12	.667		5.85	40		45.85	66
9620	1-1/4" diameter		9	.889		10.05	53.50		63.55	90.50
9630	1-1/2" diameter		7	1.143		11.90	68.50		80.40	115
9640	2" diameter		6	1.333		13.65	80		93.65	134
9650	2-1/2" diameter		6	1.333		39	80		119	162
9660	3" diameter		5	1.600		38.50	96		134.50	186
9670	3-1/2" diameter		4	2		59	120		179	244
9680	4" diameter	↓	3	2.667	↓	67	160		227	315
9690	PVC-tee fitting & cover									
9700	1/2"	1 Elec	14	.571	Ea.	6.30	34.50		40.80	58
9710	3/4"		13	.615		6.70	37		43.70	62.50
9720	1"		10	.800		6.90	48		54.90	79
9730	1-1/4"		9	.889		13	53.50		66.50	94
9740	1-1/2"		8	1		13.90	60		73.90	105
9750	2"	↓	7	1.143		22	68.50		90.50	126
9760	PVC-reducers, 3/4" x 1/2" diameter					1.04			1.04	1.14
9770	1" x 1/2" diameter					2.80			2.80	3.08
9780	1" x 3/4" diameter					3.13			3.13	3.44
9790	1-1/4" x 3/4" diameter					2.75			2.75	3.03
9800	1-1/4" x 1" diameter					4.29			4.29	4.72
9810	1-1/2" x 1-1/4" diameter					3.56			3.56	3.92
9820	2" x 1-1/4" diameter					4.46			4.46	4.91
9830	2-1/2" x 2" diameter					14.85			14.85	16.35
9840	3" x 2" diameter					19.55			19.55	21.50
9850	4" x 3" diameter					24			24	26.50
9860	Cement, quart					15.60			15.60	17.15
9870	Gallon					71.50			71.50	79
9880	Heat bender, to 6" diameter				↓	1,500			1,500	1,675
9900	Add to labor for higher elevated installation									
9905	10' to 14.5' high, add						10%			
9910	15' to 20' high, add						20%			
9920	20' to 25' high, add						25%			
9930	25' to 30' high, add						35%			
9940	30' to 35' high, add						40%			
9950	35' to 40' high, add						50%			
9960	Over 40' high, add						55%			
9995	Do not include labor when adding couplings to a fitting installation	R260533-30								

26 05 Common Work Results for Electrical

26 05 33 – Raceway and Boxes for Electrical Systems

26 05 33.14 Conduit

26 05 33.14 Conduit		Crew	Daily Output	Labor-Hours	Unit	Material	2019 Bare Costs Labor	Equipment	Total	Total Incl O&P
0010	**CONDUIT** To 10' high, includes 11 couplings per 100'									
0200	Electric metallic tubing, 1/2" diameter	1 Elec	435	.018	L.F.	.62	1.10		1.72	2.33
0220	3/4" diameter		253	.032		.96	1.90		2.86	3.89
0240	1" diameter		207	.039		1.57	2.32		3.89	5.20
0260	1-1/4" diameter		173	.046		2.67	2.78		5.45	7.10
0280	1-1/2" diameter		153	.052		3.07	3.14		6.21	8.05
0300	2" diameter		130	.062		3.39	3.70		7.09	9.25
0320	2-1/2" diameter		92	.087		3.59	5.20		8.79	11.75
0340	3" diameter	2 Elec	148	.108		4.30	6.50		10.80	14.45
0360	3-1/2" diameter		134	.119		5.80	7.15		12.95	17.10
0380	4" diameter		114	.140		10.95	8.45		19.40	24.50
0500	Steel rigid galvanized, 1/2" diameter	1 Elec	146	.055		1.54	3.29		4.83	6.60
0520	3/4" diameter		125	.064		3.97	3.84		7.81	10.10
0540	1" diameter		93	.086		6.35	5.15		11.50	14.70
0560	1-1/4" diameter		88	.091		3.60	5.45		9.05	12.10
0580	1-1/2" diameter		80	.100		6.50	6		12.50	16.10
0600	2" diameter		65	.123		8	7.40		15.40	19.85
0620	2-1/2" diameter		48	.167		10.15	10		20.15	26
0640	3" diameter	2 Elec	64	.250		11.75	15		26.75	35.50
0660	3-1/2" diameter		60	.267		14.85	16		30.85	40.50
0680	4" diameter		52	.308		17.25	18.50		35.75	46.50
0700	5" diameter		50	.320		31	19.20		50.20	63
0720	6" diameter		48	.333		43	20		63	77
1000	Steel intermediate conduit (IMC), 1/2" diameter R260533-25	1 Elec	155	.052		1.01	3.10		4.11	5.75
1010	3/4" diameter		130	.062		1.19	3.70		4.89	6.80
1020	1" diameter		100	.080		1.79	4.80		6.59	9.10
1030	1-1/4" diameter		93	.086		2.38	5.15		7.53	10.30
1040	1-1/2" diameter		85	.094		3.17	5.65		8.82	11.95
1050	2" diameter		70	.114		4.01	6.85		10.86	14.65
1060	2-1/2" diameter		53	.151		7.70	9.05		16.75	22
1070	3" diameter	2 Elec	80	.200		10.15	12		22.15	29
1080	3-1/2" diameter		70	.229		12.15	13.75		25.90	34
1090	4" diameter		60	.267		8.85	16		24.85	34

26 05 33.15 Conduit Nipples

26 05 33.15 Conduit Nipples		Crew	Daily Output	Labor-Hours	Unit	Material	2019 Bare Costs Labor	Equipment	Total	Total Incl O&P
0010	**CONDUIT NIPPLES** With locknuts and bushings									
0100	Aluminum, 1/2" diameter, close	1 Elec	36	.222	Ea.	6.10	13.35		19.45	26.50
0120	1-1/2" long		36	.222		6.15	13.35		19.50	26.50
0140	2" long		36	.222		7.20	13.35		20.55	28
0160	2-1/2" long		36	.222		8.95	13.35		22.30	30
0180	3" long		36	.222		7.80	13.35		21.15	28.50
0200	3-1/2" long		36	.222		6.45	13.35		19.80	27
0220	4" long		36	.222		6.55	13.35		19.90	27
0240	5" long		36	.222		8.80	13.35		22.15	29.50
0260	6" long		36	.222		8.95	13.35		22.30	30
0280	8" long		36	.222		9.15	13.35		22.50	30
0300	10" long		36	.222		9.70	13.35		23.05	30.50
0320	12" long		36	.222		16.35	13.35		29.70	38
0340	3/4" diameter, close		32	.250		8.40	15		23.40	31.50
0360	1-1/2" long		32	.250		8.45	15		23.45	32
0380	2" long		32	.250		8.55	15		23.55	32
0400	2-1/2" long		32	.250		9.90	15		24.90	33.50
0420	3" long		32	.250		10.05	15		25.05	33.50

26 05 33.15 Conduit Nipples		Crew	Daily Output	Labor-Hours	Unit	Material	2019 Bare Costs Labor	Equipment	Total	Total Incl O&P
0440	3-1/2" long	1 Elec	32	.250	Ea.	14.10	15		29.10	38
0460	4" long		32	.250		10.65	15		25.65	34
0480	5" long		32	.250		11.30	15		26.30	35
0500	6" long		32	.250		12.65	15		27.65	36.50
0520	8" long		32	.250		11.70	15		26.70	35.50
0540	10" long		32	.250		12.95	15		27.95	37
0560	12" long		32	.250		16.40	15		31.40	40.50
0580	1" diameter, close		27	.296		11.90	17.80		29.70	39.50
0600	2" long		27	.296		17.75	17.80		35.55	46
0620	2-1/2" long		27	.296		12.60	17.80		30.40	40.50
0640	3" long		27	.296		14.25	17.80		32.05	42
0660	3-1/2" long		27	.296		22	17.80		39.80	50.50
0680	4" long		27	.296		15.95	17.80		33.75	44
0700	5" long		27	.296		16.50	17.80		34.30	44.50
0720	6" long		27	.296		16.25	17.80		34.05	44.50
0740	8" long		27	.296		24	17.80		41.80	53
0760	10" long		27	.296		42	17.80		59.80	72.50
0780	12" long		27	.296		41	17.80		58.80	71.50
0800	1-1/4" diameter, close		23	.348		19	21		40	52
0820	2" long		23	.348		26.50	21		47.50	60
0840	2-1/2" long		23	.348		17.55	21		38.55	50.50
0860	3" long		23	.348		21.50	21		42.50	55
0880	3-1/2" long		23	.348		21.50	21		42.50	54.50
0900	4" long		23	.348		19.25	21		40.25	52
0920	5" long		23	.348		24	21		45	57.50
0940	6" long		23	.348		27.50	21		48.50	61
0960	8" long		23	.348		41	21		62	76
0980	10" long		23	.348		27	21		48	60.50
1000	12" long		23	.348		33.50	21		54.50	67.50
1020	1-1/2" diameter, close		20	.400		25.50	24		49.50	64
1040	2" long		20	.400		27	24		51	65.50
1060	2-1/2" long		20	.400		27	24		51	66
1080	3" long		20	.400		27	24		51	65.50
1100	3-1/2" long		20	.400		26.50	24		50.50	65.50
1120	4" long		20	.400		27.50	24		51.50	66.50
1140	5" long		20	.400		32.50	24		56.50	71.50
1160	6" long		20	.400		33.50	24		57.50	72.50
1180	8" long		20	.400		33	24		57	72.50
1200	10" long		20	.400		38.50	24		62.50	78
1220	12" long		20	.400		45	24		69	85.50
1240	2" diameter, close		18	.444		39	26.50		65.50	83
1260	2-1/2" long		18	.444		38	26.50		64.50	82
1280	3" long		18	.444		40.50	26.50		67	84.50
1300	3-1/2" long		18	.444		41	26.50		67.50	85
1320	4" long		18	.444		42	26.50		68.50	86
1340	5" long		18	.444		45.50	26.50		72	90
1360	6" long		18	.444		42.50	26.50		69	86.50
1380	8" long		18	.444		51.50	26.50		78	96.50
1400	10" long		18	.444		59.50	26.50		86	106
1420	12" long		18	.444		70	26.50		96.50	117
1440	2-1/2" diameter, close		15	.533		75.50	32		107.50	131
1460	3" long		15	.533		70	32		102	125
1480	3-1/2" long		15	.533		72	32		104	127

For customer support on your Electrical Costs with RSMeans data, call 800.448.8182.

149

26 05 33.15 Conduit Nipples		Crew	Daily Output	Labor-Hours	Unit	Material	2019 Bare Costs Labor	Equipment	Total	Total Incl O&P
1500	4" long	1 Elec	15	.533	Ea.	81	32		113	138
1520	5" long		15	.533		102	32		134	160
1540	6" long		15	.533		77	32		109	133
1560	8" long		15	.533		81	32		113	137
1580	10" long		15	.533		101	32		133	160
1600	12" long		15	.533		91.50	32		123.50	148
1620	3" diameter, close		12	.667		95.50	40		135.50	165
1640	3" long		12	.667		97	40		137	167
1660	3-1/2" long		12	.667		118	40		158	189
1680	4" long		12	.667		118	40		158	190
1700	5" long		12	.667		95.50	40		135.50	165
1720	6" long		12	.667		103	40		143	173
1740	8" long		12	.667		119	40		159	191
1760	10" long		12	.667		147	40		187	222
1780	12" long		12	.667		145	40		185	219
1800	3-1/2" diameter, close		11	.727		184	43.50		227.50	267
1820	4" long		11	.727		209	43.50		252.50	295
1840	5" long		11	.727		192	43.50		235.50	276
1860	6" long		11	.727		201	43.50		244.50	286
1880	8" long		11	.727		188	43.50		231.50	271
1900	10" long		11	.727		212	43.50		255.50	298
1920	12" long		11	.727		284	43.50		327.50	380
1940	4" diameter, close		9	.889		215	53.50		268.50	315
1960	4" long		9	.889		225	53.50		278.50	325
1980	5" long		9	.889		226	53.50		279.50	330
2000	6" long		9	.889		270	53.50		323.50	375
2020	8" long		9	.889		254	53.50		307.50	360
2040	10" long		9	.889		277	53.50		330.50	385
2060	12" long		9	.889		252	53.50		305.50	355
2080	5" diameter, close		7	1.143		455	68.50		523.50	600
2100	5" long		7	1.143		540	68.50		608.50	695
2120	6" long		7	1.143		530	68.50		598.50	680
2140	8" long		7	1.143		555	68.50		623.50	710
2160	10" long		7	1.143		480	68.50		548.50	630
2180	12" long		7	1.143		495	68.50		563.50	645
2200	6" diameter, close		6	1.333		695	80		775	885
2220	5" long		6	1.333		750	80		830	945
2240	6" long		6	1.333		690	80		770	880
2260	8" long		6	1.333		810	80		890	1,000
2280	10" long		6	1.333		860	80		940	1,075
2300	12" long		6	1.333		865	80		945	1,075
2320	Rigid galvanized steel, 1/2" diameter, close		32	.250		2.54	15		17.54	25.50
2340	1-1/2" long		32	.250		2.73	15		17.73	25.50
2360	2" long		32	.250		2.85	15		17.85	25.50
2380	2-1/2" long		32	.250		2.95	15		17.95	26
2400	3" long		32	.250		3.07	15		18.07	26
2420	3-1/2" long		32	.250		3.19	15		18.19	26
2440	4" long		32	.250		3.32	15		18.32	26
2460	5" long		32	.250		3.50	15		18.50	26.50
2480	6" long		32	.250		3.88	15		18.88	27
2500	8" long		32	.250		5.50	15		20.50	28.50
2520	10" long		32	.250		6.40	15		21.40	29.50
2540	12" long		32	.250		6.75	15		21.75	30

26 05 33.15 Conduit Nipples		Crew	Daily Output	Labor-Hours	Unit	Material	2019 Bare Costs Labor	Equipment	Total	Total Incl O&P
2560	3/4" diameter, close	1 Elec	27	.296	Ea.	3.37	17.80		21.17	30
2580	2" long		27	.296		3.59	17.80		21.39	30.50
2600	2-1/2" long		27	.296		3.73	17.80		21.53	30.50
2620	3" long		27	.296		3.86	17.80		21.66	31
2640	3-1/2" long		27	.296		3.93	17.80		21.73	31
2660	4" long		27	.296		4.16	17.80		21.96	31
2680	5" long		27	.296		4.43	17.80		22.23	31.50
2700	6" long		27	.296		4.79	17.80		22.59	32
2720	8" long		27	.296		6.45	17.80		24.25	33.50
2740	10" long		27	.296		7.80	17.80		25.60	35
2760	12" long		27	.296		7.95	17.80		25.75	35.50
2780	1" diameter, close		23	.348		5.50	21		26.50	37
2800	2" long		23	.348		5.65	21		26.65	37
2820	2-1/2" long		23	.348		5.80	21		26.80	37.50
2840	3" long		23	.348		6.05	21		27.05	37.50
2860	3-1/2" long		23	.348		6.40	21		27.40	38
2880	4" long		23	.348		6.55	21		27.55	38
2900	5" long		23	.348		6.90	21		27.90	38.50
2920	6" long		23	.348		7.20	21		28.20	39
2940	8" long		23	.348		9.15	21		30.15	41
2960	10" long		23	.348		10.75	21		31.75	43
2980	12" long		23	.348		11.65	21		32.65	44
3000	1-1/4" diameter, close		20	.400		7.75	24		31.75	44.50
3020	2" long		20	.400		7.95	24		31.95	45
3040	3" long		20	.400		8.40	24		32.40	45.50
3060	3-1/2" long		20	.400		8.70	24		32.70	45.50
3080	4" long		20	.400		8.95	24		32.95	46
3100	5" long		20	.400		9.50	24		33.50	46.50
3120	6" long		20	.400		10	24		34	47
3140	8" long		20	.400		12.75	24		36.75	50
3160	10" long		20	.400		14.65	24		38.65	52
3180	12" long		20	.400		16	24		40	53.50
3200	1-1/2" diameter, close		18	.444		11.10	26.50		37.60	52
3220	2" long		18	.444		11.25	26.50		37.75	52.50
3240	2-1/2" long		18	.444		11.65	26.50		38.15	53
3260	3" long		18	.444		11.85	26.50		38.35	53
3280	3-1/2" long		18	.444		12.35	26.50		38.85	53.50
3300	4" long		18	.444		12.70	26.50		39.20	54
3320	5" long		18	.444		13.30	26.50		39.80	54.50
3340	6" long		18	.444		14.35	26.50		40.85	56
3360	8" long		18	.444		17.35	26.50		43.85	59
3380	10" long		18	.444		19.25	26.50		45.75	61
3400	12" long		18	.444		20	26.50		46.50	62
3420	2" diameter, close		16	.500		14.45	30		44.45	61
3440	2-1/2" long		16	.500		14.95	30		44.95	61.50
3460	3" long		16	.500		15.60	30		45.60	62
3480	3-1/2" long		16	.500		16.20	30		46.20	63
3500	4" long		16	.500		16.70	30		46.70	63.50
3520	5" long		16	.500		17.65	30		47.65	64.50
3540	6" long		16	.500		18.50	30		48.50	65.50
3560	8" long		16	.500		22	30		52	69
3580	10" long		16	.500		24	30		54	71.50
3600	12" long		16	.500		25.50	30		55.50	73

For customer support on your Electrical Costs with RSMeans data, call 800.448.8182.

151

26 05 33.15 Conduit Nipples	Crew	Daily Output	Labor-Hours	Unit	Material	2019 Bare Costs Labor	Equipment	Total	Total Incl O&P
3620	2-1/2" diameter, close	1 Elec	13	.615	Ea.	48.50	37		85.50	108
3640	3" long		13	.615		52	37		89	112
3660	3-1/2" long		13	.615		53.50	37		90.50	114
3680	4" long		13	.615		54.50	37		91.50	115
3700	5" long		13	.615		56.50	37		93.50	117
3720	6" long		13	.615		58	37		95	119
3740	8" long		13	.615		63	37		100	124
3760	10" long		13	.615		66	37		103	128
3780	12" long		13	.615		70	37		107	132
3800	3" diameter, close		12	.667		46	40		86	110
3820	3" long		12	.667		52	40		92	117
3900	3-1/2" long		12	.667		47.50	40		87.50	112
3920	4" long		12	.667		49	40		89	114
3940	5" long		12	.667		51	40		91	116
3960	6" long		12	.667		53.50	40		93.50	119
3980	8" long		12	.667		59	40		99	125
4000	10" long		12	.667		63	40		103	129
4020	12" long		12	.667		69.50	40		109.50	136
4040	3-1/2" diameter, close		10	.800		79.50	48		127.50	159
4060	4" long		10	.800		82.50	48		130.50	163
4080	5" long		10	.800		85	48		133	165
4100	6" long		10	.800		88	48		136	168
4120	8" long		10	.800		93	48		141	174
4140	10" long		10	.800		98.50	48		146.50	180
4160	12" long		10	.800		104	48		152	187
4180	4" diameter, close		8	1		120	60		180	222
4200	4" long		8	1		122	60		182	225
4220	5" long		8	1		124	60		184	227
4240	6" long		8	1		126	60		186	229
4260	8" long		8	1		130	60		190	233
4280	10" long		8	1		134	60		194	238
4300	12" long		8	1		139	60		199	243
4320	5" diameter, close		6	1.333		198	80		278	335
4340	5" long		6	1.333		210	80		290	350
4360	6" long		6	1.333		215	80		295	355
4380	8" long		6	1.333		218	80		298	360
4400	10" long		6	1.333		232	80		312	375
4420	12" long		6	1.333		248	80		328	390
4440	6" diameter, close		5	1.600		298	96		394	475
4460	5" long		5	1.600		315	96		411	495
4480	6" long		5	1.600		325	96		421	500
4500	8" long		5	1.600		335	96		431	510
4520	10" long		5	1.600		350	96		446	530
4540	12" long		5	1.600		360	96		456	540
4560	Plastic coated, 40 mil thick, 1/2" diameter, 2" long		32	.250		19.95	15		34.95	44.50
4580	2-1/2" long		32	.250		21	15		36	45.50
4600	3" long		32	.250		21	15		36	45.50
4680	3-1/2" long		32	.250		20.50	15		35.50	45.50
4700	4" long		32	.250		23	15		38	47.50
4720	5" long		32	.250		23	15		38	47.50
4740	6" long		32	.250		21.50	15		36.50	46
4760	8" long		32	.250		26.50	15		41.50	51.50
4780	10" long		32	.250		29	15		44	54

26 05 33.15 Conduit Nipples	Crew	Daily Output	Labor-Hours	Unit	Material	2019 Bare Costs Labor	Equipment	Total	Total Incl O&P
4800	12" long	1 Elec	32	.250	Ea.	36.50	15		51.50	62.50
4820	3/4" diameter, 2" long		26	.308		21	18.50		39.50	50.50
4840	2-1/2" long		26	.308		21	18.50		39.50	51
4860	3" long		26	.308		23.50	18.50		42	53
4880	3-1/2" long		26	.308		22	18.50		40.50	51.50
4900	4" long		26	.308		25	18.50		43.50	55
4920	5" long		26	.308		23	18.50		41.50	52.50
4940	6" long		26	.308		34.50	18.50		53	65.50
4960	8" long		26	.308		27.50	18.50		46	57.50
4980	10" long		26	.308		28	18.50		46.50	58.50
5000	12" long		26	.308		26	18.50		44.50	56
5020	1" diameter, 2" long		22	.364		24	22		46	59
5040	2-1/2" long		22	.364		22	22		44	57
5060	3" long		22	.364		25.50	22		47.50	61
5080	3-1/2" long		22	.364		29	22		51	64.50
5100	4" long		22	.364		25.50	22		47.50	60.50
5120	5" long		22	.364		25	22		47	59.50
5140	6" long		22	.364		29.50	22		51.50	65
5160	8" long		22	.364		31	22		53	67
5180	10" long		22	.364		33.50	22		55.50	69.50
5200	12" long		22	.364		32	22		54	67.50
5220	1-1/4" diameter, 2" long		18	.444		29	26.50		55.50	72
5240	2-1/2" long		18	.444		32	26.50		58.50	75
5260	3" long		18	.444		31	26.50		57.50	74
5280	3-1/2" long		18	.444		33.50	26.50		60	77
5300	4" long		18	.444		38.50	26.50		65	82.50
5320	5" long		18	.444		36	26.50		62.50	79.50
5340	6" long		18	.444		31.50	26.50		58	75
5360	8" long		18	.444		37.50	26.50		64	81.50
5380	10" long		18	.444		42.50	26.50		69	86.50
5400	12" long		18	.444		44	26.50		70.50	88
5420	1-1/2" diameter, 2" long		16	.500		34	30		64	82.50
5440	2-1/2" long		16	.500		36.50	30		66.50	85
5460	3" long		16	.500		35	30		65	83.50
5480	3-1/2" long		16	.500		38.50	30		68.50	87
5500	4" long		16	.500		34	30		64	82.50
5520	5" long		16	.500		34.50	30		64.50	83
5540	6" long		16	.500		41	30		71	90
5560	8" long		16	.500		45.50	30		75.50	95.50
5580	10" long		16	.500		56	30		86	107
5600	12" long		16	.500		57.50	30		87.50	108
5620	2" diameter, 2-1/2" long		14	.571		39.50	34.50		74	94.50
5640	3" long		14	.571		41	34.50		75.50	96
5660	3-1/2" long		14	.571		44.50	34.50		79	100
5680	4" long		14	.571		50.50	34.50		85	107
5700	5" long		14	.571		48	34.50		82.50	104
5720	6" long		14	.571		50.50	34.50		85	107
5740	8" long		14	.571		57.50	34.50		92	114
5760	10" long		14	.571		72.50	34.50		107	131
5780	12" long		14	.571		77.50	34.50		112	136
5800	2-1/2" diameter, 3-1/2" long		12	.667		96.50	40		136.50	166
5820	4" long		12	.667		99.50	40		139.50	170
5840	5" long		12	.667		112	40		152	184

26 05 Common Work Results for Electrical

26 05 33 – Raceway and Boxes for Electrical Systems

26 05 33.15 Conduit Nipples

		Crew	Daily Output	Labor-Hours	Unit	Material	2019 Bare Costs Labor	Equipment	Total	Total Incl O&P
5860	6" long	1 Elec	12	.667	Ea.	103	40		143	174
5880	8" long		12	.667		128	40		168	201
5900	10" long		12	.667		143	40		183	217
5920	12" long		12	.667		154	40		194	230
5940	3" diameter, 3-1/2" long		11	.727		102	43.50		145.50	178
5960	4" long		11	.727		102	43.50		145.50	178
5980	5" long		11	.727		112	43.50		155.50	188
6000	6" long		11	.727		119	43.50		162.50	196
6020	8" long		11	.727		145	43.50		188.50	225
6040	10" long		11	.727		153	43.50		196.50	234
6060	12" long		11	.727		183	43.50		226.50	266
6080	3-1/2" diameter, 4" long		9	.889		152	53.50		205.50	247
6100	5" long		9	.889		153	53.50		206.50	249
6120	6" long		9	.889		147	53.50		200.50	241
6140	8" long		9	.889		186	53.50		239.50	284
6160	10" long		9	.889		210	53.50		263.50	310
6180	12" long		9	.889		234	53.50		287.50	335
6200	4" diameter, 4" long		7.50	1.067		201	64		265	315
6220	5" long		7.50	1.067		203	64		267	320
6240	6" long		7.50	1.067		223	64		287	340
6260	8" long		7.50	1.067		241	64		305	360
6280	10" long		7.50	1.067		300	64		364	425
6300	12" long		7.50	1.067		310	64		374	435
6320	5" diameter, 5" long		5.50	1.455		169	87.50		256.50	315
6340	6" long		5.50	1.455		264	87.50		351.50	420
6360	8" long		5.50	1.455		273	87.50		360.50	430
6380	10" long		5.50	1.455		287	87.50		374.50	445
6400	12" long		5.50	1.455		315	87.50		402.50	475
6420	6" diameter, 5" long		4.50	1.778		370	107		477	570
6440	6" long		4.50	1.778		385	107		492	580
6460	8" long		4.50	1.778		405	107		512	605
6480	10" long		4.50	1.778		420	107		527	620
6500	12" long		4.50	1.778		445	107		552	650

26 05 33.16 Boxes for Electrical Systems

		Crew	Daily Output	Labor-Hours	Unit	Material	2019 Bare Costs Labor	Equipment	Total	Total Incl O&P
0010	**BOXES FOR ELECTRICAL SYSTEMS** R260533-65									
0020	Pressed steel, octagon, 4"	1 Elec	20	.400	Ea.	3.46	24		27.46	40
0040	For Romex or BX		20	.400		5.55	24		29.55	42
0050	For Romex or BX, with bracket		20	.400		8.25	24		32.25	45
0060	Covers, blank		64	.125		.87	7.50		8.37	12.15
0100	Extension rings		40	.200		4.46	12		16.46	23
0150	Square, 4"		20	.400		5.20	24		29.20	41.50
0160	For Romex or BX		20	.400		11.90	24		35.90	49
0170	For Romex or BX, with bracket		20	.400		9.05	24		33.05	46
0200	Extension rings		40	.200		4.59	12		16.59	23
0220	2-1/8" deep, 1" KO		20	.400		2.89	24		26.89	39
0250	Covers, blank		64	.125		.81	7.50		8.31	12.10
0260	Raised device		64	.125		1.43	7.50		8.93	12.75
0300	Plaster rings		64	.125		1.53	7.50		9.03	12.90
0350	Square, 4-11/16"		20	.400		4.56	24		28.56	41
0370	2-1/8" deep, 3/4" to 1-1/4" KO		20	.400		5.60	24		29.60	42
0400	Extension rings		40	.200		10.30	12		22.30	29
0450	Covers, blank		53	.151		1.16	9.05		10.21	14.80

26 05 33.16 Boxes for Electrical Systems	Crew	Daily Output	Labor-Hours	Unit	Material	2019 Bare Costs Labor	2019 Bare Costs Equipment	Total	Total Incl O&P
0460	Raised device	1 Elec	53	.151	Ea.	6.85	9.05		15.90	21
0500	Plaster rings		53	.151		6.95	9.05		16	21
0550	Handy box		27	.296		2.41	17.80		20.21	29
0560	Covers, device		64	.125		1.01	7.50		8.51	12.30
0600	Extension rings		54	.148		3.16	8.90		12.06	16.75
0650	Switchbox		27	.296		5.60	17.80		23.40	32.50
0660	Romex or BX		27	.296		7.60	17.80		25.40	35
0670	with bracket		27	.296		6.75	17.80		24.55	34
0680	Partition, metal		27	.296		3.34	17.80		21.14	30
0700	Masonry, 1 gang, 2-1/2" deep		27	.296		8.10	17.80		25.90	35.50
0710	3-1/2" deep		27	.296		8	17.80		25.80	35.50
0750	2 gang, 2-1/2" deep		20	.400		17.20	24		41.20	55
0760	3-1/2" deep		20	.400		11.40	24		35.40	48.50
0800	3 gang, 2-1/2" deep		13	.615		17.45	37		54.45	74
0850	4 gang, 2-1/2" deep		10	.800		24	48		72	98
0860	5 gang, 2-1/2" deep		9	.889		36.50	53.50		90	120
0870	6 gang, 2-1/2" deep		8	1		61.50	60		121.50	157
0880	Masonry thru-the-wall, 1 gang, 4" block		16	.500		32.50	30		62.50	81
0890	6" block		16	.500		45.50	30		75.50	95.50
0900	8" block		16	.500		54.50	30		84.50	105
0920	2 gang, 6" block		16	.500		68.50	30		98.50	121
0940	Bar hanger with 3/8" stud, for wood and masonry boxes		53	.151		5.35	9.05		14.40	19.40
0950	Concrete, set flush, 4" deep		20	.400		8.60	24		32.60	45.50
1000	Plate with 3/8" stud		80	.100		9.95	6		15.95	19.90
1100	Concrete, floor, 1 gang		5.30	1.509		123	90.50		213.50	271
1150	2 gang		4	2		181	120		301	380
1200	3 gang		2.70	2.963		264	178		442	555
1250	For duplex receptacle, pedestal mounted, add		24	.333		129	20		149	172
1270	Flush mounted, add		27	.296		29.50	17.80		47.30	59
1300	For telephone, pedestal mounted, add		30	.267		103	16		119	138
1350	Carpet flange, 1 gang		53	.151		52	9.05		61.05	71
1400	Cast, 1 gang, FS (2" deep), 1/2" hub		12	.667		23.50	40		63.50	85.50
1410	3/4" hub		12	.667		27	40		67	89.50
1420	FD (2-11/16" deep), 1/2" hub		12	.667		18.95	40		58.95	80.50
1430	3/4" hub		12	.667		24	40		64	85.50
1450	2 gang, FS, 1/2" hub		10	.800		47.50	48		95.50	124
1460	3/4" hub		10	.800		40.50	48		88.50	116
1470	FD, 1/2" hub		10	.800		57.50	48		105.50	135
1480	3/4" hub		10	.800		60	48		108	138
1500	3 gang, FS, 3/4" hub		9	.889		79.50	53.50		133	167
1510	Switch cover, 1 gang, FS		64	.125		6.75	7.50		14.25	18.65
1520	2 gang		53	.151		10.35	9.05		19.40	25
1530	Duplex receptacle cover, 1 gang, FS		64	.125		6.50	7.50		14	18.35
1540	2 gang, FS		53	.151		10.85	9.05		19.90	25.50
1542	Weatherproof blank cover, 1 gang		64	.125		1.22	7.50		8.72	12.55
1544	2 gang		53	.151		2.50	9.05		11.55	16.25
1550	Weatherproof switch cover, 1 gang		64	.125		5.50	7.50		13	17.25
1554	2 gang		53	.151		9.75	9.05		18.80	24
1600	Weatherproof receptacle cover, 1 gang		64	.125		7.80	7.50		15.30	19.75
1604	2 gang		53	.151		7.75	9.05		16.80	22
1620	Weatherproof receptacle cover, tamper resistant, 1 gang		58	.138		13.60	8.30		21.90	27.50
1624	2 gang		48	.167		26.50	10		36.50	44
1750	FSC, 1 gang, 1/2" hub		11	.727		24	43.50		67.50	91.50

26 05 33 – Raceway and Boxes for Electrical Systems

26 05 33.16 Boxes for Electrical Systems

		Crew	Daily Output	Labor-Hours	Unit	Material	2019 Bare Costs Labor	Equipment	Total	Total Incl O&P
1760	3/4" hub	1 Elec	11	.727	Ea.	33.50	43.50		77	102
1770	2 gang, 1/2" hub		9	.889		44	53.50		97.50	128
1780	3/4" hub		9	.889		60	53.50		113.50	146
1790	FDC, 1 gang, 1/2" hub		11	.727		27	43.50		70.50	95
1800	3/4" hub		11	.727		41.50	43.50		85	111
1810	2 gang, 1/2" hub		9	.889		56.50	53.50		110	142
1820	3/4" hub		9	.889		64.50	53.50		118	151
1850	Weatherproof in-use cover, 1 gang		64	.125		24	7.50		31.50	37.50
1870	2 gang		53	.151		28.50	9.05		37.55	45
2000	Poke-thru fitting, fire rated, for 3-3/4" floor		6.80	1.176		147	70.50		217.50	266
2040	For 7" floor		6.80	1.176		172	70.50		242.50	294
2100	Pedestal, 15 amp, duplex receptacle & blank plate		5.25	1.524		139	91.50		230.50	290
2120	Duplex receptacle and telephone plate		5.25	1.524		139	91.50		230.50	290
2140	Pedestal, 20 amp, duplex recept. & phone plate		5	1.600		140	96		236	297
2160	Telephone plate, both sides		5.25	1.524		132	91.50		223.50	283
2200	Abandonment plate		32	.250		41.50	15		56.50	68.50

26 05 33.17 Outlet Boxes, Plastic

		Crew	Daily Output	Labor-Hours	Unit	Material	2019 Bare Costs Labor	Equipment	Total	Total Incl O&P
0010	**OUTLET BOXES, PLASTIC**									
0050	4" diameter, round with 2 mounting nails	1 Elec	25	.320	Ea.	3.09	19.20		22.29	32
0100	Bar hanger mounted		25	.320		5.35	19.20		24.55	34.50
0200	4", square with 2 mounting nails		25	.320		5.30	19.20		24.50	34.50
0300	Plaster ring		64	.125		2.02	7.50		9.52	13.40
0400	Switch box with 2 mounting nails, 1 gang		30	.267		3.10	16		19.10	27.50
0500	2 gang		25	.320		4.21	19.20		23.41	33
0600	3 gang		20	.400		5	24		29	41.50
0700	Old work box		30	.267		4.31	16		20.31	28.50
1400	PVC, FSS, 1 gang, 1/2" hub		14	.571		12.90	34.50		47.40	65
1410	3/4" hub		14	.571		10.75	34.50		45.25	63
1420	FD, 1 gang for variable terminations		14	.571		7.95	34.50		42.45	59.50
1450	FS, 2 gang for variable terminations		12	.667		11.20	40		51.20	72
1480	Weatherproof blank cover, FS, 1 gang		64	.125		3.68	7.50		11.18	15.25
1500	2 gang		53	.151		4.84	9.05		13.89	18.80
1510	Weatherproof switch cover, FS, 1 gang		64	.125		8.05	7.50		15.55	20
1520	2 gang		53	.151		15.25	9.05		24.30	30.50
1530	Weatherproof duplex receptacle cover, FS, 1 gang		64	.125		11.95	7.50		19.45	24.50
1540	2 gang		53	.151		13.40	9.05		22.45	28.50
1750	FSC, 1 gang, 1/2" hub		13	.615		8.80	37		45.80	64.50
1760	3/4" hub		13	.615		9.60	37		46.60	65.50
1770	FSC, 2 gang, 1/2" hub		11	.727		16.55	43.50		60.05	83
1780	3/4" hub		11	.727		14.55	43.50		58.05	81
1790	FDC, 1 gang, 1/2" hub		13	.615		10.35	37		47.35	66.50
1800	3/4" hub		13	.615		10.40	37		47.40	66.50
1810	Weatherproof, T box w/3 holes		14	.571		10.30	34.50		44.80	62.50
1820	4" diameter round w/5 holes		14	.571		9.85	34.50		44.35	62
1850	In-use cover, 1 gang		64	.125		9.50	7.50		17	21.50
1870	2 gang		53	.151		12.40	9.05		21.45	27

26 05 33.18 Pull Boxes

		Crew	Daily Output	Labor-Hours	Unit	Material	2019 Bare Costs Labor	Equipment	Total	Total Incl O&P
0010	**PULL BOXES** R260533-70									
0100	Steel, pull box, NEMA 1, type SC, 6" W x 6" H x 4" D	1 Elec	8	1	Ea.	10	60		70	101
0180	8" W x 6" H x 4" D		8	1		23	60		83	115
0200	8" W x 8" H x 4" D		8	1		12.45	60		72.45	103
0210	10" W x 10" H x 4" D		7	1.143		27	68.50		95.50	132

26 05 33 – Raceway and Boxes for Electrical Systems

26 05 33.18 Pull Boxes		Crew	Daily Output	Labor-Hours	Unit	Material	2019 Bare Costs Labor	Equipment	Total	Total Incl O&P
0220	12" W x 12" H x 4" D	1 Elec	6.50	1.231	Ea.	25	74		99	138
0230	15" W x 15" H x 4" D		5.20	1.538		47.50	92.50		140	191
0240	18" W x 18" H x 4" D		4.40	1.818		36.50	109		145.50	203
0250	6" W x 6" H x 6" D		8	1		15.05	60		75.05	106
0260	8" W x 8" H x 6" D		7.50	1.067		15.05	64		79.05	112
0270	10" W x 10" H x 6" D		5.50	1.455		19.50	87.50		107	152
0300	10" W x 12" H x 6" D		5.30	1.509		28.50	90.50		119	167
0310	12" W x 12" H x 6" D		5.20	1.538		24.50	92.50		117	165
0320	15" W x 15" H x 6" D		4.60	1.739		38.50	104		142.50	199
0330	18" W x 18" H x 6" D		4.20	1.905		46	114		160	222
0340	24" W x 24" H x 6" D		3.20	2.500		91	150		241	325
0350	12" W x 12" H x 8" D		5	1.600		31.50	96		127.50	178
0360	15" W x 15" H x 8" D		4.50	1.778		45.50	107		152.50	209
0370	18" W x 18" H x 8" D		4	2		59	120		179	244
0380	24" W x 18" H x 6" D		3.70	2.162		125	130		255	330
0400	16" W x 20" H x 8" D		4	2		80.50	120		200.50	268
0500	20" W x 24" H x 8" D		3.20	2.500		113	150		263	350
0510	24" W x 24" H x 8" D		3	2.667		102	160		262	350
0600	24" W x 36" H x 8" D		2.70	2.963		151	178		329	430
0610	30" W x 30" H x 8" D		2.70	2.963		200	178		378	485
0620	36" W x 36" H x 8" D		2	4		210	240		450	590
0630	24" W x 24" H x 10" D		2.50	3.200		212	192		404	520
0650	Pull box, hinged, NEMA 1, 6" W x 6" H x 4" D		8	1		18.45	60		78.45	110
0660	8" W x 8" H x 4" D		8	1		24.50	60		84.50	117
0670	10" W x 10" H x 4" D		7	1.143		39	68.50		107.50	145
0680	12" W x 12" H x 4" D		6	1.333		41.50	80		121.50	165
0690	15" W x 15" H x 4" D		5.20	1.538		47	92.50		139.50	190
0700	18" W x 18" H x 4" D		4.40	1.818		45.50	109		154.50	213
0710	6" W x 6" H x 6" D		8	1		22	60		82	114
0720	8" W x 8" H x 6" D		7.50	1.067		29.50	64		93.50	128
0730	10" W x 10" H x 6" D		5.50	1.455		38	87.50		125.50	172
0740	12" W x 12" H x 6" D		5.20	1.538		49	92.50		141.50	192
0800	12" W x 16" H x 6" D		4.70	1.702		50	102		152	208
0820	18" W x 18" H x 6" D		4.20	1.905		84.50	114		198.50	264
1000	20" W x 20" H x 6" D		3.60	2.222		119	133		252	330
1010	24" W x 24" H x 6" D		3.20	2.500		177	150		327	420
1020	12" W x 12" H x 8" D		5	1.600		79	96		175	230
1030	15" W x 15" H x 8" D		4.50	1.778		85.50	107		192.50	253
1040	18" W x 18" H x 8" D		4	2		120	120		240	310
1200	20" W x 20" H x 8" D		3.20	2.500		150	150		300	390
1210	24" W x 24" H x 8" D		3	2.667		169	160		329	425
1220	30" W x 30" H x 8" D		2.70	2.963		244	178		422	535
1400	24" W x 36" H x 8" D		2.70	2.963		237	178		415	525
1600	24" W x 42" H x 8" D		2	4		350	240		590	745
1610	36" W x 36" H x 8" D		2	4		325	240		565	715
2100	Pull box, NEMA 3R, type SC, raintight & weatherproof									
2150	6" L x 6" W x 6" D	1 Elec	10	.800	Ea.	16.85	48		64.85	90
2200	8" L x 6" W x 6" D		8	1		24.50	60		84.50	117
2250	10" L x 6" W x 6" D		7	1.143		32	68.50		100.50	138
2300	12" L x 12" W x 6" D		5	1.600		70	96		166	220
2350	16" L x 16" W x 6" D		4.50	1.778		90.50	107		197.50	259
2400	20" L x 20" W x 6" D		4	2		84.50	120		204.50	272
2450	24" L x 18" W x 8" D		3	2.667		150	160		310	405

For customer support on your Electrical Costs with RSMeans data, call 800.448.8182.

157

26 05 33.18 Pull Boxes		Crew	Daily Output	Labor-Hours	Unit	Material	2019 Bare Costs Labor	Equipment	Total	Total Incl O&P
2500	24" L x 24" W x 10" D	1 Elec	2.50	3.200	Ea.	305	192		497	620
2550	30" L x 24" W x 12" D		2	4		385	240		625	785
2600	36" L x 36" W x 12" D	↓	1.50	5.333	↓	660	320		980	1,200
2800	Cast iron, pull boxes for surface mounting									
3000	NEMA 4, watertight & dust tight									
3050	6" L x 6" W x 6" D	1 Elec	4	2	Ea.	269	120		389	475
3100	8" L x 6" W x 6" D		3.20	2.500		385	150		535	645
3150	10" L x 6" W x 6" D		2.50	3.200		425	192		617	755
3200	12" L x 12" W x 6" D		2.30	3.478		775	209		984	1,175
3250	16" L x 16" W x 6" D		1.30	6.154		730	370		1,100	1,350
3300	20" L x 20" W x 6" D		.80	10		850	600		1,450	1,825
3350	24" L x 18" W x 8" D		.70	11.429		2,575	685		3,260	3,875
3400	24" L x 24" W x 10" D		.50	16		4,700	960		5,660	6,600
3450	30" L x 24" W x 12" D		.40	20		6,325	1,200		7,525	8,775
3500	36" L x 36" W x 12" D		.20	40		6,975	2,400		9,375	11,300
3510	NEMA 4 clamp cover, 6" L x 6" W x 4" D		4	2		251	120		371	455
3520	8" L x 6" W x 4" D	↓	4	2	↓	315	120		435	525
4000	NEMA 7, explosion proof									
4050	6" L x 6" W x 6" D	1 Elec	2	4	Ea.	695	240		935	1,125
4100	8" L x 6" W x 6" D		1.80	4.444		1,100	267		1,367	1,625
4150	10" L x 6" W x 6" D		1.60	5		1,275	300		1,575	1,850
4200	12" L x 12" W x 6" D		1	8		2,325	480		2,805	3,300
4250	16" L x 14" W x 6" D		.60	13.333		3,475	800		4,275	5,025
4300	18" L x 18" W x 8" D		.50	16		6,600	960		7,560	8,675
4350	24" L x 18" W x 8" D		.40	20		7,725	1,200		8,925	10,300
4400	24" L x 24" W x 10" D		.30	26.667		11,400	1,600		13,000	15,000
4450	30" L x 24" W x 12" D		.20	40		14,000	2,400		16,400	19,000
5000	NEMA 9, dust tight 6" L x 6" W x 6" D		3.20	2.500		400	150		550	665
5050	8" L x 6" W x 6" D		2.70	2.963		485	178		663	800
5100	10" L x 6" W x 6" D		2	4		640	240		880	1,075
5150	12" L x 12" W x 6" D		1.60	5		1,225	300		1,525	1,800
5200	16" L x 16" W x 6" D		1	8		2,025	480		2,505	2,950
5250	18" L x 18" W x 8" D		.70	11.429		3,175	685		3,860	4,500
5300	24" L x 18" W x 8" D		.60	13.333		4,475	800		5,275	6,125
5350	24" L x 24" W x 10" D		.40	20		6,050	1,200		7,250	8,450
5400	30" L x 24" W x 12" D	↓	.30	26.667	↓	9,075	1,600		10,675	12,400
6000	J.I.C. wiring boxes, NEMA 12, dust tight & drip tight									
6050	6" L x 8" W x 4" D	1 Elec	10	.800	Ea.	41	48		89	117
6100	8" L x 10" W x 4" D		8	1		51	60		111	146
6150	12" L x 14" W x 6" D		5.30	1.509		84.50	90.50		175	228
6200	14" L x 16" W x 6" D		4.70	1.702		104	102		206	267
6250	16" L x 20" W x 6" D		4.40	1.818		320	109		429	520
6300	24" L x 30" W x 6" D		3.20	2.500		470	150		620	745
6350	24" L x 30" W x 8" D		2.90	2.759		510	166		676	805
6400	24" L x 36" W x 8" D		2.70	2.963		360	178		538	660
6450	24" L x 42" W x 8" D		2.30	3.478		605	209		814	975
6500	24" L x 48" W x 8" D	↓	2	4	↓	665	240		905	1,100

26 05 33.23 Wireway

		Crew	Daily Output	Labor-Hours	Unit	Material	2019 Bare Costs Labor	Equipment	Total	Total Incl O&P
0010	**WIREWAY** to 10' high R260533-60									
0020	For higher elevations, see Section 26 05 36.40									
0100	NEMA 1, screw cover w/fittings and supports, 2-1/2" x 2-1/2"	1 Elec	45	.178	L.F.	12.30	10.70		23	29.50
0200	4" x 4"	"	40	.200	↓	13.25	12		25.25	32.50

26 05 33 – Raceway and Boxes for Electrical Systems

26 05 33.23 Wireway	Crew	Daily Output	Labor-Hours	Unit	Material	2019 Bare Costs Labor	Equipment	Total	Total Incl O&P
0400	6" x 6"	2 Elec	60	.267	L.F.	21	16		37	47
0600	8" x 8"		40	.400		27.50	24		51.50	66.50
0620	10" x 10"		30	.533		47	32		79	99.50
0640	12" x 12"		20	.800		54.50	48		102.50	131
0800	Elbows, 90°, 2-1/2"	1 Elec	24	.333	Ea.	32.50	20		52.50	65.50
1000	4"		20	.400		36	24		60	76
1200	6"		18	.444		41	26.50		67.50	85
1400	8"		16	.500		66.50	30		96.50	119
1420	10"		12	.667		88	40		128	157
1440	12"		10	.800		117	48		165	201
1500	Elbows, 45°, 2-1/2"		24	.333		29.50	20		49.50	62.50
1510	4"		20	.400		39.50	24		63.50	79.50
1520	6"		18	.444		43.50	26.50		70	88
1530	8"		16	.500		66.50	30		96.50	119
1540	10"		12	.667		84	40		124	152
1550	12"		10	.800		169	48		217	258
1600	"T" box, 2-1/2"		18	.444		37.50	26.50		64	81
1800	4"		16	.500		47	30		77	96.50
2000	6"		14	.571		52	34.50		86.50	108
2200	8"		12	.667		93	40		133	162
2220	10"		10	.800		114	48		162	198
2240	12"		8	1		144	60		204	249
2300	Cross, 2-1/2"		16	.500		41.50	30		71.50	90.50
2310	4"		14	.571		50.50	34.50		85	107
2320	6"		12	.667		62.50	40		102.50	129
2400	Panel adapter, 2-1/2"		24	.333		11.30	20		31.30	42.50
2600	4"		20	.400		10.60	24		34.60	47.50
2800	6"		18	.444		11.35	26.50		37.85	52.50
3000	8"		16	.500		15.55	30		45.55	62
3020	10"		14	.571		29	34.50		63.50	83
3040	12"		12	.667		37	40		77	100
3200	Reducer, 4" to 2-1/2"		24	.333		11.80	20		31.80	43
3400	6" to 4"		20	.400		25	24		49	63.50
3600	8" to 6"		18	.444		30.50	26.50		57	73.50
3620	10" to 8"		16	.500		43.50	30		73.50	93
3640	12" to 10"		14	.571		55	34.50		89.50	112
3780	End cap, 2-1/2"		24	.333		4.90	20		24.90	35.50
3800	4"		20	.400		5.80	24		29.80	42.50
4000	6"		18	.444		6.90	26.50		33.40	47.50
4200	8"		16	.500		9.15	30		39.15	55
4220	10"		14	.571		14.85	34.50		49.35	67.50
4240	12"		12	.667		20	40		60	81.50
4300	U-connector, 2-1/2"		200	.040		4.25	2.40		6.65	8.25
4320	4"		200	.040		5.15	2.40		7.55	9.25
4340	6"		180	.044		6.45	2.67		9.12	11.10
4360	8"		170	.047		11.05	2.83		13.88	16.35
4380	10"		150	.053		15.55	3.20		18.75	22
4400	12"		130	.062		20.50	3.70		24.20	28
4420	Hanger, 2-1/2"		100	.080		12.50	4.80		17.30	21
4430	4"		100	.080		12.50	4.80		17.30	21
4440	6"		80	.100		13.60	6		19.60	24
4450	8"		65	.123		24.50	7.40		31.90	38
4460	10"		50	.160		44.50	9.60		54.10	63.50

26 05 33.23 Wireway		Crew	Daily Output	Labor-Hours	Unit	Material	2019 Bare Costs Labor	Equipment	Total	Total Incl O&P
4470	12"	1 Elec	40	.200	Ea.	65	12		77	89.50
4475	NEMA 3R, screw cover w/fittings and supports, 4" x 4"		36	.222	L.F.	16.40	13.35		29.75	38
4480	6" x 6"	2 Elec	55	.291		18.20	17.45		35.65	46
4485	8" x 8"		36	.444		32	26.50		58.50	75.50
4490	12" x 12"		18	.889		57.50	53.50		111	143
4500	Hinged cover, with fittings and supports, 2-1/2" x 2-1/2"	1 Elec	60	.133		17.75	8		25.75	31.50
4520	4" x 4"	"	45	.178		13.85	10.70		24.55	31
4540	6" x 6"	2 Elec	80	.200		21	12		33	41
4560	8" x 8"		60	.267		37	16		53	65
4580	10" x 10"		50	.320		48	19.20		67.20	81.50
4600	12" x 12"		24	.667		72	40		112	139
4700	Elbows 90°, 2-1/2" x 2-1/2"	1 Elec	32	.250	Ea.	46.50	15		61.50	73.50
4720	4"		27	.296		77.50	17.80		95.30	112
4730	6"		23	.348		53	21		74	89
4740	8"		18	.444		72.50	26.50		99	120
4750	10"		14	.571		111	34.50		145.50	173
4760	12"		12	.667		180	40		220	258
4800	Tee box, hinged cover, 2-1/2" x 2-1/2"		23	.348		62.50	21		83.50	100
4810	4"		20	.400		89.50	24		113.50	135
4820	6"		18	.444		60.50	26.50		87	107
4830	8"		16	.500		163	30		193	224
4840	10"		12	.667		164	40		204	240
4860	12"		10	.800		204	48		252	296
4880	Cross box, hinged cover, 2-1/2" x 2-1/2"		18	.444		61.50	26.50		88	108
4900	4"		16	.500		88.50	30		118.50	143
4920	6"		13	.615		104	37		141	170
4940	8"		11	.727		150	43.50		193.50	230
4960	10"		10	.800		242	48		290	340
4980	12"		9	.889		234	53.50		287.50	335
5000	NEMA 12, hinged cover, 2-1/2" x 2-1/2"		40	.200	L.F.	21.50	12		33.50	42
5020	4" x 4"		35	.229		24.50	13.75		38.25	47.50
5040	6" x 6"	2 Elec	60	.267		38	16		54	65.50
5060	8" x 8"	"	50	.320		50	19.20		69.20	83.50
5120	Elbows 90°, flanged, 2-1/2" x 2-1/2"	1 Elec	23	.348	Ea.	54.50	21		75.50	91
5140	4"		20	.400		69	24		93	112
5160	6"		18	.444		87.50	26.50		114	137
5180	8"		15	.533		127	32		159	188
5240	Tee box, flanged, 2-1/2" x 2-1/2"		18	.444		75	26.50		101.50	123
5260	4"		16	.500		87.50	30		117.50	142
5280	6"		15	.533		116	32		148	175
5300	8"		13	.615		172	37		209	245
5360	Cross box, flanged, 2-1/2" x 2-1/2"		15	.533		153	32		185	216
5380	4"		13	.615		125	37		162	193
5400	6"		12	.667		163	40		203	239
5420	8"		10	.800		206	48		254	299
5480	Flange gasket, 2-1/2"		160	.050		2.65	3		5.65	7.40
5500	4"		80	.100		3.67	6		9.67	13
5520	6"		53	.151		5.20	9.05		14.25	19.20
5530	8"		40	.200		6.55	12		18.55	25

26 05 33.25 Conduit Fittings for Rigid Galvanized Steel	Crew	Daily Output	Labor-Hours	Unit	Material	2019 Bare Costs Labor	Equipment	Total	Total Incl O&P
0010	**CONDUIT FITTINGS FOR RIGID GALVANIZED STEEL**								
0050	Standard, locknuts, 1/2" diameter				Ea.	.21			.21	.23
0100	3/4" diameter					.42			.42	.46
0300	1" diameter					.69			.69	.76
0500	1-1/4" diameter					.84			.84	.92
0700	1-1/2" diameter					1.45			1.45	1.60
1000	2" diameter					1.96			1.96	2.16
1030	2-1/2" diameter					4.69			4.69	5.15
1050	3" diameter					6.10			6.10	6.70
1070	3-1/2" diameter					12.05			12.05	13.25
1100	4" diameter					15.35			15.35	16.90
1110	5" diameter					33.50			33.50	36.50
1120	6" diameter					47			47	51.50
1130	Bushings, plastic, 1/2" diameter	1 Elec	40	.200		.08	12		12.08	18
1150	3/4" diameter		32	.250		.09	15		15.09	22.50
1170	1" diameter		28	.286		.19	17.15		17.34	25.50
1200	1-1/4" diameter		24	.333		.23	20		20.23	30.50
1230	1-1/2" diameter		18	.444		.32	26.50		26.82	40.50
1250	2" diameter		15	.533		.54	32		32.54	48.50
1270	2-1/2" diameter		13	.615		.99	37		37.99	56
1300	3" diameter		12	.667		1.24	40		41.24	61
1330	3-1/2" diameter		11	.727		1.36	43.50		44.86	66.50
1350	4" diameter		9	.889		1.61	53.50		55.11	81.50
1360	5" diameter		7	1.143		10.45	68.50		78.95	114
1370	6" diameter		5	1.600		42	96		138	190
1390	Steel, 1/2" diameter		40	.200		.63	12		12.63	18.60
1400	3/4" diameter		32	.250		.70	15		15.70	23.50
1430	1" diameter		28	.286		1.21	17.15		18.36	27
1450	Steel insulated, 1-1/4" diameter		24	.333		5.55	20		25.55	36
1470	1-1/2" diameter		18	.444		6.45	26.50		32.95	47
1500	2" diameter		15	.533		7.05	32		39.05	56
1530	2-1/2" diameter		13	.615		13.70	37		50.70	70
1550	3" diameter		12	.667		17.70	40		57.70	79
1570	3-1/2" diameter		11	.727		25.50	43.50		69	93
1600	4" diameter		9	.889		28.50	53.50		82	111
1610	5" diameter		7	1.143		59.50	68.50		128	168
1620	6" diameter		5	1.600		176	96		272	335
1630	Sealing locknuts, 1/2" diameter		40	.200		1.78	12		13.78	19.85
1650	3/4" diameter		32	.250		1.45	15		16.45	24
1670	1" diameter		28	.286		3.91	17.15		21.06	30
1700	1-1/4" diameter		24	.333		4.98	20		24.98	35.50
1730	1-1/2" diameter		18	.444		6.30	26.50		32.80	47
1750	2" diameter		15	.533		8.10	32		40.10	57
1760	Grounding bushing, insulated, 1/2" diameter		32	.250		4.95	15		19.95	28
1770	3/4" diameter		28	.286		6	17.15		23.15	32
1780	1" diameter		20	.400		8.95	24		32.95	46
1800	1-1/4" diameter		18	.444		10.40	26.50		36.90	51.50
1830	1-1/2" diameter		16	.500		12.15	30		42.15	58.50
1850	2" diameter		13	.615		12.60	37		49.60	69
1870	2-1/2" diameter		12	.667		25.50	40		65.50	87.50
1900	3" diameter		11	.727		27.50	43.50		71	95
1930	3-1/2" diameter		9	.889		26	53.50		79.50	108

For customer support on your Electrical Costs with RSMeans data, call 800.448.8182.

161

26 05 33.25 Conduit Fittings for Rigid Galvanized Steel		Crew	Daily Output	Labor-Hours	Unit	Material	2019 Bare Costs Labor	Equipment	Total	Total Incl O&P
1950	4" diameter	1 Elec	8	1	Ea.	51	60		111	146
1960	5" diameter		6	1.333		76.50	80		156.50	203
1970	6" diameter		4	2		112	120		232	300
1990	Coupling, with set screw, 1/2" diameter		50	.160		4.89	9.60		14.49	19.75
2000	3/4" diameter		40	.200		6.15	12		18.15	24.50
2030	1" diameter		35	.229		9.35	13.75		23.10	31
2050	1-1/4" diameter		28	.286		11.70	17.15		28.85	38.50
2070	1-1/2" diameter		23	.348		13.45	21		34.45	46
2090	2" diameter		20	.400		30	24		54	69
2100	2-1/2" diameter		18	.444		64.50	26.50		91	111
2110	3" diameter		15	.533		82.50	32		114.50	139
2120	3-1/2" diameter		12	.667		175	40		215	252
2130	4" diameter		10	.800		172	48		220	261
2140	5" diameter		9	.889		230	53.50		283.50	335
2150	6" diameter		8	1		405	60		465	535
2160	Box connector with set screw, plain, 1/2" diameter		70	.114		2.08	6.85		8.93	12.55
2170	3/4" diameter		60	.133		4.50	8		12.50	16.90
2180	1" diameter		50	.160		7.10	9.60		16.70	22
2190	Insulated, 1-1/4" diameter		40	.200		9.65	12		21.65	28.50
2200	1-1/2" diameter		30	.267		14.40	16		30.40	40
2210	2" diameter		20	.400		28	24		52	66.50
2220	2-1/2" diameter		18	.444		80	26.50		106.50	128
2230	3" diameter		15	.533		131	32		163	192
2240	3-1/2" diameter		12	.667		194	40		234	273
2250	4" diameter		10	.800		164	48		212	253
2260	5" diameter		9	.889		395	53.50		448.50	510
2270	6" diameter		8	1		345	60		405	470
2280	LB, LR or LL fittings & covers, 1/2" diameter		16	.500		6.95	30		36.95	52.50
2290	3/4" diameter		13	.615		8.20	37		45.20	64
2300	1" diameter		11	.727		12.80	43.50		56.30	79
2330	1-1/4" diameter		8	1		29.50	60		89.50	122
2350	1-1/2" diameter		6	1.333		23	80		103	144
2370	2" diameter		5	1.600		58.50	96		154.50	208
2380	2-1/2" diameter		4	2		121	120		241	310
2390	3" diameter		3.50	2.286		166	137		303	390
2400	3-1/2" diameter		3	2.667		174	160		334	430
2410	4" diameter		2.50	3.200		190	192		382	495
2420	T fittings, with cover, 1/2" diameter		12	.667		9.05	40		49.05	69.50
2430	3/4" diameter		11	.727		10.05	43.50		53.55	76
2440	1" diameter		9	.889		16.55	53.50		70.05	97.50
2450	1-1/4" diameter		6	1.333		19.75	80		99.75	141
2470	1-1/2" diameter		5	1.600		42.50	96		138.50	190
2500	2" diameter		4	2		37.50	120		157.50	221
2510	2-1/2" diameter		3.50	2.286		88	137		225	300
2520	3" diameter		3	2.667		114	160		274	365
2530	3-1/2" diameter		2.50	3.200		205	192		397	510
2540	4" diameter		2	4		189	240		429	570
2550	Nipples chase, plain, 1/2" diameter		40	.200		.38	12		12.38	18.30
2560	3/4" diameter		32	.250		1.32	15		16.32	24
2570	1" diameter		28	.286		2.08	17.15		19.23	28
2600	Insulated, 1-1/4" diameter		24	.333		7.30	20		27.30	38
2630	1-1/2" diameter		18	.444		11.60	26.50		38.10	53
2650	2" diameter		15	.533		12.85	32		44.85	62

26 05 33 – Raceway and Boxes for Electrical Systems

26 05 33.25 Conduit Fittings for Rigid Galvanized Steel	Crew	Daily Output	Labor-Hours	Unit	Material	2019 Bare Costs Labor	Equipment	Total	Total Incl O&P
2660	2-1/2" diameter	1 Elec	12	.667	Ea.	39.50	40		79.50	103
2670	3" diameter		10	.800		38.50	48		86.50	114
2680	3-1/2" diameter		9	.889		54	53.50		107.50	139
2690	4" diameter		8	1		79.50	60		139.50	177
2700	5" diameter		7	1.143		279	68.50		347.50	405
2710	6" diameter		6	1.333		425	80		505	590
2720	Nipples offset, plain, 1/2" diameter		40	.200		6.35	12		18.35	25
2730	3/4" diameter		32	.250		6.65	15		21.65	30
2740	1" diameter		24	.333		8.10	20		28.10	39
2750	Insulated, 1-1/4" diameter		20	.400		6.90	24		30.90	43.50
2760	1-1/2" diameter		18	.444		8.35	26.50		34.85	49
2770	2" diameter		16	.500		11.75	30		41.75	58
2780	3" diameter		14	.571		78	34.50		112.50	137
2850	Coupling, expansion, 1/2" diameter		12	.667		46.50	40		86.50	111
2880	3/4" diameter		10	.800		52	48		100	129
2900	1" diameter		8	1		81	60		141	179
2920	1-1/4" diameter		6.40	1.250		89	75		164	210
2940	1-1/2" diameter		5.30	1.509		115	90.50		205.50	261
2960	2" diameter		4.60	1.739		166	104		270	340
2980	2-1/2" diameter		3.60	2.222		292	133		425	520
3000	3" diameter		3	2.667		350	160		510	630
3020	3-1/2" diameter		2.80	2.857		530	172		702	835
3040	4" diameter		2.40	3.333		615	200		815	975
3060	5" diameter		2	4		1,025	240		1,265	1,475
3080	6" diameter		1.80	4.444		1,675	267		1,942	2,250
3100	Expansion deflection, 1/2" diameter		12	.667		258	40		298	345
3120	3/4" diameter		12	.667		290	40		330	380
3140	1" diameter		10	.800		325	48		373	425
3160	1-1/4" diameter		6.40	1.250		545	75		620	710
3180	1-1/2" diameter		5.30	1.509		585	90.50		675.50	780
3200	2" diameter		4.60	1.739		765	104		869	995
3220	2-1/2" diameter		3.60	2.222		885	133		1,018	1,175
3240	3" diameter		3	2.667		1,350	160		1,510	1,725
3260	3-1/2" diameter		2.80	2.857		1,150	172		1,322	1,525
3280	4" diameter		2.40	3.333		1,300	200		1,500	1,750
3300	5" diameter		2	4		2,675	240		2,915	3,300
3320	6" diameter		1.80	4.444		3,850	267		4,117	4,625
3340	Ericson, 1/2" diameter		16	.500		3.34	30		33.34	48.50
3360	3/4" diameter		14	.571		4.05	34.50		38.55	55.50
3380	1" diameter		11	.727		6.40	43.50		49.90	72
3400	1-1/4" diameter		8	1		13.85	60		73.85	105
3420	1-1/2" diameter		7	1.143		15.70	68.50		84.20	119
3440	2" diameter		5	1.600		30.50	96		126.50	177
3460	2-1/2" diameter		4	2		60.50	120		180.50	246
3480	3" diameter		3.50	2.286		104	137		241	320
3500	3-1/2" diameter		3	2.667		120	160		280	370
3520	4" diameter		2.70	2.963		145	178		323	425
3540	5" diameter		2.50	3.200		288	192		480	600
3560	6" diameter		2.30	3.478		395	209		604	745
3580	Split, 1/2" diameter		32	.250		3.81	15		18.81	26.50
3600	3/4" diameter		27	.296		4.81	17.80		22.61	32
3620	1" diameter		20	.400		11	24		35	48
3640	1-1/4" diameter		16	.500		13.35	30		43.35	59.50

For customer support on your Electrical Costs with RSMeans data, call 800.448.8182.

163

26 05 33.25 Conduit Fittings for Rigid Galvanized Steel	Crew	Daily Output	Labor-Hours	Unit	Material	2019 Bare Costs Labor	Equipment	Total	Total Incl O&P
3660	1-1/2" diameter	1 Elec	14	.571	Ea.	14.30	34.50		48.80	67
3680	2" diameter		12	.667		28	40		68	90.50
3700	2-1/2" diameter		10	.800		90	48		138	171
3720	3" diameter		9	.889		135	53.50		188.50	229
3740	3-1/2" diameter		8	1		194	60		254	305
3760	4" diameter		7	1.143		165	68.50		233.50	283
3780	5" diameter		6	1.333		380	80		460	540
3800	6" diameter		5	1.600		500	96		596	695
4600	Reducing bushings, 3/4" to 1/2" diameter		54	.148		2.50	8.90		11.40	16
4620	1" to 3/4" diameter		46	.174		3.29	10.45		13.74	19.20
4640	1-1/4" to 1" diameter		40	.200		4.42	12		16.42	23
4660	1-1/2" to 1-1/4" diameter		36	.222		5.85	13.35		19.20	26.50
4680	2" to 1-1/2" diameter		32	.250		11.60	15		26.60	35.50
4740	2-1/2" to 2" diameter		30	.267		13.90	16		29.90	39.50
4760	3" to 2-1/2" diameter		28	.286		19.30	17.15		36.45	46.50
4800	Through-wall seal, 1/2" diameter		8	1		67.50	60		127.50	164
4820	3/4" diameter		7.50	1.067		275	64		339	395
4840	1" diameter		6.50	1.231		275	74		349	415
4860	1-1/4" diameter		5.50	1.455		290	87.50		377.50	450
4880	1-1/2" diameter		5	1.600		400	96		496	585
4900	2" diameter		4.20	1.905		281	114		395	480
4920	2-1/2" diameter		3.50	2.286		560	137		697	820
4940	3" diameter		3	2.667		600	160		760	900
4960	3-1/2" diameter		2.50	3.200		905	192		1,097	1,275
4980	4" diameter		2	4		860	240		1,100	1,300
5000	5" diameter		1.50	5.333		790	320		1,110	1,350
5020	6" diameter	▼	1	8	▼	790	480		1,270	1,575
5100	Cable supports, 2 or more wires									
5120	1-1/2" diameter	1 Elec	8	1	Ea.	185	60		245	294
5140	2" diameter		6	1.333		202	80		282	340
5160	2-1/2" diameter		4	2		218	120		338	420
5180	3" diameter		3.50	2.286		280	137		417	515
5200	3-1/2" diameter		2.60	3.077		405	185		590	720
5220	4" diameter		2	4		460	240		700	865
5240	5" diameter		1.50	5.333		630	320		950	1,175
5260	6" diameter		1	8		1,025	480		1,505	1,850
5280	Service entrance cap, 1/2" diameter		16	.500		4.90	30		34.90	50.50
5300	3/4" diameter		13	.615		5.85	37		42.85	61.50
5320	1" diameter		10	.800		4.79	48		52.79	77
5340	1-1/4" diameter		8	1		5.30	60		65.30	95.50
5360	1-1/2" diameter		6.50	1.231		10.45	74		84.45	122
5380	2" diameter		5.50	1.455		22	87.50		109.50	154
5400	2-1/2" diameter		4	2		77	120		197	264
5420	3" diameter		3.40	2.353		78.50	141		219.50	297
5440	3-1/2" diameter		3	2.667		125	160		285	375
5460	4" diameter		2.70	2.963		167	178		345	450
5750	90° pull elbows steel, female, 1/2" diameter		16	.500		7.90	30		37.90	53.50
5760	3/4" diameter		13	.615		8.75	37		45.75	64.50
5780	1" diameter		11	.727		14.20	43.50		57.70	80.50
5800	1-1/4" diameter		8	1		26	60		86	118
5820	1-1/2" diameter		6	1.333		34.50	80		114.50	157
5840	2" diameter	▼	5	1.600	▼	56.50	96		152.50	205
6000	Explosion proof, flexible coupling									

26 05 Common Work Results for Electrical

26 05 33 – Raceway and Boxes for Electrical Systems

26 05 33.25 Conduit Fittings for Rigid Galvanized Steel	Crew	Daily Output	Labor-Hours	Unit	Material	2019 Bare Costs Labor	Equipment	Total	Total Incl O&P
6010	1/2" diameter, 4" long	1 Elec	12	.667	Ea.	150	40		190	225
6020	6" long		12	.667		129	40		169	202
6050	12" long		12	.667		171	40		211	248
6070	18" long		12	.667		207	40		247	288
6090	24" long		12	.667		325	40		365	420
6110	30" long		12	.667		294	40		334	385
6130	36" long		12	.667		315	40		355	405
6140	3/4" diameter, 4" long		10	.800		165	48		213	254
6150	6" long		10	.800		154	48		202	241
6180	12" long		10	.800		202	48		250	294
6200	18" long		10	.800		265	48		313	365
6220	24" long		10	.800		320	48		368	420
6240	30" long		10	.800		300	48		348	400
6260	36" long		10	.800		375	48		423	485
6270	1" diameter, 6" long		8	1		299	60		359	420
6300	12" long		8	1		425	60		485	560
6320	18" long		8	1		430	60		490	565
6340	24" long		8	1		655	60		715	810
6360	30" long		8	1		790	60		850	955
6380	36" long		8	1		775	60		835	945
6390	1-1/4" diameter, 12" long		6.40	1.250		475	75		550	630
6410	18" long		6.40	1.250		615	75		690	790
6430	24" long		6.40	1.250		930	75		1,005	1,125
6450	30" long		6.40	1.250		825	75		900	1,025
6470	36" long		6.40	1.250		1,600	75		1,675	1,850
6480	1-1/2" diameter, 12" long		5.30	1.509		635	90.50		725.50	830
6500	18" long		5.30	1.509		820	90.50		910.50	1,025
6520	24" long		5.30	1.509		1,000	90.50		1,090.50	1,225
6540	30" long		5.30	1.509		1,575	90.50		1,665.50	1,850
6560	36" long		5.30	1.509		1,250	90.50		1,340.50	1,500
6570	2" diameter, 12" long		4.60	1.739		825	104		929	1,050
6590	18" long		4.60	1.739		1,100	104		1,204	1,375
6610	24" long		4.60	1.739		1,200	104		1,304	1,475
6630	30" long		4.60	1.739		1,625	104		1,729	1,925
6650	36" long		4.60	1.739		1,550	104		1,654	1,875
7000	Close up plug, 1/2" diameter, explosion proof		40	.200		2.69	12		14.69	21
7010	3/4" diameter		32	.250		3.10	15		18.10	26
7020	1" diameter		28	.286		3.67	17.15		20.82	29.50
7030	1-1/4" diameter		24	.333		3.88	20		23.88	34.50
7040	1-1/2" diameter		18	.444		5.40	26.50		31.90	46
7050	2" diameter		15	.533		9.40	32		41.40	58.50
7060	2-1/2" diameter		13	.615		13.90	37		50.90	70.50
7070	3" diameter		12	.667		21.50	40		61.50	83
7080	3-1/2" diameter		11	.727		24.50	43.50		68	92
7090	4" diameter		9	.889		34.50	53.50		88	118
7091	Elbow female, 45°, 1/2"		16	.500		14.70	30		44.70	61
7092	3/4"		13	.615		20	37		57	77
7093	1"		11	.727		25.50	43.50		69	93
7094	1-1/4"		8	1		28.50	60		88.50	121
7095	1-1/2"		6	1.333		31	80		111	153
7096	2"		5	1.600		36.50	96		132.50	184
7097	2-1/2"		4.50	1.778		105	107		212	275
7098	3"		4.20	1.905		118	114		232	300

For customer support on your Electrical Costs with RSMeans data, call 800.448.8182.

165

26 05 33.25 Conduit Fittings for Rigid Galvanized Steel	Crew	Daily Output	Labor-Hours	Unit	Material	2019 Bare Costs Labor	Equipment	Total	Total Incl O&P
7099	3-1/2"	1 Elec	4	2	Ea.	163	120		283	360
7100	4"		3.80	2.105		224	126		350	435
7101	90°, 1/2"		16	.500		13.65	30		43.65	60
7102	3/4"		13	.615		14.80	37		51.80	71.50
7103	1"		11	.727		19.95	43.50		63.45	87
7104	1-1/4"		8	1		32	60		92	125
7105	1-1/2"		6	1.333		45.50	80		125.50	169
7106	2"		5	1.600		106	96		202	260
7107	2-1/2"		4.50	1.778		160	107		267	335
7110	Elbows 90°, long male & female, 1/2" diameter, explosion proof		16	.500		19.95	30		49.95	67
7120	3/4" diameter		13	.615		22	37		59	79
7130	1" diameter		11	.727		38	43.50		81.50	107
7140	1-1/4" diameter		8	1		51.50	60		111.50	147
7150	1-1/2" diameter		6	1.333		42	80		122	165
7160	2" diameter		5	1.600		57.50	96		153.50	206
7170	Capped elbow, 1/2" diameter, explosion proof		11	.727		22	43.50		65.50	89.50
7180	3/4" diameter		8	1		27	60		87	119
7190	1" diameter		6	1.333		29.50	80		109.50	152
7200	1-1/4" diameter		5	1.600		70	96		166	220
7210	Pulling elbow, 1/2" diameter, explosion proof		11	.727		114	43.50		157.50	191
7220	3/4" diameter		8	1		174	60		234	282
7230	1" diameter		6	1.333		234	80		314	375
7240	1-1/4" diameter		5	1.600		280	96		376	455
7250	1-1/2" diameter		5	1.600		325	96		421	505
7260	2" diameter		4	2		375	120		495	590
7270	2-1/2" diameter		3.50	2.286		840	137		977	1,125
7280	3" diameter		3	2.667		1,275	160		1,435	1,650
7290	3-1/2" diameter		2.50	3.200		1,725	192		1,917	2,175
7300	4" diameter		2.20	3.636		1,575	218		1,793	2,050
7310	LB conduit body, 1/2" diameter		11	.727		54.50	43.50		98	125
7320	3/4" diameter		8	1		76.50	60		136.50	174
7330	T conduit body, 1/2" diameter		9	.889		57	53.50		110.50	143
7340	3/4" diameter		6	1.333		80.50	80		160.50	208
7350	Explosion proof, round box w/cover, 3 threaded hubs, 1/2" diameter		8	1		68.50	60		128.50	165
7351	3/4" diameter		8	1		68	60		128	165
7352	1" diameter		7.50	1.067		72	64		136	175
7353	1-1/4" diameter		7	1.143		97.50	68.50		166	209
7354	1-1/2" diameter		7	1.143		258	68.50		326.50	385
7355	2" diameter		6	1.333		288	80		368	435
7356	Round box w/cover & mtng flange, 3 threaded hubs, 1/2" diameter		8	1		63	60		123	159
7357	3/4" diameter		8	1		54	60		114	149
7358	4 threaded hubs, 1" diameter		7	1.143		77.50	68.50		146	187
7400	Unions, 1/2" diameter		20	.400		13.10	24		37.10	50.50
7410	3/4" - 1/2" diameter		16	.500		20.50	30		50.50	68
7420	3/4" diameter		16	.500		18.20	30		48.20	65
7430	1" diameter		14	.571		33	34.50		67.50	87
7440	1-1/4" diameter		12	.667		51.50	40		91.50	116
7450	1-1/2" diameter		10	.800		63.50	48		111.50	141
7460	2" diameter		8.50	.941		84	56.50		140.50	177
7480	2-1/2" diameter		8	1		130	60		190	232
7490	3" diameter		7	1.143		192	68.50		260.50	315
7500	3-1/2" diameter		6	1.333		355	80		435	510
7510	4" diameter		5	1.600		335	96		431	515

26 05 33.25 Conduit Fittings for Rigid Galvanized Steel	Crew	Daily Output	Labor-Hours	Unit	Material	2019 Bare Costs Labor	Equipment	Total	Total Incl O&P
7680	Reducer, 3/4" to 1/2"	1 Elec	54	.148	Ea.	4.28	8.90		13.18	17.95
7690	1" to 1/2"		46	.174		5.95	10.45		16.40	22
7700	1" to 3/4"		46	.174		6	10.45		16.45	22
7710	1-1/4" to 3/4"		40	.200		8.70	12		20.70	27.50
7720	1-1/4" to 1"		40	.200		9.45	12		21.45	28.50
7730	1-1/2" to 1"		36	.222		14.90	13.35		28.25	36.50
7740	1-1/2" to 1-1/4"		36	.222		14.65	13.35		28	36
7750	2" to 3/4"		32	.250		21	15		36	45.50
7760	2" to 1-1/4"		32	.250		18.90	15		33.90	43.50
7770	2" to 1-1/2"		32	.250		25	15		40	50
7780	2-1/2" to 1-1/2"		30	.267		19.90	16		35.90	46
7790	3" to 2"		30	.267		22	16		38	48.50
7800	3-1/2" to 2-1/2"		28	.286		64.50	17.15		81.65	96.50
7810	4" to 3"		28	.286		75.50	17.15		92.65	109
7820	Sealing fitting, vertical/horizontal, 1/2" diameter		14.50	.552		19.65	33		52.65	71
7830	3/4" diameter		13.30	.602		23	36		59	79.50
7840	1" diameter		11.40	.702		29.50	42		71.50	95.50
7850	1-1/4" diameter		10	.800		30	48		78	105
7860	1-1/2" diameter		8.80	.909		54	54.50		108.50	141
7870	2" diameter		8	1		61.50	60		121.50	157
7880	2-1/2" diameter		6.70	1.194		74.50	71.50		146	189
7890	3" diameter		5.70	1.404		106	84.50		190.50	243
7900	3-1/2" diameter		4.70	1.702		293	102		395	475
7910	4" diameter		4	2		525	120		645	760
7920	Sealing hubs, 1" by 1-1/2"		12	.667		28.50	40		68.50	90.50
7930	1-1/4" by 2"		10	.800		50	48		98	127
7940	1-1/2" by 2"		9	.889		70	53.50		123.50	157
7950	2" by 2-1/2"		8	1		96.50	60		156.50	196
7960	3" by 4"		7	1.143		162	68.50		230.50	280
7970	4" by 5"		6	1.333		440	80		520	605
7980	Drain, 1/2"		32	.250		108	15		123	142
7990	Breather, 1/2"		32	.250		108	15		123	142
7991	Explosion proof sealant compound, hub, fittings, 60 min set time, 1lb. pail		48	.167		20.50	10		30.50	37.50
7992	5 lb. pail		48	.167		60.50	10		70.50	81.50
7993	2 oz. tube		50	.160		65	9.60		74.60	85.50
7994	6 oz. tube		50	.160		112	9.60		121.60	138
7995	2.0 oz. cartridge		50	.160		43.50	9.60		53.10	62.50
7996	6.0 oz. cartridge		50	.160		83.50	9.60		93.10	106
7997	2.0 oz. box		50	.160		27	9.60		36.60	44
7998	8.0 oz. box		50	.160		83.50	9.60		93.10	106
7999	1.0 lb. box		50	.160		139	9.60		148.60	167
8000	Plastic coated 40 mil thick									
8010	LB, LR or LL conduit body w/cover, 1/2" diameter	1 Elec	13	.615	Ea.	47	37		84	107
8020	3/4" diameter		11	.727		71.50	43.50		115	144
8030	1" diameter		8	1		72	60		132	169
8040	1-1/4" diameter		6	1.333		104	80		184	234
8050	1-1/2" diameter		5	1.600		128	96		224	284
8060	2" diameter		4.50	1.778		188	107		295	365
8070	2-1/2" diameter		4	2		370	120		490	585
8080	3" diameter		3.50	2.286		425	137		562	670
8090	3-1/2" diameter		3	2.667		760	160		920	1,075
8100	4" diameter		2.50	3.200		860	192		1,052	1,225
8150	T conduit body with cover, 1/2" diameter		11	.727		58	43.50		101.50	129

26 05 33.25 Conduit Fittings for Rigid Galvanized Steel	Crew	Daily Output	Labor-Hours	Unit	Material	2019 Bare Costs Labor	Equipment	Total	Total Incl O&P
8160	3/4" diameter	1 Elec	9	.889	Ea.	74.50	53.50		128	162
8170	1" diameter		6	1.333		83	80		163	211
8180	1-1/4" diameter		5	1.600		139	96		235	296
8190	1-1/2" diameter		4.50	1.778		144	107		251	315
8200	2" diameter		4	2		208	120		328	410
8210	2-1/2" diameter		3.50	2.286		385	137		522	625
8220	3" diameter		3	2.667		520	160		680	815
8230	3-1/2" diameter		2.50	3.200		795	192		987	1,150
8240	4" diameter		2	4		880	240		1,120	1,325
8300	FS conduit body, 1 gang, 3/4" diameter		11	.727		73.50	43.50		117	146
8310	1" diameter		10	.800		58	48		106	136
8350	2 gang, 3/4" diameter		9	.889		103	53.50		156.50	193
8360	1" diameter		8	1		136	60		196	240
8400	Duplex receptacle cover		64	.125		48	7.50		55.50	63.50
8410	Switch cover		64	.125		60.50	7.50		68	77.50
8420	Switch, vaportight cover		53	.151		176	9.05		185.05	207
8430	Blank, cover		64	.125		37	7.50		44.50	52
8520	FSC conduit body, 1 gang, 3/4" diameter		10	.800		68	48		116	147
8530	1" diameter		9	.889		69.50	53.50		123	156
8550	2 gang, 3/4" diameter		8	1		116	60		176	218
8560	1" diameter		7	1.143		162	68.50		230.50	281
8590	Conduit hubs, 1/2" diameter		18	.444		36	26.50		62.50	79.50
8600	3/4" diameter		16	.500		41	30		71	90.50
8610	1" diameter		14	.571		51.50	34.50		86	108
8620	1-1/4" diameter		12	.667		70	40		110	137
8630	1-1/2" diameter		10	.800		72.50	48		120.50	152
8640	2" diameter		8.80	.909		97	54.50		151.50	189
8650	2-1/2" diameter		8.50	.941		152	56.50		208.50	252
8660	3" diameter		8	1		205	60		265	315
8670	3-1/2" diameter		7.50	1.067		310	64		374	440
8680	4" diameter		7	1.143		325	68.50		393.50	460
8690	5" diameter		6	1.333		380	80		460	540
8700	Plastic coated 40 mil thick									
8710	Pipe strap, stamped 1 hole, 1/2" diameter	1 Elec	470	.017	Ea.	12.25	1.02		13.27	15.05
8720	3/4" diameter		440	.018		11.30	1.09		12.39	14.05
8730	1" diameter		400	.020		12.80	1.20		14	15.90
8740	1-1/4" diameter		355	.023		20.50	1.35		21.85	24.50
8750	1-1/2" diameter		320	.025		22	1.50		23.50	26.50
8760	2" diameter		266	.030		28	1.81		29.81	33.50
8770	2-1/2" diameter		200	.040		32.50	2.40		34.90	39.50
8780	3" diameter		133	.060		44	3.61		47.61	54
8790	3-1/2" diameter		110	.073		88.50	4.37		92.87	104
8800	4" diameter		90	.089		86.50	5.35		91.85	103
8810	5" diameter		70	.114		134	6.85		140.85	157
8840	Clamp back spacers, 3/4" diameter		440	.018		17.10	1.09		18.19	20.50
8850	1" diameter		400	.020		23.50	1.20		24.70	27.50
8860	1-1/4" diameter		355	.023		25.50	1.35		26.85	30
8870	1-1/2" diameter		320	.025		39.50	1.50		41	45.50
8880	2" diameter		266	.030		57.50	1.81		59.31	66
8900	3" diameter		133	.060		80	3.61		83.61	93.50
8920	4" diameter		90	.089		74.50	5.35		79.85	90
8950	Touch-up plastic coating, spray, 12 oz.					67			67	73.50
8960	Sealing fittings, 1/2" diameter	1 Elec	11	.727		68.50	43.50		112	141

26 05 33.25 Conduit Fittings for Rigid Galvanized Steel

		Crew	Daily Output	Labor-Hours	Unit	Material	2019 Bare Costs Labor	Equipment	Total	Total Incl O&P
8970	3/4" diameter	1 Elec	9	.889	Ea.	88.50	53.50		142	177
8980	1" diameter		7.50	1.067		75.50	64		139.50	179
8990	1-1/4" diameter		6.50	1.231		103	74		177	224
9000	1-1/2" diameter		5.50	1.455		146	87.50		233.50	290
9010	2" diameter		4.80	1.667		161	100		261	325
9020	2-1/2" diameter		4	2		266	120		386	470
9030	3" diameter		3.50	2.286		299	137		436	535
9040	3-1/2" diameter		3	2.667		630	160		790	935
9050	4" diameter		2.50	3.200		1,100	192		1,292	1,500
9060	5" diameter		1.70	4.706		2,125	283		2,408	2,750
9070	Unions, 1/2" diameter		18	.444		52	26.50		78.50	97
9080	3/4" diameter		15	.533		59	32		91	113
9090	1" diameter		13	.615		72.50	37		109.50	135
9100	1-1/4" diameter		11	.727		120	43.50		163.50	197
9110	1-1/2" diameter		9.50	.842		159	50.50		209.50	251
9120	2" diameter		8	1		194	60		254	305
9130	2-1/2" diameter		7.50	1.067		278	64		342	400
9140	3" diameter		6.80	1.176		320	70.50		390.50	455
9150	3-1/2" diameter		5.80	1.379		430	83		513	595
9160	4" diameter		4.80	1.667		580	100		680	790
9170	5" diameter		4	2		955	120		1,075	1,225

26 05 33.30 Electrical Nonmetallic Tubing (ENT)

		Crew	Daily Output	Labor-Hours	Unit	Material	2019 Bare Costs Labor	Equipment	Total	Total Incl O&P
0010	**ELECTRICAL NONMETALLIC TUBING (ENT)**									
0050	Flexible, 1/2" diameter	1 Elec	270	.030	L.F.	.53	1.78		2.31	3.23
0100	3/4" diameter		230	.035		.76	2.09		2.85	3.96
0200	1" diameter		145	.055		1.38	3.31		4.69	6.45
0210	1-1/4" diameter		125	.064		1.26	3.84		5.10	7.15
0220	1-1/2" diameter		100	.080		2.05	4.80		6.85	9.40
0230	2" diameter		75	.107		2.52	6.40		8.92	12.30
0300	Connectors, to outlet box, 1/2" diameter		230	.035	Ea.	1.07	2.09		3.16	4.30
0310	3/4" diameter		210	.038		1.88	2.29		4.17	5.50
0320	1" diameter		200	.040		2.37	2.40		4.77	6.20
0400	Couplings, to conduit, 1/2" diameter		145	.055		.84	3.31		4.15	5.85
0410	3/4" diameter		130	.062		1.10	3.70		4.80	6.70
0420	1" diameter		125	.064		2.14	3.84		5.98	8.10

26 05 33.33 Raceway/Boxes for Utility Substations

		Crew	Daily Output	Labor-Hours	Unit	Material	2019 Bare Costs Labor	Equipment	Total	Total Incl O&P
0010	**RACEWAY/BOXES FOR UTILITY SUBSTATIONS**									
7000	Conduit, conductors, and insulators									
7100	Conduit, metallic	R-11	560	.100	Lb.	3.12	5.70	1.12	9.94	13.20
7110	Non-metallic	"	800	.070	"	9.45	3.99	.78	14.22	17.20
7190	See Section 26 05 33									
7200	Wire and cable	R-11	700	.080	Lb.	9.95	4.56	.89	15.40	18.75
7290	See Section 26 05 19									

26 05 33.35 Flexible Metallic Conduit

		Crew	Daily Output	Labor-Hours	Unit	Material	2019 Bare Costs Labor	Equipment	Total	Total Incl O&P
0010	**FLEXIBLE METALLIC CONDUIT**									
0050	Steel, 3/8" diameter	1 Elec	200	.040	L.F.	.37	2.40		2.77	3.99
0100	1/2" diameter		200	.040		.45	2.40		2.85	4.08
0200	3/4" diameter		160	.050		.57	3		3.57	5.10
0250	1" diameter		100	.080		1.30	4.80		6.10	8.60
0300	1-1/4" diameter		70	.114		1.37	6.85		8.22	11.75
0350	1-1/2" diameter		50	.160		2.93	9.60		12.53	17.55
0370	2" diameter		40	.200		3.62	12		15.62	22

For customer support on your Electrical Costs with RSMeans data, call 800.448.8182.

169

26 05 33.35 Flexible Metallic Conduit		Crew	Daily Output	Labor-Hours	Unit	Material	2019 Bare Costs Labor	Equipment	Total	Total Incl O&P
0380	2-1/2" diameter	1 Elec	30	.267	L.F.	3.44	16		19.44	28
0390	3" diameter	2 Elec	50	.320		5.90	19.20		25.10	35
0400	3-1/2" diameter		40	.400		6.80	24		30.80	43.50
0410	4" diameter		30	.533		7.65	32		39.65	56.50
0420	Connectors, plain, 3/8" diameter	1 Elec	100	.080	Ea.	1.78	4.80		6.58	9.10
0430	1/2" diameter		80	.100		1.89	6		7.89	11.05
0440	3/4" diameter		70	.114		2.18	6.85		9.03	12.65
0450	1" diameter		50	.160		5	9.60		14.60	19.85
0452	1-1/4" diameter		45	.178		6.10	10.70		16.80	22.50
0454	1-1/2" diameter		40	.200		8.80	12		20.80	27.50
0456	2" diameter		28	.286		12.45	17.15		29.60	39
0458	2-1/2" diameter		25	.320		22	19.20		41.20	52.50
0460	3" diameter		20	.400		30.50	24		54.50	69.50
0462	3-1/2" diameter		16	.500		90	30		120	144
0464	4" diameter		13	.615		100	37		137	165
0490	Insulated, 1" diameter		40	.200		8	12		20	26.50
0500	1-1/4" diameter		40	.200		11.60	12		23.60	30.50
0550	1-1/2" diameter		32	.250		26	15		41	51
0600	2" diameter		23	.348		24.50	21		45.50	58
0610	2-1/2" diameter		20	.400		55.50	24		79.50	97
0620	3" diameter		17	.471		113	28.50		141.50	166
0630	3-1/2" diameter		13	.615		264	37		301	345
0640	4" diameter		10	.800		267	48		315	365
0650	Connectors 90°, plain, 3/8" diameter		80	.100		2.56	6		8.56	11.75
0660	1/2" diameter		60	.133		4.96	8		12.96	17.40
0700	3/4" diameter		50	.160		7.10	9.60		16.70	22
0750	1" diameter		40	.200		11.50	12		23.50	30.50
0790	Insulated, 1" diameter		40	.200		17.20	12		29.20	37
0800	1-1/4" diameter		30	.267		24	16		40	50.50
0850	1-1/2" diameter		23	.348		44	21		65	79.50
0900	2" diameter		18	.444		70	26.50		96.50	117
0910	2-1/2" diameter		16	.500		153	30		183	214
0920	3" diameter		14	.571		216	34.50		250.50	289
0930	3-1/2" diameter		11	.727		540	43.50		583.50	660
0940	4" diameter		8	1		670	60		730	825
0960	Couplings, to flexible conduit, 1/2" diameter		50	.160		.92	9.60		10.52	15.35
0970	3/4" diameter		40	.200		1.58	12		13.58	19.65
0980	1" diameter		35	.229		2.58	13.75		16.33	23.50
0990	1-1/4" diameter		28	.286		5.80	17.15		22.95	32
1000	1-1/2" diameter		23	.348		8.05	21		29.05	40
1010	2" diameter		20	.400		16	24		40	53.50
1020	2-1/2" diameter		18	.444		36.50	26.50		63	80
1030	3" diameter		15	.533		92	32		124	149
1032	Aluminum, 3/8" diameter		210	.038	L.F.	.41	2.29		2.70	3.86
1034	1/2" diameter		210	.038		.61	2.29		2.90	4.08
1036	3/4" diameter		165	.048		.87	2.91		3.78	5.30
1038	1" diameter		105	.076		1.31	4.58		5.89	8.30
1040	1-1/4" diameter		75	.107		1.62	6.40		8.02	11.35
1042	1-1/2" diameter		53	.151		3.20	9.05		12.25	17
1044	2" diameter		42	.190		4.09	11.45		15.54	21.50
1046	2-1/2" diameter		32	.250		4.88	15		19.88	28
1048	3" diameter	2 Elec	53	.302		7.15	18.15		25.30	35
1050	3-1/2" diameter		42	.381		6.40	23		29.40	41

For customer support on your Electrical Costs with RSMeans data, call 800.448.8182.

26 05 33.35 Flexible Metallic Conduit		Crew	Daily Output	Labor-Hours	Unit	Material	2019 Bare Costs Labor	Equipment	Total	Total Incl O&P
1052	4" diameter	2 Elec	32	.500	L.F.	10.25	30		40.25	56.50
1070	Sealtite, 3/8" diameter	1 Elec	140	.057		.66	3.43		4.09	5.85
1080	1/2" diameter		140	.057		.89	3.43		4.32	6.10
1090	3/4" diameter		100	.080		1.45	4.80		6.25	8.75
1100	1" diameter		70	.114		2.38	6.85		9.23	12.85
1200	1-1/4" diameter		50	.160		2.87	9.60		12.47	17.50
1300	1-1/2" diameter		40	.200		3.01	12		15.01	21
1400	2" diameter		30	.267		3.83	16		19.83	28
1410	2-1/2" diameter		27	.296		7.60	17.80		25.40	35
1420	3" diameter	2 Elec	50	.320		8.35	19.20		27.55	37.50
1440	4" diameter	"	30	.533		19.80	32		51.80	70
1490	Connectors, plain, 3/8" diameter	1 Elec	70	.114	Ea.	2.28	6.85		9.13	12.75
1500	1/2" diameter		70	.114		2.77	6.85		9.62	13.30
1700	3/4" diameter		50	.160		4.53	9.60		14.13	19.35
1900	1" diameter		40	.200		6.85	12		18.85	25.50
1910	Insulated, 1" diameter		40	.200		9.05	12		21.05	28
2000	1-1/4" diameter		32	.250		10.25	15		25.25	34
2100	1-1/2" diameter		27	.296		22	17.80		39.80	50.50
2200	2" diameter		20	.400		42.50	24		66.50	83
2210	2-1/2" diameter		15	.533		104	32		136	162
2220	3" diameter		12	.667		145	40		185	220
2240	4" diameter		8	1		202	60		262	310
2290	Connectors, 90°, 3/8" diameter		70	.114		4.71	6.85		11.56	15.45
2300	1/2" diameter		70	.114		6.85	6.85		13.70	17.80
2400	3/4" diameter		50	.160		6.75	9.60		16.35	22
2600	1" diameter		40	.200		13.30	12		25.30	32.50
2790	Insulated, 1" diameter		40	.200		10.60	12		22.60	29.50
2800	1-1/4" diameter		32	.250		15.20	15		30.20	39
3000	1-1/2" diameter		27	.296		23	17.80		40.80	52
3100	2" diameter		20	.400		26.50	24		50.50	65
3110	2-1/2" diameter		14	.571		126	34.50		160.50	190
3120	3" diameter		11	.727		340	43.50		383.50	440
3140	4" diameter		7	1.143		170	68.50		238.50	289
4300	Coupling sealtite to rigid, 1/2" diameter		20	.400		3.64	24		27.64	40
4500	3/4" diameter		18	.444		5	26.50		31.50	45.50
4800	1" diameter		14	.571		6.55	34.50		41.05	58.50
4900	1-1/4" diameter		12	.667		14.90	40		54.90	76
5000	1-1/2" diameter		11	.727		21	43.50		64.50	88
5100	2" diameter		10	.800		33	48		81	108
5110	2-1/2" diameter		9.50	.842		124	50.50		174.50	213
5120	3" diameter		9	.889		152	53.50		205.50	247
5130	3-1/2" diameter		9	.889		180	53.50		233.50	279
5140	4" diameter		8.50	.941		209	56.50		265.50	315

26 05 33.95 Cutting and Drilling

0010	**CUTTING AND DRILLING**									
0100	Hole drilling to 10' high, concrete wall									
0110	8" thick, 1/2" pipe size	R-31	12	.667	Ea.	.24	40	3.88	44.12	64
0120	3/4" pipe size		12	.667		.24	40	3.88	44.12	64
0130	1" pipe size		9.50	.842		.39	50.50	4.90	55.79	81.50
0140	1-1/4" pipe size		9.50	.842		.39	50.50	4.90	55.79	81.50
0150	1-1/2" pipe size		9.50	.842		.39	50.50	4.90	55.79	81.50
0160	2" pipe size		4.40	1.818		.52	109	10.60	120.12	175

For customer support on your Electrical Costs with RSMeans data, call 800.448.8182.

171

26 05 33 – Raceway and Boxes for Electrical Systems

26 05 33.95 Cutting and Drilling	Crew	Daily Output	Labor-Hours	Unit	Material	2019 Bare Costs Labor	Equipment	Total	Total Incl O&P
0170	2-1/2" pipe size	R-31	4.40	1.818	Ea.	.52	109	10.60	120.12	175
0180	3" pipe size		4.40	1.818		.52	109	10.60	120.12	175
0190	3-1/2" pipe size		3.30	2.424		.68	146	14.10	160.78	233
0200	4" pipe size		3.30	2.424		.68	146	14.10	160.78	233
0500	12" thick, 1/2" pipe size		9.40	.851		.37	51	4.95	56.32	82.50
0520	3/4" pipe size		9.40	.851		.37	51	4.95	56.32	82.50
0540	1" pipe size		7.30	1.096		.59	66	6.40	72.99	106
0560	1-1/4" pipe size		7.30	1.096		.59	66	6.40	72.99	106
0570	1-1/2" pipe size		7.30	1.096		.59	66	6.40	72.99	106
0580	2" pipe size		3.60	2.222		.79	133	12.95	146.74	214
0590	2-1/2" pipe size		3.60	2.222		.79	133	12.95	146.74	214
0600	3" pipe size		3.60	2.222		.79	133	12.95	146.74	214
0610	3-1/2" pipe size		2.80	2.857		1.02	172	16.65	189.67	275
0630	4" pipe size		2.50	3.200		1.02	192	18.60	211.62	310
0650	16" thick, 1/2" pipe size		7.60	1.053		.49	63	6.15	69.64	102
0670	3/4" pipe size		7	1.143		.49	68.50	6.65	75.64	110
0690	1" pipe size		6	1.333		.79	80	7.75	88.54	128
0710	1-1/4" pipe size		5.50	1.455		.79	87.50	8.45	96.74	140
0730	1-1/2" pipe size		5.50	1.455		.79	87.50	8.45	96.74	140
0750	2" pipe size		3	2.667		1.05	160	15.50	176.55	257
0770	2-1/2" pipe size		2.70	2.963		1.05	178	17.25	196.30	285
0790	3" pipe size		2.50	3.200		1.05	192	18.60	211.65	310
0810	3-1/2" pipe size		2.30	3.478		1.36	209	20	230.36	335
0830	4" pipe size		2	4		1.36	240	23.50	264.86	385
0850	20" thick, 1/2" pipe size		6.40	1.250		.61	75	7.30	82.91	121
0870	3/4" pipe size		6	1.333		.61	80	7.75	88.36	128
0890	1" pipe size		5	1.600		.98	96	9.30	106.28	154
0910	1-1/4" pipe size		4.80	1.667		.98	100	9.70	110.68	161
0930	1-1/2" pipe size		4.60	1.739		.98	104	10.10	115.08	168
0950	2" pipe size		2.70	2.963		1.31	178	17.25	196.56	285
0970	2-1/2" pipe size		2.40	3.333		1.31	200	19.40	220.71	320
0990	3" pipe size		2.20	3.636		1.31	218	21	240.31	350
1010	3-1/2" pipe size		2	4		1.70	240	23.50	265.20	385
1030	4" pipe size		1.70	4.706		1.70	283	27.50	312.20	450
1050	24" thick, 1/2" pipe size		5.50	1.455		.73	87.50	8.45	96.68	140
1070	3/4" pipe size		5.10	1.569		.73	94	9.15	103.88	152
1090	1" pipe size		4.30	1.860		1.18	112	10.85	124.03	180
1110	1-1/4" pipe size		4	2		1.18	120	11.65	132.83	193
1130	1-1/2" pipe size		4	2		1.18	120	11.65	132.83	193
1150	2" pipe size		2.40	3.333		1.57	200	19.40	220.97	320
1170	2-1/2" pipe size		2.20	3.636		1.57	218	21	240.57	350
1190	3" pipe size		2	4		1.57	240	23.50	265.07	385
1210	3-1/2" pipe size		1.80	4.444		2.04	267	26	295.04	430
1230	4" pipe size		1.50	5.333		2.04	320	31	353.04	515
1500	Brick wall, 8" thick, 1/2" pipe size		18	.444		.24	26.50	2.59	29.33	43
1520	3/4" pipe size		18	.444		.24	26.50	2.59	29.33	43
1540	1" pipe size		13.30	.602		.39	36	3.50	39.89	58.50
1560	1-1/4" pipe size		13.30	.602		.39	36	3.50	39.89	58.50
1580	1-1/2" pipe size		13.30	.602		.39	36	3.50	39.89	58.50
1600	2" pipe size		5.70	1.404		.52	84.50	8.15	93.17	136
1620	2-1/2" pipe size		5.70	1.404		.52	84.50	8.15	93.17	136
1640	3" pipe size		5.70	1.404		.52	84.50	8.15	93.17	136
1660	3-1/2" pipe size		4.40	1.818		.68	109	10.60	120.28	175

26 05 33 – Raceway and Boxes for Electrical Systems

26 05 33.95 Cutting and Drilling	Crew	Daily Output	Labor-Hours	Unit	Material	2019 Bare Costs Labor	Equipment	Total	Total Incl O&P
1680	4" pipe size	R-31	4	2	Ea.	.68	120	11.65	132.33	193
1700	12" thick, 1/2" pipe size		14.50	.552		.37	33	3.21	36.58	53.50
1720	3/4" pipe size		14.50	.552		.37	33	3.21	36.58	53.50
1740	1" pipe size		11	.727		.59	43.50	4.23	48.32	70.50
1760	1-1/4" pipe size		11	.727		.59	43.50	4.23	48.32	70.50
1780	1-1/2" pipe size		11	.727		.59	43.50	4.23	48.32	70.50
1800	2" pipe size		5	1.600		.79	96	9.30	106.09	154
1820	2-1/2" pipe size		5	1.600		.79	96	9.30	106.09	154
1840	3" pipe size		5	1.600		.79	96	9.30	106.09	154
1860	3-1/2" pipe size		3.80	2.105		1.02	126	12.25	139.27	204
1880	4" pipe size		3.30	2.424		1.02	146	14.10	161.12	234
1900	16" thick, 1/2" pipe size		12.30	.650		.49	39	3.79	43.28	63
1920	3/4" pipe size		12.30	.650		.49	39	3.79	43.28	63
1940	1" pipe size		9.30	.860		.79	51.50	5	57.29	83.50
1960	1-1/4" pipe size		9.30	.860		.79	51.50	5	57.29	83.50
1980	1-1/2" pipe size		9.30	.860		.79	51.50	5	57.29	83.50
2000	2" pipe size		4.40	1.818		1.05	109	10.60	120.65	176
2010	2-1/2" pipe size		4.40	1.818		1.05	109	10.60	120.65	176
2030	3" pipe size		4.40	1.818		1.05	109	10.60	120.65	176
2050	3-1/2" pipe size		3.30	2.424		1.36	146	14.10	161.46	234
2070	4" pipe size		3	2.667		1.36	160	15.50	176.86	258
2090	20" thick, 1/2" pipe size		10.70	.748		.61	45	4.35	49.96	72.50
2110	3/4" pipe size		10.70	.748		.61	45	4.35	49.96	72.50
2130	1" pipe size		8	1		.98	60	5.80	66.78	97
2150	1-1/4" pipe size		8	1		.98	60	5.80	66.78	97
2170	1-1/2" pipe size		8	1		.98	60	5.80	66.78	97
2190	2" pipe size		4	2		1.31	120	11.65	132.96	193
2210	2-1/2" pipe size		4	2		1.31	120	11.65	132.96	193
2230	3" pipe size		4	2		1.31	120	11.65	132.96	193
2250	3-1/2" pipe size		3	2.667		1.70	160	15.50	177.20	258
2270	4" pipe size		2.70	2.963		1.70	178	17.25	196.95	286
2290	24" thick, 1/2" pipe size		9.40	.851		.73	51	4.95	56.68	83
2310	3/4" pipe size		9.40	.851		.73	51	4.95	56.68	83
2330	1" pipe size		7.10	1.127		1.18	67.50	6.55	75.23	110
2350	1-1/4" pipe size		7.10	1.127		1.18	67.50	6.55	75.23	110
2370	1-1/2" pipe size		7.10	1.127		1.18	67.50	6.55	75.23	110
2390	2" pipe size		3.60	2.222		1.57	133	12.95	147.52	215
2410	2-1/2" pipe size		3.60	2.222		1.57	133	12.95	147.52	215
2430	3" pipe size		3.60	2.222		1.57	133	12.95	147.52	215
2450	3-1/2" pipe size		2.80	2.857		2.04	172	16.65	190.69	277
2470	4" pipe size	▼	2.50	3.200	▼	2.04	192	18.60	212.64	310
3000	Knockouts to 8' high, metal boxes & enclosures									
3020	With hole saw, 1/2" pipe size	1 Elec	53	.151	Ea.		9.05		9.05	13.50
3040	3/4" pipe size		47	.170			10.20		10.20	15.25
3050	1" pipe size		40	.200			12		12	17.90
3060	1-1/4" pipe size		36	.222			13.35		13.35	19.90
3070	1-1/2" pipe size		32	.250			15		15	22.50
3080	2" pipe size		27	.296			17.80		17.80	26.50
3090	2-1/2" pipe size		20	.400			24		24	36
4010	3" pipe size		16	.500			30		30	45
4030	3-1/2" pipe size		13	.615			37		37	55
4050	4" pipe size	▼	11	.727	▼		43.50		43.50	65

26 05 36.10 Cable Tray Ladder Type		Crew	Daily Output	Labor-Hours	Unit	Material	2019 Bare Costs Labor	Equipment	Total	Total Incl O&P
0010	**CABLE TRAY LADDER TYPE** w/ftngs. & supports, 4" dp., to 15' elev.									
0100	For higher elevations, see Section 26 05 36.40									
0160	Galvanized steel tray R260536-10									
0170	4" rung spacing, 6" wide	2 Elec	98	.163	L.F.	13.35	9.80		23.15	29.50
0180	9" wide R260536-11		92	.174		14.65	10.45		25.10	31.50
0200	12" wide		86	.186		16.15	11.15		27.30	34.50
0400	18" wide		82	.195		18.70	11.70		30.40	38
0600	24" wide		78	.205		21.50	12.30		33.80	42
0650	30" wide		68	.235		27.50	14.15		41.65	51
0800	6" rung spacing, 6" wide		100	.160		12.15	9.60		21.75	28
0850	9" wide		94	.170		13.35	10.20		23.55	30
0860	12" wide		88	.182		14.05	10.90		24.95	32
0870	18" wide		84	.190		16.05	11.45		27.50	34.50
0880	24" wide		80	.200		17.95	12		29.95	37.50
0890	30" wide		70	.229		21	13.75		34.75	44
0910	9" rung spacing, 6" wide		102	.157		11.35	9.40		20.75	26.50
0920	9" wide		98	.163		11.65	9.80		21.45	27.50
0930	12" wide		94	.170		13.40	10.20		23.60	30
0940	18" wide		90	.178		15.10	10.70		25.80	32.50
0950	24" wide		86	.186		17.20	11.15		28.35	35.50
0960	30" wide		80	.200		18.85	12		30.85	38.50
0980	12" rung spacing, 6" wide		106	.151		11.15	9.05		20.20	26
0990	9" wide		104	.154		11.45	9.25		20.70	26.50
1000	12" wide		100	.160		12.05	9.60		21.65	27.50
1010	18" wide		96	.167		13	10		23	29.50
1020	24" wide		94	.170		13.90	10.20		24.10	30.50
1030	30" wide		88	.182		15.85	10.90		26.75	34
1041	18" rung spacing, 6" wide		108	.148		11.10	8.90		20	25.50
1042	9" wide		106	.151		11.25	9.05		20.30	26
1043	12" wide		102	.157		12	9.40		21.40	27.50
1044	18" wide		98	.163		12.55	9.80		22.35	28.50
1045	24" wide		96	.167		13	10		23	29
1046	30" wide		90	.178		13.65	10.70		24.35	31
1050	Elbows horiz. 9" rung spacing, 90°, 12" radius, 6" wide		9.60	1.667	Ea.	51.50	100		151.50	206
1060	9" wide		8.40	1.905		55	114		169	232
1070	12" wide		7.60	2.105		59.50	126		185.50	255
1080	18" wide		6.20	2.581		78.50	155		233.50	315
1090	24" wide		5.40	2.963		87.50	178		265.50	360
1100	30" wide		4.80	3.333		116	200		316	425
1120	90°, 24" radius, 6" wide		9.20	1.739		107	104		211	273
1130	9" wide		8	2		110	120		230	300
1140	12" wide		7.20	2.222		113	133		246	325
1150	18" wide		5.80	2.759		122	166		288	380
1160	24" wide		5	3.200		132	192		324	430
1170	30" wide		4.40	3.636		143	218		361	480
1190	90°, 36" radius, 6" wide		8.80	1.818		117	109		226	292
1200	9" wide		7.60	2.105		124	126		250	325
1210	12" wide		6.80	2.353		130	141		271	355
1220	18" wide		5.40	2.963		149	178		327	430
1230	24" wide		4.60	3.478		163	209		372	490
1240	30" wide		4	4		196	240		436	575
1260	45°, 12" radius, 6" wide		13.20	1.212		40	73		113	153

26 05 36 – Cable Trays for Electrical Systems

26 05 36.10 Cable Tray Ladder Type		Crew	Daily Output	Labor-Hours	Unit	Material	2019 Bare Costs Labor	Equipment	Total	Total Incl O&P
1270	9" wide	2 Elec	11	1.455	Ea.	41.50	87.50		129	176
1280	12" wide		9.60	1.667		43	100		143	197
1290	18" wide		7.60	2.105		45.50	126		171.50	239
1300	24" wide		6.20	2.581		58.50	155		213.50	296
1310	30" wide		5.40	2.963		75	178		253	350
1330	45°, 24" radius, 6" wide		12.80	1.250		50.50	75		125.50	168
1340	9" wide		10.60	1.509		53.50	90.50		144	194
1350	12" wide		9.20	1.739		60.50	104		164.50	223
1360	18" wide		7.20	2.222		62	133		195	267
1370	24" wide		5.80	2.759		75	166		241	330
1380	30" wide		5	3.200		89.50	192		281.50	385
1400	45°, 36" radius, 6" wide		12.40	1.290		72.50	77.50		150	196
1410	9" wide		10.20	1.569		76.50	94		170.50	226
1420	12" wide		8.80	1.818		78	109		187	249
1430	18" wide		6.80	2.353		84.50	141		225.50	305
1440	24" wide		5.40	2.963		89.50	178		267.50	365
1450	30" wide	↓	4.60	3.478	↓	116	209		325	435
1470	Elbows horizontal, 4" rung spacing, use 9" rung x 1.50									
1480	6" rung spacing, use 9" rung x 1.20									
1490	12" rung spacing, use 9" rung x 0.93									
1500	Elbows vertical, 9" rung spacing, 90°, 12" radius, 6" wide	2 Elec	9.60	1.667	Ea.	71	100		171	227
1510	9" wide		8.40	1.905		72.50	114		186.50	251
1520	12" wide		7.60	2.105		75	126		201	272
1530	18" wide		6.20	2.581		78	155		233	315
1540	24" wide		5.40	2.963		81.50	178		259.50	355
1550	30" wide		4.80	3.333		87.50	200		287.50	395
1570	24" radius, 6" wide		9.20	1.739		100	104		204	266
1580	9" wide		8	2		103	120		223	293
1590	12" wide		7.20	2.222		106	133		239	315
1600	18" wide		5.80	2.759		113	166		279	370
1610	24" wide		5	3.200		119	192		311	420
1620	30" wide		4.40	3.636		124	218		342	460
1640	36" radius, 6" wide		8.80	1.818		135	109		244	310
1650	9" wide		7.60	2.105		139	126		265	340
1660	12" wide		6.80	2.353		141	141		282	365
1670	18" wide		5.40	2.963		153	178		331	435
1680	24" wide		4.60	3.478		155	209		364	480
1690	30" wide	↓	4	4	↓	175	240		415	555
1710	Elbows vertical, 4" rung spacing, use 9" rung x 1.25									
1720	6" rung spacing, use 9" rung x 1.15									
1730	12" rung spacing, use 9" rung x 0.90									
1740	Tee horizontal, 9" rung spacing, 12" radius, 6" wide	2 Elec	5	3.200	Ea.	110	192		302	410
1750	9" wide		4.60	3.478		117	209		326	440
1760	12" wide		4.40	3.636		124	218		342	460
1770	18" wide		4	4		141	240		381	515
1780	24" wide		3.60	4.444		166	267		433	585
1790	30" wide		3.40	4.706		199	283		482	640
1810	24" radius, 6" wide		4.60	3.478		171	209		380	500
1820	9" wide		4.20	3.810		179	229		408	535
1830	12" wide		4	4		185	240		425	565
1840	18" wide		3.60	4.444		200	267		467	620
1850	24" wide		3.20	5		218	300		518	690
1860	30" wide	↓	3	5.333		236	320		556	740

26 05 36.10 Cable Tray Ladder Type		Crew	Daily Output	Labor-Hours	Unit	Material	2019 Bare Costs Labor	Equipment	Total	Total Incl O&P
1880	36" radius, 6" wide	2 Elec	4.20	3.810	Ea.	255	229		484	620
1890	9" wide		3.80	4.211		266	253		519	670
1900	12" wide		3.60	4.444		283	267		550	710
1910	18" wide		3.20	5		305	300		605	785
1920	24" wide		2.80	5.714		345	345		690	885
1930	30" wide		2.60	6.154		375	370		745	960
1980	Tee vertical, 9" rung spacing, 12" radius, 6" wide		5.40	2.963		186	178		364	470
1990	9" wide		5.20	3.077		188	185		373	485
2000	12" wide		5	3.200		189	192		381	495
2010	18" wide		4.60	3.478		191	209		400	520
2020	24" wide		4.40	3.636		196	218		414	540
2030	30" wide		4	4		208	240		448	590
2050	24" radius, 6" wide		5	3.200		335	192		527	655
2060	9" wide		4.80	3.333		340	200		540	675
2070	12" wide		4.60	3.478		345	209		554	690
2080	18" wide		4.20	3.810		355	229		584	730
2090	24" wide		4	4		365	240		605	760
2100	30" wide		3.60	4.444		375	267		642	810
2120	36" radius, 6" wide		4.60	3.478		625	209		834	1,000
2130	9" wide		4.40	3.636		635	218		853	1,025
2140	12" wide		4.20	3.810		645	229		874	1,050
2150	18" wide		3.80	4.211		650	253		903	1,100
2160	24" wide		3.60	4.444		660	267		927	1,125
2170	30" wide		3.20	5		685	300		985	1,200
2190	Tee, 4" rung spacing, use 9" rung x 1.30									
2200	6" rung spacing, use 9" rung x 1.20									
2210	12" rung spacing, use 9" rung x 0.90									
2220	Cross horizontal, 9" rung spacing, 12" radius, 6" wide	2 Elec	4	4	Ea.	135	240		375	510
2230	9" wide		3.80	4.211		142	253		395	530
2240	12" wide		3.60	4.444		155	267		422	570
2250	18" wide		3.40	4.706		169	283		452	605
2260	24" wide		3	5.333		188	320		508	685
2270	30" wide		2.80	5.714		227	345		572	760
2290	24" radius, 6" wide		3.60	4.444		263	267		530	690
2300	9" wide		3.40	4.706		271	283		554	720
2310	12" wide		3.20	5		279	300		579	755
2320	18" wide		3	5.333		298	320		618	805
2330	24" wide		2.60	6.154		315	370		685	900
2340	30" wide		2.40	6.667		335	400		735	965
2360	36" radius, 6" wide		3.20	5		365	300		665	855
2370	9" wide		3	5.333		380	320		700	900
2380	12" wide		2.80	5.714		400	345		745	950
2390	18" wide		2.60	6.154		440	370		810	1,025
2400	24" wide		2.20	7.273		490	435		925	1,175
2410	30" wide		2	8		530	480		1,010	1,300
2430	Cross horizontal, 4" rung spacing, use 9" rung x 1.30									
2440	6" rung spacing, use 9" rung x 1.20									
2450	12" rung spacing, use 9" rung x 0.90									
2460	Reducer, 9" to 6" wide tray	2 Elec	13	1.231	Ea.	75	74		149	193
2470	12" to 9" wide tray		12	1.333		76.50	80		156.50	204
2480	18" to 12" wide tray		10.40	1.538		77	92.50		169.50	223
2490	24" to 18" wide tray		9	1.778		78.50	107		185.50	245
2500	30" to 24" wide tray		8	2		80	120		200	267

26 05 36 – Cable Trays for Electrical Systems

26 05 36.10 Cable Tray Ladder Type	Crew	Daily Output	Labor-Hours	Unit	Material	2019 Bare Costs Labor	Equipment	Total	Total Incl O&P
2510	36" to 30" wide tray	2 Elec	7	2.286	Ea.	89.50	137		226.50	305
2511	Reducer, 18" to 6" wide tray		10.40	1.538		80	92.50		172.50	226
2512	24" to 12" wide tray		9	1.778		81.50	107		188.50	249
2513	30" to 18" wide tray		8	2		84.50	120		204.50	272
2514	30" to 12" wide tray		8	2		88	120		208	276
2515	36" to 24" wide tray		7	2.286		91	137		228	305
2516	36" to 18" wide tray		7	2.286		91	137		228	305
2517	36" to 12" wide tray		7	2.286		91	137		228	305
2520	Dropout or end plate, 6" wide		32	.500		8.35	30		38.35	54
2530	9" wide		28	.571		9.85	34.50		44.35	62
2540	12" wide		26	.615		10.85	37		47.85	67
2550	18" wide		22	.727		12.70	43.50		56.20	79
2560	24" wide		20	.800		15.35	48		63.35	88.50
2570	30" wide		18	.889		16.75	53.50		70.25	98
2590	Tray connector		48	.333		14.85	20		34.85	46.50
3200	Aluminum tray, 4" deep, 6" rung spacing, 6" wide		134	.119	L.F.	14.30	7.15		21.45	26.50
3210	9" wide		128	.125		15.15	7.50		22.65	28
3220	12" wide		124	.129		16	7.75		23.75	29
3230	18" wide		114	.140		17.85	8.45		26.30	32.50
3240	24" wide		106	.151		21	9.05		30.05	36.50
3250	30" wide		100	.160		23	9.60		32.60	39.50
3270	9" rung spacing, 6" wide		140	.114		11.60	6.85		18.45	23
3280	9" wide		134	.119		12.20	7.15		19.35	24
3290	12" wide		130	.123		12.45	7.40		19.85	24.50
3300	18" wide		122	.131		14.50	7.90		22.40	27.50
3310	24" wide		116	.138		16.85	8.30		25.15	31
3320	30" wide		108	.148		18.55	8.90		27.45	34
3340	12" rung spacing, 6" wide		146	.110		12.35	6.60		18.95	23.50
3350	9" wide		140	.114		12.50	6.85		19.35	24
3360	12" wide		134	.119		12.85	7.15		20	25
3370	18" wide		128	.125		13.70	7.50		21.20	26.50
3380	24" wide		124	.129		15.15	7.75		22.90	28.50
3390	30" wide		114	.140		16.25	8.45		24.70	30.50
3401	18" rung spacing, 6" wide		150	.107		12.25	6.40		18.65	23
3402	9" wide tray		144	.111		12.45	6.65		19.10	23.50
3403	12" wide tray		140	.114		12.85	6.85		19.70	24.50
3404	18" wide tray		134	.119		13.60	7.15		20.75	25.50
3405	24" wide tray		130	.123		14.65	7.40		22.05	27
3406	30" wide tray		120	.133		15.15	8		23.15	28.50
3410	Elbows horiz., 9" rung spacing, 90°, 12" radius, 6" wide		9.60	1.667	Ea.	55	100		155	210
3420	9" wide		8.40	1.905		56.50	114		170.50	233
3430	12" wide		7.60	2.105		67	126		193	263
3440	18" wide		6.20	2.581		75.50	155		230.50	315
3450	24" wide		5.40	2.963		93.50	178		271.50	370
3460	30" wide		4.80	3.333		99	200		299	410
3480	24" radius, 6" wide		9.20	1.739		98	104		202	264
3490	9" wide		8	2		102	120		222	291
3500	12" wide		7.20	2.222		107	133		240	315
3510	18" wide		5.80	2.759		116	166		282	375
3520	24" wide		5	3.200		128	192		320	425
3530	30" wide		4.40	3.636		140	218		358	480
3550	90°, 36" radius, 6" wide		8.80	1.818		107	109		216	281
3560	9" wide		7.60	2.105		110	126		236	310

26 05 36.10 Cable Tray Ladder Type		Crew	Daily Output	Labor-Hours	Unit	Material	2019 Bare Costs Labor	2019 Bare Costs Equipment	Total	Total Incl O&P
3570	12" wide	2 Elec	6.80	2.353	Ea.	124	141		265	345
3580	18" wide		5.40	2.963		138	178		316	415
3590	24" wide		4.60	3.478		156	209		365	480
3600	30" wide		4	4		172	240		412	550
3620	45°, 12" radius, 6" wide		13.20	1.212		36.50	73		109.50	149
3630	9" wide		11	1.455		37.50	87.50		125	172
3640	12" wide		9.60	1.667		39	100		139	192
3650	18" wide		7.60	2.105		46	126		172	240
3660	24" wide		6.20	2.581		57.50	155		212.50	294
3670	30" wide		5.40	2.963		66	178		244	340
3690	45°, 24" radius, 6" wide		12.80	1.250		48.50	75		123.50	165
3700	9" wide		10.60	1.509		55	90.50		145.50	196
3710	12" wide		9.20	1.739		56.50	104		160.50	218
3720	18" wide		7.20	2.222		60.50	133		193.50	266
3730	24" wide		5.80	2.759		70	166		236	325
3740	30" wide		5	3.200		83.50	192		275.50	380
3760	45°, 36" radius, 6" wide		12.40	1.290		66	77.50		143.50	189
3770	9" wide		10.20	1.569		67	94		161	215
3780	12" wide		8.80	1.818		68.50	109		177.50	239
3790	18" wide		6.80	2.353		75	141		216	294
3800	24" wide		5.40	2.963		92.50	178		270.50	365
3810	30" wide	▼	4.60	3.478	▼	98	209		307	420
3830	Elbows horizontal, 4" rung spacing, use 9" rung x 1.50									
3840	6" rung spacing, use 9" rung x 1.20									
3850	12" rung spacing, use 9" rung x 0.93									
3860	Elbows vertical, 9" rung spacing, 90°, 12" radius, 6" wide	2 Elec	9.60	1.667	Ea.	76.50	100		176.50	233
3870	9" wide		8.40	1.905		80.50	114		194.50	260
3880	12" wide		7.60	2.105		82	126		208	279
3890	18" wide		6.20	2.581		84.50	155		239.50	325
3900	24" wide		5.40	2.963		90	178		268	365
3910	30" wide		4.80	3.333		94	200		294	400
3930	24" radius, 6" wide		9.20	1.739		83.50	104		187.50	248
3940	9" wide		8	2		86	120		206	274
3950	12" wide		7.20	2.222		88.50	133		221.50	297
3960	18" wide		5.80	2.759		94	166		260	350
3970	24" wide		5	3.200		99.50	192		291.50	395
3980	30" wide		4.40	3.636		105	218		323	440
4000	36" radius, 6" wide		8.80	1.818		129	109		238	305
4010	9" wide		7.60	2.105		132	126		258	335
4020	12" wide		6.80	2.353		140	141		281	365
4030	18" wide		5.40	2.963		145	178		323	425
4040	24" wide		4.60	3.478		150	209		359	475
4050	30" wide	▼	4	4	▼	163	240		403	540
4070	Elbows vertical, 4" rung spacing, use 9" rung x 1.25									
4080	6" rung spacing, use 9" rung x 1.15									
4090	12" rung spacing, use 9" rung x 0.90									
4100	Tee horizontal, 9" rung spacing, 12" radius, 6" wide	2 Elec	5	3.200	Ea.	85	192		277	380
4110	9" wide		4.60	3.478		91.50	209		300.50	410
4120	12" wide		4.40	3.636		95	218		313	430
4130	18" wide		4.20	3.810		110	229		339	460
4140	24" wide		4	4		121	240		361	495
4150	30" wide		3.60	4.444		141	267		408	555
4170	24" radius, 6" wide	▼	4.60	3.478		145	209		354	470

178

26 05 36 – Cable Trays for Electrical Systems

26 05 36.10 Cable Tray Ladder Type	Crew	Daily Output	Labor-Hours	Unit	Material	2019 Bare Costs Labor	Equipment	Total	Total Incl O&P
4180	9" wide	2 Elec	4.20	3.810	Ea.	153	229		382	510
4190	12" wide		4	4		157	240		397	535
4200	18" wide		3.80	4.211		172	253		425	565
4210	24" wide		3.60	4.444		188	267		455	605
4220	30" wide		3.20	5		204	300		504	675
4240	36" radius, 6" wide		4.20	3.810		206	229		435	565
4250	9" wide		3.80	4.211		211	253		464	605
4260	12" wide		3.60	4.444		231	267		498	655
4270	18" wide		3.40	4.706		243	283		526	685
4280	24" wide		3.20	5		269	300		569	745
4290	30" wide		2.80	5.714		340	345		685	885
4310	Tee vertical, 9" rung spacing, 12" radius, 6" wide		5.40	2.963		191	178		369	475
4320	9" wide		5.20	3.077		197	185		382	495
4330	12" wide		5	3.200		198	192		390	505
4340	18" wide		4.60	3.478		203	209		412	535
4350	24" wide		4.40	3.636		219	218		437	565
4360	30" wide		4.20	3.810		223	229		452	585
4380	24" radius, 6" wide		5	3.200		325	192		517	640
4390	9" wide		4.80	3.333		330	200		530	660
4400	12" wide		4.60	3.478		335	209		544	675
4410	18" wide		4.20	3.810		340	229		569	715
4420	24" wide		4	4		350	240		590	745
4430	30" wide		3.80	4.211		360	253		613	770
4450	36" radius, 6" wide		4.60	3.478		710	209		919	1,100
4460	9" wide		4.40	3.636		725	218		943	1,125
4470	12" wide		4.20	3.810		755	229		984	1,175
4480	18" wide		3.80	4.211		760	253		1,013	1,225
4490	24" wide		3.60	4.444		780	267		1,047	1,250
4500	30" wide	▼	3.40	4.706	▼	790	283		1,073	1,300
4520	Tees, 4" rung spacing, use 9" rung x 1.30									
4530	6" rung spacing, use 9" rung x 1.20									
4540	12" rung spacing, use 9" rung x 0.90									
4550	Cross horizontal, 9" rung spacing, 12" radius, 6" wide	2 Elec	4.40	3.636	Ea.	118	218		336	455
4560	9" wide		4.20	3.810		132	229		361	485
4570	12" wide		4	4		138	240		378	510
4580	18" wide		3.60	4.444		161	267		428	575
4590	24" wide		3.40	4.706		167	283		450	605
4600	30" wide		3	5.333		196	320		516	695
4620	24" radius, 6" wide		4	4		227	240		467	610
4630	9" wide		3.80	4.211		235	253		488	635
4640	12" wide		3.60	4.444		242	267		509	665
4650	18" wide		3.20	5		254	300		554	730
4660	24" wide		3	5.333		277	320		597	785
4670	30" wide		2.60	6.154		296	370		666	875
4690	36" radius, 6" wide		3.60	4.444		269	267		536	695
4700	9" wide		3.40	4.706		279	283		562	725
4710	12" wide		3.20	5		296	300		596	775
4720	18" wide		2.80	5.714		315	345		660	855
4730	24" wide		2.60	6.154		390	370		760	980
4740	30" wide	▼	2.20	7.273	▼	435	435		870	1,125
4760	Cross horizontal, 4" rung spacing, use 9" rung x 1.30									
4770	6" rung spacing, use 9" rung x 1.20									
4780	12" rung spacing, use 9" rung x 0.90									

26 05 36.10 Cable Tray Ladder Type	Crew	Daily Output	Labor-Hours	Unit	Material	2019 Bare Costs Labor	Equipment	Total	Total Incl O&P
4790	Reducer, 9" to 6" wide tray	2 Elec	16	1	Ea.	63	60		123	159
4800	12" to 9" wide tray		14	1.143		66	68.50		134.50	175
4810	18" to 12" wide tray		12.40	1.290		66	77.50		143.50	189
4820	24" to 18" wide tray		10.60	1.509		67	90.50		157.50	209
4830	30" to 24" wide tray		9.20	1.739		68.50	104		172.50	232
4840	36" to 30" wide tray		8	2		72.50	120		192.50	259
4841	Reducer, 18" to 6" wide tray		12.40	1.290		66	77.50		143.50	189
4842	24" to 12" wide tray		10.60	1.509		67	90.50		157.50	209
4843	30" to 18" wide tray		9.20	1.739		68.50	104		172.50	232
4844	30" to 12" wide tray		9.20	1.739		68.50	104		172.50	232
4845	36" to 24" wide tray		8	2		72.50	120		192.50	259
4846	36" to 18" wide tray		8	2		72.50	120		192.50	259
4847	36" to 12" wide tray		8	2		72.50	120		192.50	259
4850	Dropout or end plate, 6" wide		32	.500		8.25	30		38.25	54
4860	9" wide tray		28	.571		8.90	34.50		43.40	61
4870	12" wide tray		26	.615		9.60	37		46.60	65.50
4880	18" wide tray		22	.727		12.95	43.50		56.45	79.50
4890	24" wide tray		20	.800		15.15	48		63.15	88
4900	30" wide tray		18	.889		18	53.50		71.50	99.50
4920	Tray connector		48	.333		12.10	20		32.10	43.50
8000	Elbow 36" radius horiz., 60°, 6" wide tray		10.60	1.509		91.50	90.50		182	236
8010	9" wide tray		9	1.778		99.50	107		206.50	268
8020	12" wide tray		7.80	2.051		102	123		225	296
8030	18" wide tray		6.20	2.581		109	155		264	350
8040	24" wide tray		5	3.200		128	192		320	425
8050	30" wide tray		4.40	3.636		130	218		348	470
8060	30°, 6" wide tray		14	1.143		70	68.50		138.50	179
8070	9" wide tray		11.40	1.404		74	84.50		158.50	208
8080	12" wide tray		9.80	1.633		75	98		173	229
8090	18" wide tray		7.40	2.162		79.50	130		209.50	281
8100	24" wide tray		5.80	2.759		97	166		263	355
8110	30" wide tray		4.80	3.333		99.50	200		299.50	410
8120	Adjustable, 6" wide tray		12.40	1.290		92.50	77.50		170	218
8130	9" wide tray		10.20	1.569		98	94		192	249
8140	12" wide tray		8.80	1.818		99.50	109		208.50	272
8150	18" wide tray		6.80	2.353		109	141		250	330
8160	24" wide tray		5.40	2.963		115	178		293	390
8170	30" wide tray		4.60	3.478		129	209		338	450
8180	Wye 36" radius horiz., 45°, 6" wide tray		4.60	3.478		101	209		310	420
8190	9" wide tray		4.40	3.636		106	218		324	440
8200	12" wide tray		4.20	3.810		110	229		339	460
8210	18" wide tray		3.80	4.211		125	253		378	510
8220	24" wide tray		3.60	4.444		140	267		407	555
8230	30" wide tray		3.40	4.706		156	283		439	590
8240	Elbow 36" radius vert. in/outside, 60°, 6" wide tray		10.60	1.509		101	90.50		191.50	246
8250	9" wide tray		9	1.778		105	107		212	274
8260	12" wide tray		7.80	2.051		106	123		229	300
8270	18" wide tray		6.20	2.581		110	155		265	350
8280	24" wide tray		5	3.200		117	192		309	415
8290	30" wide tray		4.40	3.636		125	218		343	460
8300	45°, 6" wide tray		12.40	1.290		91	77.50		168.50	216
8310	9" wide tray		10.20	1.569		96.50	94		190.50	247
8320	12" wide tray		8.80	1.818		98	109		207	271

26 05 36 – Cable Trays for Electrical Systems

26 05 36.10 Cable Tray Ladder Type	Crew	Daily Output	Labor-Hours	Unit	Material	2019 Bare Costs Labor	Equipment	Total	Total Incl O&P
8330	18" wide tray	2 Elec	6.80	2.353	Ea.	101	141		242	320
8340	24" wide tray		5.40	2.963		106	178		284	380
8350	30" wide tray		4.60	3.478		109	209		318	430
8360	30°, 6" wide tray		14	1.143		79.50	68.50		148	190
8370	9" wide tray		11.40	1.404		81	84.50		165.50	215
8380	12" wide tray		9.80	1.633		86.50	98		184.50	242
8390	18" wide tray		7.40	2.162		88	130		218	291
8400	24" wide tray		5.80	2.759		90	166		256	345
8410	30" wide tray		4.80	3.333		95	200		295	405
8660	Adjustable, 6" wide tray		12.40	1.290		91	77.50		168.50	216
8670	9" wide tray		10.20	1.569		96.50	94		190.50	247
8680	12" wide tray		8.80	1.818		98	109		207	271
8690	18" wide tray		6.80	2.353		109	141		250	330
8700	24" wide tray		5.40	2.963		115	178		293	390
8710	30" wide tray		4.60	3.478		128	209		337	450
8720	Cross, vertical, 24" radius, 6" wide tray		3.60	4.444		665	267		932	1,125
8730	9" wide tray		3.40	4.706		675	283		958	1,175
8740	12" wide tray		3.20	5		695	300		995	1,225
8750	18" wide tray		2.80	5.714		725	345		1,070	1,300
8760	24" wide tray		2.60	6.154		750	370		1,120	1,375
8770	30" wide tray		2.20	7.273		795	435		1,230	1,525
9200	Splice plate	1 Elec	48	.167	Pr.	12.65	10		22.65	29
9210	Expansion joint		48	.167		15.15	10		25.15	31.50
9220	Horizontal hinged		48	.167		12.65	10		22.65	29
9230	Vertical hinged		48	.167		15.15	10		25.15	31.50
9240	Ladder hanger, vertical		28	.286	Ea.	3.79	17.15		20.94	29.50
9250	Ladder to channel connector		24	.333		44	20		64	78.50
9260	Ladder to box connector, 30" wide		19	.421		44	25.50		69.50	86
9270	24" wide		20	.400		44	24		68	84.50
9280	18" wide		21	.381		42.50	23		65.50	81
9290	12" wide		22	.364		40	22		62	76
9300	9" wide		23	.348		35.50	21		56.50	70
9310	6" wide		24	.333		31	20		51	64.50
9320	Ladder floor flange		24	.333		28.50	20		48.50	61.50
9330	Cable roller for tray, 30" wide		10	.800		251	48		299	350
9340	24" wide		11	.727		213	43.50		256.50	300
9350	18" wide		12	.667		203	40		243	284
9360	12" wide		13	.615		160	37		197	231
9370	9" wide		14	.571		142	34.50		176.50	207
9380	6" wide		15	.533		115	32		147	175
9390	Pulley, single wheel		12	.667		262	40		302	350
9400	Triple wheel		10	.800		525	48		573	645
9440	Nylon cable tie, 14" long		80	.100		.57	6		6.57	9.60
9450	Ladder, hold down clamp		60	.133		8.05	8		16.05	21
9460	Cable clamp		60	.133		8.15	8		16.15	21
9470	Wall bracket, 30" wide tray		19	.421		46.50	25.50		72	88.50
9480	24" wide tray		20	.400		36	24		60	75.50
9490	18" wide tray		21	.381		32.50	23		55.50	69.50
9500	12" wide tray		22	.364		30	22		52	65.50
9510	9" wide tray		23	.348		26.50	21		47.50	60
9520	6" wide tray		24	.333		26	20		46	58.50

26 05 36.20 Cable Tray Solid Bottom		Crew	Daily Output	Labor-Hours	Unit	Material	2019 Bare Costs Labor	Equipment	Total	Total Incl O&P
0010	**CABLE TRAY SOLID BOTTOM** w/ftngs. & supports, 3" deep, to 15' high									
0200	For higher elevations, see Section 26 05 36.40									
0220	Galvanized steel, tray, 6" wide	R260536-10	2 Elec	120	.133	L.F.	10.60	8		18.60	23.50
0240	12" wide			100	.160		13.60	9.60		23.20	29.50
0260	18" wide			70	.229		16.55	13.75		30.30	38.50
0280	24" wide	R260536-11		60	.267		19.35	16		35.35	45.50
0300	30" wide			50	.320		23	19.20		42.20	54
0340	Elbow horizontal, 90°, 12" radius, 6" wide			9.60	1.667	Ea.	80.50	100		180.50	238
0360	12" wide			6.80	2.353		90	141		231	310
0370	18" wide			5.40	2.963		103	178		281	380
0380	24" wide			4.40	3.636		125	218		343	465
0390	30" wide			3.80	4.211		150	253		403	540
0420	24" radius, 6" wide			9.20	1.739		115	104		219	282
0440	12" wide			6.40	2.500		133	150		283	370
0450	18" wide			5	3.200		155	192		347	455
0460	24" wide			4	4		180	240		420	560
0470	30" wide			3.40	4.706		211	283		494	650
0500	36" radius, 6" wide			8.80	1.818		167	109		276	345
0520	12" wide			6	2.667		189	160		349	445
0530	18" wide			4.60	3.478		231	209		440	565
0540	24" wide			3.60	4.444		249	267		516	675
0550	30" wide			3	5.333		300	320		620	810
0580	Elbow vertical, 90°, 12" radius, 6" wide			9.60	1.667		97.50	100		197.50	256
0600	12" wide			6.80	2.353		106	141		247	325
0610	18" wide			5.40	2.963		118	178		296	395
0620	24" wide			4.40	3.636		122	218		340	460
0630	30" wide			3.80	4.211		130	253		383	520
0670	24" radius, 6" wide			9.20	1.739		138	104		242	310
0690	12" wide			6.40	2.500		152	150		302	390
0700	18" wide			5	3.200		164	192		356	470
0710	24" wide			4	4		178	240		418	555
0720	30" wide			3.40	4.706		189	283		472	630
0750	36" radius, 6" wide			8.80	1.818		189	109		298	370
0770	12" wide			6.60	2.424		211	146		357	450
0780	18" wide			4.60	3.478		231	209		440	565
0790	24" wide			3.60	4.444		253	267		520	680
0800	30" wide			3	5.333		276	320		596	785
0840	Tee horizontal, 12" radius, 6" wide			5	3.200		113	192		305	410
0860	12" wide			4	4		122	240		362	495
0870	18" wide			3.40	4.706		146	283		429	580
0880	24" wide			2.80	5.714		163	345		508	690
0890	30" wide			2.60	6.154		186	370		556	755
0940	24" radius, 6" wide			4.60	3.478		175	209		384	505
0960	12" wide			3.60	4.444		205	267		472	625
0970	18" wide			3	5.333		228	320		548	730
0980	24" wide			2.40	6.667		310	400		710	935
0990	30" wide			2.20	7.273		335	435		770	1,025
1020	36" radius, 6" wide			4.20	3.810		276	229		505	645
1040	12" wide			3.20	5		310	300		610	790
1050	18" wide			2.60	6.154		335	370		705	920
1060	24" wide			2.20	7.273		430	435		865	1,125
1070	30" wide			2	8		470	480		950	1,225

26 05 36 – Cable Trays for Electrical Systems

26 05 36.20 Cable Tray Solid Bottom		Crew	Daily Output	Labor-Hours	Unit	Material	2019 Bare Costs Labor	Equipment	Total	Total Incl O&P
1100	Tee vertical, 12" radius, 6" wide	2 Elec	5	3.200	Ea.	186	192		378	490
1120	12" wide		4	4		194	240		434	575
1130	18" wide		3.60	4.444		197	267		464	615
1140	24" wide		3.40	4.706		215	283		498	655
1150	30" wide		3	5.333		231	320		551	735
1180	24" radius, 6" wide		4.60	3.478		274	209		483	610
1200	12" wide		3.60	4.444		287	267		554	715
1210	18" wide		3.20	5		300	300		600	780
1220	24" wide		3	5.333		315	320		635	830
1230	30" wide		2.60	6.154		345	370		715	930
1260	36" radius, 6" wide		4.20	3.810		430	229		659	810
1280	12" wide		3.20	5		435	300		735	930
1290	18" wide		2.80	5.714		470	345		815	1,025
1300	24" wide		2.60	6.154		485	370		855	1,075
1310	30" wide		2.20	7.273		565	435		1,000	1,275
1340	Cross horizontal, 12" radius, 6" wide		4	4		135	240		375	510
1360	12" wide		3.40	4.706		147	283		430	580
1370	18" wide		2.80	5.714		172	345		517	700
1380	24" wide		2.40	6.667		194	400		594	810
1390	30" wide		2	8		215	480		695	950
1420	24" radius, 6" wide		3.60	4.444		246	267		513	670
1440	12" wide		3	5.333		276	320		596	785
1450	18" wide		2.40	6.667		310	400		710	935
1460	24" wide		2	8		395	480		875	1,150
1470	30" wide		1.80	8.889		430	535		965	1,275
1500	36" radius, 6" wide		3.20	5		405	300		705	895
1520	12" wide		2.60	6.154		445	370		815	1,025
1530	18" wide		2	8		485	480		965	1,250
1540	24" wide		1.80	8.889		570	535		1,105	1,425
1550	30" wide		1.60	10		620	600		1,220	1,575
1580	Drop out or end plate, 6" wide		32	.500		16.45	30		46.45	63
1600	12" wide		26	.615		20.50	37		57.50	77.50
1610	18" wide		22	.727		22	43.50		65.50	89
1620	24" wide		20	.800		26	48		74	100
1630	30" wide		18	.889		29	53.50		82.50	112
1660	Reducer, 12" to 6" wide		12	1.333		79	80		159	206
1680	18" to 12" wide		10.60	1.509		80.50	90.50		171	224
1700	18" to 6" wide		10.60	1.509		80.50	90.50		171	224
1720	24" to 18" wide		9.20	1.739		83.50	104		187.50	248
1740	24" to 12" wide		9.20	1.739		83.50	104		187.50	248
1760	30" to 24" wide		8	2		90	120		210	278
1780	30" to 18" wide		8	2		90	120		210	278
1800	30" to 12" wide		8	2		90	120		210	278
1820	36" to 30" wide		7.20	2.222		93	133		226	300
1840	36" to 24" wide		7.20	2.222		93	133		226	300
1860	36" to 18" wide		7.20	2.222		93	133		226	300
1880	36" to 12" wide		7.20	2.222		93	133		226	300
2000	Aluminum tray, 6" wide		150	.107	L.F.	10.60	6.40		17	21
2020	12" wide		130	.123		13.90	7.40		21.30	26.50
2030	18" wide		100	.160		17.45	9.60		27.05	33.50
2040	24" wide		90	.178		22	10.70		32.70	40
2050	30" wide		70	.229		26	13.75		39.75	49
2080	Elbow horizontal, 90°, 12" radius, 6" wide		9.60	1.667	Ea.	85.50	100		185.50	243

26 05 36 – Cable Trays for Electrical Systems

26 05 36.20 Cable Tray Solid Bottom	Crew	Daily Output	Labor-Hours	Unit	Material	2019 Bare Costs Labor	Equipment	Total	Total Incl O&P
2100	12" wide	2 Elec	7.60	2.105	Ea.	96	126		222	295
2110	18" wide		6.80	2.353		118	141		259	340
2120	24" wide		5.80	2.759		134	166		300	395
2130	30" wide		5	3.200		164	192		356	465
2160	24" radius, 6" wide		9.20	1.739		122	104		226	290
2180	12" wide		7.20	2.222		143	133		276	355
2190	18" wide		6.40	2.500		164	150		314	405
2200	24" wide		5.40	2.963		202	178		380	485
2210	30" wide		4.60	3.478		232	209		441	565
2240	36" radius, 6" wide		8.80	1.818		179	109		288	360
2260	12" wide		6.80	2.353		215	141		356	450
2270	18" wide		6	2.667		256	160		416	520
2280	24" wide		5	3.200		274	192		466	585
2290	30" wide		4.20	3.810		310	229		539	680
2320	Elbow vertical, 90°, 12" radius, 6" wide		9.60	1.667		101	100		201	260
2340	12" wide		7.60	2.105		105	126		231	305
2350	18" wide		6.80	2.353		119	141		260	340
2360	24" wide		5.80	2.759		126	166		292	385
2370	30" wide		5	3.200		133	192		325	435
2400	24" radius, 6" wide		9.20	1.739		143	104		247	315
2420	12" wide		7.20	2.222		156	133		289	370
2430	18" wide		6.40	2.500		168	150		318	410
2440	24" wide		5.40	2.963		182	178		360	465
2450	30" wide		4.60	3.478		193	209		402	525
2480	36" radius, 6" wide		8.80	1.818		189	109		298	370
2500	12" wide		6.80	2.353		213	141		354	445
2510	18" wide		6	2.667		232	160		392	495
2520	24" wide		5	3.200		241	192		433	550
2530	30" wide		4.20	3.810		266	229		495	635
2560	Tee horizontal, 12" radius, 6" wide		5	3.200		132	192		324	430
2580	12" wide		4.40	3.636		160	218		378	500
2590	18" wide		4	4		179	240		419	555
2600	24" wide		3.60	4.444		201	267		468	620
2610	30" wide		3	5.333		232	320		552	735
2640	24" radius, 6" wide		4.60	3.478		218	209		427	550
2660	12" wide		4	4		245	240		485	630
2670	18" wide		3.60	4.444		277	267		544	705
2680	24" wide		3	5.333		340	320		660	855
2690	30" wide		2.40	6.667		385	400		785	1,025
2720	36" radius, 6" wide		4.20	3.810		355	229		584	730
2740	12" wide		3.60	4.444		385	267		652	825
2750	18" wide		3.20	5		445	300		745	940
2760	24" wide		2.60	6.154		520	370		890	1,125
2770	30" wide		2	8		570	480		1,050	1,350
2800	Tee vertical, 12" radius, 6" wide		5	3.200		188	192		380	495
2820	12" wide		4.40	3.636		192	218		410	535
2830	18" wide		4.20	3.810		200	229		429	560
2840	24" wide		4	4		213	240		453	595
2850	30" wide		3.60	4.444		232	267		499	655
2880	24" radius, 6" wide		4.60	3.478		277	209		486	615
2900	12" wide		4	4		293	240		533	685
2910	18" wide		3.80	4.211		310	253		563	720
2920	24" wide		3.60	4.444		340	267		607	775

26 05 36 – Cable Trays for Electrical Systems

26 05 36.20 Cable Tray Solid Bottom		Crew	Daily Output	Labor-Hours	Unit	Material	2019 Bare Costs Labor	Equipment	Total	Total Incl O&P
2930	30" wide	2 Elec	3.20	5	Ea.	375	300		675	860
2960	36" radius, 6" wide		4.20	3.810		435	229		664	815
2980	12" wide		3.40	4.706		455	283		738	920
2990	18" wide		3.40	4.706		480	283		763	945
3000	24" wide		3.20	5		495	300		795	990
3010	30" wide		2.80	5.714		535	345		880	1,100
3040	Cross horizontal, 12" radius, 6" wide		4.40	3.636		168	218		386	510
3060	12" wide		4	4		189	240		429	570
3070	18" wide		3.40	4.706		215	283		498	655
3080	24" wide		2.80	5.714		245	345		590	780
3090	30" wide		2.60	6.154		277	370		647	855
3120	24" radius, 6" wide		4	4		293	240		533	685
3140	12" wide		3.60	4.444		340	267		607	775
3150	18" wide		3	5.333		370	320		690	885
3160	24" wide		2.40	6.667		445	400		845	1,075
3170	30" wide		2.20	7.273		495	435		930	1,200
3200	36" radius, 6" wide		3.60	4.444		510	267		777	965
3220	12" wide		3.20	5		540	300		840	1,050
3230	18" wide		2.60	6.154		600	370		970	1,200
3240	24" wide		2	8		655	480		1,135	1,425
3250	30" wide		1.80	8.889		705	535		1,240	1,575
3280	Dropout, or end plate, 6" wide		32	.500		16.85	30		46.85	63.50
3300	12" wide		26	.615		19.10	37		56.10	76
3310	18" wide		22	.727		23	43.50		66.50	90
3320	24" wide		20	.800		25	48		73	99
3330	30" wide		18	.889		29.50	53.50		83	112
3380	Reducer, 12" to 6" wide		14	1.143		82.50	68.50		151	193
3400	18" to 12" wide		12	1.333		85.50	80		165.50	213
3420	18" to 6" wide		12	1.333		85.50	80		165.50	213
3440	24" to 18" wide		10.60	1.509		90.50	90.50		181	235
3460	24" to 12" wide		10.60	1.509		90.50	90.50		181	235
3480	30" to 24" wide		9.20	1.739		98.50	104		202.50	264
3500	30" to 18" wide		9.20	1.739		98.50	104		202.50	264
3520	30" to 12" wide		9.20	1.739		98.50	104		202.50	264
3540	36" to 30" wide		8	2		104	120		224	293
3560	36" to 24" wide		8	2		105	120		225	295
3580	36" to 18" wide		8	2		105	120		225	295
3600	36" to 12" wide		8	2		108	120		228	297

26 05 36.30 Cable Tray Trough

0010	**CABLE TRAY TROUGH** vented, w/ftngs. & supports, 6" deep, to 10' high									
0020	For higher elevations, see Section 26 05 36.40									
0200	Galvanized steel, tray, 6" wide	R260536-10	2 Elec	90	.178	L.F.	13.75	10.70		24.45	31
0240	12" wide			80	.200		15.55	12		27.55	35
0260	18" wide			70	.229		22	13.75		35.75	45
0280	24" wide	R260536-11		60	.267		48.50	16		64.50	77
0300	30" wide			50	.320		48	19.20		67.20	81
0340	Elbow horizontal, 90°, 12" radius, 6" wide			7.60	2.105	Ea.	110	126		236	310
0360	12" wide			5.60	2.857		131	172		303	400
0370	18" wide			4.40	3.636		151	218		369	490
0380	24" wide			3.60	4.444		186	267		453	605
0390	30" wide			3.20	5		210	300		510	680
0420	24" radius, 6" wide			7.20	2.222		163	133		296	380

26 05 36 – Cable Trays for Electrical Systems

26 05 36.30 Cable Tray Trough		Crew	Daily Output	Labor-Hours	Unit	Material	2019 Bare Costs Labor	Equipment	Total	Total Incl O&P
0440	12" wide	2 Elec	5.20	3.077	Ea.	188	185		373	485
0450	18" wide		4	4		226	240		466	610
0460	24" wide		3.20	5		256	300		556	730
0470	30" wide		2.80	5.714		295	345		640	835
0500	36" radius, 6" wide		6.80	2.353		216	141		357	450
0520	12" wide		4.80	3.333		245	200		445	570
0530	18" wide		3.60	4.444		294	267		561	725
0540	24" wide		2.80	5.714		305	345		650	845
0550	30" wide		2.40	6.667		340	400		740	970
0580	Elbow vertical, 90°, 12" radius, 6" wide		7.60	2.105		122	126		248	325
0600	12" wide		5.60	2.857		136	172		308	405
0610	18" wide		4.40	3.636		138	218		356	475
0620	24" wide		3.60	4.444		158	267		425	575
0630	30" wide		3.20	5		162	300		462	630
0660	24" radius, 6" wide		7.20	2.222		175	133		308	390
0680	12" wide		5.20	3.077		185	185		370	480
0690	18" wide		4	4		206	240		446	585
0700	24" wide		3.20	5		213	300		513	685
0710	30" wide		2.80	5.714		238	345		583	770
0740	36" radius, 6" wide		6.80	2.353		236	141		377	470
0760	12" wide		4.80	3.333		252	200		452	575
0770	18" wide		3.60	4.444		279	267		546	705
0780	24" wide		2.80	5.714		299	345		644	840
0790	30" wide		2.40	6.667		305	400		705	930
0820	Tee horizontal, 12" radius, 6" wide		4	4		143	240		383	515
0840	12" wide		3.20	5		160	300		460	625
0850	18" wide		2.80	5.714		180	345		525	710
0860	24" wide		2.40	6.667		209	400		609	825
0870	30" wide		2.20	7.273		238	435		673	910
0900	24" radius, 6" wide		3.60	4.444		236	267		503	660
0920	12" wide		2.80	5.714		259	345		604	795
0930	18" wide		2.40	6.667		289	400		689	915
0940	24" wide		2	8		375	480		855	1,125
0950	30" wide		1.80	8.889		405	535		940	1,250
0980	36" radius, 6" wide		3.20	5		340	300		640	825
1000	12" wide		2.40	6.667		390	400		790	1,025
1010	18" wide		2	8		425	480		905	1,175
1020	24" wide		1.60	10		510	600		1,110	1,450
1030	30" wide		1.40	11.429		540	685		1,225	1,625
1060	Tee vertical, 12" radius, 6" wide		4	4		223	240		463	605
1080	12" wide		3.20	5		227	300		527	700
1090	18" wide		3	5.333		233	320		553	735
1100	24" wide		2.80	5.714		246	345		591	780
1110	30" wide		2.60	6.154		261	370		631	835
1140	24" radius, 6" wide		3.60	4.444		300	267		567	730
1160	12" wide		2.80	5.714		315	345		660	860
1170	18" wide		2.60	6.154		330	370		700	915
1180	24" wide		2.40	6.667		355	400		755	985
1190	30" wide		2.20	7.273		375	435		810	1,050
1220	36" radius, 6" wide		3.20	5		490	300		790	990
1240	12" wide		2.40	6.667		500	400		900	1,150
1250	18" wide		2.20	7.273		520	435		955	1,225
1260	24" wide		2	8		550	480		1,030	1,325

26 05 36.30 Cable Tray Trough		Crew	Daily Output	Labor-Hours	Unit	Material	2019 Bare Costs Labor	Equipment	Total	Total Incl O&P
1270	30" wide	2 Elec	1.80	8.889	Ea.	640	535		1,175	1,500
1300	Cross horizontal, 12" radius, 6" wide		3.20	5		176	300		476	645
1320	12" wide		2.80	5.714		178	345		523	705
1330	18" wide		2.40	6.667		199	400		599	815
1340	24" wide		2	8		209	480		689	945
1350	30" wide		1.80	8.889		235	535		770	1,050
1380	24" radius, 6" wide		2.80	5.714		278	345		623	815
1400	12" wide		2.40	6.667		291	400		691	915
1410	18" wide		2	8		320	480		800	1,075
1420	24" wide		1.60	10		405	600		1,005	1,350
1430	30" wide		1.40	11.429		450	685		1,135	1,525
1460	36" radius, 6" wide		2.40	6.667		475	400		875	1,125
1480	12" wide		2	8		490	480		970	1,250
1490	18" wide		1.60	10		500	600		1,100	1,450
1500	24" wide		1.20	13.333		605	800		1,405	1,875
1510	30" wide		1	16		665	960		1,625	2,150
1540	Dropout or end plate, 6" wide		26	.615		21.50	37		58.50	78.50
1560	12" wide		22	.727		26	43.50		69.50	93.50
1580	18" wide		20	.800		28.50	48		76.50	103
1600	24" wide		18	.889		33	53.50		86.50	116
1620	30" wide		16	1		35	60		95	128
1660	Reducer, 12" to 6" wide		9.40	1.702		85.50	102		187.50	248
1680	18" to 12" wide		8.40	1.905		89	114		203	269
1700	18" to 6" wide		8.40	1.905		89	114		203	269
1720	24" to 18" wide		7.20	2.222		95	133		228	305
1740	24" to 12" wide		7.20	2.222		95	133		228	305
1760	30" to 24" wide		6.40	2.500		98.50	150		248.50	330
1780	30" to 18" wide		6.40	2.500		98.50	150		248.50	330
1800	30" to 12" wide		6.40	2.500		98.50	150		248.50	330
1820	36" to 30" wide		5.80	2.759		105	166		271	365
1840	36" to 24" wide		5.80	2.759		105	166		271	365
1860	36" to 18" wide		5.80	2.759		107	166		273	365
1880	36" to 12" wide		5.80	2.759		107	166		273	365
2000	Aluminum, tray, vented, 6" wide		120	.133	L.F.	14.45	8		22.45	28
2010	9" wide		110	.145		16.35	8.75		25.10	31
2020	12" wide		100	.160		21.50	9.60		31.10	38
2030	18" wide		90	.178		22	10.70		32.70	40
2040	24" wide		80	.200		26	12		38	46.50
2050	30" wide		70	.229		34.50	13.75		48.25	58.50
2080	Elbow horiz., 90°, 12" radius, 6" wide		7.60	2.105	Ea.	90.50	126		216.50	289
2090	9" wide		7	2.286		98	137		235	315
2100	12" wide		6.20	2.581		104	155		259	345
2110	18" wide		5.60	2.857		122	172		294	390
2120	24" wide		4.60	3.478		141	209		350	465
2130	30" wide		4	4		174	240		414	550
2160	24" radius, 6" wide		7.20	2.222		131	133		264	345
2180	12" wide		5.80	2.759		154	166		320	415
2190	18" wide		5.20	3.077		176	185		361	470
2200	24" wide		4.20	3.810		201	229		430	560
2210	30" wide		3.60	4.444		230	267		497	655
2240	36" radius, 6" wide		6.80	2.353		189	141		330	420
2260	12" wide		5.40	2.963		214	178		392	500
2270	18" wide		4.80	3.333		244	200		444	565

26 05 36 – Cable Trays for Electrical Systems

26 05 36.30 Cable Tray Trough	Crew	Daily Output	Labor-Hours	Unit	Material	2019 Bare Costs Labor	Equipment	Total	Total Incl O&P
2280	24" wide	2 Elec	3.80	4.211	Ea.	267	253		520	670
2290	30" wide		3.40	4.706		315	283		598	765
2320	Elbow vertical, 90°, 12" radius, 6" wide		7.60	2.105		110	126		236	310
2330	9" wide		7	2.286		117	137		254	335
2340	12" wide		6.20	2.581		118	155		273	360
2350	18" wide		5.60	2.857		122	172		294	390
2360	24" wide		4.60	3.478		135	209		344	460
2370	30" wide		4	4		141	240		381	515
2400	24" radius, 6" wide		7.20	2.222		148	133		281	360
2420	12" wide		5.80	2.759		161	166		327	425
2430	18" wide		5.20	3.077		175	185		360	470
2440	24" wide		4.20	3.810		177	229		406	535
2450	30" wide		3.60	4.444		187	267		454	605
2480	36" radius, 6" wide		6.80	2.353		187	141		328	415
2500	12" wide		5.40	2.963		200	178		378	485
2510	18" wide		4.80	3.333		225	200		425	545
2520	24" wide		3.80	4.211		242	253		495	640
2530	30" wide		3.40	4.706		267	283		550	715
2560	Tee horizontal, 12" radius, 6" wide		4	4		137	240		377	510
2570	9" wide		3.80	4.211		141	253		394	530
2580	12" wide		3.60	4.444		154	267		421	570
2590	18" wide		3.20	5		177	300		477	645
2600	24" wide		2.80	5.714		204	345		549	735
2610	30" wide		2.40	6.667		219	400		619	835
2640	24" radius, 6" wide		3.60	4.444		209	267		476	630
2660	12" wide		3.20	5		238	300		538	710
2670	18" wide		2.80	5.714		262	345		607	800
2680	24" wide		2.40	6.667		325	400		725	955
2690	30" wide		2	8		350	480		830	1,100
2720	36" radius, 6" wide		3.20	5		335	300		635	820
2740	12" wide		2.80	5.714		385	345		730	935
2750	18" wide		2.40	6.667		440	400		840	1,075
2760	24" wide		2	8		510	480		990	1,275
2770	30" wide		1.60	10		545	600		1,145	1,500
2800	Tee vertical, 12" radius, 6" wide		4	4		192	240		432	570
2810	9" wide		3.80	4.211		194	253		447	590
2820	12" wide		3.60	4.444		208	267		475	630
2830	18" wide		3.40	4.706		211	283		494	655
2840	24" wide		3.20	5		214	300		514	685
2850	30" wide		3	5.333		225	320		545	725
2880	24" radius, 6" wide		3.60	4.444		272	267		539	700
2900	12" wide		3.20	5		294	300		594	775
2910	18" wide		3	5.333		315	320		635	825
2920	24" wide		2.80	5.714		335	345		680	875
2930	30" wide		2.60	6.154		360	370		730	945
2960	36" radius, 6" wide		3.20	5		415	300		715	905
2980	12" wide		2.80	5.714		440	345		785	990
2990	18" wide		2.60	6.154		450	370		820	1,050
3000	24" wide		2.40	6.667		485	400		885	1,125
3010	30" wide		2.20	7.273		525	435		960	1,225
3040	Cross horizontal, 12" radius, 6" wide		3.60	4.444		171	267		438	590
3050	9" wide		3.40	4.706		184	283		467	620
3060	12" wide		3.20	5		189	300		489	660

26 05 36.30 Cable Tray Trough	Crew	Daily Output	Labor-Hours	Unit	Material	2019 Bare Costs Labor	Equipment	Total	Total Incl O&P
3070	18" wide	2 Elec	2.80	5.714	Ea.	198	345		543	725
3080	24" wide		2.40	6.667		220	400		620	835
3090	30" wide		2.20	7.273		282	435		717	960
3120	24" radius, 6" wide		3.20	5		315	300		615	795
3140	12" wide		2.80	5.714		335	345		680	880
3150	18" wide		2.40	6.667		360	400		760	995
3160	24" wide		2	8		420	480		900	1,175
3170	30" wide		1.80	8.889		450	535		985	1,300
3200	36" radius, 6" wide		2.80	5.714		545	345		890	1,100
3220	12" wide		2.40	6.667		570	400		970	1,225
3230	18" wide		2	8		620	480		1,100	1,400
3240	24" wide		1.60	10		720	600		1,320	1,675
3250	30" wide		1.40	11.429		780	685		1,465	1,875
3280	Dropout, or end plate, 6" wide		26	.615		17.45	37		54.45	74
3300	12" wide		22	.727		20.50	43.50		64	87.50
3310	18" wide		20	.800		25	48		73	99
3320	24" wide		18	.889		30	53.50		83.50	113
3330	30" wide		16	1		32	60		92	125
3370	Reducer, 9" to 6" wide		12	1.333		85.50	80		165.50	213
3380	12" to 6" wide		11.40	1.404		90	84.50		174.50	225
3390	12" to 9" wide		11.40	1.404		90	84.50		174.50	225
3400	18" to 12" wide		9.60	1.667		96	100		196	255
3420	18" to 6" wide		9.60	1.667		96	100		196	255
3430	18" to 9" wide		9.60	1.667		96	100		196	255
3440	24" to 18" wide		8.40	1.905		103	114		217	284
3460	24" to 12" wide		8.40	1.905		103	114		217	284
3470	24" to 9" wide		8.40	1.905		104	114		218	285
3475	24" to 6" wide		8.40	1.905		104	114		218	285
3480	30" to 24" wide		7.20	2.222		105	133		238	315
3500	30" to 18" wide		7.20	2.222		105	133		238	315
3520	30" to 12" wide		7.20	2.222		108	133		241	320
3540	36" to 30" wide		6.40	2.500		109	150		259	345
3560	36" to 24" wide		6.40	2.500		109	150		259	345
3580	36" to 18" wide		6.40	2.500		109	150		259	345
3600	36" to 12" wide		6.40	2.500		109	150		259	345
3610	Elbow horizontal, 60°, 12" radius, 6" wide		7.80	2.051		74.50	123		197.50	266
3620	9" wide		7.20	2.222		81.50	133		214.50	289
3630	12" wide		6.40	2.500		90	150		240	325
3640	18" wide		5.80	2.759		103	166		269	360
3650	24" wide		4.80	3.333		118	200		318	430
3680	Elbow horizontal, 45°, 12" radius, 6" wide		8	2		68	120		188	254
3690	9" wide		7.40	2.162		68	130		198	269
3700	12" wide		6.60	2.424		72	146		218	296
3710	18" wide		6	2.667		81	160		241	330
3720	24" wide		5	3.200		94	192		286	390
3750	Elbow horizontal, 30°, 12" radius, 6" wide		8.20	1.951		56	117		173	237
3760	9" wide		7.60	2.105		59	126		185	254
3770	12" wide		6.80	2.353		64.50	141		205.50	282
3780	18" wide		6.20	2.581		68	155		223	305
3790	24" wide		5.20	3.077		74.50	185		259.50	360
3820	Elbow vertical, 60° in/outside, 12" radius, 6" wide		7.80	2.051		92.50	123		215.50	286
3830	9" wide		7.20	2.222		93.50	133		226.50	300
3840	12" wide		6.40	2.500		95	150		245	330

26 05 36 – Cable Trays for Electrical Systems

26 05 36.30 Cable Tray Trough		Crew	Daily Output	Labor-Hours	Unit	Material	2019 Bare Costs Labor	2019 Bare Costs Equipment	Total	Total Incl O&P
3850	18" wide	2 Elec	5.80	2.759	Ea.	99	166		265	355
3860	24" wide		4.80	3.333		103	200		303	410
3890	Elbow vertical, 45° in/outside, 12" radius, 6" wide		8	2		79.50	120		199.50	267
3900	9" wide		7.40	2.162		76.50	130		206.50	278
3910	12" wide		6.60	2.424		81.50	146		227.50	305
3920	18" wide		6	2.667		84	160		244	330
3930	24" wide		5	3.200		92	192		284	390
3960	Elbow vertical, 30° in/outside, 12" radius, 6" wide		8.20	1.951		64	117		181	246
3970	9" wide		7.60	2.105		68	126		194	264
3980	12" wide		6.80	2.353		69.50	141		210.50	287
3990	18" wide		6.20	2.581		72	155		227	310
4000	24" wide		5.20	3.077		73	185		258	355
4250	Reducer, left or right hand, 24" to 18" wide		8.40	1.905		97.50	114		211.50	278
4260	24" to 12" wide		8.40	1.905		97.50	114		211.50	278
4270	24" to 9" wide		8.40	1.905		97.50	114		211.50	278
4280	24" to 6" wide		8.40	1.905		99	114		213	280
4290	18" to 12" wide		9.60	1.667		91	100		191	249
4300	18" to 9" wide		9.60	1.667		91	100		191	249
4310	18" to 6" wide		9.60	1.667		91	100		191	249
4320	12" to 9" wide		11.40	1.404		87.50	84.50		172	222
4330	12" to 6" wide		11.40	1.404		87.50	84.50		172	222
4340	9" to 6" wide		12	1.333		84.50	80		164.50	212
4350	Splice plate	1 Elec	48	.167		8.10	10		18.10	24
4360	Splice plate, expansion joint		48	.167		8.30	10		18.30	24
4370	Splice plate, hinged, horizontal		48	.167		6.65	10		16.65	22.50
4380	Vertical		48	.167		9.50	10		19.50	25.50
4390	Trough, hanger, vertical		28	.286		26.50	17.15		43.65	55
4400	Box connector, 24" wide		20	.400		35.50	24		59.50	75
4410	18" wide		21	.381		30	23		53	67
4420	12" wide		22	.364		29	22		51	64
4430	9" wide		23	.348		27.50	21		48.50	61
4440	6" wide		24	.333		26	20		46	58.50
4450	Floor flange		24	.333		27.50	20		47.50	60
4460	Hold down clamp		60	.133		3.13	8		11.13	15.40
4520	Wall bracket, 24" wide tray		20	.400		32	24		56	71
4530	18" wide tray		21	.381		31	23		54	68
4540	12" wide tray		22	.364		16.55	22		38.55	50.50
4550	9" wide tray		23	.348		14.75	21		35.75	47.50
4560	6" wide tray		24	.333		13.45	20		33.45	45
5000	Cable channel aluminum, vented, 1-1/4" deep, 4" wide, straight		80	.100	L.F.	7.45	6		13.45	17.15
5010	Elbow horizontal, 36" radius, 90°		5	1.600	Ea.	136	96		232	293
5020	60°		5.50	1.455		104	87.50		191.50	245
5030	45°		6	1.333		85	80		165	213
5040	30°		6.50	1.231		70.50	74		144.50	188
5050	Adjustable		6	1.333		67.50	80		147.50	193
5060	Elbow vertical, 36" radius, 90°		5	1.600		146	96		242	305
5070	60°		5.50	1.455		113	87.50		200.50	254
5080	45°		6	1.333		92.50	80		172.50	221
5090	30°		6.50	1.231		82.50	74		156.50	201
5100	Adjustable		6	1.333		67.50	80		147.50	193
5110	Splice plate, hinged, horizontal		48	.167		9.70	10		19.70	25.50
5120	Splice plate, hinged, vertical		48	.167		13.80	10		23.80	30
5130	Hanger, vertical		28	.286		15.20	17.15		32.35	42.50

26 05 36 – Cable Trays for Electrical Systems

26 05 36.30 Cable Tray Trough

		Crew	Daily Output	Labor-Hours	Unit	Material	2019 Bare Costs Labor	Equipment	Total	Total Incl O&P
5140	Single	1 Elec	28	.286	Ea.	23	17.15		40.15	51
5150	Double		20	.400		23.50	24		47.50	62
5160	Channel to box connector		24	.333		31.50	20		51.50	64.50
5170	Hold down clip		80	.100		4.39	6		10.39	13.80
5180	Wall bracket, single		28	.286		10.75	17.15		27.90	37.50
5190	Double		20	.400		13.80	24		37.80	51
5200	Cable roller		16	.500		169	30		199	231
5210	Splice plate		48	.167		5.20	10		15.20	20.50

26 05 36.36 Cable Trays for Utility Substations

		Crew	Daily Output	Labor-Hours	Unit	Material	2019 Bare Costs Labor	Equipment	Total	Total Incl O&P
0010	**CABLE TRAYS FOR UTILITY SUBSTATIONS**									
7700	Cable tray	R-11	40	1.400	L.F.	17.20	80	15.65	112.85	155
7790	See Section 26 05 36									

26 05 36.40 Cable Tray, Covers and Dividers

		Crew	Daily Output	Labor-Hours	Unit	Material	2019 Bare Costs Labor	Equipment	Total	Total Incl O&P
0010	**CABLE TRAY, COVERS AND DIVIDERS** To 10' high R260536-10									
0011	For higher elevations, see lines 9900 – 9960									
0100	Covers, ventilated galv. steel, straight, 6" wide tray size	2 Elec	520	.031	L.F.	8.20	1.85		10.05	11.75
0200	9" wide tray size		460	.035		8.75	2.09		10.84	12.75
0300	12" wide tray size R260536-11		400	.040		15	2.40		17.40	20
0400	18" wide tray size		300	.053		37.50	3.20		40.70	46.50
0500	24" wide tray size		220	.073		59	4.37		63.37	71.50
0600	30" wide tray size		180	.089		25.50	5.35		30.85	36
1000	Elbow horizontal, 90°, 12" radius, 6" wide tray size		150	.107	Ea.	39	6.40		45.40	52.50
1020	9" wide tray size		128	.125		43.50	7.50		51	58.50
1040	12" wide tray size		108	.148		46	8.90		54.90	64
1060	18" wide tray size		84	.190		63	11.45		74.45	86.50
1080	24" wide tray size		66	.242		76.50	14.55		91.05	106
1100	30" wide tray size		60	.267		97.50	16		113.50	131
1160	24" radius, 6" wide tray size		136	.118		65	7.05		72.05	82
1180	9" wide tray size		116	.138		66.50	8.30		74.80	85.50
1200	12" wide tray size		96	.167		74	10		84	96.50
1220	18" wide tray size		76	.211		91.50	12.65		104.15	120
1240	24" wide tray size		60	.267		111	16		127	147
1260	30" wide tray size		52	.308		146	18.50		164.50	189
1320	36" radius, 6" wide tray size		120	.133		97	8		105	119
1340	9" wide tray size		104	.154		107	9.25		116.25	132
1360	12" wide tray size		84	.190		114	11.45		125.45	143
1380	18" wide tray size		72	.222		145	13.35		158.35	180
1400	24" wide tray size		52	.308		173	18.50		191.50	218
1420	30" wide tray size		46	.348		206	21		227	257
1480	Elbow horizontal, 45°, 12" radius, 6" wide tray size		150	.107		29.50	6.40		35.90	42
1500	9" wide tray size		128	.125		35.50	7.50		43	50
1520	12" wide tray size		108	.148		38.50	8.90		47.40	56
1540	18" wide tray size		88	.182		47	10.90		57.90	68
1560	24" wide tray size		76	.211		55	12.65		67.65	79.50
1580	30" wide tray size		66	.242		64.50	14.55		79.05	92.50
1640	24" radius, 6" wide tray size		136	.118		41	7.05		48.05	55.50
1660	9" wide tray size		116	.138		46	8.30		54.30	63
1680	12" wide tray size		96	.167		51.50	10		61.50	71.50
1700	18" wide tray size		80	.200		60	12		72	84
1720	24" wide tray size		70	.229		68	13.75		81.75	95.50
1740	30" wide tray size		60	.267		86.50	16		102.50	119
1800	36" radius, 6" wide tray size		120	.133		60.50	8		68.50	78.50

26 05 36 – Cable Trays for Electrical Systems

26 05 36.40 Cable Tray, Covers and Dividers	Crew	Daily Output	Labor-Hours	Unit	Material	2019 Bare Costs Labor	Equipment	Total	Total Incl O&P
1820	9" wide tray size	2 Elec	104	.154	Ea.	68	9.25		77.25	88.50
1840	12" wide tray size		84	.190		75	11.45		86.45	99.50
1860	18" wide tray size		76	.211		85.50	12.65		98.15	113
1880	24" wide tray size		62	.258		105	15.50		120.50	138
1900	30" wide tray size		52	.308		114	18.50		132.50	153
1960	Elbow vertical, 90°, 12" radius, 6" wide tray size		150	.107		34.50	6.40		40.90	47.50
1980	9" wide tray size		128	.125		35.50	7.50		43	50
2000	12" wide tray size		108	.148		38	8.90		46.90	55.50
2020	18" wide tray size		88	.182		43.50	10.90		54.40	64
2040	24" wide tray size		68	.235		44	14.15		58.15	69.50
2060	30" wide tray size		60	.267		48.50	16		64.50	77.50
2120	24" radius, 6" wide tray size		136	.118		42.50	7.05		49.55	57.50
2140	9" wide tray size		116	.138		47	8.30		55.30	64
2160	12" wide tray size		96	.167		48.50	10		58.50	68.50
2180	18" wide tray size		80	.200		64.50	12		76.50	89
2200	24" wide tray size		62	.258		72.50	15.50		88	103
2220	30" wide tray size		52	.308		79	18.50		97.50	115
2280	36" radius, 6" wide tray size		120	.133		48	8		56	64.50
2300	9" wide tray size		104	.154		61	9.25		70.25	81
2320	12" wide tray size		84	.190		68	11.45		79.45	91.50
2340	18" wide tray size		76	.211		85.50	12.65		98.15	113
2350	24" wide tray size		54	.296		95	17.80		112.80	132
2360	30" wide tray size		46	.348		114	21		135	156
2400	Tee horizontal, 12" radius, 6" wide tray size		92	.174		63	10.45		73.45	84.50
2410	9" wide tray size		80	.200		64.50	12		76.50	89
2420	12" wide tray size		68	.235		73	14.15		87.15	102
2430	18" wide tray size		60	.267		88.50	16		104.50	122
2440	24" wide tray size		52	.308		111	18.50		129.50	151
2460	30" wide tray size		36	.444		131	26.50		157.50	184
2500	24" radius, 6" wide tray size		88	.182		101	10.90		111.90	127
2510	9" wide tray size		76	.211		107	12.65		119.65	136
2520	12" wide tray size		64	.250		110	15		125	144
2530	18" wide tray size		56	.286		141	17.15		158.15	181
2540	24" wide tray size		48	.333		216	20		236	267
2560	30" wide tray size		32	.500		246	30		276	315
2600	36" radius, 6" wide tray size		84	.190		171	11.45		182.45	206
2610	9" wide tray size		72	.222		175	13.35		188.35	212
2620	12" wide tray size		60	.267		194	16		210	238
2630	18" wide tray size		52	.308		227	18.50		245.50	278
2640	24" wide tray size		44	.364		299	22		321	365
2660	30" wide tray size		28	.571		325	34.50		359.50	405
2700	Cross horizontal, 12" radius, 6" wide tray size		68	.235		93.50	14.15		107.65	124
2710	9" wide tray size		64	.250		99	15		114	132
2720	12" wide tray size		60	.267		111	16		127	146
2730	18" wide tray size		52	.308		131	18.50		149.50	172
2740	24" wide tray size		36	.444		154	26.50		180.50	209
2760	30" wide tray size		30	.533		175	32		207	240
2800	24" radius, 6" wide tray size		64	.250		168	15		183	208
2810	9" wide tray size		60	.267		182	16		198	225
2820	12" wide tray size		56	.286		197	17.15		214.15	242
2830	18" wide tray size		48	.333		235	20		255	289
2840	24" wide tray size		32	.500		286	30		316	360
2860	30" wide tray size		26	.615		315	37		352	405

26 05 36 – Cable Trays for Electrical Systems

26 05 36.40 Cable Tray, Covers and Dividers		Crew	Daily Output	Labor-Hours	Unit	Material	2019 Bare Costs Labor	Equipment	Total	Total Incl O&P
2900	36" radius, 6" wide tray size	2 Elec	60	.267	Ea.	287	16		303	340
2910	9" wide tray size		56	.286		286	17.15		303.15	340
2920	12" wide tray size		52	.308		310	18.50		328.50	375
2930	18" wide tray size		44	.364		350	22		372	420
2940	24" wide tray size		28	.571		440	34.50		474.50	535
2960	30" wide tray size		22	.727		475	43.50		518.50	585
3000	Reducer, 9" to 6" wide tray size		128	.125		40	7.50		47.50	55
3010	12" to 6" wide tray size		108	.148		42	8.90		50.90	59.50
3020	12" to 9" wide tray size		108	.148		42	8.90		50.90	59.50
3030	18" to 12" wide tray size		88	.182		44.50	10.90		55.40	65.50
3050	18" to 6" wide tray size		88	.182		44.50	10.90		55.40	65.50
3060	24" to 18" wide tray size		80	.200		62.50	12		74.50	87
3070	24" to 12" wide tray size		80	.200		57	12		69	81
3090	30" to 24" wide tray size		70	.229		66	13.75		79.75	93
3100	30" to 18" wide tray size		70	.229		66	13.75		79.75	93
3110	30" to 12" wide tray size		70	.229		57	13.75		70.75	83.50
3140	36" to 30" wide tray size		64	.250		72	15		87	102
3150	36" to 24" wide tray size		64	.250		72	15		87	102
3160	36" to 18" wide tray size		64	.250		72	15		87	102
3170	36" to 12" wide tray size		64	.250	▼	72	15		87	102
3250	Covers, aluminum, straight, 6" wide tray size		520	.031	L.F.	4.30	1.85		6.15	7.50
3270	9" wide tray size		460	.035		5.20	2.09		7.29	8.80
3290	12" wide tray size		400	.040		6.20	2.40		8.60	10.40
3310	18" wide tray size		320	.050		8.20	3		11.20	13.50
3330	24" wide tray size		260	.062		10.25	3.70		13.95	16.75
3350	30" wide tray size		200	.080	▼	11.30	4.80		16.10	19.55
3400	Elbow horizontal, 90°, 12" radius, 6" wide tray size		150	.107	Ea.	32	6.40		38.40	45
3410	9" wide tray size		128	.125		33.50	7.50		41	48
3420	12" wide tray size		108	.148		36.50	8.90		45.40	53.50
3430	18" wide tray size		88	.182		47	10.90		57.90	68.50
3440	24" wide tray size		70	.229		59.50	13.75		73.25	86
3460	30" wide tray size		64	.250		71.50	15		86.50	101
3500	24" radius, 6" wide tray size		136	.118		41.50	7.05		48.55	56
3510	9" wide tray size		116	.138		51	8.30		59.30	68.50
3520	12" wide tray size		96	.167		56.50	10		66.50	77
3530	18" wide tray size		80	.200		66.50	12		78.50	91.50
3540	24" wide tray size		64	.250		82.50	15		97.50	113
3560	30" wide tray size		56	.286		101	17.15		118.15	137
3600	36" radius, 6" wide tray size		120	.133		72	8		80	91.50
3610	9" wide tray size		104	.154		79	9.25		88.25	100
3620	12" wide tray size		84	.190		89.50	11.45		100.95	116
3630	18" wide tray size		76	.211		106	12.65		118.65	136
3640	24" wide tray size		56	.286		130	17.15		147.15	169
3660	30" wide tray size		50	.320		149	19.20		168.20	193
3700	Elbow horizontal, 45°, 12" radius, 6" wide tray size		150	.107		23.50	6.40		29.90	35.50
3710	9" wide tray size		128	.125		24	7.50		31.50	37.50
3720	12" wide tray size		108	.148		27	8.90		35.90	43.50
3730	18" wide tray size		88	.182		31	10.90		41.90	51
3740	24" wide tray size		80	.200		34.50	12		46.50	56
3760	30" wide tray size		70	.229		43.50	13.75		57.25	68.50
3800	24" radius, 6" wide tray size		136	.118		26.50	7.05		33.55	39.50
3810	9" wide tray size		116	.138		32.50	8.30		40.80	48.50
3820	12" wide tray size		96	.167	▼	34	10		44	52.50

26 05 36 – Cable Trays for Electrical Systems

26 05 36.40 Cable Tray, Covers and Dividers		Crew	Daily Output	Labor-Hours	Unit	Material	2019 Bare Costs Labor	Equipment	Total	Total Incl O&P
3830	18" wide tray size	2 Elec	80	.200	Ea.	41.50	12		53.50	63.50
3840	24" wide tray size		72	.222		51	13.35		64.35	76
3860	30" wide tray size		64	.250		59	15		74	87.50
3900	36" radius, 6" wide tray size		120	.133		45.50	8		53.50	62
3910	9" wide tray size		104	.154		49.50	9.25		58.75	68.50
3920	12" wide tray size		84	.190		52	11.45		63.45	74.50
3930	18" wide tray size		76	.211		62.50	12.65		75.15	88
3940	24" wide tray size		64	.250		76	15		91	106
3960	30" wide tray size		56	.286		86.50	17.15		103.65	121
3970	36" wide tray size		50	.320		99	19.20		118.20	138
4000	Elbow vertical, 90°, 12" radius, 6" wide tray size		150	.107		27	6.40		33.40	39.50
4010	9" wide tray size		128	.125		27.50	7.50		35	41
4020	12" wide tray size		108	.148		30	8.90		38.90	46.50
4030	18" wide tray size		88	.182		34.50	10.90		45.40	54.50
4040	24" wide tray size		70	.229		36	13.75		49.75	60
4060	30" wide tray size		64	.250		37	15		52	63
4070	36" wide tray size		54	.296		45.50	17.80		63.30	76.50
4100	24" radius, 6" wide tray size		136	.118		31	7.05		38.05	45
4110	9" wide tray size		116	.138		34	8.30		42.30	50
4120	12" wide tray size		96	.167		37	10		47	55.50
4130	18" wide tray size		80	.200		44.50	12		56.50	67
4140	24" wide tray size		64	.250		48.50	15		63.50	76
4160	30" wide tray size		56	.286		59	17.15		76.15	90.50
4170	36" wide tray size		48	.333		64	20		84	101
4200	36" radius, 6" wide tray size		120	.133		34.50	8		42.50	50
4210	9" wide tray size		104	.154		42.50	9.25		51.75	61
4220	12" wide tray size		84	.190		49.50	11.45		60.95	71.50
4230	18" wide tray size		76	.211		60	12.65		72.65	85
4240	24" wide tray size		56	.286		70.50	17.15		87.65	104
4260	30" wide tray size		50	.320		89.50	19.20		108.70	127
4270	36" wide tray size		44	.364		96	22		118	139
4300	Tee horizontal, 12" radius, 6" wide tray size		108	.148		44.50	8.90		53.40	62.50
4310	9" wide tray size		88	.182		47	10.90		57.90	68.50
4320	12" wide tray size		80	.200		52.50	12		64.50	76
4330	18" wide tray size		68	.235		62.50	14.15		76.65	89.50
4340	24" wide tray size		56	.286		79	17.15		96.15	113
4360	30" wide tray size		44	.364		93	22		115	135
4370	36" wide tray size		36	.444		112	26.50		138.50	163
4400	24" radius, 6" wide tray size		96	.167		72	10		82	94.50
4410	9" wide tray size		80	.200		76	12		88	101
4420	12" wide tray size		72	.222		85	13.35		98.35	113
4430	18" wide tray size		60	.267		99.50	16		115.50	133
4440	24" wide tray size		48	.333		159	20		179	204
4460	30" wide tray size		40	.400		175	24		199	228
4470	36" wide tray size		32	.500		195	30		225	260
4500	36" radius, 6" wide tray size		88	.182		127	10.90		137.90	156
4510	9" wide tray size		72	.222		130	13.35		143.35	163
4520	12" wide tray size		64	.250		142	15		157	180
4530	18" wide tray size		56	.286		162	17.15		179.15	204
4540	24" wide tray size		44	.364		205	22		227	259
4560	30" wide tray size		36	.444		231	26.50		257.50	294
4570	36" wide tray size		28	.571		260	34.50		294.50	340
4600	Cross horizontal, 12" radius, 6" wide tray size		80	.200		68	12		80	92.50

26 05 Common Work Results for Electrical

26 05 36 – Cable Trays for Electrical Systems

26 05 36.40 Cable Tray, Covers and Dividers		Crew	Daily Output	Labor-Hours	Unit	Material	2019 Bare Costs Labor	Equipment	Total	Total Incl O&P
4610	9" wide tray size	2 Elec	72	.222	Ea.	72	13.35		85.35	99
4620	12" wide tray size		64	.250		80	15		95	111
4630	18" wide tray size		56	.286		97	17.15		114.15	133
4640	24" wide tray size		48	.333		112	20		132	153
4660	30" wide tray size		40	.400		133	24		157	183
4670	36" wide tray size		32	.500		153	30		183	213
4700	24" radius, 6" wide tray size		72	.222		126	13.35		139.35	158
4710	9" wide tray size		64	.250		135	15		150	171
4720	12" wide tray size		56	.286		145	17.15		162.15	185
4730	18" wide tray size		48	.333		169	20		189	216
4740	24" wide tray size		40	.400		201	24		225	257
4760	30" wide tray size		32	.500		235	30		265	305
4770	36" wide tray size		24	.667		255	40		295	340
4800	36" radius, 6" wide tray size		64	.250		205	15		220	249
4810	9" wide tray size		56	.286		214	17.15		231.15	262
4820	12" wide tray size		50	.320		242	19.20		261.20	295
4830	18" wide tray size		44	.364		266	22		288	325
4840	24" wide tray size		36	.444		330	26.50		356.50	400
4860	30" wide tray size		28	.571		380	34.50		414.50	465
4870	36" wide tray size		22	.727		410	43.50		453.50	520
4900	Reducer, 9" to 6" wide tray size		128	.125		34.50	7.50		42	48.50
4910	12" to 6" wide tray size		108	.148		35.50	8.90		44.40	52.50
4920	12" to 9" wide tray size		108	.148		35.50	8.90		44.40	52.50
4930	18" to 12" wide tray size		88	.182		39	10.90		49.90	59.50
4950	18" to 6" wide tray size		88	.182		39	10.90		49.90	59.50
4960	24" to 18" wide tray size		80	.200		50.50	12		62.50	73.50
4970	24" to 12" wide tray size		80	.200		43.50	12		55.50	65.50
4990	30" to 24" wide tray size		70	.229		52.50	13.75		66.25	78
5000	30" to 18" wide tray size		70	.229		52.50	13.75		66.25	78
5010	30" to 12" wide tray size		70	.229		52.50	13.75		66.25	78
5040	36" to 30" wide tray size		64	.250		58	15		73	86.50
5050	36" to 24" wide tray size		64	.250		58	15		73	86.50
5060	36" to 18" wide tray size		64	.250		58	15		73	86.50
5070	36" to 12" wide tray size		64	.250		58	15		73	86.50
5710	Tray cover hold down clamp	1 Elec	60	.133		9.30	8		17.30	22
8000	Divider strip, straight, galvanized, 3" deep		200	.040	L.F.	4.99	2.40		7.39	9.10
8020	4" deep		180	.044		6.10	2.67		8.77	10.70
8040	6" deep		160	.050		8	3		11	13.30
8060	Aluminum, straight, 3" deep		210	.038		4.99	2.29		7.28	8.90
8080	4" deep		190	.042		6.20	2.53		8.73	10.55
8100	6" deep		170	.047		7.90	2.83		10.73	12.90
8110	Divider strip, vertical fitting, 3" deep									
8120	12" radius, galvanized, 30°	1 Elec	28	.286	Ea.	24	17.15		41.15	52
8140	45°		27	.296		30	17.80		47.80	60
8160	60°		26	.308		32	18.50		50.50	63
8180	90°		25	.320		40	19.20		59.20	72.50
8200	Aluminum, 30°		29	.276		16	16.55		32.55	42
8220	45°		28	.286		18.90	17.15		36.05	46.50
8240	60°		27	.296		22	17.80		39.80	50.50
8260	90°		26	.308		27.50	18.50		46	58
8280	24" radius, galvanized, 30°		25	.320		36	19.20		55.20	68.50
8300	45°		24	.333		38.50	20		58.50	72
8320	60°		23	.348		48.50	21		69.50	84

26 05 36.40 Cable Tray, Covers and Dividers	Crew	Daily Output	Labor-Hours	Unit	Material	2019 Bare Costs Labor	2019 Bare Costs Equipment	Total	Total Incl O&P
8340	90°	1 Elec	22	.364	Ea.	65.50	22		87.50	105
8360	Aluminum, 30°		26	.308		25	18.50		43.50	55
8380	45°		25	.320		29	19.20		48.20	60
8400	60°		24	.333		36	20		56	69.50
8420	90°		23	.348		49	21		70	85
8440	36" radius, galvanized, 30°		22	.364		45.50	22		67.50	83
8460	45°		21	.381		53.50	23		76.50	92.50
8480	60°		20	.400		62.50	24		86.50	105
8500	90°		19	.421		86.50	25.50		112	133
8520	Aluminum, 30°		23	.348		38	21		59	73
8540	45°		22	.364		49	22		71	86.50
8560	60°		21	.381		63.50	23		86.50	104
8570	90°		20	.400		83	24		107	128
8590	Divider strip, vertical fitting, 4" deep									
8600	12" radius, galvanized, 30°	1 Elec	27	.296	Ea.	28.50	17.80		46.30	58
8610	45°		26	.308		33.50	18.50		52	64
8620	60°		25	.320		37.50	19.20		56.70	69.50
8630	90°		24	.333		46	20		66	80.50
8640	Aluminum, 30°		28	.286		21.50	17.15		38.65	49
8650	45°		27	.296		24.50	17.80		42.30	53.50
8660	60°		26	.308		28	18.50		46.50	58.50
8670	90°		25	.320		33.50	19.20		52.70	65
8680	24" radius, galvanized, 30°		24	.333		46	20		66	80.50
8690	45°		23	.348		57.50	21		78.50	94
8700	60°		22	.364		65	22		87	104
8710	90°		21	.381		86.50	23		109.50	129
8720	Aluminum, 30°		25	.320		33.50	19.20		52.70	65
8730	45°		24	.333		40.50	20		60.50	74.50
8740	60°		23	.348		48	21		69	83.50
8750	90°		22	.364		65.50	22		87.50	105
8760	36" radius, galvanized, 30°		23	.348		54.50	21		75.50	91
8770	45°		22	.364		62	22		84	101
8780	60°		21	.381		78	23		101	120
8790	90°		20	.400		104	24		128	150
8800	Aluminum, 30°		24	.333		52	20		72	87.50
8810	45°		23	.348		66.50	21		87.50	105
8820	60°		22	.364		81.50	22		103.50	122
8830	90°		21	.381		103	23		126	147
8840	Divider strip, vertical fitting, 6" deep									
8850	12" radius, galvanized, 30°	1 Elec	24	.333	Ea.	33	20		53	66.50
8860	45°		23	.348		37	21		58	72
8870	60°		22	.364		42.50	22		64.50	79.50
8880	90°		21	.381		53	23		76	92.50
8890	Aluminum, 30°		25	.320		24.50	19.20		43.70	55.50
8900	45°		24	.333		29	20		49	62
8910	60°		23	.348		30.50	21		51.50	64.50
8920	90°		22	.364		36	22		58	72
8930	24" radius, galvanized, 30°		23	.348		46	21		67	81.50
8940	45°		22	.364		57.50	22		79.50	95.50
8950	60°		21	.381		65.50	23		88.50	107
8960	90°		20	.400		86.50	24		110.50	131
8970	Aluminum, 30°		24	.333		34	20		54	67.50
8980	45°		23	.348		46.50	21		67.50	82

26 05 Common Work Results for Electrical

26 05 36 – Cable Trays for Electrical Systems

26 05 36.40 Cable Tray, Covers and Dividers

		Crew	Daily Output	Labor-Hours	Unit	Material	2019 Bare Costs Labor	Equipment	Total	Total Incl O&P
8990	60°	1 Elec	22	.364	Ea.	51	22		73	88.50
9000	90°		21	.381		69.50	23		92.50	111
9010	36" radius, galvanized, 30°		22	.364		53.50	22		75.50	91
9020	45°		21	.381		65.50	23		88.50	107
9030	60°		20	.400		86.50	24		110.50	131
9040	90°		19	.421		114	25.50		139.50	163
9050	Aluminum, 30°		23	.348		52.50	21		73.50	89
9060	45°		22	.364		70.50	22		92.50	111
9070	60°		21	.381		83.50	23		106.50	126
9080	90°		20	.400		100	24		124	146
9120	Divider strip, horizontal fitting, galvanized, 3" deep		33	.242		32.50	14.55		47.05	57.50
9130	4" deep		30	.267		36.50	16		52.50	64
9140	6" deep		27	.296		46.50	17.80		64.30	78
9150	Aluminum, 3" deep		35	.229		24.50	13.75		38.25	47
9160	4" deep		32	.250		26.50	15		41.50	52
9170	6" deep		29	.276		35.50	16.55		52.05	63.50
9300	Divider strip protector		300	.027	L.F.	2.95	1.60		4.55	5.65
9310	Fastener, ladder tray				Ea.	.52			.52	.57
9320	Trough or solid bottom tray				"	.35			.35	.39
9900	Add to labor for higher elevated installation									
9905	10' to 14.5' high, add						10%			
9910	15' to 20' high, add						20%			
9920	20' to 25' high, add						25%			
9930	25' to 30' high, add						35%			
9940	30' to 35' high, add						40%			
9950	35' to 40' high, add						50%			
9960	Over 40' high, add						55%			

26 05 39 – Underfloor Raceways for Electrical Systems

26 05 39.30 Conduit In Concrete Slab

		Crew	Daily Output	Labor-Hours	Unit	Material	2019 Bare Costs Labor	Equipment	Total	Total Incl O&P
0010	**CONDUIT IN CONCRETE SLAB** Including terminations, R260539-30									
0020	fittings and supports									
3230	PVC, schedule 40, 1/2" diameter	1 Elec	270	.030	L.F.	.54	1.78		2.32	3.24
3250	3/4" diameter		230	.035		.58	2.09		2.67	3.76
3270	1" diameter		200	.040		.77	2.40		3.17	4.43
3300	1-1/4" diameter		170	.047		1.04	2.83		3.87	5.35
3330	1-1/2" diameter		140	.057		1.27	3.43		4.70	6.50
3350	2" diameter		120	.067		1.58	4		5.58	7.70
3370	2-1/2" diameter		90	.089		2.60	5.35		7.95	10.80
3400	3" diameter	2 Elec	160	.100		3.42	6		9.42	12.70
3430	3-1/2" diameter		120	.133		4.43	8		12.43	16.85
3440	4" diameter		100	.160		4.71	9.60		14.31	19.55
3450	5" diameter		80	.200		7.40	12		19.40	26
3460	6" diameter		60	.267		9.40	16		25.40	34.50
3530	Sweeps, 1" diameter, 30" radius	1 Elec	32	.250	Ea.	9.45	15		24.45	33
3550	1-1/4" diameter		24	.333		9.30	20		29.30	40.50
3570	1-1/2" diameter		21	.381		10.10	23		33.10	45
3600	2" diameter		18	.444		11.10	26.50		37.60	52
3630	2-1/2" diameter		14	.571		16.65	34.50		51.15	69.50
3650	3" diameter		10	.800		66.50	48		114.50	145
3670	3-1/2" diameter		8	1		63.50	60		123.50	159
3700	4" diameter		7	1.143		25	68.50		93.50	130
3710	5" diameter		6	1.333		42.50	80		122.50	166

For customer support on your Electrical Costs with RSMeans data, call 800.448.8182.

197

26 05 Common Work Results for Electrical

26 05 39 – Underfloor Raceways for Electrical Systems

26 05 39.30 Conduit In Concrete Slab

		Crew	Daily Output	Labor-Hours	Unit	Material	2019 Bare Costs Labor	Equipment	Total	Total Incl O&P
3730	Couplings, 1/2" diameter				Ea.	.13			.13	.14
3750	3/4" diameter					.16			.16	.18
3770	1" diameter					.25			.25	.28
3800	1-1/4" diameter					.37			.37	.41
3830	1-1/2" diameter					.45			.45	.50
3850	2" diameter					.64			.64	.70
3870	2-1/2" diameter					1.18			1.18	1.30
3900	3" diameter					2.12			2.12	2.33
3930	3-1/2" diameter					2.66			2.66	2.93
3950	4" diameter					2.70			2.70	2.97
3960	5" diameter					5.70			5.70	6.30
3970	6" diameter					10.55			10.55	11.60
4030	End bells, 1" diameter, PVC	1 Elec	60	.133		2.73	8		10.73	14.95
4050	1-1/4" diameter		53	.151		4.32	9.05		13.37	18.25
4100	1-1/2" diameter		48	.167		2.89	10		12.89	18.15
4150	2" diameter		34	.235		4.48	14.15		18.63	26
4170	2-1/2" diameter		27	.296		4.51	17.80		22.31	31.50
4200	3" diameter		20	.400		5.75	24		29.75	42.50
4250	3-1/2" diameter		16	.500		7.05	30		37.05	53
4300	4" diameter		14	.571		7.10	34.50		41.60	59
4310	5" diameter		12	.667		10.60	40		50.60	71
4320	6" diameter		9	.889	▼	12.40	53.50		65.90	93
4350	Rigid galvanized steel, 1/2" diameter		200	.040	L.F.	2.02	2.40		4.42	5.80
4400	3/4" diameter		170	.047		4.46	2.83		7.29	9.10
4450	1" diameter		130	.062		6.90	3.70		10.60	13.10
4500	1-1/4" diameter		110	.073		4.23	4.37		8.60	11.15
4600	1-1/2" diameter		100	.080		7.25	4.80		12.05	15.15
4800	2" diameter	▼	90	.089	▼	8.95	5.35		14.30	17.80

26 05 39.40 Conduit In Trench

		Crew	Daily Output	Labor-Hours	Unit	Material	2019 Bare Costs Labor	Equipment	Total	Total Incl O&P
0010	**CONDUIT IN TRENCH** Includes terminations and fittings R260539-40									
0020	Does not include excavation or backfill, see Section 31 23 16									
0200	Rigid galvanized steel, 2" diameter	1 Elec	150	.053	L.F.	8.60	3.20		11.80	14.25
0400	2-1/2" diameter	"	100	.080		11	4.80		15.80	19.25
0600	3" diameter	2 Elec	160	.100		12.85	6		18.85	23
0800	3-1/2" diameter		140	.114		16.75	6.85		23.60	28.50
1000	4" diameter		100	.160		19.50	9.60		29.10	36
1200	5" diameter		80	.200		37	12		49	58.50
1400	6" diameter	▼	60	.267	▼	53	16		69	82.50

26 05 43 – Underground Ducts and Raceways for Electrical Systems

26 05 43.10 Trench Duct

		Crew	Daily Output	Labor-Hours	Unit	Material	2019 Bare Costs Labor	Equipment	Total	Total Incl O&P
0010	**TRENCH DUCT** Steel with cover									
0020	Standard adjustable, depths to 4"									
0100	Straight, single compartment, 9" wide	2 Elec	40	.400	L.F.	251	24		275	310
0200	12" wide		32	.500		325	30		355	405
0400	18" wide		26	.615		365	37		402	455
0600	24" wide		22	.727		425	43.50		468.50	535
0700	27" wide		21	.762		213	46		259	305
0800	30" wide		20	.800		227	48		275	320
1000	36" wide		16	1		260	60		320	375
1020	Two compartment, 9" wide		38	.421		120	25.50		145.50	170
1030	12" wide		30	.533		141	32		173	203
1040	18" wide	▼	24	.667		196	40		236	276

For customer support on your Electrical Costs with RSMeans data, call 800.448.8182.

26 05 43.10 Trench Duct		Crew	Daily Output	Labor- Hours	Unit	Material	2019 Bare Costs Labor	Equipment	Total	Total Incl O&P
1050	24" wide	2 Elec	20	.800	L.F.	244	48		292	340
1060	30" wide		18	.889		310	53.50		363.50	425
1070	36" wide		14	1.143		345	68.50		413.50	480
1090	Three compartment, 9" wide		36	.444		150	26.50		176.50	205
1100	12" wide		28	.571		173	34.50		207.50	242
1110	18" wide		22	.727		211	43.50		254.50	297
1120	24" wide		18	.889		264	53.50		317.50	370
1130	30" wide		16	1		325	60		385	450
1140	36" wide		12	1.333		375	80		455	530
1200	Horizontal elbow, 9" wide		5.40	2.963	Ea.	410	178		588	715
1400	12" wide		4.60	3.478		440	209		649	795
1600	18" wide		4	4		575	240		815	995
1800	24" wide		3.20	5		815	300		1,115	1,350
1900	27" wide		3	5.333		975	320		1,295	1,550
2000	30" wide		2.60	6.154		1,075	370		1,445	1,750
2200	36" wide		2.40	6.667		1,425	400		1,825	2,175
2220	Two compartment, 9" wide		3.80	4.211		660	253		913	1,100
2230	12" wide		3	5.333		730	320		1,050	1,275
2240	18" wide		2.40	6.667		880	400		1,280	1,550
2250	24" wide		2	8		1,075	480		1,555	1,900
2260	30" wide		1.80	8.889		1,425	535		1,960	2,375
2270	36" wide		1.60	10		1,725	600		2,325	2,800
2290	Three compartment, 9" wide		3.60	4.444		660	267		927	1,125
2300	12" wide		2.80	5.714		740	345		1,085	1,325
2310	18" wide		2.20	7.273		900	435		1,335	1,650
2320	24" wide		1.80	8.889		1,150	535		1,685	2,050
2330	30" wide		1.60	10		1,475	600		2,075	2,525
2350	36" wide		1.40	11.429		1,800	685		2,485	3,000
2400	Vertical elbow, 9" wide		5.40	2.963		177	178		355	460
2600	12" wide		4.60	3.478		155	209		364	480
2800	18" wide		4	4		178	240		418	555
3000	24" wide		3.20	5		221	300		521	695
3100	27" wide		3	5.333		230	320		550	735
3200	30" wide		2.60	6.154		244	370		614	820
3400	36" wide		2.40	6.667		268	400		668	890
3600	Cross, 9" wide		4	4		670	240		910	1,100
3800	12" wide		3.20	5		710	300		1,010	1,225
4000	18" wide		2.60	6.154		850	370		1,220	1,475
4200	24" wide		2.20	7.273		1,050	435		1,485	1,800
4300	27" wide		2.20	7.273		1,200	435		1,635	1,975
4400	30" wide		2	8		1,325	480		1,805	2,200
4600	36" wide		1.80	8.889		1,675	535		2,210	2,625
4620	Two compartment, 9" wide		3.80	4.211		670	253		923	1,100
4630	12" wide		3	5.333		700	320		1,020	1,250
4640	18" wide		2.40	6.667		850	400		1,250	1,525
4650	24" wide		2	8		1,075	480		1,555	1,925
4660	30" wide		1.80	8.889		1,425	535		1,960	2,375
4670	36" wide		1.60	10		1,725	600		2,325	2,800
4690	Three compartment, 9" wide		3.60	4.444		675	267		942	1,150
4700	12" wide		2.80	5.714		775	345		1,120	1,375
4710	18" wide		2.20	7.273		920	435		1,355	1,650
4720	24" wide		1.80	8.889		1,150	535		1,685	2,050
4730	30" wide		1.60	10		1,500	600		2,100	2,550

26 05 43.10 Trench Duct		Crew	Daily Output	Labor-Hours	Unit	Material	2019 Bare Costs Labor	Equipment	Total	Total Incl O&P
4740	36" wide	2 Elec	1.40	11.429	Ea.	1,850	685		2,535	3,050
4800	End closure, 9" wide		14.40	1.111		40	66.50		106.50	144
5000	12" wide		12	1.333		46	80		126	170
5200	18" wide		10	1.600		70	96		166	220
5400	24" wide		8	2		92.50	120		212.50	281
5500	27" wide		7	2.286		107	137		244	325
5600	30" wide		6.60	2.424		154	146		300	385
5800	36" wide		5.80	2.759		138	166		304	400
6000	Tees, 9" wide		4	4		390	240		630	790
6200	12" wide		3.60	4.444		450	267		717	895
6400	18" wide		3.20	5		580	300		880	1,075
6600	24" wide		3	5.333		830	320		1,150	1,400
6700	27" wide		2.80	5.714		920	345		1,265	1,500
6800	30" wide		2.60	6.154		1,075	370		1,445	1,750
7000	36" wide		2	8		1,425	480		1,905	2,275
7020	Two compartment, 9" wide		3.80	4.211		450	253		703	870
7030	12" wide		3.40	4.706		485	283		768	955
7040	18" wide		3	5.333		650	320		970	1,200
7050	24" wide		2.80	5.714		875	345		1,220	1,475
7060	30" wide		2.40	6.667		1,200	400		1,600	1,925
7070	36" wide		1.90	8.421		1,500	505		2,005	2,400
7090	Three compartment, 9" wide		3.60	4.444		515	267		782	970
7100	12" wide		3.20	5		540	300		840	1,050
7110	18" wide		2.80	5.714		685	345		1,030	1,275
7120	24" wide		2.60	6.154		945	370		1,315	1,600
7130	30" wide		2.20	7.273		1,225	435		1,660	2,000
7140	36" wide		1.80	8.889		1,575	535		2,110	2,525
7200	Riser, and cabinet connector, 9" wide		5.40	2.963		170	178		348	450
7400	12" wide		4.60	3.478		198	209		407	530
7600	18" wide		4	4		205	240		445	585
7800	24" wide		3.20	5		295	300		595	775
7900	27" wide		3	5.333		300	320		620	810
8000	30" wide		2.60	6.154		340	370		710	925
8200	36" wide	▼	2	8	▼	395	480		875	1,150
8400	Insert assembly, cell to conduit adapter, 1-1/4"	1 Elec	16	.500		67.50	30		97.50	120
8500	Adjustable partition	"	320	.025	L.F.	24	1.50		25.50	28.50
8600	Depth of duct over 4", per 1", add					10.90			10.90	12
8700	Support post	1 Elec	240	.033		25	2		27	30.50
8800	Cover double tile trim, 2 sides					38.50			38.50	42.50
8900	4 sides					110			110	121
9160	Trench duct 3-1/2" x 4-1/2", add					10.35			10.35	11.40
9170	Trench duct 4" x 5", add					10.35			10.35	11.40
9200	For carpet trim, add					35			35	38.50
9210	For double carpet trim, add				▼	107			107	118

26 05 43.20 Underfloor Duct

		Crew	Daily Output	Labor-Hours	Unit	Material	2019 Bare Costs Labor	Equipment	Total	Total Incl O&P
0010	**UNDERFLOOR DUCT** R260543-50									
0100	Duct, 1-3/8" x 3-1/8" blank, standard	2 Elec	160	.100	L.F.	12.85	6		18.85	23
0200	1-3/8" x 7-1/4" blank, super duct		120	.133		30	8		38	45
0400	7/8" or 1-3/8" insert type, 24" OC, 1-3/8" x 3-1/8", std.		140	.114		20	6.85		26.85	32.50
0600	1-3/8" x 7-1/4", super duct	▼	100	.160	▼	35	9.60		44.60	53
0800	Junction box, single duct, 1 level, 3-1/8"	1 Elec	4	2	Ea.	450	120		570	670
0820	3-1/8" x 7-1/4"	↓	4	2	↓	550	120		670	785

For customer support on your Electrical Costs with RSMeans data, call 800.448.8182.

26 05 Common Work Results for Electrical

26 05 43 – Underground Ducts and Raceways for Electrical Systems

26 05 43.20 Underfloor Duct		Crew	Daily Output	Labor-Hours	Unit	Material	2019 Bare Costs Labor	Equipment	Total	Total Incl O&P
0840	2 level, 3-1/8" upper & lower	1 Elec	3.20	2.500	Ea.	515	150		665	795
0860	3-1/8" upper, 7-1/4" lower		2.70	2.963		510	178		688	830
0880	Carpet pan for above		80	.100		370	6		376	420
0900	Terrazzo pan for above		67	.119		880	7.15		887.15	975
1000	Junction box, single duct, 1 level, 7-1/4"		2.70	2.963		520	178		698	840
1020	2 level, 7-1/4" upper & lower		2.70	2.963		595	178		773	920
1040	2 duct, two 3-1/8" upper & lower		3.20	2.500		800	150		950	1,100
1200	1 level, 2 duct, 3-1/8"		3.20	2.500		590	150		740	870
1220	Carpet pan for above boxes		80	.100		365	6		371	410
1240	Terrazzo pan for above boxes		67	.119		850	7.15		857.15	945
1260	Junction box, 1 level, two 3-1/8" x one 3-1/8" + one 7-1/4"		2.30	3.478		1,000	209		1,209	1,400
1280	2 level, two 3-1/8" upper, one 3-1/8" + one 7-1/4" lower		2	4		1,100	240		1,340	1,550
1300	Carpet pan for above boxes		80	.100		365	6		371	410
1320	Terrazzo pan for above boxes		67	.119		850	7.15		857.15	945
1400	Junction box, 1 level, 2 duct, 7-1/4"		2.30	3.478		1,425	209		1,634	1,875
1420	Two 3-1/8" + one 7-1/4"		2	4		1,500	240		1,740	2,000
1440	Carpet pan for above		80	.100		365	6		371	410
1460	Terrazzo pan for above		67	.119		850	7.15		857.15	945
1580	Junction box, 1 level, one 3-1/8" + one 7-1/4" x same		2.30	3.478		1,000	209		1,209	1,400
1600	Triple duct, 3-1/8"		2.30	3.478		1,000	209		1,209	1,400
1700	Junction box, 1 level, one 3-1/8" + two 7-1/4"		2	4		1,675	240		1,915	2,200
1720	Carpet pan for above		80	.100		365	6		371	410
1740	Terrazzo pan for above		67	.119		850	7.15		857.15	945
1800	Insert to conduit adapter, 3/4" & 1"		32	.250		37	15		52	63
2000	Support, single cell		27	.296		54.50	17.80		72.30	86.50
2200	Super duct		16	.500		55.50	30		85.50	106
2400	Double cell		16	.500		55.50	30		85.50	106
2600	Triple cell		11	.727		65	43.50		108.50	136
2800	Vertical elbow, standard duct		10	.800		98.50	48		146.50	180
3000	Super duct		8	1		98.50	60		158.50	198
3200	Cabinet connector, standard duct		32	.250		78.50	15		93.50	109
3400	Super duct		27	.296		77	17.80		94.80	111
3600	Conduit adapter, 1" to 1-1/4"		32	.250		74	15		89	104
3800	2" to 1-1/4"		27	.296		88.50	17.80		106.30	124
4000	Outlet, low tension (tele, computer, etc.)		8	1		104	60		164	205
4200	High tension, receptacle (120 volt)		8	1		107	60		167	207
4300	End closure, standard duct		160	.050		4.28	3		7.28	9.20
4310	Super duct		160	.050		8.10	3		11.10	13.40
4350	Elbow, horiz., standard duct		26	.308		253	18.50		271.50	305
4360	Super duct		26	.308		242	18.50		260.50	295
4380	Elbow, offset, standard duct		26	.308		98.50	18.50		117	136
4390	Super duct		26	.308		100	18.50		118.50	139
4400	Marker screw assembly for inserts		50	.160		18.70	9.60		28.30	35
4410	Y take off, standard duct		26	.308		154	18.50		172.50	197
4420	Super duct		26	.308		154	18.50		172.50	197
4430	Box opening plug, standard duct		160	.050		18.40	3		21.40	24.50
4440	Super duct		160	.050		18.40	3		21.40	24.50
4450	Sleeve coupling, standard duct		160	.050		49	3		52	58.50
4460	Super duct		160	.050		50	3		53	59.50
4470	Conduit adapter, standard duct, 3/4"		32	.250		77	15		92	107
4480	1" or 1-1/4"		32	.250		76	15		91	106
4500	1-1/2"		32	.250		74	15		89	104

26 05 80 – Wiring Connections

26 05 80.10 Motor Connections	Crew	Daily Output	Labor-Hours	Unit	Material	2019 Bare Costs Labor	Equipment	Total	Total Incl O&P
0010 **MOTOR CONNECTIONS**									
0020 Flexible conduit and fittings, 115 volt, 1 phase, up to 1 HP motor	1 Elec	8	1	Ea.	5.70	60		65.70	96
0050 2 HP motor R260580-75		6.50	1.231		10.15	74		84.15	121
0100 3 HP motor		5.50	1.455		9.05	87.50		96.55	140
0110 230 volt, 3 phase, 3 HP motor		6.78	1.180		6.80	71		77.80	114
0112 5 HP motor		5.47	1.463		5.85	88		93.85	137
0114 7-1/2 HP motor		4.61	1.735		9.35	104		113.35	165
0120 10 HP motor		4.20	1.905		18.25	114		132.25	191
0150 15 HP motor		3.30	2.424		18.25	146		164.25	237
0200 25 HP motor		2.70	2.963		28	178		206	296
0400 50 HP motor		2.20	3.636		55.50	218		273.50	385
0600 100 HP motor		1.50	5.333		118	320		438	610
1500 460 volt, 5 HP motor, 3 phase		8	1		6.15	60		66.15	96.50
1520 10 HP motor		8	1		6.15	60		66.15	96.50
1530 25 HP motor		6	1.333		10.95	80		90.95	131
1540 30 HP motor		6	1.333		10.95	80		90.95	131
1550 40 HP motor		5	1.600		17.65	96		113.65	162
1560 50 HP motor		5	1.600		22.50	96		118.50	168
1570 60 HP motor		3.80	2.105		24	126		150	216
1580 75 HP motor		3.50	2.286		31	137		168	240
1590 100 HP motor		2.50	3.200		48	192		240	340
1600 125 HP motor		2	4		56	240		296	420
1610 150 HP motor		1.80	4.444		59	267		326	465
1620 200 HP motor		1.50	5.333		96.50	320		416.50	585
2005 460 volt, 5 HP motor, 3 phase, w/sealtite		8	1		8.55	60		68.55	99
2010 10 HP motor		8	1		8.55	60		68.55	99
2015 25 HP motor		6	1.333		16.95	80		96.95	138
2020 30 HP motor		6	1.333		16.95	80		96.95	138
2025 40 HP motor		5	1.600		29	96		125	175
2030 50 HP motor		5	1.600		32.50	96		128.50	179
2035 60 HP motor		3.80	2.105		36	126		162	229
2040 75 HP motor		3.50	2.286		42	137		179	251
2045 100 HP motor		2.50	3.200		75	192		267	370
2055 150 HP motor		1.80	4.444		78	267		345	485
2060 200 HP motor		1.50	5.333		269	320		589	775

26 05 90 – Residential Applications

26 05 90.10 Residential Wiring

	Crew	Daily Output	Labor-Hours	Unit	Material	2019 Bare Costs Labor	Equipment	Total	Total Incl O&P
0010 **RESIDENTIAL WIRING**									
0020 20' avg. runs and #14/2 wiring incl. unless otherwise noted									
1000 Service & panel, includes 24' SE-AL cable, service eye, meter,									
1010 Socket, panel board, main bkr., ground rod, 15 or 20 amp									
1020 1-pole circuit breakers, and misc. hardware									
1100 100 amp, with 10 branch breakers	1 Elec	1.19	6.723	Ea.	325	405		730	960
1110 With PVC conduit and wire		.92	8.696		360	520		880	1,175
1120 With RGS conduit and wire		.73	10.959		575	660		1,235	1,600
1150 150 amp, with 14 branch breakers		1.03	7.767		845	465		1,310	1,625
1170 With PVC conduit and wire		.82	9.756		915	585		1,500	1,875
1180 With RGS conduit and wire		.67	11.940		1,225	715		1,940	2,425
1200 200 amp, with 18 branch breakers	2 Elec	1.80	8.889		1,000	535		1,535	1,900
1220 With PVC conduit and wire		1.46	10.959		1,075	660		1,735	2,150
1230 With RGS conduit and wire		1.24	12.903		1,450	775		2,225	2,750
1800 Lightning surge suppressor	1 Elec	32	.250		75	15		90	105

For customer support on your Electrical Costs with RSMeans data, call 800.448.8182.

26 05 90.10 Residential Wiring	Crew	Daily Output	Labor-Hours	Unit	Material	2019 Bare Costs Labor	Equipment	Total	Total Incl O&P
2000	Switch devices									
2100	Single pole, 15 amp, ivory, with a 1-gang box, cover plate,									
2110	Type NM (Romex) cable	1 Elec	17.10	.468	Ea.	15.50	28		43.50	59
2120	Type MC cable		14.30	.559		26	33.50		59.50	78.50
2130	EMT & wire		5.71	1.401		37.50	84		121.50	167
2150	3-way, #14/3, type NM cable		14.55	.550		9.80	33		42.80	60.50
2170	Type MC cable		12.31	.650		23.50	39		62.50	84
2180	EMT & wire		5	1.600		31.50	96		127.50	178
2200	4-way, #14/3, type NM cable		14.55	.550		18.50	33		51.50	70
2220	Type MC cable		12.31	.650		32.50	39		71.50	93.50
2230	EMT & wire		5	1.600		40.50	96		136.50	188
2250	S.P., 20 amp, #12/2, type NM cable		13.33	.600		11.30	36		47.30	66.50
2270	Type MC cable		11.43	.700		21	42		63	85.50
2280	EMT & wire		4.85	1.649		35.50	99		134.50	187
2290	S.P. rotary dimmer, 600 W, no wiring		17	.471		31.50	28.50		60	76.50
2300	S.P. rotary dimmer, 600 W, type NM cable		14.55	.550		35.50	33		68.50	88.50
2320	Type MC cable		12.31	.650		46	39		85	109
2330	EMT & wire		5	1.600		58.50	96		154.50	207
2350	3-way rotary dimmer, type NM cable		13.33	.600		23.50	36		59.50	79.50
2370	Type MC cable		11.43	.700		34	42		76	100
2380	EMT & wire		4.85	1.649		46.50	99		145.50	199
2400	Interval timer wall switch, 20 amp, 1-30 min., #12/2									
2410	Type NM cable	1 Elec	14.55	.550	Ea.	58.50	33		91.50	114
2420	Type MC cable		12.31	.650		64.50	39		103.50	129
2430	EMT & wire		5	1.600		82.50	96		178.50	234
2500	Decorator style									
2510	S.P., 15 amp, type NM cable	1 Elec	17.10	.468	Ea.	20.50	28		48.50	65
2520	Type MC cable		14.30	.559		31.50	33.50		65	84.50
2530	EMT & wire		5.71	1.401		42.50	84		126.50	173
2550	3-way, #14/3, type NM cable		14.55	.550		15	33		48	66
2570	Type MC cable		12.31	.650		29	39		68	90
2580	EMT & wire		5	1.600		37	96		133	184
2600	4-way, #14/3, type NM cable		14.55	.550		23.50	33		56.50	75.50
2620	Type MC cable		12.31	.650		37.50	39		76.50	99.50
2630	EMT & wire		5	1.600		45.50	96		141.50	193
2650	S.P., 20 amp, #12/2, type NM cable		13.33	.600		16.55	36		52.55	72
2670	Type MC cable		11.43	.700		26	42		68	91
2680	EMT & wire		4.85	1.649		41	99		140	193
2700	S.P., slide dimmer, type NM cable		17.10	.468		37	28		65	83
2720	Type MC cable		14.30	.559		47.50	33.50		81	103
2730	EMT & wire		5.71	1.401		60	84		144	192
2750	S.P., touch dimmer, type NM cable		17.10	.468		48.50	28		76.50	95
2770	Type MC cable		14.30	.559		59	33.50		92.50	115
2780	EMT & wire		5.71	1.401		71.50	84		155.50	205
2800	3-way touch dimmer, type NM cable		13.33	.600		49.50	36		85.50	109
2820	Type MC cable		11.43	.700		60.50	42		102.50	129
2830	EMT & wire		4.85	1.649		73	99		172	228
3000	Combination devices									
3100	S.P. switch/15 amp recpt., ivory, 1-gang box, plate									
3110	Type NM cable	1 Elec	11.43	.700	Ea.	23	42		65	87.50
3120	Type MC cable		10	.800		33.50	48		81.50	109
3130	EMT & wire		4.40	1.818		46	109		155	214
3150	S.P. switch/pilot light, type NM cable		11.43	.700		24	42		66	89

For customer support on your Electrical Costs with RSMeans data, call 800.448.8182.

203

26 05 90.10 Residential Wiring	Crew	Daily Output	Labor-Hours	Unit	Material	2019 Bare Costs Labor	Equipment	Total	Total Incl O&P
3170	Type MC cable	1 Elec	10	.800	Ea.	35	48		83	110
3180	EMT & wire		4.43	1.806		47.50	108		155.50	214
3190	2-S.P. switches, 2-#14/2, no wiring		14	.571		13.25	34.50		47.75	65.50
3200	2-S.P. switches, 2-#14/2, type NM cables		10	.800		25	48		73	99
3220	Type MC cable		8.89	.900		40	54		94	125
3230	EMT & wire		4.10	1.951		48	117		165	228
3250	3-way switch/15 amp recpt., #14/3, type NM cable		10	.800		30.50	48		78.50	105
3270	Type MC cable		8.89	.900		44.50	54		98.50	130
3280	EMT & wire		4.10	1.951		52.50	117		169.50	233
3300	2-3 way switches, 2-#14/3, type NM cables		8.89	.900		39	54		93	124
3320	Type MC cable		8	1		60.50	60		120.50	156
3330	EMT & wire		4	2		59.50	120		179.50	244
3350	S.P. switch/20 amp recpt., #12/2, type NM cable		10	.800		39	48		87	114
3370	Type MC cable		8.89	.900		44.50	54		98.50	130
3380	EMT & wire	▼	4.10	1.951	▼	63	117		180	245
3400	Decorator style									
3410	S.P. switch/15 amp recpt., type NM cable	1 Elec	11.43	.700	Ea.	28	42		70	93.50
3420	Type MC cable		10	.800		38.50	48		86.50	114
3430	EMT & wire		4.40	1.818		51	109		160	220
3450	S.P. switch/pilot light, type NM cable		11.43	.700		29.50	42		71.50	95
3470	Type MC cable		10	.800		40	48		88	116
3480	EMT & wire		4.40	1.818		52.50	109		161.50	221
3500	2-S.P. switches, 2-#14/2, type NM cables		10	.800		30	48		78	105
3520	Type MC cable		8.89	.900		45	54		99	130
3530	EMT & wire		4.10	1.951		53	117		170	234
3550	3-way/15 amp recpt., #14/3, type NM cable		10	.800		36	48		84	111
3570	Type MC cable		8.89	.900		49.50	54		103.50	135
3580	EMT & wire		4.10	1.951		57.50	117		174.50	239
3650	2-3 way switches, 2-#14/3, type NM cables		8.89	.900		44.50	54		98.50	129
3670	Type MC cable		8	1		66	60		126	162
3680	EMT & wire		4	2		64.50	120		184.50	250
3700	S.P. switch/20 amp recpt., #12/2, type NM cable		10	.800		44	48		92	120
3720	Type MC cable		8.89	.900		50	54		104	136
3730	EMT & wire	▼	4.10	1.951	▼	68.50	117		185.50	250
4000	Receptacle devices									
4010	Duplex outlet, 15 amp recpt., ivory, 1-gang box, plate									
4015	Type NM cable	1 Elec	14.55	.550	Ea.	8.70	33		41.70	59
4020	Type MC cable		12.31	.650		19.30	39		58.30	79
4030	EMT & wire		5.33	1.501		30.50	90		120.50	168
4050	With #12/2, type NM cable		12.31	.650		9.85	39		48.85	69
4070	Type MC cable		10.67	.750		19.40	45		64.40	88.50
4080	EMT & wire		4.71	1.699		34	102		136	190
4100	20 amp recpt., #12/2, type NM cable		12.31	.650		17.85	39		56.85	77.50
4120	Type MC cable		10.67	.750		27.50	45		72.50	97
4130	EMT & wire	▼	4.71	1.699	▼	42	102		144	199
4140	For GFI see Section 26 05 90.10 line 4300 below									
4150	Decorator style, 15 amp recpt., type NM cable	1 Elec	14.55	.550	Ea.	13.95	33		46.95	65
4170	Type MC cable		12.31	.650		24.50	39		63.50	85
4180	EMT & wire		5.33	1.501		36	90		126	174
4200	With #12/2, type NM cable		12.31	.650		15.10	39		54.10	74.50
4220	Type MC cable		10.67	.750		24.50	45		69.50	94
4230	EMT & wire		4.71	1.699		39.50	102		141.50	196
4250	20 amp recpt., #12/2, type NM cable	▼	12.31	.650	▼	23	39		62	83.50

26 05 90.10 Residential Wiring		Crew	Daily Output	Labor-Hours	Unit	Material	2019 Bare Costs Labor	Equipment	Total	Total Incl O&P
4270	Type MC cable	1 Elec	10.67	.750	Ea.	32.50	45		77.50	103
4280	EMT & wire		4.71	1.699		47.50	102		149.50	204
4300	GFI, 15 amp recpt., type NM cable		12.31	.650		21	39		60	81.50
4320	Type MC cable		10.67	.750		32	45		77	102
4330	EMT & wire		4.71	1.699		43	102		145	200
4350	GFI with #12/2, type NM cable		10.67	.750		22.50	45		67.50	91.50
4370	Type MC cable		9.20	.870		32	52		84	113
4380	EMT & wire		4.21	1.900		46.50	114		160.50	222
4400	20 amp recpt., #12/2, type NM cable		10.67	.750		49	45		94	121
4420	Type MC cable		9.20	.870		58.50	52		110.50	142
4430	EMT & wire		4.21	1.900		73	114		187	251
4500	Weather-proof cover for above receptacles, add	↓	32	.250	↓	2.06	15		17.06	25
4550	Air conditioner outlet, 20 amp-240 volt recpt.									
4560	30' of #12/2, 2 pole circuit breaker									
4570	Type NM cable	1 Elec	10	.800	Ea.	60	48		108	138
4580	Type MC cable		9	.889		71	53.50		124.50	158
4590	EMT & wire		4	2		85	120		205	272
4600	Decorator style, type NM cable		10	.800		65	48		113	143
4620	Type MC cable		9	.889		76	53.50		129.50	163
4630	EMT & wire	↓	4	2	↓	89.50	120		209.50	278
4650	Dryer outlet, 30 amp-240 volt recpt., 20' of #10/3									
4660	2 pole circuit breaker									
4670	Type NM cable	1 Elec	6.41	1.248	Ea.	54	75		129	172
4680	Type MC cable		5.71	1.401		63.50	84		147.50	196
4690	EMT & wire	↓	3.48	2.299	↓	76	138		214	290
4700	Range outlet, 50 amp-240 volt recpt., 30' of #8/3									
4710	Type NM cable	1 Elec	4.21	1.900	Ea.	82.50	114		196.50	261
4720	Type MC cable		4	2		127	120		247	320
4730	EMT & wire		2.96	2.703		107	162		269	360
4750	Central vacuum outlet, type NM cable		6.40	1.250		53.50	75		128.50	171
4770	Type MC cable		5.71	1.401		67.50	84		151.50	201
4780	EMT & wire	↓	3.48	2.299	↓	88	138		226	305
4800	30 amp-110 volt locking recpt., #10/2 circ. bkr.									
4810	Type NM cable	1 Elec	6.20	1.290	Ea.	63	77.50		140.50	185
4820	Type MC cable		5.40	1.481		80	89		169	221
4830	EMT & wire	↓	3.20	2.500	↓	99	150		249	335
4900	Low voltage outlets									
4910	Telephone recpt., 20' of 4/C phone wire	1 Elec	26	.308	Ea.	8.55	18.50		27.05	37
4920	TV recpt., 20' of RG59U coax wire, F type connector	"	16	.500	"	17.10	30		47.10	64
4950	Door bell chime, transformer, 2 buttons, 60' of bellwire									
4970	Economy model	1 Elec	11.50	.696	Ea.	55	42		97	123
4980	Custom model		11.50	.696		105	42		147	179
4990	Luxury model, 3 buttons	↓	9.50	.842		188	50.50		238.50	283
6000	Lighting outlets									
6050	Wire only (for fixture), type NM cable	1 Elec	32	.250	Ea.	5.80	15		20.80	29
6070	Type MC cable		24	.333		11.65	20		31.65	43
6080	EMT & wire		10	.800		22	48		70	96
6100	Box (4"), and wire (for fixture), type NM cable		25	.320		14.85	19.20		34.05	45
6120	Type MC cable		20	.400		20.50	24		44.50	59
6130	EMT & wire	↓	11	.727	↓	31.50	43.50		75	99.50
6200	Fixtures (use with line 6050 or 6100 above)									
6210	Canopy style, economy grade	1 Elec	40	.200	Ea.	22.50	12		34.50	43
6220	Custom grade	↓	40	.200	↓	52.50	12		64.50	75.50

26 05 90 – Residential Applications

26 05 90.10 Residential Wiring	Crew	Daily Output	Labor-Hours	Unit	Material	2019 Bare Costs Labor	Equipment	Total	Total Incl O&P
6250	Dining room chandelier, economy grade	1 Elec	19	.421	Ea.	83	25.50		108.50	129
6260	Custom grade		19	.421		335	25.50		360.50	405
6270	Luxury grade		15	.533		1,250	32		1,282	1,425
6310	Kitchen fixture (fluorescent), economy grade		30	.267		71.50	16		87.50	103
6320	Custom grade		25	.320		149	19.20		168.20	193
6350	Outdoor, wall mounted, economy grade		30	.267		29.50	16		45.50	56.50
6360	Custom grade		30	.267		120	16		136	156
6370	Luxury grade		25	.320		243	19.20		262.20	296
6410	Outdoor PAR floodlights, 1 lamp, 150 watt		20	.400		26.50	24		50.50	65.50
6420	2 lamp, 150 watt each		20	.400		45	24		69	85.50
6425	Motion sensing, 2 lamp, 150 watt each		20	.400		110	24		134	157
6430	For infrared security sensor, add		32	.250		102	15		117	135
6450	Outdoor, quartz-halogen, 300 watt flood		20	.400		38.50	24		62.50	78
6600	Recessed downlight, round, pre-wired, 50 or 75 watt trim		30	.267		66	16		82	96.50
6610	With shower light trim		30	.267		90	16		106	123
6620	With wall washer trim		28	.286		92	17.15		109.15	127
6630	With eye-ball trim		28	.286		77	17.15		94.15	111
6700	Porcelain lamp holder		40	.200		3	12		15	21
6710	With pull switch		40	.200		11.30	12		23.30	30.50
6750	Fluorescent strip, 2-20 watt tube, wrap around diffuser, 24"		24	.333		48	20		68	82.50
6760	1-34 watt tube, 48"		24	.333		137	20		157	181
6770	2-34 watt tubes, 48"		20	.400		158	24		182	209
6800	Bathroom heat lamp, 1-250 watt		28	.286		31.50	17.15		48.65	60.50
6810	2-250 watt lamps		28	.286		76.50	17.15		93.65	110
6820	For timer switch, see Section 26 05 90.10 line 2400									
6900	Outdoor post lamp, incl. post, fixture, 35' of #14/2									
6910	Type NM cable	1 Elec	3.50	2.286	Ea.	299	137		436	535
6920	Photo-eye, add		27	.296		28.50	17.80		46.30	57.50
6950	Clock dial time switch, 24 hr., w/enclosure, type NM cable		11.43	.700		78.50	42		120.50	149
6970	Type MC cable		11	.727		89	43.50		132.50	163
6980	EMT & wire		4.85	1.649		100	99		199	259
7000	Alarm systems									
7050	Smoke detectors, box, #14/3, type NM cable	1 Elec	14.55	.550	Ea.	34.50	33		67.50	87.50
7070	Type MC cable		12.31	.650		44.50	39		83.50	107
7080	EMT & wire		5	1.600		52.50	96		148.50	201
7090	For relay output to security system, add					11.15			11.15	12.25
8000	Residential equipment									
8050	Disposal hook-up, incl. switch, outlet box, 3' of flex									
8060	20 amp-1 pole circ. bkr., and 25' of #12/2									
8070	Type NM cable	1 Elec	10	.800	Ea.	28	48		76	103
8080	Type MC cable		8	1		38.50	60		98.50	132
8090	EMT & wire		5	1.600		56	96		152	205
8100	Trash compactor or dishwasher hook-up, incl. outlet box,									
8110	3' of flex, 15 amp-1 pole circ. bkr., and 25' of #14/2									
8120	Type NM cable	1 Elec	10	.800	Ea.	15.60	48		63.60	88.50
8130	Type MC cable		8	1		27.50	60		87.50	120
8140	EMT & wire		5	1.600		42.50	96		138.50	190
8150	Hot water sink dispenser hook-up, use line 8100									
8200	Vent/exhaust fan hook-up, type NM cable	1 Elec	32	.250	Ea.	5.80	15		20.80	29
8220	Type MC cable		24	.333		11.65	20		31.65	43
8230	EMT & wire		10	.800		22	48		70	96
8250	Bathroom vent fan, 50 CFM (use with above hook-up)									
8260	Economy model	1 Elec	15	.533	Ea.	19.45	32		51.45	69.50

For customer support on your Electrical Costs with RSMeans data, call 800.448.8182.

26 05 90 – Residential Applications

26 05 90.10 Residential Wiring		Crew	Daily Output	Labor-Hours	Unit	Material	2019 Bare Costs Labor	Equipment	Total	Total Incl O&P
8270	Low noise model	1 Elec	15	.533	Ea.	45.50	32		77.50	98
8280	Custom model	↓	12	.667	↓	120	40		160	192
8300	Bathroom or kitchen vent fan, 110 CFM									
8310	Economy model	1 Elec	15	.533	Ea.	67	32		99	122
8320	Low noise model	"	15	.533	"	93.50	32		125.50	151
8350	Paddle fan, variable speed (w/o lights)									
8360	Economy model (AC motor)	1 Elec	10	.800	Ea.	131	48		179	216
8362	With light kit		10	.800		171	48		219	260
8370	Custom model (AC motor)		10	.800		340	48		388	445
8372	With light kit		10	.800		380	48		428	490
8380	Luxury model (DC motor)		8	1		330	60		390	450
8382	With light kit		8	1		370	60		430	495
8390	Remote speed switch for above, add	↓	12	.667	↓	34	40		74	96.50
8500	Whole house exhaust fan, ceiling mount, 36", variable speed									
8510	Remote switch, incl. shutters, 20 amp-1 pole circ. bkr.									
8520	30' of #12/2, type NM cable	1 Elec	4	2	Ea.	1,325	120		1,445	1,650
8530	Type MC cable		3.50	2.286		1,350	137		1,487	1,675
8540	EMT & wire	↓	3	2.667	↓	1,375	160		1,535	1,750
8600	Whirlpool tub hook-up, incl. timer switch, outlet box									
8610	3' of flex, 20 amp-1 pole GFI circ. bkr.									
8620	30' of #12/2, type NM cable	1 Elec	5	1.600	Ea.	128	96		224	283
8630	Type MC cable		4.20	1.905		135	114		249	320
8640	EMT & wire	↓	3.40	2.353	↓	150	141		291	375
8650	Hot water heater hook-up, incl. 1-2 pole circ. bkr., box;									
8660	3' of flex, 20' of #10/2, type NM cable	1 Elec	5	1.600	Ea.	28.50	96		124.50	174
8670	Type MC cable		4.20	1.905		41.50	114		155.50	217
8680	EMT & wire	↓	3.40	2.353	↓	49.50	141		190.50	266
9000	Heating/air conditioning									
9050	Furnace/boiler hook-up, incl. firestat, local on-off switch									
9060	Emergency switch, and 40' of type NM cable	1 Elec	4	2	Ea.	53	120		173	237
9070	Type MC cable		3.50	2.286		68	137		205	280
9080	EMT & wire	↓	1.50	5.333	↓	92	320		412	580
9100	Air conditioner hook-up, incl. local 60 amp disc. switch									
9110	3' sealtite, 40 amp, 2 pole circuit breaker									
9130	40' of #8/2, type NM cable	1 Elec	3.50	2.286	Ea.	170	137		307	390
9140	Type MC cable		3	2.667		235	160		395	495
9150	EMT & wire	↓	1.30	6.154	↓	218	370		588	790
9200	Heat pump hook-up, 1-40 & 1-100 amp 2 pole circ. bkr.									
9210	Local disconnect switch, 3' sealtite									
9220	40' of #8/2 & 30' of #3/2									
9230	Type NM cable	1 Elec	1.30	6.154	Ea.	505	370		875	1,100
9240	Type MC cable		1.08	7.407		555	445		1,000	1,275
9250	EMT & wire	↓	.94	8.511	↓	520	510		1,030	1,350
9500	Thermostat hook-up, using low voltage wire									
9520	Heating only, 25' of #18-3	1 Elec	24	.333	Ea.	6.20	20		26.20	37
9530	Heating/cooling, 25' of #18-4	"	20	.400	"	8.15	24		32.15	45

26 09 13 – Electrical Power Monitoring

26 09 13.10 Switchboard Instruments

		Crew	Daily Output	Labor-Hours	Unit	Material	2019 Bare Costs Labor	2019 Bare Costs Equipment	Total	Total Incl O&P
0010	**SWITCHBOARD INSTRUMENTS** 3 phase, 4 wire R260913-80									
0100	AC indicating, ammeter & switch	1 Elec	8	1	Ea.	2,900	60		2,960	3,300
0200	Voltmeter & switch		8	1		2,925	60		2,985	3,300
0300	Wattmeter		8	1		4,125	60		4,185	4,650
0400	AC recording, ammeter		4	2		7,350	120		7,470	8,275
0500	Voltmeter		4	2		7,350	120		7,470	8,275
0600	Ground fault protection, zero sequence		2.70	2.963		6,500	178		6,678	7,425
0700	Ground return path		2.70	2.963		6,500	178		6,678	7,425
0800	3 current transformers, 5 to 800 amp		2	4		3,025	240		3,265	3,675
0900	1,000 to 1,500 amp		1.30	6.154		4,350	370		4,720	5,325
1200	2,000 to 4,000 amp		1	8		5,125	480		5,605	6,375
1300	Fused potential transformer, maximum 600 volt	▼	8	1	▼	1,150	60		1,210	1,350

26 09 13.20 Voltage Monitor Systems

		Crew	Daily Output	Labor-Hours	Unit	Material	2019 Bare Costs Labor	2019 Bare Costs Equipment	Total	Total Incl O&P
0010	**VOLTAGE MONITOR SYSTEMS** (test equipment)									
0100	AC voltage monitor system, 120/240 V, one-channel				Ea.	2,725			2,725	2,975
0110	Modem adapter					340			340	375
0120	Add-on detector only					1,425			1,425	1,575
0150	AC voltage remote monitor sys., 3 channel, 120, 230, or 480 V					4,925			4,925	5,425
0160	With internal modem					5,200			5,200	5,725
0170	Combination temperature and humidity probe					765			765	840
0180	Add-on detector only					3,575			3,575	3,925
0190	With internal modem				▼	3,900			3,900	4,275

26 09 13.30 Smart Metering

			Crew	Daily Output	Labor-Hours	Unit	Material	2019 Bare Costs Labor	2019 Bare Costs Equipment	Total	Total Incl O&P
0010	**SMART METERING**, In panel										
0100	Single phase, 120/208 volt, 100 amp	G	1 Elec	8.78	.911	Ea.	385	54.50		439.50	505
0120	200 amp	G		8.78	.911		390	54.50		444.50	505
0200	277 volt, 100 amp	G		8.78	.911		415	54.50		469.50	535
0220	200 amp	G		8.78	.911		400	54.50		454.50	520
1100	Three phase, 120/208 volt, 100 amp	G		4.69	1.706		645	102		747	865
1120	200 amp	G		4.69	1.706		850	102		952	1,100
1130	400 amp	G		4.69	1.706		850	102		952	1,100
1140	800 amp	G		4.69	1.706		765	102		867	995
1150	1,600 amp	G		4.69	1.706		775	102		877	1,000
1200	277/480 volt, 100 amp	G		4.69	1.706		915	102		1,017	1,150
1220	200 amp	G		4.69	1.706		795	102		897	1,025
1230	400 amp	G		4.69	1.706		820	102		922	1,050
1240	800 amp	G		4.69	1.706		825	102		927	1,075
1250	1,600 amp	G		4.69	1.706		830	102		932	1,075
2000	Data recorder, 8 meters	G		10.97	.729		1,450	44		1,494	1,675
2100	16 meters	G	▼	8.53	.938		3,600	56.50		3,656.50	4,050
3000	Software package, per meter, basic	G					257			257	282
3100	Premium	G				▼	680			680	745

26 09 23 – Lighting Control Devices

26 09 23.10 Energy Saving Lighting Devices

			Crew	Daily Output	Labor-Hours	Unit	Material	2019 Bare Costs Labor	2019 Bare Costs Equipment	Total	Total Incl O&P
0010	**ENERGY SAVING LIGHTING DEVICES**										
0100	Occupancy sensors, passive infrared ceiling mounted	G	1 Elec	7	1.143	Ea.	75	68.50		143.50	185
0110	Ultrasonic ceiling mounted	G		7	1.143		99.50	68.50		168	211
0120	Dual technology ceiling mounted	G		6.50	1.231		138	74		212	261
0150	Automatic wall switches	G		24	.333		74.50	20		94.50	112
0160	Daylighting sensor, manual control, ceiling mounted	G		7	1.143		158	68.50		226.50	275
0170	Remote and dimming control with remote controller	G		6.50	1.231		196	74		270	325
0200	Passive infrared ceiling mounted	G	▼	6.50	1.231		32.50	74		106.50	146

26 09 Instrumentation and Control for Electrical Systems

26 09 23 – Lighting Control Devices

26 09 23.10 Energy Saving Lighting Devices

		Crew	Daily Output	Labor-Hours	Unit	Material	2019 Bare Costs Labor	Equipment	Total	Total Incl O&P
0400	Remote power pack	G	1 Elec	10	.800	Ea.	32.50	48		80.50	108
0450	Photoelectric control, S.P.S.T. 120 V	G		8	1		21	60		81	113
0500	S.P.S.T. 208 V/277 V	G		8	1		26	60		86	118
0550	D.P.S.T. 120 V	G		6	1.333		208	80		288	350
0600	D.P.S.T. 208 V/277 V	G		6	1.333		227	80		307	370
0650	S.P.D.T. 208 V/277 V	G		6	1.333		229	80		309	370
0660	Daylight level sensor, wall mounted, on/off or dimming	G		8	1		197	60		257	305

26 09 26 – Lighting Control Panelboards

26 09 26.10 Lighting Control Relay Panel

		Crew	Daily Output	Labor-Hours	Unit	Material	2019 Bare Costs Labor	Equipment	Total	Total Incl O&P
0010	**LIGHTING CONTROL RELAY PANEL** with timeclock										
0100	4 Relay	G	1 Elec	2.50	3.200	Ea.	1,025	192		1,217	1,400
0110	8 Relay	G		2.30	3.478		1,375	209		1,584	1,825
0120	16 Relay	G		1.80	4.444		1,600	267		1,867	2,175
0130	24 Relay	G		1.50	5.333		2,300	320		2,620	3,000
0140	48 Relay	G		1	8		1,650	480		2,130	2,550
0200	Room Controller, switching only										
0210	1 Relay	G	1 Elec	3	2.667	Ea.	1,425	160		1,585	1,825
0220	2 Relay	G		3	2.667		1,350	160		1,510	1,750
0230	3 Relay	G		3	2.667		1,650	160		1,810	2,075
0240	Dimming										
0250	1 Relay	G	1 Elec	3	2.667	Ea.	1,275	160		1,435	1,650
0260	2 Relay	G		3	2.667		1,350	160		1,510	1,725
0270	3 Relay	G		3	2.667		1,650	160		1,810	2,075

26 09 36 – Modular Dimming Controls

26 09 36.13 Manual Modular Dimming Controls

		Crew	Daily Output	Labor-Hours	Unit	Material	2019 Bare Costs Labor	Equipment	Total	Total Incl O&P
0010	**MANUAL MODULAR DIMMING CONTROLS**										
2000	Lighting control module	G	1 Elec	2	4	Ea.	305	240		545	700

26 12 Medium-Voltage Transformers

26 12 19 – Pad-Mounted, Liquid-Filled, Medium-Voltage Transformers

26 12 19.10 Transformer, Oil-Filled

		Crew	Daily Output	Labor-Hours	Unit	Material	2019 Bare Costs Labor	Equipment	Total	Total Incl O&P
0010	**TRANSFORMER, OIL-FILLED** primary delta or Y, R262213-60										
0050	Pad mounted 5 kV or 15 kV, with taps, 277/480 V secondary, 3 phase										
0100	150 kVA		R-3	.65	30.769	Ea.	9,975	1,850	200	12,025	14,000
0110	225 kVA			.55	36.364		14,600	2,175	236	17,011	19,600
0200	300 kVA			.45	44.444		14,400	2,650	288	17,338	20,100
0300	500 kVA			.40	50		20,000	2,975	325	23,300	26,800
0400	750 kVA			.38	52.632		25,900	3,150	340	29,390	33,600
0500	1,000 kVA			.26	76.923		30,700	4,600	500	35,800	41,100
0600	1,500 kVA			.23	86.957		36,400	5,200	565	42,165	48,500
0700	2,000 kVA			.20	100		46,000	5,975	650	52,625	60,000
0710	2,500 kVA			.19	105		55,500	6,275	685	62,460	71,000
0720	3,000 kVA			.17	118		67,000	7,025	765	74,790	85,500
0800	3,750 kVA			.16	125		89,000	7,475	810	97,285	109,500
1990	Pole mounted distribution type, single phase										
2000	13.8 kV primary, 120/240 V secondary, 10 kVA		R-15	7.45	6.443	Ea.	1,150	380	38	1,568	1,850
2010	50 kVA			3.70	12.973		2,250	760	76.50	3,086.50	3,700
2020	100 kVA			2.75	17.455		3,950	1,025	103	5,078	6,000
2030	167 kVA			2.15	22.326		6,600	1,300	132	8,032	9,350
2900	2,400 V primary, 120/240 V secondary, 10 kVA			7.45	6.443		1,075	380	38	1,493	1,775

For customer support on your Electrical Costs with RSMeans data, call 800.448.8182.

209

26 12 Medium-Voltage Transformers

26 12 19 – Pad-Mounted, Liquid-Filled, Medium-Voltage Transformers

26 12 19.10 Transformer, Oil-Filled

		Crew	Daily Output	Labor-Hours	Unit	Material	2019 Bare Costs Labor	Equipment	Total	Total Incl O&P
2910	15 kVA	R-15	6.70	7.164	Ea.	1,275	420	42.50	1,737.50	2,075
2920	25 kVA		6	8		1,650	470	47	2,167	2,550
2930	37.5 kVA		4.30	11.163		2,025	655	66	2,746	3,275
2940	50 kVA		3.70	12.973		2,300	760	76.50	3,136.50	3,775
2950	75 kVA		3	16		3,475	940	94.50	4,509.50	5,325
2960	100 kVA		2.75	17.455		3,550	1,025	103	4,678	5,550

26 12 19.20 Transformer, Liquid-Filled

		Crew	Daily Output	Labor-Hours	Unit	Material	2019 Bare Costs Labor	Equipment	Total	Total Incl O&P
0010	**TRANSFORMER, LIQUID-FILLED** Pad mounted									
0020	5 kV or 15 kV primary, 277/480 volt secondary, 3 phase									
0050	225 kVA	R-3	.55	36.364	Ea.	14,800	2,175	236	17,211	19,700
0100	300 kVA		.45	44.444		17,500	2,650	288	20,438	23,600
0200	500 kVA		.40	50		22,200	2,975	325	25,500	29,200
0250	750 kVA		.38	52.632		28,600	3,150	340	32,090	36,600
0300	1,000 kVA		.26	76.923		33,200	4,600	500	38,300	44,000
0350	1,500 kVA		.23	86.957		38,800	5,200	565	44,565	51,000
0400	2,000 kVA		.20	100		48,000	5,975	650	54,625	62,500
0450	2,500 kVA		.19	105		57,000	6,275	685	63,960	73,000

26 13 Medium-Voltage Switchgear

26 13 16 – Medium-Voltage Fusible Interrupter Switchgear

26 13 16.10 Switchgear

		Crew	Daily Output	Labor-Hours	Unit	Material	2019 Bare Costs Labor	Equipment	Total	Total Incl O&P
0010	**SWITCHGEAR**, Incorporate switch with cable connections, transformer,									
0100	& low voltage section									
0200	Load interrupter switch, 600 amp, 2 position									
0300	NEMA 1, 4.8 kV, 300 kVA & below w/CLF fuses	R-3	.40	50	Ea.	21,900	2,975	325	25,200	28,900
0400	400 kVA & above w/CLF fuses		.38	52.632		24,200	3,150	340	27,690	31,700
0500	Non fusible		.41	48.780		18,300	2,925	315	21,540	24,900
0600	13.8 kV, 300 kVA & below w/CLF fuses		.38	52.632		29,400	3,150	340	32,890	37,400
0700	400 kVA & above w/CLF fuses		.36	55.556		29,400	3,325	360	33,085	37,600
0800	Non fusible		.40	50		21,100	2,975	325	24,400	28,000
0900	Cable lugs for 2 feeders 4.8 kV or 13.8 kV	1 Elec	8	1		705	60		765	865
1000	Pothead, one 3 conductor or three 1 conductor		4	2		3,400	120		3,520	3,900
1100	Two 3 conductor or six 1 conductor		2	4		6,700	240		6,940	7,700
1200	Key interlocks		8	1		785	60		845	950
1300	Lightning arresters, distribution class (no charge)									
1400	Intermediate class or line type 4.8 kV	1 Elec	2.70	2.963	Ea.	3,800	178		3,978	4,450
1500	13.8 kV		2	4		5,050	240		5,290	5,900
1600	Station class, 4.8 kV		2.70	2.963		6,525	178		6,703	7,450
1700	13.8 kV		2	4		11,200	240		11,440	12,700
1800	Transformers, 4,800 volts to 480/277 volts, 75 kVA	R-3	.68	29.412		19,800	1,750	191	21,741	24,600
1900	112.5 kVA		.65	30.769		24,200	1,850	200	26,250	29,600
2000	150 kVA		.57	35.088		27,600	2,100	228	29,928	33,700
2100	225 kVA		.48	41.667		31,700	2,500	270	34,470	38,800
2200	300 kVA		.41	48.780		35,400	2,925	315	38,640	43,600
2300	500 kVA		.36	55.556		46,500	3,325	360	50,185	56,500
2400	750 kVA		.29	68.966		53,000	4,125	450	57,575	64,500
2500	13,800 volts to 480/277 volts, 75 kVA		.61	32.787		27,900	1,950	213	30,063	33,900
2600	112.5 kVA		.55	36.364		37,100	2,175	236	39,511	44,300
2700	150 kVA		.49	40.816		37,400	2,450	265	40,115	45,100
2800	225 kVA		.41	48.780		43,200	2,925	315	46,440	52,000
2900	300 kVA		.37	54.054		44,100	3,225	350	47,675	53,500

26 13 Medium-Voltage Switchgear

26 13 16 – Medium-Voltage Fusible Interrupter Switchgear

26 13 16.10 Switchgear		Crew	Daily Output	Labor-Hours	Unit	Material	2019 Bare Costs Labor	Equipment	Total	Total Incl O&P
3000	500 kVA	R-3	.31	64.516	Ea.	48,800	3,850	420	53,070	59,500
3100	750 kVA		.26	76.923		53,500	4,600	500	58,600	66,500
3200	Forced air cooling & temperature alarm	1 Elec	1	8		4,325	480		4,805	5,475
3300	Low voltage components									
3400	Maximum panel height 49-1/2", single or twin row									
3500	Breaker heights, type FA or FH, 6"									
3600	type KA or KH, 8"									
3700	type LA, 11"									
3800	type MA, 14"									
3900	Breakers, 2 pole, 15 to 60 amp, type FA	1 Elec	5.60	1.429	Ea.	745	86		831	945
4000	70 to 100 amp, type FA		4.20	1.905		935	114		1,049	1,200
4100	15 to 60 amp, type FH		5.60	1.429		1,125	86		1,211	1,375
4200	70 to 100 amp, type FH		4.20	1.905		1,350	114		1,464	1,675
4300	125 to 225 amp, type KA		3.40	2.353		1,375	141		1,516	1,700
4400	125 to 225 amp, type KH		3.40	2.353		2,500	141		2,641	2,950
4500	125 to 400 amp, type LA		2.50	3.200		2,550	192		2,742	3,075
4600	125 to 600 amp, type MA		1.80	4.444		5,700	267		5,967	6,650
4700	700 & 800 amp, type MA		1.50	5.333		7,175	320		7,495	8,375
4800	3 pole, 15 to 60 amp, type FA		5.30	1.509		900	90.50		990.50	1,125
4900	70 to 100 amp, type FA		4	2		1,100	120		1,220	1,400
5000	15 to 60 amp, type FH		5.30	1.509		1,475	90.50		1,565.50	1,750
5100	70 to 100 amp, type FH		4	2		1,700	120		1,820	2,025
5200	125 to 225 amp, type KA		3.20	2.500		2,125	150		2,275	2,550
5300	125 to 225 amp, type KH		3.20	2.500		3,850	150		4,000	4,450
5400	125 to 400 amp, type LA		2.30	3.478		3,125	209		3,334	3,725
5500	125 to 600 amp, type MA		1.60	5		5,225	300		5,525	6,200
5600	700 & 800 amp, type MA		1.30	6.154		6,725	370		7,095	7,925

26 22 Low-Voltage Transformers

26 22 13 – Low-Voltage Distribution Transformers

26 22 13.10 Transformer, Dry-Type

		Crew	Daily Output	Labor-Hours	Unit	Material	2019 Bare Costs Labor	Equipment	Total	Total Incl O&P
0010	**TRANSFORMER, DRY-TYPE** R262213-10									
0050	Single phase, 240/480 volt primary, 120/240 volt secondary									
0100	1 kVA R262213-60	1 Elec	2	4	Ea.	395	240		635	795
0300	2 kVA		1.60	5		600	300		900	1,100
0500	3 kVA		1.40	5.714		750	345		1,095	1,325
0700	5 kVA		1.20	6.667		1,225	400		1,625	1,950
0900	7.5 kVA	2 Elec	2.20	7.273		1,400	435		1,835	2,175
1100	10 kVA		1.60	10		1,725	600		2,325	2,800
1300	15 kVA		1.20	13.333		2,125	800		2,925	3,525
1500	25 kVA		1	16		2,600	960		3,560	4,275
1700	37.5 kVA		.80	20		2,825	1,200		4,025	4,900
1900	50 kVA		.70	22.857		3,425	1,375		4,800	5,800
2100	75 kVA		.65	24.615		4,425	1,475		5,900	7,075
2110	100 kVA	R-3	.90	22.222		5,775	1,325	144	7,244	8,475
2120	167 kVA	"	.80	25		9,575	1,500	162	11,237	12,900
2190	480 V primary, 120/240 V secondary, nonvent., 15 kVA	2 Elec	1.20	13.333		1,625	800		2,425	3,000
2200	25 kVA		.90	17.778		2,400	1,075		3,475	4,250
2210	37 kVA		.75	21.333		2,850	1,275		4,125	5,050
2220	50 kVA		.65	24.615		3,375	1,475		4,850	5,925
2230	75 kVA		.60	26.667		4,475	1,600		6,075	7,325

For customer support on your Electrical Costs with RSMeans data, call 800.448.8182.

211

26 22 13.10 Transformer, Dry-Type		Crew	Daily Output	Labor-Hours	Unit	Material	2019 Bare Costs Labor	Equipment	Total	Total Incl O&P
2240	100 kVA	2 Elec	.50	32	Ea.	5,825	1,925		7,750	9,300
2250	Low operating temperature (80°C), 25 kVA		1	16		4,375	960		5,335	6,225
2260	37 kVA		.80	20		4,700	1,200		5,900	6,975
2270	50 kVA		.70	22.857		6,125	1,375		7,500	8,800
2280	75 kVA		.65	24.615		9,750	1,475		11,225	12,900
2290	100 kVA	↓	.55	29.091	↓	10,200	1,750		11,950	13,800
2300	3 phase, 480 volt primary, 120/208 volt secondary									
2310	Ventilated, 3 kVA	1 Elec	1	8	Ea.	1,125	480		1,605	1,975
2700	6 kVA		.80	10		1,275	600		1,875	2,300
2900	9 kVA	↓	.70	11.429		1,275	685		1,960	2,425
3100	15 kVA	2 Elec	1.10	14.545		1,875	875		2,750	3,375
3300	30 kVA		.90	17.778		1,800	1,075		2,875	3,575
3500	45 kVA		.80	20		1,925	1,200		3,125	3,900
3700	75 kVA	↓	.70	22.857		2,575	1,375		3,950	4,875
3900	112.5 kVA	R-3	.90	22.222		4,225	1,325	144	5,694	6,775
4100	150 kVA		.85	23.529		5,150	1,400	153	6,703	7,950
4300	225 kVA		.65	30.769		9,000	1,850	200	11,050	12,900
4500	300 kVA		.55	36.364		9,350	2,175	236	11,761	13,800
4700	500 kVA		.45	44.444		18,800	2,650	288	21,738	25,000
4800	750 kVA		.35	57.143		22,800	3,425	370	26,595	30,600
4820	1,000 kVA	↓	.32	62.500		26,800	3,725	405	30,930	35,500
4850	K-4 rated, 15 kVA	2 Elec	1.10	14.545		3,525	875		4,400	5,200
4855	30 kVA		.90	17.778		5,725	1,075		6,800	7,900
4860	45 kVA		.80	20		6,950	1,200		8,150	9,450
4865	75 kVA	↓	.70	22.857		10,000	1,375		11,375	13,100
4870	112.5 kVA	R-3	.90	22.222		12,700	1,325	144	14,169	16,100
4875	150 kVA		.85	23.529		16,700	1,400	153	18,253	20,700
4880	225 kVA		.65	30.769		23,400	1,850	200	25,450	28,700
4885	300 kVA		.55	36.364		30,700	2,175	236	33,111	37,300
4890	500 kVA	↓	.45	44.444		42,900	2,650	288	45,838	51,500
4900	K-13 rated, 15 kVA	2 Elec	1.10	14.545		4,025	875		4,900	5,725
4905	30 kVA		.90	17.778		6,000	1,075		7,075	8,200
4910	45 kVA		.80	20		7,225	1,200		8,425	9,750
4915	75 kVA	↓	.70	22.857		11,200	1,375		12,575	14,400
4920	112.5 kVA	R-3	.90	22.222		14,400	1,325	144	15,869	17,900
4925	150 kVA		.85	23.529		24,000	1,400	153	25,553	28,700
4930	225 kVA		.65	30.769		26,100	1,850	200	28,150	31,700
4935	300 kVA		.55	36.364		34,100	2,175	236	36,511	41,000
4940	500 kVA	↓	.45	44.444	↓	57,500	2,650	288	60,438	68,000
5020	480 volt primary, 120/208 volt secondary									
5030	Nonventilated, 15 kVA	2 Elec	1.10	14.545	Ea.	3,275	875		4,150	4,900
5040	30 kVA		.80	20		4,775	1,200		5,975	7,050
5050	45 kVA		.70	22.857		5,950	1,375		7,325	8,600
5060	75 kVA	↓	.65	24.615		10,100	1,475		11,575	13,400
5070	112.5 kVA	R-3	.85	23.529		19,900	1,400	153	21,453	24,100
5081	150 kVA		.85	23.529		25,200	1,400	153	26,753	30,000
5090	225 kVA		.60	33.333		22,100	2,000	216	24,316	27,500
5100	300 kVA		.50	40		21,500	2,400	260	24,160	27,500
5200	Low operating temperature (80°C), 30 kVA	2 Elec	.90	17.778		6,200	1,075		7,275	8,400
5210	45 kVA		.80	20		6,175	1,200		7,375	8,600
5220	75 kVA	↓	.70	22.857		8,975	1,375		10,350	11,900
5230	112.5 kVA	R-3	.90	22.222		11,700	1,325	144	13,169	15,000
5240	150 kVA	↓	.85	23.529		15,000	1,400	153	16,553	18,800

26 22 13 – Low-Voltage Distribution Transformers

26 22 13.10 Transformer, Dry-Type

		Crew	Daily Output	Labor-Hours	Unit	Material	2019 Bare Costs Labor	Equipment	Total	Total Incl O&P
5250	225 kVA	R-3	.65	30.769	Ea.	20,400	1,850	200	22,450	25,500
5260	300 kVA		.55	36.364		29,100	2,175	236	31,511	35,500
5270	500 kVA		.45	44.444		36,800	2,650	288	39,738	44,800
5380	3 phase, 5 kV primary, 277/480 volt secondary									
5400	High voltage, 112.5 kVA	R-3	.85	23.529	Ea.	17,900	1,400	153	19,453	22,000
5410	150 kVA		.65	30.769		20,900	1,850	200	22,950	26,000
5420	225 kVA		.55	36.364		24,600	2,175	236	27,011	30,600
5430	300 kVA		.45	44.444		31,500	2,650	288	34,438	38,900
5440	500 kVA		.35	57.143		40,600	3,425	370	44,395	50,000
5450	750 kVA		.32	62.500		61,000	3,725	405	65,130	73,000
5460	1,000 kVA		.30	66.667		71,000	3,975	435	75,410	85,000
5470	1,500 kVA		.27	74.074		83,000	4,425	480	87,905	98,000
5480	2,000 kVA		.25	80		97,500	4,775	520	102,795	114,500
5490	2,500 kVA		.20	100		109,500	5,975	650	116,125	130,000
5500	3,000 kVA		.18	111		143,500	6,625	720	150,845	168,500
5590	15 kV primary, 277/480 volt secondary									
5600	High voltage, 112.5 kVA	R-3	.85	23.529	Ea.	29,200	1,400	153	30,753	34,400
5610	150 kVA		.65	30.769		34,200	1,850	200	36,250	40,600
5620	225 kVA		.55	36.364		37,700	2,175	236	40,111	45,000
5630	300 kVA		.45	44.444		44,500	2,650	288	47,438	53,000
5640	500 kVA		.35	57.143		53,500	3,425	370	57,295	64,500
5650	750 kVA		.32	62.500		70,500	3,725	405	74,630	83,500
5660	1,000 kVA		.30	66.667		80,500	3,975	435	84,910	95,000
5670	1,500 kVA		.27	74.074		92,500	4,425	480	97,405	109,000
5680	2,000 kVA		.25	80		103,000	4,775	520	108,295	120,500
5690	2,500 kVA		.20	100		119,000	5,975	650	125,625	140,500
5700	3,000 kVA		.18	111		142,000	6,625	720	149,345	166,500
6000	2400 volt primary, 480 volt secondary, 300 kVA		.45	44.444		31,500	2,650	288	34,438	38,900
6010	500 kVA		.35	57.143		40,600	3,425	370	44,395	50,000
6020	750 kVA		.32	62.500		56,000	3,725	405	60,130	67,500
9300	Energy efficient transformer 3 ph									
9400	15 kVA, 480VAC delta, 208Y/120VAC	R-3	1.50	13.333	Ea.	1,975	795	86.50	2,856.50	3,475
9405	30 kVA, 480VAC delta, 208Y/120VAC		1.50	13.333		2,525	795	86.50	3,406.50	4,075
9410	45 kVA, 480VAC delta, 208Y/120VAC		1.50	13.333		3,050	795	86.50	3,931.50	4,650
9415	75 kVA, 480VAC delta, 208Y/120VAC		1.25	16		4,500	955	104	5,559	6,500
9420	112 kVA, 480VAC delta, 208Y/120VAC		1.25	16		6,400	955	104	7,459	8,600
9425	150 kVA, 480VAC delta, 208Y/120VAC		1.25	16		15,400	955	104	16,459	18,400
9430	225 kVA, 480VAC delta, 208Y/120VAC		1.25	16		10,700	955	104	11,759	13,200
9435	300 kVA, 480VAC delta, 208Y/120VAC		1.25	16		17,500	955	104	18,559	20,800

26 22 13.20 Isolating Panels

		Crew	Daily Output	Labor-Hours	Unit	Material	2019 Bare Costs Labor	Equipment	Total	Total Incl O&P
0010	**ISOLATING PANELS** used with isolating transformers									
0020	For hospital applications									
0100	Critical care area, 8 circuit, 3 kVA	1 Elec	.58	13.793	Ea.	6,800	830		7,630	8,700
0200	5 kVA		.54	14.815		6,950	890		7,840	8,975
0400	7.5 kVA		.52	15.385		6,075	925		7,000	8,050
0600	10 kVA		.44	18.182		8,075	1,100		9,175	10,500
0800	Operating room power & lighting, 8 circuit, 3 kVA		.58	13.793		4,775	830		5,605	6,475
1000	5 kVA		.54	14.815		5,250	890		6,140	7,100
1200	7.5 kVA		.52	15.385		4,025	925		4,950	5,800
1400	10 kVA		.44	18.182		5,200	1,100		6,300	7,325
1600	X-ray systems, 15 kVA, 90 amp		.44	18.182		14,800	1,100		15,900	17,900
1800	25 kVA, 125 amp		.36	22.222		15,400	1,325		16,725	18,900

For customer support on your Electrical Costs with RSMeans data, call 800.448.8182.

213

26 22 Low-Voltage Transformers

26 22 13 – Low-Voltage Distribution Transformers

26 22 13.30 Isolating Transformer

		Crew	Daily Output	Labor-Hours	Unit	Material	2019 Bare Costs Labor	Equipment	Total	Total Incl O&P
0010	**ISOLATING TRANSFORMER**									
0100	Single phase, 120/240 volt primary, 120/240 volt secondary									
0200	0.50 kVA R262213-60	1 Elec	4	2	Ea.	395	120		515	615
0400	1 kVA		2	4		935	240		1,175	1,375
0600	2 kVA		1.60	5		1,500	300		1,800	2,100
0800	3 kVA		1.40	5.714		830	345		1,175	1,425
1000	5 kVA		1.20	6.667		1,100	400		1,500	1,800
1200	7.5 kVA		1.10	7.273		1,375	435		1,810	2,175
1400	10 kVA		.80	10		1,775	600		2,375	2,850
1600	15 kVA		.60	13.333		2,250	800		3,050	3,675
1800	25 kVA		.50	16		3,250	960		4,210	5,000
1810	37.5 kVA	2 Elec	.80	20		5,375	1,200		6,575	7,700
1820	75 kVA	"	.65	24.615		8,075	1,475		9,550	11,100
1830	3 phase, 120/240 V primary, 120/240 V secondary, 112.5 kVA	R-3	.90	22.222		10,600	1,325	144	12,069	13,700
1840	150 kVA		.85	23.529		13,400	1,400	153	14,953	17,000
1850	225 kVA		.65	30.769		18,700	1,850	200	20,750	23,500
1860	300 kVA		.55	36.364		24,800	2,175	236	27,211	30,800
1870	500 kVA		.45	44.444		41,400	2,650	288	44,338	49,800
1880	750 kVA		.35	57.143		41,800	3,425	370	45,595	51,500

26 22 13.90 Transformer Handling

		Crew	Daily Output	Labor-Hours	Unit	Material	2019 Bare Costs Labor	Equipment	Total	Total Incl O&P
0010	**TRANSFORMER HANDLING** Add to normal labor cost in restricted areas									
5000	Transformers									
5150	15 kVA, approximately 200 pounds	2 Elec	2.70	5.926	Ea.		355		355	530
5160	25 kVA, approximately 300 pounds		2.50	6.400			385		385	575
5170	37.5 kVA, approximately 400 pounds		2.30	6.957			420		420	625
5180	50 kVA, approximately 500 pounds		2	8			480		480	715
5190	75 kVA, approximately 600 pounds		1.80	8.889			535		535	795
5200	100 kVA, approximately 700 pounds		1.60	10			600		600	895
5210	112.5 kVA, approximately 800 pounds	3 Elec	2.20	10.909			655		655	975
5220	125 kVA, approximately 900 pounds		2	12			720		720	1,075
5230	150 kVA, approximately 1,000 pounds		1.80	13.333			800		800	1,200
5240	167 kVA, approximately 1,200 pounds		1.60	15			900		900	1,350
5250	200 kVA, approximately 1,400 pounds		1.40	17.143			1,025		1,025	1,525
5260	225 kVA, approximately 1,600 pounds		1.30	18.462			1,100		1,100	1,650
5270	250 kVA, approximately 1,800 pounds		1.10	21.818			1,300		1,300	1,950
5280	300 kVA, approximately 2,000 pounds		1	24			1,450		1,450	2,150
5290	500 kVA, approximately 3,000 pounds		.75	32			1,925		1,925	2,875
5300	600 kVA, approximately 3,500 pounds		.67	35.821			2,150		2,150	3,200
5310	750 kVA, approximately 4,000 pounds		.60	40			2,400		2,400	3,575
5320	1,000 kVA, approximately 5,000 pounds		.50	48			2,875		2,875	4,300

26 22 16 – Low-Voltage Buck-Boost Transformers

26 22 16.10 Buck-Boost Transformer

		Crew	Daily Output	Labor-Hours	Unit	Material	2019 Bare Costs Labor	Equipment	Total	Total Incl O&P
0010	**BUCK-BOOST TRANSFORMER** R262213-60									
0100	Single phase, 120/240 V primary, 12/24 V secondary									
0200	0.10 kVA	1 Elec	8	1	Ea.	105	60		165	205
0400	0.25 kVA		5.70	1.404		147	84.50		231.50	288
0600	0.50 kVA		4	2		200	120		320	400
0800	0.75 kVA		3.10	2.581		262	155		417	520
1000	1.0 kVA		2	4		277	240		517	665
1200	1.5 kVA		1.80	4.444		370	267		637	805
1400	2.0 kVA		1.60	5		510	300		810	1,000

26 22 Low-Voltage Transformers

26 22 16 – Low-Voltage Buck-Boost Transformers

26 22 16.10 Buck-Boost Transformer	Crew	Daily Output	Labor-Hours	Unit	Material	2019 Bare Costs Labor	Equipment	Total	Total Incl O&P
1600	3.0 kVA	1 Elec	1.40	5.714	Ea.	595	345		940	1,150
1800	5.0 kVA		1.20	6.667		1,100	400		1,500	1,825
2000	3 phase, 240 V primary, 208/120 V secondary, 15 kVA	2 Elec	2.40	6.667		2,275	400		2,675	3,100
2200	30 kVA		1.60	10		2,550	600		3,150	3,700
2400	45 kVA		1.40	11.429		3,050	685		3,735	4,400
2600	75 kVA		1.20	13.333		3,700	800		4,500	5,275
2800	112.5 kVA	R-3	1.40	14.286		4,600	855	92.50	5,547.50	6,450
3000	150 kVA		1.10	18.182		6,125	1,075	118	7,318	8,500
3200	225 kVA		1	20		7,975	1,200	130	9,305	10,700
3400	300 kVA		.90	22.222		10,800	1,325	144	12,269	14,000

26 24 Switchboards and Panelboards

26 24 13 – Switchboards

26 24 13.10 Incoming Switchboards

26 24 13.10 Incoming Switchboards		Crew	Daily Output	Labor-Hours	Unit	Material	2019 Bare Costs Labor	Equipment	Total	Total Incl O&P
0010	**INCOMING SWITCHBOARDS** main service section	R262419-84								
0100	Aluminum bus bars, not including CT's or PT's									
0200	No main disconnect, includes CT compartment									
0300	120/208 volt, 4 wire, 600 amp	2 Elec	1	16	Ea.	3,950	960		4,910	5,775
0400	800 amp		.88	18.182		3,950	1,100		5,050	5,975
0500	1,000 amp		.80	20		4,750	1,200		5,950	7,025
0600	1,200 amp		.72	22.222		4,750	1,325		6,075	7,225
0700	1,600 amp		.66	24.242		4,750	1,450		6,200	7,400
0800	2,000 amp		.62	25.806		5,125	1,550		6,675	7,925
1000	3,000 amp		.56	28.571		6,750	1,725		8,475	10,000
1200	277/480 volt, 4 wire, 600 amp		1	16		3,950	960		4,910	5,775
1300	800 amp		.88	18.182		3,950	1,100		5,050	5,975
1400	1,000 amp		.80	20		5,025	1,200		6,225	7,325
1500	1,200 amp		.72	22.222		5,025	1,325		6,350	7,525
1600	1,600 amp		.66	24.242		5,025	1,450		6,475	7,700
1700	2,000 amp		.62	25.806		5,125	1,550		6,675	7,925
1800	3,000 amp		.56	28.571		6,750	1,725		8,475	10,000
1900	4,000 amp		.52	30.769		8,375	1,850		10,225	12,000
2000	Fused switch & CT compartment									
2100	120/208 volt, 4 wire, 400 amp	2 Elec	1.12	14.286	Ea.	2,650	860		3,510	4,175
2200	600 amp		.94	17.021		3,125	1,025		4,150	4,975
2300	800 amp		.84	19.048		10,700	1,150		11,850	13,500
2400	1,200 amp		.68	23.529		13,900	1,425		15,325	17,400
2500	277/480 volt, 4 wire, 400 amp		1.14	14.035		3,525	845		4,370	5,125
2600	600 amp		.94	17.021		5,950	1,025		6,975	8,075
2700	800 amp		.84	19.048		10,700	1,150		11,850	13,500
2800	1,200 amp		.68	23.529		13,900	1,425		15,325	17,400
2900	Pressure switch & CT compartment									
3000	120/208 volt, 4 wire, 800 amp	2 Elec	.80	20	Ea.	9,625	1,200		10,825	12,400
3100	1,200 amp		.66	24.242		18,600	1,450		20,050	22,700
3200	1,600 amp		.62	25.806		19,800	1,550		21,350	24,100
3300	2,000 amp		.56	28.571		21,000	1,725		22,725	25,700
3310	2,500 amp		.50	32		25,900	1,925		27,825	31,400
3320	3,000 amp		.44	36.364		34,900	2,175		37,075	41,700
3330	4,000 amp		.40	40		44,800	2,400		47,200	53,000
3340	120/208 volt, 4 wire, 800 amp, with ground fault		.80	20		15,900	1,200		17,100	19,300
3350	1,200 amp, with ground fault		.66	24.242		20,500	1,450		21,950	24,800

26 24 13 – Switchboards

26 24 13.10 Incoming Switchboards

		Crew	Daily Output	Labor-Hours	Unit	Material	2019 Bare Costs Labor	Equipment	Total	Total Incl O&P
3360	1,600 amp, with ground fault	2 Elec	.62	25.806	Ea.	22,300	1,550		23,850	26,800
3370	2,000 amp, with ground fault		.56	28.571		24,100	1,725		25,825	29,100
3400	277/480 volt, 4 wire, 800 amp, with ground fault		.80	20		15,900	1,200		17,100	19,300
3600	1,200 amp, with ground fault		.66	24.242		20,500	1,450		21,950	24,800
4000	1,600 amp, with ground fault		.62	25.806		22,300	1,550		23,850	26,800
4200	2,000 amp, with ground fault		.56	28.571		24,100	1,725		25,825	29,100
4400	Circuit breaker, molded case & CT compartment									
4600	3 pole, 4 wire, 600 amp	2 Elec	.94	17.021	Ea.	8,275	1,025		9,300	10,700
4800	800 amp		.84	19.048		9,925	1,150		11,075	12,600
5000	1,200 amp		.68	23.529		13,500	1,425		14,925	17,000
5100	Copper bus bars, not incl. CT's or PT's, add, minimum					15%				

26 24 13.20 In Plant Distribution Switchboards

		Crew	Daily Output	Labor-Hours	Unit	Material	2019 Bare Costs Labor	Equipment	Total	Total Incl O&P
0010	**IN PLANT DISTRIBUTION SWITCHBOARDS**									
0100	Main lugs only, to 600 volt, 3 pole, 3 wire, 200 amp	2 Elec	1.20	13.333	Ea.	1,250	800		2,050	2,575
0110	400 amp		1.20	13.333		1,250	800		2,050	2,550
0120	600 amp		1.20	13.333		1,275	800		2,075	2,600
0130	800 amp		1.08	14.815		1,400	890		2,290	2,850
0140	1,200 amp		.92	17.391		1,675	1,050		2,725	3,400
0150	1,600 amp		.86	18.605		2,225	1,125		3,350	4,100
0160	2,000 amp		.82	19.512		2,450	1,175		3,625	4,450
0250	To 480 volt, 3 pole, 4 wire, 200 amp		1.20	13.333		1,075	800		1,875	2,400
0260	400 amp		1.20	13.333		1,250	800		2,050	2,575
0270	600 amp		1.20	13.333		1,350	800		2,150	2,700
0280	800 amp		1.08	14.815		1,475	890		2,365	2,950
0290	1,200 amp		.92	17.391		1,825	1,050		2,875	3,550
0300	1,600 amp		.86	18.605		2,075	1,125		3,200	3,975
0310	2,000 amp		.82	19.512		2,425	1,175		3,600	4,425
0400	Main circuit breaker, to 600 volt, 3 pole, 3 wire, 200 amp		1.20	13.333		2,950	800		3,750	4,450
0410	400 amp		1.14	14.035		2,975	845		3,820	4,525
0420	600 amp		1.10	14.545		3,750	875		4,625	5,425
0430	800 amp		1.04	15.385		6,275	925		7,200	8,275
0440	1,200 amp		.88	18.182		8,225	1,100		9,325	10,700
0450	1,600 amp		.84	19.048		13,200	1,150		14,350	16,200
0460	2,000 amp		.80	20		14,100	1,200		15,300	17,300
0550	277/480 volt, 3 pole, 4 wire, 200 amp		1.20	13.333		3,125	800		3,925	4,625
0560	400 amp		1.14	14.035		3,125	845		3,970	4,675
0570	600 amp		1.10	14.545		3,900	875		4,775	5,600
0580	800 amp		1.04	15.385		6,575	925		7,500	8,600
0590	1,200 amp		.88	18.182		8,500	1,100		9,600	11,000
0600	1,600 amp		.84	19.048		13,200	1,150		14,350	16,200
0610	2,000 amp		.80	20		14,100	1,200		15,300	17,300
0700	Main fusible switch w/fuse, 208/240 volt, 3 pole, 3 wire, 200 amp		1.20	13.333		3,275	800		4,075	4,800
0710	400 amp		1.14	14.035		3,275	845		4,120	4,850
0720	600 amp		1.10	14.545		3,975	875		4,850	5,675
0730	800 amp		1.04	15.385		8,125	925		9,050	10,300
0740	1,200 amp		.88	18.182		9,450	1,100		10,550	12,000
0800	120/208, 120/240 volt, 3 pole, 4 wire, 200 amp		1.20	13.333		2,900	800		3,700	4,400
0810	400 amp		1.14	14.035		2,900	845		3,745	4,450
0820	600 amp		1.10	14.545		3,800	875		4,675	5,475
0830	800 amp		1.04	15.385		6,025	925		6,950	8,000
0840	1,200 amp		.88	18.182		7,000	1,100		8,100	9,325
0900	480 or 600 volt, 3 pole, 3 wire, 200 amp		1.20	13.333		3,150	800		3,950	4,650

For customer support on your Electrical Costs with RSMeans data, call 800.448.8182.

26 24 Switchboards and Panelboards

26 24 13 – Switchboards

26 24 13.20 In Plant Distribution Switchboards	Crew	Daily Output	Labor-Hours	Unit	Material	2019 Bare Costs Labor	Equipment	Total	Total Incl O&P
0910	400 amp	2 Elec	1.14	14.035	Ea.	3,150	845		3,995	4,725
0920	600 amp		1.10	14.545		3,750	875		4,625	5,425
0930	800 amp		1.04	15.385		5,800	925		6,725	7,750
0940	1,200 amp		.88	18.182		6,725	1,100		7,825	9,025
1000	277 or 480 volt, 3 pole, 4 wire, 200 amp		1.20	13.333		3,250	800		4,050	4,775
1010	400 amp		1.14	14.035		3,275	845		4,120	4,850
1020	600 amp		1.10	14.545		3,925	875		4,800	5,625
1030	800 amp		1.04	15.385		6,025	925		6,950	8,025
1040	1,200 amp		.88	18.182		7,000	1,100		8,100	9,325
1120	1,600 amp		.76	21.053		12,800	1,275		14,075	16,000
1130	2,000 amp		.68	23.529		16,800	1,425		18,225	20,600
1150	Pressure switch, bolted, 3 pole, 208/240 volt, 3 wire, 800 amp		.96	16.667		9,475	1,000		10,475	11,900
1160	1,200 amp		.80	20		12,100	1,200		13,300	15,100
1170	1,600 amp		.76	21.053		13,800	1,275		15,075	17,100
1180	2,000 amp		.68	23.529		15,700	1,425		17,125	19,400
1200	120/208 or 120/240 volt, 3 pole, 4 wire, 800 amp		.96	16.667		7,475	1,000		8,475	9,725
1210	1,200 amp		.80	20		8,750	1,200		9,950	11,400
1220	1,600 amp		.76	21.053		13,800	1,275		15,075	17,100
1230	2,000 amp		.68	23.529		15,700	1,425		17,125	19,400
1300	480 or 600 volt, 3 wire, 800 amp		.96	16.667		9,475	1,000		10,475	11,900
1310	1,200 amp		.80	20		13,300	1,200		14,500	16,400
1320	1,600 amp		.76	21.053		14,900	1,275		16,175	18,300
1330	2,000 amp		.68	23.529		16,800	1,425		18,225	20,600
1400	277/480 volt, 4 wire, 800 amp		.96	16.667		9,475	1,000		10,475	11,900
1410	1,200 amp		.80	20		13,300	1,200		14,500	16,400
1420	1,600 amp		.76	21.053		14,900	1,275		16,175	18,300
1430	2,000 amp		.68	23.529		16,800	1,425		18,225	20,600
1500	Main ground fault protector, 1,200-2,000 amp		5.40	2.963		3,150	178		3,328	3,750
1600	Busway connection, 200 amp		5.40	2.963		480	178		658	790
1610	400 amp		4.60	3.478		480	209		689	835
1620	600 amp		4	4		480	240		720	885
1630	800 amp		3.20	5		480	300		780	975
1640	1,200 amp		2.60	6.154		480	370		850	1,075
1650	1,600 amp		2.40	6.667		990	400		1,390	1,675
1660	2,000 amp		2	8		990	480		1,470	1,800
1700	Shunt trip for remote operation, 200 amp		8	2		625	120		745	870
1710	400 amp		8	2		1,000	120		1,120	1,275
1720	600 amp		8	2		1,175	120		1,295	1,475
1730	800 amp		8	2		1,525	120		1,645	1,850
1740	1,200-2,000 amp		8	2		3,225	120		3,345	3,725
1800	Motor operated main breaker, 200 amp		8	2		3,125	120		3,245	3,600
1810	400 amp		8	2		3,125	120		3,245	3,600
1820	600 amp		8	2		3,200	120		3,320	3,675
1830	800 amp		8	2		3,250	120		3,370	3,725
1840	1,200-2,000 amp		8	2		3,275	120		3,395	3,775
1900	Current/potential transformer metering compartment, 200-800 amp		5.40	2.963		2,125	178		2,303	2,600
1940	1,200 amp		5.40	2.963		5,825	178		6,003	6,700
1950	1,600-2,000 amp		5.40	2.963		9,425	178		9,603	10,700
2000	With watt meter, 200-800 amp		4	4		8,000	240		8,240	9,150
2040	1,200 amp		4	4		9,950	240		10,190	11,300
2050	1,600-2,000 amp		4	4		9,875	240		10,115	11,300
2100	Split bus, 60-200 amp	1 Elec	5.30	1.509		189	90.50		279.50	345
2130	400 amp	2 Elec	4.60	3.478		330	209		539	670

26 24 13 – Switchboards

26 24 13.20 In Plant Distribution Switchboards

		Crew	Daily Output	Labor-Hours	Unit	Material	2019 Bare Costs Labor	Equipment	Total	Total Incl O&P
2140	600 amp	2 Elec	3.60	4.444	Ea.	400	267		667	840
2150	800 amp		2.60	6.154		510	370		880	1,100
2170	1,200 amp		2	8		580	480		1,060	1,350
2250	Contactor control, 60 amp	1 Elec	2	4		1,250	240		1,490	1,725
2260	100 amp		1.50	5.333		1,400	320		1,720	2,000
2270	200 amp		1	8		2,100	480		2,580	3,025
2280	400 amp	2 Elec	1	16		6,375	960		7,335	8,450
2290	600 amp		.84	19.048		7,150	1,150		8,300	9,575
2300	800 amp		.72	22.222		8,475	1,325		9,800	11,300
2500	Modifier, two distribution sections, add		.80	20		2,875	1,200		4,075	4,950
2520	Three distribution sections, add		.40	40		5,550	2,400		7,950	9,675
2560	Auxiliary pull section, 20", add		2	8		1,050	480		1,530	1,900
2580	24", add		1.80	8.889		1,050	535		1,585	1,975
2600	30", add		1.60	10		1,050	600		1,650	2,075
2620	36", add		1.40	11.429		1,275	685		1,960	2,425
2640	Dog house, 12", add		2.40	6.667		218	400		618	835
2660	18", add		2	8		435	480		915	1,200
3000	Transition section between switchboard and transformer									
3050	or motor control center, 4 wire alum. bus, 600 amp	2 Elec	1.14	14.035	Ea.	2,175	845		3,020	3,625
3100	800 amp		1	16		2,425	960		3,385	4,100
3150	1,000 amp		.88	18.182		2,725	1,100		3,825	4,625
3200	1,200 amp		.80	20		3,000	1,200		4,200	5,100
3250	1,600 amp		.72	22.222		3,550	1,325		4,875	5,900
3300	2,000 amp		.66	24.242		4,075	1,450		5,525	6,650
3350	2,500 amp		.62	25.806		4,750	1,550		6,300	7,525
3400	3,000 amp		.56	28.571		5,475	1,725		7,200	8,575
4000	Weatherproof construction, per vertical section		1.76	9.091		2,475	545		3,020	3,550

26 24 13.30 Distribution Switchboards Section

		Crew	Daily Output	Labor-Hours	Unit	Material	2019 Bare Costs Labor	Equipment	Total	Total Incl O&P
0010	**DISTRIBUTION SWITCHBOARDS SECTION** R262419-80									
0100	Aluminum bus bars, not including breakers									
0160	Subfeed lug-rated at 60 amp	2 Elec	1.30	12.308	Ea.	1,175	740		1,915	2,375
0170	100 amp		1.26	12.698		1,525	765		2,290	2,850
0180	200 amp		1.20	13.333		1,250	800		2,050	2,600
0190	400 amp		1.10	14.545		1,425	875		2,300	2,850
0195	120/208 or 277/480 volt, 4 wire, 400 amp		1.10	14.545		1,425	875		2,300	2,850
0200	600 amp		1	16		1,500	960		2,460	3,075
0300	800 amp		.88	18.182		1,900	1,100		3,000	3,725
0400	1,000 amp		.80	20		2,500	1,200		3,700	4,550
0500	1,200 amp		.72	22.222		2,950	1,325		4,275	5,250
0600	1,600 amp		.66	24.242		3,975	1,450		5,425	6,525
0700	2,000 amp		.62	25.806		4,825	1,550		6,375	7,625
0800	2,500 amp		.60	26.667		5,875	1,600		7,475	8,875
0900	3,000 amp		.56	28.571		6,850	1,725		8,575	10,100
0950	4,000 amp		.52	30.769		7,875	1,850		9,725	11,400

26 24 13.40 Switchboards Feeder Section

		Crew	Daily Output	Labor-Hours	Unit	Material	2019 Bare Costs Labor	Equipment	Total	Total Incl O&P
0010	**SWITCHBOARDS FEEDER SECTION** group mounted devices R262419-82									
0030	Circuit breakers									
0160	FA frame, 15 to 60 amp, 240 volt, 1 pole	1 Elec	8	1	Ea.	121	60		181	223
0170	2 pole		7	1.143		370	68.50		438.50	510
0180	3 pole		5.30	1.509		565	90.50		655.50	760
0210	480 volt, 1 pole		8	1		163	60		223	270
0220	2 pole		7	1.143		460	68.50		528.50	605

26 24 Switchboards and Panelboards

26 24 13 – Switchboards

26 24 13.40 Switchboards Feeder Section	Crew	Daily Output	Labor-Hours	Unit	Material	2019 Bare Costs Labor	Equipment	Total	Total Incl O&P
0230	3 pole	1 Elec	5.30	1.509	Ea.	590	90.50		680.50	780
0260	600 volt, 2 pole		7	1.143		755	68.50		823.50	930
0270	3 pole		5.30	1.509		520	90.50		610.50	710
0280	FA frame, 70 to 100 amp, 240 volt, 1 pole		7	1.143		219	68.50		287.50	345
0310	2 pole		5	1.600		585	96		681	790
0320	3 pole		4	2		570	120		690	805
0330	480 volt, 1 pole		7	1.143		291	68.50		359.50	420
0360	2 pole		5	1.600		585	96		681	790
0370	3 pole		4	2		695	120		815	945
0380	600 volt, 2 pole		5	1.600		665	96		761	880
0410	3 pole		4	2		835	120		955	1,100
0420	KA frame, 70 to 225 amp		3.20	2.500		1,525	150		1,675	1,900
0430	LA frame, 125 to 400 amp		2.30	3.478		3,500	209		3,709	4,175
0460	MA frame, 450 to 600 amp		1.60	5		6,150	300		6,450	7,225
0470	700 to 800 amp		1.30	6.154		7,975	370		8,345	9,325
0480	MAL frame, 1,000 amp		1	8		8,300	480		8,780	9,850
0490	PA frame, 1,200 amp		.80	10		16,800	600		17,400	19,400
0500	Branch circuit, fusible switch, 600 volt, double 30/30 amp		4	2		835	120		955	1,100
0550	60/60 amp		3.20	2.500		860	150		1,010	1,175
0600	100/100 amp		2.70	2.963		1,075	178		1,253	1,475
0650	Single, 30 amp		5.30	1.509		685	90.50		775.50	890
0700	60 amp		4.70	1.702		800	102		902	1,025
0750	100 amp		4	2		1,150	120		1,270	1,450
0800	200 amp		2.70	2.963		1,325	178		1,503	1,750
0850	400 amp		2.30	3.478		2,425	209		2,634	2,975
0900	600 amp		1.80	4.444		2,975	267		3,242	3,675
0950	800 amp		1.30	6.154		5,025	370		5,395	6,075
1000	1,200 amp		.80	10		5,725	600		6,325	7,200
1080	Branch circuit, circuit breakers, high interrupting capacity									
1100	60 amp, 240, 480 or 600 volt, 1 pole	1 Elec	8	1	Ea.	475	60		535	610
1120	2 pole		7	1.143		1,175	68.50		1,243.50	1,400
1140	3 pole		5.30	1.509		650	90.50		740.50	850
1150	100 amp, 240, 480 or 600 volt, 1 pole		7	1.143		525	68.50		593.50	680
1160	2 pole		5	1.600		1,350	96		1,446	1,650
1180	3 pole		4	2		705	120		825	955
1200	225 amp, 240, 480 or 600 volt, 2 pole		3.50	2.286		2,225	137		2,362	2,650
1220	3 pole		3.20	2.500		2,200	150		2,350	2,625
1240	400 amp, 240, 480 or 600 volt, 2 pole		2.50	3.200		3,250	192		3,442	3,850
1260	3 pole		2.30	3.478		2,975	209		3,184	3,550
1280	600 amp, 240, 480 or 600 volt, 2 pole		1.80	4.444		5,025	267		5,292	5,925
1300	3 pole		1.60	5		3,625	300		3,925	4,450
1320	800 amp, 240, 480 or 600 volt, 2 pole		1.50	5.333		4,775	320		5,095	5,725
1340	3 pole		1.30	6.154		5,425	370		5,795	6,525
1360	1,000 amp, 240, 480 or 600 volt, 2 pole		1.10	7.273		6,650	435		7,085	7,975
1380	3 pole		1	8		5,875	480		6,355	7,200
1400	1,200 amp, 240, 480 or 600 volt, 2 pole		.90	8.889		6,450	535		6,985	7,900
1420	3 pole		.80	10		7,650	600		8,250	9,325
1700	Fusible switch, 240 V, 60 amp, 2 pole		3.20	2.500		420	150		570	690
1720	3 pole		3	2.667		580	160		740	880
1740	100 amp, 2 pole		2.70	2.963		430	178		608	735
1760	3 pole		2.50	3.200		650	192		842	1,000
1780	200 amp, 2 pole		2	4		900	240		1,140	1,350
1800	3 pole		1.90	4.211		1,050	253		1,303	1,525

For customer support on your Electrical Costs with RSMeans data, call 800.448.8182.

219

26 24 Switchboards and Panelboards

26 24 13 – Switchboards

26 24 13.40 Switchboards Feeder Section

		Crew	Daily Output	Labor-Hours	Unit	Material	2019 Bare Costs Labor	2019 Bare Costs Equipment	Total	Total Incl O&P
1820	400 amp, 2 pole	1 Elec	1.50	5.333	Ea.	1,675	320		1,995	2,325
1840	3 pole		1.30	6.154		2,100	370		2,470	2,875
1860	600 amp, 2 pole		1	8		2,400	480		2,880	3,350
1880	3 pole		.90	8.889		2,950	535		3,485	4,025
1900	240-600 V, 800 amp, 2 pole		.70	11.429		5,325	685		6,010	6,875
1920	3 pole		.60	13.333		6,525	800		7,325	8,375
2000	600 V, 60 amp, 2 pole		3.20	2.500		640	150		790	930
2040	100 amp, 2 pole		2.70	2.963		660	178		838	990
2080	200 amp, 2 pole		2	4		1,150	240		1,390	1,600
2120	400 amp, 2 pole		1.50	5.333		2,275	320		2,595	2,975
2160	600 amp, 2 pole		1	8		2,750	480		3,230	3,750
2500	Branch circuit, circuit breakers, 60 amp, 600 volt, 3 pole		5.30	1.509		460	90.50		550.50	640
2520	240, 480 or 600 volt, 1 pole		8	1		99	60		159	199
2540	240 volt, 2 pole		7	1.143		169	68.50		237.50	288
2560	480 or 600 volt, 2 pole		7	1.143		360	68.50		428.50	495
2580	240 volt, 3 pole		5.30	1.509		410	90.50		500.50	585
2600	480 volt, 3 pole		5.30	1.509		480	90.50		570.50	665
2620	100 amp, 600 volt, 2 pole		5	1.600		460	96		556	650
2640	3 pole		4	2		580	120		700	820
2660	480 volt, 2 pole		5	1.600		395	96		491	580
2680	240 volt, 2 pole		5	1.600		226	96		322	390
2700	3 pole		4	2		365	120		485	580
2720	480 volt, 3 pole		4	2		535	120		655	765
2740	225 amp, 240, 480 or 600 volt, 2 pole		3.50	2.286		660	137		797	930
2760	3 pole		3.20	2.500		695	150		845	990
2780	400 amp, 240, 480 or 600 volt, 2 pole		2.50	3.200		1,400	192		1,592	1,800
2800	3 pole		2.30	3.478		1,625	209		1,834	2,075
2820	600 amp, 240 or 480 volt, 2 pole		1.80	4.444		2,250	267		2,517	2,900
2840	3 pole		1.60	5		2,775	300		3,075	3,500
2860	800 amp, 240, 480 or 600 volt, 2 pole		1.50	5.333		3,400	320		3,720	4,225
2880	3 pole		1.30	6.154		3,975	370		4,345	4,925
2900	1,000 amp, 240, 480 or 600 volt, 2 pole		1.10	7.273		4,175	435		4,610	5,250
2920	480 or 600 volt, 3 pole		1	8		4,800	480		5,280	6,000
2940	1,200 amp, 240, 480 or 600 volt, 2 pole		.90	8.889		5,900	535		6,435	7,275
2960	3 pole		.80	10		6,425	600		7,025	7,975
2980	600 volt, 3 pole		.80	10		6,450	600		7,050	8,000

26 24 16 – Panelboards

26 24 16.10 Load Centers

		Crew	Daily Output	Labor-Hours	Unit	Material	2019 Bare Costs Labor	2019 Bare Costs Equipment	Total	Total Incl O&P
0010	**LOAD CENTERS** (residential type)　R262416-50									
0100	3 wire, 120/240 V, 1 phase, including 1 pole plug-in breakers									
0200	100 amp main lugs, indoor, 8 circuits	1 Elec	1.40	5.714	Ea.	98	345		443	620
0300	12 circuits		1.20	6.667		123	400		523	730
0400	Rainproof, 8 circuits		1.40	5.714		126	345		471	650
0500	12 circuits		1.20	6.667		151	400		551	760
0600	200 amp main lugs, indoor, 16 circuits	R-1A	1.80	8.889		203	480		683	940
0700	20 circuits		1.50	10.667		202	575		777	1,075
0800	24 circuits		1.30	12.308		253	665		918	1,275
0900	30 circuits		1.20	13.333		289	720		1,009	1,400
1000	40 circuits		.80	20		405	1,075		1,480	2,075
1200	Rainproof, 16 circuits		1.80	8.889		245	480		725	985
1300	20 circuits		1.50	10.667		285	575		860	1,175
1400	24 circuits		1.30	12.308		335	665		1,000	1,350

26 24 16 – Panelboards

26 24 16.10 Load Centers

		Crew	Daily Output	Labor-Hours	Unit	Material	2019 Bare Costs Labor	Equipment	Total	Total Incl O&P
1500	30 circuits	R-1A	1.20	13.333	Ea.	395	720		1,115	1,500
1600	40 circuits		.80	20		510	1,075		1,585	2,175
1800	400 amp main lugs, indoor, 42 circuits		.72	22.222		1,125	1,200		2,325	3,050
1900	Rainproof, 42 circuits	↓	.72	22.222	↓	1,625	1,200		2,825	3,600
2200	Plug in breakers, 20 amp, 1 pole, 4 wire, 120/208 volts									
2210	125 amp main lugs, indoor, 12 circuits	1 Elec	1.20	6.667	Ea.	180	400		580	795
2300	18 circuits		.80	10		256	600		856	1,175
2400	Rainproof, 12 circuits		1.20	6.667		225	400		625	845
2500	18 circuits	↓	.80	10		310	600		910	1,250
2600	200 amp main lugs, indoor, 24 circuits	R-1A	1.30	12.308		330	665		995	1,350
2700	30 circuits		1.20	13.333		380	720		1,100	1,500
2800	36 circuits		1	16		365	865		1,230	1,700
2900	42 circuits		.80	20		495	1,075		1,570	2,175
3000	Rainproof, 24 circuits		1.30	12.308		268	665		933	1,300
3100	30 circuits		1.20	13.333		300	720		1,020	1,400
3200	36 circuits		1	16		455	865		1,320	1,800
3300	42 circuits		.80	20		615	1,075		1,690	2,300
3500	400 amp main lugs, indoor, 42 circuits		.72	22.222		1,250	1,200		2,450	3,175
3600	Rainproof, 42 circuits	↓	.72	22.222	↓	1,350	1,200		2,550	3,275
3700	Plug-in breakers, 20 amp, 1 pole, 3 wire, 120/240 volts									
3800	100 amp main breaker, indoor, 12 circuits	1 Elec	1.20	6.667	Ea.	155	400		555	765
3900	18 circuits	"	.80	10		191	600		791	1,100
4000	200 amp main breaker, indoor, 20 circuits	R-1A	1.50	10.667		279	575		854	1,175
4200	24 circuits		1.30	12.308		305	665		970	1,325
4300	30 circuits		1.20	13.333		340	720		1,060	1,450
4400	40 circuits		.90	17.778		400	960		1,360	1,875
4500	Rainproof, 20 circuits		1.50	10.667		355	575		930	1,250
4600	24 circuits		1.30	12.308		390	665		1,055	1,425
4700	30 circuits		1.20	13.333		470	720		1,190	1,600
4800	40 circuits		.90	17.778		595	960		1,555	2,075
5000	400 amp main breaker, indoor, 42 circuits		.72	22.222		2,950	1,200		4,150	5,050
5100	Rainproof, 42 circuits	↓	.72	22.222	↓	3,225	1,200		4,425	5,350
5300	Plug in breakers, 20 amp, 1 pole, 4 wire, 120/208 volts									
5400	200 amp main breaker, indoor, 30 circuits	R-1A	1.20	13.333	Ea.	390	720		1,110	1,500
5500	42 circuits		.80	20		470	1,075		1,545	2,150
5600	Rainproof, 30 circuits		1.20	13.333		800	720		1,520	1,950
5700	42 circuits		.80	20		535	1,075		1,610	2,225

26 24 16.20 Panelboard and Load Center Circuit Breakers

		Crew	Daily Output	Labor-Hours	Unit	Material	2019 Bare Costs Labor	Equipment	Total	Total Incl O&P
0010	**PANELBOARD AND LOAD CENTER CIRCUIT BREAKERS** R262416-50									
0050	Bolt-on, 10,000 amp I.C., 120 volt, 1 pole									
0100	15-50 amp	1 Elec	10	.800	Ea.	18.90	48		66.90	92.50
0200	60 amp		8	1		19.05	60		79.05	111
0300	70 amp	↓	8	1	↓	28	60		88	120
0350	240 volt, 2 pole									
0400	15-50 amp	1 Elec	8	1	Ea.	50.50	60		110.50	145
0500	60 amp		7.50	1.067		33.50	64		97.50	133
0600	80-100 amp		5	1.600		114	96		210	268
0700	3 pole, 15-60 amp		6.20	1.290		137	77.50		214.50	267
0800	70 amp		5	1.600		178	96		274	340
0900	80-100 amp		3.60	2.222		189	133		322	405
1000	22,000 amp I.C., 240 volt, 2 pole, 70-225 amp		2.70	2.963		560	178		738	880
1100	3 pole, 70-225 amp	↓	2.30	3.478	↓	270	209		479	605

For customer support on your Electrical Costs with RSMeans data, call 800.448.8182.

221

26 24 16.20 Panelboard and Load Center Circuit Breakers	Crew	Daily Output	Labor-Hours	Unit	Material	2019 Bare Costs Labor	Equipment	Total	Total Incl O&P
1200	14,000 amp I.C., 277 volts, 1 pole, 15-30 amp	1 Elec	8	1	Ea.	35.50	60		95.50	129
1300	22,000 amp I.C., 480 volts, 2 pole, 70-225 amp		2.70	2.963		495	178		673	810
1400	3 pole, 70-225 amp		2.30	3.478		915	209		1,124	1,300
2000	Plug-in panel or load center, 120/240 volt, to 60 amp, 1 pole		12	.667		6.15	40		46.15	66.50
2004	Circuit breaker, 120/240 volt, 20 A, 1 pole with NM cable		6.50	1.231		11	74		85	122
2006	30 A, 1 pole with NM cable		6.50	1.231		11	74		85	122
2010	2 pole		9	.889		25.50	53.50		79	108
2014	50 A, 2 pole with NM cable		5.50	1.455		30.50	87.50		118	164
2020	3 pole		7.50	1.067		84	64		148	188
2030	100 amp, 2 pole		6	1.333		103	80		183	232
2040	3 pole		4.50	1.778		115	107		222	286
2050	150-200 amp, 2 pole		3	2.667		257	160		417	520
2060	Plug-in tandem, 120/240 V, 2-15 A, 1 pole		11	.727		20.50	43.50		64	87.50
2070	1-15 A & 1-20 A		11	.727		17.30	43.50		60.80	84
2080	2-20 A		11	.727		20.50	43.50		64	87.50
2082	Arc fault circuit interrupter, 120/240 V, 1-15 A & 1-20 A, 1 pole		11	.727		58	43.50		101.50	129
2100	High interrupting capacity, 120/240 volt, plug-in, 30 amp, 1 pole		12	.667		25.50	40		65.50	87.50
2110	60 amp, 2 pole		9	.889		24.50	53.50		78	107
2120	3 pole		7.50	1.067		274	64		338	395
2130	100 amp, 2 pole		6	1.333		138	80		218	271
2140	3 pole		4.50	1.778		390	107		497	590
2150	125 amp, 2 pole		3	2.667		760	160		920	1,075
2200	Bolt-on, 30 amp, 1 pole		10	.800		50.50	48		98.50	127
2210	60 amp, 2 pole		7.50	1.067		110	64		174	217
2220	3 pole		6.20	1.290		273	77.50		350.50	415
2230	100 amp, 2 pole		5	1.600		267	96		363	435
2240	3 pole		3.60	2.222		395	133		528	635
2300	Ground fault, 240 volt, 30 amp, 1 pole		7	1.143		90.50	68.50		159	202
2310	2 pole		6	1.333		170	80		250	305
2350	Key operated, 240 volt, 1 pole, 30 amp		7	1.143		121	68.50		189.50	235
2360	Switched neutral, 240 volt, 30 amp, 2 pole		6	1.333		39.50	80		119.50	162
2370	3 pole		5.50	1.455		65.50	87.50		153	202
2400	Shunt trip, for 240 volt breaker, 60 amp, 1 pole		4	2		70	120		190	256
2410	2 pole		3.50	2.286		70	137		207	282
2420	3 pole		3	2.667		70	160		230	315
2430	100 amp, 2 pole		3	2.667		70	160		230	315
2440	3 pole		2.50	3.200		70	192		262	365
2450	150 amp, 2 pole		2	4		202	240		442	580
2500	Auxiliary switch, for 240 volt breaker, 60 amp, 1 pole		4	2		77.50	120		197.50	264
2510	2 pole		3.50	2.286		77.50	137		214.50	290
2520	3 pole		3	2.667		77.50	160		237.50	325
2530	100 amp, 2 pole		3	2.667		77.50	160		237.50	325
2540	3 pole		2.50	3.200		77.50	192		269.50	370
2550	150 amp, 2 pole		2	4		106	240		346	475
2600	Panel or load center, 277/480 volt, plug-in, 30 amp, 1 pole		12	.667		62	40		102	128
2610	60 amp, 2 pole		9	.889		187	53.50		240.50	286
2620	3 pole		7.50	1.067		274	64		338	395
2650	Bolt-on, 60 amp, 2 pole		7.50	1.067		187	64		251	300
2660	3 pole		6.20	1.290		274	77.50		351.50	415
2700	H line, 277/480 volt, 30 amp, 1 pole		8	1		59	60		119	155
2710	60 amp, 2 pole		7.50	1.067		208	64		272	325
2720	3 pole		6.20	1.290		325	77.50		402.50	470
2730	100 amp, 1 pole		7.50	1.067		138	64		202	248

26 24 16 – Panelboards

26 24 16.20 Panelboard and Load Center Circuit Breakers	Crew	Daily Output	Labor-Hours	Unit	Material	2019 Bare Costs Labor	Equipment	Total	Total Incl O&P
2740	2 pole	1 Elec	5	1.600	Ea.	325	96		421	500
2750	3 pole		3.50	2.286		380	137		517	625
2800	High interrupting capacity, 277/480 volt, plug-in, 30 amp, 1 pole		12	.667		360	40		400	455
2810	60 amp, 2 pole		9	.889		555	53.50		608.50	695
2820	3 pole		7	1.143		650	68.50		718.50	815
2830	Bolt-on, 30 amp, 1 pole		8	1		360	60		420	485
2840	60 amp, 2 pole		7.50	1.067		605	64		669	760
2850	3 pole		6.20	1.290		680	77.50		757.50	860
2900	I-line, 30 amp, 1 pole		8	1		360	60		420	485
2910	60 amp, 2 pole		7.50	1.067		510	64		574	655
2920	3 pole		6.20	1.290		570	77.50		647.50	740
2930	100 amp, 1 pole		7.50	1.067		585	64		649	735
2940	2 pole		5	1.600		485	96		581	680
2950	3 pole		3.60	2.222		635	133		768	900
2960	Shunt trip, 277/480 volt breaker, remote oper., 30 amp, 1 pole		4	2		380	120		500	600
2970	60 amp, 2 pole		3.50	2.286		560	137		697	820
2980	3 pole		3	2.667		635	160		795	940
2990	100 amp, 1 pole		3.50	2.286		455	137		592	705
3000	2 pole		3	2.667		635	160		795	940
3010	3 pole		2.50	3.200		580	192		772	920
3050	Under voltage trip, 277/480 volt breaker, 30 amp, 1 pole		4	2		380	120		500	600
3060	60 amp, 2 pole		3.50	2.286		560	137		697	820
3070	3 pole		3	2.667		635	160		795	940
3080	100 amp, 1 pole		3.50	2.286		455	137		592	705
3090	2 pole		3	2.667		635	160		795	940
3100	3 pole		2.50	3.200		580	192		772	920
3150	Motor operated, 277/480 volt breaker, 30 amp, 1 pole		4	2		680	120		800	930
3160	60 amp, 2 pole		3.50	2.286		715	137		852	995
3170	3 pole		3	2.667		945	160		1,105	1,300
3180	100 amp, 1 pole		3.50	2.286		755	137		892	1,025
3190	2 pole		3	2.667		945	160		1,105	1,300
3200	3 pole		2.50	3.200		830	192		1,022	1,200
3250	Panelboard spacers, per pole		40	.200		3.68	12		15.68	22
5110	NEMA 1 enclosure only, 600V, 3 p, 14k AIC, 100A	▼	2.20	3.636	▼	126	218		344	465

26 24 16.30 Panelboards Commercial Applications

		Crew	Daily Output	Labor-Hours	Unit	Material	2019 Bare Costs Labor	Equipment	Total	Total Incl O&P
0010	**PANELBOARDS COMMERCIAL APPLICATIONS** R262416-50									
0050	NQOD, w/20 amp 1 pole bolt-on circuit breakers									
0100	3 wire, 120/240 volts, 100 amp main lugs									
0150	10 circuits	1 Elec	1	8	Ea.	925	480		1,405	1,750
0200	14 circuits		.88	9.091		1,025	545		1,570	1,950
0250	18 circuits		.75	10.667		1,125	640		1,765	2,175
0300	20 circuits		.65	12.308		1,250	740		1,990	2,475
0350	225 amp main lugs, 24 circuits	2 Elec	1.20	13.333		1,425	800		2,225	2,750
0400	30 circuits		.90	17.778		1,625	1,075		2,700	3,400
0450	36 circuits		.80	20		1,850	1,200		3,050	3,850
0500	38 circuits		.72	22.222		2,025	1,325		3,350	4,225
0550	42 circuits		.66	24.242		2,075	1,450		3,525	4,475
0600	4 wire, 120/208 volts, 100 amp main lugs, 12 circuits	1 Elec	1	8		990	480		1,470	1,825
0650	16 circuits		.75	10.667		1,150	640		1,790	2,200
0700	20 circuits		.65	12.308		1,300	740		2,040	2,550
0750	24 circuits		.60	13.333		1,250	800		2,050	2,575
0800	30 circuits	▼	.53	15.094		1,625	905		2,530	3,150

For customer support on your Electrical Costs with RSMeans data, call 800.448.8182.

223

26 24 16.30 Panelboards Commercial Applications	Crew	Daily Output	Labor-Hours	Unit	Material	2019 Bare Costs Labor	Equipment	Total	Total Incl O&P
0850 225 amp main lugs, 32 circuits	2 Elec	.90	17.778	Ea.	1,825	1,075		2,900	3,600
0900 34 circuits		.84	19.048		1,875	1,150		3,025	3,750
0950 36 circuits		.80	20		1,900	1,200		3,100	3,900
1000 42 circuits		.68	23.529		2,150	1,425		3,575	4,475
1010 400 amp main lugs, 42 circs		.68	23.529		2,150	1,425		3,575	4,475
1040 225 amp main lugs, NEMA 7, 12 circuits		1	16		5,775	960		6,735	7,775
1100 24 circuits		.40	40		6,975	2,400		9,375	11,300
1200 NEHB, w/20 amp, 1 pole bolt-on circuit breakers									
1250 4 wire, 277/480 volts, 100 amp main lugs, 12 circuits	1 Elec	.88	9.091	Ea.	1,375	545		1,920	2,325
1300 20 circuits	"	.60	13.333		2,025	800		2,825	3,425
1350 225 amp main lugs, 24 circuits	2 Elec	.90	17.778		2,300	1,075		3,375	4,125
1400 30 circuits		.80	20		2,750	1,200		3,950	4,825
1448 32 circuits		4.90	3.265		3,200	196		3,396	3,825
1450 36 circuits		.72	22.222		3,200	1,325		4,525	5,525
1500 42 circuits		.60	26.667		3,650	1,600		5,250	6,425
1510 225 amp main lugs, NEMA 7, 12 circuits		.90	17.778		6,050	1,075		7,125	8,250
1590 24 circuits		.30	53.333		8,075	3,200		11,275	13,700
1600 NQOD panel, w/20 amp, 1 pole, circuit breakers									
1650 3 wire, 120/240 volt with main circuit breaker									
1700 100 amp main, 12 circuits	1 Elec	.80	10	Ea.	1,225	600		1,825	2,250
1750 20 circuits	"	.60	13.333		1,550	800		2,350	2,925
1800 225 amp main, 30 circuits	2 Elec	.68	23.529		2,925	1,425		4,350	5,325
1801 225 amp main, 32 circuits		5	3.200		2,925	192		3,117	3,500
1850 42 circuits		.52	30.769		3,375	1,850		5,225	6,475
1900 400 amp main, 30 circuits		.54	29.630		4,050	1,775		5,825	7,100
1950 42 circuits		.50	32		4,525	1,925		6,450	7,850
2000 4 wire, 120/208 volts with main circuit breaker									
2050 100 amp main, 24 circuits	1 Elec	.47	17.021	Ea.	1,875	1,025		2,900	3,600
2100 30 circuits	"	.40	20		2,050	1,200		3,250	4,050
2200 225 amp main, 32 circuits	2 Elec	.72	22.222		3,400	1,325		4,725	5,750
2250 42 circuits		.56	28.571		3,850	1,725		5,575	6,775
2300 400 amp main, 42 circuits		.48	33.333		5,000	2,000		7,000	8,475
2350 600 amp main, 42 circuits		.40	40		7,375	2,400		9,775	11,700
2400 NEHB, with 20 amp, 1 pole circuit breaker									
2450 4 wire, 277/480 volts with main circuit breaker									
2500 100 amp main, 24 circuits	1 Elec	.42	19.048	Ea.	2,650	1,150		3,800	4,625
2550 30 circuits	"	.38	21.053		3,100	1,275		4,375	5,275
2600 225 amp main, 30 circuits	2 Elec	.72	22.222		3,900	1,325		5,225	6,275
2650 42 circuits		.56	28.571		4,800	1,725		6,525	7,825
2700 400 amp main, 42 circuits		.46	34.783		5,750	2,100		7,850	9,450
2750 600 amp main, 42 circuits		.38	42.105		7,875	2,525		10,400	12,500
2900 Note: the following line items don't include branch circuit breakers									
2910 For branch circuit breakers information, see Section 26 24 16.20									
3010 Main lug, no main breaker, 240 volt, 1 pole, 3 wire, 100 amp	1 Elec	2.30	3.478	Ea.	595	209		804	965
3020 225 amp	2 Elec	2.40	6.667		715	400		1,115	1,375
3030 400 amp	"	1.80	8.889		1,050	535		1,585	1,950
3060 3 pole, 3 wire, 100 amp	1 Elec	2.30	3.478		645	209		854	1,025
3070 225 amp	2 Elec	2.40	6.667		755	400		1,155	1,425
3080 400 amp		1.80	8.889		1,125	535		1,660	2,050
3090 600 amp		1.60	10		1,325	600		1,925	2,350
3110 3 pole, 4 wire, 100 amp	1 Elec	2.30	3.478		710	209		919	1,100
3120 225 amp	2 Elec	2.40	6.667		890	400		1,290	1,575
3130 400 amp		1.80	8.889		1,125	535		1,660	2,050

26 24 16 – Panelboards

26 24 16.30 Panelboards Commercial Applications

		Crew	Daily Output	Labor-Hours	Unit	Material	2019 Bare Costs Labor	Equipment	Total	Total Incl O&P
3140	600 amp	2 Elec	1.60	10	Ea.	1,325	600		1,925	2,350
3160	480 volt, 3 pole, 3 wire, 100 amp	1 Elec	2.30	3.478		845	209		1,054	1,250
3170	225 amp	2 Elec	2.40	6.667		1,025	400		1,425	1,725
3180	400 amp		1.80	8.889		1,425	535		1,960	2,375
3190	600 amp		1.60	10		1,600	600		2,200	2,650
3210	277/480 volt, 3 pole, 4 wire, 100 amp	1 Elec	2.30	3.478		825	209		1,034	1,225
3220	225 amp	2 Elec	2.40	6.667		1,000	400		1,400	1,700
3230	400 amp		1.80	8.889		1,400	535		1,935	2,325
3240	600 amp		1.60	10		1,550	600		2,150	2,600
3260	Main circuit breaker, 240 volt, 1 pole, 3 wire, 100 amp	1 Elec	2	4		800	240		1,040	1,250
3270	225 amp	2 Elec	2	8		1,625	480		2,105	2,525
3280	400 amp	"	1.60	10		2,550	600		3,150	3,725
3310	3 pole, 3 wire, 100 amp	1 Elec	2	4		895	240		1,135	1,350
3320	225 amp	2 Elec	2	8		1,875	480		2,355	2,775
3330	400 amp		1.60	10		2,925	600		3,525	4,125
3360	120/208 volt, 3 pole, 4 wire, 100 amp		4	4		915	240		1,155	1,350
3370	225 amp		2	8		1,875	480		2,355	2,775
3380	400 amp		1.60	10		2,925	600		3,525	4,125
3410	480 volt, 3 pole, 3 wire, 100 amp	1 Elec	2	4		1,275	240		1,515	1,750
3420	225 amp	2 Elec	2	8		2,200	480		2,680	3,150
3430	400 amp		1.60	10		3,275	600		3,875	4,525
3460	277/480 volt, 3 pole, 4 wire, 100 amp		4	4		1,250	240		1,490	1,725
3470	225 amp		2	8		2,150	480		2,630	3,075
3480	400 amp		1.60	10		3,300	600		3,900	4,525
3510	Main circuit breaker, HIC, 240 volt, 1 pole, 3 wire, 100 amp	1 Elec	2	4		1,225	240		1,465	1,700
3520	225 amp	2 Elec	2	8		3,200	480		3,680	4,225
3530	400 amp	"	1.60	10		4,275	600		4,875	5,600
3560	3 pole, 3 wire, 100 amp	1 Elec	2	4		1,350	240		1,590	1,850
3570	225 amp	2 Elec	2	8		3,600	480		4,080	4,675
3580	400 amp	"	1.60	10		4,750	600		5,350	6,125
3610	120/208 volt, 3 pole, 4 wire, 100 amp	1 Elec	2	4		1,350	240		1,590	1,850
3620	225 amp	2 Elec	2	8		3,600	480		4,080	4,675
3630	400 amp	"	1.60	10		4,750	600		5,350	6,125
3660	480 volt, 3 pole, 3 wire, 100 amp	1 Elec	2	4		2,025	240		2,265	2,575
3670	225 amp	2 Elec	2	8		4,000	480		4,480	5,125
3680	400 amp	"	1.60	10		5,100	600		5,700	6,500
3710	277/480 volt, 3 pole, 4 wire, 100 amp	1 Elec	2	4		1,925	240		2,165	2,475
3720	225 amp	2 Elec	2	8		3,800	480		4,280	4,925
3730	400 amp	"	1.60	10		5,050	600		5,650	6,450
3760	Main circuit breaker, shunt trip, 100 amp	1 Elec	1.20	6.667		1,000	400		1,400	1,700
3770	225 amp	2 Elec	1.60	10		2,325	600		2,925	3,450
3780	400 amp	"	1.40	11.429		3,350	685		4,035	4,725

26 24 19 – Motor-Control Centers

26 24 19.20 Motor Control Center Components

		Crew	Daily Output	Labor-Hours	Unit	Material	2019 Bare Costs Labor	Equipment	Total	Total Incl O&P
0010	**MOTOR CONTROL CENTER COMPONENTS** R262419-60									
0100	Starter, size 1, FVNR, NEMA 1, type A, fusible	1 Elec	2.70	2.963	Ea.	1,650	178		1,828	2,100
0120	Circuit breaker		2.70	2.963		1,800	178		1,978	2,250
0140	Type B, fusible		2.70	2.963		1,825	178		2,003	2,275
0160	Circuit breaker		2.70	2.963		1,975	178		2,153	2,450
0180	NEMA 12, type A, fusible		2.60	3.077		1,700	185		1,885	2,125
0200	Circuit breaker		2.60	3.077		1,825	185		2,010	2,300
0220	Type B, fusible		2.60	3.077		1,850	185		2,035	2,325

For customer support on your Electrical Costs with RSMeans data, call 800.448.8182.

225

26 24 19.20 Motor Control Center Components		Crew	Daily Output	Labor-Hours	Unit	Material	2019 Bare Costs		Total	Total Incl O&P
							Labor	Equipment		
0240	Circuit breaker	1 Elec	2.60	3.077	Ea.	2,025	185		2,210	2,500
0300	Starter, size 1, FVR, NEMA 1, type A, fusible		2	4		2,375	240		2,615	2,975
0320	Circuit breaker		2	4		2,375	240		2,615	2,975
0340	Type B, fusible		2	4		2,625	240		2,865	3,225
0360	Circuit breaker		2	4		2,625	240		2,865	3,225
0380	NEMA 12, type A, fusible		1.90	4.211		2,400	253		2,653	3,025
0400	Circuit breaker		1.90	4.211		2,400	253		2,653	3,025
0420	Type B, fusible		1.90	4.211		2,650	253		2,903	3,300
0440	Circuit breaker	▼	1.90	4.211	▼	2,650	253		2,903	3,300
0490	Starter size 1, 2 speed, separate winding									
0500	NEMA 1, type A, fusible	1 Elec	2.60	3.077	Ea.	3,150	185		3,335	3,725
0520	Circuit breaker		2.60	3.077		3,150	185		3,335	3,725
0540	Type B, fusible		2.60	3.077		3,450	185		3,635	4,075
0560	Circuit breaker		2.60	3.077		3,450	185		3,635	4,075
0580	NEMA 12, type A, fusible		2.50	3.200		3,200	192		3,392	3,800
0600	Circuit breaker		2.50	3.200		3,200	192		3,392	3,800
0620	Type B, fusible		2.50	3.200		3,500	192		3,692	4,125
0640	Circuit breaker	▼	2.50	3.200	▼	3,500	192		3,692	4,125
0650	Starter size 1, 2 speed, consequent pole									
0660	NEMA 1, type A, fusible	1 Elec	2.60	3.077	Ea.	3,150	185		3,335	3,725
0680	Circuit breaker		2.60	3.077		3,150	185		3,335	3,725
0700	Type B, fusible		2.60	3.077		3,450	185		3,635	4,075
0720	Circuit breaker		2.60	3.077		3,450	185		3,635	4,075
0740	NEMA 12, type A, fusible		2.50	3.200		3,200	192		3,392	3,800
0760	Circuit breaker		2.50	3.200		3,200	192		3,392	3,800
0780	Type B, fusible		2.50	3.200		3,500	192		3,692	4,100
0800	Circuit breaker	▼	2.50	3.200	▼	3,500	192		3,692	4,125
0810	Starter size 1, 2 speed, space only									
0820	NEMA 1, type A, fusible	1 Elec	16	.500	Ea.	705	30		735	820
0840	Circuit breaker		16	.500		705	30		735	820
0860	Type B, fusible		16	.500		705	30		735	820
0880	Circuit breaker		16	.500		705	30		735	820
0900	NEMA 12, type A, fusible		15	.533		735	32		767	855
0920	Circuit breaker		15	.533		735	32		767	855
0940	Type B, fusible		15	.533		735	32		767	855
0960	Circuit breaker	▼	15	.533		735	32		767	855
1100	Starter size 2, FVNR, NEMA 1, type A, fusible	2 Elec	4	4		1,875	240		2,115	2,425
1120	Circuit breaker		4	4		2,050	240		2,290	2,625
1140	Type B, fusible		4	4		2,075	240		2,315	2,625
1160	Circuit breaker		4	4		2,250	240		2,490	2,825
1180	NEMA 12, type A, fusible		3.80	4.211		1,900	253		2,153	2,475
1200	Circuit breaker		3.80	4.211		2,075	253		2,328	2,675
1220	Type B, fusible		3.80	4.211		2,075	253		2,328	2,675
1240	Circuit breaker		3.80	4.211		2,275	253		2,528	2,875
1300	FVR, NEMA 1, type A, fusible		3.20	5		3,175	300		3,475	3,950
1320	Circuit breaker		3.20	5		3,175	300		3,475	3,950
1340	Type B, fusible		3.20	5		3,500	300		3,800	4,300
1360	Circuit breaker		3.20	5		3,500	300		3,800	4,300
1380	NEMA type 12, type A, fusible		3	5.333		3,225	320		3,545	4,025
1400	Circuit breaker		3	5.333		3,225	320		3,545	4,025
1420	Type B, fusible		3	5.333		3,550	320		3,870	4,375
1440	Circuit breaker	▼	3	5.333		3,550	320		3,870	4,375
1490	Starter size 2, 2 speed, separate winding									

26 24 19 – Motor-Control Centers

26 24 19.20 Motor Control Center Components	Crew	Daily Output	Labor-Hours	Unit	Material	2019 Bare Costs Labor	Equipment	Total	Total Incl O&P
1500	NEMA 1, type A, fusible	2 Elec	3.80	4.211	Ea.	3,550	253		3,803	4,275
1520	Circuit breaker		3.80	4.211		3,550	253		3,803	4,275
1540	Type B, fusible		3.80	4.211		3,900	253		4,153	4,675
1560	Circuit breaker		3.80	4.211		3,900	253		4,153	4,675
1570	NEMA 12, type A, fusible		3.60	4.444		3,600	267		3,867	4,375
1580	Circuit breaker		3.60	4.444		3,600	267		3,867	4,375
1600	Type B, fusible		3.60	4.444		3,950	267		4,217	4,750
1620	Circuit breaker	↓	3.60	4.444	↓	3,950	267		4,217	4,750
1630	Starter size 2, 2 speed, consequent pole									
1640	NEMA 1, type A, fusible	2 Elec	3.80	4.211	Ea.	4,025	253		4,278	4,800
1660	Circuit breaker		3.80	4.211		4,025	253		4,278	4,800
1680	Type B, fusible		3.80	4.211		4,250	253		4,503	5,050
1700	Circuit breaker		3.80	4.211		4,250	253		4,503	5,050
1720	NEMA 12, type A, fusible		3.80	4.211		4,050	253		4,303	4,825
1740	Circuit breaker		3.60	4.444		4,075	267		4,342	4,900
1760	Type B, fusible		3.60	4.444		4,325	267		4,592	5,150
1780	Circuit breaker	↓	3.60	4.444	↓	4,325	267		4,592	5,150
1830	Starter size 2, autotransformer									
1840	NEMA 1, type A, fusible	2 Elec	3.40	4.706	Ea.	6,275	283		6,558	7,325
1860	Circuit breaker		3.40	4.706		6,425	283		6,708	7,500
1880	Type B, fusible		3.40	4.706		6,850	283		7,133	7,975
1900	Circuit breaker		3.40	4.706		6,850	283		7,133	7,975
1920	NEMA 12, type A, fusible		3.20	5		6,400	300		6,700	7,500
1940	Circuit breaker		3.20	5		6,400	300		6,700	7,500
1960	Type B, fusible		3.20	5		6,975	300		7,275	8,125
1980	Circuit breaker	↓	3.20	5	↓	6,975	300		7,275	8,125
2030	Starter size 2, space only									
2040	NEMA 1, type A, fusible	1 Elec	16	.500	Ea.	705	30		735	820
2060	Circuit breaker		16	.500		705	30		735	820
2080	Type B, fusible		16	.500		705	30		735	820
2100	Circuit breaker		16	.500		705	30		735	820
2120	NEMA 12, type A, fusible		15	.533		735	32		767	855
2140	Circuit breaker		15	.533		735	32		767	855
2160	Type B, fusible		15	.533		735	32		767	855
2180	Circuit breaker	↓	15	.533		735	32		767	855
2300	Starter size 3, FVNR, NEMA 1, type A, fusible	2 Elec	2	8		3,575	480		4,055	4,650
2320	Circuit breaker		2	8		3,175	480		3,655	4,225
2340	Type B, fusible		2	8		3,925	480		4,405	5,050
2360	Circuit breaker		2	8		3,500	480		3,980	4,575
2380	NEMA 12, type A, fusible		1.90	8.421		3,650	505		4,155	4,750
2400	Circuit breaker		1.90	8.421		3,225	505		3,730	4,300
2420	Type B, fusible		1.90	8.421		4,075	505		4,580	5,225
2440	Circuit breaker		1.90	8.421		3,550	505		4,055	4,675
2500	Starter size 3, FVR, NEMA 1, type A, fusible		1.60	10		4,875	600		5,475	6,250
2520	Circuit breaker		1.60	10		4,650	600		5,250	6,025
2540	Type B, fusible		1.60	10		5,300	600		5,900	6,750
2560	Circuit breaker		1.60	10		5,100	600		5,700	6,500
2580	NEMA 12, type A, fusible		1.50	10.667		4,975	640		5,615	6,425
2600	Circuit breaker		1.50	10.667		4,750	640		5,390	6,175
2620	Type B, fusible		1.50	10.667		5,400	640		6,040	6,900
2640	Circuit breaker	↓	1.50	10.667	↓	5,600	640		6,240	7,125
2690	Starter size 3, 2 speed, separate winding									
2700	NEMA 1, type A, fusible	2 Elec	2	8	Ea.	5,600	480		6,080	6,900

26 24 19.20 Motor Control Center Components	Crew	Daily Output	Labor-Hours	Unit	Material	2019 Bare Costs Labor	Equipment	Total	Total Incl O&P
2720	Circuit breaker	2 Elec	2	8	Ea.	4,950	480		5,430	6,175
2740	Type B, fusible		2	8		6,150	480		6,630	7,475
2760	Circuit breaker		2	8		5,425	480		5,905	6,675
2780	NEMA 12, type A, fusible		1.90	8.421		5,725	505		6,230	7,050
2800	Circuit breaker		1.90	8.421		5,050	505		5,555	6,300
2820	Type B, fusible		1.90	8.421		6,250	505		6,755	7,625
2840	Circuit breaker		1.90	8.421		5,525	505		6,030	6,825
2850	Starter size 3, 2 speed, consequent pole									
2860	NEMA 1, type A, fusible	2 Elec	2	8	Ea.	6,250	480		6,730	7,600
2880	Circuit breaker		2	8		5,600	480		6,080	6,875
2900	Type B, fusible		2	8		6,800	480		7,280	8,200
2920	Circuit breaker		2	8		6,075	480		6,555	7,400
2940	NEMA 12, type A, fusible		1.90	8.421		6,375	505		6,880	7,750
2960	Circuit breaker		1.90	8.421		5,700	505		6,205	7,025
2980	Type B, fusible		1.90	8.421		6,900	505		7,405	8,350
3000	Circuit breaker		1.90	8.421		6,175	505		6,680	7,525
3100	Starter size 3, autotransformer, NEMA 1, type A, fusible		1.60	10		7,850	600		8,450	9,550
3120	Circuit breaker		1.60	10		7,775	600		8,375	9,475
3140	Type B, fusible		1.60	10		8,500	600		9,100	10,200
3160	Circuit breaker		1.60	10		8,500	600		9,100	10,200
3180	NEMA 12, type A, fusible		1.50	10.667		7,925	640		8,565	9,675
3200	Circuit breaker		1.50	10.667		7,925	640		8,565	9,675
3220	Type B, fusible		1.50	10.667		8,650	640		9,290	10,500
3240	Circuit breaker		1.50	10.667		8,650	640		9,290	10,500
3260	Starter size 3, space only, NEMA 1, type A, fusible	1 Elec	15	.533		1,200	32		1,232	1,375
3280	Circuit breaker		15	.533		940	32		972	1,075
3300	Type B, fusible		15	.533		1,200	32		1,232	1,375
3320	Circuit breaker		15	.533		940	32		972	1,075
3340	NEMA 12, type A, fusible		14	.571		1,275	34.50		1,309.50	1,450
3360	Circuit breaker		14	.571		995	34.50		1,029.50	1,150
3380	Type B, fusible		14	.571		1,275	34.50		1,309.50	1,450
3400	Circuit breaker		14	.571		995	34.50		1,029.50	1,150
3500	Starter size 4, FVNR, NEMA 1, type A, fusible	2 Elec	1.60	10		4,650	600		5,250	6,025
3520	Circuit breaker		1.60	10		4,225	600		4,825	5,550
3540	Type B, fusible		1.60	10		5,100	600		5,700	6,500
3560	Circuit breaker		1.60	10		4,650	600		5,250	6,025
3580	NEMA 12, type A, fusible		1.50	10.667		4,775	640		5,415	6,200
3600	Circuit breaker		1.50	10.667		4,300	640		4,940	5,675
3620	Type B, fusible		1.50	10.667		5,200	640		5,840	6,675
3640	Circuit breaker		1.50	10.667		4,725	640		5,365	6,150
3700	Starter size 4, FVR, NEMA 1, type A, fusible		1.20	13.333		6,375	800		7,175	8,200
3720	Circuit breaker		1.20	13.333		5,750	800		6,550	7,525
3740	Type B, fusible		1.20	13.333		6,950	800		7,750	8,850
3760	Circuit breaker		1.20	13.333		6,325	800		7,125	8,150
3780	NEMA 12, type A, fusible		1.16	13.793		6,525	830		7,355	8,400
3800	Circuit breaker		1.16	13.793		5,825	830		6,655	7,625
3820	Type B, fusible		1.16	13.793		5,875	830		6,705	7,700
3840	Circuit breaker		1.16	13.793		6,400	830		7,230	8,275
3890	Starter size 4, 2 speed, separate windings									
3900	NEMA 1, type A, fusible	2 Elec	1.60	10	Ea.	8,050	600		8,650	9,750
3920	Circuit breaker		1.60	10		6,025	600		6,625	7,550
3940	Type B, fusible		1.60	10		8,825	600		9,425	10,600
3960	Circuit breaker		1.60	10		6,625	600		7,225	8,175

26 24 Switchboards and Panelboards

26 24 19 – Motor-Control Centers

26 24 19.20 Motor Control Center Components	Crew	Daily Output	Labor-Hours	Unit	Material	2019 Bare Costs Labor	Equipment	Total	Total Incl O&P
3980	NEMA 12, type A, fusible	2 Elec	1.50	10.667	Ea.	8,175	640		8,815	9,950
4000	Circuit breaker		1.50	10.667		6,150	640		6,790	7,700
4020	Type B, fusible		1.50	10.667		8,975	640		9,615	10,800
4040	Circuit breaker		1.50	10.667		6,725	640		7,365	8,350
4050	Starter size 4, 2 speed, consequent pole									
4060	NEMA 1, type A, fusible	2 Elec	1.60	10	Ea.	9,400	600		10,000	11,200
4080	Circuit breaker		1.60	10		6,975	600		7,575	8,575
4100	Type B, fusible		1.60	10		10,300	600		10,900	12,300
4120	Circuit breaker		1.60	10		7,650	600		8,250	9,325
4140	NEMA 12, type A, fusible		1.50	10.667		9,525	640		10,165	11,500
4160	Circuit breaker		1.50	10.667		7,075	640		7,715	8,750
4180	Type B, fusible		1.50	10.667		10,500	640		11,140	12,500
4200	Circuit breaker		1.50	10.667		7,750	640		8,390	9,475
4300	Starter size 4, autotransformer, NEMA 1, type A, fusible		1.30	12.308		9,250	740		9,990	11,300
4320	Circuit breaker		1.30	12.308		9,325	740		10,065	11,400
4340	Type B, fusible		1.30	12.308		10,200	740		10,940	12,300
4360	Circuit breaker		1.30	12.308		10,200	740		10,940	12,300
4380	NEMA 12, type A, fusible		1.24	12.903		9,400	775		10,175	11,500
4400	Circuit breaker		1.24	12.903		9,450	775		10,225	11,600
4420	Type B, fusible		1.24	12.903		10,300	775		11,075	12,600
4440	Circuit breaker		1.24	12.903		10,300	775		11,075	12,500
4500	Starter size 4, space only, NEMA 1, type A, fusible	1 Elec	14	.571		1,675	34.50		1,709.50	1,900
4520	Circuit breaker		14	.571		1,200	34.50		1,234.50	1,375
4540	Type B, fusible		14	.571		1,675	34.50		1,709.50	1,900
4560	Circuit breaker		14	.571		1,200	34.50		1,234.50	1,375
4580	NEMA 12, type A, fusible		13	.615		1,775	37		1,812	2,000
4600	Circuit breaker		13	.615		1,275	37		1,312	1,450
4620	Type B, fusible		13	.615		1,775	37		1,812	2,000
4640	Circuit breaker		13	.615		1,275	37		1,312	1,450
4800	Starter size 5, FVNR, NEMA 1, type A, fusible	2 Elec	1	16		9,625	960		10,585	12,000
4820	Circuit breaker		1	16		6,825	960		7,785	8,925
4840	Type B, fusible		1	16		10,600	960		11,560	13,000
4860	Circuit breaker		1	16		7,500	960		8,460	9,675
4880	NEMA 12, type A, fusible		.96	16.667		9,750	1,000		10,750	12,200
4900	Circuit breaker		.96	16.667		6,950	1,000		7,950	9,150
4920	Type B, fusible		.96	16.667		10,700	1,000		11,700	13,300
4940	Circuit breaker		.96	16.667		7,650	1,000		8,650	9,900
5000	Starter size 5, FVR, NEMA 1, type A, fusible		.80	20		14,200	1,200		15,400	17,400
5020	Circuit breaker		.80	20		11,100	1,200		12,300	14,000
5040	Type B, fusible		.80	20		15,500	1,200		16,700	18,900
5060	Circuit breaker		.80	20		12,200	1,200		13,400	15,200
5080	NEMA 12, type A, fusible		.76	21.053		14,400	1,275		15,675	17,800
5100	Circuit breaker		.76	21.053		11,300	1,275		12,575	14,300
5120	Type B, fusible		.76	21.053		15,800	1,275		17,075	19,300
5140	Circuit breaker		.76	21.053		12,400	1,275		13,675	15,500
5190	Starter size 5, 2 speed, separate windings									
5200	NEMA 1, type A, fusible	2 Elec	1	16	Ea.	19,500	960		20,460	22,900
5220	Circuit breaker		1	16		14,600	960		15,560	17,400
5240	Type B, fusible		1	16		21,400	960		22,360	25,000
5260	Circuit breaker		1	16		16,000	960		16,960	19,000
5280	NEMA 12, type A, fusible		.96	16.667		19,800	1,000		20,800	23,300
5300	Circuit breaker		.96	16.667		14,700	1,000		15,700	17,700
5320	Type B, fusible		.96	16.667		21,700	1,000		22,700	25,400

For customer support on your Electrical Costs with RSMeans data, call 800.448.8182.

229

26 24 19.20 Motor Control Center Components	Crew	Daily Output	Labor-Hours	Unit	Material	2019 Bare Costs Labor	Equipment	Total	Total Incl O&P
5340	Circuit breaker	2 Elec	.96	16.667	Ea.	16,200	1,000		17,200	19,300
5400	Starter size 5, autotransformer, NEMA 1, type A, fusible		.70	22.857		16,000	1,375		17,375	19,700
5420	Circuit breaker		.70	22.857		13,300	1,375		14,675	16,700
5440	Type B, fusible		.70	22.857		17,500	1,375		18,875	21,400
5460	Circuit breaker		.70	22.857		14,600	1,375		15,975	18,100
5480	NEMA 12, type A, fusible		.68	23.529		16,200	1,425		17,625	19,900
5500	Circuit breaker		.68	23.529		13,400	1,425		14,825	16,900
5520	Type B, fusible		.68	23.529		17,700	1,425		19,125	21,600
5540	Circuit breakers		.68	23.529		14,700	1,425		16,125	18,300
5600	Starter size 5, space only, NEMA 1, type A, fusible	1 Elec	12	.667		2,150	40		2,190	2,425
5620	Circuit breaker		12	.667		2,150	40		2,190	2,425
5640	Type B, fusible		12	.667		2,150	40		2,190	2,425
5660	Circuit breaker		12	.667		1,400	40		1,440	1,600
5680	NEMA 12, type A, fusible		11	.727		2,275	43.50		2,318.50	2,575
5700	Circuit breaker		11	.727		1,500	43.50		1,543.50	1,725
5720	Type B, fusible		11	.727		2,275	43.50		2,318.50	2,575
5740	Circuit breaker		11	.727		1,500	43.50		1,543.50	1,725
5800	Fuse, light contactor NEMA 1, type A, 30 amp		2.70	2.963		1,525	178		1,703	1,950
5820	60 amp		2	4		1,725	240		1,965	2,250
5840	100 amp		1	8		3,275	480		3,755	4,325
5860	200 amp		.80	10		7,600	600		8,200	9,250
5880	Type B, 30 amp		2.70	2.963		1,650	178		1,828	2,100
5900	60 amp		2	4		1,850	240		2,090	2,375
5920	100 amp		1	8		3,600	480		4,080	4,675
5940	200 amp	2 Elec	1.60	10		8,325	600		8,925	10,000
5960	NEMA 12, type A, 30 amp	1 Elec	2.60	3.077		1,550	185		1,735	1,975
5980	60 amp		1.90	4.211		1,750	253		2,003	2,300
6000	100 amp		.95	8.421		3,325	505		3,830	4,425
6020	200 amp	2 Elec	1.50	10.667		7,700	640		8,340	9,425
6040	Type B, 30 amp	1 Elec	2.60	3.077		1,675	185		1,860	2,125
6060	60 amp		1.90	4.211		1,875	253		2,128	2,450
6080	100 amp		.95	8.421		3,675	505		4,180	4,775
6100	200 amp	2 Elec	1.50	10.667		8,450	640		9,090	10,300
6200	Circuit breaker, light contactor NEMA 1, type A, 30 amp	1 Elec	2.70	2.963		1,650	178		1,828	2,100
6220	60 amp		2	4		1,875	240		2,115	2,425
6240	100 amp		1	8		2,900	480		3,380	3,925
6260	200 amp	2 Elec	1.60	10		6,300	600		6,900	7,825
6280	Type B, 30 amp	1 Elec	2.70	2.963		1,825	178		2,003	2,275
6300	60 amp		2	4		2,075	240		2,315	2,650
6320	100 amp		1	8		3,175	480		3,655	4,200
6340	200 amp	2 Elec	1.60	10		6,875	600		7,475	8,450
6360	NEMA 12, type A, 30 amp	1 Elec	2.60	3.077		1,675	185		1,860	2,125
6380	60 amp		1.90	4.211		1,900	253		2,153	2,475
6400	100 amp		.95	8.421		2,950	505		3,455	4,000
6420	200 amp	2 Elec	1.50	10.667		6,325	640		6,965	7,900
6440	Type B, 30 amp	1 Elec	2.60	3.077		1,975	185		2,160	2,450
6460	60 amp		1.90	4.211		2,100	253		2,353	2,700
6480	100 amp		.95	8.421		3,225	505		3,730	4,300
6500	200 amp	2 Elec	1.50	10.667		6,950	640		7,590	8,600
6600	Fusible switch, NEMA 1, type A, 30 amp	1 Elec	5.30	1.509		1,100	90.50		1,190.50	1,325
6620	60 amp		5	1.600		1,175	96		1,271	1,450
6640	100 amp		4	2		1,300	120		1,420	1,600
6660	200 amp		3.20	2.500		2,250	150		2,400	2,700

26 24 19 – Motor-Control Centers

26 24 19.20 Motor Control Center Components	Crew	Daily Output	Labor-Hours	Unit	Material	2019 Bare Costs Labor	Equipment	Total	Total Incl O&P
6680	400 amp	2 Elec	4.60	3.478	Ea.	5,275	209		5,484	6,100
6700	600 amp		3.20	5		5,700	300		6,000	6,725
6720	800 amp		2.60	6.154		15,700	370		16,070	17,800
6740	NEMA 12, type A, 30 amp	1 Elec	5.20	1.538		1,125	92.50		1,217.50	1,400
6760	60 amp		4.90	1.633		1,200	98		1,298	1,475
6780	100 amp		3.90	2.051		1,325	123		1,448	1,625
6800	200 amp		3.10	2.581		2,300	155		2,455	2,750
6820	400 amp	2 Elec	4.40	3.636		5,375	218		5,593	6,250
6840	600 amp		3	5.333		5,875	320		6,195	6,925
6860	800 amp		2.40	6.667		15,800	400		16,200	18,000
6900	Circuit breaker, NEMA 1, type A, 30 amp	1 Elec	5.30	1.509		1,000	90.50		1,090.50	1,225
6920	60 amp		5	1.600		1,000	96		1,096	1,250
6940	100 amp		4	2		1,000	120		1,120	1,275
6960	225 amp		3.20	2.500		1,800	150		1,950	2,200
6980	400 amp	2 Elec	4.60	3.478		3,425	209		3,634	4,075
7000	600 amp		3.20	5		4,000	300		4,300	4,850
7020	800 amp		2.60	6.154		8,375	370		8,745	9,775
7040	NEMA 12, type A, 30 amp	1 Elec	5.20	1.538		1,050	92.50		1,142.50	1,300
7060	60 amp		4.90	1.633		1,050	98		1,148	1,300
7080	100 amp		3.90	2.051		1,050	123		1,173	1,325
7100	225 amp		3.10	2.581		1,825	155		1,980	2,250
7120	400 amp	2 Elec	4.40	3.636		3,500	218		3,718	4,150
7140	600 amp		3	5.333		4,075	320		4,395	4,950
7160	800 amp		2.40	6.667		8,550	400		8,950	10,000
7300	Incoming line, main lug only, 600 amp, alum., NEMA 1		1.60	10		1,275	600		1,875	2,300
7320	NEMA 12		1.50	10.667		1,300	640		1,940	2,400
7340	Copper, NEMA 1		1.60	10		1,350	600		1,950	2,375
7360	800 amp, alum., NEMA 1		1.50	10.667		3,200	640		3,840	4,475
7380	NEMA 12		1.40	11.429		3,250	685		3,935	4,600
7400	Copper, NEMA 1		1.50	10.667		3,325	640		3,965	4,625
7420	1,200 amp, copper, NEMA 1		1.40	11.429		3,450	685		4,135	4,825
7440	Incoming line, fusible switch, 400 amp, alum., NEMA 1		1.20	13.333		4,400	800		5,200	6,050
7460	NEMA 12		1.10	14.545		4,500	875		5,375	6,250
7480	Copper, NEMA 1		1.20	13.333		4,475	800		5,275	6,125
7500	600 amp, alum., NEMA 1		1.10	14.545		5,375	875		6,250	7,225
7520	NEMA 12		1	16		5,475	960		6,435	7,450
7540	Copper, NEMA 1		1.10	14.545		5,450	875		6,325	7,300
7560	Incoming line, circuit breaker, 225 amp, alum., NEMA 1		1.20	13.333		2,300	800		3,100	3,725
7580	NEMA 12		1.10	14.545		2,350	875		3,225	3,875
7600	Copper, NEMA 1		1.20	13.333		2,375	800		3,175	3,800
7620	400 amp, alum., NEMA 1		1.20	13.333		3,425	800		4,225	4,975
7640	NEMA 12		1.10	14.545		3,500	875		4,375	5,125
7660	Copper, NEMA 1		1.20	13.333		3,500	800		4,300	5,050
7680	600 amp, alum., NEMA 1		1.10	14.545		4,000	875		4,875	5,700
7700	NEMA 12		1	16		4,075	960		5,035	5,900
7720	Copper, NEMA 1		1.10	14.545		4,075	875		4,950	5,775
7740	800 amp, copper, NEMA 1		.90	17.778		8,375	1,075		9,450	10,800
7760	Incoming line, for copper bus, add					126			126	139
7780	For 65,000 amp bus bracing, add					188			188	207
7800	For NEMA 3R enclosure, add					5,900			5,900	6,500
7820	For NEMA 12 enclosure, add					162			162	179
7840	For 1/4" x 1" ground bus, add	1 Elec	16	.500		105	30		135	160
7860	For 1/4" x 2" ground bus, add	"	12	.667		105	40		145	175

26 24 19.20 Motor Control Center Components

		Crew	Daily Output	Labor-Hours	Unit	Material	2019 Bare Costs Labor	2019 Bare Costs Equipment	Total	Total Incl O&P
7900	Main rating basic section, alum., NEMA 1, 800 amp	2 Elec	1.40	11.429	Ea.	345	685		1,030	1,400
7920	1,200 amp	"	1.20	13.333		675	800		1,475	1,950
7940	For copper bus, add					505			505	555
7960	For 65,000 amp bus bracing, add					340			340	375
7980	For NEMA 3R enclosure, add					5,900			5,900	6,500
8000	For NEMA 12, enclosure, add					162			162	179
8020	For 1/4" x 1" ground bus, add	1 Elec	16	.500		105	30		135	160
8040	For 1/4" x 2" ground bus, add		12	.667		105	40		145	175
8060	Unit devices, pilot light, standard		16	.500		90.50	30		120.50	145
8080	Pilot light, push to test		16	.500		126	30		156	184
8100	Pilot light, standard, and push button		12	.667		217	40		257	298
8120	Pilot light, push to test, and push button		12	.667		253	40		293	340
8140	Pilot light, standard, and select switch		12	.667		217	40		257	298
8160	Pilot light, push to test, and select switch		12	.667		253	40		293	340

26 24 19.30 Motor Control Center

		Crew	Daily Output	Labor-Hours	Unit	Material	2019 Bare Costs Labor	2019 Bare Costs Equipment	Total	Total Incl O&P
0010	**MOTOR CONTROL CENTER** Consists of starters & structures R262419-60									
0050	Starters, class 1, type B, comb. MCP, FVNR, with									
0100	control transformer, 10 HP, size 1, 12" high	1 Elec	2.70	2.963	Ea.	1,800	178		1,978	2,250
0200	25 HP, size 2, 18" high	2 Elec	4	4		2,050	240		2,290	2,625
0300	50 HP, size 3, 24" high		2	8		3,175	480		3,655	4,225
0350	75 HP, size 4, 24" high		1.60	10		4,225	600		4,825	5,550
0400	100 HP, size 4, 30" high		1.40	11.429		5,575	685		6,260	7,150
0500	200 HP, size 5, 48" high		1	16		8,325	960		9,285	10,600
0600	400 HP, size 6, 72" high		.80	20		17,100	1,200		18,300	20,600
0800	Structures, 600 amp, 22,000 rms, takes any									
0900	combination of starters up to 72" high	2 Elec	1.60	10	Ea.	1,950	600		2,550	3,025
1000	Back to back, 72" front & 66" back	"	1.20	13.333		2,625	800		3,425	4,075
1100	For copper bus, add per structure					277			277	305
1200	For NEMA 12, add per structure					160			160	176
1300	For 42,000 rms, add per structure					227			227	250
1400	For 100,000 rms, size 1 & 2, add					710			710	780
1500	Size 3, add					1,150			1,150	1,250
1600	Size 4, add					925			925	1,025
1700	For pilot lights, add per starter	1 Elec	16	.500		124	30		154	182
1800	For push button, add per starter		16	.500		124	30		154	182
1900	For auxiliary contacts, add per starter		16	.500		202	30		232	268

26 24 19.40 Motor Starters and Controls

		Crew	Daily Output	Labor-Hours	Unit	Material	2019 Bare Costs Labor	2019 Bare Costs Equipment	Total	Total Incl O&P
0010	**MOTOR STARTERS AND CONTROLS** R262419-65									
0050	Magnetic, FVNR, with enclosure and heaters, 480 volt									
0080	2 HP, size 00	1 Elec	3.50	2.286	Ea.	199	137		336	425
0100	5 HP, size 0		2.30	3.478		290	209		499	630
0200	10 HP, size 1		1.60	5		266	300		566	740
0300	25 HP, size 2 R263413-33	2 Elec	2.20	7.273		500	435		935	1,200
0400	50 HP, size 3		1.80	8.889		815	535		1,350	1,700
0500	100 HP, size 4		1.20	13.333		1,800	800		2,600	3,175
0600	200 HP, size 5		.90	17.778		4,225	1,075		5,300	6,250
0610	400 HP, size 6		.80	20		18,100	1,200		19,300	21,700
0620	NEMA 7, 5 HP, size 0	1 Elec	1.60	5		1,350	300		1,650	1,925
0630	10 HP, size 1	"	1.10	7.273		1,400	435		1,835	2,200
0640	25 HP, size 2	2 Elec	1.80	8.889		2,275	535		2,810	3,300
0650	50 HP, size 3		1.20	13.333		3,425	800		4,225	4,950
0660	100 HP, size 4		.90	17.778		5,500	1,075		6,575	7,675

232

For customer support on your Electrical Costs with RSMeans data, call 800.448.8182.

26 24 19 – Motor-Control Centers

26 24 19.40 Motor Starters and Controls		Crew	Daily Output	Labor-Hours	Unit	Material	2019 Bare Costs Labor	Equipment	Total	Total Incl O&P
0670	200 HP, size 5	2 Elec	.50	32	Ea.	13,200	1,925		15,125	17,400
0700	Combination, with motor circuit protectors, 5 HP, size 0	1 Elec	1.80	4.444		990	267		1,257	1,500
0800	10 HP, size 1	"	1.30	6.154		1,025	370		1,395	1,675
0900	25 HP, size 2	2 Elec	2	8		1,425	480		1,905	2,300
1000	50 HP, size 3		1.32	12.121		2,100	730		2,830	3,375
1200	100 HP, size 4	▼	.80	20		4,500	1,200		5,700	6,750
1220	NEMA 7, 5 HP, size 0	1 Elec	1.30	6.154		3,050	370		3,420	3,925
1230	10 HP, size 1	"	1	8		3,125	480		3,605	4,175
1240	25 HP, size 2	2 Elec	1.32	12.121		4,175	730		4,905	5,675
1250	50 HP, size 3		.80	20		6,900	1,200		8,100	9,400
1260	100 HP, size 4		.60	26.667		10,800	1,600		12,400	14,200
1270	200 HP, size 5	▼	.40	40		23,400	2,400		25,800	29,300
1400	Combination, with fused switch, 5 HP, size 0	1 Elec	1.80	4.444		595	267		862	1,050
1600	10 HP, size 1	"	1.30	6.154		640	370		1,010	1,250
1800	25 HP, size 2	2 Elec	2	8		1,025	480		1,505	1,875
2000	50 HP, size 3		1.32	12.121		1,750	730		2,480	3,000
2200	100 HP, size 4	▼	.80	20		3,050	1,200		4,250	5,175
2610	NEMA 4, with start-stop push button, size 1	1 Elec	1.30	6.154		1,600	370		1,970	2,325
2620	Size 2	2 Elec	2	8		2,150	480		2,630	3,075
2630	Size 3		1.32	12.121		3,400	730		4,130	4,825
2640	Size 4	▼	.80	20	▼	5,225	1,200		6,425	7,550
2650	NEMA 4, FVNR, including control transformer									
2660	Size 1	2 Elec	2.60	6.154	Ea.	1,550	370		1,920	2,250
2670	Size 2		2	8		2,200	480		2,680	3,150
2680	Size 3		1.32	12.121		3,525	730		4,255	4,950
2690	Size 4	▼	.80	20		5,425	1,200		6,625	7,775
2710	Magnetic, FVR, control circuit transformer, NEMA 1, size 1	1 Elec	1.30	6.154		910	370		1,280	1,550
2720	Size 2	2 Elec	2	8		1,450	480		1,930	2,325
2730	Size 3		1.32	12.121		2,225	730		2,955	3,500
2740	Size 4	▼	.80	20		4,825	1,200		6,025	7,100
2760	NEMA 4, size 1	1 Elec	1.10	7.273		1,300	435		1,735	2,100
2770	Size 2	2 Elec	1.60	10		2,100	600		2,700	3,225
2780	Size 3		1.20	13.333		3,175	800		3,975	4,700
2790	Size 4	▼	.70	22.857		6,525	1,375		7,900	9,225
2820	NEMA 12, size 1	1 Elec	1.10	7.273		1,075	435		1,510	1,825
2830	Size 2	2 Elec	1.60	10		1,700	600		2,300	2,775
2840	Size 3		1.20	13.333		2,675	800		3,475	4,150
2850	Size 4	▼	.70	22.857		5,500	1,375		6,875	8,100
2870	Combination FVR, fused, w/control XFMR & PB, NEMA 1, size 1	1 Elec	1	8		1,575	480		2,055	2,450
2880	Size 2	2 Elec	1.50	10.667		2,250	640		2,890	3,400
2890	Size 3		1.10	14.545		3,325	875		4,200	4,950
2900	Size 4	▼	.70	22.857		6,750	1,375		8,125	9,475
2910	NEMA 4, size 1	1 Elec	.90	8.889		2,325	535		2,860	3,375
2920	Size 2	2 Elec	1.40	11.429		3,400	685		4,085	4,750
2930	Size 3		1	16		5,350	960		6,310	7,325
2940	Size 4	▼	.60	26.667		8,900	1,600		10,500	12,200
2950	NEMA 12, size 1	1 Elec	1	8		1,775	480		2,255	2,675
2960	Size 2	2 Elec	1.40	11.429		2,500	685		3,185	3,800
2970	Size 3		1	16		3,700	960		4,660	5,475
2980	Size 4		.60	26.667		7,325	1,600		8,925	10,500
3010	Manual, single phase, w/pilot, 1 pole, 120 V, NEMA 1	1 Elec	6.40	1.250		79.50	75		154.50	200
3020	NEMA 4		4	2		365	120		485	585
3030	2 pole, 120/240 V, NEMA 1		6.40	1.250		112	75		187	235

26 24 19 – Motor-Control Centers

26 24 19.40 Motor Starters and Controls	Crew	Daily Output	Labor-Hours	Unit	Material	2019 Bare Costs Labor	Equipment	Total	Total Incl O&P
3040	NEMA 4	1 Elec	4	2	Ea.	299	120		419	510
3041	3 phase, 3 pole, 600 V, NEMA 1		5.50	1.455		246	87.50		333.50	400
3042	NEMA 4		3.50	2.286		405	137		542	655
3043	NEMA 12	▼	3.50	2.286		272	137		409	505
3070	Auxiliary contact, normally open				▼	91.50			91.50	100
3500	Magnetic FVNR with NEMA 12, enclosure & heaters, 480 volt									
3600	5 HP, size 0	1 Elec	2.20	3.636	Ea.	240	218		458	590
3700	10 HP, size 1	"	1.50	5.333		360	320		680	880
3800	25 HP, size 2	2 Elec	2	8		675	480		1,155	1,450
3900	50 HP, size 3		1.60	10		1,050	600		1,650	2,050
4000	100 HP, size 4		1	16		2,475	960		3,435	4,150
4100	200 HP, size 5	▼	.80	20		5,975	1,200		7,175	8,375
4200	Combination, with motor circuit protectors, 5 HP, size 0	1 Elec	1.70	4.706		795	283		1,078	1,300
4300	10 HP, size 1	"	1.20	6.667		825	400		1,225	1,500
4400	25 HP, size 2	2 Elec	1.80	8.889		1,225	535		1,760	2,150
4500	50 HP, size 3		1.20	13.333		2,025	800		2,825	3,425
4600	100 HP, size 4	▼	.74	21.622		4,550	1,300		5,850	6,925
4700	Combination, with fused switch, 5 HP, size 0	1 Elec	1.70	4.706		715	283		998	1,200
4800	10 HP, size 1	"	1.20	6.667		745	400		1,145	1,425
4900	25 HP, size 2	2 Elec	1.80	8.889		1,125	535		1,660	2,050
5000	50 HP, size 3		1.20	13.333		1,825	800		2,625	3,200
5100	100 HP, size 4	▼	.74	21.622	▼	3,700	1,300		5,000	5,975
5200	Factory installed controls, adders to size 0 thru 5									
5300	Start-stop push button	1 Elec	32	.250	Ea.	50.50	15		65.50	78
5400	Hand-off-auto-selector switch		32	.250		50.50	15		65.50	78
5500	Pilot light		32	.250		94.50	15		109.50	127
5600	Start-stop-pilot		32	.250		145	15		160	182
5700	Auxiliary contact, NO or NC		32	.250		69	15		84	98.50
5800	NO-NC		32	.250		138	15		153	175
5810	Magnetic FVR, NEMA 7, w/heaters, size 1		.66	12.121		3,050	730		3,780	4,425
5830	Size 2	2 Elec	1.10	14.545		5,100	875		5,975	6,925
5840	Size 3		.70	22.857		8,200	1,375		9,575	11,100
5850	Size 4	▼	.60	26.667		9,450	1,600		11,050	12,800
5860	Combination w/circuit breakers, heaters, control XFMR PB, size 1	1 Elec	.60	13.333		1,675	800		2,475	3,025
5870	Size 2	2 Elec	.80	20		2,150	1,200		3,350	4,150
5880	Size 3		.50	32		2,850	1,925		4,775	6,000
5890	Size 4	▼	.40	40		5,650	2,400		8,050	9,775
5900	Manual, 240 volt, 0.75 HP motor	1 Elec	4	2		53.50	120		173.50	238
5910	2 HP motor		4	2		145	120		265	340
6000	Magnetic, 240 volt, 1 or 2 pole, 0.75 HP motor		4	2		201	120		321	400
6020	2 HP motor		4	2		222	120		342	425
6040	5 HP motor		3	2.667		320	160		480	590
6060	10 HP motor		2.30	3.478		790	209		999	1,175
6100	3 pole, 0.75 HP motor		3	2.667		199	160		359	460
6120	5 HP motor		2.30	3.478		270	209		479	605
6140	10 HP motor		1.60	5		510	300		810	1,000
6160	15 HP motor		1.60	5		510	300		810	1,000
6180	20 HP motor	▼	1.10	7.273		830	435		1,265	1,550
6200	25 HP motor	2 Elec	2.20	7.273		790	435		1,225	1,525
6210	30 HP motor		1.80	8.889		790	535		1,325	1,675
6220	40 HP motor		1.80	8.889		1,750	535		2,285	2,725
6230	50 HP motor		1.80	8.889		1,750	535		2,285	2,725
6240	60 HP motor		1.20	13.333		4,100	800		4,900	5,700

234

For customer support on your Electrical Costs with RSMeans data, call 800.448.8182.

26 24 19.40 Motor Starters and Controls		Crew	Daily Output	Labor-Hours	Unit	Material	2019 Bare Costs Labor	Equipment	Total	Total Incl O&P
6250	75 HP motor	2 Elec	1.20	13.333	Ea.	4,100	800		4,900	5,700
6260	100 HP motor		1.20	13.333		4,100	800		4,900	5,700
6270	125 HP motor		.90	17.778		11,500	1,075		12,575	14,300
6280	150 HP motor		.90	17.778		11,500	1,075		12,575	14,300
6290	200 HP motor	▼	.90	17.778		11,500	1,075		12,575	14,300
6400	Starter & nonfused disconnect, 240 volt, 1-2 pole,0 .75 HP motor	1 Elec	2	4		269	240		509	655
6410	2 HP motor		2	4		291	240		531	680
6420	5 HP motor		1.80	4.444		385	267		652	825
6430	10 HP motor		1.40	5.714		885	345		1,230	1,475
6440	3 pole, 0.75 HP motor		1.60	5		267	300		567	745
6450	5 HP motor		1.40	5.714		340	345		685	880
6460	10 HP motor		1.10	7.273		600	435		1,035	1,300
6470	15 HP motor	▼	1	8		600	480		1,080	1,375
6480	20 HP motor	2 Elec	1.50	10.667		1,050	640		1,690	2,100
6490	25 HP motor		1.50	10.667		1,000	640		1,640	2,050
6500	30 HP motor		1.30	12.308		1,000	740		1,740	2,200
6510	40 HP motor		1.24	12.903		2,125	775		2,900	3,500
6520	50 HP motor		1.12	14.286		2,125	860		2,985	3,625
6530	60 HP motor		.90	17.778		4,475	1,075		5,550	6,525
6540	75 HP motor		.76	21.053		4,475	1,275		5,750	6,800
6550	100 HP motor		.70	22.857		4,475	1,375		5,850	6,975
6560	125 HP motor		.60	26.667		12,400	1,600		14,000	16,000
6570	150 HP motor		.52	30.769		12,400	1,850		14,250	16,400
6580	200 HP motor	▼	.50	32		12,400	1,925		14,325	16,500
6600	Starter & fused disconnect, 240 volt, 1-2 pole, 0.75 HP motor	1 Elec	2	4		287	240		527	675
6610	2 HP motor		2	4		310	240		550	700
6620	5 HP motor		1.80	4.444		405	267		672	845
6630	10 HP motor		1.40	5.714		935	345		1,280	1,525
6640	3 pole, 0.75 HP motor		1.60	5		285	300		585	765
6650	5 HP motor		1.40	5.714		355	345		700	900
6660	10 HP motor		1.10	7.273		655	435		1,090	1,375
6690	15 HP motor	▼	1	8		655	480		1,135	1,425
6700	20 HP motor	2 Elec	1.60	10		1,075	600		1,675	2,075
6710	25 HP motor		1.60	10		1,025	600		1,625	2,050
6720	30 HP motor		1.40	11.429		1,025	685		1,710	2,175
6730	40 HP motor		1.20	13.333		2,275	800		3,075	3,725
6740	50 HP motor		1.20	13.333		2,275	800		3,075	3,725
6750	60 HP motor		.90	17.778		4,625	1,075		5,700	6,700
6760	75 HP motor		.90	17.778		4,625	1,075		5,700	6,700
6770	100 HP motor		.70	22.857		4,625	1,375		6,000	7,150
6780	125 HP motor		.54	29.630	▼	12,900	1,775		14,675	16,900
6790	Combination starter & nonfusible disconnect									
6800	240 volt, 1-2 pole, 0.75 HP motor	1 Elec	2	4	Ea.	660	240		900	1,100
6810	2 HP motor		2	4		510	240		750	920
6820	5 HP motor		1.50	5.333		645	320		965	1,200
6830	10 HP motor		1.20	6.667		985	400		1,385	1,675
6840	3 pole, 0.75 HP motor		1.80	4.444		600	267		867	1,075
6850	5 HP motor		1.30	6.154		635	370		1,005	1,250
6860	10 HP motor		1	8		985	480		1,465	1,800
6870	15 HP motor	▼	1	8		985	480		1,465	1,800
6880	20 HP motor	2 Elec	1.32	12.121		1,625	730		2,355	2,850
6890	25 HP motor		1.32	12.121		1,625	730		2,355	2,850
6900	30 HP motor	▼	1.32	12.121	▼	1,625	730		2,355	2,850

26 24 Switchboards and Panelboards

26 24 19 – Motor-Control Centers

26 24 19.40 Motor Starters and Controls	Crew	Daily Output	Labor-Hours	Unit	Material	2019 Bare Costs Labor	Equipment	Total	Total Incl O&P
6910	40 HP motor	2 Elec	.80	20	Ea.	3,100	1,200		4,300	5,225
6920	50 HP motor		.80	20		3,100	1,200		4,300	5,225
6930	60 HP motor		.70	22.857		6,950	1,375		8,325	9,675
6940	75 HP motor		.70	22.857		6,950	1,375		8,325	9,675
6950	100 HP motor		.70	22.857		6,950	1,375		8,325	9,675
6960	125 HP motor		.60	26.667		18,300	1,600		19,900	22,500
6970	150 HP motor		.60	26.667		18,300	1,600		19,900	22,500
6980	200 HP motor		.60	26.667		18,300	1,600		19,900	22,500
6990	Combination starter and fused disconnect									
7000	240 volt, 1-2 pole, 0.75 HP motor	1 Elec	2	4	Ea.	635	240		875	1,050
7010	2 HP motor		2	4		635	240		875	1,050
7020	5 HP motor		1.50	5.333		665	320		985	1,200
7030	10 HP motor		1.20	6.667		1,025	400		1,425	1,725
7040	3 pole, 0.75 HP motor		1.80	4.444		635	267		902	1,100
7050	5 HP motor		1.30	6.154		665	370		1,035	1,275
7060	10 HP motor		1	8		1,025	480		1,505	1,850
7070	15 HP motor		1	8		1,025	480		1,505	1,850
7080	20 HP motor	2 Elec	1.32	12.121		1,725	730		2,455	2,975
7090	25 HP motor		1.32	12.121		1,725	730		2,455	2,975
7100	30 HP motor		1.32	12.121		1,725	730		2,455	2,975
7110	40 HP motor		.80	20		3,275	1,200		4,475	5,400
7120	50 HP motor		.80	20		3,275	1,200		4,475	5,400
7130	60 HP motor		.80	20		7,300	1,200		8,500	9,825
7140	75 HP motor		.70	22.857		7,300	1,375		8,675	10,100
7150	100 HP motor		.70	22.857		7,300	1,375		8,675	10,100
7160	125 HP motor		.70	22.857		19,300	1,375		20,675	23,300
7170	150 HP motor		.60	26.667		19,300	1,600		20,900	23,600
7180	200 HP motor		.60	26.667		19,300	1,600		20,900	23,600
7190	Combination starter & circuit breaker disconnect									
7200	240 volt, 1-2 pole, 0.75 HP motor	1 Elec	2	4	Ea.	630	240		870	1,050
7210	2 HP motor		2	4		630	240		870	1,050
7220	5 HP motor		1.50	5.333		660	320		980	1,200
7230	10 HP motor		1.20	6.667		1,000	400		1,400	1,700
7240	3 pole, 0.75 HP motor		1.80	4.444		645	267		912	1,100
7250	5 HP motor		1.30	6.154		680	370		1,050	1,300
7260	10 HP motor		1	8		1,025	480		1,505	1,850
7270	15 HP motor		1	8		1,025	480		1,505	1,850
7280	20 HP motor	2 Elec	1.32	12.121		1,750	730		2,480	3,000
7290	25 HP motor		1.32	12.121		1,750	730		2,480	3,000
7300	30 HP motor		1.32	12.121		1,750	730		2,480	3,000
7310	40 HP motor		.80	20		3,800	1,200		5,000	5,975
7320	50 HP motor		.80	20		3,800	1,200		5,000	5,975
7330	60 HP motor		.80	20		8,750	1,200		9,950	11,500
7340	75 HP motor		.70	22.857		8,750	1,375		10,125	11,700
7350	100 HP motor		.70	22.857		8,750	1,375		10,125	11,700
7360	125 HP motor		.70	22.857		19,000	1,375		20,375	23,000
7370	150 HP motor		.60	26.667		19,000	1,600		20,600	23,300
7380	200 HP motor		.60	26.667		19,000	1,600		20,600	23,300
7400	Magnetic FVNR with enclosure & heaters, 2 pole,									
7410	230 volt, 1 HP, size 00	1 Elec	4	2	Ea.	181	120		301	380
7420	2 HP, size 0		4	2		202	120		322	400
7430	3 HP, size 1		3	2.667		232	160		392	495
7440	5 HP, size 1P		3	2.667		297	160		457	565

26 24 19.40 Motor Starters and Controls		Crew	Daily Output	Labor-Hours	Unit	Material	2019 Bare Costs Labor	Equipment	Total	Total Incl O&P
7450	115 volt, 1/3 HP, size 00	1 Elec	4	2	Ea.	181	120		301	380
7460	1 HP, size 0		4	2		202	120		322	400
7470	2 HP, size 1		3	2.667		232	160		392	495
7480	3 HP, size 1P		3	2.667		288	160		448	555
7500	3 pole, 480 volt, 600 HP, size 7	2 Elec	.70	22.857		15,300	1,375		16,675	18,900
7590	Magnetic FVNR with heater, NEMA 1									
7600	600 volt, 3 pole, 5 HP motor	1 Elec	2.30	3.478	Ea.	241	209		450	575
7610	10 HP motor	"	1.60	5		270	300		570	745
7620	25 HP motor	2 Elec	2.20	7.273		510	435		945	1,200
7630	30 HP motor		1.80	8.889		830	535		1,365	1,700
7640	40 HP motor		1.80	8.889		830	535		1,365	1,700
7650	50 HP motor		1.80	8.889		830	535		1,365	1,700
7660	60 HP motor		1.20	13.333		1,850	800		2,650	3,225
7670	75 HP motor		1.20	13.333		2,225	800		3,025	3,650
7680	100 HP motor		1.20	13.333		2,950	800		3,750	4,450
7690	125 HP motor		.90	17.778		4,200	1,075		5,275	6,225
7700	150 HP motor		.90	17.778		4,400	1,075		5,475	6,450
7710	200 HP motor		.90	17.778		4,550	1,075		5,625	6,625
7750	Starter & nonfused disconnect, 600 volt, 3 pole, 5 HP motor	1 Elec	1.40	5.714		320	345		665	860
7760	10 HP motor	"	1.10	7.273		350	435		785	1,025
7770	25 HP motor	2 Elec	1.50	10.667		585	640		1,225	1,600
7780	30 HP motor		1.30	12.308		1,050	740		1,790	2,250
7790	40 HP motor		1.30	12.308		1,050	740		1,790	2,250
7800	50 HP motor		1.30	12.308		1,050	740		1,790	2,250
7810	60 HP motor		.92	17.391		2,175	1,050		3,225	3,925
7820	75 HP motor		.92	17.391		2,550	1,050		3,600	4,350
7830	100 HP motor		.84	19.048		3,275	1,150		4,425	5,325
7840	125 HP motor		.70	22.857		4,725	1,375		6,100	7,225
7850	150 HP motor		.70	22.857		4,900	1,375		6,275	7,450
7860	200 HP motor		.60	26.667		5,075	1,600		6,675	7,975
7870	Starter & fused disconnect, 600 volt, 3 pole, 5 HP motor	1 Elec	1.40	5.714		385	345		730	935
7880	10 HP motor	"	1.10	7.273		415	435		850	1,100
7890	25 HP motor	2 Elec	1.50	10.667		655	640		1,295	1,675
7900	30 HP starters		1.30	12.308		1,000	740		1,740	2,200
7910	40 HP motor		1.30	12.308		1,000	740		1,740	2,200
7920	50 HP motor		1.30	12.308		1,000	740		1,740	2,200
7930	60 HP motor		.92	17.391		2,150	1,050		3,200	3,925
7940	75 HP motorP		.92	17.391		2,525	1,050		3,575	4,325
7950	100 HP motor		.84	19.048		3,275	1,150		4,425	5,300
7960	125 HP motor		.70	22.857		4,650	1,375		6,025	7,175
7970	150 HP motor		.70	22.857		4,850	1,375		6,225	7,375
7980	200 HP motor		.60	26.667		5,000	1,600		6,600	7,900
7990	Combination starter and nonfusible disconnect									
8000	600 volt, 3 pole, 5 HP motor	1 Elec	1.80	4.444	Ea.	685	267		952	1,150
8010	10 HP motor	"	1.30	6.154		810	370		1,180	1,450
8020	25 HP motor	2 Elec	2	8		1,125	480		1,605	1,950
8030	30 HP motor		1.32	12.121		1,700	730		2,430	2,925
8040	40 HP motor		1.32	12.121		1,700	730		2,430	2,925
8050	50 HP motor		1.32	12.121		2,075	730		2,805	3,375
8060	60 HP motor		.80	20		3,225	1,200		4,425	5,325
8070	75 HP motor		.80	20		3,225	1,200		4,425	5,325
8080	100 HP motor		.80	20		4,125	1,200		5,325	6,325
8090	125 HP motor		.70	22.857		7,150	1,375		8,525	9,925

26 24 Switchboards and Panelboards

26 24 19 – Motor-Control Centers

26 24 19.40 Motor Starters and Controls	Crew	Daily Output	Labor-Hours	Unit	Material	2019 Bare Costs Labor	Equipment	Total	Total Incl O&P
8100	150 HP motor	2 Elec	.70	22.857	Ea.	7,250	1,375		8,625	10,000
8110	200 HP motor	▼	.70	22.857	▼	7,925	1,375		9,300	10,800
8140	Combination starter and fused disconnect									
8150	600 volt, 3 pole, 5 HP motor	1 Elec	1.80	4.444	Ea.	635	267		902	1,100
8160	10 HP motor	"	1.30	6.154		690	370		1,060	1,300
8170	25 HP motor	2 Elec	2	8		1,125	480		1,605	1,950
8180	30 HP motor		1.32	12.121		1,750	730		2,480	3,000
8190	40 HP motor		1.32	12.121		1,900	730		2,630	3,150
8200	50 HP motor		1.32	12.121		1,900	730		2,630	3,150
8210	60 HP motor		.80	20		3,300	1,200		4,500	5,425
8220	75 HP motor		.80	20		3,300	1,200		4,500	5,425
8230	100 HP motor		.80	20		3,300	1,200		4,500	5,425
8240	125 HP motor		.70	22.857		7,300	1,375		8,675	10,100
8250	150 HP motor		.70	22.857		7,300	1,375		8,675	10,100
8260	200 HP motor	▼	.70	22.857	▼	7,300	1,375		8,675	10,100
8290	Combination starter & circuit breaker disconnect									
8300	600 volt, 3 pole, 5 HP motor	1 Elec	1.80	4.444	Ea.	940	267		1,207	1,425
8310	10 HP motor	"	1.30	6.154		915	370		1,285	1,550
8320	25 HP motor	2 Elec	2	8		1,375	480		1,855	2,225
8330	30 HP motor		1.32	12.121		1,750	730		2,480	3,000
8340	40 HP motor		1.32	12.121		1,750	730		2,480	3,000
8350	50 HP motor		1.32	12.121		1,825	730		2,555	3,075
8360	60 HP motor		.80	20		3,800	1,200		5,000	5,975
8370	75 HP motor		.80	20		3,800	1,200		5,000	5,975
8380	100 HP motor		.80	20		3,950	1,200		5,150	6,150
8390	125 HP motor		.70	22.857		8,750	1,375		10,125	11,700
8400	150 HP motor		.70	22.857		8,350	1,375		9,725	11,200
8410	200 HP motor	▼	.70	22.857	▼	9,150	1,375		10,525	12,200
8430	Starter & circuit breaker disconnect									
8440	600 volt, 3 pole, 5 HP motor	1 Elec	1.40	5.714	Ea.	755	345		1,100	1,350
8450	10 HP motor	"	1.10	7.273		785	435		1,220	1,525
8460	25 HP motor	2 Elec	1.50	10.667		1,025	640		1,665	2,075
8470	30 HP motor		1.30	12.308		1,450	740		2,190	2,700
8480	40 HP motor		1.30	12.308		1,450	740		2,190	2,700
8490	50 HP motor		1.30	12.308		1,450	740		2,190	2,700
8500	60 HP motor		.92	17.391		3,475	1,050		4,525	5,375
8510	75 HP motor		.92	17.391		3,875	1,050		4,925	5,800
8520	100 HP motor		.84	19.048		4,600	1,150		5,750	6,750
8530	125 HP motor		.70	22.857		5,850	1,375		7,225	8,500
8540	150 HP motor		.70	22.857		6,050	1,375		7,425	8,700
8550	200 HP motor	▼	.60	26.667		7,375	1,600		8,975	10,500
8900	240 volt, 1-2 pole, 0.75 HP motor	1 Elec	2	4		715	240		955	1,150
8910	2 HP motor		2	4		740	240		980	1,175
8920	5 HP motor		1.80	4.444		835	267		1,102	1,325
8930	10 HP motor		1.40	5.714		1,400	345		1,745	2,050
8950	3 pole, 0.75 HP motor		1.60	5		715	300		1,015	1,225
8970	5 HP motor		1.40	5.714		785	345		1,130	1,375
8980	10 HP motor		1.10	7.273		1,125	435		1,560	1,900
8990	15 HP motor	▼	1	8		1,125	480		1,605	1,975
9100	20 HP motor	2 Elec	1.50	10.667		1,550	640		2,190	2,650
9110	25 HP motor		1.50	10.667		1,500	640		2,140	2,600
9120	30 HP motor		1.30	12.308		1,500	740		2,240	2,750
9130	40 HP motor	▼	1.24	12.903	▼	3,400	775		4,175	4,875

238

26 24 Switchboards and Panelboards

26 24 19 – Motor-Control Centers

26 24 19.40 Motor Starters and Controls		Crew	Daily Output	Labor-Hours	Unit	Material	2019 Bare Costs Labor	Equipment	Total	Total Incl O&P
9140	50 HP motor	2 Elec	1.12	14.286	Ea.	3,400	860		4,260	5,000
9150	60 HP motor		.90	17.778		6,900	1,075		7,975	9,200
9160	75 HP motor		.76	21.053		6,900	1,275		8,175	9,475
9170	100 HP motor		.70	22.857		6,900	1,375		8,275	9,650
9180	125 HP motor		.60	26.667		15,600	1,600		17,200	19,500
9190	150 HP motor		.52	30.769		15,600	1,850		17,450	19,900
9200	200 HP motor		.50	32		15,600	1,925		17,525	20,000

26 25 Low-Voltage Enclosed Bus Assemblies

26 25 13 – Low-Voltage Busways

26 25 13.10 Aluminum Bus Duct

26 25 13.10 Aluminum Bus Duct		Crew	Daily Output	Labor-Hours	Unit	Material	2019 Bare Costs Labor	Equipment	Total	Total Incl O&P
0010	**ALUMINUM BUS DUCT** 10 ft. long R262513-10									
0050	Indoor 3 pole 4 wire, plug-in, straight section, 225 amp	2 Elec	44	.364	L.F.	175	22		197	225
0100	400 amp		36	.444		220	26.50		246.50	282
0150	600 amp		32	.500		305	30		335	380
0200	800 amp		26	.615		355	37		392	445
0250	1,000 amp		24	.667		445	40		485	550
0270	1,200 amp		23	.696		545	42		587	660
0300	1,350 amp		22	.727		283	43.50		326.50	375
0310	1,600 amp		18	.889		300	53.50		353.50	410
0320	2,000 amp		16	1		395	60		455	525
0330	2,500 amp		14	1.143		450	68.50		518.50	595
0340	3,000 amp		12	1.333		545	80		625	720
0350	Feeder, 600 amp		34	.471		111	28.50		139.50	164
0400	800 amp		28	.571		425	34.50		459.50	515
0450	1,000 amp		26	.615		515	37		552	620
0455	1,200 amp		25	.640		645	38.50		683.50	770
0500	1,350 amp		24	.667		231	40		271	315
0550	1,600 amp		20	.800		890	48		938	1,050
0600	2,000 amp		18	.889		1,075	53.50		1,128.50	1,275
0620	2,500 amp		14	1.143		1,325	68.50		1,393.50	1,575
0630	3,000 amp		12	1.333		1,550	80		1,630	1,825
0640	4,000 amp		10	1.600		665	96		761	875
0650	Elbow, 225 amp		4.40	3.636	Ea.	835	218		1,053	1,250
0700	400 amp		3.80	4.211		845	253		1,098	1,300
0750	600 amp		3.40	4.706		850	283		1,133	1,350
0800	800 amp		3	5.333		880	320		1,200	1,450
0850	1,000 amp		2.80	5.714		1,125	345		1,470	1,750
0870	1,200 amp		2.70	5.926		1,400	355		1,755	2,075
0900	1,350 amp		2.60	6.154		1,250	370		1,620	1,925
0950	1,600 amp		2.40	6.667		1,400	400		1,800	2,125
1000	2,000 amp		2	8		1,525	480		2,005	2,400
1020	2,500 amp		1.80	8.889		1,800	535		2,335	2,800
1030	3,000 amp		1.60	10		2,100	600		2,700	3,200
1040	4,000 amp		1.40	11.429		3,375	685		4,060	4,750
1100	Cable tap box end, 225 amp		3.60	4.444		1,750	267		2,017	2,325
1150	400 amp		3.20	5		1,975	300		2,275	2,600
1200	600 amp		2.60	6.154		1,525	370		1,895	2,225
1250	800 amp		2.20	7.273		1,850	435		2,285	2,675
1300	1,000 amp		2	8		3,800	480		4,280	4,900
1320	1,200 amp		2	8		1,850	480		2,330	2,750

26 25 13 – Low-Voltage Busways

26 25 13.10 Aluminum Bus Duct		Crew	Daily Output	Labor-Hours	Unit	Material	2019 Bare Costs Labor	Equipment	Total	Total Incl O&P
1350	1,350 amp	2 Elec	1.60	10	Ea.	1,500	600		2,100	2,550
1400	1,600 amp		1.40	11.429		1,675	685		2,360	2,875
1450	2,000 amp		1.20	13.333		1,800	800		2,600	3,200
1460	2,500 amp		1	16		2,350	960		3,310	4,025
1470	3,000 amp		.80	20		2,625	1,200		3,825	4,675
1480	4,000 amp		.60	26.667		3,550	1,600		5,150	6,300
1500	Switchboard stub, 225 amp		5.80	2.759		1,625	166		1,791	2,050
1550	400 amp		5.40	2.963		1,650	178		1,828	2,100
1600	600 amp		4.60	3.478		1,850	209		2,059	2,325
1650	800 amp		4	4		1,725	240		1,965	2,250
1700	1,000 amp		3.20	5		1,725	300		2,025	2,350
1720	1,200 amp		3.10	5.161		1,775	310		2,085	2,400
1750	1,350 amp		3	5.333		1,675	320		1,995	2,325
1800	1,600 amp		2.60	6.154		1,800	370		2,170	2,525
1850	2,000 amp		2.40	6.667		1,800	400		2,200	2,600
1860	2,500 amp		2.20	7.273		1,975	435		2,410	2,825
1870	3,000 amp		2	8		2,175	480		2,655	3,125
1880	4,000 amp		1.80	8.889		2,400	535		2,935	3,450
1890	Tee fittings, 225 amp		3.20	5		875	300		1,175	1,425
1900	400 amp		2.80	5.714		875	345		1,220	1,475
1950	600 amp		2.60	6.154		875	370		1,245	1,525
2000	800 amp		2.40	6.667		930	400		1,330	1,625
2050	1,000 amp		2.20	7.273		985	435		1,420	1,725
2070	1,200 amp		2.10	7.619		1,150	460		1,610	1,950
2100	1,350 amp		2	8		1,600	480		2,080	2,475
2150	1,600 amp		1.60	10		1,925	600		2,525	3,025
2200	2,000 amp		1.20	13.333		2,125	800		2,925	3,550
2220	2,500 amp		1	16		2,550	960		3,510	4,225
2230	3,000 amp		.80	20		2,900	1,200		4,100	5,000
2240	4,000 amp		.60	26.667		4,825	1,600		6,425	7,725
2300	Wall flange, 600 amp		20	.800		262	48		310	360
2310	800 amp		16	1		262	60		322	380
2320	1,000 amp		13	1.231		262	74		336	400
2325	1,200 amp		12	1.333		262	80		342	405
2330	1,350 amp		10.80	1.481		262	89		351	420
2340	1,600 amp		9	1.778		262	107		369	445
2350	2,000 amp		8	2		262	120		382	465
2360	2,500 amp		6.60	2.424		262	146		408	505
2370	3,000 amp		5.40	2.963		262	178		440	555
2380	4,000 amp		4	4		385	240		625	785
2390	5,000 amp		3	5.333		385	320		705	905
2400	Vapor barrier		8	2		475	120		595	705
2420	Roof flange kit		4	4		775	240		1,015	1,225
2600	Expansion fitting, 225 amp		10	1.600		1,375	96		1,471	1,650
2610	400 amp		8	2		1,375	120		1,495	1,675
2620	600 amp		6	2.667		1,375	160		1,535	1,750
2630	800 amp		4.60	3.478		1,550	209		1,759	2,000
2640	1,000 amp		4	4		1,750	240		1,990	2,275
2650	1,350 amp		3.60	4.444		2,050	267		2,317	2,650
2660	1,600 amp		3.20	5		2,450	300		2,750	3,125
2670	2,000 amp		2.80	5.714		2,725	345		3,070	3,500
2680	2,500 amp		2.40	6.667		3,300	400		3,700	4,225
2690	3,000 amp		2	8		3,775	480		4,255	4,900

26 25 13.10 Aluminum Bus Duct		Crew	Daily Output	Labor-Hours	Unit	Material	2019 Bare Costs Labor	Equipment	Total	Total Incl O&P
2700	4,000 amp	2 Elec	1.60	10	Ea.	4,975	600		5,575	6,375
2800	Reducer nonfused, 400 amp		8	2		860	120		980	1,125
2810	600 amp		6	2.667		860	160		1,020	1,175
2820	800 amp		4.60	3.478		1,025	209		1,234	1,425
2830	1,000 amp		4	4		1,200	240		1,440	1,675
2840	1,350 amp		3.60	4.444		1,575	267		1,842	2,150
2850	1,600 amp		3.20	5		2,125	300		2,425	2,800
2860	2,000 amp		2.80	5.714		2,450	345		2,795	3,200
2870	2,500 amp		2.40	6.667		3,075	400		3,475	4,000
2880	3,000 amp		2	8		3,575	480		4,055	4,650
2890	4,000 amp		1.60	10		4,775	600		5,375	6,150
2950	Reducer fuse included, 225 amp		4.40	3.636		2,525	218		2,743	3,100
2960	400 amp		4.20	3.810		2,575	229		2,804	3,175
2970	600 amp		3.60	4.444		3,025	267		3,292	3,725
2980	800 amp		3.20	5		4,825	300		5,125	5,750
2990	1,000 amp		3	5.333		5,500	320		5,820	6,525
3000	1,200 amp		2.80	5.714		5,500	345		5,845	6,550
3010	1,600 amp		2.20	7.273		12,600	435		13,035	14,500
3020	2,000 amp		1.80	8.889		14,000	535		14,535	16,100
3100	Reducer circuit breaker, 225 amp		4.40	3.636		2,475	218		2,693	3,050
3110	400 amp		4.20	3.810		3,025	229		3,254	3,675
3120	600 amp		3.60	4.444		4,300	267		4,567	5,125
3130	800 amp		3.20	5		5,025	300		5,325	6,000
3140	1,000 amp		3	5.333		5,700	320		6,020	6,750
3150	1,200 amp		2.80	5.714		6,875	345		7,220	8,050
3160	1,600 amp		2.20	7.273		10,100	435		10,535	11,900
3170	2,000 amp		1.80	8.889		11,100	535		11,635	13,000
3250	Reducer circuit breaker, 75,000 AIC, 225 amp		4.40	3.636		3,875	218		4,093	4,600
3260	400 amp		4.20	3.810		3,875	229		4,104	4,625
3270	600 amp		3.60	4.444		5,200	267		5,467	6,125
3280	800 amp		3.20	5		5,700	300		6,000	6,725
3290	1,000 amp		3	5.333		9,175	320		9,495	10,600
3300	1,200 amp		2.80	5.714		9,175	345		9,520	10,600
3310	1,600 amp		2.20	7.273		10,100	435		10,535	11,900
3320	2,000 amp		1.80	8.889		11,100	535		11,635	13,000
3400	Reducer circuit breaker CLF 225 amp		4.40	3.636		4,000	218		4,218	4,725
3410	400 amp		4.20	3.810		4,750	229		4,979	5,575
3420	600 amp		3.60	4.444		7,100	267		7,367	8,200
3430	800 amp		3.20	5		7,400	300		7,700	8,600
3440	1,000 amp		3	5.333		7,725	320		8,045	8,975
3450	1,200 amp		2.80	5.714		10,100	345		10,445	11,600
3460	1,600 amp		2.20	7.273		10,100	435		10,535	11,900
3470	2,000 amp		1.80	8.889		11,100	535		11,635	13,000
3550	Ground bus added to bus duct, 225 amp		320	.050	L.F.	33	3		36	40.50
3560	400 amp		320	.050		33	3		36	40.50
3570	600 amp		280	.057		33	3.43		36.43	41
3580	800 amp		240	.067		33	4		37	42
3590	1,000 amp		200	.080		33	4.80		37.80	43
3600	1,350 amp		180	.089		33	5.35		38.35	44
3610	1,600 amp		160	.100		33	6		39	45
3620	2,000 amp		160	.100		33	6		39	45
3630	2,500 amp		140	.114		33	6.85		39.85	46.50
3640	3,000 amp		120	.133		33	8		41	48

26 25 13.10 Aluminum Bus Duct		Crew	Daily Output	Labor-Hours	Unit	Material	2019 Bare Costs Labor	Equipment	Total	Total Incl O&P
3650	4,000 amp	2 Elec	100	.160	L.F.	33	9.60		42.60	50.50
3810	High short circuit, 400 amp		36	.444		207	26.50		233.50	268
3820	600 amp		32	.500		370	30		400	450
3830	800 amp		26	.615		430	37		467	525
3840	1,000 amp		24	.667		535	40		575	650
3850	1,350 amp		22	.727		185	43.50		228.50	268
3860	1,600 amp		18	.889		213	53.50		266.50	315
3870	2,000 amp		16	1		250	60		310	365
3880	2,500 amp		14	1.143		415	68.50		483.50	560
3890	3,000 amp		12	1.333		470	80		550	640
3920	Cross, 225 amp		5.60	2.857	Ea.	1,325	172		1,497	1,700
3930	400 amp		4.60	3.478		1,325	209		1,534	1,750
3940	600 amp		4	4		1,325	240		1,565	1,800
3950	800 amp		3.40	4.706		1,375	283		1,658	1,950
3960	1,000 amp		3	5.333		1,450	320		1,770	2,075
3970	1,350 amp		2.80	5.714		2,350	345		2,695	3,100
3980	1,600 amp		2.20	7.273		2,800	435		3,235	3,725
3990	2,000 amp		1.80	8.889		3,075	535		3,610	4,175
4000	2,500 amp		1.60	10		3,625	600		4,225	4,900
4010	3,000 amp		1.20	13.333		4,175	800		4,975	5,800
4020	4,000 amp		1	16		6,375	960		7,335	8,450
4040	Cable tap box center, 225 amp		3.60	4.444		2,750	267		3,017	3,425
4050	400 amp		3.20	5		3,125	300		3,425	3,875
4060	600 amp		2.60	6.154		4,000	370		4,370	4,950
4070	800 amp		2.20	7.273		4,375	435		4,810	5,475
4080	1,000 amp		2	8		5,600	480		6,080	6,875
4090	1,350 amp		1.60	10		1,425	600		2,025	2,450
4100	1,600 amp		1.40	11.429		1,600	685		2,285	2,775
4110	2,000 amp		1.20	13.333		1,825	800		2,625	3,225
4120	2,500 amp		1	16		2,225	960		3,185	3,875
4130	3,000 amp		.80	20		2,450	1,200		3,650	4,500
4140	4,000 amp		.60	26.667		3,275	1,600		4,875	6,000
4500	Weatherproof 3 pole 4 wire, feeder, 600 amp		30	.533	L.F.	133	32		165	194
4520	800 amp		24	.667		155	40		195	231
4540	1,000 amp		22	.727		177	43.50		220.50	260
4550	1,200 amp		21	.762		245	46		291	340
4560	1,350 amp		20	.800		277	48		325	375
4580	1,600 amp		17	.941		320	56.50		376.50	440
4600	2,000 amp		16	1		375	60		435	505
4620	2,500 amp		12	1.333		490	80		570	655
4640	3,000 amp		10	1.600		555	96		651	755
4660	4,000 amp		8	2		800	120		920	1,050
5000	Indoor 3 pole, 3 wire, feeder, 600 amp		40	.400		102	24		126	148
5010	800 amp		32	.500		350	30		380	430
5020	1,000 amp		30	.533		390	32		422	480
5025	1,200 amp		29	.552		535	33		568	640
5030	1,350 amp		28	.571		176	34.50		210.50	244
5040	1,600 amp		24	.667		745	40		785	880
5050	2,000 amp		20	.800		895	48		943	1,050
5060	2,500 amp		16	1		1,075	60		1,135	1,300
5070	3,000 amp		14	1.143		1,250	68.50		1,318.50	1,475
5080	4,000 amp		12	1.333		480	80		560	650
5200	Plug-in type, 225 amp		50	.320		138	19.20		157.20	181

26 25 Low-Voltage Enclosed Bus Assemblies

26 25 13 – Low-Voltage Busways

26 25 13.10 Aluminum Bus Duct		Crew	Daily Output	Labor-Hours	Unit	Material	2019 Bare Costs Labor	Equipment	Total	Total Incl O&P
5210	400 amp	2 Elec	42	.381	L.F.	169	23		192	220
5220	600 amp		36	.444		208	26.50		234.50	269
5230	800 amp		30	.533		290	32		322	370
5240	1,000 amp		28	.571		325	34.50		359.50	410
5245	1,200 amp		27	.593		233	35.50		268.50	310
5250	1,350 amp		26	.615		185	37		222	258
5260	1,600 amp		20	.800		213	48		261	305
5270	2,000 amp		18	.889		250	53.50		303.50	355
5280	2,500 amp		16	1		340	60		400	465
5290	3,000 amp		14	1.143		405	68.50		473.50	545
5300	4,000 amp		12	1.333		490	80		570	660
5330	High short circuit, 400 amp		42	.381		201	23		224	256
5340	600 amp		36	.444		248	26.50		274.50	315
5350	800 amp		30	.533		360	32		392	445
5360	1,000 amp		28	.571		405	34.50		439.50	495
5370	1,350 amp		26	.615		185	37		222	258
5380	1,600 amp		20	.800		213	48		261	305
5390	2,000 amp		18	.889		250	53.50		303.50	355
5400	2,500 amp		16	1		390	60		450	520
5410	3,000 amp		14	1.143		405	68.50		473.50	545
5440	Elbow, 225 amp		5	3.200	Ea.	645	192		837	995
5450	400 amp		4.40	3.636		645	218		863	1,025
5460	600 amp		4	4		645	240		885	1,075
5470	800 amp		3.40	4.706		690	283		973	1,175
5480	1,000 amp		3.20	5		710	300		1,010	1,225
5485	1,200 amp		3.10	5.161		755	310		1,065	1,300
5490	1,350 amp		3	5.333		715	320		1,035	1,275
5500	1,600 amp		2.80	5.714		1,100	345		1,445	1,700
5510	2,000 amp		2.40	6.667		1,200	400		1,600	1,925
5520	2,500 amp		2	8		1,475	480		1,955	2,350
5530	3,000 amp		1.80	8.889		1,750	535		2,285	2,725
5540	4,000 amp		1.60	10		2,350	600		2,950	3,475
5560	Tee fittings, 225 amp		3.60	4.444		785	267		1,052	1,275
5570	400 amp		3.20	5		785	300		1,085	1,325
5580	600 amp		3	5.333		785	320		1,105	1,350
5590	800 amp		2.80	5.714		840	345		1,185	1,425
5600	1,000 amp		2.60	6.154		870	370		1,240	1,500
5605	1,200 amp		2.50	6.400		925	385		1,310	1,600
5610	1,350 amp		2.40	6.667		1,275	400		1,675	2,025
5620	1,600 amp		1.80	8.889		1,525	535		2,060	2,475
5630	2,000 amp		1.40	11.429		1,700	685		2,385	2,875
5640	2,500 amp		1.20	13.333		2,100	800		2,900	3,500
5650	3,000 amp		1	16		2,475	960		3,435	4,150
5660	4,000 amp		.70	22.857		3,625	1,375		5,000	6,025
5680	Cross, 225 amp		6.40	2.500		1,275	150		1,425	1,625
5690	400 amp		5.40	2.963		1,275	178		1,453	1,675
5700	600 amp		4.60	3.478		1,275	209		1,484	1,700
5710	800 amp		4	4		1,350	240		1,590	1,825
5720	1,000 amp		3.60	4.444		1,375	267		1,642	1,925
5730	1,350 amp		3.20	5		2,050	300		2,350	2,725
5740	1,600 amp		2.60	6.154		2,400	370		2,770	3,200
5750	2,000 amp		2.20	7.273		2,625	435		3,060	3,550
5760	2,500 amp		1.80	8.889		3,175	535		3,710	4,300

For customer support on your Electrical Costs with RSMeans data, call 800.448.8182.

243

26 25 13.10 Aluminum Bus Duct	Crew	Daily Output	Labor-Hours	Unit	Material	2019 Bare Costs Labor	Equipment	Total	Total Incl O&P
5770	3,000 amp	2 Elec	1.40	11.429	Ea.	3,800	685		4,485	5,200
5780	4,000 amp		1.20	13.333		5,275	800		6,075	7,000
5800	Expansion fitting, 225 amp		11.60	1.379		1,000	83		1,083	1,225
5810	400 amp		9.20	1.739		1,000	104		1,104	1,250
5820	600 amp		7	2.286		1,000	137		1,137	1,300
5830	800 amp		5.20	3.077		1,175	185		1,360	1,575
5840	1,000 amp		4.60	3.478		1,300	209		1,509	1,725
5850	1,350 amp		4.20	3.810		1,450	229		1,679	1,950
5860	1,600 amp		3.60	4.444		1,800	267		2,067	2,375
5870	2,000 amp		3.20	5		2,125	300		2,425	2,800
5880	2,500 amp		2.80	5.714		2,400	345		2,745	3,150
5890	3,000 amp		2.40	6.667		2,800	400		3,200	3,700
5900	4,000 amp		1.80	8.889		3,850	535		4,385	5,050
5940	Reducer, nonfused, 400 amp		9.20	1.739		740	104		844	965
5950	600 amp		7	2.286		740	137		877	1,025
5960	800 amp		5.20	3.077		775	185		960	1,125
5970	1,000 amp		4.60	3.478		955	209		1,164	1,350
5980	1,350 amp		4.20	3.810		1,425	229		1,654	1,900
5990	1,600 amp		3.60	4.444		1,600	267		1,867	2,175
6000	2,000 amp		3.20	5		1,875	300		2,175	2,525
6010	2,500 amp		2.80	5.714		2,400	345		2,745	3,150
6020	3,000 amp		2.20	7.273		2,825	435		3,260	3,750
6030	4,000 amp		1.80	8.889		3,850	535		4,385	5,050
6050	Reducer, fuse included, 225 amp		5	3.200		1,925	192		2,117	2,400
6060	400 amp		4.80	3.333		2,575	200		2,775	3,125
6070	600 amp		4.20	3.810		3,275	229		3,504	3,975
6080	800 amp		3.60	4.444		5,100	267		5,367	6,000
6090	1,000 amp		3.40	4.706		5,550	283		5,833	6,525
6100	1,350 amp		3.20	5		10,100	300		10,400	11,700
6110	1,600 amp		2.60	6.154		12,000	370		12,370	13,800
6120	2,000 amp		2	8		13,900	480		14,380	16,000
6160	Reducer, circuit breaker, 225 amp		5	3.200		2,375	192		2,567	2,900
6170	400 amp		4.80	3.333		2,925	200		3,125	3,500
6180	600 amp		4.20	3.810		4,175	229		4,404	4,950
6190	800 amp		3.60	4.444		4,875	267		5,142	5,775
6200	1,000 amp		3.40	4.706		5,550	283		5,833	6,525
6210	1,350 amp		3.20	5		6,700	300		7,000	7,825
6220	1,600 amp		2.60	6.154		10,000	370		10,370	11,600
6230	2,000 amp		2	8		10,900	480		11,380	12,700
6270	Cable tap box center, 225 amp		4.20	3.810		1,025	229		1,254	1,475
6280	400 amp		3.60	4.444		1,025	267		1,292	1,525
6290	600 amp		3	5.333		1,025	320		1,345	1,600
6300	800 amp		2.60	6.154		1,100	370		1,470	1,775
6310	1,000 amp		2.40	6.667		1,175	400		1,575	1,900
6320	1,350 amp		1.80	8.889		1,225	535		1,760	2,150
6330	1,600 amp		1.60	10		1,375	600		1,975	2,425
6340	2,000 amp		1.40	11.429		1,575	685		2,260	2,775
6350	2,500 amp		1.20	13.333		1,975	800		2,775	3,375
6360	3,000 amp		1	16		2,250	960		3,210	3,900
6370	4,000 amp		.70	22.857		2,650	1,375		4,025	4,975
6390	Cable tap box end, 225 amp		4.20	3.810		620	229		849	1,025
6400	400 amp		3.60	4.444		620	267		887	1,075
6410	600 amp		3	5.333		620	320		940	1,150

26 25 13.10 Aluminum Bus Duct

		Crew	Daily Output	Labor-Hours	Unit	Material	2019 Bare Costs Labor	Equipment	Total	Total Incl O&P
6420	800 amp	2 Elec	2.60	6.154	Ea.	680	370		1,050	1,300
6430	1,000 amp		2.40	6.667		730	400		1,130	1,400
6435	1,200 amp		2.10	7.619		795	460		1,255	1,550
6440	1,350 amp		1.80	8.889		770	535		1,305	1,650
6450	1,600 amp		1.60	10		870	600		1,470	1,850
6460	2,000 amp		1.40	11.429		985	685		1,670	2,100
6470	2,500 amp		1.20	13.333		1,150	800		1,950	2,450
6480	3,000 amp		1	16		1,350	960		2,310	2,900
6490	4,000 amp		.70	22.857		1,625	1,375		3,000	3,825
7000	Weatherproof 3 pole 3 wire, feeder, 600 amp		34	.471	L.F.	131	28.50		159.50	186
7020	800 amp		28	.571		145	34.50		179.50	210
7040	1,000 amp		26	.615		155	37		192	226
7050	1,200 amp		25	.640		182	38.50		220.50	258
7060	1,350 amp		24	.667		211	40		251	292
7080	1,600 amp		20	.800		244	48		292	340
7100	2,000 amp		18	.889		288	53.50		341.50	395
7120	2,500 amp		14	1.143		400	68.50		468.50	540
7140	3,000 amp		12	1.333		475	80		555	645
7160	4,000 amp		10	1.600		575	96		671	780

26 25 13.20 Bus Duct

		Crew	Daily Output	Labor-Hours	Unit	Material	2019 Bare Costs Labor	Equipment	Total	Total Incl O&P
0010	**BUS DUCT** 100 amp and less, aluminum or copper, plug-in									
0080	Bus duct, 3 pole 3 wire, 100 amp	1 Elec	42	.190	L.F.	71.50	11.45		82.95	95.50
0110	Elbow		4	2	Ea.	132	120		252	325
0120	Tee		2	4		187	240		427	565
0130	Wall flange		8	1		27	60		87	120
0140	Ground kit		16	.500		61.50	30		91.50	113
0180	3 pole 4 wire, 100 amp		40	.200	L.F.	81.50	12		93.50	107
0200	Cable tap box		3.10	2.581	Ea.	254	155		409	510
0300	End closure		16	.500		34	30		64	82
0400	Elbow		4	2		212	120		332	410
0500	Tee		2	4		310	240		550	700
0600	Hangers		10	.800		20.50	48		68.50	94
0700	Circuit breakers, 15 to 50 amp, 1 pole		8	1		585	60		645	730
0800	15 to 60 amp, 2 pole		6.70	1.194		380	71.50		451.50	525
0900	3 pole		5.30	1.509		455	90.50		545.50	635
1000	60 to 100 amp, 1 pole		6.70	1.194		610	71.50		681.50	775
1100	70 to 100 amp, 2 pole		5.30	1.509		1,250	90.50		1,340.50	1,500
1200	3 pole		4.50	1.778		1,675	107		1,782	2,000
1220	Switch, nonfused, 3 pole, 4 wire		8	1		202	60		262	310
1240	Fused, 3 fuses, 4 wire, 30 amp		8	1		360	60		420	490
1260	60 amp		5.30	1.509		420	90.50		510.50	600
1280	100 amp		4.50	1.778		575	107		682	795
1300	Plug, fusible, 3 pole 250 volt, 30 amp		5.30	1.509		515	90.50		605.50	705
1310	60 amp		5.30	1.509		595	90.50		685.50	790
1320	100 amp		4.50	1.778		795	107		902	1,025
1330	3 pole 480 volt, 30 amp		5.30	1.509		525	90.50		615.50	715
1340	60 amp		5.30	1.509		570	90.50		660.50	760
1350	100 amp		4.50	1.778		820	107		927	1,075
1360	Circuit breaker, 3 pole 250 volt, 60 amp		5.30	1.509		935	90.50		1,025.50	1,150
1370	3 pole 480 volt, 100 amp		4.50	1.778		935	107		1,042	1,175
2000	Bus duct, 2 wire, 250 volt, 30 amp		60	.133	L.F.	7.75	8		15.75	20.50
2100	60 amp		50	.160		7.75	9.60		17.35	23

26 25 Low-Voltage Enclosed Bus Assemblies

26 25 13 – Low-Voltage Busways

26 25 13.20 Bus Duct		Crew	Daily Output	Labor-Hours	Unit	Material	2019 Bare Costs Labor	Equipment	Total	Total Incl O&P
2200	300 volt, 30 amp	1 Elec	60	.133	L.F.	7.75	8		15.75	20.50
2300	60 amp		50	.160		7.75	9.60		17.35	23
2400	3 wire, 250 volt, 30 amp		60	.133		10.55	8		18.55	23.50
2500	60 amp		50	.160		10.35	9.60		19.95	26
2600	480/277 volt, 30 amp		60	.133		10.55	8		18.55	23.50
2700	60 amp		50	.160		10.55	9.60		20.15	26
2750	End feed, 300 volt 2 wire max. 30 amp		6	1.333	Ea.	73.50	80		153.50	200
2800	60 amp		5.50	1.455		73.50	87.50		161	211
2850	30 amp miniature		6	1.333		73.50	80		153.50	200
2900	3 wire, 30 amp		6	1.333		92.50	80		172.50	221
2950	60 amp		5.50	1.455		92.50	87.50		180	232
3000	30 amp miniature		6	1.333		92.50	80		172.50	221
3050	Center feed, 300 volt 2 wire, 30 amp		6	1.333		102	80		182	231
3100	60 amp		5.50	1.455		102	87.50		189.50	242
3150	3 wire, 30 amp		6	1.333		116	80		196	246
3200	60 amp		5.50	1.455		116	87.50		203.50	257
3220	Elbow, 30 amp		6	1.333		39.50	80		119.50	163
3240	60 amp		5.50	1.455		39.50	87.50		127	174
3260	End cap		40	.200		9	12		21	28
3280	Strength beam, 10'		15	.533		24	32		56	74.50
3300	Hanger		24	.333		5.10	20		25.10	35.50
3320	Tap box, nonfusible		6.30	1.270		76.50	76.50		153	198
3340	Fusible switch 30 amp, 1 fuse		6	1.333		465	80		545	630
3360	2 fuse		6	1.333		485	80		565	655
3380	3 fuse		6	1.333		520	80		600	690
3400	Circuit breaker handle on cover, 1 pole		6	1.333		63	80		143	188
3420	2 pole		6	1.333		540	80		620	715
3440	3 pole		6	1.333		665	80		745	855
3460	Circuit breaker external operhandle, 1 pole		6	1.333		74	80		154	201
3480	2 pole		6	1.333		695	80		775	885
3500	3 pole		6	1.333		815	80		895	1,025
3520	Terminal plug only		16	.500		103	30		133	158
3540	Terminal with receptacle		16	.500		111	30		141	167
3560	Fixture plug		16	.500		75	30		105	128
4000	Copper bus duct, lighting, 2 wire 300 volt, 20 amp		70	.114	L.F.	6.95	6.85		13.80	17.90
4020	35 amp		60	.133		6.95	8		14.95	19.60
4040	50 amp		55	.145		6.95	8.75		15.70	20.50
4060	60 amp		50	.160		6.95	9.60		16.55	22
4080	3 wire 300 volt, 20 amp		70	.114		6.85	6.85		13.70	17.75
4100	35 amp		60	.133		6.65	8		14.65	19.25
4120	50 amp		55	.145		6.85	8.75		15.60	20.50
4140	60 amp		50	.160		6.85	9.60		16.45	22
4160	Feeder in box, end, 1 circuit		6	1.333	Ea.	93.50	80		173.50	222
4180	2 circuit		5.50	1.455		97.50	87.50		185	237
4200	Center, 1 circuit		6	1.333		129	80		209	261
4220	2 circuit		5.50	1.455		134	87.50		221.50	277
4240	End cap		40	.200		15.80	12		27.80	35.50
4260	Hanger, surface mount		24	.333		9.45	20		29.45	40.50
4280	Coupling		40	.200		12	12		24	31

26 25 13.30 Copper Bus Duct	Crew	Daily Output	Labor-Hours	Unit	Material	2019 Bare Costs Labor	Equipment	Total	Total Incl O&P
0010 **COPPER BUS DUCT**									
0100 Weatherproof 3 pole 4 wire, feeder duct, 600 amp	2 Elec	24	.667	L.F.	320	40		360	410
0110 800 amp		18	.889		470	53.50		523.50	595
0120 1,000 amp		17	.941		490	56.50		546.50	620
0125 1,200 amp		16.50	.970		570	58		628	710
0130 1,350 amp		16	1		590	60		650	740
0140 1,600 amp		12	1.333		640	80		720	825
0150 2,000 amp		10	1.600		705	96		801	920
0160 2,500 amp		7	2.286		780	137		917	1,075
0170 3,000 amp		5	3.200		1,075	192		1,267	1,450
0180 4,000 amp		3.60	4.444		1,700	267		1,967	2,250
0200 Indoor 3 pole 4 wire, plug-in, bus duct high short circuit, 400 amp		32	.500		450	30		480	540
0210 600 amp		26	.615		515	37		552	620
0220 800 amp		20	.800		750	48		798	895
0230 1,000 amp		18	.889		875	53.50		928.50	1,050
0240 1,350 amp		16	1		475	60		535	615
0250 1,600 amp		12	1.333		540	80		620	715
0260 2,000 amp		10	1.600		685	96		781	895
0270 2,500 amp		8	2		845	120		965	1,100
0280 3,000 amp		6	2.667		945	160		1,105	1,300
0310 Cross, 225 amp		3	5.333	Ea.	3,000	320		3,320	3,775
0320 400 amp		2.80	5.714		3,000	345		3,345	3,800
0330 600 amp		2.60	6.154		3,000	370		3,370	3,850
0340 800 amp		2.20	7.273		3,250	435		3,685	4,225
0350 1,000 amp		2	8		3,625	480		4,105	4,725
0360 1,350 amp		1.80	8.889		4,125	535		4,660	5,350
0370 1,600 amp		1.70	9.412		4,600	565		5,165	5,900
0380 2,000 amp		1.60	10		7,525	600		8,125	9,200
0390 2,500 amp		1.40	11.429		9,125	685		9,810	11,000
0400 3,000 amp		1.20	13.333		9,775	800		10,575	12,000
0410 4,000 amp		1	16		12,800	960		13,760	15,400
0430 Expansion fitting, 225 amp		5.40	2.963		1,825	178		2,003	2,275
0440 400 amp		4.60	3.478		2,075	209		2,284	2,575
0450 600 amp		4	4		2,550	240		2,790	3,175
0460 800 amp		3.40	4.706		3,000	283		3,283	3,725
0470 1,000 amp		3	5.333		3,450	320		3,770	4,275
0480 1,350 amp		2.80	5.714		3,750	345		4,095	4,625
0490 1,600 amp		2.60	6.154		5,250	370		5,620	6,325
0500 2,000 amp		2.20	7.273		6,025	435		6,460	7,275
0510 2,500 amp		1.80	8.889		7,350	535		7,885	8,875
0520 3,000 amp		1.60	10		8,075	600		8,675	9,775
0530 4,000 amp		1.20	13.333		10,400	800		11,200	12,600
0550 Reducer nonfused, 225 amp		5.40	2.963		1,925	178		2,103	2,375
0560 400 amp		4.60	3.478		1,925	209		2,134	2,400
0570 600 amp		4	4		1,925	240		2,165	2,450
0580 800 amp		3.40	4.706		2,325	283		2,608	2,975
0590 1,000 amp		3	5.333		2,775	320		3,095	3,550
0600 1,350 amp		2.80	5.714		3,950	345		4,295	4,850
0610 1,600 amp		2.60	6.154		4,525	370		4,895	5,550
0620 2,000 amp		2.20	7.273		5,450	435		5,885	6,650
0630 2,500 amp		1.80	8.889		6,875	535		7,410	8,375
0640 3,000 amp		1.60	10		7,550	600		8,150	9,200

26 25 13.30 Copper Bus Duct		Crew	Daily Output	Labor-Hours	Unit	Material	2019 Bare Costs Labor	Equipment	Total	Total Incl O&P
0650	4,000 amp	2 Elec	1.20	13.333	Ea.	9,825	800		10,625	12,000
0670	Reducer fuse included, 225 amp		4.40	3.636		4,100	218		4,318	4,825
0680	400 amp		4.20	3.810		5,125	229		5,354	6,000
0690	600 amp		3.60	4.444		6,325	267		6,592	7,350
0700	800 amp		3.20	5		8,925	300		9,225	10,300
0710	1,000 amp		3	5.333		11,200	320		11,520	12,800
0720	1,350 amp		2.80	5.714		18,200	345		18,545	20,500
0730	1,600 amp		2.20	7.273		23,200	435		23,635	26,200
0740	2,000 amp		1.80	8.889		26,100	535		26,635	29,500
0790	Reducer, circuit breaker, 225 amp		4.40	3.636		5,175	218		5,393	6,025
0800	400 amp		4.20	3.810		6,050	229		6,279	7,000
0810	600 amp		3.60	4.444		8,600	267		8,867	9,850
0820	800 amp		3.20	5		10,100	300		10,400	11,600
0830	1,000 amp		3	5.333		11,400	320		11,720	13,100
0840	1,350 amp		2.80	5.714		12,800	345		13,145	14,600
0850	1,600 amp		2.20	7.273		19,000	435		19,435	21,600
0860	2,000 amp		1.80	8.889		20,800	535		21,335	23,700
0910	Cable tap box, center, 225 amp		3.20	5		2,025	300		2,325	2,675
0920	400 amp		2.60	6.154		2,025	370		2,395	2,775
0930	600 amp		2.20	7.273		2,025	435		2,460	2,875
0940	800 amp		2	8		2,250	480		2,730	3,200
0950	1,000 amp		1.60	10		2,425	600		3,025	3,575
0960	1,350 amp		1.40	11.429		2,750	685		3,435	4,075
0970	1,600 amp		1.20	13.333		3,075	800		3,875	4,600
0980	2,000 amp		1	16		3,700	960		4,660	5,500
1040	2,500 amp		.80	20		4,400	1,200		5,600	6,625
1060	3,000 amp		.60	26.667		4,725	1,600		6,325	7,575
1080	4,000 amp		.40	40		5,950	2,400		8,350	10,100
1800	Weatherproof 3 pole 3 wire, feeder duct, 600 amp		28	.571	L.F.	266	34.50		300.50	345
1820	800 amp		22	.727		325	43.50		368.50	420
1840	1,000 amp		20	.800		360	48		408	465
1850	1,200 amp		19	.842		375	50.50		425.50	485
1860	1,350 amp		18	.889		535	53.50		588.50	670
1880	1,600 amp		14	1.143		610	68.50		678.50	770
1900	2,000 amp		12	1.333		780	80		860	980
1920	2,500 amp		8	2		975	120		1,095	1,250
1940	3,000 amp		6	2.667		1,100	160		1,260	1,450
1960	4,000 amp		4	4		1,450	240		1,690	1,950
2000	Indoor 3 pole 3 wire, feeder duct, 600 amp		32	.500		222	30		252	289
2010	800 amp		26	.615		585	37		622	700
2020	1,000 amp		24	.667		300	40		340	390
2025	1,200 amp		22	.727		395	43.50		438.50	500
2030	1,350 amp		20	.800		445	48		493	560
2040	1,600 amp		16	1		1,075	60		1,135	1,275
2050	2,000 amp		14	1.143		1,400	68.50		1,468.50	1,625
2060	2,500 amp		10	1.600		810	96		906	1,025
2070	3,000 amp		8	2		890	120		1,010	1,150
2080	4,000 amp		6	2.667		1,200	160		1,360	1,575
2090	5,000 amp		5	3.200		1,475	192		1,667	1,900
2200	Indoor 3 pole 3 wire, bus duct plug-in, 225 amp		46	.348		173	21		194	222
2210	400 amp		36	.444		266	26.50		292.50	330
2220	600 amp		30	.533		330	32		362	415
2230	800 amp		24	.667		490	40		530	595

26 25 13.30 Copper Bus Duct		Crew	Daily Output	Labor-Hours	Unit	Material	2019 Bare Costs Labor	Equipment	Total	Total Incl O&P
2240	1,000 amp	2 Elec	20	.800	L.F.	530	48		578	650
2250	1,350 amp		18	.889		800	53.50		853.50	960
2260	1,600 amp		14	1.143		850	68.50		918.50	1,025
2270	2,000 amp		12	1.333		685	80		765	870
2280	2,500 amp		10	1.600		845	96		941	1,075
2290	3,000 amp		8	2		945	120		1,065	1,225
2330	High short circuit, 400 amp		36	.444		335	26.50		361.50	410
2340	600 amp		30	.533		400	32		432	490
2350	800 amp		24	.667		585	40		625	705
2360	1,000 amp		20	.800		635	48		683	765
2370	1,350 amp		18	.889		955	53.50		1,008.50	1,125
2380	1,600 amp		14	1.143		1,025	68.50		1,093.50	1,225
2390	2,000 amp		12	1.333		685	80		765	870
2400	2,500 amp		10	1.600		845	96		941	1,075
2410	3,000 amp		8	2		945	120		1,065	1,225
2440	Elbows, 225 amp		4.60	3.478	Ea.	1,425	209		1,634	1,850
2450	400 amp		4.20	3.810		1,425	229		1,654	1,900
2460	600 amp		3.60	4.444		1,425	267		1,692	1,950
2470	800 amp		3.20	5		1,525	300		1,825	2,125
2480	1,000 amp		3	5.333		1,600	320		1,920	2,225
2485	1,200 amp		2.90	5.517		1,525	330		1,855	2,175
2490	1,350 amp		2.80	5.714		1,825	345		2,170	2,500
2500	1,600 amp		2.60	6.154		1,950	370		2,320	2,700
2510	2,000 amp		2	8		2,250	480		2,730	3,200
2520	2,500 amp		1.80	8.889		3,400	535		3,935	4,550
2530	3,000 amp		1.60	10		3,650	600		4,250	4,900
2540	4,000 amp		1.40	11.429		4,675	685		5,360	6,150
2560	Tee fittings, 225 amp		2.80	5.714		1,500	345		1,845	2,150
2570	400 amp		2.40	6.667		1,500	400		1,900	2,250
2580	600 amp		2	8		1,500	480		1,980	2,375
2590	800 amp		1.80	8.889		1,625	535		2,160	2,600
2600	1,000 amp		1.60	10		1,825	600		2,425	2,900
2605	1,200 amp		1.50	10.667		1,950	640		2,590	3,075
2610	1,350 amp		1.40	11.429		2,150	685		2,835	3,375
2620	1,600 amp		1.20	13.333		2,500	800		3,300	3,950
2630	2,000 amp		1	16		4,150	960		5,110	5,975
2640	2,500 amp		.70	22.857		4,850	1,375		6,225	7,400
2650	3,000 amp		.60	26.667		5,250	1,600		6,850	8,175
2660	4,000 amp		.50	32		6,775	1,925		8,700	10,300
2680	Cross, 225 amp		3.60	4.444		2,275	267		2,542	2,900
2690	400 amp		3.20	5		2,275	300		2,575	2,950
2700	600 amp		3	5.333		2,275	320		2,595	2,975
2710	800 amp		2.60	6.154		2,725	370		3,095	3,550
2720	1,000 amp		2.40	6.667		2,850	400		3,250	3,725
2730	1,350 amp		2.20	7.273		3,425	435		3,860	4,425
2740	1,600 amp		2	8		3,700	480		4,180	4,775
2750	2,000 amp		1.80	8.889		5,850	535		6,385	7,225
2760	2,500 amp		1.60	10		6,800	600		7,400	8,375
2770	3,000 amp		1.40	11.429		7,300	685		7,985	9,050
2780	4,000 amp		1	16		9,325	960		10,285	11,700
2800	Expansion fitting, 225 amp		6.40	2.500		1,925	150		2,075	2,350
2810	400 amp		5.40	2.963		4,875	178		5,053	5,625
2820	600 amp		4.60	3.478		1,925	209		2,134	2,425

26 25 13.30 Copper Bus Duct		Crew	Daily Output	Labor-Hours	Unit	Material	2019 Bare Costs Labor	Equipment	Total	Total Incl O&P
2830	800 amp	2 Elec	4	4	Ea.	2,300	240		2,540	2,875
2840	1,000 amp		3.60	4.444		2,525	267		2,792	3,175
2850	1,350 amp		3.20	5		2,925	300		3,225	3,675
2860	1,600 amp		3	5.333		3,200	320		3,520	4,000
2870	2,000 amp		2.60	6.154		3,975	370		4,345	4,925
2880	2,500 amp		2.20	7.273		5,550	435		5,985	6,750
2890	3,000 amp		1.80	8.889		6,075	535		6,610	7,475
2900	4,000 amp		1.40	11.429		7,750	685		8,435	9,550
2920	Reducer nonfused, 225 amp		6.40	2.500		1,575	150		1,725	1,950
2930	400 amp		5.40	2.963		1,575	178		1,753	2,000
2940	600 amp		4.60	3.478		1,575	209		1,784	2,025
2950	800 amp		4	4		1,825	240		2,065	2,375
2960	1,000 amp		3.60	4.444		2,050	267		2,317	2,675
2970	1,350 amp		3.20	5		2,550	300		2,850	3,250
2980	1,600 amp		3	5.333		2,900	320		3,220	3,650
2990	2,000 amp		2.60	6.154		3,475	370		3,845	4,375
3000	2,500 amp		2.20	7.273		5,075	435		5,510	6,225
3010	3,000 amp		1.80	8.889		5,525	535		6,060	6,900
3020	4,000 amp		1.40	11.429		7,200	685		7,885	8,950
3040	Reducer fuse included, 225 amp		5	3.200		3,725	192		3,917	4,375
3050	400 amp		4.80	3.333		4,950	200		5,150	5,750
3060	600 amp		4.20	3.810		5,900	229		6,129	6,850
3070	800 amp		3.60	4.444		8,400	267		8,667	9,650
3080	1,000 amp		3.40	4.706		9,950	283		10,233	11,400
3090	1,350 amp		3.20	5		9,475	300		9,775	10,900
3100	1,600 amp		2.60	6.154		22,100	370		22,470	24,900
3110	2,000 amp		2	8		25,000	480		25,480	28,200
3160	Reducer circuit breaker, 225 amp		5	3.200		4,750	192		4,942	5,500
3170	400 amp		4.80	3.333		5,825	200		6,025	6,700
3180	600 amp		4.20	3.810		8,375	229		8,604	9,550
3190	800 amp		3.60	4.444		9,775	267		10,042	11,100
3200	1,000 amp		3.40	4.706		11,100	283		11,383	12,600
3210	1,350 amp		3.20	5		12,600	300		12,900	14,300
3220	1,600 amp		2.60	6.154		18,700	370		19,070	21,200
3230	2,000 amp		2	8		20,500	480		20,980	23,200
3280	3 pole, 3 wire, cable tap box center, 225 amp		3.60	4.444		2,300	267		2,567	2,950
3290	400 amp		3	5.333		2,300	320		2,620	3,025
3300	600 amp		2.60	6.154		2,300	370		2,670	3,100
3310	800 amp		2.40	6.667		2,575	400		2,975	3,450
3320	1,000 amp		1.80	8.889		2,800	535		3,335	3,875
3330	1,350 amp		1.60	10		3,425	600		4,025	4,675
3340	1,600 amp		1.40	11.429		3,825	685		4,510	5,225
3350	2,000 amp		1.20	13.333		4,400	800		5,200	6,025
3360	2,500 amp		1	16		5,225	960		6,185	7,175
3370	3,000 amp		.70	22.857		5,650	1,375		7,025	8,275
3380	4,000 amp		.50	32		7,175	1,925		9,100	10,800
3400	Cable tap box end, 225 amp		3.60	4.444		1,250	267		1,517	1,775
3410	400 amp		3	5.333		1,250	320		1,570	1,850
3420	600 amp		2.60	6.154		1,400	370		1,770	2,100
3430	800 amp		2.40	6.667		1,400	400		1,800	2,150
3440	1,000 amp		1.80	8.889		1,550	535		2,085	2,500
3445	1,200 amp		1.70	9.412		1,550	565		2,115	2,550
3450	1,350 amp		1.60	10		1,825	600		2,425	2,925

26 25 13 – Low-Voltage Busways

26 25 13.30 Copper Bus Duct		Crew	Daily Output	Labor-Hours	Unit	Material	2019 Bare Costs Labor	Equipment	Total	Total Incl O&P
3460	1,600 amp	2 Elec	1.40	11.429	Ea.	2,100	685		2,785	3,350
3470	2,000 amp		1.20	13.333		2,275	800		3,075	3,700
3480	2,500 amp		1	16		2,725	960		3,685	4,425
3490	3,000 amp		.70	22.857		2,925	1,375		4,300	5,275
3500	4,000 amp		.50	32		3,700	1,925		5,625	6,950
4600	Plug-in, fusible switch w/3 fuses, 3 pole, 250 volt, 30 amp	1 Elec	4	2		465	120		585	695
4610	60 amp		3.60	2.222		610	133		743	870
4620	100 amp		2.70	2.963		870	178		1,048	1,225
4630	200 amp	2 Elec	3.20	5		1,450	300		1,750	2,050
4640	400 amp		1.40	11.429		3,800	685		4,485	5,225
4650	600 amp		.90	17.778		5,275	1,075		6,350	7,400
4700	4 pole, 120/208 volt, 30 amp	1 Elec	3.90	2.051		635	123		758	885
4710	60 amp		3.50	2.286		685	137		822	960
4720	100 amp		2.60	3.077		955	185		1,140	1,325
4730	200 amp	2 Elec	3	5.333		1,600	320		1,920	2,250
4740	400 amp		1.30	12.308		3,775	740		4,515	5,250
4750	600 amp		.80	20		5,275	1,200		6,475	7,600
4800	3 pole, 480 volt, 30 amp	1 Elec	4	2		480	120		600	705
4810	60 amp		3.60	2.222		505	133		638	755
4820	100 amp		2.70	2.963		850	178		1,028	1,200
4830	200 amp	2 Elec	3.20	5		1,475	300		1,775	2,050
4840	400 amp		1.40	11.429		3,400	685		4,085	4,775
4850	600 amp		.90	17.778		4,850	1,075		5,925	6,925
4860	800 amp		.66	24.242		17,500	1,450		18,950	21,400
4870	1,000 amp		.60	26.667		18,300	1,600		19,900	22,600
4880	1,200 amp		.50	32		19,500	1,925		21,425	24,400
4890	1,600 amp		.44	36.364		21,100	2,175		23,275	26,600
4900	4 pole, 277/480 volt, 30 amp	1 Elec	3.90	2.051		690	123		813	940
4910	60 amp		3.50	2.286		735	137		872	1,025
4920	100 amp		2.60	3.077		1,075	185		1,260	1,450
4930	200 amp	2 Elec	3	5.333		2,125	320		2,445	2,825
4940	400 amp		1.30	12.308		4,025	740		4,765	5,525
4950	600 amp		.80	20		5,500	1,200		6,700	7,850
5050	800 amp		.60	26.667		18,100	1,600		19,700	22,300
5060	1,000 amp		.56	28.571		20,900	1,725		22,625	25,600
5070	1,200 amp		.48	33.333		21,100	2,000		23,100	26,200
5080	1,600 amp		.42	38.095		22,900	2,300		25,200	28,600
5150	Fusible with starter, 3 pole 250 volt, 30 amp	1 Elec	3.50	2.286		3,425	137		3,562	3,975
5160	60 amp		3.20	2.500		3,625	150		3,775	4,225
5170	100 amp		2.50	3.200		4,100	192		4,292	4,775
5180	200 amp	2 Elec	2.80	5.714		6,775	345		7,120	7,950
5200	3 pole 480 volt, 30 amp	1 Elec	3.50	2.286		3,425	137		3,562	3,975
5210	60 amp		3.20	2.500		3,625	150		3,775	4,225
5220	100 amp		2.50	3.200		4,100	192		4,292	4,775
5230	200 amp	2 Elec	2.80	5.714		6,775	345		7,120	7,950
5300	Fusible with contactor, 3 pole 250 volt, 30 amp	1 Elec	3.50	2.286		3,350	137		3,487	3,875
5310	60 amp		3.20	2.500		4,250	150		4,400	4,900
5320	100 amp		2.50	3.200		5,950	192		6,142	6,825
5330	200 amp	2 Elec	2.80	5.714		6,800	345		7,145	8,000
5400	3 pole 480 volt, 30 amp	1 Elec	3.50	2.286		3,600	137		3,737	4,150
5410	60 amp		3.20	2.500		5,075	150		5,225	5,800
5420	100 amp		2.50	3.200		6,950	192		7,142	7,925
5430	200 amp	2 Elec	2.80	5.714		7,125	345		7,470	8,350

26 25 Low-Voltage Enclosed Bus Assemblies

26 25 13 – Low-Voltage Busways

26 25 13.30 Copper Bus Duct		Crew	Daily Output	Labor-Hours	Unit	Material	2019 Bare Costs Labor	Equipment	Total	Total Incl O&P
5450	Fusible with capacitor, 3 pole 250 volt, 30 amp	1 Elec	3	2.667	Ea.	8,675	160		8,835	9,800
5460	60 amp		2	4		10,100	240		10,340	11,500
5500	3 pole 480 volt, 30 amp		3	2.667		7,275	160		7,435	8,250
5510	60 amp		2	4		9,075	240		9,315	10,400
5600	Circuit breaker, 3 pole, 250 volt, 60 amp		4.50	1.778		655	107		762	880
5610	100 amp		3.20	2.500		805	150		955	1,100
5650	4 pole, 120/208 volt, 60 amp		4.40	1.818		740	109		849	980
5660	100 amp		3.10	2.581		875	155		1,030	1,200
5700	3 pole, 4 wire 277/480 volt, 60 amp		4.30	1.860		995	112		1,107	1,275
5710	100 amp	▼	3	2.667		1,100	160		1,260	1,475
5720	225 amp	2 Elec	3.20	5		2,450	300		2,750	3,150
5730	400 amp		1.20	13.333		5,100	800		5,900	6,800
5740	600 amp		.96	16.667		6,850	1,000		7,850	9,025
5750	700 amp		.60	26.667		8,675	1,600		10,275	12,000
5760	800 amp		.60	26.667		8,675	1,600		10,275	12,000
5770	900 amp		.54	29.630		11,400	1,775		13,175	15,300
5780	1,000 amp		.54	29.630		11,400	1,775		13,175	15,300
5790	1,200 amp	▼	.42	38.095		13,700	2,300		16,000	18,500
5810	Circuit breaker w/HIC fuses, 3 pole 480 volt, 60 amp	1 Elec	4.40	1.818		1,250	109		1,359	1,550
5820	100 amp	"	3.10	2.581		1,375	155		1,530	1,725
5830	225 amp	2 Elec	3.40	4.706		4,375	283		4,658	5,250
5840	400 amp		1.40	11.429		6,950	685		7,635	8,675
5850	600 amp		1	16		7,050	960		8,010	9,175
5860	700 amp		.64	25		9,350	1,500		10,850	12,600
5870	800 amp		.64	25		9,350	1,500		10,850	12,600
5880	900 amp		.56	28.571		20,200	1,725		21,925	24,800
5890	1,000 amp	▼	.56	28.571		20,200	1,725		21,925	24,800
5950	3 pole 4 wire 277/480 volt, 60 amp	1 Elec	4.30	1.860		1,250	112		1,362	1,550
5960	100 amp	"	3	2.667		1,375	160		1,535	1,750
5970	225 amp	2 Elec	3	5.333		4,375	320		4,695	5,300
5980	400 amp		1.10	14.545		6,950	875		7,825	8,950
5990	600 amp		.94	17.021		7,050	1,025		8,075	9,275
6000	700 amp		.58	27.586		9,350	1,650		11,000	12,800
6010	800 amp		.58	27.586		9,350	1,650		11,000	12,800
6020	900 amp		.52	30.769		20,200	1,850		22,050	25,000
6030	1,000 amp		.52	30.769		20,200	1,850		22,050	25,000
6040	1,200 amp	▼	.40	40		20,200	2,400		22,600	25,800
6100	Circuit breaker with starter, 3 pole 250 volt, 60 amp	1 Elec	3.20	2.500		1,900	150		2,050	2,325
6110	100 amp	"	2.50	3.200		2,625	192		2,817	3,175
6120	225 amp	2 Elec	3	5.333		3,775	320		4,095	4,625
6130	3 pole 480 volt, 60 amp	1 Elec	3.20	2.500		1,900	150		2,050	2,325
6140	100 amp	"	2.50	3.200		2,625	192		2,817	3,175
6150	225 amp	2 Elec	3	5.333		3,250	320		3,570	4,050
6200	Circuit breaker with contactor, 3 pole 250 volt, 60 amp	1 Elec	3.20	2.500		1,800	150		1,950	2,200
6210	100 amp	"	2.50	3.200		2,450	192		2,642	2,975
6220	225 amp	2 Elec	3	5.333		3,500	320		3,820	4,325
6250	3 pole 480 volt, 60 amp	1 Elec	3.20	2.500		1,800	150		1,950	2,200
6260	100 amp	"	2.50	3.200		2,450	192		2,642	2,975
6270	225 amp	2 Elec	3	5.333		3,375	320		3,695	4,200
6300	Circuit breaker with capacitor, 3 pole 250 volt, 60 amp	1 Elec	2	4		9,775	240		10,015	11,200
6310	3 pole 480 volt, 60 amp		2	4		10,200	240		10,440	11,600
6400	Add control transformer with pilot light to above starter		16	.500		560	30		590	665
6410	Switch, fusible, mechanically held contactor optional	▼	16	.500	▼	1,475	30		1,505	1,675

26 25 Low-Voltage Enclosed Bus Assemblies

26 25 13 – Low-Voltage Busways

26 25 13.30 Copper Bus Duct	Crew	Daily Output	Labor-Hours	Unit	Material	2019 Bare Costs Labor	Equipment	Total	Total Incl O&P
6430	Circuit breaker, mechanically held contactor optional	1 Elec	16	.500	Ea.	1,475	30		1,505	1,675
6450	Ground neutralizer, 3 pole	↓	16	.500	↓	68.50	30		98.50	121

26 25 13.40 Copper Bus Duct

	26 25 13.40 Copper Bus Duct	Crew	Daily Output	Labor-Hours	Unit	Material	Labor	Equipment	Total	Total Incl O&P
0010	**COPPER BUS DUCT** 10' long									
0050	Indoor 3 pole 4 wire, plug-in, straight section, 225 amp	2 Elec	40	.400	L.F.	222	24		246	280
1000	400 amp		32	.500		375	30		405	460
1500	600 amp		26	.615		430	37		467	525
2400	800 amp		20	.800		625	48		673	755
2450	1,000 amp		18	.889		305	53.50		358.50	415
2470	1,200 amp		17	.941		395	56.50		451.50	520
2500	1,350 amp		16	1		415	60		475	550
2510	1,600 amp		12	1.333		470	80		550	640
2520	2,000 amp		10	1.600		595	96		691	800
2530	2,500 amp		8	2		735	120		855	990
2540	3,000 amp		6	2.667		840	160		1,000	1,175
2550	Feeder, 600 amp		28	.571		204	34.50		238.50	276
2600	800 amp		22	.727		755	43.50		798.50	895
2700	1,000 amp		20	.800		895	48		943	1,050
2750	1,200 amp		19	.842		1,075	50.50		1,125.50	1,250
2800	1,350 amp		18	.889		390	53.50		443.50	505
2900	1,600 amp		14	1.143		1,450	68.50		1,518.50	1,700
3000	2,000 amp		12	1.333		1,775	80		1,855	2,075
3010	2,500 amp		8	2		710	120		830	960
3020	3,000 amp		6	2.667		815	160		975	1,125
3030	4,000 amp		4	4		1,075	240		1,315	1,525
3040	5,000 amp		2	8	↓	1,325	480		1,805	2,175
3100	Elbows, 225 amp		4	4	Ea.	1,550	240		1,790	2,050
3200	400 amp		3.60	4.444		1,550	267		1,817	2,100
3300	600 amp		3.20	5		1,550	300		1,850	2,150
3400	800 amp		2.80	5.714		1,675	345		2,020	2,350
3500	1,000 amp		2.60	6.154		1,875	370		2,245	2,625
3550	1,200 amp		2.50	6.400		1,875	385		2,260	2,650
3600	1,350 amp		2.40	6.667		1,900	400		2,300	2,675
3700	1,600 amp		2.20	7.273		2,075	435		2,510	2,925
3800	2,000 amp		1.80	8.889		2,550	535		3,085	3,600
3810	2,500 amp		1.60	10		4,025	600		4,625	5,325
3820	3,000 amp		1.40	11.429		4,400	685		5,085	5,850
3830	4,000 amp		1.20	13.333		5,675	800		6,475	7,450
3840	5,000 amp		1	16		9,200	960		10,160	11,500
4000	End box, 225 amp		34	.471		170	28.50		198.50	229
4100	400 amp		32	.500		189	30		219	253
4200	600 amp		28	.571		189	34.50		223.50	259
4300	800 amp		26	.615		189	37		226	263
4400	1,000 amp		24	.667		179	40		219	257
4410	1,200 amp		23	.696		189	42		231	271
4500	1,350 amp		22	.727		180	43.50		223.50	263
4600	1,600 amp		20	.800		180	48		228	270
4700	2,000 amp		18	.889		221	53.50		274.50	325
4710	2,500 amp		16	1		221	60		281	335
4720	3,000 amp		14	1.143		208	68.50		276.50	330
4730	4,000 amp		12	1.333		252	80		332	395
4740	5,000 amp		10	1.600		252	96		348	420

26 25 13.40 Copper Bus Duct		Crew	Daily Output	Labor-Hours	Unit	Material	2019 Bare Costs Labor	Equipment	Total	Total Incl O&P
4800	Cable tap box end, 225 amp	2 Elec	3.20	5	Ea.	1,200	300		1,500	1,775
5000	400 amp		2.60	6.154		1,175	370		1,545	1,850
5100	600 amp		2.20	7.273		1,600	435		2,035	2,400
5200	800 amp		2	8		1,725	480		2,205	2,625
5300	1,000 amp		1.60	10		1,700	600		2,300	2,775
5350	1,200 amp		1.50	10.667		2,025	640		2,665	3,175
5400	1,350 amp		1.40	11.429		2,250	685		2,935	3,525
5500	1,600 amp		1.20	13.333		2,525	800		3,325	3,975
5600	2,000 amp		1	16		2,825	960		3,785	4,525
5610	2,500 amp		.80	20		3,125	1,200		4,325	5,250
5620	3,000 amp		.60	26.667		3,650	1,600		5,250	6,425
5630	4,000 amp		.40	40		4,175	2,400		6,575	8,150
5640	5,000 amp		.20	80		4,975	4,800		9,775	12,700
5700	Switchboard stub, 225 amp		5.40	2.963		1,250	178		1,428	1,650
5800	400 amp		4.60	3.478		1,325	209		1,534	1,750
5900	600 amp		4	4		1,400	240		1,640	1,875
6000	800 amp		3.20	5		1,675	300		1,975	2,300
6100	1,000 amp		3	5.333		1,950	320		2,270	2,625
6150	1,200 amp		2.80	5.714		2,275	345		2,620	3,025
6200	1,350 amp		2.60	6.154		2,400	370		2,770	3,200
6300	1,600 amp		2.40	6.667		2,700	400		3,100	3,575
6400	2,000 amp		2	8		3,275	480		3,755	4,350
6410	2,500 amp		1.80	8.889		4,000	535		4,535	5,200
6420	3,000 amp		1.60	10		4,425	600		5,025	5,775
6430	4,000 amp		1.40	11.429		5,775	685		6,460	7,375
6440	5,000 amp		1.20	13.333		7,075	800		7,875	9,000
6490	Tee fittings, 225 amp		2.40	6.667		2,225	400		2,625	3,050
6500	400 amp		2	8		2,225	480		2,705	3,175
6600	600 amp		1.80	8.889		2,225	535		2,760	3,250
6700	800 amp		1.60	10		2,300	600		2,900	3,425
6750	1,000 amp		1.40	11.429		2,575	685		3,260	3,875
6770	1,200 amp		1.30	12.308		2,900	740		3,640	4,300
6800	1,350 amp		1.20	13.333		3,050	800		3,850	4,575
7000	1,600 amp		1	16		3,475	960		4,435	5,250
7100	2,000 amp		.80	20		4,125	1,200		5,325	6,350
7110	2,500 amp		.60	26.667		5,075	1,600		6,675	7,975
7120	3,000 amp		.50	32		5,575	1,925		7,500	9,000
7130	4,000 amp		.40	40		7,200	2,400		9,600	11,500
7140	5,000 amp		.20	80		8,525	4,800		13,325	16,600
7200	Plug-in fusible switches w/3 fuses, 600 volt, 3 pole, 30 amp	1 Elec	4	2		870	120		990	1,150
7300	60 amp		3.60	2.222		980	133		1,113	1,275
7400	100 amp		2.70	2.963		1,325	178		1,503	1,725
7500	200 amp	2 Elec	3.20	5		2,325	300		2,625	3,025
7600	400 amp		1.40	11.429		6,525	685		7,210	8,200
7700	600 amp		.90	17.778		7,850	1,075		8,925	10,300
7800	800 amp		.66	24.242		12,700	1,450		14,150	16,100
7900	1,200 amp		.50	32		23,800	1,925		25,725	29,100
7910	1,600 amp		.44	36.364		22,600	2,175		24,775	28,100
8000	Plug-in circuit breakers, molded case, 15 to 50 amp	1 Elec	4.40	1.818		825	109		934	1,075
8100	70 to 100 amp	"	3.10	2.581		915	155		1,070	1,225
8200	150 to 225 amp	2 Elec	3.40	4.706		2,475	283		2,758	3,150
8300	250 to 400 amp		1.40	11.429		4,350	685		5,035	5,825
8400	500 to 600 amp		1	16		5,875	960		6,835	7,900

26 25 13.40 Copper Bus Duct		Crew	Daily Output	Labor-Hours	Unit	Material	2019 Bare Costs Labor	Equipment	Total	Total Incl O&P
8500	700 to 800 amp	2 Elec	.64	25	Ea.	7,250	1,500		8,750	10,200
8600	900 to 1,000 amp		.56	28.571		10,400	1,725		12,125	14,000
8700	1,200 amp		.44	36.364		12,500	2,175		14,675	17,000
8720	1,400 amp		.40	40		17,500	2,400		19,900	22,900
8730	1,600 amp		.40	40		19,200	2,400		21,600	24,700
8750	Circuit breakers, with current limiting fuse, 15 to 50 amp	1 Elec	4.40	1.818		1,650	109		1,759	2,000
8760	70 to 100 amp	"	3.10	2.581		1,950	155		2,105	2,375
8770	150 to 225 amp	2 Elec	3.40	4.706		4,200	283		4,483	5,050
8780	250 to 400 amp		1.40	11.429		6,500	685		7,185	8,175
8790	500 to 600 amp		1	16		7,475	960		8,435	9,650
8800	700 to 800 amp		.64	25		12,300	1,500		13,800	15,800
8810	900 to 1,000 amp		.56	28.571		14,000	1,725		15,725	18,000
8850	Combination starter FVNR, fusible switch, NEMA size 0, 30 amp	1 Elec	2	4		2,225	240		2,465	2,800
8860	NEMA size 1, 60 amp		1.80	4.444		2,350	267		2,617	3,000
8870	NEMA size 2, 100 amp		1.30	6.154		2,975	370		3,345	3,800
8880	NEMA size 3, 200 amp	2 Elec	2	8		4,725	480		5,205	5,925
8900	Circuit breaker, NEMA size 0, 30 amp	1 Elec	2	4		2,275	240		2,515	2,875
8910	NEMA size 1, 60 amp		1.80	4.444		2,375	267		2,642	3,000
8920	NEMA size 2, 100 amp		1.30	6.154		3,400	370		3,770	4,300
8930	NEMA size 3, 200 amp	2 Elec	2	8		4,325	480		4,805	5,475
8950	Combination contactor, fusible switch, NEMA size 0, 30 amp	1 Elec	2	4		1,300	240		1,540	1,775
8960	NEMA size 1, 60 amp		1.80	4.444		1,325	267		1,592	1,875
8970	NEMA size 2, 100 amp		1.30	6.154		1,975	370		2,345	2,725
8980	NEMA size 3, 200 amp	2 Elec	2	8		2,300	480		2,780	3,250
9000	Circuit breaker, NEMA size 0, 30 amp	1 Elec	2	4		1,500	240		1,740	2,000
9010	NEMA size 1, 60 amp		1.80	4.444		1,525	267		1,792	2,075
9020	NEMA size 2, 100 amp		1.30	6.154		2,350	370		2,720	3,150
9030	NEMA size 3, 200 amp	2 Elec	2	8		2,875	480		3,355	3,875
9050	Control transformer for above, NEMA size 0, 30 amp	1 Elec	8	1		252	60		312	365
9060	NEMA size 1, 60 amp		8	1		252	60		312	365
9070	NEMA size 2, 100 amp		7	1.143		350	68.50		418.50	485
9080	NEMA size 3, 200 amp	2 Elec	14	1.143		490	68.50		558.50	640
9100	Comb. fusible switch & lighting control, electrically held, 30 amp	1 Elec	2	4		1,025	240		1,265	1,475
9110	60 amp		1.80	4.444		1,475	267		1,742	2,025
9120	100 amp		1.30	6.154		1,900	370		2,270	2,650
9130	200 amp	2 Elec	2	8		4,700	480		5,180	5,900
9150	Mechanically held, 30 amp	1 Elec	2	4		1,300	240		1,540	1,775
9160	60 amp		1.80	4.444		1,925	267		2,192	2,525
9170	100 amp		1.30	6.154		2,475	370		2,845	3,275
9180	200 amp	2 Elec	2	8		5,050	480		5,530	6,275
9200	Ground bus added to bus duct, 225 amp		320	.050	L.F.	44.50	3		47.50	53.50
9210	400 amp		240	.067		44.50	4		48.50	55
9220	600 amp		240	.067		44.50	4		48.50	55
9230	800 amp		160	.100		52	6		58	66.50
9240	1,000 amp		160	.100		59.50	6		65.50	74.50
9250	1,350 amp		140	.114		85	6.85		91.85	104
9260	1,600 amp		120	.133		92	8		100	113
9270	2,000 amp		110	.145		120	8.75		128.75	145
9280	2,500 amp		100	.160		148	9.60		157.60	177
9290	3,000 amp		90	.178		167	10.70		177.70	199
9300	4,000 amp		80	.200		220	12		232	260
9310	5,000 amp		70	.229		267	13.75		280.75	315
9320	High short circuit bracing, add					18.60			18.60	20.50

26 25 Low-Voltage Enclosed Bus Assemblies

26 25 13 – Low-Voltage Busways

26 25 13.60 Copper or Aluminum Bus Duct Fittings	Crew	Daily Output	Labor-Hours	Unit	Material	2019 Bare Costs Labor	Equipment	Total	Total Incl O&P
0010 **COPPER OR ALUMINUM BUS DUCT FITTINGS**									
0100 Flange, wall, with vapor barrier, 225 amp	2 Elec	6.20	2.581	Ea.	810	155		965	1,125
0110 400 amp		6	2.667		790	160		950	1,100
0120 600 amp		5.80	2.759		810	166		976	1,150
0130 800 amp		5.40	2.963		810	178		988	1,150
0140 1,000 amp		5	3.200		810	192		1,002	1,175
0145 1,200 amp		4.80	3.333		810	200		1,010	1,200
0150 1,350 amp		4.60	3.478		810	209		1,019	1,200
0160 1,600 amp		4.20	3.810		810	229		1,039	1,225
0170 2,000 amp		4	4		810	240		1,050	1,250
0180 2,500 amp		3.60	4.444		810	267		1,077	1,300
0190 3,000 amp		3.20	5		810	300		1,110	1,350
0200 4,000 amp		2.60	6.154		800	370		1,170	1,425
0300 Roof, 225 amp		6.20	2.581		940	155		1,095	1,250
0310 400 amp		6	2.667		940	160		1,100	1,275
0320 600 amp		5.80	2.759		940	166		1,106	1,275
0330 800 amp		5.40	2.963		940	178		1,118	1,300
0340 1,000 amp		5	3.200		940	192		1,132	1,300
0345 1,200 amp		4.80	3.333		940	200		1,140	1,325
0350 1,350 amp		4.60	3.478		940	209		1,149	1,325
0360 1,600 amp		4.20	3.810		940	229		1,169	1,375
0370 2,000 amp		4	4		940	240		1,180	1,375
0380 2,500 amp		3.60	4.444		940	267		1,207	1,425
0390 3,000 amp		3.20	5		940	300		1,240	1,475
0400 4,000 amp		2.60	6.154		940	370		1,310	1,575
0420 Support, floor mounted, 225 amp		20	.800		169	48		217	258
0430 400 amp		20	.800		169	48		217	258
0440 600 amp		18	.889		169	53.50		222.50	266
0450 800 amp		16	1		169	60		229	276
0460 1,000 amp		13	1.231		169	74		243	296
0465 1,200 amp		11.80	1.356		169	81.50		250.50	305
0470 1,350 amp		10.60	1.509		169	90.50		259.50	320
0480 1,600 amp		9.20	1.739		169	104		273	340
0490 2,000 amp		8	2		169	120		289	365
0500 2,500 amp		6.40	2.500		169	150		319	410
0510 3,000 amp		5.40	2.963		169	178		347	450
0520 4,000 amp		4	4		169	240		409	545
0540 Weather stop, 225 amp		12	1.333		525	80		605	700
0550 400 amp		10	1.600		525	96		621	725
0560 600 amp		9	1.778		525	107		632	740
0570 800 amp		8	2		525	120		645	760
0580 1,000 amp		6.40	2.500		525	150		675	805
0585 1,200 amp		5.90	2.712		525	163		688	825
0590 1,350 amp		5.40	2.963		525	178		703	845
0600 1,600 amp		4.60	3.478		525	209		734	890
0610 2,000 amp		4	4		525	240		765	940
0620 2,500 amp		3.20	5		525	300		825	1,025
0630 3,000 amp		2.60	6.154		525	370		895	1,125
0640 4,000 amp		2	8		525	480		1,005	1,300
0660 End closure, 225 amp		34	.471		195	28.50		223.50	257
0670 400 amp		32	.500		195	30		225	260
0680 600 amp		28	.571		195	34.50		229.50	266

26 25 Low-Voltage Enclosed Bus Assemblies

26 25 13 – Low-Voltage Busways

26 25 13.60 Copper or Aluminum Bus Duct Fittings		Crew	Daily Output	Labor-Hours	Unit	Material	2019 Bare Costs Labor	Equipment	Total	Total Incl O&P
0690	800 amp	2 Elec	26	.615	Ea.	170	37		207	242
0700	1,000 amp		24	.667		195	40		235	275
0705	1,200 amp		23	.696		168	42		210	248
0710	1,350 amp		22	.727		168	43.50		211.50	250
0720	1,600 amp		20	.800		195	48		243	287
0730	2,000 amp		18	.889		276	53.50		329.50	385
0740	2,500 amp		16	1		253	60		313	370
0750	3,000 amp		14	1.143		253	68.50		321.50	380
0760	4,000 amp		12	1.333		249	80		329	395
0780	Switchboard stub, 3 pole 3 wire, 225 amp		6	2.667		1,000	160		1,160	1,375
0790	400 amp		5.20	3.077		1,000	185		1,185	1,400
0800	600 amp		4.60	3.478		1,000	209		1,209	1,425
0810	800 amp		3.60	4.444		1,225	267		1,492	1,750
0820	1,000 amp		3.40	4.706		1,425	283		1,708	2,000
0825	1,200 amp		3.20	5		1,425	300		1,725	2,025
0830	1,350 amp		3	5.333		1,725	320		2,045	2,350
0840	1,600 amp		2.80	5.714		2,025	345		2,370	2,725
0850	2,000 amp		2.40	6.667		2,350	400		2,750	3,200
0860	2,500 amp		2	8		2,875	480		3,355	3,875
0870	3,000 amp		1.80	8.889		3,275	535		3,810	4,400
0880	4,000 amp		1.60	10		4,225	600		4,825	5,550
0890	5,000 amp		1.40	11.429		5,200	685		5,885	6,750
0900	3 pole 4 wire, 225 amp		5.40	2.963		1,250	178		1,428	1,650
0910	400 amp		4.60	3.478		1,250	209		1,459	1,675
0920	600 amp		4	4		1,250	240		1,490	1,725
0930	800 amp		3.20	5		1,525	300		1,825	2,125
0940	1,000 amp		3	5.333		1,775	320		2,095	2,425
0950	1,350 amp		2.60	6.154		2,275	370		2,645	3,050
0960	1,600 amp		2.40	6.667		2,575	400		2,975	3,425
0970	2,000 amp		2	8		3,100	480		3,580	4,150
0980	2,500 amp		1.80	8.889		3,800	535		4,335	4,975
0990	3,000 amp		1.60	10		4,425	600		5,025	5,775
1000	4,000 amp		1.40	11.429		5,750	685		6,435	7,350
1050	Service head, weatherproof, 3 pole 3 wire, 225 amp		3	5.333		1,725	320		2,045	2,375
1060	400 amp		2.80	5.714		1,725	345		2,070	2,400
1070	600 amp		2.60	6.154		1,725	370		2,095	2,450
1080	800 amp		2.40	6.667		1,925	400		2,325	2,725
1090	1,000 amp		2	8		2,075	480		2,555	3,025
1100	1,350 amp		1.80	8.889		2,700	535		3,235	3,775
1110	1,600 amp		1.60	10		3,025	600		3,625	4,225
1120	2,000 amp		1.40	11.429		3,675	685		4,360	5,075
1130	2,500 amp		1.20	13.333		4,400	800		5,200	6,025
1140	3,000 amp		.90	17.778		5,125	1,075		6,200	7,225
1150	4,000 amp		.70	22.857		6,500	1,375		7,875	9,200
1200	3 pole 4 wire, 225 amp		2.60	6.154		1,925	370		2,295	2,675
1210	400 amp		2.40	6.667		1,925	400		2,325	2,725
1220	600 amp		2.20	7.273		1,925	435		2,360	2,775
1230	800 amp		2	8		2,225	480		2,705	3,175
1240	1,000 amp		1.70	9.412		2,575	565		3,140	3,675
1250	1,350 amp		1.50	10.667		3,050	640		3,690	4,300
1260	1,600 amp		1.40	11.429		3,325	685		4,010	4,675
1270	2,000 amp		1.20	13.333		4,450	800		5,250	6,100
1280	2,500 amp		1	16		5,475	960		6,435	7,450

For customer support on your Electrical Costs with RSMeans data, call 800.448.8182.

257

26 25 13.60 Copper or Aluminum Bus Duct Fittings	Crew	Daily Output	Labor-Hours	Unit	Material	2019 Bare Costs Labor	Equipment	Total	Total Incl O&P
1290	3,000 amp	2 Elec	.80	20	Ea.	6,425	1,200		7,625	8,850
1300	4,000 amp		.60	26.667		8,325	1,600		9,925	11,600
1350	Flanged end, 3 pole 3 wire, 225 amp		6	2.667		935	160		1,095	1,275
1360	400 amp		5.20	3.077		935	185		1,120	1,300
1370	600 amp		4.60	3.478		935	209		1,144	1,325
1380	800 amp		3.60	4.444		1,025	267		1,292	1,550
1390	1,000 amp		3.40	4.706		1,175	283		1,458	1,700
1395	1,200 amp		3.20	5		1,300	300		1,600	1,875
1400	1,350 amp		3	5.333		1,400	320		1,720	2,000
1410	1,600 amp		2.80	5.714		1,575	345		1,920	2,250
1420	2,000 amp		2.40	6.667		1,875	400		2,275	2,650
1430	2,500 amp		2	8		2,175	480		2,655	3,125
1440	3,000 amp		1.80	8.889		2,475	535		3,010	3,525
1450	4,000 amp		1.60	10		3,100	600		3,700	4,300
1500	3 pole 4 wire, 225 amp		5.40	2.963		1,075	178		1,253	1,450
1510	400 amp		4.60	3.478		1,075	209		1,284	1,475
1520	600 amp		4	4		1,075	240		1,315	1,525
1530	800 amp		3.20	5		1,275	300		1,575	1,850
1540	1,000 amp		3	5.333		1,425	320		1,745	2,025
1545	1,200 amp		2.80	5.714		1,600	345		1,945	2,250
1550	1,350 amp		2.60	6.154		1,725	370		2,095	2,450
1560	1,600 amp		2.40	6.667		1,975	400		2,375	2,775
1570	2,000 amp		2	8		2,325	480		2,805	3,275
1580	2,500 amp		1.80	8.889		2,775	535		3,310	3,850
1590	3,000 amp		1.60	10		3,175	600		3,775	4,375
1600	4,000 amp		1.40	11.429		4,100	685		4,785	5,525
1650	Hanger, standard, 225 amp		64	.250		26.50	15		41.50	51.50
1660	400 amp		48	.333		26.50	20		46.50	59
1670	600 amp		40	.400		26.50	24		50.50	65
1680	800 amp		32	.500		26.50	30		56.50	74
1690	1,000 amp		24	.667		26.50	40		66.50	88.50
1695	1,200 amp		22	.727		26.50	43.50		70	94
1700	1,350 amp		20	.800		26.50	48		74.50	101
1710	1,600 amp		20	.800		26.50	48		74.50	101
1720	2,000 amp		18	.889		26.50	53.50		80	109
1730	2,500 amp		16	1		26.50	60		86.50	119
1740	3,000 amp		16	1		26.50	60		86.50	119
1750	4,000 amp		16	1		26.50	60		86.50	119
1800	Spring type, 225 amp		16	1		88	60		148	186
1810	400 amp		14	1.143		88	68.50		156.50	199
1820	600 amp		14	1.143		88	68.50		156.50	199
1830	800 amp		14	1.143		88	68.50		156.50	199
1840	1,000 amp		14	1.143		88	68.50		156.50	199
1845	1,200 amp		14	1.143		88	68.50		156.50	199
1850	1,350 amp		14	1.143		88	68.50		156.50	199
1860	1,600 amp		12	1.333		88	80		168	216
1870	2,000 amp		12	1.333		88	80		168	216
1880	2,500 amp		12	1.333		88	80		168	216
1890	3,000 amp		10	1.600		88	96		184	240
1900	4,000 amp		10	1.600		88	96		184	240

26 25 13 – Low-Voltage Busways

26 25 13.70 Feedrail	Crew	Daily Output	Labor-Hours	Unit	Material	2019 Bare Costs Labor	Equipment	Total	Total Incl O&P
0010 **FEEDRAIL**, 12' mounting									
0050 Trolley busway, 3 pole									
0100 300 volt 60 amp, plain, 10' lengths	1 Elec	50	.160	L.F.	21	9.60		30.60	38
0300 Door track		50	.160		36	9.60		45.60	54
0500 Curved track	↓	30	.267	↓	15.10	16		31.10	40.50
0700 Coupling				Ea.	12			12	13.20
0900 Center feed	1 Elec	5.30	1.509		47.50	90.50		138	187
1100 End feed		5.30	1.509		48.50	90.50		139	189
1300 Hanger set		24	.333	↓	2.82	20		22.82	33
3000 600 volt 100 amp, plain, 10' lengths		35	.229	L.F.	60	13.75		73.75	86
3300 Door track	↓	35	.229	"	90.50	13.75		104.25	120
3700 Coupling				Ea.	52.50			52.50	57.50
4000 End cap	1 Elec	40	.200		67	12		79	92
4200 End feed		4	2		235	120		355	435
4500 Trolley, 600 volt, 20 amp		5.30	1.509		245	90.50		335.50	405
4700 30 amp		5.30	1.509		245	90.50		335.50	405
4900 Duplex, 40 amp		4	2		885	120		1,005	1,150
5000 60 amp		4	2		890	120		1,010	1,150
5300 Fusible, 20 amp		4	2		500	120		620	730
5500 30 amp		4	2		500	120		620	730
5900 300 volt, 20 amp		5.30	1.509		234	90.50		324.50	395
6000 30 amp		5.30	1.509		310	90.50		400.50	475
6300 Fusible, 20 amp		4.70	1.702		350	102		452	540
6500 30 amp		4.70	1.702	↓	490	102		592	695
7300 Busway, 250 volt 50 amp, 2 wire	↓	70	.114	L.F.	20	6.85		26.85	33
7330 Coupling				Ea.	44			44	48.50
7340 Center feed	1 Elec	6	1.333		500	80		580	670
7350 End feed		6	1.333		112	80		192	242
7360 End cap		40	.200		29	12		41	50
7370 Hanger set		24	.333	↓	2.82	20		22.82	33
7400 125/250 volt 50 amp, 3 wire		60	.133	L.F.	20.50	8		28.50	34.50
7430 Coupling		6	1.333	Ea.	63	80		143	189
7440 Center feed		6	1.333		500	80		580	670
7450 End feed		6	1.333		111	80		191	241
7460 End cap		40	.200		28.50	12		40.50	49.50
7470 Hanger set		24	.333		3.85	20		23.85	34
7480 Trolley, 250 volt, 2 pole, 20 amp		6	1.333		37	80		117	160
7490 30 amp		6	1.333		37	80		117	160
7500 125/250 volt, 3 pole, 20 amp		6	1.333		37.50	80		117.50	160
7510 30 amp	↓	6	1.333		37.50	80		117.50	160
8000 Cleaning tools, 300 volt, dust remover					109			109	120
8100 Bus bar cleaner					206			206	226
8300 600 volt, dust remover, 60 amp					300			300	330
8400 100 amp					655			655	720
8600 Bus bar cleaner, 60 amp					680			680	750
8700 100 amp				↓	800			800	885

259

For customer support on your Electrical Costs with RSMeans data, call 800.448.8182.

26 27 13.10 Meter Centers and Sockets	Crew	Daily Output	Labor-Hours	Unit	Material	2019 Bare Costs Labor	Equipment	Total	Total Incl O&P
0010 **METER CENTERS AND SOCKETS**									
0100 Sockets, single position, 4 terminal, 100 amp	1 Elec	3.20	2.500	Ea.	45	150		195	274
0200 150 amp		2.30	3.478		63.50	209		272.50	380
0300 200 amp		1.90	4.211		98	253		351	485
0400 Transformer rated, 20 amp		3.20	2.500		163	150		313	405
0500 Double position, 4 terminal, 100 amp		2.80	2.857		215	172		387	490
0600 150 amp		2.10	3.810		287	229		516	655
0700 200 amp		1.70	4.706		555	283		838	1,025
0800 Trans-socket, 13 terminal, 3 CT mounts, 400 amp		1	8		1,225	480		1,705	2,075
0900 800 amp	2 Elec	1.20	13.333		1,375	800		2,175	2,700
1100 Meter centers and sockets, three phase, single pos, 7 terminal, 100 amp	1 Elec	2.80	2.857		118	172		290	385
1200 200 amp		2.10	3.810		228	229		457	590
1400 400 amp		1.70	4.706		840	283		1,123	1,350
2000 Meter center, main fusible switch, 1P 3W 120/240 V									
2030 400 amp	2 Elec	1.60	10	Ea.	670	600		1,270	1,625
2040 600 amp		1.10	14.545		1,050	875		1,925	2,450
2050 800 amp		.90	17.778		1,250	1,075		2,325	2,950
2060 Rainproof 1P 3W 120/240 V, 400 A		1.60	10		1,675	600		2,275	2,750
2070 600 amp		1.10	14.545		2,925	875		3,800	4,500
2080 800 amp		.90	17.778		4,575	1,075		5,650	6,625
2100 3P 4W 120/208 V, 400 amp		1.60	10		560	600		1,160	1,500
2110 600 amp		1.10	14.545		1,150	875		2,025	2,575
2120 800 amp		.90	17.778		1,725	1,075		2,800	3,500
2130 Rainproof 3P 4W 120/208 V, 400 amp		1.60	10		1,925	600		2,525	3,000
2140 600 amp		1.10	14.545		3,575	875		4,450	5,250
2150 800 amp		.90	17.778		6,650	1,075		7,725	8,925
2170 Main circuit breaker, 1P 3W 120/240 V									
2180 400 amp	2 Elec	1.60	10	Ea.	1,425	600		2,025	2,475
2190 600 amp		1.10	14.545		1,775	875		2,650	3,250
2200 800 amp		.90	17.778		1,850	1,075		2,925	3,625
2210 1,000 amp		.80	20		3,150	1,200		4,350	5,275
2220 1,200 amp		.76	21.053		3,225	1,275		4,500	5,425
2230 1,600 amp		.68	23.529		18,600	1,425		20,025	22,600
2240 Rainproof 1P 3W 120/240 V, 400 amp		1.60	10		3,000	600		3,600	4,200
2250 600 amp		1.10	14.545		4,025	875		4,900	5,750
2260 800 amp		.90	17.778		4,700	1,075		5,775	6,775
2270 1,000 amp		.80	20		6,500	1,200		7,700	8,950
2280 1,200 amp		.76	21.053		8,750	1,275		10,025	11,500
2300 3P 4W 120/208 V, 400 amp		1.60	10		3,325	600		3,925	4,550
2310 600 amp		1.10	14.545		4,800	875		5,675	6,575
2320 800 amp		.90	17.778		5,725	1,075		6,800	7,900
2330 1,000 amp		.80	20		7,500	1,200		8,700	10,100
2340 1,200 amp		.76	21.053		9,600	1,275		10,875	12,500
2350 1,600 amp		.68	23.529		19,600	1,425		21,025	23,600
2360 Rainproof 3P 4W 120/208 V, 400 amp		1.60	10		3,425	600		4,025	4,650
2370 600 amp		1.10	14.545		4,800	875		5,675	6,575
2380 800 amp		.90	17.778		5,725	1,075		6,800	7,900
2390 1,000 amp		.76	21.053		7,500	1,275		8,775	10,200
2400 1,200 amp		.68	23.529		9,600	1,425		11,025	12,700
2420 Main lugs terminal box, 1P 3W 120/240 V									
2430 800 amp	2 Elec	.94	17.021	Ea.	485	1,025		1,510	2,050
2440 1,200 amp		.72	22.222		535	1,325		1,860	2,600
2450 Rainproof 1P 3W 120/240 V, 225 amp		2.40	6.667		385	400		785	1,025

26 27 13 – Electricity Metering

26 27 13.10 Meter Centers and Sockets	Crew	Daily Output	Labor-Hours	Unit	Material	2019 Bare Costs Labor	Equipment	Total	Total Incl O&P
2460	800 amp	2 Elec	.94	17.021	Ea.	565	1,025		1,590	2,150
2470	1,200 amp		.72	22.222		1,125	1,325		2,450	3,250
2500	3P 4W 120/208 V, 800 amp		.94	17.021		625	1,025		1,650	2,225
2510	1,200 amp		.72	22.222		1,250	1,325		2,575	3,375
2520	Rainproof 3P 4W 120/208 V, 225 amp		2.40	6.667		385	400		785	1,025
2530	800 amp		.94	17.021		625	1,025		1,650	2,225
2540	1,200 amp		.72	22.222		1,250	1,325		2,575	3,375
2590	Basic meter device									
2600	1P 3W 120/240 V 4 jaw 125A sockets, 3 meter	2 Elec	1	16	Ea.	360	960		1,320	1,825
2610	4 meter		.90	17.778		445	1,075		1,520	2,100
2620	5 meter		.80	20		530	1,200		1,730	2,375
2630	6 meter		.60	26.667		555	1,600		2,155	3,025
2640	7 meter		.56	28.571		1,425	1,725		3,150	4,100
2650	8 meter		.52	30.769		1,550	1,850		3,400	4,450
2660	10 meter		.48	33.333		1,925	2,000		3,925	5,100
2680	Rainproof 1P 3W 120/240 V 4 jaw 125A sockets									
2690	3 meter	2 Elec	1	16	Ea.	645	960		1,605	2,125
2700	4 meter		.90	17.778		770	1,075		1,845	2,450
2710	6 meter		.60	26.667		1,100	1,600		2,700	3,625
2720	7 meter		.56	28.571		1,425	1,725		3,150	4,100
2730	8 meter		.52	30.769		1,550	1,850		3,400	4,450
2750	1P 3W 120/240 V 4 jaw sockets									
2760	with 125A circuit breaker, 3 meter	2 Elec	1	16	Ea.	1,200	960		2,160	2,750
2770	4 meter		.90	17.778		1,525	1,075		2,600	3,275
2780	5 meter		.80	20		1,900	1,200		3,100	3,900
2790	6 meter		.60	26.667		2,225	1,600		3,825	4,850
2800	7 meter		.56	28.571		2,725	1,725		4,450	5,550
2810	8 meter		.52	30.769		3,050	1,850		4,900	6,100
2820	10 meter		.48	33.333		3,800	2,000		5,800	7,175
2830	Rainproof 1P 3W 120/240 V 4 jaw sockets									
2840	with 125A circuit breaker, 3 meter	2 Elec	1	16	Ea.	1,200	960		2,160	2,750
2850	4 meter		.90	17.778		1,525	1,075		2,600	3,275
2870	6 meter		.60	26.667		2,225	1,600		3,825	4,850
2880	7 meter		.56	28.571		2,725	1,725		4,450	5,550
2890	8 meter		.52	30.769		3,050	1,850		4,900	6,100
2920	1P 3W on 3P 4W 120/208 V system 5 jaw									
2930	125A sockets, 3 meter	2 Elec	1	16	Ea.	645	960		1,605	2,125
2940	4 meter		.90	17.778		770	1,075		1,845	2,450
2950	5 meter		.80	20		965	1,200		2,165	2,875
2960	6 meter		.60	26.667		1,100	1,600		2,700	3,625
2970	7 meter		.56	28.571		1,425	1,725		3,150	4,100
2980	8 meter		.52	30.769		1,550	1,850		3,400	4,450
2990	10 meter		.48	33.333		1,925	2,000		3,925	5,100
3000	Rainproof 1P 3W on 3P 4W 120/208 V system									
3020	5 jaw 125A sockets, 3 meter	2 Elec	1	16	Ea.	645	960		1,605	2,125
3030	4 meter		.90	17.778		770	1,075		1,845	2,450
3050	6 meter		.60	26.667		1,100	1,600		2,700	3,625
3060	7 meter		.56	28.571		1,425	1,725		3,150	4,100
3070	8 meter		.52	30.769		1,550	1,850		3,400	4,450
3090	1P 3W on 3P 4W 120/208 V system 5 jaw sockets									
3100	With 125A circuit breaker, 3 meter	2 Elec	1	16	Ea.	1,200	960		2,160	2,750
3110	4 meter		.90	17.778		1,525	1,075		2,600	3,275
3120	5 meter		.80	20		1,900	1,200		3,100	3,900

26 27 13.10 Meter Centers and Sockets	Crew	Daily Output	Labor-Hours	Unit	Material	2019 Bare Costs Labor	Equipment	Total	Total Incl O&P
3130	6 meter	2 Elec	.60	26.667	Ea.	2,225	1,600		3,825	4,850
3140	7 meter		.56	28.571		2,725	1,725		4,450	5,550
3150	8 meter		.52	30.769		3,050	1,850		4,900	6,100
3160	10 meter		.48	33.333		3,800	2,000		5,800	7,175
3170	Rainproof 1P 3W on 3P 4W 120/208 V system									
3180	5 jaw sockets w/125A circuit breaker, 3 meter	2 Elec	1	16	Ea.	1,200	960		2,160	2,750
3190	4 meter		.90	17.778		1,525	1,075		2,600	3,275
3210	6 meter		.60	26.667		2,225	1,600		3,825	4,850
3220	7 meter		.56	28.571		2,725	1,725		4,450	5,550
3230	8 meter		.52	30.769		3,050	1,850		4,900	6,100
3250	1P 3W 120/240 V 4 jaw sockets									
3260	with 200A circuit breaker, 3 meter	2 Elec	1	16	Ea.	1,800	960		2,760	3,425
3270	4 meter		.90	17.778		2,450	1,075		3,525	4,300
3290	6 meter		.60	26.667		3,625	1,600		5,225	6,375
3300	7 meter		.56	28.571		4,250	1,725		5,975	7,225
3310	8 meter		.56	28.571		4,900	1,725		6,625	7,925
3330	Rainproof 1P 3W 120/240 V 4 jaw sockets									
3350	with 200A circuit breaker, 3 meter	2 Elec	1	16	Ea.	1,800	960		2,760	3,425
3360	4 meter		.90	17.778		2,450	1,075		3,525	4,300
3380	6 meter		.60	26.667		3,625	1,600		5,225	6,375
3390	7 meter		.56	28.571		4,250	1,725		5,975	7,225
3400	8 meter		.52	30.769		4,900	1,850		6,750	8,125
3420	1P 3W on 3P 4W 120/208 V 5 jaw sockets									
3430	with 200A circuit breaker, 3 meter	2 Elec	1	16	Ea.	1,800	960		2,760	3,425
3440	4 meter		.90	17.778		2,450	1,075		3,525	4,300
3460	6 meter		.60	26.667		3,625	1,600		5,225	6,400
3470	7 meter		.56	28.571		4,250	1,725		5,975	7,225
3480	8 meter		.52	30.769		4,900	1,850		6,750	8,125
3500	Rainproof 1P 3W on 3P 4W 120/208 V 5 jaw socket									
3510	with 200A circuit breaker, 3 meter	2 Elec	1	16	Ea.	1,800	960		2,760	3,425
3520	4 meter		.90	17.778		2,450	1,075		3,525	4,300
3540	6 meter		.60	26.667		3,625	1,600		5,225	6,375
3550	7 meter		.56	28.571		4,250	1,725		5,975	7,225
3560	8 meter		.52	30.769		4,900	1,850		6,750	8,125
3600	Automatic circuit closing, add					70			70	77
3610	Manual circuit closing, add					80			80	88
3650	Branch meter device									
3660	3P 4W 208/120 or 240/120 V 7 jaw sockets									
3670	with 200A circuit breaker, 2 meter	2 Elec	.90	17.778	Ea.	3,025	1,075		4,100	4,925
3680	3 meter		.80	20		4,525	1,200		5,725	6,800
3690	4 meter		.70	22.857		6,050	1,375		7,425	8,700
3700	Main circuit breaker 42,000 rms, 400 amp		1.60	10		1,925	600		2,525	3,000
3710	600 amp		1.10	14.545		2,575	875		3,450	4,150
3720	800 amp		.90	17.778		3,275	1,075		4,350	5,200
3730	Rainproof main circ. breaker 42,000 rms, 400 amp		1.60	10		2,650	600		3,250	3,800
3740	600 amp		1.10	14.545		4,250	875		5,125	5,975
3750	800 amp		.90	17.778		5,725	1,075		6,800	7,900
3760	Main circuit breaker 65,000 rms, 400 amp		1.60	10		3,600	600		4,200	4,850
3770	600 amp		1.10	14.545		5,050	875		5,925	6,850
3780	800 amp		.90	17.778		5,725	1,075		6,800	7,900
3790	1,000 amp		.80	20		7,500	1,200		8,700	10,100
3800	1,200 amp		.76	21.053		9,600	1,275		10,875	12,500
3810	1,600 amp		.68	23.529		19,600	1,425		21,025	23,600

For customer support on your Electrical Costs with RSMeans data, call 800.448.8182.

26 27 Low-Voltage Distribution Equipment

26 27 13 – Electricity Metering

26 27 13.10 Meter Centers and Sockets

		Crew	Daily Output	Labor-Hours	Unit	Material	2019 Bare Costs Labor	Equipment	Total	Total Incl O&P
3820	Rainproof main circ. breaker 65,000 rms, 400 amp	2 Elec	1.60	10	Ea.	3,600	600		4,200	4,850
3830	600 amp		1.10	14.545		5,050	875		5,925	6,850
3840	800 amp		.90	17.778		5,700	1,075		6,775	7,875
3850	1,000 amp		.80	20		7,500	1,200		8,700	10,100
3860	1,200 amp		.76	21.053		9,600	1,275		10,875	12,500
3880	Main circuit breaker 100,000 rms, 400 amp		1.60	10		3,600	600		4,200	4,850
3890	600 amp		1.10	14.545		5,050	875		5,925	6,850
3900	800 amp		.90	17.778		5,925	1,075		7,000	8,125
3910	Rainproof main circ. breaker 100,000 rms, 400 amp		1.60	10		3,600	600		4,200	4,850
3920	600 amp		1.10	14.545		5,050	875		5,925	6,850
3930	800 amp		.90	17.778		5,925	1,075		7,000	8,125
3940	Main lugs terminal box, 800 amp		.94	17.021		625	1,025		1,650	2,225
3950	1,600 amp		.72	22.222		2,525	1,325		3,850	4,775
3960	Rainproof, 800 amp		.94	17.021		625	1,025		1,650	2,225
3970	1,600 amp	▼	.72	22.222	▼	2,525	1,325		3,850	4,775

26 27 16 – Electrical Cabinets and Enclosures

26 27 16.10 Cabinets

			Crew	Daily Output	Labor-Hours	Unit	Material	2019 Bare Costs Labor	Equipment	Total	Total Incl O&P
0010	**CABINETS**	R260533-70									
7000	Cabinets, current transformer										
7050	Single door, 24" H x 24" W x 10" D	R262716-40	1 Elec	1.60	5	Ea.	168	300		468	635
7100	30" H x 24" W x 10" D			1.30	6.154		177	370		547	745
7150	36" H x 24" W x 10" D			1.10	7.273		425	435		860	1,125
7200	30" H x 30" W x 10" D			1	8		245	480		725	985
7250	36" H x 30" W x 10" D			.90	8.889		320	535		855	1,150
7300	36" H x 36" W x 10" D			.80	10		325	600		925	1,250
7500	Double door, 48" H x 36" W x 10" D			.60	13.333		600	800		1,400	1,850
7550	24" H x 24" W x 12" D		▼	1	8	▼	192	480		672	925
8000	NEMA 12, double door, floor mounted										
8020	54" H x 42" W x 8" D		2 Elec	6	2.667	Ea.	900	160		1,060	1,225
8040	60" H x 48" W x 8" D			5.40	2.963		1,200	178		1,378	1,600
8060	60" H x 48" W x 10" D			5.40	2.963		1,250	178		1,428	1,625
8080	60" H x 60" W x 10" D			5	3.200		1,375	192		1,567	1,800
8100	72" H x 60" W x 10" D			4	4		1,625	240		1,865	2,150
8120	72" H x 72" W x 10" D			3.40	4.706		1,800	283		2,083	2,400
8140	60" H x 48" W x 12" D			3.40	4.706		1,250	283		1,533	1,800
8160	60" H x 60" W x 12" D			3.20	5		1,425	300		1,725	2,025
8180	72" H x 60" W x 12" D			3	5.333		2,275	320		2,595	2,975
8200	72" H x 72" W x 12" D			3	5.333		2,500	320		2,820	3,225
8220	60" H x 48" W x 16" D			3.20	5		1,300	300		1,600	1,900
8240	72" H x 72" W x 16" D			2.60	6.154		1,900	370		2,270	2,650
8260	60" H x 48" W x 20" D			3	5.333		1,450	320		1,770	2,050
8280	72" H x 72" W x 20" D			2.20	7.273		2,050	435		2,485	2,900
8300	60" H x 48" W x 24" D			2.60	6.154		1,525	370		1,895	2,225
8320	72" H x 72" W x 24" D		▼	2	8	▼	2,175	480		2,655	3,100
8340	Pushbutton enclosure, oiltight										
8360	3-1/2" H x 3-1/4" W x 2-3/4" D, for 1 P.B.		1 Elec	12	.667	Ea.	39.50	40		79.50	103
8380	5-3/4" H x 3-1/4" W x 2-3/4" D, for 2 P.B.			11	.727		45	43.50		88.50	115
8400	8" H x 3-1/4" W x 2-3/4" D, for 3 P.B.			10.50	.762		50.50	46		96.50	124
8420	10-1/4" H x 3-1/4" W x 2-3/4" D, for 4 P.B.			10.50	.762		56	46		102	130
8460	12-1/2" H x 3-1/4" W x 3" D, for 5 P.B.			9	.889		93.50	53.50		147	183
8480	9-1/2" H x 6-1/4" W x 3" D, for 6 P.B.			8.50	.941		72.50	56.50		129	165
8500	9-1/2" H x 8-1/2" W x 3" D, for 9 P.B.			8	1		80.50	60		140.50	178

For customer support on your Electrical Costs with RSMeans data, call 800.448.8182.

263

26 27 16.10 Cabinets		Crew	Daily Output	Labor-Hours	Unit	Material	2019 Bare Costs Labor	Equipment	Total	Total Incl O&P
8510	11-3/4" H x 8-1/2" W x 3" D, for 12 P.B.	1 Elec	7	1.143	Ea.	88	68.50		156.50	199
8520	11-3/4" H x 10-3/4" W x 3" D, for 16 P.B.		6.50	1.231		100	74		174	220
8540	14" H x 10-3/4" W x 3" D, for 20 P.B.		5	1.600		147	96		243	305
8560	14" H x 13" W x 3" D, for 25 P.B.		4.50	1.778		160	107		267	335
8580	Sloping front pushbutton enclosures									
8600	3-1/2" H x 7-3/4" W x 4-7/8" D, for 3 P.B.	1 Elec	10	.800	Ea.	67.50	48		115.50	146
8620	7-1/4" H x 8-1/2" W x 6-3/4" D, for 6 P.B.		8	1		98.50	60		158.50	198
8640	9-1/2" H x 8-1/2" W x 7-7/8" D, for 9 P.B.		7	1.143		117	68.50		185.50	231
8660	11-1/4" H x 8-1/2" W x 9" D, for 12 P.B.		5	1.600		133	96		229	289
8680	11-3/4" H x 10" W x 9" D, for 16 P.B.		5	1.600		146	96		242	305
8700	11-3/4" H x 13" W x 9" D, for 20 P.B.		5	1.600		168	96		264	330
8720	14" H x 13" W x 10-1/8" D, for 25 P.B.		4.50	1.778		190	107		297	370
8740	Pedestals, not including P.B. enclosure or base									
8760	Straight column 4" x 4"	1 Elec	4.50	1.778	Ea.	178	107		285	355
8780	6" x 6"		4	2		278	120		398	485
8800	Angled column 4" x 4"		4.50	1.778		205	107		312	385
8820	6" x 6"		4	2		310	120		430	520
8840	Pedestal, base 18" x 18"		10	.800		105	48		153	188
8860	24" x 24"		9	.889		242	53.50		295.50	345
8900	Electronic rack enclosures									
8920	72" H x 19" W x 24" D	1 Elec	1.50	5.333	Ea.	2,025	320		2,345	2,700
8940	72" H x 23" W x 24" D		1.50	5.333		2,250	320		2,570	2,950
8960	72" H x 19" W x 30" D		1.30	6.154		2,350	370		2,720	3,150
8980	72" H x 19" W x 36" D		1.20	6.667		2,800	400		3,200	3,675
9000	72" H x 23" W x 36" D		1.20	6.667		3,050	400		3,450	3,950
9020	NEMA 12 & 4 enclosure panels									
9040	12" x 24"	1 Elec	20	.400	Ea.	47	24		71	88
9060	16" x 12"		20	.400		31.50	24		55.50	71
9080	20" x 16"		20	.400		46.50	24		70.50	87
9100	20" x 20"		19	.421		56	25.50		81.50	99
9120	24" x 20"		18	.444		72	26.50		98.50	120
9140	24" x 24"		17	.471		83	28.50		111.50	134
9160	30" x 20"		16	.500		108	30		138	163
9180	30" x 24"		16	.500		103	30		133	158
9200	36" x 24"		15	.533		127	32		159	188
9220	36" x 30"		15	.533		161	32		193	225
9240	42" x 24"		15	.533		146	32		178	208
9260	42" x 30"		14	.571		188	34.50		222.50	258
9280	42" x 36"		14	.571		227	34.50		261.50	300
9300	48" x 24"		14	.571		98	34.50		132.50	159
9320	48" x 30"		14	.571		127	34.50		161.50	191
9340	48" x 36"		13	.615		244	37		281	325
9360	60" x 36"		12	.667		310	40		350	400
9400	Wiring trough steel JIC, clamp cover									
9490	4" x 4", 12" long	1 Elec	12	.667	Ea.	62.50	40		102.50	129
9510	24" long		10	.800		79	48		127	158
9530	36" long		8	1		107	60		167	207
9540	48" long		7	1.143		114	68.50		182.50	228
9550	60" long		6	1.333		144	80		224	278
9560	6" x 6", 12" long		11	.727		90	43.50		133.50	164
9580	24" long		9	.889		117	53.50		170.50	209
9600	36" long		7	1.143		135	68.50		203.50	250
9610	48" long		6	1.333		177	80		257	315

26 27 Low-Voltage Distribution Equipment

26 27 16 – Electrical Cabinets and Enclosures

26 27 16.10 Cabinets

		Crew	Daily Output	Labor-Hours	Unit	Material	2019 Bare Costs Labor	Equipment	Total	Total Incl O&P
9620	60" long	1 Elec	5	1.600	Ea.	208	96		304	370

26 27 16.20 Cabinets and Enclosures

		Crew	Daily Output	Labor-Hours	Unit	Material	2019 Bare Costs Labor	Equipment	Total	Total Incl O&P
0010	**CABINETS AND ENCLOSURES** Nonmetallic									
0080	Enclosures fiberglass NEMA 4X									
0100	Wall mount, quick release latch door, 20" H x 16" W x 6" D	1 Elec	4.80	1.667	Ea.	725	100		825	950
0110	20" H x 20" W x 6" D		4.50	1.778		860	107		967	1,100
0120	24" H x 20" W x 6" D		4.20	1.905		920	114		1,034	1,175
0130	20" H x 16" W x 8" D		4.50	1.778		825	107		932	1,075
0140	20" H x 20" W x 8" D		4.20	1.905		910	114		1,024	1,175
0150	24" H x 24" W x 8" D		3.80	2.105		1,025	126		1,151	1,325
0160	30" H x 24" W x 8" D		3.20	2.500		1,150	150		1,300	1,475
0170	36" H x 30" W x 8" D		3	2.667		1,575	160		1,735	2,000
0180	20" H x 16" W x 10" D		3.50	2.286		950	137		1,087	1,250
0190	20" H x 20" W x 10" D		3.20	2.500		1,075	150		1,225	1,400
0200	24" H x 20" W x 10" D		3	2.667		1,100	160		1,260	1,475
0210	30" H x 24" W x 10" D		2.80	2.857		1,250	172		1,422	1,600
0220	20" H x 16" W x 12" D		3	2.667		1,025	160		1,185	1,375
0230	20" H x 20" W x 12" D		2.80	2.857		1,075	172		1,247	1,425
0240	24" H x 24" W x 12" D		2.60	3.077		1,200	185		1,385	1,600
0250	30" H x 24" W x 12" D		2.40	3.333		1,400	200		1,600	1,850
0260	36" H x 30" W x 12" D		2.20	3.636		1,825	218		2,043	2,325
0270	36" H x 36" W x 12" D		2.10	3.810		2,000	229		2,229	2,550
0280	48" H x 36" W x 12" D		2	4		2,375	240		2,615	2,975
0290	60" H x 36" W x 12" D		1.80	4.444		1,650	267		1,917	2,200
0300	30" H x 24" W x 16" D		1.40	5.714		1,550	345		1,895	2,200
0310	48" H x 36" W x 16" D		1.20	6.667		2,575	400		2,975	3,425
0320	60" H x 36" W x 16" D		1	8		2,900	480		3,380	3,900
0480	Freestanding, one door, 72" H x 25" W x 25" D		.80	10		2,825	600		3,425	4,025
0490	Two doors with two panels, 72" H x 49" W x 24" D		.50	16		7,050	960		8,010	9,200
0500	Floor stand kits, for NEMA 4 & 12, 20" W or more, 6" H x 8" D		24	.333		112	20		132	153
0510	6" H x 10" D		24	.333		122	20		142	165
0520	6" H x 12" D		24	.333		138	20		158	182
0530	6" H x 18" D		24	.333		169	20		189	216
0540	12" H x 8" D		22	.364		142	22		164	189
0550	12" H x 10" D		22	.364		244	22		266	300
0560	12" H x 12" D		22	.364		286	22		308	350
0570	12" H x 16" D		22	.364		186	22		208	238
0580	12" H x 18" D		22	.364		202	22		224	256
0590	12" H x 20" D		22	.364		205	22		227	258
0600	18" H x 8" D		20	.400		179	24		203	233
0610	18" H x 10" D		20	.400		184	24		208	238
0620	18" H x 12" D		20	.400		195	24		219	251
0630	18" H x 16" D		20	.400		217	24		241	275
0640	24" H x 8" D		16	.500		208	30		238	274
0650	24" H x 10" D		16	.500		217	30		247	284
0660	24" H x 12" D		16	.500		232	30		262	300
0670	24" H x 16" D		16	.500		258	30		288	330
0680	Small, screw cover, 5-1/2" H x 4" W x 4-15/16" D		12	.667		116	40		156	187
0690	7-1/2" H x 4" W x 4-15/16" D		12	.667		76	40		116	144
0700	7-1/2" H x 6" W x 5-3/16" D		10	.800		142	48		190	229
0710	9-1/2" H x 6" W x 5-11/16" D		10	.800		148	48		196	235
0720	11-1/2" H x 8" W x 6-11/16" D		8	1		223	60		283	335

265

26 27 16.20 Cabinets and Enclosures		Crew	Daily Output	Labor-Hours	Unit	Material	2019 Bare Costs Labor	Equipment	Total	Total Incl O&P
0730	13-1/2" H x 10" W x 7-3/16" D	1 Elec	7	1.143	Ea.	259	68.50		327.50	385
0740	15-1/2" H x 12" W x 8-3/16" D		6	1.333		335	80		415	485
0750	17-1/2" H x 14" W x 8-11/16" D		5	1.600		395	96		491	580
0760	Screw cover with window, 6" H x 4" W x 5" D		12	.667		177	40		217	255
0770	8" H x 4" W x 5" D		11	.727		180	43.50		223.50	263
0780	8" H x 6" W x 5" D		11	.727		284	43.50		327.50	375
0790	10" H x 6" W x 6" D		10	.800		267	48		315	365
0800	12" H x 8" W x 7" D		8	1		345	60		405	470
0810	14" H x 10" W x 7" D		7	1.143		465	68.50		533.50	610
0820	16" H x 12" W x 8" D		6	1.333		490	80		570	655
0830	18" H x 14" W x 9" D		5	1.600		585	96		681	790
0840	Quick-release latch cover, 5-1/2" H x 4" W x 5" D		12	.667		151	40		191	226
0850	7-1/2" H x 4" W x 5" D		12	.667		138	40		178	212
0860	7-1/2" H x 6" W x 5-1/4" D		10	.800		181	48		229	271
0870	9-1/2" H x 6" W x 5-3/4" D		10	.800		110	48		158	193
0880	11-1/2" H x 8" W x 6-3/4" D		8	1		177	60		237	284
0890	13-1/2" H x 10" W x 7-1/4" D		7	1.143		196	68.50		264.50	320
0900	15-1/2" H x 12" W x 8-1/4" D		6	1.333		440	80		520	605
0910	17-1/2" H x 14" W x 8-3/4" D		5	1.600		530	96		626	725
0920	Pushbutton, 1 hole 5-1/2" H x 4" W x 4-15/16" D		12	.667		440	40		480	545
0930	2 hole 7-1/2" H x 4" W x 4-15/16" D		11	.727		620	43.50		663.50	745
0940	4 hole 7-1/2" H x 6" W x 5-3/16" D		10.50	.762		430	46		476	540
0950	6 hole 9-1/2" H x 6" W x 5-11/16" D		9	.889		520	53.50		573.50	650
0960	8 hole 11-1/2" H x 8" W x 6-11/16" D		8.50	.941		665	56.50		721.50	815
0970	12 hole 13-1/2" H x 10" W x 7-3/16" D		8	1		875	60		935	1,050
0980	20 hole 15-1/2" H x 12" W x 8-3/16" D		5	1.600		835	96		931	1,075
0990	30 hole 17-1/2" H x 14" W x 8-11/16" D	▼	4.50	1.778	▼	990	107		1,097	1,250
1450	Enclosures polyester NEMA 4X									
1460	Small, screw cover,									
1500	3-15/16" H x 3-15/16" W x 3-1/16" D	1 Elec	12	.667	Ea.	64.50	40		104.50	131
1510	5-3/16" H x 3-5/16" W x 3-1/16" D		12	.667		63.50	40		103.50	130
1520	5-7/8" H x 3-7/8" W x 4-3/16" D		12	.667		67	40		107	133
1530	5-7/8" H x 5-7/8" W x 4-3/16" D		12	.667		75.50	40		115.50	143
1540	7-5/8" H x 3-5/16" W x 3-1/16" D		12	.667		69.50	40		109.50	136
1550	10-3/16" H x 3-5/16" W x 3-1/16" D		10	.800		82.50	48		130.50	162
1560	Clear cover, 3-15/16" H x 3-15/16" W x 2-7/8" D		12	.667		73	40		113	140
1570	5-3/16" H x 3-5/16" W x 2-7/8" D		12	.667		77.50	40		117.50	145
1580	5-7/8" H x 3-7/8" W x 4" D		12	.667		96.50	40		136.50	166
1590	5-7/8" H x 5-7/8" W x 4" D		12	.667		119	40		159	191
1600	7-5/8" H x 3-5/16" W x 2-7/8" D		12	.667		87	40		127	155
1610	10-3/16" H x 3-5/16" W x 2-7/8" D		10	.800		109	48		157	192
1620	Pushbutton, 1 hole, 5-5/16" H x 3-5/16" W x 3-1/16" D		12	.667		56.50	40		96.50	122
1630	2 hole, 7-5/8" H x 3-5/16" W x 3-1/8" D		11	.727		63.50	43.50		107	135
1640	3 hole, 10-3/16" H x 3-5/16" W x 3-1/16" D		10.50	.762		75.50	46		121.50	152
8000	Wireway fiberglass, straight sect. screwcover, 12" L, 4" W x 4" D		40	.200		218	12		230	257
8010	6" W x 6" D		30	.267		298	16		314	355
8020	24" L, 4" W x 4" D		20	.400		284	24		308	345
8030	6" W x 6" D		15	.533		435	32		467	530
8040	36" L, 4" W x 4" D		13.30	.602		345	36		381	430
8050	6" W x 6" D		10	.800		550	48		598	675
8060	48" L, 4" W x 4" D		10	.800		400	48		448	510
8070	6" W x 6" D		7.50	1.067		650	64		714	810
8080	60" L, 4" W x 4" D		8	1		535	60		595	680

26 27 16 – Electrical Cabinets and Enclosures

26 27 16.20 Cabinets and Enclosures		Crew	Daily Output	Labor-Hours	Unit	Material	2019 Bare Costs Labor	Equipment	Total	Total Incl O&P
8090	6" W x 6" D	1 Elec	6	1.333	Ea.	760	80		840	960
8100	Elbow, 90°, 4" W x 4" D		20	.400		221	24		245	279
8110	6" W x 6" D		18	.444		445	26.50		471.50	530
8120	Elbow, 45°, 4" W x 4" D		20	.400		223	24		247	282
8130	6" W x 6" D		18	.444		435	26.50		461.50	520
8140	Tee, 4" W x 4" D		16	.500		280	30		310	350
8150	6" W x 6" D		14	.571		510	34.50		544.50	610
8160	Cross, 4" W x 4" D		14	.571		310	34.50		344.50	390
8170	6" W x 6" D		12	.667		700	40		740	830
8180	Cut-off fitting, w/flange & adhesive, 4" W x 4" D		18	.444		172	26.50		198.50	229
8190	6" W x 6" D		16	.500		345	30		375	425
8200	Flexible ftng., hvy. neoprene coated nylon, 4" W x 4" D		20	.400		310	24		334	380
8210	6" W x 6" D		18	.444		470	26.50		496.50	555
8220	Closure plate, fiberglass, 4" W x 4" D		20	.400		61	24		85	103
8230	6" W x 6" D		18	.444		70	26.50		96.50	117
8240	Box connector, stainless steel type 304, 4" W x 4" D		20	.400		108	24		132	155
8250	6" W x 6" D		18	.444		120	26.50		146.50	172
8260	Hanger, 4" W x 4" D		100	.080		26.50	4.80		31.30	36
8270	6" W x 6" D		80	.100		32.50	6		38.50	44.50
8280	Straight tube section fiberglass, 4" W x 4" D, 12" long		40	.200		192	12		204	229
8290	24" long		20	.400		258	24		282	320
8300	36" long		13.30	.602		220	36		256	296
8310	48" long		10	.800		274	48		322	370
8320	60" long		8	1		350	60		410	475
8330	120" long		4	2		540	120		660	775

26 27 19 – Multi-Outlet Assemblies

26 27 19.10 Wiring Duct

		Crew	Daily Output	Labor-Hours	Unit	Material	Labor	Equipment	Total	Total Incl O&P
0010	**WIRING DUCT** Plastic									
1250	PVC, snap-in slots, adhesive backed									
1270	1-1/2" W x 2" H	2 Elec	120	.133	L.F.	4.94	8		12.94	17.40
1280	1-1/2" W x 3" H		120	.133		7.20	8		15.20	19.85
1290	1-1/2" W x 4" H		120	.133		7.05	8		15.05	19.70
1300	2" W x 1" H		120	.133		4.44	8		12.44	16.85
1310	2" W x 1-1/2" H		120	.133		4.81	8		12.81	17.25
1320	2" W x 2" H		120	.133		4.83	8		12.83	17.25
1340	2" W x 3" H		120	.133		6.30	8		14.30	18.90
1350	2" W x 4" H		120	.133		7.60	8		15.60	20.50
1360	2-1/2" W x 3" H		120	.133		6.80	8		14.80	19.40
1370	3" W x 1" H		110	.145		5.05	8.75		13.80	18.60
1390	3" W x 2" H		110	.145		6	8.75		14.75	19.65
1400	3" W x 3" H		110	.145		7.25	8.75		16	21
1410	3" W x 4" H		110	.145		9.10	8.75		17.85	23
1420	3" W x 5" H		110	.145		11.95	8.75		20.70	26
1430	4" W x 1-1/2" H		100	.160		7.30	9.60		16.90	22.50
1440	4" W x 2" H		100	.160		7.05	9.60		16.65	22
1450	4" W x 3" H		100	.160		7.95	9.60		17.55	23
1460	4" W x 4" H		100	.160		9.60	9.60		19.20	25
1470	4" W x 5" H		100	.160		13.50	9.60		23.10	29
1550	Cover, 1-1/2" W		200	.080		1	4.80		5.80	8.25
1560	2" W		200	.080		1.15	4.80		5.95	8.40
1570	2-1/2" W		200	.080		1.54	4.80		6.34	8.85
1580	3" W		200	.080		1.74	4.80		6.54	9.05

For customer support on your Electrical Costs with RSMeans data, call 800.448.8182.

267

26 27 Low-Voltage Distribution Equipment

26 27 19 – Multi-Outlet Assemblies

26 27 19.10 Wiring Duct	Crew	Daily Output	Labor-Hours	Unit	Material	2019 Bare Costs Labor	Equipment	Total	Total Incl O&P
1590	4" W	2 Elec	200	.080	L.F.	2.11	4.80		6.91	9.45

26 27 23 – Indoor Service Poles

26 27 23.40 Surface Raceway

		Crew	Daily Output	Labor-Hours	Unit	Material	2019 Bare Costs Labor	Equipment	Total	Total Incl O&P
0010	**SURFACE RACEWAY**									
0090	Metal, straight section									
0100	No. 500	1 Elec	100	.080	L.F.	1.27	4.80		6.07	8.55
0110	No. 700		100	.080		1.43	4.80		6.23	8.70
0200	No. 1000		90	.089		1.46	5.35		6.81	9.55
0400	No. 1500, small pancake		90	.089		2.20	5.35		7.55	10.35
0600	No. 2000, base & cover, blank		90	.089		2.24	5.35		7.59	10.40
0610	Receptacle, 6" OC		40	.200		21	12		33	41
0620	12" OC		44	.182		13.75	10.90		24.65	31.50
0630	18" OC		46	.174		9.65	10.45		20.10	26.50
0650	30" OC		50	.160		5.20	9.60		14.80	20
0660	60" OC		50	.160		4.81	9.60		14.41	19.65
0670	No. 2400, base & cover, blank		80	.100		1.96	6		7.96	11.10
0680	Receptacle, 6" OC		42	.190		42	11.45		53.45	63
0690	12" OC		53	.151		29.50	9.05		38.55	46
0700	18" OC		55	.145		27	8.75		35.75	42.50
0710	24" OC		57	.140		9.95	8.45		18.40	23.50
0720	30" OC		59	.136		6.80	8.15		14.95	19.60
0730	60" OC		61	.131		5.45	7.90		13.35	17.75
0800	No. 3000, base & cover, blank		75	.107		4.38	6.40		10.78	14.35
0810	Receptacle, 6" OC		45	.178		38	10.70		48.70	57.50
0820	12" OC		62	.129		21.50	7.75		29.25	35
0830	18" OC		64	.125		17.30	7.50		24.80	30
0840	24" OC		66	.121		12.95	7.30		20.25	25
0850	30" OC		68	.118		11.65	7.05		18.70	23.50
0860	60" OC		70	.114		8.55	6.85		15.40	19.65
1000	No. 4000, base & cover, blank		65	.123		7.45	7.40		14.85	19.25
1010	Receptacle, 6" OC		41	.195		51	11.70		62.70	73.50
1020	12" OC		52	.154		31.50	9.25		40.75	49
1030	18" OC		54	.148		26	8.90		34.90	42
1040	24" OC		56	.143		21	8.60		29.60	36.50
1050	30" OC		58	.138		19.15	8.30		27.45	33.50
1060	60" OC		60	.133		14.70	8		22.70	28
1200	No. 6000, base & cover, blank		50	.160		12.70	9.60		22.30	28.50
1210	Receptacle, 6" OC		30	.267		62	16		78	92.50
1220	12" OC		37	.216		41	13		54	65
1230	18" OC		39	.205		35	12.30		47.30	57
1240	24" OC		41	.195		28.50	11.70		40.20	49
1250	30" OC		43	.186		27	11.15		38.15	46.50
1260	60" OC		45	.178		21	10.70		31.70	39
2400	Fittings, elbows, No. 500		40	.200	Ea.	2.29	12		14.29	20.50
2800	Elbow cover, No. 2000		40	.200		3.67	12		15.67	22
2880	Tee, No. 500		42	.190		4.23	11.45		15.68	21.50
2900	No. 2000		27	.296		12.20	17.80		30	40
3000	Switch box, No. 500		16	.500		14.75	30		44.75	61
3400	Telephone outlet, No. 1500		16	.500		17.90	30		47.90	64.50
3600	Junction box, No. 1500		16	.500		11.30	30		41.30	57.50
3800	Plugmold wired sections, No. 2000									
4000	1 circuit, 6 outlets, 3' long	1 Elec	8	1	Ea.	38.50	60		98.50	132

268

26 27 23.40 Surface Raceway	Crew	Daily Output	Labor-Hours	Unit	Material	2019 Bare Costs Labor	Equipment	Total	Total Incl O&P
4100	2 circuits, 8 outlets, 6' long	1 Elec	5.30	1.509	Ea.	61.50	90.50		152	203
4110	Tele-power pole, alum, w/2 recept, 10'		4	2		260	120		380	465
4120	12'		3.85	2.078		262	125		387	475
4130	15'		3.70	2.162		355	130		485	585
4140	Steel, w/2 recept, 10'		4	2		165	120		285	360
4150	One phone fitting, 10'		4	2		152	120		272	345
4160	Alum, 4 outlets, 10'		3.70	2.162		350	130		480	580
4300	Overhead distribution systems, 125 volt									
4800	No. 2000, entrance end fitting	1 Elec	20	.400	Ea.	6.85	24		30.85	43.50
5000	Blank end fitting		40	.200		2.27	12		14.27	20.50
5200	Supporting clip		40	.200		1.34	12		13.34	19.35
5800	No. 3000, entrance end fitting		20	.400		9.30	24		33.30	46.50
6000	Blank end fitting		40	.200		2.64	12		14.64	21
6020	Internal elbow		20	.400		12.20	24		36.20	49.50
6030	External elbow		20	.400		17.25	24		41.25	55
6040	Device bracket		53	.151		4.14	9.05		13.19	18.05
6400	Hanger clamp		32	.250		6.30	15		21.30	29.50
7000	No. 4000 base		90	.089	L.F.	4.44	5.35		9.79	12.85
7200	Divider		100	.080	"	.86	4.80		5.66	8.10
7400	Entrance end fitting		16	.500	Ea.	25.50	30		55.50	73
7600	Blank end fitting		40	.200		6.25	12		18.25	25
7610	Recpt. & tele. cover		53	.151		10.65	9.05		19.70	25
7620	External elbow		16	.500		35	30		65	83.50
7630	Coupling		53	.151		5.30	9.05		14.35	19.35
7640	Divider clip & coupling		80	.100		1.05	6		7.05	10.10
7650	Panel connector		16	.500		22	30		52	69.50
7800	Take off connector		16	.500		73.50	30		103.50	126
8000	No. 6000, take off connector		16	.500		87	30		117	141
8100	Take off fitting		16	.500		64.50	30		94.50	116
8200	Hanger clamp		32	.250		15.70	15		30.70	40
8230	Coupling					10.20			10.20	11.20
8240	One gang device plate	1 Elec	53	.151		8.35	9.05		17.40	22.50
8250	Two gang device plate		40	.200		10.25	12		22.25	29
8260	Blank end fitting		40	.200		8.90	12		20.90	27.50
8270	Combination elbow		14	.571		36	34.50		70.50	91
8300	Panel connector		16	.500		15.90	30		45.90	62.50
8500	Chan-L-Wire system installed in 1-5/8" x 1-5/8" strut. Strut									
8600	not incl., 30 amp, 4 wire, 3 phase	1 Elec	200	.040	L.F.	5.10	2.40		7.50	9.20
8700	Junction box		8	1	Ea.	34	60		94	127
8800	Insulating end cap		40	.200		9.05	12		21.05	28
8900	Strut splice plate		40	.200		11.95	12		23.95	31
9000	Tap		40	.200		23.50	12		35.50	44
9100	Fixture hanger		60	.133		10.30	8		18.30	23.50
9200	Pulling tool					91.50			91.50	100
9300	Non-metallic, straight section									
9310	7/16" x 7/8", base & cover, blank	1 Elec	160	.050	L.F.	1.73	3		4.73	6.40
9320	Base & cover w/adhesive		160	.050		1.62	3		4.62	6.25
9340	7/16" x 1-5/16", base & cover, blank		145	.055		1.90	3.31		5.21	7.05
9350	Base & cover w/adhesive		145	.055		2.23	3.31		5.54	7.40
9370	11/16" x 2-1/4", base & cover, blank		130	.062		2.68	3.70		6.38	8.45
9380	Base & cover w/adhesive		130	.062		3.10	3.70		6.80	8.90
9385	1-11/16" x 5-1/4", two compartment base & cover w/screws		80	.100		10.30	6		16.30	20.50
9400	Fittings, elbows, 7/16" x 7/8"		50	.160	Ea.	1.94	9.60		11.54	16.50

26 27 23 – Indoor Service Poles

26 27 23.40 Surface Raceway

		Crew	Daily Output	Labor-Hours	Unit	Material	2019 Bare Costs Labor	2019 Bare Costs Equipment	Total	Total Incl O&P
9410	7/16" x 1-5/16"	1 Elec	45	.178	Ea.	2.01	10.70		12.71	18.15
9420	11/16" x 2-1/4"		40	.200		2.17	12		14.17	20.50
9425	1-11/16" x 5-1/4"		28	.286		14.05	17.15		31.20	41
9430	Tees, 7/16" x 7/8"		35	.229		2.47	13.75		16.22	23
9440	7/16" x 1-5/16"		32	.250		2.53	15		17.53	25.50
9450	11/16" x 2-1/4"		30	.267		2.67	16		18.67	27
9455	1-11/16" x 5-1/4"		24	.333		22.50	20		42.50	54.50
9460	Cover clip, 7/16" x 7/8"		80	.100		.50	6		6.50	9.50
9470	7/16" x 1-5/16"		72	.111		.45	6.65		7.10	10.45
9480	11/16" x 2-1/4"		64	.125		.95	7.50		8.45	12.25
9484	1-11/16" x 5-1/4"		42	.190		3.28	11.45		14.73	20.50
9486	Wire clip, 1-11/16" x 5-1/4"		68	.118		.48	7.05		7.53	11.10
9490	Blank end, 7/16" x 7/8"		50	.160		.86	9.60		10.46	15.30
9500	7/16" x 1-5/16"		45	.178		.93	10.70		11.63	16.95
9510	11/16" x 2-1/4"		40	.200		1.37	12		13.37	19.40
9515	1-11/16" x 5-1/4"		38	.211		5.75	12.65		18.40	25
9520	Round fixture box, 5.5" diam. x 1"		25	.320		11.25	19.20		30.45	41
9530	Device box, 1 gang		30	.267		5.20	16		21.20	29.50
9540	2 gang		25	.320		7.60	19.20		26.80	37

26 27 26 – Wiring Devices

26 27 26.10 Low Voltage Switching

		Crew	Daily Output	Labor-Hours	Unit	Material	2019 Bare Costs Labor	2019 Bare Costs Equipment	Total	Total Incl O&P
0010	**LOW VOLTAGE SWITCHING**									
3600	Relays, 120 V or 277 V standard	1 Elec	12	.667	Ea.	46	40		86	110
3800	Flush switch, standard		40	.200		11.60	12		23.60	30.50
4000	Interchangeable		40	.200		15.20	12		27.20	34.50
4100	Surface switch, standard		40	.200		8.25	12		20.25	27
4200	Transformer 115 V to 25 V		12	.667		129	40		169	202
4400	Master control, 12 circuit, manual		4	2		129	120		249	320
4500	25 circuit, motorized		4	2		143	120		263	335
4600	Rectifier, silicon		12	.667		46	40		86	110
4800	Switchplates, 1 gang, 1, 2 or 3 switch, plastic		80	.100		5.05	6		11.05	14.50
5000	Stainless steel		80	.100		11.40	6		17.40	21.50
5400	2 gang, 3 switch, stainless steel		53	.151		23	9.05		32.05	39
5500	4 switch, plastic		53	.151		10.50	9.05		19.55	25
5600	2 gang, 4 switch, stainless steel		53	.151		22	9.05		31.05	38
5700	6 switch, stainless steel		53	.151		43.50	9.05		52.55	61.50
5800	3 gang, 9 switch, stainless steel		32	.250		66	15		81	95
5900	Receptacle, triple, 1 return, 1 feed		26	.308		43.50	18.50		62	75.50
6000	2 feed		20	.400		43.50	24		67.50	84
6100	Relay gang boxes, flush or surface, 6 gang		5.30	1.509		91	90.50		181.50	235
6200	12 gang		4.70	1.702		107	102		209	271
6400	18 gang		4	2		117	120		237	310
6500	Frame, to hold up to 6 relays		12	.667		88.50	40		128.50	157
7200	Control wire, 2 conductor		6.30	1.270	C.L.F.	31.50	76.50		108	149
7400	3 conductor		5	1.600		39	96		135	186
7600	19 conductor		2.50	3.200		315	192		507	630
7800	26 conductor		2	4		430	240		670	830
8000	Weatherproof, 3 conductor		5	1.600		81.50	96		177.50	233

26 27 26.20 Wiring Devices Elements

			Crew	Daily Output	Labor-Hours	Unit	Material	2019 Bare Costs Labor	2019 Bare Costs Equipment	Total	Total Incl O&P
0010	**WIRING DEVICES ELEMENTS**	R262726-90									
0200	Toggle switch, quiet type, single pole, 15 amp		1 Elec	40	.200	Ea.	.47	12		12.47	18.40
0500	20 amp			27	.296		2.96	17.80		20.76	30

270

For customer support on your Electrical Costs with RSMeans data, call 800.448.8182.

26 27 26 – Wiring Devices

26 27 26.20 Wiring Devices Elements		Crew	Daily Output	Labor-Hours	Unit	Material	2019 Bare Costs Labor	Equipment	Total	Total Incl O&P
0510	30 amp	1 Elec	23	.348	Ea.	26.50	21		47.50	60.50
0530	Lock handle, 20 amp		27	.296		28.50	17.80		46.30	58
0540	Security key, 20 amp		26	.308		87	18.50		105.50	123
0550	Rocker, 15 amp		40	.200		2.35	12		14.35	20.50
0560	20 amp		27	.296		9.25	17.80		27.05	36.50
0600	3 way, 15 amp		23	.348		1.90	21		22.90	33
0800	20 amp		18	.444		3.68	26.50		30.18	44
0810	30 amp		9	.889		24	53.50		77.50	106
0830	Lock handle, 20 amp		18	.444		33	26.50		59.50	76.50
0840	Security key, 20 amp		17	.471		122	28.50		150.50	176
0850	Rocker, 15 amp		23	.348		8.20	21		29.20	40
0860	20 amp		18	.444		12.55	26.50		39.05	54
0900	4 way, 15 amp		15	.533		10.90	32		42.90	60
1000	20 amp		11	.727		43	43.50		86.50	113
1020	Lock handle, 20 amp		11	.727		38	43.50		81.50	107
1030	Rocker, 15 amp		15	.533		8.90	32		40.90	58
1040	20 amp		11	.727		17.80	43.50		61.30	84.50
1100	Toggle switch, quiet type, double pole, 15 amp		15	.533		12.45	32		44.45	61.50
1200	20 amp		11	.727		17.25	43.50		60.75	84
1210	30 amp		9	.889		31.50	53.50		85	114
1230	Lock handle, 20 amp		11	.727		31.50	43.50		75	100
1250	Security key, 20 amp		10	.800		93	48		141	174
1420	Toggle switch quiet type, 1 pole, 2 throw center off, 15 amp		23	.348		41.50	21		62.50	77
1440	20 amp		18	.444		77	26.50		103.50	125
1460	2 pole, 2 throw center off, lock handle, 20 amp		11	.727		78.50	43.50		122	152
1480	1 pole, momentary contact, 15 amp		23	.348		17.85	21		38.85	50.50
1500	20 amp		18	.444		28	26.50		54.50	71
1520	Momentary contact, lock handle, 20 amp		18	.444		31	26.50		57.50	74
1650	Dimmer switch, 120 volt, incandescent, 600 watt, 1 pole G		16	.500		23	30		53	70
1700	600 watt, 3 way G		12	.667		10.95	40		50.95	71.50
1750	1,000 watt, 1 pole G		16	.500		40.50	30		70.50	89.50
1800	1,000 watt, 3 way G		12	.667		73.50	40		113.50	141
2000	1,500 watt, 1 pole G		11	.727		91.50	43.50		135	166
2100	2,000 watt, 1 pole G		8	1		131	60		191	234
2110	Fluorescent, 600 watt G		15	.533		110	32		142	169
2120	1,000 watt G		15	.533		134	32		166	195
2130	1,500 watt G		10	.800		248	48		296	345
2160	Explosion proof, toggle switch, wall, single pole 20 amp		5.30	1.509		230	90.50		320.50	390
2180	Receptacle, single outlet, 20 amp		5.30	1.509		540	90.50		630.50	730
2190	30 amp		4	2		860	120		980	1,125
2290	60 amp		2.50	3.200		780	192		972	1,150
2360	Plug, 20 amp		16	.500		181	30		211	244
2370	30 amp		12	.667		330	40		370	425
2380	60 amp		8	1		525	60		585	670
2410	Furnace, thermal cutoff switch with plate		26	.308		16.75	18.50		35.25	46
2460	Receptacle, duplex, 120 volt, grounded, 15 amp		40	.200		1.51	12		13.51	19.55
2470	20 amp		27	.296		9.55	17.80		27.35	37
2480	Ground fault interrupting, 15 amp		27	.296		14	17.80		31.80	42
2482	20 amp		27	.296		40.50	17.80		58.30	71
2486	Clock receptacle, 15 amp		40	.200		30	12		42	50.50
2490	Dryer, 30 amp		15	.533		4.39	32		36.39	53
2500	Range, 50 amp		11	.727		10.80	43.50		54.30	77
2530	Surge suppressor receptacle, duplex, 20 amp		27	.296		50.50	17.80		68.30	82

271

26 27 26.20 Wiring Devices Elements	Crew	Daily Output	Labor-Hours	Unit	Material	2019 Bare Costs Labor	Equipment	Total	Total Incl O&P
2532	Quad, 20 amp	1 Elec	20	.400	Ea.	87	24		111	132
2540	Isolated ground receptacle, duplex, 20 amp		27	.296		23.50	17.80		41.30	52.50
2542	Quad, 20 amp		20	.400		30	24		54	69
2550	Simplex, 20 amp		27	.296		29	17.80		46.80	58.50
2560	Simplex, 30 amp	↓	15	.533		34.50	32		66.50	86
2570	Cable reel w/receptacle 50' w/3#12, 120 V, 20 A	2 Elec	2.67	5.999		1,075	360		1,435	1,750
2600	Wall plates, stainless steel, 1 gang	1 Elec	80	.100		2.58	6		8.58	11.80
2800	2 gang		53	.151		4.37	9.05		13.42	18.30
3000	3 gang		32	.250		11.05	15		26.05	34.50
3100	4 gang		27	.296		11.10	17.80		28.90	38.50
3110	Brown plastic, 1 gang		80	.100		.38	6		6.38	9.35
3120	2 gang		53	.151		.75	9.05		9.80	14.35
3130	3 gang		32	.250		1.11	15		16.11	23.50
3140	4 gang		27	.296		3	17.80		20.80	30
3150	Brushed brass, 1 gang		80	.100		5.60	6		11.60	15.15
3160	Anodized aluminum, 1 gang		80	.100		3.04	6		9.04	12.30
3170	Switch cover, weatherproof, 1 gang		60	.133		5.35	8		13.35	17.80
3180	Vandal proof lock, 1 gang		60	.133		14.75	8		22.75	28
3200	Lampholder, keyless		26	.308		17.45	18.50		35.95	46.50
3400	Pullchain with receptacle		22	.364		21	22		43	55.50
3500	Pilot light, neon with jewel		27	.296		9.30	17.80		27.10	37
3600	Receptacle, 20 amp, 250 volt, NEMA 6		27	.296		23.50	17.80		41.30	52
3620	277 volt NEMA 7		27	.296		24	17.80		41.80	53
3640	125/250 volt NEMA 10		27	.296		22	17.80		39.80	51
3680	125/250 volt NEMA 14		25	.320		28	19.20		47.20	59
3700	3 pole, 250 volt NEMA 15		25	.320		32.50	19.20		51.70	64.50
3720	120/208 volt NEMA 18		25	.320		32	19.20		51.20	63.50
3740	30 amp, 125 volt NEMA 5		15	.533		19.65	32		51.65	69.50
3760	250 volt NEMA 6		15	.533		23	32		55	73
3780	277 volt NEMA 7		15	.533		27.50	32		59.50	78
3820	125/250 volt NEMA 14		14	.571		67	34.50		101.50	125
3840	3 pole, 250 volt NEMA 15		14	.571		57	34.50		91.50	114
3880	50 amp, 125 volt NEMA 5		11	.727		32	43.50		75.50	101
3900	250 volt NEMA 6		11	.727		27	43.50		70.50	94.50
3920	277 volt NEMA 7		11	.727		34.50	43.50		78	103
3960	125/250 volt NEMA 14		10	.800		86	48		134	166
3980	3 pole, 250 volt NEMA 15		10	.800		85.50	48		133.50	166
4020	60 amp, 125/250 volt, NEMA 14		8	1		88	60		148	187
4040	3 pole, 250 volt NEMA 15		8	1		93.50	60		153.50	193
4060	120/208 volt NEMA 18		8	1		99	60		159	199
4100	Receptacle locking, 20 amp, 125 volt NEMA L5		27	.296		22	17.80		39.80	51
4120	250 volt NEMA L6		27	.296		22.50	17.80		40.30	51.50
4140	277 volt NEMA L7		27	.296		23	17.80		40.80	52
4150	3 pole, 250 volt, NEMA L11		27	.296		25	17.80		42.80	54
4160	20 amp, 480 volt NEMA L8		27	.296		25.50	17.80		43.30	54.50
4180	600 volt NEMA L9		27	.296		36.50	17.80		54.30	66.50
4200	125/250 volt NEMA L10		27	.296		28	17.80		45.80	57
4230	125/250 volt NEMA L14		25	.320		27.50	19.20		46.70	58.50
4280	250 volt NEMA L15		25	.320		34.50	19.20		53.70	66.50
4300	480 volt NEMA L16		25	.320		26.50	19.20		45.70	57.50
4320	3 phase, 120/208 volt NEMA L18		25	.320		39.50	19.20		58.70	72
4340	277/480 volt NEMA L19		25	.320		38	19.20		57.20	70.50
4360	347/600 volt NEMA L20	↓	25	.320	↓	31	19.20		50.20	63

26 27 26 – Wiring Devices

26 27 26.20 Wiring Devices Elements	Crew	Daily Output	Labor-Hours	Unit	Material	2019 Bare Costs Labor	Equipment	Total	Total Incl O&P
4380	120/208 volt NEMA L21	1 Elec	23	.348	Ea.	34	21		55	68.50
4400	277/480 volt NEMA L22		23	.348		37	21		58	71.50
4420	347/600 volt NEMA L23		23	.348		47	21		68	83
4440	30 amp, 125 volt NEMA L5		15	.533		31	32		63	82
4460	250 volt NEMA L6		15	.533		34	32		66	85
4480	277 volt NEMA L7		15	.533		34	32		66	85.50
4500	480 volt NEMA L8		15	.533		32.50	32		64.50	84
4520	600 volt NEMA L9		15	.533		40.50	32		72.50	93
4540	125/250 volt NEMA L10		15	.533		39	32		71	91
4560	3 phase, 250 volt NEMA L11		15	.533		27.50	32		59.50	78
4620	125/250 volt NEMA L14		14	.571		46.50	34.50		81	103
4640	250 volt NEMA L15		14	.571		50.50	34.50		85	107
4660	480 volt NEMA L16		14	.571		46.50	34.50		81	103
4680	600 volt NEMA L17		14	.571		45.50	34.50		80	101
4700	120/208 volt NEMA L18		14	.571		60.50	34.50		95	118
4720	277/480 volt NEMA L19		14	.571		62	34.50		96.50	120
4740	347/600 volt NEMA L20		14	.571		53	34.50		87.50	110
4760	120/208 volt NEMA L21		13	.615		53	37		90	113
4780	277/480 volt NEMA L22		13	.615		54	37		91	115
4800	347/600 volt NEMA L23		13	.615		64	37		101	126
4840	Receptacle, corrosion resistant, 15 or 20 amp, 125 volt NEMA L5		27	.296		35.50	17.80		53.30	65.50
4860	250 volt NEMA L6		27	.296		20	17.80		37.80	48.50
4870	Receptacle box assembly, cast aluminum, 60A, 4P5W, 3P, 120/208V, NEMA PR4		22	.364		650	22		672	750
4875	100A, 4P5W, 3P, 347/600V, NEMA PR4		22	.364		795	22		817	910
4880	200A, 4P5W, 3P, 347/600V, NEMA PR4		22	.364		1,725	22		1,747	1,925
4900	Receptacle, cover plate, phenolic plastic, NEMA 5 & 6		80	.100		.65	6		6.65	9.65
4910	NEMA 7-23		80	.100		.76	6		6.76	9.80
4920	Stainless steel, NEMA 5 & 6		80	.100		2.74	6		8.74	11.95
4930	NEMA 7-23		80	.100		3.65	6		9.65	12.95
4940	Brushed brass NEMA 5 & 6		80	.100		5.45	6		11.45	14.95
4950	NEMA 7-23		80	.100		5.90	6		11.90	15.45
4960	Anodized aluminum, NEMA 5 & 6		80	.100		3.54	6		9.54	12.85
4970	NEMA 7-23		80	.100		5.85	6		11.85	15.40
4980	Weatherproof NEMA 7-23		60	.133		43.50	8		51.50	59.50
5000	Duplex receptacle, combo 15A/125V, 3 wire w/2-5V 0.7A, port USB, AL		27	.296		17.40	17.80		35.20	45.50
5002	15A/125V, 3 wire w/2-5V 0.7A, port USB, BK		27	.296		18.20	17.80		36	46.50
5004	15A/125V, 3 wire w/2-5V 0.7A, port USB, LA		27	.296		18.30	17.80		36.10	46.50
5006	15A/125V, 3 wire w/2-5V 0.7A, port USB, IV		27	.296		27	17.80		44.80	56.50
5008	15A/125V, 3 wire w/2-5V 0.7A, port USB, WH		27	.296		23.50	17.80		41.30	52
5010	Duplex receptacle, combo 15A/125V, 3 wire w/2-5V 2.1A, port USB, AL		27	.296		50.50	17.80		68.30	82
5012	15A/125V, 3 wire w/2-5V 2.1A, port USB, BK		27	.296		52	17.80		69.80	84
5014	15A/125V, 3 wire w/2-5V 2.1A, port USB, LA		27	.296		46	17.80		63.80	77
5016	15A/125V, 3 wire w/2-5V 2.1A, port USB, IV		27	.296		46	17.80		63.80	77
5018	15A/125V, 3 wire w/2-5V 2.1A, port USB, WH		27	.296		31	17.80		48.80	60.50
5020	Duplex receptacle, combo 20A/125V, 3 wire w/2-5V 2.1A, port USB, AL		27	.296		51	17.80		68.80	82.50
5022	20A/125V, 3 wire w/2-5V 2.1A, port USB, BK		27	.296		57.50	17.80		75.30	89.50
5024	20A/125V, 3 wire w/2-5V 2.1A, port USB, LA		27	.296		35	17.80		52.80	65
5026	20A/125V, 3 wire w/2-5V 2.1A, port USB, IV		27	.296		34.50	17.80		52.30	64.50
5028	20A/125V, 3 wire w/2-5V 2.1A, port USB, WH		27	.296		35.50	17.80		53.30	65.50
5100	Plug, 20 amp, 250 volt NEMA 6		30	.267		17.80	16		33.80	43.50
5110	277 volt NEMA 7		30	.267		19.35	16		35.35	45.50
5120	3 pole, 120/250 volt NEMA 10		26	.308		22	18.50		40.50	52
5130	125/250 volt NEMA 14		26	.308		50	18.50		68.50	82.50

26 27 26.20 Wiring Devices Elements	Crew	Daily Output	Labor-Hours	Unit	Material	2019 Bare Costs Labor	Equipment	Total	Total Incl O&P
5140	250 volt NEMA 15	1 Elec	26	.308	Ea.	49	18.50		67.50	81
5150	120/208 volt NEMA 8		26	.308		51.50	18.50		70	84
5160	30 amp, 125 volt NEMA 5		13	.615		51.50	37		88.50	112
5170	250 volt NEMA 6		13	.615		54	37		91	114
5180	277 volt NEMA 7		13	.615		46.50	37		83.50	106
5190	125/250 volt NEMA 14		13	.615		56	37		93	117
5200	3 pole, 250 volt NEMA 15		12	.667		60	40		100	126
5210	50 amp, 125 volt NEMA 5		9	.889		68.50	53.50		122	155
5220	250 volt NEMA 6		9	.889		72	53.50		125.50	159
5230	277 volt NEMA 7		9	.889		76.50	53.50		130	164
5240	125/250 volt NEMA 14		9	.889		81	53.50		134.50	169
5250	3 pole, 250 volt NEMA 15		8	1		71	60		131	168
5260	60 amp, 125/250 volt NEMA 14		7	1.143		76	68.50		144.50	186
5270	3 pole, 250 volt NEMA 15		7	1.143		81.50	68.50		150	192
5280	120/208 volt NEMA 18		7	1.143		92	68.50		160.50	203
5300	Plug angle, 20 amp, 250 volt NEMA 6		30	.267		26.50	16		42.50	53.50
5310	30 amp, 125 volt NEMA 5		13	.615		47.50	37		84.50	107
5320	250 volt NEMA 6		13	.615		50.50	37		87.50	111
5330	277 volt NEMA 7		13	.615		63	37		100	125
5340	125/250 volt NEMA 14		13	.615		61.50	37		98.50	123
5350	3 pole, 250 volt NEMA 15		12	.667		68	40		108	134
5360	50 amp, 125 volt NEMA 5		9	.889		67.50	53.50		121	154
5370	250 volt NEMA 6		9	.889		52.50	53.50		106	137
5380	277 volt NEMA 7		9	.889		78	53.50		131.50	166
5390	125/250 volt NEMA 14		9	.889		73.50	53.50		127	161
5400	3 pole, 250 volt NEMA 15		8	1		81	60		141	179
5410	60 amp, 125/250 volt NEMA 14		7	1.143		86	68.50		154.50	197
5420	3 pole, 250 volt NEMA 15		7	1.143		87.50	68.50		156	199
5430	120/208 volt NEMA 18		7	1.143		91	68.50		159.50	202
5500	Plug, locking, 20 amp, 125 volt NEMA L5		30	.267		17.45	16		33.45	43
5510	250 volt NEMA L6		30	.267		17.50	16		33.50	43.50
5520	277 volt NEMA L7		30	.267		17.85	16		33.85	43.50
5530	480 volt NEMA L8		30	.267		19.40	16		35.40	45.50
5540	600 volt NEMA L9		30	.267		23	16		39	49.50
5550	3 pole, 125/250 volt NEMA L10		26	.308		23.50	18.50		42	53.50
5560	250 volt NEMA L11		26	.308		23.50	18.50		42	53.50
5570	480 volt NEMA L12		26	.308		27.50	18.50		46	57.50
5580	125/250 volt NEMA L14		26	.308		26	18.50		44.50	56.50
5590	250 volt NEMA L15		26	.308		28.50	18.50		47	59
5600	480 volt NEMA L16		26	.308		30.50	18.50		49	61.50
5610	4 pole, 120/208 volt NEMA L18		24	.333		35	20		55	68.50
5620	277/480 volt NEMA L19		24	.333		39.50	20		59.50	73.50
5630	347/600 volt NEMA L20		24	.333		32.50	20		52.50	66
5640	120/208 volt NEMA L21		24	.333		34	20		54	67.50
5650	277/480 volt NEMA L22		24	.333		36.50	20		56.50	70.50
5660	347/600 volt NEMA L23		24	.333		40.50	20		60.50	74.50
5670	30 amp, 125 volt NEMA L5		13	.615		26.50	37		63.50	84
5680	250 volt NEMA L6		13	.615		28	37		65	86
5690	277 volt NEMA L7		13	.615		27	37		64	85
5700	480 volt NEMA L8		13	.615		29	37		66	87
5710	600 volt NEMA L9		13	.615		29	37		66	87
5720	3 pole, 125/250 volt NEMA L10		11	.727		25.50	43.50		69	93
5730	250 volt NEMA L11		11	.727		25.50	43.50		69	93

For customer support on your Electrical Costs with RSMeans data, call 800.448.8182.

26 27 Low-Voltage Distribution Equipment

26 27 26 – Wiring Devices

26 27 26.20 Wiring Devices Elements	Crew	Daily Output	Labor-Hours	Unit	Material	2019 Bare Costs Labor	Equipment	Total	Total Incl O&P
5760	125/250 volt NEMA L14	1 Elec	11	.727	Ea.	35.50	43.50		79	105
5770	250 volt NEMA L15		11	.727		36	43.50		79.50	105
5780	480 volt NEMA L16		11	.727		38	43.50		81.50	107
5790	600 volt NEMA L17		11	.727		39	43.50		82.50	108
5800	4 pole, 120/208 volt NEMA L18		10	.800		49.50	48		97.50	126
5810	120/208 volt NEMA L19		10	.800		49	48		97	126
5820	347/600 volt NEMA L20		10	.800		51.50	48		99.50	128
5830	120/208 volt NEMA L21		10	.800		38	48		86	113
5840	277/480 volt NEMA L22		10	.800		41.50	48		89.50	118
5850	347/600 volt NEMA L23		10	.800		48.50	48		96.50	125
6000	Connector, 20 amp, 250 volt NEMA 6		30	.267		30.50	16		46.50	57.50
6010	277 volt NEMA 7		30	.267		32	16		48	59
6020	3 pole, 120/250 volt NEMA 10		26	.308		36	18.50		54.50	67
6030	125/250 volt NEMA 14		26	.308		36	18.50		54.50	67
6040	250 volt NEMA 15		26	.308		36	18.50		54.50	67.50
6050	120/208 volt NEMA 18		26	.308		39.50	18.50		58	71
6060	30 amp, 125 volt NEMA 5		13	.615		62	37		99	124
6070	250 volt NEMA 6		13	.615		62	37		99	124
6080	277 volt NEMA 7		13	.615		62	37		99	124
6110	50 amp, 125 volt NEMA 5		9	.889		85.50	53.50		139	174
6120	250 volt NEMA 6		9	.889		85.50	53.50		139	174
6130	277 volt NEMA 7		9	.889		85.50	53.50		139	174
6200	Connector, locking, 20 amp, 125 volt NEMA L5		30	.267		26.50	16		42.50	53.50
6210	250 volt NEMA L6		30	.267		26.50	16		42.50	53
6220	277 volt NEMA L7		30	.267		26	16		42	52.50
6230	480 volt NEMA L8		30	.267		33	16		49	60
6240	600 volt NEMA L9		30	.267		38	16		54	66
6250	3 pole, 125/250 volt NEMA L10		26	.308		35.50	18.50		54	66.50
6260	250 volt NEMA L11		26	.308		35.50	18.50		54	66.50
6280	125/250 volt NEMA L14		26	.308		36	18.50		54.50	67
6290	250 volt NEMA L15		26	.308		36	18.50		54.50	67
6300	480 volt NEMA L16		26	.308		40.50	18.50		59	72
6310	4 pole, 120/208 volt NEMA L18		24	.333		49.50	20		69.50	84.50
6320	277/480 volt NEMA L19		24	.333		49	20		69	84
6330	347/600 volt NEMA L20		24	.333		50.50	20		70.50	85.50
6340	120/208 volt NEMA L21		24	.333		53	20		73	88.50
6350	277/480 volt NEMA L22		24	.333		65	20		85	101
6360	347/600 volt NEMA L23		24	.333		67.50	20		87.50	105
6370	30 amp, 125 volt NEMA L5		13	.615		52	37		89	112
6380	250 volt NEMA L6		13	.615		52.50	37		89.50	113
6390	277 volt NEMA L7		13	.615		57	37		94	118
6400	480 volt NEMA L8		13	.615		57	37		94	118
6410	600 volt NEMA L9		13	.615		64.50	37		101.50	126
6420	3 pole, 125/250 volt NEMA L10		11	.727		68.50	43.50		112	141
6430	250 volt NEMA L11		11	.727		68.50	43.50		112	141
6460	125/250 volt NEMA L14		11	.727		72	43.50		115.50	145
6470	250 volt NEMA L15		11	.727		72.50	43.50		116	145
6480	480 volt NEMA L16		11	.727		79.50	43.50		123	153
6490	600 volt NEMA L17		11	.727		80.50	43.50		124	154
6500	4 pole, 120/208 volt NEMA L18		10	.800		97	48		145	179
6510	120/208 volt NEMA L19		10	.800		87.50	48		135.50	168
6520	347/600 volt NEMA L20		10	.800		98.50	48		146.50	180
6530	120/208 volt NEMA L21		10	.800		71	48		119	150

275

26 27 26.20 Wiring Devices Elements	Crew	Daily Output	Labor-Hours	Unit	Material	2019 Bare Costs Labor	Equipment	Total	Total Incl O&P
6540	277/480 volt NEMA L22	1 Elec	10	.800	Ea.	80.50	48		128.50	160
6550	347/600 volt NEMA L23		10	.800		87.50	48		135.50	168
7000	Receptacle computer, 250 volt, 15 amp, 3 pole 4 wire		8	1		83.50	60		143.50	182
7010	20 amp, 2 pole 3 wire		8	1		112	60		172	214
7020	30 amp, 2 pole 3 wire		6.50	1.231		173	74		247	300
7030	30 amp, 3 pole 4 wire		6.50	1.231		186	74		260	315
7040	60 amp, 3 pole 4 wire		4.50	1.778		305	107		412	495
7050	100 amp, 3 pole 4 wire		3	2.667		385	160		545	665
7100	Connector computer, 250 volt, 15 amp, 3 pole 4 wire		27	.296		146	17.80		163.80	188
7110	20 amp, 2 pole 3 wire		27	.296		149	17.80		166.80	191
7120	30 amp, 2 pole 3 wire		15	.533		196	32		228	263
7130	30 amp, 3 pole 4 wire		15	.533		223	32		255	293
7140	60 amp, 3 pole 4 wire		8	1		385	60		445	515
7150	100 amp, 3 pole 4 wire		4	2		565	120		685	805
7200	Plug, computer, 250 volt, 15 amp, 3 pole 4 wire		27	.296		131	17.80		148.80	171
7210	20 amp, 2 pole 3 wire		27	.296		133	17.80		150.80	173
7220	30 amp, 2 pole 3 wire		15	.533		191	32		223	259
7230	30 amp, 3 pole 4 wire		15	.533		206	32		238	275
7240	60 amp, 3 pole 4 wire		8	1		310	60		370	430
7250	100 amp, 3 pole 4 wire		4	2		380	120		500	595
7300	Connector adapter to flexible conduit, 1/2"		60	.133		3.87	8		11.87	16.20
7310	3/4"		50	.160		4.50	9.60		14.10	19.30
7320	1-1/4"		30	.267		14	16		30	39.50
7330	1-1/2"		23	.348		18.20	21		39.20	51
8100	Pin/sleeve, 20A, 480V, DSN1, male inlet, 3 pole, 3 phase, 4 wire		16	.500		48	30		78	98
8110	20A, 125V, DSN20, male inlet, 3 pole, 3 phase, 3 wire		16	.500		39.50	30		69.50	88.50
8120	60A, 480V, DS60, male inlet, 3 pole, 3 phase, 3 wire		16	.500		315	30		345	395
8130	Pin/sleeve, 20A, 480V, DSN10, female RCPT, 3P, 3 phase, 4 wire		16	.500		92.50	30		122.50	147
8140	30A, 480V, DS30, female RCPT, 3P, 3 phase, 4 wire		16	.500		315	30		345	395
8500	Wiring device terminal strip, 2 pole, 12-2AWG, 300VAC, CSA, 600VAC, 10A		36	.222		3.30	13.35		16.65	23.50
8505	4 pole, 12-2AWG, 300VAC, CSA, 600VAC, 10A		36	.222		4.66	13.35		18.01	25
8510	6 pole, 12-2AWG, 300VAC, CSA, 600VAC, 10A		36	.222		5.25	13.35		18.60	25.50
8515	8 pole, 12-2AWG, 300VAC, CSA, 600VAC, 10A		30	.267		5.85	16		21.85	30.50
8520	10 pole, 12-2AWG, 300VAC, CSA, 600VAC, 10A		30	.267		6.45	16		22.45	31
8525	12 pole, 12-2AWG, 300VAC, CSA, 600VAC, 10A		30	.267		7.15	16		23.15	32
8530	Wiring device terminal strip jumper, 10A		36	.222		.82	13.35		14.17	21
8535	Wiring device terminal strip, 2 pole, 12-22AWG, 300VAC, CSA, 600VAC, 20A		34	.235		3.58	14.15		17.73	25
8540	4 pole, 12-22AWG, 300VAC, CSA, 600VAC, 20A		34	.235		4.88	14.15		19.03	26.50
8545	6 pole, 12-22AWG, 300VAC, CSA, 600VAC, 20A		34	.235		5.55	14.15		19.70	27
8550	8 pole, 12-22AWG, 300VAC, CSA, 600VAC, 20A		30	.267		6.20	16		22.20	31
8555	10 pole, 12-22AWG, 300VAC, CSA, 600VAC, 20A		30	.267		6.85	16		22.85	31.50
8560	12 pole, 12-22AWG, 300VAC, CSA, 600VAC, 20A		30	.267		7.50	16		23.50	32.50
8565	Wiring device terminal strip jumper, 20A		34	.235		1	14.15		15.15	22
8570	Wiring device terminal shorting type no cover, 4 pole, 6-10AWG, 600VAC, 50A		30	.267		19.95	16		35.95	46
8575	6 pole, 6-10AWG, 600VAC, 50A		30	.267		24.50	16		40.50	51
8580	8 pole, 6-10AWG, 600VAC, 50A		30	.267		29.50	16		45.50	56.50
8585	12 pole, 6-10AWG, 600VAC, 50A		30	.267		32.50	16		48.50	60
8590	Wiring device terminal 2 shorting pin 1/2 cover, 2 pole, 8-4AWG, 600V, 45A		30	.267		22	16		38	48
8595	4 pole, 8-4AWG, 600V, 45A		30	.267		27.50	16		43.50	54.50
8600	6 pole, 8-4AWG, 600V, 45A		30	.267		34	16		50	61.50
8605	8 pole, 8-4AWG, 600V, 45A		30	.267		46.50	16		62.50	75
8610	12 pole, 8-4AWG, 600V, 45A		30	.267		55.50	16		71.50	85
8615	Wiring device terminal 2 shorting pin W/cover, 4 pole, 8-4AWG, 600V, 45A		30	.267		35	16		51	62.50

26 27 Low-Voltage Distribution Equipment

26 27 26 – Wiring Devices

26 27 26.20 Wiring Devices Elements

		Crew	Daily Output	Labor-Hours	Unit	Material	2019 Bare Costs Labor	Equipment	Total	Total Incl O&P
8620	6 pole, 8-4AWG, 600V, 45A	1 Elec	30	.267	Ea.	51	16		67	80.50
8625	8 pole, 8-4AWG, 600V, 45A		30	.267		56	16		72	85.50
8630	12 pole, 8-4AWG, 600V, 45A	↓	30	.267	↓	66.50	16		82.50	97.50

26 27 33 – Power Distribution Units

26 27 33.10 Power Distribution Unit Cabinet

		Crew	Daily Output	Labor-Hours	Unit	Material	2019 Bare Costs Labor	Equipment	Total	Total Incl O&P
0010	**POWER DISTRIBUTION UNIT CABINET**									
0100	Power distribution unit, single cabinet, 50 kVA output, 480/208 input	1 Elec	2	4	Ea.	9,125	240		9,365	10,400
0110	75 kVA output, 480/208 input		2	4		9,925	240		10,165	11,300
0120	100 kVA output, 480/208 input		2	4		10,800	240		11,040	12,300
0130	125 kVA output, 480/208 input		1.50	5.333		11,900	320		12,220	13,600
0140	150 kVA output, 480/208 input		1.50	5.333		14,400	320		14,720	16,400
0150	200 kVA output, 480/208 input		1	8		15,900	480		16,380	18,200
0160	225 kVA output, 480/208 input	↓	1	8	↓	16,800	480		17,280	19,200

26 27 33.20 Power Distribution Unit

		Crew	Daily Output	Labor-Hours	Unit	Material	2019 Bare Costs Labor	Equipment	Total	Total Incl O&P
0010	**POWER DISTRIBUTION UNIT**									
0050	3PH, 60 kVA, PDU									
0100	208V-208V/120V	2 Elec	2	8	Ea.	14,000	480		14,480	16,100
0110	480V-208V/120V		2	8		13,400	480		13,880	15,500
0120	600V-208V/120V	↓	2	8	↓	15,200	480		15,680	17,400
0125	3PH, 80 kVA, PDU									
0130	208V-208V/120V	2 Elec	2	8	Ea.	27,700	480		28,180	31,100
0140	480V-208V/120V		2	8		28,000	480		28,480	31,500
0150	600V-208V/120V		2	8		27,700	480		28,180	31,200
0200	3PH, 15 kVA, 480-208/120V, 208-208/120V	3	5.333		11,600	320		11,920	13,200
0210	30 kVA, 400-208/120V, 208-208/120V	3	5.333		19,100	320		19,420	21,500
0250	50 kVA, 480-208/120V, 208-208/120V	2.50	6.400		10,400	385		10,785	12,100
0260	75 kVA, 480-208/120V, 208-208/120V	2.50	6.400		11,800	385		12,185	13,600
0270	100 kVA, 480-208/120V, 208-208/120V	2	8		12,700	480		13,180	14,600
0280	125 kVA, 480-208/120V, 208-208/120V	2	8		13,600	480		14,080	15,700
0290	150 kVA, 480-208/120V, 208-208/120V	2	8		12,400	480		12,880	14,400
0300	200 kVA, 480-208/120V, 208-208/120V	1.50	10.667		13,700	640		14,340	16,100
0310	225 kVA, 480-208/120V, 208-208/120V	↓	1.50	10.667	↓	14,500	640		15,140	16,900

26 27 73 – Door Chimes

26 27 73.10 Doorbell System

		Crew	Daily Output	Labor-Hours	Unit	Material	2019 Bare Costs Labor	Equipment	Total	Total Incl O&P
0010	**DOORBELL SYSTEM**, incl. transformer, button & signal									
0100	6" bell	1 Elec	4	2	Ea.	141	120		261	335
0200	Buzzer		4	2		111	120		231	300
1000	Door chimes, 2 notes		16	.500		31	30		61	79.50
1020	with ambient light		12	.667		123	40		163	195
1100	Tube type, 3 tube system		12	.667		231	40		271	315
1180	4 tube system		10	.800		495	48		543	615
1900	For transformer & button, add		5	1.600		16.25	96		112.25	161
3000	For push button only		24	.333		.95	20		20.95	31
3200	Bell transformer	↓	16	.500	↓	22.50	30		52.50	70

For customer support on your Electrical Costs with RSMeans data, call 800.448.8182.

277

26 28 13 – Fuses

26 28 13.10 Fuse Elements	Crew	Daily Output	Labor-Hours	Unit	Material	2019 Bare Costs Labor	Equipment	Total	Total Incl O&P
0010 **FUSE ELEMENTS**									
0020 Cartridge, nonrenewable									
0050 250 volt, 30 amp	1 Elec	50	.160	Ea.	2.26	9.60		11.86	16.85
0100 60 amp		50	.160		4.11	9.60		13.71	18.85
0150 100 amp		40	.200		16.30	12		28.30	36
0200 200 amp		36	.222		39	13.35		52.35	63
0250 400 amp		30	.267		107	16		123	141
0300 600 amp		24	.333		187	20		207	235
0400 600 volt, 30 amp		40	.200		10.70	12		22.70	29.50
0450 60 amp		40	.200		16.25	12		28.25	36
0500 100 amp		36	.222		33.50	13.35		46.85	57
0550 200 amp		30	.267		69	16		85	100
0600 400 amp		24	.333		143	20		163	187
0650 600 amp		20	.400		170	24		194	223
0800 Dual element, time delay, 250 volt, 30 amp		50	.160		9.75	9.60		19.35	25
0840 50 amp		50	.160		11.05	9.60		20.65	26.50
0850 60 amp		50	.160		11.40	9.60		21	27
0900 100 amp		40	.200		38.50	12		50.50	60
0950 200 amp		36	.222		83.50	13.35		96.85	111
1000 400 amp		30	.267		119	16		135	155
1050 600 amp		24	.333		196	20		216	246
1300 600 volt, 15 to 30 amp		40	.200		22	12		34	42
1350 35 to 60 amp		40	.200		31	12		43	52
1400 70 to 100 amp		36	.222		65	13.35		78.35	91.50
1450 110 to 200 amp		30	.267		130	16		146	167
1500 225 to 400 amp		24	.333		277	20		297	335
1550 600 amp		20	.400		360	24		384	430
1800 Class RK1, high capacity, 250 volt, 30 amp		50	.160		9.85	9.60		19.45	25
1850 60 amp		50	.160		18	9.60		27.60	34
1900 100 amp		40	.200		25	12		37	45.50
1950 200 amp		36	.222		73	13.35		86.35	100
2000 400 amp		30	.267		188	16		204	230
2050 600 amp		24	.333		272	20		292	330
2200 600 volt, 30 amp		40	.200		11.80	12		23.80	31
2250 60 amp		40	.200		19.25	12		31.25	39
2300 100 amp		36	.222		79	13.35		92.35	107
2350 200 amp		30	.267		103	16		119	137
2400 400 amp		24	.333		218	20		238	269
2450 600 amp		20	.400		400	24		424	475
2700 Class J, current limiting, 250 or 600 volt, 30 amp		40	.200		20.50	12		32.50	40.50
2750 60 amp		40	.200		32.50	12		44.50	54
2800 100 amp		36	.222		66.50	13.35		79.85	93
2850 200 amp		30	.267		96.50	16		112.50	130
2900 400 amp		24	.333		267	20		287	325
2950 600 amp		20	.400		345	24		369	415
3100 Class L, current limiting, 250 or 600 volt, 601 to 1,200 amp		16	.500		775	30		805	900
3150 1,500 to 1,600 amp		13	.615		765	37		802	895
3200 1,800 to 2,000 amp		10	.800		1,500	48		1,548	1,725
3250 2,500 amp		10	.800		1,275	48		1,323	1,475
3300 3,000 amp		8	1		2,075	60		2,135	2,375
3350 3,500 to 4,000 amp		8	1		2,075	60		2,135	2,400
3400 4,500 to 5,000 amp		6.70	1.194		1,825	71.50		1,896.50	2,100
3450 6,000 amp		5.70	1.404		5,050	84.50		5,134.50	5,700

26 28 Low-Voltage Circuit Protective Devices

26 28 13 - Fuses

26 28 13.10 Fuse Elements

		Crew	Daily Output	Labor-Hours	Unit	Material	2019 Bare Costs Labor	Equipment	Total	Total Incl O&P
3600	Plug, 120 volt, 1 to 10 amp	1 Elec	50	.160	Ea.	3.92	9.60		13.52	18.65
3650	15 to 30 amp		50	.160		4.05	9.60		13.65	18.80
3700	Dual element 0.3 to 14 amp		50	.160		5.65	9.60		15.25	20.50
3750	15 to 30 amp		50	.160		6.05	9.60		15.65	21
3800	Fustat, 120 volt, 15 to 30 amp		50	.160		5.85	9.60		15.45	21
3850	0.3 to 14 amp		50	.160		5.70	9.60		15.30	20.50
3900	Adapters 0.3 to 10 amp		50	.160		5.40	9.60		15	20.50
3950	15 to 30 amp		50	.160		8.10	9.60		17.70	23.50
4000	F-frame current limiting fuse, 14 to 2 AWG, 3 ampere, 3P, aluminum terminal		40	.200		1,825	12		1,837	2,025
4010	7 ampere, 3P, aluminum terminal		40	.200		1,750	12		1,762	1,950
4020	15 AMP, 3P, aluminum terminal		40	.200		2,075	12		2,087	2,325
4030	30 AMP, 3P, aluminum terminal		40	.200		1,650	12		1,662	1,850
4040	50 AMP, 3P, aluminum terminal		40	.200		1,900	12		1,912	2,125
4050	F-frame current limiting fuse, 1 to 4/0 AWG, 100 AMP, 3P, aluminum terminal		40	.200		1,800	12		1,812	2,000
4060	150 AMP, 3P, aluminum terminal		40	.200		2,075	12		2,087	2,300

26 28 16 - Enclosed Switches and Circuit Breakers

26 28 16.10 Circuit Breakers

		Crew	Daily Output	Labor-Hours	Unit	Material	2019 Bare Costs Labor	Equipment	Total	Total Incl O&P
0010	**CIRCUIT BREAKERS** (in enclosure)									
0100	Enclosed (NEMA 1), 600 volt, 3 pole, 30 amp	1 Elec	3.20	2.500	Ea.	515	150		665	790
0200	60 amp		2.80	2.857		620	172		792	940
0400	100 amp		2.30	3.478		710	209		919	1,100
0500	200 amp		1.50	5.333		1,800	320		2,120	2,475
0600	225 amp		1.50	5.333		1,650	320		1,970	2,275
0700	400 amp	2 Elec	1.60	10		2,800	600		3,400	4,000
0800	600 amp		1.20	13.333		4,075	800		4,875	5,675
1000	800 amp		.94	17.021		5,300	1,025		6,325	7,350
1200	1,000 amp		.84	19.048		6,700	1,150		7,850	9,050
1220	1,200 amp		.80	20		8,550	1,200		9,750	11,200
1240	1,600 amp		.72	22.222		15,800	1,325		17,125	19,400
1260	2,000 amp		.64	25		17,100	1,500		18,600	21,100
1400	1,200 amp with ground fault		.80	20		13,900	1,200		15,100	17,100
1600	1,600 amp with ground fault		.72	22.222		17,900	1,325		19,225	21,700
1800	2,000 amp with ground fault		.64	25		19,200	1,500		20,700	23,400
2000	Disconnect, 240 volt 3 pole, 5 HP motor	1 Elec	3.20	2.500		435	150		585	705
2020	10 HP motor		3.20	2.500		435	150		585	705
2040	15 HP motor		2.80	2.857		435	172		607	735
2060	20 HP motor		2.30	3.478		530	209		739	890
2080	25 HP motor		2.30	3.478		530	209		739	890
2100	30 HP motor		2.30	3.478		530	209		739	890
2120	40 HP motor		2	4		905	240		1,145	1,350
2140	50 HP motor		1.50	5.333		905	320		1,225	1,475
2160	60 HP motor		1.50	5.333		2,075	320		2,395	2,775
2180	75 HP motor	2 Elec	2	8		2,075	480		2,555	3,025
2200	100 HP motor		1.60	10		2,075	600		2,675	3,200
2220	125 HP motor		1.60	10		2,075	600		2,675	3,200
2240	150 HP motor		1.20	13.333		4,075	800		4,875	5,675
2260	200 HP motor		1.20	13.333		5,300	800		6,100	7,025
2300	Enclosed (NEMA 7), explosion proof, 600 volt 3 pole, 50 amp	1 Elec	2.30	3.478		1,850	209		2,059	2,350
2350	100 amp		1.50	5.333		1,925	320		2,245	2,600
2400	150 amp		1	8		4,650	480		5,130	5,850
2450	250 amp	2 Elec	1.60	10		5,825	600		6,425	7,300
2500	400 amp	"	1.20	13.333		6,450	800		7,250	8,275

26 28 16.13 Circuit Breakers

	Crew	Daily Output	Labor-Hours	Unit	Material	2019 Bare Costs Labor	Equipment	Total	Total Incl O&P
0010 CIRCUIT BREAKERS									
0100 Circuit breaker current limiter, 225 to 400, ampere, 1 pole	1 Elec	48	.167	Ea.	800	10		810	895
0200 Circuit breaker current limiter, 200 to 500, ampere, 1 pole, thermal magnet		46	.174		820	10.45		830.45	915
0300 300 to 500, ampere, 1 pole, thermal magnet		46	.174		1,225	10.45		1,235.45	1,375
0400 Circuit breaker current limiter, 600 to 1000, ampere, 1 pole, tri pac	▼	46	.174	▼	1,600	10.45		1,610.45	1,775

26 28 16.20 Safety Switches

	Crew	Daily Output	Labor-Hours	Unit	Material	2019 Bare Costs Labor	Equipment	Total	Total Incl O&P
0010 SAFETY SWITCHES R262816-80									
0100 General duty 240 volt, 3 pole NEMA 1, fusible, 30 amp	1 Elec	3.20	2.500	Ea.	85.50	150		235.50	320
0200 60 amp		2.30	3.478		145	209		354	470
0300 100 amp		1.90	4.211		246	253		499	645
0400 200 amp	▼	1.30	6.154		535	370		905	1,125
0500 400 amp	2 Elec	1.80	8.889		1,400	535		1,935	2,325
0600 600 amp	"	1.20	13.333		2,300	800		3,100	3,750
0610 Nonfusible, 30 amp	1 Elec	3.20	2.500		68.50	150		218.50	299
0650 60 amp		2.30	3.478		93	209		302	415
0700 100 amp		1.90	4.211		212	253		465	610
0750 200 amp		1.30	6.154		385	370		755	970
0800 400 amp	2 Elec	1.80	8.889		880	535		1,415	1,750
0850 600 amp	"	1.20	13.333	▼	1,550	800		2,350	2,900
1100 Heavy duty, 600 volt, 3 pole NEMA 1 nonfused									
1110 30 amp	1 Elec	3.20	2.500	Ea.	78	150		228	310
1500 60 amp		2.30	3.478		214	209		423	545
1700 100 amp		1.90	4.211		330	253		583	740
1900 200 amp	▼	1.30	6.154		505	370		875	1,100
2100 400 amp	2 Elec	1.80	8.889		1,125	535		1,660	2,025
2300 600 amp		1.20	13.333		2,025	800		2,825	3,425
2500 800 amp		.94	17.021		3,425	1,025		4,450	5,300
2700 1,200 amp	▼	.80	20	▼	3,350	1,200		4,550	5,475
2900 Heavy duty, 240 volt, 3 pole NEMA 1 fusible									
2910 30 amp	1 Elec	3.20	2.500	Ea.	87.50	150		237.50	320
3000 60 amp		2.30	3.478		145	209		354	470
3300 100 amp		1.90	4.211		228	253		481	625
3500 200 amp	▼	1.30	6.154		390	370		760	975
3700 400 amp	2 Elec	1.80	8.889		990	535		1,525	1,900
3900 600 amp		1.20	13.333		1,750	800		2,550	3,125
4100 800 amp		.94	17.021		6,100	1,025		7,125	8,225
4300 1,200 amp	▼	.80	20		5,175	1,200		6,375	7,475
4340 2 pole fusible, 30 amp	1 Elec	3.50	2.286		65	137		202	277
4350 600 volt, 3 pole, fusible, 30 amp		3.20	2.500		145	150		295	385
4380 60 amp		2.30	3.478		176	209		385	505
4400 100 amp		1.90	4.211		315	253		568	720
4420 200 amp		1.30	6.154		450	370		820	1,050
4440 400 amp	2 Elec	1.80	8.889		1,250	535		1,785	2,175
4450 600 amp		1.20	13.333		1,975	800		2,775	3,375
4460 800 amp		.94	17.021		6,100	1,025		7,125	8,225
4480 1,200 amp	▼	.80	20	▼	5,175	1,200		6,375	7,475
4500 240 volt 3 pole NEMA 3R (no hubs), fusible									
4510 30 amp	1 Elec	3.10	2.581	Ea.	153	155		308	400
4700 60 amp		2.20	3.636		256	218		474	605
4900 100 amp		1.80	4.444		345	267		612	775
5100 200 amp	▼	1.20	6.667		470	400		870	1,125
5300 400 amp	2 Elec	1.60	10	▼	1,100	600		1,700	2,100

280

For customer support on your Electrical Costs with RSMeans data, call 800.448.8182.

26 28 Low-Voltage Circuit Protective Devices

26 28 16 – Enclosed Switches and Circuit Breakers

26 28 16.20 Safety Switches		Crew	Daily Output	Labor-Hours	Unit	Material	2019 Bare Costs Labor	Equipment	Total	Total Incl O&P
5500	600 amp	2 Elec	1	16	Ea.	2,200	960		3,160	3,850
5510	Heavy duty, 600 volt, 3 pole 3 ph. NEMA 3R fusible, 30 amp	1 Elec	3.10	2.581		245	155		400	500
5520	60 amp		2.20	3.636		310	218		528	665
5530	100 amp		1.80	4.444		435	267		702	880
5540	200 amp		1.20	6.667		630	400		1,030	1,300
5550	400 amp	2 Elec	1.60	10	▼	1,525	600		2,125	2,575
5700	600 volt, 3 pole NEMA 3R nonfused									
5710	30 amp	1 Elec	3.10	2.581	Ea.	204	155		359	455
5900	60 amp		2.20	3.636		365	218		583	725
6100	100 amp		1.80	4.444		505	267		772	960
6300	200 amp	▼	1.20	6.667		645	400		1,045	1,300
6500	400 amp	2 Elec	1.60	10		1,475	600		2,075	2,525
6700	600 amp	"	1	16		2,175	960		3,135	3,800
6900	600 volt, 6 pole NEMA 3R nonfused, 30 amp	1 Elec	2.70	2.963		900	178		1,078	1,250
7100	60 amp		2	4		1,675	240		1,915	2,175
7300	100 amp		1.50	5.333		2,275	320		2,595	2,975
7500	200 amp	▼	1.20	6.667	▼	3,150	400		3,550	4,075
7600	600 volt, 3 pole NEMA 7 explosion proof nonfused									
7610	30 amp	1 Elec	2.20	3.636	Ea.	1,425	218		1,643	1,900
7620	60 amp		1.80	4.444		1,725	267		1,992	2,300
7630	100 amp		1.20	6.667		2,475	400		2,875	3,325
7640	200 amp		.80	10		5,275	600		5,875	6,700
7710	600 volt 6 pole, NEMA 3R fusible, 30 amp		2.70	2.963		1,275	178		1,453	1,675
7900	60 amp		2	4		2,025	240		2,265	2,575
8100	100 amp		1.50	5.333		1,600	320		1,920	2,250
8110	240 volt 3 pole, NEMA 12 fusible, 30 amp		3.10	2.581		300	155		455	560
8120	60 amp		2.20	3.636		385	218		603	745
8130	100 amp		1.80	4.444		690	267		957	1,150
8140	200 amp	▼	1.20	6.667		595	400		995	1,250
8150	400 amp	2 Elec	1.60	10		2,275	600		2,875	3,400
8160	600 amp	"	1	16		2,400	960		3,360	4,050
8180	600 volt 3 pole, NEMA 12 fusible, 30 amp	1 Elec	3.10	2.581		300	155		455	560
8190	60 amp		2.20	3.636		365	218		583	725
8200	100 amp		1.80	4.444		510	267		777	960
8210	200 amp		1.20	6.667		730	400		1,130	1,400
8220	400 amp	2 Elec	1.60	10		2,325	600		2,925	3,475
8230	600 amp	"	1	16		3,075	960		4,035	4,800
8240	600 volt 3 pole, NEMA 12 nonfused, 30 amp	1 Elec	3.10	2.581		187	155		342	435
8250	60 amp		2.20	3.636		350	218		568	710
8260	100 amp		1.80	4.444		535	267		802	990
8270	200 amp	▼	1.20	6.667		475	400		875	1,125
8280	400 amp	2 Elec	1.60	10		1,900	600		2,500	2,975
8290	600 amp	"	1	16		2,450	960		3,410	4,100
8310	600 volt, 3 pole NEMA 4 fusible, 30 amp	1 Elec	3	2.667		660	160		820	965
8320	60 amp		2.20	3.636		775	218		993	1,175
8330	100 amp		1.80	4.444		1,550	267		1,817	2,125
8340	200 amp	▼	1.20	6.667		2,025	400		2,425	2,850
8350	400 amp	2 Elec	1.60	10		4,075	600		4,675	5,400
8360	600 volt 3 pole NEMA 4 nonfused, 30 amp	1 Elec	3	2.667		850	160		1,010	1,175
8370	60 amp		2.20	3.636		1,000	218		1,218	1,425
8380	100 amp		1.80	4.444		2,100	267		2,367	2,725
8390	200 amp	▼	1.20	6.667		1,975	400		2,375	2,775
8400	400 amp	2 Elec	1.60	10	▼	5,475	600		6,075	6,925

26 28 Low-Voltage Circuit Protective Devices

26 28 16 – Enclosed Switches and Circuit Breakers

26 28 16.20 Safety Switches

		Crew	Daily Output	Labor-Hours	Unit	Material	2019 Bare Costs Labor	Equipment	Total	Total Incl O&P
8490	Motor starters, manual, single phase, NEMA 1	1 Elec	6.40	1.250	Ea.	84.50	75		159.50	205
8500	NEMA 4		4	2		239	120		359	440
8700	NEMA 7		4	2		300	120		420	510
8900	NEMA 1 with pilot		6.40	1.250		110	75		185	233
8920	3 pole, NEMA 1, 230/460 volt, 5 HP, size 0		3.50	2.286		202	137		339	425
8940	10 HP, size 1		2	4		240	240		480	625
9010	Disc. switch, 600 volt 3 pole fusible, 30 amp, to 10 HP motor		3.20	2.500		345	150		495	605
9050	60 amp, to 30 HP motor		2.30	3.478		795	209		1,004	1,175
9070	100 amp, to 60 HP motor		1.90	4.211		795	253		1,048	1,250
9100	200 amp, to 125 HP motor	▼	1.30	6.154		1,200	370		1,570	1,875
9110	400 amp, to 200 HP motor	2 Elec	1.80	8.889	▼	3,025	535		3,560	4,125

26 28 16.40 Time Switches

		Crew	Daily Output	Labor-Hours	Unit	Material	2019 Bare Costs Labor	Equipment	Total	Total Incl O&P
0010	**TIME SWITCHES**									
0100	Single pole, single throw, 24 hour dial	1 Elec	4	2	Ea.	130	120		250	320
0200	24 hour dial with reserve power		3.60	2.222		590	133		723	850
0300	Astronomic dial		3.60	2.222		235	133		368	455
0400	Astronomic dial with reserve power		3.30	2.424		895	146		1,041	1,200
0500	7 day calendar dial		3.30	2.424		191	146		337	425
0600	7 day calendar dial with reserve power		3.20	2.500		222	150		372	470
0700	Photo cell 2,000 watt		8	1		26	60		86	118
1080	Load management device, 4 loads		2	4		1,100	240		1,340	1,550
1100	8 loads	▼	1	8	▼	2,275	480		2,755	3,225

26 28 16.50 Meter Socket Entry Hub

		Crew	Daily Output	Labor-Hours	Unit	Material	2019 Bare Costs Labor	Equipment	Total	Total Incl O&P
0010	**METER SOCKET ENTRY HUB**									
0100	Meter socket entry hub closing cap, 3/4 to 4 inch	1 Elec	40	.200	Ea.	2.16	12		14.16	20.50
0110	Meter socket entry hub, 3/4 conduit		40	.200		17.55	12		29.55	37
0120	1 inch conduit		40	.200		19.75	12		31.75	39.50
0130	1.25 inch conduit		40	.200		18.70	12		30.70	38.50
0140	1.50 inch conduit		40	.200		18.95	12		30.95	39
0150	2 inch conduit		40	.200		18.25	12		30.25	38
0160	2.50 inch conduit		40	.200		62	12		74	86
0170	3 inch conduit	▼	40	.200	▼	86	12		98	112

26 29 Low-Voltage Controllers

26 29 13 – Enclosed Controllers

26 29 13.10 Contactors, AC

		Crew	Daily Output	Labor-Hours	Unit	Material	2019 Bare Costs Labor	Equipment	Total	Total Incl O&P
0010	**CONTACTORS, AC** Enclosed (NEMA 1)									
0050	Lighting, 600 volt 3 pole, electrically held									
0100	20 amp	1 Elec	4	2	Ea.	380	120		500	595
0200	30 amp		3.60	2.222		254	133		387	480
0300	60 amp		3	2.667		545	160		705	840
0400	100 amp		2.50	3.200		835	192		1,027	1,200
0500	200 amp	▼	1.40	5.714		2,125	345		2,470	2,825
0600	300 amp	2 Elec	1.60	10		4,325	600		4,925	5,650
0800	600 volt 3 pole, mechanically held, 30 amp	1 Elec	3.60	2.222		605	133		738	865
0900	60 amp		3	2.667		1,225	160		1,385	1,600
1000	75 amp		2.80	2.857		1,500	172		1,672	1,900
1100	100 amp		2.50	3.200		1,550	192		1,742	2,000
1200	150 amp		2	4		3,525	240		3,765	4,225
1300	200 amp	▼	1.40	5.714	▼	5,300	345		5,645	6,325

26 29 13 – Enclosed Controllers

26 29 13.10 Contactors, AC

		Crew	Daily Output	Labor-Hours	Unit	Material	2019 Bare Costs Labor	Equipment	Total	Total Incl O&P
1500	Magnetic with auxiliary contact, size 00, 9 amp	1 Elec	4	2	Ea.	183	120		303	380
1600	Size 0, 18 amp		4	2		218	120		338	420
1700	Size 1, 27 amp		3.60	2.222		248	133		381	470
1800	Size 2, 45 amp		3	2.667		460	160		620	745
1900	Size 3, 90 amp		2.50	3.200		740	192		932	1,100
2000	Size 4, 135 amp		2.30	3.478		1,700	209		1,909	2,175
2100	Size 5, 270 amp	2 Elec	1.80	8.889		3,575	535		4,110	4,725
2200	Size 6, 540 amp		1.20	13.333		10,400	800		11,200	12,700
2300	Size 7, 810 amp		1	16		14,000	960		14,960	16,900
2310	Size 8, 1,215 amp		.80	20		21,900	1,200		23,100	25,800
2500	Magnetic, 240 volt, 1-2 pole, 0.75 HP motor	1 Elec	4	2		147	120		267	340
2520	2 HP motor		3.60	2.222		198	133		331	415
2540	5 HP motor		2.50	3.200		400	192		592	725
2560	10 HP motor		1.40	5.714		660	345		1,005	1,225
2600	240 volt or less, 3 pole, 0.75 HP motor		4	2		147	120		267	340
2620	5 HP motor		3.60	2.222		183	133		316	400
2640	10 HP motor		3.60	2.222		212	133		345	430
2660	15 HP motor		2.50	3.200		425	192		617	750
2700	25 HP motor		2.50	3.200		425	192		617	750
2720	30 HP motor	2 Elec	2.80	5.714		705	345		1,050	1,300
2740	40 HP motor		2.80	5.714		705	345		1,050	1,300
2760	50 HP motor		1.60	10		705	600		1,305	1,675
2800	75 HP motor		1.60	10		1,650	600		2,250	2,725
2820	100 HP motor		1	16		1,650	960		2,610	3,250
2860	150 HP motor		1	16		3,525	960		4,485	5,325
2880	200 HP motor		1	16		3,525	960		4,485	5,325
3000	600 volt, 3 pole, 5 HP motor	1 Elec	4	2		183	120		303	380
3020	10 HP motor		3.60	2.222		212	133		345	430
3040	25 HP motor		3	2.667		425	160		585	705
3100	50 HP motor		2.50	3.200		705	192		897	1,075
3160	100 HP motor	2 Elec	2.80	5.714		1,650	345		1,995	2,325
3220	200 HP motor	"	1.60	10		3,525	600		4,125	4,800

26 29 13.20 Control Stations

		Crew	Daily Output	Labor-Hours	Unit	Material	2019 Bare Costs Labor	Equipment	Total	Total Incl O&P
0010	**CONTROL STATIONS**									
0050	NEMA 1, heavy duty, stop/start	1 Elec	8	1	Ea.	146	60		206	250
0100	Stop/start, pilot light		6.20	1.290		199	77.50		276.50	335
0200	Hand/off/automatic		6.20	1.290		108	77.50		185.50	235
0400	Stop/start/reverse		5.30	1.509		196	90.50		286.50	350
0500	NEMA 7, heavy duty, stop/start		6	1.333		540	80		620	715
0600	Stop/start, pilot light		4	2		655	120		775	905
0700	NEMA 7 or 9, 1 element		6	1.333		440	80		520	600
0800	2 element		6	1.333		565	80		645	740
0900	3 element		4	2		1,125	120		1,245	1,425
0910	Selector switch, 2 position		6	1.333		440	80		520	605
0920	3 position		4	2		440	120		560	665
0930	Oiltight, 1 element		8	1		110	60		170	211
0940	2 element		6.20	1.290		153	77.50		230.50	284
0950	3 element		5.30	1.509		142	90.50		232.50	292
0960	Selector switch, 2 position		6.20	1.290		111	77.50		188.50	238
0970	3 position		5.30	1.509		116	90.50		206.50	263

For customer support on your Electrical Costs with RSMeans data, call 800.448.8182.

283

26 29 13.30 Control Switches

		Crew	Daily Output	Labor-Hours	Unit	Material	2019 Bare Costs Labor	Equipment	Total	Total Incl O&P
0010	**CONTROL SWITCHES** Field installed									
6000	Push button 600 V 10A, momentary contact									
6150	Standard operator with colored button	1 Elec	34	.235	Ea.	18.55	14.15		32.70	41.50
6160	With single block 1NO 1NC		18	.444		39	26.50		65.50	83
6170	With double block 2NO 2NC		15	.533		59.50	32		91.50	114
6180	Std operator w/mushroom button 1-9/16" diam.		34	.235		38	14.15		52.15	63
6190	Std operator w/mushroom button 2-1/4" diam.									
6200	With single block 1NO 1NC	1 Elec	18	.444	Ea.	54.50	26.50		81	100
6210	With double block 2NO 2NC		15	.533		73.50	32		105.50	129
6500	Maintained contact, selector operator		34	.235		54.50	14.15		68.65	81
6510	With single block 1NO 1NC		18	.444		73.50	26.50		100	121
6520	With double block 2NO 2NC		15	.533		92	32		124	149
6560	Spring-return selector operator		34	.235		54.50	14.15		68.65	81
6570	With single block 1NO 1NC		18	.444		73.50	26.50		100	121
6580	With double block 2NO 2NC		15	.533		92	32		124	149
6620	Transformer operator w/illuminated									
6630	button 6 V #12 lamp	1 Elec	32	.250	Ea.	92	15		107	124
6640	With single block 1NO 1NC w/guard		16	.500		111	30		141	167
6650	With double block 2NO 2NC w/guard		13	.615		130	37		167	198
6690	Combination operator		34	.235		54.50	14.15		68.65	81
6700	With single block 1NO 1NC		18	.444		73.50	26.50		100	121
6710	With double block 2NO 2NC		15	.533		92	32		124	149
9000	Indicating light unit, full voltage									
9010	110-125 V front mount	1 Elec	32	.250	Ea.	73.50	15		88.50	103
9020	130 V resistor type		32	.250		58	15		73	86.50
9030	6 V transformer type		32	.250		67	15		82	96.50

26 29 13.40 Relays

		Crew	Daily Output	Labor-Hours	Unit	Material	2019 Bare Costs Labor	Equipment	Total	Total Incl O&P
0010	**RELAYS** Enclosed (NEMA 1)									
0050	600 volt AC, 1 pole, 12 amp	1 Elec	5.30	1.509	Ea.	86.50	90.50		177	230
0100	2 pole, 12 amp		5	1.600		86.50	96		182.50	238
0200	4 pole, 10 amp		4.50	1.778		115	107		222	286
0500	250 volt DC, 1 pole, 15 amp		5.30	1.509		113	90.50		203.50	260
0600	2 pole, 10 amp		5	1.600		107	96		203	261
0700	4 pole, 4 amp		4.50	1.778		143	107		250	315

26 29 23 – Variable-Frequency Motor Controllers

26 29 23.10 Variable Frequency Drives/Adj. Frequency Drives

			Crew	Daily Output	Labor-Hours	Unit	Material	2019 Bare Costs Labor	Equipment	Total	Total Incl O&P
0010	**VARIABLE FREQUENCY DRIVES/ADJ. FREQUENCY DRIVES**										
0100	Enclosed (NEMA 1), 460 volt, for 3 HP motor size	G	1 Elec	.80	10	Ea.	1,975	600		2,575	3,075
0110	5 HP motor size	G		.80	10		2,175	600		2,775	3,275
0120	7.5 HP motor size	G		.67	11.940		2,550	715		3,265	3,875
0130	10 HP motor size	G		.67	11.940		2,800	715		3,515	4,175
0140	15 HP motor size	G	2 Elec	.89	17.978		3,625	1,075		4,700	5,600
0150	20 HP motor size	G		.89	17.978		4,450	1,075		5,525	6,500
0160	25 HP motor size	G		.67	23.881		5,250	1,425		6,675	7,925
0170	30 HP motor size	G		.67	23.881		6,550	1,425		7,975	9,350
0180	40 HP motor size	G		.67	23.881		7,550	1,425		8,975	10,500
0190	50 HP motor size	G		.53	30.189		10,100	1,825		11,925	13,800
0200	60 HP motor size	G	R-3	.56	35.714		11,800	2,125	232	14,157	16,300
0210	75 HP motor size	G		.56	35.714		13,900	2,125	232	16,257	18,700
0220	100 HP motor size	G		.50	40		17,400	2,400	260	20,060	23,000
0230	125 HP motor size	G		.50	40		18,900	2,400	260	21,560	24,700

26 29 Low-Voltage Controllers

26 29 23 – Variable-Frequency Motor Controllers

26 29 23.10 Variable Frequency Drives/Adj. Frequency Drives		Crew	Daily Output	Labor-Hours	Unit	Material	2019 Bare Costs Labor	Equipment	Total	Total Incl O&P
0240	150 HP motor size	G R-3	.50	40	Ea.	23,900	2,400	260	26,560	30,200
0250	200 HP motor size	G ▼	.42	47.619		26,100	2,850	310	29,260	33,300
1100	Custom-engineered, 460 volt, for 3 HP motor size	G 1 Elec	.56	14.286		3,100	860		3,960	4,675
1110	5 HP motor size	G	.56	14.286		3,075	860		3,935	4,650
1120	7.5 HP motor size	G	.47	17.021		3,250	1,025		4,275	5,100
1130	10 HP motor size	G ▼	.47	17.021		3,400	1,025		4,425	5,275
1140	15 HP motor size	G 2 Elec	.62	25.806		4,250	1,550		5,800	6,975
1150	20 HP motor size	G	.62	25.806		4,500	1,550		6,050	7,250
1160	25 HP motor size	G	.47	34.043		5,625	2,050		7,675	9,225
1170	30 HP motor size	G	.47	34.043		6,975	2,050		9,025	10,700
1180	40 HP motor size	G	.47	34.043		8,250	2,050		10,300	12,200
1190	50 HP motor size	G ▼	.37	43.243		9,450	2,600		12,050	14,300
1200	60 HP motor size	G R-3	.39	51.282		14,300	3,075	335	17,710	20,600
1210	75 HP motor size	G	.39	51.282		15,100	3,075	335	18,510	21,500
1220	100 HP motor size	G	.35	57.143		16,600	3,425	370	20,395	23,700
1230	125 HP motor size	G	.35	57.143		17,800	3,425	370	21,595	25,000
1240	150 HP motor size	G	.35	57.143		20,400	3,425	370	24,195	27,900
1250	200 HP motor size	G ▼	.29	68.966	▼	26,500	4,125	450	31,075	35,700
2000	For complex & special design systems to meet specific									
2010	requirements, obtain quote from vendor.									

26 31 Photovoltaic Collectors

26 31 13 – Photovoltaics

26 31 13.50 Solar Energy - Photovoltaics

			Crew	Daily Output	Labor-Hours	Unit	Material	2019 Bare Costs Labor	Equipment	Total	Total Incl O&P
0010	**SOLAR ENERGY - PHOTOVOLTAICS**										
0220	Alt. energy source, photovoltaic module, 6 watt, 15 V	G	1 Elec	8	1	Ea.	51	60		111	146
0230	10 watt, 16.3 V	G		8	1		117	60		177	219
0240	20 watt, 14.5 V	G		8	1		180	60		240	288
0250	36 watt, 17 V	G		8	1		163	60		223	270
0260	55 watt, 17 V	G		8	1		241	60		301	355
0270	75 watt, 17 V	G		8	1		395	60		455	525
0280	130 watt, 33 V	G		8	1		620	60		680	775
0290	140 watt, 33 V	G		8	1		480	60		540	620
0300	150 watt, 33 V	G		8	1		400	60		460	530
0310	DC to AC inverter for, 12 V, 2,000 watt	G		4	2		1,750	120		1,870	2,100
0320	12 V, 2,500 watt	G		4	2		1,100	120		1,220	1,375
0330	24 V, 2,500 watt	G		4	2		1,625	120		1,745	1,950
0340	12 V, 3,000 watt	G		3	2.667		1,500	160		1,660	1,900
0350	24 V, 3,000 watt	G		3	2.667		2,350	160		2,510	2,850
0360	24 V, 4,000 watt	G		2	4		3,375	240		3,615	4,050
0370	48 V, 4,000 watt	G		2	4		2,825	240		3,065	3,450
0380	48 V, 5,500 watt	G		2	4		3,000	240		3,240	3,650
0390	PV components, combiner box, 10 lug, NEMA 3R enclosure	G		4	2		271	120		391	475
0400	Fuse, 15 A for combiner box	G		40	.200		22	12		34	42
0410	Battery charger controller w/temperature sensor	G		4	2		495	120		615	725
0420	Digital readout panel, displays hours, volts, amps, etc.	G		4	2		215	120		335	415
0430	Deep cycle solar battery, 6 V, 180 Ah (C/20)	G		8	1		315	60		375	440
0440	Battery interconn, 15" AWG #2/0, sealed w/copper ring lugs	G		16	.500		18.30	30		48.30	65
0442	Battery interconn, 24" AWG #2/0, sealed w/copper ring lugs	G		16	.500		23.50	30		53.50	71
0444	Battery interconn, 60" AWG #2/0, sealed w/copper ring lugs	G		16	.500		56	30		86	107
0446	Batt temp computer probe, RJ11 jack, 15' cord	G	▼	16	.500	▼	22.50	30		52.50	69.50

26 31 Photovoltaic Collectors

26 31 13 – Photovoltaics

26 31 13.50 Solar Energy - Photovoltaics

	Crew	Daily Output	Labor-Hours	Unit	Material	2019 Bare Costs Labor	Equipment	Total	Total Incl O&P
0450	System disconnect, DC 175 amp circuit breaker **G**	1 Elec	8	1	Ea.	229	60		289	340
0460	Conduit box for inverter **G**		8	1		63	60		123	159
0470	Low voltage disconnect **G**		8	1		62	60		122	158
0480	Vented battery enclosure, wood **G**	1 Carp	2	4		277	207		484	620
0490	PV rack system, roof, non-penetrating ballast, 1 panel **G**	R-1A	30.50	.525		970	28.50		998.50	1,125
0500	Penetrating surface mount, on steel framing, 1 panel		5.13	3.122		57	169		226	315
0510	On wood framing, 1 panel		11	1.455		56	78.50		134.50	179
0520	With standoff, 1 panel		11	1.455		65	78.50		143.50	189
0530	Ground, ballast, fixed, 3 panel		20.50	.780		1,025	42		1,067	1,200
0540	4 panel		20.50	.780		1,400	42		1,442	1,625
0550	5 panel		20.50	.780		1,850	42		1,892	2,125
0560	6 panel		20.50	.780		2,150	42		2,192	2,425
0570	Adjustable, 3 panel		20.50	.780		1,100	42		1,142	1,275
0580	4 panel		20.50	.780		1,500	42		1,542	1,725
0590	5 panel		20.50	.780		1,975	42		2,017	2,250
0600	6 panel		20.50	.780		2,425	42		2,467	2,750
0605	Top of Pole, see 26 56 13.10 3800+, 5400+ & 33 71 16.33 7400+ for poles									
0610	Passive tracking, 1 panel	R-1A	20.50	.780	Ea.	645	42		687	770
0620	2 panel		20.50	.780		1,300	42		1,342	1,500
0630	3 panel		20.50	.780		1,850	42		1,892	2,125
0640	4 panel		20.50	.780		2,000	42		2,042	2,275
0650	6 panel		20.50	.780		2,125	42		2,167	2,425
0660	8 panel		20.50	.780		3,150	42		3,192	3,550
1020	Photovoltaic module, Thin film	1 Elec	8	1		180	60		240	288
1040	Photovoltaic module, Cadium telluride		8	1		241	60		301	355
1060	Photovoltaic module, Polycrystalline		8	1		180	60		240	288
1080	Photovoltaic module, Monocrystalline		8	1		620	60		680	775

26 32 Packaged Generator Assemblies

26 32 13 – Engine Generators

26 32 13.13 Diesel-Engine-Driven Generator Sets

	Crew	Daily Output	Labor-Hours	Unit	Material	2019 Bare Costs Labor	Equipment	Total	Total Incl O&P
0010	**DIESEL-ENGINE-DRIVEN GENERATOR SETS**									
2000	Diesel engine, including battery, charger,									
2010	muffler, & day tank, 30 kW	R-3	.55	36.364	Ea.	11,200	2,175	236	13,611	15,800
2100	50 kW		.42	47.619		20,300	2,850	310	23,460	26,900
2110	60 kW		.39	51.282		21,200	3,075	335	24,610	28,200
2200	75 kW		.35	57.143		23,600	3,425	370	27,395	31,500
2300	100 kW		.31	64.516		27,500	3,850	420	31,770	36,500
2400	125 kW		.29	68.966		28,900	4,125	450	33,475	38,400
2500	150 kW		.26	76.923		42,900	4,600	500	48,000	54,500
2501	Generator set, dsl eng in alum encl, incl btry, chgr, muf & day tank, 150 kW		.26	76.923		41,800	4,600	500	46,900	53,500
2600	175 kW		.25	80		45,300	4,775	520	50,595	57,500
2700	200 kW		.24	83.333		45,300	4,975	540	50,815	58,000
2800	250 kW		.23	86.957		48,700	5,200	565	54,465	62,000
2850	275 kW		.22	90.909		53,500	5,425	590	59,515	68,000
2900	300 kW		.22	90.909		54,000	5,425	590	60,015	68,000
3000	350 kW		.20	100		63,500	5,975	650	70,125	79,500
3100	400 kW		.19	105		72,500	6,275	685	79,460	90,000
3200	500 kW		.18	111		98,500	6,625	720	105,845	119,000
3220	600 kW		.17	118		119,500	7,025	765	127,290	143,000
3230	650 kW	R-13	.38	111		157,500	6,400	620	164,520	184,000

26 32 Packaged Generator Assemblies

26 32 13 - Engine Generators

26 32 13.13 Diesel-Engine-Driven Generator Sets

	26 32 13.13 Diesel-Engine-Driven Generator Sets	Crew	Daily Output	Labor-Hours	Unit	Material	2019 Bare Costs Labor	Equipment	Total	Total Incl O&P
3240	750 kW	R-13	.38	111	Ea.	148,000	6,400	620	155,020	173,500
3250	800 kW		.36	117		155,500	6,750	655	162,905	182,500
3260	900 kW		.31	135		181,500	7,850	765	190,115	212,000
3270	1,000 kW	↓	.27	156	↓	190,000	9,025	875	199,900	223,500

26 32 13.16 Gas-Engine-Driven Generator Sets

		Crew	Daily Output	Labor-Hours	Unit	Material	Labor	Equipment	Total	Total Incl O&P
0010	**GAS-ENGINE-DRIVEN GENERATOR SETS**									
0020	Gas or gasoline operated, includes battery,									
0050	charger, & muffler									
0200	7.5 kW	R-3	.83	24.096	Ea.	8,700	1,450	156	10,306	11,900
0300	11.5 kW		.71	28.169		12,300	1,675	183	14,158	16,300
0400	20 kW		.63	31.746		14,500	1,900	206	16,606	19,100
0460	25 kW, steel enclosure		.63	31.746		21,200	1,900	206	23,306	26,400
0470	27 kW, aluminum enclosure		.63	31.746		33,000	1,900	206	35,106	39,400
0480	30 kW, steel enclosure		.63	31.746		29,100	1,900	206	31,206	35,100
0485	120/208 V, 30kW		.63	31.746		24,000	1,900	206	26,106	29,500
0500	35 kW		.55	36.364		17,300	2,175	236	19,711	22,500
0510	45 kW, steel enclosure		.50	40		34,400	2,400	260	37,060	41,800
0515	48 kW, aluminum enclosure		.50	40		49,500	2,400	260	52,160	58,500
0520	60 kW	↓	.50	40		22,800	2,400	260	25,460	29,000
0550	70 kW, aluminum enclosure	R-13	.46	91.304		60,000	5,300	515	65,815	74,500
0600	80 kW		.40	105		28,400	6,075	590	35,065	41,000
0700	100 kW		.33	127		31,100	7,375	715	39,190	46,000
0800	125 kW		.28	150		63,500	8,700	845	73,045	84,000
0850	130 kW, aluminum enclosure		.28	150		91,000	8,700	845	100,545	114,000
0875	150 kW, aluminum enclosure		.25	168		92,000	9,725	945	102,670	117,000
0900	185 kW	↓	.25	168	↓	84,000	9,725	945	94,670	108,000

26 33 Battery Equipment

26 33 19 - Battery Units

26 33 19.10 Battery Units

		Crew	Daily Output	Labor-Hours	Unit	Material	Labor	Equipment	Total	Total Incl O&P
0010	**BATTERY UNITS**									
0500	Salt water battery, 2.1 kWA, 24V, 750W, 30A, wired in crate, IP22 rated	1 Elec	16.25	.492	Ea.	970	29.50		999.50	1,125
0510	2.2 kWh, 48V, 800W, 17A, wired in crate, IP22 rated		16.25	.492		1,200	29.50		1,229.50	1,375
0520	25.9 kWh,48V, 11700W, 240A, wired in crate, IP2X rated		16.25	.492		14,600	29.50		14,629.50	16,000
0700	Nickel iron nife battery, deep cycle, 100Ah, 12V, 10 cells		14.25	.561		1,050	33.50		1,083.50	1,200
0710	24V, 10 cells		14.25	.561		2,225	33.50		2,258.50	2,500
0720	48V, 10 cells		14.25	.561		4,150	33.50		4,183.50	4,625
0730	Nickel iron nife battery, deep cycle, 300Ah, 12V, 10 cells		12.25	.653		3,050	39		3,089	3,425
0740	24V, 10 cells		12.25	.653		6,125	39		6,164	6,775
0750	48V, 10 cells	↓	12.25	.653	↓	12,800	39		12,839	14,200

26 33 43 - Battery Chargers

26 33 43.55 Electric Vehicle Charging

			Crew	Daily Output	Labor-Hours	Unit	Material	Labor	Equipment	Total	Total Incl O&P
0010	**ELECTRIC VEHICLE CHARGING**										
0020	Level 2, wall mounted										
2200	Heavy duty	G	R-1A	15.36	1.042	Ea.	2,000	56.50		2,056.50	2,275
2210	with RFID	G		12.29	1.302		3,375	70.50		3,445.50	3,825
2300	Free standing, single connector	G		10.24	1.563		2,200	84.50		2,284.50	2,550
2310	with RFID	G		8.78	1.822		4,475	98.50		4,573.50	5,075
2320	Double connector	G		7.68	2.083		4,075	113		4,188	4,650
2330	with RFID	G	↓	6.83	2.343	↓	7,475	127		7,602	8,425

For customer support on your Electrical Costs with RSMeans data, call 800.448.8182.

287

26 33 Battery Equipment

26 33 53 – Static Uninterruptible Power Supply

26 33 53.10 Uninterruptible Power Supply/Conditioner Trans.		Crew	Daily Output	Labor-Hours	Unit	Material	2019 Bare Costs Labor	Equipment	Total	Total Incl O&P
0010	**UNINTERRUPTIBLE POWER SUPPLY/CONDITIONER TRANSFORMERS**									
0100	Volt. regulating, isolating transf., w/invert. & 10 min. battery pack									
0110	Single-phase, 120 V, 0.35 kVA · R263353-80	1 Elec	2.29	3.493	Ea.	1,200	210		1,410	1,650
0120	0.5 kVA		2	4		1,250	240		1,490	1,725
0130	For additional 55 min. battery, add to 0.35 kVA		2.29	3.493		705	210		915	1,100
0140	Add to 0.5 kVA		1.14	7.018		740	420		1,160	1,450
0150	Single-phase, 120 V, 0.75 kVA		.80	10		1,575	600		2,175	2,650
0160	1.0 kVA		.80	10		2,250	600		2,850	3,400
0170	1.5 kVA	2 Elec	1.14	14.035		3,900	845		4,745	5,525
0180	2 kVA	"	.89	17.978		4,225	1,075		5,300	6,250
0190	3 kVA	R-3	.63	31.746		5,200	1,900	206	7,306	8,775
0200	5 kVA		.42	47.619		7,425	2,850	310	10,585	12,800
0210	7.5 kVA		.33	60.606		9,475	3,625	395	13,495	16,200
0220	10 kVA		.28	71.429		12,200	4,275	465	16,940	20,300
0230	15 kVA		.22	90.909		15,600	5,425	590	21,615	26,000
0240	3 phase, 120/208 V input 120/208 V output, 20 kVA, incl 17 min. battery		.21	95.238		28,700	5,700	620	35,020	40,800
0242	30 kVA, incl 11 min. battery		.20	100		32,100	5,975	650	38,725	44,900
0250	40 kVA, incl 15 min. battery		.20	100		42,900	5,975	650	49,525	57,000
0260	480 V input 277/480 V output, 60 kVA, incl 6 min. battery		.19	105		48,200	6,275	685	55,160	63,000
0262	80 kVA, incl 4 min. battery		.19	105		57,500	6,275	685	64,460	73,000
0400	For additional 34 min./15 min. battery, add to 40 kVA		.71	28.011		15,000	1,675	182	16,857	19,200
0600	For complex & special design systems to meet specific									
0610	requirements, obtain quote from vendor									

26 35 Power Filters and Conditioners

26 35 13 – Capacitors

26 35 13.10 Capacitors Indoor

		Crew	Daily Output	Labor-Hours	Unit	Material	2019 Bare Costs Labor	Equipment	Total	Total Incl O&P
0010	**CAPACITORS INDOOR**									
0020	240 volts, single & 3 phase, 0.5 kVAR	1 Elec	2.70	2.963	Ea.	575	178		753	900
0100	1.0 kVAR		2.70	2.963		695	178		873	1,025
0150	2.5 kVAR		2	4		785	240		1,025	1,225
0200	5.0 kVAR		1.80	4.444		800	267		1,067	1,275
0250	7.5 kVAR		1.60	5		1,500	300		1,800	2,100
0300	10 kVAR		1.50	5.333		1,625	320		1,945	2,250
0350	15 kVAR		1.30	6.154		2,025	370		2,395	2,775
0400	20 kVAR		1.10	7.273		2,450	435		2,885	3,350
0450	25 kVAR		1	8		2,875	480		3,355	3,875
1000	480 volts, single & 3 phase, 1 kVAR		2.70	2.963		525	178		703	840
1050	2 kVAR		2.70	2.963		605	178		783	930
1100	5 kVAR		2	4		760	240		1,000	1,200
1150	7.5 kVAR		2	4		820	240		1,060	1,275
1200	10 kVAR		2	4		1,075	240		1,315	1,550
1250	15 kVAR		2	4		1,325	240		1,565	1,825
1300	20 kVAR		1.60	5		1,400	300		1,700	2,000
1350	30 kVAR		1.50	5.333		1,650	320		1,970	2,300
1400	40 kVAR		1.20	6.667		2,125	400		2,525	2,925
1450	50 kVAR		1.10	7.273		2,475	435		2,910	3,375
2000	600 volts, single & 3 phase, 1 kVAR		2.70	2.963		540	178		718	860
2050	2 kVAR		2.70	2.963		620	178		798	950
2100	5 kVAR		2	4		760	240		1,000	1,200

26 35 13 – Capacitors

26 35 13.10 Capacitors Indoor	Crew	Daily Output	Labor-Hours	Unit	Material	2019 Bare Costs Labor	Equipment	Total	Total Incl O&P
2150	7.5 kVAR	1 Elec	2	4	Ea.	820	240		1,060	1,275
2200	10 kVAR		2	4		1,150	240		1,390	1,625
2250	15 kVAR		1.60	5		1,450	300		1,750	2,050
2300	20 kVAR		1.60	5		1,425	300		1,725	2,025
2350	25 kVAR		1.50	5.333		1,600	320		1,920	2,250
2400	35 kVAR		1.40	5.714		1,950	345		2,295	2,650
2450	50 kVAR	↓	1.30	6.154	↓	2,500	370		2,870	3,300

26 35 26 – Harmonic Filters

26 35 26.10 Computer Isolation Transformer

		Crew	Daily Output	Labor-Hours	Unit	Material	Labor	Equipment	Total	Total Incl O&P
0010	**COMPUTER ISOLATION TRANSFORMER**									
0100	Computer grade									
0110	Single-phase, 120/240 V, 0.5 kVA	1 Elec	4	2	Ea.	440	120		560	665
0120	1.0 kVA		2.67	2.996		630	180		810	960
0130	2.5 kVA		2	4		955	240		1,195	1,400
0140	5 kVA	↓	1.14	7.018	↓	1,100	420		1,520	1,825

26 35 26.20 Computer Regulator Transformer

		Crew	Daily Output	Labor-Hours	Unit	Material	Labor	Equipment	Total	Total Incl O&P
0010	**COMPUTER REGULATOR TRANSFORMER**									
0100	Ferro-resonant, constant voltage, variable transformer									
0110	Single-phase, 240 V, 0.5 kVA	1 Elec	2.67	2.996	Ea.	530	180		710	850
0120	1.0 kVA		2	4		725	240		965	1,150
0130	2.0 kVA		1	8		1,250	480		1,730	2,100
0210	Plug-in unit 120 V, 0.14 kVA		8	1		305	60		365	425
0220	0.25 kVA		8	1		355	60		415	485
0230	0.5 kVA		8	1		530	60		590	670
0240	1.0 kVA		5.33	1.501		725	90		815	935
0250	2.0 kVA	↓	4	2	↓	1,250	120		1,370	1,550

26 35 26.30 Power Conditioner Transformer

		Crew	Daily Output	Labor-Hours	Unit	Material	Labor	Equipment	Total	Total Incl O&P
0010	**POWER CONDITIONER TRANSFORMER**									
0100	Electronic solid state, buck-boost, transformer, w/tap switch									
0110	Single-phase, 115 V, 3.0 kVA, + or - 3% accuracy	2 Elec	1.60	10	Ea.	2,850	600		3,450	4,050
0120	208, 220, 230, or 240 V, 5.0 kVA, + or - 1.5% accuracy	3 Elec	1.60	15		3,700	900		4,600	5,425
0130	5.0 kVA, + or - 6% accuracy	2 Elec	1.14	14.035		3,300	845		4,145	4,875
0140	7.5 kVA, + or - 1.5% accuracy	3 Elec	1.50	16		4,675	960		5,635	6,575
0150	7.5 kVA, + or - 6% accuracy		1.60	15		3,900	900		4,800	5,650
0160	10.0 kVA, + or - 1.5% accuracy		1.33	18.045		6,250	1,075		7,325	8,500
0170	10.0 kVA, + or - 6% accuracy	↓	1.41	17.021	↓	5,325	1,025		6,350	7,375

26 35 26.40 Transient Voltage Suppressor Transformer

		Crew	Daily Output	Labor-Hours	Unit	Material	Labor	Equipment	Total	Total Incl O&P
0010	**TRANSIENT VOLTAGE SUPPRESSOR TRANSFORMER**									
0110	Single-phase, 120 V, 1.8 kVA	1 Elec	4	2	Ea.	1,500	120		1,620	1,825
0120	3.6 kVA		4	2		2,350	120		2,470	2,750
0130	7.2 kVA		3.20	2.500		2,800	150		2,950	3,325
0150	240 V, 3.6 kVA		4	2		3,225	120		3,345	3,725
0160	7.2 kVA		4	2		3,800	120		3,920	4,350
0170	14.4 kVA		3.20	2.500		4,925	150		5,075	5,625
0210	Plug-in unit, 120 V, 1.8 kVA	↓	8	1	↓	995	60		1,055	1,200

26 35 53 – Voltage Regulators

26 35 53.10 Automatic Voltage Regulators

		Crew	Daily Output	Labor-Hours	Unit	Material	Labor	Equipment	Total	Total Incl O&P
0010	**AUTOMATIC VOLTAGE REGULATORS**									
0100	Computer grade, solid state, variable transf. volt. regulator									
0110	Single-phase, 120 V, 8.6 kVA	2 Elec	1.33	12.030	Ea.	6,425	720		7,145	8,150
0120	17.3 kVA	↓	1.14	14.035	↓	7,600	845		8,445	9,600

26 35 53 – Voltage Regulators

26 35 53.10 Automatic Voltage Regulators

		Crew	Daily Output	Labor-Hours	Unit	Material	2019 Bare Costs Labor	Equipment	Total	Total Incl O&P
0130	208/240 V, 7.5/8.6 kVA	2 Elec	1.33	12.030	Ea.	6,425	720		7,145	8,150
0140	13.5/15.6 kVA		1.33	12.030		7,600	720		8,320	9,425
0150	27.0/31.2 kVA		1.14	14.035		9,575	845		10,420	11,800
0210	Two-phase, single control, 208/240 V, 15.0/17.3 kVA		1.14	14.035		7,600	845		8,445	9,600
0220	Individual phase control, 15.0/17.3 kVA	▼	1.14	14.035		7,600	845		8,445	9,600
0230	30.0/34.6 kVA	3 Elec	1.33	18.045		9,575	1,075		10,650	12,100
0310	Three-phase single control, 208/240 V, 26/30 kVA	2 Elec	1	16		7,600	960		8,560	9,775
0320	380/480 V, 24/30 kVA	"	1	16		7,600	960		8,560	9,775
0330	43/54 kVA	3 Elec	1.33	18.045		13,800	1,075		14,875	16,800
0340	Individual phase control, 208 V, 26 kVA	"	1.33	18.045		7,600	1,075		8,675	9,975
0350	52 kVA	R-3	.91	21.978		9,575	1,325	143	11,043	12,600
0360	340/480 V, 24/30 kVA	2 Elec	1	16		7,600	960		8,560	9,775
0370	43/54 kVA	"	1	16		9,575	960		10,535	11,900
0380	48/60 kVA	3 Elec	1.33	18.045		13,900	1,075		14,975	16,900
0390	86/108 kVA	R-3	.91	21.978	▼	15,400	1,325	143	16,868	19,100
0500	Standard grade, solid state, variable transformer volt. regulator									
0510	Single-phase, 115 V, 2.3 kVA	1 Elec	2.29	3.493	Ea.	2,250	210		2,460	2,800
0520	4.2 kVA		2	4		3,600	240		3,840	4,300
0530	6.6 kVA		1.14	7.018		4,400	420		4,820	5,475
0540	13.0 kVA	▼	1.14	7.018		7,600	420		8,020	8,975
0550	16.6 kVA	2 Elec	1.23	13.008		8,975	780		9,755	11,100
0610	230 V, 8.3 kVA		1.33	12.030		7,600	720		8,320	9,425
0620	21.4 kVA		1.23	13.008		8,975	780		9,755	11,100
0630	29.9 kVA		1.23	13.008		8,975	780		9,755	11,100
0710	460 V, 9.2 kVA		1.33	12.030		7,600	720		8,320	9,425
0720	20.7 kVA	▼	1.23	13.008		8,975	780		9,755	11,100
0810	Three-phase, 230 V, 13.1 kVA	3 Elec	1.41	17.021		7,600	1,025		8,625	9,875
0820	19.1 kVA		1.41	17.021		8,975	1,025		10,000	11,400
0830	25.1 kVA	▼	1.23	19.512		8,975	1,175		10,150	11,600
0840	57.8 kVA	R-3	.95	21.053		16,300	1,250	137	17,687	19,900
0850	74.9 kVA	"	.91	21.978		16,300	1,325	143	17,768	20,000
0910	460 V, 14.3 kVA	3 Elec	1.41	17.021		7,600	1,025		8,625	9,875
0920	19.1 kVA		1.41	17.021		8,975	1,025		10,000	11,400
0930	27.9 kVA	▼	1.23	19.512		8,975	1,175		10,150	11,600
0940	59.8 kVA	R-3	1	20		16,300	1,200	130	17,630	19,800
0950	79.7 kVA		.95	21.053		18,200	1,250	137	19,587	22,000
0960	118 kVA	▼	.95	21.053	▼	19,200	1,250	137	20,587	23,100
1000	Laboratory grade, precision, electronic voltage regulator									
1110	Single-phase, 115 V, 0.5 kVA	1 Elec	2.29	3.493	Ea.	1,575	210		1,785	2,050
1120	1.0 kVA		2	4		1,675	240		1,915	2,200
1130	3.0 kVA	▼	.80	10		2,350	600		2,950	3,475
1140	6.0 kVA	2 Elec	1.46	10.959		4,350	660		5,010	5,750
1150	10.0 kVA	3 Elec	1	24		5,650	1,450		7,100	8,375
1160	15.0 kVA	"	1.50	16		6,450	960		7,410	8,525
1210	230 V, 3.0 kVA	1 Elec	.80	10		2,650	600		3,250	3,825
1220	6.0 kVA	2 Elec	1.46	10.959		4,450	660		5,110	5,850
1230	10.0 kVA	3 Elec	1.71	14.035		5,900	845		6,745	7,750
1240	15.0 kVA	"	1.60	15		6,675	900		7,575	8,700

26 35 53.30 Transient Suppressor/Voltage Regulator

		Crew	Daily Output	Labor-Hours	Unit	Material	2019 Bare Costs Labor	Equipment	Total	Total Incl O&P
0010	**TRANSIENT SUPPRESSOR/VOLTAGE REGULATOR** (without isolation)									
0110	Single-phase, 115 V, 1.0 kVA	1 Elec	2.67	2.996	Ea.	1,175	180		1,355	1,575
0120	2.0 kVA	▼	2.29	3.493	▼	1,600	210		1,810	2,100

26 35 Power Filters and Conditioners

26 35 53 – Voltage Regulators

26 35 53.30 Transient Suppressor/Voltage Regulator	Crew	Daily Output	Labor-Hours	Unit	Material	2019 Bare Costs Labor	Equipment	Total	Total Incl O&P
0130	4.0 kVA	1 Elec	2.13	3.756	Ea.	1,975	226		2,201	2,500
0140	220 V, 1.0 kVA		2.67	2.996		1,175	180		1,355	1,575
0150	2.0 kVA		2.29	3.493		1,600	210		1,810	2,100
0160	4.0 kVA		2.13	3.756		2,075	226		2,301	2,625
0210	Plug-in unit, 120 V, 1.0 kVA		8	1		1,125	60		1,185	1,325
0220	2.0 kVA		8	1		1,575	60		1,635	1,850

26 36 Transfer Switches

26 36 13 – Manual Transfer Switches

26 36 13.10 Non-Automatic Transfer Switches

		Crew	Daily Output	Labor-Hours	Unit	Material	2019 Bare Costs Labor	Equipment	Total	Total Incl O&P
0010	**NON-AUTOMATIC TRANSFER SWITCHES** enclosed									
0100	Manual operated, 480 volt 3 pole, 30 amp	1 Elec	2.30	3.478	Ea.	1,025	209		1,234	1,425
0150	60 amp		1.90	4.211		2,025	253		2,278	2,625
0200	100 amp		1.30	6.154		3,625	370		3,995	4,550
0250	200 amp	2 Elec	2	8		4,475	480		4,955	5,650
0300	400 amp		1.60	10		8,625	600		9,225	10,400
0350	600 amp		1	16		5,375	960		6,335	7,325
1000	250 volt 3 pole, 30 amp	1 Elec	2.30	3.478		1,125	209		1,334	1,550
1100	60 amp		1.90	4.211		1,700	253		1,953	2,225
1150	100 amp		1.30	6.154		3,075	370		3,445	3,925
1200	200 amp	2 Elec	2	8		4,175	480		4,655	5,325
1300	600 amp	"	1	16		9,550	960		10,510	11,900
1500	Electrically operated, 480 volt 3 pole, 60 amp	1 Elec	1.90	4.211		1,475	253		1,728	2,000
1600	100 amp	"	1.30	6.154		1,475	370		1,845	2,175
1650	200 amp	2 Elec	2	8		2,450	480		2,930	3,425
1700	400 amp		1.60	10		3,425	600		4,025	4,675
1750	600 amp		1	16		4,925	960		5,885	6,850
2000	250 volt 3 pole, 30 amp	1 Elec	2.30	3.478		1,800	209		2,009	2,275
2050	60 amp	"	1.90	4.211		2,150	253		2,403	2,750
2150	200 amp	2 Elec	2	8		3,550	480		4,030	4,625
2200	400 amp		1.60	10		4,950	600		5,550	6,350
2250	600 amp		1	16		7,150	960		8,110	9,275
2500	NEMA 3R, 480 volt 3 pole, 60 amp	1 Elec	1.80	4.444		2,550	267		2,817	3,200
2550	100 amp	"	1.20	6.667		4,225	400		4,625	5,250
2600	200 amp	2 Elec	1.80	8.889		5,750	535		6,285	7,125
2650	400 amp	"	1.40	11.429		3,750	685		4,435	5,150
2800	NEMA 3R, 250 volt 3 pole solid state, 100 amp	1 Elec	1.20	6.667		3,925	400		4,325	4,900
2850	150 amp	2 Elec	1.80	8.889		5,200	535		5,735	6,525
2900	250 volt 2 pole solid state, 100 amp	1 Elec	1.30	6.154		3,800	370		4,170	4,750
2950	150 amp	2 Elec	2	8		5,075	480		5,555	6,300

26 36 23 – Automatic Transfer Switches

26 36 23.10 Automatic Transfer Switch Devices

		Crew	Daily Output	Labor-Hours	Unit	Material	2019 Bare Costs Labor	Equipment	Total	Total Incl O&P
0010	**AUTOMATIC TRANSFER SWITCH DEVICES** R263623-60									
0015	Switches, enclosed 120/240 volt, 2 pole, 30 amp	1 Elec	2.40	3.333	Ea.	1,900	200		2,100	2,375
0020	70 amp		2	4		1,900	240		2,140	2,425
0030	100 amp		1.35	5.926		1,900	355		2,255	2,600
0040	225 amp	2 Elec	2.10	7.619		2,800	460		3,260	3,750
0050	400 amp		1.70	9.412		4,250	565		4,815	5,525
0060	600 amp		1.06	15.094		9,375	905		10,280	11,700
0070	800 amp		.84	19.048		11,000	1,150		12,150	13,800

For customer support on your Electrical Costs with RSMeans data, call 800.448.8182.

291

26 36 Transfer Switches

26 36 23 – Automatic Transfer Switches

26 36 23.10 Automatic Transfer Switch Devices	Crew	Daily Output	Labor-Hours	Unit	Material	2019 Bare Costs Labor	Equipment	Total	Total Incl O&P
0100	Switches, enclosed 480 volt, 3 pole, 30 amp	1 Elec	2.30	3.478	Ea.	3,375	209		3,584	4,025
0200	60 amp		1.90	4.211		3,375	253		3,628	4,100
0300	100 amp		1.30	6.154		3,375	370		3,745	4,275
0400	150 amp	2 Elec	2.40	6.667		4,225	400		4,625	5,225
0500	225 amp		2	8		5,325	480		5,805	6,575
0600	260 amp		2	8		6,150	480		6,630	7,500
0700	400 amp		1.60	10		7,950	600		8,550	9,650
0800	600 amp		1	16		11,100	960		12,060	13,600
0900	800 amp		.80	20		13,000	1,200		14,200	16,100
1000	1,000 amp		.76	21.053		16,900	1,275		18,175	20,500
1100	1,200 amp		.70	22.857		23,100	1,375		24,475	27,500
1200	1,600 amp		.60	26.667		26,300	1,600		27,900	31,300
1300	2,000 amp		.50	32		29,400	1,925		31,325	35,300
1600	Accessories, time delay on engine starting					258			258	284
1700	Adjustable time delay on retransfer					258			258	284
1800	Shunt trips for customer connections					460			460	505
1900	Maintenance select switch					105			105	115
2000	Auxiliary contact when normal fails					122			122	134
2100	Pilot light-emergency					105			105	115
2200	Pilot light-normal					105			105	115
2300	Auxiliary contact-closed on normal					121			121	133
2400	Auxiliary contact-closed on emergency					121			121	133
2500	Emergency source sensing, frequency relay					535			535	585

26 41 Facility Lightning Protection

26 41 13 – Lightning Protection for Structures

26 41 13.13 Lightning Protection for Buildings

		Crew	Daily Output	Labor-Hours	Unit	Material	2019 Bare Costs Labor	Equipment	Total	Total Incl O&P
0010	**LIGHTNING PROTECTION FOR BUILDINGS**									
0200	Air terminals & base, copper									
0400	3/8" diameter x 10" (to 75' high)	1 Elec	8	1	Ea.	24	60		84	116
0500	1/2" diameter x 12" (over 75' high)		8	1		28	60		88	120
0520	1/2" diameter x 24"		7.30	1.096		37.50	66		103.50	139
0540	1/2" diameter x 60"		6.70	1.194		68.50	71.50		140	183
1000	Aluminum, 1/2" diameter x 12" (to 75' high)		8	1		19.50	60		79.50	111
1020	1/2" diameter x 24"		7.30	1.096		21.50	66		87.50	122
1040	1/2" diameter x 60"		6.70	1.194		28	71.50		99.50	138
1100	5/8" diameter x 12" (over 75' high)		8	1		20.50	60		80.50	112
2000	Cable, copper, 220 lb. per thousand ft. (to 75' high)		320	.025	L.F.	3.05	1.50		4.55	5.60
2100	375 lb. per thousand ft. (over 75' high)		230	.035		5.55	2.09		7.64	9.20
2500	Aluminum, 101 lb. per thousand ft. (to 75' high)		280	.029		.95	1.72		2.67	3.61
2600	199 lb. per thousand ft. (over 75' high)		240	.033		1.40	2		3.40	4.53
3000	Arrester, 175 volt AC to ground		8	1	Ea.	146	60		206	251
3100	650 volt AC to ground		6.70	1.194	"	148	71.50		219.50	269
4000	Air terminals, copper									
4010	3/8" x 10"	1 Elec	50	.160	Ea.	7.70	9.60		17.30	23
4020	1/2" x 12"		38	.211		11.40	12.65		24.05	31.50
4030	5/8" x 12"		33	.242		16.10	14.55		30.65	39.50
4040	1/2" x 24"		30	.267		21	16		37	47
4050	1/2" x 60"		19	.421		52.50	25.50		78	95
4060	Air terminals, aluminum									
4070	3/8" x 10"	1 Elec	50	.160	Ea.	3.12	9.60		12.72	17.80

26 41 13.13 Lightning Protection for Buildings	Crew	Daily Output	Labor-Hours	Unit	Material	2019 Bare Costs Labor	Equipment	Total	Total Incl O&P
4080	1/2" x 12"	1 Elec	38	.211	Ea.	3.41	12.65		16.06	22.50
4090	5/8" x 12"		33	.242		4.37	14.55		18.92	26.50
4100	1/2" x 24"		30	.267		5.30	16		21.30	30
4110	1/2" x 60"	↓	19	.421	↓	12.05	25.50		37.55	51
4200	Air terminal bases, copper									
4210	Adhesive bases, 1/2"	1 Elec	15	.533	Ea.	18.05	32		50.05	68
4215	Adhesive base, lightning air terminal					18.05			18.05	19.85
4220	Bolted bases	1 Elec	9	.889		16.45	53.50		69.95	97.50
4230	Hinged bases		7	1.143		27	68.50		95.50	132
4240	Side mounted bases		9	.889		16.30	53.50		69.80	97.50
4250	Offset point support		9	.889		27.50	53.50		81	110
4260	Concealed base assembly		5	1.600		83	96		179	235
4270	Tee connector base		13	.615		47	37		84	107
4280	Tripod base, 36" for 60" air terminal		7	1.143		39	68.50		107.50	145
4290	Intermediate base, 1/2"	↓	15	.533	↓	30	32		62	81
4300	Air terminal bases, aluminum									
4310	Adhesive bases, 1/2"	1 Elec	15	.533	Ea.	9.15	32		41.15	58
4320	Bolted bases		9	.889		16.10	53.50		69.60	97
4330	Hinged bases		7	1.143		19.05	68.50		87.55	123
4340	Side mounted bases		9	.889		11.85	53.50		65.35	92.50
4350	Offset point support		9	.889		13.30	53.50		66.80	94
4360	Concealed base assembly		5	1.600		42	96		138	190
4370	Tee connector base		13	.615		14.65	37		51.65	71
4380	Tripod base, 36" for 60" air terminal		7	1.143		29.50	68.50		98	135
4390	Intermediate base, 1/2"	↓	15	.533	↓	13.70	32		45.70	63
4400	Connector cable, copper									
4410	Through wall connector	1 Elec	7	1.143	Ea.	84	68.50		152.50	195
4420	Through roof connector		5	1.600		103	96		199	257
4430	Beam connector		7	1.143		27	68.50		95.50	132
4440	Double bolt connector		38	.211		11.70	12.65		24.35	31.50
4450	Bar, copper	↓	40	.200	↓	12.95	12		24.95	32
4500	Connector cable, aluminum									
4510	Through wall connector	1 Elec	7	1.143	Ea.	33.50	68.50		102	139
4520	Through roof connector		5	1.600		36.50	96		132.50	183
4530	Beam connector		7	1.143		18.30	68.50		86.80	122
4540	Double bolt connector	↓	38	.211	↓	6.20	12.65		18.85	25.50
4600	Bonding plates, copper									
4610	I-beam, 8" square	1 Elec	7	1.143	Ea.	21.50	68.50		90	126
4620	Purlin, 8" square		7	1.143		26.50	68.50		95	131
4630	Large heavy duty, 16" square		5	1.600		32	96		128	178
4640	Aluminum, I-beam, 8" square		7	1.143		16.60	68.50		85.10	120
4650	Purlin, 8" square		7	1.143		16.55	68.50		85.05	120
4660	Large heavy duty, 16" square		5	1.600		18	96		114	163
4700	Cable support, copper, loop fastener		38	.211		1.11	12.65		13.76	20
4710	Adhesive cable holder		38	.211		1.45	12.65		14.10	20.50
4720	Aluminum, loop fastener		38	.211		.72	12.65		13.37	19.65
4730	Adhesive cable holder		38	.211		.78	12.65		13.43	19.70
4740	Bonding strap, copper, 3/4" x 9-1/2"		15	.533		11.40	32		43.40	60.50
4750	Aluminum, 3/4" x 9-1/2"		15	.533		4.85	32		36.85	53.50
4760	Swivel adapter, copper, 1/2"		73	.110		17.15	6.60		23.75	28.50
4770	Aluminum, 1/2"	↓	73	.110	↓	9.15	6.60		15.75	19.85

26 51 Interior Lighting

26 51 13 – Interior Lighting Fixtures, Lamps, and Ballasts

26 51 13.10 Fixture Hangers

		Crew	Daily Output	Labor-Hours	Unit	Material	2019 Bare Costs Labor	Equipment	Total	Total Incl O&P
0010	**FIXTURE HANGERS**									
0220	Box hub cover	1 Elec	32	.250	Ea.	4.14	15		19.14	27
0240	Canopy		12	.667		9.75	40		49.75	70.50
0260	Connecting block		40	.200		2.79	12		14.79	21
0280	Cushion hanger		16	.500		23.50	30		53.50	71
0300	Box hanger, with mounting strap		8	1		10	60		70	101
0320	Connecting block		40	.200		2.39	12		14.39	20.50
0340	Flexible, 1/2" diameter, 4" long		12	.667		16.45	40		56.45	77.50
0360	6" long		12	.667		17.85	40		57.85	79
0380	8" long		12	.667		19.80	40		59.80	81.50
0400	10" long		12	.667		21	40		61	83
0420	12" long		12	.667		22	40		62	84
0440	15" long		12	.667		23.50	40		63.50	85
0460	18" long		12	.667		27.50	40		67.50	89.50
0480	3/4" diameter, 4" long		10	.800		21	48		69	94.50
0500	6" long		10	.800		23.50	48		71.50	97
0520	8" long		10	.800		23.50	48		71.50	97
0540	10" long		10	.800		26	48		74	100
0560	12" long		10	.800		28	48		76	102
0580	15" long		10	.800		31.50	48		79.50	106
0600	18" long		10	.800		34.50	48		82.50	110

26 51 13.40 Interior HID Fixtures

		Crew	Daily Output	Labor-Hours	Unit	Material	2019 Bare Costs Labor	Equipment	Total	Total Incl O&P
0010	**INTERIOR HID FIXTURES** Incl. lamps and mounting hardware									
0700	High pressure sodium, recessed, round, 70 watt	1 Elec	3.50	2.286	Ea.	550	137		687	805
0720	100 watt		3.50	2.286		620	137		757	890
0740	150 watt		3.20	2.500		715	150		865	1,025
0760	Square, 70 watt		3.60	2.222		600	133		733	860
0780	100 watt		3.60	2.222		625	133		758	885
0820	250 watt		3	2.667		870	160		1,030	1,200
0840	1,000 watt	2 Elec	4.80	3.333		1,675	200		1,875	2,125
0860	Surface, round, 70 watt	1 Elec	3	2.667		840	160		1,000	1,175
0880	100 watt		3	2.667		860	160		1,020	1,200
0900	150 watt		2.70	2.963		910	178		1,088	1,275
0920	Square, 70 watt		3	2.667		810	160		970	1,125
0940	100 watt		3	2.667		850	160		1,010	1,175
0980	250 watt		2.50	3.200		705	192		897	1,050
1040	Pendent, round, 70 watt		3	2.667		860	160		1,020	1,175
1060	100 watt		3	2.667		845	160		1,005	1,175
1080	150 watt		2.70	2.963		840	178		1,018	1,200
1100	Square, 70 watt		3	2.667		920	160		1,080	1,250
1120	100 watt		3	2.667		930	160		1,090	1,275
1140	150 watt		2.70	2.963		960	178		1,138	1,325
1160	250 watt		2.50	3.200		1,300	192		1,492	1,700
1180	400 watt		2.40	3.333		1,350	200		1,550	1,800
1220	Wall, round, 70 watt		3	2.667		765	160		925	1,075
1240	100 watt		3	2.667		775	160		935	1,100
1260	150 watt		2.70	2.963		810	178		988	1,150
1300	Square, 70 watt		3	2.667		830	160		990	1,150
1320	100 watt		3	2.667		850	160		1,010	1,175
1340	150 watt		2.70	2.963		895	178		1,073	1,250
1360	250 watt		2.50	3.200		920	192		1,112	1,275
1380	400 watt	2 Elec	4.80	3.333		1,125	200		1,325	1,550

26 51 Interior Lighting

26 51 13 – Interior Lighting Fixtures, Lamps, and Ballasts

26 51 13.40 Interior HID Fixtures		Crew	Daily Output	Labor-Hours	Unit	Material	2019 Bare Costs Labor	Equipment	Total	Total Incl O&P
1400	1,000 watt	2 Elec	3.60	4.444	Ea.	1,650	267		1,917	2,225
1500	Metal halide, recessed, round, 175 watt	1 Elec	3.40	2.353		435	141		576	690
1520	250 watt	"	3.20	2.500		545	150		695	825
1540	400 watt	2 Elec	5.80	2.759		825	166		991	1,150
1580	Square, 175 watt	1 Elec	3.40	2.353		440	141		581	695
1640	Surface, round, 175 watt		2.90	2.759		595	166		761	900
1660	250 watt	↓	2.70	2.963		1,025	178		1,203	1,400
1680	400 watt	2 Elec	4.80	3.333		1,150	200		1,350	1,550
1720	Square, 175 watt	1 Elec	2.90	2.759		645	166		811	955
1800	Pendent, round, 175 watt		2.90	2.759		805	166		971	1,125
1820	250 watt	↓	2.70	2.963		1,125	178		1,303	1,500
1840	400 watt	2 Elec	4.80	3.333		1,225	200		1,425	1,650
1880	Square, 175 watt	1 Elec	2.90	2.759		515	166		681	815
1900	250 watt	"	2.70	2.963		735	178		913	1,075
1920	400 watt	2 Elec	4.80	3.333		1,250	200		1,450	1,650
1980	Wall, round, 175 watt	1 Elec	2.90	2.759		965	166		1,131	1,325
2000	250 watt	"	2.70	2.963		970	178		1,148	1,350
2020	400 watt	2 Elec	4.80	3.333		1,025	200		1,225	1,425
2060	Square, 175 watt	1 Elec	2.90	2.759		635	166		801	945
2080	250 watt	"	2.70	2.963		725	178		903	1,050
2100	400 watt	2 Elec	4.80	3.333		1,025	200		1,225	1,425
2800	High pressure sodium, vaporproof, recessed, 70 watt	1 Elec	3.50	2.286		765	137		902	1,050
2820	100 watt		3.50	2.286		785	137		922	1,075
2840	150 watt		3.20	2.500		805	150		955	1,100
2900	Surface, 70 watt		3	2.667		865	160		1,025	1,200
2920	100 watt		3	2.667		905	160		1,065	1,225
2940	150 watt		2.70	2.963		940	178		1,118	1,300
3000	Pendent, 70 watt		3	2.667		855	160		1,015	1,175
3020	100 watt		3	2.667		880	160		1,040	1,200
3040	150 watt		2.70	2.963		935	178		1,113	1,300
3100	Wall, 70 watt		3	2.667		920	160		1,080	1,275
3120	100 watt		3	2.667		960	160		1,120	1,300
3140	150 watt		2.70	2.963		1,000	178		1,178	1,375
3200	Metal halide, vaporproof, recessed, 175 watt		3.40	2.353		670	141		811	945
3220	250 watt	↓	3.20	2.500		720	150		870	1,025
3240	400 watt	2 Elec	5.80	2.759		905	166		1,071	1,250
3260	1,000 watt	"	4.80	3.333		1,650	200		1,850	2,125
3280	Surface, 175 watt	1 Elec	2.90	2.759		1,100	166		1,266	1,475
3300	250 watt	"	2.70	2.963		1,025	178		1,203	1,400
3320	400 watt	2 Elec	4.80	3.333		1,275	200		1,475	1,700
3340	1,000 watt	"	3.60	4.444		1,875	267		2,142	2,450
3360	Pendent, 175 watt	1 Elec	2.90	2.759		1,075	166		1,241	1,425
3380	250 watt	"	2.70	2.963		1,050	178		1,228	1,450
3400	400 watt	2 Elec	4.80	3.333		1,275	200		1,475	1,700
3420	1,000 watt	"	3.60	4.444		2,050	267		2,317	2,650
3440	Wall, 175 watt	1 Elec	2.90	2.759		1,125	166		1,291	1,500
3460	250 watt	"	2.70	2.963		1,125	178		1,303	1,500
3480	400 watt	2 Elec	4.80	3.333		1,325	200		1,525	1,750
3500	1,000 watt	"	3.60	4.444	↓	2,125	267		2,392	2,725

For customer support on your Electrical Costs with RSMeans data, call 800.448.8182.

295

26 51 13.50 Interior Lighting Fixtures		Crew	Daily Output	Labor-Hours	Unit	Material	2019 Bare Costs Labor	Equipment	Total	Total Incl O&P
0010	**INTERIOR LIGHTING FIXTURES** Including lamps, mounting R265113-40									
0030	hardware and connections									
0100	Fluorescent, C.W. lamps, troffer, recess mounted in grid, RS									
0130	Grid ceiling mount									
0200	Acrylic lens, 1' W x 4' L, two 40 watt	1 Elec	5.70	1.404	Ea.	51.50	84.50		136	183
0210	1' W x 4' L, three 40 watt		5.40	1.481		58.50	89		147.50	197
0300	2' W x 2' L, two U40 watt		5.70	1.404		55	84.50		139.50	187
0400	2' W x 4' L, two 40 watt		5.30	1.509		54	90.50		144.50	195
0500	2' W x 4' L, three 40 watt		5	1.600		59.50	96		155.50	209
0600	2' W x 4' L, four 40 watt		4.70	1.702		62.50	102		164.50	222
0700	4' W x 4' L, four 40 watt	2 Elec	6.40	2.500		283	150		433	535
0800	4' W x 4' L, six 40 watt		6.20	2.581		298	155		453	560
0900	4' W x 4' L, eight 40 watt		5.80	2.759		335	166		501	615
0910	Acrylic lens, 1' W x 4' L, two 32 watt T8 G	1 Elec	5.70	1.404		66.50	84.50		151	199
0930	2' W x 2' L, two U32 watt T8 G		5.70	1.404		96	84.50		180.50	232
0940	2' W x 4' L, two 32 watt T8 G		5.30	1.509		76	90.50		166.50	219
0950	2' W x 4' L, three 32 watt T8 G		5	1.600		70	96		166	220
0960	2' W x 4' L, four 32 watt T8 G		4.70	1.702		74.50	102		176.50	235
1000	Surface mounted, RS									
1030	Acrylic lens with hinged & latched door frame									
1100	1' W x 4' L, two 40 watt	1 Elec	7	1.143	Ea.	66	68.50		134.50	175
1110	1' W x 4' L, three 40 watt		6.70	1.194		68.50	71.50		140	182
1200	2' W x 2' L, two U40 watt		7	1.143		71	68.50		139.50	180
1300	2' W x 4' L, two 40 watt		6.20	1.290		80.50	77.50		158	205
1400	2' W x 4' L, three 40 watt		5.70	1.404		83	84.50		167.50	218
1500	2' W x 4' L, four 40 watt		5.30	1.509		85	90.50		175.50	229
1501	2' W x 4' L, six 40 watt T8		5.20	1.538		85	92.50		177.50	232
1600	4' W x 4' L, four 40 watt	2 Elec	7.20	2.222		420	133		553	660
1700	4' W x 4' L, six 40 watt		6.60	2.424		455	146		601	715
1800	4' W x 4' L, eight 40 watt		6.20	2.581		475	155		630	750
1900	2' W x 8' L, four 40 watt		6.40	2.500		167	150		317	410
2000	2' W x 8' L, eight 40 watt		6.20	2.581		180	155		335	430
2010	Acrylic wrap around lens									
2020	6" W x 4' L, one 40 watt	1 Elec	8	1	Ea.	68.50	60		128.50	165
2030	6" W x 8' L, two 40 watt	2 Elec	8	2		75	120		195	262
2040	11" W x 4' L, two 40 watt	1 Elec	7	1.143		45.50	68.50		114	153
2050	11" W x 8' L, four 40 watt	2 Elec	6.60	2.424		75.50	146		221.50	300
2060	16" W x 4' L, four 40 watt	1 Elec	5.30	1.509		75.50	90.50		166	218
2070	16" W x 8' L, eight 40 watt	2 Elec	6.40	2.500		165	150		315	405
2080	2' W x 2' L, two U40 watt	1 Elec	7	1.143		92.50	68.50		161	204
2100	Strip fixture									
2130	Surface mounted									
2200	4' long, one 40 watt, RS	1 Elec	8.50	.941	Ea.	31	56.50		87.50	119
2300	4' long, two 40 watt, RS		8	1		43.50	60		103.50	138
2310	4' long, two 32 watt T8, RS G		8	1		65.50	60		125.50	162
2400	4' long, one 40 watt, SL		8	1		52	60		112	147
2500	4' long, two 40 watt, SL		7	1.143		71	68.50		139.50	180
2580	8' long, one 60 watt T8, SL G	2 Elec	13.40	1.194		112	71.50		183.50	230
2590	8' long, two 60 watt T8, SL G		12.40	1.290		92	77.50		169.50	217
2600	8' long, one 75 watt, SL		13.40	1.194		55	71.50		126.50	168
2700	8' long, two 75 watt, SL		12.40	1.290		66.50	77.50		144	189
2800	4' long, two 60 watt, HO	1 Elec	6.70	1.194		107	71.50		178.50	225

26 51 13 – Interior Lighting Fixtures, Lamps, and Ballasts

26 51 13.50 Interior Lighting Fixtures		Crew	Daily Output	Labor-Hours	Unit	Material	2019 Bare Costs Labor	Equipment	Total	Total Incl O&P
2810	4' long, two 54 watt, T5HO [G]	1 Elec	6.70	1.194	Ea.	175	71.50		246.50	299
2900	8' long, two 110 watt, HO	2 Elec	10.60	1.509		110	90.50		200.50	256
2910	4' long, two 115 watt, VHO	1 Elec	6.50	1.231		144	74		218	268
2920	8' long, two 215 watt, VHO	2 Elec	10.40	1.538		156	92.50		248.50	310
2950	High bay pendent mounted, 16" W x 4' L, four 54 watt, T5HO [G]		8.90	1.798		277	108		385	465
2952	2' W x 4' L, six 54 watt, T5HO [G]		8.50	1.882		305	113		418	505
2954	2' W x 4' L, six 32 watt, T8 [G]		8.50	1.882		179	113		292	365
3000	Strip, pendent mounted, industrial, white porcelain enamel									
3100	4' long, two 40 watt, RS	1 Elec	5.70	1.404	Ea.	52.50	84.50		137	184
3110	4' long, two 32 watt T8, RS [G]		5.70	1.404		73	84.50		157.50	207
3200	4' long, two 60 watt, HO		5	1.600		83	96		179	234
3290	8' long, two 60 watt T8, SL [G]	2 Elec	8.80	1.818		118	109		227	292
3300	8' long, two 75 watt, SL		8.80	1.818		98.50	109		207.50	271
3400	8' long, two 110 watt, HO		8	2		126	120		246	315
3410	Acrylic finish, 4' long, two 40 watt, RS	1 Elec	5.70	1.404		86.50	84.50		171	221
3420	4' long, two 60 watt, HO		5	1.600		161	96		257	320
3430	4' long, two 115 watt, VHO		4.80	1.667		208	100		308	380
3440	8' long, two 75 watt, SL	2 Elec	8.80	1.818		170	109		279	350
3450	8' long, two 110 watt, HO		8	2		193	120		313	390
3460	8' long, two 215 watt, VHO		7.60	2.105		273	126		399	490
3470	Troffer, air handling, 2' W x 4' L with four 32 watt T8 [G]	1 Elec	4	2		117	120		237	305
3480	2' W x 2' L with two U32 watt T8 [G]		5.50	1.455		112	87.50		199.50	253
3490	Air connector insulated, 5" diameter		20	.400		67.50	24		91.50	111
3500	6" diameter		20	.400		69	24		93	112
3502	Troffer, direct/indirect, 2' W x 4' L with two 32 W T8 [G]		5.30	1.509		284	90.50		374.50	445
3510	Troffer parabolic lay-in, 1' W x 4' L with one 32 W T8 [G]		5.70	1.404		118	84.50		202.50	256
3520	1' W x 4' L with two 32 W T8 [G]		5.30	1.509		140	90.50		230.50	289
3525	2' W x 2' L with two U32 W T8 [G]		5.70	1.404		119	84.50		203.50	257
3530	2' W x 4' L with three 32 W T8 [G]		5	1.600		131	96		227	287
3531	Intr fxtr, fluor, troffer prismatic lay-in, 2' W x 4'l w/three 32 W T8		5	1.600		153	96		249	310
3535	Downlight, recess mounted [G]		8	1		148	60		208	252
3540	Wall wash reflector, recess mounted [G]		8	1		110	60		170	211
3550	Direct/indirect, 4' long, stl., pendent mtd. [G]		5	1.600		162	96		258	320
3560	4' long, alum., pendent mtd. [G]		5	1.600		335	96		431	515
3565	Prefabricated cove, 4' long, stl. continuous row [G]		5	1.600		203	96		299	365
3570	4' long, alum. continuous row [G]		5	1.600		345	96		441	525
3580	Wet location, recess mounted, 2' W x 4' L with two 32 watt T8 [G]		5.30	1.509		238	90.50		328.50	395
3590	Pendent mounted, 2' W x 4' L with two 32 watt T8 [G]		5.70	1.404		380	84.50		464.50	545
4000	Induction lamp, integral ballast, ceiling mounted									
4110	High bay, aluminum reflector, 160 watt	1 Elec	3.20	2.500	Ea.	860	150		1,010	1,175
4120	320 watt	"	3	2.667		1,550	160		1,710	1,950
4130	480 watt	2 Elec	5.80	2.759		2,350	166		2,516	2,825
4150	Low bay, aluminum reflector, 250 watt	1 Elec	3.20	2.500		930	150		1,080	1,250
4170	Garage, aluminum reflector, 80 watt		3.60	2.222		840	133		973	1,125
4180	Vandalproof, aluminum reflector, 100 watt		3.20	2.500		670	150		820	960
4220	Metal halide, integral ballast, ceiling, recess mounted									
4230	prismatic glass lens, floating door									
4240	2' W x 2' L, 250 watt	1 Elec	3.20	2.500	Ea.	263	150		413	515
4250	2' W x 2' L, 400 watt	2 Elec	5.80	2.759		365	166		531	650
4260	Surface mounted, 2' W x 2' L, 250 watt	1 Elec	2.70	2.963		345	178		523	640
4270	400 watt	2 Elec	4.80	3.333		405	200		605	745
4280	High bay, aluminum reflector,									
4290	Single unit, 400 watt	2 Elec	4.60	3.478	Ea.	410	209		619	765

26 51 13 – Interior Lighting Fixtures, Lamps, and Ballasts

26 51 13.50 Interior Lighting Fixtures	Crew	Daily Output	Labor-Hours	Unit	Material	2019 Bare Costs Labor	Equipment	Total	Total Incl O&P
4300	Single unit, 1,000 watt	2 Elec	4	4	Ea.	595	240		835	1,000
4310	Twin unit, 400 watt		3.20	5		825	300		1,125	1,350
4320	Low bay, aluminum reflector, 250W DX lamp	1 Elec	3.20	2.500		365	150		515	625
4330	400 watt lamp	2 Elec	5	3.200		525	192		717	865
4340	High pressure sodium integral ballast ceiling, recess mounted									
4350	prismatic glass lens, floating door									
4360	2' W x 2' L, 150 watt lamp	1 Elec	3.20	2.500	Ea.	385	150		535	650
4370	2' W x 2' L, 400 watt lamp	2 Elec	5.80	2.759		465	166		631	755
4380	Surface mounted, 2' W x 2' L, 150 watt lamp	1 Elec	2.70	2.963		470	178		648	785
4390	400 watt lamp	2 Elec	4.80	3.333		530	200		730	880
4400	High bay, aluminum reflector,									
4410	Single unit, 400 watt lamp	2 Elec	4.60	3.478	Ea.	380	209		589	730
4430	Single unit, 1,000 watt lamp	"	4	4		550	240		790	960
4440	Low bay, aluminum reflector, 150 watt lamp	1 Elec	3.20	2.500		330	150		480	585
4445	High bay H.I.D. quartz restrike	"	16	.500		164	30		194	225
4450	Incandescent, high hat can, round alzak reflector, prewired									
4470	100 watt	1 Elec	8	1	Ea.	68	60		128	164
4480	150 watt		8	1		104	60		164	205
4500	300 watt		6.70	1.194		241	71.50		312.50	370
4520	Round with reflector and baffles, 150 watt		8	1		51.50	60		111.50	146
4540	Round with concentric louver, 150 watt PAR		8	1		78.50	60		138.50	176
4600	Square glass lens with metal trim, prewired									
4630	100 watt	1 Elec	6.70	1.194	Ea.	55.50	71.50		127	168
4680	150 watt		6.70	1.194		98	71.50		169.50	215
4700	200 watt		6.70	1.194		98	71.50		169.50	215
4800	300 watt		5.70	1.404		145	84.50		229.50	286
4810	500 watt		5	1.600		286	96		382	460
4900	Ceiling/wall, surface mounted, metal cylinder, 75 watt		10	.800		52.50	48		100.50	130
4920	150 watt		10	.800		81	48		129	161
4930	300 watt		8	1		168	60		228	274
5000	500 watt		6.70	1.194		365	71.50		436.50	505
5010	Square, 100 watt		8	1		113	60		173	214
5020	150 watt		8	1		124	60		184	227
5030	300 watt		7	1.143		340	68.50		408.50	470
5040	500 watt		6	1.333		345	80		425	500
5200	Ceiling, surface mounted, opal glass drum									
5300	8", one 60 watt lamp	1 Elec	10	.800	Ea.	66.50	48		114.50	145
5400	10", two 60 watt lamps		8	1		74	60		134	171
5500	12", four 60 watt lamps		6.70	1.194		104	71.50		175.50	222
5510	Pendent, round, 100 watt		8	1		113	60		173	214
5520	150 watt		8	1		123	60		183	225
5530	300 watt		6.70	1.194		171	71.50		242.50	295
5540	500 watt		5.50	1.455		320	87.50		407.50	485
5550	Square, 100 watt		6.70	1.194		152	71.50		223.50	274
5560	150 watt		6.70	1.194		158	71.50		229.50	280
5570	300 watt		5.70	1.404		230	84.50		314.50	380
5580	500 watt		5	1.600		310	96		406	490
5600	Wall, round, 100 watt		8	1		65.50	60		125.50	162
5620	300 watt		8	1		142	60		202	246
5630	500 watt		6.70	1.194		380	71.50		451.50	520
5640	Square, 100 watt		8	1		103	60		163	204
5650	150 watt		8	1		105	60		165	206
5660	300 watt		7	1.143		166	68.50		234.50	284

26 51 13.50 Interior Lighting Fixtures		Crew	Daily Output	Labor-Hours	Unit	Material	2019 Bare Costs Labor	Equipment	Total	Total Incl O&P
5670	500 watt	1 Elec	6	1.333	Ea.	279	80		359	425
6010	Vapor tight, incandescent, ceiling mounted, 200 watt		6.20	1.290		76.50	77.50		154	201
6020	Recessed, 200 watt		6.70	1.194		124	71.50		195.50	244
6030	Pendent, 200 watt		6.70	1.194		76.50	71.50		148	192
6040	Wall, 200 watt		8	1		80.50	60		140.50	178
6100	Fluorescent, surface mounted, 2 lamps, 4' L, RS, 40 watt		3.20	2.500		116	150		266	350
6110	Industrial, 2 lamps, 4' L in tandem, 430 MA		2.20	3.636		213	218		431	560
6130	2 lamps, 4' L, 800 MA		1.90	4.211		181	253		434	575
6160	Pendent, indust, 2 lamps, 4' L in tandem, 430 MA		1.90	4.211		247	253		500	645
6170	2 lamps, 4' L, 430 MA		2.30	3.478		166	209		375	490
6180	2 lamps, 4' L, 800 MA		1.70	4.706		209	283		492	650
6850	Vandalproof, surface mounted, fluorescent, two 32 watt T8 [G]		3.20	2.500		253	150		403	500
6860	Incandescent, one 150 watt		8	1		90	60		150	189
6900	Mirror light, fluorescent, RS, acrylic enclosure, two 40 watt		8	1		113	60		173	214
6910	One 40 watt		8	1		97	60		157	197
6920	One 20 watt		12	.667		83.50	40		123.50	151
7000	Low bay, aluminum reflector, 70 watt, high pressure sodium		4	2		276	120		396	485
7010	250 watt		3.20	2.500		365	150		515	630
7020	400 watt	2 Elec	5	3.200		385	192		577	705
7500	Ballast replacement, by weight of ballast, to 15' high									
7520	Indoor fluorescent, less than 2 lb.	1 Elec	10	.800	Ea.	24	48		72	98
7540	Two 40W, watt reducer, 2 to 5 lb.		9.40	.851		72	51		123	156
7560	Two F96 slimline, over 5 lb.		8	1		112	60		172	213
7580	Vaportite ballast, less than 2 lb.		9.40	.851		24	51		75	103
7600	2 lb. to 5 lb.		8.90	.899		72	54		126	160
7620	Over 5 lb.		7.60	1.053		112	63		175	218
7630	Electronic ballast for two tubes		8	1		43.50	60		103.50	137
7640	Dimmable ballast one-lamp [G]		8	1		108	60		168	209
7650	Dimmable ballast two-lamp [G]		7.60	1.053		113	63		176	219
7660	Dimmable ballast three-lamp		6.90	1.159		130	69.50		199.50	247
7690	Emergency ballast (factory installed in fixture)					181			181	199
7990	Decorator									
8000	Pendent RLM in colors, shallow dome, 12" diam., 100 watt	1 Elec	8	1	Ea.	80.50	60		140.50	178
8010	Regular dome, 12" diam., 100 watt		8	1		83	60		143	181
8020	16" diam., 200 watt		7	1.143		84.50	68.50		153	195
8030	18" diam., 300 watt		6	1.333		93	80		173	222
8100	Picture framing light		16	.500		96	30		126	151
8150	Miniature low voltage, recessed, pinhole		8	1		129	60		189	232
8160	Star		8	1		133	60		193	237
8170	Adjustable cone		8	1		166	60		226	273
8180	Eyeball		8	1		106	60		166	206
8190	Cone		8	1		130	60		190	232
8200	Coilex baffle		8	1		123	60		183	225
8210	Surface mounted, adjustable cylinder		8	1		127	60		187	230
8250	Chandeliers, incandescent									
8260	24" diam. x 42" high, 6 light candle	1 Elec	6	1.333	Ea.	430	80		510	590
8270	24" diam. x 42" high, 6 light candle w/glass shade		6	1.333		430	80		510	595
8280	17" diam. x 12" high, 8 light w/glass panels		8	1		264	60		324	380
8300	27" diam. x 29" high, 10 light bohemian lead crystal		4	2		450	120		570	670
8310	21" diam. x 9" high, 6 light sculptured ice crystal		8	1		420	60		480	550
8500	Accent lights, on floor or edge, 0.5 W low volt incandescent									
8520	incl. transformer & fastenings, based on 100' lengths									
8550	Lights in clear tubing, 12" OC	1 Elec	230	.035	L.F.	8.45	2.09		10.54	12.40

26 51 13 – Interior Lighting Fixtures, Lamps, and Ballasts

26 51 13.50 Interior Lighting Fixtures		Crew	Daily Output	Labor-Hours	Unit	Material	2019 Bare Costs Labor	Equipment	Total	Total Incl O&P
8560	6" OC	1 Elec	160	.050	L.F.	11.05	3		14.05	16.65
8570	4" OC		130	.062		16.90	3.70		20.60	24
8580	3" OC		125	.064		18.75	3.84		22.59	26.50
8590	2" OC		100	.080		27.50	4.80		32.30	37
8600	Carpet, lights both sides 6" OC, in alum. extrusion		270	.030		25.50	1.78		27.28	30.50
8610	In bronze extrusion		270	.030		29	1.78		30.78	34.50
8620	Carpet-bare floor, lights 18" OC, in alum. extrusion		270	.030		20.50	1.78		22.28	25
8630	In bronze extrusion		270	.030		24	1.78		25.78	29
8640	Carpet edge-wall, lights 6" OC in alum. extrusion		270	.030		25.50	1.78		27.28	30.50
8650	In bronze extrusion		270	.030		29	1.78		30.78	34.50
8660	Bare floor, lights 18" OC, in alum. extrusion		300	.027		20.50	1.60		22.10	25
8670	In bronze extrusion		300	.027		24	1.60		25.60	29
8680	Bare floor conduit, alum. extrusion		300	.027		6.75	1.60		8.35	9.85
8690	In bronze extrusion		300	.027	▼	13.55	1.60		15.15	17.30
8700	Step edge to 36", lights 6" OC, in alum. extrusion		100	.080	Ea.	68	4.80		72.80	82
8710	In bronze extrusion		100	.080		71	4.80		75.80	85
8720	Step edge to 54", lights 6" OC, in alum. extrusion		100	.080		102	4.80		106.80	119
8730	In bronze extrusion		100	.080		108	4.80		112.80	126
8740	Step edge to 72", lights 6" OC, in alum. extrusion		100	.080		136	4.80		140.80	157
8750	In bronze extrusion		100	.080		149	4.80		153.80	170
8760	Connector, male		32	.250		2.52	15		17.52	25.50
8770	Female with pigtail		32	.250		5.30	15		20.30	28.50
8780	Clamps		400	.020		.49	1.20		1.69	2.33
8790	Transformers, 50 watt		8	1		92.50	60		152.50	192
8800	250 watt		4	2		294	120		414	505
8810	1,000 watt	▼	2.70	2.963	▼	335	178		513	630

26 51 13.55 Interior LED Fixtures

				Crew	Daily Output	Labor-Hours	Unit	Material	2019 Bare Costs Labor	Equipment	Total	Total Incl O&P
0010	**INTERIOR LED FIXTURES** Incl. lamps and mounting hardware											
0100	Downlight, recess mounted, 7.5" diameter, 25 watt		G	1 Elec	8	1	Ea.	335	60		395	460
0120	10" diameter, 36 watt		G		8	1		360	60		420	485
0160	cylinder, 10 watts		G		8	1		104	60		164	204
0180	20 watts		G		8	1		129	60		189	232
0900	Interior LED fixts, troffer, recess mounted, 2' x 2', 3,500K				8.50	.941		126	56.50		182.50	224
0910	2' x 2', 4,000K				8.50	.941		130	56.50		186.50	228
1000	Troffer, recess mounted, 2' x 4', 3,200 lumens		G		5.30	1.509		138	90.50		228.50	286
1010	4,800 lumens		G		5	1.600		178	96		274	340
1020	6,400 lumens		G		4.70	1.702		190	102		292	360
1100	Troffer retrofit lamp, 38 watt		G		21	.381		69.50	23		92.50	111
1110	60 watt		G		20	.400		142	24		166	192
1120	100 watt		G		18	.444		206	26.50		232.50	267
1200	Troffer, volumetric recess mounted, 2' x 2'		G		5.70	1.404		320	84.50		404.50	475
2000	Strip, surface mounted, one light bar 4' long, 3,500 K		G		8.50	.941		310	56.50		366.50	425
2010	5,000 K		G		8	1		276	60		336	395
2020	Two light bar 4' long, 5,000 K		G		7	1.143		430	68.50		498.50	575
3000	Linear, suspended mounted, one light bar 4' long, 37 watt		G	▼	6.70	1.194		168	71.50		239.50	291
3010	One light bar 8' long, 74 watt		G	2 Elec	12.20	1.311		310	79		389	460
3020	Two light bar 4' long, 74 watt		G	1 Elec	5.70	1.404		335	84.50		419.50	490
3030	Two light bar 8' long, 148 watt		G	2 Elec	8.80	1.818		380	109		489	580
4000	High bay, surface mounted, round, 150 watts		G		5.41	2.959		425	178		603	735
4010	2 bars, 164 watts		G		5.41	2.959		385	178		563	690
4020	3 bars, 246 watts		G		5.01	3.197		520	192		712	855
4030	4 bars, 328 watts		G	▼	4.60	3.478		750	209		959	1,125

26 51 Interior Lighting

26 51 13 – Interior Lighting Fixtures, Lamps, and Ballasts

26 51 13.55 Interior LED Fixtures

		Crew	Daily Output	Labor-Hours	Unit	Material	2019 Bare Costs Labor	Equipment	Total	Total Incl O&P
4040	5 bars, 410 watts	G 3 Elec	4.20	5.716	Ea.	840	345		1,185	1,425
4050	6 bars, 492 watts	G	3.80	6.324		940	380		1,320	1,600
4060	7 bars, 574 watts	G	3.39	7.075		1,025	425		1,450	1,750
4070	8 bars, 656 watts	G	2.99	8.029		1,050	480		1,530	1,875
5000	Track, lighthead, 6 watt	G 1 Elec	32	.250		55.50	15		70.50	83.50
5010	9 watt	G "	32	.250		63	15		78	92
6000	Garage, surface mounted, 103 watts	G 2 Elec	6.50	2.462		990	148		1,138	1,325
6100	pendent mounted, 80 watts	G	6.50	2.462		720	148		868	1,000
6200	95 watts	G	6.50	2.462		855	148		1,003	1,150
6300	125 watts	G	6.50	2.462		905	148		1,053	1,225
7000	Downlight, recess mtd., low profile, 4" diam., 9W, 3,000K	1 Elec	8.50	.941		24.50	56.50		81	112
7010	4,000K		8.50	.941		23.50	56.50		80	110
7020	5,000K		8.50	.941		24	56.50		80.50	111

26 51 13.70 Residential Fixtures

		Crew	Daily Output	Labor-Hours	Unit	Material	2019 Bare Costs Labor	Equipment	Total	Total Incl O&P
0010	**RESIDENTIAL FIXTURES**									
0400	Fluorescent, interior, surface, circline, 32 watt & 40 watt	1 Elec	20	.400	Ea.	165	24		189	218
0500	2' x 2', two U-tube 32 watt T8		8	1		130	60		190	232
0700	Shallow under cabinet, two 20 watt		16	.500		70.50	30		100.50	123
0900	Wall mounted, 4' L, two 32 watt T8, with baffle		10	.800		164	48		212	252
2000	Incandescent, exterior lantern, wall mounted, 60 watt		16	.500		61	30		91	112
2100	Post light, 150 W, with 7' post		4	2		268	120		388	475
2500	Lamp holder, weatherproof with 150 W PAR		16	.500		35	30		65	83.50
2550	With reflector and guard		12	.667		69.50	40		109.50	136
2600	Interior pendent, globe with shade, 150 W		20	.400		204	24		228	261

26 51 13.90 Ballast, Replacement HID

		Crew	Daily Output	Labor-Hours	Unit	Material	2019 Bare Costs Labor	Equipment	Total	Total Incl O&P
0010	**BALLAST, REPLACEMENT HID**									
7510	Multi-tap 120/208/240/277 V									
7550	High pressure sodium, 70 watt	1 Elec	10	.800	Ea.	176	48		224	266
7560	100 watt		9.40	.851		126	51		177	215
7570	150 watt		9	.889		198	53.50		251.50	298
7580	250 watt		8.50	.941		294	56.50		350.50	410
7590	400 watt		7	1.143		260	68.50		328.50	390
7600	1,000 watt		6	1.333		279	80		359	425
7610	Metal halide, 175 watt		8	1		92	60		152	191
7620	250 watt		8	1		121	60		181	223
7630	400 watt		7	1.143		150	68.50		218.50	267
7640	1,000 watt		6	1.333		245	80		325	390
7650	1,500 watt		5	1.600		293	96		389	465

26 51 19 – LED Interior Lighting

26 51 19.10 LED Interior Lighting

		Crew	Daily Output	Labor-Hours	Unit	Material	2019 Bare Costs Labor	Equipment	Total	Total Incl O&P
0010	**LED INTERIOR LIGHTING**									
7000	Interior LED fixts, tape rope kit, 120V, 3' strip, 4W, w/5' cord, 3,000K	1 Elec	24	.333	Ea.	8.45	20		28.45	39.50
7010	5,000K		24	.333		12.10	20		32.10	43.50
7020	Interior LED fixts, tape rope kit, 120V, 6' strip, 8W, w/5' cord, 3,000K		23	.348		15.40	21		36.40	48
7030	5,000K		23	.348		21	21		42	54
7040	Interior LED fixts, tape rope kit, 120V, 13' strip, 15W, w/5' cord, 3,000K		22	.364		28.50	22		50.50	64
7050	5,000K		22	.364		28.50	22		50.50	64
7060	Interior LED fixts, tape rope kit, 120V, 20' strip, 22W, w/5' cord, 3,000K		21	.381		41.50	23		64.50	79.50
7070	5,000K		21	.381		41.50	23		64.50	79.50
7080	Interior LED fixts, tape rope kit, 120V, 33' strip, 37W, w/5' cord, 3,000K		20	.400		61	24		85	103
7090	5,000K		20	.400		61	24		85	103
7100	Interior LED fixts, tape rope kit, 120V, 50' strip, 55W, w/5' cord, 3,000K		19	.421		92	25.50		117.50	139

For customer support on your Electrical Costs with RSMeans data, call 800.448.8182.

301

26 51 Interior Lighting

26 51 19 – LED Interior Lighting

	26 51 19.10 LED Interior Lighting	Crew	Daily Output	Labor-Hours	Unit	Material	2019 Bare Costs Labor	Equipment	Total	Total Incl O&P
7110	5,000K	1 Elec	19	.421	Ea.	132	25.50		157.50	183
7120	Interior LED tape light, starstrand, dimmable, 12V, super star, 2", 3,000K		25	.320		11.30	19.20		30.50	41
7130	Ultra star, 2", 3,000K		25	.320		14.15	19.20		33.35	44
7140	Rrainbow star, 2", 3,000K		25	.320		11.30	19.20		30.50	41
7150	Interior LED tape light, starstrand, dimmable, 24V, elite star, 4", 35K		25	.320		9.90	19.20		29.10	39.50
7160	12", 35K		25	.320		22.50	19.20		41.70	53.50
7170	60", 35K		23	.348		99	21		120	140
7180	120", 35K		21	.381		139	23		162	187
7190	240", 35K		18	.444		274	26.50		300.50	340
7200	Interior LED tape lighting channel , 36", AL flex, starstrand, elite star		48	.167		21	10		31	38

26 52 Safety Lighting

26 52 13 – Emergency and Exit Lighting

26 52 13.10 Emergency Lighting and Battery Units

		Crew	Daily Output	Labor-Hours	Unit	Material	2019 Bare Costs Labor	Equipment	Total	Total Incl O&P
0010	**EMERGENCY LIGHTING AND BATTERY UNITS**									
0300	Emergency light units, battery operated									
0350	Twin sealed beam light, 25 W, 6 V each									
0500	Lead battery operated	1 Elec	4	2	Ea.	144	120		264	335
0700	Nickel cadmium battery operated		4	2		345	120		465	555
0780	Additional remote mount, sealed beam, 25 W 6 V		26.70	.300		29.50	18		47.50	59.50
0790	Twin sealed beam light, 25 W 6 V each		26.70	.300		62.50	18		80.50	95.50
0900	Self-contained fluorescent lamp pack		10	.800		191	48		239	282

26 52 13.16 Exit Signs

		Crew	Daily Output	Labor-Hours	Unit	Material	2019 Bare Costs Labor	Equipment	Total	Total Incl O&P
0010	**EXIT SIGNS**									
0080	Exit light ceiling or wall mount, incandescent, single face	1 Elec	8	1	Ea.	72.50	60		132.50	170
0100	Double face		6.70	1.194		51	71.50		122.50	163
0120	Explosion proof		3.80	2.105		560	126		686	810
0150	Fluorescent, single face		8	1		84	60		144	182
0160	Double face		6.70	1.194		76.50	71.50		148	191
0200	LED standard, single face [G]		8	1		51	60		111	146
0220	Double face [G]		6.70	1.194		53	71.50		124.50	165
0230	LED vandal-resistant, single face [G]		7.27	1.100		217	66		283	335
0240	LED w/battery unit, single face [G]		4.40	1.818		189	109		298	370
0260	Double face [G]		4	2		217	120		337	420
0262	LED w/battery unit, vandal-resistant, single face [G]		4.40	1.818		256	109		365	445
0270	Combination emergency light units and exit sign		4	2		184	120		304	380
0290	LED retrofit kits [G]		60	.133		50.50	8		58.50	68
1500	Exit sign, 12 V, 1 face, remote end mounted (Type 602A1)	R-19	18	1.111		70.50	67		137.50	177
1780	With emergency battery, explosion proof	"	7.70	2.597		4,650	156		4,806	5,350

26 54 Classified Location Lighting

26 54 13 - Incandescent Classified Location Lighting

26 54 13.20 Explosion Proof

26 54 13.20 Explosion Proof		Crew	Daily Output	Labor-Hours	Unit	Material	2019 Bare Costs Labor	2019 Bare Costs Equipment	Total	Total Incl O&P
0010	**EXPLOSION PROOF**, incl. lamps, mounting hardware and connections									
6310	Metal halide with ballast, ceiling, surface mounted, 175 watt	1 Elec	2.90	2.759	Ea.	1,350	166		1,516	1,750
6320	250 watt	"	2.70	2.963		1,625	178		1,803	2,075
6330	400 watt	2 Elec	4.80	3.333		1,750	200		1,950	2,225
6340	Ceiling, pendent mounted, 175 watt	1 Elec	2.60	3.077		1,300	185		1,485	1,700
6350	250 watt	"	2.40	3.333		1,575	200		1,775	2,025
6360	400 watt	2 Elec	4.20	3.810		1,675	229		1,904	2,200
6370	Wall, surface mounted, 175 watt	1 Elec	2.90	2.759		1,450	166		1,616	1,850
6380	250 watt	"	2.70	2.963		1,725	178		1,903	2,175
6390	400 watt	2 Elec	4.80	3.333		1,850	200		2,050	2,325
6400	High pressure sodium, ceiling surface mounted, 70 watt	1 Elec	3	2.667		2,375	160		2,535	2,875
6410	100 watt		3	2.667		2,475	160		2,635	2,950
6420	150 watt		2.70	2.963		2,550	178		2,728	3,075
6430	Pendent mounted, 70 watt		2.70	2.963		2,275	178		2,453	2,775
6440	100 watt		2.70	2.963		2,350	178		2,528	2,875
6450	150 watt		2.40	3.333		2,025	200		2,225	2,525
6460	Wall mounted, 70 watt		3	2.667		2,550	160		2,710	3,050
6470	100 watt		3	2.667		2,625	160		2,785	3,125
6480	150 watt		2.70	2.963		2,700	178		2,878	3,250
6510	Incandescent, ceiling mounted, 200 watt		4	2		1,450	120		1,570	1,775
6520	Pendent mounted, 200 watt		3.50	2.286		1,250	137		1,387	1,575
6530	Wall mounted, 200 watt		4	2		1,450	120		1,570	1,750
6600	Fluorescent, RS, 4' long, ceiling mounted, two 40 watt		2.70	2.963		4,700	178		4,878	5,450
6610	Three 40 watt		2.20	3.636		6,825	218		7,043	7,825
6620	Four 40 watt		1.90	4.211		8,750	253		9,003	10,000
6630	Pendent mounted, two 40 watt		2.30	3.478		5,450	209		5,659	6,300
6640	Three 40 watt		1.90	4.211		7,750	253		8,003	8,900
6650	Four 40 watt		1.70	4.706		10,200	283		10,483	11,600

26 55 Special Purpose Lighting

26 55 33 - Hazard Warning Lighting

26 55 33.10 Warning Beacons

		Crew	Daily Output	Labor-Hours	Unit	Material	Labor	Equipment	Total	Total Incl O&P
0010	**WARNING BEACONS**									
0015	Surface mount with colored or clear lens									
0100	Rotating beacon									
0110	120V, 40 watt halogen	1 Elec	3.50	2.286	Ea.	107	137		244	320
0120	24V, 20 watt halogen	"	3.50	2.286	"	283	137		420	515
0200	Steady beacon									
0210	120V, 40 watt halogen	1 Elec	3.50	2.286	Ea.	109	137		246	325
0220	24V, 20 watt		3.50	2.286		112	137		249	330
0230	12V DC, incandescent		3.50	2.286		104	137		241	320
0300	Flashing beacon									
0310	120V, 40 watt halogen	1 Elec	3.50	2.286	Ea.	106	137		243	320
0320	24V, 20 watt halogen		3.50	2.286		107	137		244	320
0410	12V DC with two 6V lantern batteries		7	1.143		108	68.50		176.50	221

26 55 59 - Display Lighting

26 55 59.10 Track Lighting

		Crew	Daily Output	Labor-Hours	Unit	Material	Labor	Equipment	Total	Total Incl O&P
0010	**TRACK LIGHTING**									
0080	Track, 1 circuit, 4' section	1 Elec	6.70	1.194	Ea.	43	71.50		114.50	155
0100	8' section	2 Elec	10.60	1.509		71	90.50		161.50	213

For customer support on your Electrical Costs with RSMeans data, call 800.448.8182.

303

26 55 Special Purpose Lighting

26 55 59 – Display Lighting

26 55 59.10 Track Lighting

		Crew	Daily Output	Labor-Hours	Unit	Material	2019 Bare Costs Labor	Equipment	Total	Total Incl O&P
0200	12' section	2 Elec	8.80	1.818	Ea.	118	109		227	292
0300	3 circuits, 4' section	1 Elec	6.70	1.194		129	71.50		200.50	249
0400	8' section	2 Elec	10.60	1.509		133	90.50		223.50	281
0500	12' section	"	8.80	1.818		160	109		269	340
1000	Feed kit, surface mounting	1 Elec	16	.500		14.95	30		44.95	61.50
1100	End cover		24	.333		8.25	20		28.25	39
1200	Feed kit, stem mounting, 1 circuit		16	.500		51	30		81	101
1300	3 circuit		16	.500		51	30		81	101
2000	Electrical joiner, for continuous runs, 1 circuit		32	.250		39.50	15		54.50	66
2100	3 circuit		32	.250		83	15		98	114
2200	Fixtures, spotlight, 75 W PAR halogen		16	.500		51.50	30		81.50	102
2210	50 W MR16 halogen		16	.500		173	30		203	235
3000	Wall washer, 250 W tungsten halogen		16	.500		133	30		163	191
3100	Low voltage, 25/50 W, 1 circuit		16	.500		134	30		164	192
3120	3 circuit		16	.500		197	30		227	262

26 55 61 – Theatrical Lighting

26 55 61.10 Lights

		Crew	Daily Output	Labor-Hours	Unit	Material	2019 Bare Costs Labor	Equipment	Total	Total Incl O&P
0010	**LIGHTS**									
2000	Lights, border, quartz, reflector, vented,									
2100	colored or white	1 Elec	20	.400	L.F.	186	24		210	241
2500	Spotlight, follow spot, with transformer, 2,100 watt	"	4	2	Ea.	3,500	120		3,620	4,025
2600	For no transformer, deduct					985			985	1,075
3000	Stationary spot, fresnel quartz, 6" lens	1 Elec	4	2		199	120		319	400
3100	8" lens		4	2		250	120		370	455
3500	Ellipsoidal quartz, 1,000 watt, 6" lens		4	2		370	120		490	590
3600	12" lens		4	2		665	120		785	910
4000	Strobe light, 1 to 15 flashes per second, quartz		3	2.667		855	160		1,015	1,175
4500	Color wheel, portable, five hole, motorized		4	2		222	120		342	425

26 55 63 – Detention Lighting

26 55 63.10 Detention Lighting Fixtures

		Crew	Daily Output	Labor-Hours	Unit	Material	2019 Bare Costs Labor	Equipment	Total	Total Incl O&P
0010	**DETENTION LIGHTING FIXTURES**									
1000	Surface mounted, cold rolled steel, 14 ga., 1' x 4'	1 Elec	5.40	1.481	Ea.	810	89		899	1,025
1010	2' x 2'		5.40	1.481		850	89		939	1,075
1020	2' x 4'		5	1.600		1,000	96		1,096	1,250
1100	12 ga., 1' x 4'		5.40	1.481		1,050	89		1,139	1,300
1110	2' x 2'		5.40	1.481		850	89		939	1,075
1120	2' x 4'		5	1.600		1,100	96		1,196	1,350
3000	Fluorescent vandal resistant light fixture, high abuse troffer		5.40	1.481		221	89		310	375
3010	50" L x 8" W x 4" H		5.40	1.481		204	89		293	355
3020	32W		5.40	1.481		280	89		369	445
3030	T8/32W		5.40	1.481		293	89		382	455
3040	No lamp		5.60	1.429		335	86		421	500
3050	Lamp included		5.40	1.481		385	89		474	560
3060	16 ga. cold rolled steel		5.40	1.481		267	89		356	425
3070	16 ga. cold rolled steel, ceiling/wall mount		5.40	1.481		285	89		374	450
3080	w/lamp, white		5.40	1.481		217	89		306	370
3090	wo/lamp, bronze		5.60	1.429		222	86		308	370
3100	w/lamp 6" L x 4-1/2" W x 8-1/2" H		5.40	1.481		77.50	89		166.50	218
3110	w/lamp 5-7/8" W x 4-1/2" D x 8-1/2" H clear		5.40	1.481		51.50	89		140.50	190
3120	w/lamp white		5.40	1.481		490	89		579	675
3130	Tall wallpack, w/lamp, bronze		5.40	1.481		129	89		218	274
3140	Tall wallpack, w/lamp, white		5.40	1.481		131	89		220	277

26 56 13.10 Lighting Poles	Crew	Daily Output	Labor-Hours	Unit	Material	2019 Bare Costs Labor	2019 Bare Costs Equipment	Total	Total Incl O&P
0010 LIGHTING POLES									
0100 Exterior, light poles, concrete, 30' above 5' below, 13.5" Base, 5.5" Tip	2 Elec	4.60	3.478	Ea.	1,575	209		1,784	2,025
0110 39' above 6' below, 15.5" Base, 5.25" Tip		4.60	3.478		1,850	209		2,059	2,325
0120 43' above 7' below, 17.25" Base, 6.5" Tip		4.60	3.478		1,875	209		2,084	2,350
0130 43' above 7' below, 19.5" Base, 8.25" Tip		4.60	3.478		1,900	209		2,109	2,375
2800 Light poles, anchor base									
2820 not including concrete bases									
2840 Aluminum pole, 8' high	1 Elec	4	2	Ea.	810	120		930	1,075
2850 10' high		4	2		865	120		985	1,125
2860 12' high		3.80	2.105		895	126		1,021	1,175
2870 14' high		3.40	2.353		925	141		1,066	1,225
2880 16' high		3	2.667		1,025	160		1,185	1,375
3000 20' high	R-3	2.90	6.897		1,075	410	45	1,530	1,875
3200 30' high		2.60	7.692		2,050	460	50	2,560	3,025
3400 35' high		2.30	8.696		2,250	520	56.50	2,826.50	3,300
3600 40' high		2	10		2,750	595	65	3,410	3,975
3800 Bracket arms, 1 arm	1 Elec	8	1		139	60		199	243
4000 2 arms		8	1		282	60		342	400
4200 3 arms		5.30	1.509		430	90.50		520.50	610
4400 4 arms		4.80	1.667		575	100		675	780
4500 Steel pole, galvanized, 8' high		3.80	2.105		675	126		801	930
4510 10' high		3.70	2.162		705	130		835	970
4520 12' high		3.40	2.353		760	141		901	1,050
4530 14' high		3.10	2.581		810	155		965	1,125
4540 16' high		2.90	2.759		855	166		1,021	1,200
4550 18' high		2.70	2.963		905	178		1,083	1,250
4600 20' high	R-3	2.60	7.692		1,050	460	50	1,560	1,900
4800 30' high		2.30	8.696		1,525	520	56.50	2,101.50	2,525
5000 35' high		2.20	9.091		1,650	545	59	2,254	2,700
5200 40' high		1.70	11.765		2,025	705	76.50	2,806.50	3,375
5400 Bracket arms, 1 arm	1 Elec	8	1		212	60		272	325
5600 2 arms		8	1		310	60		370	430
5800 3 arms		5.30	1.509		225	90.50		315.50	380
6000 4 arms		5.30	1.509		325	90.50		415.50	490
6100 Fiberglass pole, 1 or 2 fixtures, 20' high	R-3	4	5		800	299	32.50	1,131.50	1,350
6200 30' high		3.60	5.556		1,050	330	36	1,416	1,675
6300 35' high		3.20	6.250		1,650	375	40.50	2,065.50	2,425
6400 40' high		2.80	7.143		1,925	425	46.50	2,396.50	2,800
6420 Wood pole, 4-1/2" x 5-1/8", 8' high	1 Elec	6	1.333		385	80		465	545
6430 10' high		6	1.333		445	80		525	610
6440 12' high		5.70	1.404		560	84.50		644.50	740
6450 15' high		5	1.600		655	96		751	865
6460 20' high		4	2		795	120		915	1,050
6461 Light poles, anchor base, w/o conc base, pwdr ct stl, 16' H	2 Elec	3.10	5.161		855	310		1,165	1,400
6462 20' high	R-3	2.90	6.897		1,050	410	45	1,505	1,825
6463 30' high		2.30	8.696		1,525	520	56.50	2,101.50	2,525
6464 35' high		2.40	8.333		1,650	500	54	2,204	2,625
6465 25' high		2.70	7.407		1,050	440	48	1,538	1,875
7300 Transformer bases, not including concrete bases									
7320 Maximum pole size, steel, 40' high	1 Elec	2	4	Ea.	1,700	240		1,940	2,225
7340 Cast aluminum, 30' high		3	2.667		895	160		1,055	1,225
7350 40' high		2.50	3.200		1,375	192		1,567	1,775
8000 Line cover protective devices									

For customer support on your Electrical Costs with RSMeans data, call 800.448.8182.

305

26 56 Exterior Lighting

26 56 13 – Lighting Poles and Standards

26 56 13.10 Lighting Poles

	26 56 13.10 Lighting Poles	Crew	Daily Output	Labor-Hours	Unit	Material	2019 Bare Costs Labor	Equipment	Total	Total Incl O&P
8010	line cover									
8015	Refer to 26 01 02.20 for labor adjustment factors as they apply									
8100	MVLC 1" W x 1-1/2" H, 14/5'	1 Elec	60	.133	Ea.	4.61	8		12.61	17
8110	8'		58	.138		5.15	8.30		13.45	18
8120	MVLC 1" W x 1-1/2" H, 18/5'		60	.133		5.85	8		13.85	18.40
8130	8'		58	.138		6.50	8.30		14.80	19.50
8140	MVLC 1" W x 1-1/2" H, 38/5'		60	.133		10.50	8		18.50	23.50
8150	8'		58	.138		12.65	8.30		20.95	26.50

26 56 19 – LED Exterior Lighting

26 56 19.55 Roadway LED Luminaire

	26 56 19.55 Roadway LED Luminaire		Crew	Daily Output	Labor-Hours	Unit	Material	2019 Bare Costs Labor	Equipment	Total	Total Incl O&P
0010	**ROADWAY LED LUMINAIRE**										
0100	LED fixture, 72 LEDs, 120 V AC or 12 V DC, equal to 60 watt	G	1 Elec	2.70	2.963	Ea.	585	178		763	910
0110	108 LEDs, 120 V AC or 12 V DC, equal to 90 watt	G		2.70	2.963		690	178		868	1,025
0120	144 LEDs, 120 V AC or 12 V DC, equal to 120 watt	G		2.70	2.963		850	178		1,028	1,200
0130	252 LEDs, 120 V AC or 12 V DC, equal to 210 watt	G	2 Elec	4.40	3.636		1,175	218		1,393	1,625
0140	Replaces high pressure sodium fixture, 75 watt	G	1 Elec	2.70	2.963		400	178		578	705
0150	125 watt	G		2.70	2.963		455	178		633	765
0160	150 watt	G		2.70	2.963		535	178		713	855
0170	175 watt	G		2.70	2.963		790	178		968	1,125
0180	200 watt	G		2.70	2.963		750	178		928	1,100
0190	250 watt	G	2 Elec	4.40	3.636		860	218		1,078	1,275
0200	320 watt	G		4.40	3.636		940	218		1,158	1,350
2000	Exterior fixtures, LED roadway, type 2, 50W, neutral slipfitter bronze			5.20	3.077		580	185		765	910
2010	3, 50W, neutral slipfitter bronze			5.20	3.077		570	185		755	905
2020	4, 50W, neutral slipfitter bronze			5.20	3.077		580	185		765	910
2030	Exterior fixtures, LED roadway, type 2, 78W, neutral slipfitter bronze			5.20	3.077		585	185		770	915
2040	3, 78W, neutral slipfitter bronze			5.20	3.077		555	185		740	885
2050	4, 78W, neutral slipfitter bronze			5.20	3.077		625	185		810	960
2060	Exterior fixtures, LED roadway, type 2, 105W, neutral slipfitter bronze			5.20	3.077		740	185		925	1,100
2070	3, 105W, neutral slipfitter bronze			5.20	3.077		690	185		875	1,025
2080	4, 105W, neutral slipfitter bronze			5.20	3.077		725	185		910	1,075
2090	Exterior fixtures, LED roadway, type 2, 125W, neutral slipfitter bronze			5.20	3.077		685	185		870	1,025

26 56 19.60 Parking LED Lighting

	26 56 19.60 Parking LED Lighting		Crew	Daily Output	Labor-Hours	Unit	Material	2019 Bare Costs Labor	Equipment	Total	Total Incl O&P
0010	**PARKING LED LIGHTING**										
0100	Round pole mounting, 88 lamp watts	G	1 Elec	2	4	Ea.	1,225	240		1,465	1,700

26 56 21 – HID Exterior Lighting

26 56 21.20 Roadway Luminaire

	26 56 21.20 Roadway Luminaire	Crew	Daily Output	Labor-Hours	Unit	Material	2019 Bare Costs Labor	Equipment	Total	Total Incl O&P
0010	**ROADWAY LUMINAIRE**									
2650	Roadway area luminaire, low pressure sodium, 135 watt	1 Elec	2	4	Ea.	890	240		1,130	1,350
2700	180 watt	"	2	4		950	240		1,190	1,400
2750	Metal halide, 400 watt	2 Elec	4.40	3.636		700	218		918	1,100
2760	1,000 watt		4	4		785	240		1,025	1,225
2780	High pressure sodium, 400 watt		4.40	3.636		835	218		1,053	1,250
2790	1,000 watt		4	4		945	240		1,185	1,400

26 56 23 – Area Lighting

26 56 23.10 Exterior Fixtures

	26 56 23.10 Exterior Fixtures	Crew	Daily Output	Labor-Hours	Unit	Material	2019 Bare Costs Labor	Equipment	Total	Total Incl O&P
0010	**EXTERIOR FIXTURES** With lamps									
0200	Wall mounted, incandescent, 100 watt	1 Elec	8	1	Ea.	53.50	60		113.50	149
0400	Quartz, 500 watt		5.30	1.509		61.50	90.50		152	203
0420	1,500 watt		4.20	1.905		100	114		214	281
1100	Wall pack, low pressure sodium, 35 watt		4	2		202	120		322	400

26 56 Exterior Lighting

26 56 23 – Area Lighting

26 56 23.10 Exterior Fixtures		Crew	Daily Output	Labor-Hours	Unit	Material	2019 Bare Costs Labor	2019 Bare Costs Equipment	Total	Total Incl O&P
1150	55 watt	1 Elec	4	2	Ea.	240	120		360	445
1160	High pressure sodium, 70 watt		4	2		200	120		320	400
1170	150 watt		4	2		217	120		337	415
1175	High pressure sodium, 250 watt		4	2		217	120		337	415
1180	Metal halide, 175 watt		4	2		221	120		341	420
1190	250 watt		4	2		250	120		370	455
1195	400 watt		4	2		410	120		530	630
1250	Induction lamp, 40 watt		4	2		470	120		590	695
1260	80 watt		4	2		560	120		680	795
1278	LED, poly lens, 26 watt		4	2		310	120		430	520
1280	110 watt		4	2		830	120		950	1,100
1500	LED, glass lens, 13 watt		4	2		310	120		430	520

26 56 23.55 Exterior LED Fixtures

			Crew	Daily Output	Labor-Hours	Unit	Material	Labor	Equipment	Total	Total Incl O&P
0010	**EXTERIOR LED FIXTURES**										
0100	Wall mounted, indoor/outdoor, 12 watt	G	1 Elec	10	.800	Ea.	212	48		260	305
0110	32 watt	G		10	.800		475	48		523	595
0120	66 watt	G		10	.800		490	48		538	610
0200	outdoor, 110 watt	G		10	.800		1,200	48		1,248	1,400
0210	220 watt	G		10	.800		1,800	48		1,848	2,050
0300	modular, type IV, 120 V, 50 lamp watts	G		9	.889		1,175	53.50		1,228.50	1,375
0310	101 lamp watts	G		9	.889		1,325	53.50		1,378.50	1,525
0320	126 lamp watts	G		9	.889		1,650	53.50		1,703.50	1,900
0330	202 lamp watts	G		9	.889		1,925	53.50		1,978.50	2,200
0340	240 V, 50 lamp watts	G		8	1		1,250	60		1,310	1,475
0350	101 lamp watts	G		8	1		1,375	60		1,435	1,600
0360	126 lamp watts	G		8	1		1,450	60		1,510	1,700
0370	202 lamp watts	G		8	1		1,925	60		1,985	2,225
0400	wall pack, glass, 13 lamp watts	G		4	2		455	120		575	680
0410	poly w/photocell, 26 lamp watts	G		4	2		360	120		480	580
0420	50 lamp watts	G		4	2		780	120		900	1,050
0430	replacement, 40 watts	G		4	2		390	120		510	610
0440	60 watts	G		4	2		400	120		520	620

26 56 26 – Landscape Lighting

26 56 26.20 Landscape Fixtures

			Crew	Daily Output	Labor-Hours	Unit	Material	Labor	Equipment	Total	Total Incl O&P
0010	**LANDSCAPE FIXTURES**										
7380	Landscape recessed uplight, incl. housing, ballast, transformer										
7390	& reflector, not incl. conduit, wire, trench										
7420	Incandescent, 250 watt		1 Elec	5	1.600	Ea.	635	96		731	845
7440	Quartz, 250 watt			5	1.600		605	96		701	810
7460	500 watt			4	2		620	120		740	865

26 56 26.50 Landscape LED Fixtures

			Crew	Daily Output	Labor-Hours	Unit	Material	Labor	Equipment	Total	Total Incl O&P
0010	**LANDSCAPE LED FIXTURES**										
0100	12 volt alum bullet hooded-BLK		1 Elec	5	1.600	Ea.	97	96		193	250
0200	12 volt alum bullet hooded-BRZ			5	1.600		97	96		193	250
0300	12 volt alum bullet hooded-GRN			5	1.600		97	96		193	250
1000	12 volt alum large bullet hooded-BLK			5	1.600		71.50	96		167.50	222
1100	12 volt alum large bullet hooded-BRZ			5	1.600		71.50	96		167.50	222
1200	12 volt alum large bullet hooded-GRN			5	1.600		71.50	96		167.50	222
2000	12 volt large bullet landscape light fixture			5	1.600		71.50	96		167.50	222
2100	12 volt alum light large bullet			5	1.600		71.50	96		167.50	222
2200	12 volt alum bullet light			5	1.600		71.50	96		167.50	222

For customer support on your Electrical Costs with RSMeans data, call 800.448.8182.

307

26 56 33.10 Walkway Luminaire

	26 56 33.10 Walkway Luminaire	Crew	Daily Output	Labor-Hours	Unit	Material	2019 Bare Costs Labor	Equipment	Total	Total Incl O&P
0010	**WALKWAY LUMINAIRE**									
6500	Bollard light, lamp & ballast, 42" high with polycarbonate lens									
6800	Metal halide, 175 watt	1 Elec	3	2.667	Ea.	985	160		1,145	1,325
6900	High pressure sodium, 70 watt		3	2.667		995	160		1,155	1,350
7000	100 watt		3	2.667		995	160		1,155	1,350
7100	150 watt		3	2.667		970	160		1,130	1,325
7200	Incandescent, 150 watt		3	2.667		715	160		875	1,025
7810	Walkway luminaire, square 16", metal halide 250 watt		2.70	2.963		760	178		938	1,100
7820	High pressure sodium, 70 watt		3	2.667		875	160		1,035	1,200
7830	100 watt		3	2.667		890	160		1,050	1,225
7840	150 watt		3	2.667		890	160		1,050	1,225
7850	200 watt		3	2.667		895	160		1,055	1,225
7910	Round 19", metal halide, 250 watt		2.70	2.963		1,100	178		1,278	1,475
7920	High pressure sodium, 70 watt		3	2.667		1,200	160		1,360	1,575
7930	100 watt		3	2.667		1,200	160		1,360	1,575
7940	150 watt		3	2.667		1,225	160		1,385	1,600
7950	250 watt		2.70	2.963		1,275	178		1,453	1,700
8000	Sphere 14" opal, incandescent, 200 watt		4	2		350	120		470	565
8020	Sphere 18" opal, incandescent, 300 watt		3.50	2.286		425	137		562	670
8040	Sphere 16" clear, high pressure sodium, 70 watt		3	2.667		735	160		895	1,050
8050	100 watt		3	2.667		785	160		945	1,100
8100	Cube 16" opal, incandescent, 300 watt		3.50	2.286		470	137		607	725
8120	High pressure sodium, 70 watt		3	2.667		690	160		850	995
8130	100 watt		3	2.667		705	160		865	1,025
8230	Lantern, high pressure sodium, 70 watt		3	2.667		610	160		770	910
8240	100 watt		3	2.667		655	160		815	965
8250	150 watt		3	2.667		615	160		775	915
8260	250 watt		2.70	2.963		860	178		1,038	1,200
8270	Incandescent, 300 watt		3.50	2.286		455	137		592	705
8330	Reflector 22" w/globe, high pressure sodium, 70 watt		3	2.667		590	160		750	885
8340	100 watt		3	2.667		595	160		755	895
8350	150 watt		3	2.667		605	160		765	905
8360	250 watt		2.70	2.963		770	178		948	1,125

26 56 33.55 Walkway LED Luminaire

	26 56 33.55 Walkway LED Luminaire		Crew	Daily Output	Labor-Hours	Unit	Material	2019 Bare Costs Labor	Equipment	Total	Total Incl O&P
0010	**WALKWAY LED LUMINAIRE**										
0100	Pole mounted, 86 watts, 4,350 lumens	G	1 Elec	3	2.667	Ea.	1,450	160		1,610	1,850
0110	4,630 lumens	G		3	2.667		2,375	160		2,535	2,850
0120	80 watts, 4,000 lumens	G		3	2.667		2,125	160		2,285	2,600
0600	LED bollard pole, 6" diam. x 36" H, cast alum, surf. mtd., 8W, 6,000K, 120V			8.50	.941		385	56.50		441.50	505
0610	6" diam. x 42" H, cast alum, surf. mtd., 12W, 5,100K,120V			8.50	.941		550	56.50		606.50	690
0620	6" diam. x 42" H, cast alum, surf. mtd., 18W, 5,100K,120V			8.50	.941		640	56.50		696.50	790
0630	6" diam. x 42" H, cast alum, surf. mtd., 24W, 5,100K,120V			8.50	.941		505	56.50		561.50	640
0640	9.5" square x 42" H, cast resin, surf. mtd., A19, 9W, 120V			8.50	.941		605	56.50		661.50	750
0650	9.75" diam. x 42" H, cast resin, side mount, dome, 9W, 120V			8.50	.941		690	56.50		746.50	845
0660	9.75" diam. x 43" H, cast resin, burial base, dome, 9W, 120V			8.50	.941		690	56.50		746.50	845
0670	8" diam. x 36" H, concrete security base, w/rebar, 8W, HID			8	1		1,025	60		1,085	1,225
0680	10" diam. x 49" H, concrete base, w/Sch 40 Steel pipe, 8W			6.75	1.185		1,150	71		1,221	1,375
0690	12" diam. x 48" H, concrete base, w/Sch 40 Steel pipe, 8W			6.75	1.185		1,325	71		1,396	1,550
0700	8" diam. x 36" H, concrete, cast aluminum dome, 90 degree, 4.1W			7	1.143		1,025	68.50		1,093.50	1,225
0710	8" diam. x 36" H, concrete, cast aluminum dome, 180 degree, 4.1W			7	1.143		1,275	68.50		1,343.50	1,500
0720	8" diam. x 48" H, concrete, cast aluminum dome, 360 degree, 4.1W			7	1.143		1,550	68.50		1,618.50	1,800

26 56 Exterior Lighting

26 56 36 – Flood Lighting

26 56 36.20 Floodlights		Crew	Daily Output	Labor-Hours	Unit	Material	2019 Bare Costs Labor	Equipment	Total	Total Incl O&P
0010	**FLOODLIGHTS** with ballast and lamp,									
1290	floor mtd, mount with swivel bracket									
1300	Induction lamp, 40 watt	1 Elec	3	2.667	Ea.	555	160		715	850
1310	80 watt		3	2.667		740	160		900	1,050
1320	150 watt		3	2.667		1,350	160		1,510	1,750
1400	Pole mounted, pole not included									
1950	Metal halide, 175 watt	1 Elec	2.70	2.963	Ea.	216	178		394	505
2000	400 watt	2 Elec	4.40	3.636		223	218		441	570
2200	1,000 watt		4	4		830	240		1,070	1,275
2210	1,500 watt		3.70	4.324		430	260		690	855
2250	Low pressure sodium, 55 watt	1 Elec	2.70	2.963		465	178		643	775
2270	90 watt		2	4		595	240		835	1,025
2290	180 watt		2	4		680	240		920	1,100
2340	High pressure sodium, 70 watt		2.70	2.963		251	178		429	540
2360	100 watt		2.70	2.963		258	178		436	550
2380	150 watt		2.70	2.963		310	178		488	605
2400	400 watt	2 Elec	4.40	3.636		325	218		543	680
2600	1,000 watt	"	4	4		600	240		840	1,025
9005	Solar powered floodlight, w/motion det, incl. batt pack for cloudy days G	1 Elec	8	1		60.50	60		120.50	156
9020	Replacement battery pack G	"	8	1		20	60		80	112

26 56 36.55 LED Floodlights

		Crew	Daily Output	Labor-Hours	Unit	Material	Labor	Equipment	Total	Total Incl O&P
0010	**LED FLOODLIGHTS** with ballast and lamp,									
0020	Pole mounted, pole not included									
0100	11 watt G	1 Elec	4	2	Ea.	435	120		555	660
0110	46 watt G		4	2		1,475	120		1,595	1,800
0120	90 watt G		4	2		2,100	120		2,220	2,500
0130	288 watt G		4	2		1,900	120		2,020	2,275

26 61 Lighting Systems and Accessories

26 61 13 – Lighting Accessories

26 61 13.30 Fixture Whips

		Crew	Daily Output	Labor-Hours	Unit	Material	Labor	Equipment	Total	Total Incl O&P
0010	**FIXTURE WHIPS**									
0080	3/8" Greenfield, 2 connectors, 6' long									
0100	TFFN wire, three #18	1 Elec	32	.250	Ea.	7.75	15		22.75	31
0150	Four #18		28	.286		8.35	17.15		25.50	34.50
0200	Three #16		32	.250		7.95	15		22.95	31.50
0250	Four #16		28	.286		8.70	17.15		25.85	35
0300	THHN wire, three #14		32	.250		5.55	15		20.55	28.50
0350	Four #14		28	.286		7.90	17.15		25.05	34
0360	Three #12		32	.250		10.55	15		25.55	34

26 61 23 – Lamps Applications

26 61 23.10 Lamps

		Crew	Daily Output	Labor-Hours	Unit	Material	Labor	Equipment	Total	Total Incl O&P
0010	**LAMPS**									
0080	Fluorescent, rapid start, cool white, 2' long, 20 watt	1 Elec	1	8	C	375	480		855	1,125
0100	4' long, 40 watt		.90	8.889		248	535		783	1,075
0120	3' long, 30 watt		.90	8.889		460	535		995	1,300
0125	3' long, 25 watt energy saver R265723-10 G		.90	8.889		1,300	535		1,835	2,225
0150	U-40 watt		.80	10		1,150	600		1,750	2,150
0155	U-34 watt energy saver G		.80	10		11.80	600		611.80	910
0170	4' long, 34 watt energy saver G		.90	8.889		795	535		1,330	1,675

26 61 23.10 Lamps		Crew	Daily Output	Labor-Hours	Unit	Material	2019 Bare Costs Labor	Equipment	Total	Total Incl O&P
0176	2' long, T8, 17 watt energy saver	G	1 Elec	1	8	C	325	480		805	1,075
0178	3' long, T8, 25 watt energy saver	G		.90	8.889		355	535		890	1,175
0180	4' long, T8, 32 watt energy saver	G		.90	8.889		216	535		751	1,025
0200	Slimline, 4' long, 40 watt			.90	8.889		1,050	535		1,585	1,950
0210	4' long, 30 watt energy saver	G		.90	8.889		1,050	535		1,585	1,950
0300	8' long, 75 watt			.80	10		1,025	600		1,625	2,025
0350	8' long, 60 watt energy saver			.80	10		395	600		995	1,325
0400	High output, 4' long, 60 watt			.90	8.889		560	535		1,095	1,400
0410	8' long, 95 watt energy saver	G		.80	10		610	600		1,210	1,575
0500	8' long, 110 watt			.80	10		610	600		1,210	1,575
0512	2' long, T5, 14 watt energy saver	G		1	8		154	480		634	885
0514	3' long, T5, 21 watt energy saver	G		.90	8.889		174	535		709	985
0516	4' long, T5, 28 watt energy saver	G		.90	8.889		174	535		709	985
0517	4' long, T5, 54 watt energy saver	G		.90	8.889		535	535		1,070	1,375
0520	Very high output, 4' long, 110 watt			.90	8.889		1,525	535		2,060	2,475
0525	8' long, 195 watt energy saver	G		.70	11.429		1,275	685		1,960	2,425
0550	8' long, 215 watt			.70	11.429		1,275	685		1,960	2,425
0554	Full spectrum, 4' long, 60 watt			.90	8.889		660	535		1,195	1,525
0556	6' long, 85 watt			.90	8.889		780	535		1,315	1,650
0558	8' long, 110 watt			.80	10		1,975	600		2,575	3,075
0560	Twin tube compact lamp	G		.90	8.889		435	535		970	1,275
0570	Double twin tube compact lamp	G		.80	10		950	600		1,550	1,950
0600	Mercury vapor, mogul base, deluxe white, 100 watt			.30	26.667		5,150	1,600		6,750	8,075
0650	175 watt			.30	26.667		3,150	1,600		4,750	5,850
0700	250 watt			.30	26.667		5,025	1,600		6,625	7,925
0800	400 watt			.30	26.667		5,050	1,600		6,650	7,950
0900	1,000 watt			.20	40		13,200	2,400		15,600	18,100
1000	Metal halide, mogul base, 175 watt			.30	26.667		1,050	1,600		2,650	3,550
1100	250 watt			.30	26.667		1,650	1,600		3,250	4,225
1200	400 watt			.30	26.667		2,025	1,600		3,625	4,625
1300	1,000 watt			.20	40		3,300	2,400		5,700	7,225
1320	1,000 watt, 125,000 initial lumens			.20	40		15,400	2,400		17,800	20,500
1330	1,500 watt			.20	40		3,575	2,400		5,975	7,500
1350	High pressure sodium, 70 watt			.30	26.667		1,575	1,600		3,175	4,150
1360	100 watt			.30	26.667		1,550	1,600		3,150	4,100
1370	150 watt			.30	26.667		1,625	1,600		3,225	4,175
1380	250 watt			.30	26.667		2,150	1,600		3,750	4,775
1400	400 watt			.30	26.667		1,600	1,600		3,200	4,175
1450	1,000 watt			.20	40		4,300	2,400		6,700	8,300
1500	Low pressure sodium, 35 watt			.30	26.667		15,900	1,600		17,500	19,900
1550	55 watt			.30	26.667		18,100	1,600		19,700	22,400
1600	90 watt			.30	26.667		6,275	1,600		7,875	9,325
1650	135 watt			.20	40		26,400	2,400		28,800	32,600
1700	180 watt			.20	40		39,500	2,400		41,900	47,100
1750	Quartz line, clear, 500 watt			1.10	7.273		630	435		1,065	1,350
1760	1,500 watt			.20	40		1,750	2,400		4,150	5,500
1762	Spot, MR 16, 50 watt			1.30	6.154		1,050	370		1,420	1,725
1770	Tungsten halogen, T4, 400 watt			1.10	7.273		3,850	435		4,285	4,900
1775	T3, 1,200 watt			.30	26.667		4,850	1,600		6,450	7,750
1778	PAR 30, 50 watt			1.30	6.154		1,100	370		1,470	1,750
1780	PAR 38, 90 watt			1.30	6.154		9,325	370		9,695	10,900
1800	Incandescent, interior, A21, 100 watt			1.60	5		2,700	300		3,000	3,425
1900	A21, 150 watt			1.60	5		16,100	300		16,400	18,200

26 61 23 – Lamps Applications

26 61 23.10 Lamps

26 61 23.10 Lamps	Crew	Daily Output	Labor-Hours	Unit	Material	2019 Bare Costs Labor	Equipment	Total	Total Incl O&P
2000	A23, 200 watt	1 Elec	1.60	5	C	320	300		620	805
2200	PS 35, 300 watt		1.60	5		965	300		1,265	1,525
2210	PS 35, 500 watt		1.60	5		1,375	300		1,675	1,950
2230	PS 52, 1,000 watt		1.30	6.154		2,525	370		2,895	3,325
2240	PS 52, 1,500 watt		1.30	6.154		6,600	370		6,970	7,825
2300	R30, 75 watt		1.30	6.154		700	370		1,070	1,325
2400	R40, 100 watt		1.30	6.154		700	370		1,070	1,325
2500	Exterior, PAR 38, 75 watt		1.30	6.154		1,825	370		2,195	2,575
2600	PAR 38, 150 watt		1.30	6.154		2,050	370		2,420	2,800
2700	PAR 46, 200 watt		1.10	7.273		3,750	435		4,185	4,775
2800	PAR 56, 300 watt		1.10	7.273		2,575	435		3,010	3,475
3000	Guards, fluorescent lamp, 4' long		100	.080	Ea.	13.45	4.80		18.25	22
3200	8' long		90	.089	"	27	5.35		32.35	37.50

26 61 23.55 LED Lamps

26 61 23.55 LED Lamps	Crew	Daily Output	Labor-Hours	Unit	Material	2019 Bare Costs Labor	Equipment	Total	Total Incl O&P
0010	**LED LAMPS**									
0100	LED lamp, interior, shape A60, equal to 60 W [G]	1 Elec	160	.050	Ea.	18.85	3		21.85	25
0110	7 W LED decorative c, ca, f, g shape		160	.050		33	3		36	41
0120	12V mini lamp LED		160	.050		5.30	3		8.30	10.35
0200	Globe frosted A60, equal to 60 W [G]		160	.050		11.25	3		14.25	16.85
0205	LED lamp, interior, globe		160	.050		15.60	3		18.60	21.50
0210	2.2 W LED LMP		160	.050		64	3		67	75
0220	2.2 W LED replacement decorative lamp		160	.050		8.40	3		11.40	13.75
0230	3.5 W LED replacement decorative lamp		160	.050		10.40	3		13.40	15.95
0240	4.5 W 120V LED replacement decorative lamp		160	.050		11.35	3		14.35	16.95
0250	4.5 W 120V LED, 2700k replacement decorative lamp		160	.050		9.80	3		12.80	15.25
0260	4.9 W 120V LED, 2700k replacement decorative lamp candelabra base		160	.050		9.70	3		12.70	15.20
0270	4.9 W 120V LED, 3000k replacement decorative lamp candelabra base		160	.050		9.10	3		12.10	14.50
0280	5 W LED PAR 20 parabolic reflector lamp FL		140	.057		58	3.43		61.43	69
0300	Globe earth, equal to 100 W [G]		140	.057		27	3.43		30.43	35
0305	7 W LED reflector lamp 3000k DIM		140	.057		56	3.43		59.43	67
0310	10 W omni LED warm white light bulb E26 medium base 120 volt card		140	.057		16.20	3.43		19.63	23
0315	7 W LED reflector lamp WFL 2700k DIM		140	.057		56	3.43		59.43	67
0320	9 W omni LED warm white light bulb E26 medium base 120 volt card		140	.057		8.75	3.43		12.18	14.75
0500	8 W LED, A19 lamp, equal to 40 W		140	.057		5.70	3.43		9.13	11.40
0505	9 W LED, A19 lamp, 5000K dimmable, equal to 60 W		140	.057		6.95	3.43		10.38	12.75
0510	10.5 W LED, A19 lamp, dimmable, equal to 60 W		140	.057		8.60	3.43		12.03	14.60
0515	10 W LED, A19 lamp, frosted, dimmable, equal to 60 W		140	.057		13	3.43		16.43	19.40
0520	10 W LED, A19 lamp, omni-directional, dimmable, equal to 60 W		140	.057		7.60	3.43		11.03	13.45
0525	12 W LED, A19 lamp, dimmable, equal to 75 W		140	.057		10.45	3.43		13.88	16.60
1100	MR16, 3 W, replacement of halogen lamp 25 W [G]		130	.062		18.90	3.70		22.60	26.50
1200	6 W replacement of halogen lamp 45 W [G]		130	.062		19.85	3.70		23.55	27.50
2100	10 W, PAR20, equal to 60 W [G]		130	.062		26	3.70		29.70	34.50
2200	15 W, PAR30, equal to 100 W [G]		130	.062		46	3.70		49.70	56
2210	50 W LED lamp		130	.062		555	3.70		558.70	615
2220	11 Watt reflector dimmable warm white LED light bulb with medium base		130	.062		40	3.70		43.70	49.50
2221	12 W A-Line LED lamp DIM		130	.062		75.50	3.70		79.20	88.50
2225	13 Watt reflector LED warm white e26 with medium base 120 volt box		130	.062		38	3.70		41.70	47
2226	13 W br30 LED lamp		130	.062		65.50	3.70		69.20	77.50
2227	15 W 120V br30 inc LED lamp, 2700k		130	.062		61.50	3.70		65.20	73
2228	15 W 120V br30 inc LED lamp, 4000k		130	.062		58	3.70		61.70	69.50
2230	3 Watt dimmable warm white decorative LED lamp with medium base		160	.050		11.85	3		14.85	17.55
2240	.43 W night light LED daylight bulb E12 candelabra base 120 volt 2 pack		160	.050		4.57	3		7.57	9.55

For customer support on your Electrical Costs with RSMeans data, call 800.448.8182.

311

26 61 23 – Lamps Applications

26 61 23.55 LED Lamps		Crew	Daily Output	Labor-Hours	Unit	Material	2019 Bare Costs Labor	2019 Bare Costs Equipment	Total	Total Incl O&P
2250	15 W omni-directional LED warm white e26 medium base 120 volt box	1 Elec	160	.050	Ea.	39	3		42	47.50
2251	11 W omni-directional LED warm white e26 medium base 120 volt box		160	.050		27	3		30	34
2252	10 W omni-directional LED warm white e26 medium base 120 volt box		160	.050		27	3		30	34
2253	7 W omni A19 LED warm white e26 medium base 120 volt box		160	.050		19.80	3		22.80	26.50
2255	8 PAR 20 parabolic reflector 2700 LED lamp		160	.050		18.15	3		21.15	24.50
2256	8 PAR 20 parabolic reflector 3000 LED lamp		160	.050		18.35	3		21.35	24.50
2260	12 W PAR 38 120V LED 15 degree directional lamp		160	.050		196	3		199	220
2270	12 W PAR 38 120V LED 25 degree directional lamp		160	.050		30.50	3		33.50	38
2280	12 W PAR 38 120V LED 40 degree directional lamp		160	.050		196	3		199	220
2285	16 W PAR 38 120V 2700k LED parabolic reflector lamp		160	.050		29.50	3		32.50	37
2290	16 W PAR 38 120V 3000k LED parabolic reflector lamp		160	.050		22	3		25	28.50
3000	15 W PAR 30 LED daylight E26 medium base 120V box		160	.050		58	3		61	68.50
3100	17 W LED 3000k PAR 38 100 W replacement		160	.050		45.50	3		48.50	54.50
3110	17 W LED PAR 38 100 W replacement		160	.050		34	3		37	42
3120	17 W LED PAR 38 100 W replacement parabolic reflector		160	.050		35.50	3		38.50	43.50
3130	24 W LED T8 PW straight fluorescent lamp		8.89	.900		279	54		333	385
3135	Linear fluorescent LED lamp 120V		8.89	.900		128	54		182	222
3200	30 W LED 2700K recessed 8055E PAR 38 high power		160	.050		129	3		132	146
3210	30 W LED 4200K 120 degree 8055E PAR 38 high power		160	.050		129	3		132	146
3220	30 W LED 5700K 8055E PAR 38 high power		160	.050		131	3		134	148
3230	50 W LED 2700K 8045M PAR 38 277V high power		160	.050		330	3		333	365
3240	50 W LED 4200K 8045M PAR 38 277V high power retro fit		160	.050		175	3		178	196
3250	50 W LED 5700K 8045M PAR 38 277V high power retro fit		160	.050		330	3		333	365
8000	10 W LED PAR 30/fl 10 pk		160	.050		93	3		96	106
8010	Gen 3 PAR 30 15 W short neck power LED 120 VAC E26 80 +cri 300k dimm		160	.050		12.55	3		15.55	18.30
8020	12 PAR 30 2700K parabolic reflector LED lamp		160	.050		42	3		45	50.50
8030	12 PAR 30 3000K parabolic reflector LED lamp		160	.050		36	3		39	44
8040	3500K LED advantage T8 9 W 800LM 2ft linear 2 BD frosted		69	.116		10.40	6.95		17.35	22
8050	3500K LED litespan T8 9 W 900LM 2ft linear frosted		69	.116		12.80	6.95		19.75	24.50
8060	4000K LED advantage T8 9 W 800LM 2ft linear 2BD frosted		69	.116		12.15	6.95		19.10	24
8070	4000K LED litespan T8 9 W 900LM 2ft linear frosted		69	.116		13.40	6.95		20.35	25
8080	5000K LED advantage T8 9 W 800LM 2ft linear 2BD frosted		69	.116		10.40	6.95		17.35	22
8090	5000K LED litespan T8 9 W 900LM 2ft linear frosted		69	.116		11.45	6.95		18.40	23
8100	3500K LED advantage T8 18 W 1600LM 4ft linear 2 BD frosted		69	.116		10.05	6.95		17	21.50
8105	18 W LED 4ft T8 4000K frost 1600L linear lamp		69	.116		17.90	6.95		24.85	30
8108	18 W LED 48 inch T8 4100K 1890LM linear lamp		69	.116		55.50	6.95		62.45	71.50
8110	3500K LED advantage T8 18 W 1800LM 4ft linear 2 BD frosted		69	.116		28.50	6.95		35.45	42
8120	4000K LED advantage T8 18 W 1600LM 4ft linear 2 BD frosted		69	.116		20	6.95		26.95	32.50
8130	4000K LED litespan T8 18 W 1800LM 4ft linear frosted		69	.116		20.50	6.95		27.45	33
8140	5000K LED advantage T8 18 W 1600LM 4ft linear 2BD frosted		69	.116		12.80	6.95		19.75	24.50
8150	5000K LED litespan T8 18 W 1800LM 4ft linear 2BD frosted		69	.116		20.50	6.95		27.45	33
8200	16.5 T8 3000 IF-6U U-shape fluorescent LED lamp		65	.123		21	7.40		28.40	34
8210	16.5 T8 3500 IF-6U U-shape fluorescent LED lamp		65	.123		21	7.40		28.40	34
8220	16.5 T8 4000 IF-6U U-shape fluorescent LED lamp		65	.123		20.50	7.40		27.90	33.50
8230	16.5 T8 5000 IF-6U U-shape fluorescent LED lamp		65	.123		21	7.40		28.40	34
8240	18 W 6 inch T8 4100K U-shape frosted fluorescent LED lamp		60	.133		11.30	8		19.30	24.50
8250	18 W 6 inch T8 5000K U-shape frosted fluorescent LED lamp		60	.133		11.30	8		19.30	24.50
8260	Circular 12 W linear fluorescent LED 2700K MOD lamp		62.40	.128		292	7.70		299.70	330
8270	Circular 12 W linear fluorescent LED 3000K MOD lamp		62.40	.128		292	7.70		299.70	330
8280	Circular 12 W linear fluorescent LED 3500K MOD lamp		62.40	.128		264	7.70		271.70	305
8290	Circular 18 W linear fluorescent LED 3000K MOD lamp		62.40	.128		360	7.70		367.70	405
8300	Circular 18 W linear fluorescent LED 3500K MOD lamp		62.40	.128		400	7.70		407.70	450
8310	Circular 18 W linear fluorescent LED 5000K MOD lamp		62.40	.128		360	7.70		367.70	405

26 61 Lighting Systems and Accessories

26 61 23 – Lamps Applications

26 61 23.55 LED Lamps		Crew	Daily Output	Labor-Hours	Unit	Material	2019 Bare Costs Labor	Equipment	Total	Total Incl O&P
8320	Circular 18 W linear fluorescent LED 4100K MOD lamp	1 Elec	62.40	.128	Ea.	385	7.70		392.70	430

26 71 Electrical Machines

26 71 13 – Motors Applications

26 71 13.10 Handling

		Crew	Daily Output	Labor-Hours	Unit	Material	Labor	Equipment	Total	Total Incl O&P
0010	**HANDLING** Add to normal labor cost for restricted areas									
5000	Motors									
5100	1/2 HP, 23 pounds	1 Elec	4	2	Ea.		120		120	179
5110	3/4 HP, 28 pounds		4	2			120		120	179
5120	1 HP, 33 pounds		4	2			120		120	179
5130	1-1/2 HP, 44 pounds		3.20	2.500			150		150	224
5140	2 HP, 56 pounds		3	2.667			160		160	239
5150	3 HP, 71 pounds		2.30	3.478			209		209	310
5160	5 HP, 82 pounds		1.90	4.211			253		253	375
5170	7-1/2 HP, 124 pounds		1.50	5.333			320		320	480
5180	10 HP, 144 pounds		1.20	6.667			400		400	595
5190	15 HP, 185 pounds		1	8			480		480	715
5200	20 HP, 214 pounds	2 Elec	1.50	10.667			640		640	955
5210	25 HP, 266 pounds		1.40	11.429			685		685	1,025
5220	30 HP, 310 pounds		1.20	13.333			800		800	1,200
5230	40 HP, 400 pounds		1	16			960		960	1,425
5240	50 HP, 450 pounds		.90	17.778			1,075		1,075	1,600
5250	75 HP, 680 pounds		.80	20			1,200		1,200	1,800
5260	100 HP, 870 pounds	3 Elec	1	24			1,450		1,450	2,150
5270	125 HP, 940 pounds		.80	30			1,800		1,800	2,700
5280	150 HP, 1,200 pounds		.70	34.286			2,050		2,050	3,075
5290	175 HP, 1,300 pounds		.60	40			2,400		2,400	3,575
5300	200 HP, 1,400 pounds		.50	48			2,875		2,875	4,300

26 71 13.20 Motors

		Crew	Daily Output	Labor-Hours	Unit	Material	Labor	Equipment	Total	Total Incl O&P
0010	**MOTORS** 230/460 V, 60 HZ R263413-30									
0050	Dripproof, premium efficiency, 1.15 service factor									
0060	1,800 RPM, 1/4 HP	1 Elec	5.33	1.501	Ea.	270	90		360	430
0070	1/3 HP		5.33	1.501		240	90		330	400
0080	1/2 HP		5.33	1.501		206	90		296	360
0090	3/4 HP		5.33	1.501		285	90		375	450
0100	1 HP		4.50	1.778		305	107		412	495
0150	2 HP		4.50	1.778		370	107		477	565
0200	3 HP		4.50	1.778		760	107		867	995
0250	5 HP		4.50	1.778		610	107		717	830
0300	7.5 HP		4.20	1.905		1,100	114		1,214	1,400
0350	10 HP		4	2		1,350	120		1,470	1,675
0400	15 HP		3.20	2.500		1,800	150		1,950	2,200
0450	20 HP	2 Elec	5.20	3.077		2,125	185		2,310	2,625
0500	25 HP		5	3.200		2,750	192		2,942	3,300
0550	30 HP		4.80	3.333		3,025	200		3,225	3,625
0600	40 HP		4	4		3,950	240		4,190	4,700
0650	50 HP		3.20	5		4,100	300		4,400	4,950
0700	60 HP		2.80	5.714		4,175	345		4,520	5,100
0750	75 HP		2.40	6.667		5,050	400		5,450	6,175
0800	100 HP	3 Elec	2.70	8.889		6,525	535		7,060	7,975
0850	125 HP		2.10	11.429		6,825	685		7,510	8,550

For customer support on your Electrical Costs with RSMeans data, call 800.448.8182.

313

26 71 13.20 Motors

		Crew	Daily Output	Labor-Hours	Unit	Material	2019 Bare Costs Labor	Equipment	Total	Total Incl O&P
0900	150 HP	3 Elec	1.80	13.333	Ea.	10,300	800		11,100	12,500
0950	200 HP	↓	1.50	16		12,600	960		13,560	15,200
1000	1,200 RPM, 1 HP	1 Elec	4.50	1.778		490	107		597	700
1050	2 HP		4.50	1.778		545	107		652	760
1100	3 HP		4.50	1.778		750	107		857	985
1150	5 HP		4.50	1.778		975	107		1,082	1,225
1200	3,600 RPM, 2 HP		4.50	1.778		575	107		682	790
1250	3 HP		4.50	1.778		590	107		697	810
1300	5 HP	↓	4.50	1.778	↓	555	107		662	770
1350	Totally enclosed, premium efficiency 1.15 service factor									
1360	1,800 RPM, 1/4 HP	1 Elec	5.33	1.501	Ea.	400	90		490	575
1370	1/3 HP		5.33	1.501		320	90		410	490
1380	1/2 HP		5.33	1.501		375	90		465	545
1390	3/4 HP		5.33	1.501		415	90		505	595
1400	1 HP		4.50	1.778		640	107		747	860
1450	2 HP		4.50	1.778		735	107		842	970
1500	3 HP		4.50	1.778		725	107		832	960
1550	5 HP		4.50	1.778		800	107		907	1,050
1600	7.5 HP		4.20	1.905		1,100	114		1,214	1,375
1650	10 HP		4	2		1,425	120		1,545	1,750
1700	15 HP	↓	3.20	2.500		2,250	150		2,400	2,700
1750	20 HP	2 Elec	5.20	3.077		2,400	185		2,585	2,900
1800	25 HP		5	3.200		3,600	192		3,792	4,250
1850	30 HP		4.80	3.333		3,700	200		3,900	4,375
1900	40 HP		4	4		4,050	240		4,290	4,800
1950	50 HP		3.20	5		4,475	300		4,775	5,375
2000	60 HP		2.80	5.714		5,975	345		6,320	7,050
2050	75 HP	↓	2.40	6.667		7,750	400		8,150	9,125
2100	100 HP	3 Elec	2.70	8.889		9,075	535		9,610	10,800
2150	125 HP		2.10	11.429		12,000	685		12,685	14,200
2200	150 HP		1.80	13.333		13,900	800		14,700	16,400
2250	200 HP	↓	1.50	16		17,500	960		18,460	20,700
2300	1,200 RPM, 1 HP	1 Elec	4.50	1.778		495	107		602	705
2350	2 HP		4.50	1.778		520	107		627	730
2400	3 HP		4.50	1.778		750	107		857	985
2450	5 HP		4.50	1.778		970	107		1,077	1,225
2500	3,600 RPM, 2 HP		4.50	1.778		435	107		542	635
2550	3 HP		4.50	1.778		575	107		682	790
2600	5 HP	↓	4.50	1.778	↓	675	107		782	900

26 71 13.40 Motors Explosion Proof

		Crew	Daily Output	Labor-Hours	Unit	Material	2019 Bare Costs Labor	Equipment	Total	Total Incl O&P
0010	**MOTORS EXPLOSION PROOF**, 208-230/460 V, 60 HZ									
0020	1,800 RPM, 1/4 HP	1 Elec	5	1.600	Ea.	660	96		756	870
0030	1/3 HP		5	1.600		278	96		374	450
0040	1/2 HP		5	1.600		450	96		546	640
0050	3/4 HP		5.33	1.501		360	90		450	530
0060	1 HP		4.20	1.905		580	114		694	805
0070	2 HP		4.20	1.905		600	114		714	830
0080	3 HP		4.20	1.905		930	114		1,044	1,200
0090	5 HP		4.20	1.905		910	114		1,024	1,175
0100	7.5 HP		4	2		1,075	120		1,195	1,350
0110	10 HP		3.70	2.162		1,325	130		1,455	1,675
0120	15 HP	↓	3.20	2.500		1,750	150		1,900	2,150

26 71 Electrical Machines

26 71 13 - Motors Applications

26 71 13.40 Motors Explosion Proof	Crew	Daily Output	Labor-Hours	Unit	Material	2019 Bare Costs Labor	Equipment	Total	Total Incl O&P
0130	20 HP	2 Elec	5	3.200	Ea.	2,175	192		2,367	2,650
0140	25 HP		4.80	3.333		2,600	200		2,800	3,175
0150	30 HP		4.40	3.636		3,025	218		3,243	3,650
0160	40 HP		3.80	4.211		4,425	253		4,678	5,225
0170	50 HP		3.20	5		4,075	300		4,375	4,925
0180	60 HP		2.60	6.154		7,700	370		8,070	9,025
0190	75 HP		2	8		9,475	480		9,955	11,100
0200	100 HP	3 Elec	2.50	9.600		12,300	575		12,875	14,400
0210	125 HP		2.20	10.909		15,700	655		16,355	18,300
0220	150 HP		1.60	15		18,600	900		19,500	21,900
0230	200 HP		1.20	20		22,900	1,200		24,100	27,000
1000	1,200 RPM, 1 HP	1 Elec	4.20	1.905		480	114		594	700
1010	2 HP		4.20	1.905		555	114		669	780
1020	3 HP		4.20	1.905		755	114		869	1,000
1030	5 HP		4.20	1.905		1,025	114		1,139	1,300
2000	3,600 RPM, 1/4 HP		5	1.600		415	96		511	600
2010	1/3 HP		5	1.600		435	96		531	620
2020	1/2 HP		5	1.600		325	96		421	505
2030	3/4 HP		5	1.600		620	96		716	830
2040	1 HP		4.20	1.905		705	114		819	945
2050	2 HP		4.20	1.905		855	114		969	1,100
2060	3 HP		4.20	1.905		935	114		1,049	1,200
2070	5 HP		4.20	1.905		1,250	114		1,364	1,550
2080	7.5 HP		4	2		1,425	120		1,545	1,725
2090	10 HP		3.70	2.162		1,650	130		1,780	2,000
2100	15 HP		3.20	2.500		2,225	150		2,375	2,675
2110	20 HP	2 Elec	5	3.200		2,750	192		2,942	3,300
2120	25 HP		4.80	3.333		3,400	200		3,600	4,025
2130	30 HP		4.40	3.636		3,975	218		4,193	4,675
2140	40 HP		3.80	4.211		4,975	253		5,228	5,850
2150	50 HP		3.20	5		4,225	300		4,525	5,100
2160	60 HP		2.60	6.154		7,775	370		8,145	9,100
2170	75 HP		2	8		9,500	480		9,980	11,100
2180	100 HP	3 Elec	2.50	9.600		12,900	575		13,475	15,000
2190	125 HP		2.20	10.909		16,300	655		16,955	18,900
2200	150 HP		1.60	15		19,400	900		20,300	22,800
2210	200 HP		1.20	20		24,700	1,200		25,900	29,000
2220	250 HP		1.20	20		18,800	1,200		20,000	22,500

For customer support on your Electrical Costs with RSMeans data, call 800.448.8182.

315

Division Notes

	CREW	DAILY OUTPUT	LABOR-HOURS	UNIT	BARE COSTS				TOTAL INCL O&P
					MAT.	LABOR	EQUIP.	TOTAL	

Estimating Tips

27 20 00 Data Communications
27 30 00 Voice Communications
27 40 00 Audio-Video Communications

- When estimating material costs for special systems, it is always prudent to obtain manufacturers' quotations for equipment prices and special installation requirements that may affect the total cost.

- For cost modifications for elevated tray installation, add the percentages to labor according to the height of the installation and only to the quantities exceeding the different height levels, not to the total tray quantities. Refer to subdivision 26 01 02.20 for labor adjustment factors.

- Do not overlook the costs for equipment used in the installation. If scissor lifts and boom lifts are available in the field, contractors may use them in lieu of the proposed ladders and rolling staging.

Reference Numbers

Reference numbers are shown at the beginning of some major classifications. These numbers refer to related items in the Reference Section. The reference information may be an estimating procedure, an alternate pricing method, or technical information.

Note: Not all subdivisions listed here necessarily appear. ■

Did you know?

RSMeans data is available through our online application:

- Search for costs by keyword
- Leverage the most up-to-date data
- Build and export estimates

Try it free
rsmeans.com/2019freetrial

27 01 Operation and Maintenance of Communications Systems

27 01 30 – Operation and Maintenance of Voice Communications

27 01 30.51 Operation and Maintenance of Voice Equipment

		Crew	Daily Output	Labor-Hours	Unit	Material	2019 Bare Costs Labor	2019 Bare Costs Equipment	Total	Total Incl O&P
0010	**OPERATION AND MAINTENANCE OF VOICE EQUIPMENT**									
3400	Remove and replace (reinstall), speaker	1 Elec	6	1.333	Ea.		80		80	119

27 05 Common Work Results for Communications

27 05 05 – Selective Demolition for Communications

27 05 05.20 Electrical Demolition, Communications

		Crew	Daily Output	Labor-Hours	Unit	Material	2019 Bare Costs Labor	2019 Bare Costs Equipment	Total	Total Incl O&P
0010	**ELECTRICAL DEMOLITION, COMMUNICATIONS** R024119-10									
0100	Fiber optics									
0120	Cable R260105-30	1 Elec	2400	.003	L.F.		.20		.20	.30
0160	Multi-channel rack enclosure		6	1.333	Ea.		80		80	119
0180	Patch panel	↓	18	.444	"		26.50		26.50	40
0200	Communication cables & fittings									
0220	Voice/data outlet	1 Elec	140	.057	Ea.		3.43		3.43	5.10
0240	Telephone cable		2800	.003	L.F.		.17		.17	.26
0260	Phone jack		135	.059	Ea.		3.56		3.56	5.30
0280	Data jack		135	.059	"		3.56		3.56	5.30
0300	High performance cable, 2 pair		3000	.003	L.F.		.16		.16	.24
0320	4 pair		2100	.004			.23		.23	.34
0340	25 pair		900	.009	↓		.53		.53	.80
0400	Terminal cabinet	↓	5	1.600	Ea.		96		96	143
1000	Nurse call system									
1020	Station	1 Elec	24	.333	Ea.		20		20	30
1040	Standard call button		24	.333			20		20	30
1060	Corridor dome light or zone indicator	↓	24	.333			20		20	30
1080	Master control station	2 Elec	2	8	↓		480		480	715

27 05 05.30 Electrical Demolition, Sound and Video

		Crew	Daily Output	Labor-Hours	Unit	Material	2019 Bare Costs Labor	2019 Bare Costs Equipment	Total	Total Incl O&P
0010	**ELECTRICAL DEMOLITION, SOUND & VIDEO** R024119-10									
0100	Cables									
0120	TV antenna lead-in cable R260105-30	1 Elec	2100	.004	L.F.		.23		.23	.34
0140	Sound cable		2400	.003			.20		.20	.30
0160	Microphone cable		2400	.003			.20		.20	.30
0180	Coaxial cable		2400	.003	↓		.20		.20	.30
0200	Doorbell system, not including wires, cables, and conduit		16	.500	Ea.		30		30	45
0220	Door chime or devices	↓	36	.222	"		13.35		13.35	19.90
0300	Public address system, not including wires, cables, and conduit									
0320	Conventional office	1 Elec	16	.500	Speaker		30		30	45
0340	Conventional industrial	"	8	1	"		60		60	89.50
0352	PA cabinet/panel	2 Elec	5	3.200	Ea.		192		192	287
0400	Sound system, not including wires, cables, and conduit									
0410	Components	1 Elec	24	.333	Ea.		20		20	30
0412	Speaker		24	.333			20		20	30
0416	Volume control		24	.333			20		20	30
0420	Intercom, master station		6	1.333			80		80	119
0440	Remote station		24	.333			20		20	30
0460	Intercom outlets		24	.333			20		20	30
0480	Handset	↓	12	.667	↓		40		40	59.50
0500	Emergency call system, not including wires, cables, and conduit									
0520	Annunciator	1 Elec	4	2	Ea.		120		120	179
0540	Devices	"	16	.500			30		30	45
0600	Master door, buzzer type unit	2 Elec	1.60	10	↓		600		600	895
0800	TV system, not including wires, cables, and conduit									

318

27 05 Common Work Results for Communications

27 05 05 – Selective Demolition for Communications

27 05 05.30 Electrical Demolition, Sound and Video

		Crew	Daily Output	Labor-Hours	Unit	Material	2019 Bare Costs Labor	Equipment	Total	Total Incl O&P
0820	Master TV antenna system, per outlet	1 Elec	39	.205	Outlet		12.30		12.30	18.40
0840	School application, per outlet		16	.500	"		30		30	45
0860	Amplifier		12	.667	Ea.		40		40	59.50
0880	Antenna	↓	6	1.333	"		80		80	119
0900	One camera & one monitor	2 Elec	7.80	2.051	Total		123		123	184
0920	One camera	1 Elec	8	1	Ea.		60		60	89.50

27 05 29 – Hangers and Supports for Communications Systems

27 05 29.10 Cable Support

		Crew	Daily Output	Labor-Hours	Unit	Material	2019 Bare Costs Labor	Equipment	Total	Total Incl O&P
0010	**CABLE SUPPORT**									
0110	J-hook, single tier, single sided, 1" diam.	1 Elec	68.57	.117	Ea.	2.16	7		9.16	12.85
0120	1-1/2" diam.		67.80	.118		2.72	7.10		9.82	13.55
0130	2" diam.		67.30	.119		3.18	7.15		10.33	14.15
0140	4" diam.		66.80	.120		6.20	7.20		13.40	17.55
0330	Double tier, single sided, 2" diam.		67	.119		11.45	7.15		18.60	23.50
0340	4" diam.		66.20	.121		15	7.25		22.25	27.50
0430	Double sided, 2" diam.		65.80	.122		17.55	7.30		24.85	30
0440	4" diam.		65.40	.122		27	7.35		34.35	40.50
0530	Triple tier, single sided, 2" diam.		66.40	.120		14.80	7.25		22.05	27
0540	4" diam.		65.85	.121		21.50	7.30		28.80	35
0630	Double sided, 2" diam.		65.10	.123		21	7.40		28.40	34
0640	4" diam.	↓	64.70	.124	↓	32	7.45		39.45	46.50

27 11 Communications Equipment Room Fittings

27 11 16 – Communications Cabinets, Racks, Frames and Enclosures

27 11 16.10 Public Phone

		Crew	Daily Output	Labor-Hours	Unit	Material	2019 Bare Costs Labor	Equipment	Total	Total Incl O&P
0010	**PUBLIC PHONE**									
7600	Telephone with wood backboard									
7620	Single door, 12" H x 12" W x 4" D	1 Elec	5.30	1.509	Ea.	94	90.50		184.50	239
7650	18" H x 12" W x 4" D		4.70	1.702		107	102		209	271
7700	24" H x 12" W x 4" D		4.20	1.905		183	114		297	370
7720	18" H x 18" W x 4" D		4.20	1.905		153	114		267	340
7750	24" H x 18" W x 4" D		4	2		221	120		341	420
7780	36" H x 36" W x 4" D		3.60	2.222		212	133		345	430
7800	24" H x 24" W x 6" D		3.60	2.222		290	133		423	520
7820	30" H x 24" W x 6" D		3.20	2.500		320	150		470	580
7850	30" H x 30" W x 6" D		2.70	2.963		395	178		573	700
7880	36" H x 30" W x 6" D		2.50	3.200		430	192		622	755
7900	48" H x 36" W x 6" D		2.20	3.636		770	218		988	1,175
7920	Double door, 48" H x 36" W x 6" D	↓	2	4	↓	1,125	240		1,365	1,575

27 11 16.20 Rack Mount Cabinet

		Crew	Daily Output	Labor-Hours	Unit	Material	2019 Bare Costs Labor	Equipment	Total	Total Incl O&P
0010	**RACK MOUNT CABINET**									
0100	80" H x 24" W x 40" D									
0110	No sides	1 Elec	9.60	.833	Ea.	1,400	50		1,450	1,600
0120	One side		8	1		1,475	60		1,535	1,725
0130	Two sides	↓	7.86	1.018	↓	1,600	61		1,661	1,875
0200	80" H x 24" W x 42" D									
0210	No sides	1 Elec	9.60	.833	Ea.	1,425	50		1,475	1,650
0220	One side		8	1		1,475	60		1,535	1,725
0230	Two sides	↓	7.86	1.018	↓	1,475	61		1,536	1,725
0300	80" H x 24" W x 48" D									

27 11 Communications Equipment Room Fittings

27 11 16 – Communications Cabinets, Racks, Frames and Enclosures

27 11 16.20 Rack Mount Cabinet	Crew	Daily Output	Labor-Hours	Unit	Material	2019 Bare Costs Labor	Equipment	Total	Total Incl O&P
0310	No sides	1 Elec	9.60	.833	Ea.	1,625	50		1,675	1,850
0320	One side		8	1		1,700	60		1,760	1,950
0330	Two sides	↓	7.86	1.018	↓	1,725	61		1,786	2,000
0400	80" H x 30" W x 40" D									
0410	No sides	1 Elec	9.60	.833	Ea.	1,450	50		1,500	1,650
0420	One side		8	1		1,475	60		1,535	1,725
0430	Two sides	↓	7.86	1.018	↓	1,525	61		1,586	1,775
0500	80" H x 30" W x 42" D									
0510	No sides	1 Elec	9.60	.833	Ea.	1,450	50		1,500	1,675
0520	One side		8	1		1,475	60		1,535	1,725
0530	Two sides	↓	7.86	1.018	↓	1,450	61		1,511	1,700
0600	80" H x 30" W x 48" D									
0610	No sides	1 Elec	9.60	.833	Ea.	1,550	50		1,600	1,775
0620	One side		8	1		1,600	60		1,660	1,850
0630	Two sides	↓	7.86	1.018	↓	1,650	61		1,711	1,925
0700	80" H x 32" W x 32" D									
0710	No sides	1 Elec	9.60	.833	Ea.	1,600	50		1,650	1,850
0720	One side		8	1		1,625	60		1,685	1,900
0730	Two sides	↓	7.86	1.018	↓	1,650	61		1,711	1,925
0800	80" H x 32" W x 40" D									
0810	No sides	1 Elec	9.60	.833	Ea.	1,725	50		1,775	1,975
0820	One side		8	1		1,775	60		1,835	2,050
0830	Two sides	↓	7.86	1.018	↓	1,800	61		1,861	2,075
0900	80" H x 32" W x 42" D									
0910	No sides	1 Elec	9.60	.833	Ea.	1,800	50		1,850	2,050
0920	One side		8	1		1,825	60		1,885	2,100
0930	Two sides	↓	7.86	1.018	↓	1,825	61		1,886	2,125
1000	80" H x 32" W x 48" D									
1010	No sides	1 Elec	9.60	.833	Ea.	1,850	50		1,900	2,125
1020	One side		8	1		1,925	60		1,985	2,225
1030	Two sides	↓	7.86	1.018	↓	1,350	61		1,411	1,575
1200	Freestanding cabinet with front and rear doors									
1300	84" H x 19" W x 24" D, gray									
1310	No sides	1 Elec	9.60	.833	Ea.	1,625	50		1,675	1,875
1320	One side		9.60	.833		2,525	50		2,575	2,850
1330	Two sides	↓	9.60	.833	↓	2,525	50		2,575	2,850
1400	84" H x 19" W x 24" D, white									
1410	No sides	1 Elec	9.60	.833	Ea.	2,125	50		2,175	2,425
1420	One side		9.60	.833		2,525	50		2,575	2,850
1430	Two sides	↓	9.60	.833	↓	2,525	50		2,575	2,850
1500	84" H x 19" W x 24" D, black									
1510	No sides	1 Elec	9.60	.833	Ea.	1,625	50		1,675	1,850
1520	One side		9.60	.833		2,400	50		2,450	2,725
1530	Two sides		9.60	.833		2,175	50		2,225	2,475
1600	84" H x 28" W x 34" D front glass, rear louvered door	↓	9.60	.833	↓	1,450	50		1,500	1,675

27 11 19 – Communications Termination Blocks and Patch Panels

27 11 19.10 Termination Blocks and Patch Panels

0010	**TERMINATION BLOCKS AND PATCH PANELS**									
2960	Patch panel, RJ-45/110 type, 24 ports	2 Elec	6	2.667	Ea.	189	160		349	445
3000	48 ports	3 Elec	6	4		315	240		555	705
3040	96 ports	"	4	6		490	360		850	1,075
3100	Punch down termination per port	1 Elec	107	.075	↓		4.49		4.49	6.70

320

For customer support on your Electrical Costs with RSMeans data, call 800.448.8182.

27 13 23 – Communications Optical Fiber Backbone Cabling

27 13 23.13 Communications Optical Fiber		Crew	Daily Output	Labor-Hours	Unit	Material	2019 Bare Costs Labor	Equipment	Total	Total Incl O&P
0010	**COMMUNICATIONS OPTICAL FIBER**									
0040	Specialized tools & techniques cause installation costs to vary.									
0070	Fiber optic, cable, bulk simplex, single mode R271323-40	1 Elec	8	1	C.L.F.	23	60		83	115
0080	Multi mode		8	1		29.50	60		89.50	122
0090	4 strand, single mode		7.34	1.090		37.50	65.50		103	139
0095	Multi mode		7.34	1.090		50	65.50		115.50	153
0100	12 strand, single mode		6.67	1.199		75.50	72		147.50	191
0105	Multi mode		6.67	1.199		99.50	72		171.50	216
0150	Jumper				Ea.	36			36	39.50
0200	Pigtail					34			34	37.50
0300	Connector	1 Elec	24	.333		25.50	20		45.50	58
0350	Finger splice		32	.250		37.50	15		52.50	64
0400	Transceiver (low cost bi-directional)		8	1		440	60		500	575
0450	Rack housing, 4 rack spaces, 12 panels (144 fibers)		2	4		610	240		850	1,025
0500	Patch panel, 12 ports		6	1.333		300	80		380	450
1000	Cable, 62.5 microns, direct burial, 4 fiber	R-15	1200	.040	L.F.	.90	2.35	.24	3.49	4.76
1020	Indoor, 2 fiber	R-19	1000	.020		.44	1.20		1.64	2.28
1040	Outdoor, aerial/duct	"	1670	.012		.65	.72		1.37	1.80
1060	50 microns, direct burial, 8 fiber	R-22	4000	.009		1.27	.51		1.78	2.16
1080	12 fiber		4000	.009		2.05	.51		2.56	3.02
1100	Indoor, 12 fiber		759	.049		1.99	2.70		4.69	6.20
1120	Connectors, 62.5 micron cable, transmission	R-19	40	.500	Ea.	14.15	30		44.15	60.50
1140	Cable splice		40	.500		14.90	30		44.90	61.50
1160	125 micron cable, transmission		16	1.250		14.90	75		89.90	128
1180	Receiver, 1.2 mile range		20	1		261	60		321	375
1200	1.9 mile range		20	1		221	60		281	335
1220	6.2 mile range		5	4		278	241		519	665
1240	Transmitter, 1.2 mile range		20	1		287	60		347	405
1260	1.9 mile range		20	1		279	60		339	395
1280	6.2 mile range		5	4		415	241		656	815
1300	Modem, 1.2 mile range		5	4		172	241		413	550
1320	6.2 mile range		5	4		315	241		556	705
1340	1.9 mile range, 12 channel		5	4		2,000	241		2,241	2,550
1360	Repeater, 1.2 mile range		10	2		370	120		490	585
1380	1.9 mile range		10	2		475	120		595	700
1400	6.2 mile range		5	4		910	241		1,151	1,350
1420	1.2 mile range, digital		5	4		435	241		676	840
1440	Transceiver, 1.9 mile range		5	4		385	241		626	780
1460	1.2 mile range, digital		5	4		395	241		636	795
1480	Cable enclosure, interior NEMA 13		7	2.857		144	172		316	415
1500	Splice w/enclosure encapsulant		16	1.250		221	75		296	355
1510	1/4", black, non-metallic, flexible, tube, liquid tight		8	2.500	C.L.F.	150	150		300	390
1520	3/8"		8	2.500		90.50	150		240.50	325
1530	1/2"		7.34	2.725		125	164		289	385
1540	3/4"		7.34	2.725		225	164		389	495
1550	1"		7.34	2.725		330	164		494	605
1560	1-1/4"		6.67	2.999		520	180		700	840
1570	1-1/2"		6.67	2.999		570	180		750	900
1580	2"		6.35	3.150		860	189		1,049	1,225
2040	Fiber optic cable, 48 strand, single mode, steel armor, material		1.67	11.976	M.L.F.	4,650	720		5,370	6,175

27 15 01 – Communications Horizontal Cabling Applications

27 15 01.19 Fire Alarm Communications Conductors & Cables	Crew	Daily Output	Labor-Hours	Unit	Material	2019 Bare Costs Labor	Equipment	Total	Total Incl O&P
0010 **FIRE ALARM COMMUNICATIONS CONDUCTORS AND CABLES**									
1500 Fire alarm FEP teflon 150 V to 200°C									
1550 #22, 1 pair	1 Elec	10	.800	C.L.F.	97	48		145	179
1600 2 pair		8	1		95.50	60		155.50	195
1650 4 pair		7	1.143		350	68.50		418.50	485
1700 6 pair		6	1.333		425	80		505	585
1750 8 pair		5.50	1.455		710	87.50		797.50	910
1800 10 pair		5	1.600		655	96		751	865
1850 #18, 1 pair	2 Elec	16	1		68.50	60		128.50	165
1900 2 pair		13	1.231		221	74		295	355
1950 4 pair		9.60	1.667		205	100		305	375
2000 6 pair		8	2		435	120		555	655
2050 8 pair		7	2.286		320	137		457	560
2100 10 pair		6	2.667		480	160		640	765

27 15 01.23 Audio-Video Communications Horizontal Cabling

	Crew	Daily Output	Labor-Hours	Unit	Material	2019 Bare Costs Labor	Equipment	Total	Total Incl O&P
0010 **AUDIO-VIDEO COMMUNICATIONS HORIZONTAL CABLING**									
3000 S-video cable, MD4, 1M Pro interconnects, 3'	1 Elec	38	.211	Ea.	34	12.65		46.65	56.50
3010 6'		38	.211		31	12.65		43.65	53
3020 9'		38	.211		45	12.65		57.65	68.50
3030 16'		36	.222		48.50	13.35		61.85	73.50
3040 22'		36	.222		57	13.35		70.35	82.50
3300 HDMI cable, coaxial, HDMI/HDMI, Non plenum, 3', 24 kt terminations		40	.200		14.05	12		26.05	33.50
3310 6', 24 kt terminations		40	.200		16.60	12		28.60	36
3320 HDMI cable, coaxial, HDMI/HDMI metal, Non plenum, 3', 24 kt terminations		40	.200		16.05	12		28.05	35.50
3330 6', 24 kt terminations		40	.200		19.65	12		31.65	39.50
3340 15', 24 kt terminations		38	.211		27.50	12.65		40.15	49
3350 25', 24 kt terminations		38	.211		35.50	12.65		48.15	58.50

27 15 10 – Special Communications Cabling

27 15 10.23 Sound and Video Cables and Fittings

	Crew	Daily Output	Labor-Hours	Unit	Material	2019 Bare Costs Labor	Equipment	Total	Total Incl O&P
0010 **SOUND AND VIDEO CABLES & FITTINGS**									
0900 TV antenna lead-in, 300 ohm, #20-2 conductor	1 Elec	7	1.143	C.L.F.	22	68.50		90.50	126
0950 Coaxial, feeder outlet		7	1.143		20.50	68.50		89	125
1000 Coaxial, main riser		6	1.333		13.50	80		93.50	134
1100 Sound, shielded with drain, #22-2 conductor		8	1		8.35	60		68.35	98.50
1150 #22-3 conductor		7.50	1.067		12.40	64		76.40	109
1200 #22-4 conductor		6.50	1.231		19.90	74		93.90	132
1250 Nonshielded, #22-2 conductor		10	.800		10.25	48		58.25	83
1300 #22-3 conductor		9	.889		20.50	53.50		74	102
1350 #22-4 conductor		8	1		22.50	60		82.50	115
1400 Microphone cable		8	1		30	60		90	123

27 15 13 – Communications Copper Horizontal Cabling

27 15 13.13 Communication Cables and Fittings

	Crew	Daily Output	Labor-Hours	Unit	Material	2019 Bare Costs Labor	Equipment	Total	Total Incl O&P
0010 **COMMUNICATION CABLES AND FITTINGS**									
2200 Telephone twisted, PVC insulation, #22-2 conductor	1 Elec	10	.800	C.L.F.	10.30	48		58.30	83
2250 #22-3 conductor		9	.889		12.70	53.50		66.20	93.50
2300 #22-4 conductor		8	1		15	60		75	106
2350 #18-2 conductor		9	.889		21	53.50		74.50	103
2370 Telephone jack, eight pins		32	.250	Ea.	3.48	15		18.48	26.50
5000 High performance unshielded twisted pair (UTP)									
5100 Cable, category 3, #24, 2 pair solid, PVC jacket R271513-75	1 Elec	10	.800	C.L.F.	9.65	48		57.65	82
5200 4 pair solid, PVC jacket		7	1.143		13.65	68.50		82.15	117

27 15 13 – Communications Copper Horizontal Cabling

27 15 13.13 Communication Cables and Fittings		Crew	Daily Output	Labor-Hours	Unit	Material	2019 Bare Costs Labor	Equipment	Total	Total Incl O&P
5300	25 pair solid, PVC jacket	1 Elec	3	2.667	C.L.F.	81	160		241	330
5400	2 pair solid, plenum		10	.800		11.15	48		59.15	84
5500	4 pair solid, plenum		7	1.143		13.20	68.50		81.70	117
5600	25 pair solid, plenum		3	2.667		114	160		274	365
5700	4 pair stranded, PVC jacket		7	1.143		12	68.50		80.50	115
7000	Category 5, #24, 4 pair solid, PVC jacket		7	1.143		10.35	68.50		78.85	113
7100	4 pair solid, plenum		7	1.143		14.45	68.50		82.95	118
7200	4 pair stranded, PVC jacket		7	1.143		21.50	68.50		90	126
7210	Category 5e, #24, 4 pair solid, PVC jacket		7	1.143		15.05	68.50		83.55	119
7212	4 pair solid, plenum		7	1.143		21.50	68.50		90	126
7214	4 pair stranded, PVC jacket		7	1.143		22	68.50		90.50	126
7240	Category 6, #24, 4 pair solid, PVC jacket		7	1.143		18.55	68.50		87.05	123
7242	4 pair solid, plenum		7	1.143		13.70	68.50		82.20	117
7244	4 pair stranded, PVC jacket		7	1.143		14.10	68.50		82.60	118
7300	Connector, RJ45, category 5		80	.100	Ea.	.77	6		6.77	9.80
7302	Shielded RJ45, category 5		72	.111		1.75	6.65		8.40	11.90
7310	Jack, UTP RJ45, category 3		72	.111		.43	6.65		7.08	10.40
7312	Category 5		65	.123		5.80	7.40		13.20	17.40
7314	Category 5e		65	.123		4.92	7.40		12.32	16.45
7316	Category 6		65	.123		3.21	7.40		10.61	14.60
7322	Jack, shielded RJ45, category 5		60	.133		6.45	8		14.45	19.05
7324	Category 5e		60	.133		6.45	8		14.45	19.05
7326	Category 6		60	.133		7	8		15	19.65
7400	Voice/data expansion module, category 5e		8	1		47	60		107	142
7401	Modular jack, cat 6 keystone, RJ45, office white		16.50	.485		6	29		35	50
7402	RJ45, green		16.50	.485		5.90	29		34.90	50
7404	RJ45, grey		16.50	.485		7.35	29		36.35	51.50
7408	RJ45, orange		16.50	.485		6.05	29		35.05	50
7420	5M, RJ45, 568B, cord set		16.50	.485		32	29		61	79
7422	3M, RJ45, 568B, cord set		16.50	.485		33	29		62	79.50
7424	1M, RJ45, 568B, cord set		16.50	.485		25	29		54	71
7450	M12 TO RJ45, 2M ultra-loch D-mode, cord set		15.50	.516		37.50	31		68.50	88
8000	Multipair unshielded non-plenum cable, 150 V PVC jacket									
8002	#22, 2 pair	1 Elec	8.40	.952	C.L.F.	15.40	57		72.40	102
8003	3 pair		8	1		24.50	60		84.50	117
8004	4 pair		7.30	1.096		53.50	66		119.50	157
8006	6 pair		6.20	1.290		35	77.50		112.50	155
8008	8 pair		5.70	1.404		74	84.50		158.50	207
8010	10 pair		5.30	1.509		62.50	90.50		153	204
8012	12 pair		4.20	1.905		156	114		270	345
8015	15 pair		3.80	2.105		200	126		326	410
8020	20 pair	2 Elec	6.60	2.424		237	146		383	475
8025	25 pair		6	2.667		274	160		434	540
8030	30 pair		5.60	2.857		450	172		622	750
8040	40 pair		5.20	3.077		425	185		610	740
8050	50 pair		4.90	3.265		595	196		791	945
8100	Multipair unshielded non-plenum cable, 300 V PVC jacket									
8102	#20, 2 pair	1 Elec	7.30	1.096	C.L.F.	21.50	66		87.50	122
8103	3 pair		6.70	1.194		29	71.50		100.50	139
8104	4 pair		6	1.333		33.50	80		113.50	156
8106	6 pair		5.30	1.509		65	90.50		155.50	207
8108	8 pair		4.40	1.818		80	109		189	251
8110	10 pair	2 Elec	7.30	2.192		87.50	132		219.50	293

27 15 13 – Communications Copper Horizontal Cabling

27 15 13.13 Communication Cables and Fittings

		Crew	Daily Output	Labor-Hours	Unit	Material	2019 Bare Costs Labor	Equipment	Total	Total Incl O&P
8112	12 pair	2 Elec	6.20	2.581	C.L.F.	107	155		262	350
8115	15 pair	↓	5.90	2.712		134	163		297	390
8201	#18, 1 pair	1 Elec	8	1		75	60		135	172
8202	2 pair		6.50	1.231		77	74		151	195
8203	3 pair		5.60	1.429		116	86		202	255
8204	4 pair		5	1.600		151	96		247	310
8206	6 pair		4.40	1.818		247	109		356	435
8208	8 pair	↓	4	2		242	120		362	445
8215	15 pair	2 Elec	5	3.200	↓	640	192		832	990
8300	Multipair shielded non-plenum cable, 300 V PVC jacket									
8303	#22, 3 pair	1 Elec	6.70	1.194	C.L.F.	57	71.50		128.50	170
8306	6 pair		5.70	1.404		104	84.50		188.50	241
8309	9 pair	↓	5	1.600		140	96		236	297
8312	12 pair	2 Elec	8	2		400	120		520	620
8315	15 pair		7.30	2.192		505	132		637	755
8317	17 pair		6.80	2.353		595	141		736	865
8319	19 pair		6.40	2.500		595	150		745	880
8327	27 pair	↓	5.90	2.712		1,025	163		1,188	1,375
8402	#20, 2 pair	1 Elec	6.70	1.194		94	71.50		165.50	211
8403	3 pair		6.20	1.290		107	77.50		184.50	233
8406	6 pair		4.40	1.818		268	109		377	460
8409	9 pair	↓	3.60	2.222		277	133		410	505
8412	12 pair	2 Elec	5.90	2.712		500	163		663	800
8415	15 pair	"	5.70	2.807		485	169		654	780
8502	#18, 2 pair	1 Elec	6.20	1.290		108	77.50		185.50	235
8503	3 pair		5.30	1.509		170	90.50		260.50	320
8504	4 pair		4.70	1.702		198	102		300	370
8506	6 pair		4	2		315	120		435	525
8509	9 pair	↓	3.40	2.353		480	141		621	740
8515	15 pair	2 Elec	4.80	3.333	↓	810	200		1,010	1,200

27 15 33 – Communications Coaxial Horizontal Cabling

27 15 33.10 Coaxial Cable and Fittings

		Crew	Daily Output	Labor-Hours	Unit	Material	2019 Bare Costs Labor	Equipment	Total	Total Incl O&P
0010	**COAXIAL CABLE & FITTINGS**									
3500	Coaxial connectors, 50 ohm impedance quick disconnect									
3540	BNC plug, for RG A/U #58 cable	1 Elec	42	.190	Ea.	2.92	11.45		14.37	20.50
3550	RG A/U #59 cable		42	.190		2.89	11.45		14.34	20
3560	RG A/U #62 cable		42	.190		2.92	11.45		14.37	20.50
3600	BNC jack, for RG A/U #58 cable		42	.190		3.35	11.45		14.80	20.50
3610	RG A/U #59 cable		42	.190		3.53	11.45		14.98	21
3620	RG A/U #62 cable		42	.190		3.46	11.45		14.91	21
3660	BNC panel jack, for RG A/U #58 cable		40	.200		8.05	12		20.05	27
3670	RG A/U #59 cable		40	.200		6.60	12		18.60	25
3680	RG A/U #62 cable		40	.200		6.60	12		18.60	25
3720	BNC bulkhead jack, for RG A/U #58 cable		40	.200		5.95	12		17.95	24.50
3730	RG A/U #59 cable		40	.200		5.95	12		17.95	24.50
3740	RG A/U #62 cable		40	.200	↓	5.95	12		17.95	24.50
3850	Coaxial cable, RG A/U #58, 50 ohm		8	1	C.L.F.	49	60		109	144
3860	RG A/U #59, 75 ohm		8	1		39	60		99	133
3870	RG A/U #62, 93 ohm		8	1		45.50	60		105.50	140
3875	RG 6/U, 75 ohm		8	1		36.50	60		96.50	130
3950	Fire rated, RG A/U #58, 50 ohm		8	1		94	60		154	193
3960	RG A/U #59, 75 ohm		8	1		124	60		184	226

27 15 Communications Horizontal Cabling

27 15 33 – Communications Coaxial Horizontal Cabling

27 15 33.10 Coaxial Cable and Fittings	Crew	Daily Output	Labor-Hours	Unit	Material	2019 Bare Costs Labor	2019 Bare Costs Equipment	Total	Total Incl O&P
3970	RG A/U #62, 93 ohm	1 Elec	8	1	C.L.F.	109	60		169	210

27 15 43 – Communications Faceplates and Connectors

27 15 43.13 Communication Outlets

		Crew	Daily Output	Labor-Hours	Unit	Material	Labor	Equipment	Total	Total Incl O&P
0010	**COMMUNICATION OUTLETS**									
0100	Voice/data devices not included									
0120	Voice/data outlets, single opening	1 Elec	48	.167	Ea.	6.50	10		16.50	22
0140	Two jack openings		48	.167		2.26	10		12.26	17.45
0160	One jack & one 3/4" round opening		48	.167		7.35	10		17.35	23
0180	One jack & one twinaxial opening		48	.167		7.60	10		17.60	23.50
0200	One jack & one connector cabling opening		48	.167		7.35	10		17.35	23
0220	Two 3/8" coaxial openings		48	.167		7.35	10		17.35	23
0300	Data outlets, single opening		48	.167		7.35	10		17.35	23
0320	One 25-pin subminiature opening		48	.167		7.35	10		17.35	23
1000	Voice/data wall plate, plastic, 1 gang, 1-port		72	.111		2.33	6.65		8.98	12.50
1020	2-port		72	.111		2.80	6.65		9.45	13.05
1040	3-port		72	.111		2.41	6.65		9.06	12.60
1060	4-port		72	.111		2.87	6.65		9.52	13.10
1080	6-port		72	.111		2.41	6.65		9.06	12.60
1100	2 gang, 6-port		48	.167		6.95	10		16.95	22.50
1120	Voice/data wall plate, stainless steel, 1 gang, 1-port		72	.111		8.20	6.65		14.85	18.95
1140	2-port		72	.111		7.50	6.65		14.15	18.20
1160	3-port		72	.111		7.55	6.65		14.20	18.25
1180	4-port		72	.111		7.55	6.65		14.20	18.30
1200	2 gang, 6-port	▼	48	.167	▼	11.85	10		21.85	28

27 21 Data Communications Network Equipment

27 21 23 – Data Communications Switches and Hubs

27 21 23.10 Switching and Routing Equipment

		Crew	Daily Output	Labor-Hours	Unit	Material	Labor	Equipment	Total	Total Incl O&P
0010	**SWITCHING AND ROUTING EQUIPMENT**									
1100	Network hub, dual speed, 24 ports, includes cabinet	3 Elec	.66	36.364	Ea.	865	2,175		3,040	4,200
1300	Network switch, 50/60 HZ, 8 port, multi-platform, analog KVM	1 Elec	6.85	1.168		1,100	70		1,170	1,300
1310	16 port, multi-platform, analog KVM		6	1.333		1,450	80		1,530	1,700
1320	Network switch, 0x2x16,CAT5, analog KVM		6	1.333		1,100	80		1,180	1,325
1330	2x1x16, digital KVM, w/VM		16	.500		3,475	30		3,505	3,875
1340	2x1x32, digital KVM, w/VM		16	.500		4,150	30		4,180	4,600
1350	8x1x32, digital KVM, w/VM		16	.500		4,775	30		4,805	5,325
1500	KVM,1 RMU 16 port, no keyboard		16	.500		1,575	30		1,605	1,800
1510	KVM,1 RMU, 17" LCD, 16 port, US keyboard		15.15	.528		3,050	31.50		3,081.50	3,400
1520	KVM,1 RMU, 17" LCD, 16 port, UK keyboard		15.15	.528		3,050	31.50		3,081.50	3,400
2000	10/100/1000 Mbps, 24 ports		14	.571		590	34.50		624.50	695
2040	10/100/1000 Mbps, 48 ports		12	.667		1,300	40		1,340	1,475
2050	10/100/Mbps, 5 port, industrial ethernet type		32	.250		365	15		380	425
2060	10/100/Mbps, 6 port, industrial ethernet type		32	.250		1,300	15		1,315	1,450
2070	10/100/1000 Mbps, 10 port, X307-3		28	.286		3,075	17.15		3,092.15	3,425
2080	10/100/1000 Mbps, 10 port, X308-2		28	.286		2,850	17.15		2,867.15	3,150
3000	10/100/1000 Mbps, 20 port, front ports		20	.400		4,475	24		4,499	4,950
3010	10/100/1000 Mbps, 20 port, rear ports		20	.400		4,475	24		4,499	4,950
3040	KVM, 10/100/1000/10000 Mbps, 28 port, rear ports		16	.500		9,275	30		9,305	10,200
3050	KVM, 10/100/1000/10000 Mbps, 28 port, front ports	▼	16	.500		10,900	30		10,930	12,000
3070	KVM, 10/100/1000/10000 Mbps, 52 port, front ports	2 Elec	12	1.333	▼	11,400	80		11,480	12,700

27 21 Data Communications Network Equipment

27 21 23 – Data Communications Switches and Hubs

27 21 23.10 Switching and Routing Equipment	Crew	Daily Output	Labor-Hours	Unit	Material	2019 Bare Costs Labor	Equipment	Total	Total Incl O&P
3080	KVM, 10/100/1000/10000 Mbps, 52 port, rear ports	2 Elec	12	1.333	Ea.	11,400	80		11,480	12,700

27 32 Voice Communications Terminal Equipment

27 32 26 – Ring-Down Emergency Telephones

27 32 26.10 Emergency Phone Stations

	27 32 26.10 Emergency Phone Stations	Crew	Daily Output	Labor-Hours	Unit	Material	Labor	Equipment	Total	Total Incl O&P
0010	**EMERGENCY PHONE STATIONS**									
0100	Hands free emergency speaker phone w/LED strobe	1 Elec	7.22	1.108	Ea.	865	66.50		931.50	1,050
0110	Wall/pole mounted emergency phone station w/LED strobe	"	4.60	1.739		1,525	104		1,629	1,825
2000	Call station, 30", pole mounted emergency phone, 120V/ VoIP, blue beacon	2 Elec	5	3.200		1,100	192		1,292	1,475
2010	Call station, 36"		5	3.200		1,575	192		1,767	2,025
2020	Call station, 5'8"		4.80	3.333		2,675	200		2,875	3,250
2030	Call station, 9'		4.80	3.333		2,700	200		2,900	3,275

27 32 36 – TTY Equipment

27 32 36.10 TTY Telephone Equipment

	27 32 36.10 TTY Telephone Equipment	Crew	Daily Output	Labor-Hours	Unit	Material	Labor	Equipment	Total	Total Incl O&P
0010	**TTY TELEPHONE EQUIPMENT**									
1620	Telephone, TTY, compact, pocket type				Ea.	305			305	335
1630	Advanced, desk type	2 Elec	20	.800		515	48		563	640
1640	Full-featured public, wall type	"	4	4		685	240		925	1,125

27 41 Audio-Video Systems

27 41 33 – Master Antenna Television Systems

27 41 33.10 TV Systems

	27 41 33.10 TV Systems	Crew	Daily Output	Labor-Hours	Unit	Material	Labor	Equipment	Total	Total Incl O&P
0010	**TV SYSTEMS,** not including rough-in wires, cables & conduits									
0100	Master TV antenna system									
0200	VHF reception & distribution, 12 outlets	1 Elec	6	1.333	Outlet	118	80		198	249
0400	30 outlets		10	.800		240	48		288	335
0600	100 outlets		13	.615		286	37		323	370
0800	VHF & UHF reception & distribution, 12 outlets		6	1.333		251	80		331	395
1000	30 outlets		10	.800		169	48		217	258
1200	100 outlets		13	.615		172	37		209	245
1400	School and deluxe systems, 12 outlets		2.40	3.333		330	200		530	665
1600	30 outlets		4	2		291	120		411	500
1800	80 outlets		5.30	1.509		279	90.50		369.50	440
1900	Amplifier		4	2	Ea.	830	120		950	1,100
5000	Antenna, small		6	1.333		55.50	80		135.50	180
5100	Large		4	2		233	120		353	435
5110	Rotor unit		8	1		64	60		124	160
5120	Single booster		8	1		31	60		91	124
5130	Antenna pole, 10'		3.20	2.500		19.70	150		169.70	246
6100	Satellite TV system	2 Elec	1	16		2,175	960		3,135	3,800
6110	Dish, mesh, 10' diam.	"	2.40	6.667		1,725	400		2,125	2,500
6111	Two way RF/IF tapeoff	1 Elec	36	.222		3.59	13.35		16.94	24
6112	Two way RF/IF splitter		24	.333		13.80	20		33.80	45
6113	Line amplifier		24	.333		13.15	20		33.15	44.50
6114	Line splitters		36	.222		4.59	13.35		17.94	25
6115	Line multi switches		8	1		660	60		720	815
6120	Motor unit		2.40	3.333		263	200		463	590
7000	Home theater, widescreen, 42", high definition, TV		10.24	.781		400	47		447	510
7050	Flat wall mount bracket		10.24	.781		68.50	47		115.50	146

27 41 Audio-Video Systems

27 41 33 – Master Antenna Television Systems

27 41 33.10 TV Systems	Crew	Daily Output	Labor-Hours	Unit	Material	2019 Bare Costs Labor	Equipment	Total	Total Incl O&P
7100	7 channel home theater receiver	1 Elec	10.24	.781	Ea.	385	47		432	495
7200	Home theater speakers		10.24	.781	Set	216	47		263	305
7300	Home theater programmable remote		10.24	.781	Ea.	250	47		297	345
8000	Main video splitter		4	2		3,275	120		3,395	3,775
8010	Video distribution units		4	2		141	120		261	335

27 42 Electronic Digital Systems

27 42 13 – Point of Sale Systems

27 42 13.10 Bar Code Scanner

		Crew	Daily Output	Labor-Hours	Unit	Material	Labor	Equipment	Total	Total Incl O&P
0010	**BAR CODE SCANNER**									
0100	CR3600, w/B4 battery, handle	1 Elec	6	1.333	Outlet	910	80		990	1,125
0110	Palm		12	.667		860	40		900	1,000
0120	Batch, B4 batt, handle, chgstn, 3' USB		9	.889		860	53.50		913.50	1,025
0130	Palm, chgstn, 3' USB		12	.667		840	40		880	985
0140	CR2500, w/H2 handle, no cable		10	.800		640	48		688	775
0150	Core unit, palm, no cable		16	.500		570	30		600	675

27 51 Distributed Audio-Video Communications Systems

27 51 16 – Public Address Systems

27 51 16.10 Public Address System

		Crew	Daily Output	Labor-Hours	Unit	Material	Labor	Equipment	Total	Total Incl O&P
0010	**PUBLIC ADDRESS SYSTEM**									
0100	Conventional, office	1 Elec	5.33	1.501	Speaker	165	90		255	315
0200	Industrial	"	2.70	2.963	"	320	178		498	615

27 51 19 – Sound Masking Systems

27 51 19.10 Sound System

		Crew	Daily Output	Labor-Hours	Unit	Material	Labor	Equipment	Total	Total Incl O&P
0010	**SOUND SYSTEM**, not including rough-in wires, cables & conduits									
0100	Components, projector outlet	1 Elec	8	1	Ea.	58.50	60		118.50	154
0200	Microphone		4	2		98	120		218	287
0400	Speakers, ceiling or wall		8	1		144	60		204	248
0600	Trumpets		4	2		267	120		387	475
0800	Privacy switch		8	1		107	60		167	207
1000	Monitor panel		4	2		475	120		595	705
1200	Antenna, AM/FM		4	2		134	120		254	325
1400	Volume control		8	1		57	60		117	152
1600	Amplifier, 250 W		1	8		1,225	480		1,705	2,075
1800	Cabinets		1	8		1,025	480		1,505	1,850
2000	Intercom, 30 station capacity, master station	2 Elec	2	8		2,325	480		2,805	3,275
2020	10 station capacity	"	4	4		1,375	240		1,615	1,875
2200	Remote station	1 Elec	8	1		199	60		259	310
2400	Intercom outlets		8	1		117	60		177	219
2600	Handset		4	2		390	120		510	605
2800	Emergency call system, 12 zones, annunciator		1.30	6.154		1,175	370		1,545	1,825
3000	Bell		5.30	1.509		121	90.50		211.50	268
3200	Light or relay		8	1		60.50	60		120.50	156
3400	Transformer		4	2		266	120		386	470
3600	House telephone, talking station		1.60	5		570	300		870	1,075
3800	Press to talk, release to listen		5.30	1.509		133	90.50		223.50	281
4000	System-on button					79.50			79.50	87.50
4200	Door release	1 Elec	4	2		142	120		262	335

27 51 Distributed Audio-Video Communications Systems

27 51 19 – Sound Masking Systems

27 51 19.10 Sound System		Crew	Daily Output	Labor-Hours	Unit	Material	2019 Bare Costs Labor	Equipment	Total	Total Incl O&P
4400	Combination speaker and microphone	1 Elec	8	1	Ea.	242	60		302	355
4600	Termination box		3.20	2.500		76	150		226	310
4800	Amplifier or power supply		5.30	1.509		875	90.50		965.50	1,100
5000	Vestibule door unit		16	.500	Name	161	30		191	222
5200	Strip cabinet		27	.296	Ea.	305	17.80		322.80	360
5400	Directory		16	.500		143	30		173	202
6000	Master door, button buzzer type, 100 unit	2 Elec	.54	29.630		1,450	1,775		3,225	4,250
6020	200 unit		.30	53.333		2,750	3,200		5,950	7,800
6040	300 unit		.20	80		4,175	4,800		8,975	11,800
6060	Transformer	1 Elec	8	1		37	60		97	131
6080	Door opener		5.30	1.509		53.50	90.50		144	194
6100	Buzzer with door release and plate		4	2		53.50	120		173.50	238
6200	Intercom type, 100 unit	2 Elec	.54	29.630		1,800	1,775		3,575	4,650
6220	200 unit		.30	53.333		3,525	3,200		6,725	8,650
6240	300 unit		.20	80		5,325	4,800		10,125	13,000
6260	Amplifier	1 Elec	2	4		266	240		506	650
6280	Speaker with door release	"	4	2		79.50	120		199.50	267

27 52 Healthcare Communications and Monitoring Systems

27 52 23 – Nurse Call/Code Blue Systems

27 52 23.10 Nurse Call Systems

		Crew	Daily Output	Labor-Hours	Unit	Material	Labor	Equipment	Total	Total Incl O&P
0010	**NURSE CALL SYSTEMS**									
0100	Single bedside call station	1 Elec	8	1	Ea.	171	60		231	278
0200	Ceiling speaker station		8	1		87	60		147	186
0400	Emergency call station		8	1		90	60		150	189
0600	Pillow speaker		8	1		223	60		283	335
0800	Double bedside call station		4	2		180	120		300	375
1000	Duty station		4	2		131	120		251	325
1200	Standard call button		8	1		126	60		186	229
1400	Lights, corridor, dome or zone indicator		8	1		60.50	60		120.50	156
1600	Master control station for 20 stations	2 Elec	.65	24.615	Total	3,625	1,475		5,100	6,175

27 53 Distributed Systems

27 53 13 – Clock Systems

27 53 13.50 Clock Equipments

		Crew	Daily Output	Labor-Hours	Unit	Material	Labor	Equipment	Total	Total Incl O&P
0010	**CLOCK EQUIPMENTS**, not including wires & conduits									
0100	Time system components, master controller	1 Elec	.33	24.242	Ea.	1,475	1,450		2,925	3,775
0200	Program bell		8	1		110	60		170	210
0400	Combination clock & speaker		3.20	2.500		198	150		348	440
0600	Frequency generator		2	4		2,400	240		2,640	3,000
0800	Job time automatic stamp recorder		4	2		560	120		680	795
1600	Master time clock system, clocks & bells, 20 room	4 Elec	.20	160		5,425	9,600		15,025	20,300
1800	50 room	"	.08	400		11,400	24,000		35,400	48,300
1900	Time clock	1 Elec	3.20	2.500		385	150		535	650
2000	100 cards in & out, 1 color					9.35			9.35	10.25
2200	2 colors					8.75			8.75	9.60
2800	Metal rack for 25 cards	1 Elec	7	1.143		46	68.50		114.50	153
4000	Wireless time systems component, master controller	"	2	4		635	240		875	1,050
4010	For transceiver and antenna, see Section 28 39 10.10									

27 53 13.50 Clock Equipments	Crew	Daily Output	Labor-Hours	Unit	Material	2019 Bare Costs Labor	Equipment	Total	Total Incl O&P
4100	Wireless analog clock w/battery operated	1 Elec	8	1	Ea.	110	60		170	211
4200	Wireless digital clock w/battery operated	"	8	1	"	420	60		480	550

For customer support on your Electrical Costs with RSMeans data, call 800.448.8182.

329

Division Notes

		CREW	DAILY OUTPUT	LABOR-HOURS	UNIT	BARE COSTS				TOTAL INCL O&P
						MAT.	LABOR	EQUIP.	TOTAL	

Estimating Tips

- When estimating material costs for electronic safety and security systems, it is always prudent to obtain manufacturers' quotations for equipment prices and special installation requirements that may affect the total cost.

- Fire alarm systems consist of control panels, annunciator panels, batteries with rack, charger, and fire alarm actuating and indicating devices. Some fire alarm systems include speakers, telephone lines, door closer controls, and other components. Be careful not to overlook the costs related to installation for these items. Also be aware of costs for integrated automation instrumentation and terminal devices, control equipment, control wiring, and programming. Insurance underwriters may have specific requirements for the type of materials to be installed or design requirements based on the hazard to be protected. Local jurisdictions may have requirements not covered by code. It is advisable to be aware of any special conditions.

- Security equipment includes items such as CCTV, access control, and other detection and identification systems to perform alert and alarm functions. Be sure to consider the costs related to installation for this security equipment, such as for integrated automation instrumentation and terminal devices, control equipment, control wiring, and programming.

Reference Numbers

Reference numbers are shown at the beginning of some major classifications. These numbers refer to related items in the Reference Section. The reference information may be an estimating procedure, an alternate pricing method, or technical information.

Did you know?

RSMeans data is available through our online application:

- Search for costs by keyword
- Leverage the most up-to-date data
- Build and export estimates

Try it free
rsmeans.com/2019freetrial

28 01 Operation and Maint. of Electronic Safety and Security

28 01 80 – Operation and Maintenance of Fire Detection and Alarm

28 01 80.51 Maintenance & Admin. of Fire Detection & Alarm	Crew	Daily Output	Labor-Hours	Unit	Material	2019 Bare Costs Labor	2019 Bare Costs Equipment	Total	Total Incl O&P
0010 **MAINTENANCE AND ADMINISTRATION OF FIRE DETECTION AND ALARM**									
3300 Remove and replace (reinstall), fire alarm device	1 Elec	5.33	1.501	Ea.		90		90	134

28 05 Common Work Results for Electronic Safety and Security

28 05 05 – Selective Demolition for Electronic Safety and Security

28 05 05.10 Safety and Security Demolition

		Crew	Daily Output	Labor-Hours	Unit	Material	2019 Bare Costs Labor	2019 Bare Costs Equipment	Total	Total Incl O&P
0010	**SAFETY AND SECURITY DEMOLITION**									
1050	Finger print/card reader	1 Elec	8	1	Ea.		60		60	89.50
1051	Electrical dml, CA 3000 card reader sys w/25 users		.31	25.600			1,525		1,525	2,300
1060	Video camera		8	1			60		60	89.50
1070	Motion detector, multi-channel		7	1.143			68.50		68.50	102
1090	Infrared detector		7	1.143			68.50		68.50	102
1130	Video monitor, 19"		8	1			60		60	89.50
1210	Fire alarm horn and strobe light		16	.500			30		30	45
1215	Remote fire alarm indicator light		23	.348			21		21	31
1220	Flame detector		23	.348			21		21	31
1230	Duct detector		10	.800			48		48	71.50
1240	Smoke detector		23	.348			21		21	31
1241	Fire alarm duct smoke remote test station		21.60	.370			22		22	33
1250	Fire alarm pull station		23	.348			21		21	31
1260	Fire alarm control panel, 4 to 8 zone	2 Elec	4	4			240		240	360
1270	12 to 16 zone	"	2.67	5.993			360		360	535
1280	Fire alarm annunciation panel, 4 to 8 zone	1 Elec	6	1.333			80		80	119
1290	12 to 16 zone	2 Elec	6	2.667			160		160	239
1300	Electrical dml, tamper switch	1 Elec	38	.211			12.65		12.65	18.85
1310	Electrical dml, flow switch	"	38	.211			12.65		12.65	18.85

28 05 19 – Storage Appliances for Electronic Safety and Security

28 05 19.11 Digital Video Recorder (DVR)

		Crew	Daily Output	Labor-Hours	Unit	Material	2019 Bare Costs Labor	2019 Bare Costs Equipment	Total	Total Incl O&P
0010	**DIGITAL VIDEO RECORDERS**									
0100	Pentaplex hybrid, internet protocol, and hard drive									
0200	4 channel	1 Elec	1.33	6.015	Ea.	1,650	360		2,010	2,350
0300	8 channel		1	8		3,675	480		4,155	4,775
0400	16 channel		1	8		3,725	480		4,205	4,825

28 15 Access Control Hardware Devices

28 15 11 – Integrated Credential Readers and Field Entry Management

28 15 11.11 Standard Card Readers

		Crew	Daily Output	Labor-Hours	Unit	Material	2019 Bare Costs Labor	2019 Bare Costs Equipment	Total	Total Incl O&P
0010	**STANDARD CARD READERS**									
0015	Card key access									
0020	Computerized system, processor, proximity reader and cards									
0030	Does not include door hardware, lockset or wiring									
0040	Card key system for 1 door				Ea.	1,425			1,425	1,575
0060	Card key system for 2 doors					2,150			2,150	2,375
0080	Card key system for 4 doors					2,525			2,525	2,775
0100	Processor for card key access system					1,000			1,000	1,100
0160	Magnetic lock for electric access, 600 pound holding force					177			177	195
0170	Magnetic lock for electric access, 1200 pound holding force					177			177	195
0200	Proximity card reader					161			161	177
0210	Entrance card systems									

28 15 Access Control Hardware Devices

28 15 11 – Integrated Credential Readers and Field Entry Management

28 15 11.11 Standard Card Readers

		Crew	Daily Output	Labor-Hours	Unit	Material	2019 Bare Costs Labor	Equipment	Total	Total Incl O&P
0300	Entrance card, barium ferrite				Ea.	4.95			4.95	5.45
0320	Credential					8.15			8.15	8.95
0340	Proximity					3.65			3.65	4.02
0360	Weigand					8.30			8.30	9.15
0700	Entrance card reader, barium ferrite	R-19	4	5		350	300		650	835
0720	Credential		4	5		212	300		512	685
0740	Proximity		4	5		300	300		600	780
0760	Weigand		4	5		495	300		795	995
0850	Scanner, eye retina	↓	4	5		2,650	300		2,950	3,375
0900	Gate opener, cantilever	R-18	3	8.667		2,100	455		2,555	2,975
0910	Light duty		3	8.667		1,750	455		2,205	2,600
0912	Custom	↓	3	8.667		2,200	455		2,655	3,100
0950	Switch, tamper	R-19	6	3.333		37	201		238	340
1100	Accessories, electric door strike/bolt		5	4		78.50	241		319.50	445
1120	Electromagnetic lock		5	4		147	241		388	520
1140	Keypad for card reader	↓	4	5	↓	126	300		426	590

28 15 11.13 Keypads

		Crew	Daily Output	Labor-Hours	Unit	Material	2019 Bare Costs Labor	Equipment	Total	Total Incl O&P
0010	**KEYPADS**									
0340	Digital keypad, int./ext., basic, excl. striker/power/wiring	1 Elec	3	2.667	Ea.	108	160		268	360
0350	Lockset, mechanical push-button type, complete, incl. hardware	1 Carp	4	2	"	415	103		518	610

28 15 11.15 Biometric Identity Devices

		Crew	Daily Output	Labor-Hours	Unit	Material	2019 Bare Costs Labor	Equipment	Total	Total Incl O&P
0010	**BIOMETRIC IDENTITY DEVICES**									
0200	Fingerprint scanner unit, excl. striker/power supply	1 Elec	4	2	Ea.	1,650	120		1,770	1,975
0210	Fingerprint scanner unit, for computer keyboard access		8	1		1,250	60		1,310	1,475
0220	Hand geometry scanner, mem of 512 users, excl. striker/power		3	2.667		2,375	160		2,535	2,875
0230	Memory upgrade for, adds 9,700 user profiles		8	1		345	60		405	470
0240	Adds 32,500 user profiles		8	1		680	60		740	840
0250	Prison type, memory of 256 users, excl. striker/power		3	2.667		2,675	160		2,835	3,175
0260	Memory upgrade for, adds 3,300 user profiles		8	1		330	60		390	455
0270	Adds 9,700 user profiles		8	1		520	60		580	665
0280	Adds 27,900 user profiles		8	1		710	60		770	870
0290	All weather, mem of 512 users, excl. striker/power		3	2.667		4,225	160		4,385	4,900
0300	Facial & fingerprint scanner, combination unit, excl. striker/power		3	2.667		925	160		1,085	1,275
0310	Access for, for initial setup, excl. striker/power	↓	3	2.667	↓	1,250	160		1,410	1,625

28 15 11.19 Security Access Control Accessories

		Crew	Daily Output	Labor-Hours	Unit	Material	2019 Bare Costs Labor	Equipment	Total	Total Incl O&P
0010	**SECURITY ACCESS CONTROL ACCESSORIES**									
0360	Scanner/reader access, power supply/transf, 110V to 12/24V	1 Elec	4	2	Ea.	231	120		351	435
0370	Elec/mag strikers, 12/24V		4	2		99	120		219	288
0380	1 hour battery backup power supply	↓	4	2		49.50	120		169.50	234
0390	Deadbolt, digital, batt-operated, indoor/outdoor, complete, incl. hardware	1 Carp	4	2	↓	380	103		483	570

For customer support on your Electrical Costs with RSMeans data, call 800.448.8182.

333

28 18 Security Access Detection Equipment

28 18 11 – Security Access Metal Detectors

28 18 11.13 Security Access Metal Detectors

28 18 11.13 Security Access Metal Detectors	Crew	Daily Output	Labor-Hours	Unit	Material	2019 Bare Costs Labor	Equipment	Total	Total Incl O&P
0010 **SECURITY ACCESS METAL DETECTORS**									
0240 Metal detector, hand-held, wand type, unit only				Ea.	129			129	138
0250 Metal detector, walk through portal type, single zone	1 Elec	2	4		3,300	240		3,540	4,000
0260 Multi-zone	"	2	4		4,500	240		4,740	5,300

28 18 13 – Security Access X-Ray Equipment

28 18 13.16 Security Access X-Ray Equipment

	Crew	Daily Output	Labor-Hours	Unit	Material	2019 Bare Costs Labor	Equipment	Total	Total Incl O&P
0010 **SECURITY ACCESS X-RAY EQUIPMENT**									
0290 X-ray machine, desk top, for mail/small packages/letters	1 Elec	4	2	Ea.	3,450	120		3,570	3,950
0300 Conveyor type, incl. monitor		2	4		15,800	240		16,040	17,800
0310 Includes additional features		2	4		27,700	240		27,940	30,800
0320 X-ray machine, large unit, for airports, incl. monitor	2 Elec	1	16		39,600	960		40,560	44,900
0330 Full console	"	.50	32		68,000	1,925		69,925	77,500

28 18 15 – Security Access Explosive Detection Equipment

28 18 15.23 Security Access Explosive Detection Equipment

	Crew	Daily Output	Labor-Hours	Unit	Material	2019 Bare Costs Labor	Equipment	Total	Total Incl O&P
0010 **SECURITY ACCESS EXPLOSIVE DETECTION EQUIPMENT**									
0270 Explosives detector, walk through portal type	1 Elec	2	4	Ea.	43,700	240		43,940	48,500
0280 Hand-held, battery operated				"				25,500	28,100

28 23 Video Management System

28 23 13 – Video Management System Interfaces

28 23 13.10 Closed Circuit Television System

	Crew	Daily Output	Labor-Hours	Unit	Material	2019 Bare Costs Labor	Equipment	Total	Total Incl O&P
0010 **CLOSED CIRCUIT TELEVISION SYSTEM**									
2000 Surveillance, one station (camera & monitor)	2 Elec	2.60	6.154	Total	720	370		1,090	1,350
2200 For additional camera stations, add	1 Elec	2.70	2.963	Ea.	315	178		493	610
2400 Industrial quality, one station (camera & monitor)	2 Elec	2.60	6.154	Total	1,700	370		2,070	2,400
2600 For additional camera stations, add	1 Elec	2.70	2.963	Ea.	675	178		853	1,000
2610 For low light, add		2.70	2.963		505	178		683	825
2620 For very low light, add		2.70	2.963		3,300	178		3,478	3,900
2800 For weatherproof camera station, add		1.30	6.154		540	370		910	1,150
3000 For pan and tilt, add		1.30	6.154		1,600	370		1,970	2,300
3200 For zoom lens - remote control, add		2	4		1,850	240		2,090	2,400
3400 Extended zoom lens		2	4		5,075	240		5,315	5,925
3410 For automatic iris for low light, add		2	4		1,025	240		1,265	1,475
3600 Educational TV studio, basic 3 camera system, black & white,									
3800 electrical & electronic equip. only	4 Elec	.80	40	Total	8,450	2,400		10,850	12,900
4000 Full console		.28	114		25,100	6,875		31,975	37,800
4100 As above, but color system		.28	114		52,000	6,875		58,875	67,000
4120 Full console		.12	267		193,000	16,000		209,000	236,000
4200 For film chain, black & white, add	1 Elec	1	8	Ea.	13,000	480		13,480	15,000
4250 Color, add		.25	32		8,900	1,925		10,825	12,700
4400 For video recorders, add		1	8		2,525	480		3,005	3,500
4600 Premium	4 Elec	.40	80		16,300	4,800		21,100	25,200

28 23 23 – Video Surveillance Systems Infrastructure

28 23 23.50 Video Surveillance Equipments

	Crew	Daily Output	Labor-Hours	Unit	Material	2019 Bare Costs Labor	Equipment	Total	Total Incl O&P
0010 **VIDEO SURVEILLANCE EQUIPMENTS**									
0200 Video cameras, wireless, hidden in exit signs, clocks, etc., incl. receiver	1 Elec	3	2.667	Ea.	171	160		331	425
0210 Accessories for video recorder, single camera		3	2.667		175	160		335	430
0220 For multiple cameras		3	2.667		1,675	160		1,835	2,100
0230 Video cameras, wireless, for under vehicle searching, complete		2	4		15,100	240		15,340	17,000

28 23 Video Management System

28 23 23 – Video Surveillance Systems Infrastructure

28 23 23.50 Video Surveillance Equipments	Crew	Daily Output	Labor-Hours	Unit	Material	2019 Bare Costs Labor	Equipment	Total	Total Incl O&P
0234 Master monitor station, 3 doors x 5 color monitor with tilt feature	1 Elec	2	4	Ea.	935	240		1,175	1,375
0400 Internet protocol network camera, day/night, color & power supply		2.60	3.077		1,250	185		1,435	1,650
0500 Monitor, color flat screen, liquid crystal display (LCD), 15"		2.70	2.963		535	178		713	855
0520 17"		2.70	2.963		445	178		623	750
0540 19"		2.70	2.963		590	178		768	910

28 31 Intrusion Detection

28 31 16 – Intrusion Detection Systems Infrastructure

28 31 16.50 Intrusion Detection

		Crew	Daily Output	Labor-Hours	Unit	Material	2019 Bare Costs Labor	Equipment	Total	Total Incl O&P
0010	INTRUSION DETECTION, not including wires & conduits									
0100	Burglar alarm, battery operated, mechanical trigger	1 Elec	4	2	Ea.	281	120		401	490
0200	Electrical trigger		4	2		335	120		455	550
0400	For outside key control, add		8	1		87	60		147	185
0600	For remote signaling circuitry, add		8	1		144	60		204	249
0800	Card reader, flush type, standard		2.70	2.963		815	178		993	1,150
1000	Multi-code		2.70	2.963		1,225	178		1,403	1,600
1010	Card reader, proximity type		2.70	2.963		300	178		478	595
1200	Door switches, hinge switch		5.30	1.509		64.50	90.50		155	206
1400	Magnetic switch		5.30	1.509		105	90.50		195.50	251
1600	Exit control locks, horn alarm		4	2		219	120		339	420
1800	Flashing light alarm		4	2		247	120		367	450
2000	Indicating panels, 1 channel		2.70	2.963		247	178		425	535
2200	10 channel	2 Elec	3.20	5		1,075	300		1,375	1,625
2400	20 channel		2	8		2,500	480		2,980	3,475
2600	40 channel		1.14	14.035		4,525	845		5,370	6,225
2800	Ultrasonic motion detector, 12 V	1 Elec	2.30	3.478		200	209		409	530
3000	Infrared photoelectric detector		4	2		135	120		255	325
3200	Passive infrared detector		4	2		238	120		358	440
3400	Glass break alarm switch		8	1		91	60		151	190
3420	Switchmats, 30" x 5'		5.30	1.509		105	90.50		195.50	251
3440	30" x 25'		4	2		206	120		326	405
3460	Police connect panel		4	2		285	120		405	495
3480	Telephone dialer		5.30	1.509		390	90.50		480.50	565
3500	Alarm bell		4	2		106	120		226	295
3520	Siren		4	2		148	120		268	340
3540	Microwave detector, 10' to 200'		2	4		590	240		830	1,000
3560	10' to 350'		2	4		1,775	240		2,015	2,300

28 33 Security Monitoring and Control

28 33 11 – Electronic Structural Monitoring Systems

28 33 11.10 Load Cell Units

		Crew	Daily Output	Labor-Hours	Unit	Material	2019 Bare Costs Labor	Equipment	Total	Total Incl O&P
0010	LOAD CELL UNITS									
0100	Load cell, stainless steel, 0.5/1 ton rated, mounting unit	2 Elec	1.33	12.030	Ea.	635	720		1,355	1,775
0110	2/3 to 5/5 ton rated, mounting unit		1.20	13.333		680	800		1,480	1,950
0120	0.5/1/2/3.5/5 ton, self aligning bottom base		1.66	9.639		211	580		791	1,100
0130	Combo mounting unit, 10/25 ton rated		.86	18.605		1,775	1,125		2,900	3,625
0140	40/60 ton rated		.66	24.242		2,875	1,450		4,325	5,325
0150	60kg/3m, rated		1.84	8.696		1,500	520		2,020	2,425
0160	130kg/3m, rated		1.55	10.323		1,425	620		2,045	2,500

For customer support on your Electrical Costs with RSMeans data, call 800.448.8182.

335

28 33 Security Monitoring and Control

28 33 11 – Electronic Structural Monitoring Systems

28 33 11.10 Load Cell Units	Crew	Daily Output	Labor-Hours	Unit	Material	2019 Bare Costs Labor	Equipment	Total	Total Incl O&P
0170	280 kg, rated	2 Elec	1.22	13.115	Ea.	1,425	790		2,215	2,750

28 46 Fire Detection and Alarm

28 46 11 – Fire Sensors and Detectors

28 46 11.21 Carbon-Monoxide Detection Sensors

		Crew	Daily Output	Labor-Hours	Unit	Material	2019 Bare Costs Labor	Equipment	Total	Total Incl O&P
0010	**CARBON-MONOXIDE DETECTION SENSORS**									
8400	Smoke and carbon monoxide alarm battery operated photoelectric low profile	1 Elec	24	.333	Ea.	59	20		79	95
8410	low profile photoelectric battery powered		24	.333		36.50	20		56.50	70
8420	photoelectric low profile sealed lithium		24	.333		66.50	20		86.50	104
8430	Photoelectric low profile sealed lithium smoke and CO with voice combo		24	.333		41.50	20		61.50	75.50
8500	Carbon monoxide sensor, wall mount 1Mod 1 relay output smoke & heat		24	.333		505	20		525	585
8510	1Mod, 1relay no display, smoke & heat		24	.333		490	20		510	570
8530	1Mod 1 relay no HW logo smoke & heat		24	.333		530	20		550	615
8700	Carbon monoxide detector, battery operated, wall mounted		16	.500		46	30		76	95.50
8710	Hardwired, wall and ceiling mounted		8	1		90	60		150	189
8720	Duct mounted		8	1		350	60		410	475
8730	Continuous air monitoring system, PC based, incl. remote sensors, min.	2 Elec	.25	64		31,600	3,850		35,450	40,500
8740	Maximum	"	.10	160		51,000	9,600		60,600	71,000

28 46 11.27 Other Sensors

		Crew	Daily Output	Labor-Hours	Unit	Material	2019 Bare Costs Labor	Equipment	Total	Total Incl O&P
0010	**OTHER SENSORS**									
5200	Smoke detector, ceiling type	1 Elec	6.20	1.290	Ea.	123	77.50		200.50	251
5240	Smoke detector, addressable type		6	1.333		227	80		307	370
5400	Duct type		3.20	2.500		294	150		444	550
5420	Duct addressable type		3.20	2.500		525	150		675	800
8300	Smoke alarm with integrated strobe light 120 V, 16DB 60 fpm flash rate		16	.500		112	30		142	168
8310	Photoelectric smoke detector with strobe 120 V, 90 DB ceiling mount		12	.667		205	40		245	285
8320	120 V, 90 DB wall mount		12	.667		189	40		229	267
8330	120 V, 9 VDC, 90 DB ceiling mount		12	.667		199	40		239	278
8340	120 V, 9 VDC backup 90 DB wall mount		12	.667		199	40		239	279
8440	Fire alarm beam detector, motorized reflective, infrared optical beam	2 Elec	1.60	10		775	600		1,375	1,750
8450	Fire alarm beam detector, motorized reflective, IR/UV optical beam	"	1.60	10		1,025	600		1,625	2,025

28 46 11.50 Fire and Heat Detectors

		Crew	Daily Output	Labor-Hours	Unit	Material	2019 Bare Costs Labor	Equipment	Total	Total Incl O&P
0010	**FIRE & HEAT DETECTORS**									
5000	Detector, rate of rise	1 Elec	8	1	Ea.	52	60		112	147
5010	Heat addressable type		7.25	1.103		231	66.50		297.50	355
5100	Fixed temp fire alarm		7	1.143		45.50	68.50		114	152
5200	125 V SRF MT-135F		7.25	1.103		31	66.50		97.50	134
5300	MT-195F		7.25	1.103		33	66.50		99.50	136
5400	104/85 DB, indoor/outdoor, ceiling/wall, red		10.25	.780		69	47		116	146
5410	Ceiling/wall, white		10.25	.780		67.50	47		114.50	145
5420	102/98 DB, lumin 110cd, indoor/wall, red		10.25	.780		99.50	47		146.50	179
5430	Lumin15/75cd, indoor/wall, red		10.25	.780		107	47		154	187
5440	HI/LO DB, 24 V, fire marking, red		10.25	.780		74.50	47		121.50	152

28 46 21 – Fire Alarm

28 46 21.50 Alarm Panels and Devices

		Crew	Daily Output	Labor-Hours	Unit	Material	2019 Bare Costs Labor	Equipment	Total	Total Incl O&P
0010	**ALARM PANELS AND DEVICES**, not including wires & conduits									
2200	Intercom remote station	1 Elec	8	1	Ea.	77	60		137	175
2400	Intercom outlet		8	1		57.50	60		117.50	153
2600	Sound system, intercom handset		8	1		460	60		520	595
3590	Signal device, beacon		8	1		212	60		272	325

28 46 Fire Detection and Alarm

28 46 21 – Fire Alarm

28 46 21.50 Alarm Panels and Devices	Crew	Daily Output	Labor-Hours	Unit	Material	2019 Bare Costs Labor	Equipment	Total	Total Incl O&P
3594	2 zone	1 Elec	1.35	5.926	Ea.	310	355		665	870
3600	4 zone	2 Elec	2	8		380	480		860	1,125
3610	3 zone		2	8		620	480		1,100	1,400
3800	8 zone		1	16		960	960		1,920	2,475
3810	5 zone		1.50	10.667		660	640		1,300	1,675
3900	10 zone		1.25	12.800		910	770		1,680	2,150
4000	12 zone		.67	23.988		2,175	1,450		3,625	4,550
4020	Alarm device, tamper, flow	1 Elec	8	1		227	60		287	340
4025	Fire alarm, loop expander card		16	.500		805	30		835	930
4050	Actuating device		8	1		345	60		405	470
4160	Alarm control panel, addressable w/o voice, up to 200 points	2 Elec	1.14	13.998		4,600	840		5,440	6,300
4170	addressable w/voice, up to 400 points	"	.73	22.008		9,700	1,325		11,025	12,700
4175	Addressable interface device	1 Elec	7.25	1.103		129	66.50		195.50	241
4200	Battery and rack		4	2		390	120		510	610
4400	Automatic charger		8	1		525	60		585	665
4600	Signal bell		8	1		135	60		195	239
4610	Fire alarm signal bell 10" red 20-24 V P		8	1		149	60		209	254
4800	Trouble buzzer or manual station		8	1		87.50	60		147.50	186
5425	Duct smoke and heat detector 2 wire		8	1		118	60		178	220
5430	Fire alarm duct detector controller		3	2.667		255	160		415	520
5435	Fire alarm duct detector sensor kit		8	1		71	60		131	168
5440	Remote test station for smoke detector duct type		5.30	1.509		60	90.50		150.50	202
5460	Remote fire alarm indicator light		5.30	1.509		22.50	90.50		113	160
5600	Strobe and horn		5.30	1.509		140	90.50		230.50	289
5610	Strobe and horn (ADA type)		5.30	1.509		168	90.50		258.50	320
5620	Visual alarm (ADA type)		6.70	1.194		117	71.50		188.50	236
5800	Fire alarm horn		6.70	1.194		55.50	71.50		127	168
6000	Door holder, electro-magnetic		4	2		91.50	120		211.50	280
6200	Combination holder and closer		3.20	2.500		126	150		276	365
6600	Drill switch		8	1		380	60		440	510
6800	Master box		2.70	2.963		6,575	178		6,753	7,500
7000	Break glass station		8	1		58	60		118	153
7010	Break glass station, addressable		7.25	1.103		168	66.50		234.50	284
7800	Remote annunciator, 8 zone lamp		1.80	4.444		213	267		480	635
8000	12 zone lamp	2 Elec	2.60	6.154		415	370		785	1,000
8200	16 zone lamp	"	2.20	7.273		375	435		810	1,050

28 47 Mass Notification

28 47 12 – Notification Systems

28 47 12.10 Mass Notification System

		Crew	Daily Output	Labor-Hours	Unit	Material	Labor	Equipment	Total	Total Incl O&P
0010	**MASS NOTIFICATION SYSTEM**									
0100	Wireless command center, 10,000 devices	2 Elec	1.33	12.030	Ea.	2,325	720		3,045	3,625
0200	Option, email notification					2,025			2,025	2,225
0210	Remote device supervision & monitor					2,575			2,575	2,825
0300	Antenna VHF or UHF, for medium range	1 Elec	4	2		87.50	120		207.50	276
0310	For high-power transmitter		2	4		1,100	240		1,340	1,575
0400	Transmitter, 25 watt		4	2		1,725	120		1,845	2,075
0410	40 watt		2.66	3.008		2,000	181		2,181	2,475
0420	100 watt		1.33	6.015		6,500	360		6,860	7,700
0500	Wireless receiver/control module for speaker		8	1		282	60		342	400
0600	Desktop paging controller, stand alone		4	2		845	120		965	1,100

28 47 Mass Notification

28 47 12 – Notification Systems

28 47 12.10 Mass Notification System	Crew	Daily Output	Labor-Hours	Unit	Material	2019 Bare Costs Labor	Equipment	Total	Total Incl O&P
0700 Auxiliary level inputs	1 Elec	8.36	.957	Ea.	99.50	57.50		157	196

28 52 Detention Security Systems

28 52 11 – Detention Monitoring and Control Systems

28 52 11.10 Detention Control Systems

		Crew	Daily Output	Labor-Hours	Unit	Material	2019 Bare Costs Labor	Equipment	Total	Total Incl O&P
0010	**DETENTION CONTROL SYSTEMS**									
1000	Desk top control systems for 10 doors, 10 intercoms, and 10 lights	2 Elec	.30	53.333	Ea.	63,000	3,200		66,200	74,000
1020	Push button control panel systems for 10 doors, 10 intercoms, and 10 lights	"	.32	50	"	56,500	3,000		59,500	66,500

Estimating Tips
31 05 00 Common Work Results for Earthwork

- Estimating the actual cost of performing earthwork requires careful consideration of the variables involved. This includes items such as type of soil, whether water will be encountered, dewatering, whether banks need bracing, disposal of excavated earth, and length of haul to fill or spoil sites, etc. If the project has large quantities of cut or fill, consider raising or lowering the site to reduce costs, while paying close attention to the effect on site drainage and utilities.

- If the project has large quantities of fill, creating a borrow pit on the site can significantly lower the costs.

- It is very important to consider what time of year the project is scheduled for completion. Bad weather can create large cost overruns from dewatering, site repair, and lost productivity from cold weather.

Reference Numbers

Reference numbers are shown at the beginning of some major classifications. These numbers refer to related items in the Reference Section. The reference information may be an estimating procedure, an alternate pricing method, or technical information.

Note: Not all subdivisions listed here necessarily appear. ■

Did you know?

RSMeans data is available through our online application:

- Search for costs by keyword
- Leverage the most up-to-date data
- Build and export estimates

Try it free
rsmeans.com/2019freetrial

31 23 Excavation and Fill

31 23 16 – Excavation

31 23 16.13 Excavating, Trench

	Crew	Daily Output	Labor-Hours	Unit	Material	2019 Bare Costs Labor	2019 Bare Costs Equipment	Total	Total Incl O&P
0010 **EXCAVATING, TRENCH**									
0011 Or continuous footing									
0020 Common earth with no sheeting or dewatering included									
0050 1' to 4' deep, 3/8 C.Y. excavator	B-11C	150	.107	B.C.Y.		5.15	2.13	7.28	10.05
0060 1/2 C.Y. excavator	B-11M	200	.080			3.85	1.94	5.79	7.95
0062 3/4 C.Y. excavator	B-12F	270	.059			2.92	2.52	5.44	7.20
1400 By hand with pick and shovel 2' to 6' deep, light soil	1 Clab	8	1			41		41	62
1500 Heavy soil	"	4	2			82		82	124
3000 Backfill trench, F.E. loader, wheel mtd., 1 C.Y. bucket									
3020 Minimal haul	B-10R	400	.030	L.C.Y.		1.51	.74	2.25	3.09
5020 Loam & sandy clay with no sheeting or dewatering included									
5050 1' to 4' deep, 3/8 C.Y. tractor loader/backhoe	B-11C	162	.099	B.C.Y.		4.75	1.98	6.73	9.30
5060 1/2 C.Y. excavator	B-11M	216	.074			3.56	1.80	5.36	7.30
5070 3/4 C.Y. excavator	B-12F	292	.055			2.70	2.33	5.03	6.65
6020 Sand & gravel with no sheeting or dewatering included									
6050 1' to 4' deep, 3/8 C.Y. excavator	B-11C	165	.097	B.C.Y.		4.66	1.94	6.60	9.15
6060 1/2 C.Y. excavator	B-11M	220	.073			3.50	1.76	5.26	7.20
6070 3/4 C.Y. excavator	B-12F	297	.054			2.65	2.29	4.94	6.50
7020 Dense hard clay with no sheeting or dewatering included									
7050 1' to 4' deep, 3/8 C.Y. excavator	B-11C	132	.121	B.C.Y.		5.85	2.43	8.28	11.45
7060 1/2 C.Y. excavator	B-11M	176	.091			4.37	2.20	6.57	9
7070 3/4 C.Y. excavator	B-12F	238	.067			3.31	2.86	6.17	8.15

31 23 16.14 Excavating, Utility Trench

	Crew	Daily Output	Labor-Hours	Unit	Material	2019 Bare Costs Labor	2019 Bare Costs Equipment	Total	Total Incl O&P
0010 **EXCAVATING, UTILITY TRENCH**									
0011 Common earth									
0050 Trenching with chain trencher, 12 HP, operator walking									
0100 4" wide trench, 12" deep	B-53	800	.010	L.F.		.52	.08	.60	.87
0150 18" deep		750	.011			.55	.08	.63	.92
0200 24" deep		700	.011			.59	.09	.68	.99
0300 6" wide trench, 12" deep		650	.012			.64	.10	.74	1.06
0350 18" deep		600	.013			.69	.11	.80	1.15
0400 24" deep		550	.015			.75	.12	.87	1.26
0450 36" deep		450	.018			.92	.14	1.06	1.54
0600 8" wide trench, 12" deep		475	.017			.87	.13	1	1.46
0650 18" deep		400	.020			1.03	.16	1.19	1.72
0700 24" deep		350	.023			1.18	.18	1.36	1.97
0750 36" deep		300	.027			1.38	.21	1.59	2.30
1000 Backfill by hand including compaction, add									
1050 4" wide trench, 12" deep	A-1G	800	.010	L.F.		.41	.07	.48	.69
1100 18" deep		530	.015			.62	.10	.72	1.05
1150 24" deep		400	.020			.82	.13	.95	1.39
1300 6" wide trench, 12" deep		540	.015			.61	.10	.71	1.03
1350 18" deep		405	.020			.81	.13	.94	1.37
1400 24" deep		270	.030			1.22	.20	1.42	2.06
1450 36" deep		180	.044			1.82	.29	2.11	3.08
1600 8" wide trench, 12" deep		400	.020			.82	.13	.95	1.39
1650 18" deep		265	.030			1.24	.20	1.44	2.09
1700 24" deep		200	.040			1.64	.27	1.91	2.77
1750 36" deep		135	.059			2.43	.39	2.82	4.11
2000 Chain trencher, 40 HP operator riding									
2050 6" wide trench and backfill, 12" deep	B-54	1200	.007	L.F.		.34	.29	.63	.84
2100 18" deep		1000	.008			.41	.35	.76	1

31 23 16.14 Excavating, Utility Trench

		Crew	Daily Output	Labor-Hours	Unit	Material	2019 Bare Costs Labor	Equipment	Total	Total Incl O&P
2150	24" deep	B-54	975	.008	L.F.		.42	.36	.78	1.03
2200	36" deep		900	.009			.46	.39	.85	1.12
2250	48" deep		750	.011			.55	.46	1.01	1.34
2300	60" deep		650	.012			.64	.54	1.18	1.54
2400	8" wide trench and backfill, 12" deep		1000	.008			.41	.35	.76	1
2450	18" deep		950	.008			.43	.37	.80	1.05
2500	24" deep		900	.009			.46	.39	.85	1.12
2550	36" deep		800	.010			.52	.44	.96	1.26
2600	48" deep		650	.012			.64	.54	1.18	1.54
2700	12" wide trench and backfill, 12" deep		975	.008			.42	.36	.78	1.03
2750	18" deep		860	.009			.48	.41	.89	1.17
2800	24" deep		800	.010			.52	.44	.96	1.26
2850	36" deep		725	.011			.57	.48	1.05	1.39
3000	16" wide trench and backfill, 12" deep		835	.010			.49	.42	.91	1.20
3050	18" deep		750	.011			.55	.46	1.01	1.34
3100	24" deep	▼	700	.011	▼		.59	.50	1.09	1.44
3200	Compaction with vibratory plate, add								35%	35%
5100	Hand excavate and trim for pipe bells after trench excavation									
5200	8" pipe	1 Clab	155	.052	L.F.		2.12		2.12	3.21
5300	18" pipe	"	130	.062	"		2.53		2.53	3.82

31 23 16.15 Excavating, Utility Trench, Plow

		Crew	Daily Output	Labor-Hours	Unit	Material	2019 Bare Costs Labor	Equipment	Total	Total Incl O&P
0010	**EXCAVATING, UTILITY TRENCH, PLOW**									
0100	Single cable, plowed into fine material	B-54C	3800	.004	L.F.		.20	.29	.49	.62
0200	Two cable		3200	.005			.24	.35	.59	.74
0300	Single cable, plowed into coarse material	▼	2000	.008	▼		.38	.56	.94	1.19

31 23 16.16 Structural Excavation for Minor Structures

		Crew	Daily Output	Labor-Hours	Unit	Material	2019 Bare Costs Labor	Equipment	Total	Total Incl O&P
0010	**STRUCTURAL EXCAVATION FOR MINOR STRUCTURES**									
0015	Hand, pits to 6' deep, sandy soil	1 Clab	8	1	B.C.Y.		41		41	62
0100	Heavy soil or clay		4	2			82		82	124
0300	Pits 6' to 12' deep, sandy soil		5	1.600			65.50		65.50	99.50
0500	Heavy soil or clay		3	2.667			109		109	166
0700	Pits 12' to 18' deep, sandy soil		4	2			82		82	124
0900	Heavy soil or clay	▼	2	4			164		164	248
6030	Common earth, hydraulic backhoe, 1/2 C.Y. bucket	B-12E	55	.291			14.35	8.05	22.40	30.50
6035	3/4 C.Y. bucket	B-12F	90	.178			8.75	7.55	16.30	21.50
6040	1 C.Y. bucket	B-12A	108	.148			7.30	6.95	14.25	18.65
6050	1-1/2 C.Y. bucket	B-12B	144	.111			5.45	6.30	11.75	15.20
6060	2 C.Y. bucket	B-12C	200	.080			3.94	5.40	9.34	11.90
6070	Sand and gravel, 3/4 C.Y. bucket	B-12F	100	.160			7.90	6.80	14.70	19.35
6080	1 C.Y. bucket	B-12A	120	.133			6.55	6.25	12.80	16.75
6090	1-1/2 C.Y. bucket	B-12B	160	.100			4.93	5.70	10.63	13.65
6100	2 C.Y. bucket	B-12C	220	.073			3.58	4.90	8.48	10.80
6110	Clay, till, or blasted rock, 3/4 C.Y. bucket	B-12F	80	.200			9.85	8.50	18.35	24
6120	1 C.Y. bucket	B-12A	95	.168			8.30	7.90	16.20	21
6130	1-1/2 C.Y. bucket	B-12B	130	.123			6.05	7	13.05	16.85
6140	2 C.Y. bucket	B-12C	175	.091			4.50	6.15	10.65	13.60
6230	Sandy clay & loam, hydraulic backhoe, 1/2 C.Y. bucket	B-12E	60	.267			13.15	7.40	20.55	28
6235	3/4 C.Y. bucket	B-12F	98	.163			8.05	6.95	15	19.75
6240	1 C.Y. bucket	B-12A	116	.138			6.80	6.45	13.25	17.35
6250	1-1/2 C.Y. bucket	B-12B	156	.103			5.05	5.80	10.85	14

31 23 19 – Dewatering

31 23 19.20 Dewatering Systems	Crew	Daily Output	Labor-Hours	Unit	Material	2019 Bare Costs Labor	Equipment	Total	Total Incl O&P
0010 **DEWATERING SYSTEMS**									
0020 Excavate drainage trench, 2' wide, 2' deep	B-11C	90	.178	C.Y.		8.55	3.56	12.11	16.75
0100 2' wide, 3' deep, with backhoe loader	"	135	.119			5.70	2.37	8.07	11.20
0200 Excavate sump pits by hand, light soil	1 Clab	7.10	1.127			46.50		46.50	70
0300 Heavy soil	"	3.50	2.286			94		94	142
0500 Pumping 8 hrs., attended 2 hrs./day, incl. 20 L.F.									
0550 of suction hose & 100 L.F. discharge hose									
0600 2" diaphragm pump used for 8 hrs.	B-10H	4	3	Day		151	19.55	170.55	249
0650 4" diaphragm pump used for 8 hrs.	B-10I	4	3			151	37.50	188.50	268
0800 8 hrs. attended, 2" diaphragm pump	B-10H	1	12			605	78	683	995
0900 3" centrifugal pump	B-10J	1	12			605	84.50	689.50	1,000
1000 4" diaphragm pump	B-10I	1	12			605	149	754	1,075
1100 6" centrifugal pump	B-10K	1	12			605	320	925	1,250

31 23 23 – Fill

31 23 23.13 Backfill

	Crew	Daily Output	Labor-Hours	Unit	Material	Labor	Equipment	Total	Total Incl O&P
0010 **BACKFILL**									
0015 By hand, no compaction, light soil	1 Clab	14	.571	L.C.Y.		23.50		23.50	35.50
0100 Heavy soil		11	.727	"		30		30	45
0300 Compaction in 6" layers, hand tamp, add to above		20.60	.388	E.C.Y.		15.95		15.95	24
0400 Roller compaction operator walking, add	B-10A	100	.120			6.05	1.82	7.87	11.10
0500 Air tamp, add	B-9D	190	.211			8.75	1.42	10.17	14.75
0600 Vibrating plate, add	A-1D	60	.133			5.45	.52	5.97	8.85
0800 Compaction in 12" layers, hand tamp, add to above	1 Clab	34	.235			9.65		9.65	14.60
0900 Roller compaction operator walking, add	B-10A	150	.080			4.03	1.21	5.24	7.40
1000 Air tamp, add	B-9	285	.140			5.80	.82	6.62	9.70
1100 Vibrating plate, add	A-1E	90	.089			3.65	.45	4.10	6
1300 Dozer backfilling, bulk, up to 300' haul, no compaction	B-10B	1200	.010	L.C.Y.		.50	1.08	1.58	1.94
1400 Air tamped, add	B-11B	80	.200	E.C.Y.		9.25	3.59	12.84	17.90
1600 Compacting backfill, 6" to 12" lifts, vibrating roller	B-10C	800	.015			.76	2.14	2.90	3.49
1700 Sheepsfoot roller	B-10D	750	.016			.81	2.29	3.10	3.73
1900 Dozer backfilling, trench, up to 300' haul, no compaction	B-10B	900	.013	L.C.Y.		.67	1.43	2.10	2.59
2000 Air tamped, add	B-11B	80	.200	E.C.Y.		9.25	3.59	12.84	17.90
2200 Compacting backfill, 6" to 12" lifts, vibrating roller	B-10C	700	.017			.86	2.45	3.31	3.99
2300 Sheepsfoot roller	B-10D	650	.018			.93	2.65	3.58	4.31

31 23 23.16 Fill By Borrow and Utility Bedding

	Crew	Daily Output	Labor-Hours	Unit	Material	Labor	Equipment	Total	Total Incl O&P
0010 **FILL BY BORROW AND UTILITY BEDDING**									
0049 Utility bedding, for pipe & conduit, not incl. compaction									
0050 Crushed or screened bank run gravel	B-6	150	.160	L.C.Y.	21	7.15	2.13	30.28	36.50
0100 Crushed stone 3/4" to 1/2"		150	.160		27.50	7.15	2.13	36.78	43
0200 Sand, dead or bank		150	.160		18.10	7.15	2.13	27.38	33
0500 Compacting bedding in trench	A-1D	90	.089	E.C.Y.		3.65	.35	4	5.90
0610 See Section 31 23 23.20 for hauling mileage add.									

31 23 23.20 Hauling

	Crew	Daily Output	Labor-Hours	Unit	Material	Labor	Equipment	Total	Total Incl O&P
0010 **HAULING**									
0011 Excavated or borrow, loose cubic yards									
0012 no loading equipment, including hauling, waiting, loading/dumping									
0013 time per cycle (wait, load, travel, unload or dump & return)									
0014 8 C.Y. truck, 15 MPH avg., cycle 0.5 miles, 10 min. wait/ld./uld.	B-34A	320	.025	L.C.Y.		1.18	1.06	2.24	2.94
0016 cycle 1 mile		272	.029			1.38	1.25	2.63	3.45
0018 cycle 2 miles		208	.038			1.81	1.63	3.44	4.52
0020 cycle 4 miles		144	.056			2.61	2.36	4.97	6.50

31 23 23.20 Hauling		Crew	Daily Output	Labor-Hours	Unit	Material	2019 Bare Costs Labor	Equipment	Total	Total Incl O&P
0022	cycle 6 miles	B-34A	112	.071	L.C.Y.		3.36	3.03	6.39	8.40
0024	cycle 8 miles		88	.091			4.27	3.86	8.13	10.70
0026	20 MPH avg., cycle 0.5 mile		336	.024			1.12	1.01	2.13	2.79
0028	cycle 1 mile		296	.027			1.27	1.15	2.42	3.17
0030	cycle 2 miles		240	.033			1.57	1.41	2.98	3.92
0032	cycle 4 miles		176	.045			2.14	1.93	4.07	5.35
0034	cycle 6 miles		136	.059			2.76	2.50	5.26	6.90
0036	cycle 8 miles		112	.071			3.36	3.03	6.39	8.40
0044	25 MPH avg., cycle 4 miles		192	.042			1.96	1.77	3.73	4.90
0046	cycle 6 miles		160	.050			2.35	2.12	4.47	5.90
0048	cycle 8 miles		128	.063			2.94	2.65	5.59	7.35
0050	30 MPH avg., cycle 4 miles		216	.037			1.74	1.57	3.31	4.35
0052	cycle 6 miles		176	.045			2.14	1.93	4.07	5.35
0054	cycle 8 miles		144	.056			2.61	2.36	4.97	6.50
0114	15 MPH avg., cycle 0.5 mile, 15 min. wait/ld./uld.		224	.036			1.68	1.52	3.20	4.20
0116	cycle 1 mile		200	.040			1.88	1.70	3.58	4.70
0118	cycle 2 miles		168	.048			2.24	2.02	4.26	5.60
0120	cycle 4 miles		120	.067			3.13	2.83	5.96	7.85
0122	cycle 6 miles		96	.083			3.92	3.54	7.46	9.80
0124	cycle 8 miles		80	.100			4.70	4.25	8.95	11.75
0126	20 MPH avg., cycle 0.5 mile		232	.034			1.62	1.46	3.08	4.05
0128	cycle 1 mile		208	.038			1.81	1.63	3.44	4.52
0130	cycle 2 miles		184	.043			2.04	1.85	3.89	5.10
0132	cycle 4 miles		144	.056			2.61	2.36	4.97	6.50
0134	cycle 6 miles		112	.071			3.36	3.03	6.39	8.40
0136	cycle 8 miles		96	.083			3.92	3.54	7.46	9.80
0144	25 MPH avg., cycle 4 miles		152	.053			2.47	2.23	4.70	6.20
0146	cycle 6 miles		128	.063			2.94	2.65	5.59	7.35
0148	cycle 8 miles		112	.071			3.36	3.03	6.39	8.40
0150	30 MPH avg., cycle 4 miles		168	.048			2.24	2.02	4.26	5.60
0152	cycle 6 miles		144	.056			2.61	2.36	4.97	6.50
0154	cycle 8 miles		120	.067			3.13	2.83	5.96	7.85
0214	15 MPH avg., cycle 0.5 mile, 20 min. wait/ld./uld.		176	.045			2.14	1.93	4.07	5.35
0216	cycle 1 mile		160	.050			2.35	2.12	4.47	5.90
0218	cycle 2 miles		136	.059			2.76	2.50	5.26	6.90
0220	cycle 4 miles		104	.077			3.62	3.27	6.89	9.05
0222	cycle 6 miles		88	.091			4.27	3.86	8.13	10.70
0224	cycle 8 miles		72	.111			5.20	4.72	9.92	13.05
0226	20 MPH avg., cycle 0.5 mile		176	.045			2.14	1.93	4.07	5.35
0228	cycle 1 mile		168	.048			2.24	2.02	4.26	5.60
0230	cycle 2 miles		144	.056			2.61	2.36	4.97	6.50
0232	cycle 4 miles		120	.067			3.13	2.83	5.96	7.85
0234	cycle 6 miles		96	.083			3.92	3.54	7.46	9.80
0236	cycle 8 miles		88	.091			4.27	3.86	8.13	10.70
0244	25 MPH avg., cycle 4 miles		128	.063			2.94	2.65	5.59	7.35
0246	cycle 6 miles		112	.071			3.36	3.03	6.39	8.40
0248	cycle 8 miles		96	.083			3.92	3.54	7.46	9.80
0250	30 MPH avg., cycle 4 miles		136	.059			2.76	2.50	5.26	6.90
0252	cycle 6 miles		120	.067			3.13	2.83	5.96	7.85
0254	cycle 8 miles		104	.077			3.62	3.27	6.89	9.05
0314	15 MPH avg., cycle 0.5 mile, 25 min. wait/ld./uld.		144	.056			2.61	2.36	4.97	6.50
0316	cycle 1 mile		128	.063			2.94	2.65	5.59	7.35
0318	cycle 2 miles		112	.071			3.36	3.03	6.39	8.40

31 23 23 – Fill

31 23 23.20 Hauling		Crew	Daily Output	Labor-Hours	Unit	Material	2019 Bare Costs Labor	2019 Bare Costs Equipment	Total	Total Incl O&P
0320	cycle 4 miles	B-34A	96	.083	L.C.Y.		3.92	3.54	7.46	9.80
0322	cycle 6 miles		80	.100			4.70	4.25	8.95	11.75
0324	cycle 8 miles		64	.125			5.90	5.30	11.20	14.70
0326	20 MPH avg., cycle 0.5 mile		144	.056			2.61	2.36	4.97	6.50
0328	cycle 1 mile		136	.059			2.76	2.50	5.26	6.90
0330	cycle 2 miles		120	.067			3.13	2.83	5.96	7.85
0332	cycle 4 miles		104	.077			3.62	3.27	6.89	9.05
0334	cycle 6 miles		88	.091			4.27	3.86	8.13	10.70
0336	cycle 8 miles		80	.100			4.70	4.25	8.95	11.75
0344	25 MPH avg., cycle 4 miles		112	.071			3.36	3.03	6.39	8.40
0346	cycle 6 miles		96	.083			3.92	3.54	7.46	9.80
0348	cycle 8 miles		88	.091			4.27	3.86	8.13	10.70
0350	30 MPH avg., cycle 4 miles		112	.071			3.36	3.03	6.39	8.40
0352	cycle 6 miles		104	.077			3.62	3.27	6.89	9.05
0354	cycle 8 miles		96	.083			3.92	3.54	7.46	9.80
0414	15 MPH avg., cycle 0.5 mile, 30 min. wait/ld./uld.		120	.067			3.13	2.83	5.96	7.85
0416	cycle 1 mile		112	.071			3.36	3.03	6.39	8.40
0418	cycle 2 miles		96	.083			3.92	3.54	7.46	9.80
0420	cycle 4 miles		80	.100			4.70	4.25	8.95	11.75
0422	cycle 6 miles		72	.111			5.20	4.72	9.92	13.05
0424	cycle 8 miles		64	.125			5.90	5.30	11.20	14.70
0426	20 MPH avg., cycle 0.5 mile		120	.067			3.13	2.83	5.96	7.85
0428	cycle 1 mile		112	.071			3.36	3.03	6.39	8.40
0430	cycle 2 miles		104	.077			3.62	3.27	6.89	9.05
0432	cycle 4 miles		88	.091			4.27	3.86	8.13	10.70
0434	cycle 6 miles		80	.100			4.70	4.25	8.95	11.75
0436	cycle 8 miles		72	.111			5.20	4.72	9.92	13.05
0444	25 MPH avg., cycle 4 miles		96	.083			3.92	3.54	7.46	9.80
0446	cycle 6 miles		88	.091			4.27	3.86	8.13	10.70
0448	cycle 8 miles		80	.100			4.70	4.25	8.95	11.75
0450	30 MPH avg., cycle 4 miles		96	.083			3.92	3.54	7.46	9.80
0452	cycle 6 miles		88	.091			4.27	3.86	8.13	10.70
0454	cycle 8 miles		80	.100			4.70	4.25	8.95	11.75
0514	15 MPH avg., cycle 0.5 mile, 35 min. wait/ld./uld.		104	.077			3.62	3.27	6.89	9.05
0516	cycle 1 mile		96	.083			3.92	3.54	7.46	9.80
0518	cycle 2 miles		88	.091			4.27	3.86	8.13	10.70
0520	cycle 4 miles		72	.111			5.20	4.72	9.92	13.05
0522	cycle 6 miles		64	.125			5.90	5.30	11.20	14.70
0524	cycle 8 miles		56	.143			6.70	6.05	12.75	16.75
0526	20 MPH avg., cycle 0.5 mile		104	.077			3.62	3.27	6.89	9.05
0528	cycle 1 mile		96	.083			3.92	3.54	7.46	9.80
0530	cycle 2 miles		96	.083			3.92	3.54	7.46	9.80
0532	cycle 4 miles		80	.100			4.70	4.25	8.95	11.75
0534	cycle 6 miles		72	.111			5.20	4.72	9.92	13.05
0536	cycle 8 miles		64	.125			5.90	5.30	11.20	14.70
0544	25 MPH avg., cycle 4 miles		88	.091			4.27	3.86	8.13	10.70
0546	cycle 6 miles		80	.100			4.70	4.25	8.95	11.75
0548	cycle 8 miles		72	.111			5.20	4.72	9.92	13.05
0550	30 MPH avg., cycle 4 miles		88	.091			4.27	3.86	8.13	10.70
0552	cycle 6 miles		80	.100			4.70	4.25	8.95	11.75
0554	cycle 8 miles		72	.111			5.20	4.72	9.92	13.05
1014	12 C.Y. truck, cycle 0.5 mile, 15 MPH avg., 15 min. wait/ld./uld.	B-34B	336	.024			1.12	1.69	2.81	3.54
1016	cycle 1 mile		300	.027			1.25	1.89	3.14	3.97

344

31 23 23.20 Hauling		Crew	Daily Output	Labor-Hours	Unit	Material	2019 Bare Costs Labor	Equipment	Total	Total Incl O&P
1018	cycle 2 miles	B-34B	252	.032	L.C.Y.		1.49	2.25	3.74	4.73
1020	cycle 4 miles		180	.044			2.09	3.15	5.24	6.60
1022	cycle 6 miles		144	.056			2.61	3.94	6.55	8.25
1024	cycle 8 miles		120	.067			3.13	4.73	7.86	9.90
1025	cycle 10 miles		96	.083			3.92	5.90	9.82	12.40
1026	20 MPH avg., cycle 0.5 mile		348	.023			1.08	1.63	2.71	3.42
1028	cycle 1 mile		312	.026			1.21	1.82	3.03	3.81
1030	cycle 2 miles		276	.029			1.36	2.05	3.41	4.31
1032	cycle 4 miles		216	.037			1.74	2.63	4.37	5.50
1034	cycle 6 miles		168	.048			2.24	3.38	5.62	7.10
1036	cycle 8 miles		144	.056			2.61	3.94	6.55	8.25
1038	cycle 10 miles		120	.067			3.13	4.73	7.86	9.90
1040	25 MPH avg., cycle 4 miles		228	.035			1.65	2.49	4.14	5.20
1042	cycle 6 miles		192	.042			1.96	2.95	4.91	6.20
1044	cycle 8 miles		168	.048			2.24	3.38	5.62	7.10
1046	cycle 10 miles		144	.056			2.61	3.94	6.55	8.25
1050	30 MPH avg., cycle 4 miles		252	.032			1.49	2.25	3.74	4.73
1052	cycle 6 miles		216	.037			1.74	2.63	4.37	5.50
1054	cycle 8 miles		180	.044			2.09	3.15	5.24	6.60
1056	cycle 10 miles		156	.051			2.41	3.63	6.04	7.65
1060	35 MPH avg., cycle 4 miles		264	.030			1.42	2.15	3.57	4.50
1062	cycle 6 miles		228	.035			1.65	2.49	4.14	5.20
1064	cycle 8 miles		204	.039			1.84	2.78	4.62	5.85
1066	cycle 10 miles		180	.044			2.09	3.15	5.24	6.60
1068	cycle 20 miles		120	.067			3.13	4.73	7.86	9.90
1069	cycle 30 miles		84	.095			4.48	6.75	11.23	14.20
1070	cycle 40 miles		72	.111			5.20	7.90	13.10	16.50
1072	40 MPH avg., cycle 6 miles		240	.033			1.57	2.36	3.93	4.96
1074	cycle 8 miles		216	.037			1.74	2.63	4.37	5.50
1076	cycle 10 miles		192	.042			1.96	2.95	4.91	6.20
1078	cycle 20 miles		120	.067			3.13	4.73	7.86	9.90
1080	cycle 30 miles		96	.083			3.92	5.90	9.82	12.40
1082	cycle 40 miles		72	.111			5.20	7.90	13.10	16.50
1084	cycle 50 miles		60	.133			6.25	9.45	15.70	19.85
1094	45 MPH avg., cycle 8 miles		216	.037			1.74	2.63	4.37	5.50
1096	cycle 10 miles		204	.039			1.84	2.78	4.62	5.85
1098	cycle 20 miles		132	.061			2.85	4.30	7.15	9
1100	cycle 30 miles		108	.074			3.48	5.25	8.73	11.05
1102	cycle 40 miles		84	.095			4.48	6.75	11.23	14.20
1104	cycle 50 miles		72	.111			5.20	7.90	13.10	16.50
1106	50 MPH avg., cycle 10 miles		216	.037			1.74	2.63	4.37	5.50
1108	cycle 20 miles		144	.056			2.61	3.94	6.55	8.25
1110	cycle 30 miles		108	.074			3.48	5.25	8.73	11.05
1112	cycle 40 miles		84	.095			4.48	6.75	11.23	14.20
1114	cycle 50 miles		72	.111			5.20	7.90	13.10	16.50
1214	15 MPH avg., cycle 0.5 mile, 20 min. wait/ld./uld.		264	.030			1.42	2.15	3.57	4.50
1216	cycle 1 mile		240	.033			1.57	2.36	3.93	4.96
1218	cycle 2 miles		204	.039			1.84	2.78	4.62	5.85
1220	cycle 4 miles		156	.051			2.41	3.63	6.04	7.65
1222	cycle 6 miles		132	.061			2.85	4.30	7.15	9
1224	cycle 8 miles		108	.074			3.48	5.25	8.73	11.05
1225	cycle 10 miles		96	.083			3.92	5.90	9.82	12.40
1226	20 MPH avg., cycle 0.5 mile		264	.030			1.42	2.15	3.57	4.50

31 23 Excavation and Fill

31 23 23 – Fill

31 23 23.20 Hauling		Crew	Daily Output	Labor-Hours	Unit	Material	2019 Bare Costs Labor	Equipment	Total	Total Incl O&P
1228	cycle 1 mile	B-34B	252	.032	L.C.Y.		1.49	2.25	3.74	4.73
1230	cycle 2 miles		216	.037			1.74	2.63	4.37	5.50
1232	cycle 4 miles		180	.044			2.09	3.15	5.24	6.60
1234	cycle 6 miles		144	.056			2.61	3.94	6.55	8.25
1236	cycle 8 miles		132	.061			2.85	4.30	7.15	9
1238	cycle 10 miles		108	.074			3.48	5.25	8.73	11.05
1240	25 MPH avg., cycle 4 miles		192	.042			1.96	2.95	4.91	6.20
1242	cycle 6 miles		168	.048			2.24	3.38	5.62	7.10
1244	cycle 8 miles		144	.056			2.61	3.94	6.55	8.25
1246	cycle 10 miles		132	.061			2.85	4.30	7.15	9
1250	30 MPH avg., cycle 4 miles		204	.039			1.84	2.78	4.62	5.85
1252	cycle 6 miles		180	.044			2.09	3.15	5.24	6.60
1254	cycle 8 miles		156	.051			2.41	3.63	6.04	7.65
1256	cycle 10 miles		144	.056			2.61	3.94	6.55	8.25
1260	35 MPH avg., cycle 4 miles		216	.037			1.74	2.63	4.37	5.50
1262	cycle 6 miles		192	.042			1.96	2.95	4.91	6.20
1264	cycle 8 miles		168	.048			2.24	3.38	5.62	7.10
1266	cycle 10 miles		156	.051			2.41	3.63	6.04	7.65
1268	cycle 20 miles		108	.074			3.48	5.25	8.73	11.05
1269	cycle 30 miles		72	.111			5.20	7.90	13.10	16.50
1270	cycle 40 miles		60	.133			6.25	9.45	15.70	19.85
1272	40 MPH avg., cycle 6 miles		192	.042			1.96	2.95	4.91	6.20
1274	cycle 8 miles		180	.044			2.09	3.15	5.24	6.60
1276	cycle 10 miles		156	.051			2.41	3.63	6.04	7.65
1278	cycle 20 miles		108	.074			3.48	5.25	8.73	11.05
1280	cycle 30 miles		84	.095			4.48	6.75	11.23	14.20
1282	cycle 40 miles		72	.111			5.20	7.90	13.10	16.50
1284	cycle 50 miles		60	.133			6.25	9.45	15.70	19.85
1294	45 MPH avg., cycle 8 miles		180	.044			2.09	3.15	5.24	6.60
1296	cycle 10 miles		168	.048			2.24	3.38	5.62	7.10
1298	cycle 20 miles		120	.067			3.13	4.73	7.86	9.90
1300	cycle 30 miles		96	.083			3.92	5.90	9.82	12.40
1302	cycle 40 miles		72	.111			5.20	7.90	13.10	16.50
1304	cycle 50 miles		60	.133			6.25	9.45	15.70	19.85
1306	50 MPH avg., cycle 10 miles		180	.044			2.09	3.15	5.24	6.60
1308	cycle 20 miles		132	.061			2.85	4.30	7.15	9
1310	cycle 30 miles		96	.083			3.92	5.90	9.82	12.40
1312	cycle 40 miles		84	.095			4.48	6.75	11.23	14.20
1314	cycle 50 miles		72	.111			5.20	7.90	13.10	16.50
1414	15 MPH avg., cycle 0.5 mile, 25 min. wait/ld./uld.		204	.039			1.84	2.78	4.62	5.85
1416	cycle 1 mile		192	.042			1.96	2.95	4.91	6.20
1418	cycle 2 miles		168	.048			2.24	3.38	5.62	7.10
1420	cycle 4 miles		132	.061			2.85	4.30	7.15	9
1422	cycle 6 miles		120	.067			3.13	4.73	7.86	9.90
1424	cycle 8 miles		96	.083			3.92	5.90	9.82	12.40
1425	cycle 10 miles		84	.095			4.48	6.75	11.23	14.20
1426	20 MPH avg., cycle 0.5 mile		216	.037			1.74	2.63	4.37	5.50
1428	cycle 1 mile		204	.039			1.84	2.78	4.62	5.85
1430	cycle 2 miles		180	.044			2.09	3.15	5.24	6.60
1432	cycle 4 miles		156	.051			2.41	3.63	6.04	7.65
1434	cycle 6 miles		132	.061			2.85	4.30	7.15	9
1436	cycle 8 miles		120	.067			3.13	4.73	7.86	9.90
1438	cycle 10 miles		96	.083			3.92	5.90	9.82	12.40

31 23 Excavation and Fill

31 23 23 – Fill

31 23 23.20 Hauling

		Crew	Daily Output	Labor-Hours	Unit	Material	2019 Bare Costs Labor	2019 Bare Costs Equipment	Total	Total Incl O&P
1440	25 MPH avg., cycle 4 miles	B-34B	168	.048	L.C.Y.		2.24	3.38	5.62	7.10
1442	cycle 6 miles		144	.056			2.61	3.94	6.55	8.25
1444	cycle 8 miles		132	.061			2.85	4.30	7.15	9
1446	cycle 10 miles		108	.074			3.48	5.25	8.73	11.05
1450	30 MPH avg., cycle 4 miles		168	.048			2.24	3.38	5.62	7.10
1452	cycle 6 miles		156	.051			2.41	3.63	6.04	7.65
1454	cycle 8 miles		132	.061			2.85	4.30	7.15	9
1456	cycle 10 miles		120	.067			3.13	4.73	7.86	9.90
1460	35 MPH avg., cycle 4 miles		180	.044			2.09	3.15	5.24	6.60
1462	cycle 6 miles		156	.051			2.41	3.63	6.04	7.65
1464	cycle 8 miles		144	.056			2.61	3.94	6.55	8.25
1466	cycle 10 miles		132	.061			2.85	4.30	7.15	9
1468	cycle 20 miles		96	.083			3.92	5.90	9.82	12.40
1469	cycle 30 miles		72	.111			5.20	7.90	13.10	16.50
1470	cycle 40 miles		60	.133			6.25	9.45	15.70	19.85
1472	40 MPH avg., cycle 6 miles		168	.048			2.24	3.38	5.62	7.10
1474	cycle 8 miles		156	.051			2.41	3.63	6.04	7.65
1476	cycle 10 miles		144	.056			2.61	3.94	6.55	8.25
1478	cycle 20 miles		96	.083			3.92	5.90	9.82	12.40
1480	cycle 30 miles		84	.095			4.48	6.75	11.23	14.20
1482	cycle 40 miles		60	.133			6.25	9.45	15.70	19.85
1484	cycle 50 miles		60	.133			6.25	9.45	15.70	19.85
1494	45 MPH avg., cycle 8 miles		156	.051			2.41	3.63	6.04	7.65
1496	cycle 10 miles		144	.056			2.61	3.94	6.55	8.25
1498	cycle 20 miles		108	.074			3.48	5.25	8.73	11.05
1500	cycle 30 miles		84	.095			4.48	6.75	11.23	14.20
1502	cycle 40 miles		72	.111			5.20	7.90	13.10	16.50
1504	cycle 50 miles		60	.133			6.25	9.45	15.70	19.85
1506	50 MPH avg., cycle 10 miles		156	.051			2.41	3.63	6.04	7.65
1508	cycle 20 miles		120	.067			3.13	4.73	7.86	9.90
1510	cycle 30 miles		96	.083			3.92	5.90	9.82	12.40
1512	cycle 40 miles		72	.111			5.20	7.90	13.10	16.50
1514	cycle 50 miles		60	.133			6.25	9.45	15.70	19.85
1614	15 MPH avg., cycle 0.5 mile, 30 min. wait/ld./uld.		180	.044			2.09	3.15	5.24	6.60
1616	cycle 1 mile		168	.048			2.24	3.38	5.62	7.10
1618	cycle 2 miles		144	.056			2.61	3.94	6.55	8.25
1620	cycle 4 miles		120	.067			3.13	4.73	7.86	9.90
1622	cycle 6 miles		108	.074			3.48	5.25	8.73	11.05
1624	cycle 8 miles		84	.095			4.48	6.75	11.23	14.20
1625	cycle 10 miles		84	.095			4.48	6.75	11.23	14.20
1626	20 MPH avg., cycle 0.5 mile		180	.044			2.09	3.15	5.24	6.60
1628	cycle 1 mile		168	.048			2.24	3.38	5.62	7.10
1630	cycle 2 miles		156	.051			2.41	3.63	6.04	7.65
1632	cycle 4 miles		132	.061			2.85	4.30	7.15	9
1634	cycle 6 miles		120	.067			3.13	4.73	7.86	9.90
1636	cycle 8 miles		108	.074			3.48	5.25	8.73	11.05
1638	cycle 10 miles		96	.083			3.92	5.90	9.82	12.40
1640	25 MPH avg., cycle 4 miles		144	.056			2.61	3.94	6.55	8.25
1642	cycle 6 miles		132	.061			2.85	4.30	7.15	9
1644	cycle 8 miles		108	.074			3.48	5.25	8.73	11.05
1646	cycle 10 miles		108	.074			3.48	5.25	8.73	11.05
1650	30 MPH avg., cycle 4 miles		144	.056			2.61	3.94	6.55	8.25
1652	cycle 6 miles		132	.061			2.85	4.30	7.15	9

347

31 23 23.20 Hauling	Crew	Daily Output	Labor-Hours	Unit	Material	2019 Bare Costs Labor	2019 Bare Costs Equipment	Total	Total Incl O&P
1654	cycle 8 miles	B-34B	120	.067	L.C.Y.		3.13	4.73	7.86	9.90
1656	cycle 10 miles		108	.074			3.48	5.25	8.73	11.05
1660	35 MPH avg., cycle 4 miles		156	.051			2.41	3.63	6.04	7.65
1662	cycle 6 miles		144	.056			2.61	3.94	6.55	8.25
1664	cycle 8 miles		132	.061			2.85	4.30	7.15	9
1666	cycle 10 miles		120	.067			3.13	4.73	7.86	9.90
1668	cycle 20 miles		84	.095			4.48	6.75	11.23	14.20
1669	cycle 30 miles		72	.111			5.20	7.90	13.10	16.50
1670	cycle 40 miles		60	.133			6.25	9.45	15.70	19.85
1672	40 MPH avg., cycle 6 miles		144	.056			2.61	3.94	6.55	8.25
1674	cycle 8 miles		132	.061			2.85	4.30	7.15	9
1676	cycle 10 miles		120	.067			3.13	4.73	7.86	9.90
1678	cycle 20 miles		96	.083			3.92	5.90	9.82	12.40
1680	cycle 30 miles		72	.111			5.20	7.90	13.10	16.50
1682	cycle 40 miles		60	.133			6.25	9.45	15.70	19.85
1684	cycle 50 miles		48	.167			7.85	11.80	19.65	25
1694	45 MPH avg., cycle 8 miles		144	.056			2.61	3.94	6.55	8.25
1696	cycle 10 miles		132	.061			2.85	4.30	7.15	9
1698	cycle 20 miles		96	.083			3.92	5.90	9.82	12.40
1700	cycle 30 miles		84	.095			4.48	6.75	11.23	14.20
1702	cycle 40 miles		60	.133			6.25	9.45	15.70	19.85
1704	cycle 50 miles		60	.133			6.25	9.45	15.70	19.85
1706	50 MPH avg., cycle 10 miles		132	.061			2.85	4.30	7.15	9
1708	cycle 20 miles		108	.074			3.48	5.25	8.73	11.05
1710	cycle 30 miles		84	.095			4.48	6.75	11.23	14.20
1712	cycle 40 miles		72	.111			5.20	7.90	13.10	16.50
1714	cycle 50 miles		60	.133			6.25	9.45	15.70	19.85
2000	Hauling, 8 C.Y. truck, small project cost per hour	B-34A	8	1	Hr.		47	42.50	89.50	118
2100	12 C.Y. truck	B-34B	8	1			47	71	118	149
2150	16.5 C.Y. truck	B-34C	8	1			47	77.50	124.50	157
2175	18 C.Y. 8 wheel truck	B-34I	8	1			47	88	135	168
2200	20 C.Y. truck	B-34D	8	1			47	79.50	126.50	159
2300	Grading at dump, or embankment if required, by dozer	B-10B	1000	.012	L.C.Y.		.61	1.29	1.90	2.33
2310	Spotter at fill or cut, if required	1 Clab	8	1	Hr.		41		41	62
9014	18 C.Y. truck, 8 wheels,15 min. wait/ld./uld.,15 MPH, cycle 0.5 mi.	B-34I	504	.016	L.C.Y.		.75	1.40	2.15	2.66
9016	cycle 1 mile		450	.018			.84	1.57	2.41	2.98
9018	cycle 2 miles		378	.021			.99	1.86	2.85	3.55
9020	cycle 4 miles		270	.030			1.39	2.61	4	4.97
9022	cycle 6 miles		216	.037			1.74	3.26	5	6.20
9024	cycle 8 miles		180	.044			2.09	3.92	6.01	7.45
9025	cycle 10 miles		144	.056			2.61	4.90	7.51	9.35
9026	20 MPH avg., cycle 0.5 mile		522	.015			.72	1.35	2.07	2.57
9028	cycle 1 mile		468	.017			.80	1.51	2.31	2.87
9030	cycle 2 miles		414	.019			.91	1.70	2.61	3.24
9032	cycle 4 miles		324	.025			1.16	2.18	3.34	4.14
9034	cycle 6 miles		252	.032			1.49	2.80	4.29	5.35
9036	cycle 8 miles		216	.037			1.74	3.26	5	6.20
9038	cycle 10 miles		180	.044			2.09	3.92	6.01	7.45
9040	25 MPH avg., cycle 4 miles		342	.023			1.10	2.06	3.16	3.92
9042	cycle 6 miles		288	.028			1.31	2.45	3.76	4.66
9044	cycle 8 miles		252	.032			1.49	2.80	4.29	5.35
9046	cycle 10 miles		216	.037			1.74	3.26	5	6.20
9050	30 MPH avg., cycle 4 miles		378	.021			.99	1.86	2.85	3.55

31 23 23.20 Hauling		Crew	Daily Output	Labor-Hours	Unit	Material	2019 Bare Costs Labor	Equipment	Total	Total Incl O&P
9052	cycle 6 miles	B-34I	324	.025	L.C.Y.		1.16	2.18	3.34	4.14
9054	cycle 8 miles		270	.030			1.39	2.61	4	4.97
9056	cycle 10 miles		234	.034			1.61	3.01	4.62	5.75
9060	35 MPH avg., cycle 4 miles		396	.020			.95	1.78	2.73	3.39
9062	cycle 6 miles		342	.023			1.10	2.06	3.16	3.92
9064	cycle 8 miles		288	.028			1.31	2.45	3.76	4.66
9066	cycle 10 miles		270	.030			1.39	2.61	4	4.97
9068	cycle 20 miles		162	.049			2.32	4.35	6.67	8.30
9070	cycle 30 miles		126	.063			2.98	5.60	8.58	10.65
9072	cycle 40 miles		90	.089			4.18	7.85	12.03	14.90
9074	40 MPH avg., cycle 6 miles		360	.022			1.04	1.96	3	3.72
9076	cycle 8 miles		324	.025			1.16	2.18	3.34	4.14
9078	cycle 10 miles		288	.028			1.31	2.45	3.76	4.66
9080	cycle 20 miles		180	.044			2.09	3.92	6.01	7.45
9082	cycle 30 miles		144	.056			2.61	4.90	7.51	9.35
9084	cycle 40 miles		108	.074			3.48	6.55	10.03	12.45
9086	cycle 50 miles		90	.089			4.18	7.85	12.03	14.90
9094	45 MPH avg., cycle 8 miles		324	.025			1.16	2.18	3.34	4.14
9096	cycle 10 miles		306	.026			1.23	2.30	3.53	4.38
9098	cycle 20 miles		198	.040			1.90	3.56	5.46	6.80
9100	cycle 30 miles		144	.056			2.61	4.90	7.51	9.35
9102	cycle 40 miles		126	.063			2.98	5.60	8.58	10.65
9104	cycle 50 miles		108	.074			3.48	6.55	10.03	12.45
9106	50 MPH avg., cycle 10 miles		324	.025			1.16	2.18	3.34	4.14
9108	cycle 20 miles		216	.037			1.74	3.26	5	6.20
9110	cycle 30 miles		162	.049			2.32	4.35	6.67	8.30
9112	cycle 40 miles		126	.063			2.98	5.60	8.58	10.65
9114	cycle 50 miles		108	.074			3.48	6.55	10.03	12.45
9214	20 min. wait/ld./uld.,15 MPH, cycle 0.5 mi.		396	.020			.95	1.78	2.73	3.39
9216	cycle 1 mile		360	.022			1.04	1.96	3	3.72
9218	cycle 2 miles		306	.026			1.23	2.30	3.53	4.38
9220	cycle 4 miles		234	.034			1.61	3.01	4.62	5.75
9222	cycle 6 miles		198	.040			1.90	3.56	5.46	6.80
9224	cycle 8 miles		162	.049			2.32	4.35	6.67	8.30
9225	cycle 10 miles		144	.056			2.61	4.90	7.51	9.35
9226	20 MPH avg., cycle 0.5 mile		396	.020			.95	1.78	2.73	3.39
9228	cycle 1 mile		378	.021			.99	1.86	2.85	3.55
9230	cycle 2 miles		324	.025			1.16	2.18	3.34	4.14
9232	cycle 4 miles		270	.030			1.39	2.61	4	4.97
9234	cycle 6 miles		216	.037			1.74	3.26	5	6.20
9236	cycle 8 miles		198	.040			1.90	3.56	5.46	6.80
9238	cycle 10 miles		162	.049			2.32	4.35	6.67	8.30
9240	25 MPH avg., cycle 4 miles		288	.028			1.31	2.45	3.76	4.66
9242	cycle 6 miles		252	.032			1.49	2.80	4.29	5.35
9244	cycle 8 miles		216	.037			1.74	3.26	5	6.20
9246	cycle 10 miles		198	.040			1.90	3.56	5.46	6.80
9250	30 MPH avg., cycle 4 miles		306	.026			1.23	2.30	3.53	4.38
9252	cycle 6 miles		270	.030			1.39	2.61	4	4.97
9254	cycle 8 miles		234	.034			1.61	3.01	4.62	5.75
9256	cycle 10 miles		216	.037			1.74	3.26	5	6.20
9260	35 MPH avg., cycle 4 miles		324	.025			1.16	2.18	3.34	4.14
9262	cycle 6 miles		288	.028			1.31	2.45	3.76	4.66
9264	cycle 8 miles		252	.032			1.49	2.80	4.29	5.35

31 23 23.20 Hauling		Crew	Daily Output	Labor-Hours	Unit	Material	Labor	Equipment	Total	Total Incl O&P
								2019 Bare Costs		
9266	cycle 10 miles	B-34I	234	.034	L.C.Y.		1.61	3.01	4.62	5.75
9268	cycle 20 miles		162	.049			2.32	4.35	6.67	8.30
9270	cycle 30 miles		108	.074			3.48	6.55	10.03	12.45
9272	cycle 40 miles		90	.089			4.18	7.85	12.03	14.90
9274	40 MPH avg., cycle 6 miles		288	.028			1.31	2.45	3.76	4.66
9276	cycle 8 miles		270	.030			1.39	2.61	4	4.97
9278	cycle 10 miles		234	.034			1.61	3.01	4.62	5.75
9280	cycle 20 miles		162	.049			2.32	4.35	6.67	8.30
9282	cycle 30 miles		126	.063			2.98	5.60	8.58	10.65
9284	cycle 40 miles		108	.074			3.48	6.55	10.03	12.45
9286	cycle 50 miles		90	.089			4.18	7.85	12.03	14.90
9294	45 MPH avg., cycle 8 miles		270	.030			1.39	2.61	4	4.97
9296	cycle 10 miles		252	.032			1.49	2.80	4.29	5.35
9298	cycle 20 miles		180	.044			2.09	3.92	6.01	7.45
9300	cycle 30 miles		144	.056			2.61	4.90	7.51	9.35
9302	cycle 40 miles		108	.074			3.48	6.55	10.03	12.45
9304	cycle 50 miles		90	.089			4.18	7.85	12.03	14.90
9306	50 MPH avg., cycle 10 miles		270	.030			1.39	2.61	4	4.97
9308	cycle 20 miles		198	.040			1.90	3.56	5.46	6.80
9310	cycle 30 miles		144	.056			2.61	4.90	7.51	9.35
9312	cycle 40 miles		126	.063			2.98	5.60	8.58	10.65
9314	cycle 50 miles		108	.074			3.48	6.55	10.03	12.45
9414	25 min. wait/ld./uld.,15 MPH, cycle 0.5 mi.		306	.026			1.23	2.30	3.53	4.38
9416	cycle 1 mile		288	.028			1.31	2.45	3.76	4.66
9418	cycle 2 miles		252	.032			1.49	2.80	4.29	5.35
9420	cycle 4 miles		198	.040			1.90	3.56	5.46	6.80
9422	cycle 6 miles		180	.044			2.09	3.92	6.01	7.45
9424	cycle 8 miles		144	.056			2.61	4.90	7.51	9.35
9425	cycle 10 miles		126	.063			2.98	5.60	8.58	10.65
9426	20 MPH avg., cycle 0.5 mile		324	.025			1.16	2.18	3.34	4.14
9428	cycle 1 mile		306	.026			1.23	2.30	3.53	4.38
9430	cycle 2 miles		270	.030			1.39	2.61	4	4.97
9432	cycle 4 miles		234	.034			1.61	3.01	4.62	5.75
9434	cycle 6 miles		198	.040			1.90	3.56	5.46	6.80
9436	cycle 8 miles		180	.044			2.09	3.92	6.01	7.45
9438	cycle 10 miles		144	.056			2.61	4.90	7.51	9.35
9440	25 MPH avg., cycle 4 miles		252	.032			1.49	2.80	4.29	5.35
9442	cycle 6 miles		216	.037			1.74	3.26	5	6.20
9444	cycle 8 miles		198	.040			1.90	3.56	5.46	6.80
9446	cycle 10 miles		180	.044			2.09	3.92	6.01	7.45
9450	30 MPH avg., cycle 4 miles		252	.032			1.49	2.80	4.29	5.35
9452	cycle 6 miles		234	.034			1.61	3.01	4.62	5.75
9454	cycle 8 miles		198	.040			1.90	3.56	5.46	6.80
9456	cycle 10 miles		180	.044			2.09	3.92	6.01	7.45
9460	35 MPH avg., cycle 4 miles		270	.030			1.39	2.61	4	4.97
9462	cycle 6 miles		234	.034			1.61	3.01	4.62	5.75
9464	cycle 8 miles		216	.037			1.74	3.26	5	6.20
9466	cycle 10 miles		198	.040			1.90	3.56	5.46	6.80
9468	cycle 20 miles		144	.056			2.61	4.90	7.51	9.35
9470	cycle 30 miles		108	.074			3.48	6.55	10.03	12.45
9472	cycle 40 miles		90	.089			4.18	7.85	12.03	14.90
9474	40 MPH avg., cycle 6 miles		252	.032			1.49	2.80	4.29	5.35
9476	cycle 8 miles		234	.034			1.61	3.01	4.62	5.75

31 23 23.20 Hauling		Crew	Daily Output	Labor-Hours	Unit	Material	2019 Bare Costs Labor	Equipment	Total	Total Incl O&P
9478	cycle 10 miles	B-34I	216	.037	L.C.Y.		1.74	3.26	5	6.20
9480	cycle 20 miles		144	.056			2.61	4.90	7.51	9.35
9482	cycle 30 miles		126	.063			2.98	5.60	8.58	10.65
9484	cycle 40 miles		90	.089			4.18	7.85	12.03	14.90
9486	cycle 50 miles		90	.089			4.18	7.85	12.03	14.90
9494	45 MPH avg., cycle 8 miles		234	.034			1.61	3.01	4.62	5.75
9496	cycle 10 miles		216	.037			1.74	3.26	5	6.20
9498	cycle 20 miles		162	.049			2.32	4.35	6.67	8.30
9500	cycle 30 miles		126	.063			2.98	5.60	8.58	10.65
9502	cycle 40 miles		108	.074			3.48	6.55	10.03	12.45
9504	cycle 50 miles		90	.089			4.18	7.85	12.03	14.90
9506	50 MPH avg., cycle 10 miles		234	.034			1.61	3.01	4.62	5.75
9508	cycle 20 miles		180	.044			2.09	3.92	6.01	7.45
9510	cycle 30 miles		144	.056			2.61	4.90	7.51	9.35
9512	cycle 40 miles		108	.074			3.48	6.55	10.03	12.45
9514	cycle 50 miles		90	.089			4.18	7.85	12.03	14.90
9614	30 min. wait/ld./uld.,15 MPH, cycle 0.5 mi.		270	.030			1.39	2.61	4	4.97
9616	cycle 1 mile		252	.032			1.49	2.80	4.29	5.35
9618	cycle 2 miles		216	.037			1.74	3.26	5	6.20
9620	cycle 4 miles		180	.044			2.09	3.92	6.01	7.45
9622	cycle 6 miles		162	.049			2.32	4.35	6.67	8.30
9624	cycle 8 miles		126	.063			2.98	5.60	8.58	10.65
9625	cycle 10 miles		126	.063			2.98	5.60	8.58	10.65
9626	20 MPH avg., cycle 0.5 mile		270	.030			1.39	2.61	4	4.97
9628	cycle 1 mile		252	.032			1.49	2.80	4.29	5.35
9630	cycle 2 miles		234	.034			1.61	3.01	4.62	5.75
9632	cycle 4 miles		198	.040			1.90	3.56	5.46	6.80
9634	cycle 6 miles		180	.044			2.09	3.92	6.01	7.45
9636	cycle 8 miles		162	.049			2.32	4.35	6.67	8.30
9638	cycle 10 miles		144	.056			2.61	4.90	7.51	9.35
9640	25 MPH avg., cycle 4 miles		216	.037			1.74	3.26	5	6.20
9642	cycle 6 miles		198	.040			1.90	3.56	5.46	6.80
9644	cycle 8 miles		180	.044			2.09	3.92	6.01	7.45
9646	cycle 10 miles		162	.049			2.32	4.35	6.67	8.30
9650	30 MPH avg., cycle 4 miles		216	.037			1.74	3.26	5	6.20
9652	cycle 6 miles		198	.040			1.90	3.56	5.46	6.80
9654	cycle 8 miles		180	.044			2.09	3.92	6.01	7.45
9656	cycle 10 miles		162	.049			2.32	4.35	6.67	8.30
9660	35 MPH avg., cycle 4 miles		234	.034			1.61	3.01	4.62	5.75
9662	cycle 6 miles		216	.037			1.74	3.26	5	6.20
9664	cycle 8 miles		198	.040			1.90	3.56	5.46	6.80
9666	cycle 10 miles		180	.044			2.09	3.92	6.01	7.45
9668	cycle 20 miles		126	.063			2.98	5.60	8.58	10.65
9670	cycle 30 miles		108	.074			3.48	6.55	10.03	12.45
9672	cycle 40 miles		90	.089			4.18	7.85	12.03	14.90
9674	40 MPH avg., cycle 6 miles		216	.037			1.74	3.26	5	6.20
9676	cycle 8 miles		198	.040			1.90	3.56	5.46	6.80
9678	cycle 10 miles		180	.044			2.09	3.92	6.01	7.45
9680	cycle 20 miles		144	.056			2.61	4.90	7.51	9.35
9682	cycle 30 miles		108	.074			3.48	6.55	10.03	12.45
9684	cycle 40 miles		90	.089			4.18	7.85	12.03	14.90
9686	cycle 50 miles		72	.111			5.20	9.80	15	18.60
9694	45 MPH avg., cycle 8 miles		216	.037			1.74	3.26	5	6.20

31 23 Excavation and Fill

31 23 23 – Fill

31 23 23.20 Hauling		Crew	Daily Output	Labor-Hours	Unit	Material	2019 Bare Costs Labor	Equipment	Total	Total Incl O&P
9696	cycle 10 miles	B-34I	198	.040	L.C.Y.		1.90	3.56	5.46	6.80
9698	cycle 20 miles		144	.056			2.61	4.90	7.51	9.35
9700	cycle 30 miles		126	.063			2.98	5.60	8.58	10.65
9702	cycle 40 miles		108	.074			3.48	6.55	10.03	12.45
9704	cycle 50 miles		90	.089			4.18	7.85	12.03	14.90
9706	50 MPH avg., cycle 10 miles		198	.040			1.90	3.56	5.46	6.80
9708	cycle 20 miles		162	.049			2.32	4.35	6.67	8.30
9710	cycle 30 miles		126	.063			2.98	5.60	8.58	10.65
9712	cycle 40 miles		108	.074			3.48	6.55	10.03	12.45
9714	cycle 50 miles		90	.089			4.18	7.85	12.03	14.90

31 23 23.23 Compaction

31 23 23.23 Compaction		Crew	Daily Output	Labor-Hours	Unit	Material	2019 Bare Costs Labor	Equipment	Total	Total Incl O&P
0010	**COMPACTION**									
7000	Walk behind, vibrating plate 18" wide, 6" lifts, 2 passes	A-1D	200	.040	E.C.Y.		1.64	.16	1.80	2.65
7020	3 passes		185	.043			1.78	.17	1.95	2.88
7040	4 passes		140	.057			2.35	.22	2.57	3.79

Estimating Tips

32 01 00 Operations and Maintenance of Exterior Improvements

- Recycling of asphalt pavement is becoming very popular and is an alternative to removal and replacement. It can be a good value engineering proposal if removed pavement can be recycled, either at the project site or at another site that is reasonably close to the project site. Sections on repair of flexible and rigid pavement are included.

32 10 00 Bases, Ballasts, and Paving

- When estimating paving, keep in mind the project schedule. Also note that prices for asphalt and concrete are generally higher in the cold seasons. Lines for pavement markings, including tactile warning systems and fence lines, are included.

32 90 00 Planting

- The timing of planting and guarantee specifications often dictate the costs for establishing tree and shrub growth and a stand of grass or ground cover. Establish the work performance schedule to coincide with the local planting season. Maintenance and growth guarantees can add 20–100% to the total landscaping cost and can be contractually cumbersome. The cost to replace trees and shrubs can be as high as 5% of the total cost, depending on the planting zone, soil conditions, and time of year.

Reference Numbers

Reference numbers are shown at the beginning of some major classifications. These numbers refer to related items in the Reference Section. The reference information may be an estimating procedure, an alternate pricing method, or technical information.

Note: Not all subdivisions listed here necessarily appear. ∎

Did you know?

RSMeans data is available through our online application:

- Search for costs by keyword
- Leverage the most up-to-date data
- Build and export estimates

Try it free
rsmeans.com/2019freetrial

32 31 Fences and Gates

32 31 13 – Chain Link Fences and Gates

32 31 13.30 Fence, Chain Link, Gates and Posts	Crew	Daily Output	Labor-Hours	Unit	Material	2019 Bare Costs Labor	Equipment	Total	Total Incl O&P
0010 **FENCE, CHAIN LINK, GATES & POSTS**									
0011 (1/3 post length in ground)									
0013 For Concrete, See Section 03 31									
6710 Corner post, galv. steel, 4" OD, set in conc., 4'	B-80	65	.492	Ea.	63	22.50	9.70	95.20	113
6715 6'		63	.508		97.50	23	10.05	130.55	153
6720 7'		61	.525		115	24	10.35	149.35	173
6725 8'		65	.492		180	22.50	9.70	212.20	242
7900 Auger fence post hole, 3' deep, medium soil, by hand	1 Clab	30	.267			10.95		10.95	16.55
7925 By machine	B-80	175	.183			8.30	3.61	11.91	16.45
7950 Rock, with jackhammer	B-9	32	1.250			52	7.35	59.35	86.50
7975 With rock drill	B-47C	65	.246			11.40	23	34.40	42

Estimating Tips

33 10 00 Water Utilities
33 30 00 Sanitary Sewerage Utilities
33 40 00 Storm Drainage Utilities

- Never assume that the water, sewer, and drainage lines will go in at the early stages of the project. Consider the site access needs before dividing the site in half with open trenches, loose pipe, and machinery obstructions. Always inspect the site to establish that the site drawings are complete. Check off all existing utilities on your drawings as you locate them. Be especially careful with underground utilities because appurtenances are sometimes buried during regrading or repaving operations. If you find any discrepancies, mark up the site plan for further research. Differing site conditions can be very costly if discovered later in the project.

- See also Section 33 01 00 for restoration of pipe where removal/replacement may be undesirable. Use of new types of piping materials can reduce the overall project cost. Owners/design engineers should consider the installing contractor as a valuable source of current information on utility products and local conditions that could lead to significant cost savings.

Reference Numbers

Reference numbers are shown at the beginning of some major classifications. These numbers refer to related items in the Reference Section. The reference information may be an estimating procedure, an alternate pricing method, or technical information.

Note: Not all subdivisions listed here necessarily appear. ∎

33 05 07 – Trenchless Installation of Utility Piping

33 05 07.13 Utility Directional Drilling	Crew	Daily Output	Labor-Hours	Unit	Material	2019 Bare Costs Labor	Equipment	Total	Total Incl O&P
0010 **UTILITY DIRECTIONAL DRILLING**									
0011 Excluding access and splice pits (if required) and conduit (required)									
0012 Drilled hole diameters shown should be 50% larger than conduit									
0013 Assume access to H2O & removal of spoil/drilling mud as non-hazard material									
0014 Actual production rates adj. to account for risk factors for each soil type									
0100 Sand, silt, clay, common earth									
0110 Mobilization or demobilization	B-82A	1	32	Ea.		1,475	2,275	3,750	4,725
0120 6" diameter		156	.205	L.F.		9.50	14.65	24.15	30.50
0130 12" diameter		78	.410			19	29.50	48.50	60.50
0140 18" diameter		52	.615			28.50	44	72.50	91.50
0150 24" diameter		39	.821			38	58.50	96.50	122
0200 Hard clay, cobble, random boulders									
0210 Mobilization or demobilization	B-82B	1	32	Ea.		1,475	2,550	4,025	5,025
0220 6" diameter		124	.258	L.F.		11.95	20.50	32.45	40.50
0230 12" diameter		50	.640			29.50	51	80.50	101
0240 18" diameter		30	1.067			49.50	85	134.50	168
0250 24" diameter		25	1.280			59.50	102	161.50	202
0260 30" diameter		20	1.600			74	127	201	252
0270 36" diameter		15	2.133			99	170	269	335
0300 Hard rock (solid bed, 24,000+ psi)									
0310 Mobilization or demobilization	B-82C	1	32	Ea.		1,475	2,825	4,300	5,350
0320 6" diameter		28	1.143	L.F.		53	101	154	191
0330 12" diameter		12	2.667			124	236	360	445
0340 18" diameter		8	4			185	355	540	670
0350 24" diameter		7	4.571			212	405	617	765
0360 30" diameter		6	5.333			247	475	722	890
0370 36" diameter		4.50	7.111			330	630	960	1,200

33 05 07.23 Utility Boring and Jacking

	Crew	Daily Output	Labor-Hours	Unit	Material	Labor	Equipment	Total	Total Incl O&P
0010 **UTILITY BORING AND JACKING**									
0011 Casing only, 100' minimum,									
0020 not incl. jacking pits or dewatering									
0100 Roadwork, 1/2" thick wall, 24" diameter casing	B-42	20	3.200	L.F.	119	148	56	323	420
0200 36" diameter		16	4		214	185	70.50	469.50	595
0300 48" diameter		15	4.267		287	197	75	559	700
0500 Railroad work, 24" diameter		15	4.267		119	197	75	391	515
0600 36" diameter		14	4.571		214	211	80.50	505.50	645
0700 48" diameter		12	5.333		287	247	93.50	627.50	795
0900 For ledge, add								20%	20%
1000 Small diameter boring, 3", sandy soil	B-82	900	.018		21.50	.82	.09	22.41	25
1040 Rocky soil	"	500	.032		21.50	1.48	.16	23.14	26
1100 Prepare jacking pits, incl. mobilization & demobilization, minimum				Ea.				3,225	3,700
1101 Maximum				"				22,000	25,500

33 05 07.36 Microtunneling

	Crew	Daily Output	Labor-Hours	Unit	Material	Labor	Equipment	Total	Total Incl O&P
0010 **MICROTUNNELING**									
0011 Not including excavation, backfill, shoring,									
0020 or dewatering, average 50'/day, slurry method									
0100 24" to 48" outside diameter, minimum				L.F.				965	965
0110 Adverse conditions, add				"				500	500
1000 Rent microtunneling machine, average monthly lease				Month				97,500	107,000
1010 Operating technician				Day				630	705
1100 Mobilization and demobilization, minimum				Job				41,200	45,900
1110 Maximum				"				445,500	490,500

33 05 Common Work Results for Utilities

33 05 61 – Concrete Manholes

33 05 61.10 Storm Drainage Manholes, Frames and Covers

		Crew	Daily Output	Labor-Hours	Unit	Material	2019 Bare Costs Labor	Equipment	Total	Total Incl O&P
0010	**STORM DRAINAGE MANHOLES, FRAMES & COVERS**									
0020	Excludes footing, excavation, backfill (See line items for frame & cover)									
0050	Brick, 4' inside diameter, 4' deep	D-1	1	16	Ea.	590	730		1,320	1,775
0100	6' deep		.70	22.857		835	1,050		1,885	2,525
0150	8' deep		.50	32		1,075	1,475		2,550	3,400
0200	For depths over 8', add		4	4	V.L.F.	87.50	183		270.50	375
0400	Concrete blocks (radial), 4' ID, 4' deep		1.50	10.667	Ea.	415	490		905	1,200
0500	6' deep		1	16		550	730		1,280	1,725
0600	8' deep		.70	22.857		685	1,050		1,735	2,350
0700	For depths over 8', add		5.50	2.909	V.L.F.	70	133		203	280
0800	Concrete, cast in place, 4' x 4', 8" thick, 4' deep	C-14H	2	24	Ea.	555	1,200	13.20	1,768.20	2,450
0900	6' deep		1.50	32		795	1,600	17.60	2,412.60	3,325
1000	8' deep		1	48		1,150	2,425	26.50	3,601.50	4,950
1100	For depths over 8', add		8	6	V.L.F.	127	300	3.30	430.30	600
1110	Precast, 4' ID, 4' deep	B-22	4.10	7.317	Ea.	865	350	47.50	1,262.50	1,525
1120	6' deep		3	10		1,075	480	65	1,620	1,975
1130	8' deep		2	15		1,200	720	97.50	2,017.50	2,525
1140	For depths over 8', add		16	1.875	V.L.F.	128	90.50	12.15	230.65	291
1150	5' ID, 4' deep	B-6	3	8	Ea.	1,750	355	107	2,212	2,575
1160	6' deep		2	12		2,075	535	160	2,770	3,250
1170	8' deep		1.50	16		2,575	715	213	3,503	4,150
1180	For depths over 8', add		12	2	V.L.F.	290	89	26.50	405.50	485
1190	6' ID, 4' deep		2	12	Ea.	2,475	535	160	3,170	3,700
1200	6' deep		1.50	16		3,025	715	213	3,953	4,625
1210	8' deep		1	24		3,725	1,075	320	5,120	6,075
1220	For depths over 8', add		8	3	V.L.F.	405	134	40	579	690
1250	Slab tops, precast, 8" thick									
1300	4' diameter manhole	B-6	8	3	Ea.	282	134	40	456	555
1400	5' diameter manhole		7.50	3.200		460	143	42.50	645.50	765
1500	6' diameter manhole		7	3.429		715	153	45.50	913.50	1,075
3800	Steps, heavyweight cast iron, 7" x 9"	1 Bric	40	.200		18.75	10.20		28.95	36
3900	8" x 9"		40	.200		22.50	10.20		32.70	40.50
3928	12" x 10-1/2"		40	.200		27.50	10.20		37.70	45.50
4000	Standard sizes, galvanized steel		40	.200		25	10.20		35.20	43
4100	Aluminum		40	.200		29	10.20		39.20	47.50
4150	Polyethylene		40	.200		31	10.20		41.20	49.50

33 05 63 – Concrete Vaults and Chambers

33 05 63.13 Precast Concrete Utility Structures

		Crew	Daily Output	Labor-Hours	Unit	Material	2019 Bare Costs Labor	Equipment	Total	Total Incl O&P
0010	**PRECAST CONCRETE UTILITY STRUCTURES**, 6" thick									
0050	5' x 10' x 6' high, ID	B-13	2	28	Ea.	1,800	1,250	291	3,341	4,200
0100	6' x 10' x 6' high, ID		2	28		1,875	1,250	291	3,416	4,275
0150	5' x 12' x 6' high, ID		2	28		1,975	1,250	291	3,516	4,400
0200	6' x 12' x 6' high, ID		1.80	31.111		2,225	1,400	325	3,950	4,900
0250	6' x 13' x 6' high, ID		1.50	37.333		2,925	1,675	390	4,990	6,150
0300	8' x 14' x 7' high, ID		1	56		3,150	2,500	580	6,230	7,925

33 05 97 – Identification and Signage for Utilities

33 05 97.05 Utility Connection

		Crew	Daily Output	Labor-Hours	Unit	Material	2019 Bare Costs Labor	Equipment	Total	Total Incl O&P
0010	**UTILITY CONNECTION**									
0020	Water, sanitary, stormwater, gas, single connection	B-14	1	48	Ea.	3,050	2,075	320	5,445	6,825
0030	Telecommunication	"	3	16	"	435	690	107	1,232	1,650

For customer support on your Electrical Costs with RSMeans data, call 800.448.8182.

357

33 05 Common Work Results for Utilities

33 05 97 – Identification and Signage for Utilities

33 05 97.10 Utility Accessories	Crew	Daily Output	Labor-Hours	Unit	Material	2019 Bare Costs Labor	Equipment	Total	Total Incl O&P
0010 **UTILITY ACCESSORIES**									
0400 Underground tape, detectable, reinforced, alum. foil core, 2"	1 Clab	150	.053	C.L.F.	9	2.19		11.19	13.20
0500 6"	"	140	.057	"	36	2.35		38.35	43

33 42 Stormwater Conveyance

33 42 33 – Stormwater Curbside Drains and Inlets

33 42 33.13 Catch Basins

	Crew	Daily Output	Labor-Hours	Unit	Material	2019 Bare Costs Labor	Equipment	Total	Total Incl O&P
0010 **CATCH BASINS**									
0011 Not including footing & excavation									
1600 Frames & grates, C.I., 24" square, 500 lb.	B-6	7.80	3.077	Ea.	355	137	41	533	640
1700 26" D shape, 600 lb.		7	3.429		640	153	45.50	838.50	985
1800 Light traffic, 18" diameter, 100 lb.		10	2.400		211	107	32	350	430
1900 24" diameter, 300 lb.		8.70	2.759		202	123	37	362	450
2000 36" diameter, 900 lb.		5.80	4.138		660	184	55	899	1,075
2100 Heavy traffic, 24" diameter, 400 lb.		7.80	3.077		273	137	41	451	550
2200 36" diameter, 1,150 lb.		3	8		885	355	107	1,347	1,625
2300 Mass. State standard, 26" diameter, 475 lb.		7	3.429		680	153	45.50	878.50	1,025
2400 30" diameter, 620 lb.		7	3.429		345	153	45.50	543.50	660
2500 Watertight, 24" diameter, 350 lb.		7.80	3.077		365	137	41	543	655
2600 26" diameter, 500 lb.		7	3.429		430	153	45.50	628.50	755
2700 32" diameter, 575 lb.	▼	6	4	▼	950	178	53.50	1,181.50	1,375
2800 3 piece cover & frame, 10" deep,									
2900 1,200 lb., for heavy equipment	B-6	3	8	Ea.	1,075	355	107	1,537	1,825
3000 Raised for paving 1-1/4" to 2" high									
3100 4 piece expansion ring									
3200 20" to 26" diameter	1 Clab	3	2.667	Ea.	184	109		293	370
3300 30" to 36" diameter	"	3	2.667	"	255	109		364	445
3320 Frames and covers, existing, raised for paving, 2", including									
3340 row of brick, concrete collar, up to 12" wide frame	B-6	18	1.333	Ea.	52	59.50	17.80	129.30	166
3360 20" to 26" wide frame		11	2.182		78.50	97.50	29	205	266
3380 30" to 36" wide frame	▼	9	2.667		97.50	119	35.50	252	325
3400 Inverts, single channel brick	D-1	3	5.333	Ea.	106	244		350	485
3500 Concrete		5	3.200		127	146		273	365
3600 Triple channel, brick		2	8		170	365		535	745
3700 Concrete	▼	3	5.333	▼	156	244		400	540

33 71 Electrical Utility Transmission and Distribution

33 71 13 – Electrical Utility Towers

33 71 13.23 Steel Electrical Utility Towers

	Crew	Daily Output	Labor-Hours	Unit	Material	2019 Bare Costs Labor	Equipment	Total	Total Incl O&P
0010 **STEEL ELECTRICAL UTILITY TOWERS**									
0100 Excavation and backfill, earth	R-5	135.38	.650	C.Y.		34	8.85	42.85	61
0105 Rock		21.46	4.101	"		215	56	271	385
0200 Steel footings (grillage) in earth		3.91	22.506	Ton	2,175	1,175	305	3,655	4,500
0205 In rock		3.20	27.500	"	2,175	1,450	375	4,000	4,975
0300 Rock anchors		5.87	14.991	Ea.	585	785	204	1,574	2,050
0400 Concrete foundations	▼	12.85	6.848	C.Y.	156	360	93.50	609.50	815
0490 See also Section 03 30 53.40									
0500 Towers: material handling and spotting	R-7	22.56	2.128	Ton		90.50	6.85	97.35	146
0540 Steel tower erection	R-5	7.65	11.503	▼	2,125	605	157	2,887	3,400

33 71 13 – Electrical Utility Towers

33 71 13.23 Steel Electrical Utility Towers		Crew	Daily Output	Labor-Hours	Unit	Material	2019 Bare Costs Labor	Equipment	Total	Total Incl O&P
0550	Lace and box	R-5	7.10	12.394	Ton	2,125	650	169	2,944	3,475
0560	Painting total structure	↓	1.47	59.864	Ea.	495	3,125	815	4,435	6,175
0570	Disposal of surplus material	R-7	20.87	2.300	Mile		97.50	7.40	104.90	157
0600	Special towers: material handling and spotting	"	12.31	3.899	Ton		166	12.55	178.55	267
0640	Special steel structure erection	R-6	6.52	13.497	↓	2,650	705	505	3,860	4,525
0650	Special steel lace and box	"	6.29	13.990	↓	2,650	735	525	3,910	4,575
0670	Disposal of surplus material	R-7	7.87	6.099	Mile		259	19.65	278.65	415

33 71 13.80 Transmission Line Right of Way

		Crew	Daily Output	Labor-Hours	Unit	Material	2019 Bare Costs Labor	Equipment	Total	Total Incl O&P
0010	**TRANSMISSION LINE RIGHT OF WAY**									
0100	Clearing right of way	B-87	6.67	5.997	Acre		315	455	770	970
0200	Restoration & seeding	B-10D	4	3	"	460	151	430	1,041	1,200

33 71 16 – Electrical Utility Poles

33 71 16.23 Steel Electrical Utility Poles

33 71 16.23 Steel Electrical Utility Poles		Crew	Daily Output	Labor-Hours	Unit	Material	2019 Bare Costs Labor	Equipment	Total	Total Incl O&P
0010	**STEEL ELECTRICAL UTILITY POLES**									
0100	Steel, square, tapered shaft, 20'	R-15A	6	8	Ea.	450	420	47	917	1,175
0120	25'		5.35	8.972		450	470	53	973	1,250
0140	30'		4.80	10		510	525	59	1,094	1,400
0160	35'		4.35	11.034		540	580	65	1,185	1,525
0180	40'		4	12		755	630	71	1,456	1,850
0300	Galvanized, round, tapered shaft, 40'		4	12		2,175	630	71	2,876	3,425
0320	45'		3.70	12.973		2,525	680	76.50	3,281.50	3,875
0340	50'		3.45	13.913		2,675	730	82	3,487	4,125
0360	55'		3	16		2,925	840	94.50	3,859.50	4,575
0380	60'	↓	2.65	18.113		3,125	950	107	4,182	5,000
0400	65'	R-13	2.20	19.091		4,050	1,100	107	5,257	6,225
0420	70'		1.90	22.105		4,875	1,275	124	6,274	7,425
0440	75'		1.70	24.706		6,400	1,425	139	7,964	9,325
0460	80'		1.50	28		5,775	1,625	158	7,558	8,950
0480	85'		1.40	30		6,600	1,750	169	8,519	10,000
0500	90'		1.25	33.600		6,475	1,950	189	8,614	10,200
0520	95'		1.15	36.522		7,225	2,125	206	9,556	11,300
0540	100'	↓	1	42		8,025	2,425	236	10,686	12,700
0700	12 sided, tapered shaft, 40'	R-15A	4	12		1,300	630	71	2,001	2,450
0720	45'		3.70	12.973		1,525	680	76.50	2,281.50	2,775
0740	50'		3.45	13.913		1,750	730	82	2,562	3,125
0760	55'		3	16		2,000	840	94.50	2,934.50	3,550
0780	60'	↓	2.65	18.113		1,950	950	107	3,007	3,675
0800	65'	R-13	2.20	19.091		2,425	1,100	107	3,632	4,450
0820	70'		1.90	22.105		2,650	1,275	124	4,049	4,950
0840	75'		1.70	24.706		2,875	1,425	139	4,439	5,450
0860	80'		1.50	28		4,625	1,625	158	6,408	7,675
0880	85'		1.40	30		7,900	1,750	169	9,819	11,500
0900	90'		1.25	33.600		10,300	1,950	189	12,439	14,500
0920	95'		1.15	36.522		13,000	2,125	206	15,331	17,700
0940	100'		1	42		15,800	2,425	236	18,461	21,300
0960	105'	↓	.90	46.667		16,200	2,700	263	19,163	22,100
0980	Ladder clips					270			270	297
1000	Galvanized steel, round, tapered, w/one 6' arm, 20'	R-15A	9.20	5.217		1,075	273	31	1,379	1,625
1020	25'		7.65	6.275		1,150	330	37	1,517	1,800
1040	30'		6.55	7.328		1,300	385	43	1,728	2,050
1060	35'		5.75	8.348		1,375	435	49.50	1,859.50	2,200
1080	40'	↓	5.15	9.320	↓	1,475	490	55	2,020	2,425

33 71 16.23 Steel Electrical Utility Poles	Crew	Daily Output	Labor-Hours	Unit	Material	2019 Bare Costs Labor	2019 Bare Costs Equipment	Total	Total Incl O&P
1200	Two 6' arms, 20'	R-15A	6.55	7.328	Ea.	1,375	385	43	1,803	2,125
1220	25'		5.75	8.348		1,400	435	49.50	1,884.50	2,250
1240	30'		5.15	9.320		1,425	490	55	1,970	2,375
1260	35'		4.60	10.435		1,600	545	61.50	2,206.50	2,650
1280	40'		4.20	11.429		1,825	600	67.50	2,492.50	3,000
1400	One 12' truss arm, 25'		7.65	6.275		1,550	330	37	1,917	2,225
1420	30'		6.55	7.328		1,325	385	43	1,753	2,100
1440	35'		5.75	8.348		1,500	435	49.50	1,984.50	2,350
1460	40'		5.15	9.320		1,950	490	55	2,495	2,925
1480	45'		4.60	10.435		2,075	545	61.50	2,681.50	3,175
1600	Two 12' truss arms, 25'		5.75	8.348		1,675	435	49.50	2,159.50	2,525
1620	30'		5.15	9.320		1,625	490	55	2,170	2,600
1640	35'		4.60	10.435		1,875	545	61.50	2,481.50	2,950
1660	40'		4.20	11.429		2,250	600	67.50	2,917.50	3,450
1680	45'		3.85	12.468		2,025	655	73.50	2,753.50	3,275
3400	Galvanized steel, tapered, 10'		15.50	3.097		435	162	18.25	615.25	740
3420	12'		11.50	4.174		515	219	24.50	758.50	925
3440	14'		9.20	5.217		380	273	31	684	860
3460	16'		8.45	5.680		395	298	33.50	726.50	915
3480	18'		7.65	6.275		405	330	37	772	980
3500	20'		7.10	6.761		390	355	40	785	1,000
5200	Galvanized steel, bracket type, 2", 6' long		18.50	2.595		166	136	15.30	317.30	405
5220	8' long		15.50	3.097		205	162	18.25	385.25	490
5240	10' long		12.50	3.840		345	201	22.50	568.50	705
6000	Digging holes in earth, average	R-5	25.14	3.500			183	48	231	330
6010	In rock, average	"	4.51	19.512			1,025	266	1,291	1,825
6020	Formed plate pole structure									
6030	Material handling and spotting	R-7	2.40	20	Ea.		850	64.50	914.50	1,375
6040	Erect steel plate pole	R-5	1.95	45.128		11,300	2,375	615	14,290	16,600
6050	Guys, anchors and hardware for pole, in earth		7.04	12.500		685	655	171	1,511	1,925
6060	In rock		17.96	4.900		820	257	67	1,144	1,350
6070	Foundations for line poles									
6080	Excavation, in earth	R-5	135.38	.650	C.Y.		34	8.85	42.85	61
6090	In rock		20	4.400			231	60	291	410
6110	Concrete foundations		11	8		156	420	109	685	920

33 71 16.33 Wood Electrical Utility Poles

		Crew	Daily Output	Labor-Hours	Unit	Material	2019 Bare Costs Labor	2019 Bare Costs Equipment	Total	Total Incl O&P
0010	**WOOD ELECTRICAL UTILITY POLES**									
0011	Excludes excavation, backfill and cast-in-place concrete									
1020	12" Ponderosa Pine Poles treated 0.40 ACQ, 16'	R-3	3.20	6.250	Ea.	890	375	40.50	1,305.50	1,575
5000	Wood, class 3 yellow pine, penta-treated, 25'	R-15A	8.60	5.581		295	292	33	620	800
5020	30'		7.70	6.234		345	325	37	707	910
5040	35'		5.80	8.276		425	435	49	909	1,175
5060	40'		5.30	9.057		515	475	53.50	1,043.50	1,350
5080	45'		4.70	10.213		555	535	60.50	1,150.50	1,475
5100	50'		4.20	11.429		680	600	67.50	1,347.50	1,725
5120	55'		3.80	12.632		790	660	74.50	1,524.50	1,950
5140	60'		3.50	13.714		1,050	720	81	1,851	2,325
5160	65'		3.20	15		1,350	785	88.50	2,223.50	2,750
5180	70'		3	16		1,900	840	94.50	2,834.50	3,450
5200	75'		2.80	17.143		2,400	900	101	3,401	4,100
6000	Wood, class 1 type C, CCA/ACA-treated, 25'		8.60	5.581		298	292	33	623	805
6020	30'		7.70	6.234		360	325	37	722	930

360

For customer support on your Electrical Costs with RSMeans data, call 800.448.8182.

33 71 Electrical Utility Transmission and Distribution

33 71 16 – Electrical Utility Poles

33 71 16.33 Wood Electrical Utility Poles

		Crew	Daily Output	Labor-Hours	Unit	Material	2019 Bare Costs Labor	Equipment	Total	Total Incl O&P
6040	35'	R-15A	5.80	8.276	Ea.	500	435	49	984	1,250
6060	40'		5.30	9.057		605	475	53.50	1,133.50	1,425
6080	45'		4.70	10.213		710	535	60.50	1,305.50	1,650
6100	50'		4.20	11.429		805	600	67.50	1,472.50	1,850
6120	55'	↓	3.80	12.632		900	660	74.50	1,634.50	2,050
6200	Wood, class 3 Douglas Fir, penta-treated, 20'	R-3	3.10	6.452		194	385	42	621	835
6400	25' R337119-30		2.90	6.897		246	410	45	701	935
6600	30'		2.60	7.692		340	460	50	850	1,125
6800	35'		2.40	8.333		495	500	54	1,049	1,350
7000	40'		2.30	8.696		645	520	56.50	1,221.50	1,550
7200	45'	↓	1.70	11.765	↓	880	705	76.50	1,661.50	2,100
7400	Cross arms with hardware & insulators									
7600	4' long	1 Elec	2.50	3.200	Ea.	151	192		343	455
7800	5' long		2.40	3.333		167	200		367	485
8000	6' long	↓	2.20	3.636	↓	174	218		392	515
9000	Disposal of pole & hardware surplus material	R-7	20.87	2.300	Mile		97.50	7.40	104.90	157
9100	Disposal of crossarms & hardware surplus material	"	40	1.200	"		51	3.86	54.86	82

33 71 19 – Electrical Underground Ducts and Manholes

33 71 19.15 Underground Ducts and Manholes

		Crew	Daily Output	Labor-Hours	Unit	Material	2019 Bare Costs Labor	Equipment	Total	Total Incl O&P
0010	**UNDERGROUND DUCTS AND MANHOLES**									
0011	Not incl. excavation, backfill, or concrete in slab and duct bank									
1000	Direct burial R337119-30									
1010	PVC, schedule 40, w/coupling, 1/2" diameter	1 Elec	340	.024	L.F.	.33	1.41		1.74	2.47
1020	3/4" diameter		290	.028		.40	1.66		2.06	2.91
1030	1" diameter		260	.031		.66	1.85		2.51	3.49
1040	1-1/2" diameter		210	.038		.98	2.29		3.27	4.49
1050	2" diameter	↓	180	.044		1.27	2.67		3.94	5.40
1060	3" diameter	2 Elec	240	.067		2.34	4		6.34	8.50
1070	4" diameter		160	.100		3.11	6		9.11	12.35
1080	5" diameter		120	.133		4.62	8		12.62	17.05
1090	6" diameter	↓	90	.178	↓	6.10	10.70		16.80	22.50
1110	Elbows, 1/2" diameter	1 Elec	48	.167	Ea.	.62	10		10.62	15.65
1120	3/4" diameter		38	.211		.68	12.65		13.33	19.60
1130	1" diameter		32	.250		1.30	15		16.30	24
1140	1-1/2" diameter		21	.381		3.01	23		26.01	37.50
1150	2" diameter		16	.500		3.97	30		33.97	49.50
1160	3" diameter		12	.667		13.15	40		53.15	74
1170	4" diameter		9	.889		23	53.50		76.50	105
1180	5" diameter		8	1		35.50	60		95.50	129
1190	6" diameter		5	1.600		36	96		132	183
1210	Adapters, 1/2" diameter		52	.154		.29	9.25		9.54	14.10
1220	3/4" diameter		43	.186		.48	11.15		11.63	17.20
1230	1" diameter		39	.205		.48	12.30		12.78	18.95
1240	1-1/2" diameter		35	.229		.80	13.75		14.55	21.50
1250	2" diameter		26	.308		1.53	18.50		20.03	29
1260	3" diameter		20	.400		3.11	24		27.11	39.50
1270	4" diameter		14	.571		5.15	34.50		39.65	56.50
1280	5" diameter		12	.667		10.30	40		50.30	71
1290	6" diameter		9	.889		19.80	53.50		73.30	102
1340	Bell end & cap, 1-1/2" diameter		35	.229		4.50	13.75		18.25	25.50
1350	Bell end & plug, 2" diameter		26	.308		6.10	18.50		24.60	34.50
1360	3" diameter		20	.400	↓	7.85	24		31.85	44.50

For customer support on your Electrical Costs with RSMeans data, call 800.448.8182.

361

33 71 19.15 Underground Ducts and Manholes	Crew	Daily Output	Labor-Hours	Unit	Material	2019 Bare Costs Labor	Equipment	Total	Total Incl O&P
1370	4" diameter	1 Elec	14	.571	Ea.	9.40	34.50		43.90	61.50
1380	5" diameter		12	.667		14.45	40		54.45	75.50
1390	6" diameter		9	.889		16.70	53.50		70.20	98
1450	Base spacer, 2" diameter		56	.143		1.21	8.60		9.81	14.15
1460	3" diameter		46	.174		1.61	10.45		12.06	17.35
1470	4" diameter		41	.195		1.77	11.70		13.47	19.45
1480	5" diameter		37	.216		1.98	13		14.98	21.50
1490	6" diameter		34	.235		2.35	14.15		16.50	23.50
1550	Intermediate spacer, 2" diameter		60	.133		1.19	8		9.19	13.25
1560	3" diameter		46	.174		1.72	10.45		12.17	17.50
1570	4" diameter		41	.195		1.54	11.70		13.24	19.20
1580	5" diameter		37	.216		1.82	13		14.82	21.50
1590	6" diameter		34	.235		2.45	14.15		16.60	23.50
4010	PVC, schedule 80, w/coupling, 1/2" diameter		215	.037	L.F.	.91	2.23		3.14	4.33
4020	3/4" diameter		180	.044		1.24	2.67		3.91	5.35
4030	1" diameter		145	.055		1.69	3.31		5	6.80
4040	1-1/2" diameter		120	.067		2.78	4		6.78	9
4050	2" diameter		100	.080		3.91	4.80		8.71	11.45
4060	3" diameter	2 Elec	130	.123		7.90	7.40		15.30	19.75
4070	4" diameter		90	.178		11.30	10.70		22	28.50
4080	5" diameter		70	.229		17.40	13.75		31.15	39.50
4090	6" diameter		50	.320		22.50	19.20		41.70	53.50
4110	Elbows, 1/2" diameter	1 Elec	29	.276	Ea.	2.41	16.55		18.96	27
4120	3/4" diameter		23	.348		5.55	21		26.55	37
4130	1" diameter		20	.400		6.60	24		30.60	43.50
4140	1-1/2" diameter		16	.500		11	30		41	57
4150	2" diameter		12	.667		14.85	40		54.85	76
4160	3" diameter		9	.889		37	53.50		90.50	121
4170	4" diameter		7	1.143		66	68.50		134.50	175
4180	5" diameter		6	1.333		177	80		257	315
4190	6" diameter		4	2		193	120		313	390
4210	Adapter, 1/2" diameter		39	.205		.29	12.30		12.59	18.70
4220	3/4" diameter		33	.242		.48	14.55		15.03	22
4230	1" diameter		29	.276		.48	16.55		17.03	25
4240	1-1/2" diameter		26	.308		.80	18.50		19.30	28.50
4250	2" diameter		23	.348		1.53	21		22.53	32.50
4260	3" diameter		18	.444		3.11	26.50		29.61	43.50
4270	4" diameter		13	.615		5.15	37		42.15	60.50
4280	5" diameter		11	.727		10.30	43.50		53.80	76.50
4290	6" diameter		8	1		19.80	60		79.80	112
4310	Bell end & cap, 1-1/2" diameter		26	.308		4.50	18.50		23	32.50
4320	Bell end & plug, 2" diameter		23	.348		6.10	21		27.10	38
4330	3" diameter		18	.444		7.85	26.50		34.35	48.50
4340	4" diameter		13	.615		9.40	37		46.40	65.50
4350	5" diameter		11	.727		14.45	43.50		57.95	81
4360	6" diameter		8	1		16.70	60		76.70	108
4370	Base spacer, 2" diameter		42	.190		1.21	11.45		12.66	18.40
4380	3" diameter		33	.242		1.61	14.55		16.16	23.50
4390	4" diameter		29	.276		1.77	16.55		18.32	26.50
4400	5" diameter		26	.308		1.98	18.50		20.48	29.50
4410	6" diameter		25	.320		2.35	19.20		21.55	31
4420	Intermediate spacer, 2" diameter		45	.178		1.19	10.70		11.89	17.25
4430	3" diameter		34	.235		1.72	14.15		15.87	23

33 71 Electrical Utility Transmission and Distribution

33 71 19 – Electrical Underground Ducts and Manholes

33 71 19.15 Underground Ducts and Manholes

		Crew	Daily Output	Labor-Hours	Unit	Material	2019 Bare Costs Labor	Equipment	Total	Total Incl O&P
4440	4" diameter	1 Elec	31	.258	Ea.	1.54	15.50		17.04	24.50
4450	5" diameter		28	.286		1.82	17.15		18.97	27.50
4460	6" diameter		25	.320		2.45	19.20		21.65	31

33 71 19.17 Electric and Telephone Underground

		Crew	Daily Output	Labor-Hours	Unit	Material	2019 Bare Costs Labor	Equipment	Total	Total Incl O&P
0010	**ELECTRIC AND TELEPHONE UNDERGROUND**									
0011	Not including excavation									
0200	backfill and cast in place concrete									
0250	For bedding, see Section 31 23 23.16 R337119-30									
0400	Hand holes, precast concrete, with concrete cover									
0600	2' x 2' x 3' deep	R-3	2.40	8.333	Ea.	505	500	54	1,059	1,350
0800	3' x 3' x 3' deep		1.90	10.526		655	630	68.50	1,353.50	1,725
1000	4' x 4' x 4' deep		1.40	14.286		1,450	855	92.50	2,397.50	2,975
1200	Manholes, precast with iron racks & pulling irons, C.I. frame									
1400	and cover, 4' x 6' x 7' deep	B-13	2	28	Ea.	2,925	1,250	291	4,466	5,425
1600	6' x 8' x 7' deep		1.90	29.474		3,275	1,325	305	4,905	5,925
1800	6' x 10' x 7' deep		1.80	31.111		3,675	1,400	325	5,400	6,500
4200	Underground duct, banks ready for concrete fill, min. of 7.5"									
4400	between conduits, center to center									
4580	PVC, type EB, 1 @ 2" diameter	2 Elec	480	.033	L.F.	1.12	2		3.12	4.22
4600	2 @ 2" diameter		240	.067		2.24	4		6.24	8.40
4800	4 @ 2" diameter		120	.133		4.48	8		12.48	16.90
4900	1 @ 3" diameter		400	.040		1.82	2.40		4.22	5.60
5000	2 @ 3" diameter		200	.080		3.64	4.80		8.44	11.15
5200	4 @ 3" diameter		100	.160		7.30	9.60		16.90	22.50
5300	1 @ 4" diameter		320	.050		1.86	3		4.86	6.50
5400	2 @ 4" diameter		160	.100		3.71	6		9.71	13.05
5600	4 @ 4" diameter		80	.200		7.45	12		19.45	26
5800	6 @ 4" diameter		54	.296		11.15	17.80		28.95	39
5810	1 @ 5" diameter		260	.062		2.06	3.70		5.76	7.75
5820	2 @ 5" diameter		130	.123		4.12	7.40		11.52	15.60
5840	4 @ 5" diameter		70	.229		8.25	13.75		22	29.50
5860	6 @ 5" diameter		50	.320		12.35	19.20		31.55	42
5870	1 @ 6" diameter		200	.080		3.56	4.80		8.36	11.05
5880	2 @ 6" diameter		100	.160		7.10	9.60		16.70	22
5900	4 @ 6" diameter		50	.320		14.20	19.20		33.40	44
5920	6 @ 6" diameter		30	.533		21.50	32		53.50	71.50
6200	Rigid galvanized steel, 2 @ 2" diameter		180	.089		17.25	5.35		22.60	27
6400	4 @ 2" diameter		90	.178		34.50	10.70		45.20	54
6800	2 @ 3" diameter		100	.160		25.50	9.60		35.10	42.50
7000	4 @ 3" diameter		50	.320		51	19.20		70.20	84.50
7200	2 @ 4" diameter		70	.229		38	13.75		51.75	62.50
7400	4 @ 4" diameter		34	.471		76	28.50		104.50	126
7600	6 @ 4" diameter		22	.727		114	43.50		157.50	191
7620	2 @ 5" diameter		60	.267		71.50	16		87.50	103
7640	4 @ 5" diameter		30	.533		143	32		175	205
7660	6 @ 5" diameter		18	.889		214	53.50		267.50	315
7680	2 @ 6" diameter		40	.400		101	24		125	147
7700	4 @ 6" diameter		20	.800		203	48		251	295
7720	6 @ 6" diameter		14	1.143		305	68.50		373.50	435
7800	For cast-in-place concrete, add									
7810	Under 1 C.Y.	C-6	16	3	C.Y.	198	128	3.33	329.33	415
7820	1 C.Y. to 5 C.Y.		19.20	2.500		179	107	2.78	288.78	360

For customer support on your Electrical Costs with RSMeans data, call 800.448.8182.

363

33 71 19 – Electrical Underground Ducts and Manholes

33 71 19.17 Electric and Telephone Underground	Crew	Daily Output	Labor-Hours	Unit	Material	2019 Bare Costs Labor	Equipment	Total	Total Incl O&P
7830	Over 5 C.Y.	C-6	24	2	C.Y.	148	85.50	2.22	235.72	294
7850	For reinforcing rods, add									
7860	#4 to #7	2 Rodm	1.10	14.545	Ton	1,025	800		1,825	2,325
7870	#8 to #14	"	1.50	10.667	"	1,025	585		1,610	2,000
8000	Fittings, PVC type EB, elbow, 2" diameter	1 Elec	16	.500	Ea.	16.25	30		46.25	63
8200	3" diameter		14	.571		17.75	34.50		52.25	70.50
8400	4" diameter		12	.667		39	40		79	103
8420	5" diameter		10	.800		124	48		172	209
8440	6" diameter		9	.889		173	53.50		226.50	271
8500	Coupling, 2" diameter					.82			.82	.90
8600	3" diameter					4.10			4.10	4.51
8700	4" diameter					4.34			4.34	4.77
8720	5" diameter					8.35			8.35	9.15
8740	6" diameter					23			23	25
8800	Adapter, 2" diameter	1 Elec	26	.308		.98	18.50		19.48	28.50
9000	3" diameter		20	.400		3.02	24		27.02	39.50
9200	4" diameter		16	.500		3.90	30		33.90	49.50
9220	5" diameter		13	.615		9.65	37		46.65	65.50
9240	6" diameter		10	.800		12.15	48		60.15	85
9400	End bell, 2" diameter		16	.500		2.20	30		32.20	47.50
9600	3" diameter		14	.571		3.75	34.50		38.25	55
9800	4" diameter		12	.667		4.65	40		44.65	64.50
9810	5" diameter		10	.800		7.85	48		55.85	80
9820	6" diameter		8	1		9.80	60		69.80	100
9830	5° angle coupling, 2" diameter		26	.308		7.90	18.50		26.40	36
9840	3" diameter		20	.400		27	24		51	65.50
9850	4" diameter		16	.500		18.55	30		48.55	65.50
9860	5" diameter		13	.615		16.20	37		53.20	73
9870	6" diameter		10	.800		25.50	48		73.50	99.50
9880	Expansion joint, 2" diameter		16	.500		32.50	30		62.50	81
9890	3" diameter		18	.444		68	26.50		94.50	115
9900	4" diameter		12	.667		100	40		140	170
9910	5" diameter		10	.800		184	48		232	275
9920	6" diameter		8	1		170	60		230	277
9930	Heat bender, 2" diameter					415			415	455
9940	6" diameter					1,500			1,500	1,675
9950	Cement, quart					15.60			15.60	17.15
9960	Nylon polyethylene pull rope, 1/4"	2 Elec	2000	.008	L.F.	.17	.48		.65	.91

33 71 23 – Insulators and Fittings

33 71 23.16 Post Insulators

		Crew	Daily Output	Labor-Hours	Unit	Material	2019 Bare Costs Labor	Equipment	Total	Total Incl O&P
0010	**POST INSULATORS**									
7400	Insulators, pedestal type	R-11	112	.500	Ea.		28.50	5.60	34.10	48.50
7490	See also line 33 71 39.13 1000									

33 71 26 – Transmission and Distribution Equipment

33 71 26.13 Capacitor Banks

		Crew	Daily Output	Labor-Hours	Unit	Material	2019 Bare Costs Labor	Equipment	Total	Total Incl O&P
0010	**CAPACITOR BANKS**									
1300	Station capacitors									
1350	Synchronous, 13 to 26 kV	R-11	3.11	18.006	MVAR	7,625	1,025	201	8,851	10,100
1360	46 kV		3.33	16.817		9,725	960	188	10,873	12,300
1370	69 kV		3.81	14.698		9,575	840	164	10,579	11,900
1380	161 kV		6.51	8.602		8,950	490	96	9,536	10,700
1390	500 kV		10.37	5.400		7,775	310	60.50	8,145.50	9,075

33 71 Electrical Utility Transmission and Distribution

33 71 26 – Transmission and Distribution Equipment

33 71 26.13 Capacitor Banks

		Crew	Daily Output	Labor-Hours	Unit	Material	2019 Bare Costs Labor	2019 Bare Costs Equipment	Total	Total Incl O&P
1450	Static, 13 to 26 kV	R-11	3.11	18.006	MVAR	6,475	1,025	201	7,701	8,850
1460	46 kV		3.01	18.605		8,175	1,050	208	9,433	10,800
1470	69 kV		3.81	14.698		7,925	840	164	8,929	10,200
1480	161 kV		6.51	8.602		7,375	490	96	7,961	8,950
1490	500 kV		10.37	5.400	↓	6,725	310	60.50	7,095.50	7,925
1600	Voltage regulators, 13 to 26 kV	↓	.75	74.667	Ea.	289,000	4,250	835	294,085	325,500

33 71 26.23 Current Transformers

		Crew	Daily Output	Labor-Hours	Unit	Material	2019 Bare Costs Labor	2019 Bare Costs Equipment	Total	Total Incl O&P
0010	**CURRENT TRANSFORMERS**									
4050	Current transformers, 13 to 26 kV	R-11	14	4	Ea.	3,550	228	44.50	3,822.50	4,325
4060	46 kV		9.33	6.002		10,400	340	67	10,807	12,000
4070	69 kV		7	8		10,800	455	89.50	11,344.50	12,600
4080	161 kV	↓	1.87	29.947	↓	34,900	1,700	335	36,935	41,300

33 71 26.26 Potential Transformers

		Crew	Daily Output	Labor-Hours	Unit	Material	2019 Bare Costs Labor	2019 Bare Costs Equipment	Total	Total Incl O&P
0010	**POTENTIAL TRANSFORMERS**									
4100	Potential transformers, 13 to 26 kV	R-11	11.20	5	Ea.	5,075	285	56	5,416	6,075
4110	46 kV		8	7		10,500	400	78.50	10,978.50	12,200
4120	69 kV		6.22	9.003		11,100	515	101	11,716	13,100
4130	161 kV		2.24	25		23,900	1,425	280	25,605	28,700
4140	500 kV	↓	1.40	40	↓	71,500	2,275	445	74,220	82,500

33 71 39 – High-Voltage Wiring

33 71 39.13 Overhead High-Voltage Wiring

		Crew	Daily Output	Labor-Hours	Unit	Material	2019 Bare Costs Labor	2019 Bare Costs Equipment	Total	Total Incl O&P
0010	**OVERHEAD HIGH-VOLTAGE WIRING** R337116-60									
0100	Conductors, primary circuits									
0110	Material handling and spotting	R-5	9.78	8.998	W.Mile		470	123	593	845
0120	For river crossing, add		11	8			420	109	529	750
0150	Conductors, per wire, 210 to 636 kcmil		1.96	44.898		13,000	2,350	610	15,960	18,500
0160	795 to 954 kcmil		1.87	47.059		26,200	2,475	640	29,315	33,300
0170	1,000 to 1,600 kcmil		1.47	59.864		45,900	3,125	815	49,840	56,000
0180	Over 1,600 kcmil		1.35	65.185		64,500	3,425	890	68,815	76,500
0200	For river crossing, add, 210 to 636 kcmil		1.24	70.968			3,725	970	4,695	6,675
0220	795 to 954 kcmil		1.09	80.734			4,225	1,100	5,325	7,550
0230	1,000 to 1,600 kcmil		.97	90.722			4,750	1,225	5,975	8,500
0240	Over 1,600 kcmil	↓	.87	101	↓		5,300	1,375	6,675	9,500
0300	Joints and dead ends	R-8	6	8	Ea.	1,800	425	44	2,269	2,675
0400	Sagging	R-5	7.33	12.001	W.Mile		630	164	794	1,125
0500	Clipping, per structure, 69 kV	R-10	9.60	5	Ea.		283	62	345	495
0510	161 kV		5.33	9.006			510	111	621	885
0520	345 to 500 kV	↓	2.53	18.972			1,075	234	1,309	1,850
0600	Make and install jumpers, per structure, 69 kV	R-8	3.20	15		490	795	82.50	1,367.50	1,825
0620	161 kV		1.20	40		975	2,125	220	3,320	4,525
0640	345 to 500 kV	↓	.32	150		1,650	7,950	825	10,425	14,700
0700	Spacers	R-10	68.57	.700		97.50	39.50	8.65	145.65	176
0720	For river crossings, add	"	60	.800	↓		45.50	9.90	55.40	79
0800	Installing pulling line (500 kV only)	R-9	1.45	44.138	W.Mile	850	2,175	182	3,207	4,425
0810	Disposal of surplus material, high voltage conductors	R-7	6.96	6.897	Mile		293	22	315	470
0820	With trailer mounted reel stands	"	13.71	3.501	"		149	11.25	160.25	239
0900	Insulators and hardware, primary circuits									
0920	Material handling and spotting, 69 kV	R-7	480	.100	Ea.		4.25	.32	4.57	6.85
0930	161 kV		685.71	.070			2.97	.23	3.20	4.78
0950	345 to 500 kV	↓	960	.050			2.12	.16	2.28	3.42
1000	Disk insulators, 69 kV	R-5	880	.100		99.50	5.25	1.36	106.11	118
1020	161 kV	↓	977.78	.090	↓	114	4.72	1.23	119.95	133

For customer support on your Electrical Costs with RSMeans data, call 800.448.8182.

365

33 71 39.13 Overhead High-Voltage Wiring	Crew	Daily Output	Labor-Hours	Unit	Material	2019 Bare Costs Labor	Equipment	Total	Total Incl O&P
1040	345 to 500 kV	R-5	1100	.080	Ea.	114	4.19	1.09	119.28	133
1060	See Section 33 71 23.16 for pin or pedestal insulator									
1100	Install disk insulator at river crossing, add									
1110	69 kV	R-5	586.67	.150	Ea.		7.85	2.05	9.90	14.05
1120	161 kV		880	.100			5.25	1.36	6.61	9.40
1140	345 to 500 kV	↓	880	.100	↓		5.25	1.36	6.61	9.40
1150	Disposal of surplus material, high voltage insulators	R-7	41.74	1.150	Mile		49	3.70	52.70	78.50
1300	Overhead ground wire installation									
1320	Material handling and spotting	R-7	5.65	8.496	W.Mile		360	27.50	387.50	580
1340	Overhead ground wire	R-5	1.76	50		3,500	2,625	680	6,805	8,550
1350	At river crossing, add	↓	1.17	75.214	↓		3,950	1,025	4,975	7,050
1360	Disposal of surplus material, grounding wire	↓	41.74	2.108	Mile		110	29	139	198
1400	Installing conductors, underbuilt circuits									
1420	Material handling and spotting	R-7	5.65	8.496	W.Mile		360	27.50	387.50	580
1440	Conductors, per wire, 210 to 636 kcmil	R-5	1.96	44.898		13,000	2,350	610	15,960	18,500
1450	795 to 954 kcmil		1.87	47.059		26,200	2,475	640	29,315	33,300
1460	1,000 to 1,600 kcmil		1.47	59.864		45,900	3,125	815	49,840	56,000
1470	Over 1,600 kcmil	↓	1.35	65.185	↓	64,500	3,425	890	68,815	76,500
1500	Joints and dead ends	R-8	6	8	Ea.	1,800	425	44	2,269	2,675
1550	Sagging	R-5	8.80	10	W.Mile		525	136	661	940
1600	Clipping, per structure, 69 kV	R-10	9.60	5	Ea.		283	62	345	495
1620	161 kV		5.33	9.006			510	111	621	885
1640	345 to 500 kV	↓	2.53	18.972			1,075	234	1,309	1,850
1700	Making and installing jumpers, per structure, 69 kV	R-8	5.87	8.177		490	435	45	970	1,225
1720	161 kV		.96	50		975	2,650	276	3,901	5,350
1740	345 to 500 kV	↓	.32	150	↓	1,650	7,950	825	10,425	14,700
1800	Spacers	R-10	96	.500		97.50	28.50	6.20	132.20	156
1810	Disposal of surplus material, conductors & hardware	R-7	6.96	6.897	Mile		293	22	315	470
2000	Insulators and hardware for underbuilt circuits									
2100	Material handling and spotting	R-7	1200	.040	Ea.		1.70	.13	1.83	2.73
2150	Disk insulators, 69 kV	R-8	600	.080		99.50	4.25	.44	104.19	116
2160	161 kV		686	.070		114	3.71	.39	118.10	131
2170	345 to 500 kV	↓	800	.060	↓	114	3.18	.33	117.51	130
2180	Disposal of surplus material, insulators & hardware	R-7	41.74	1.150	Mile		49	3.70	52.70	78.50
2300	Sectionalizing switches, 69 kV	R-5	1.26	69.841	Ea.	25,600	3,650	955	30,205	34,700
2310	161 kV		.80	110		28,900	5,775	1,500	36,175	42,100
2500	Protective devices	↓	5.50	16	↓	8,375	840	218	9,433	10,700
2600	Clearance poles, 8 poles per mile									
2650	In earth, 69 kV	R-5	1.16	75.862	Mile	6,875	3,975	1,025	11,875	14,700
2660	161 kV	"	.64	138		11,300	7,200	1,875	20,375	25,400
2670	345 to 500 kV	R-6	.48	183		13,600	9,600	6,875	30,075	36,900
2800	In rock, 69 kV	R-5	.69	128	↓	6,875	6,675	1,750	15,300	19,500
2820	161 kV	"	.35	251		11,300	13,200	3,425	27,925	36,100
2840	345 to 500 kV	R-6	.24	367	↓	13,600	19,200	13,700	46,500	59,000

33 72 Utility Substations

33 72 26 – Substation Bus Assemblies

33 72 26.13 Aluminum Substation Bus Assemblies	Crew	Daily Output	Labor-Hours	Unit	Material	2019 Bare Costs Labor	Equipment	Total	Total Incl O&P
0010 **ALUMINUM SUBSTATION BUS ASSEMBLIES**									
7300 Bus	R-11	590	.095	Lb.	3.38	5.40	1.06	9.84	13

33 72 33 – Control House Equipment

33 72 33.43 Substation Backup Batteries

	Crew	Daily Output	Labor-Hours	Unit	Material	Labor	Equipment	Total	Total Incl O&P
0010 **SUBSTATION BACKUP BATTERIES**									
9120 Battery chargers	R-11	11.20	5	Ea.	4,325	285	56	4,666	5,250
9200 Control batteries	"	14	4	K.A.H.	540	228	44.50	812.50	985

33 73 Utility Transformers

33 73 23 – Dry-Type Utility Transformers

33 73 23.20 Primary Transformers

	Crew	Daily Output	Labor-Hours	Unit	Material	Labor	Equipment	Total	Total Incl O&P
0010 **PRIMARY TRANSFORMERS**									
1000 Main conversion equipment									
1050 Power transformers, 13 to 26 kV	R-11	1.72	32.558	MVA	27,300	1,850	365	29,515	33,300
1060 46 kV		3.50	16		25,600	915	179	26,694	29,800
1070 69 kV		3.11	18.006		21,800	1,025	201	23,026	25,700
1080 110 kV		3.29	17.021		20,800	970	190	21,960	24,500
1090 161 kV		4.31	12.993		19,300	740	145	20,185	22,500
1100 500 kV		7	8	↓	19,200	455	89.50	19,744.50	21,900
1200 Grounding transformers	↓	3.11	18.006	Ea.	121,000	1,025	201	122,226	134,500

33 75 High-Voltage Switchgear and Protection Devices

33 75 13 – Air High-Voltage Circuit Breaker

33 75 13.13 High-Voltage Circuit Breaker, Air

	Crew	Daily Output	Labor-Hours	Unit	Material	Labor	Equipment	Total	Total Incl O&P
0010 **HIGH-VOLTAGE CIRCUIT BREAKER, AIR**									
2100 Air circuit breakers, 13 to 26 kV	R-11	.56	100	Ea.	68,000	5,700	1,125	74,825	84,500
2110 161 kV	"	.24	233	"	290,500	13,300	2,600	306,400	342,500

33 75 16 – Oil High-Voltage Circuit Breaker

33 75 16.13 Oil Circuit Breaker

	Crew	Daily Output	Labor-Hours	Unit	Material	Labor	Equipment	Total	Total Incl O&P
0010 **OIL CIRCUIT BREAKER**									
2000 Power circuit breakers									
2050 Oil circuit breakers, 13 to 26 kV	R-11	1.12	50	Ea.	65,500	2,850	560	68,910	77,000
2060 46 kV		.75	74.667		95,000	4,250	835	100,085	112,000
2070 69 kV		.45	124		220,500	7,100	1,400	229,000	254,500
2080 161 kV		.16	350		335,500	20,000	3,925	359,425	403,000
2090 500 kV	↓	.06	933	↓	1,253,500	53,000	10,400	1,316,900	1,470,000

33 75 19 – Gas High-Voltage Circuit Breaker

33 75 19.13 Gas Circuit Breaker

	Crew	Daily Output	Labor-Hours	Unit	Material	Labor	Equipment	Total	Total Incl O&P
0010 **GAS CIRCUIT BREAKER**									
2000 Power circuit breakers									
2150 Gas circuit breakers, 13 to 26 kV	R-11	.56	100	Ea.	254,000	5,700	1,125	260,825	289,500
2160 161 kV		.08	700		333,500	39,900	7,825	381,225	435,000
2170 500 kV	↓	.04	1400	↓	1,173,500	80,000	15,700	1,269,200	1,427,000

33 75 High-Voltage Switchgear and Protection Devices

33 75 23 – Vacuum High-Voltage Circuit Breaker

33 75 23.13 Vacuum Circuit Breaker

33 75 23.13 Vacuum Circuit Breaker	Crew	Daily Output	Labor-Hours	Unit	Material	2019 Bare Costs Labor	Equipment	Total	Total Incl O&P
0010 **VACUUM CIRCUIT BREAKER**									
2000 Power circuit breakers									
2200 Vacuum circuit breakers, 13 to 26 kV	R-11	.56	100	Ea.	57,500	5,700	1,125	64,325	73,500

33 75 36 – High-Voltage Utility Fuses

33 75 36.13 Fuses

	Crew	Daily Output	Labor-Hours	Unit	Material	Labor	Equipment	Total	Total Incl O&P
0010 **FUSES**									
8000 Protective equipment									
8250 Fuses, 13 to 26 kV	R-11	18.67	2.999	Ea.	2,375	171	33.50	2,579.50	2,925
8260 46 kV		11.20	5		2,725	285	56	3,066	3,475
8270 69 kV		8	7		2,875	400	78.50	3,353.50	3,825
8280 161 kV	↓	4.67	11.991	↓	3,575	685	134	4,394	5,100

33 75 39 – High-Voltage Surge Arresters

33 75 39.13 Surge Arresters

	Crew	Daily Output	Labor-Hours	Unit	Material	Labor	Equipment	Total	Total Incl O&P
0010 **SURGE ARRESTERS**									
8000 Protective equipment									
8050 Lightning arresters, 13 to 26 kV	R-11	18.67	2.999	Ea.	1,775	171	33.50	1,979.50	2,275
8060 46 kV		14	4		4,750	228	44.50	5,022.50	5,625
8070 69 kV		11.20	5		6,100	285	56	6,441	7,200
8080 161 kV		5.60	10		8,475	570	112	9,157	10,300
8090 500 kV	↓	1.40	40	↓	28,900	2,275	445	31,620	35,700

33 75 43 – Shunt Reactors

33 75 43.13 Reactors

	Crew	Daily Output	Labor-Hours	Unit	Material	Labor	Equipment	Total	Total Incl O&P
0010 **REACTORS**									
8000 Protective equipment									
8150 Reactors and resistors, 13 to 26 kV	R-11	28	2	Ea.	4,225	114	22.50	4,361.50	4,850
8160 46 kV		4.31	12.993		12,700	740	145	13,585	15,200
8170 69 kV		2.80	20		20,200	1,150	224	21,574	24,200
8180 161 kV		2.24	25		23,000	1,425	280	24,705	27,700
8190 500 kV	↓	.08	700	↓	90,000	39,900	7,825	137,725	167,000

33 75 53 – High-Voltage Switches

33 75 53.13 Switches

	Crew	Daily Output	Labor-Hours	Unit	Material	Labor	Equipment	Total	Total Incl O&P
0010 **SWITCHES**									
3000 Disconnecting switches									
3050 Gang operated switches									
3060 Manual operation, 13 to 26 kV	R-11	1.65	33.939	Ea.	16,000	1,925	380	18,305	20,900
3070 46 kV		1.12	50		27,000	2,850	560	30,410	34,700
3080 69 kV		.80	70		29,800	4,000	785	34,585	39,500
3090 161 kV		.56	100		35,700	5,700	1,125	42,525	49,100
3100 500 kV		.14	400		96,000	22,800	4,475	123,275	144,500
3110 Motor operation, 161 kV		.51	110		53,500	6,275	1,225	61,000	69,500
3120 500 kV		.28	200		145,500	11,400	2,225	159,125	180,000
3250 Circuit switches, 161 kV	↓	.41	137	↓	105,000	7,800	1,525	114,325	129,000
3300 Single pole switches									
3350 Disconnecting switches, 13 to 26 kV	R-11	28	2	Ea.	14,400	114	22.50	14,536.50	16,000
3360 46 kV		8	7		25,300	400	78.50	25,778.50	28,500
3370 69 kV		5.60	10		27,600	570	112	28,282	31,300
3380 161 kV		2.80	20		102,500	1,150	224	103,874	115,000
3390 500 kV		.22	255		292,000	14,500	2,850	309,350	346,000
3450 Grounding switches, 46 kV		5.60	10		37,900	570	112	38,582	42,700
3460 69 kV		3.73	15.013		39,100	855	168	40,123	44,500

33 75 High-Voltage Switchgear and Protection Devices

33 75 53 – High-Voltage Switches

33 75 53.13 Switches

		Crew	Daily Output	Labor-Hours	Unit	Material	2019 Bare Costs Labor	2019 Bare Costs Equipment	Total	Total Incl O&P
3470	161 kV	R-11	2.24	25	Ea.	41,700	1,425	280	43,405	48,300
3480	500 kV	↓	.62	90.323	↓	52,000	5,150	1,000	58,150	66,000

33 78 Substation Converter Stations

33 78 33 – Converter Stations

33 78 33.46 Substation Converter Stations

		Crew	Daily Output	Labor-Hours	Unit	Material	2019 Bare Costs Labor	2019 Bare Costs Equipment	Total	Total Incl O&P
0010	**SUBSTATION CONVERTER STATIONS**									
9000	Station service equipment									
9100	Conversion equipment									
9110	Station service transformers	R-11	5.60	10	Ea.	105,000	570	112	105,682	116,500

33 79 Site Grounding

33 79 83 – Site Grounding Conductors

33 79 83.13 Grounding Wire, Bar, and Rod

		Crew	Daily Output	Labor-Hours	Unit	Material	2019 Bare Costs Labor	2019 Bare Costs Equipment	Total	Total Incl O&P
0010	**GROUNDING WIRE, BAR, AND ROD**									
7500	Grounding systems	R-11	280	.200	Lb.	11.55	11.40	2.24	25.19	32

33 81 Communications Structures

33 81 13 – Communications Transmission Towers

33 81 13.10 Radio Towers

		Crew	Daily Output	Labor-Hours	Unit	Material	2019 Bare Costs Labor	2019 Bare Costs Equipment	Total	Total Incl O&P
0010	**RADIO TOWERS**									
0020	Guyed, 50' H, 40 lb. section, 70 MPH basic wind speed	2 Sswk	1	16	Ea.	2,775	895		3,670	4,475
0100	Wind load 90 MPH basic wind speed	"	1	16		3,375	895		4,270	5,150
0300	190' high, 40 lb. section, wind load 70 MPH basic wind speed	K-2	.33	72.727		9,450	3,875	745	14,070	17,300
0400	200' high, 70 lb. section, wind load 90 MPH basic wind speed		.33	72.727		15,600	3,875	745	20,220	24,100
0600	300' high, 70 lb. section, wind load 70 MPH basic wind speed		.20	120		24,700	6,375	1,225	32,300	38,600
0700	270' high, 90 lb. section, wind load 90 MPH basic wind speed		.20	120		26,900	6,375	1,225	34,500	41,000
0800	400' high, 100 lb. section, wind load 70 MPH basic wind speed		.14	171		37,600	9,125	1,750	48,475	57,500
0900	Self-supporting, 60' high, wind load 70 MPH basic wind speed		.80	30		4,400	1,600	310	6,310	7,675
0910	60' high, wind load 90 MPH basic wind speed		.45	53.333		5,225	2,825	545	8,595	10,800
1000	120' high, wind load 70 MPH basic wind speed		.40	60		9,725	3,200	615	13,540	16,400
1200	190' high, wind load 90 MPH basic wind speed	↓	.20	120	↓	28,100	6,375	1,225	35,700	42,300
2000	For states west of Rocky Mountains, add for shipping					10%				

Division Notes

	CREW	DAILY OUTPUT	LABOR-HOURS	UNIT	BARE COSTS				TOTAL INCL O&P
					MAT.	LABOR	EQUIP.	TOTAL	

Estimating Tips

34 11 00 Rail Tracks

This subdivision includes items that may involve either repair of existing or construction of new railroad tracks. Additional preparation work, such as the roadbed earthwork, would be found in Division 31. Additional new construction siding and turnouts are found in Subdivision 34 72. Maintenance of railroads is found under 34 01 23 Operation and Maintenance of Railways.

34 40 00 Traffic Signals

This subdivision includes traffic signal systems. Other traffic control devices such as traffic signs are found in Subdivision 10 14 53 Traffic Signage.

34 70 00 Vehicle Barriers

This subdivision includes security vehicle barriers, guide and guard rails, crash barriers, and delineators. The actual maintenance and construction of concrete and asphalt pavement are found in Division 32.

Reference Numbers

Reference numbers are shown at the beginning of some major classifications. These numbers refer to related items in the Reference Section. The reference information may be an estimating procedure, an alternate pricing method, or technical information.

Note: Not all subdivisions listed here necessarily appear. ■

Did you know?

RSMeans data is available through our online application:

■ Search for costs by keyword

■ Leverage the most up-to-date data

■ Build and export estimates

Try it free
rsmeans.com/2019freetrial

34 43 Airfield Signaling and Control Equipment

34 43 13 – Airfield Signals and Lighting

34 43 13.16 Airfield Runway and Taxiway Inset Lighting

	34 43 13.16 Airfield Runway and Taxiway Inset Lighting	Crew	Daily Output	Labor-Hours	Unit	Material	2019 Bare Costs Labor	Equipment	Total	Total Incl O&P
0010	**AIRFIELD RUNWAY AND TAXIWAY INSET LIGHTING**									
0100	Runway centerline, bidir., semi-flush, 200 W, w/shallow insert base	R-22	12.40	3.006	Ea.	2,050	165		2,215	2,500
0110	for mounting in base housing		18.60	2.004		1,075	110		1,185	1,375
0120	Flush, 200 W, w/shallow insert base		12.40	3.006		2,025	165		2,190	2,475
0130	for mounting in base housing		18.64	2		1,050	110		1,160	1,350
0140	45 W, for mounting in base housing		18.64	2		1,050	110		1,160	1,350
0150	Touchdown zone light, unidirectional, 200 W, w/shallow insert base		12.40	3.006		1,875	165		2,040	2,300
0160	115 W		12.40	3.006		1,750	165		1,915	2,175
0170	Bidirectional, 62 W		12.40	3.006		1,875	165		2,040	2,325
0180	Unidirectional, 200 W, for mounting in base housing		18.60	2.004		835	110		945	1,075
0190	115 W		18.60	2.004		895	110		1,005	1,150
0200	Bidirectional, 62 W		18.60	2.004		1,000	110		1,110	1,275
0210	Runway edge & threshold light, bidir., 200 W, for base housing		9.36	3.983		1,125	219		1,344	1,550
0220	300 W		9.36	3.983		1,575	219		1,794	2,050
0230	499 W		7.44	5.011		2,625	275		2,900	3,275
0240	Threshold & approach light, unidir., 200 W, for base housing		9.36	3.983		640	219		859	1,025
0250	499 W		7.44	5.011		780	275		1,055	1,275
0260	Runway edge, bidirectional, 2-115 W, for base housing		12.40	3.006		1,700	165		1,865	2,100
0270	2-185 W		12.40	3.006		1,875	165		2,040	2,325
0280	Runway threshold & end, bidir., 2-115 W, for base housing		12.40	3.006		1,875	165		2,040	2,300
0370	45 W, flush, for mounting in base housing		18.64	2		910	110		1,020	1,175
0380	115 W		18.64	2		930	110		1,040	1,200
0400	Transformer, 45 W		37.20	1.002		200	55		255	300
0410	65 W		37.20	1.002		205	55		260	305
0420	100 W		37.20	1.002		248	55		303	355
0430	200 W, 6.6/6.6 A		37.20	1.002		249	55		304	355
0440	20/6.6 A		24.80	1.503		266	82.50		348.50	415
0450	500 W		24.80	1.503		350	82.50		432.50	510
0460	Base housing, 24" deep, 12" diameter		18.64	2		565	110		675	785
0470	15" diameter		18.64	2		1,225	110		1,335	1,525
0480	Connection kit for 1 conductor cable		24.80	1.503		40	82.50		122.50	167
0500	Elevated runway marker, high intensity quartz, edge, 115 W		14.88	2.505		455	138		593	710
0510	175 W		14.88	2.505		470	138		608	720
0520	Threshold, 115 W		14.88	2.505		510	138		648	765
0530	Edge, 200 W, w/base and transformer		11.85	3.146		1,250	173		1,423	1,625
0540	Elevated runway light, quartz, runway edge, 45 W		14.88	2.505		265	138		403	495
0550	Threshold, 115 W		14.88	2.505		445	138		583	695
0560	Taxiway, 30 W		14.88	2.505		246	138		384	475
0570	Fixture, 200 W, w/base and transformer		11.85	3.146		1,250	173		1,423	1,625
0580	Elevated threshold marker, quartz, 115 W		14.88	2.505		555	138		693	815
0590	175 W		14.88	2.505		555	138		693	815

34 43 23 – Weather Observation Equipment

34 43 23.16 Airfield Wind Cones

		Crew	Daily Output	Labor-Hours	Unit	Material	2019 Bare Costs Labor	Equipment	Total	Total Incl O&P
0010	**AIRFIELD WIND CONES**									
1200	Wind cone, 12' lighted assembly, rigid, w/obstruction light	R-21	1.36	24.118	Ea.	10,800	1,450	49	12,299	14,100
1210	Without obstruction light		1.52	21.579		9,075	1,300	44	10,419	11,900
1220	Unlighted assembly, w/obstruction light		1.68	19.524		9,875	1,175	39.50	11,089.50	12,700
1230	Without obstruction light		1.84	17.826		9,075	1,075	36	10,186	11,600
1240	Wind cone slip fitter, 2-1/2" pipe		21.84	1.502		120	90	3.05	213.05	270
1250	Wind cone sock, 12' x 3', cotton		6.56	5		655	300	10.15	965.15	1,175
1260	Nylon		6.56	5		675	300	10.15	985.15	1,200

Estimating Tips

- When estimating costs for the installation of electrical power generation equipment, factors to review include access to the job site, access and setting up at the installation site, required connections, uncrating pads, anchors, leveling, final assembly of the components, and temporary protection from physical damage, such as environmental exposure.

- Be aware of the costs of equipment supports, concrete pads, and vibration isolators. Cross-reference them against other trades' specifications. Also, review site and structural drawings for items that must be included in the estimates.

- It is important to include items that are not documented in the plans and specifications but must be priced. These items include, but are not limited to, testing, dust protection, roof penetration, core drilling concrete floors and walls, patching, cleanup, and final adjustments. Add a contingency or allowance for utility company fees for power hookups, if needed.

- The project size and scope of electrical power generation equipment will have a significant impact on cost. The intent of RSMeans cost data is to provide a benchmark cost so that owners, engineers, and electrical contractors will have a comfortable number with which to start a project. Additionally, there are many websites available to use for research and to obtain a vendor's quote to finalize costs.

Reference Numbers

Reference numbers are shown at the beginning of some major classifications. These numbers refer to related items in the Reference Section. The reference information may be an estimating procedure, an alternate pricing method, or technical information.

Did you know?

RSMeans data is available through our online application:

- Search for costs by keyword
- Leverage the most up-to-date data
- Build and export estimates

Try it free
rsmeans.com/2019freetrial

48 13 Hydroelectric Plant Electrical Power Generation Equipment

48 13 13 – Hydroelectric Power Plant Water Turbines

48 13 13.10 Water Turbines and Components	Crew	Daily Output	Labor-Hours	Unit	Material	2019 Bare Costs Labor	Equipment	Total	Total Incl O&P
0010 **WATER TURBINES & COMPONENTS**									
0500 Water turbine, 100-500 watt, 3P, 48VDC, 54"W x 24"H, 260LBS	2 Elec	2	8	Ea.	10,100	480		10,580	11,800
0600 Water turbine, 1500-2000 watt, 3P, 130-200VAC to DC, 95LBS	"	3	5.333	"	3,975	320		4,295	4,850

48 15 Wind Energy Electrical Power Generation Equipment

48 15 13 – Wind Turbines

48 15 13.50 Wind Turbines and Components

	Crew	Daily Output	Labor-Hours	Unit	Material	2019 Bare Costs Labor	Equipment	Total	Total Incl O&P
0010 **WIND TURBINES & COMPONENTS**									
0500 Complete system, grid connected									
0501 Wind turbine marine rated, 160 W, 12 V, DC	2 Elec	2.20	7.273	Ea.	1,150	435		1,585	1,925
0510 160 W, 24 V, DC		2.20	7.273		1,175	435		1,610	1,950
0520 160 W, 48 V, DC		2.20	7.273		1,150	435		1,585	1,925
1000 20 kW, 31' diam., incl. labor & material G				System				49,900	49,900
1001 Wind turbine, 300 W, 12 V, DC	2 Elec	2.20	7.273	Ea.	860	435		1,295	1,600
1011 300 W, 24 V, DC		2.20	7.273		875	435		1,310	1,625
1020 300 W, 48 V, DC		2.20	7.273		860	435		1,295	1,600
1030 Wind turbine, 400 W, 12 V, DC		2.20	7.273		860	435		1,295	1,600
1040 400 W, 24 V, DC		2.20	7.273		860	435		1,295	1,600
1050 400 W, 48 V, DC		2.20	7.273		875	435		1,310	1,625
2000 2.4 kW, 12' diam., incl. labor & material G				System				18,000	18,000
2900 Component system									
3200 1,000 W, 9' diam. G	1 Elec	2.05	3.902	Ea.	1,700	234		1,934	2,225
3300 Wind turbine/charge control, 2,900-7,900 W, 5.5" H x 11' W, 180 lb.	2 Elec	2	8		16,300	480		16,780	18,600
3320 Wind turbine/charge control, 4,000-10,800 W, 5.5" H x 13' W, 240 lb.	"	2	8		27,000	480		27,480	30,500
3400 Mounting hardware									
3401 Wind turbine tower, 2 stage steel, 8' H x 6' W	2 Elec	2	8	Ea.	2,050	480		2,530	2,975
3500 30' guyed tower kit G	2 Clab	5.12	3.125		405	128		533	640
3505 3' galvanized helical earth screw G	1 Clab	8	1		54.50	41		95.50	122
3510 Attic mount kit G	1 Rofc	2.56	3.125		216	141		357	470
3520 Roof mount kit G	1 Clab	3.41	2.346		264	96.50		360.50	435
4000 Anemometer, standard AC or 9V, 2-digit LCD, wood case, +/-2% of Input	2 Elec	2	8		171	480		651	905
8900 Equipment									
9100 DC to AC inverter for, 48 V, 4,000 W G	1 Elec	2	4	Ea.	2,300	240		2,540	2,875

48 18 Fuel Cell Electrical Power Generation Equipment

48 18 13 – Electrical Power Generation Fuel Cells

48 18 13.10 Fuel Cells

	Crew	Daily Output	Labor-Hours	Unit	Material	2019 Bare Costs Labor	Equipment	Total	Total Incl O&P
0010 **FUEL CELLS**									
2001 Complete system, natural gas, installed, 200 kW G				System				1,150,000	1,150,000
2002 400 kW G								2,168,000	2,168,000
2003 600 kW G								3,060,000	3,060,000
2004 800 kW G								4,000,000	4,000,000
2005 1,000 kW G								5,005,000	5,005,000
2010 Comm type, for battery charging, uses hydrogen, 100 W, 12 V G				Ea.	2,375			2,375	2,625
2020 500 W, 12 V G					5,375			5,375	5,925
2030 1,000 W, 48 V G					8,025			8,025	8,825
2040 Small, for demonstration or battery charging units, 12 V G					163			163	179
2050 Spare anode set for unit G					40			40	44

48 18 Fuel Cell Electrical Power Generation Equipment

48 18 13 – Electrical Power Generation Fuel Cells

48 18 13.10 Fuel Cells		Crew	Daily Output	Labor-Hours	Unit	Material	2019 Bare Costs Labor	Equipment	Total	Total Incl O&P
2060	3 W, uses hydrogen gas	G				Ea.	296	-		296	325
2070	6 W, uses hydrogen gas	G					570			570	625
2080	10 W, uses hydrogen gas	G					1,375			1,375	1,525

Assemblies Section

Table of Contents

RSMeans data: Assemblies— How They Work

Assemblies estimating provides a fast and reasonably accurate way to develop construction costs. An assembly is the grouping of individual work items—with appropriate quantities— to provide a cost for a major construction component in a convenient unit of measure.

An assemblies estimate is often used during early stages of design development to compare

the cost impact of various design alternatives on total building cost.

Assemblies estimates are also used as an efficient tool to verify construction estimates.

Assemblies estimates do not require a completed design or detailed drawings. Instead, they are based on the general size of the structure and

other known parameters of the project. The degree of accuracy of an assemblies estimate is generally within +/- 15%.

Most assemblies consist of three major elements: a graphic, the system components, and the cost data itself. The **Graphic** is a visual representation showing the typical appearance of the assembly

① Unique 12-character Identifier

Our assemblies are identified by a **unique 12-character identifier**. The assemblies are numbered using UNIFORMAT II, ASTM Standard E1557. The first 5 characters represent this system to Level 3. The last 7 characters represent further breakdown in order to arrange items in understandable groups of similar tasks. Line numbers are consistent across all of our publications, so a line number in any assemblies data set will always refer to the same item.

② Reference Box

Information is available in the Reference Section to assist the estimator with estimating procedures, alternate pricing methods, and additional technical information.

The **Reference Box** indicates the exact location of this information in the Reference Section. The "R" stands for "reference," and the remaining characters are the line numbers.

③ Narrative Descriptions

Our assemblies descriptions appear in two formats: narrative and table. **Narrative descriptions** are shown in a hierarchical structure to make them readable. In order to read a complete description, read up through the indents to the top of the section. Include everything that is above and to the left that is not contradicted by information below.

D20 Plumbing

D2020 Domestic Water Distribution

Installation includes piping and fittings within 10' of heater. Electric water heaters do not require venting.

1 Kilowatt hour will raise:			
Gallons of Water	Degrees F	Gallons of Water	Degrees F
4.1	100°	6.8	60°
4.5	90°	8.2	50°
5.1	80°	10.0	40°
5.9	70°		

System Components			COST EACH		
	QUANTITY	UNIT	MAT.	INST.	TOTAL
SYSTEM D2020 210 1780 **ELECTRIC WATER HEATER, RESIDENTIAL, 100°F RISE** **10 GALLON TANK, 7 GPH**					
Water heater, residential electric, glass lined tank, 10 gal.	1.000	Ea.	485	330	815
Copper tubing, type L, solder joint, hanger 10' OC 1/2" diam.	30.000	L.F.	111.60	280.50	392.10
Wrought copper 90° elbow for solder joints 1/2" diam.	4.000	Ea.	5.28	152	157.28
Wrought copper Tee for solder joints, 1/2" diam.	2.000	Ea.	5	117	122
Union, wrought copper, 1/2" diam.	2.000	Ea.	45	80	125
Valve, gate, bronze, 125 lb, NRS, soldered 1/2" diam.	2.000	Ea.	131	63	194
Relief valve, bronze, press & temp, self-close, 3/4" IPS	1.000	Ea.	231	27	258
Wrought copper adapter, CTS to MPT 3/4" IPS	1.000	Ea.	6.35	44.50	50.85
Copper tubing, type L, solder joints, 3/4" diam.	1.000	L.F.	4.82	9.95	14.77
Wrought copper 90° elbow for solder joints 3/4" diam.	1.000	Ea.	2.64	40	42.64
TOTAL			1,027.69	1,143.95	2,171.64

D2020 210	Electric Water Heaters - Residential Systems	COST EACH		
		MAT.	INST.	TOTAL
1760	Electric water heater, residential, 100°F rise			
1780	10 gallon tank, 7 GPH	1,025	1,150	2,175
1820	20 gallon tank, 7 GPH	1,275	1,225	2,500
1860	30 gallon tank, 7 GPH	1,900	1,275	3,175
1900	40 gallon tank, 8 GPH	2,150	1,400	3,550
1940	52 gallon tank, 10 GPH	2,325	1,425	3,750
1980	66 gallon tank, 13 GPH	3,625	1,625	5,250
2020	80 gallon tank, 16 GPH	3,875	1,700	5,575
2060	120 gallon tank, 23 GPH	5,500	1,975	7,475

(Reference box at line 1820/1860: R224000 -20)

For supplemental customizable square foot estimating forms, visit: **RSMeans.com/2019books** in question. It is frequently accompanied by additional explanatory technical information describing the class of items. The **System Components** is a listing of the individual tasks that make up the assembly, including the quantity and unit of measure for each item, along with the cost of material and installation. The **Assemblies** data below lists prices for other similar systems with dimensional and/or size variations.

All of our assemblies costs represent the cost for the installing contractor. An allowance for profit has been added to all material, labor, and equipment rental costs. A markup for labor burdens, including workers' compensation, fixed overhead, and business overhead, is included with installation costs.

The information in RSMeans cost data represents a "national average" cost. This data should be modified to the project location using the **City Cost Indexes** or **Location Factors** tables found in the Reference Section.

D20 Plumbing

D2020 Domestic Water Distribution

Installation includes piping and fittings within 10' of heater. Electric water heaters do not require venting.

1 Kilowatt hour will raise:			
Gallons of Water	Degrees F	Gallons of Water	Degrees F
4.1	100°	6.8	60°
4.5	90°	8.2	50°
5.1	80°	10.0	40°
5.9	70°		

④ Unit of Measure

All RSMeans data: Assemblies include a typical **Unit of Measure** used for estimating that item. For instance, for continuous footings or foundation walls the unit is linear feet (L.F.). For spread footings the unit is each (Ea.). The estimator needs to take special care that the unit in the data matches the unit in the takeoff. Abbreviations and unit conversions can be found in the Reference Section.

⑤ System Components

System components are listed separately to detail what is included in the development of the total system price.

System Components ⑤	QUANTITY	UNIT	COST EACH		
			MAT.	INST.	TOTAL
SYSTEM D2020 210 1780					
ELECTRIC WATER HEATER, RESIDENTIAL, 100°F RISE					
10 GALLON TANK, 7 GPH					
Water heater, residential electric, glass lined tank, 10 gal.	1.000	Ea.	485	330	815
Copper tubing, type L, solder joint, hanger 10' OC 1/2" diam.	30.000	L.F.	111.60	280.50	392.10
Wrought copper 90° elbow for solder joints 1/2" diam.	4.000	Ea.	5.28	152	157.28
Wrought copper Tee for solder joints, 1/2" diam.	2.000	Ea.	5	117	122
Union, wrought copper, 1/2" diam.	2.000	Ea.	45	80	125
Valve, gate, bronze, 125 lb, NRS, soldered 1/2" diam.	2.000	Ea.	131	63	194
Relief valve, bronze, press & temp, self-close, 3/4" IPS	1.000	Ea.	231	27	258
Wrought copper adapter, CTS to MPT 3/4" IPS	1.000	Ea.	6.35	44.50	50.85
Copper tubing, type L, solder joints, 3/4" diam.	1.000	L.F.	4.82	9.95	14.77
Wrought copper 90° elbow for solder joints 3/4" diam.	1.000	Ea.	2.64	40	42.64
TOTAL			1,027.69	1,143.95	2,171.64

D2020 210	Electric Water Heaters - Residential Systems		COST EACH		
			MAT.	INST.	TOTAL
1760	Electric water heater, residential, 100°F rise				
1780	10 gallon tank, 7 GPH		1,025	1,150	2,175
1820	20 gallon tank, 7 GPH	R224000	1,275	1,225	2,500
1860	30 gallon tank, 7 GPH	-20	1,900	1,275	3,175
1900	40 gallon tank, 8 GPH		2,150	1,400	3,550
1940	52 gallon tank, 10 GPH		2,325	1,425	3,750
1980	66 gallon tank, 13 GPH		3,625	1,625	5,250
2020	80 gallon tank, 16 GPH		3,875	1,700	5,575
2060	120 gallon tank, 23 GPH		5,500	1,975	7,475

Sample Estimate

This sample demonstrates the elements of an estimate, including a tally of the RSMeans data lines. Published assemblies costs include all markups for labor burden and profit for the installing contractor. This estimate adds a summary of the markups applied by a general contractor on the installing contractor's work. These figures represent the total cost to the owner. The location factor with RSMeans data is applied at the bottom of the estimate to adjust the cost of the work to a specific location.

Project Name:	Interior Fit-out, ABC Office			
Location:	Anywhere, USA		Date: 1/1/2019	STD
Assembly Number	**Description**	**Qty.**	**Unit**	**Subtotal**
① C1010 124 1200	Wood partition, 2 x 4 @ 16" OC w/5/8" FR gypsum board	560	S.F.	$2,889.60
C1020 114 1800	Metal door & frame, flush hollow core, 3'-0" x 7'-0"	2	Ea.	$2,560.00
C3010 230 0080	Painting, brushwork, primer & 2 coats	1,120	S.F.	$1,433.60
C3020 410 0140	Carpet, tufted, nylon, roll goods, 12' wide, 26 oz	240	S.F.	$885.60
C3030 210 6000	Acoustic ceilings, 24" x 48" tile, tee grid suspension	200	S.F.	$1,288.00
D5020 125 0560	Receptacles incl plate, box, conduit, wire, 20 A duplex	8	Ea.	$2,388.00
D5020 125 0720	Light switch incl plate, box, conduit, wire, 20 A single pole	2	Ea.	$583.00
D5020 210 0560	Fluorescent fixtures, recess mounted, 20 per 1000 SF	200	S.F.	$2,270.00
	Assembly Subtotal			**$14,297.80**
	Sales Tax @ **②**		5 %	$ 357.45
	General Requirements @ **③**		7 %	$ 1,000.85
	Subtotal A			**$15,656.09**
	GC Overhead @ **④**		5 %	$ 782.80
	Subtotal B			**$16,438.90**
	GC Profit @ **⑤**		5 %	$ 821.94
	Subtotal C			**$17,260.84**
	Adjusted by Location Factor **⑥**	113.9		$ 19,660.10
	Architects Fee @ **⑦**		8 %	$ 1,572.81
	Contingency @ **⑧**		15 %	$ 2,949.01
	Project Total Cost			**$ 24,181.92**

This estimate is based on an interactive spreadsheet. You are free to download it and adjust it to your methodology. A copy of this spreadsheet is available at **RSMeans.com/2019books.**

① Work Performed

The body of the estimate shows the RSMeans data selected, including line numbers, a brief description of each item, its takeoff quantity and unit, and the total installed cost, including the installing contractor's overhead and profit.

② Sales Tax

If the work is subject to state or local sales taxes, the amount must be added to the estimate. In a conceptual estimate it can be assumed that one half of the total represents material costs. Therefore, apply the sales tax rate to 50% of the assembly subtotal.

③ General Requirements

This item covers project-wide needs provided by the general contractor. These items vary by project but may include temporary facilities and utilities, security, testing, project cleanup, etc. In assemblies estimates a percentage is used—typically between 5% and 15% of project cost.

④ General Contractor Overhead

This entry represents the general contractor's markup on all work to cover project administration costs.

⑤ General Contractor Profit

This entry represents the GC's profit on all work performed. The value included here can vary widely by project and is influenced by the GC's perception of the project's financial risk and market conditions.

⑥ Location Factor

RSMeans published data are based on national average costs. If necessary, adjust the total cost of the project using a location factor from the "Location Factor" table or the "City Cost Indexes" table found in the Reference Section. Use location factors if the work is general, covering the work of multiple trades. If the work is by a single trade (e.g., masonry) use the more specific data found in the City Cost Indexes.

To adjust costs by location factors, multiply the base cost by the factor and divide by 100.

⑦ Architect's Fee

If appropriate, add the design cost to the project estimate. These fees vary based on project complexity and size. Typical design and engineering fees can be found in the Reference Section.

⑧ Contingency

A factor for contingency may be added to any estimate to represent the cost of unknowns that may occur between the time that the estimate is performed and the time the project is constructed. The amount of the allowance will depend on the stage of design at which the estimate is done, as well as the contractor's assessment of the risk involved.

D2020 Domestic Water Distribution

Installation includes piping and fittings within 10′ of heater. Electric water heaters do not require venting.

1 Kilowatt hour will raise:			
Gallons of Water	Degrees F	Gallons of Water	Degrees F
4.1	100°	6.8	60°
4.5	90°	8.2	50°
5.1	80°	10.0	40°
5.9	70°		

System Components	QUANTITY	UNIT	COST EACH		
			MAT.	INST.	TOTAL
SYSTEM D2020 210 1780					
ELECTRIC WATER HEATER, RESIDENTIAL, 100°F RISE					
10 GALLON TANK, 7 GPH					
Water heater, residential electric, glass lined tank, 10 gal.	1.000	Ea.	485	330	815
Copper tubing, type L, solder joint, hanger 10′ OC 1/2″ diam.	30.000	L.F.	111.60	280.50	392.10
Wrought copper 90° elbow for solder joints 1/2″ diam.	4.000	Ea.	5.28	152	157.28
Wrought copper Tee for solder joints, 1/2″ diam.	2.000	Ea.	5	117	122
Union, wrought copper, 1/2″ diam.	2.000	Ea.	45	80	125
Valve, gate, bronze, 125 lb, NRS, soldered 1/2″ diam.	2.000	Ea.	131	63	194
Relief valve, bronze, press & temp, self-close, 3/4″ IPS	1.000	Ea.	231	27	258
Wrought copper adapter, CTS to MPT 3/4″ IPS	1.000	Ea.	6.35	44.50	50.85
Copper tubing, type L, solder joints, 3/4″ diam.	1.000	L.F.	4.82	9.95	14.77
Wrought copper 90° elbow for solder joints 3/4″ diam.	1.000	Ea.	2.64	40	42.64
TOTAL			1,027.69	1,143.95	2,171.64

D2020 210	Electric Water Heaters - Residential Systems	COST EACH		
		MAT.	INST.	TOTAL
1760	Electric water heater, residential, 100°F rise			
1780	10 gallon tank, 7 GPH	1,025	1,150	2,175
1820	20 gallon tank, 7 GPH	1,275	1,225	2,500
1860	30 gallon tank, 7 GPH R224000 -20	1,900	1,275	3,175
1900	40 gallon tank, 8 GPH	2,150	1,400	3,550
1940	52 gallon tank, 10 GPH	2,325	1,425	3,750
1980	66 gallon tank, 13 GPH	3,625	1,625	5,250
2020	80 gallon tank, 16 GPH	3,875	1,700	5,575
2060	120 gallon tank, 23 GPH	5,500	1,975	7,475

D2020 Domestic Water Distribution

Systems below include piping and fittings within 10' of heater. Electric water heaters do not require venting.

System Components	QUANTITY	UNIT	COST EACH MAT.	COST EACH INST.	COST EACH TOTAL
SYSTEM D2020 240 1820					
ELECTRIC WATER HEATER, COMMERCIAL, 100°F RISE					
50 GALLON TANK, 9 KW, 37 GPH					
Water heater, commercial, electric, 50 Gal, 9 KW, 37 GPH	1.000	Ea.	8,275	420	8,695
Copper tubing, type L, solder joint, hanger 10' OC, 3/4" diam	34.000	L.F.	163.88	338.30	502.18
Wrought copper 90° elbow for solder joints 3/4" diam	5.000	Ea.	13.20	200	213.20
Wrought copper Tee for solder joints, 3/4" diam	2.000	Ea.	13.30	126	139.30
Wrought copper union for soldered joints, 3/4" diam.	2.000	Ea.	56	84	140
Valve, gate, bronze, 125 lb, NRS, soldered 3/4" diam	2.000	Ea.	150	76	226
Relief valve, bronze, press & temp, self-close, 3/4" IPS	1.000	Ea.	231	27	258
Wrought copper adapter, copper tubing to male, 3/4" IPS	1.000	Ea.	6.35	44.50	50.85
TOTAL			8,908.73	1,315.80	10,224.53

D2020 240	Electric Water Heaters - Commercial Systems		COST EACH MAT.	COST EACH INST.	COST EACH TOTAL
1800	Electric water heater, commercial, 100°F rise				
1820	50 gallon tank, 9 KW 37 GPH		8,900	1,325	10,225
1860	80 gal, 12 KW 49 GPH		10,700	1,625	12,325
1900	36 KW 147 GPH	R224000 -10	14,700	1,750	16,450
1940	120 gal, 36 KW 147 GPH		16,600	1,900	18,500
1980	150 gal, 120 KW 490 GPH	R224000 -20	52,000	2,025	54,025
2020	200 gal, 120 KW 490 GPH		51,000	2,100	53,100
2060	250 gal, 150 KW 615 GPH		56,500	2,450	58,950
2100	300 gal, 180 KW 738 GPH		64,500	2,575	67,075
2140	350 gal, 30 KW 123 GPH		48,700	2,775	51,475
2180	180 KW 738 GPH		68,000	2,775	70,775
2220	500 gal, 30 KW 123 GPH		59,500	3,250	62,750
2260	240 KW 984 GPH		100,500	3,250	103,750
2300	700 gal, 30 KW 123 GPH		77,000	3,725	80,725
2340	300 KW 1230 GPH		114,500	3,725	118,225
2380	1000 gal, 60 KW 245 GPH		93,000	5,175	98,175
2420	480 KW 1970 GPH		151,000	5,200	156,200
2460	1500 gal, 60 KW 245 GPH		127,500	6,425	133,925
2500	480 KW 1970 GPH		187,000	6,425	193,425

D3020 Heat Generating Systems

Boiler

Baseboard Radiation

Small Electric Boiler
System Considerations:
1. Terminal units are fin tube baseboard radiation rated at 720 BTU/hr with 200° water temperature or 820 BTU/hr steam.
2. Primary use being for residential or smaller supplementary areas, the floor levels are based on 7-1/2' ceiling heights.
3. All distribution piping is copper for boilers through 205 MBH. All piping for larger systems is steel pipe.

System Components	QUANTITY	UNIT	COST EACH		
			MAT.	INST.	TOTAL
SYSTEM D3020 102 1120					
SMALL HEATING SYSTEM, HYDRONIC, ELECTRIC BOILER					
1,480 S.F., 61 MBH, STEAM, 1 FLOOR					
Boiler, electric steam, std cntrls, trim, ftngs and valves, 18 KW, 61.4 MBH	1.000	Ea.	5,335	1,925	7,260
Copper tubing type L, solder joint, hanger 10'OC, 1-1/4" diam	160.000	L.F.	2,048	2,088	4,136
Radiation, 3/4" copper tube w/alum fin baseboard pkg 7" high	60.000	L.F.	534	1,440	1,974
Rough in baseboard panel or fin tube with valves & traps	10.000	Set	2,840	7,250	10,090
Pipe covering, calcium silicate w/cover, 1" wall 1-1/4" diam	160.000	L.F.	740.80	1,240	1,980.80
Low water cut-off, quick hookup, in gage glass tappings	1.000	Ea.	340	48	388
TOTAL			11,837.80	13,991	25,828.80
COST PER S.F.			8	9.45	17.45

D3020 102	Small Heating Systems, Hydronic, Electric Boilers	COST PER S.F.		
		MAT.	INST.	TOTAL
1100	Small heating systems, hydronic, electric boilers			
1120	Steam, 1 floor, 1480 S.F., 61 M.B.H.	7.97	9.46	17.43
1160	3,000 S.F., 123 M.B.H.	6.15	8.30	14.45
1200	5,000 S.F., 205 M.B.H.	5.05	7.65	12.70
1240	2 floors, 12,400 S.F., 512 M.B.H.	4.10	7.60	11.70
1280	3 floors, 24,800 S.F., 1023 M.B.H.	4.69	7.50	12.19
1320	34,750 S.F., 1,433 M.B.H.	4.31	7.30	11.61
1360	Hot water, 1 floor, 1,000 S.F., 41 M.B.H.	13.40	5.30	18.70
1400	2,500 S.F., 103 M.B.H.	9.20	9.45	18.65
1440	2 floors, 4,850 S.F., 205 M.B.H.	8.55	11.35	19.90
1480	3 floors, 9,700 S.F., 410 M.B.H.	9.05	11.75	20.80

D30 HVAC

D3020 Heat Generating Systems

Boiler

Unit Heater

Large Electric Boiler System Considerations:

1. Terminal units are all unit heaters of the same size. Quantities are varied to accommodate total requirements.
2. All air is circulated through the heaters a minimum of three times per hour.
3. As the capacities are adequate for commercial use, floor levels are based on 10' ceiling heights.
4. All distribution piping is black steel pipe.

System Components	QUANTITY	UNIT	COST EACH		
			MAT.	INST.	TOTAL
SYSTEM D3020 104 1240					
LARGE HEATING SYSTEM, HYDRONIC, ELECTRIC BOILER					
9,280 S.F., 135 KW, 461 MBH, 1 FLOOR					
Boiler, electric hot water, std ctrls, trim, ftngs, valves, 135 KW, 461 MBH	1.000	Ea.	12,487.50	4,200	16,687.50
Expansion tank, painted steel, 60 Gal capacity ASME	1.000	Ea.	5,225	230	5,455
Circulating pump, CI, close cpld, 50 GPM, 2 HP, 2" pipe conn	1.000	Ea.	3,325	455	3,780
Unit heater, 1 speed propeller, horizontal, 200° EWT, 72.7 MBH	7.000	Ea.	6,755	1,757	8,512
Unit heater piping hookup with controls	7.000	Set	5,985	11,375	17,360
Pipe, steel, black, schedule 40, welded, 2-1/2" diam	380.000	L.F.	6,270	11,517.80	17,787.80
Pipe covering, calcium silicate w/cover, 1" wall, 2-1/2" diam	380.000	L.F.	2,185	3,021	5,206
TOTAL			42,232.50	32,555.80	74,788.30
COST PER S.F.			4.55	3.51	8.06

D3020 104	Large Heating Systems, Hydronic, Electric Boilers	COST PER S.F.		
		MAT.	INST.	TOTAL
1230	Large heating systems, hydronic, electric boilers			
1240	9,280 S.F., 135 K.W., 461 M.B.H., 1 floor	4.55	3.51	8.06
1280	14,900 S.F., 240 K.W., 820 M.B.H., 2 floors	5.95	5.65	11.60
1320	18,600 S.F., 296 K.W., 1,010 M.B.H., 3 floors	5.90	6.15	12.05
1360	26,100 S.F., 420 K.W., 1,432 M.B.H., 4 floors	5.80	6.05	11.85
1400	39,100 S.F., 666 K.W., 2,273 M.B.H., 4 floors	4.95	5.05	10
1440	57,700 S.F., 900 K.W., 3,071 M.B.H., 5 floors	4.71	4.96	9.67
1480	111,700 S.F., 1,800 K.W., 6,148 M.B.H., 6 floors	4.15	4.26	8.41
1520	149,000 S.F., 2,400 K.W., 8,191 M.B.H., 8 floors	4.13	4.26	8.39
1560	223,300 S.F., 3,600 K.W., 12,283 M.B.H., 14 floors	4.39	4.83	9.22

D40 Fire Protection

D4090 Other Fire Protection Systems

General: Automatic fire protection (suppression) systems other than water sprinklers may be desired for special environments, high risk areas, isolated locations or unusual hazards. Some typical applications would include:

Paint dip tanks
Securities vaults
Electronic data processing
Tape and data storage
Transformer rooms
Spray booths
Petroleum storage
High rack storage

Piping and wiring costs are dependent on the individual application and must be added to the component costs shown below.

All areas are assumed to be open.

D4090 910	Fire Suppression Unit Components	COST EACH		
		MAT.	INST.	TOTAL
0020	Detectors with brackets			
0040	Fixed temperature heat detector	50	102	152
0060	Rate of temperature rise detector	57.50	89.50	147
0080	Ion detector (smoke) detector	135	116	251
0200	Extinguisher agent			
0240	200 lb FM200, container	7,375	340	7,715
0280	75 lb carbon dioxide cylinder	1,475	227	1,702
0320	Dispersion nozzle			
0340	FM200 1-1/2" dispersion nozzle	210	54	264
0380	Carbon dioxide 3" x 5" dispersion nozzle	169	42	211
0420	Control station			
0440	Single zone control station with batteries	1,075	715	1,790
0470	Multizone (4) control station with batteries	3,225	1,425	4,650
0490				
0500	Electric mechanical release	1,075	370	1,445
0520				
0550	Manual pull station	91.50	126	217.50
0570				
0640	Battery standby power 10" x 10" x 17"	430	179	609
0700				
0740	Bell signalling device	149	89.50	238.50

D4090 920	FM200 Systems	COST PER C.F.		
		MAT.	INST.	TOTAL
0820	Average FM200 system, minimum			1.97
0840	Maximum			3.92

388

D5010 Electrical Service/Distribution

System Components	QUANTITY	UNIT	COST PER L.F.		
			MAT.	INST.	TOTAL
SYSTEM D5010 110 0200					
HIGH VOLTAGE CABLE, NEUTRAL AND CONDUIT INCLUDED, COPPER #2, 5 kV					
Shielded cable, no splice/termn, copper, XLP shielding, 5 kV, #2	.030	C.L.F.	8.85	10.80	19.65
Wire, 600 volt, type THW, copper, stranded, #8	.010	C.L.F.	.37	.90	1.27
Rigid galv steel conduit to 15' H, 2" diam, w/term, ftng & support	1.000	L.F.	10.95	15.95	26.90
TOTAL			20.17	27.65	47.82

D5010 110	High Voltage Shielded Conductors	COST PER L.F.		
		MAT.	INST.	TOTAL
0200	High voltage cable, neutral & conduit included, copper #2, 5 kV	20	27.50	47.50
0240	Copper #1, 5 kV	24.50	32.50	57
0280	15 kV	25	32.50	57.50
0320	Copper 1/0, 5 kV	27.50	41	68.50
0360	15 kV	29.50	41	70.50
0400	25 kV	35	41.50	76.50
0440	35 kV	36.50	42	78.50
0480	Copper 2/0, 5 kV	30.50	42	72.50
0520	15 kV	32	42	74
0560	25 kV	37	42.50	79.50
0600	35 kV	43.50	47.50	91
0640	Copper 4/0, 5 kV	39.50	47.50	87
0680	15 kV	41	47.50	88.50
0720	25 kV	46.50	48.50	95
0760	35 kV	50	49.50	99.50
0800	Copper 250 kcmil, 5 kV	43	48.50	91.50
0840	15 kV	43	48.50	91.50
0880	25 kV	52.50	49.50	102
0920	35 kV	58.50	54	112.50
0960	Copper 350 kcmil, 5 kV	53.50	54	107.50
1000	15 kV	53	54	107
1040	25 kV	62.50	55.50	118
1080	35 kV	65.50	57	122.50
1120	Copper 500 kcmil, 5 kV	83	68	151
1160	15 kV	80	68	148
1200	25 kV	89	69.50	158.50
1240	35 kV	93.50	71.50	165

389

System Components	QUANTITY	UNIT	COST EACH		
			MAT.	INST.	TOTAL
SYSTEM D5010 120 0220					
SERVICE INSTALLATION, INCLUDES BREAKERS, METERING, 20' CONDUIT & WIRE					
3 PHASE, 4 WIRE, 60 A					
Wire, 600 volt, copper type XHHW, stranded #6	.600	C.L.F.	33.60	66	99.60
Rigid galvanized steel conduit, 3/4″ including fittings	20.000	L.F.	104	179	283
Service entrance cap, 3/4″ diameter	1.000	Ea.	6.40	55	61.40
Conduit LB fitting with cover, 3/4″ diameter	1.000	Ea.	9	55	64
Meter socket, three phase, 100 A	1.000	Ea.	129	256	385
Safety switches, heavy duty, 240 volt, 3 pole NEMA 1 fusible, 60 amp	1.000	Ea.	160	310	470
Grounding, wire ground bare armored, #8-1 conductor	.200	C.L.F.	15.10	72	87.10
Grounding, clamp, bronze, 3/4″ diameter	1.000	Ea.	6.15	22.50	28.65
Grounding, rod, copper clad, 8' long, 3/4″ diameter	1.000	Ea.	39	135	174
Wireway w/fittings, 2-1/2″ x 2-1/2″	1.000	L.F.	13.50	15.95	29.45
TOTAL			515.75	1,166.45	1,682.20

D5010 120	Overhead Electric Service, 3 Phase - 4 Wire	COST EACH		
		MAT.	INST.	TOTAL
0200	Service installation, includes breakers, metering, 20' conduit & wire			
0220	3 phase, 4 wire, 120/208 volts, 60 A	515	1,175	1,690
0240	100 A	690	1,325	2,015
0245	100 A w/circuit breaker	1,475	1,625	3,100
0280	200 A	1,175	1,825	3,000
0285	200 A w/circuit breaker	3,175	2,300	5,475
0320	400 A	2,375	3,625	6,000
0325	400 A w/circuit breaker	5,475	4,525	10,000
0360	600 A	4,000	5,375	9,375
0365	600 A, w/switchboard	8,350	6,800	15,150
0400	800 A	6,150	6,425	12,575
0405	800 A, w/switchboard	10,500	8,050	18,550
0440	1000 A	7,575	7,900	15,475
0445	1000 A, w/switchboard	12,800	9,700	22,500
0480	1200 A	10,000	9,025	19,025
0485	1200 A, w/groundfault switchboard	32,600	11,200	43,800
0520	1600 A	12,600	11,800	24,400
0525	1600 A, w/groundfault switchboard	37,100	14,100	51,200

D5010 Electrical Service/Distribution

D5010 120	Overhead Electric Service, 3 Phase - 4 Wire	COST EACH		
		MAT.	INST.	TOTAL
0560	2000 A	16,500	14,300	30,800
0565	2000 A, w/groundfault switchboard	44,100	17,600	61,700
0610	1 phase, 3 wire, 120/240 volts, 100 A (no safety switch)	160	645	805
0615	100 A w/load center	330	1,250	1,580
0620	200 A	370	885	1,255
0625	200 A w/load center	705	1,875	2,580

For customer support on your Electrical Costs with RSMeans data, call 800.448.8182.

D50 Electrical

D5010 Electrical Service/Distribution

Underground service conductor

Safety Switch

Meter and service equipment

Ground rod

System Components			COST EACH		
	QUANTITY	UNIT	MAT.	INST.	TOTAL
SYSTEM D5010 130 1000					
2000 AMP UNDERGROUND ELECTRIC SERVICE WITH SAFETY SWITCH					
INCLUDING EXCAVATION, BACKFILL AND COMPACTION					
Excavate Trench	44.440	B.C.Y.		319.08	319.08
4 inch conduit bank	108.000	L.F.	1,323	2,862	4,185
4 inch fitting	8.000	Ea.	344	476	820
4 inch bells	4.000	Ea.	20.40	238	258.40
Concrete material	16.580	C.Y.	2,254.88		2,254.88
Concrete placement	16.580	C.Y.		430.91	430.91
Backfill trench	33.710	L.C.Y.		104.16	104.16
Compact fill material in trench	25.930	E.C.Y.		136.91	136.91
Dispose of excess fill material on-site	24.070	L.C.Y.		297.75	297.75
500 kcmil power cable	18.000	C.L.F.	15,840	8,100	23,940
Wire, 600 volt, type THW, copper, stranded, 1/0	6.000	C.L.F.	1,398	1,302	2,700
Saw cutting, concrete walls, plain, per inch of depth	64.000	L.F.	3.20	496	499.20
Meter centers and sockets, single pos, 4 terminal, 400 amp	5.000	Ea.	4,625	2,100	6,725
Safety switch, 400 amp	5.000	Ea.	5,500	3,975	9,475
Ground rod clamp	6.000	Ea.	36.90	135	171.90
Wireway	1.000	L.F.	13.50	15.95	29.45
600 volt stranded copper wire, 1/0	1.200	C.L.F.	279.60	260.40	540
Flexible metallic conduit	120.000	L.F.	75.60	537.60	613.20
Ground rod	6.000	Ea.	327	1,074	1,401
TOTAL			32,041.08	22,860.76	54,901.84

D5010 130	Underground Electric Service		COST EACH		
			MAT.	INST.	TOTAL
0950	Underground electric service including excavation, backfill and compaction				
1000	3 phase, 4 wire, 277/480 volts, 2000 A		32,000	22,900	54,900
1050	2000 A w/groundfault switchboard		58,500	25,500	84,000
1100	1600 A		25,200	18,400	43,600
1150	1600 A w/groundfault switchboard		49,700	20,700	70,400
1200	1200 A		19,500	14,700	34,200
1250	1200 A w/groundfault switchboard		42,100	16,800	58,900
1400	800 A		12,800	12,100	24,900

D5010 Electrical Service/Distribution

D5010 130	Underground Electric Service	COST EACH		
		MAT.	INST.	TOTAL
1450	800 A w/switchboard	17,200	13,800	31,000
1500	600 A	8,400	11,100	19,500
1550	600 A w/switchboard	12,800	12,600	25,400
1600	1 phase, 3 wire, 120/240 volts, 200 A	2,975	3,775	6,750
1650	200 A w/load center	3,325	4,775	8,100
1700	100 A	2,400	3,075	5,475
1750	100 A w/load center	2,725	4,075	6,800

D50 Electrical

D5010 Electrical Service/Distribution

System Components			COST PER L.F.		
	QUANTITY	UNIT	MAT.	INST.	TOTAL
SYSTEM D5010 230 0200					
FEEDERS, INCLUDING STEEL CONDUIT & WIRE, 60 A					
Rigid galvanized steel conduit, 3/4", including fittings	1.000	L.F.	5.20	8.95	14.15
Wire 600 volt, type XHHW copper stranded #6	.030	C.L.F.	1.68	3.30	4.98
Wire 600 volt, type XHHW copper stranded #8	.010	C.L.F.	.34	.90	1.24
TOTAL			7.22	13.15	20.37

D5010 230	Feeder Installation	COST PER L.F.		
		MAT.	INST.	TOTAL
0200	Feeder installation 600 V, including RGS conduit and XHHW wire, 60 A	7.20	13.15	20.35
0240	100 A	8.85	17.15	26
0280	200 A	22.50	26	48.50
0320	400 A	45	52	97
0360	600 A	73	84	157
0400	800 A	94.50	100	194.50
0440	1000 A	117	132	249
0480	1200 A	146	168	314
0520	1600 A	189	200	389
0560	2000 A	234	263	497
1200	Branch installation 600 V, including EMT conduit and THW wire, 15 A	1.22	6.40	7.62
1240	20 A	1.49	6.80	8.29
1280	30 A	1.83	7.10	8.93
1320	50 A	3.57	9.85	13.42
1360	65 A	3.87	10.45	14.32
1400	85 A	6.35	12.10	18.45
1440	100 A	7.40	12.35	19.75
1480	130 A	8.70	14.40	23.10
1520	150 A	11.75	16.55	28.30
1560	200 A	12.75	18.90	31.65

D50 Electrical

D5010 Electrical Service/Distribution

System Components	QUANTITY	UNIT	COST EACH		
			MAT.	INST.	TOTAL
SYSTEM D5010 240 0240					
SWITCHGEAR INSTALLATION, INCL SWBD, PANELS & CIRC BREAKERS, 600 A					
Switchboards, 120/208 V, 600 amp	1.000	Ea.	4,350	1,425	5,775
Aluminum bus bars, 120/208 V, 600 amp	1.000	Ea.	1,650	1,425	3,075
Feeder section, circuit breakers, KA frame, 70 to 225 amp	1.000	Ea.	1,675	224	1,899
Feeder section, circuit breakers, LA frame, 125 to 400 amp	1.000	Ea.	3,875	310	4,185
TOTAL			11,550	3,384	14,934

D5010 240	Switchgear	COST PER EACH		
		MAT.	INST.	TOTAL
0190	Switchgear installation, including switchboard, panels, & circuit breaker			
0200	120/208 V, 3 phase, 400 A	8,825	3,025	11,850
0240	600 A	11,600	3,375	14,975
0280	800 A	14,900	3,925	18,825
0300	1000 A	18,400	4,375	22,775
0320	1200 A	19,300	4,950	24,250
0360	1600 A	31,400	5,700	37,100
0400	2000 A	37,900	6,175	44,075
0500	277/480 V, 3 phase, 400 A	11,900	5,625	17,525
0520	600 A	16,500	6,450	22,950
0540	800 A	19,100	7,175	26,275
0560	1000 A	24,200	7,975	32,175
0580	1200 A	26,100	8,950	35,050
0600	1600 A	38,800	9,825	48,625
0620	2000 A	48,600	10,800	59,400

395

D5010 Electrical Service/Distribution

System Components			COST EACH		
	QUANTITY	UNIT	MAT.	INST.	TOTAL
SYSTEM D5010 250 1020					
PANELBOARD INSTALLATION, INCLUDES PANELBOARD, CONDUCTOR, & CONDUIT					
Conduit, galvanized steel, 1-1/4"dia.	37.000	L.F.	196.10	442.15	638.25
Panelboards, NQOD, 4 wire, 120/208 volts, 100 amp main, 24 circuits	1.000	Ea.	2,075	1,525	3,600
Wire, 600 volt, type THW, copper, stranded, #3	1.480	C.L.F.	182.04	211.64	393.68
TOTAL			2,453.14	2,178.79	4,631.93

D5010 250	Panelboard	COST EACH		
		MAT.	INST.	TOTAL
0900	Panelboards, NQOD, 4 wire, 120/208 volts w/conductor & conduit			
1000	100 A, 0 stories, 0' horizontal	2,075	1,525	3,600
1020	1 stories, 25' horizontal	2,450	2,175	4,625
1040	5 stories, 50' horizontal	3,200	3,475	6,675
1060	10 stories, 75' horizontal	4,075	4,975	9,050
1080	225A, 0 stories, 0' horizontal	3,750	2,000	5,750
2000	1 stories, 25' horizontal	4,800	3,175	7,975
2020	5 stories, 50' horizontal	6,875	5,500	12,375
2040	10 stories, 75' horizontal	9,300	8,200	17,500
2060	400A, 0 stories, 0' horizontal	5,500	2,975	8,475
2080	1 stories, 25' horizontal	6,925	4,900	11,825
3000	5 stories, 50' horizontal	9,775	8,675	18,450
3020	10 stories, 75' horizontal	16,400	17,000	33,400
3040	600 A, 0 stories, 0' horizontal	8,125	3,575	11,700
3060	1 stories, 25' horizontal	10,500	6,700	17,200
3080	5 stories, 50' horizontal	15,100	12,800	27,900
4000	10 stories, 75' horizontal	26,700	25,300	52,000
4010	Panelboards, NEHB, 4 wire, 277/480 volts w/conductor, conduit, & safety switch			
4020	100 A, 0 stories, 0' horizontal, includes safety switch	3,175	2,075	5,250
4040	1 stories, 25' horizontal	3,550	2,725	6,275
4060	5 stories, 50' horizontal	4,300	4,025	8,325
4080	10 stories, 75' horizontal	5,175	5,525	10,700
5000	225 A, 0 stories, 0' horizontal	4,700	2,550	7,250
5020	1 stories, 25' horizontal	5,750	3,725	9,475
5040	5 stories, 50' horizontal	7,825	6,050	13,875
5060	10 stories, 75' horizontal	10,200	8,750	18,950

D5010 Electrical Service/Distribution

D5010 250	Panelboard	COST EACH		
		MAT.	INST.	TOTAL
5080	400 A, 0 stories, 0' horizontal	7,425	3,925	11,350
6000	1 stories, 25' horizontal	8,850	5,825	14,675
6020	5 stories, 50' horizontal	11,700	9,625	21,325
6040	10 stories, 75' horizontal	18,300	17,900	36,200
6060	600 A, 0 stories, 0' horizontal	10,600	4,975	15,575
6080	1 stories, 25' horizontal	12,900	8,100	21,000
7000	5 stories, 50' horizontal	17,500	14,200	31,700
7020	10 stories, 75' horizontal	29,200	26,700	55,900

Duplex Receptacle

System Components			COST PER S.F.		
	QUANTITY	UNIT	MAT.	INST.	TOTAL
SYSTEM D5020 110 0200					
RECEPTACLES INCL. PLATE, BOX, CONDUIT, WIRE & TRANS. WHEN REQUIRED					
2.5 PER 1000 S.F., .3 WATTS PER S.F.					
Steel intermediate conduit, (IMC) 1/2" diam	167.000	L.F.	.31	1.19	1.50
Wire 600V type THWN-THHN, copper solid #12	3.340	C.L.F.	.04	.22	.26
Wiring device, receptacle, duplex, 120V grounded, 15 amp	2.500	Ea.		.04	.04
Wall plate, 1 gang, brown plastic	2.500	Ea.		.02	.02
Steel outlet box 4" square	2.500	Ea.	.01	.09	.10
Steel outlet box 4" plaster rings	2.500	Ea.		.03	.03
TOTAL			.36	1.59	1.95

D5020 110	Receptacle (by Wattage)	COST PER S.F.		
		MAT.	INST.	TOTAL
0190	Receptacles include plate, box, conduit, wire & transformer when required			
0200	2.5 per 1000 S.F., .3 watts per S.F.	.36	1.59	1.95
0240	With transformer	.46	1.67	2.13
0280	4 per 1000 S.F., .5 watts per S.F.	.42	1.85	2.27
0320	With transformer	.57	1.98	2.55
0360	5 per 1000 S.F., .6 watts per S.F.	.14	.74	.88
0400	With transformer	.69	2.35	3.04
0440	8 per 1000 S.F., .9 watts per S.F.	.53	2.43	2.96
0480	With transformer	.84	2.69	3.53
0520	10 per 1000 S.F., 1.2 watts per S.F.	.57	2.64	3.21
0560	With transformer	.96	2.96	3.92
0600	16.5 per 1000 S.F., 2.0 watts per S.F.	.67	3.30	3.97
0640	With transformer	1.31	3.83	5.14
0680	20 per 1000 S.F., 2.4 watts per S.F.	.70	3.60	4.30
0720	With transformer	1.47	4.24	5.71

D5020 Lighting and Branch Wiring

Underfloor Receptacle System

Description: Table D5020 115 includes installed costs of raceways and copper wire from panel to and including receptacle.

National Electrical Code prohibits use of undercarpet system in residential, school or hospital buildings. Can only be used with carpet squares.

Low density = (1) Outlet per 259 S.F. of floor area.

High density = (1) Outlet per 127 S.F. of floor area.

System Components	QUANTITY	UNIT	COST PER S.F.		
			MAT.	INST.	TOTAL
SYSTEM D5020 115 0200					
RECEPTACLE SYSTEMS, UNDERFLOOR DUCT, 5′ ON CENTER, LOW DENSITY					
Underfloor duct 3-1/8″ x 7/8″ w/insert 24″ on center	.190	L.F.	4.18	1.95	6.13
Underfloor duct junction box, single duct, 3-1/8″	.003	Ea.	1.47	.54	2.01
Underfloor junction box carpet pan	.003	Ea.	1.23	.03	1.26
Underfloor duct outlet, high tension receptacle	.004	Ea.	.47	.36	.83
Wire 600V type THWN-THHN copper solid #12	.010	C.L.F.	.11	.65	.76
Vertical elbow for underfloor duct, 3-1/8″, included					
Underfloor duct conduit adapter, 2″ x 1-1/4″, included					
TOTAL			7.46	3.53	10.99

D5020 115	Receptacles, Floor	COST PER S.F.		
		MAT.	INST.	TOTAL
0200	Receptacle systems, underfloor duct, 5′ on center, low density	7.45	3.53	10.98
0240	High density	8.05	4.54	12.59
0280	7′ on center, low density	5.90	3.03	8.93
0320	High density	6.50	4.04	10.54
0400	Poke thru fittings, low density	1.33	1.71	3.04
0440	High density	2.65	3.42	6.07
0520	Telepoles, using Romex, low density	1.57	1.07	2.64
0560	High density	3.13	2.12	5.25
0600	Using EMT, low density	1.65	1.40	3.05
0640	High density	3.31	2.81	6.12
0720	Conduit system with floor boxes, low density	1.45	1.20	2.65
0760	High density	2.92	2.41	5.33
0840	Undercarpet power system, 3 conductor with 5 conductor feeder, low density	1.48	.42	1.90
0880	High density	2.98	.87	3.85

D5020 Lighting and Branch Wiring

Duplex Receptacle   **Wall Switch**

System Components	QUANTITY	UNIT	COST PER S.F.		
			MAT.	INST.	TOTAL
SYSTEM D5020 120 0520					
RECEPTACLES AND WALL SWITCHES					
4 RECEPTACLES PER 400 S.F.					
Steel intermediate conduit, (IMC), 1/2" diam	.220	L.F.	.41	1.57	1.98
Wire, 600 volt, type THWN-THHN, copper, solid #12	.005	C.L.F.	.05	.33	.38
Steel outlet box 4" square	.010	Ea.	.06	.36	.42
Steel outlet box, 4" square, plaster rings	.010	Ea.	.02	.11	.13
Receptacle, duplex, 120 volt grounded, 15 amp	.010	Ea.	.02	.18	.20
Wall plate, 1 gang, brown plastic	.010	Ea.		.09	.09
TOTAL			.56	2.64	3.20

D5020 120	Receptacles & Wall Switches	COST PER S.F.		
		MAT.	INST.	TOTAL
0520	Receptacles and wall switches, 400 S.F., 4 receptacles	.56	2.64	3.20
0560	6 receptacles	.63	3.08	3.71
0600	8 receptacles	.69	3.58	4.27
0640	1 switch	.12	.60	.72
0680	600 S.F., 6 receptacles	.56	2.64	3.20
0720	8 receptacles	.61	3	3.61
0760	10 receptacles	.68	3.35	4.03
0800	2 switches	.16	.78	.94
0840	1000 S.F., 10 receptacles	.55	2.63	3.18
0880	12 receptacles	.57	2.77	3.34
0920	14 receptacles	.60	3	3.60
0960	2 switches	.08	.43	.51
1000	1600 S.F., 12 receptacles	.48	2.34	2.82
1040	14 receptacles	.53	2.55	3.08
1080	16 receptacles	.55	2.63	3.18
1120	4 switches	.09	.51	.60
1160	2000 S.F., 14 receptacles	.51	2.41	2.92
1200	16 receptacles	.50	2.41	2.91
1240	18 receptacles	.53	2.55	3.08
1280	4 switches	.08	.43	.51
1320	3000 S.F., 12 receptacles	.41	1.84	2.25
1360	18 receptacles	.48	2.27	2.75
1400	24 receptacles	.50	2.41	2.91
1440	6 switches	.08	.43	.51
1480	3600 S.F., 20 receptacles	.46	2.20	2.66
1520	24 receptacles	.47	2.27	2.74
1560	28 receptacles	.49	2.34	2.83
1600	8 switches	.09	.49	.58
1640	4000 S.F., 16 receptacles	.41	1.84	2.25
1680	24 receptacles	.48	2.27	2.75
1720	30 receptacles	.48	2.34	2.82
1760	8 switches	.08	.43	.51
1800	5000 S.F., 20 receptacles	.41	1.84	2.25
1840	26 receptacles	.48	2.19	2.67
1880	30 receptacles	.48	2.27	2.75
1920	10 switches	.08	.43	.51

D5020 Lighting and Branch Wiring

Duplex Receptacle

Wall Switch

System Components	QUANTITY	UNIT	COST PER EACH		
			MAT.	INST.	TOTAL
SYSTEM D5020 125 0520					
RECEPTACLES AND WALL SWITCHES, RECEPTICLE DUPLEX 120 V GROUNDED, 15 A					
Electric metallic tubing conduit, (EMT), 3/4" diam	22.000	L.F.	28.60	121	149.60
Wire, 600 volt, type THWN-THHN, copper, solid #12	.630	C.L.F.	6.80	40.95	47.75
Steel outlet box 4" square	1.000	Ea.	5.70	36	41.70
Steel outlet box, 4" square, plaster rings	1.000	Ea.	1.68	11.20	12.88
Receptacle, duplex, 120 volt grounded, 15 amp	1.000	Ea.	1.66	17.90	19.56
Wall plate, 1 gang, brown plastic	1.000	Ea.	.42	8.95	9.37
TOTAL			44.86	236	280.86

D5020 125	Receptacles & Switches by Each	COST PER EACH		
		MAT.	INST.	TOTAL
0460	Receptacles & Switches, with box, plate, 3/4" EMT conduit & wire			
0520	Receptacle duplex 120 V grounded, 15 A	45	236	281
0560	20 A	53.50	245	298.50
0600	Receptacle duplex ground fault interrupting, 15 A	58.50	245	303.50
0640	20 A	87.50	245	332.50
0680	Toggle switch single, 15 A	43.50	236	279.50
0720	20 A	46.50	245	291.50
0760	3 way switch, 15 A	45.50	249	294.50
0800	20 A	47.50	258	305.50
0840	4 way switch, 15 A	55	266	321
0880	20 A	90.50	283	373.50

D50 Electrical

D5020　Lighting and Branch Wiring

Description: Table D5020 130 includes the cost for switch, plate, box, conduit in slab or EMT exposed and copper wire. Add 20% for exposed conduit.

No power required for switches.

Federal energy guidelines recommend the maximum lighting area controlled per switch shall not exceed 1000 S.F. and that areas over 500 S.F. shall be so controlled that total illumination can be reduced by at least 50%.

System Components	QUANTITY	UNIT	COST PER S.F.		
			MAT.	INST.	TOTAL
SYSTEM D5020 130 0360					
WALL SWITCHES, 5.0 PER 1000 S.F.					
Steel, intermediate conduit (IMC), 1/2" diameter	88.000	L.F.	.16	.63	.79
Wire, 600V type THWN-THHN, copper solid #12	1.710	C.L.F.	.02	.11	.13
Toggle switch, single pole, 15 amp	5.000	Ea.		.09	.09
Wall plate, 1 gang, brown plastic	5.000	Ea.		.04	.04
Steel outlet box 4" plaster rings	5.000	Ea.	.03	.18	.21
Plaster rings	5.000	Ea.	.01	.06	.07
TOTAL			.22	1.11	1.33

D5020　130	Wall Switch by Sq. Ft.	COST PER S.F.		
		MAT.	INST.	TOTAL
0200	Wall switches, 1.0 per 1000 S.F.	.05	.27	.32
0240	1.2 per 1000 S.F.	.06	.27	.33
0280	2.0 per 1000 S.F.	.08	.41	.49
0320	2.5 per 1000 S.F.	.09	.51	.60
0360	5.0 per 1000 S.F.	.22	1.11	1.33
0400	10.0 per 1000 S.F.	.46	2.24	2.70

D5020 Lighting and Branch Wiring

System D5020 135 includes all wiring and connections.

System Components	QUANTITY	UNIT	COST PER S.F.		
			MAT.	INST.	TOTAL
SYSTEM D5020 135 0200					
MISCELLANEOUS POWER, TO .5 WATTS					
Steel intermediate conduit, (IMC) 1/2" diam	15.000	L.F.	.03	.11	.14
Wire 600V type THWN-THHN, copper solid #12	.325	C.L.F.		.02	.02
TOTAL			.03	.13	.16

D5020 135	Miscellaneous Power	COST PER S.F.		
		MAT.	INST.	TOTAL
0200	Miscellaneous power, to .5 watts	.03	.13	.16
0240	.8 watts	.04	.18	.22
0280	1 watt	.06	.25	.31
0320	1.2 watts	.07	.29	.36
0360	1.5 watts	.08	.34	.42
0400	1.8 watts	.10	.39	.49
0440	2 watts	.11	.46	.57
0480	2.5 watts	.13	.57	.70
0520	3 watts	.17	.66	.83

D5020 Lighting and Branch Wiring

System D5020 140 includes all wiring and connections for central air conditioning units.

System Components	QUANTITY	UNIT	COST PER S.F.		
			MAT.	INST.	TOTAL
SYSTEM D5020 140 0200					
CENTRAL AIR CONDITIONING POWER, 1 WATT					
Steel intermediate conduit, 1/2" diam.	.030	L.F.	.06	.21	.27
Wire 600V type THWN-THHN, copper solid #12	.001	C.L.F.	.01	.07	.08
TOTAL			.07	.28	.35

D5020 140	Central A. C. Power (by Wattage)	COST PER S.F.		
		MAT.	INST.	TOTAL
0200	Central air conditioning power, 1 watt	.07	.28	.35
0220	2 watts	.08	.33	.41
0240	3 watts	.15	.48	.63
0280	4 watts	.16	.48	.64
0320	6 watts	.24	.64	.88
0360	8 watts	.34	.70	1.04
0400	10 watts	.46	.81	1.27

D5020 Lighting and Branch Wiring

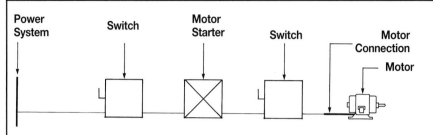

Power System — Switch — Motor Starter — Switch — Motor Connection — Motor

System D5020 145 installed cost of motor wiring using 50' of rigid conduit and copper wire. **Cost and setting of motor not included.**

System Components	QUANTITY	UNIT	COST EACH		
			MAT.	INST.	TOTAL
SYSTEM D5020 145 0200					
MOTOR INST., SINGLE PHASE, 115V, TO AND INCLUDING 1/3 HP MOTOR SIZE					
Wire 600V type THWN-THHN, copper solid #12	1.250	C.L.F.	13.50	81.25	94.75
Steel intermediate conduit, (IMC) 1/2" diam.	50.000	L.F.	93	357.50	450.50
Magnetic FVNR, 115V, 1/3 HP, size 00 starter	1.000	Ea.	199	179	378
Safety switch, fused, heavy duty, 240V 2P 30 amp	1.000	Ea.	71.50	205	276.50
Safety switch, non fused, heavy duty, 600V, 3 phase, 30 A	1.000	Ea.	85.50	224	309.50
Flexible metallic conduit, Greenfield 1/2" diam.	1.500	L.F.	.75	5.37	6.12
Connectors for flexible metallic conduit Greenfield 1/2" diam.	1.000	Ea.	2.08	8.95	11.03
Coupling for Greenfield to conduit 1/2" diam. flexible metallic conduit	1.000	Ea.	1.01	14.35	15.36
Fuse cartridge nonrenewable, 250V 30 amp	1.000	Ea.	2.49	14.35	16.84
TOTAL			468.83	1,089.77	1,558.60

D5020 145	Motor Installation	COST EACH		
		MAT.	INST.	TOTAL
0200	Motor installation, single phase, 115V, 1/3 HP motor size	470	1,100	1,570
0240	1 HP motor size	490	1,100	1,590
0280	2 HP motor size	540	1,150	1,690
0320	3 HP motor size	620	1,175	1,795
0360	230V, 1 HP motor size	470	1,100	1,570
0400	2 HP motor size	510	1,100	1,610
0440	3 HP motor size	575	1,175	1,750
0520	Three phase, 200V, 1-1/2 HP motor size	550	1,225	1,775
0560	3 HP motor size	650	1,325	1,975
0600	5 HP motor size	655	1,475	2,130
0640	7-1/2 HP motor size	685	1,500	2,185
0680	10 HP motor size	1,125	1,875	3,000
0720	15 HP motor size	1,550	2,100	3,650
0760	20 HP motor size	1,950	2,400	4,350
0800	25 HP motor size	2,025	2,425	4,450
0840	30 HP motor size	3,125	2,850	5,975
0880	40 HP motor size	3,875	3,375	7,250
0920	50 HP motor size	6,825	3,925	10,750
0960	60 HP motor size	7,100	4,150	11,250
1000	75 HP motor size	8,775	4,750	13,525
1040	100 HP motor size	25,100	5,600	30,700
1080	125 HP motor size	25,600	6,150	31,750
1120	150 HP motor size	29,200	7,250	36,450
1160	200 HP motor size	29,800	8,575	38,375
1240	230V, 1-1/2 HP motor size	525	1,200	1,725
1280	3 HP motor size	625	1,300	1,925
1320	5 HP motor size	630	1,450	2,080
1360	7-1/2 HP motor size	630	1,450	2,080
1400	10 HP motor size	985	1,775	2,760
1440	15 HP motor size	1,200	1,950	3,150
1480	20 HP motor size	1,825	2,350	4,175
1520	25 HP motor size	1,950	2,400	4,350

D5020 Lighting and Branch Wiring

D5020 145	Motor Installation	COST EACH		
		MAT.	INST.	TOTAL
1560	30 HP motor size	1,975	2,425	4,400
1600	40 HP motor size	3,725	3,300	7,025
1640	50 HP motor size	3,925	3,475	7,400
1680	60 HP motor size	6,850	3,950	10,800
1720	75 HP motor size	7,975	4,475	12,450
1760	100 HP motor size	8,875	5,000	13,875
1800	125 HP motor size	26,000	5,775	31,775
1840	150 HP motor size	26,600	6,575	33,175
1880	200 HP motor size	27,500	7,325	34,825
1960	460V, 2 HP motor size	655	1,200	1,855
2000	5 HP motor size	755	1,325	2,080
2040	10 HP motor size	725	1,450	2,175
2080	15 HP motor size	995	1,675	2,670
2120	20 HP motor size	1,050	1,775	2,825
2160	25 HP motor size	1,200	1,875	3,075
2200	30 HP motor size	1,550	2,025	3,575
2240	40 HP motor size	1,900	2,175	4,075
2280	50 HP motor size	2,125	2,400	4,525
2320	60 HP motor size	3,250	2,825	6,075
2360	75 HP motor size	3,750	3,125	6,875
2400	100 HP motor size	4,150	3,500	7,650
2440	125 HP motor size	7,075	3,950	11,025
2480	150 HP motor size	8,725	4,425	13,150
2520	200 HP motor size	9,675	5,000	14,675
2600	575V, 2 HP motor size	655	1,200	1,855
2640	5 HP motor size	755	1,325	2,080
2680	10 HP motor size	725	1,450	2,175
2720	20 HP motor size	995	1,675	2,670
2760	25 HP motor size	1,050	1,775	2,825
2800	30 HP motor size	1,550	2,025	3,575
2840	50 HP motor size	1,650	2,100	3,750
2880	60 HP motor size	3,225	2,825	6,050
2920	75 HP motor size	3,250	2,825	6,075
2960	100 HP motor size	3,750	3,125	6,875
3000	125 HP motor size	6,825	3,900	10,725
3040	150 HP motor size	7,075	3,950	11,025
3080	200 HP motor size	8,775	4,500	13,275

D5020 Lighting and Branch Wiring

System Components	QUANTITY	UNIT	COST PER L.F.		
			MAT.	INST.	TOTAL
SYSTEM D5020 155 0200					
MOTOR FEEDER SYSTEMS, SINGLE PHASE, UP TO 115V, 1HP OR 230V, 2HP					
Steel intermediate conduit, (IMC) 1/2" diam	1.000	L.F.	1.86	7.15	9.01
Wire 600V type THWN-THHN, copper solid #12	.020	C.L.F.	.22	1.30	1.52
TOTAL			2.08	8.45	10.53

D5020 155	Motor Feeder	COST PER L.F.		
		MAT.	INST.	TOTAL
0200	Motor feeder systems, single phase, feed up to 115V 1HP or 230V 2 HP	2.08	8.45	10.53
0240	115V 2HP, 230V 3HP	2.18	8.60	10.78
0280	115V 3HP	2.48	8.95	11.43
0360	Three phase, feed to 200V 3HP, 230V 5HP, 460V 10HP, 575V 10HP	2.18	9.10	11.28
0440	200V 5HP, 230V 7.5HP, 460V 15HP, 575V 20HP	2.34	9.30	11.64
0520	200V 10HP, 230V 10HP, 460V 30HP, 575V 30HP	2.79	9.85	12.64
0600	200V 15HP, 230V 15HP, 460V 40HP, 575V 50HP	3.86	11.25	15.11
0680	200V 20HP, 230V 25HP, 460V 50HP, 575V 60HP	5.10	14.30	19.40
0760	200V 25HP, 230V 30HP, 460V 60HP, 575V 75HP	6.05	14.55	20.60
0840	200V 30HP	6.70	15	21.70
0920	230V 40HP, 460V 75HP, 575V 100HP	9.05	16.40	25.45
1000	200V 40HP	10.50	17.55	28.05
1080	230V 50HP, 460V 100HP, 575V 125HP	12.45	19.35	31.80
1160	200V 50HP, 230V 60HP, 460V 125HP, 575V 150HP	14.45	20.50	34.95
1240	200V 60HP, 460V 150HP	18.45	24	42.45
1320	230V 75HP, 575V 200HP	21	25	46
1400	200V 75HP	23.50	25.50	49
1480	230V 100HP, 460V 200HP	29.50	30	59.50
1560	200V 100HP	41	37.50	78.50
1640	230V 125HP	41	37.50	78.50
1720	200V 125HP, 230V 150HP	46.50	46.50	93
1800	200V 150HP	58	50.50	108.50
1880	200V 200HP	74.50	53.50	128
1960	230V 200HP	68.50	55.50	124

D5020 Lighting and Branch Wiring

Starters are full voltage, type NEMA 1 for general purpose indoor application with motor overload protection and include mounting and wire connections.

System Components	QUANTITY	UNIT	COST EACH		
			MAT.	INST.	TOTAL
SYSTEM D5020 160 0200					
MAGNETIC STARTER, SIZE 00 TO 1/3 HP, 1 PHASE 115V OR 1 HP 230V					
Magnetic starter, size 00, to 1/3 HP, 1 phase, 115V or 1 HP 230V	1.000	Ea.	199	179	378
TOTAL			199	179	378

D5020 160	Magnetic Starter	COST EACH		
		MAT.	INST.	TOTAL
0200	Magnetic starter, size 00, to 1/3 HP, 1 phase, 115V or 1 HP 230V	199	179	378
0280	Size 00, to 1-1/2 HP, 3 phase, 200-230V or 2 HP 460-575V	219	205	424
0360	Size 0, to 1 HP, 1 phase, 115V or 2 HP 230V	222	179	401
0440	Size 0, to 3 HP, 3 phase, 200-230V or 5 HP 460-575V	320	310	630
0520	Size 1, to 2 HP, 1 phase, 115V or 3 HP 230V	255	239	494
0600	Size 1, to 7-1/2 HP, 3 phase, 200-230V or 10 HP 460-575V	292	450	742
0680	Size 2, to 10 HP, 3 phase, 200V, 15 HP-230V or 25 HP 460-575V	550	650	1,200
0760	Size 3, to 25 HP, 3 phase, 200V, 30 HP-230V or 50 HP 460-575V	895	795	1,690
0840	Size 4, to 40 HP, 3 phase, 200V, 50 HP-230V or 100 HP 460-575V	1,975	1,200	3,175
0920	Size 5, to 75 HP, 3 phase, 200V, 100 HP-230V or 200 HP 460-575V	4,650	1,600	6,250
1000	Size 6, to 150 HP, 3 phase, 200V, 200 HP-230V or 400 HP 460-575V	19,900	1,800	21,700

D50 Electrical

D5020 Lighting and Branch Wiring

Safety switches are type NEMA 1 for general purpose indoor application, and include time delay fuses, insulation and wire terminations.

System Components	QUANTITY	UNIT	COST EACH		
			MAT.	INST.	TOTAL
SYSTEM D5020 165 0200					
SAFETY SWITCH, 30A FUSED, 1 PHASE, 115V OR 230V					
Safety switch fused, hvy duty, 240V 2p 30 amp	1.000	Ea.	71.50	205	276.50
Fuse, dual element time delay 250V, 30 amp	2.000	Ea.	21.40	28.70	50.10
TOTAL			92.90	233.70	326.60

D5020 165	Safety Switches	COST EACH		
		MAT.	INST.	TOTAL
0200	Safety switch, 30 A fused, 1 phase, 2 HP 115 V or 3 HP, 230 V	93	234	327
0280	3 phase, 5 HP, 200 V or 7 1/2 HP, 230 V	129	267	396
0360	15 HP, 460 V or 20 HP, 575 V	232	278	510
0440	60 A fused, 3 phase, 15 HP 200 V or 15 HP 230 V	198	355	553
0520	30 HP 460 V or 40 HP 575 V	296	365	661
0600	100 A fused, 3 phase, 20 HP 200 V or 25 HP 230 V	375	430	805
0680	50 HP 460 V or 60 HP 575 V	560	435	995
0760	200 A fused, 3 phase, 50 HP 200 V or 60 HP 230 V	700	610	1,310
0840	125 HP 460 V or 150 HP 575 V	925	620	1,545
0920	400 A fused, 3 phase, 100 HP 200 V or 125 HP 230 V	1,500	865	2,365
1000	250 HP 460 V or 350 HP 575 V	2,300	885	3,185
1020	600 A fused, 3 phase, 150 HP 200 V or 200 HP 230 V	2,575	1,300	3,875
1040	400 HP 460 V	3,350	1,300	4,650

D5020 Lighting and Branch Wiring

Straight Connector

Angle Connector

Flexible Conduit

Table below includes costs for the flexible conduit. Not included are wire terminations and testing motor for correct rotation.

System Components	QUANTITY	UNIT	COST EACH		
			MAT.	INST.	TOTAL
SYSTEM D5020 170 0200					
MOTOR CONNECTIONS, SINGLE PHASE, 115V/230V UP TO 1 HP					
Motor connection, flexible conduit & fittings, 1 HP motor 115V	1.000	Ea.	6.25	89.50	95.75
TOTAL			6.25	89.50	95.75

D5020 170	Motor Connections	COST EACH		
		MAT.	INST.	TOTAL
0200	Motor connections, single phase, 115/230V, up to 1 HP	6.25	89.50	95.75
0240	Up to 3 HP	11.15	110	121.15
0280	Three phase, 200/230/460/575V, up to 3 HP	7.50	106	113.50
0320	Up to 5 HP	6.40	131	137.40
0360	Up to 7-1/2 HP	10.25	155	165.25
0400	Up to 10 HP	20	171	191
0440	Up to 15 HP	20	217	237
0480	Up to 25 HP	30.50	265	295.50
0520	Up to 50 HP	61.50	325	386.50
0560	Up to 100 HP	130	480	610

D5020 Lighting and Branch Wiring

Manual Starter

Magnetic Starter

Induction Motor

For 230/460 Volt A.C., 3 phase, 60 cycle ball bearing squirrel cage induction motors, NEMA Class B standard line. Installation included.

No conduit, wire, or terminations included.

System Components	QUANTITY	UNIT	COST EACH		
			MAT.	INST.	TOTAL
SYSTEM D5020 175 0220					
MOTOR, DRIPPROOF CLASS B INSULATION, 1.15 SERVICE FACTOR, WITH STARTER					
1 H.P., 1200 RPM WITH MANUAL STARTER					
Motor, dripproof, class B insul, 1.15 serv fact, 1,200 RPM, 1 HP	1.000	Ea.	540	159	699
Motor starter, manual, 3 phase, 1 HP motor	1.000	Ea.	222	205	427
TOTAL			762	364	1,126

D5020 175	Motor & Starter	COST EACH		
		MAT.	INST.	TOTAL
0190	Motor, dripproof, premium efficient, 1.15 service factor			
0200	1 HP, 1200 RPM, motor only	540	159	699
0220	With manual starter	760	365	1,125
0240	With magnetic starter	760	365	1,125
0260	1800 RPM, motor only	335	159	494
0280	With manual starter	555	365	920
0300	With magnetic starter	555	365	920
0320	2 HP, 1200 RPM, motor only	600	159	759
0340	With manual starter	820	365	1,185
0360	With magnetic starter	920	470	1,390
0380	1800 RPM, motor only	405	159	564
0400	With manual starter	625	365	990
0420	With magnetic starter	725	470	1,195
0440	3600 RPM, motor only	630	159	789
0460	With manual starter	850	365	1,215
0480	With magnetic starter	950	470	1,420
0500	3 HP, 1200 RPM, motor only	825	159	984
0520	With manual starter	1,050	365	1,415
0540	With magnetic starter	1,150	470	1,620
0560	1800 RPM, motor only	835	159	994
0580	With manual starter	1,050	365	1,415
0600	With magnetic starter	1,150	470	1,620
0620	3600 RPM, motor only	650	159	809
0640	With manual starter	870	365	1,235
0660	With magnetic starter	970	470	1,440
0680	5 HP, 1200 RPM, motor only	1,075	159	1,234
0700	With manual starter	1,350	520	1,870
0720	With magnetic starter	1,375	610	1,985
0740	1800 RPM, motor only	670	159	829
0760	With manual starter	935	520	1,455
0780	With magnetic starter	960	610	1,570
0800	3600 RPM, motor only	610	159	769

D5020 Lighting and Branch Wiring

D5020 175	Motor & Starter	COST EACH		
		MAT.	INST.	TOTAL
0820	With manual starter	875	520	1,395
0840	With magnetic starter	900	610	1,510
0860	7.5 HP, 1800 RPM, motor only	1,225	171	1,396
0880	With manual starter	1,500	530	2,030
0900	With magnetic starter	1,775	820	2,595
0920	10 HP, 1800 RPM, motor only	1,500	179	1,679
0940	With manual starter	1,775	540	2,315
0960	With magnetic starter	2,050	830	2,880
0980	15 HP, 1800 RPM, motor only	1,975	224	2,199
1000	With magnetic starter	2,525	875	3,400
1040	20 HP, 1800 RPM, motor only	2,350	276	2,626
1060	With magnetic starter	3,250	1,075	4,325
1100	25 HP, 1800 RPM, motor only	3,025	287	3,312
1120	With magnetic starter	3,925	1,075	5,000
1160	30 HP, 1800 RPM, motor only	3,325	299	3,624
1180	With magnetic starter	4,225	1,100	5,325
1220	40 HP, 1800 RPM, motor only	4,350	360	4,710
1240	With magnetic starter	6,325	1,550	7,875
1280	50 HP, 1800 RPM, motor only	4,500	450	4,950
1300	With magnetic starter	6,475	1,650	8,125
1340	60 HP, 1800 RPM, motor only	4,600	510	5,110
1360	With magnetic starter	9,250	2,100	11,350
1400	75 HP, 1800 RPM, motor only	5,575	595	6,170
1420	With magnetic starter	10,200	2,200	12,400
1460	100 HP, 1800 RPM, motor only	7,175	795	7,970
1480	With magnetic starter	11,800	2,400	14,200
1520	125 HP, 1800 RPM, motor only	7,525	1,025	8,550
1540	With magnetic starter	27,400	2,825	30,225
1580	150 HP, 1800 RPM, motor only	11,300	1,200	12,500
1600	With magnetic starter	31,200	3,000	34,200
1640	200 HP, 1800 RPM, motor only	13,800	1,425	15,225
1660	With magnetic starter	33,700	3,225	36,925
1680	Totally encl, premium efficient, 1.0 ser. fac., 1HP, 1200 RPM, motor only	545	159	704
1700	With manual starter	765	365	1,130
1720	With magnetic starter	765	365	1,130
1740	1800 RPM, motor only	700	159	859
1760	With manual starter	920	365	1,285
1780	With magnetic starter	920	365	1,285
1800	2 HP, 1200 RPM, motor only	570	159	729
1820	With manual starter	790	365	1,155
1840	With magnetic starter	890	470	1,360
1860	1800 RPM, motor only	810	159	969
1880	With manual starter	1,025	365	1,390
1900	With magnetic starter	1,125	470	1,595
1920	3600 RPM, motor only	475	159	634
1940	With manual starter	695	365	1,060
1960	With magnetic starter	795	470	1,265
1980	3 HP, 1200 RPM, motor only	825	159	984
2000	With manual starter	1,050	365	1,415
2020	With magnetic starter	1,150	470	1,620
2040	1800 RPM, motor only	800	159	959
2060	With manual starter	1,025	365	1,390
2080	With magnetic starter	1,125	470	1,595
2100	3600 RPM, motor only	630	159	789
2120	With manual starter	850	365	1,215
2140	With magnetic starter	950	470	1,420
2160	5 HP, 1200 RPM, motor only	1,075	159	1,234
2180	With manual starter	1,350	520	1,870

D50 Electrical

D5020 Lighting and Branch Wiring

D5020 175	Motor & Starter	COST EACH		
		MAT.	INST.	TOTAL
2200	With magnetic starter	1,375	610	1,985
2220	1800 RPM, motor only	880	159	1,039
2240	With manual starter	1,150	520	1,670
2260	With magnetic starter	1,175	610	1,785
2280	3600 RPM, motor only	740	159	899
2300	With manual starter	1,000	520	1,520
2320	With magnetic starter	1,025	610	1,635
2340	7.5 HP, 1800 RPM, motor only	1,200	171	1,371
2360	With manual starter	1,475	530	2,005
2380	With magnetic starter	1,750	820	2,570
2400	10 HP, 1800 RPM, motor only	1,575	179	1,754
2420	With manual starter	1,850	540	2,390
2440	With magnetic starter	2,125	830	2,955
2460	15 HP, 1800 RPM, motor only	2,475	224	2,699
2480	With magnetic starter	3,025	875	3,900
2500	20 HP, 1800 RPM, motor only	2,625	276	2,901
2520	With magnetic starter	3,525	1,075	4,600
2540	25 HP, 1800 RPM, motor only	3,975	287	4,262
2560	With magnetic starter	4,875	1,075	5,950
2580	30 HP, 1800 RPM, motor only	4,075	299	4,374
2600	With magnetic starter	4,975	1,100	6,075
2620	40 HP, 1800 RPM, motor only	4,450	360	4,810
2640	With magnetic starter	6,425	1,550	7,975
2660	50 HP, 1800 RPM, motor only	4,925	450	5,375
2680	With magnetic starter	6,900	1,650	8,550
2700	60 HP, 1800 RPM, motor only	6,550	510	7,060
2720	With magnetic starter	11,200	2,100	13,300
2740	75 HP, 1800 RPM, motor only	8,525	595	9,120
2760	With magnetic starter	13,200	2,200	15,400
2780	100 HP, 1800 RPM, motor only	9,975	795	10,770
2800	With magnetic starter	14,600	2,400	17,000
2820	125 HP, 1800 RPM, motor only	13,200	1,025	14,225
2840	With magnetic starter	33,100	2,825	35,925
2860	150 HP, 1800 RPM, motor only	15,200	1,200	16,400
2880	With magnetic starter	35,100	3,000	38,100
2900	200 HP, 1800 RPM, motor only	19,300	1,425	20,725
2920	With magnetic starter	39,200	3,225	42,425

For customer support on your Electrical Costs with RSMeans data, call 800.448.8182.

D50 Electrical

D5020 Lighting and Branch Wiring

General: The cost of the lighting portion of the electrical costs is dependent upon:
1. The footcandle requirement of the proposed building.
2. The type of fixtures required.
3. The ceiling heights of the building.
4. Reflectance value of ceilings, walls and floors.
5. Fixture efficiencies and spacing vs. mounting height ratios.

Footcandle Requirements: See Table D5020-204 for Footcandle and Watts per S.F. determination.

Table D5020-201   IESNA* Recommended Illumination Levels in Footcandles

Commercial Buildings			Industrial Buildings		
Type	Description	Footcandles	Type	Description	Footcandles
Bank	Lobby	50	Assembly Areas	Rough bench & machine work	50
	Customer Areas	70		Medium bench & machine work	100
	Teller Stations	150		Fine bench & machine work	500
	Accounting Areas	150	Inspection Areas	Ordinary	50
Offices	Routine Work	100		Difficult	100
	Accounting	150		Highly Difficult	200
	Drafting	200	Material Handling	Loading	20
	Corridors, Halls, Washrooms	30		Stock Picking	30
Schools	Reading or Writing	70		Packing, Wrapping	50
	Drafting, Labs, Shops	100	Stairways	Service Areas	20
	Libraries	70	Washrooms	Service Areas	20
	Auditoriums, Assembly	15	Storage Areas	Inactive	5
	Auditoriums, Exhibition	30		Active, Rough, Bulky	10
Stores	Circulation Areas	30		Active, Medium	20
	Stock Rooms	30		Active, Fine	50
	Merchandise Areas, Service	100	Garages	Active Traffic Areas	20
	Self-Service Areas	200		Service & Repair	100

*IESNA - Illuminating Engineering Society of North America

414

Table D5020-202 General Lighting Loads by Occupancies

Type of Occupancy	Unit Load per S.F. (Watts)
Armories and Auditoriums	1
Banks	5
Barber Shops and Beauty Parlors	3
Churches	1
Clubs	2
Court Rooms	2
*Dwelling Units	3
Garages — Commercial (storage)	½
Hospitals	2
*Hotels and Motels, including apartment houses without provisions for cooking by tenants	2
Industrial Commercial (Loft) Buildings	2
Lodge Rooms	1½
Office Buildings	5
Restaurants	2
Schools	3
Stores	3
Warehouses (storage)	¼
*In any of the above occupancies except one-family dwellings and individual dwelling units of multi-family dwellings:	
Assembly Halls and Auditoriums	1
Halls, Corridors, Closets	½
Storage Spaces	¼

Table D5020-203 Lighting Limit (Connected Load) for Listed Occupancies: New Building Proposed Energy Conservation Guideline

Type of Use	Maximum Watts per S.F.
Interior	
Category A: Classrooms, office areas, automotive mechanical areas, museums, conference rooms, drafting rooms, clerical areas, laboratories, merchandising areas, kitchens, examining rooms, book stacks, athletic facilities.	3.00
Category B: Auditoriums, waiting areas, spectator areas, restrooms, dining areas, transportation terminals, working corridors in prisons and hospitals, book storage areas, active inventory storage, hospital bedrooms, hotel and motel bedrooms, enclosed shopping mall concourse areas, stairways.	1.00
Category C: Corridors, lobbies, elevators, inactive storage areas.	0.50
Category D: Indoor parking.	0.25
Exterior	
Category E: Building perimeter: wall-wash, facade, canopy.	5.00 (per linear foot)
Category F: Outdoor parking.	0.10

Table D5020-204 Procedure for Calculating Footcandles and Watts Per Square Foot

1. Initial footcandles = No. of fixtures × lamps per fixture × lumens per lamp × coefficient of utilization ÷ square feet
2. Maintained footcandles = initial footcandles × maintenance factor
3. Watts per square foot = No. of fixtures × lamps × (lamp watts + ballast watts) ÷ square feet

Example: To find footcandles and watts per S.F. for an office 20′ x 20′ with 11 fluorescent fixtures each having 4–40 watt C.W. lamps.

Based on good reflectance and clean conditions:

Lumens per lamp = 40 watt cool white at 3150 lumens per lamp RD5020-250, Table RD5020-251

Coefficient of utilization = .42 (varies from .62 for light colored areas to .27 for dark)

Maintenance factor = .75 (varies from .80 for clean areas with good maintenance to .50 for poor)

Ballast loss = 8 watts per lamp. (Varies with manufacturer. See manufacturers' catalog.)

1. Initial footcandles:

$$\frac{11 \times 4 \times 3150 \times .42}{400} = \frac{58,212}{400} = 145 \text{ footcandles}$$

2. Maintained footcandles:

$$145 \times .75 = 109 \text{ footcandles}$$

3. Watts per S.F.

$$\frac{11 \times 4 \ (40 + 8)}{400} = \frac{2,112}{400} = 5.3 \text{ watts per S.F.}$$

Table D5020-205 Approximate Watts Per Square Foot for Popular Fixture Types

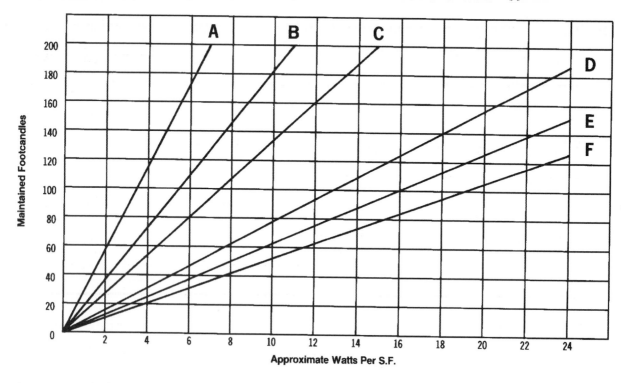

Due to the many variables involved, use for preliminary estimating only:
a. Fluorescent – industrial System D5020 208
b. Fluorescent – lens unit System D5020 208 Fixture types B & C
c. Fluorescent – louvered unit
d. Incandescent – open reflector System D5020 214, Type D
e. Incandescent – lens unit System D5020 214, Type A
f. Incandescent – down light System D5020 214, Type B

416

D50 Electrical

D5020 Lighting and Branch Wiring

A. Strip Fixture

B. Surface Mounted

C. Recessed

D. Pendent Mounted

Design Assumptions:

1. A 100 footcandle average maintained level of illumination.
2. Ceiling heights range from 9' to 11'.
3. Average reflectance values are assumed for ceilings, walls and floors.
4. Cool white (CW) fluorescent lamps with 3150 lumens for 40 watt lamps and 6300 lumens for 8' slimline lamps.
5. Four 40 watt lamps per 4' fixture and two 8' lamps per 8' fixture.
6. Average fixture efficiency values and spacing to mounting height ratios.
7. Installation labor is average U.S. rate as of January 1.

System Components	QUANTITY	UNIT	COST PER S.F.		
			MAT.	INST.	TOTAL
SYSTEM D5020 208 0520					
FLUORESCENT FIXTURES MOUNTED 9'-11" ABOVE FLOOR, 100 FC					
TYPE A, 8 FIXTURES PER 400 S.F.					
Conduit, steel intermediate, 1/2" diam.	.185	L.F.	.34	1.32	1.66
Wire, 600V, type THWN-THHN, copper, solid, #12	.004	C.L.F.	.04	.24	.28
Fluorescent strip fixture 8' long, surface mounted, two 75W SL	.020	Ea.	1.46	2.32	3.78
Steel outlet box 4" concrete	.020	Ea.	.19	.72	.91
Steel outlet box plate with stud, 4" concrete	.020	Ea.	.22	.18	.40
Fixture hangers, flexible, 1/2" diameter, 4" long	.040	Ea.	.72	2.38	3.10
Fixture whip, THHN wire, three #12, 3/8" Greenfield	.040	Ea.	.46	.90	1.36
TOTAL			**3.43**	**8.06**	**11.49**

D5020 208	Fluorescent Fixtures (by Type)	COST PER S.F.		
		MAT.	INST.	TOTAL
0520	Fluorescent fixtures, type A, 8 fixtures per 400 S.F.	3.43	8.05	11.48
0560	11 fixtures per 600 S.F.	3.14	7.35	10.49
0600	17 fixtures per 1000 S.F.	2.91	6.85	9.76
0640	23 fixtures per 1600 S.F.	2.44	5.75	8.19
0680	28 fixtures per 2000 S.F.	2.40	5.65	8.05
0720	41 fixtures per 3000 S.F.	2.36	5.55	7.91
0800	53 fixtures per 4000 S.F.	2.27	5.35	7.62
0840	64 fixtures per 5000 S.F.	2.23	5.25	7.48
0880	Type B, 11 fixtures per 400 S.F.	5.30	11.45	16.75
0920	15 fixtures per 600 S.F.	4.76	10.35	15.11
0960	24 fixtures per 1000 S.F.	4.60	9.95	14.55
1000	35 fixtures per 1600 S.F.	4.19	9.10	13.29
1040	42 fixtures per 2000 S.F.	4.01	8.70	12.71
1080	61 fixtures per 3000 S.F.	3.86	8.35	12.21
1160	80 fixtures per 4000 S.F.	2.76	5.30	8.06
1200	98 fixtures per 5000 S.F.	3.77	8.15	11.92
1240	Type C, 11 fixtures per 400 S.F.	4.59	11.95	16.54
1280	14 fixtures per 600 S.F.	3.85	10.05	13.90
1320	23 fixtures per 1000 S.F.	3.80	9.95	13.75
1360	34 fixtures per 1600 S.F.	3.51	9.15	12.66
1400	43 fixtures per 2000 S.F.	3.61	9.40	13.01
1440	63 fixtures per 3000 S.F.	3.48	9.05	12.53
1520	81 fixtures per 4000 S.F.	3.35	8.70	12.05
1560	101 fixtures per 5000 S.F.	3.31	8.65	11.96
1600	Type D, 8 fixtures per 400 S.F.	4.13	9	13.13
1640	12 fixtures per 600 S.F.	4.13	9	13.13
1680	19 fixtures per 1000 S.F.	3.94	8.55	12.49
1720	27 fixtures per 1600 S.F.	3.51	7.65	11.16
1760	34 fixtures per 2000 S.F.	3.51	7.65	11.16
1800	48 fixtures per 3000 S.F.	3.32	7.20	10.52
1880	64 fixtures per 4000 S.F.	3.32	7.20	10.52
1920	79 fixtures per 5000 S.F.	3.31	7.20	10.51

D5020 Lighting and Branch Wiring

Type C. Recessed, mounted on grid ceiling suspension system, 2' x 4', four 40 watt lamps, acrylic prismatic diffusers.

5.3 watts per S.F. for 100 footcandles.

3 watts per S.F. for 57 footcandles.

System Components	QUANTITY	UNIT	COST PER S.F.		
			MAT.	INST.	TOTAL
SYSTEM D5020 210 0200					
FLUORESCENT FIXTURES RECESS MOUNTED IN CEILING					
1 WATT PER S.F., 20 FC, 5 FIXTURES PER 1000 S.F.					
Steel intermediate conduit, (IMC) 1/2″ diam.	.128	L.F.	.24	.92	1.16
Wire, 600 volt, type THW, copper, solid, #12	.003	C.L.F.	.03	.20	.23
Fluorescent fixture, recessed, 2'x 4', four 40W, w/lens, for grid ceiling	.005	Ea.	.34	.77	1.11
Steel outlet box 4″ square	.005	Ea.	.05	.18	.23
Fixture whip, Greenfield w/#12 THHN wire	.005	Ea.	.05	.04	.09
TOTAL			.71	2.11	2.82

D5020 210	Fluorescent Fixtures (by Wattage)	COST PER S.F.		
		MAT.	INST.	TOTAL
0190	Fluorescent fixtures recess mounted in ceiling			
0195	T12, standard 40 watt lamps			
0200	1 watt per S.F., 20 FC, 5 fixtures @40 watts per 1000 S.F.	.71	2.11	2.82
0240	2 watt per S.F., 40 FC, 10 fixtures @40 watt per 1000 S.F.	1.42	4.14	5.56
0280	3 watt per S.F., 60 FC, 15 fixtures @40 watt per 1000 S.F	2.13	6.25	8.38
0320	4 watt per S.F., 80 FC, 20 fixtures @40 watt per 1000 S.F.	2.84	8.25	11.09
0400	5 watt per S.F., 100 FC, 25 fixtures @40 watt per 1000 S.F.	3.55	10.40	13.95
0450	T8, energy saver 32 watt lamps			
0500	0.8 watt per S.F., 20 FC, 5 fixtures @32 watt per 1000 S.F.	.78	2.11	2.89
0520	1.6 watt per S.F., 40 FC, 10 fixtures @32 watt per 1000 S.F.	1.55	4.14	5.69
0540	2.4 watt per S.F., 60 FC, 15 fixtures @ 32 watt per 1000 S.F	2.32	6.25	8.57
0560	3.2 watt per S.F., 80 FC, 20 fixtures @32 watt per 1000 S.F.	3.10	8.25	11.35
0580	4 watt per S.F., 100 FC, 25 fixtures @32 watt per 1000 S.F.	3.88	10.40	14.28

D5020 Lighting and Branch Wiring

Type A. Recessed wide distribution reflector with flat glass lens 150 W.

Maximum spacing = 1.2 x mounting height.

13 watts per S.F. for 100 footcandles.

Type B. Recessed reflector down light with baffles 150 W.

Maximum spacing = 0.8 x mounting height.

18 watts per S.F. for 100 footcandles.

Type C. Recessed PAR–38 flood lamp with concentric louver 150 W.

Maximum spacing = 0.5 x mounting height.

19 watts per S.F. for 100 footcandles.

Type D. Recessed R–40 flood lamp with reflector skirt.

Maximum spacing = 0.7 x mounting height.

15 watts per S.F. for 100 footcandles.

System Components	QUANTITY	UNIT	COST PER S.F.		
			MAT.	INST.	TOTAL
SYSTEM D5020 214 0400					
INCANDESCENT FIXTURE RECESS MOUNTED, 100 FC					
TYPE A, 34 FIXTURES PER 400 S.F.					
Steel intermediate conduit, (IMC) 1/2" diam	1.060	L.F.	1.97	7.58	9.55
Wire, 600V, type THWN-THHN, copper, solid, #12	.033	C.L.F.	.36	2.15	2.51
Steel outlet box 4" square	.085	Ea.	.80	3.06	3.86
Fixture whip, Greenfield w/#12 THHN wire	.085	Ea.	.93	.76	1.69
Incandescent fixture, recessed, w/lens, prewired, square trim, 200W	.085	Ea.	9.18	9.10	18.28
TOTAL			13.24	22.65	35.89

D5020 214	Incandescent Fixture (by Type)	COST PER S.F.		
		MAT.	INST.	TOTAL
0380	Incandescent fixture recess mounted, 100 FC			
0400	Type A, 34 fixtures per 400 S.F.	13.25	22.50	35.75
0440	49 fixtures per 600 S.F.	12.90	22.50	35.40
0480	63 fixtures per 800 S.F.	12.55	22	34.55
0520	90 fixtures per 1200 S.F.	12.05	21.50	33.55
0560	116 fixtures per 1600 S.F.	11.85	21.50	33.35
0600	143 fixtures per 2000 S.F.	11.75	21.50	33.25
0640	Type B, 47 fixtures per 400 S.F.	12.15	28.50	40.65
0680	66 fixtures per 600 S.F.	11.60	28	39.60
0720	88 fixtures per 800 S.F.	11.60	28	39.60
0760	127 fixtures per 1200 S.F.	11.35	27.50	38.85
0800	160 fixtures per 1600 S.F.	11.20	27.50	38.70
0840	206 fixtures per 2000 S.F.	11.15	27.50	38.65
0880	Type C, 51 fixtures per 400 S.F.	16.70	30	46.70
0920	74 fixtures per 600 S.F.	16.20	29.50	45.70
0960	97 fixtures per 800 S.F.	16	29	45
1000	142 fixtures per 1200 S.F.	15.75	29	44.75
1040	186 fixtures per 1600 S.F.	15.55	28.50	44.05
1080	230 fixtures per 2000 S.F.	15.50	28.50	44
1120	Type D, 39 fixtures per 400 S.F.	15.75	23.50	39.25
1160	57 fixtures per 600 S.F.	15.35	23	38.35
1200	75 fixtures per 800 S.F.	15.25	23	38.25
1240	109 fixtures per 1200 S.F.	14.90	23	37.90
1280	143 fixtures per 1600 S.F.	14.60	22.50	37.10
1320	176 fixtures per 2000 S.F.	14.50	22.50	37

D50 Electrical

D5020 Lighting and Branch Wiring

Type A. Recessed, wide distribution reflector with flat glass lens.

150 watt inside frost—2500 lumens per lamp.

PS–25 extended service lamp.

Maximum spacing = 1.2 x mounting height.

13 watts per S.F. for 100 footcandles.

System Components	QUANTITY	UNIT	COST PER S.F.		
			MAT.	INST.	TOTAL
SYSTEM D5020 216 0200					
INCANDESCENT FIXTURE RECESS MOUNTED, TYPE A					
1 WATT PER S.F., 8 FC, 6 FIXT PER 1000 S.F.					
Steel intermediate conduit, (IMC) 1/2" diam	.091	L.F.	.17	.65	.82
Wire, 600V, type THWN-THHN, copper, solid, #12	.002	C.L.F.	.02	.13	.15
Incandescent fixture, recessed, w/lens, prewired, square trim, 200W	.006	Ea.	.65	.64	1.29
Steel outlet box 4" square	.006	Ea.	.06	.22	.28
Fixture whip, Greenfield w/#12 THHN wire	.006	Ea.	.07	.05	.12
TOTAL			.97	1.69	2.66

D5020 216	Incandescent Fixture (by Wattage)	COST PER S.F.		
		MAT.	INST.	TOTAL
0190	Incandescent fixture recess mounted, type A			
0200	1 watt per S.F., 8 FC, 6 fixtures per 1000 S.F.	.97	1.69	2.66
0240	2 watt per S.F., 16 FC, 12 fixtures per 1000 S.F.	1.92	3.38	5.30
0280	3 watt per S.F., 24 FC, 18 fixtures, per 1000 S.F.	2.87	5	7.87
0320	4 watt per S.F., 32 FC, 24 fixtures per 1000 S.F.	3.84	6.70	10.54
0400	5 watt per S.F., 40 FC, 30 fixtures per 1000 S.F.	4.80	8.40	13.20

For customer support on your Electrical Costs with RSMeans data, call 800.448.8182.

D5020 Lighting and Branch Wiring

High Bay Fluorescent Fixtures
Four T5 HO/54 watt lamps

System Components	QUANTITY	UNIT	COST PER S.F.		
			MAT.	INST.	TOTAL
SYSTEM D5020 218 0200					
FLUORESCENT HIGH BAY FIXTURE, 4 LAMP, 8′-10′ ABOVE WORK PLANE					
0.5 WATT/S.F., 29 FC, 2 FIXTURES/1000 S.F.					
Steel intermediate conduit, (IMC) 1/2″ diam	.100	L.F.	.19	.72	.91
Wire, 600V, type THWN-THHN, copper, solid, #10	.002	C.L.F.	.03	.14	.17
Steel outlet box 4″ concrete	.002	Ea.	.02	.07	.09
Steel outlet box plate with stud, 4″ concrete	.002	Ea.	.02	.02	.04
Flourescent , hi bay, 4 lamp, T-5 fixture	.002	Ea.	.61	.32	.93
TOTAL			.87	1.27	2.14

D5020 218	Fluorescent Fixture, High Bay, 8′-10′ (by Wattage)	COST PER S.F.		
		MAT.	INST.	TOTAL
0190	Fluorescent high bay-4 lamp fixture, 8′-10′ above work plane			
0200	.5 watt/SF, 29 FC, 2 fixtures per 1000 S.F.	.87	1.27	2.14
0400	1 watt/SF, 59 FC, 4 fixtures per 1000 S.F.	1.71	2.44	4.15
0600	1.5 watt/SF, 103 FC, 7 fixtures per 1000 S.F.	2.89	3.80	6.69
0800	2 watt/SF, 133 FC, 9 fixtures per 1000 S.F.	3.74	5	8.74
1000	2.5 watt/SF, 162 FC, 11 fixtures per 1000 S.F.	4.58	6.25	10.83

D5020 Lighting and Branch Wiring

HIGH BAY FIXTURES

B. Metal halide 400 watt

C. High pressure sodium 400 watt

E. Metal halide 1000 watt

F. High pressure sodium 1000 watt

G. Metal halide 1000 watt
125,000 lumen lamp

System Components	QUANTITY	UNIT	COST PER S.F.		
			MAT.	INST.	TOTAL
SYSTEM D5020 220 0880					
HIGH INTENSITY DISCHARGE FIXTURE, 8'-10' ABOVE WORK PLANE, 100 FC					
TYPE B, 8 FIXTURES PER 900 S.F.					
Steel intermediate conduit, (IMC) 1/2" diam	.460	L.F.	.86	3.29	4.15
Wire, 600V, type THWN-THHN, copper, solid, #10	.009	C.L.F.	.14	.64	.78
Steel outlet box 4" concrete	.009	Ea.	.09	.32	.41
Steel outlet box plate with stud 4" concrete	.009	Ea.	.10	.08	.18
Metal halide, hi bay, aluminum reflector, 400 W lamp	.009	Ea.	4.10	2.79	6.89
TOTAL			5.29	7.12	12.41

D5020 220	H.I.D. Fixture, High Bay, 8'-10' (by Type)	COST PER S.F.		
		MAT.	INST.	TOTAL
0500	High intensity discharge fixture, 8'-10' above work plane, 100 FC			
0880	Type B, 8 fixtures per 900 S.F.	5.30	7.10	12.40
0920	15 fixtures per 1800 S.F.	4.83	6.85	11.68
0960	24 fixtures per 3000 S.F.	4.83	6.85	11.68
1000	31 fixtures per 4000 S.F.	4.82	6.85	11.67
1040	38 fixtures per 5000 S.F.	4.82	6.85	11.67
1080	60 fixtures per 8000 S.F.	4.82	6.85	11.67
1120	72 fixtures per 10000 S.F.	4.32	6.35	10.67
1160	115 fixtures per 16000 S.F.	4.32	6.35	10.67
1200	230 fixtures per 32000 S.F.	4.32	6.35	10.67
1240	Type C, 4 fixtures per 900 S.F.	2.50	4.35	6.85
1280	8 fixtures per 1800 S.F.	2.51	4.37	6.88
1320	13 fixtures per 3000 S.F.	2.50	4.35	6.85
1360	17 fixtures per 4000 S.F.	2.50	4.35	6.85
1400	21 fixtures per 5000 S.F.	2.42	4	6.42
1440	33 fixtures per 8000 S.F.	2.40	3.93	6.33
1480	40 fixtures per 10000 S.F.	2.34	3.71	6.05
1520	63 fixtures per 16000 S.F.	2.34	3.71	6.05
1560	126 fixtures per 32000 S.F.	2.34	3.71	6.05

D5020 Lighting and Branch Wiring

HIGH BAY FIXTURES

B. Metal halide 400 watt

C. High pressure sodium 400 watt

E. Metal halide 1000 watt

F. High pressure sodium 1000 watt

G. Metal halide 1000 watt
125,000 lumen lamp

System Components	QUANTITY	UNIT	COST PER S.F.		
			MAT.	INST.	TOTAL
SYSTEM D5020 222 0240					
HIGH INTENSITY DISCHARGE FIXTURE, 8'-10' ABOVE WORK PLANE					
1 WATT/S.F., TYPE B, 29 FC, 2 FIXTURES/1000 S.F.					
Steel intermediate conduit, (IMC) 1/2" diam	.100	L.F.	.19	.72	.91
Wire, 600V, type THWN-THHN, copper, solid, #10	.002	C.L.F.	.03	.14	.17
Steel outlet box 4" concrete	.002	Ea.	.02	.07	.09
Steel outlet box plate with stud, 4" concrete	.002	Ea.	.02	.02	.04
Metal halide, hi bay, aluminum reflector, 400 W lamp	.002	Ea.	.91	.62	1.53
TOTAL			1.17	1.57	2.74

D5020 222	H.I.D. Fixture, High Bay, 8'-10' (by Wattage)	COST PER S.F.		
		MAT.	INST.	TOTAL
0190	High intensity discharge fixture, 8'-10' above work plane			
0240	1 watt/S.F., type B, 29 FC, 2 fixtures/1000 S.F.	1.17	1.57	2.74
0280	Type C, 54 FC, 2 fixtures/1000 S.F.	1.01	1.19	2.20
0400	2 watt/S.F., type B, 59 FC, 4 fixtures/1000 S.F.	2.31	3.04	5.35
0440	Type C, 108 FC, 4 fixtures/1000 S.F.	1.99	2.31	4.30
0560	3 watt/S.F., type B, 103 FC, 7 fixtures/1000 S.F.	3.94	4.84	8.78
0600	Type C, 189 FC, 6 fixtures/1000 S.F.	3.21	2.98	6.19
0720	4 watt/S.F., type B, 133 FC, 9 fixtures/1000 S.F.	5.10	6.35	11.45
0760	Type C, 243 FC, 9 fixtures/1000 S.F.	4.46	5.15	9.61
0880	5 watt/S.F., type B, 162 FC, 11 fixtures/1000 S.F.	6.25	7.85	14.10
0920	Type C, 297 FC, 11 fixtures/1000 S.F.	5.45	6.40	11.85

D5020 Lighting and Branch Wiring

HIGH BAY FIXTURES

B. Metal halide 400 watt

C. High pressure sodium 400 watt

E. Metal halide 1000 watt

F. High pressure sodium 1000 watt

G. Metal halide 1000 watt
125,000 lumen lamp

System Components			COST PER S.F.		
	QUANTITY	UNIT	MAT.	INST.	TOTAL
SYSTEM D5020 224 1240					
HIGH INTENSITY DISCHARGE FIXTURE, 16' ABOVE WORK PLANE, 100 FC					
TYPE C, 5 FIXTURES PER 900 S.F.					
Steel intermediate conduit, (IMC) 1/2" diam	.260	L.F.	.48	1.86	2.34
Wire, 600V, type THWN-THHN, copper, solid, #10	.007	C.L.F.	.11	.50	.61
Steel outlet box 4" concrete	.006	Ea.	.06	.22	.28
Steel outlet box plate with stud, 4" concrete	.006	Ea.	.07	.05	.12
High pressure sodium, hi bay, aluminum reflector, 400 W lamp	.006	Ea.	2.52	1.86	4.38
TOTAL			3.24	4.49	7.73

D5020 224	H.I.D. Fixture, High Bay, 16' (by Type)	COST PER S.F.		
		MAT.	INST.	TOTAL
0510	High intensity discharge fixture, 16' above work plane, 100 FC			
1240	Type C, 5 fixtures per 900 S.F.	3.24	4.49	7.73
1280	9 fixtures per 1800 S.F.	2.94	4.70	7.64
1320	15 fixtures per 3000 S.F.	2.94	4.70	7.64
1360	18 fixtures per 4000 S.F.	2.81	4.20	7.01
1400	22 fixtures per 5000 S.F.	2.81	4.20	7.01
1440	36 fixtures per 8000 S.F.	2.81	4.20	7.01
1480	42 fixtures per 10,000 S.F.	2.50	4.35	6.85
1520	65 fixtures per 16,000 S.F.	2.50	4.35	6.85
1600	Type G, 4 fixtures per 900 S.F.	4.07	7.10	11.17
1640	6 fixtures per 1800 S.F.	3.39	6.65	10.04
1720	9 fixtures per 4000 S.F.	2.77	6.45	9.22
1760	11 fixtures per 5000 S.F.	2.77	6.45	9.22
1840	21 fixtures per 10,000 S.F.	2.60	5.80	8.40
1880	33 fixtures per 16,000 S.F.	2.60	5.80	8.40

D5020 Lighting and Branch Wiring

HIGH BAY FIXTURES

B. Metal halide 400 watt

C. High pressure sodium 400 watt

E. Metal halide 1000 watt

F. High pressure sodium 1000 watt

G. Metal halide 1000 watt
125,000 lumen lamp

System Components	QUANTITY	UNIT	COST PER S.F. MAT.	COST PER S.F. INST.	COST PER S.F. TOTAL
SYSTEM D5020 226 0240					
HIGH INTENSITY DISCHARGE FIXTURE, 16' ABOVE WORK PLANE					
1 WATT/S.F., TYPE E, 42 FC, 1 FIXTURE/1000 S.F.					
Steel intermediate conduit, (IMC) 1/2" diam	.160	L.F.	.30	1.14	1.44
Wire, 600V, type THWN-THHN, copper, solid, #10	.003	C.L.F.	.05	.21	.26
Steel outlet box 4" concrete	.001	Ea.	.01	.04	.05
Steel outlet box plate with stud, 4" concrete	.001	Ea.	.01	.01	.02
Metal halide, hi bay, aluminum reflector, 1000 W lamp	.001	Ea.	.65	.36	1.01
TOTAL			1.02	1.76	2.78

D5020 226	H.I.D. Fixture, High Bay, 16' (by Wattage)	COST PER S.F. MAT.	COST PER S.F. INST.	COST PER S.F. TOTAL
0190	High intensity discharge fixture, 16' above work plane			
0240	1 watt/S.F., type E, 42 FC, 1 fixture/1000 S.F.	1.02	1.76	2.78
0280	Type G, 52 FC, 1 fixture/1000 S.F.	1.02	1.76	2.78
0320	Type C, 54 FC, 2 fixture/1000 S.F.	1.20	1.96	3.16
0440	2 watt/S.F., type E, 84 FC, 2 fixture/1000 S.F.	2.05	3.60	5.65
0480	Type G, 105 FC, 2 fixture/1000 S.F.	2.05	3.60	5.65
0520	Type C, 108 FC, 4 fixture/1000 S.F.	2.40	3.94	6.34
0640	3 watt/S.F., type E, 126 FC, 3 fixture/1000 S.F.	3.06	5.35	8.41
0680	Type G, 157 FC, 3 fixture/1000 S.F.	3.06	5.35	8.41
0720	Type C, 162 FC, 6 fixture/1000 S.F.	3.60	5.90	9.50
0840	4 watt/S.F., type E, 168 FC, 4 fixture/1000 S.F.	4.10	7.20	11.30
0880	Type G, 210 FC, 4 fixture/1000 S.F.	4.10	7.20	11.30
0920	Type C, 243 FC, 9 fixture/1000 S.F.	5.25	8.25	13.50
1040	5 watt/S.F., type E, 210 FC, 5 fixture/1000 S.F.	5.10	8.95	14.05
1080	Type G, 262 FC, 5 fixture/1000 S.F.	5.10	8.95	14.05
1120	Type C, 297 FC, 11 fixture/1000 S.F.	6.45	10.20	16.65

D5020 Lighting and Branch Wiring

HIGH BAY FIXTURES

B. Metal halide 400 watt

C. High pressure sodium 400 watt

E. Metal halide 1000 watt

F. High pressure sodium 1000 watt

G. Metal halide 1000 watt
125,000 lumen lamp

System Components	QUANTITY	UNIT	COST PER S.F. MAT.	COST PER S.F. INST.	COST PER S.F. TOTAL
SYSTEM D5020 228 1240					
HIGH INTENSITY DISCHARGE FIXTURE, 20′ ABOVE WORK PLANE, 100 FC					
TYPE C, 6 FIXTURES PER 900 S.F.					
Steel intermediate conduit, (IMC) 1/2″ diam	.350	L.F.	.65	2.50	3.15
Wire, 600V, type THWN-THHN, copper, solid, #10.	.011	C.L.F.	.18	.79	.97
Steel outlet box 4″ concrete	.007	Ea.	.07	.25	.32
Steel outlet box plate with stud, 4″ concrete	.007	Ea.	.08	.06	.14
High pressure sodium, hi bay, aluminum reflector, 400 W lamp	.007	Ea.	2.94	2.17	5.11
TOTAL			3.92	5.77	9.69

D5020 228	H.I.D. Fixture, High Bay, 20′ (by Type)	COST PER S.F. MAT.	COST PER S.F. INST.	COST PER S.F. TOTAL
0510	High intensity discharge fixture 20′ above work plane, 100 FC			
1240	Type C, 6 fixtures per 900 S.F.	3.92	5.75	9.67
1280	10 fixtures per 1800 S.F.	3.48	5.40	8.88
1320	16 fixtures per 3000 S.F.	3.06	5.20	8.26
1360	20 fixtures per 4000 S.F.	3.06	5.20	8.26
1400	24 fixtures per 5000 S.F.	3.06	5.20	8.26
1440	38 fixtures per 8000 S.F.	3	4.99	7.99
1520	68 fixtures per 16000 S.F.	2.80	5.50	8.30
1560	132 fixtures per 32000 S.F.	2.80	5.50	8.30
1600	Type G, 4 fixtures per 900 S.F.	4.03	7	11.03
1640	6 fixtures per 1800 S.F.	3.38	6.65	10.03
1680	7 fixtures per 3000 S.F.	3.32	6.45	9.77
1720	10 fixtures per 4000 S.F.	3.30	6.35	9.65
1760	11 fixtures per 5000 S.F.	2.86	6.80	9.66
1800	18 fixtures per 8000 S.F.	2.86	6.80	9.66
1840	22 fixtures per 10000 S.F.	2.86	6.80	9.66
1880	34 fixtures per 16000 S.F.	2.80	6.60	9.40
1920	66 fixtures per 32000 S.F.	2.65	6.05	8.70

D5020 Lighting and Branch Wiring

HIGH BAY FIXTURES

B. Metal halide 400 watt

C. High pressure sodium 400 watt

E. Metal halide 1000 watt

F. High pressure sodium 1000 watt

G. Metal halide 1000 watt
125,000 lumen lamp

System Components	QUANTITY	UNIT	COST PER S.F. MAT.	COST PER S.F. INST.	COST PER S.F. TOTAL
SYSTEM D5020 230 0240					
HIGH INTENSITY DISCHARGE FIXTURE, 20' ABOVE WORK PLANE					
1 WATT/S.F., TYPE E, 40 FC, 1 FIXTURE 1000 S.F.					
Steel intermediate conduit, (IMC) 1/2" diam	.160	L.F.	.30	1.14	1.44
Wire, 600V, type THWN-THHN, copper, solid, #10	.005	C.L.F.	.08	.36	.44
Steel outlet box 4" concrete	.001	Ea.	.01	.04	.05
Steel outlet box plate with stud, 4" concrete	.001	Ea.	.01	.01	.02
Metal halide, hi bay, aluminum reflector, 1000 W lamp	.001	Ea.	.65	.36	1.01
TOTAL			1.05	1.91	2.96

D5020 230	H.I.D. Fixture, High Bay, 20' (by Wattage)	COST PER S.F. MAT.	COST PER S.F. INST.	COST PER S.F. TOTAL
0190	High intensity discharge fixture, 20' above work plane			
0240	1 watt/S.F., type E, 40 FC, 1 fixture/1000 S.F.	1.05	1.91	2.96
0280	Type G, 50 FC, 1 fixture/1000 S.F.	1.05	1.91	2.96
0320	Type C, 52 FC, 2 fixtures/1000 S.F.	1.25	2.20	3.45
0440	2 watt/S.F., type E, 81 FC, 2 fixtures/1000 S.F.	2.10	3.82	5.92
0480	Type G, 101 FC, 2 fixtures/1000 S.F.	2.10	3.82	5.92
0520	Type C, 104 FC, 4 fixtures/1000 S.F.	2.49	4.32	6.81
0640	3 watt/S.F., type E, 121 FC, 3 fixtures/1000 S.F.	3.14	5.70	8.84
0680	Type G, 151 FC, 3 fixtures/1000 S.F.	3.14	5.70	8.84
0720	Type C, 155 FC, 6 fixtures/1000 S.F.	3.76	6.50	10.26
0840	4 watt/S.F., type E, 161 FC, 4 fixtures/1000 S.F.	4.19	7.65	11.84
0880	Type G, 202 FC, 4 fixtures/1000 S.F.	4.19	7.65	11.84
0920	Type C, 233 FC, 9 fixtures/1000 S.F.	5.45	9	14.45
1040	5 watt/S.F., type E, 202 FC, 5 fixtures/1000 S.F.	5.25	9.55	14.80
1080	Type G, 252 FC, 5 fixtures/1000 S.F.	5.25	9.55	14.80
1120	Type C, 285 FC, 11 fixtures/1000 S.F.	6.70	11.20	17.90

HIGH BAY FIXTURES

B. Metal halide 400 watt

C. High pressure sodium 400 watt

E. Metal halide 1000 watt

F. High pressure sodium 1000 watt

G. Metal halide 1000 watt
125,000 lumen lamp

System Components	QUANTITY	UNIT	COST PER S.F.		
			MAT.	INST.	TOTAL
SYSTEM D5020 232 1240					
HIGH INTENSITY DISCHARGE FIXTURE, 30′ ABOVE WORK PLANE, 100 FC					
TYPE F, 4 FIXTURES PER 900 S.F.					
Steel intermediate conduit, (IMC) 1/2″ diam	.580	L.F.	1.08	4.15	5.23
Wire, 600V, type THWN-THHN, copper, solid, #10	.018	C.L.F.	.29	1.29	1.58
Steel outlet box 4″ concrete	.004	Ea.	.04	.14	.18
Steel outlet box plate with stud, 4″ concrete	.004	Ea.	.04	.04	.08
High pressure sodium, hi bay, aluminum, 1000 W lamp	.004	Ea.	2.40	1.44	3.84
TOTAL			3.85	7.06	10.91

D5020 232	H.I.D. Fixture, High Bay, 30′ (by Type)	COST PER S.F.		
		MAT.	INST.	TOTAL
0510	High intensity discharge fixture, 30′ above work plane, 100 FC			
1240	Type F, 4 fixtures per 900 S.F.	3.85	7.05	10.90
1280	6 fixtures per 1800 S.F.	3.25	6.75	10
1320	8 fixtures per 3000 S.F.	3.25	6.75	10
1360	9 fixtures per 4000 S.F.	2.65	6.40	9.05
1400	10 fixtures per 5000 S.F.	2.65	6.40	9.05
1440	17 fixtures per 8000 S.F.	2.65	6.40	9.05
1480	18 fixtures per 10,000 S.F.	2.52	5.90	8.42
1520	27 fixtures per 16,000 S.F.	2.52	5.90	8.42
1560	52 fixtures per 32000 S.F.	2.50	5.80	8.30
1600	Type G, 4 fixtures per 900 S.F.	4.16	7.50	11.66
1640	6 fixtures per 1800 S.F.	3.41	6.80	10.21
1680	9 fixtures per 3000 S.F.	3.36	6.60	9.96
1720	11 fixtures per 4000 S.F.	3.30	6.35	9.65
1760	13 fixtures per 5000 S.F.	3.30	6.35	9.65
1800	21 fixtures per 8000 S.F.	3.30	6.35	9.65
1840	23 fixtures per 10,000 S.F.	2.81	6.75	9.56
1880	36 fixtures per 16,000 S.F.	2.81	6.75	9.56
1920	70 fixtures per 32,000 S.F.	2.81	6.75	9.56

D5020 Lighting and Branch Wiring

HIGH BAY FIXTURES

B. Metal halide 400 watt

C. High pressure sodium 400 watt

E. Metal halide 1000 watt

F. High pressure sodium 1000 watt

G. Metal halide 1000 watt
125,000 lumen lamp

System Components	QUANTITY	UNIT	COST PER S.F. MAT.	COST PER S.F. INST.	COST PER S.F. TOTAL
SYSTEM D5020 234 0240					
HIGH INTENSITY DISCHARGE FIXTURE, 30' ABOVE WORK PLANE					
1 WATT/S.F., TYPE E, 37 FC, 1 FIXTURE/1000 S.F.					
Steel intermediate conduit, (IMC) 1/2" diam	.196	L.F.	.36	1.40	1.76
Wire, 600V type THWN-THHN, copper, solid, #10	.006	C.L.F.	.10	.43	.53
Steel outlet box 4" concrete	.001	Ea.	.01	.04	.05
Steel outlet box plate with stud, 4" concrete	.001	Ea.	.01	.01	.02
Metal halide, hi bay, aluminum reflector, 1000 W lamp	.001	Ea.	.65	.36	1.01
TOTAL			1.13	2.24	3.37

D5020 234	H.I.D. Fixture, High Bay, 30' (by Wattage)	COST PER S.F. MAT.	COST PER S.F. INST.	COST PER S.F. TOTAL
0190	High intensity discharge fixture, 30' above work plane			
0240	1 watt/S.F., type E, 37 FC, 1 fixture/1000 S.F.	1.13	2.24	3.37
0280	Type G, 45 FC., 1 fixture/1000 S.F.	1.13	2.24	3.37
0320	Type F, 50 FC, 1 fixture/1000 S.F.	.96	1.76	2.72
0440	2 watt/S.F., type E, 74 FC, 2 fixtures/1000 S.F.	2.26	4.47	6.73
0480	Type G, 92 FC, 2 fixtures/1000 S.F.	2.26	4.47	6.73
0520	Type F, 100 FC, 2 fixtures/1000 S.F.	1.93	3.57	5.50
0640	3 watt/S.F., type E, 110 FC, 3 fixtures/1000 S.F.	3.41	6.80	10.21
0680	Type G, 138 FC, 3 fixtures/1000 S.F.	3.41	6.80	10.21
0720	Type F, 150 FC, 3 fixtures/1000 S.F.	2.90	5.30	8.20
0840	4 watt/S.F., type E, 148 FC, 4 fixtures/1000 S.F.	4.54	9	13.54
0880	Type G, 185 FC, 4 fixtures/1000 S.F.	4.54	9	13.54
0920	Type F, 200 FC, 4 fixtures/1000 S.F.	3.87	7.15	11.02
1040	5 watt/S.F., type E, 185 FC, 5 fixtures/1000 S.F.	5.65	11.25	16.90
1080	Type G, 230 FC, 5 fixtures/1000 S.F.	5.65	11.25	16.90
1120	Type F, 250 FC, 5 fixtures/1000 S.F.	4.83	8.90	13.73

LOW BAY FIXTURES

J. Metal halide 250 watt

K. High pressure sodium 150 watt

System Components	QUANTITY	UNIT	COST PER S.F.		
			MAT.	INST.	TOTAL
SYSTEM D5020 236 0920					
HIGH INTENSITY DISCHARGE FIXTURE, 8'-10' ABOVE WORK PLANE, 50 FC					
TYPE J, 13 FIXTURES PER 1800 S.F.					
Steel intermediate conduit, (IMC) 1/2" diam	.550	L.F.	1.02	3.93	4.95
Wire, 600V, type THWN-THHN, copper, solid, #10	.012	C.L.F.	.19	.86	1.05
Steel outlet box 4" concrete	.007	Ea.	.07	.25	.32
Steel outlet box plate with stud, 4" concrete	.007	Ea.	.08	.06	.14
Metal halide, lo bay, aluminum reflector, 250 W DX lamp	.007	Ea.	2.80	1.57	4.37
TOTAL			4.16	6.67	10.83

D5020 236	H.I.D. Fixture, Low Bay, 8'-10' (by Type)	COST PER S.F.		
		MAT.	INST.	TOTAL
0510	High intensity discharge fixture, 8'-10' above work plane, 50 FC			
0880	Type J, 7 fixtures per 900 S.F.	4.54	6.75	11.29
0920	13 fixtures per 1800 S.F.	4.16	6.65	10.81
0960	21 fixtures per 3000 S.F.	4.16	6.65	10.81
1000	28 fixtures per 4000 S.F.	4.16	6.65	10.81
1040	35 fixtures per 5000 S.F.	4.16	6.65	10.81
1120	62 fixtures per 10,000 S.F.	3.74	6.40	10.14
1160	99 fixtures per 16,000 S.F.	3.74	6.40	10.14
1200	199 fixtures per 32,000 S.F.	3.74	6.40	10.14
1240	Type K, 9 fixtures per 900 S.F.	4.58	5.75	10.33
1280	16 fixtures per 1800 S.F.	4.23	5.55	9.78
1320	26 fixtures per 3000 S.F.	4.21	5.50	9.71
1360	31 fixtures per 4000 S.F.	3.87	5.35	9.22
1400	39 fixtures per 5000 S.F.	3.87	5.35	9.22
1440	62 fixtures per 8000 S.F.	3.87	5.35	9.22
1480	78 fixtures per 10,000 S.F.	3.87	5.35	9.22
1520	124 fixtures per 16,000 S.F.	3.83	5.20	9.03
1560	248 fixtures per 32,000 S.F.	3.67	5.80	9.47

D5020 Lighting and Branch Wiring

LOW BAY FIXTURES

J. Metal halide 250 watt

K. High pressure sodium 150 watt

System Components	QUANTITY	UNIT	COST PER S.F. MAT.	COST PER S.F. INST.	COST PER S.F. TOTAL
SYSTEM D5020 238 0240					
HIGH INTENSITY DISCHARGE FIXTURE, 8'-10' ABOVE WORK PLANE					
1 WATT/S.F., TYPE J, 30 FC, 4 FIXTURES/1000 S.F.					
Steel intermediate conduit, (IMC) 1/2" diam	.280	L.F.	.52	2	2.52
Wire, 600V, type THWN-THHN, copper, solid, #10	.008	C.L.F.	.13	.57	.70
Steel outlet box 4" concrete	.004	Ea.	.04	.14	.18
Steel outlet box plate with stud, 4" concrete	.004	Ea.	.04	.04	.08
Metal halide, lo bay, aluminum reflector, 250 W DX lamp	.004	Ea.	1.60	.90	2.50
TOTAL			2.33	3.65	5.98

D5020 238	H.I.D. Fixture, Low Bay, 8'-10' (by Wattage)	COST PER S.F. MAT.	COST PER S.F. INST.	COST PER S.F. TOTAL
0190	High intensity discharge fixture, 8'-10' above work plane			
0240	1 watt/S.F., type J, 30 FC, 4 fixtures/1000 S.F.	2.33	3.65	5.98
0280	Type K, 29 FC, 5 fixtures/1000 S.F.	2.39	3.27	5.66
0400	2 watt/S.F., type J, 52 FC, 7 fixtures/1000 S.F.	4.16	6.70	10.86
0440	Type K, 63 FC, 11 fixtures/1000 S.F.	5.15	6.90	12.05
0560	3 watt/S.F., type J, 81 FC, 11 fixtures/1000 S.F.	6.40	10.10	16.50
0600	Type K, 92 FC, 16 fixtures/1000 S.F.	7.55	10.15	17.70
0720	4 watt/S.F., type J, 103 FC, 14 fixtures/1000 S.F.	8.30	13.30	21.60
0760	Type K, 127 FC, 22 fixtures/1000 S.F.	10.35	13.80	24.15
0880	5 watt/S.F., type J, 133 FC, 18 fixtures/1000 S.F.	10.60	16.80	27.40
0920	Type K, 155 FC, 27 fixtures/1000 S.F.	12.75	17.05	29.80

D5020 Lighting and Branch Wiring

LOW BAY FIXTURES

J. Metal halide 250 watt

K. High pressure sodium 150 watt

System Components	QUANTITY	UNIT	COST PER S.F.		
			MAT.	INST.	TOTAL
SYSTEM D5020 240 0880					
HIGH INTENSITY DISCHARGE FIXTURE, 16' ABOVE WORK PLANE, 50 FC					
TYPE J, 9 FIXTURES PER 900 S.F.					
Steel intermediate conduit, (IMC) 1/2" diam	.630	L.F.	1.17	4.50	5.67
Wire, 600V type, THWN-THHN, copper, solid, #10	.012	C.L.F.	.19	.86	1.05
Steel outlet box 4" concrete	.010	Ea.	.09	.36	.45
Steel outlet box plate with stud, 4" concrete	.010	Ea.	.11	.09	.20
Metal halide, lo bay, aluminum reflector, 250 W DX lamp	.010	Ea.	4	2.24	6.24
TOTAL			5.56	8.05	13.61

D5020 240	H.I.D. Fixture, Low Bay, 16' (by Type)	COST PER S.F.		
		MAT.	INST.	TOTAL
0510	High intensity discharge fixture, 16' above work plane, 50 FC			
0880	Type J, 9 fixtures per 900 S.F.	5.55	8.05	13.60
0920	14 fixtures per 1800 S.F.	4.75	7.60	12.35
0960	24 fixtures per 3000 S.F.	4.79	7.75	12.54
1000	32 fixtures per 4000 S.F.	4.79	7.75	12.54
1040	35 fixtures per 5000 S.F.	4.39	7.55	11.94
1080	56 fixtures per 8000 S.F.	4.39	7.55	11.94
1120	70 fixtures per 10,000 S.F.	4.38	7.55	11.93
1160	111 fixtures per 16,000 S.F.	4.39	7.55	11.94
1200	222 fixtures per 32,000 S.F.	4.39	7.55	11.94
1240	Type K, 11 fixtures per 900 S.F.	5.65	7.55	13.20
1280	20 fixtures per 1800 S.F.	5.20	6.95	12.15
1320	29 fixtures per 3000 S.F.	4.85	6.85	11.70
1360	39 fixtures per 4000 S.F.	4.83	6.75	11.58
1400	44 fixtures per 5000 S.F.	4.50	6.65	11.15
1440	62 fixtures per 8000 S.F.	4.50	6.65	11.15
1480	87 fixtures per 10,000 S.F.	4.50	6.65	11.15
1520	138 fixtures per 16,000 S.F.	4.50	6.65	11.15

D5020 Lighting and Branch Wiring

LOW BAY FIXTURES

J. Metal halide 250 watt

K. High pressure sodium 150 watt

System Components	QUANTITY	UNIT	COST PER S.F.		
			MAT.	INST.	TOTAL
SYSTEM D5020 242 0240					
HIGH INTENSITY DISCHARGE FIXTURE, 16' ABOVE WORK PLANE					
1 WATT/S.F., TYPE J, 28 FC, 4 FIXTURES/1000 S.F.					
Steel intermediate conduit, (IMC) 1/2" diam	.328	L.F.	.61	2.35	2.96
Wire, 600V, type THWN-THHN, copper, solid, #10	.010	C.L.F.	.16	.72	.88
Steel outlet box 4" concrete	.004	Ea.	.04	.14	.18
Steel outlet box plate with stud, 4" concrete	.004	Ea.	.04	.04	.08
Metal halide, lo bay, aluminum reflector, 250 W DX lamp	.004	Ea.	1.60	.90	2.50
TOTAL			2.45	4.15	6.60

D5020 242	H.I.D. Fixture, Low Bay, 16' (by Wattage)	COST PER S.F.		
		MAT.	INST.	TOTAL
0190	High intensity discharge fixture, mounted 16' above work plane			
0240	1 watt/S.F., type J, 28 FC, 4 fixt./1000 S.F.	2.45	4.15	6.60
0280	Type K, 27 FC, 5 fixt./1000 S.F.	2.73	4.63	7.36
0400	2 watt/S.F., type J, 48 FC, 7 fixt/1000 S.F.	4.49	8	12.49
0440	Type K, 58 FC, 11 fixt/1000 S.F.	5.80	9.45	15.25
0560	3 watt/S.F., type J, 75 FC, 11 fixt/1000 S.F.	6.95	12.15	19.10
0600	Type K, 85 FC, 16 fixt/1000 S.F.	8.55	14.10	22.65
0720	4 watt/S.F., type J, 95 FC, 14 fixt/1000 S.F.	8.95	16	24.95
0760	Type K, 117 FC, 22 fixt/1000 S.F.	11.65	18.95	30.60
0880	5 watt/S.F., type J, 122 FC, 18 fixt/1000 S.F.	11.45	20	31.45
0920	Type K, 143 FC, 27 fixt/1000 S.F.	14.40	23.50	37.90

D5020 Lighting and Branch Wiring

Daylight Dimming System

System Components			COST PER S.F.		
	QUANTITY	UNIT	MAT.	INST.	TOTAL
SYSTEM D5020 290 0800					
DAYLIGHT DIMMING CONTROL SYSTEM					
5 FIXTURES PER 1000 S.F.					
Tray cable, type TC, copper #16-4 conductor	.300	C.L.F.	.02	.03	.05
Wire, 600 volt, type THWN-THHN, copper, solid, #12	.320	C.L.F.		.02	.02
Conduit (EMT) , to 10' H, incl 2 termn,2 elb&11 bm clp per 100', 3/4"	10.000	L.F.	.01	.06	.07
Cabinet, hinged, steel, NEMA 1, 12"W x 12"H x 4"D	.200	Ea.	.01	.02	.03
Lighting control module	.200	Ea.	.07	.07	.14
Dimmable ballast three-lamp	5.000	Ea.	.72	.52	1.24
Daylight level sensor, wall mounted, on/off or dimming	.200	Ea.	.04	.02	.06
Automatic wall switches	.200	Ea.	.02	.01	.03
Remote power pack	.100	Ea.		.01	.01
TOTAL			.89	.76	1.65

D5020 290	Daylight Dimming System	COST PER S.F.		
		MAT.	INST.	TOTAL
0500	Daylight Dimming System (no fixtures or fixture power)			
0800	5 fixtures per 1000 S.F.	.89	.76	1.65
1000	10 fixtures per 1000 S.F.	1.76	1.43	3.19
2000	12 fixtures per 1000 S.F.	2.10	1.70	3.80
3000	15 fixtures per 1000 S.F.	2.63	2.11	4.74
4000	20 fixtures per 1000 S.F.	3.50	2.79	6.29
5000	25 fixtures per 1000 S.F.	4.38	3.47	7.85
6000	50 fixtures per 1000 S.F.	8.75	6.85	15.60

D5020 Lighting and Branch Wiring

Lighting AC Control System

System Components	QUANTITY	UNIT	COST PER S.F.		
			MAT.	INST.	TOTAL
SYSTEM D5020 295 0800					
LIGHTING ON/OFF CONTROL SYSTEM					
5 FIXTURES PER 1000 S.F.					
Tray cable, type TC, copper #16-4 conductor	.600	C.L.F.	.03	.06	.09
Wire, 600 volt, type THWN-THHN, copper, solid, #12	.320	C.L.F.		.02	.02
Conduit (EMT) , to 10' H, incl 2 termn,2 elb&11 bm clp per 100', 3/4"	10.000	L.F.	.01	.06	.07
Cabinet, hinged, steel, NEMA 1, 12"W x 12"H x 4"D	.400	Ea.	.02	.05	.07
Relays, 120 V or 277 V standard	.800	Ea.	.04	.05	.09
24 hour dial with reserve power	.400	Ea.	.26	.08	.34
Lighting control module	.200	Ea.	.07	.07	.14
Occupancy sensors, passive infrared ceiling mounted	.400	Ea.	.03	.04	.07
Automatic wall switches	.400	Ea.	.03	.01	.04
Remote power pack	.200	Ea.	.01	.01	.02
TOTAL			.50	.45	.95

D5020 295	Lighting On/Off Control System	COST PER S.F.		
		MAT.	INST.	TOTAL
0500	Includes occupancy and time switching (no fixtures or fixture power)			
0800	5 fixtures per 1000 SF	.50	.45	.95
1000	10 fixtures per 1000 S.F.	.57	.52	1.09
2000	12 fixtures per 1000 S.F.	.68	.62	1.30
3000	15 fixtures per 1000 S.F.	.85	.75	1.60
4000	20 fixtures per 1000 S.F.	1.14	.98	2.12
5000	25 fixtures per 1000 S.F.	1.41	1.20	2.61
6000	50 fixtures per 1000 S.F.	2.82	2.30	5.12

D5030 Communications and Security

Description: System below includes telephone fitting installed. Does not include cable.

When poke thru fittings and telepoles are used for power, they can also be used for telephones at a negligible additional cost.

System Components	QUANTITY	UNIT	COST PER S.F.		
			MAT.	INST.	TOTAL
SYSTEM D5030 310 0200					
TELEPHONE SYSTEMS, UNDERFLOOR DUCT, 5' ON CENTER, LOW DENSITY					
Underfloor duct 7-1/4" w/insert 2' O.C. 1-3/8" x 7-1/4" super duct	.190	L.F.	7.32	2.73	10.05
Vertical elbow for underfloor superduct, 7-1/4", included					
Underfloor duct conduit adapter, 2" x 1-1/4", included					
Underfloor duct junction box, single duct, 7-1/4" x 3 1/8"	.003	Ea.	1.82	.54	2.36
Underfloor junction box carpet pan	.003	Ea.	1.23	.03	1.26
Underfloor duct outlet, low tension	.004	Ea.	.46	.36	.82
TOTAL			10.83	3.66	14.49

D5030 310	Telephone Systems	COST PER S.F.		
		MAT.	INST.	TOTAL
0200	Telephone systems, underfloor duct, 5' on center, low density	10.85	3.66	14.51
0240	5' on center, high density	11.30	4.02	15.32
0280	7' on center, low density	8.65	3.04	11.69
0320	7' on center, high density	9.10	3.40	12.50
0400	Poke thru fittings, low density	1.25	1.29	2.54
0440	High density	2.51	2.57	5.08
0520	Telepoles, low density	1.54	.78	2.32
0560	High density	3.08	1.55	4.63
0640	Conduit system with floor boxes, low density	1.64	1.28	2.92
0680	High density	3.26	2.52	5.78
1020	Telephone wiring for offices & laboratories, 8 jacks/M.S.F.	.39	1.97	2.36

D5030 810	Security & Detection Systems	COST EACH		
		MAT.	INST.	TOTAL
0200	Security system, head end equipment	30,000	4,775	34,775
0240	Security system, door/window contact biased, box, conduit & cable	181	540	721
0280	Security system, door/window contact balanced, box, conduit & cable	124	540	664
0440	Security system, proximity card reader, box, conduit & cable	450	715	1,165
1600	Card control entrance system, to 6 zone, including hardware for 100 doors	43,100	33,200	76,300

D5030 Communications and Security

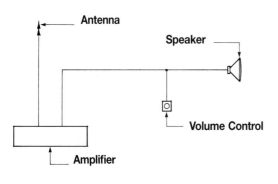

Sound System Includes AM–FM antenna, outlets, rigid conduit, and copper wire.
Fire Detection System Includes pull stations, signals, smoke and heat detectors, rigid conduit, and copper wire.
Intercom System Includes master and remote stations, rigid conduit, and copper wire.
Master Clock System Includes clocks, bells, rigid conduit, and copper wire.
Master TV Antenna Includes antenna, VHF–UHF reception and distribution, rigid conduit, and copper wire.

System Components	QUANTITY	UNIT	COST EACH		
			MAT.	INST.	TOTAL
SYSTEM D5030 910 0220					
SOUND SYSTEM, INCLUDES OUTLETS, BOXES, CONDUIT & WIRE					
Steel intermediate conduit, (IMC) 1/2" diam	1200.000	L.F.	2,232	8,580	10,812
Wire sound shielded w/drain, #22-2 conductor	15.500	C.L.F.	141.83	1,387.25	1,529.08
Sound system speakers ceiling or wall	12.000	Ea.	1,896	1,074	2,970
Sound system volume control	12.000	Ea.	750	1,074	1,824
Sound system amplifier, 250 W	1.000	Ea.	1,350	715	2,065
Sound system antenna, AM FM	1.000	Ea.	147	179	326
Sound system monitor panel	1.000	Ea.	525	179	704
Sound system cabinet	1.000	Ea.	1,125	715	1,840
Steel outlet box 4" square	12.000	Ea.	68.40	432	500.40
Steel outlet box 4" plaster rings	12.000	Ea.	20.16	134.40	154.56
TOTAL			8,255.39	14,469.65	22,725.04

D5030 910	Communication & Alarm Systems	COST EACH		
		MAT.	INST.	TOTAL
0200	Communication & alarm systems, includes outlets, boxes, conduit & wire			
0210	Sound system, 6 outlets	5,800	9,050	14,850
0220	12 outlets	8,250	14,500	22,750
0240	30 outlets	14,500	27,300	41,800
0280	100 outlets	43,300	91,500	134,800
0320	Fire detection systems, non-addressable, 12 detectors	3,350	7,500	10,850
0360	25 detectors	5,950	12,700	18,650
0400	50 detectors	11,700	25,200	36,900
0440	100 detectors	22,000	45,800	67,800
0450	Addressable type, 12 detectors	5,025	7,550	12,575
0452	25 detectors	9,150	12,800	21,950
0454	50 detectors	17,500	25,100	42,600
0456	100 detectors	34,100	46,100	80,200
0458	Fire alarm control panel, 8 zone, excluding wire and conduit	1,050	1,425	2,475
0459	12 zone	2,400	2,150	4,550
0460	Fire alarm command center, addressable without voice, excl. wire & conduit	5,050	1,250	6,300
0462	Addressable with voice	10,700	1,975	12,675
0480	Intercom systems, 6 stations	4,200	6,175	10,375
0520	12 stations	7,875	12,300	20,175
0560	25 stations	13,300	23,700	37,000
0600	50 stations	26,000	44,600	70,600
0640	100 stations	51,500	87,000	138,500
0680	Master clock systems, 6 rooms	4,500	10,100	14,600
0720	12 rooms	7,200	17,200	24,400
0760	20 rooms	10,100	24,400	34,500
0800	30 rooms	16,700	44,900	61,600

D50 Electrical

D5030 Communications and Security

D5030 910	Communication & Alarm Systems	COST EACH		
		MAT.	**INST.**	**TOTAL**
0840	50 rooms	27,200	75,500	102,700
0880	100 rooms	53,000	150,000	203,000
0920	Master TV antenna systems, 6 outlets	2,300	6,375	8,675
0960	12 outlets	4,250	11,900	16,150
1000	30 outlets	14,300	27,500	41,800
1040	100 outlets	52,500	89,500	142,000

D5030 920	Data Communication	COST PER M.S.F.		
		MAT.	**INST.**	**TOTAL**
0100	Data communication system, incl data/voice outlets, boxes, conduit & cable			
0102	Data and voice system, 2 data/voice outlets per 1000 S.F.	144	530	674
0104	4 data/voice outlets per 1000 S.F.	271	1,050	1,321
0106	6 data/voice outlets per 1000 S.F.	380	1,550	1,930
0110	8 data/voice outlets per 1000 S.F.	490	2,025	2,515

D5090 Other Electrical Systems

Muffler **Sleeve**

Transfer Switch

Engine

Charger

Battery

Description: System below tabulates the installed cost for generators by kW. Included in costs are battery, charger, muffler, and transfer switch.

No conduit, wire, or terminations included.

System Components	QUANTITY	UNIT	COST PER kW		
			MAT.	INST.	TOTAL
SYSTEM D5090 210 0200 GENERATOR SET, INCL. BATTERY, CHARGER, MUFFLER & TRANSFER SWITCH GAS/GASOLINE OPER., 3 PHASE, 4 WIRE, 277/480V, 7.5 kW					
Generator set, gas or gasoline operated, 3 ph 4 W, 277/480 V, 7.5 kW	.133	Ea.	1,276.67	309.60	1,586.27
TOTAL			1,276.67	309.60	1,586.27

D5090 210	Generators (by kW)	COST PER kW		
		MAT.	INST.	TOTAL
0190	Generator sets, include battery, charger, muffler & transfer switch			
0200	Gas/gasoline operated, 3 phase, 4 wire, 277/480 volt, 7.5 kW	1,275	310	1,585
0240	11.5 kW	1,175	238	1,413
0280	20 kW	800	152	952
0320	35 kW	545	100	645
0360	80 kW	390	122	512
0400	100 kW	340	118	458
0440	125 kW	560	111	671
0480	185 kW	500	84	584
0560	Diesel engine with fuel tank, 30 kW	410	117	527
0600	50 kW	445	92	537
0640	75 kW	345	73.50	418.50
0680	100 kW	305	62	367
0720	125 kW	254	53	307
0760	150 kW	315	49.50	364.50
0800	175 kW	285	44.50	329.50
0840	200 kW	250	40	290
0880	250 kW	214	33.50	247.50
0920	300 kW	197	29	226
0960	350 kW	200	27.50	227.50
1000	400 kW	200	25.50	225.50
1040	500 kW	217	21.50	238.50
1200	750 kW	217	13.65	230.65
1400	1000 kW	209	14.45	223.45

439

For customer support on your Electrical Costs with RSMeans data, call 800.448.8182.

PV power system, stand alone, AC and DC loads

(Diagram labels: PV modules, Combiner box 15amp pull out fuses, PV array disconnect 50 amp 2-pole, added lightning protection, Battery disconnect 175 amp, DC to AC inverter, To AC loads, AC load center, DC loads, Low voltage disconnect, C40 charge controller 40 amp, 48 V, PV Charge battery temp sensor, Batteries, DC load center, Grounding rod)

System Components	QUANTITY	UNIT	COST EACH		
			MAT.	INST.	TOTAL
SYSTEM D5090 420 0100					
PHOTOVOLTAIC POWER SYSTEM, STAND ALONE					
Alternative energy sources, photovoltaic module, 75 watt, 17 V	12.000	Ea.	5,220	1,074	6,294
PV rack system, roof, penetrating surface mount, on wood framing, 1 panel	12.000	Ea.	738	1,404	2,142
DC to AC inverter for, 24 V, 2,500 watt	1.000	Ea.	1,775	179	1,954
PV components, combiner box, 10 lug, NEMA 3R enclosure	1.000	Ea.	298	179	477
Alternative energy sources, PV components, fuse, 15 A for combiner box	10.000	Ea.	240	179	419
Battery charger controller w/temperature sensor	1.000	Ea.	545	179	724
Digital readout panel, displays hours, volts, amps, etc.	1.000	Ea.	236	179	415
Deep cycle solar battery, 6 V, 180 Ah (C/20)	4.000	Ea.	1,400	358	1,758
Battery interconnection, 15" AWG #2/0, sealed w/copper ring lugs	3.000	Ea.	60	135	195
Battery interconn, 24" AWG #2/0, sealed w/copper ring lugs	2.000	Ea.	52	90	142
Battery interconn, 60" AWG #2/0, sealed w/copper ring lugs	2.000	Ea.	123	90	213
Batt temp computer probe, RJ11 jack, 15' cord	1.000	Ea.	24.50	45	69.50
System disconnect, DC 175 amp circuit breaker	1.000	Ea.	252	89.50	341.50
Conduit box for inverter	1.000	Ea.	69	89.50	158.50
Low voltage disconnect	1.000	Ea.	68.50	89.50	158
Vented battery enclosure, wood	1.000	Ea.	305	315	620
Grounding, rod, copper clad, 8' long, 5/8" diameter	1.000	Ea.	24.50	130	154.50
Grounding, clamp, bronze, 5/8" dia	1.000	Ea.	5.45	22.50	27.95
Bare copper wire, stranded, #8	1.000	C.L.F.	47.50	65	112.50
Wire, 600 volt, type THW, copper, stranded, #12	3.600	C.L.F.	54.18	234	288.18
Wire, 600 volt, type THW, copper, stranded, #10	1.050	C.L.F.	24.68	75.08	99.76
Cond to 10' H,incl 2 termn,2 elb&11 bm CLP per 100',galv stl,1/2" dia	120.000	L.F.	303.60	954	1,257.60
Conduit,to 10' H,incl 2 termn,2 elb&11 bm clp per 100',(EMT), 1" dia	30.000	L.F.	62.70	187.50	250.20
Lightning surge suppressor	1.000	Ea.	82.50	22.50	105
General duty 240 volt, 2 pole, nonfusible, NEMA 3R, 60 amp	1.000	Ea.	131	310	441
Load centers, 3 wire, 120/240V, 100 amp main lugs, indoor, 8 circuits	2.000	Ea.	216	1,020	1,236
Circuit breaker, Plug-in, 120/240 volt, to 60 amp, 1 pole	5.000	Ea.	33.75	297.50	331.25
Fuses, dual element, time delay, 250 volt, 50 amp	2.000	Ea.	24.30	28.70	53
TOTAL			12,416.16	8,021.28	20,437.44

D50 Electrical

D5090 Other Electrical Systems

D5090 420	Photovoltaic Power System, Stand Alone	COST EACH		
		MAT.	INST.	TOTAL
0050	24 V capacity, 900 W (~102 SF)			
0100	Roof mounted, on wood framing	12,400	8,025	20,425
0150	with standoff	12,500	8,025	20,525
0200	Ground placement, ballast, fixed	17,300	8,550	25,850
0250	Adjustable	18,000	8,550	26,550
0300	Top of pole, passive tracking	17,300	8,550	25,850
0350	1.8 kW (~204 SF)			
0400	Roof mounted, on wood framing	20,100	12,700	32,800
0450	with standoff	20,300	12,700	33,000
0500	Ground placement, ballast, fixed	29,000	12,000	41,000
0550	Adjustable	30,300	12,000	42,300
0600	Top of pole, passive tracking	30,000	12,000	42,000
0650	3.2 kW (~408 SF)			
0700	Roof mounted, on wood framing	36,500	21,900	58,400
0750	with standoff	37,000	21,900	58,900
0800	Ground placement, ballast, fixed	53,500	18,700	72,200
0850	Adjustable	56,000	18,700	74,700
0900	Top of pole, passive tracking	55,500	18,600	74,100

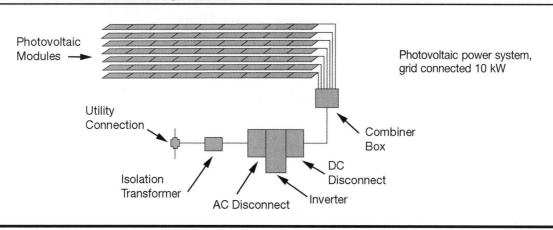

Photovoltaic Modules →

Photovoltaic power system, grid connected 10 kW

Combiner Box

Utility Connection

DC Disconnect

Isolation Transformer

AC Disconnect

Inverter

System Components	QUANTITY	UNIT	COST EACH		
			MAT.	INST.	TOTAL
SYSTEM D5090 430 0100					
PHOTOVOLTAIC POWER SYSTEM, GRID CONNECTED 10 kW					
Photovoltaic module, 150 watt, 33 V	60.000	Ea.	26,400	5,370	31,770
PV rack system, roof, non-penetrating ballast, 1 panel	60.000	Ea.	64,500	2,550	67,050
DC to AC inverter for, 24 V, 4,000 watt	3.000	Ea.	11,100	1,080	12,180
Combiner box 10 lug, NEMA 3R	1.000	Ea.	298	179	477
15 amp fuses	10.000	Ea.	240	179	419
Safety switch, 60 amp	1.000	Ea.	281	325	606
Safety switch, 100 amp	1.000	Ea.	480	400	880
Utility connection, 3 pole breaker	1.000	Ea.	300	116	416
Fuse, 60 A	3.000	Ea.	37.65	43.05	80.70
Fuse, 100 A	3.000	Ea.	126	53.70	179.70
10 kVA isolation tranformer	1.000	Ea.	1,950	895	2,845
EMT Conduit w/fittings & support	1.000	L.F.	376.20	1,125	1,501.20
RGS Conduit w/fittings & support	267.000	L.F.	2,069.25	2,950.35	5,019.60
Enclosure 24" x 24" x 10", NEMA 4	1.000	Ea.	5,175	1,425	6,600
Wire, 600 volt, type THWN-THHN, copper, stranded, #6	8.000	C.L.F.	424	880	1,304
Wire, 600 volt, copper type XLPE-USE(RHW), stranded, #12	18.000	C.L.F.	360	1,170	1,530
Grounding, bare copper wire stranded, 4/0	2.000	C.L.F.	970	504	1,474
Grounding, exothermic weld, 4/0 wire to building steel	4.000	Ea.	41.20	408	449.20
Grounding, brazed connections, #6 wire	2.000	Ea.	35.50	119	154.50
Insulated ground wire, copper, #6	.400	C.L.F.	21.20	44	65.20
TOTAL			115,185	19,816.10	135,001.10

D5090 430	Photovoltaic Power System, Grid Connected	COST EACH		
		MAT.	INST.	TOTAL
0050	10 kW			
0100	Roof mounted, non-penetrating ballast	115,000	19,800	134,800
0200	Penetrating surface mount, on steel framing	54,500	32,400	86,900
0300	on wood framing	54,500	24,300	78,800
0400	with standoff	55,000	24,300	79,300
0500	Ground placement, ballast, fixed	75,000	19,700	94,700
0600	adjustable	78,500	19,700	98,200
0700	Top of pole, passive tracking	77,000	20,100	97,100
1050	20 kW			
1100	Roof mounted, non-penetrating ballast	223,000	33,200	256,200
1200	Penetrating surface mount, on steel framing	101,500	58,500	160,000
1300	on wood framing	101,500	42,100	143,600
1400	with standoff	102,500	42,100	144,600
1500	Ground placement, ballast, fixed	142,500	31,900	174,400
1600	adjustable	149,000	31,900	180,900
1700	Top of pole, passive tracking	146,500	32,800	179,300

D5090 Other Electrical Systems

System Components	QUANTITY	UNIT	COST EACH		
			MAT.	INST.	TOTAL
SYSTEM D5090 480 1100					
ELECTRICAL, SINGLE PHASE, 1 METER					
#18 twisted shielded pair in 1/2" EMT conduit	.050	C.L.F.	7.85	11.95	19.80
Wire, 600 volt, type THW, copper, solid, #12	.150	C.L.F.	1.60	9.75	11.35
Conduit (EMT) , to 10' H, incl 2 termn,2 elb&11 bm clp per 100', 3/4"	5.000	L.F.	6.50	27.50	34
Outlet boxes, pressed steel, handy box	1.000	Ea.	2.65	26.50	29.15
Outlet boxes, pressed steel, handy box, covers, device	1.000	Ea.	1.11	11.20	12.31
Wiring devices, receptacle, duplex, 120 volt, ground, 20 amp	1.000	Ea.	10.50	26.50	37
Single phase, 277 volt, 200 amp	1.000	Ea.	440	81.50	521.50
Data recorder, 8 meters	1.000	Ea.	1,600	65.50	1,665.50
Software package, per meter, premium	1.000	Ea.	745		745
TOTAL			2,815.21	260.40	3,075.61

D5090 480	Energy Monitoring Systems	COST EACH		
		MAT.	INST.	TOTAL
1000	Electrical			
1100	Single phase, 1 meter	2,825	260	3,085
1110	4 meters	7,175	1,975	9,150
1120	8 meters	14,400	4,050	18,450
1200	Three phase, 1 meter	3,225	330	3,555
1210	5 meters	10,900	3,100	14,000
1220	10 meters	22,100	6,325	28,425
1230	25 meters	52,500	12,600	65,100
2000	Mechanical			
2100	BTU, 1 meter	3,550	935	4,485
2110	w/1 duct sensor	3,750	1,175	4,925
2120	& 1 space sensor	4,250	1,325	5,575
2130	& 5 space sensors	6,275	1,950	8,225
2140	& 10 space sensors	11,200	2,775	13,975
2200	BTU, 3 meters	7,425	2,675	10,100
2210	w/3 duct sensors	8,050	3,375	11,425
2220	& 3 space sensors	9,550	3,850	13,400
2230	& 15 space sensors	15,600	5,750	21,350
2240	& 30 space sensors	25,600	8,150	33,750
9000	Front end display	515	140	655
9100	Computer workstation	1,300	1,850	3,150

D5090 Other Electrical Systems

Low Density and Medium Density Baseboard Radiation

The costs shown in Table below are based on the following system considerations:

1. The heat loss per square foot is based on approximately 34 BTU/hr. per S.F. of floor or 10 watts per S.F. of floor.
2. Baseboard radiation is based on the low watt density type rated 187 watts per L.F. and the medium density type rated 250 watts per L.F.
3. Thermostat is not included.
4. Wiring costs include branch circuit wiring.

System Components	QUANTITY	UNIT	COST PER S.F.		
			MAT.	INST.	TOTAL
SYSTEM D5090 510 1000					
ELECTRIC BASEBOARD RADIATION, LOW DENSITY, 900 S.F., 31 MBH, 9 kW					
Electric baseboard radiator, 5' long 935 watt	.011	Ea.	.56	1.39	1.95
Steel intermediate conduit, (IMC) 1/2" diam	.170	L.F.	.32	1.22	1.54
Wire 600 volt, type THW, copper, solid, #12	.005	C.L.F.	.05	.33	.38
TOTAL			.93	2.94	3.87

D5090 510	Electric Baseboard Radiation (Low Density)	COST PER S.F.		
		MAT.	INST.	TOTAL
1000	Electric baseboard radiation, low density, 900 S.F., 31 MBH, 9 kW	.93	2.94	3.87
1200	1500 S.F., 51 MBH, 15 kW	.91	2.86	3.77
1400	2100 S.F., 72 MBH, 21 kW	.82	2.56	3.38
1600	3000 S.F., 102 MBH, 30 kW	.74	2.33	3.07
2000	Medium density, 900 S.F., 31 MBH, 9 kW	.83	2.68	3.51
2200	1500 S.F., 51 MBH, 15 kW	.81	2.60	3.41
2400	2100 S.F., 72 MBH, 21 kW	.77	2.43	3.20
2600	3000 S.F., 102 MBH, 30 kW	.69	2.21	2.90

D5090 Other Electrical Systems

Commercial Duty Baseboard Radiation
The costs shown in Table below are based on the following system considerations:

1. The heat loss per square foot is based on approximately 41 BTU/hr. per S.F. of floor or 12 watts per S.F.
2. The baseboard radiation is of the commercial duty type rated 250 watts per L.F. served by 277 volt, single phase power.
3. Thermostat is not included.
4. Wiring costs include branch circuit wiring.

System Components	QUANTITY	UNIT	COST PER S.F.		
			MAT.	INST.	TOTAL
SYSTEM D5090 520 1000					
ELECTRIC BASEBOARD RADIATION, MEDIUM DENSITY, 1230 S.F., 51 MBH, 15 kW					
Electric baseboard radiator, 5' long	.013	Ea.	.66	1.64	2.30
Steel intermediate conduit, (IMC) 1/2" diam	.154	L.F.	.29	1.10	1.39
Wire 600 volt, type THW, copper, solid, #12	.004	C.L.F.	.04	.26	.30
TOTAL			.99	3	3.99

D5090 520	Electric Baseboard Radiation (Medium Density)	COST PER S.F.		
		MAT.	INST.	TOTAL
1000	Electric baseboard radiation, medium density, 1230 SF, 51 MBH, 15 kW	.99	3	3.99
1200	2500 S.F. floor area, 106 MBH, 31 kW	.92	2.81	3.73
1400	3700 S.F. floor area, 157 MBH, 46 kW	.90	2.73	3.63
1600	4800 S.F. floor area, 201 MBH, 59 kW	.83	2.53	3.36
1800	11,300 S.F. floor area, 464 MBH, 136 kW	.78	2.33	3.11
2000	30,000 S.F. floor area, 1229 MBH, 360 kW	.77	2.27	3.04

G1030 Site Earthwork

Trenching Systems are shown on a cost per linear foot basis. The systems include: excavation; backfill and removal of spoil; and compaction for various depths and trench bottom widths. The backfill has been reduced to accommodate a pipe of suitable diameter and bedding.

The slope for trench sides varies from none to 1:1.

The Expanded System Listing shows Trenching Systems that range from 2' to 12' in width. Depths range from 2' to 25'.

System Components				COST PER L.F.		
		QUANTITY	UNIT	EQUIP.	LABOR	TOTAL
SYSTEM G1030 805 1310						
TRENCHING, COMMON EARTH, NO SLOPE, 2' WIDE, 2' DP, 3/8 C.Y. BUCKET						
	Excavation, trench, hyd. backhoe, track mtd., 3/8 C.Y. bucket	.148	B.C.Y.	.35	1.14	1.49
	Backfill and load spoil, from stockpile	.153	L.C.Y.	.13	.35	.48
	Compaction by vibrating plate, 6" lifts, 4 passes	.118	E.C.Y.	.03	.42	.45
	Remove excess spoil, 8 C.Y. dump truck, 2 mile roundtrip	.040	L.C.Y.	.12	.19	.31
	TOTAL			.63	2.10	2.73

G1030 805	Trenching Common Earth	COST PER L.F.		
		EQUIP.	LABOR	TOTAL
1310	Trenching, common earth, no slope, 2' wide, 2' deep, 3/8 C.Y. bucket	.63	2.10	2.73
1320	3' deep, 3/8 C.Y. bucket	.90	3.15	4.05
1330	4' deep, 3/8 C.Y. bucket	1.16	4.20	5.36
1340	6' deep, 3/8 C.Y. bucket	1.61	5.45	7.06
1350	8' deep, 1/2 C.Y. bucket	2.14	7.25	9.39
1360	10' deep, 1 C.Y. bucket	3.51	8.60	12.11
1400	4' wide, 2' deep, 3/8 C.Y. bucket	1.40	4.16	5.56
1410	3' deep, 3/8 C.Y. bucket	1.94	6.25	8.19
1420	4' deep, 1/2 C.Y. bucket	2.40	7	9.40
1430	6' deep, 1/2 C.Y. bucket	3.97	11.25	15.22
1440	8' deep, 1/2 C.Y. bucket	6.65	14.55	21.20
1450	10' deep, 1 C.Y. bucket	8.10	18.05	26.15
1460	12' deep, 1 C.Y. bucket	10.45	23	33.45
1470	15' deep, 1-1/2 C.Y. bucket	8.80	20.50	29.30
1480	18' deep, 2-1/2 C.Y. bucket	12.40	29	41.40
1520	6' wide, 6' deep, 5/8 C.Y. bucket w/trench box	8.45	16.85	25.30
1530	8' deep, 3/4 C.Y. bucket	11.25	22	33.25
1540	10' deep, 1 C.Y. bucket	10.85	23	33.85
1550	12' deep, 1-1/2 C.Y. bucket	11.50	24.50	36
1560	16' deep, 2-1/2 C.Y. bucket	15.85	30.50	46.35
1570	20' deep, 3-1/2 C.Y. bucket	21	36.50	57.50
1580	24' deep, 3-1/2 C.Y. bucket	25	44	69
1640	8' wide, 12' deep, 1-1/2 C.Y. bucket w/trench box	15.95	31	46.95
1650	15' deep, 1-1/2 C.Y. bucket	21	41	62
1660	18' deep, 2-1/2 C.Y. bucket	22.50	41	63.50
1680	24' deep, 3-1/2 C.Y. bucket	33.50	57	90.50
1730	10' wide, 20' deep, 3-1/2 C.Y. bucket w/trench box	27.50	54	81.50
1740	24' deep, 3-1/2 C.Y. bucket	40	65	105
1780	12' wide, 20' deep, 3-1/2 C.Y. bucket w/trench box	42.50	68.50	111
1790	25' deep, bucket	52.50	87.50	140
1800	1/2 to 1 slope, 2' wide, 2' deep, 3/8 C.Y. bucket	.90	3.15	4.05
1810	3' deep, 3/8 C.Y. bucket	1.50	5.50	7

G1030 Site Earthwork

G1030 805	Trenching Common Earth	COST PER L.F.		
		EQUIP.	LABOR	TOTAL
1820	4' deep, 3/8 C.Y. bucket	2.25	8.40	10.65
1840	6' deep, 3/8 C.Y. bucket	3.90	13.65	17.55
1860	8' deep, 1/2 C.Y. bucket	6.20	22	28.20
1880	10' deep, 1 C.Y. bucket	12.10	30.50	42.60
2300	4' wide, 2' deep, 3/8 C.Y. bucket	1.67	5.20	6.87
2310	3' deep, 3/8 C.Y. bucket	2.56	8.65	11.21
2320	4' deep, 1/2 C.Y. bucket	3.43	10.65	14.08
2340	6' deep, 1/2 C.Y. bucket	6.75	20	26.75
2360	8' deep, 1/2 C.Y. bucket	12.95	29.50	42.45
2380	10' deep, 1 C.Y. bucket	18	41.50	59.50
2400	12' deep, 1 C.Y. bucket	23	56	79
2430	15' deep, 1-1/2 C.Y. bucket	25	60	85
2460	18' deep, 2-1/2 C.Y. bucket	43.50	93.50	137
2840	6' wide, 6' deep, 5/8 C.Y. bucket w/trench box	12.55	25	37.55
2860	8' deep, 3/4 C.Y. bucket	18.30	37.50	55.80
2880	10' deep, 1 C.Y. bucket	17.35	38	55.35
2900	12' deep, 1-1/2 C.Y. bucket	22	50	72
2940	16' deep, 2-1/2 C.Y. bucket	36.50	73	109.50
2980	20' deep, 3-1/2 C.Y. bucket	52.50	98	150.50
3020	24' deep, 3-1/2 C.Y. bucket	74	134	208
3100	8' wide, 12' deep, 1-1/2 C.Y. bucket w/trench box	27	57	84
3120	15' deep, 1-1/2 C.Y. bucket	39.50	82	121.50
3140	18' deep, 2-1/2 C.Y. bucket	50	97.50	147.50
3180	24' deep, 3-1/2 C.Y. bucket	83	147	230
3270	10' wide, 20' deep, 3-1/2 C.Y. bucket w/trench box	53.50	113	166.50
3280	24' deep, 3-1/2 C.Y. bucket	91.50	161	252.50
3370	12' wide, 20' deep, 3-1/2 C.Y. bucket w/trench box	76.50	131	207.50
3380	25' deep, 3-1/2 C.Y. bucket	107	188	295
3500	1 to 1 slope, 2' wide, 2' deep, 3/8 C.Y. bucket	1.16	4.21	5.37
3520	3' deep, 3/8 C.Y. bucket	3.24	9.60	12.84
3540	4' deep, 3/8 C.Y. bucket	3.34	12.65	15.99
3560	6' deep, 1/2 C.Y. bucket	3.89	13.65	17.54
3580	8' deep, 1/2 C.Y. bucket	7.75	27.50	35.25
3600	10' deep, 1 C.Y. bucket	20.50	52	72.50
3800	4' wide, 2' deep, 3/8 C.Y. bucket	1.94	6.25	8.19
3820	3' deep, 3/8 C.Y. bucket	3.17	11	14.17
3840	4' deep, 1/2 C.Y. bucket	4.44	14.30	18.74
3860	6' deep, 1/2 C.Y. bucket	9.50	28.50	38
3880	8' deep, 1/2 C.Y. bucket	19.30	44	63.30
3900	10' deep, 1 C.Y. bucket	28	64.50	92.50
3920	12' deep, 1 C.Y. bucket	41	93.50	134.50
3940	15' deep, 1-1/2 C.Y. bucket	41	99.50	140.50
3960	18' deep, 2-1/2 C.Y. bucket	60	129	189
4030	6' wide, 6' deep, 5/8 C.Y. bucket w/trench box	16.55	33.50	50.05
4040	8' deep, 3/4 C.Y. bucket	24	46	70
4050	10' deep, 1 C.Y. bucket	25	56	81
4060	12' deep, 1-1/2 C.Y. bucket	33.50	76	109.50
4070	16' deep, 2-1/2 C.Y. bucket	57	115	172
4080	20' deep, 3-1/2 C.Y. bucket	84.50	160	244.50
4090	24' deep, 3-1/2 C.Y. bucket	123	225	348
4500	8' wide, 12' deep, 1-1/2 C.Y. bucket w/trench box	38	82.50	120.50
4550	15' deep, 1-1/2 C.Y. bucket	57.50	124	181.50
4600	18' deep, 2-1/2 C.Y. bucket	76	151	227
4650	24' deep, 3-1/2 C.Y. bucket	132	238	370
4800	10' wide, 20' deep, 3-1/2 C.Y. bucket w/trench box	79.50	172	251.50
4850	24' deep, 3-1/2 C.Y. bucket	141	251	392
4950	12' wide, 20' deep, 3-1/2 C.Y. bucket w/trench box	111	194	305
4980	25' deep, 3-1/2 C.Y. bucket	160	284	444

For customer support on your Electrical Costs with RSMeans data, call 800.448.8182.

Trenching Systems are shown on a cost per linear foot basis. The systems include: excavation; backfill and removal of spoil; and compaction for various depths and trench bottom widths. The backfill has been reduced to accommodate a pipe of suitable diameter and bedding.

The slope for trench sides varies from none to 1:1.

The Expanded System Listing shows Trenching Systems that range from 2' to 12' in width. Depths range from 2' to 25'.

System Components	QUANTITY	UNIT	COST PER L.F.		
			EQUIP.	LABOR	TOTAL
SYSTEM G1030 806 1310					
TRENCHING, LOAM & SANDY CLAY, NO SLOPE, 2' WIDE, 2' DP, 3/8 C.Y. BUCKET					
Excavation, trench, hyd. backhoe, track mtd., 3/8 C.Y. bucket	.148	B.C.Y.	.32	1.06	1.38
Backfill and load spoil, from stockpile	.165	L.C.Y.	.14	.38	.52
Compaction by vibrating plate 18" wide, 6" lifts, 4 passes	.118	E.C.Y.	.03	.42	.45
Remove excess spoil, 8 C.Y. dump truck, 2 mile roundtrip	.042	L.C.Y.	.13	.20	.33
TOTAL			.62	2.06	2.68

G1030 806	Trenching Loam & Sandy Clay	COST PER L.F.		
		EQUIP.	LABOR	TOTAL
1310	Trenching, loam & sandy clay, no slope, 2' wide, 2' deep, 3/8 C.Y. bucket	.62	2.06	2.68
1320	3' deep, 3/8 C.Y. bucket	.97	3.37	4.34
1330	4' deep, 3/8 C.Y. bucket	1.14	4.11	5.25
1340	6' deep, 3/8 C.Y. bucket	1.75	4.90	6.65
1350	8' deep, 1/2 C.Y. bucket	2.33	6.45	8.80
1360	10' deep, 1 C.Y. bucket	2.53	6.90	9.45
1400	4' wide, 2' deep, 3/8 C.Y. bucket	1.39	4.09	5.50
1410	3' deep, 3/8 C.Y. bucket	1.92	6.15	8.05
1420	4' deep, 1/2 C.Y. bucket	2.39	6.90	9.30
1430	6' deep, 1/2 C.Y. bucket	4.26	10.15	14.40
1440	8' deep, 1/2 C.Y. bucket	6.50	14.35	21
1450	10' deep, 1 C.Y. bucket	6.15	14.70	21
1460	12' deep, 1 C.Y. bucket	7.70	18.30	26
1470	15' deep, 1-1/2 C.Y. bucket	9.15	21.50	30.50
1480	18' deep, 2-1/2 C.Y. bucket	10.85	23.50	34.50
1520	6' wide, 6' deep, 5/8 C.Y. bucket w/trench box	8.15	16.55	24.50
1530	8' deep, 3/4 C.Y. bucket	10.85	21.50	32.50
1540	10' deep, 1 C.Y. bucket	10	22	32
1550	12' deep, 1-1/2 C.Y. bucket	11.25	24.50	36
1560	16' deep, 2-1/2 C.Y. bucket	15.50	31	46.50
1570	20' deep, 3-1/2 C.Y. bucket	19.30	36.50	56
1580	24' deep, 3-1/2 C.Y. bucket	24	44.50	68.50
1640	8' wide, 12' deep, 1-1/4 C.Y. bucket w/trench box	15.75	31	47
1650	15' deep, 1-1/2 C.Y. bucket	19.30	39.50	59
1660	18' deep, 2-1/2 C.Y. bucket	23.50	44.50	68
1680	24' deep, 3-1/2 C.Y. bucket	32.50	57.50	90
1730	10' wide, 20' deep, 3-1/2 C.Y. bucket w/trench box	33	58	91
1740	24' deep, 3-1/2 C.Y. bucket	41	71.50	113
1780	12' wide, 20' deep, 3-1/2 C.Y. bucket w/trench box	39.50	69	109
1790	25' deep, 3-1/2 C.Y. bucket	51.50	88.50	140
1800	1/2:1 slope, 2' wide, 2' deep, 3/8 C.Y. bucket	.88	3.08	3.96
1810	3' deep, 3/8 C.Y. bucket	1.47	5.40	6.85
1820	4' deep, 3/8 C.Y. bucket	2.21	8.25	10.45
1840	6' deep, 3/8 C.Y. bucket	4.23	12.25	16.50

G1030 Site Earthwork

G1030 806	Trenching Loam & Sandy Clay	COST PER L.F.		
		EQUIP.	LABOR	TOTAL
1860	8' deep, 1/2 C.Y. bucket	6.75	19.55	26.50
1880	10' deep, 1 C.Y. bucket	8.65	24.50	33
2300	4' wide, 2' deep, 3/8 C.Y. bucket	1.65	5.10	6.75
2310	3' deep, 3/8 C.Y. bucket	2.52	8.45	10.95
2320	4' deep, 1/2 C.Y. bucket	3.39	10.45	13.85
2340	6' deep, 1/2 C.Y. bucket	7.20	18	25
2360	8' deep, 1/2 C.Y. bucket	12.60	29	41.50
2380	10' deep, 1 C.Y. bucket	13.55	33.50	47
2400	12' deep, 1 C.Y. bucket	22.50	56	78.50
2430	15' deep, 1-1/2 C.Y. bucket	26	62	88
2460	18' deep, 2-1/2 C.Y. bucket	43	95	138
2840	6' wide, 6' deep, 5/8 C.Y. bucket w/trench box	12	25	37
2860	8' deep, 3/4 C.Y. bucket	17.60	36.50	54
2880	10' deep, 1 C.Y. bucket	18	41	59
2900	12' deep, 1-1/2 C.Y. bucket	22	50.50	72.50
2940	16' deep, 2-1/2 C.Y. bucket	35.50	74	110
2980	20' deep, 3-1/2 C.Y. bucket	51	99.50	151
3020	24' deep, 3-1/2 C.Y. bucket	72	136	208
3100	8' wide, 12' deep, 1-1/2 C.Y. bucket w/trench box	26.50	57	83.50
3120	15' deep, 1-1/2 C.Y. bucket	36	80	116
3140	18' deep, 2-1/2 C.Y. bucket	49	99	148
3180	24' deep, 3-1/2 C.Y. bucket	80.50	150	231
3270	10' wide, 20' deep, 3-1/2 C.Y. bucket w/trench box	64.50	121	186
3280	24' deep, 3-1/2 C.Y. bucket	89	163	252
3320	12' wide, 20' deep, 3-1/2 C.Y. bucket w/trench box	71.50	132	204
3380	25' deep, 3-1/2 C.Y. bucket w/trench box	97	177	274
3500	1:1 slope, 2' wide, 2' deep, 3/8 C.Y. bucket	1.14	4.12	5.25
3520	3' deep, 3/8 C.Y. bucket	2.07	7.70	9.75
3540	4' deep, 3/8 C.Y. bucket	3.26	12.35	15.60
3560	6' deep, 1/2 C.Y. bucket	4.23	12.25	16.50
3580	8' deep, 1/2 C.Y. bucket	11.15	32.50	43.50
3600	10' deep, 1 C.Y. bucket	14.75	42	57
3800	4' wide, 2' deep, 3/8 C.Y. bucket	1.92	6.15	8.05
3820	3' deep, 1/2 C.Y. bucket	3.11	10.75	13.85
3840	4' deep, 1/2 C.Y. bucket	4.38	14.05	18.45
3860	6' deep, 1/2 C.Y. bucket	10.20	26	36
3880	8' deep, 1/2 C.Y. bucket	18.75	43.50	62.50
3900	10' deep, 1 C.Y. bucket	21	52.50	73.50
3920	12' deep, 1 C.Y. bucket	30	74.50	105
3940	15' deep, 1-1/2 C.Y. bucket	42.50	103	146
3960	18' deep, 2-1/2 C.Y. bucket	59	131	190
4030	6' wide, 6' deep, 5/8 C.Y. bucket w/trench box	15.80	33.50	49.50
4040	8' deep, 3/4 C.Y. bucket	24.50	51.50	76
4050	10' deep, 1 C.Y. bucket	26	60.50	86.50
4060	12' deep, 1-1/2 C.Y. bucket	33	76	109
4070	16' deep, 2-1/2 C.Y. bucket	55.50	117	173
4080	20' deep, 3-1/2 C.Y. bucket	82.50	163	246
4090	24' deep, 3-1/2 C.Y. bucket	120	228	350
4500	8' wide, 12' deep, 1-1/4 C.Y. bucket w/trench box	37.50	82.50	120
4550	15' deep, 1-1/2 C.Y. bucket	53	120	173
4600	18' deep, 2-1/2 C.Y. bucket	74.50	153	228
4650	24' deep, 3-1/2 C.Y. bucket	128	241	370
4800	10' wide, 20' deep, 3-1/2 C.Y. bucket w/trench box	96.50	184	281
4850	24' deep, 3-1/2 C.Y. bucket	137	255	390
4950	12' wide, 20' deep, 3-1/2 C.Y. bucket w/trench box	103	195	298
4980	25' deep, 3-1/2 C.Y. bucket	155	288	445

For customer support on your Electrical Costs with RSMeans data, call 800.448.8182.

451

G1030 Site Earthwork

Trenching Systems are shown on a cost per linear foot basis. The systems include: excavation; backfill and removal of spoil; and compaction for various depths and trench bottom widths. The backfill has been reduced to accommodate a pipe of suitable diameter and bedding.

The slope for trench sides varies from none to 1:1.

The Expanded System Listing shows Trenching Systems that range from 2' to 12' in width. Depths range from 2' to 25'.

System Components	QUANTITY	UNIT	COST PER L.F. EQUIP.	COST PER L.F. LABOR	COST PER L.F. TOTAL
SYSTEM G1030 807 1310					
TRENCHING, SAND & GRAVEL, NO SLOPE, 2' WIDE, 2' DEEP, 3/8 C.Y. BUCKET					
Excavation, trench, hyd. backhoe, track mtd., 3/8 C.Y. bucket	.148	B.C.Y.	.32	1.04	1.36
Backfill and load spoil, from stockpile	.140	L.C.Y.	.11	.32	.43
Compaction by vibrating plate 18" wide, 6" lifts, 4 passes	.118	E.C.Y.	.03	.42	.45
Remove excess spoil, 8 C.Y. dump truck, 2 mile roundtrip	.035	L.C.Y.	.11	.17	.28
TOTAL			.57	1.95	2.52

G1030 807	Trenching Sand & Gravel	COST PER L.F. EQUIP.	COST PER L.F. LABOR	COST PER L.F. TOTAL
1310	Trenching, sand & gravel, no slope, 2' wide, 2' deep, 3/8 C.Y. bucket	.57	1.95	2.52
1320	3' deep, 3/8 C.Y. bucket	.91	3.23	4.14
1330	4' deep, 3/8 C.Y. bucket	1.06	3.89	4.95
1340	6' deep, 3/8 C.Y. bucket	1.65	4.67	6.30
1350	8' deep, 1/2 C.Y. bucket	2.19	6.20	8.40
1360	10' deep, 1 C.Y. bucket	2.35	6.50	8.85
1400	4' wide, 2' deep, 3/8 C.Y. bucket	1.27	3.84	5.10
1410	3' deep, 3/8 C.Y. bucket	1.78	5.80	7.60
1420	4' deep, 1/2 C.Y. bucket	2.20	6.50	8.70
1430	6' deep, 1/2 C.Y. bucket	4.05	9.70	13.75
1440	8' deep, 1/2 C.Y. bucket	6.10	13.65	19.75
1450	10' deep, 1 C.Y. bucket	5.75	13.90	19.65
1460	12' deep, 1 C.Y. bucket	7.25	17.30	24.50
1470	15' deep, 1-1/2 C.Y. bucket	8.60	20	28.50
1480	18' deep, 2-1/2 C.Y. bucket	10.25	22	32.50
1520	6' wide, 6' deep, 5/8 C.Y. bucket w/trench box	7.70	15.65	23.50
1530	8' deep, 3/4 C.Y. bucket	10.30	20.50	31
1540	10' deep, 1 C.Y. bucket	9.45	20.50	30
1550	12' deep, 1-1/2 C.Y. bucket	10.60	23	33.50
1560	16' deep, 2 C.Y. bucket	14.80	29	44
1570	20' deep, 3-1/2 C.Y. bucket	18.25	34.50	53
1580	24' deep, 3-1/2 C.Y. bucket	23	42	65
1640	8' wide, 12' deep, 1-1/2 C.Y. bucket w/trench box	14.80	29.50	44.50
1650	15' deep, 1-1/2 C.Y. bucket	18	37.50	55.50
1660	18' deep, 2-1/2 C.Y. bucket	22	42	64
1680	24' deep, 3-1/2 C.Y. bucket	31	54	85
1730	10' wide, 20' deep, 3-1/2 C.Y. bucket w/trench box	31	54.50	85.50
1740	24' deep, 3-1/2 C.Y. bucket	38.50	67	106
1780	12' wide, 20' deep, 3-1/2 C.Y. bucket w/trench box	37.50	64.50	102
1790	25' deep, 3-1/2 C.Y. bucket	48.50	83	132
1800	1/2:1 slope, 2' wide, 2' deep, 3/8 C.Y. bucket	.82	2.92	3.74
1810	3' deep, 3/8 C.Y. bucket	1.38	5.10	6.50
1820	4' deep, 3/8 C.Y. bucket	2.05	7.85	9.90
1840	6' deep, 3/8 C.Y. bucket	4.03	11.70	15.75

G10 Site Preparation

G1030 Site Earthwork

G1030 807	Trenching Sand & Gravel	COST PER L.F.		
		EQUIP.	LABOR	TOTAL
1860	8' deep, 1/2 C.Y. bucket	6.40	18.65	25
1880	10' deep, 1 C.Y. bucket	8.05	23	31
2300	4' wide, 2' deep, 3/8 C.Y. bucket	1.53	4.82	6.35
2310	3' deep, 3/8 C.Y. bucket	2.33	8	10.35
2320	4' deep, 1/2 C.Y. bucket	3.14	9.90	13.05
2340	6' deep, 1/2 C.Y. bucket	6.90	17.25	24
2360	8' deep, 1/2 C.Y. bucket	11.95	27.50	39.50
2380	10' deep, 1 C.Y. bucket	12.80	32	45
2400	12' deep, 1 C.Y. bucket	21.50	53	74.50
2430	15' deep, 1-1/2 C.Y. bucket	24.50	58.50	83
2460	18' deep, 2-1/2 C.Y. bucket	40.50	89.50	130
2840	6' wide, 6' deep, 5/8 C.Y. bucket w/trench box	11.35	24	35.50
2860	8' deep, 3/4 C.Y. bucket	16.75	35	52
2880	10' deep, 1 C.Y. bucket	17.05	39	56
2900	12' deep, 1-1/2 C.Y. bucket	21	47.50	68.50
2940	16' deep, 2 C.Y. bucket	34	69.50	104
2980	20' deep, 3-1/2 C.Y. bucket	48.50	93.50	142
3020	24' deep, 3-1/2 C.Y. bucket	68.50	128	197
3100	8' wide, 12' deep, 1-1/4 C.Y. bucket w/trench box	25	54	79
3120	15' deep, 1-1/2 C.Y. bucket	34	75.50	110
3140	18' deep, 2-1/2 C.Y. bucket	46.50	93	140
3180	24' deep, 3-1/2 C.Y. bucket	76.50	140	217
3270	10' wide, 20' deep, 3-1/2 C.Y. bucket w/trench box	61	113	174
3280	24' deep, 3-1/2 C.Y. bucket	84	153	237
3370	12' wide, 20' deep, 3-1/2 C.Y. bucket w/trench box	68	126	194
3380	25' deep, 3-1/2 C.Y. bucket	98	177	275
3500	1:1 slope, 2' wide, 2' deep, 3/8 C.Y. bucket	1.76	4.89	6.65
3520	3' deep, 3/8 C.Y. bucket	1.93	7.35	9.30
3540	4' deep, 3/8 C.Y. bucket	3.04	11.75	14.80
3560	6' deep, 3/8 C.Y. bucket	4.03	11.70	15.75
3580	8' deep, 1/2 C.Y. bucket	10.60	31	41.50
3600	10' deep, 1 C.Y. bucket	13.75	39.50	53.50
3800	4' wide, 2' deep, 3/8 C.Y. bucket	1.78	5.80	7.60
3820	3' deep, 3/8 C.Y. bucket	2.89	10.20	13.10
3840	4' deep, 1/2 C.Y. bucket	4.07	13.30	17.35
3860	6' deep, 1/2 C.Y. bucket	9.75	25	35
3880	8' deep, 1/2 C.Y. bucket	17.80	41.50	59.50
3900	10' deep, 1 C.Y. bucket	19.80	50	70
3920	12' deep, 1 C.Y. bucket	28.50	70.50	99
3940	15' deep, 1-1/2 C.Y. bucket	40.50	97	138
3960	18' deep, 2-1/2 C.Y. bucket	56	123	179
4030	6' wide, 6' deep, 5/8 C.Y. bucket w/trench box	15	32	47
4040	8' deep, 3/4 C.Y. bucket	23	49	72
4050	10' deep, 1 C.Y. bucket	24.50	57	81.50
4060	12' deep, 1-1/2 C.Y. bucket	31	71.50	103
4070	16' deep, 2 C.Y. bucket	53.50	110	164
4080	20' deep, 3-1/2 C.Y. bucket	78.50	153	232
4090	24' deep, 3-1/2 C.Y. bucket	114	214	330
4500	8' wide, 12' deep, 1-1/2 C.Y. bucket w/trench box	35.50	78	114
4550	15' deep, 1-1/2 C.Y. bucket	50	113	163
4600	18' deep, 2-1/2 C.Y. bucket	70.50	144	215
4650	24' deep, 3-1/2 C.Y. bucket	122	227	350
4800	10' wide, 20' deep, 3-1/2 C.Y. bucket w/trench box	91	173	264
4850	24' deep, 3-1/2 C.Y. bucket	130	239	370
4950	12' wide, 20' deep, 3-1/2 C.Y. bucket w/trench box	97.50	183	281
4980	25' deep, 3-1/2 C.Y. bucket	147	270	415

G4010 Electrical Distribution

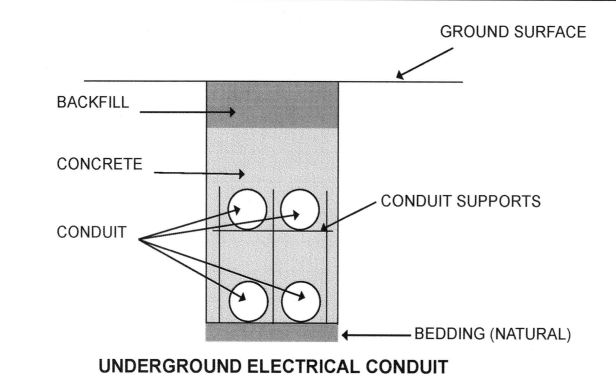

UNDERGROUND ELECTRICAL CONDUIT

System Components			COST EACH		
	QUANTITY	UNIT	MAT.	INST.	TOTAL
SYSTEM G4010 312 1000					
2000 AMP UNDERGROUND SERVICE INCLUDING COMMON EARTH EXCAVATION,					
CONCRETE, BACKFILL AND COMPACTION					
Excavate Trench	44.440	B.C.Y.		319.08	319.08
4 inch conduit bank	108.000	L.F.	1,323	2,862	4,185
4 inch fitting	8.000	Ea.	344	476	820
4 inch bells	4.000	Ea.	20.40	238	258.40
Concrete material	16.580	C.Y.	2,254.88		2,254.88
Concrete placement	16.580	C.Y.		430.91	430.91
Backfill trench	33.710	L.C.Y.		104.16	104.16
Compact fill material in trench	25.930	E.C.Y.		136.91	136.91
Dispose of excess fill material on-site	24.070	L.C.Y.		297.75	297.75
500 kcmil power cable	18.000	C.L.F.	15,840	8,100	23,940
Wire, 600 volt stranded copper wire, type THW, 1/0	6.000	C.L.F.	1,398	1,302	2,700
TOTAL			21,180.28	14,266.81	35,447.09

G4010 312	Underground Power Feed	COST EACH		
		MAT.	INST.	TOTAL
0900	Underground electrical			
0950	Including common earth excavation, concrete, backfill and compaction			
1000	2000 Amp service, 100' length, 4' depth	21,200	14,300	35,500
1100	1600 Amp service	16,400	11,000	27,400
1200	1200 Amp service	13,000	9,125	22,125
1300	1000 Amp service	9,675	8,325	18,000
1400	800 Amp service	8,375	7,850	16,225
1500	600 Amp service	6,150	7,275	13,425

G4010 Electrical Distribution

System Components	QUANTITY	UNIT	COST L.F.		
			MAT.	INST.	TOTAL
SYSTEM G4010 320 1012					
UNDERGROUND ELECTRICAL CONDUIT, 2″ DIAMETER, INCLUDING EXCAVATION,					
CONCRETE, BEDDING, BACKFILL AND COMPACTION					
Excavate Trench	.150	B.C.Y.		1.08	1.08
Base spacer	.200	Ea.	.27	2.56	2.83
Conduit	1.000	L.F.	1.40	3.98	5.38
Concrete material	.050	C.Y.	6.80		6.80
Concrete placement	.050	C.Y.		1.30	1.30
Backfill trench	.120	L.C.Y.		.37	.37
Compact fill material in trench	.100	E.C.Y.		.53	.53
Dispose of excess fill material on-site	.120	L.C.Y.		1.48	1.48
Marking tape	.010	C.L.F.	.40	.04	.44
TOTAL			8.87	11.34	20.21

G4010 320	Underground Electrical Conduit	COST L.F.		
		MAT.	INST.	TOTAL
0950	Underground electrical conduit, including common earth excavation,			
0952	Concrete, bedding, backfill and compaction			
1012	2″ dia. Schedule 40 PVC, 2′ deep	8.85	11.35	20.20
1014	4′ deep	8.85	16.10	24.95
1016	6′ deep	8.85	20.50	29.35
1022	2 @ 2″ dia. Schedule 40 PVC, 2′ deep	10.55	17.90	28.45
1024	4′ deep	10.55	23	33.55
1026	6′ deep	10.55	27	37.55
1032	3 @ 2″ dia. Schedule 40 PVC, 2′ deep	12.20	24.50	36.70
1034	4′ deep	12.20	29.50	41.70
1036	6′ deep	12.20	33.50	45.70
1042	4 @ 2″ dia. Schedule 40 PVC, 2′ deep	16.55	30.50	47.05
1044	4′ deep	16.55	35	51.55
1046	6′ deep	16.55	40	56.55

G4010 Electrical Distribution

G4010 320	Underground Electrical Conduit	COST L.F.		
		MAT.	INST.	TOTAL
1062	6 @ 2" dia. Schedule 40 PVC, 2' deep	19.90	43.50	63.40
1064	4' deep	19.90	48	67.90
1066	6' deep	19.90	53	72.90
1082	8 @ 2" dia. Schedule 40 PVC, 2' deep	23	56.50	79.50
1084	4' deep	23	61	84
1086	6' deep	23	65.50	88.50
1092	9 @ 2" dia. Schedule 40 PVC, 2' deep	25	62.50	87.50
1094	4' deep	27.50	67	94.50
1096	6' deep	27.50	72	99.50
1212	3" dia. Schedule 40 PVC, 2' deep	11.50	13.95	25.45
1214	4' deep	11.50	18.70	30.20
1216	6' deep	11.50	23.50	35
1222	2 @ 3" dia. Schedule 40 PVC, 2' deep	13.05	23	36.05
1224	4' deep	13.05	27.50	40.55
1226	6' deep	13.05	32	45.05
1232	3 @ 3" dia. Schedule 40 PVC, 2' deep	15.95	32	47.95
1234	4' deep	15.95	37	52.95
1236	6' deep	15.95	41	56.95
1242	4 @ 3" dia. Schedule 40 PVC, 2' deep	23	41	64
1244	4' deep	23	45.50	68.50
1246	6' deep	23	50.50	73.50
1262	6 @ 3" dia. Schedule 40 PVC, 2' deep	27.50	59	86.50
1264	4' deep	27.50	63.50	91
1266	6' deep	27.50	68	95.50
1282	8 @ 3" dia. Schedule 40 PVC, 2' deep	33.50	77.50	111
1284	4' deep	33.50	82	115.50
1286	6' deep	33.50	87	120.50
1292	9 @ 3" dia. Schedule 40 PVC, 2' deep	39	87	126
1294	4' deep	40.50	91	131.50
1296	6' deep	40.50	95	135.50
1312	4" dia. Schedule 40 PVC, 2' deep	12.35	17.10	29.45
1314	4' deep	12.35	22	34.35
1316	6' deep	12.35	26.50	38.85
1322	2 @ 4" dia. Schedule 40 PVC, 2' deep	16.20	29.50	45.70
1324	4' deep	16.20	34.50	50.70
1326	6' deep	16.20	39	55.20
1332	3 @ 4" dia. Schedule 40 PVC, 2' deep	18.65	42	60.65
1334	4' deep	18.65	47	65.65
1336	6' deep	18.65	51	69.65
1342	4 @ 4" dia. Schedule 40 PVC, 2' deep	28	54.50	82.50
1344	4' deep	28	59	87
1346	6' deep	28	64	92
1362	6 @ 4" dia. Schedule 40 PVC, 2' deep	34	79.50	113.50
1364	4' deep	34	83.50	117.50
1366	6' deep	34	88.50	122.50
1382	8 @ 4" dia. Schedule 40 PVC, 2' deep	43	105	148
1384	4' deep	43	110	153
1386	6' deep	43	116	159
1392	9 @ 4" dia. Schedule 40 PVC, 2' deep	48	118	166
1394	4' deep	48	121	169
1396	6' deep	48	125	173
1412	5" dia. Schedule 40 PVC, 2' deep	14.10	20.50	34.60
1414	4' deep	14.10	25	39.10
1416	6' deep	14.10	30	44.10
1422	2 @ 5" dia. Schedule 40 PVC, 2' deep	19.65	36.50	56.15
1424	4' deep	19.65	40.50	60.15
1426	6' deep	19.65	45.50	65.15
1432	3 @ 5" dia. Schedule 40 PVC, 2' deep	25	52	77

G4010 Electrical Distribution

G4010 320	Underground Electrical Conduit	COST L.F.		
		MAT.	INST.	TOTAL
1434	4' deep	25	57.50	82.50
1436	6' deep	25	62.50	87.50
1442	4 @ 5" dia. Schedule 40 PVC, 2' deep	34.50	67.50	102
1444	4' deep	34.50	72.50	107
1446	6' deep	34.50	77	111.50
1462	6 @ 5" dia. Schedule 40 PVC, 2' deep	44.50	99	143.50
1464	4' deep	44.50	104	148.50
1466	6' deep	44.50	109	153.50
1482	8 @ 5" dia. Schedule 40 PVC, 2' deep	58	132	190
1484	4' deep	58	139	197
1486	6' deep	58	144	202
1492	9 @ 5" dia. Schedule 40 PVC, 2' deep	65	148	213
1494	4' deep	65	151	216
1496	6' deep	65	156	221
1512	6" dia. Schedule 40 PVC, 2' deep	17.15	25	42.15
1514	4' deep	17.15	29.50	46.65
1516	6' deep	17.15	34.50	51.65
1522	2 @ 6" dia. Schedule 40 PVC, 2' deep	23	45	68
1524	4' deep	23	49	72
1526	6' deep	23	54	77
1532	3 @ 6" dia. Schedule 40 PVC, 2' deep	31.50	65.50	97
1534	4' deep	31.50	71.50	103
1536	6' deep	31.50	77	108.50
1542	4 @ 6" dia. Schedule 40 PVC, 2' deep	41.50	84.50	126
1544	4' deep	41.50	89.50	131
1546	6' deep	41.50	93.50	135
1562	6 @ 6" dia. Schedule 40 PVC, 2' deep	57.50	126	183.50
1564	4' deep	57.50	131	188.50
1566	6' deep	57.50	137	194.50
1582	8 @ 6" dia. Schedule 40 PVC, 2' deep	76	167	243
1584	4' deep	76	174	250
1586	6' deep	76	182	258
1592	9 @ 6" dia. Schedule 40 PVC, 2' deep	84.50	188	272.50
1594	4' deep	84.50	191	275.50
1596	6' deep	84.50	197	281.50

G4020 Site Lighting

The Site Lighting System includes the complete unit from foundation to electrical fixtures. Each system includes: excavation; concrete base; backfill by hand; compaction with a plate compacter; pole of specified material; all fixtures; and lamps.

The Expanded System Listing shows Site Lighting Systems that use one of two types of lamps: high pressure sodium; and metal halide. Systems are listed for 400-watt and 1000-watt lamps. Pole height varies from 20′ to 40′. There are four types of poles possibly listed: aluminum, fiberglass, steel and wood.

System Components			COST EACH		
	QUANTITY	UNIT	MAT.	INST.	TOTAL
SYSTEM G4020 110 5820					
SITE LIGHTING, METAL HALIDE, 400 WATT, ALUMINUM POLE, 20′ HIGH					
Excavating, by hand, pits to 6′ deep, heavy soil	2.368	C.Y.		293.63	293.63
Concrete in place incl. forms and reinf. stl. spread footings under 1 CY	.465	C.Y.	98.58	138.12	236.70
Backfill	1.903	C.Y.		85.64	85.64
Compact, vibrating plate	1.903	C.Y.		11.42	11.42
Aluminum light pole, 20′, no concrete base	1.000	Ea.	1,200	664	1,864
Road area luminaire wire	.300	C.L.F.	9.30	26.85	36.15
Roadway area luminaire, metal halide 400 W	1.000	Ea.	770	325	1,095
TOTAL			2,077.88	1,544.66	3,622.54

G4020 110	Site Lighting	COST EACH		
		MAT.	INST.	TOTAL
2320	Site lighting, high pressure sodium, 400 watt, aluminum pole, 20′ high	2,225	1,550	3,775
2340	30′ high	3,325	1,875	5,200
2360	40′ high	4,100	2,425	6,525
2520	Fiberglass pole, 20′ high	1,900	1,350	3,250
2540	30′ high	2,200	1,700	3,900
2560	40′ high	3,200	2,150	5,350
2720	Steel pole, 20′ high	2,175	1,625	3,800
2740	30′ high	2,750	2,000	4,750
2760	40′ high	3,325	2,600	5,925
2920	Wood pole, 20′ high	1,225	1,500	2,725
2940	30′ high	1,425	1,875	3,300
2960	40′ high	1,775	2,325	4,100
3120	1000 watt, aluminum pole, 20′ high	2,350	1,575	3,925
3140	30′ high	3,450	1,925	5,375
3160	40′ high	4,250	2,475	6,725
3320	Fiberglass pole, 20′ high	2,050	1,375	3,425
3340	30′ high	2,325	1,725	4,050
3360	40′ high	3,350	2,200	5,550
3420	Steel pole, 20′ high	2,300	1,650	3,950
3440	30′ high	2,875	2,025	4,900

G4020 Site Lighting

G4020 110	Site Lighting	COST EACH		
		MAT.	INST.	TOTAL
3460	40' high	3,475	2,625	6,100
3520	Wood pole, 20' high	1,375	1,550	2,925
3540	30' high	1,550	1,925	3,475
3560	40' high	1,925	2,350	4,275
5820	Metal halide, 400 watt, aluminum pole, 20' high	2,075	1,550	3,625
5840	30' high	3,175	1,875	5,050
5860	40' high	3,950	2,425	6,375
6120	Fiberglass pole, 20' high	1,750	1,350	3,100
6140	30' high	2,050	1,700	3,750
6160	40' high	3,050	2,150	5,200
6320	Steel pole, 20' high	2,025	1,625	3,650
6340	30' high	2,600	2,000	4,600
6360	40' high	3,175	2,600	5,775
7620	1000 watt, aluminum pole, 20' high	2,175	1,575	3,750
7640	30' high	3,275	1,925	5,200
7660	40' high	4,050	2,475	6,525
7900	Fiberglass pole, 20' high	1,850	1,375	3,225
7920	30' high	2,150	1,725	3,875
7940	40' high	3,150	2,200	5,350
8120	Steel pole, 20' high	2,125	1,650	3,775
8140	30' high	2,700	2,025	4,725
8160	40' high	3,275	2,625	5,900

Table G4020 210 Procedure for Calculating Floodlights Required for Various Footcandles
Poles should not be spaced more than 4 times the fixture mounting height for good light distribution.

Estimating Chart
Select Lamp type.

Determine total square feet.

Chart will show quantity of fixtures to provide 1 footcandle initial, at intersection of lines. Multiply fixture quantity by desired footcandle level.

Chart based on use of wide beam luminaires in an area whose dimensions are large compared to mounting height and is approximate only.

To maintain 1 footcandle over a large area use these watts per square foot:

Incandescent	0.15
Metal Halide	0.032
Mercury Vapor	0.05
High Pressure Sodium	0.024

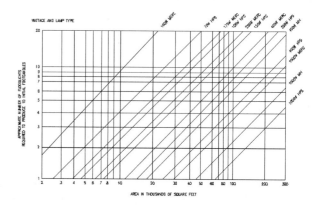

System Components			COST EACH		
	QUANTITY	UNIT	MAT.	INST.	TOTAL
SYSTEM G4020 210 0200					
LIGHT POLES, ALUMINUM, 20' HIGH, 1 ARM BRACKET					
Aluminum light pole, 20', no concrete base	1.000	Ea.	1,200	664	1,864
Bracket arm for aluminum light pole	1.000	Ea.	153	89.50	242.50
Excavation by hand, pits to 6' deep, heavy soil or clay	2.368	C.Y.		293.63	293.63
Footing, concrete incl forms, reinforcing, spread, under 1 C.Y.	.465	C.Y.	98.58	138.12	236.70
Backfill by hand	1.903	C.Y.		85.64	85.64
Compaction vibrating plate	1.903	C.Y.		11.42	11.42
TOTAL			1,451.58	1,282.31	2,733.89

G4020 210	Light Pole (Installed)	COST EACH		
		MAT.	INST.	TOTAL
0200	Light pole, aluminum, 20' high, 1 arm bracket	1,450	1,275	2,725
0240	2 arm brackets	1,600	1,275	2,875
0280	3 arm brackets	1,775	1,325	3,100
0320	4 arm brackets	1,925	1,350	3,275
0360	30' high, 1 arm bracket	2,550	1,625	4,175
0400	2 arm brackets	2,700	1,625	4,325
0440	3 arm brackets	2,875	1,650	4,525
0480	4 arm brackets	3,025	1,675	4,700
0680	40' high, 1 arm bracket	3,325	2,150	5,475
0720	2 arm brackets	3,475	2,150	5,625
0760	3 arm brackets	3,650	2,200	5,850
0800	4 arm brackets	3,800	2,225	6,025
0840	Steel, 20' high, 1 arm bracket	1,475	1,350	2,825
0880	2 arm brackets	1,600	1,350	2,950
0920	3 arm brackets	1,500	1,400	2,900
0960	4 arm brackets	1,600	1,400	3,000
1000	30' high, 1 arm bracket	2,050	1,725	3,775
1040	2 arm brackets	2,175	1,725	3,900
1080	3 arm brackets	2,075	1,775	3,850
1120	4 arm brackets	2,175	1,775	3,950
1320	40' high, 1 arm bracket	2,625	2,325	4,950
1360	2 arm brackets	2,750	2,325	5,075
1400	3 arm brackets	2,650	2,350	5,000
1440	4 arm brackets	2,750	2,350	5,100

Reference Section

All the reference information is in one section, making it easy to find what you need to know and easy to use the data set on a daily basis. This section is visually identified by a vertical black bar on the page edges.

In this Reference Section, we've included Equipment Rental Costs, a listing of rental and operating costs; Crew Listings, a full listing of all crews, equipment, and their costs; Historical Cost Indexes for cost comparisons over time; City Cost Indexes and Location Factors for adjusting costs to the region you are in; Reference Tables, where you will find explanations, estimating information and procedures, and technical data; and an explanation of all the Abbreviations in the data set.

Table of Contents

Estimating Tips

- This section contains the average costs to rent and operate hundreds of pieces of construction equipment. This is useful information when one is estimating the time and material requirements of any particular operation in order to establish a unit or total cost. Bare equipment costs shown on a unit cost line include, not only rental, but also operating costs for equipment under normal use.

Rental Costs

- Equipment rental rates are obtained from the following industry sources throughout North America: contractors, suppliers, dealers, manufacturers, and distributors.

- Rental rates vary throughout the country, with larger cities generally having lower rates. Lease plans for new equipment are available for periods in excess of six months, with a percentage of payments applying toward purchase.

- Monthly rental rates vary from 2% to 5% of the purchase price of the equipment depending on the anticipated life of the equipment and its wearing parts.

- Weekly rental rates are about 1/3 of the monthly rates, and daily rental rates are about 1/3 of the weekly rate.

- Rental rates can also be treated as reimbursement costs for contractor-owned equipment. Owned equipment costs include depreciation, loan payments, interest, taxes, insurance, storage, and major repairs.

Operating Costs

- The operating costs include parts and labor for routine servicing, such as the repair and replacement of pumps, filters, and worn lines. Normal operating expendables, such as fuel, lubricants, tires, and electricity (where applicable), are also included.

- Extraordinary operating expendables with highly variable wear patterns, such as diamond bits and blades, are excluded. These costs can be found as material costs in the Unit Price section.

- The hourly operating costs listed do not include the operator's wages.

Equipment Cost/Day

- Any power equipment required by a crew is shown in the Crew Listings with a daily cost.

- This daily cost of equipment needed by a crew includes both the rental cost and the operating cost and is based on dividing the weekly rental rate by 5 (the number of working days in the week), then adding the hourly operating cost multiplied by 8 (the number of hours in a day). This "Equipment Cost/Day" is shown in the far right column of the Equipment Rental section.

- If equipment is needed for only one or two days, it is best to develop your own cost by including components for daily rent and hourly operating costs. This is important when the listed Crew for a task does not contain the equipment needed, such as a crane for lifting mechanical heating/cooling equipment up onto a roof.

- If the quantity of work is less than the crew's Daily Output shown for a Unit Price line item that includes a bare unit equipment cost, the recommendation is to estimate one day's rental cost and operating cost for equipment shown in the Crew Listing for that line item.

- Please note, in some cases the equipment description in the crew is followed by a time period in parenthesis. For example: (daily) or (monthly). In these cases the equipment cost/day is calculated by adding the rental cost per time period to the hourly operating cost multiplied by 8.

Mobilization, Demobilization Costs

- The cost to move construction equipment from an equipment yard or rental company to the job site and back again is not included in equipment rental costs listed in the Reference Section. It is also not included in the bare equipment cost of any Unit Price line item or in any equipment costs shown in the Crew Listings.

- Mobilization (to the site) and demobilization (from the site) costs can be found in the Unit Price section.

- If a piece of equipment is already at the job site, it is not appropriate to utilize mobilization or demobilization costs again in an estimate. ■

01 54 33 | Equipment Rental

			UNIT	HOURLY OPER. COST	RENT PER DAY	RENT PER WEEK	RENT PER MONTH	EQUIPMENT COST/DAY	
10	0010	**CONCRETE EQUIPMENT RENTAL** without operators	R015433 -10						10
	0200	Bucket, concrete lightweight, 1/2 C.Y.	Ea.	.88	24.50	74	222	21.80	
	0300	1 C.Y.		.98	28.50	85.50	257	24.95	
	0400	1-1/2 C.Y.		1.24	38.50	115	345	32.90	
	0500	2 C.Y.		1.34	47	141	425	38.90	
	0580	8 C.Y.		6.49	265	795	2,375	210.90	
	0600	Cart, concrete, self-propelled, operator walking, 10 C.F.		2.86	58.50	176	530	58.10	
	0700	Operator riding, 18 C.F.		4.82	98.50	296	890	97.75	
	0800	Conveyer for concrete, portable, gas, 16" wide, 26' long		10.64	130	390	1,175	163.10	
	0900	46' long		11.02	155	465	1,400	181.15	
	1000	56' long		11.18	162	485	1,450	186.45	
	1100	Core drill, electric, 2-1/2 H.P., 1" to 8" bit diameter		1.57	56.50	170	510	46.55	
	1150	11 H.P., 8" to 18" cores		5.40	115	345	1,025	112.20	
	1200	Finisher, concrete floor, gas, riding trowel, 96" wide		9.66	148	445	1,325	166.30	
	1300	Gas, walk-behind, 3 blade, 36" trowel		2.04	24.50	73.50	221	31	
	1400	4 blade, 48" trowel		3.07	28	84.50	254	41.50	
	1500	Float, hand-operated (Bull float), 48" wide		.08	13.85	41.50	125	8.95	
	1570	Curb builder, 14 H.P., gas, single screw		14.04	285	855	2,575	283.30	
	1590	Double screw		15.04	340	1,025	3,075	325.35	
	1600	Floor grinder, concrete and terrazzo, electric, 22" path		3.04	190	570	1,700	138.30	
	1700	Edger, concrete, electric, 7" path		1.18	56.50	170	510	43.45	
	1750	Vacuum pick-up system for floor grinders, wet/dry		1.62	98.50	296	890	72.15	
	1800	Mixer, powered, mortar and concrete, gas, 6 C.F., 18 H.P.		7.42	127	380	1,150	135.35	
	1900	10 C.F., 25 H.P.		9.00	150	450	1,350	162	
	2000	16 C.F.		9.36	173	520	1,550	178.85	
	2100	Concrete, stationary, tilt drum, 2 C.Y.		7.23	243	730	2,200	203.85	
	2120	Pump, concrete, truck mounted, 4" line, 80' boom		29.88	1,000	3,025	9,075	844.05	
	2140	5" line, 110' boom		37.46	1,375	4,150	12,500	1,130	
	2160	Mud jack, 50 C.F. per hr.		6.45	123	370	1,100	125.60	
	2180	225 C.F. per hr.		8.55	145	435	1,300	155.40	
	2190	Shotcrete pump rig, 12 C.Y./hr.		13.97	218	655	1,975	242.75	
	2200	35 C.Y./hr.		15.80	242	725	2,175	271.40	
	2600	Saw, concrete, manual, gas, 18 H.P.		5.54	47.50	143	430	72.90	
	2650	Self-propelled, gas, 30 H.P.		7.90	78.50	235	705	110.15	
	2675	V-groove crack chaser, manual, gas, 6 H.P.		1.64	19	57	171	24.50	
	2700	Vibrators, concrete, electric, 60 cycle, 2 H.P.		.47	9.15	27.50	82.50	9.25	
	2800	3 H.P.		.56	11.65	35	105	11.50	
	2900	Gas engine, 5 H.P.		1.54	16.35	49	147	22.15	
	3000	8 H.P.		2.08	16.50	49.50	149	26.55	
	3050	Vibrating screed, gas engine, 8 H.P.		2.81	95	285	855	79.45	
	3120	Concrete transit mixer, 6 x 4, 250 H.P., 8 C.Y., rear discharge		50.72	615	1,850	5,550	775.80	
	3200	Front discharge		58.88	740	2,225	6,675	916.05	
	3300	6 x 6, 285 H.P., 12 C.Y., rear discharge		58.15	710	2,125	6,375	890.20	
	3400	Front discharge		60.59	750	2,250	6,750	934.70	
20	0010	**EARTHWORK EQUIPMENT RENTAL** without operators	R015433 -10						20
	0040	Aggregate spreader, push type, 8' to 12' wide	Ea.	2.60	27.50	82.50	248	37.30	
	0045	Tailgate type, 8' wide		2.55	33	99	297	40.15	
	0055	Earth auger, truck mounted, for fence & sign posts, utility poles		13.85	460	1,375	4,125	385.85	
	0060	For borings and monitoring wells		42.65	690	2,075	6,225	756.20	
	0070	Portable, trailer mounted		2.30	33	99	297	38.20	
	0075	Truck mounted, for caissons, water wells		85.39	2,900	8,725	26,200	2,428	
	0080	Horizontal boring machine, 12" to 36" diameter, 45 H.P.		22.77	188	565	1,700	295.15	
	0090	12" to 48" diameter, 65 H.P.		31.25	330	985	2,950	447	
	0095	Auger, for fence posts, gas engine, hand held		.45	6.35	19	57	7.40	
	0100	Excavator, diesel hydraulic, crawler mounted, 1/2 C.Y. cap.		21.72	450	1,350	4,050	443.75	
	0120	5/8 C.Y. capacity		29.04	590	1,775	5,325	587.30	
	0140	3/4 C.Y. capacity		32.66	700	2,100	6,300	681.25	
	0150	1 C.Y. capacity		41.21	700	2,100	6,300	749.70	

01 54 33 | Equipment Rental

		UNIT	HOURLY OPER. COST	RENT PER DAY	RENT PER WEEK	RENT PER MONTH	EQUIPMENT COST/DAY
0200	1-1/2 C.Y. capacity	Ea.	48.59	865	2,600	7,800	908.70
0300	2 C.Y. capacity		56.58	1,050	3,125	9,375	1,078
0320	2-1/2 C.Y. capacity		82.64	1,300	3,900	11,700	1,441
0325	3-1/2 C.Y. capacity		120.12	2,150	6,475	19,400	2,256
0330	4-1/2 C.Y. capacity		151.63	2,750	8,275	24,800	2,868
0335	6 C.Y. capacity		192.39	3,400	10,200	30,600	3,579
0340	7 C.Y. capacity		175.20	3,225	9,650	29,000	3,332
0342	Excavator attachments, bucket thumbs		3.40	250	750	2,250	177.20
0345	Grapples		3.14	215	645	1,925	154.10
0346	Hydraulic hammer for boom mounting, 4,000 ft. lb.		13.48	375	1,125	3,375	332.85
0347	5,000 ft. lb.		15.95	465	1,400	4,200	407.60
0348	8,000 ft. lb.		23.54	675	2,025	6,075	593.35
0349	12,000 ft. lb.		25.72	800	2,400	7,200	685.75
0350	Gradall type, truck mounted, 3 ton @ 15' radius, 5/8 C.Y.		43.44	860	2,575	7,725	862.55
0370	1 C.Y. capacity		59.40	1,175	3,550	10,700	1,185
0400	Backhoe-loader, 40 to 45 H.P., 5/8 C.Y. capacity		11.90	207	620	1,850	219.20
0450	45 H.P. to 60 H.P., 3/4 C.Y. capacity		18.02	293	880	2,650	320.20
0460	80 H.P., 1-1/4 C.Y. capacity		20.36	375	1,125	3,375	387.85
0470	112 H.P., 1-1/2 C.Y. capacity		32.99	590	1,775	5,325	618.90
0482	Backhoe-loader attachment, compactor, 20,000 lb.		6.44	150	450	1,350	141.50
0485	Hydraulic hammer, 750 ft. lb.		3.68	103	310	930	91.45
0486	Hydraulic hammer, 1,200 ft. lb.		6.55	198	595	1,775	171.40
0500	Brush chipper, gas engine, 6" cutter head, 35 H.P.		9.17	110	330	990	139.35
0550	Diesel engine, 12" cutter head, 130 H.P.		23.67	335	1,000	3,000	389.40
0600	15" cutter head, 165 H.P.		26.59	400	1,200	3,600	452.70
0750	Bucket, clamshell, general purpose, 3/8 C.Y.		1.40	41	123	370	35.80
0800	1/2 C.Y.		1.52	50.50	152	455	42.55
0850	3/4 C.Y.		1.64	57.50	173	520	47.75
0900	1 C.Y.		1.70	62.50	188	565	51.20
0950	1-1/2 C.Y.		2.79	87	261	785	74.50
1000	2 C.Y.		2.92	95	285	855	80.40
1010	Bucket, dragline, medium duty, 1/2 C.Y.		.82	24.50	74	222	21.35
1020	3/4 C.Y.		.78	26	78	234	21.85
1030	1 C.Y.		.80	27	81.50	245	22.70
1040	1-1/2 C.Y.		1.26	41.50	125	375	35.10
1050	2 C.Y.		1.29	45	135	405	37.35
1070	3 C.Y.		2.08	66	198	595	56.25
1200	Compactor, manually guided 2-drum vibratory smooth roller, 7.5 H.P.		7.22	207	620	1,850	181.75
1250	Rammer/tamper, gas, 8"		2.21	46.50	140	420	45.65
1260	15"		2.63	53.50	160	480	53.05
1300	Vibratory plate, gas, 18" plate, 3,000 lb. blow		2.13	23.50	70.50	212	31.10
1350	21" plate, 5,000 lb. blow		2.62	32.50	98	294	40.55
1370	Curb builder/extruder, 14 H.P., gas, single screw		14.03	283	850	2,550	282.25
1390	Double screw		15.04	340	1,025	3,075	325.30
1500	Disc harrow attachment, for tractor		.47	79.50	239	715	51.60
1810	Feller buncher, shearing & accumulating trees, 100 H.P.		39.20	815	2,450	7,350	803.55
1860	Grader, self-propelled, 25,000 lb.		33.35	760	2,275	6,825	721.80
1910	30,000 lb.		32.85	660	1,975	5,925	657.85
1920	40,000 lb.		51.89	1,275	3,825	11,500	1,180
1930	55,000 lb.		66.93	1,650	4,925	14,800	1,520
1950	Hammer, pavement breaker, self-propelled, diesel, 1,000 to 1,250 lb.		28.40	475	1,425	4,275	512.20
2000	1,300 to 1,500 lb.		42.80	975	2,925	8,775	927.40
2050	Pile driving hammer, steam or air, 4,150 ft. lb. @ 225 bpm		12.15	565	1,700	5,100	437.20
2100	8,750 ft. lb. @ 145 bpm		14.35	810	2,425	7,275	599.75
2150	15,000 ft. lb. @ 60 bpm		14.68	840	2,525	7,575	622.40
2200	24,450 ft. lb. @ 111 bpm		15.69	935	2,800	8,400	685.50
2250	Leads, 60' high for pile driving hammers up to 20,000 ft. lb.		3.67	85.50	256	770	80.55
2300	90' high for hammers over 20,000 ft. lb.		5.45	152	455	1,375	134.60

20 ... **20**

For customer support on your Electrical Costs with RSMeans data, call 800.448.8182.

465

01 54 33 | Equipment Rental

		UNIT	HOURLY OPER. COST	RENT PER DAY	RENT PER WEEK	RENT PER MONTH	EQUIPMENT COST/DAY	
20	2350	Diesel type hammer, 22,400 ft. lb.	Ea.	17.82	475	1,425	4,275	427.55	20
	2400	41,300 ft. lb.		25.68	600	1,800	5,400	565.45	
	2450	141,000 ft. lb.		41.33	950	2,850	8,550	900.65	
	2500	Vib. elec. hammer/extractor, 200 kW diesel generator, 34 H.P.		41.37	690	2,075	6,225	745.95	
	2550	80 H.P.		73.03	1,000	3,000	9,000	1,184	
	2600	150 H.P.		135.14	1,925	5,775	17,300	2,236	
	2800	Log chipper, up to 22" diameter, 600 H.P.		46.26	665	2,000	6,000	770.10	
	2850	Logger, for skidding & stacking logs, 150 H.P.		43.53	835	2,500	7,500	848.25	
	2860	Mulcher, diesel powered, trailer mounted		18.04	220	660	1,975	276.35	
	2900	Rake, spring tooth, with tractor		14.72	360	1,075	3,225	332.75	
	3000	Roller, vibratory, tandem, smooth drum, 20 H.P.		7.81	150	450	1,350	152.45	
	3050	35 H.P.		10.13	252	755	2,275	232.05	
	3100	Towed type vibratory compactor, smooth drum, 50 H.P.		25.27	365	1,100	3,300	422.15	
	3150	Sheepsfoot, 50 H.P.		25.64	375	1,125	3,375	430.15	
	3170	Landfill compactor, 220 H.P.		70.01	1,575	4,750	14,300	1,510	
	3200	Pneumatic tire roller, 80 H.P.		12.92	390	1,175	3,525	338.35	
	3250	120 H.P.		19.39	640	1,925	5,775	540.10	
	3300	Sheepsfoot vibratory roller, 240 H.P.		62.21	1,375	4,125	12,400	1,323	
	3320	340 H.P.		83.82	2,075	6,250	18,800	1,921	
	3350	Smooth drum vibratory roller, 75 H.P.		23.34	635	1,900	5,700	566.70	
	3400	125 H.P.		27.61	715	2,150	6,450	650.90	
	3410	Rotary mower, brush, 60", with tractor		18.79	350	1,050	3,150	360.30	
	3420	Rototiller, walk-behind, gas, 5 H.P.		2.14	89	267	800	70.50	
	3422	8 H.P.		2.81	103	310	930	84.50	
	3440	Scrapers, towed type, 7 C.Y. capacity		6.44	123	370	1,100	125.50	
	3450	10 C.Y. capacity		7.20	165	495	1,475	156.65	
	3500	15 C.Y. capacity		7.40	190	570	1,700	173.20	
	3525	Self-propelled, single engine, 14 C.Y. capacity		133.29	2,575	7,725	23,200	2,611	
	3550	Dual engine, 21 C.Y. capacity		141.37	2,350	7,050	21,200	2,541	
	3600	31 C.Y. capacity		187.85	3,500	10,500	31,500	3,603	
	3640	44 C.Y. capacity		232.68	4,500	13,500	40,500	4,561	
	3650	Elevating type, single engine, 11 C.Y. capacity		61.87	1,125	3,350	10,100	1,165	
	3700	22 C.Y. capacity		114.62	2,325	6,950	20,900	2,307	
	3710	Screening plant, 110 H.P. w/5' x 10' screen		21.13	385	1,150	3,450	399.05	
	3720	5' x 16' screen		26.68	490	1,475	4,425	508.40	
	3850	Shovel, crawler-mounted, front-loading, 7 C.Y. capacity		218.65	3,800	11,400	34,200	4,029	
	3855	12 C.Y. capacity		336.90	5,275	15,800	47,400	5,855	
	3860	Shovel/backhoe bucket, 1/2 C.Y.		2.69	70.50	212	635	63.95	
	3870	3/4 C.Y.		2.67	79.50	238	715	68.90	
	3880	1 C.Y.		2.76	88	264	790	74.85	
	3890	1-1/2 C.Y.		2.95	103	310	930	85.65	
	3910	3 C.Y.		3.44	140	420	1,250	111.50	
	3950	Stump chipper, 18" deep, 30 H.P.		6.90	212	635	1,900	182.20	
	4110	Dozer, crawler, torque converter, diesel 80 H.P.		25.25	450	1,350	4,050	472	
	4150	105 H.P.		34.34	560	1,675	5,025	609.70	
	4200	140 H.P.		41.27	840	2,525	7,575	835.20	
	4260	200 H.P.		63.16	1,300	3,925	11,800	1,290	
	4310	300 H.P.		80.73	1,975	5,925	17,800	1,831	
	4360	410 H.P.		106.74	2,350	7,075	21,200	2,269	
	4370	500 H.P.		133.38	2,900	8,725	26,200	2,812	
	4380	700 H.P.		230.17	5,300	15,900	47,700	5,021	
	4400	Loader, crawler, torque conv., diesel, 1-1/2 C.Y., 80 H.P.		29.54	590	1,775	5,325	591.30	
	4450	1-1/2 to 1-3/4 C.Y., 95 H.P.		30.27	675	2,025	6,075	647.15	
	4510	1-3/4 to 2-1/4 C.Y., 130 H.P.		47.76	1,025	3,100	9,300	1,002	
	4530	2-1/2 to 3-1/4 C.Y., 190 H.P.		57.79	1,225	3,700	11,100	1,202	
	4560	3-1/2 to 5 C.Y., 275 H.P.		71.41	1,475	4,450	13,400	1,461	
	4610	Front end loader, 4WD, articulated frame, diesel, 1 to 1-1/4 C.Y., 70 H.P.		16.62	273	820	2,450	297	
	4620	1-1/2 to 1-3/4 C.Y., 95 H.P.		20.00	320	965	2,900	352.95	

01 54 33 | Equipment Rental

		UNIT	HOURLY OPER. COST	RENT PER DAY	RENT PER WEEK	RENT PER MONTH	EQUIPMENT COST/DAY	
20								**20**
4650	1-3/4 to 2 C.Y., 130 H.P.	Ea.	21.07	385	1,150	3,450	398.55	
4710	2-1/2 to 3-1/2 C.Y., 145 H.P.		29.53	490	1,475	4,425	531.20	
4730	3 to 4-1/2 C.Y., 185 H.P.		32.09	515	1,550	4,650	566.70	
4760	5-1/4 to 5-3/4 C.Y., 270 H.P.		53.19	915	2,750	8,250	975.55	
4810	7 to 9 C.Y., 475 H.P.		91.17	1,750	5,275	15,800	1,784	
4870	9 to 11 C.Y., 620 H.P.		131.92	2,625	7,850	23,600	2,625	
4880	Skid-steer loader, wheeled, 10 C.F., 30 H.P. gas		9.57	163	490	1,475	174.55	
4890	1 C.Y., 78 H.P., diesel		18.44	410	1,225	3,675	392.50	
4892	Skid-steer attachment, auger		.75	137	410	1,225	87.95	
4893	Backhoe		.74	118	355	1,075	76.90	
4894	Broom		.71	123	370	1,100	79.65	
4895	Forks		.15	26.50	80	240	17.25	
4896	Grapple		.72	107	320	960	69.80	
4897	Concrete hammer		1.05	177	530	1,600	114.40	
4898	Tree spade		.60	100	300	900	64.80	
4899	Trencher		.65	108	325	975	70.20	
4900	Trencher, chain, boom type, gas, operator walking, 12 H.P.		4.18	50	150	450	63.40	
4910	Operator riding, 40 H.P.		16.69	360	1,075	3,225	348.50	
5000	Wheel type, diesel, 4' deep, 12" wide		68.71	935	2,800	8,400	1,110	
5100	6' deep, 20" wide		87.58	2,150	6,475	19,400	1,996	
5150	Chain type, diesel, 5' deep, 8" wide		16.30	350	1,050	3,150	340.40	
5200	Diesel, 8' deep, 16" wide		89.66	1,875	5,600	16,800	1,837	
5202	Rock trencher, wheel type, 6" wide x 18" deep		47.12	700	2,100	6,300	797	
5206	Chain type, 18" wide x 7' deep		104.70	3,075	9,200	27,600	2,678	
5210	Tree spade, self-propelled		13.69	390	1,175	3,525	344.55	
5250	Truck, dump, 2-axle, 12 ton, 8 C.Y. payload, 220 H.P.		23.95	247	740	2,225	339.60	
5300	Three axle dump, 16 ton, 12 C.Y. payload, 400 H.P.		44.63	350	1,050	3,150	567.05	
5310	Four axle dump, 25 ton, 18 C.Y. payload, 450 H.P.		50.00	510	1,525	4,575	705	
5350	Dump trailer only, rear dump, 16-1/2 C.Y.		5.74	147	440	1,325	133.95	
5400	20 C.Y.		6.20	165	495	1,475	148.60	
5450	Flatbed, single axle, 1-1/2 ton rating		19.05	71.50	214	640	195.20	
5500	3 ton rating		23.12	102	305	915	245.95	
5550	Off highway rear dump, 25 ton capacity		62.86	1,425	4,250	12,800	1,353	
5600	35 ton capacity		67.10	1,550	4,675	14,000	1,472	
5610	50 ton capacity		84.12	1,775	5,325	16,000	1,738	
5620	65 ton capacity		89.83	1,925	5,800	17,400	1,879	
5630	100 ton capacity		121.61	2,850	8,550	25,700	2,683	
6000	Vibratory plow, 25 H.P., walking		6.79	60.50	181	545	90.55	
40	0010	**GENERAL EQUIPMENT RENTAL** without operators	R015433 -10						**40**
0020	Aerial lift, scissor type, to 20' high, 1200 lb. capacity, electric		Ea.	3.49	50.50	151	455	58.15	
0030	To 30' high, 1,200 lb. capacity		3.78	67.50	202	605	70.70	
0040	Over 30' high, 1,500 lb. capacity		5.15	122	365	1,100	114.20	
0070	Articulating boom, to 45' high, 500 lb. capacity, diesel	R015433 -15	9.95	275	825	2,475	244.60	
0075	To 60' high, 500 lb. capacity		13.70	510	1,525	4,575	414.60	
0080	To 80' high, 500 lb. capacity		16.10	560	1,675	5,025	463.80	
0085	To 125' high, 500 lb. capacity		18.40	775	2,325	6,975	612.20	
0100	Telescoping boom to 40' high, 500 lb. capacity, diesel		11.27	320	965	2,900	283.15	
0105	To 45' high, 500 lb. capacity		12.54	315	945	2,825	289.35	
0110	To 60' high, 500 lb. capacity		16.41	535	1,600	4,800	451.25	
0115	To 80' high, 500 lb. capacity		21.33	835	2,500	7,500	670.65	
0120	To 100' high, 500 lb. capacity		28.80	860	2,575	7,725	745.40	
0125	To 120' high, 500 lb. capacity		29.25	840	2,525	7,575	739	
0195	Air compressor, portable, 6.5 CFM, electric		.91	12.65	38	114	14.85	
0196	Gasoline		.65	18	54	162	16.05	
0200	Towed type, gas engine, 60 CFM		9.46	51.50	155	465	106.70	
0300	160 CFM		10.51	53	159	475	115.85	
0400	Diesel engine, rotary screw, 250 CFM		12.12	118	355	1,075	167.95	
0500	365 CFM		16.05	142	425	1,275	213.35	

For customer support on your Electrical Costs with RSMeans data, call 800.448.8182.

467

01 54 33 | Equipment Rental

		UNIT	HOURLY OPER. COST	RENT PER DAY	RENT PER WEEK	RENT PER MONTH	EQUIPMENT COST/DAY	
40	0550	450 CFM	Ea.	20.01	177	530	1,600	266.05	40
	0600	600 CFM		34.20	240	720	2,150	417.60	
	0700	750 CFM		34.72	252	755	2,275	428.80	
	0930	Air tools, breaker, pavement, 60 lb.		.57	10.35	31	93	10.75	
	0940	80 lb.		.57	10.65	32	96	10.90	
	0950	Drills, hand (jackhammer), 65 lb.		.67	17.35	52	156	15.80	
	0960	Track or wagon, swing boom, 4" drifter		54.83	940	2,825	8,475	1,004	
	0970	5" drifter		63.49	1,100	3,325	9,975	1,173	
	0975	Track mounted quarry drill, 6" diameter drill		102.22	1,625	4,875	14,600	1,793	
	0980	Dust control per drill		1.04	25	75.50	227	23.45	
	0990	Hammer, chipping, 12 lb.		.60	27	81	243	21	
	1000	Hose, air with couplings, 50' long, 3/4" diameter		.07	10	30	90	6.55	
	1100	1" diameter		.08	13	39	117	8.40	
	1200	1-1/2" diameter		.22	35	105	315	22.75	
	1300	2" diameter		.24	40	120	360	25.90	
	1400	2-1/2" diameter		.36	56.50	170	510	36.85	
	1410	3" diameter		.42	37.50	112	335	25.75	
	1450	Drill, steel, 7/8" x 2'		.08	12.50	37.50	113	8.20	
	1460	7/8" x 6'		.12	19	57	171	12.30	
	1520	Moil points		.03	4.73	14.20	42.50	3.05	
	1525	Pneumatic nailer w/accessories		.48	31	92.50	278	22.35	
	1530	Sheeting driver for 60 lb. breaker		.04	7.50	22.50	67.50	4.85	
	1540	For 90 lb. breaker		.13	10.15	30.50	91.50	7.15	
	1550	Spade, 25 lb.		.50	7.15	21.50	64.50	8.30	
	1560	Tamper, single, 35 lb.		.59	39.50	119	355	28.55	
	1570	Triple, 140 lb.		.89	59.50	179	535	42.90	
	1580	Wrenches, impact, air powered, up to 3/4" bolt		.43	12.85	38.50	116	11.10	
	1590	Up to 1-1/4" bolt		.58	23	69.50	209	18.50	
	1600	Barricades, barrels, reflectorized, 1 to 99 barrels		.03	5.50	16.55	49.50	3.55	
	1610	100 to 200 barrels		.02	4.27	12.80	38.50	2.75	
	1620	Barrels with flashers, 1 to 99 barrels		.03	6.20	18.55	55.50	4	
	1630	100 to 200 barrels		.03	4.95	14.85	44.50	3.20	
	1640	Barrels with steady burn type C lights		.05	8.15	24.50	73.50	5.30	
	1650	Illuminated board, trailer mounted, with generator		3.29	135	405	1,225	107.30	
	1670	Portable barricade, stock, with flashers, 1 to 6 units		.03	6.15	18.50	55.50	4	
	1680	25 to 50 units		.03	5.75	17.25	52	3.70	
	1685	Butt fusion machine, wheeled, 1.5 HP electric, 2" - 8" diameter pipe		2.64	203	610	1,825	143.10	
	1690	Tracked, 20 HP diesel, 4" - 12" diameter pipe		11.27	525	1,575	4,725	405.15	
	1695	83 HP diesel, 8" - 24" diameter pipe		51.47	2,525	7,575	22,700	1,927	
	1700	Carts, brick, gas engine, 1,000 lb. capacity		2.95	69.50	208	625	65.20	
	1800	1,500 lb., 7-1/2' lift		2.92	65	195	585	62.40	
	1822	Dehumidifier, medium, 6 lb./hr., 150 CFM		1.19	74	222	665	53.95	
	1824	Large, 18 lb./hr., 600 CFM		2.20	140	420	1,250	101.60	
	1830	Distributor, asphalt, trailer mounted, 2,000 gal., 38 H.P. diesel		11.03	340	1,025	3,075	293.20	
	1840	3,000 gal., 38 H.P. diesel		12.90	365	1,100	3,300	323.25	
	1850	Drill, rotary hammer, electric		1.12	27	80.50	242	25.05	
	1860	Carbide bit, 1-1/2" diameter, add to electric rotary hammer		.03	5.10	15.30	46	3.30	
	1865	Rotary, crawler, 250 H.P.		136.18	2,225	6,650	20,000	2,419	
	1870	Emulsion sprayer, 65 gal., 5 H.P. gas engine		2.78	103	310	930	84.20	
	1880	200 gal., 5 H.P. engine		7.25	173	520	1,550	161.95	
	1900	Floor auto-scrubbing machine, walk-behind, 28" path		5.64	365	1,100	3,300	265.15	
	1930	Floodlight, mercury vapor, or quartz, on tripod, 1,000 watt		.46	22	66.50	200	16.95	
	1940	2,000 watt		.59	27.50	82	246	21.15	
	1950	Floodlights, trailer mounted with generator, one - 300 watt light		3.56	76	228	685	74.05	
	1960	two 1000 watt lights		4.50	84.50	254	760	86.80	
	2000	four 300 watt lights		4.26	96.50	290	870	92.05	
	2005	Foam spray rig, incl. box trailer, compressor, generator, proportioner		25.54	515	1,550	4,650	514.30	
	2015	Forklift, pneumatic tire, rough terr., straight mast, 5,000 lb., 12' lift, gas		18.65	212	635	1,900	276.20	

01 54 33 | Equipment Rental

		UNIT	HOURLY OPER. COST	RENT PER DAY	RENT PER WEEK	RENT PER MONTH	EQUIPMENT COST/DAY	
40	2025	8,000 lb, 12' lift	Ea.	22.75	350	1,050	3,150	392	**40**
	2030	5,000 lb., 12' lift, diesel		15.45	237	710	2,125	265.60	
	2035	8,000 lb, 12' lift, diesel		16.75	268	805	2,425	295	
	2045	All terrain, telescoping boom, diesel, 5,000 lb., 10' reach, 19' lift		17.25	410	1,225	3,675	383	
	2055	6,600 lb., 29' reach, 42' lift		21.10	400	1,200	3,600	408.80	
	2065	10,000 lb., 31' reach, 45' lift		23.10	465	1,400	4,200	464.75	
	2070	Cushion tire, smooth floor, gas, 5,000 lb. capacity		8.25	76.50	230	690	112	
	2075	8,000 lb. capacity		11.37	89	267	800	144.30	
	2085	Diesel, 5,000 lb. capacity		7.75	83.50	250	750	112	
	2090	12,000 lb. capacity		12.05	130	390	1,175	174.40	
	2095	20,000 lb. capacity		17.25	150	450	1,350	228.05	
	2100	Generator, electric, gas engine, 1.5 kW to 3 kW		2.58	10.15	30.50	91.50	26.70	
	2200	5 kW		3.22	12.65	38	114	33.35	
	2300	10 kW		5.93	33.50	101	305	67.65	
	2400	25 kW		7.41	84.50	254	760	110.05	
	2500	Diesel engine, 20 kW		9.21	74	222	665	118.05	
	2600	50 kW		15.95	96.50	289	865	185.35	
	2700	100 kW		28.60	133	400	1,200	308.80	
	2800	250 kW		54.35	255	765	2,300	587.85	
	2850	Hammer, hydraulic, for mounting on boom, to 500 ft. lb.		2.90	90.50	271	815	77.40	
	2860	1,000 ft. lb.		4.61	135	405	1,225	117.85	
	2900	Heaters, space, oil or electric, 50 MBH		1.46	8	24	72	16.50	
	3000	100 MBH		2.72	11.85	35.50	107	28.85	
	3100	300 MBH		7.92	40	120	360	87.40	
	3150	500 MBH		13.16	45	135	405	132.25	
	3200	Hose, water, suction with coupling, 20' long, 2" diameter		.02	3.13	9.40	28	2.05	
	3210	3" diameter		.03	4.10	12.30	37	2.70	
	3220	4" diameter		.03	4.80	14.40	43	3.10	
	3230	6" diameter		.11	17	51	153	11.10	
	3240	8" diameter		.27	45	135	405	29.15	
	3250	Discharge hose with coupling, 50' long, 2" diameter		.01	1.33	4	12	.90	
	3260	3" diameter		.01	2.22	6.65	19.95	1.40	
	3270	4" diameter		.02	3.47	10.40	31	2.25	
	3280	6" diameter		.06	8.65	26	78	5.70	
	3290	8" diameter		.24	40	120	360	25.90	
	3295	Insulation blower		.83	6	18.05	54	10.25	
	3300	Ladders, extension type, 16' to 36' long		.18	31.50	95	285	20.45	
	3400	40' to 60' long		.64	112	335	1,000	72.10	
	3405	Lance for cutting concrete		2.21	64	192	575	56.05	
	3407	Lawn mower, rotary, 22", 5 H.P.		1.06	25	75	225	23.45	
	3408	48" self-propelled		2.90	127	380	1,150	99.20	
	3410	Level, electronic, automatic, with tripod and leveling rod		1.05	61.50	185	555	45.40	
	3430	Laser type, for pipe and sewer line and grade		2.18	153	460	1,375	109.40	
	3440	Rotating beam for interior control		.90	62	186	560	44.40	
	3460	Builder's optical transit, with tripod and rod		.10	17	51	153	11	
	3500	Light towers, towable, with diesel generator, 2,000 watt		4.27	98	294	880	92.95	
	3600	4,000 watt		4.51	102	305	915	97.10	
	3700	Mixer, powered, plaster and mortar, 6 C.F., 7 H.P.		2.06	21	62.50	188	28.95	
	3800	10 C.F., 9 H.P.		2.24	33.50	101	305	38.15	
	3850	Nailer, pneumatic		.48	32.50	97	291	23.25	
	3900	Paint sprayers complete, 8 CFM		.85	59.50	179	535	42.60	
	4000	17 CFM		1.60	107	320	960	76.80	
	4020	Pavers, bituminous, rubber tires, 8' wide, 50 H.P., diesel		32.02	550	1,650	4,950	586.20	
	4030	10' wide, 150 H.P.		95.91	1,875	5,650	17,000	1,897	
	4050	Crawler, 8' wide, 100 H.P., diesel		87.85	2,000	5,975	17,900	1,898	
	4060	10' wide, 150 H.P.		104.28	2,275	6,825	20,500	2,199	
	4070	Concrete paver, 12' to 24' wide, 250 H.P.		87.88	1,625	4,900	14,700	1,683	
	4080	Placer-spreader-trimmer, 24' wide, 300 H.P.		117.87	2,475	7,425	22,300	2,428	

01 54 33 | Equipment Rental

		UNIT	HOURLY OPER. COST	RENT PER DAY	RENT PER WEEK	RENT PER MONTH	EQUIPMENT COST/DAY	
40	4100	Pump, centrifugal gas pump, 1-1/2" diam., 65 GPM	Ea.	3.93	52.50	157	470	62.90	40
	4200	2" diameter, 130 GPM		4.99	64.50	194	580	78.75	
	4300	3" diameter, 250 GPM		5.13	63.50	191	575	79.25	
	4400	6" diameter, 1,500 GPM		22.31	197	590	1,775	296.45	
	4500	Submersible electric pump, 1-1/4" diameter, 55 GPM		.40	17.65	53	159	13.80	
	4600	1-1/2" diameter, 83 GPM		.45	20.50	61	183	15.75	
	4700	2" diameter, 120 GPM		1.65	25	75.50	227	28.30	
	4800	3" diameter, 300 GPM		3.04	45	135	405	51.35	
	4900	4" diameter, 560 GPM		14.80	163	490	1,475	216.35	
	5000	6" diameter, 1,590 GPM		22.14	200	600	1,800	297.15	
	5100	Diaphragm pump, gas, single, 1-1/2" diameter		1.13	57.50	172	515	43.45	
	5200	2" diameter		3.99	70.50	212	635	74.30	
	5300	3" diameter		4.06	74.50	224	670	77.30	
	5400	Double, 4" diameter		6.05	155	465	1,400	141.40	
	5450	Pressure washer, 5 GPM, 3,000 psi		3.88	53.50	160	480	63.05	
	5460	7 GPM, 3,000 psi		4.95	63	189	565	77.40	
	5500	Trash pump, self-priming, gas, 2" diameter		3.83	23.50	71	213	44.80	
	5600	Diesel, 4" diameter		6.70	94.50	284	850	110.40	
	5650	Diesel, 6" diameter		16.90	165	495	1,475	234.20	
	5655	Grout pump		18.75	272	815	2,450	313	
	5700	Salamanders, L.P. gas fired, 100,000 BTU		2.89	14.15	42.50	128	31.65	
	5705	50,000 BTU		1.67	11.65	35	105	20.35	
	5720	Sandblaster, portable, open top, 3 C.F. capacity		.60	26.50	79.50	239	20.70	
	5730	6 C.F. capacity		1.01	39.50	118	355	31.65	
	5740	Accessories for above		.14	23	69	207	14.90	
	5750	Sander, floor		.77	19.35	58	174	17.75	
	5760	Edger		.52	15.85	47.50	143	13.65	
	5800	Saw, chain, gas engine, 18" long		1.76	22.50	67.50	203	27.55	
	5900	Hydraulic powered, 36" long		.78	67.50	202	605	46.65	
	5950	60" long		.78	69.50	208	625	47.85	
	6000	Masonry, table mounted, 14" diameter, 5 H.P.		1.32	81	243	730	59.20	
	6050	Portable cut-off, 8 H.P.		1.82	32.50	97.50	293	34	
	6100	Circular, hand held, electric, 7-1/4" diameter		.23	4.93	14.80	44.50	4.80	
	6200	12" diameter		.24	7.85	23.50	70.50	6.60	
	6250	Wall saw, w/hydraulic power, 10 H.P.		3.30	33.50	101	305	46.60	
	6275	Shot blaster, walk-behind, 20" wide		4.75	272	815	2,450	201	
	6280	Sidewalk broom, walk-behind		2.25	80	240	720	65.95	
	6300	Steam cleaner, 100 gallons per hour		3.35	80	240	720	74.80	
	6310	200 gallons per hour		4.35	96.50	290	870	92.75	
	6340	Tar kettle/pot, 400 gallons		16.53	76.50	230	690	178.20	
	6350	Torch, cutting, acetylene-oxygen, 150' hose, excludes gases		.45	14.85	44.50	134	12.50	
	6360	Hourly operating cost includes tips and gas		20.98				167.85	
	6410	Toilet, portable chemical		.13	22.50	67	201	14.45	
	6420	Recycle flush type		.16	27.50	83	249	17.90	
	6430	Toilet, fresh water flush, garden hose,		.19	33	99	297	21.35	
	6440	Hoisted, non-flush, for high rise		.16	27	81	243	17.45	
	6465	Tractor, farm with attachment		17.42	300	900	2,700	319.35	
	6480	Trailers, platform, flush deck, 2 axle, 3 ton capacity		1.69	21.50	65	195	26.55	
	6500	25 ton capacity		6.25	138	415	1,250	133	
	6600	40 ton capacity		8.06	197	590	1,775	182.50	
	6700	3 axle, 50 ton capacity		8.75	218	655	1,975	200.95	
	6800	75 ton capacity		11.11	290	870	2,600	262.90	
	6810	Trailer mounted cable reel for high voltage line work		5.90	282	845	2,525	216.20	
	6820	Trailer mounted cable tensioning rig		11.71	560	1,675	5,025	428.65	
	6830	Cable pulling rig		73.99	3,125	9,375	28,100	2,467	
	6850	Portable cable/wire puller, 8,000 lb. max. pulling capacity		3.71	205	615	1,850	152.70	
	6900	Water tank trailer, engine driven discharge, 5,000 gallons		7.18	153	460	1,375	149.45	
	6925	10,000 gallons		9.78	208	625	1,875	203.25	

01 54 33 | Equipment Rental

		Description	UNIT	HOURLY OPER. COST	RENT PER DAY	RENT PER WEEK	RENT PER MONTH	EQUIPMENT COST/DAY	
40	6950	Water truck, off highway, 6,000 gallons	Ea.	71.97	810	2,425	7,275	1,061	40
	7010	Tram car for high voltage line work, powered, 2 conductor		6.90	152	455	1,375	146.20	
	7020	Transit (builder's level) with tripod		.10	17	51	153	11	
	7030	Trench box, 3,000 lb., 6' x 8'		.56	93.50	281	845	60.70	
	7040	7,200 lb., 6' x 20'		.72	120	360	1,075	77.75	
	7050	8,000 lb., 8' x 16'		1.08	180	540	1,625	116.65	
	7060	9,500 lb., 8' x 20'		1.21	225	675	2,025	144.65	
	7065	11,000 lb., 8' x 24'		1.27	212	635	1,900	137.15	
	7070	12,000 lb., 10' x 20'		1.49	255	765	2,300	164.95	
	7100	Truck, pickup, 3/4 ton, 2 wheel drive		9.26	59.50	179	535	109.90	
	7200	4 wheel drive		9.51	76	228	685	121.70	
	7250	Crew carrier, 9 passenger		12.70	88	264	790	154.35	
	7290	Flat bed truck, 20,000 lb. GVW		15.31	128	385	1,150	199.45	
	7300	Tractor, 4 x 2, 220 H.P.		22.32	208	625	1,875	303.55	
	7410	330 H.P.		32.43	285	855	2,575	430.40	
	7500	6 x 4, 380 H.P.		36.20	330	990	2,975	487.60	
	7600	450 H.P.		44.36	400	1,200	3,600	594.90	
	7610	Tractor, with A frame, boom and winch, 225 H.P.		24.81	283	850	2,550	368.50	
	7620	Vacuum truck, hazardous material, 2,500 gallons		12.83	300	900	2,700	282.65	
	7625	5,000 gallons		13.06	425	1,275	3,825	359.50	
	7650	Vacuum, HEPA, 16 gallon, wet/dry		.85	17.15	51.50	155	17.10	
	7655	55 gallon, wet/dry		.78	26.50	80	240	22.25	
	7660	Water tank, portable		.73	155	465	1,400	98.85	
	7690	Sewer/catch basin vacuum, 14 C.Y., 1,500 gallons		17.37	640	1,925	5,775	523.95	
	7700	Welder, electric, 200 amp		3.83	16	48	144	40.20	
	7800	300 amp		5.56	19.35	58	174	56.10	
	7900	Gas engine, 200 amp		8.97	23	69	207	85.60	
	8000	300 amp		10.16	24	72	216	95.65	
	8100	Wheelbarrow, any size		.06	10.35	31	93	6.70	
	8200	Wrecking ball, 4,000 lb.		2.51	72	216	650	63.25	
50	0010	**HIGHWAY EQUIPMENT RENTAL** without operators							50
	0050	Asphalt batch plant, portable drum mixer, 100 ton/hr.	Ea.	88.67	1,500	4,475	13,400	1,604	
	0060	200 ton/hr.		102.29	1,600	4,775	14,300	1,773	
	0070	300 ton/hr.		120.22	1,875	5,600	16,800	2,082	
	0100	Backhoe attachment, long stick, up to 185 H.P., 10-1/2' long		.37	24.50	74	222	17.75	
	0140	Up to 250 H.P., 12' long		.41	27.50	83	249	19.90	
	0180	Over 250 H.P., 15' long		.57	37.50	113	340	27.15	
	0200	Special dipper arm, up to 100 H.P., 32' long		1.16	77	231	695	55.50	
	0240	Over 100 H.P., 33' long		1.45	96.50	290	870	69.60	
	0280	Catch basin/sewer cleaning truck, 3 ton, 9 C.Y., 1,000 gal.		35.50	410	1,225	3,675	528.95	
	0300	Concrete batch plant, portable, electric, 200 C.Y./hr.		24.26	540	1,625	4,875	519.05	
	0520	Grader/dozer attachment, ripper/scarifier, rear mounted, up to 135 H.P.		3.16	61.50	184	550	62.10	
	0540	Up to 180 H.P.		4.15	92.50	278	835	88.80	
	0580	Up to 250 H.P.		5.87	148	445	1,325	135.95	
	0700	Pvmt. removal bucket, for hyd. excavator, up to 90 H.P.		2.17	56.50	169	505	51.10	
	0740	Up to 200 H.P.		2.31	72	216	650	61.70	
	0780	Over 200 H.P.		2.53	88.50	265	795	73.20	
	0900	Aggregate spreader, self-propelled, 187 H.P.		50.75	715	2,150	6,450	836	
	1000	Chemical spreader, 3 C.Y.		3.18	46.50	139	415	53.20	
	1900	Hammermill, traveling, 250 H.P.		67.55	2,250	6,750	20,300	1,890	
	2000	Horizontal borer, 3" diameter, 13 H.P. gas driven		5.43	57.50	172	515	77.85	
	2150	Horizontal directional drill, 20,000 lb. thrust, 78 H.P. diesel		27.66	685	2,050	6,150	631.30	
	2160	30,000 lb. thrust, 115 H.P.		34.00	1,050	3,125	9,375	897	
	2170	50,000 lb. thrust, 170 H.P.		48.74	1,325	3,975	11,900	1,185	
	2190	Mud trailer for HDD, 1,500 gallons, 175 H.P., gas		25.58	157	470	1,400	298.65	
	2200	Hydromulcher, diesel, 3,000 gallon, for truck mounting		17.48	255	765	2,300	292.90	
	2300	Gas, 600 gallon		7.52	107	320	960	124.15	
	2400	Joint & crack cleaner, walk behind, 25 H.P.		3.17	52.50	158	475	56.95	

R015433 -10

01 54 33 | Equipment Rental

		UNIT	HOURLY OPER. COST	RENT PER DAY	RENT PER WEEK	RENT PER MONTH	EQUIPMENT COST/DAY	
50	**2500**	Filler, trailer mounted, 400 gallons, 20 H.P.	Ea.	8.37	220	660	1,975	198.95	**50**
	3000	Paint striper, self-propelled, 40 gallon, 22 H.P.		6.78	163	490	1,475	152.25	
	3100	120 gallon, 120 H.P.		19.28	410	1,225	3,675	399.25	
	3200	Post drivers, 6" I-Beam frame, for truck mounting		12.45	390	1,175	3,525	334.60	
	3400	Road sweeper, self-propelled, 8' wide, 90 H.P.		36.01	690	2,075	6,225	703.10	
	3450	Road sweeper, vacuum assisted, 4 C.Y., 220 gallons		58.45	650	1,950	5,850	857.60	
	4000	Road mixer, self-propelled, 130 H.P.		46.37	800	2,400	7,200	851	
	4100	310 H.P.		75.23	2,100	6,275	18,800	1,857	
	4220	Cold mix paver, incl. pug mill and bitumen tank, 165 H.P.		95.25	2,250	6,725	20,200	2,107	
	4240	Pavement brush, towed		3.44	96.50	290	870	85.50	
	4250	Paver, asphalt, wheel or crawler, 130 H.P., diesel		94.52	2,200	6,600	19,800	2,076	
	4300	Paver, road widener, gas, 1' to 6', 67 H.P.		46.80	940	2,825	8,475	939.40	
	4400	Diesel, 2' to 14', 88 H.P.		56.55	1,125	3,350	10,100	1,122	
	4600	Slipform pavers, curb and gutter, 2 track, 75 H.P.		58.01	1,225	3,650	11,000	1,194	
	4700	4 track, 165 H.P.		35.79	815	2,450	7,350	776.35	
	4800	Median barrier, 215 H.P.		58.61	1,300	3,900	11,700	1,249	
	4901	Trailer, low bed, 75 ton capacity		10.74	273	820	2,450	249.90	
	5000	Road planer, walk behind, 10" cutting width, 10 H.P.		2.46	33.50	100	300	39.70	
	5100	Self-propelled, 12" cutting width, 64 H.P.		8.28	117	350	1,050	136.25	
	5120	Traffic line remover, metal ball blaster, truck mounted, 115 H.P.		46.70	785	2,350	7,050	843.60	
	5140	Grinder, truck mounted, 115 H.P.		51.04	810	2,425	7,275	893.35	
	5160	Walk-behind, 11 H.P.		3.57	54.50	164	490	61.35	
	5200	Pavement profiler, 4' to 6' wide, 450 H.P.		217.23	3,425	10,300	30,900	3,798	
	5300	8' to 10' wide, 750 H.P.		332.58	4,525	13,600	40,800	5,381	
	5400	Roadway plate, steel, 1" x 8' x 20'		.09	14.65	44	132	9.50	
	5600	Stabilizer, self-propelled, 150 H.P.		41.26	640	1,925	5,775	715.05	
	5700	310 H.P.		76.41	1,700	5,075	15,200	1,626	
	5800	Striper, truck mounted, 120 gallon paint, 460 H.P.		48.89	490	1,475	4,425	686.10	
	5900	Thermal paint heating kettle, 115 gallons		7.73	26.50	80	240	77.85	
	6000	Tar kettle, 330 gallon, trailer mounted		12.31	60	180	540	134.50	
	7000	Tunnel locomotive, diesel, 8 to 12 ton		29.85	600	1,800	5,400	598.75	
	7005	Electric, 10 ton		29.33	685	2,050	6,150	644.70	
	7010	Muck cars, 1/2 C.Y. capacity		2.30	26	77.50	233	33.95	
	7020	1 C.Y. capacity		2.52	33.50	101	305	40.35	
	7030	2 C.Y. capacity		2.66	37.50	113	340	43.90	
	7040	Side dump, 2 C.Y. capacity		2.88	46.50	140	420	51.05	
	7050	3 C.Y. capacity		3.86	51.50	154	460	61.70	
	7060	5 C.Y. capacity		5.64	66.50	199	595	84.90	
	7100	Ventilating blower for tunnel, 7-1/2 H.P.		2.14	51	153	460	47.75	
	7110	10 H.P.		2.43	53.50	160	480	51.40	
	7120	20 H.P.		3.55	69.50	208	625	70	
	7140	40 H.P.		6.16	91.50	275	825	104.25	
	7160	60 H.P.		8.71	98.50	295	885	128.70	
	7175	75 H.P.		10.40	153	460	1,375	175.20	
	7180	200 H.P.		20.85	300	905	2,725	347.75	
	7800	Windrow loader, elevating		54.10	1,350	4,025	12,100	1,238	
60	**0010**	**LIFTING AND HOISTING EQUIPMENT RENTAL** without operators R015433 -10							**60**
	0150	Crane, flatbed mounted, 3 ton capacity	Ea.	14.45	195	585	1,750	232.65	
	0200	Crane, climbing, 106' jib, 6,000 lb. capacity, 410 fpm R312316		39.84	1,775	5,350	16,100	1,389	
	0300	101' jib, 10,250 lb. capacity, 270 fpm -45		46.57	2,275	6,800	20,400	1,733	
	0500	Tower, static, 130' high, 106' jib, 6,200 lb. capacity at 400 fpm		45.29	2,075	6,200	18,600	1,602	
	0520	Mini crawler spider crane, up to 24" wide, 1,990 lb. lifting capacity		12.54	535	1,600	4,800	420.30	
	0525	Up to 30" wide, 6,450 lb. lifting capacity		14.57	635	1,900	5,700	496.55	
	0530	Up to 52" wide, 6,680 lb. lifting capacity		23.17	775	2,325	6,975	650.40	
	0535	Up to 55" wide, 8,920 lb. lifting capacity		25.87	860	2,575	7,725	722	
	0540	Up to 66" wide, 13,350 lb. lifting capacity		35.03	1,325	4,000	12,000	1,080	
	0600	Crawler mounted, lattice boom, 1/2 C.Y., 15 tons at 12' radius		37.07	950	2,850	8,550	866.60	
	0700	3/4 C.Y., 20 tons at 12' radius		50.57	1,200	3,625	10,900	1,130	

01 54 33 | Equipment Rental

		UNIT	HOURLY OPER. COST	RENT PER DAY	RENT PER WEEK	RENT PER MONTH	EQUIPMENT COST/DAY
0800	1 C.Y., 25 tons at 12' radius	Ea.	67.62	1,400	4,225	12,700	1,386
0900	1-1/2 C.Y., 40 tons at 12' radius		66.51	1,425	4,300	12,900	1,392
1000	2 C.Y., 50 tons at 12' radius		89.03	2,200	6,600	19,800	2,032
1100	3 C.Y., 75 tons at 12' radius		75.49	1,875	5,650	17,000	1,734
1200	100 ton capacity, 60' boom		86.17	1,975	5,950	17,900	1,879
1300	165 ton capacity, 60' boom		106.42	2,275	6,850	20,600	2,221
1400	200 ton capacity, 70' boom		138.63	3,175	9,550	28,700	3,019
1500	350 ton capacity, 80' boom		182.75	4,075	12,200	36,600	3,902
1600	Truck mounted, lattice boom, 6 x 4, 20 tons at 10' radius		39.88	915	2,750	8,250	869
1700	25 tons at 10' radius		42.86	1,400	4,175	12,500	1,178
1800	8 x 4, 30 tons at 10' radius		45.67	1,475	4,425	13,300	1,250
1900	40 tons at 12' radius		48.70	1,550	4,625	13,900	1,315
2000	60 tons at 15' radius		53.85	1,650	4,950	14,900	1,421
2050	82 tons at 15' radius		59.60	1,775	5,350	16,100	1,547
2100	90 tons at 15' radius		66.59	1,950	5,825	17,500	1,698
2200	115 tons at 15' radius		75.12	2,175	6,525	19,600	1,906
2300	150 tons at 18' radius		81.33	2,275	6,850	20,600	2,021
2350	165 tons at 18' radius		87.30	2,425	7,275	21,800	2,153
2400	Truck mounted, hydraulic, 12 ton capacity		29.59	390	1,175	3,525	471.75
2500	25 ton capacity		36.46	485	1,450	4,350	581.70
2550	33 ton capacity		50.82	900	2,700	8,100	946.60
2560	40 ton capacity		49.62	900	2,700	8,100	937
2600	55 ton capacity		53.94	915	2,750	8,250	981.50
2700	80 ton capacity		75.94	1,475	4,400	13,200	1,487
2720	100 ton capacity		75.18	1,550	4,675	14,000	1,536
2740	120 ton capacity		103.12	1,825	5,500	16,500	1,925
2760	150 ton capacity		110.25	2,050	6,125	18,400	2,107
2800	Self-propelled, 4 x 4, with telescoping boom, 5 ton		15.18	230	690	2,075	259.45
2900	12-1/2 ton capacity		21.48	335	1,000	3,000	371.85
3000	15 ton capacity		34.52	535	1,600	4,800	596.20
3050	20 ton capacity		24.09	660	1,975	5,925	587.75
3100	25 ton capacity		36.80	615	1,850	5,550	664.40
3150	40 ton capacity		45.03	650	1,950	5,850	750.25
3200	Derricks, guy, 20 ton capacity, 60' boom, 75' mast		22.81	435	1,300	3,900	442.50
3300	100' boom, 115' mast		36.15	750	2,250	6,750	739.20
3400	Stiffleg, 20 ton capacity, 70' boom, 37' mast		25.48	565	1,700	5,100	543.85
3500	100' boom, 47' mast		39.43	910	2,725	8,175	860.50
3550	Helicopter, small, lift to 1,250 lb. maximum, w/pilot		99.44	3,525	10,600	31,800	2,916
3600	Hoists, chain type, overhead, manual, 3/4 ton		.15	.33	1	3	1.35
3900	10 ton		.79	6	18	54	9.90
4000	Hoist and tower, 5,000 lb. cap., portable electric, 40' high		5.14	252	755	2,275	192.10
4100	For each added 10' section, add		.12	19.65	59	177	12.75
4200	Hoist and single tubular tower, 5,000 lb. electric, 100' high		6.98	350	1,050	3,150	265.80
4300	For each added 6'-6" section, add		.21	34.50	103	310	22.25
4400	Hoist and double tubular tower, 5,000 lb., 100' high		7.59	385	1,150	3,450	290.75
4500	For each added 6'-6" section, add		.23	37.50	113	340	24.40
4550	Hoist and tower, mast type, 6,000 lb., 100' high		8.26	400	1,200	3,600	306.10
4570	For each added 10' section, add		.13	23	69	207	14.85
4600	Hoist and tower, personnel, electric, 2,000 lb., 100' @ 125 fpm		17.55	1,075	3,200	9,600	780.40
4700	3,000 lb., 100' @ 200 fpm		20.08	1,200	3,625	10,900	885.65
4800	3,000 lb., 150' @ 300 fpm		22.28	1,350	4,075	12,200	993.30
4900	4,000 lb., 100' @ 300 fpm		23.05	1,375	4,150	12,500	1,014
5000	6,000 lb., 100' @ 275 fpm		24.77	1,450	4,350	13,100	1,068
5100	For added heights up to 500', add	L.F.	.01	1.67	5	15	1.10
5200	Jacks, hydraulic, 20 ton	Ea.	.05	2	6	18	1.60
5500	100 ton		.40	12	36	108	10.40
6100	Jacks, hydraulic, climbing w/50' jackrods, control console, 30 ton cap.		2.17	145	435	1,300	104.40
6150	For each added 10' jackrod section, add		.05	3.33	10	30	2.40

01 54 33 | Equipment Rental

		UNIT	HOURLY OPER. COST	RENT PER DAY	RENT PER WEEK	RENT PER MONTH	EQUIPMENT COST/DAY	
60	6300	50 ton capacity	Ea.	3.49	232	695	2,075	166.95	**60**
	6350	For each added 10' jackrod section, add		.06	4	12	36	2.90	
	6500	125 ton capacity		9.13	610	1,825	5,475	438.05	
	6550	For each added 10' jackrod section, add		.62	41	123	370	29.55	
	6600	Cable jack, 10 ton capacity with 200' cable		1.82	122	365	1,100	87.60	
	6650	For each added 50' of cable, add	▼	.22	14.65	44	132	10.55	
70	0010	**WELLPOINT EQUIPMENT RENTAL** without operators	R015433 -10						**70**
	0020	Based on 2 months rental							
	0100	Combination jetting & wellpoint pump, 60 H.P. diesel	Ea.	15.72	360	1,075	3,225	340.75	
	0200	High pressure gas jet pump, 200 H.P., 300 psi	"	33.93	305	920	2,750	455.45	
	0300	Discharge pipe, 8" diameter	L.F.	.01	.58	1.74	5.20	.40	
	0350	12" diameter		.01	.83	2.50	7.50	.60	
	0400	Header pipe, flows up to 150 GPM, 4" diameter		.01	.53	1.59	4.77	.40	
	0500	400 GPM, 6" diameter		.01	.62	1.86	5.60	.45	
	0600	800 GPM, 8" diameter		.01	.83	2.50	7.50	.60	
	0700	1,500 GPM, 10" diameter		.01	.88	2.64	7.90	.65	
	0800	2,500 GPM, 12" diameter		.03	1.70	5.10	15.30	1.20	
	0900	4,500 GPM, 16" diameter		.03	2.18	6.55	19.65	1.55	
	0950	For quick coupling aluminum and plastic pipe, add	▼	.03	2.27	6.80	20.50	1.65	
	1100	Wellpoint, 25' long, with fittings & riser pipe, 1-1/2" or 2" diameter	Ea.	.07	4.52	13.55	40.50	3.25	
	1200	Wellpoint pump, diesel powered, 4" suction, 20 H.P.		7.02	207	620	1,850	180.20	
	1300	6" suction, 30 H.P.		9.42	257	770	2,300	229.35	
	1400	8" suction, 40 H.P.		12.76	350	1,050	3,150	312.10	
	1500	10" suction, 75 H.P.		18.83	410	1,225	3,675	395.65	
	1600	12" suction, 100 H.P.		27.32	660	1,975	5,925	613.55	
	1700	12" suction, 175 H.P.	▼	39.09	725	2,175	6,525	747.75	
80	0010	**MARINE EQUIPMENT RENTAL** without operators	R015433 -10						**80**
	0200	Barge, 400 ton, 30' wide x 90' long	Ea.	17.69	1,150	3,475	10,400	836.50	
	0240	800 ton, 45' wide x 90' long		22.21	1,425	4,275	12,800	1,033	
	2000	Tugboat, diesel, 100 H.P.		29.66	230	690	2,075	375.25	
	2040	250 H.P.		57.58	415	1,250	3,750	710.70	
	2080	380 H.P.		125.36	1,250	3,750	11,300	1,753	
	3000	Small work boat, gas, 16-foot, 50 H.P.		11.38	46.50	139	415	118.85	
	4000	Large, diesel, 48-foot, 200 H.P.	▼	74.90	1,325	3,975	11,900	1,394	

Crew No.	Bare Costs		Incl. Subs O&P		Cost Per Labor-Hour	

Crew A-1

Crew A-1	Hr.	Daily	Hr.	Daily	Bare Costs	Incl. O&P
1 Building Laborer	$41.05	$328.40	$62.10	$496.80	$41.05	$62.10
1 Concrete Saw, Gas Manual		72.90		80.19	9.11	10.02
8 L.H., Daily Totals		$401.30		$576.99	$50.16	$72.12

Crew A-1A	Hr.	Daily	Hr.	Daily	Bare Costs	Incl. O&P
1 Skilled Worker	$53.40	$427.20	$81.25	$650.00	$53.40	$81.25
1 Shot Blaster, 20"		201.00		221.10	25.13	27.64
8 L.H., Daily Totals		$628.20		$871.10	$78.53	$108.89

Crew A-1B	Hr.	Daily	Hr.	Daily	Bare Costs	Incl. O&P
1 Building Laborer	$41.05	$328.40	$62.10	$496.80	$41.05	$62.10
1 Concrete Saw		110.15		121.17	13.77	15.15
8 L.H., Daily Totals		$438.55		$617.97	$54.82	$77.25

Crew A-1C	Hr.	Daily	Hr.	Daily	Bare Costs	Incl. O&P
1 Building Laborer	$41.05	$328.40	$62.10	$496.80	$41.05	$62.10
1 Chain Saw, Gas, 18"		27.55		30.31	3.44	3.79
8 L.H., Daily Totals		$355.95		$527.11	$44.49	$65.89

Crew A-1D	Hr.	Daily	Hr.	Daily	Bare Costs	Incl. O&P
1 Building Laborer	$41.05	$328.40	$62.10	$496.80	$41.05	$62.10
1 Vibrating Plate, Gas, 18"		31.10		34.21	3.89	4.28
8 L.H., Daily Totals		$359.50		$531.01	$44.94	$66.38

Crew A-1E	Hr.	Daily	Hr.	Daily	Bare Costs	Incl. O&P
1 Building Laborer	$41.05	$328.40	$62.10	$496.80	$41.05	$62.10
1 Vibrating Plate, Gas, 21"		40.55		44.60	5.07	5.58
8 L.H., Daily Totals		$368.95		$541.40	$46.12	$67.68

Crew A-1F	Hr.	Daily	Hr.	Daily	Bare Costs	Incl. O&P
1 Building Laborer	$41.05	$328.40	$62.10	$496.80	$41.05	$62.10
1 Rammer/Tamper, Gas, 8"		45.65		50.22	5.71	6.28
8 L.H., Daily Totals		$374.05		$547.01	$46.76	$68.38

Crew A-1G	Hr.	Daily	Hr.	Daily	Bare Costs	Incl. O&P
1 Building Laborer	$41.05	$328.40	$62.10	$496.80	$41.05	$62.10
1 Rammer/Tamper, Gas, 15"		53.05		58.35	6.63	7.29
8 L.H., Daily Totals		$381.45		$555.15	$47.68	$69.39

Crew A-1H	Hr.	Daily	Hr.	Daily	Bare Costs	Incl. O&P
1 Building Laborer	$41.05	$328.40	$62.10	$496.80	$41.05	$62.10
1 Exterior Steam Cleaner		74.80		82.28	9.35	10.29
8 L.H., Daily Totals		$403.20		$579.08	$50.40	$72.39

Crew A-1J	Hr.	Daily	Hr.	Daily	Bare Costs	Incl. O&P
1 Building Laborer	$41.05	$328.40	$62.10	$496.80	$41.05	$62.10
1 Cultivator, Walk-Behind, 5 H.P.		70.50		77.55	8.81	9.69
8 L.H., Daily Totals		$398.90		$574.35	$49.86	$71.79

Crew A-1K	Hr.	Daily	Hr.	Daily	Bare Costs	Incl. O&P
1 Building Laborer	$41.05	$328.40	$62.10	$496.80	$41.05	$62.10
1 Cultivator, Walk-Behind, 8 H.P.		84.50		92.95	10.56	11.62
8 L.H., Daily Totals		$412.90		$589.75	$51.61	$73.72

Crew A-1M	Hr.	Daily	Hr.	Daily	Bare Costs	Incl. O&P
1 Building Laborer	$41.05	$328.40	$62.10	$496.80	$41.05	$62.10
1 Snow Blower, Walk-Behind		65.95		72.55	8.24	9.07
8 L.H., Daily Totals		$394.35		$569.35	$49.29	$71.17

Crew A-2	Hr.	Daily	Hr.	Daily	Bare Costs	Incl. O&P
2 Laborers	$41.05	$656.80	$62.10	$993.60	$42.53	$64.23
1 Truck Driver (light)	45.50	364.00	68.50	548.00		
1 Flatbed Truck, Gas, 1.5 Ton		195.20		214.72	8.13	8.95
24 L.H., Daily Totals		$1216.00		$1756.32	$50.67	$73.18

Crew A-2A	Hr.	Daily	Hr.	Daily	Bare Costs	Incl. O&P
2 Laborers	$41.05	$656.80	$62.10	$993.60	$42.53	$64.23
1 Truck Driver (light)	45.50	364.00	68.50	548.00		
1 Flatbed Truck, Gas, 1.5 Ton		195.20		214.72		
1 Concrete Saw		110.15		121.17	12.72	14.00
24 L.H., Daily Totals		$1326.15		$1877.48	$55.26	$78.23

Crew A-2B	Hr.	Daily	Hr.	Daily	Bare Costs	Incl. O&P
1 Truck Driver (light)	$45.50	$364.00	$68.50	$548.00	$45.50	$68.50
1 Flatbed Truck, Gas, 1.5 Ton		195.20		214.72	24.40	26.84
8 L.H., Daily Totals		$559.20		$762.72	$69.90	$95.34

Crew A-3A	Hr.	Daily	Hr.	Daily	Bare Costs	Incl. O&P
1 Equip. Oper. (light)	$51.65	$413.20	$77.55	$620.40	$51.65	$77.55
1 Pickup Truck, 4x4, 3/4 Ton		121.70		133.87	15.21	16.73
8 L.H., Daily Totals		$534.90		$754.27	$66.86	$94.28

Crew A-3B	Hr.	Daily	Hr.	Daily	Bare Costs	Incl. O&P
1 Equip. Oper. (medium)	$55.10	$440.80	$82.70	$661.60	$51.05	$76.72
1 Truck Driver (heavy)	47.00	376.00	70.75	566.00		
1 Dump Truck, 12 C.Y., 400 H.P.		567.05		623.76		
1 F.E. Loader, W.M., 2.5 C.Y.		531.20		584.32	68.64	75.50
16 L.H., Daily Totals		$1915.05		$2435.68	$119.69	$152.23

Crew A-3C	Hr.	Daily	Hr.	Daily	Bare Costs	Incl. O&P
1 Equip. Oper. (light)	$51.65	$413.20	$77.55	$620.40	$51.65	$77.55
1 Loader, Skid Steer, 78 H.P.		392.50		431.75	49.06	53.97
8 L.H., Daily Totals		$805.70		$1052.15	$100.71	$131.52

Crew A-3D	Hr.	Daily	Hr.	Daily	Bare Costs	Incl. O&P
1 Truck Driver (light)	$45.50	$364.00	$68.50	$548.00	$45.50	$68.50
1 Pickup Truck, 4x4, 3/4 Ton		121.70		133.87		
1 Flatbed Trailer, 25 Ton		133.00		146.30	31.84	35.02
8 L.H., Daily Totals		$618.70		$828.17	$77.34	$103.52

Crew A-3E	Hr.	Daily	Hr.	Daily	Bare Costs	Incl. O&P
1 Equip. Oper. (crane)	$57.45	$459.60	$86.25	$690.00	$52.23	$78.50
1 Truck Driver (heavy)	47.00	376.00	70.75	566.00		
1 Pickup Truck, 4x4, 3/4 Ton		121.70		133.87	7.61	8.37
16 L.H., Daily Totals		$957.30		$1389.87	$59.83	$86.87

Crew A-3F	Hr.	Daily	Hr.	Daily	Bare Costs	Incl. O&P
1 Equip. Oper. (crane)	$57.45	$459.60	$86.25	$690.00	$52.23	$78.50
1 Truck Driver (heavy)	47.00	376.00	70.75	566.00		
1 Pickup Truck, 4x4, 3/4 Ton		121.70		133.87		
1 Truck Tractor, 6x4, 380 H.P.		487.60		536.36		
1 Lowbed Trailer, 75 Ton		249.90		274.89	53.70	59.07
16 L.H., Daily Totals		$1694.80		$2201.12	$105.93	$137.57

Crew A-3G

Crew No.	Bare Costs Hr.	Bare Costs Daily	Incl. Subs O&P Hr.	Incl. Subs O&P Daily	Cost Per Labor-Hour Bare Costs	Cost Per Labor-Hour Incl. O&P
1 Equip. Oper. (crane)	$57.45	$459.60	$86.25	$690.00	$52.23	$78.50
1 Truck Driver (heavy)	47.00	376.00	70.75	566.00		
1 Pickup Truck, 4x4, 3/4 Ton		121.70		133.87		
1 Truck Tractor, 6x4, 450 H.P.		594.90		654.39		
1 Lowbed Trailer, 75 Ton		249.90		274.89	60.41	66.45
16 L.H., Daily Totals		$1802.10		$2319.15	$112.63	$144.95

Crew A-3H

Crew No.	Bare Costs Hr.	Bare Costs Daily	Incl. Subs O&P Hr.	Incl. Subs O&P Daily	Cost Per Labor-Hour Bare Costs	Cost Per Labor-Hour Incl. O&P
1 Equip. Oper. (crane)	$57.45	$459.60	$86.25	$690.00	$57.45	$86.25
1 Hyd. Crane, 12 Ton (Daily)		708.90		779.79	88.61	97.47
8 L.H., Daily Totals		$1168.50		$1469.79	$146.06	$183.72

Crew A-3I

Crew No.	Bare Costs Hr.	Bare Costs Daily	Incl. Subs O&P Hr.	Incl. Subs O&P Daily	Cost Per Labor-Hour Bare Costs	Cost Per Labor-Hour Incl. O&P
1 Equip. Oper. (crane)	$57.45	$459.60	$86.25	$690.00	$57.45	$86.25
1 Hyd. Crane, 25 Ton (Daily)		785.85		864.43	98.23	108.05
8 L.H., Daily Totals		$1245.45		$1554.43	$155.68	$194.30

Crew A-3J

Crew No.	Bare Costs Hr.	Bare Costs Daily	Incl. Subs O&P Hr.	Incl. Subs O&P Daily	Cost Per Labor-Hour Bare Costs	Cost Per Labor-Hour Incl. O&P
1 Equip. Oper. (crane)	$57.45	$459.60	$86.25	$690.00	$57.45	$86.25
1 Hyd. Crane, 40 Ton (Daily)		1280.00		1408.00	160.00	176.00
8 L.H., Daily Totals		$1739.60		$2098.00	$217.45	$262.25

Crew A-3K

Crew No.	Bare Costs Hr.	Bare Costs Daily	Incl. Subs O&P Hr.	Incl. Subs O&P Daily	Cost Per Labor-Hour Bare Costs	Cost Per Labor-Hour Incl. O&P
1 Equip. Oper. (crane)	$57.45	$459.60	$86.25	$690.00	$53.33	$80.05
1 Equip. Oper. (oiler)	49.20	393.60	73.85	590.80		
1 Hyd. Crane, 55 Ton (Daily)		1299.00		1428.90		
1 P/U Truck, 3/4 Ton (Daily)		140.45		154.50	89.97	98.96
16 L.H., Daily Totals		$2292.65		$2864.20	$143.29	$179.01

Crew A-3L

Crew No.	Bare Costs Hr.	Bare Costs Daily	Incl. Subs O&P Hr.	Incl. Subs O&P Daily	Cost Per Labor-Hour Bare Costs	Cost Per Labor-Hour Incl. O&P
1 Equip. Oper. (crane)	$57.45	$459.60	$86.25	$690.00	$53.33	$80.05
1 Equip. Oper. (oiler)	49.20	393.60	73.85	590.80		
1 Hyd. Crane, 80 Ton (Daily)		2056.00		2261.60		
1 P/U Truck, 3/4 Ton (Daily)		140.45		154.50	137.28	151.01
16 L.H., Daily Totals		$3049.65		$3696.90	$190.60	$231.06

Crew A-3M

Crew No.	Bare Costs Hr.	Bare Costs Daily	Incl. Subs O&P Hr.	Incl. Subs O&P Daily	Cost Per Labor-Hour Bare Costs	Cost Per Labor-Hour Incl. O&P
1 Equip. Oper. (crane)	$57.45	$459.60	$86.25	$690.00	$53.33	$80.05
1 Equip. Oper. (oiler)	49.20	393.60	73.85	590.80		
1 Hyd. Crane, 100 Ton (Daily)		2179.00		2396.90		
1 P/U Truck, 3/4 Ton (Daily)		140.45		154.50	144.97	159.46
16 L.H., Daily Totals		$3172.65		$3832.20	$198.29	$239.51

Crew A-3N

Crew No.	Bare Costs Hr.	Bare Costs Daily	Incl. Subs O&P Hr.	Incl. Subs O&P Daily	Cost Per Labor-Hour Bare Costs	Cost Per Labor-Hour Incl. O&P
1 Equip. Oper. (crane)	$57.45	$459.60	$86.25	$690.00	$57.45	$86.25
1 Tower Crane (monthly)		1181.00		1299.10	147.63	162.39
8 L.H., Daily Totals		$1640.60		$1989.10	$205.07	$248.64

Crew A-3P

Crew No.	Bare Costs Hr.	Bare Costs Daily	Incl. Subs O&P Hr.	Incl. Subs O&P Daily	Cost Per Labor-Hour Bare Costs	Cost Per Labor-Hour Incl. O&P
1 Equip. Oper. (light)	$51.65	$413.20	$77.55	$620.40	$51.65	$77.55
1 A.T. Forklift, 31' reach, 45' lift		464.75		511.23	58.09	63.90
8 L.H., Daily Totals		$877.95		$1131.63	$109.74	$141.45

Crew A-3Q

Crew No.	Bare Costs Hr.	Bare Costs Daily	Incl. Subs O&P Hr.	Incl. Subs O&P Daily	Cost Per Labor-Hour Bare Costs	Cost Per Labor-Hour Incl. O&P
1 Equip. Oper. (light)	$51.65	$413.20	$77.55	$620.40	$51.65	$77.55
1 Pickup Truck, 4x4, 3/4 Ton		121.70		133.87		
1 Flatbed Trailer, 3 Ton		26.55		29.20	18.53	20.38
8 L.H., Daily Totals		$561.45		$783.48	$70.18	$97.93

Crew A-3R

Crew No.	Bare Costs Hr.	Bare Costs Daily	Incl. Subs O&P Hr.	Incl. Subs O&P Daily	Cost Per Labor-Hour Bare Costs	Cost Per Labor-Hour Incl. O&P
1 Equip. Oper. (light)	$51.65	$413.20	$77.55	$620.40	$51.65	$77.55
1 Forklift, Smooth Floor, 8,000 Lb.		144.30		158.73	18.04	19.84
8 L.H., Daily Totals		$557.50		$779.13	$69.69	$97.39

Crew A-4

Crew No.	Bare Costs Hr.	Bare Costs Daily	Incl. Subs O&P Hr.	Incl. Subs O&P Daily	Cost Per Labor-Hour Bare Costs	Cost Per Labor-Hour Incl. O&P
2 Carpenters	$51.65	$826.40	$78.15	$1250.40	$48.85	$73.68
1 Painter, Ordinary	43.25	346.00	64.75	518.00		
24 L.H., Daily Totals		$1172.40		$1768.40	$48.85	$73.68

Crew A-5

Crew No.	Bare Costs Hr.	Bare Costs Daily	Incl. Subs O&P Hr.	Incl. Subs O&P Daily	Cost Per Labor-Hour Bare Costs	Cost Per Labor-Hour Incl. O&P
2 Laborers	$41.05	$656.80	$62.10	$993.60	$41.54	$62.81
.25 Truck Driver (light)	45.50	91.00	68.50	137.00		
.25 Flatbed Truck, Gas, 1.5 Ton		48.80		53.68	2.71	2.98
18 L.H., Daily Totals		$796.60		$1184.28	$44.26	$65.79

Crew A-6

Crew No.	Bare Costs Hr.	Bare Costs Daily	Incl. Subs O&P Hr.	Incl. Subs O&P Daily	Cost Per Labor-Hour Bare Costs	Cost Per Labor-Hour Incl. O&P
1 Instrument Man	$53.40	$427.20	$81.25	$650.00	$51.52	$78.00
1 Rodman/Chainman	49.65	397.20	74.75	598.00		
1 Level, Electronic		45.40		49.94	2.84	3.12
16 L.H., Daily Totals		$869.80		$1297.94	$54.36	$81.12

Crew A-7

Crew No.	Bare Costs Hr.	Bare Costs Daily	Incl. Subs O&P Hr.	Incl. Subs O&P Daily	Cost Per Labor-Hour Bare Costs	Cost Per Labor-Hour Incl. O&P
1 Chief of Party	$65.10	$520.80	$98.10	$784.80	$56.05	$84.70
1 Instrument Man	53.40	427.20	81.25	650.00		
1 Rodman/Chainman	49.65	397.20	74.75	598.00		
1 Level, Electronic		45.40		49.94	1.89	2.08
24 L.H., Daily Totals		$1390.60		$2082.74	$57.94	$86.78

Crew A-8

Crew No.	Bare Costs Hr.	Bare Costs Daily	Incl. Subs O&P Hr.	Incl. Subs O&P Daily	Cost Per Labor-Hour Bare Costs	Cost Per Labor-Hour Incl. O&P
1 Chief of Party	$65.10	$520.80	$98.10	$784.80	$54.45	$82.21
1 Instrument Man	53.40	427.20	81.25	650.00		
2 Rodmen/Chainmen	49.65	794.40	74.75	1196.00		
1 Level, Electronic		45.40		49.94	1.42	1.56
32 L.H., Daily Totals		$1787.80		$2680.74	$55.87	$83.77

Crew A-9

Crew No.	Bare Costs Hr.	Bare Costs Daily	Incl. Subs O&P Hr.	Incl. Subs O&P Daily	Cost Per Labor-Hour Bare Costs	Cost Per Labor-Hour Incl. O&P
1 Asbestos Foreman	$57.85	$462.80	$89.30	$714.40	$57.41	$88.64
7 Asbestos Workers	57.35	3211.60	88.55	4958.80		
64 L.H., Daily Totals		$3674.40		$5673.20	$57.41	$88.64

Crew A-10A

Crew No.	Bare Costs Hr.	Bare Costs Daily	Incl. Subs O&P Hr.	Incl. Subs O&P Daily	Cost Per Labor-Hour Bare Costs	Cost Per Labor-Hour Incl. O&P
1 Asbestos Foreman	$57.85	$462.80	$89.30	$714.40	$57.52	$88.80
2 Asbestos Workers	57.35	917.60	88.55	1416.80		
24 L.H., Daily Totals		$1380.40		$2131.20	$57.52	$88.80

Crew A-10B

Crew No.	Bare Costs Hr.	Bare Costs Daily	Incl. Subs O&P Hr.	Incl. Subs O&P Daily	Cost Per Labor-Hour Bare Costs	Cost Per Labor-Hour Incl. O&P
1 Asbestos Foreman	$57.85	$462.80	$89.30	$714.40	$57.48	$88.74
3 Asbestos Workers	57.35	1376.40	88.55	2125.20		
32 L.H., Daily Totals		$1839.20		$2839.60	$57.48	$88.74

Crew A-10C

Crew No.	Bare Costs Hr.	Bare Costs Daily	Incl. Subs O&P Hr.	Incl. Subs O&P Daily	Cost Per Labor-Hour Bare Costs	Cost Per Labor-Hour Incl. O&P
3 Asbestos Workers	$57.35	$1376.40	$88.55	$2125.20	$57.35	$88.55
1 Flatbed Truck, Gas, 1.5 Ton		195.20		214.72	8.13	8.95
24 L.H., Daily Totals		$1571.60		$2339.92	$65.48	$97.50

Crews - Standard

Crew No.	Bare Costs Hr.	Daily	Incl. Subs O&P Hr.	Daily	Cost Per Labor-Hour Bare Costs	Incl. O&P
Crew A-10D	Hr.	Daily	Hr.	Daily	Bare Costs	Incl. O&P
2 Asbestos Workers	$57.35	$917.60	$88.55	$1416.80	$55.34	$84.30
1 Equip. Oper. (crane)	57.45	459.60	86.25	690.00		
1 Equip. Oper. (oiler)	49.20	393.60	73.85	590.80		
1 Hydraulic Crane, 33 Ton		946.60		1041.26	29.58	32.54
32 L.H., Daily Totals		$2717.40		$3738.86	$84.92	$116.84

Crew A-11	Hr.	Daily	Hr.	Daily	Bare Costs	Incl. O&P
1 Asbestos Foreman	$57.85	$462.80	$89.30	$714.40	$57.41	$88.64
7 Asbestos Workers	57.35	3211.60	88.55	4958.80		
2 Chip. Hammers, 12 Lb., Elec.		42.00		46.20	.66	.72
64 L.H., Daily Totals		$3716.40		$5719.40	$58.07	$89.37

Crew A-12	Hr.	Daily	Hr.	Daily	Bare Costs	Incl. O&P
1 Asbestos Foreman	$57.85	$462.80	$89.30	$714.40	$57.41	$88.64
7 Asbestos Workers	57.35	3211.60	88.55	4958.80		
1 Trk-Mtd Vac, 14 CY, 1500 Gal.		523.95		576.35		
1 Flatbed Truck, 20,000 GVW		199.45		219.40	11.30	12.43
64 L.H., Daily Totals		$4397.80		$6468.94	$68.72	$101.08

Crew A-13	Hr.	Daily	Hr.	Daily	Bare Costs	Incl. O&P
1 Equip. Oper. (light)	$51.65	$413.20	$77.55	$620.40	$51.65	$77.55
1 Trk-Mtd Vac, 14 CY, 1500 Gal.		523.95		576.35		
1 Flatbed Truck, 20,000 GVW		199.45		219.40	90.42	99.47
8 L.H., Daily Totals		$1136.60		$1416.14	$142.07	$177.02

Crew B-1	Hr.	Daily	Hr.	Daily	Bare Costs	Incl. O&P
1 Labor Foreman (outside)	$43.05	$344.40	$65.10	$520.80	$41.72	$63.10
2 Laborers	41.05	656.80	62.10	993.60		
24 L.H., Daily Totals		$1001.20		$1514.40	$41.72	$63.10

Crew B-1A	Hr.	Daily	Hr.	Daily	Bare Costs	Incl. O&P
1 Labor Foreman (outside)	$43.05	$344.40	$65.10	$520.80	$41.72	$63.10
2 Laborers	41.05	656.80	62.10	993.60		
2 Cutting Torches		25.00		27.50		
2 Sets of Gases		335.70		369.27	15.03	16.53
24 L.H., Daily Totals		$1361.90		$1911.17	$56.75	$79.63

Crew B-1B	Hr.	Daily	Hr.	Daily	Bare Costs	Incl. O&P
1 Labor Foreman (outside)	$43.05	$344.40	$65.10	$520.80	$45.65	$68.89
2 Laborers	41.05	656.80	62.10	993.60		
1 Equip. Oper. (crane)	57.45	459.60	86.25	690.00		
2 Cutting Torches		25.00		27.50		
2 Sets of Gases		335.70		369.27		
1 Hyd. Crane, 12 Ton		471.75		518.92	26.01	28.62
32 L.H., Daily Totals		$2293.25		$3120.09	$71.66	$97.50

Crew B-1C	Hr.	Daily	Hr.	Daily	Bare Costs	Incl. O&P
1 Labor Foreman (outside)	$43.05	$344.40	$65.10	$520.80	$41.72	$63.10
2 Laborers	41.05	656.80	62.10	993.60		
1 Telescoping Boom Lift, to 60'		451.25		496.38	18.80	20.68
24 L.H., Daily Totals		$1452.45		$2010.78	$60.52	$83.78

Crew B-1D	Hr.	Daily	Hr.	Daily	Bare Costs	Incl. O&P
2 Laborers	$41.05	$656.80	$62.10	$993.60	$41.05	$62.10
1 Small Work Boat, Gas, 50 H.P.		118.85		130.74		
1 Pressure Washer, 7 GPM		77.40		85.14	12.27	13.49
16 L.H., Daily Totals		$853.05		$1209.47	$53.32	$75.59

Crew B-1E	Hr.	Daily	Hr.	Daily	Bare Costs	Incl. O&P
1 Labor Foreman (outside)	$43.05	$344.40	$65.10	$520.80	$41.55	$62.85
3 Laborers	41.05	985.20	62.10	1490.40		
1 Work Boat, Diesel, 200 H.P.		1394.00		1533.40		
2 Pressure Washers, 7 GPM		154.80		170.28	48.40	53.24
32 L.H., Daily Totals		$2878.40		$3714.88	$89.95	$116.09

Crew B-1F	Hr.	Daily	Hr.	Daily	Bare Costs	Incl. O&P
2 Skilled Workers	$53.40	$854.40	$81.25	$1300.00	$49.28	$74.87
1 Laborer	41.05	328.40	62.10	496.80		
1 Small Work Boat, Gas, 50 H.P.		118.85		130.74		
1 Pressure Washer, 7 GPM		77.40		85.14	8.18	8.99
24 L.H., Daily Totals		$1379.05		$2012.68	$57.46	$83.86

Crew B-1G	Hr.	Daily	Hr.	Daily	Bare Costs	Incl. O&P
2 Laborers	$41.05	$656.80	$62.10	$993.60	$41.05	$62.10
1 Small Work Boat, Gas, 50 H.P.		118.85		130.74	7.43	8.17
16 L.H., Daily Totals		$775.65		$1124.34	$48.48	$70.27

Crew B-1H	Hr.	Daily	Hr.	Daily	Bare Costs	Incl. O&P
2 Skilled Workers	$53.40	$854.40	$81.25	$1300.00	$49.28	$74.87
1 Laborer	41.05	328.40	62.10	496.80		
1 Small Work Boat, Gas, 50 H.P.		118.85		130.74	4.95	5.45
24 L.H., Daily Totals		$1301.65		$1927.54	$54.24	$80.31

Crew B-1J	Hr.	Daily	Hr.	Daily	Bare Costs	Incl. O&P
1 Labor Foreman (inside)	$41.55	$332.40	$62.85	$502.80	$41.30	$62.48
1 Laborer	41.05	328.40	62.10	496.80		
16 L.H., Daily Totals		$660.80		$999.60	$41.30	$62.48

Crew B-1K	Hr.	Daily	Hr.	Daily	Bare Costs	Incl. O&P
1 Carpenter Foreman (inside)	$52.15	$417.20	$78.90	$631.20	$51.90	$78.53
1 Carpenter	51.65	413.20	78.15	625.20		
16 L.H., Daily Totals		$830.40		$1256.40	$51.90	$78.53

Crew B-2	Hr.	Daily	Hr.	Daily	Bare Costs	Incl. O&P
1 Labor Foreman (outside)	$43.05	$344.40	$65.10	$520.80	$41.45	$62.70
4 Laborers	41.05	1313.60	62.10	1987.20		
40 L.H., Daily Totals		$1658.00		$2508.00	$41.45	$62.70

Crew B-2A	Hr.	Daily	Hr.	Daily	Bare Costs	Incl. O&P
1 Labor Foreman (outside)	$43.05	$344.40	$65.10	$520.80	$41.72	$63.10
2 Laborers	41.05	656.80	62.10	993.60		
1 Telescoping Boom Lift, to 60'		451.25		496.38	18.80	20.68
24 L.H., Daily Totals		$1452.45		$2010.78	$60.52	$83.78

Crew B-3	Hr.	Daily	Hr.	Daily	Bare Costs	Incl. O&P
1 Labor Foreman (outside)	$43.05	$344.40	$65.10	$520.80	$45.71	$68.92
2 Laborers	41.05	656.80	62.10	993.60		
1 Equip. Oper. (medium)	55.10	440.80	82.70	661.60		
2 Truck Drivers (heavy)	47.00	752.00	70.75	1132.00		
1 Crawler Loader, 3 C.Y.		1202.00		1322.20		
2 Dump Trucks, 12 C.Y., 400 H.P.		1134.10		1247.51	48.67	53.54
48 L.H., Daily Totals		$4530.10		$5877.71	$94.38	$122.45

Crew B-3A	Hr.	Daily	Hr.	Daily	Bare Costs	Incl. O&P
4 Laborers	$41.05	$1313.60	$62.10	$1987.20	$43.86	$66.22
1 Equip. Oper. (medium)	55.10	440.80	82.70	661.60		
1 Hyd. Excavator, 1.5 C.Y.		908.70		999.57	22.72	24.99
40 L.H., Daily Totals		$2663.10		$3648.37	$66.58	$91.21

Crew No.	Bare Costs		Incl. Subs O&P		Cost Per Labor-Hour	
Crew B-3B	Hr.	Daily	Hr.	Daily	Bare Costs	Incl. O&P
2 Laborers	$41.05	$656.80	$62.10	$993.60	$46.05	$69.41
1 Equip. Oper. (medium)	55.10	440.80	82.70	661.60		
1 Truck Driver (heavy)	47.00	376.00	70.75	566.00		
1 Backhoe Loader, 80 H.P.		387.85		426.63		
1 Dump Truck, 12 C.Y., 400 H.P.		567.05		623.76	29.84	32.82
32 L.H., Daily Totals		$2428.50		$3271.59	$75.89	$102.24
Crew B-3C	Hr.	Daily	Hr.	Daily	Bare Costs	Incl. O&P
3 Laborers	$41.05	$985.20	$62.10	$1490.40	$44.56	$67.25
1 Equip. Oper. (medium)	55.10	440.80	82.70	661.60		
1 Crawler Loader, 4 C.Y.		1461.00		1607.10	45.66	50.22
32 L.H., Daily Totals		$2887.00		$3759.10	$90.22	$117.47
Crew B-4	Hr.	Daily	Hr.	Daily	Bare Costs	Incl. O&P
1 Labor Foreman (outside)	$43.05	$344.40	$65.10	$520.80	$42.38	$64.04
4 Laborers	41.05	1313.60	62.10	1987.20		
1 Truck Driver (heavy)	47.00	376.00	70.75	566.00		
1 Truck Tractor, 220 H.P.		303.55		333.90		
1 Flatbed Trailer, 40 Ton		182.50		200.75	10.13	11.14
48 L.H., Daily Totals		$2520.05		$3608.66	$52.50	$75.18
Crew B-5	Hr.	Daily	Hr.	Daily	Bare Costs	Incl. O&P
1 Labor Foreman (outside)	$43.05	$344.40	$65.10	$520.80	$45.35	$68.41
4 Laborers	41.05	1313.60	62.10	1987.20		
2 Equip. Oper. (medium)	55.10	881.60	82.70	1323.20		
1 Air Compressor, 250 cfm		167.95		184.75		
2 Breakers, Pavement, 60 lb.		21.50		23.65		
2 -50' Air Hoses, 1.5"		45.50		50.05		
1 Crawler Loader, 3 C.Y.		1202.00		1322.20	25.66	28.23
56 L.H., Daily Totals		$3976.55		$5411.85	$71.01	$96.64
Crew B-5A	Hr.	Daily	Hr.	Daily	Bare Costs	Incl. O&P
1 Labor Foreman (outside)	$43.05	$344.40	$65.10	$520.80	$45.43	$68.51
6 Laborers	41.05	1970.40	62.10	2980.80		
2 Equip. Oper. (medium)	55.10	881.60	82.70	1323.20		
1 Equip. Oper. (light)	51.65	413.20	77.55	620.40		
2 Truck Drivers (heavy)	47.00	752.00	70.75	1132.00		
1 Air Compressor, 365 cfm		213.35		234.69		
2 Breakers, Pavement, 60 lb.		21.50		23.65		
8 -50' Air Hoses, 1"		67.20		73.92		
2 Dump Trucks, 8 C.Y., 220 H.P.		679.20		747.12	10.22	11.24
96 L.H., Daily Totals		$5342.85		$7656.57	$55.65	$79.76
Crew B-5B	Hr.	Daily	Hr.	Daily	Bare Costs	Incl. O&P
1 Powderman	$53.40	$427.20	$81.25	$650.00	$50.77	$76.48
2 Equip. Oper. (medium)	55.10	881.60	82.70	1323.20		
3 Truck Drivers (heavy)	47.00	1128.00	70.75	1698.00		
1 F.E. Loader, W.M., 2.5 C.Y.		531.20		584.32		
3 Dump Trucks, 12 C.Y., 400 H.P.		1701.15		1871.27		
1 Air Compressor, 365 cfm		213.35		234.69	50.95	56.05
48 L.H., Daily Totals		$4882.50		$6361.47	$101.72	$132.53

Crew No.	Bare Costs		Incl. Subs O&P		Cost Per Labor-Hour	
Crew B-5C	Hr.	Daily	Hr.	Daily	Bare Costs	Incl. O&P
3 Laborers	$41.05	$985.20	$62.10	$1490.40	$47.36	$71.33
1 Equip. Oper. (medium)	55.10	440.80	82.70	661.60		
2 Truck Drivers (heavy)	47.00	752.00	70.75	1132.00		
1 Equip. Oper. (crane)	57.45	459.60	86.25	690.00		
1 Equip. Oper. (oiler)	49.20	393.60	73.85	590.80		
2 Dump Trucks, 12 C.Y., 400 H.P.		1134.10		1247.51		
1 Crawler Loader, 4 C.Y.		1461.00		1607.10		
1 S.P. Crane, 4x4, 25 Ton		664.40		730.84	50.93	56.02
64 L.H., Daily Totals		$6290.70		$8150.25	$98.29	$127.35
Crew B-5D	Hr.	Daily	Hr.	Daily	Bare Costs	Incl. O&P
1 Labor Foreman (outside)	$43.05	$344.40	$65.10	$520.80	$45.56	$68.71
4 Laborers	41.05	1313.60	62.10	1987.20		
2 Equip. Oper. (medium)	55.10	881.60	82.70	1323.20		
1 Truck Driver (heavy)	47.00	376.00	70.75	566.00		
1 Air Compressor, 250 cfm		167.95		184.75		
2 Breakers, Pavement, 60 lb.		21.50		23.65		
2 -50' Air Hoses, 1.5"		45.50		50.05		
1 Crawler Loader, 3 C.Y.		1202.00		1322.20		
1 Dump Truck, 12 C.Y., 400 H.P.		567.05		623.76	31.31	34.44
64 L.H., Daily Totals		$4919.60		$6601.60	$76.87	$103.15
Crew B-6	Hr.	Daily	Hr.	Daily	Bare Costs	Incl. O&P
2 Laborers	$41.05	$656.80	$62.10	$993.60	$44.58	$67.25
1 Equip. Oper. (light)	51.65	413.20	77.55	620.40		
1 Backhoe Loader, 48 H.P.		320.20		352.22	13.34	14.68
24 L.H., Daily Totals		$1390.20		$1966.22	$57.92	$81.93
Crew B-6A	Hr.	Daily	Hr.	Daily	Bare Costs	Incl. O&P
.5 Labor Foreman (outside)	$43.05	$172.20	$65.10	$260.40	$47.07	$70.94
1 Laborer	41.05	328.40	62.10	496.80		
1 Equip. Oper. (medium)	55.10	440.80	82.70	661.60		
1 Vacuum Truck, 5000 Gal.		359.50		395.45	17.98	19.77
20 L.H., Daily Totals		$1300.90		$1814.25	$65.05	$90.71
Crew B-6B	Hr.	Daily	Hr.	Daily	Bare Costs	Incl. O&P
2 Labor Foremen (outside)	$43.05	$688.80	$65.10	$1041.60	$41.72	$63.10
4 Laborers	41.05	1313.60	62.10	1987.20		
1 S.P. Crane, 4x4, 5 Ton		259.45		285.39		
1 Flatbed Truck, Gas, 1.5 Ton		195.20		214.72		
1 Butt Fusion Mach., 4"-12" diam.		405.15		445.67	17.91	19.70
48 L.H., Daily Totals		$2862.20		$3974.58	$59.63	$82.80
Crew B-6C	Hr.	Daily	Hr.	Daily	Bare Costs	Incl. O&P
2 Labor Foremen (outside)	$43.05	$688.80	$65.10	$1041.60	$41.72	$63.10
4 Laborers	41.05	1313.60	62.10	1987.20		
1 S.P. Crane, 4x4, 12 Ton		371.85		409.04		
1 Flatbed Truck, Gas, 3 Ton		245.95		270.55		
1 Butt Fusion Mach., 8"-24" diam.		1927.00		2119.70	53.02	58.32
48 L.H., Daily Totals		$4547.20		$5828.08	$94.73	$121.42
Crew B-7	Hr.	Daily	Hr.	Daily	Bare Costs	Incl. O&P
1 Labor Foreman (outside)	$43.05	$344.40	$65.10	$520.80	$43.73	$66.03
4 Laborers	41.05	1313.60	62.10	1987.20		
1 Equip. Oper. (medium)	55.10	440.80	82.70	661.60		
1 Brush Chipper, 12", 130 H.P.		389.40		428.34		
1 Crawler Loader, 3 C.Y.		1202.00		1322.20		
2 Chain Saws, Gas, 36" Long		93.30		102.63	35.10	38.61
48 L.H., Daily Totals		$3783.50		$5022.77	$78.82	$104.64

Crews - Standard

Crew No.	Bare Costs Hr.	Daily	Incl. Subs O&P Hr.	Daily	Cost Per Labor-Hour Bare Costs	Incl. O&P
Crew B-7A	**Hr.**	**Daily**	**Hr.**	**Daily**	**Bare Costs**	**Incl. O&P**
2 Laborers	$41.05	$656.80	$62.10	$993.60	$44.58	$67.25
1 Equip. Oper. (light)	51.65	413.20	77.55	620.40		
1 Rake w/Tractor		332.75		366.02		
2 Chain Saws, Gas, 18"		55.10		60.61	16.16	17.78
24 L.H., Daily Totals		$1457.85		$2040.64	$60.74	$85.03
Crew B-7B	**Hr.**	**Daily**	**Hr.**	**Daily**	**Bare Costs**	**Incl. O&P**
1 Labor Foreman (outside)	$43.05	$344.40	$65.10	$520.80	$44.19	$66.71
4 Laborers	41.05	1313.60	62.10	1987.20		
1 Equip. Oper. (medium)	55.10	440.80	82.70	661.60		
1 Truck Driver (heavy)	47.00	376.00	70.75	566.00		
1 Brush Chipper, 12", 130 H.P.		389.40		428.34		
1 Crawler Loader, 3 C.Y.		1202.00		1322.20		
2 Chain Saws, Gas, 36" Long		93.30		102.63		
1 Dump Truck, 8 C.Y., 220 H.P.		339.60		373.56	36.15	39.76
56 L.H., Daily Totals		$4499.10		$5962.33	$80.34	$106.47
Crew B-7C	**Hr.**	**Daily**	**Hr.**	**Daily**	**Bare Costs**	**Incl. O&P**
1 Labor Foreman (outside)	$43.05	$344.40	$65.10	$520.80	$44.19	$66.71
4 Laborers	41.05	1313.60	62.10	1987.20		
1 Equip. Oper. (medium)	55.10	440.80	82.70	661.60		
1 Truck Driver (heavy)	47.00	376.00	70.75	566.00		
1 Brush Chipper, 12", 130 H.P.		389.40		428.34		
1 Crawler Loader, 3 C.Y.		1202.00		1322.20		
2 Chain Saws, Gas, 36" Long		93.30		102.63		
1 Dump Truck, 12 C.Y., 400 H.P.		567.05		623.76	40.21	44.23
56 L.H., Daily Totals		$4726.55		$6212.52	$84.40	$110.94
Crew B-8	**Hr.**	**Daily**	**Hr.**	**Daily**	**Bare Costs**	**Incl. O&P**
1 Labor Foreman (outside)	$43.05	$344.40	$65.10	$520.80	$47.32	$71.26
2 Laborers	41.05	656.80	62.10	993.60		
2 Equip. Oper. (medium)	55.10	881.60	82.70	1323.20		
1 Equip. Oper. (oiler)	49.20	393.60	73.85	590.80		
2 Truck Drivers (heavy)	47.00	752.00	70.75	1132.00		
1 Hyd. Crane, 25 Ton		581.70		639.87		
1 Crawler Loader, 3 C.Y.		1202.00		1322.20		
2 Dump Trucks, 12 C.Y., 400 H.P.		1134.10		1247.51	45.59	50.15
64 L.H., Daily Totals		$5946.20		$7769.98	$92.91	$121.41
Crew B-9	**Hr.**	**Daily**	**Hr.**	**Daily**	**Bare Costs**	**Incl. O&P**
1 Labor Foreman (outside)	$43.05	$344.40	$65.10	$520.80	$41.45	$62.70
4 Laborers	41.05	1313.60	62.10	1987.20		
1 Air Compressor, 250 cfm		167.95		184.75		
2 Breakers, Pavement, 60 lb.		21.50		23.65		
2 -50' Air Hoses, 1.5"		45.50		50.05	5.87	6.46
40 L.H., Daily Totals		$1892.95		$2766.45	$47.32	$69.16
Crew B-9A	**Hr.**	**Daily**	**Hr.**	**Daily**	**Bare Costs**	**Incl. O&P**
2 Laborers	$41.05	$656.80	$62.10	$993.60	$43.03	$64.98
1 Truck Driver (heavy)	47.00	376.00	70.75	566.00		
1 Water Tank Trailer, 5000 Gal.		149.45		164.40		
1 Truck Tractor, 220 H.P.		303.55		333.90		
2 -50' Discharge Hoses, 3"		2.80		3.08	18.99	20.89
24 L.H., Daily Totals		$1488.60		$2060.98	$62.02	$85.87

Crew No.	Bare Costs Hr.	Daily	Incl. Subs O&P Hr.	Daily	Cost Per Labor-Hour Bare Costs	Incl. O&P
Crew B-9B	**Hr.**	**Daily**	**Hr.**	**Daily**	**Bare Costs**	**Incl. O&P**
2 Laborers	$41.05	$656.80	$62.10	$993.60	$43.03	$64.98
1 Truck Driver (heavy)	47.00	376.00	70.75	566.00		
2 -50' Discharge Hoses, 3"		2.80		3.08		
1 Water Tank Trailer, 5000 Gal.		149.45		164.40		
1 Truck Tractor, 220 H.P.		303.55		333.90		
1 Pressure Washer		63.05		69.36	21.62	23.78
24 L.H., Daily Totals		$1551.65		$2130.34	$64.65	$88.76
Crew B-9D	**Hr.**	**Daily**	**Hr.**	**Daily**	**Bare Costs**	**Incl. O&P**
1 Labor Foreman (outside)	$43.05	$344.40	$65.10	$520.80	$41.45	$62.70
4 Common Laborers	41.05	1313.60	62.10	1987.20		
1 Air Compressor, 250 cfm		167.95		184.75		
2 -50' Air Hoses, 1.5"		45.50		50.05		
2 Air Powered Tampers		57.10		62.81	6.76	7.44
40 L.H., Daily Totals		$1928.55		$2805.61	$48.21	$70.14
Crew B-9E	**Hr.**	**Daily**	**Hr.**	**Daily**	**Bare Costs**	**Incl. O&P**
1 Cement Finisher	$48.90	$391.20	$72.20	$577.60	$44.98	$67.15
1 Laborer	41.05	328.40	62.10	496.80		
1 Chip. Hammers, 12 Lb., Elec.		21.00		23.10	1.31	1.44
16 L.H., Daily Totals		$740.60		$1097.50	$46.29	$68.59
Crew B-10	**Hr.**	**Daily**	**Hr.**	**Daily**	**Bare Costs**	**Incl. O&P**
1 Equip. Oper. (medium)	$55.10	$440.80	$82.70	$661.60	$50.42	$75.83
.5 Laborer	41.05	164.20	62.10	248.40		
12 L.H., Daily Totals		$605.00		$910.00	$50.42	$75.83
Crew B-10A	**Hr.**	**Daily**	**Hr.**	**Daily**	**Bare Costs**	**Incl. O&P**
1 Equip. Oper. (medium)	$55.10	$440.80	$82.70	$661.60	$50.42	$75.83
.5 Laborer	41.05	164.20	62.10	248.40		
1 Roller, 2-Drum, W.B., 7.5 H.P.		181.75		199.93	15.15	16.66
12 L.H., Daily Totals		$786.75		$1109.93	$65.56	$92.49
Crew B-10B	**Hr.**	**Daily**	**Hr.**	**Daily**	**Bare Costs**	**Incl. O&P**
1 Equip. Oper. (medium)	$55.10	$440.80	$82.70	$661.60	$50.42	$75.83
.5 Laborer	41.05	164.20	62.10	248.40		
1 Dozer, 200 H.P.		1290.00		1419.00	107.50	118.25
12 L.H., Daily Totals		$1895.00		$2329.00	$157.92	$194.08
Crew B-10C	**Hr.**	**Daily**	**Hr.**	**Daily**	**Bare Costs**	**Incl. O&P**
1 Equip. Oper. (medium)	$55.10	$440.80	$82.70	$661.60	$50.42	$75.83
.5 Laborer	41.05	164.20	62.10	248.40		
1 Dozer, 200 H.P.		1290.00		1419.00		
1 Vibratory Roller, Towed, 23 Ton		422.15		464.37	142.68	156.95
12 L.H., Daily Totals		$2317.15		$2793.36	$193.10	$232.78
Crew B-10D	**Hr.**	**Daily**	**Hr.**	**Daily**	**Bare Costs**	**Incl. O&P**
1 Equip. Oper. (medium)	$55.10	$440.80	$82.70	$661.60	$50.42	$75.83
.5 Laborer	41.05	164.20	62.10	248.40		
1 Dozer, 200 H.P.		1290.00		1419.00		
1 Sheepsft. Roller, Towed		430.15		473.17	143.35	157.68
12 L.H., Daily Totals		$2325.15		$2802.17	$193.76	$233.51
Crew B-10E	**Hr.**	**Daily**	**Hr.**	**Daily**	**Bare Costs**	**Incl. O&P**
1 Equip. Oper. (medium)	$55.10	$440.80	$82.70	$661.60	$50.42	$75.83
.5 Laborer	41.05	164.20	62.10	248.40		
1 Tandem Roller, 5 Ton		152.45		167.69	12.70	13.97
12 L.H., Daily Totals		$757.45		$1077.69	$63.12	$89.81

Crew No.	Bare Costs		Incl. Subs O&P		Cost Per Labor-Hour	

Crew B-10F

Crew B-10F	Hr.	Daily	Hr.	Daily	Bare Costs	Incl. O&P
1 Equip. Oper. (medium)	$55.10	$440.80	$82.70	$661.60	$50.42	$75.83
.5 Laborer	41.05	164.20	62.10	248.40		
1 Tandem Roller, 10 Ton		232.05		255.26	19.34	21.27
12 L.H., Daily Totals		$837.05		$1165.26	$69.75	$97.10

Crew B-10G	Hr.	Daily	Hr.	Daily	Bare Costs	Incl. O&P
1 Equip. Oper. (medium)	$55.10	$440.80	$82.70	$661.60	$50.42	$75.83
.5 Laborer	41.05	164.20	62.10	248.40		
1 Sheepsfoot Roller, 240 H.P.		1323.00		1455.30	110.25	121.28
12 L.H., Daily Totals		$1928.00		$2365.30	$160.67	$197.11

Crew B-10H	Hr.	Daily	Hr.	Daily	Bare Costs	Incl. O&P
1 Equip. Oper. (medium)	$55.10	$440.80	$82.70	$661.60	$50.42	$75.83
.5 Laborer	41.05	164.20	62.10	248.40		
1 Diaphragm Water Pump, 2"		74.30		81.73		
1 -20' Suction Hose, 2"		2.05		2.25		
2 -50' Discharge Hoses, 2"		1.80		1.98	6.51	7.16
12 L.H., Daily Totals		$683.15		$995.97	$56.93	$83.00

Crew B-10I	Hr.	Daily	Hr.	Daily	Bare Costs	Incl. O&P
1 Equip. Oper. (medium)	$55.10	$440.80	$82.70	$661.60	$50.42	$75.83
.5 Laborer	41.05	164.20	62.10	248.40		
1 Diaphragm Water Pump, 4"		141.40		155.54		
1 -20' Suction Hose, 4"		3.10		3.41		
2 -50' Discharge Hoses, 4"		4.50		4.95	12.42	13.66
12 L.H., Daily Totals		$754.00		$1073.90	$62.83	$89.49

Crew B-10J	Hr.	Daily	Hr.	Daily	Bare Costs	Incl. O&P
1 Equip. Oper. (medium)	$55.10	$440.80	$82.70	$661.60	$50.42	$75.83
.5 Laborer	41.05	164.20	62.10	248.40		
1 Centrifugal Water Pump, 3"		79.25		87.17		
1 -20' Suction Hose, 3"		2.70		2.97		
2 -50' Discharge Hoses, 3"		2.80		3.08	7.06	7.77
12 L.H., Daily Totals		$689.75		$1003.23	$57.48	$83.60

Crew B-10K	Hr.	Daily	Hr.	Daily	Bare Costs	Incl. O&P
1 Equip. Oper. (medium)	$55.10	$440.80	$82.70	$661.60	$50.42	$75.83
.5 Laborer	41.05	164.20	62.10	248.40		
1 Centr. Water Pump, 6"		296.45		326.10		
1 -20' Suction Hose, 6"		11.10		12.21		
2 -50' Discharge Hoses, 6"		11.40		12.54	26.58	29.24
12 L.H., Daily Totals		$923.95		$1260.85	$77.00	$105.07

Crew B-10L	Hr.	Daily	Hr.	Daily	Bare Costs	Incl. O&P
1 Equip. Oper. (medium)	$55.10	$440.80	$82.70	$661.60	$50.42	$75.83
.5 Laborer	41.05	164.20	62.10	248.40		
1 Dozer, 80 H.P.		472.00		519.20	39.33	43.27
12 L.H., Daily Totals		$1077.00		$1429.20	$89.75	$119.10

Crew B-10M	Hr.	Daily	Hr.	Daily	Bare Costs	Incl. O&P
1 Equip. Oper. (medium)	$55.10	$440.80	$82.70	$661.60	$50.42	$75.83
.5 Laborer	41.05	164.20	62.10	248.40		
1 Dozer, 300 H.P.		1831.00		2014.10	152.58	167.84
12 L.H., Daily Totals		$2436.00		$2924.10	$203.00	$243.68

Crew B-10N	Hr.	Daily	Hr.	Daily	Bare Costs	Incl. O&P
1 Equip. Oper. (medium)	$55.10	$440.80	$82.70	$661.60	$50.42	$75.83
.5 Laborer	41.05	164.20	62.10	248.40		
1 F.E. Loader, T.M., 1.5 C.Y.		591.30		650.43	49.27	54.20
12 L.H., Daily Totals		$1196.30		$1560.43	$99.69	$130.04

Crew B-100	Hr.	Daily	Hr.	Daily	Bare Costs	Incl. O&P
1 Equip. Oper. (medium)	$55.10	$440.80	$82.70	$661.60	$50.42	$75.83
.5 Laborer	41.05	164.20	62.10	248.40		
1 F.E. Loader, T.M., 2.25 C.Y.		1002.00		1102.20	83.50	91.85
12 L.H., Daily Totals		$1607.00		$2012.20	$133.92	$167.68

Crew B-10P	Hr.	Daily	Hr.	Daily	Bare Costs	Incl. O&P
1 Equip. Oper. (medium)	$55.10	$440.80	$82.70	$661.60	$50.42	$75.83
.5 Laborer	41.05	164.20	62.10	248.40		
1 Crawler Loader, 3 C.Y.		1202.00		1322.20	100.17	110.18
12 L.H., Daily Totals		$1807.00		$2232.20	$150.58	$186.02

Crew B-10Q	Hr.	Daily	Hr.	Daily	Bare Costs	Incl. O&P
1 Equip. Oper. (medium)	$55.10	$440.80	$82.70	$661.60	$50.42	$75.83
.5 Laborer	41.05	164.20	62.10	248.40		
1 Crawler Loader, 4 C.Y.		1461.00		1607.10	121.75	133.93
12 L.H., Daily Totals		$2066.00		$2517.10	$172.17	$209.76

Crew B-10R	Hr.	Daily	Hr.	Daily	Bare Costs	Incl. O&P
1 Equip. Oper. (medium)	$55.10	$440.80	$82.70	$661.60	$50.42	$75.83
.5 Laborer	41.05	164.20	62.10	248.40		
1 F.E. Loader, W.M., 1 C.Y.		297.00		326.70	24.75	27.23
12 L.H., Daily Totals		$902.00		$1236.70	$75.17	$103.06

Crew B-10S	Hr.	Daily	Hr.	Daily	Bare Costs	Incl. O&P
1 Equip. Oper. (medium)	$55.10	$440.80	$82.70	$661.60	$50.42	$75.83
.5 Laborer	41.05	164.20	62.10	248.40		
1 F.E. Loader, W.M., 1.5 C.Y.		352.95		388.25	29.41	32.35
12 L.H., Daily Totals		$957.95		$1298.24	$79.83	$108.19

Crew B-10T	Hr.	Daily	Hr.	Daily	Bare Costs	Incl. O&P
1 Equip. Oper. (medium)	$55.10	$440.80	$82.70	$661.60	$50.42	$75.83
.5 Laborer	41.05	164.20	62.10	248.40		
1 F.E. Loader, W.M., 2.5 C.Y.		531.20		584.32	44.27	48.69
12 L.H., Daily Totals		$1136.20		$1494.32	$94.68	$124.53

Crew B-10U	Hr.	Daily	Hr.	Daily	Bare Costs	Incl. O&P
1 Equip. Oper. (medium)	$55.10	$440.80	$82.70	$661.60	$50.42	$75.83
.5 Laborer	41.05	164.20	62.10	248.40		
1 F.E. Loader, W.M., 5.5 C.Y.		975.55		1073.11	81.30	89.43
12 L.H., Daily Totals		$1580.55		$1983.11	$131.71	$165.26

Crew B-10V	Hr.	Daily	Hr.	Daily	Bare Costs	Incl. O&P
1 Equip. Oper. (medium)	$55.10	$440.80	$82.70	$661.60	$50.42	$75.83
.5 Laborer	41.05	164.20	62.10	248.40		
1 Dozer, 700 H.P.		5021.00		5523.10	418.42	460.26
12 L.H., Daily Totals		$5626.00		$6433.10	$468.83	$536.09

Crew B-10W	Hr.	Daily	Hr.	Daily	Bare Costs	Incl. O&P
1 Equip. Oper. (medium)	$55.10	$440.80	$82.70	$661.60	$50.42	$75.83
.5 Laborer	41.05	164.20	62.10	248.40		
1 Dozer, 105 H.P.		609.70		670.67	50.81	55.89
12 L.H., Daily Totals		$1214.70		$1580.67	$101.22	$131.72

Crew B-10X	Hr.	Daily	Hr.	Daily	Bare Costs	Incl. O&P
1 Equip. Oper. (medium)	$55.10	$440.80	$82.70	$661.60	$50.42	$75.83
.5 Laborer	41.05	164.20	62.10	248.40		
1 Dozer, 410 H.P.		2269.00		2495.90	189.08	207.99
12 L.H., Daily Totals		$2874.00		$3405.90	$239.50	$283.82

Crew No.	Bare Costs Hr.	Bare Costs Daily	Incl. Subs O&P Hr.	Incl. Subs O&P Daily	Cost Per Labor-Hour Bare Costs	Cost Per Labor-Hour Incl. O&P
Crew B-10Y	Hr.	Daily	Hr.	Daily	Bare Costs	Incl. O&P
1 Equip. Oper. (medium)	$55.10	$440.80	$82.70	$661.60	$50.42	$75.83
.5 Laborer	41.05	164.20	62.10	248.40		
1 Vibr. Roller, Towed, 12 Ton		566.70		623.37	47.23	51.95
12 L.H., Daily Totals		$1171.70		$1533.37	$97.64	$127.78

Crew B-11A	Hr.	Daily	Hr.	Daily	Bare Costs	Incl. O&P
1 Equipment Oper. (med.)	$55.10	$440.80	$82.70	$661.60	$48.08	$72.40
1 Laborer	41.05	328.40	62.10	496.80		
1 Dozer, 200 H.P.		1290.00		1419.00	80.63	88.69
16 L.H., Daily Totals		$2059.20		$2577.40	$128.70	$161.09

Crew B-11B	Hr.	Daily	Hr.	Daily	Bare Costs	Incl. O&P
1 Equipment Oper. (light)	$51.65	$413.20	$77.55	$620.40	$46.35	$69.83
1 Laborer	41.05	328.40	62.10	496.80		
1 Air Powered Tamper		28.55		31.41		
1 Air Compressor, 365 cfm		213.35		234.69		
2 -50' Air Hoses, 1.5"		45.50		50.05	17.96	19.76
16 L.H., Daily Totals		$1029.00		$1433.34	$64.31	$89.58

Crew B-11C	Hr.	Daily	Hr.	Daily	Bare Costs	Incl. O&P
1 Equipment Oper. (med.)	$55.10	$440.80	$82.70	$661.60	$48.08	$72.40
1 Laborer	41.05	328.40	62.10	496.80		
1 Backhoe Loader, 48 H.P.		320.20		352.22	20.01	22.01
16 L.H., Daily Totals		$1089.40		$1510.62	$68.09	$94.41

Crew B-11J	Hr.	Daily	Hr.	Daily	Bare Costs	Incl. O&P
1 Equipment Oper. (med.)	$55.10	$440.80	$82.70	$661.60	$48.08	$72.40
1 Laborer	41.05	328.40	62.10	496.80		
1 Grader, 30,000 Lbs.		657.85		723.63		
1 Ripper, Beam & 1 Shank		88.80		97.68	46.67	51.33
16 L.H., Daily Totals		$1515.85		$1979.71	$94.74	$123.73

Crew B-11K	Hr.	Daily	Hr.	Daily	Bare Costs	Incl. O&P
1 Equipment Oper. (med.)	$55.10	$440.80	$82.70	$661.60	$48.08	$72.40
1 Laborer	41.05	328.40	62.10	496.80		
1 Trencher, Chain Type, 8' D		1837.00		2020.70	114.81	126.29
16 L.H., Daily Totals		$2606.20		$3179.10	$162.89	$198.69

Crew B-11L	Hr.	Daily	Hr.	Daily	Bare Costs	Incl. O&P
1 Equipment Oper. (med.)	$55.10	$440.80	$82.70	$661.60	$48.08	$72.40
1 Laborer	41.05	328.40	62.10	496.80		
1 Grader, 30,000 Lbs.		657.85		723.63	41.12	45.23
16 L.H., Daily Totals		$1427.05		$1882.04	$89.19	$117.63

Crew B-11M	Hr.	Daily	Hr.	Daily	Bare Costs	Incl. O&P
1 Equipment Oper. (med.)	$55.10	$440.80	$82.70	$661.60	$48.08	$72.40
1 Laborer	41.05	328.40	62.10	496.80		
1 Backhoe Loader, 80 H.P.		387.85		426.63	24.24	26.66
16 L.H., Daily Totals		$1157.05		$1585.04	$72.32	$99.06

Crew B-11N	Hr.	Daily	Hr.	Daily	Bare Costs	Incl. O&P
1 Labor Foreman (outside)	$43.05	$344.40	$65.10	$520.80	$48.36	$72.78
2 Equipment Operators (med.)	55.10	881.60	82.70	1323.20		
6 Truck Drivers (heavy)	47.00	2256.00	70.75	3396.00		
1 F.E. Loader, W.M., 5.5 C.Y.		975.55		1073.11		
1 Dozer, 410 H.P.		2269.00		2495.90		
6 Dump Trucks, Off Hwy., 50 Ton		10428.00		11470.80	189.90	208.89
72 L.H., Daily Totals		$17154.55		$20279.81	$238.26	$281.66

Crew B-11Q	Hr.	Daily	Hr.	Daily	Bare Costs	Incl. O&P
1 Equipment Operator (med.)	$55.10	$440.80	$82.70	$661.60	$50.42	$75.83
.5 Laborer	41.05	164.20	62.10	248.40		
1 Dozer, 140 H.P.		835.20		918.72	69.60	76.56
12 L.H., Daily Totals		$1440.20		$1828.72	$120.02	$152.39

Crew B-11R	Hr.	Daily	Hr.	Daily	Bare Costs	Incl. O&P
1 Equipment Operator (med.)	$55.10	$440.80	$82.70	$661.60	$50.42	$75.83
.5 Laborer	41.05	164.20	62.10	248.40		
1 Dozer, 200 H.P.		1290.00		1419.00	107.50	118.25
12 L.H., Daily Totals		$1895.00		$2329.00	$157.92	$194.08

Crew B-11S	Hr.	Daily	Hr.	Daily	Bare Costs	Incl. O&P
1 Equipment Operator (med.)	$55.10	$440.80	$82.70	$661.60	$50.42	$75.83
.5 Laborer	41.05	164.20	62.10	248.40		
1 Dozer, 300 H.P.		1831.00		2014.10		
1 Ripper, Beam & 1 Shank		88.80		97.68	159.98	175.98
12 L.H., Daily Totals		$2524.80		$3021.78	$210.40	$251.82

Crew B-11T	Hr.	Daily	Hr.	Daily	Bare Costs	Incl. O&P
1 Equipment Operator (med.)	$55.10	$440.80	$82.70	$661.60	$50.42	$75.83
.5 Laborer	41.05	164.20	62.10	248.40		
1 Dozer, 410 H.P.		2269.00		2495.90		
1 Ripper, Beam & 2 Shanks		135.95		149.54	200.41	220.45
12 L.H., Daily Totals		$3009.95		$3555.45	$250.83	$296.29

Crew B-11U	Hr.	Daily	Hr.	Daily	Bare Costs	Incl. O&P
1 Equipment Operator (med.)	$55.10	$440.80	$82.70	$661.60	$50.42	$75.83
.5 Laborer	41.05	164.20	62.10	248.40		
1 Dozer, 520 H.P.		2812.00		3093.20	234.33	257.77
12 L.H., Daily Totals		$3417.00		$4003.20	$284.75	$333.60

Crew B-11V	Hr.	Daily	Hr.	Daily	Bare Costs	Incl. O&P
3 Laborers	$41.05	$985.20	$62.10	$1490.40	$41.05	$62.10
1 Roller, 2-Drum, W.B., 7.5 H.P.		181.75		199.93	7.57	8.33
24 L.H., Daily Totals		$1166.95		$1690.33	$48.62	$70.43

Crew B-11W	Hr.	Daily	Hr.	Daily	Bare Costs	Incl. O&P
1 Equipment Operator (med.)	$55.10	$440.80	$82.70	$661.60	$47.18	$71.03
1 Common Laborer	41.05	328.40	62.10	496.80		
10 Truck Drivers (heavy)	47.00	3760.00	70.75	5660.00		
1 Dozer, 200 H.P.		1290.00		1419.00		
1 Vibratory Roller, Towed, 23 Ton		422.15		464.37		
10 Dump Trucks, 8 C.Y., 220 H.P.		3396.00		3735.60	53.21	58.53
96 L.H., Daily Totals		$9637.35		$12437.37	$100.39	$129.56

Crew B-11Y	Hr.	Daily	Hr.	Daily	Bare Costs	Incl. O&P
1 Labor Foreman (outside)	$43.05	$344.40	$65.10	$520.80	$45.96	$69.30
5 Common Laborers	41.05	1642.00	62.10	2484.00		
3 Equipment Operators (med.)	55.10	1322.40	82.70	1984.80		
1 Dozer, 80 H.P.		472.00		519.20		
2 Rollers, 2-Drum, W.B., 7.5 H.P.		363.50		399.85		
4 Vibrating Plates, Gas, 21"		162.20		178.42	13.86	15.24
72 L.H., Daily Totals		$4306.50		$6087.07	$59.81	$84.54

Crew B-12A	Hr.	Daily	Hr.	Daily	Bare Costs	Incl. O&P
1 Equip. Oper. (crane)	$57.45	$459.60	$86.25	$690.00	$49.25	$74.17
1 Laborer	41.05	328.40	62.10	496.80		
1 Hyd. Excavator, 1 C.Y.		749.70		824.67	46.86	51.54
16 L.H., Daily Totals		$1537.70		$2011.47	$96.11	$125.72

Crew No.	Bare Costs		Incl. Subs O&P		Cost Per Labor-Hour	
Crew B-12B	Hr.	Daily	Hr.	Daily	Bare Costs	Incl. O&P
1 Equip. Oper. (crane)	$57.45	$459.60	$86.25	$690.00	$49.25	$74.17
1 Laborer	41.05	328.40	62.10	496.80		
1 Hyd. Excavator, 1.5 C.Y.		908.70		999.57	56.79	62.47
16 L.H., Daily Totals		$1696.70		$2186.37	$106.04	$136.65
Crew B-12C	Hr.	Daily	Hr.	Daily	Bare Costs	Incl. O&P
1 Equip. Oper. (crane)	$57.45	$459.60	$86.25	$690.00	$49.25	$74.17
1 Laborer	41.05	328.40	62.10	496.80		
1 Hyd. Excavator, 2 C.Y.		1078.00		1185.80	67.38	74.11
16 L.H., Daily Totals		$1866.00		$2372.60	$116.63	$148.29
Crew B-12D	Hr.	Daily	Hr.	Daily	Bare Costs	Incl. O&P
1 Equip. Oper. (crane)	$57.45	$459.60	$86.25	$690.00	$49.25	$74.17
1 Laborer	41.05	328.40	62.10	496.80		
1 Hyd. Excavator, 3.5 C.Y.		2256.00		2481.60	141.00	155.10
16 L.H., Daily Totals		$3044.00		$3668.40	$190.25	$229.28
Crew B-12E	Hr.	Daily	Hr.	Daily	Bare Costs	Incl. O&P
1 Equip. Oper. (crane)	$57.45	$459.60	$86.25	$690.00	$49.25	$74.17
1 Laborer	41.05	328.40	62.10	496.80		
1 Hyd. Excavator, .5 C.Y.		443.75		488.13	27.73	30.51
16 L.H., Daily Totals		$1231.75		$1674.93	$76.98	$104.68
Crew B-12F	Hr.	Daily	Hr.	Daily	Bare Costs	Incl. O&P
1 Equip. Oper. (crane)	$57.45	$459.60	$86.25	$690.00	$49.25	$74.17
1 Laborer	41.05	328.40	62.10	496.80		
1 Hyd. Excavator, .75 C.Y.		681.25		749.38	42.58	46.84
16 L.H., Daily Totals		$1469.25		$1936.18	$91.83	$121.01
Crew B-12G	Hr.	Daily	Hr.	Daily	Bare Costs	Incl. O&P
1 Equip. Oper. (crane)	$57.45	$459.60	$86.25	$690.00	$49.25	$74.17
1 Laborer	41.05	328.40	62.10	496.80		
1 Crawler Crane, 15 Ton		866.60		953.26		
1 Clamshell Bucket, .5 C.Y.		42.55		46.81	56.82	62.50
16 L.H., Daily Totals		$1697.15		$2186.86	$106.07	$136.68
Crew B-12H	Hr.	Daily	Hr.	Daily	Bare Costs	Incl. O&P
1 Equip. Oper. (crane)	$57.45	$459.60	$86.25	$690.00	$49.25	$74.17
1 Laborer	41.05	328.40	62.10	496.80		
1 Crawler Crane, 25 Ton		1386.00		1524.60		
1 Clamshell Bucket, 1 C.Y.		51.20		56.32	89.83	98.81
16 L.H., Daily Totals		$2225.20		$2767.72	$139.07	$172.98
Crew B-12I	Hr.	Daily	Hr.	Daily	Bare Costs	Incl. O&P
1 Equip. Oper. (crane)	$57.45	$459.60	$86.25	$690.00	$49.25	$74.17
1 Laborer	41.05	328.40	62.10	496.80		
1 Crawler Crane, 20 Ton		1130.00		1243.00		
1 Dragline Bucket, .75 C.Y.		21.85		24.04	71.99	79.19
16 L.H., Daily Totals		$1939.85		$2453.84	$121.24	$153.36
Crew B-12J	Hr.	Daily	Hr.	Daily	Bare Costs	Incl. O&P
1 Equip. Oper. (crane)	$57.45	$459.60	$86.25	$690.00	$49.25	$74.17
1 Laborer	41.05	328.40	62.10	496.80		
1 Gradall, 5/8 C.Y.		862.55		948.80	53.91	59.30
16 L.H., Daily Totals		$1650.55		$2135.61	$103.16	$133.48

Crew No.	Bare Costs		Incl. Subs O&P		Cost Per Labor-Hour	
Crew B-12K	Hr.	Daily	Hr.	Daily	Bare Costs	Incl. O&P
1 Equip. Oper. (crane)	$57.45	$459.60	$86.25	$690.00	$49.25	$74.17
1 Laborer	41.05	328.40	62.10	496.80		
1 Gradall, 3 Ton, 1 C.Y.		1185.00		1303.50	74.06	81.47
16 L.H., Daily Totals		$1973.00		$2490.30	$123.31	$155.64
Crew B-12L	Hr.	Daily	Hr.	Daily	Bare Costs	Incl. O&P
1 Equip. Oper. (crane)	$57.45	$459.60	$86.25	$690.00	$49.25	$74.17
1 Laborer	41.05	328.40	62.10	496.80		
1 Crawler Crane, 15 Ton		866.60		953.26		
1 F.E. Attachment, .5 C.Y.		63.95		70.34	58.16	63.98
16 L.H., Daily Totals		$1718.55		$2210.41	$107.41	$138.15
Crew B-12M	Hr.	Daily	Hr.	Daily	Bare Costs	Incl. O&P
1 Equip. Oper. (crane)	$57.45	$459.60	$86.25	$690.00	$49.25	$74.17
1 Laborer	41.05	328.40	62.10	496.80		
1 Crawler Crane, 20 Ton		1130.00		1243.00		
1 F.E. Attachment, .75 C.Y.		68.90		75.79	74.93	82.42
16 L.H., Daily Totals		$1986.90		$2505.59	$124.18	$156.60
Crew B-12N	Hr.	Daily	Hr.	Daily	Bare Costs	Incl. O&P
1 Equip. Oper. (crane)	$57.45	$459.60	$86.25	$690.00	$49.25	$74.17
1 Laborer	41.05	328.40	62.10	496.80		
1 Crawler Crane, 25 Ton		1386.00		1524.60		
1 F.E. Attachment, 1 C.Y.		74.85		82.33	91.30	100.43
16 L.H., Daily Totals		$2248.85		$2793.74	$140.55	$174.61
Crew B-12O	Hr.	Daily	Hr.	Daily	Bare Costs	Incl. O&P
1 Equip. Oper. (crane)	$57.45	$459.60	$86.25	$690.00	$49.25	$74.17
1 Laborer	41.05	328.40	62.10	496.80		
1 Crawler Crane, 40 Ton		1392.00		1531.20		
1 F.E. Attachment, 1.5 C.Y.		85.65		94.22	92.35	101.59
16 L.H., Daily Totals		$2265.65		$2812.22	$141.60	$175.76
Crew B-12P	Hr.	Daily	Hr.	Daily	Bare Costs	Incl. O&P
1 Equip. Oper. (crane)	$57.45	$459.60	$86.25	$690.00	$49.25	$74.17
1 Laborer	41.05	328.40	62.10	496.80		
1 Crawler Crane, 40 Ton		1392.00		1531.20		
1 Dragline Bucket, 1.5 C.Y.		35.10		38.61	89.19	98.11
16 L.H., Daily Totals		$2215.10		$2756.61	$138.44	$172.29
Crew B-12Q	Hr.	Daily	Hr.	Daily	Bare Costs	Incl. O&P
1 Equip. Oper. (crane)	$57.45	$459.60	$86.25	$690.00	$49.25	$74.17
1 Laborer	41.05	328.40	62.10	496.80		
1 Hyd. Excavator, 5/8 C.Y.		587.30		646.03	36.71	40.38
16 L.H., Daily Totals		$1375.30		$1832.83	$85.96	$114.55
Crew B-12S	Hr.	Daily	Hr.	Daily	Bare Costs	Incl. O&P
1 Equip. Oper. (crane)	$57.45	$459.60	$86.25	$690.00	$49.25	$74.17
1 Laborer	41.05	328.40	62.10	496.80		
1 Hyd. Excavator, 2.5 C.Y.		1441.00		1585.10	90.06	99.07
16 L.H., Daily Totals		$2229.00		$2771.90	$139.31	$173.24
Crew B-12T	Hr.	Daily	Hr.	Daily	Bare Costs	Incl. O&P
1 Equip. Oper. (crane)	$57.45	$459.60	$86.25	$690.00	$49.25	$74.17
1 Laborer	41.05	328.40	62.10	496.80		
1 Crawler Crane, 75 Ton		1734.00		1907.40		
1 F.E. Attachment, 3 C.Y.		111.50		122.65	115.34	126.88
16 L.H., Daily Totals		$2633.50		$3216.85	$164.59	$201.05

Crew B-12V

	Bare Costs Hr.	Daily	Incl. Subs O&P Hr.	Daily	Cost Per Labor-Hour Bare Costs	Incl. O&P
1 Equip. Oper. (crane)	$57.45	$459.60	$86.25	$690.00	$49.25	$74.17
1 Laborer	41.05	328.40	62.10	496.80		
1 Crawler Crane, 75 Ton		1734.00		1907.40		
1 Dragline Bucket, 3 C.Y.		56.25		61.88	111.89	123.08
16 L.H., Daily Totals		$2578.25		$3156.07	$161.14	$197.25

Crew B-12Y

	Bare Costs Hr.	Daily	Incl. Subs O&P Hr.	Daily	Cost Per Labor-Hour Bare Costs	Incl. O&P
1 Equip. Oper. (crane)	$57.45	$459.60	$86.25	$690.00	$46.52	$70.15
2 Laborers	41.05	656.80	62.10	993.60		
1 Hyd. Excavator, 3.5 C.Y.		2256.00		2481.60	94.00	103.40
24 L.H., Daily Totals		$3372.40		$4165.20	$140.52	$173.55

Crew B-12Z

	Bare Costs Hr.	Daily	Incl. Subs O&P Hr.	Daily	Cost Per Labor-Hour Bare Costs	Incl. O&P
1 Equip. Oper. (crane)	$57.45	$459.60	$86.25	$690.00	$46.52	$70.15
2 Laborers	41.05	656.80	62.10	993.60		
1 Hyd. Excavator, 2.5 C.Y.		1441.00		1585.10	60.04	66.05
24 L.H., Daily Totals		$2557.40		$3268.70	$106.56	$136.20

Crew B-13

	Bare Costs Hr.	Daily	Incl. Subs O&P Hr.	Daily	Cost Per Labor-Hour Bare Costs	Incl. O&P
1 Labor Foreman (outside)	$43.05	$344.40	$65.10	$520.80	$44.84	$67.66
4 Laborers	41.05	1313.60	62.10	1987.20		
1 Equip. Oper. (crane)	57.45	459.60	86.25	690.00		
1 Equip. Oper. (oiler)	49.20	393.60	73.85	590.80		
1 Hyd. Crane, 25 Ton		581.70		639.87	10.39	11.43
56 L.H., Daily Totals		$3092.90		$4428.67	$55.23	$79.08

Crew B-13A

	Bare Costs Hr.	Daily	Incl. Subs O&P Hr.	Daily	Cost Per Labor-Hour Bare Costs	Incl. O&P
1 Labor Foreman (outside)	$43.05	$344.40	$65.10	$520.80	$47.05	$70.89
2 Laborers	41.05	656.80	62.10	993.60		
2 Equipment Operators (med.)	55.10	881.60	82.70	1323.20		
2 Truck Drivers (heavy)	47.00	752.00	70.75	1132.00		
1 Crawler Crane, 75 Ton		1734.00		1907.40		
1 Crawler Loader, 4 C.Y.		1461.00		1607.10		
2 Dump Trucks, 8 C.Y., 220 H.P.		679.20		747.12	69.18	76.10
56 L.H., Daily Totals		$6509.00		$8231.22	$116.23	$146.99

Crew B-13B

	Bare Costs Hr.	Daily	Incl. Subs O&P Hr.	Daily	Cost Per Labor-Hour Bare Costs	Incl. O&P
1 Labor Foreman (outside)	$43.05	$344.40	$65.10	$520.80	$44.84	$67.66
4 Laborers	41.05	1313.60	62.10	1987.20		
1 Equip. Oper. (crane)	57.45	459.60	86.25	690.00		
1 Equip. Oper. (oiler)	49.20	393.60	73.85	590.80		
1 Hyd. Crane, 55 Ton		981.50		1079.65	17.53	19.28
56 L.H., Daily Totals		$3492.70		$4868.45	$62.37	$86.94

Crew B-13C

	Bare Costs Hr.	Daily	Incl. Subs O&P Hr.	Daily	Cost Per Labor-Hour Bare Costs	Incl. O&P
1 Labor Foreman (outside)	$43.05	$344.40	$65.10	$520.80	$44.84	$67.66
4 Laborers	41.05	1313.60	62.10	1987.20		
1 Equip. Oper. (crane)	57.45	459.60	86.25	690.00		
1 Equip. Oper. (oiler)	49.20	393.60	73.85	590.80		
1 Crawler Crane, 100 Ton		1879.00		2066.90	33.55	36.91
56 L.H., Daily Totals		$4390.20		$5855.70	$78.40	$104.57

Crew B-13D

	Bare Costs Hr.	Daily	Incl. Subs O&P Hr.	Daily	Cost Per Labor-Hour Bare Costs	Incl. O&P
1 Laborer	$41.05	$328.40	$62.10	$496.80	$49.25	$74.17
1 Equip. Oper. (crane)	57.45	459.60	86.25	690.00		
1 Hyd. Excavator, 1 C.Y.		749.70		824.67		
1 Trench Box		77.75		85.53	51.72	56.89
16 L.H., Daily Totals		$1615.45		$2096.99	$100.97	$131.06

Crew B-13E

	Bare Costs Hr.	Daily	Incl. Subs O&P Hr.	Daily	Cost Per Labor-Hour Bare Costs	Incl. O&P
1 Laborer	$41.05	$328.40	$62.10	$496.80	$49.25	$74.17
1 Equip. Oper. (crane)	57.45	459.60	86.25	690.00		
1 Hyd. Excavator, 1.5 C.Y.		908.70		999.57		
1 Trench Box		77.75		85.53	61.65	67.82
16 L.H., Daily Totals		$1774.45		$2271.90	$110.90	$141.99

Crew B-13F

	Bare Costs Hr.	Daily	Incl. Subs O&P Hr.	Daily	Cost Per Labor-Hour Bare Costs	Incl. O&P
1 Laborer	$41.05	$328.40	$62.10	$496.80	$49.25	$74.17
1 Equip. Oper. (crane)	57.45	459.60	86.25	690.00		
1 Hyd. Excavator, 3.5 C.Y.		2256.00		2481.60		
1 Trench Box		77.75		85.53	145.86	160.45
16 L.H., Daily Totals		$3121.75		$3753.93	$195.11	$234.62

Crew B-13G

	Bare Costs Hr.	Daily	Incl. Subs O&P Hr.	Daily	Cost Per Labor-Hour Bare Costs	Incl. O&P
1 Laborer	$41.05	$328.40	$62.10	$496.80	$49.25	$74.17
1 Equip. Oper. (crane)	57.45	459.60	86.25	690.00		
1 Hyd. Excavator, .75 C.Y.		681.25		749.38		
1 Trench Box		77.75		85.53	47.44	52.18
16 L.H., Daily Totals		$1547.00		$2021.70	$96.69	$126.36

Crew B-13H

	Bare Costs Hr.	Daily	Incl. Subs O&P Hr.	Daily	Cost Per Labor-Hour Bare Costs	Incl. O&P
1 Laborer	$41.05	$328.40	$62.10	$496.80	$49.25	$74.17
1 Equip. Oper. (crane)	57.45	459.60	86.25	690.00		
1 Gradall, 5/8 C.Y.		862.55		948.80		
1 Trench Box		77.75		85.53	58.77	64.65
16 L.H., Daily Totals		$1728.30		$2221.13	$108.02	$138.82

Crew B-13I

	Bare Costs Hr.	Daily	Incl. Subs O&P Hr.	Daily	Cost Per Labor-Hour Bare Costs	Incl. O&P
1 Laborer	$41.05	$328.40	$62.10	$496.80	$49.25	$74.17
1 Equip. Oper. (crane)	57.45	459.60	86.25	690.00		
1 Gradall, 3 Ton, 1 C.Y.		1185.00		1303.50		
1 Trench Box		77.75		85.53	78.92	86.81
16 L.H., Daily Totals		$2050.75		$2575.82	$128.17	$160.99

Crew B-13J

	Bare Costs Hr.	Daily	Incl. Subs O&P Hr.	Daily	Cost Per Labor-Hour Bare Costs	Incl. O&P
1 Laborer	$41.05	$328.40	$62.10	$496.80	$49.25	$74.17
1 Equip. Oper. (crane)	57.45	459.60	86.25	690.00		
1 Hyd. Excavator, 2.5 C.Y.		1441.00		1585.10		
1 Trench Box		77.75		85.53	94.92	104.41
16 L.H., Daily Totals		$2306.75		$2857.43	$144.17	$178.59

Crew B-13K

	Bare Costs Hr.	Daily	Incl. Subs O&P Hr.	Daily	Cost Per Labor-Hour Bare Costs	Incl. O&P
2 Equip. Opers. (crane)	$57.45	$919.20	$86.25	$1380.00	$57.45	$86.25
1 Hyd. Excavator, .75 C.Y.		681.25		749.38		
1 Hyd. Hammer, 4000 ft-lb		332.85		366.13		
1 Hyd. Excavator, .75 C.Y.		681.25		749.38	105.96	116.56
16 L.H., Daily Totals		$2614.55		$3244.89	$163.41	$202.81

Crew B-13L

	Bare Costs Hr.	Daily	Incl. Subs O&P Hr.	Daily	Cost Per Labor-Hour Bare Costs	Incl. O&P
2 Equip. Opers. (crane)	$57.45	$919.20	$86.25	$1380.00	$57.45	$86.25
1 Hyd. Excavator, 1.5 C.Y.		908.70		999.57		
1 Hyd. Hammer, 5000 ft-lb		407.60		448.36		
1 Hyd. Excavator, .75 C.Y.		681.25		749.38	124.85	137.33
16 L.H., Daily Totals		$2916.75		$3577.30	$182.30	$223.58

For customer support on your Electrical Costs with RSMeans data, call 800.448.8182.

483

Crew No.	Bare Costs Hr.	Daily	Incl. Subs O&P Hr.	Daily	Cost Per Labor-Hour Bare Costs	Incl. O&P
Crew B-13M	**Hr.**	**Daily**	**Hr.**	**Daily**	**Bare Costs**	**Incl. O&P**
2 Equip. Opers. (crane)	$57.45	$919.20	$86.25	$1380.00	$57.45	$86.25
1 Hyd. Excavator, 2.5 C.Y.		1441.00		1585.10		
1 Hyd. Hammer, 8000 ft-lb		593.35		652.68		
1 Hyd. Excavator, 1.5 C.Y.		908.70		999.57	183.94	202.33
16 L.H., Daily Totals		$3862.25		$4617.35	$241.39	$288.58
Crew B-13N	**Hr.**	**Daily**	**Hr.**	**Daily**	**Bare Costs**	**Incl. O&P**
2 Equip. Opers. (crane)	$57.45	$919.20	$86.25	$1380.00	$57.45	$86.25
1 Hyd. Excavator, 3.5 C.Y.		2256.00		2481.60		
1 Hyd. Hammer, 12,000 ft-lb		685.75		754.33		
1 Hyd. Excavator, 1.5 C.Y.		908.70		999.57	240.65	264.72
16 L.H., Daily Totals		$4769.65		$5615.49	$298.10	$350.97
Crew B-14	**Hr.**	**Daily**	**Hr.**	**Daily**	**Bare Costs**	**Incl. O&P**
1 Labor Foreman (outside)	$43.05	$344.40	$65.10	$520.80	$43.15	$65.17
4 Laborers	41.05	1313.60	62.10	1987.20		
1 Equip. Oper. (light)	51.65	413.20	77.55	620.40		
1 Backhoe Loader, 48 H.P.		320.20		352.22	6.67	7.34
48 L.H., Daily Totals		$2391.40		$3480.62	$49.82	$72.51
Crew B-14A	**Hr.**	**Daily**	**Hr.**	**Daily**	**Bare Costs**	**Incl. O&P**
1 Equip. Oper. (crane)	$57.45	$459.60	$86.25	$690.00	$51.98	$78.20
.5 Laborer	41.05	164.20	62.10	248.40		
1 Hyd. Excavator, 4.5 C.Y.		2868.00		3154.80	239.00	262.90
12 L.H., Daily Totals		$3491.80		$4093.20	$290.98	$341.10
Crew B-14B	**Hr.**	**Daily**	**Hr.**	**Daily**	**Bare Costs**	**Incl. O&P**
1 Equip. Oper. (crane)	$57.45	$459.60	$86.25	$690.00	$51.98	$78.20
.5 Laborer	41.05	164.20	62.10	248.40		
1 Hyd. Excavator, 6 C.Y.		3579.00		3936.90	298.25	328.07
12 L.H., Daily Totals		$4202.80		$4875.30	$350.23	$406.27
Crew B-14C	**Hr.**	**Daily**	**Hr.**	**Daily**	**Bare Costs**	**Incl. O&P**
1 Equip. Oper. (crane)	$57.45	$459.60	$86.25	$690.00	$51.98	$78.20
.5 Laborer	41.05	164.20	62.10	248.40		
1 Hyd. Excavator, 7 C.Y.		3332.00		3665.20	277.67	305.43
12 L.H., Daily Totals		$3955.80		$4603.60	$329.65	$383.63
Crew B-14F	**Hr.**	**Daily**	**Hr.**	**Daily**	**Bare Costs**	**Incl. O&P**
1 Equip. Oper. (crane)	$57.45	$459.60	$86.25	$690.00	$51.98	$78.20
.5 Laborer	41.05	164.20	62.10	248.40		
1 Hyd. Shovel, 7 C.Y.		4029.00		4431.90	335.75	369.32
12 L.H., Daily Totals		$4652.80		$5370.30	$387.73	$447.52
Crew B-14G	**Hr.**	**Daily**	**Hr.**	**Daily**	**Bare Costs**	**Incl. O&P**
1 Equip. Oper. (crane)	$57.45	$459.60	$86.25	$690.00	$51.98	$78.20
.5 Laborer	41.05	164.20	62.10	248.40		
1 Hyd. Shovel, 12 C.Y.		5855.00		6440.50	487.92	536.71
12 L.H., Daily Totals		$6478.80		$7378.90	$539.90	$614.91
Crew B-14J	**Hr.**	**Daily**	**Hr.**	**Daily**	**Bare Costs**	**Incl. O&P**
1 Equip. Oper. (medium)	$55.10	$440.80	$82.70	$661.60	$50.42	$75.83
.5 Laborer	41.05	164.20	62.10	248.40		
1 F.E. Loader, 8 C.Y.		1784.00		1962.40	148.67	163.53
12 L.H., Daily Totals		$2389.00		$2872.40	$199.08	$239.37

Crew No.	Bare Costs Hr.	Daily	Incl. Subs O&P Hr.	Daily	Cost Per Labor-Hour Bare Costs	Incl. O&P
Crew B-14K	**Hr.**	**Daily**	**Hr.**	**Daily**	**Bare Costs**	**Incl. O&P**
1 Equip. Oper. (medium)	$55.10	$440.80	$82.70	$661.60	$50.42	$75.83
.5 Laborer	41.05	164.20	62.10	248.40		
1 F.E. Loader, 10 C.Y.		2625.00		2887.50	218.75	240.63
12 L.H., Daily Totals		$3230.00		$3797.50	$269.17	$316.46
Crew B-15	**Hr.**	**Daily**	**Hr.**	**Daily**	**Bare Costs**	**Incl. O&P**
1 Equipment Oper. (med.)	$55.10	$440.80	$82.70	$661.60	$48.46	$72.93
.5 Laborer	41.05	164.20	62.10	248.40		
2 Truck Drivers (heavy)	47.00	752.00	70.75	1132.00		
2 Dump Trucks, 12 C.Y., 400 H.P.		1134.10		1247.51		
1 Dozer, 200 H.P.		1290.00		1419.00	86.58	95.23
28 L.H., Daily Totals		$3781.10		$4708.51	$135.04	$168.16
Crew B-16	**Hr.**	**Daily**	**Hr.**	**Daily**	**Bare Costs**	**Incl. O&P**
1 Labor Foreman (outside)	$43.05	$344.40	$65.10	$520.80	$43.04	$65.01
2 Laborers	41.05	656.80	62.10	993.60		
1 Truck Driver (heavy)	47.00	376.00	70.75	566.00		
1 Dump Truck, 12 C.Y., 400 H.P.		567.05		623.76	17.72	19.49
32 L.H., Daily Totals		$1944.25		$2704.16	$60.76	$84.50
Crew B-17	**Hr.**	**Daily**	**Hr.**	**Daily**	**Bare Costs**	**Incl. O&P**
2 Laborers	$41.05	$656.80	$62.10	$993.60	$45.19	$68.13
1 Equip. Oper. (light)	51.65	413.20	77.55	620.40		
1 Truck Driver (heavy)	47.00	376.00	70.75	566.00		
1 Backhoe Loader, 48 H.P.		320.20		352.22		
1 Dump Truck, 8 C.Y., 220 H.P.		339.60		373.56	20.62	22.68
32 L.H., Daily Totals		$2105.80		$2905.78	$65.81	$90.81
Crew B-17A	**Hr.**	**Daily**	**Hr.**	**Daily**	**Bare Costs**	**Incl. O&P**
2 Labor Foremen (outside)	$43.05	$688.80	$65.10	$1041.60	$44.12	$66.83
6 Laborers	41.05	1970.40	62.10	2980.80		
1 Skilled Worker Foreman (out)	55.40	443.20	84.25	674.00		
1 Skilled Worker	53.40	427.20	81.25	650.00		
80 L.H., Daily Totals		$3529.60		$5346.40	$44.12	$66.83
Crew B-17B	**Hr.**	**Daily**	**Hr.**	**Daily**	**Bare Costs**	**Incl. O&P**
2 Laborers	$41.05	$656.80	$62.10	$993.60	$45.19	$68.13
1 Equip. Oper. (light)	51.65	413.20	77.55	620.40		
1 Truck Driver (heavy)	47.00	376.00	70.75	566.00		
1 Backhoe Loader, 48 H.P.		320.20		352.22		
1 Dump Truck, 12 C.Y., 400 H.P.		567.05		623.76	27.73	30.50
32 L.H., Daily Totals		$2333.25		$3155.97	$72.91	$98.62
Crew B-18	**Hr.**	**Daily**	**Hr.**	**Daily**	**Bare Costs**	**Incl. O&P**
1 Labor Foreman (outside)	$43.05	$344.40	$65.10	$520.80	$41.72	$63.10
2 Laborers	41.05	656.80	62.10	993.60		
1 Vibrating Plate, Gas, 21"		40.55		44.60	1.69	1.86
24 L.H., Daily Totals		$1041.75		$1559.01	$43.41	$64.96
Crew B-19	**Hr.**	**Daily**	**Hr.**	**Daily**	**Bare Costs**	**Incl. O&P**
1 Pile Driver Foreman (outside)	$54.15	$433.20	$84.65	$677.20	$53.36	$82.15
4 Pile Drivers	52.15	1668.80	81.55	2609.60		
2 Equip. Oper. (crane)	57.45	919.20	86.25	1380.00		
1 Equip. Oper. (oiler)	49.20	393.60	73.85	590.80		
1 Crawler Crane, 40 Ton		1392.00		1531.20		
1 Lead, 90' High		134.60		148.06		
1 Hammer, Diesel, 22k ft-lb		427.55		470.31	30.53	33.59
64 L.H., Daily Totals		$5368.95		$7407.17	$83.89	$115.74

Crew No.	Bare Costs		Incl. Subs O&P		Cost Per Labor-Hour	

Crew B-19A	Hr.	Daily	Hr.	Daily	Bare Costs	Incl. O&P
1 Pile Driver Foreman (outside)	$54.15	$433.20	$84.65	$677.20	$53.36	$82.15
4 Pile Drivers	52.15	1668.80	81.55	2609.60		
2 Equip. Oper. (crane)	57.45	919.20	86.25	1380.00		
1 Equip. Oper. (oiler)	49.20	393.60	73.85	590.80		
1 Crawler Crane, 75 Ton		1734.00		1907.40		
1 Lead, 90' High		134.60		148.06		
1 Hammer, Diesel, 41k ft-lb		565.45		622.00	38.03	41.84
64 L.H., Daily Totals		$5848.85		$7935.06	$91.39	$123.99

Crew B-19B	Hr.	Daily	Hr.	Daily	Bare Costs	Incl. O&P
1 Pile Driver Foreman (outside)	$54.15	$433.20	$84.65	$677.20	$53.36	$82.15
4 Pile Drivers	52.15	1668.80	81.55	2609.60		
2 Equip. Oper. (crane)	57.45	919.20	86.25	1380.00		
1 Equip. Oper. (oiler)	49.20	393.60	73.85	590.80		
1 Crawler Crane, 40 Ton		1392.00		1531.20		
1 Lead, 90' High		134.60		148.06		
1 Hammer, Diesel, 22k ft-lb		427.55		470.31		
1 Barge, 400 Ton		836.50		920.15	43.60	47.96
64 L.H., Daily Totals		$6205.45		$8327.32	$96.96	$130.11

Crew B-19C	Hr.	Daily	Hr.	Daily	Bare Costs	Incl. O&P
1 Pile Driver Foreman (outside)	$54.15	$433.20	$84.65	$677.20	$53.36	$82.15
4 Pile Drivers	52.15	1668.80	81.55	2609.60		
2 Equip. Oper. (crane)	57.45	919.20	86.25	1380.00		
1 Equip. Oper. (oiler)	49.20	393.60	73.85	590.80		
1 Crawler Crane, 75 Ton		1734.00		1907.40		
1 Lead, 90' High		134.60		148.06		
1 Hammer, Diesel, 41k ft-lb		565.45		622.00		
1 Barge, 400 Ton		836.50		920.15	51.10	56.21
64 L.H., Daily Totals		$6685.35		$8855.20	$104.46	$138.36

Crew B-20	Hr.	Daily	Hr.	Daily	Bare Costs	Incl. O&P
1 Labor Foreman (outside)	$43.05	$344.40	$65.10	$520.80	$45.83	$69.48
1 Skilled Worker	53.40	427.20	81.25	650.00		
1 Laborer	41.05	328.40	62.10	496.80		
24 L.H., Daily Totals		$1100.00		$1667.60	$45.83	$69.48

Crew B-20A	Hr.	Daily	Hr.	Daily	Bare Costs	Incl. O&P
1 Labor Foreman (outside)	$43.05	$344.40	$65.10	$520.80	$49.44	$74.44
1 Laborer	41.05	328.40	62.10	496.80		
1 Plumber	63.15	505.20	94.75	758.00		
1 Plumber Apprentice	50.50	404.00	75.80	606.40		
32 L.H., Daily Totals		$1582.00		$2382.00	$49.44	$74.44

Crew B-21	Hr.	Daily	Hr.	Daily	Bare Costs	Incl. O&P
1 Labor Foreman (outside)	$43.05	$344.40	$65.10	$520.80	$47.49	$71.88
1 Skilled Worker	53.40	427.20	81.25	650.00		
1 Laborer	41.05	328.40	62.10	496.80		
.5 Equip. Oper. (crane)	57.45	229.80	86.25	345.00		
.5 S.P. Crane, 4x4, 5 Ton		129.72		142.70	4.63	5.10
28 L.H., Daily Totals		$1459.53		$2155.30	$52.13	$76.97

Crew B-21A	Hr.	Daily	Hr.	Daily	Bare Costs	Incl. O&P
1 Labor Foreman (outside)	$43.05	$344.40	$65.10	$520.80	$51.04	$76.80
1 Laborer	41.05	328.40	62.10	496.80		
1 Plumber	63.15	505.20	94.75	758.00		
1 Plumber Apprentice	50.50	404.00	75.80	606.40		
1 Equip. Oper. (crane)	57.45	459.60	86.25	690.00		
1 S.P. Crane, 4x4, 12 Ton		371.85		409.04	9.30	10.23
40 L.H., Daily Totals		$2413.45		$3481.03	$60.34	$87.03

Crew B-21B	Hr.	Daily	Hr.	Daily	Bare Costs	Incl. O&P
1 Labor Foreman (outside)	$43.05	$344.40	$65.10	$520.80	$44.73	$67.53
3 Laborers	41.05	985.20	62.10	1490.40		
1 Equip. Oper. (crane)	57.45	459.60	86.25	690.00		
1 Hyd. Crane, 12 Ton		471.75		518.92	11.79	12.97
40 L.H., Daily Totals		$2260.95		$3220.13	$56.52	$80.50

Crew B-21C	Hr.	Daily	Hr.	Daily	Bare Costs	Incl. O&P
1 Labor Foreman (outside)	$43.05	$344.40	$65.10	$520.80	$44.84	$67.66
4 Laborers	41.05	1313.60	62.10	1987.20		
1 Equip. Oper. (crane)	57.45	459.60	86.25	690.00		
1 Equip. Oper. (oiler)	49.20	393.60	73.85	590.80		
2 Cutting Torches		25.00		27.50		
2 Sets of Gases		335.70		369.27		
1 Lattice Boom Crane, 90 Ton		1698.00		1867.80	36.76	40.44
56 L.H., Daily Totals		$4569.90		$6053.37	$81.61	$108.10

Crew B-22	Hr.	Daily	Hr.	Daily	Bare Costs	Incl. O&P
1 Labor Foreman (outside)	$43.05	$344.40	$65.10	$520.80	$48.16	$72.84
1 Skilled Worker	53.40	427.20	81.25	650.00		
1 Laborer	41.05	328.40	62.10	496.80		
.75 Equip. Oper. (crane)	57.45	344.70	86.25	517.50		
.75 S.P. Crane, 4x4, 5 Ton		194.59		214.05	6.49	7.13
30 L.H., Daily Totals		$1639.29		$2399.15	$54.64	$79.97

Crew B-22A	Hr.	Daily	Hr.	Daily	Bare Costs	Incl. O&P
1 Labor Foreman (outside)	$43.05	$344.40	$65.10	$520.80	$47.20	$71.36
1 Skilled Worker	53.40	427.20	81.25	650.00		
2 Laborers	41.05	656.80	62.10	993.60		
1 Equipment Operator, Crane	57.45	459.60	86.25	690.00		
1 S.P. Crane, 4x4, 5 Ton		259.45		285.39		
1 Butt Fusion Mach., 4"-12" diam.		405.15		445.67	16.61	18.28
40 L.H., Daily Totals		$2552.60		$3585.46	$63.81	$89.64

Crew B-22B	Hr.	Daily	Hr.	Daily	Bare Costs	Incl. O&P
1 Labor Foreman (outside)	$43.05	$344.40	$65.10	$520.80	$47.20	$71.36
1 Skilled Worker	53.40	427.20	81.25	650.00		
2 Laborers	41.05	656.80	62.10	993.60		
1 Equip. Oper. (crane)	57.45	459.60	86.25	690.00		
1 S.P. Crane, 4x4, 5 Ton		259.45		285.39		
1 Butt Fusion Mach., 8"-24" diam.		1927.00		2119.70	54.66	60.13
40 L.H., Daily Totals		$4074.45		$5259.49	$101.86	$131.49

Crew B-22C	Hr.	Daily	Hr.	Daily	Bare Costs	Incl. O&P
1 Skilled Worker	$53.40	$427.20	$81.25	$650.00	$47.23	$71.67
1 Laborer	41.05	328.40	62.10	496.80		
1 Butt Fusion Mach., 2"-8" diam.		143.10		157.41	8.94	9.84
16 L.H., Daily Totals		$898.70		$1304.21	$56.17	$81.51

Crew B-23	Hr.	Daily	Hr.	Daily	Bare Costs	Incl. O&P
1 Labor Foreman (outside)	$43.05	$344.40	$65.10	$520.80	$41.45	$62.70
4 Laborers	41.05	1313.60	62.10	1987.20		
1 Drill Rig, Truck-Mounted		2428.00		2670.80		
1 Flatbed Truck, Gas, 3 Ton		245.95		270.55	66.85	73.53
40 L.H., Daily Totals		$4331.95		$5449.35	$108.30	$136.23

For customer support on your Electrical Costs with RSMeans data, call 800.448.8182.

Crew No.	Bare Costs Hr.	Daily	Incl. Subs O&P Hr.	Daily	Cost Per Labor-Hour Bare Costs	Incl. O&P
Crew B-23A	**Hr.**	**Daily**	**Hr.**	**Daily**	**Bare Costs**	**Incl. O&P**
1 Labor Foreman (outside)	$43.05	$344.40	$65.10	$520.80	$46.40	$69.97
1 Laborer	41.05	328.40	62.10	496.80		
1 Equip. Oper. (medium)	55.10	440.80	82.70	661.60		
1 Drill Rig, Truck-Mounted		2428.00		2670.80		
1 Pickup Truck, 3/4 Ton		109.90		120.89	105.75	116.32
24 L.H., Daily Totals		$3651.50		$4470.89	$152.15	$186.29

Crew No.	Hr.	Daily	Hr.	Daily	Bare Costs	Incl. O&P
Crew B-23B	**Hr.**	**Daily**	**Hr.**	**Daily**	**Bare Costs**	**Incl. O&P**
1 Labor Foreman (outside)	$43.05	$344.40	$65.10	$520.80	$46.40	$69.97
1 Laborer	41.05	328.40	62.10	496.80		
1 Equip. Oper. (medium)	55.10	440.80	82.70	661.60		
1 Drill Rig, Truck-Mounted		2428.00		2670.80		
1 Pickup Truck, 3/4 Ton		109.90		120.89		
1 Centr. Water Pump, 6"		296.45		326.10	118.10	129.91
24 L.H., Daily Totals		$3947.95		$4796.98	$164.50	$199.87

Crew No.	Hr.	Daily	Hr.	Daily	Bare Costs	Incl. O&P
Crew B-24	**Hr.**	**Daily**	**Hr.**	**Daily**	**Bare Costs**	**Incl. O&P**
1 Cement Finisher	$48.90	$391.20	$72.20	$577.60	$47.20	$70.82
1 Laborer	41.05	328.40	62.10	496.80		
1 Carpenter	51.65	413.20	78.15	625.20		
24 L.H., Daily Totals		$1132.80		$1699.60	$47.20	$70.82

Crew No.	Hr.	Daily	Hr.	Daily	Bare Costs	Incl. O&P
Crew B-25	**Hr.**	**Daily**	**Hr.**	**Daily**	**Bare Costs**	**Incl. O&P**
1 Labor Foreman (outside)	$43.05	$344.40	$65.10	$520.80	$45.06	$67.99
7 Laborers	41.05	2298.80	62.10	3477.60		
3 Equip. Oper. (medium)	55.10	1322.40	82.70	1984.80		
1 Asphalt Paver, 130 H.P.		2076.00		2283.60		
1 Tandem Roller, 10 Ton		232.05		255.26		
1 Roller, Pneum. Whl., 12 Ton		338.35		372.19	30.07	33.08
88 L.H., Daily Totals		$6612.00		$8894.24	$75.14	$101.07

Crew No.	Hr.	Daily	Hr.	Daily	Bare Costs	Incl. O&P
Crew B-25B	**Hr.**	**Daily**	**Hr.**	**Daily**	**Bare Costs**	**Incl. O&P**
1 Labor Foreman (outside)	$43.05	$344.40	$65.10	$520.80	$45.90	$69.22
7 Laborers	41.05	2298.80	62.10	3477.60		
4 Equip. Oper. (medium)	55.10	1763.20	82.70	2646.40		
1 Asphalt Paver, 130 H.P.		2076.00		2283.60		
2 Tandem Rollers, 10 Ton		464.10		510.51		
1 Roller, Pneum. Whl., 12 Ton		338.35		372.19	29.98	32.98
96 L.H., Daily Totals		$7284.85		$9811.09	$75.88	$102.20

Crew No.	Hr.	Daily	Hr.	Daily	Bare Costs	Incl. O&P
Crew B-25C	**Hr.**	**Daily**	**Hr.**	**Daily**	**Bare Costs**	**Incl. O&P**
1 Labor Foreman (outside)	$43.05	$344.40	$65.10	$520.80	$46.07	$69.47
3 Laborers	41.05	985.20	62.10	1490.40		
2 Equip. Oper. (medium)	55.10	881.60	82.70	1323.20		
1 Asphalt Paver, 130 H.P.		2076.00		2283.60		
1 Tandem Roller, 10 Ton		232.05		255.26	48.08	52.89
48 L.H., Daily Totals		$4519.25		$5873.26	$94.15	$122.36

Crew No.	Hr.	Daily	Hr.	Daily	Bare Costs	Incl. O&P
Crew B-25D	**Hr.**	**Daily**	**Hr.**	**Daily**	**Bare Costs**	**Incl. O&P**
1 Labor Foreman (outside)	$43.05	$344.40	$65.10	$520.80	$46.27	$69.76
3 Laborers	41.05	985.20	62.10	1490.40		
2.125 Equip. Oper. (medium)	55.10	936.70	82.70	1405.90		
.125 Truck Driver (heavy)	47.00	47.00	70.75	70.75		
.125 Truck Tractor, 6x4, 380 H.P.		60.95		67.05		
.125 Dist. Tanker, 3000 Gallon		40.41		44.45		
1 Asphalt Paver, 130 H.P.		2076.00		2283.60		
1 Tandem Roller, 10 Ton		232.05		255.26	48.19	53.01
50 L.H., Daily Totals		$4722.71		$6138.20	$94.45	$122.76

Crew No.	Bare Costs Hr.	Daily	Incl. Subs O&P Hr.	Daily	Cost Per Labor-Hour Bare Costs	Incl. O&P
Crew B-25E	**Hr.**	**Daily**	**Hr.**	**Daily**	**Bare Costs**	**Incl. O&P**
1 Labor Foreman (outside)	$43.05	$344.40	$65.10	$520.80	$46.45	$70.03
3 Laborers	41.05	985.20	62.10	1490.40		
2.250 Equip. Oper. (medium)	55.10	991.80	82.70	1488.60		
.25 Truck Driver (heavy)	47.00	94.00	70.75	141.50		
.25 Truck Tractor, 6x4, 380 H.P.		121.90		134.09		
.25 Dist. Tanker, 3000 Gallon		80.81		88.89		
1 Asphalt Paver, 130 H.P.		2076.00		2283.60		
1 Tandem Roller, 10 Ton		232.05		255.26	48.28	53.11
52 L.H., Daily Totals		$4926.16		$6403.14	$94.73	$123.14

Crew No.	Hr.	Daily	Hr.	Daily	Bare Costs	Incl. O&P
Crew B-26	**Hr.**	**Daily**	**Hr.**	**Daily**	**Bare Costs**	**Incl. O&P**
1 Labor Foreman (outside)	$43.05	$344.40	$65.10	$520.80	$45.75	$68.94
6 Laborers	41.05	1970.40	62.10	2980.80		
2 Equip. Oper. (medium)	55.10	881.60	82.70	1323.20		
1 Rodman (reinf.)	54.85	438.80	83.00	664.00		
1 Cement Finisher	48.90	391.20	72.20	577.60		
1 Grader, 30,000 Lbs.		657.85		723.63		
1 Paving Mach. & Equip.		2428.00		2670.80	35.07	38.57
88 L.H., Daily Totals		$7112.25		$9460.83	$80.82	$107.51

Crew No.	Hr.	Daily	Hr.	Daily	Bare Costs	Incl. O&P
Crew B-26A	**Hr.**	**Daily**	**Hr.**	**Daily**	**Bare Costs**	**Incl. O&P**
1 Labor Foreman (outside)	$43.05	$344.40	$65.10	$520.80	$45.75	$68.94
6 Laborers	41.05	1970.40	62.10	2980.80		
2 Equip. Oper. (medium)	55.10	881.60	82.70	1323.20		
1 Rodman (reinf.)	54.85	438.80	83.00	664.00		
1 Cement Finisher	48.90	391.20	72.20	577.60		
1 Grader, 30,000 Lbs.		657.85		723.63		
1 Paving Mach. & Equip.		2428.00		2670.80		
1 Concrete Saw		110.15		121.17	36.32	39.95
88 L.H., Daily Totals		$7222.40		$9582.00	$82.07	$108.89

Crew No.	Hr.	Daily	Hr.	Daily	Bare Costs	Incl. O&P
Crew B-26B	**Hr.**	**Daily**	**Hr.**	**Daily**	**Bare Costs**	**Incl. O&P**
1 Labor Foreman (outside)	$43.05	$344.40	$65.10	$520.80	$46.53	$70.08
6 Laborers	41.05	1970.40	62.10	2980.80		
3 Equip. Oper. (medium)	55.10	1322.40	82.70	1984.80		
1 Rodman (reinf.)	54.85	438.80	83.00	664.00		
1 Cement Finisher	48.90	391.20	72.20	577.60		
1 Grader, 30,000 Lbs.		657.85		723.63		
1 Paving Mach. & Equip.		2428.00		2670.80		
1 Concrete Pump, 110' Boom		1130.00		1243.00	43.92	48.31
96 L.H., Daily Totals		$8683.05		$11365.43	$90.45	$118.39

Crew No.	Hr.	Daily	Hr.	Daily	Bare Costs	Incl. O&P
Crew B-26C	**Hr.**	**Daily**	**Hr.**	**Daily**	**Bare Costs**	**Incl. O&P**
1 Labor Foreman (outside)	$43.05	$344.40	$65.10	$520.80	$44.82	$67.56
6 Laborers	41.05	1970.40	62.10	2980.80		
1 Equip. Oper. (medium)	55.10	440.80	82.70	661.60		
1 Rodman (reinf.)	54.85	438.80	83.00	664.00		
1 Cement Finisher	48.90	391.20	72.20	577.60		
1 Paving Mach. & Equip.		2428.00		2670.80		
1 Concrete Saw		110.15		121.17	31.73	34.90
80 L.H., Daily Totals		$6123.75		$8196.76	$76.55	$102.46

Crew No.	Hr.	Daily	Hr.	Daily	Bare Costs	Incl. O&P
Crew B-27	**Hr.**	**Daily**	**Hr.**	**Daily**	**Bare Costs**	**Incl. O&P**
1 Labor Foreman (outside)	$43.05	$344.40	$65.10	$520.80	$41.55	$62.85
3 Laborers	41.05	985.20	62.10	1490.40		
1 Berm Machine		325.30		357.83	10.17	11.18
32 L.H., Daily Totals		$1654.90		$2369.03	$51.72	$74.03

Crew No.	Bare Costs		Incl. Subs O&P		Cost Per Labor-Hour	
Crew B-28	Hr.	Daily	Hr.	Daily	Bare Costs	Incl. O&P
2 Carpenters	$51.65	$826.40	$78.15	$1250.40	$48.12	$72.80
1 Laborer	41.05	328.40	62.10	496.80		
24 L.H., Daily Totals		$1154.80		$1747.20	$48.12	$72.80
Crew B-29	Hr.	Daily	Hr.	Daily	Bare Costs	Incl. O&P
1 Labor Foreman (outside)	$43.05	$344.40	$65.10	$520.80	$44.84	$67.66
4 Laborers	41.05	1313.60	62.10	1987.20		
1 Equip. Oper. (crane)	57.45	459.60	86.25	690.00		
1 Equip. Oper. (oiler)	49.20	393.60	73.85	590.80		
1 Gradall, 5/8 C.Y.		862.55		948.80	15.40	16.94
56 L.H., Daily Totals		$3373.75		$4737.60	$60.25	$84.60
Crew B-30	Hr.	Daily	Hr.	Daily	Bare Costs	Incl. O&P
1 Equip. Oper. (medium)	$55.10	$440.80	$82.70	$661.60	$49.70	$74.73
2 Truck Drivers (heavy)	47.00	752.00	70.75	1132.00		
1 Hyd. Excavator, 1.5 C.Y.		908.70		999.57		
2 Dump Trucks, 12 C.Y., 400 H.P.		1134.10		1247.51	85.12	93.63
24 L.H., Daily Totals		$3235.60		$4040.68	$134.82	$168.36
Crew B-31	Hr.	Daily	Hr.	Daily	Bare Costs	Incl. O&P
1 Labor Foreman (outside)	$43.05	$344.40	$65.10	$520.80	$43.57	$65.91
3 Laborers	41.05	985.20	62.10	1490.40		
1 Carpenter	51.65	413.20	78.15	625.20		
1 Air Compressor, 250 cfm		167.95		184.75		
1 Sheeting Driver		7.15		7.87		
2 -50' Air Hoses, 1.5"		45.50		50.05	5.51	6.07
40 L.H., Daily Totals		$1963.40		$2879.06	$49.09	$71.98
Crew B-32	Hr.	Daily	Hr.	Daily	Bare Costs	Incl. O&P
1 Laborer	$41.05	$328.40	$62.10	$496.80	$51.59	$77.55
3 Equip. Oper. (medium)	55.10	1322.40	82.70	1984.80		
1 Grader, 30,000 Lbs.		657.85		723.63		
1 Tandem Roller, 10 Ton		232.05		255.26		
1 Dozer, 200 H.P.		1290.00		1419.00	68.12	74.93
32 L.H., Daily Totals		$3830.70		$4879.49	$119.71	$152.48
Crew B-32A	Hr.	Daily	Hr.	Daily	Bare Costs	Incl. O&P
1 Laborer	$41.05	$328.40	$62.10	$496.80	$50.42	$75.83
2 Equip. Oper. (medium)	55.10	881.60	82.70	1323.20		
1 Grader, 30,000 Lbs.		657.85		723.63		
1 Roller, Vibratory, 25 Ton		650.90		715.99	54.53	59.98
24 L.H., Daily Totals		$2518.75		$3259.63	$104.95	$135.82
Crew B-32B	Hr.	Daily	Hr.	Daily	Bare Costs	Incl. O&P
1 Laborer	$41.05	$328.40	$62.10	$496.80	$50.42	$75.83
2 Equip. Oper. (medium)	55.10	881.60	82.70	1323.20		
1 Dozer, 200 H.P.		1290.00		1419.00		
1 Roller, Vibratory, 25 Ton		650.90		715.99	80.87	88.96
24 L.H., Daily Totals		$3150.90		$3954.99	$131.29	$164.79
Crew B-32C	Hr.	Daily	Hr.	Daily	Bare Costs	Incl. O&P
1 Labor Foreman (outside)	$43.05	$344.40	$65.10	$520.80	$48.41	$72.90
2 Laborers	41.05	656.80	62.10	993.60		
3 Equip. Oper. (medium)	55.10	1322.40	82.70	1984.80		
1 Grader, 30,000 Lbs.		657.85		723.63		
1 Tandem Roller, 10 Ton		232.05		255.26		
1 Dozer, 200 H.P.		1290.00		1419.00	45.41	49.96
48 L.H., Daily Totals		$4503.50		$5897.09	$93.82	$122.86

Crew No.	Bare Costs		Incl. Subs O&P		Cost Per Labor-Hour	
Crew B-33A	Hr.	Daily	Hr.	Daily	Bare Costs	Incl. O&P
1 Equip. Oper. (medium)	$55.10	$440.80	$82.70	$661.60	$51.09	$76.81
.5 Laborer	41.05	164.20	62.10	248.40		
.25 Equip. Oper. (medium)	55.10	110.20	82.70	165.40		
1 Scraper, Towed, 7 C.Y.		125.50		138.05		
1.25 Dozers, 300 H.P.		2288.75		2517.63	172.45	189.69
14 L.H., Daily Totals		$3129.45		$3731.07	$223.53	$266.51
Crew B-33B	Hr.	Daily	Hr.	Daily	Bare Costs	Incl. O&P
1 Equip. Oper. (medium)	$55.10	$440.80	$82.70	$661.60	$51.09	$76.81
.5 Laborer	41.05	164.20	62.10	248.40		
.25 Equip. Oper. (medium)	55.10	110.20	82.70	165.40		
1 Scraper, Towed, 10 C.Y.		156.65		172.32		
1.25 Dozers, 300 H.P.		2288.75		2517.63	174.67	192.14
14 L.H., Daily Totals		$3160.60		$3765.34	$225.76	$268.95
Crew B-33C	Hr.	Daily	Hr.	Daily	Bare Costs	Incl. O&P
1 Equip. Oper. (medium)	$55.10	$440.80	$82.70	$661.60	$51.09	$76.81
.5 Laborer	41.05	164.20	62.10	248.40		
.25 Equip. Oper. (medium)	55.10	110.20	82.70	165.40		
1 Scraper, Towed, 15 C.Y.		173.20		190.52		
1.25 Dozers, 300 H.P.		2288.75		2517.63	175.85	193.44
14 L.H., Daily Totals		$3177.15		$3783.55	$226.94	$270.25
Crew B-33D	Hr.	Daily	Hr.	Daily	Bare Costs	Incl. O&P
1 Equip. Oper. (medium)	$55.10	$440.80	$82.70	$661.60	$51.09	$76.81
.5 Laborer	41.05	164.20	62.10	248.40		
.25 Equip. Oper. (medium)	55.10	110.20	82.70	165.40		
1 S.P. Scraper, 14 C.Y.		2611.00		2872.10		
.25 Dozer, 300 H.P.		457.75		503.52	219.20	241.12
14 L.H., Daily Totals		$3783.95		$4451.02	$270.28	$317.93
Crew B-33E	Hr.	Daily	Hr.	Daily	Bare Costs	Incl. O&P
1 Equip. Oper. (medium)	$55.10	$440.80	$82.70	$661.60	$51.09	$76.81
.5 Laborer	41.05	164.20	62.10	248.40		
.25 Equip. Oper. (medium)	55.10	110.20	82.70	165.40		
1 S.P. Scraper, 21 C.Y.		2541.00		2795.10		
.25 Dozer, 300 H.P.		457.75		503.52	214.20	235.62
14 L.H., Daily Totals		$3713.95		$4374.02	$265.28	$312.43
Crew B-33F	Hr.	Daily	Hr.	Daily	Bare Costs	Incl. O&P
1 Equip. Oper. (medium)	$55.10	$440.80	$82.70	$661.60	$51.09	$76.81
.5 Laborer	41.05	164.20	62.10	248.40		
.25 Equip. Oper. (medium)	55.10	110.20	82.70	165.40		
1 Elev. Scraper, 11 C.Y.		1165.00		1281.50		
.25 Dozer, 300 H.P.		457.75		503.52	115.91	127.50
14 L.H., Daily Totals		$2337.95		$2860.43	$167.00	$204.32
Crew B-33G	Hr.	Daily	Hr.	Daily	Bare Costs	Incl. O&P
1 Equip. Oper. (medium)	$55.10	$440.80	$82.70	$661.60	$51.09	$76.81
.5 Laborer	41.05	164.20	62.10	248.40		
.25 Equip. Oper. (medium)	55.10	110.20	82.70	165.40		
1 Elev. Scraper, 22 C.Y.		2307.00		2537.70		
.25 Dozer, 300 H.P.		457.75		503.52	197.48	217.23
14 L.H., Daily Totals		$3479.95		$4116.63	$248.57	$294.04

Crews - Standard

Crew No.		Bare Costs		Incl. Subs O&P		Cost Per Labor-Hour	
Crew B-33H	Hr.	Daily	Hr.	Daily	Bare Costs	Incl. O&P
.5 Laborer	$41.05	$164.20	$62.10	$248.40	$51.09	$76.81
1 Equipment Operator (med.)	55.10	440.80	82.70	661.60		
.25 Equipment Operator (med.)	55.10	110.20	82.70	165.40		
1 S.P. Scraper, 44 C.Y.		4561.00		5017.10		
.25 Dozer, 410 H.P.		567.25		623.98	366.30	402.93
14 L.H., Daily Totals		$5843.45		$6716.48	$417.39	$479.75
Crew B-33J	Hr.	Daily	Hr.	Daily	Bare Costs	Incl. O&P
1 Equipment Operator (med.)	$55.10	$440.80	$82.70	$661.60	$55.10	$82.70
1 S.P. Scraper, 14 C.Y.		2611.00		2872.10	326.38	359.01
8 L.H., Daily Totals		$3051.80		$3533.70	$381.48	$441.71
Crew B-33K	Hr.	Daily	Hr.	Daily	Bare Costs	Incl. O&P
1 Equipment Operator (med.)	$55.10	$440.80	$82.70	$661.60	$51.09	$76.81
.25 Equipment Operator (med.)	55.10	110.20	82.70	165.40		
.5 Laborer	41.05	164.20	62.10	248.40		
1 S.P. Scraper, 31 C.Y.		3603.00		3963.30		
.25 Dozer, 410 H.P.		567.25		623.98	297.88	327.66
14 L.H., Daily Totals		$4885.45		$5662.68	$348.96	$404.48
Crew B-34A	Hr.	Daily	Hr.	Daily	Bare Costs	Incl. O&P
1 Truck Driver (heavy)	$47.00	$376.00	$70.75	$566.00	$47.00	$70.75
1 Dump Truck, 8 C.Y., 220 H.P.		339.60		373.56	42.45	46.70
8 L.H., Daily Totals		$715.60		$939.56	$89.45	$117.44
Crew B-34B	Hr.	Daily	Hr.	Daily	Bare Costs	Incl. O&P
1 Truck Driver (heavy)	$47.00	$376.00	$70.75	$566.00	$47.00	$70.75
1 Dump Truck, 12 C.Y., 400 H.P.		567.05		623.76	70.88	77.97
8 L.H., Daily Totals		$943.05		$1189.76	$117.88	$148.72
Crew B-34C	Hr.	Daily	Hr.	Daily	Bare Costs	Incl. O&P
1 Truck Driver (heavy)	$47.00	$376.00	$70.75	$566.00	$47.00	$70.75
1 Truck Tractor, 6x4, 380 H.P.		487.60		536.36		
1 Dump Trailer, 16.5 C.Y.		133.95		147.35	77.69	85.46
8 L.H., Daily Totals		$997.55		$1249.70	$124.69	$156.21
Crew B-34D	Hr.	Daily	Hr.	Daily	Bare Costs	Incl. O&P
1 Truck Driver (heavy)	$47.00	$376.00	$70.75	$566.00	$47.00	$70.75
1 Truck Tractor, 6x4, 380 H.P.		487.60		536.36		
1 Dump Trailer, 20 C.Y.		148.60		163.46	79.53	87.48
8 L.H., Daily Totals		$1012.20		$1265.82	$126.53	$158.23
Crew B-34E	Hr.	Daily	Hr.	Daily	Bare Costs	Incl. O&P
1 Truck Driver (heavy)	$47.00	$376.00	$70.75	$566.00	$47.00	$70.75
1 Dump Truck, Off Hwy., 25 Ton		1353.00		1488.30	169.13	186.04
8 L.H., Daily Totals		$1729.00		$2054.30	$216.13	$256.79
Crew B-34F	Hr.	Daily	Hr.	Daily	Bare Costs	Incl. O&P
1 Truck Driver (heavy)	$47.00	$376.00	$70.75	$566.00	$47.00	$70.75
1 Dump Truck, Off Hwy., 35 Ton		1472.00		1619.20	184.00	202.40
8 L.H., Daily Totals		$1848.00		$2185.20	$231.00	$273.15
Crew B-34G	Hr.	Daily	Hr.	Daily	Bare Costs	Incl. O&P
1 Truck Driver (heavy)	$47.00	$376.00	$70.75	$566.00	$47.00	$70.75
1 Dump Truck, Off Hwy., 50 Ton		1738.00		1911.80	217.25	238.97
8 L.H., Daily Totals		$2114.00		$2477.80	$264.25	$309.73

Crew No.		Bare Costs		Incl. Subs O&P		Cost Per Labor-Hour	
Crew B-34H	Hr.	Daily	Hr.	Daily	Bare Costs	Incl. O&P
1 Truck Driver (heavy)	$47.00	$376.00	$70.75	$566.00	$47.00	$70.75
1 Dump Truck, Off Hwy., 65 Ton		1879.00		2066.90	234.88	258.36
8 L.H., Daily Totals		$2255.00		$2632.90	$281.88	$329.11
Crew B-34I	Hr.	Daily	Hr.	Daily	Bare Costs	Incl. O&P
1 Truck Driver (heavy)	$47.00	$376.00	$70.75	$566.00	$47.00	$70.75
1 Dump Truck, 18 C.Y., 450 H.P.		705.00		775.50	88.13	96.94
8 L.H., Daily Totals		$1081.00		$1341.50	$135.13	$167.69
Crew B-34J	Hr.	Daily	Hr.	Daily	Bare Costs	Incl. O&P
1 Truck Driver (heavy)	$47.00	$376.00	$70.75	$566.00	$47.00	$70.75
1 Dump Truck, Off Hwy., 100 Ton		2683.00		2951.30	335.38	368.91
8 L.H., Daily Totals		$3059.00		$3517.30	$382.38	$439.66
Crew B-34K	Hr.	Daily	Hr.	Daily	Bare Costs	Incl. O&P
1 Truck Driver (heavy)	$47.00	$376.00	$70.75	$566.00	$47.00	$70.75
1 Truck Tractor, 6x4, 450 H.P.		594.90		654.39		
1 Lowbed Trailer, 75 Ton		249.90		274.89	105.60	116.16
8 L.H., Daily Totals		$1220.80		$1495.28	$152.60	$186.91
Crew B-34L	Hr.	Daily	Hr.	Daily	Bare Costs	Incl. O&P
1 Equip. Oper. (light)	$51.65	$413.20	$77.55	$620.40	$51.65	$77.55
1 Flatbed Truck, Gas, 1.5 Ton		195.20		214.72	24.40	26.84
8 L.H., Daily Totals		$608.40		$835.12	$76.05	$104.39
Crew B-34M	Hr.	Daily	Hr.	Daily	Bare Costs	Incl. O&P
1 Equip. Oper. (light)	$51.65	$413.20	$77.55	$620.40	$51.65	$77.55
1 Flatbed Truck, Gas, 3 Ton		245.95		270.55	30.74	33.82
8 L.H., Daily Totals		$659.15		$890.95	$82.39	$111.37
Crew B-34N	Hr.	Daily	Hr.	Daily	Bare Costs	Incl. O&P
1 Truck Driver (heavy)	$47.00	$376.00	$70.75	$566.00	$51.05	$76.72
1 Equip. Oper. (medium)	55.10	440.80	82.70	661.60		
1 Truck Tractor, 6x4, 380 H.P.		487.60		536.36		
1 Flatbed Trailer, 40 Ton		182.50		200.75	41.88	46.07
16 L.H., Daily Totals		$1486.90		$1964.71	$92.93	$122.79
Crew B-34P	Hr.	Daily	Hr.	Daily	Bare Costs	Incl. O&P
1 Pipe Fitter	$63.95	$511.60	$95.95	$767.60	$54.85	$82.38
1 Truck Driver (light)	45.50	364.00	68.50	548.00		
1 Equip. Oper. (medium)	55.10	440.80	82.70	661.60		
1 Flatbed Truck, Gas, 3 Ton		245.95		270.55		
1 Backhoe Loader, 48 H.P.		320.20		352.22	23.59	25.95
24 L.H., Daily Totals		$1882.55		$2599.97	$78.44	$108.33
Crew B-34Q	Hr.	Daily	Hr.	Daily	Bare Costs	Incl. O&P
1 Pipe Fitter	$63.95	$511.60	$95.95	$767.60	$55.63	$83.57
1 Truck Driver (light)	45.50	364.00	68.50	548.00		
1 Equip. Oper. (crane)	57.45	459.60	86.25	690.00		
1 Flatbed Trailer, 25 Ton		133.00		146.30		
1 Dump Truck, 8 C.Y., 220 H.P.		339.60		373.56		
1 Hyd. Crane, 25 Ton		581.70		639.87	43.93	48.32
24 L.H., Daily Totals		$2389.50		$3165.33	$99.56	$131.89

Crew No.	Bare Costs Hr.	Bare Costs Daily	Incl. Subs O&P Hr.	Incl. Subs O&P Daily	Cost Per Labor-Hour Bare Costs	Cost Per Labor-Hour Incl. O&P
Crew B-34R	Hr.	Daily	Hr.	Daily	Bare Costs	Incl. O&P
1 Pipe Fitter	$63.95	$511.60	$95.95	$767.60	$55.63	$83.57
1 Truck Driver (light)	45.50	364.00	68.50	548.00		
1 Equip. Oper. (crane)	57.45	459.60	86.25	690.00		
1 Flatbed Trailer, 25 Ton		133.00		146.30		
1 Dump Truck, 8 C.Y., 220 H.P.		339.60		373.56		
1 Hyd. Crane, 25 Ton		581.70		639.87		
1 Hyd. Excavator, 1 C.Y.		749.70		824.67	75.17	82.68
24 L.H., Daily Totals		$3139.20		$3990.00	$130.80	$166.25
Crew B-34S	Hr.	Daily	Hr.	Daily	Bare Costs	Incl. O&P
2 Pipe Fitters	$63.95	$1023.20	$95.95	$1535.20	$58.09	$87.22
1 Truck Driver (heavy)	47.00	376.00	70.75	566.00		
1 Equip. Oper. (crane)	57.45	459.60	86.25	690.00		
1 Flatbed Trailer, 40 Ton		182.50		200.75		
1 Truck Tractor, 6x4, 380 H.P.		487.60		536.36		
1 Hyd. Crane, 80 Ton		1487.00		1635.70		
1 Hyd. Excavator, 2 C.Y.		1078.00		1185.80	101.10	111.21
32 L.H., Daily Totals		$5093.90		$6349.81	$159.18	$198.43
Crew B-34T	Hr.	Daily	Hr.	Daily	Bare Costs	Incl. O&P
2 Pipe Fitters	$63.95	$1023.20	$95.95	$1535.20	$58.09	$87.22
1 Truck Driver (heavy)	47.00	376.00	70.75	566.00		
1 Equip. Oper. (crane)	57.45	459.60	86.25	690.00		
1 Flatbed Trailer, 40 Ton		182.50		200.75		
1 Truck Tractor, 6x4, 380 H.P.		487.60		536.36		
1 Hyd. Crane, 80 Ton		1487.00		1635.70	67.41	74.15
32 L.H., Daily Totals		$4015.90		$5164.01	$125.50	$161.38
Crew B-34U	Hr.	Daily	Hr.	Daily	Bare Costs	Incl. O&P
1 Truck Driver (heavy)	$47.00	$376.00	$70.75	$566.00	$49.33	$74.15
1 Equip. Oper. (light)	51.65	413.20	77.55	620.40		
1 Truck Tractor, 220 H.P.		303.55		333.90		
1 Flatbed Trailer, 25 Ton		133.00		146.30	27.28	30.01
16 L.H., Daily Totals		$1225.75		$1666.61	$76.61	$104.16
Crew B-34V	Hr.	Daily	Hr.	Daily	Bare Costs	Incl. O&P
1 Truck Driver (heavy)	$47.00	$376.00	$70.75	$566.00	$52.03	$78.18
1 Equip. Oper. (crane)	57.45	459.60	86.25	690.00		
1 Equip. Oper. (light)	51.65	413.20	77.55	620.40		
1 Truck Tractor, 6x4, 450 H.P.		594.90		654.39		
1 Equipment Trailer, 50 Ton		200.95		221.04		
1 Pickup Truck, 4x4, 3/4 Ton		121.70		133.87	38.23	42.05
24 L.H., Daily Totals		$2166.35		$2885.70	$90.26	$120.24
Crew B-34W	Hr.	Daily	Hr.	Daily	Bare Costs	Incl. O&P
5 Truck Drivers (heavy)	$47.00	$1880.00	$70.75	$2830.00	$49.83	$74.97
2 Equip. Opers. (crane)	57.45	919.20	86.25	1380.00		
1 Equip. Oper. (mechanic)	57.55	460.40	86.40	691.20		
1 Laborer	41.05	328.40	62.10	496.80		
4 Truck Tractors, 6x4, 380 H.P.		1950.40		2145.44		
2 Equipment Trailers, 50 Ton		401.90		442.09		
2 Flatbed Trailers, 40 Ton		365.00		401.50		
1 Pickup Truck, 4x4, 3/4 Ton		121.70		133.87		
1 S.P. Crane, 4x4, 20 Ton		587.75		646.52	47.59	52.35
72 L.H., Daily Totals		$7014.75		$9167.42	$97.43	$127.33

Crew No.	Bare Costs Hr.	Bare Costs Daily	Incl. Subs O&P Hr.	Incl. Subs O&P Daily	Cost Per Labor-Hour Bare Costs	Cost Per Labor-Hour Incl. O&P
Crew B-35	Hr.	Daily	Hr.	Daily	Bare Costs	Incl. O&P
1 Labor Foreman (outside)	$43.05	$344.40	$65.10	$520.80	$51.22	$77.22
1 Skilled Worker	53.40	427.20	81.25	650.00		
1 Welder (plumber)	63.15	505.20	94.75	758.00		
1 Laborer	41.05	328.40	62.10	496.80		
1 Equip. Oper. (crane)	57.45	459.60	86.25	690.00		
1 Equip. Oper. (oiler)	49.20	393.60	73.85	590.80		
1 Welder, Electric, 300 amp		56.10		61.71		
1 Hyd. Excavator, .75 C.Y.		681.25		749.38	15.36	16.90
48 L.H., Daily Totals		$3195.75		$4517.48	$66.58	$94.11
Crew B-35A	Hr.	Daily	Hr.	Daily	Bare Costs	Incl. O&P
1 Labor Foreman (outside)	$43.05	$344.40	$65.10	$520.80	$49.76	$75.06
2 Laborers	41.05	656.80	62.10	993.60		
1 Skilled Worker	53.40	427.20	81.25	650.00		
1 Welder (plumber)	63.15	505.20	94.75	758.00		
1 Equip. Oper. (crane)	57.45	459.60	86.25	690.00		
1 Equip. Oper. (oiler)	49.20	393.60	73.85	590.80		
1 Welder, Gas Engine, 300 amp		95.65		105.22		
1 Crawler Crane, 75 Ton		1734.00		1907.40	32.67	35.94
56 L.H., Daily Totals		$4616.45		$6215.81	$82.44	$111.00
Crew B-36	Hr.	Daily	Hr.	Daily	Bare Costs	Incl. O&P
1 Labor Foreman (outside)	$43.05	$344.40	$65.10	$520.80	$47.07	$70.94
2 Laborers	41.05	656.80	62.10	993.60		
2 Equip. Oper. (medium)	55.10	881.60	82.70	1323.20		
1 Dozer, 200 H.P.		1290.00		1419.00		
1 Aggregate Spreader		37.30		41.03		
1 Tandem Roller, 10 Ton		232.05		255.26	38.98	42.88
40 L.H., Daily Totals		$3442.15		$4552.89	$86.05	$113.82
Crew B-36A	Hr.	Daily	Hr.	Daily	Bare Costs	Incl. O&P
1 Labor Foreman (outside)	$43.05	$344.40	$65.10	$520.80	$49.36	$74.30
2 Laborers	41.05	656.80	62.10	993.60		
4 Equip. Oper. (medium)	55.10	1763.20	82.70	2646.40		
1 Dozer, 200 H.P.		1290.00		1419.00		
1 Aggregate Spreader		37.30		41.03		
1 Tandem Roller, 10 Ton		232.05		255.26		
1 Roller, Pneum. Whl., 12 Ton		338.35		372.19	33.89	37.28
56 L.H., Daily Totals		$4662.10		$6248.27	$83.25	$111.58
Crew B-36B	Hr.	Daily	Hr.	Daily	Bare Costs	Incl. O&P
1 Labor Foreman (outside)	$43.05	$344.40	$65.10	$520.80	$49.07	$73.86
2 Laborers	41.05	656.80	62.10	993.60		
4 Equip. Oper. (medium)	55.10	1763.20	82.70	2646.40		
1 Truck Driver (heavy)	47.00	376.00	70.75	566.00		
1 Grader, 30,000 Lbs.		657.85		723.63		
1 F.E. Loader, Crl. 1.5 C.Y.		647.15		711.87		
1 Dozer, 300 H.P.		1831.00		2014.10		
1 Roller, Vibratory, 25 Ton		650.90		715.99		
1 Truck Tractor, 6x4, 450 H.P.		594.90		654.39		
1 Water Tank Trailer, 5000 Gal.		149.45		164.40	70.80	77.88
64 L.H., Daily Totals		$7671.65		$9711.17	$119.87	$151.74

489

Crew No.	Bare Costs Hr.	Daily	Incl. Subs O&P Hr.	Daily	Cost Per Labor-Hour Bare Costs	Incl. O&P
Crew B-36C	Hr.	Daily	Hr.	Daily	Bare Costs	Incl. O&P
1 Labor Foreman (outside)	$43.05	$344.40	$65.10	$520.80	$51.07	$76.79
3 Equip. Oper. (medium)	55.10	1322.40	82.70	1984.80		
1 Truck Driver (heavy)	47.00	376.00	70.75	566.00		
1 Grader, 30,000 Lbs.		657.85		723.63		
1 Dozer, 300 H.P.		1831.00		2014.10		
1 Roller, Vibratory, 25 Ton		650.90		715.99		
1 Truck Tractor, 6x4, 450 H.P.		594.90		654.39		
1 Water Tank Trailer, 5000 Gal.		149.45		164.40	97.10	106.81
40 L.H., Daily Totals		$5926.90		$7344.11	$148.17	$183.60

Crew B-36D	Hr.	Daily	Hr.	Daily	Bare Costs	Incl. O&P
1 Labor Foreman (outside)	$43.05	$344.40	$65.10	$520.80	$52.09	$78.30
3 Equip. Oper. (medium)	55.10	1322.40	82.70	1984.80		
1 Grader, 30,000 Lbs.		657.85		723.63		
1 Dozer, 300 H.P.		1831.00		2014.10		
1 Roller, Vibratory, 25 Ton		650.90		715.99	98.12	107.93
32 L.H., Daily Totals		$4806.55		$5959.32	$150.20	$186.23

Crew B-37	Hr.	Daily	Hr.	Daily	Bare Costs	Incl. O&P
1 Labor Foreman (outside)	$43.05	$344.40	$65.10	$520.80	$43.15	$65.17
4 Laborers	41.05	1313.60	62.10	1987.20		
1 Equip. Oper. (light)	51.65	413.20	77.55	620.40		
1 Tandem Roller, 5 Ton		152.45		167.69	3.18	3.49
48 L.H., Daily Totals		$2223.65		$3296.09	$46.33	$68.67

Crew B-37A	Hr.	Daily	Hr.	Daily	Bare Costs	Incl. O&P
2 Laborers	$41.05	$656.80	$62.10	$993.60	$42.53	$64.23
1 Truck Driver (light)	45.50	364.00	68.50	548.00		
1 Flatbed Truck, Gas, 1.5 Ton		195.20		214.72		
1 Tar Kettle, T.M.		134.50		147.95	13.74	15.11
24 L.H., Daily Totals		$1350.50		$1904.27	$56.27	$79.34

Crew B-37B	Hr.	Daily	Hr.	Daily	Bare Costs	Incl. O&P
3 Laborers	$41.05	$985.20	$62.10	$1490.40	$42.16	$63.70
1 Truck Driver (light)	45.50	364.00	68.50	548.00		
1 Flatbed Truck, Gas, 1.5 Ton		195.20		214.72		
1 Tar Kettle, T.M.		134.50		147.95	10.30	11.33
32 L.H., Daily Totals		$1678.90		$2401.07	$52.47	$75.03

Crew B-37C	Hr.	Daily	Hr.	Daily	Bare Costs	Incl. O&P
2 Laborers	$41.05	$656.80	$62.10	$993.60	$43.27	$65.30
2 Truck Drivers (light)	45.50	728.00	68.50	1096.00		
2 Flatbed Trucks, Gas, 1.5 Ton		390.40		429.44		
1 Tar Kettle, T.M.		134.50		147.95	16.40	18.04
32 L.H., Daily Totals		$1909.70		$2666.99	$59.68	$83.34

Crew B-37D	Hr.	Daily	Hr.	Daily	Bare Costs	Incl. O&P
1 Laborer	$41.05	$328.40	$62.10	$496.80	$43.27	$65.30
1 Truck Driver (light)	45.50	364.00	68.50	548.00		
1 Pickup Truck, 3/4 Ton		109.90		120.89	6.87	7.56
16 L.H., Daily Totals		$802.30		$1165.69	$50.14	$72.86

Crew B-37E	Hr.	Daily	Hr.	Daily	Bare Costs	Incl. O&P
3 Laborers	$41.05	$985.20	$62.10	$1490.40	$45.84	$69.08
1 Equip. Oper. (light)	51.65	413.20	77.55	620.40		
1 Equip. Oper. (medium)	55.10	440.80	82.70	661.60		
2 Truck Drivers (light)	45.50	728.00	68.50	1096.00		
4 Barrels w/ Flasher		16.00		17.60		
1 Concrete Saw		110.15		121.17		
1 Rotary Hammer Drill		25.05		27.56		
1 Hammer Drill Bit		3.30		3.63		
1 Loader, Skid Steer, 30 H.P.		174.55		192.01		
1 Conc. Hammer Attach.		114.40		125.84		
1 Vibrating Plate, Gas, 18"		31.10		34.21		
2 Flatbed Trucks, Gas, 1.5 Ton		390.40		429.44	15.45	16.99
56 L.H., Daily Totals		$3432.15		$4819.85	$61.29	$86.07

Crew B-37F	Hr.	Daily	Hr.	Daily	Bare Costs	Incl. O&P
3 Laborers	$41.05	$985.20	$62.10	$1490.40	$42.16	$63.70
1 Truck Driver (light)	45.50	364.00	68.50	548.00		
4 Barrels w/ Flasher		16.00		17.60		
1 Concrete Mixer, 10 C.F.		162.00		178.20		
1 Air Compressor, 60 cfm		106.70		117.37		
1 -50' Air Hose, 3/4"		6.55		7.21		
1 Spade (Chipper)		8.30		9.13		
1 Flatbed Truck, Gas, 1.5 Ton		195.20		214.72	15.46	17.01
32 L.H., Daily Totals		$1843.95		$2582.63	$57.62	$80.71

Crew B-37G	Hr.	Daily	Hr.	Daily	Bare Costs	Incl. O&P
1 Labor Foreman (outside)	$43.05	$344.40	$65.10	$520.80	$43.15	$65.17
4 Laborers	41.05	1313.60	62.10	1987.20		
1 Equip. Oper. (light)	51.65	413.20	77.55	620.40		
1 Berm Machine		325.30		357.83		
1 Tandem Roller, 5 Ton		152.45		167.69	9.95	10.95
48 L.H., Daily Totals		$2548.95		$3653.93	$53.10	$76.12

Crew B-37H	Hr.	Daily	Hr.	Daily	Bare Costs	Incl. O&P
1 Labor Foreman (outside)	$43.05	$344.40	$65.10	$520.80	$43.15	$65.17
4 Laborers	41.05	1313.60	62.10	1987.20		
1 Equip. Oper. (light)	51.65	413.20	77.55	620.40		
1 Tandem Roller, 5 Ton		152.45		167.69		
1 Flatbed Truck, Gas, 1.5 Ton		195.20		214.72		
1 Tar Kettle, T.M.		134.50		147.95	10.04	11.05
48 L.H., Daily Totals		$2553.35		$3658.76	$53.19	$76.22

Crew B-37I	Hr.	Daily	Hr.	Daily	Bare Costs	Incl. O&P
3 Laborers	$41.05	$985.20	$62.10	$1490.40	$45.84	$69.08
1 Equip. Oper. (light)	51.65	413.20	77.55	620.40		
1 Equip. Oper. (medium)	55.10	440.80	82.70	661.60		
2 Truck Drivers (light)	45.50	728.00	68.50	1096.00		
4 Barrels w/ Flasher		16.00		17.60		
1 Concrete Saw		110.15		121.17		
1 Rotary Hammer Drill		25.05		27.56		
1 Hammer Drill Bit		3.30		3.63		
1 Air Compressor, 60 cfm		106.70		117.37		
1 -50' Air Hose, 3/4"		6.55		7.21		
1 Spade (Chipper)		8.30		9.13		
1 Loader, Skid Steer, 30 H.P.		174.55		192.01		
1 Conc. Hammer Attach.		114.40		125.84		
1 Concrete Mixer, 10 C.F.		162.00		178.20		
1 Vibrating Plate, Gas, 18"		31.10		34.21		
2 Flatbed Trucks, Gas, 1.5 Ton		390.40		429.44	20.51	22.56
56 L.H., Daily Totals		$3715.70		$5131.75	$66.35	$91.64

For customer support on your Electrical Costs with RSMeans data, call 800.448.8182.

Crew No.	Bare Costs		Incl. Subs O&P		Cost Per Labor-Hour	

Crew B-37J	Hr.	Daily	Hr.	Daily	Bare Costs	Incl. O&P
1 Labor Foreman (outside)	$43.05	$344.40	$65.10	$520.80	$43.15	$65.17
4 Laborers	41.05	1313.60	62.10	1987.20		
1 Equip. Oper. (light)	51.65	413.20	77.55	620.40		
1 Air Compressor, 60 cfm		106.70		117.37		
1 -50' Air Hose, 3/4"		6.55		7.21		
2 Concrete Mixers, 10 C.F.		324.00		356.40		
2 Flatbed Trucks, Gas, 1.5 Ton		390.40		429.44		
1 Shot Blaster, 20"		201.00		221.10	21.43	23.57
48 L.H., Daily Totals		$3099.85		$4259.92	$64.58	$88.75

Crew B-37K	Hr.	Daily	Hr.	Daily	Bare Costs	Incl. O&P
1 Labor Foreman (outside)	$43.05	$344.40	$65.10	$520.80	$43.15	$65.17
4 Laborers	41.05	1313.60	62.10	1987.20		
1 Equip. Oper. (light)	51.65	413.20	77.55	620.40		
1 Air Compressor, 60 cfm		106.70		117.37		
1 -50' Air Hose, 3/4"		6.55		7.21		
2 Flatbed Trucks, Gas, 1.5 Ton		390.40		429.44		
1 Shot Blaster, 20"		201.00		221.10	14.68	16.15
48 L.H., Daily Totals		$2775.85		$3903.51	$57.83	$81.32

Crew B-38	Hr.	Daily	Hr.	Daily	Bare Costs	Incl. O&P
1 Labor Foreman (outside)	$43.05	$344.40	$65.10	$520.80	$46.38	$69.91
2 Laborers	41.05	656.80	62.10	993.60		
1 Equip. Oper. (light)	51.65	413.20	77.55	620.40		
1 Equip. Oper. (medium)	55.10	440.80	82.70	661.60		
1 Backhoe Loader, 48 H.P.		320.20		352.22		
1 Hyd. Hammer (1200 lb.)		171.40		188.54		
1 F.E. Loader, W.M., 4 C.Y.		566.70		623.37		
1 Pvmt. Rem. Bucket		61.70		67.87	28.00	30.80
40 L.H., Daily Totals		$2975.20		$4028.40	$74.38	$100.71

Crew B-39	Hr.	Daily	Hr.	Daily	Bare Costs	Incl. O&P
1 Labor Foreman (outside)	$43.05	$344.40	$65.10	$520.80	$43.15	$65.17
4 Laborers	41.05	1313.60	62.10	1987.20		
1 Equip. Oper. (light)	51.65	413.20	77.55	620.40		
1 Air Compressor, 250 cfm		167.95		184.75		
2 Breakers, Pavement, 60 lb.		21.50		23.65		
2 -50' Air Hoses, 1.5"		45.50		50.05	4.89	5.38
48 L.H., Daily Totals		$2306.15		$3386.84	$48.04	$70.56

Crew B-40	Hr.	Daily	Hr.	Daily	Bare Costs	Incl. O&P
1 Pile Driver Foreman (outside)	$54.15	$433.20	$84.65	$677.20	$53.36	$82.15
4 Pile Drivers	52.15	1668.80	81.55	2609.60		
2 Equip. Oper. (crane)	57.45	919.20	86.25	1380.00		
1 Equip. Oper. (oiler)	49.20	393.60	73.85	590.80		
1 Crawler Crane, 40 Ton		1392.00		1531.20		
1 Vibratory Hammer & Gen.		2236.00		2459.60	56.69	62.36
64 L.H., Daily Totals		$7042.80		$9248.40	$110.04	$144.51

Crew B-40B	Hr.	Daily	Hr.	Daily	Bare Costs	Incl. O&P
1 Labor Foreman (outside)	$43.05	$344.40	$65.10	$520.80	$45.48	$68.58
3 Laborers	41.05	985.20	62.10	1490.40		
1 Equip. Oper. (crane)	57.45	459.60	86.25	690.00		
1 Equip. Oper. (oiler)	49.20	393.60	73.85	590.80		
1 Lattice Boom Crane, 40 Ton		1315.00		1446.50	27.40	30.14
48 L.H., Daily Totals		$3497.80		$4738.50	$72.87	$98.72

Crew B-41	Hr.	Daily	Hr.	Daily	Bare Costs	Incl. O&P
1 Labor Foreman (outside)	$43.05	$344.40	$65.10	$520.80	$42.53	$64.28
4 Laborers	41.05	1313.60	62.10	1987.20		
.25 Equip. Oper. (crane)	57.45	114.90	86.25	172.50		
.25 Equip. Oper. (oiler)	49.20	98.40	73.85	147.70		
.25 Crawler Crane, 40 Ton		348.00		382.80	7.91	8.70
44 L.H., Daily Totals		$2219.30		$3211.00	$50.44	$72.98

Crew B-42	Hr.	Daily	Hr.	Daily	Bare Costs	Incl. O&P
1 Labor Foreman (outside)	$43.05	$344.40	$65.10	$520.80	$46.24	$70.37
4 Laborers	41.05	1313.60	62.10	1987.20		
1 Equip. Oper. (crane)	57.45	459.60	86.25	690.00		
1 Equip. Oper. (oiler)	49.20	393.60	73.85	590.80		
1 Welder	56.00	448.00	89.35	714.80		
1 Hyd. Crane, 25 Ton		581.70		639.87		
1 Welder, Gas Engine, 300 amp		95.65		105.22		
1 Horz. Boring Csg. Mch.		447.00		491.70	17.57	19.32
64 L.H., Daily Totals		$4083.55		$5740.39	$63.81	$89.69

Crew B-43	Hr.	Daily	Hr.	Daily	Bare Costs	Incl. O&P
1 Labor Foreman (outside)	$43.05	$344.40	$65.10	$520.80	$45.48	$68.58
3 Laborers	41.05	985.20	62.10	1490.40		
1 Equip. Oper. (crane)	57.45	459.60	86.25	690.00		
1 Equip. Oper. (oiler)	49.20	393.60	73.85	590.80		
1 Drill Rig, Truck-Mounted		2428.00		2670.80	50.58	55.64
48 L.H., Daily Totals		$4610.80		$5962.80	$96.06	$124.22

Crew B-44	Hr.	Daily	Hr.	Daily	Bare Costs	Incl. O&P
1 Pile Driver Foreman (outside)	$54.15	$433.20	$84.65	$677.20	$52.34	$80.68
4 Pile Drivers	52.15	1668.80	81.55	2609.60		
2 Equip. Oper. (crane)	57.45	919.20	86.25	1380.00		
1 Laborer	41.05	328.40	62.10	496.80		
1 Crawler Crane, 40 Ton		1392.00		1531.20		
1 Lead, 60' High		80.55		88.61		
1 Hammer, Diesel, 15K ft.-lbs.		622.40		684.64	32.73	36.01
64 L.H., Daily Totals		$5444.55		$7468.05	$85.07	$116.69

Crew B-45	Hr.	Daily	Hr.	Daily	Bare Costs	Incl. O&P
1 Equip. Oper. (medium)	$55.10	$440.80	$82.70	$661.60	$51.05	$76.72
1 Truck Driver (heavy)	47.00	376.00	70.75	566.00		
1 Dist. Tanker, 3000 Gallon		323.25		355.57		
1 Truck Tractor, 6x4, 380 H.P.		487.60		536.36	50.68	55.75
16 L.H., Daily Totals		$1627.65		$2119.53	$101.73	$132.47

Crew B-46	Hr.	Daily	Hr.	Daily	Bare Costs	Incl. O&P
1 Pile Driver Foreman (outside)	$54.15	$433.20	$84.65	$677.20	$46.93	$72.34
2 Pile Drivers	52.15	834.40	81.55	1304.80		
3 Laborers	41.05	985.20	62.10	1490.40		
1 Chain Saw, Gas, 36" Long		46.65		51.31	.97	1.07
48 L.H., Daily Totals		$2299.45		$3523.72	$47.91	$73.41

Crew B-47	Hr.	Daily	Hr.	Daily	Bare Costs	Incl. O&P
1 Blast Foreman (outside)	$43.05	$344.40	$65.10	$520.80	$45.25	$68.25
1 Driller	41.05	328.40	62.10	496.80		
1 Equip. Oper. (light)	51.65	413.20	77.55	620.40		
1 Air Track Drill, 4"		1004.00		1104.40		
1 Air Compressor, 600 cfm		417.60		459.36		
2 -50' Air Hoses, 3"		51.50		56.65	61.38	67.52
24 L.H., Daily Totals		$2559.10		$3258.41	$106.63	$135.77

Crew B-47A

	Bare Costs Hr.	Daily	Incl. Subs O&P Hr.	Daily	Cost Per Labor-Hour Bare Costs	Incl. O&P
1 Drilling Foreman (outside)	$43.05	$344.40	$65.10	$520.80	$49.90	$75.07
1 Equip. Oper. (heavy)	57.45	459.60	86.25	690.00		
1 Equip. Oper. (oiler)	49.20	393.60	73.85	590.80		
1 Air Track Drill, 5"		1173.00		1290.30	48.88	53.76
24 L.H., Daily Totals		$2370.60		$3091.90	$98.78	$128.83

Crew B-47C

	Bare Costs Hr.	Daily	Incl. Subs O&P Hr.	Daily	Cost Per Labor-Hour Bare Costs	Incl. O&P
1 Laborer	$41.05	$328.40	$62.10	$496.80	$46.35	$69.83
1 Equip. Oper. (light)	51.65	413.20	77.55	620.40		
1 Air Compressor, 750 cfm		428.80		471.68		
2 -50' Air Hoses, 3"		51.50		56.65		
1 Air Track Drill, 4"		1004.00		1104.40	92.77	102.05
16 L.H., Daily Totals		$2225.90		$2749.93	$139.12	$171.87

Crew B-47E

	Bare Costs Hr.	Daily	Incl. Subs O&P Hr.	Daily	Cost Per Labor-Hour Bare Costs	Incl. O&P
1 Labor Foreman (outside)	$43.05	$344.40	$65.10	$520.80	$41.55	$62.85
3 Laborers	41.05	985.20	62.10	1490.40		
1 Flatbed Truck, Gas, 3 Ton		245.95		270.55	7.69	8.45
32 L.H., Daily Totals		$1575.55		$2281.74	$49.24	$71.30

Crew B-47G

	Bare Costs Hr.	Daily	Incl. Subs O&P Hr.	Daily	Cost Per Labor-Hour Bare Costs	Incl. O&P
1 Labor Foreman (outside)	$43.05	$344.40	$65.10	$520.80	$44.20	$66.71
2 Laborers	41.05	656.80	62.10	993.60		
1 Equip. Oper. (light)	51.65	413.20	77.55	620.40		
1 Air Track Drill, 4"		1004.00		1104.40		
1 Air Compressor, 600 cfm		417.60		459.36		
2 -50' Air Hoses, 3"		51.50		56.65		
1 Gunite Pump Rig		313.00		344.30	55.82	61.40
32 L.H., Daily Totals		$3200.50		$4099.51	$100.02	$128.11

Crew B-47H

	Bare Costs Hr.	Daily	Incl. Subs O&P Hr.	Daily	Cost Per Labor-Hour Bare Costs	Incl. O&P
1 Skilled Worker Foreman (out)	$55.40	$443.20	$84.25	$674.00	$53.90	$82.00
3 Skilled Workers	53.40	1281.60	81.25	1950.00		
1 Flatbed Truck, Gas, 3 Ton		245.95		270.55	7.69	8.45
32 L.H., Daily Totals		$1970.75		$2894.55	$61.59	$90.45

Crew B-48

	Bare Costs Hr.	Daily	Incl. Subs O&P Hr.	Daily	Cost Per Labor-Hour Bare Costs	Incl. O&P
1 Labor Foreman (outside)	$43.05	$344.40	$65.10	$520.80	$46.36	$69.86
3 Laborers	41.05	985.20	62.10	1490.40		
1 Equip. Oper. (crane)	57.45	459.60	86.25	690.00		
1 Equip. Oper. (oiler)	49.20	393.60	73.85	590.80		
1 Equip. Oper. (light)	51.65	413.20	77.55	620.40		
1 Centr. Water Pump, 6"		296.45		326.10		
1 -20' Suction Hose, 6"		11.10		12.21		
1 -50' Discharge Hose, 6"		5.70		6.27		
1 Drill Rig, Truck-Mounted		2428.00		2670.80	48.95	53.85
56 L.H., Daily Totals		$5337.25		$6927.77	$95.31	$123.71

Crew B-49

	Bare Costs Hr.	Daily	Incl. Subs O&P Hr.	Daily	Cost Per Labor-Hour Bare Costs	Incl. O&P
1 Labor Foreman (outside)	$43.05	$344.40	$65.10	$520.80	$48.68	$73.84
3 Laborers	41.05	985.20	62.10	1490.40		
2 Equip. Oper. (crane)	57.45	919.20	86.25	1380.00		
2 Equip. Oper. (oilers)	49.20	787.20	73.85	1181.60		
1 Equip. Oper. (light)	51.65	413.20	77.55	620.40		
2 Pile Drivers	52.15	834.40	81.55	1304.80		
1 Hyd. Crane, 25 Ton		581.70		639.87		
1 Centr. Water Pump, 6"		296.45		326.10		
1 -20' Suction Hose, 6"		11.10		12.21		
1 -50' Discharge Hose, 6"		5.70		6.27		
1 Drill Rig, Truck-Mounted		2428.00		2670.80	37.76	41.54
88 L.H., Daily Totals		$7606.55		$10153.25	$86.44	$115.38

Crew B-50

	Bare Costs Hr.	Daily	Incl. Subs O&P Hr.	Daily	Cost Per Labor-Hour Bare Costs	Incl. O&P
2 Pile Driver Foremen (outside)	$54.15	$866.40	$84.65	$1354.40	$50.60	$77.95
6 Pile Drivers	52.15	2503.20	81.55	3914.40		
2 Equip. Oper. (crane)	57.45	919.20	86.25	1380.00		
1 Equip. Oper. (oiler)	49.20	393.60	73.85	590.80		
3 Laborers	41.05	985.20	62.10	1490.40		
1 Crawler Crane, 40 Ton		1392.00		1531.20		
1 Lead, 60' High		80.55		88.61		
1 Hammer, Diesel, 15K ft.-lbs.		622.40		684.64		
1 Air Compressor, 600 cfm		417.60		459.36		
2 -50' Air Hoses, 3"		51.50		56.65		
1 Chain Saw, Gas, 36" Long		46.65		51.31	23.31	25.64
112 L.H., Daily Totals		$8278.30		$11601.77	$73.91	$103.59

Crew B-51

	Bare Costs Hr.	Daily	Incl. Subs O&P Hr.	Daily	Cost Per Labor-Hour Bare Costs	Incl. O&P
1 Labor Foreman (outside)	$43.05	$344.40	$65.10	$520.80	$42.13	$63.67
4 Laborers	41.05	1313.60	62.10	1987.20		
1 Truck Driver (light)	45.50	364.00	68.50	548.00		
1 Flatbed Truck, Gas, 1.5 Ton		195.20		214.72	4.07	4.47
48 L.H., Daily Totals		$2217.20		$3270.72	$46.19	$68.14

Crew B-52

	Bare Costs Hr.	Daily	Incl. Subs O&P Hr.	Daily	Cost Per Labor-Hour Bare Costs	Incl. O&P
1 Carpenter Foreman (outside)	$53.65	$429.20	$81.15	$649.20	$47.48	$71.52
1 Carpenter	51.65	413.20	78.15	625.20		
3 Laborers	41.05	985.20	62.10	1490.40		
1 Cement Finisher	48.90	391.20	72.20	577.60		
.5 Rodman (reinf.)	54.85	219.40	83.00	332.00		
.5 Equip. Oper. (medium)	55.10	220.40	82.70	330.80		
.5 Crawler Loader, 3 C.Y.		601.00		661.10	10.73	11.81
56 L.H., Daily Totals		$3259.60		$4666.30	$58.21	$83.33

Crew B-53

	Bare Costs Hr.	Daily	Incl. Subs O&P Hr.	Daily	Cost Per Labor-Hour Bare Costs	Incl. O&P
1 Equip. Oper. (light)	$51.65	$413.20	$77.55	$620.40	$51.65	$77.55
1 Trencher, Chain, 12 H.P.		63.40		69.74	7.92	8.72
8 L.H., Daily Totals		$476.60		$690.14	$59.58	$86.27

Crew B-54

	Bare Costs Hr.	Daily	Incl. Subs O&P Hr.	Daily	Cost Per Labor-Hour Bare Costs	Incl. O&P
1 Equip. Oper. (light)	$51.65	$413.20	$77.55	$620.40	$51.65	$77.55
1 Trencher, Chain, 40 H.P.		348.50		383.35	43.56	47.92
8 L.H., Daily Totals		$761.70		$1003.75	$95.21	$125.47

Crew B-54A

	Bare Costs Hr.	Daily	Incl. Subs O&P Hr.	Daily	Cost Per Labor-Hour Bare Costs	Incl. O&P
.17 Labor Foreman (outside)	$43.05	$58.55	$65.10	$88.54	$53.35	$80.14
1 Equipment Operator (med.)	55.10	440.80	82.70	661.60		
1 Wheel Trencher, 67 H.P.		1110.00		1221.00	118.59	130.45
9.36 L.H., Daily Totals		$1609.35		$1971.14	$171.94	$210.59

Crew B-54B

	Bare Costs Hr.	Daily	Incl. Subs O&P Hr.	Daily	Cost Per Labor-Hour Bare Costs	Incl. O&P
.25 Labor Foreman (outside)	$43.05	$86.10	$65.10	$130.20	$52.69	$79.18
1 Equipment Operator (med.)	55.10	440.80	82.70	661.60		
1 Wheel Trencher, 150 H.P.		1996.00		2195.60	199.60	219.56
10 L.H., Daily Totals		$2522.90		$2987.40	$252.29	$298.74

Crew B-54C

	Bare Costs Hr.	Daily	Incl. Subs O&P Hr.	Daily	Cost Per Labor-Hour Bare Costs	Incl. O&P
1 Laborer	$41.05	$328.40	$62.10	$496.80	$48.08	$72.40
1 Equipment Operator (med.)	55.10	440.80	82.70	661.60		
1 Wheel Trencher, 67 H.P.		1110.00		1221.00	69.38	76.31
16 L.H., Daily Totals		$1879.20		$2379.40	$117.45	$148.71

Crews - Standard

Crew No.	Bare Costs Hr.	Daily	Incl. Subs O&P Hr.	Daily	Cost Per Labor-Hour Bare Costs	Incl. O&P
Crew B-54D	Hr.	Daily	Hr.	Daily	Bare Costs	Incl. O&P
1 Laborer	$41.05	$328.40	$62.10	$496.80	$48.08	$72.40
1 Equipment Operator (med.)	55.10	440.80	82.70	661.60		
1 Rock Trencher, 6" Width		797.00		876.70	49.81	54.79
16 L.H., Daily Totals		$1566.20		$2035.10	$97.89	$127.19
Crew B-54E	Hr.	Daily	Hr.	Daily	Bare Costs	Incl. O&P
1 Laborer	$41.05	$328.40	$62.10	$496.80	$48.08	$72.40
1 Equipment Operator (med.)	55.10	440.80	82.70	661.60		
1 Rock Trencher, 18" Width		2678.00		2945.80	167.38	184.11
16 L.H., Daily Totals		$3447.20		$4104.20	$215.45	$256.51
Crew B-55	Hr.	Daily	Hr.	Daily	Bare Costs	Incl. O&P
2 Laborers	$41.05	$656.80	$62.10	$993.60	$42.53	$64.23
1 Truck Driver (light)	45.50	364.00	68.50	548.00		
1 Truck-Mounted Earth Auger		756.20		831.82		
1 Flatbed Truck, Gas, 3 Ton		245.95		270.55	41.76	45.93
24 L.H., Daily Totals		$2022.95		$2643.97	$84.29	$110.17
Crew B-56	Hr.	Daily	Hr.	Daily	Bare Costs	Incl. O&P
1 Laborer	$41.05	$328.40	$62.10	$496.80	$46.35	$69.83
1 Equip. Oper. (light)	51.65	413.20	77.55	620.40		
1 Air Track Drill, 4"		1004.00		1104.40		
1 Air Compressor, 600 cfm		417.60		459.36		
1 -50' Air Hose, 3"		25.75		28.32	90.46	99.51
16 L.H., Daily Totals		$2188.95		$2709.28	$136.81	$169.33
Crew B-57	Hr.	Daily	Hr.	Daily	Bare Costs	Incl. O&P
1 Labor Foreman (outside)	$43.05	$344.40	$65.10	$520.80	$47.24	$71.16
2 Laborers	41.05	656.80	62.10	993.60		
1 Equip. Oper. (crane)	57.45	459.60	86.25	690.00		
1 Equip. Oper. (light)	51.65	413.20	77.55	620.40		
1 Equip. Oper. (oiler)	49.20	393.60	73.85	590.80		
1 Crawler Crane, 25 Ton		1386.00		1524.60		
1 Clamshell Bucket, 1 C.Y.		51.20		56.32		
1 Centr. Water Pump, 6"		296.45		326.10		
1 -20' Suction Hose, 6"		11.10		12.21		
20 -50' Discharge Hoses, 6"		114.00		125.40	38.72	42.60
48 L.H., Daily Totals		$4126.35		$5460.23	$85.97	$113.75
Crew B-58	Hr.	Daily	Hr.	Daily	Bare Costs	Incl. O&P
2 Laborers	$41.05	$656.80	$62.10	$993.60	$44.58	$67.25
1 Equip. Oper. (light)	51.65	413.20	77.55	620.40		
1 Backhoe Loader, 48 H.P.		320.20		352.22		
1 Small Helicopter, w/ Pilot		2916.00		3207.60	134.84	148.33
24 L.H., Daily Totals		$4306.20		$5173.82	$179.43	$215.58
Crew B-59	Hr.	Daily	Hr.	Daily	Bare Costs	Incl. O&P
1 Truck Driver (heavy)	$47.00	$376.00	$70.75	$566.00	$47.00	$70.75
1 Truck Tractor, 220 H.P.		303.55		333.90		
1 Water Tank Trailer, 5000 Gal.		149.45		164.40	56.63	62.29
8 L.H., Daily Totals		$829.00		$1064.30	$103.63	$133.04
Crew B-59A	Hr.	Daily	Hr.	Daily	Bare Costs	Incl. O&P
2 Laborers	$41.05	$656.80	$62.10	$993.60	$43.03	$64.98
1 Truck Driver (heavy)	47.00	376.00	70.75	566.00		
1 Water Tank Trailer, 5000 Gal.		149.45		164.40		
1 Truck Tractor, 220 H.P.		303.55		333.90	18.88	20.76
24 L.H., Daily Totals		$1485.80		$2057.90	$61.91	$85.75

Crew No.	Bare Costs Hr.	Daily	Incl. Subs O&P Hr.	Daily	Cost Per Labor-Hour Bare Costs	Incl. O&P
Crew B-60	Hr.	Daily	Hr.	Daily	Bare Costs	Incl. O&P
1 Labor Foreman (outside)	$43.05	$344.40	$65.10	$520.80	$47.87	$72.07
2 Laborers	41.05	656.80	62.10	993.60		
1 Equip. Oper. (crane)	57.45	459.60	86.25	690.00		
2 Equip. Oper. (light)	51.65	826.40	77.55	1240.80		
1 Equip. Oper. (oiler)	49.20	393.60	73.85	590.80		
1 Crawler Crane, 40 Ton		1392.00		1531.20		
1 Lead, 60' High		80.55		88.61		
1 Hammer, Diesel, 15K ft.-lbs.		622.40		684.64		
1 Backhoe Loader, 48 H.P.		320.20		352.22	43.13	47.44
56 L.H., Daily Totals		$5095.95		$6692.67	$91.00	$119.51
Crew B-61	Hr.	Daily	Hr.	Daily	Bare Costs	Incl. O&P
1 Labor Foreman (outside)	$43.05	$344.40	$65.10	$520.80	$43.57	$65.79
3 Laborers	41.05	985.20	62.10	1490.40		
1 Equip. Oper. (light)	51.65	413.20	77.55	620.40		
1 Cement Mixer, 2 C.Y.		203.85		224.24		
1 Air Compressor, 160 cfm		115.85		127.44	7.99	8.79
40 L.H., Daily Totals		$2062.50		$2983.27	$51.56	$74.58
Crew B-62	Hr.	Daily	Hr.	Daily	Bare Costs	Incl. O&P
2 Laborers	$41.05	$656.80	$62.10	$993.60	$44.58	$67.25
1 Equip. Oper. (light)	51.65	413.20	77.55	620.40		
1 Loader, Skid Steer, 30 H.P.		174.55		192.01	7.27	8.00
24 L.H., Daily Totals		$1244.55		$1806.01	$51.86	$75.25
Crew B-62A	Hr.	Daily	Hr.	Daily	Bare Costs	Incl. O&P
2 Laborers	$41.05	$656.80	$62.10	$993.60	$44.58	$67.25
1 Equip. Oper. (light)	51.65	413.20	77.55	620.40		
1 Loader, Skid Steer, 30 H.P.		174.55		192.01		
1 Trencher Attachment		70.20		77.22	10.20	11.22
24 L.H., Daily Totals		$1314.75		$1883.22	$54.78	$78.47
Crew B-63	Hr.	Daily	Hr.	Daily	Bare Costs	Incl. O&P
4 Laborers	$41.05	$1313.60	$62.10	$1987.20	$43.17	$65.19
1 Equip. Oper. (light)	51.65	413.20	77.55	620.40		
1 Loader, Skid Steer, 30 H.P.		174.55		192.01	4.36	4.80
40 L.H., Daily Totals		$1901.35		$2799.61	$47.53	$69.99
Crew B-63B	Hr.	Daily	Hr.	Daily	Bare Costs	Incl. O&P
1 Labor Foreman (inside)	$41.55	$332.40	$62.85	$502.80	$43.83	$66.15
2 Laborers	41.05	656.80	62.10	993.60		
1 Equip. Oper. (light)	51.65	413.20	77.55	620.40		
1 Loader, Skid Steer, 78 H.P.		392.50		431.75	12.27	13.49
32 L.H., Daily Totals		$1794.90		$2548.55	$56.09	$79.64
Crew B-64	Hr.	Daily	Hr.	Daily	Bare Costs	Incl. O&P
1 Laborer	$41.05	$328.40	$62.10	$496.80	$43.27	$65.30
1 Truck Driver (light)	45.50	364.00	68.50	548.00		
1 Power Mulcher (small)		139.35		153.29		
1 Flatbed Truck, Gas, 1.5 Ton		195.20		214.72	20.91	23.00
16 L.H., Daily Totals		$1026.95		$1412.81	$64.18	$88.30
Crew B-65	Hr.	Daily	Hr.	Daily	Bare Costs	Incl. O&P
1 Laborer	$41.05	$328.40	$62.10	$496.80	$43.27	$65.30
1 Truck Driver (light)	45.50	364.00	68.50	548.00		
1 Power Mulcher (Large)		276.35		303.99		
1 Flatbed Truck, Gas, 1.5 Ton		195.20		214.72	29.47	32.42
16 L.H., Daily Totals		$1163.95		$1563.51	$72.75	$97.72

For customer support on your Electrical Costs with RSMeans data, call 800.448.8182.

493

Crews - Standard

Crew B-66

Crew No.	Bare Costs Hr.	Bare Costs Daily	Incl. Subs O&P Hr.	Incl. Subs O&P Daily	Cost Per Labor-Hour Bare Costs	Cost Per Labor-Hour Incl. O&P
1 Equip. Oper. (light)	$51.65	$413.20	$77.55	$620.40	$51.65	$77.55
1 Loader-Backhoe, 40 H.P.		219.20		241.12	27.40	30.14
8 L.H., Daily Totals		$632.40		$861.52	$79.05	$107.69

Crew B-67

Crew No.	Bare Costs Hr.	Bare Costs Daily	Incl. Subs O&P Hr.	Incl. Subs O&P Daily	Cost Per Labor-Hour Bare Costs	Cost Per Labor-Hour Incl. O&P
1 Millwright	$54.15	$433.20	$78.70	$629.60	$52.90	$78.13
1 Equip. Oper. (light)	51.65	413.20	77.55	620.40		
1 R.T. Forklift, 5,000 Lb., diesel		265.60		292.16	16.60	18.26
16 L.H., Daily Totals		$1112.00		$1542.16	$69.50	$96.39

Crew B-67B

Crew No.	Bare Costs Hr.	Bare Costs Daily	Incl. Subs O&P Hr.	Incl. Subs O&P Daily	Cost Per Labor-Hour Bare Costs	Cost Per Labor-Hour Incl. O&P
1 Millwright Foreman (inside)	$54.65	$437.20	$79.40	$635.20	$54.40	$79.05
1 Millwright	54.15	433.20	78.70	629.60		
16 L.H., Daily Totals		$870.40		$1264.80	$54.40	$79.05

Crew B-68

Crew No.	Bare Costs Hr.	Bare Costs Daily	Incl. Subs O&P Hr.	Incl. Subs O&P Daily	Cost Per Labor-Hour Bare Costs	Cost Per Labor-Hour Incl. O&P
2 Millwrights	$54.15	$866.40	$78.70	$1259.20	$53.32	$78.32
1 Equip. Oper. (light)	51.65	413.20	77.55	620.40		
1 R.T. Forklift, 5,000 Lb., diesel		265.60		292.16	11.07	12.17
24 L.H., Daily Totals		$1545.20		$2171.76	$64.38	$90.49

Crew B-68A

Crew No.	Bare Costs Hr.	Bare Costs Daily	Incl. Subs O&P Hr.	Incl. Subs O&P Daily	Cost Per Labor-Hour Bare Costs	Cost Per Labor-Hour Incl. O&P
1 Millwright Foreman (inside)	$54.65	$437.20	$79.40	$635.20	$54.32	$78.93
2 Millwrights	54.15	866.40	78.70	1259.20		
1 Forklift, Smooth Floor, 8,000 Lb.		144.30		158.73	6.01	6.61
24 L.H., Daily Totals		$1447.90		$2053.13	$60.33	$85.55

Crew B-68B

Crew No.	Bare Costs Hr.	Bare Costs Daily	Incl. Subs O&P Hr.	Incl. Subs O&P Daily	Cost Per Labor-Hour Bare Costs	Cost Per Labor-Hour Incl. O&P
1 Millwright Foreman (inside)	$54.65	$437.20	$79.40	$635.20	$58.48	$86.50
2 Millwrights	54.15	866.40	78.70	1259.20		
2 Electricians	60.05	960.80	89.60	1433.60		
2 Plumbers	63.15	1010.40	94.75	1516.00		
1 R.T. Forklift, 5,000 Lb., gas		276.20		303.82	4.93	5.43
56 L.H., Daily Totals		$3551.00		$5147.82	$63.41	$91.93

Crew B-68C

Crew No.	Bare Costs Hr.	Bare Costs Daily	Incl. Subs O&P Hr.	Incl. Subs O&P Daily	Cost Per Labor-Hour Bare Costs	Cost Per Labor-Hour Incl. O&P
1 Millwright Foreman (inside)	$54.65	$437.20	$79.40	$635.20	$58.00	$85.61
1 Millwright	54.15	433.20	78.70	629.60		
1 Electrician	60.05	480.40	89.60	716.80		
1 Plumber	63.15	505.20	94.75	758.00		
1 R.T. Forklift, 5,000 Lb., gas		276.20		303.82	8.63	9.49
32 L.H., Daily Totals		$2132.20		$3043.42	$66.63	$95.11

Crew B-68D

Crew No.	Bare Costs Hr.	Bare Costs Daily	Incl. Subs O&P Hr.	Incl. Subs O&P Daily	Cost Per Labor-Hour Bare Costs	Cost Per Labor-Hour Incl. O&P
1 Labor Foreman (inside)	$41.55	$332.40	$62.85	$502.80	$44.75	$67.50
1 Laborer	41.05	328.40	62.10	496.80		
1 Equip. Oper. (light)	51.65	413.20	77.55	620.40		
1 R.T. Forklift, 5,000 Lb., gas		276.20		303.82	11.51	12.66
24 L.H., Daily Totals		$1350.20		$1923.82	$56.26	$80.16

Crew B-68E

Crew No.	Bare Costs Hr.	Bare Costs Daily	Incl. Subs O&P Hr.	Incl. Subs O&P Daily	Cost Per Labor-Hour Bare Costs	Cost Per Labor-Hour Incl. O&P
1 Struc. Steel Foreman (inside)	$56.50	$452.00	$90.10	$720.80	$56.10	$89.50
3 Struc. Steel Workers	56.00	1344.00	89.35	2144.40		
1 Welder	56.00	448.00	89.35	714.80		
1 Forklift, Smooth Floor, 8,000 Lb.		144.30		158.73	3.61	3.97
40 L.H., Daily Totals		$2388.30		$3738.73	$59.71	$93.47

Crew B-68F

Crew No.	Bare Costs Hr.	Bare Costs Daily	Incl. Subs O&P Hr.	Incl. Subs O&P Daily	Cost Per Labor-Hour Bare Costs	Cost Per Labor-Hour Incl. O&P
1 Skilled Worker Foreman (out)	$55.40	$443.20	$84.25	$674.00	$54.07	$82.25
2 Skilled Workers	53.40	854.40	81.25	1300.00		
1 R.T. Forklift, 5,000 Lb., gas		276.20		303.82	11.51	12.66
24 L.H., Daily Totals		$1573.80		$2277.82	$65.58	$94.91

Crew B-68G

Crew No.	Bare Costs Hr.	Bare Costs Daily	Incl. Subs O&P Hr.	Incl. Subs O&P Daily	Cost Per Labor-Hour Bare Costs	Cost Per Labor-Hour Incl. O&P
2 Structural Steel Workers	$56.00	$896.00	$89.35	$1429.60	$56.00	$89.35
1 R.T. Forklift, 5,000 Lb., gas		276.20		303.82	17.26	18.99
16 L.H., Daily Totals		$1172.20		$1733.42	$73.26	$108.34

Crew B-69

Crew No.	Bare Costs Hr.	Bare Costs Daily	Incl. Subs O&P Hr.	Incl. Subs O&P Daily	Cost Per Labor-Hour Bare Costs	Cost Per Labor-Hour Incl. O&P
1 Labor Foreman (outside)	$43.05	$344.40	$65.10	$520.80	$45.48	$68.58
3 Laborers	41.05	985.20	62.10	1490.40		
1 Equip. Oper. (crane)	57.45	459.60	86.25	690.00		
1 Equip. Oper. (oiler)	49.20	393.60	73.85	590.80		
1 Hyd. Crane, 80 Ton		1487.00		1635.70	30.98	34.08
48 L.H., Daily Totals		$3669.80		$4927.70	$76.45	$102.66

Crew B-69A

Crew No.	Bare Costs Hr.	Bare Costs Daily	Incl. Subs O&P Hr.	Incl. Subs O&P Daily	Cost Per Labor-Hour Bare Costs	Cost Per Labor-Hour Incl. O&P
1 Labor Foreman (outside)	$43.05	$344.40	$65.10	$520.80	$45.03	$67.72
3 Laborers	41.05	985.20	62.10	1490.40		
1 Equip. Oper. (medium)	55.10	440.80	82.70	661.60		
1 Concrete Finisher	48.90	391.20	72.20	577.60		
1 Curb/Gutter Paver, 2-Track		1194.00		1313.40	24.88	27.36
48 L.H., Daily Totals		$3355.60		$4563.80	$69.91	$95.08

Crew B-69B

Crew No.	Bare Costs Hr.	Bare Costs Daily	Incl. Subs O&P Hr.	Incl. Subs O&P Daily	Cost Per Labor-Hour Bare Costs	Cost Per Labor-Hour Incl. O&P
1 Labor Foreman (outside)	$43.05	$344.40	$65.10	$520.80	$45.03	$67.72
3 Laborers	41.05	985.20	62.10	1490.40		
1 Equip. Oper. (medium)	55.10	440.80	82.70	661.60		
1 Cement Finisher	48.90	391.20	72.20	577.60		
1 Curb/Gutter Paver, 4-Track		776.35		853.99	16.17	17.79
48 L.H., Daily Totals		$2937.95		$4104.39	$61.21	$85.51

Crew B-70

Crew No.	Bare Costs Hr.	Bare Costs Daily	Incl. Subs O&P Hr.	Incl. Subs O&P Daily	Cost Per Labor-Hour Bare Costs	Cost Per Labor-Hour Incl. O&P
1 Labor Foreman (outside)	$43.05	$344.40	$65.10	$520.80	$47.36	$71.36
3 Laborers	41.05	985.20	62.10	1490.40		
3 Equip. Oper. (medium)	55.10	1322.40	82.70	1984.80		
1 Grader, 30,000 Lbs.		657.85		723.63		
1 Ripper, Beam & 1 Shank		88.80		97.68		
1 Road Sweeper, S.P., 8' wide		703.10		773.41		
1 F.E. Loader, W.M., 1.5 C.Y.		352.95		388.25	32.19	35.41
56 L.H., Daily Totals		$4454.70		$5978.97	$79.55	$106.77

Crew B-70A

Crew No.	Bare Costs Hr.	Bare Costs Daily	Incl. Subs O&P Hr.	Incl. Subs O&P Daily	Cost Per Labor-Hour Bare Costs	Cost Per Labor-Hour Incl. O&P
1 Laborer	$41.05	$328.40	$62.10	$496.80	$52.29	$78.58
4 Equip. Oper. (medium)	55.10	1763.20	82.70	2646.40		
1 Grader, 40,000 Lbs.		1180.00		1298.00		
1 F.E. Loader, W.M., 2.5 C.Y.		531.20		584.32		
1 Dozer, 80 H.P.		472.00		519.20		
1 Roller, Pneum. Whl., 12 Ton		338.35		372.19	63.04	69.34
40 L.H., Daily Totals		$4613.15		$5916.90	$115.33	$147.92

Crew B-71

Crew No.	Bare Costs Hr.	Bare Costs Daily	Incl. Subs O&P Hr.	Incl. Subs O&P Daily	Cost Per Labor-Hour Bare Costs	Cost Per Labor-Hour Incl. O&P
1 Labor Foreman (outside)	$43.05	$344.40	$65.10	$520.80	$47.36	$71.36
3 Laborers	41.05	985.20	62.10	1490.40		
3 Equip. Oper. (medium)	55.10	1322.40	82.70	1984.80		
1 Pvmt. Profiler, 750 H.P.		5381.00		5919.10		
1 Road Sweeper, S.P., 8' wide		703.10		773.41		
1 F.E. Loader, W.M., 1.5 C.Y.		352.95		388.25	114.95	126.44
56 L.H., Daily Totals		$9089.05		$11076.75	$162.30	$197.80

For customer support on your Electrical Costs with RSMeans data, call 800.448.8182.

Crew No.	Bare Costs		Incl. Subs O&P		Cost Per Labor-Hour	

Crew B-72

Crew B-72	Hr.	Daily	Hr.	Daily	Bare Costs	Incl. O&P
1 Labor Foreman (outside)	$43.05	$344.40	$65.10	$520.80	$48.33	$72.78
3 Laborers	41.05	985.20	62.10	1490.40		
4 Equip. Oper. (medium)	55.10	1763.20	82.70	2646.40		
1 Pvmt. Profiler, 750 H.P.		5381.00		5919.10		
1 Hammermill, 250 H.P.		1890.00		2079.00		
1 Windrow Loader		1238.00		1361.80		
1 Mix Paver, 165 H.P.		2107.00		2317.70		
1 Roller, Pneum. Whl., 12 Ton		338.35		372.19	171.16	188.28
64 L.H., Daily Totals		$14047.15		$16707.38	$219.49	$261.05

Crew B-73

Crew B-73	Hr.	Daily	Hr.	Daily	Bare Costs	Incl. O&P
1 Labor Foreman (outside)	$43.05	$344.40	$65.10	$520.80	$50.08	$75.35
2 Laborers	41.05	656.80	62.10	993.60		
5 Equip. Oper. (medium)	55.10	2204.00	82.70	3308.00		
1 Road Mixer, 310 H.P.		1857.00		2042.70		
1 Tandem Roller, 10 Ton		232.05		255.26		
1 Hammermill, 250 H.P.		1890.00		2079.00		
1 Grader, 30,000 Lbs.		657.85		723.63		
.5 F.E. Loader, W.M., 1.5 C.Y.		176.47		194.12		
.5 Truck Tractor, 220 H.P.		151.78		166.95		
.5 Water Tank Trailer, 5000 Gal.		74.72		82.20	78.75	86.62
64 L.H., Daily Totals		$8245.08		$10366.26	$128.83	$161.97

Crew B-74

Crew B-74	Hr.	Daily	Hr.	Daily	Bare Costs	Incl. O&P
1 Labor Foreman (outside)	$43.05	$344.40	$65.10	$520.80	$49.81	$74.94
1 Laborer	41.05	328.40	62.10	496.80		
4 Equip. Oper. (medium)	55.10	1763.20	82.70	2646.40		
2 Truck Drivers (heavy)	47.00	752.00	70.75	1132.00		
1 Grader, 30,000 Lbs.		657.85		723.63		
1 Ripper, Beam & 1 Shank		88.80		97.68		
2 Stabilizers, 310 H.P.		3252.00		3577.20		
1 Flatbed Truck, Gas, 3 Ton		245.95		270.55		
1 Chem. Spreader, Towed		53.20		58.52		
1 Roller, Vibratory, 25 Ton		650.90		715.99		
1 Water Tank Trailer, 5000 Gal.		149.45		164.40		
1 Truck Tractor, 220 H.P.		303.55		333.90	84.40	92.84
64 L.H., Daily Totals		$8589.70		$10737.87	$134.21	$167.78

Crew B-75

Crew B-75	Hr.	Daily	Hr.	Daily	Bare Costs	Incl. O&P
1 Labor Foreman (outside)	$43.05	$344.40	$65.10	$520.80	$50.21	$75.54
1 Laborer	41.05	328.40	62.10	496.80		
4 Equip. Oper. (medium)	55.10	1763.20	82.70	2646.40		
1 Truck Driver (heavy)	47.00	376.00	70.75	566.00		
1 Grader, 30,000 Lbs.		657.85		723.63		
1 Ripper, Beam & 1 Shank		88.80		97.68		
2 Stabilizers, 310 H.P.		3252.00		3577.20		
1 Dist. Tanker, 3000 Gallon		323.25		355.57		
1 Truck Tractor, 6x4, 380 H.P.		487.60		536.36		
1 Roller, Vibratory, 25 Ton		650.90		715.99	97.51	107.26
56 L.H., Daily Totals		$8272.40		$10236.44	$147.72	$182.79

Crew B-76

Crew B-76	Hr.	Daily	Hr.	Daily	Bare Costs	Incl. O&P
1 Dock Builder Foreman (outside)	$54.15	$433.20	$84.65	$677.20	$53.22	$82.08
5 Dock Builders	52.15	2086.00	81.55	3262.00		
2 Equip. Oper. (crane)	57.45	919.20	86.25	1380.00		
1 Equip. Oper. (oiler)	49.20	393.60	73.85	590.80		
1 Crawler Crane, 50 Ton		2032.00		2235.20		
1 Barge, 400 Ton		836.50		920.15		
1 Hammer, Diesel, 15K ft.-lbs.		622.40		684.64		
1 Lead, 60' High		80.55		88.61		
1 Air Compressor, 600 cfm		417.60		459.36		
2 -50' Air Hoses, 3"		51.50		56.65	56.12	61.73
72 L.H., Daily Totals		$7872.55		$10354.61	$109.34	$143.81

Crew B-76A

Crew B-76A	Hr.	Daily	Hr.	Daily	Bare Costs	Incl. O&P
1 Labor Foreman (outside)	$43.05	$344.40	$65.10	$520.80	$44.37	$66.96
5 Laborers	41.05	1642.00	62.10	2484.00		
1 Equip. Oper. (crane)	57.45	459.60	86.25	690.00		
1 Equip. Oper. (oiler)	49.20	393.60	73.85	590.80		
1 Crawler Crane, 50 Ton		2032.00		2235.20		
1 Barge, 400 Ton		836.50		920.15	44.82	49.30
64 L.H., Daily Totals		$5708.10		$7440.95	$89.19	$116.26

Crew B-77

Crew B-77	Hr.	Daily	Hr.	Daily	Bare Costs	Incl. O&P
1 Labor Foreman (outside)	$43.05	$344.40	$65.10	$520.80	$42.34	$63.98
3 Laborers	41.05	985.20	62.10	1490.40		
1 Truck Driver (light)	45.50	364.00	68.50	548.00		
1 Crack Cleaner, 25 H.P.		56.95		62.65		
1 Crack Filler, Trailer Mtd.		198.95		218.85		
1 Flatbed Truck, Gas, 3 Ton		245.95		270.55	12.55	13.80
40 L.H., Daily Totals		$2195.45		$3111.24	$54.89	$77.78

Crew B-78

Crew B-78	Hr.	Daily	Hr.	Daily	Bare Costs	Incl. O&P
1 Labor Foreman (outside)	$43.05	$344.40	$65.10	$520.80	$42.13	$63.67
4 Laborers	41.05	1313.60	62.10	1987.20		
1 Truck Driver (light)	45.50	364.00	68.50	548.00		
1 Paint Striper, S.P., 40 Gallon		152.25		167.47		
1 Flatbed Truck, Gas, 3 Ton		245.95		270.55		
1 Pickup Truck, 3/4 Ton		109.90		120.89	10.59	11.64
48 L.H., Daily Totals		$2530.10		$3614.91	$52.71	$75.31

Crew B-78A

Crew B-78A	Hr.	Daily	Hr.	Daily	Bare Costs	Incl. O&P
1 Equip. Oper. (light)	$51.65	$413.20	$77.55	$620.40	$51.65	$77.55
1 Line Rem. (Metal Balls) 115 H.P.		843.60		927.96	105.45	116.00
8 L.H., Daily Totals		$1256.80		$1548.36	$157.10	$193.54

Crew B-78B

Crew B-78B	Hr.	Daily	Hr.	Daily	Bare Costs	Incl. O&P
2 Laborers	$41.05	$656.80	$62.10	$993.60	$42.23	$63.82
.25 Equip. Oper. (light)	51.65	103.30	77.55	155.10		
1 Pickup Truck, 3/4 Ton		109.90		120.89		
1 Line Rem.,11 H.P.,Walk Behind		61.35		67.48		
.25 Road Sweeper, S.P., 8' wide		175.78		193.35	19.28	21.21
18 L.H., Daily Totals		$1107.13		$1530.43	$61.51	$85.02

Crew B-78C

Crew B-78C	Hr.	Daily	Hr.	Daily	Bare Costs	Incl. O&P
1 Labor Foreman (outside)	$43.05	$344.40	$65.10	$520.80	$42.13	$63.67
4 Laborers	41.05	1313.60	62.10	1987.20		
1 Truck Driver (light)	45.50	364.00	68.50	548.00		
1 Paint Striper, T.M., 120 Gal.		686.10		754.71		
1 Flatbed Truck, Gas, 3 Ton		245.95		270.55		
1 Pickup Truck, 3/4 Ton		109.90		120.89	21.71	23.88
48 L.H., Daily Totals		$3063.95		$4202.15	$63.83	$87.54

For customer support on your Electrical Costs with RSMeans data, call 800.448.8182.

495

Crew No.	Bare Costs		Incl. Subs O&P		Cost Per Labor-Hour	
	Hr.	Daily	Hr.	Daily	Bare Costs	Incl. O&P
Crew B-78D	**Hr.**	**Daily**	**Hr.**	**Daily**	**Bare Costs**	**Incl. O&P**
2 Labor Foremen (outside)	$43.05	$688.80	$65.10	$1041.60	$41.90	$63.34
7 Laborers	41.05	2298.80	62.10	3477.60		
1 Truck Driver (light)	45.50	364.00	68.50	548.00		
1 Paint Striper, T.M., 120 Gal.		686.10		754.71		
1 Flatbed Truck, Gas, 3 Ton		245.95		270.55		
3 Pickup Trucks, 3/4 Ton		329.70		362.67		
1 Air Compressor, 60 cfm		106.70		117.37		
1 -50' Air Hose, 3/4"		6.55		7.21		
1 Breaker, Pavement, 60 lb.		10.75		11.82	17.32	19.05
80 L.H., Daily Totals		$4737.35		$6591.52	$59.22	$82.39
Crew B-78E	**Hr.**	**Daily**	**Hr.**	**Daily**	**Bare Costs**	**Incl. O&P**
2 Labor Foremen (outside)	$43.05	$688.80	$65.10	$1041.60	$41.75	$63.13
9 Laborers	41.05	2955.60	62.10	4471.20		
1 Truck Driver (light)	45.50	364.00	68.50	548.00		
1 Paint Striper, T.M., 120 Gal.		686.10		754.71		
1 Flatbed Truck, Gas, 3 Ton		245.95		270.55		
4 Pickup Trucks, 3/4 Ton		439.60		483.56		
2 Air Compressors, 60 cfm		213.40		234.74		
2 -50' Air Hoses, 3/4"		13.10		14.41		
2 Breakers, Pavement, 60 lb.		21.50		23.65	16.87	18.56
96 L.H., Daily Totals		$5628.05		$7842.42	$58.63	$81.69
Crew B-78F	**Hr.**	**Daily**	**Hr.**	**Daily**	**Bare Costs**	**Incl. O&P**
2 Labor Foremen (outside)	$43.05	$688.80	$65.10	$1041.60	$41.65	$62.99
11 Laborers	41.05	3612.40	62.10	5464.80		
1 Truck Driver (light)	45.50	364.00	68.50	548.00		
1 Paint Striper, T.M., 120 Gal.		686.10		754.71		
1 Flatbed Truck, Gas, 3 Ton		245.95		270.55		
7 Pickup Trucks, 3/4 Ton		769.30		846.23		
3 Air Compressors, 60 cfm		320.10		352.11		
3 -50' Air Hoses, 3/4"		19.65		21.61		
3 Breakers, Pavement, 60 lb.		32.25		35.48	18.51	20.36
112 L.H., Daily Totals		$6738.55		$9335.08	$60.17	$83.35
Crew B-79	**Hr.**	**Daily**	**Hr.**	**Daily**	**Bare Costs**	**Incl. O&P**
1 Labor Foreman (outside)	$43.05	$344.40	$65.10	$520.80	$42.34	$63.98
3 Laborers	41.05	985.20	62.10	1490.40		
1 Truck Driver (light)	45.50	364.00	68.50	548.00		
1 Paint Striper, T.M., 120 Gal.		686.10		754.71		
1 Heating Kettle, 115 Gallon		77.85		85.64		
1 Flatbed Truck, Gas, 3 Ton		245.95		270.55		
2 Pickup Trucks, 3/4 Ton		219.80		241.78	30.74	33.82
40 L.H., Daily Totals		$2923.30		$3911.87	$73.08	$97.80
Crew B-79A	**Hr.**	**Daily**	**Hr.**	**Daily**	**Bare Costs**	**Incl. O&P**
1.5 Equip. Oper. (light)	$51.65	$619.80	$77.55	$930.60	$51.65	$77.55
.5 Line Remov. (Grinder) 115 H.P.		446.68		491.34		
1 Line Rem. (Metal Balls) 115 H.P.		843.60		927.96	107.52	118.28
12 L.H., Daily Totals		$1910.08		$2349.90	$159.17	$195.83
Crew B-79B	**Hr.**	**Daily**	**Hr.**	**Daily**	**Bare Costs**	**Incl. O&P**
1 Laborer	$41.05	$328.40	$62.10	$496.80	$41.05	$62.10
1 Set of Gases		167.85		184.63	20.98	23.08
8 L.H., Daily Totals		$496.25		$681.43	$62.03	$85.18

Crew No.	Bare Costs		Incl. Subs O&P		Cost Per Labor-Hour	
Crew B-79C	**Hr.**	**Daily**	**Hr.**	**Daily**	**Bare Costs**	**Incl. O&P**
1 Labor Foreman (outside)	$43.05	$344.40	$65.10	$520.80	$41.97	$63.44
5 Laborers	41.05	1642.00	62.10	2484.00		
1 Truck Driver (light)	45.50	364.00	68.50	548.00		
1 Paint Striper, T.M., 120 Gal.		686.10		754.71		
1 Heating Kettle, 115 Gallon		77.85		85.64		
1 Flatbed Truck, Gas, 3 Ton		245.95		270.55		
3 Pickup Trucks, 3/4 Ton		329.70		362.67		
1 Air Compressor, 60 cfm		106.70		117.37		
1 -50' Air Hose, 3/4"		6.55		7.21		
1 Breaker, Pavement, 60 lb.		10.75		11.82	26.14	28.75
56 L.H., Daily Totals		$3814.00		$5162.76	$68.11	$92.19
Crew B-79D	**Hr.**	**Daily**	**Hr.**	**Daily**	**Bare Costs**	**Incl. O&P**
2 Labor Foremen (outside)	$43.05	$688.80	$65.10	$1041.60	$42.11	$63.65
5 Laborers	41.05	1642.00	62.10	2484.00		
1 Truck Driver (light)	45.50	364.00	68.50	548.00		
1 Paint Striper, T.M., 120 Gal.		686.10		754.71		
1 Heating Kettle, 115 Gallon		77.85		85.64		
1 Flatbed Truck, Gas, 3 Ton		245.95		270.55		
4 Pickup Trucks, 3/4 Ton		439.60		483.56		
1 Air Compressor, 60 cfm		106.70		117.37		
1 -50' Air Hose, 3/4"		6.55		7.21		
1 Breaker, Pavement, 60 lb.		10.75		11.82	24.59	27.04
64 L.H., Daily Totals		$4268.30		$5804.45	$66.69	$90.69
Crew B-79E	**Hr.**	**Daily**	**Hr.**	**Daily**	**Bare Costs**	**Incl. O&P**
2 Labor Foremen (outside)	$43.05	$688.80	$65.10	$1041.60	$41.90	$63.34
7 Laborers	41.05	2298.80	62.10	3477.60		
1 Truck Driver (light)	45.50	364.00	68.50	548.00		
1 Paint Striper, T.M., 120 Gal.		686.10		754.71		
1 Heating Kettle, 115 Gallon		77.85		85.64		
1 Flatbed Truck, Gas, 3 Ton		245.95		270.55		
5 Pickup Trucks, 3/4 Ton		549.50		604.45		
2 Air Compressors, 60 cfm		213.40		234.74		
2 -50' Air Hoses, 3/4"		13.10		14.41		
2 Breakers, Pavement, 60 lb.		21.50		23.65	22.59	24.85
80 L.H., Daily Totals		$5159.00		$7055.34	$64.49	$88.19
Crew B-80	**Hr.**	**Daily**	**Hr.**	**Daily**	**Bare Costs**	**Incl. O&P**
1 Labor Foreman (outside)	$43.05	$344.40	$65.10	$520.80	$45.31	$68.31
1 Laborer	41.05	328.40	62.10	496.80		
1 Truck Driver (light)	45.50	364.00	68.50	548.00		
1 Equip. Oper. (light)	51.65	413.20	77.55	620.40		
1 Flatbed Truck, Gas, 3 Ton		245.95		270.55		
1 Earth Auger, Truck-Mtd.		385.85		424.44	19.74	21.72
32 L.H., Daily Totals		$2081.80		$2880.98	$65.06	$90.03
Crew B-80A	**Hr.**	**Daily**	**Hr.**	**Daily**	**Bare Costs**	**Incl. O&P**
3 Laborers	$41.05	$985.20	$62.10	$1490.40	$41.05	$62.10
1 Flatbed Truck, Gas, 3 Ton		245.95		270.55	10.25	11.27
24 L.H., Daily Totals		$1231.15		$1760.94	$51.30	$73.37
Crew B-80B	**Hr.**	**Daily**	**Hr.**	**Daily**	**Bare Costs**	**Incl. O&P**
3 Laborers	$41.05	$985.20	$62.10	$1490.40	$43.70	$65.96
1 Equip. Oper. (light)	51.65	413.20	77.55	620.40		
1 Crane, Flatbed Mounted, 3 Ton		232.65		255.91	7.27	8.00
32 L.H., Daily Totals		$1631.05		$2366.72	$50.97	$73.96

Crew No.	Bare Costs		Incl. Subs O&P		Cost Per Labor-Hour	
Crew B-80C	Hr.	Daily	Hr.	Daily	Bare Costs	Incl. O&P
2 Laborers	$41.05	$656.80	$62.10	$993.60	$42.53	$64.23
1 Truck Driver (light)	45.50	364.00	68.50	548.00		
1 Flatbed Truck, Gas, 1.5 Ton		195.20		214.72		
1 Manual Fence Post Auger, Gas		7.40		8.14	8.44	9.29
24 L.H., Daily Totals		$1223.40		$1764.46	$50.98	$73.52
Crew B-81	Hr.	Daily	Hr.	Daily	Bare Costs	Incl. O&P
1 Laborer	$41.05	$328.40	$62.10	$496.80	$47.72	$71.85
1 Equip. Oper. (medium)	55.10	440.80	82.70	661.60		
1 Truck Driver (heavy)	47.00	376.00	70.75	566.00		
1 Hydromulcher, T.M., 3000 Gal.		292.90		322.19		
1 Truck Tractor, 220 H.P.		303.55		333.90	24.85	27.34
24 L.H., Daily Totals		$1741.65		$2380.49	$72.57	$99.19
Crew B-81A	Hr.	Daily	Hr.	Daily	Bare Costs	Incl. O&P
1 Laborer	$41.05	$328.40	$62.10	$496.80	$43.27	$65.30
1 Truck Driver (light)	45.50	364.00	68.50	548.00		
1 Hydromulcher, T.M., 600 Gal.		124.15		136.57		
1 Flatbed Truck, Gas, 3 Ton		245.95		270.55	23.13	25.44
16 L.H., Daily Totals		$1062.50		$1451.91	$66.41	$90.74
Crew B-82	Hr.	Daily	Hr.	Daily	Bare Costs	Incl. O&P
1 Laborer	$41.05	$328.40	$62.10	$496.80	$46.35	$69.83
1 Equip. Oper. (light)	51.65	413.20	77.55	620.40		
1 Horiz. Borer, 6 H.P.		77.85		85.64	4.87	5.35
16 L.H., Daily Totals		$819.45		$1202.84	$51.22	$75.18
Crew B-82A	Hr.	Daily	Hr.	Daily	Bare Costs	Incl. O&P
2 Laborers	$41.05	$656.80	$62.10	$993.60	$46.35	$69.83
2 Equip. Opers. (light)	51.65	826.40	77.55	1240.80		
2 Dump Trucks, 8 C.Y., 220 H.P.		679.20		747.12		
1 Flatbed Trailer, 25 Ton		133.00		146.30		
1 Horiz. Dir. Drill, 20k lb. Thrust		631.30		694.43		
1 Mud Trailer for HDD, 1500 Gal.		298.65		328.51		
1 Pickup Truck, 4x4, 3/4 Ton		121.70		133.87		
1 Flatbed Trailer, 3 Ton		26.55		29.20		
1 Loader, Skid Steer, 78 H.P.		392.50		431.75	71.34	78.47
32 L.H., Daily Totals		$3766.10		$4745.59	$117.69	$148.30
Crew B-82B	Hr.	Daily	Hr.	Daily	Bare Costs	Incl. O&P
2 Laborers	$41.05	$656.80	$62.10	$993.60	$46.35	$69.83
2 Equip. Opers. (light)	51.65	826.40	77.55	1240.80		
2 Dump Trucks, 8 C.Y., 220 H.P.		679.20		747.12		
1 Flatbed Trailer, 25 Ton		133.00		146.30		
1 Horiz. Dir. Drill, 30k lb. Thrust		897.00		986.70		
1 Mud Trailer for HDD, 1500 Gal.		298.65		328.51		
1 Pickup Truck, 4x4, 3/4 Ton		121.70		133.87		
1 Flatbed Trailer, 3 Ton		26.55		29.20		
1 Loader, Skid Steer, 78 H.P.		392.50		431.75	79.64	87.61
32 L.H., Daily Totals		$4031.80		$5037.86	$125.99	$157.43
Crew B-82C	Hr.	Daily	Hr.	Daily	Bare Costs	Incl. O&P
2 Laborers	$41.05	$656.80	$62.10	$993.60	$46.35	$69.83
2 Equip. Opers. (light)	51.65	826.40	77.55	1240.80		
2 Dump Trucks, 8 C.Y., 220 H.P.		679.20		747.12		
1 Flatbed Trailer, 25 Ton		133.00		146.30		
1 Horiz. Dir. Drill, 50k lb. Thrust		1185.00		1303.50		
1 Mud Trailer for HDD, 1500 Gal.		298.65		328.51		
1 Pickup Truck, 4x4, 3/4 Ton		121.70		133.87		
1 Flatbed Trailer, 3 Ton		26.55		29.20		
1 Loader, Skid Steer, 78 H.P.		392.50		431.75	88.64	97.51
32 L.H., Daily Totals		$4319.80		$5354.66	$134.99	$167.33
Crew B-82D	Hr.	Daily	Hr.	Daily	Bare Costs	Incl. O&P
1 Equip. Oper. (light)	$51.65	$413.20	$77.55	$620.40	$51.65	$77.55
1 Mud Trailer for HDD, 1500 Gal.		298.65		328.51	37.33	41.06
8 L.H., Daily Totals		$711.85		$948.91	$88.98	$118.61
Crew B-83	Hr.	Daily	Hr.	Daily	Bare Costs	Incl. O&P
1 Tugboat Captain	$55.10	$440.80	$82.70	$661.60	$48.08	$72.40
1 Tugboat Hand	41.05	328.40	62.10	496.80		
1 Tugboat, 250 H.P.		710.70		781.77	44.42	48.86
16 L.H., Daily Totals		$1479.90		$1940.17	$92.49	$121.26
Crew B-84	Hr.	Daily	Hr.	Daily	Bare Costs	Incl. O&P
1 Equip. Oper. (medium)	$55.10	$440.80	$82.70	$661.60	$55.10	$82.70
1 Rotary Mower/Tractor		360.30		396.33	45.04	49.54
8 L.H., Daily Totals		$801.10		$1057.93	$100.14	$132.24
Crew B-85	Hr.	Daily	Hr.	Daily	Bare Costs	Incl. O&P
3 Laborers	$41.05	$985.20	$62.10	$1490.40	$45.05	$67.95
1 Equip. Oper. (medium)	55.10	440.80	82.70	661.60		
1 Truck Driver (heavy)	47.00	376.00	70.75	566.00		
1 Telescoping Boom Lift, to 80'		670.65		737.72		
1 Brush Chipper, 12", 130 H.P.		389.40		428.34		
1 Pruning Saw, Rotary		6.60		7.26	26.67	29.33
40 L.H., Daily Totals		$2868.65		$3891.32	$71.72	$97.28
Crew B-86	Hr.	Daily	Hr.	Daily	Bare Costs	Incl. O&P
1 Equip. Oper. (medium)	$55.10	$440.80	$82.70	$661.60	$55.10	$82.70
1 Stump Chipper, S.P.		182.20		200.42	22.77	25.05
8 L.H., Daily Totals		$623.00		$862.02	$77.88	$107.75
Crew B-86A	Hr.	Daily	Hr.	Daily	Bare Costs	Incl. O&P
1 Equip. Oper. (medium)	$55.10	$440.80	$82.70	$661.60	$55.10	$82.70
1 Grader, 30,000 Lbs.		657.85		723.63	82.23	90.45
8 L.H., Daily Totals		$1098.65		$1385.23	$137.33	$173.15
Crew B-86B	Hr.	Daily	Hr.	Daily	Bare Costs	Incl. O&P
1 Equip. Oper. (medium)	$55.10	$440.80	$82.70	$661.60	$55.10	$82.70
1 Dozer, 200 H.P.		1290.00		1419.00	161.25	177.38
8 L.H., Daily Totals		$1730.80		$2080.60	$216.35	$260.07
Crew B-87	Hr.	Daily	Hr.	Daily	Bare Costs	Incl. O&P
1 Laborer	$41.05	$328.40	$62.10	$496.80	$52.29	$78.58
4 Equip. Oper. (medium)	55.10	1763.20	82.70	2646.40		
2 Feller Bunchers, 100 H.P.		1607.10		1767.81		
1 Log Chipper, 22" Tree		770.10		847.11		
1 Dozer, 105 H.P.		609.70		670.67		
1 Chain Saw, Gas, 36" Long		46.65		51.31	75.84	83.42
40 L.H., Daily Totals		$5125.15		$6480.10	$128.13	$162.00

497

For customer support on your Electrical Costs with RSMeans data, call 800.448.8182.

Crew No.	Bare Costs		Incl. Subs O&P		Cost Per Labor-Hour	

Crew B-88	Hr.	Daily	Hr.	Daily	Bare Costs	Incl. O&P
1 Laborer	$41.05	$328.40	$62.10	$496.80	$53.09	$79.76
6 Equip. Oper. (medium)	55.10	2644.80	82.70	3969.60		
2 Feller Bunchers, 100 H.P.		1607.10		1767.81		
1 Log Chipper, 22" Tree		770.10		847.11		
2 Log Skidders, 50 H.P.		1696.50		1866.15		
1 Dozer, 105 H.P.		609.70		670.67		
1 Chain Saw, Gas, 36" Long		46.65		51.31	84.47	92.91
56 L.H., Daily Totals		$7703.25		$9669.45	$137.56	$172.67

Crew B-89	Hr.	Daily	Hr.	Daily	Bare Costs	Incl. O&P
1 Equip. Oper. (light)	$51.65	$413.20	$77.55	$620.40	$48.58	$73.03
1 Truck Driver (light)	45.50	364.00	68.50	548.00		
1 Flatbed Truck, Gas, 3 Ton		245.95		270.55		
1 Concrete Saw		110.15		121.17		
1 Water Tank, 65 Gal.		98.85		108.74	28.43	31.28
16 L.H., Daily Totals		$1232.15		$1668.85	$77.01	$104.30

Crew B-89A	Hr.	Daily	Hr.	Daily	Bare Costs	Incl. O&P
1 Skilled Worker	$53.40	$427.20	$81.25	$650.00	$47.23	$71.67
1 Laborer	41.05	328.40	62.10	496.80		
1 Core Drill (Large)		112.20		123.42	7.01	7.71
16 L.H., Daily Totals		$867.80		$1270.22	$54.24	$79.39

Crew B-89B	Hr.	Daily	Hr.	Daily	Bare Costs	Incl. O&P
1 Equip. Oper. (light)	$51.65	$413.20	$77.55	$620.40	$48.58	$73.03
1 Truck Driver (light)	45.50	364.00	68.50	548.00		
1 Wall Saw, Hydraulic, 10 H.P.		46.60		51.26		
1 Generator, Diesel, 100 kW		308.80		339.68		
1 Water Tank, 65 Gal.		98.85		108.74		
1 Flatbed Truck, Gas, 3 Ton		245.95		270.55	43.76	48.14
16 L.H., Daily Totals		$1477.40		$1938.62	$92.34	$121.16

Crew B-90	Hr.	Daily	Hr.	Daily	Bare Costs	Incl. O&P
1 Labor Foreman (outside)	$43.05	$344.40	$65.10	$520.80	$45.44	$68.50
3 Laborers	41.05	985.20	62.10	1490.40		
2 Equip. Oper. (light)	51.65	826.40	77.55	1240.80		
2 Truck Drivers (heavy)	47.00	752.00	70.75	1132.00		
1 Road Mixer, 310 H.P.		1857.00		2042.70		
1 Dist. Truck, 2000 Gal.		293.20		322.52	33.60	36.96
64 L.H., Daily Totals		$5058.20		$6749.22	$79.03	$105.46

Crew B-90A	Hr.	Daily	Hr.	Daily	Bare Costs	Incl. O&P
1 Labor Foreman (outside)	$43.05	$344.40	$65.10	$520.80	$49.36	$74.30
2 Laborers	41.05	656.80	62.10	993.60		
4 Equip. Oper. (medium)	55.10	1763.20	82.70	2646.40		
2 Graders, 30,000 Lbs.		1315.70		1447.27		
1 Tandem Roller, 10 Ton		232.05		255.26		
1 Roller, Pneum. Whl., 12 Ton		338.35		372.19	33.68	37.05
56 L.H., Daily Totals		$4650.50		$6235.51	$83.04	$111.35

Crew B-90B	Hr.	Daily	Hr.	Daily	Bare Costs	Incl. O&P
1 Labor Foreman (outside)	$43.05	$344.40	$65.10	$520.80	$48.41	$72.90
2 Laborers	41.05	656.80	62.10	993.60		
3 Equip. Oper. (medium)	55.10	1322.40	82.70	1984.80		
1 Roller, Pneum. Whl., 12 Ton		338.35		372.19		
1 Road Mixer, 310 H.P.		1857.00		2042.70	45.74	50.31
48 L.H., Daily Totals		$4518.95		$5914.09	$94.14	$123.21

Crew B-90C	Hr.	Daily	Hr.	Daily	Bare Costs	Incl. O&P
1 Labor Foreman (outside)	$43.05	$344.40	$65.10	$520.80	$46.69	$70.35
4 Laborers	41.05	1313.60	62.10	1987.20		
3 Equip. Oper. (medium)	55.10	1322.40	82.70	1984.80		
3 Truck Drivers (heavy)	47.00	1128.00	70.75	1698.00		
3 Road Mixers, 310 H.P.		5571.00		6128.10	63.31	69.64
88 L.H., Daily Totals		$9679.40		$12318.90	$109.99	$139.99

Crew B-90D	Hr.	Daily	Hr.	Daily	Bare Costs	Incl. O&P
1 Labor Foreman (outside)	$43.05	$344.40	$65.10	$520.80	$45.82	$69.08
6 Laborers	41.05	1970.40	62.10	2980.80		
3 Equip. Oper. (medium)	55.10	1322.40	82.70	1984.80		
3 Truck Drivers (heavy)	47.00	1128.00	70.75	1698.00		
3 Road Mixers, 310 H.P.		5571.00		6128.10	53.57	58.92
104 L.H., Daily Totals		$10336.20		$13312.50	$99.39	$128.00

Crew B-90E	Hr.	Daily	Hr.	Daily	Bare Costs	Incl. O&P
1 Labor Foreman (outside)	$43.05	$344.40	$65.10	$520.80	$46.62	$70.26
4 Laborers	41.05	1313.60	62.10	1987.20		
3 Equip. Oper. (medium)	55.10	1322.40	82.70	1984.80		
1 Truck Driver (heavy)	47.00	376.00	70.75	566.00		
1 Road Mixer, 310 H.P.		1857.00		2042.70	25.79	28.37
72 L.H., Daily Totals		$5213.40		$7101.50	$72.41	$98.63

Crew B-91	Hr.	Daily	Hr.	Daily	Bare Costs	Incl. O&P
1 Labor Foreman (outside)	$43.05	$344.40	$65.10	$520.80	$49.07	$73.86
2 Laborers	41.05	656.80	62.10	993.60		
4 Equip. Oper. (medium)	55.10	1763.20	82.70	2646.40		
1 Truck Driver (heavy)	47.00	376.00	70.75	566.00		
1 Dist. Tanker, 3000 Gallon		323.25		355.57		
1 Truck Tractor, 6x4, 380 H.P.		487.60		536.36		
1 Aggreg. Spreader, S.P.		836.00		919.60		
1 Roller, Pneum. Whl., 12 Ton		338.35		372.19		
1 Tandem Roller, 10 Ton		232.05		255.26	34.64	38.11
64 L.H., Daily Totals		$5357.65		$7165.77	$83.71	$111.97

Crew B-91B	Hr.	Daily	Hr.	Daily	Bare Costs	Incl. O&P
1 Laborer	$41.05	$328.40	$62.10	$496.80	$48.08	$72.40
1 Equipment Oper. (med.)	55.10	440.80	82.70	661.60		
1 Road Sweeper, Vac. Assist.		857.60		943.36	53.60	58.96
16 L.H., Daily Totals		$1626.80		$2101.76	$101.68	$131.36

Crew B-91C	Hr.	Daily	Hr.	Daily	Bare Costs	Incl. O&P
1 Laborer	$41.05	$328.40	$62.10	$496.80	$43.27	$65.30
1 Truck Driver (light)	45.50	364.00	68.50	548.00		
1 Catch Basin Cleaning Truck		528.95		581.85	33.06	36.37
16 L.H., Daily Totals		$1221.35		$1626.65	$76.33	$101.67

Crew B-91D	Hr.	Daily	Hr.	Daily	Bare Costs	Incl. O&P
1 Labor Foreman (outside)	$43.05	$344.40	$65.10	$520.80	$47.52	$71.58
5 Laborers	41.05	1642.00	62.10	2484.00		
5 Equip. Oper. (medium)	55.10	2204.00	82.70	3308.00		
2 Truck Drivers (heavy)	47.00	752.00	70.75	1132.00		
1 Aggreg. Spreader, S.P.		836.00		919.60		
2 Truck Tractors, 6x4, 380 H.P.		975.20		1072.72		
2 Dist. Tankers, 3000 Gallon		646.50		711.15		
2 Pavement Brushes, Towed		171.00		188.10		
2 Rollers Pneum. Whl., 12 Ton		676.70		744.37	31.78	34.96
104 L.H., Daily Totals		$8247.80		$11080.74	$79.31	$106.55

Crews - Standard

Crew No.	Bare Costs		Incl. Subs O&P		Cost Per Labor-Hour	
	Hr.	**Daily**	**Hr.**	**Daily**	**Bare Costs**	**Incl. O&P**
Crew B-92						
1 Labor Foreman (outside)	$43.05	$344.40	$65.10	$520.80	$41.55	$62.85
3 Laborers	41.05	985.20	62.10	1490.40		
1 Crack Cleaner, 25 H.P.		56.95		62.65		
1 Air Compressor, 60 cfm		106.70		117.37		
1 Tar Kettle, T.M.		134.50		147.95		
1 Flatbed Truck, Gas, 3 Ton		245.95		270.55	17.00	18.70
32 L.H., Daily Totals		$1873.70		$2609.71	$58.55	$81.55
Crew B-93	**Hr.**	**Daily**	**Hr.**	**Daily**	**Bare Costs**	**Incl. O&P**
1 Equip. Oper. (medium)	$55.10	$440.80	$82.70	$661.60	$55.10	$82.70
1 Feller Buncher, 100 H.P.		803.55		883.90	100.44	110.49
8 L.H., Daily Totals		$1244.35		$1545.51	$155.54	$193.19
Crew B-94A	**Hr.**	**Daily**	**Hr.**	**Daily**	**Bare Costs**	**Incl. O&P**
1 Laborer	$41.05	$328.40	$62.10	$496.80	$41.05	$62.10
1 Diaphragm Water Pump, 2"		74.30		81.73		
1 -20' Suction Hose, 2"		2.05		2.25		
2 -50' Discharge Hoses, 2"		1.80		1.98	9.77	10.75
8 L.H., Daily Totals		$406.55		$582.76	$50.82	$72.85
Crew B-94B	**Hr.**	**Daily**	**Hr.**	**Daily**	**Bare Costs**	**Incl. O&P**
1 Laborer	$41.05	$328.40	$62.10	$496.80	$41.05	$62.10
1 Diaphragm Water Pump, 4"		141.40		155.54		
1 -20' Suction Hose, 4"		3.10		3.41		
2 -50' Discharge Hoses, 4"		4.50		4.95	18.63	20.49
8 L.H., Daily Totals		$477.40		$660.70	$59.67	$82.59
Crew B-94C	**Hr.**	**Daily**	**Hr.**	**Daily**	**Bare Costs**	**Incl. O&P**
1 Laborer	$41.05	$328.40	$62.10	$496.80	$41.05	$62.10
1 Centrifugal Water Pump, 3"		79.25		87.17		
1 -20' Suction Hose, 3"		2.70		2.97		
2 -50' Discharge Hoses, 3"		2.80		3.08	10.59	11.65
8 L.H., Daily Totals		$413.15		$590.02	$51.64	$73.75
Crew B-94D	**Hr.**	**Daily**	**Hr.**	**Daily**	**Bare Costs**	**Incl. O&P**
1 Laborer	$41.05	$328.40	$62.10	$496.80	$41.05	$62.10
1 Centr. Water Pump, 6"		296.45		326.10		
1 -20' Suction Hose, 6"		11.10		12.21		
2 -50' Discharge Hoses, 6"		11.40		12.54	39.87	43.86
8 L.H., Daily Totals		$647.35		$847.64	$80.92	$105.96
Crew C-1	**Hr.**	**Daily**	**Hr.**	**Daily**	**Bare Costs**	**Incl. O&P**
3 Carpenters	$51.65	$1239.60	$78.15	$1875.60	$49.00	$74.14
1 Laborer	41.05	328.40	62.10	496.80		
32 L.H., Daily Totals		$1568.00		$2372.40	$49.00	$74.14
Crew C-2	**Hr.**	**Daily**	**Hr.**	**Daily**	**Bare Costs**	**Incl. O&P**
1 Carpenter Foreman (outside)	$53.65	$429.20	$81.15	$649.20	$50.22	$75.97
4 Carpenters	51.65	1652.80	78.15	2500.80		
1 Laborer	41.05	328.40	62.10	496.80		
48 L.H., Daily Totals		$2410.40		$3646.80	$50.22	$75.97
Crew C-2A	**Hr.**	**Daily**	**Hr.**	**Daily**	**Bare Costs**	**Incl. O&P**
1 Carpenter Foreman (outside)	$53.65	$429.20	$81.15	$649.20	$49.76	$74.98
3 Carpenters	51.65	1239.60	78.15	1875.60		
1 Cement Finisher	48.90	391.20	72.20	577.60		
1 Laborer	41.05	328.40	62.10	496.80		
48 L.H., Daily Totals		$2388.40		$3599.20	$49.76	$74.98

Crew No.	Bare Costs		Incl. Subs O&P		Cost Per Labor-Hour	
	Hr.	**Daily**	**Hr.**	**Daily**	**Bare Costs**	**Incl. O&P**
Crew C-3						
1 Rodman Foreman (outside)	$56.85	$454.80	$86.05	$688.40	$51.25	$77.47
4 Rodmen (reinf.)	54.85	1755.20	83.00	2656.00		
1 Equip. Oper. (light)	51.65	413.20	77.55	620.40		
2 Laborers	41.05	656.80	62.10	993.60		
3 Stressing Equipment		31.20		34.32		
.5 Grouting Equipment		77.70		85.47	1.70	1.87
64 L.H., Daily Totals		$3388.90		$5078.19	$52.95	$79.35
Crew C-4	**Hr.**	**Daily**	**Hr.**	**Daily**	**Bare Costs**	**Incl. O&P**
1 Rodman Foreman (outside)	$56.85	$454.80	$86.05	$688.40	$55.35	$83.76
3 Rodmen (reinf.)	54.85	1316.40	83.00	1992.00		
3 Stressing Equipment		31.20		34.32	.97	1.07
32 L.H., Daily Totals		$1802.40		$2714.72	$56.33	$84.83
Crew C-4A	**Hr.**	**Daily**	**Hr.**	**Daily**	**Bare Costs**	**Incl. O&P**
2 Rodmen (reinf.)	$54.85	$877.60	$83.00	$1328.00	$54.85	$83.00
4 Stressing Equipment		41.60		45.76	2.60	2.86
16 L.H., Daily Totals		$919.20		$1373.76	$57.45	$85.86
Crew C-5	**Hr.**	**Daily**	**Hr.**	**Daily**	**Bare Costs**	**Incl. O&P**
1 Rodman Foreman (outside)	$56.85	$454.80	$86.05	$688.40	$54.70	$82.59
4 Rodmen (reinf.)	54.85	1755.20	83.00	2656.00		
1 Equip. Oper. (crane)	57.45	459.60	86.25	690.00		
1 Equip. Oper. (oiler)	49.20	393.60	73.85	590.80		
1 Hyd. Crane, 25 Ton		581.70		639.87	10.39	11.43
56 L.H., Daily Totals		$3644.90		$5265.07	$65.09	$94.02
Crew C-6	**Hr.**	**Daily**	**Hr.**	**Daily**	**Bare Costs**	**Incl. O&P**
1 Labor Foreman (outside)	$43.05	$344.40	$65.10	$520.80	$42.69	$64.28
4 Laborers	41.05	1313.60	62.10	1987.20		
1 Cement Finisher	48.90	391.20	72.20	577.60		
2 Gas Engine Vibrators		53.10		58.41	1.11	1.22
48 L.H., Daily Totals		$2102.30		$3144.01	$43.80	$65.50
Crew C-6A	**Hr.**	**Daily**	**Hr.**	**Daily**	**Bare Costs**	**Incl. O&P**
2 Cement Finishers	$48.90	$782.40	$72.20	$1155.20	$48.90	$72.20
1 Concrete Vibrator, Elec, 2 HP		9.25		10.18	.58	.64
16 L.H., Daily Totals		$791.65		$1165.38	$49.48	$72.84
Crew B-89C	**Hr.**	**Daily**	**Hr.**	**Daily**	**Bare Costs**	**Incl. O&P**
1 Cement Finisher	$48.90	$391.20	$72.20	$577.60	$48.90	$72.20
1 Masonry cut-off saw, gas		34.00		37.40	4.25	4.67
8 L.H., Daily Totals		$425.20		$615.00	$53.15	$76.88
Crew C-7	**Hr.**	**Daily**	**Hr.**	**Daily**	**Bare Costs**	**Incl. O&P**
1 Labor Foreman (outside)	$43.05	$344.40	$65.10	$520.80	$44.61	$67.15
5 Laborers	41.05	1642.00	62.10	2484.00		
1 Cement Finisher	48.90	391.20	72.20	577.60		
1 Equip. Oper. (medium)	55.10	440.80	82.70	661.60		
1 Equip. Oper. (oiler)	49.20	393.60	73.85	590.80		
2 Gas Engine Vibrators		53.10		58.41		
1 Concrete Bucket, 1 C.Y.		24.95		27.45		
1 Hyd. Crane, 55 Ton		981.50		1079.65	14.72	16.19
72 L.H., Daily Totals		$4271.55		$6000.31	$59.33	$83.34

For customer support on your Electrical Costs with RSMeans data, call 800.448.8182.

Crews - Standard

Crew C-7A	Hr.	Daily	Hr.	Daily	Bare Costs	Incl. O&P
1 Labor Foreman (outside)	$43.05	$344.40	$65.10	$520.80	$42.79	$64.64
5 Laborers	41.05	1642.00	62.10	2484.00		
2 Truck Drivers (heavy)	47.00	752.00	70.75	1132.00		
2 Conc. Transit Mixers		1869.40		2056.34	29.21	32.13
64 L.H., Daily Totals		$4607.80		$6193.14	$72.00	$96.77

Crew C-7B	Hr.	Daily	Hr.	Daily	Bare Costs	Incl. O&P
1 Labor Foreman (outside)	$43.05	$344.40	$65.10	$520.80	$44.37	$66.96
5 Laborers	41.05	1642.00	62.10	2484.00		
1 Equipment Operator, Crane	57.45	459.60	86.25	690.00		
1 Equipment Oiler	49.20	393.60	73.85	590.80		
1 Conc. Bucket, 2 C.Y.		38.90		42.79		
1 Lattice Boom Crane, 165 Ton		2153.00		2368.30	34.25	37.67
64 L.H., Daily Totals		$5031.50		$6696.69	$78.62	$104.64

Crew C-7C	Hr.	Daily	Hr.	Daily	Bare Costs	Incl. O&P
1 Labor Foreman (outside)	$43.05	$344.40	$65.10	$520.80	$44.81	$67.63
5 Laborers	41.05	1642.00	62.10	2484.00		
2 Equipment Operators (med.)	55.10	881.60	82.70	1323.20		
2 F.E. Loaders, W.M., 4 C.Y.		1133.40		1246.74	17.71	19.48
64 L.H., Daily Totals		$4001.40		$5574.74	$62.52	$87.11

Crew C-7D	Hr.	Daily	Hr.	Daily	Bare Costs	Incl. O&P
1 Labor Foreman (outside)	$43.05	$344.40	$65.10	$520.80	$43.34	$65.47
5 Laborers	41.05	1642.00	62.10	2484.00		
1 Equip. Oper. (medium)	55.10	440.80	82.70	661.60		
1 Concrete Conveyer		186.45		205.10	3.33	3.66
56 L.H., Daily Totals		$2613.65		$3871.49	$46.67	$69.13

Crew C-8	Hr.	Daily	Hr.	Daily	Bare Costs	Incl. O&P
1 Labor Foreman (outside)	$43.05	$344.40	$65.10	$520.80	$45.59	$68.36
3 Laborers	41.05	985.20	62.10	1490.40		
2 Cement Finishers	48.90	782.40	72.20	1155.20		
1 Equip. Oper. (medium)	55.10	440.80	82.70	661.60		
1 Concrete Pump (Small)		844.05		928.46	15.07	16.58
56 L.H., Daily Totals		$3396.85		$4756.45	$60.66	$84.94

Crew C-8A	Hr.	Daily	Hr.	Daily	Bare Costs	Incl. O&P
1 Labor Foreman (outside)	$43.05	$344.40	$65.10	$520.80	$44.00	$65.97
3 Laborers	41.05	985.20	62.10	1490.40		
2 Cement Finishers	48.90	782.40	72.20	1155.20		
48 L.H., Daily Totals		$2112.00		$3166.40	$44.00	$65.97

Crew C-8B	Hr.	Daily	Hr.	Daily	Bare Costs	Incl. O&P
1 Labor Foreman (outside)	$43.05	$344.40	$65.10	$520.80	$44.26	$66.82
3 Laborers	41.05	985.20	62.10	1490.40		
1 Equip. Oper. (medium)	55.10	440.80	82.70	661.60		
1 Vibrating Power Screed		79.45		87.39		
1 Roller, Vibratory, 25 Ton		650.90		715.99		
1 Dozer, 200 H.P.		1290.00		1419.00	50.51	55.56
40 L.H., Daily Totals		$3790.75		$4895.19	$94.77	$122.38

Crew C-8C	Hr.	Daily	Hr.	Daily	Bare Costs	Incl. O&P
1 Labor Foreman (outside)	$43.05	$344.40	$65.10	$520.80	$45.03	$67.72
3 Laborers	41.05	985.20	62.10	1490.40		
1 Cement Finisher	48.90	391.20	72.20	577.60		
1 Equip. Oper. (medium)	55.10	440.80	82.70	661.60		
1 Shotcrete Rig, 12 C.Y./hr		242.75		267.02		
1 Air Compressor, 160 cfm		115.85		127.44		
4 -50' Air Hoses, 1"		33.60		36.96		
4 -50' Air Hoses, 2"		103.60		113.96	10.33	11.36
48 L.H., Daily Totals		$2657.40		$3795.78	$55.36	$79.08

Crew C-8D	Hr.	Daily	Hr.	Daily	Bare Costs	Incl. O&P
1 Labor Foreman (outside)	$43.05	$344.40	$65.10	$520.80	$46.16	$69.24
1 Laborer	41.05	328.40	62.10	496.80		
1 Cement Finisher	48.90	391.20	72.20	577.60		
1 Equipment Oper. (light)	51.65	413.20	77.55	620.40		
1 Air Compressor, 250 cfm		167.95		184.75		
2 -50' Air Hoses, 1"		16.80		18.48	5.77	6.35
32 L.H., Daily Totals		$1661.95		$2418.82	$51.94	$75.59

Crew C-8E	Hr.	Daily	Hr.	Daily	Bare Costs	Incl. O&P
1 Labor Foreman (outside)	$43.05	$344.40	$65.10	$520.80	$44.46	$66.86
3 Laborers	41.05	985.20	62.10	1490.40		
1 Cement Finisher	48.90	391.20	72.20	577.60		
1 Equipment Oper. (light)	51.65	413.20	77.55	620.40		
1 Shotcrete Rig, 35 C.Y./hr		271.40		298.54		
1 Air Compressor, 250 cfm		167.95		184.75		
4 -50' Air Hoses, 1"		33.60		36.96		
4 -50' Air Hoses, 2"		103.60		113.96	12.01	13.21
48 L.H., Daily Totals		$2710.55		$3843.41	$56.47	$80.07

Crew C-9	Hr.	Daily	Hr.	Daily	Bare Costs	Incl. O&P
1 Cement Finisher	$48.90	$391.20	$72.20	$577.60	$45.66	$68.49
2 Laborers	41.05	656.80	62.10	993.60		
1 Equipment Oper. (light)	51.65	413.20	77.55	620.40		
1 Grout Pump, 50 C.F./hr.		125.60		138.16		
1 Air Compressor, 160 cfm		115.85		127.44		
2 -50' Air Hoses, 1"		16.80		18.48		
2 -50' Air Hoses, 2"		51.80		56.98	9.69	10.66
32 L.H., Daily Totals		$1771.25		$2532.66	$55.35	$79.15

Crew C-10	Hr.	Daily	Hr.	Daily	Bare Costs	Incl. O&P
1 Laborer	$41.05	$328.40	$62.10	$496.80	$46.28	$68.83
2 Cement Finishers	48.90	782.40	72.20	1155.20		
24 L.H., Daily Totals		$1110.80		$1652.00	$46.28	$68.83

Crew C-10B	Hr.	Daily	Hr.	Daily	Bare Costs	Incl. O&P
3 Laborers	$41.05	$985.20	$62.10	$1490.40	$44.19	$66.14
2 Cement Finishers	48.90	782.40	72.20	1155.20		
1 Concrete Mixer, 10 C.F.		162.00		178.20		
2 Trowels, 48" Walk-Behind		83.00		91.30	6.13	6.74
40 L.H., Daily Totals		$2012.60		$2915.10	$50.31	$72.88

Crew C-10C	Hr.	Daily	Hr.	Daily	Bare Costs	Incl. O&P
1 Laborer	$41.05	$328.40	$62.10	$496.80	$46.28	$68.83
2 Cement Finishers	48.90	782.40	72.20	1155.20		
1 Trowel, 48" Walk-Behind		41.50		45.65	1.73	1.90
24 L.H., Daily Totals		$1152.30		$1697.65	$48.01	$70.74

Crews - Standard

Crew No.	Bare Costs Hr.	Daily	Incl. Subs O&P Hr.	Daily	Cost Per Labor-Hour Bare Costs	Incl. O&P
Crew C-10D	Hr.	Daily	Hr.	Daily	Bare Costs	Incl. O&P
1 Laborer	$41.05	$328.40	$62.10	$496.80	$46.28	$68.83
2 Cement Finishers	48.90	782.40	72.20	1155.20		
1 Vibrating Power Screed		79.45		87.39		
1 Trowel, 48" Walk-Behind		41.50		45.65	5.04	5.54
24 L.H., Daily Totals		$1231.75		$1785.05	$51.32	$74.38
Crew C-10E	Hr.	Daily	Hr.	Daily	Bare Costs	Incl. O&P
1 Laborer	$41.05	$328.40	$62.10	$496.80	$46.28	$68.83
2 Cement Finishers	48.90	782.40	72.20	1155.20		
1 Vibrating Power Screed		79.45		87.39		
1 Cement Trowel, 96" Ride-On		166.30		182.93	10.24	11.26
24 L.H., Daily Totals		$1356.55		$1922.33	$56.52	$80.10
Crew C-10F	Hr.	Daily	Hr.	Daily	Bare Costs	Incl. O&P
1 Laborer	$41.05	$328.40	$62.10	$496.80	$46.28	$68.83
2 Cement Finishers	48.90	782.40	72.20	1155.20		
1 Telescoping Boom Lift, to 60'		451.25		496.38	18.80	20.68
24 L.H., Daily Totals		$1562.05		$2148.38	$65.09	$89.52
Crew C-11	Hr.	Daily	Hr.	Daily	Bare Costs	Incl. O&P
1 Struc. Steel Foreman (outside)	$58.00	$464.00	$92.50	$740.00	$55.63	$87.63
6 Struc. Steel Workers	56.00	2688.00	89.35	4288.80		
1 Equip. Oper. (crane)	57.45	459.60	86.25	690.00		
1 Equip. Oper. (oiler)	49.20	393.60	73.85	590.80		
1 Lattice Boom Crane, 150 Ton		2021.00		2223.10	28.07	30.88
72 L.H., Daily Totals		$6026.20		$8532.70	$83.70	$118.51
Crew C-12	Hr.	Daily	Hr.	Daily	Bare Costs	Incl. O&P
1 Carpenter Foreman (outside)	$53.65	$429.20	$81.15	$649.20	$51.18	$77.33
3 Carpenters	51.65	1239.60	78.15	1875.60		
1 Laborer	41.05	328.40	62.10	496.80		
1 Equip. Oper. (crane)	57.45	459.60	86.25	690.00		
1 Hyd. Crane, 12 Ton		471.75		518.92	9.83	10.81
48 L.H., Daily Totals		$2928.55		$4230.52	$61.01	$88.14
Crew C-13	Hr.	Daily	Hr.	Daily	Bare Costs	Incl. O&P
1 Struc. Steel Worker	$56.00	$448.00	$89.35	$714.80	$54.55	$85.62
1 Welder	56.00	448.00	89.35	714.80		
1 Carpenter	51.65	413.20	78.15	625.20		
1 Welder, Gas Engine, 300 amp		95.65		105.22	3.99	4.38
24 L.H., Daily Totals		$1404.85		$2160.01	$58.54	$90.00
Crew C-14	Hr.	Daily	Hr.	Daily	Bare Costs	Incl. O&P
1 Carpenter Foreman (outside)	$53.65	$429.20	$81.15	$649.20	$50.00	$75.38
5 Carpenters	51.65	2066.00	78.15	3126.00		
4 Laborers	41.05	1313.60	62.10	1987.20		
4 Rodmen (reinf.)	54.85	1755.20	83.00	2656.00		
2 Cement Finishers	48.90	782.40	72.20	1155.20		
1 Equip. Oper. (crane)	57.45	459.60	86.25	690.00		
1 Equip. Oper. (oiler)	49.20	393.60	73.85	590.80		
1 Hyd. Crane, 80 Ton		1487.00		1635.70	10.33	11.36
144 L.H., Daily Totals		$8686.60		$12490.10	$60.32	$86.74

Crew No.	Bare Costs Hr.	Daily	Incl. Subs O&P Hr.	Daily	Cost Per Labor-Hour Bare Costs	Incl. O&P
Crew C-14A	Hr.	Daily	Hr.	Daily	Bare Costs	Incl. O&P
1 Carpenter Foreman (outside)	$53.65	$429.20	$81.15	$649.20	$51.42	$77.71
16 Carpenters	51.65	6611.20	78.15	10003.20		
4 Rodmen (reinf.)	54.85	1755.20	83.00	2656.00		
2 Laborers	41.05	656.80	62.10	993.60		
1 Cement Finisher	48.90	391.20	72.20	577.60		
1 Equip. Oper. (medium)	55.10	440.80	82.70	661.60		
1 Gas Engine Vibrator		26.55		29.20		
1 Concrete Pump (Small)		844.05		928.46	4.35	4.79
200 L.H., Daily Totals		$11155.00		$16498.86	$55.77	$82.49
Crew C-14B	Hr.	Daily	Hr.	Daily	Bare Costs	Incl. O&P
1 Carpenter Foreman (outside)	$53.65	$429.20	$81.15	$649.20	$51.33	$77.49
16 Carpenters	51.65	6611.20	78.15	10003.20		
4 Rodmen (reinf.)	54.85	1755.20	83.00	2656.00		
2 Laborers	41.05	656.80	62.10	993.60		
2 Cement Finishers	48.90	782.40	72.20	1155.20		
1 Equip. Oper. (medium)	55.10	440.80	82.70	661.60		
1 Gas Engine Vibrator		26.55		29.20		
1 Concrete Pump (Small)		844.05		928.46	4.19	4.60
208 L.H., Daily Totals		$11546.20		$17076.46	$55.51	$82.10
Crew C-14C	Hr.	Daily	Hr.	Daily	Bare Costs	Incl. O&P
1 Carpenter Foreman (outside)	$53.65	$429.20	$81.15	$649.20	$49.02	$74.05
6 Carpenters	51.65	2479.20	78.15	3751.20		
2 Rodmen (reinf.)	54.85	877.60	83.00	1328.00		
4 Laborers	41.05	1313.60	62.10	1987.20		
1 Cement Finisher	48.90	391.20	72.20	577.60		
1 Gas Engine Vibrator		26.55		29.20	.24	.26
112 L.H., Daily Totals		$5517.35		$8322.41	$49.26	$74.31
Crew C-14D	Hr.	Daily	Hr.	Daily	Bare Costs	Incl. O&P
1 Carpenter Foreman (outside)	$53.65	$429.20	$81.15	$649.20	$51.17	$77.32
18 Carpenters	51.65	7437.60	78.15	11253.60		
2 Rodmen (reinf.)	54.85	877.60	83.00	1328.00		
2 Laborers	41.05	656.80	62.10	993.60		
1 Cement Finisher	48.90	391.20	72.20	577.60		
1 Equip. Oper. (medium)	55.10	440.80	82.70	661.60		
1 Gas Engine Vibrator		26.55		29.20		
1 Concrete Pump (Small)		844.05		928.46	4.35	4.79
200 L.H., Daily Totals		$11103.80		$16421.26	$55.52	$82.11
Crew C-14E	Hr.	Daily	Hr.	Daily	Bare Costs	Incl. O&P
1 Carpenter Foreman (outside)	$53.65	$429.20	$81.15	$649.20	$49.85	$75.27
2 Carpenters	51.65	826.40	78.15	1250.40		
4 Rodmen (reinf.)	54.85	1755.20	83.00	2656.00		
3 Laborers	41.05	985.20	62.10	1490.40		
1 Cement Finisher	48.90	391.20	72.20	577.60		
1 Gas Engine Vibrator		26.55		29.20	.30	.33
88 L.H., Daily Totals		$4413.75		$6652.81	$50.16	$75.60
Crew C-14F	Hr.	Daily	Hr.	Daily	Bare Costs	Incl. O&P
1 Labor Foreman (outside)	$43.05	$344.40	$65.10	$520.80	$46.51	$69.17
2 Laborers	41.05	656.80	62.10	993.60		
6 Cement Finishers	48.90	2347.20	72.20	3465.60		
1 Gas Engine Vibrator		26.55		29.20	.37	.41
72 L.H., Daily Totals		$3374.95		$5009.20	$46.87	$69.57

Crews - Standard

Crew C-14G	Hr.	Daily	Hr.	Daily	Bare Costs	Incl. O&P
1 Labor Foreman (outside)	$43.05	$344.40	$65.10	$520.80	$45.82	$68.30
2 Laborers	41.05	656.80	62.10	993.60		
4 Cement Finishers	48.90	1564.80	72.20	2310.40		
1 Gas Engine Vibrator		26.55		29.20	.47	.52
56 L.H., Daily Totals		$2592.55		$3854.01	$46.30	$68.82

Crew C-14H	Hr.	Daily	Hr.	Daily	Bare Costs	Incl. O&P
1 Carpenter Foreman (outside)	$53.65	$429.20	$81.15	$649.20	$50.29	$75.79
2 Carpenters	51.65	826.40	78.15	1250.40		
1 Rodman (reinf.)	54.85	438.80	83.00	664.00		
1 Laborer	41.05	328.40	62.10	496.80		
1 Cement Finisher	48.90	391.20	72.20	577.60		
1 Gas Engine Vibrator		26.55		29.20	.55	.61
48 L.H., Daily Totals		$2440.55		$3667.20	$50.84	$76.40

Crew C-14L	Hr.	Daily	Hr.	Daily	Bare Costs	Incl. O&P
1 Carpenter Foreman (outside)	$53.65	$429.20	$81.15	$649.20	$48.05	$72.55
6 Carpenters	51.65	2479.20	78.15	3751.20		
4 Laborers	41.05	1313.60	62.10	1987.20		
1 Cement Finisher	48.90	391.20	72.20	577.60		
1 Gas Engine Vibrator		26.55		29.20	.28	.30
96 L.H., Daily Totals		$4639.75		$6994.40	$48.33	$72.86

Crew C-14M	Hr.	Daily	Hr.	Daily	Bare Costs	Incl. O&P
1 Carpenter Foreman (outside)	$53.65	$429.20	$81.15	$649.20	$49.74	$74.94
2 Carpenters	51.65	826.40	78.15	1250.40		
1 Rodman (reinf.)	54.85	438.80	83.00	664.00		
2 Laborers	41.05	656.80	62.10	993.60		
1 Cement Finisher	48.90	391.20	72.20	577.60		
1 Equip. Oper. (medium)	55.10	440.80	82.70	661.60		
1 Gas Engine Vibrator		26.55		29.20		
1 Concrete Pump (Small)		844.05		928.46	13.60	14.96
64 L.H., Daily Totals		$4053.80		$5754.06	$63.34	$89.91

Crew C-15	Hr.	Daily	Hr.	Daily	Bare Costs	Incl. O&P
1 Carpenter Foreman (outside)	$53.65	$429.20	$81.15	$649.20	$48.08	$72.35
2 Carpenters	51.65	826.40	78.15	1250.40		
3 Laborers	41.05	985.20	62.10	1490.40		
2 Cement Finishers	48.90	782.40	72.20	1155.20		
1 Rodman (reinf.)	54.85	438.80	83.00	664.00		
72 L.H., Daily Totals		$3462.00		$5209.20	$48.08	$72.35

Crew C-16	Hr.	Daily	Hr.	Daily	Bare Costs	Incl. O&P
1 Labor Foreman (outside)	$43.05	$344.40	$65.10	$520.80	$45.59	$68.36
3 Laborers	41.05	985.20	62.10	1490.40		
2 Cement Finishers	48.90	782.40	72.20	1155.20		
1 Equip. Oper. (medium)	55.10	440.80	82.70	661.60		
1 Gunite Pump Rig		313.00		344.30		
2 -50' Air Hoses, 3/4"		13.10		14.41		
2 -50' Air Hoses, 2"		51.80		56.98	6.75	7.42
56 L.H., Daily Totals		$2930.70		$4243.69	$52.33	$75.78

Crew C-16A	Hr.	Daily	Hr.	Daily	Bare Costs	Incl. O&P
1 Laborer	$41.05	$328.40	$62.10	$496.80	$48.49	$72.30
2 Cement Finishers	48.90	782.40	72.20	1155.20		
1 Equip. Oper. (medium)	55.10	440.80	82.70	661.60		
1 Gunite Pump Rig		313.00		344.30		
2 -50' Air Hoses, 3/4"		13.10		14.41		
2 -50' Air Hoses, 2"		51.80		56.98		
1 Telescoping Boom Lift, to 60'		451.25		496.38	25.91	28.50
32 L.H., Daily Totals		$2380.75		$3225.67	$74.40	$100.80

Crew C-17	Hr.	Daily	Hr.	Daily	Bare Costs	Incl. O&P
2 Skilled Worker Foremen (out)	$55.40	$886.40	$84.25	$1348.00	$53.80	$81.85
8 Skilled Workers	53.40	3417.60	81.25	5200.00		
80 L.H., Daily Totals		$4304.00		$6548.00	$53.80	$81.85

Crew C-17A	Hr.	Daily	Hr.	Daily	Bare Costs	Incl. O&P
2 Skilled Worker Foremen (out)	$55.40	$886.40	$84.25	$1348.00	$53.85	$81.90
8 Skilled Workers	53.40	3417.60	81.25	5200.00		
.125 Equip. Oper. (crane)	57.45	57.45	86.25	86.25		
.125 Hyd. Crane, 80 Ton		185.88		204.46	2.29	2.52
81 L.H., Daily Totals		$4547.32		$6838.71	$56.14	$84.43

Crew C-17B	Hr.	Daily	Hr.	Daily	Bare Costs	Incl. O&P
2 Skilled Worker Foremen (out)	$55.40	$886.40	$84.25	$1348.00	$53.89	$81.96
8 Skilled Workers	53.40	3417.60	81.25	5200.00		
.25 Equip. Oper. (crane)	57.45	114.90	86.25	172.50		
.25 Hyd. Crane, 80 Ton		371.75		408.93		
.25 Trowel, 48" Walk-Behind		10.38		11.41	4.66	5.13
82 L.H., Daily Totals		$4801.02		$7140.84	$58.55	$87.08

Crew C-17C	Hr.	Daily	Hr.	Daily	Bare Costs	Incl. O&P
2 Skilled Worker Foremen (out)	$55.40	$886.40	$84.25	$1348.00	$53.93	$82.01
8 Skilled Workers	53.40	3417.60	81.25	5200.00		
.375 Equip. Oper. (crane)	57.45	172.35	86.25	258.75		
.375 Hyd. Crane, 80 Ton		557.63		613.39	6.72	7.39
83 L.H., Daily Totals		$5033.98		$7420.14	$60.65	$89.40

Crew C-17D	Hr.	Daily	Hr.	Daily	Bare Costs	Incl. O&P
2 Skilled Worker Foremen (out)	$55.40	$886.40	$84.25	$1348.00	$53.97	$82.06
8 Skilled Workers	53.40	3417.60	81.25	5200.00		
.5 Equip. Oper. (crane)	57.45	229.80	86.25	345.00		
.5 Hyd. Crane, 80 Ton		743.50		817.85	8.85	9.74
84 L.H., Daily Totals		$5277.30		$7710.85	$62.83	$91.80

Crew C-17E	Hr.	Daily	Hr.	Daily	Bare Costs	Incl. O&P
2 Skilled Worker Foremen (out)	$55.40	$886.40	$84.25	$1348.00	$53.80	$81.85
8 Skilled Workers	53.40	3417.60	81.25	5200.00		
1 Hyd. Jack with Rods		104.40		114.84	1.30	1.44
80 L.H., Daily Totals		$4408.40		$6662.84	$55.10	$83.29

Crew C-18	Hr.	Daily	Hr.	Daily	Bare Costs	Incl. O&P
.125 Labor Foreman (outside)	$43.05	$43.05	$65.10	$65.10	$41.27	$62.43
1 Laborer	41.05	328.40	62.10	496.80		
1 Concrete Cart, 10 C.F.		58.10		63.91	6.46	7.10
9 L.H., Daily Totals		$429.55		$625.81	$47.73	$69.53

Crew C-19	Hr.	Daily	Hr.	Daily	Bare Costs	Incl. O&P
.125 Labor Foreman (outside)	$43.05	$43.05	$65.10	$65.10	$41.27	$62.43
1 Laborer	41.05	328.40	62.10	496.80		
1 Concrete Cart, 18 C.F.		97.75		107.53	10.86	11.95
9 L.H., Daily Totals		$469.20		$669.42	$52.13	$74.38

Crew C-20	Hr.	Daily	Hr.	Daily	Bare Costs	Incl. O&P
1 Labor Foreman (outside)	$43.05	$344.40	$65.10	$520.80	$44.04	$66.31
5 Laborers	41.05	1642.00	62.10	2484.00		
1 Cement Finisher	48.90	391.20	72.20	577.60		
1 Equip. Oper. (medium)	55.10	440.80	82.70	661.60		
2 Gas Engine Vibrators		53.10		58.41		
1 Concrete Pump (Small)		844.05		928.46	14.02	15.42
64 L.H., Daily Totals		$3715.55		$5230.86	$58.06	$81.73

Crew No.	Bare Costs		Incl. Subs O&P		Cost Per Labor-Hour	
Crew C-21	Hr.	Daily	Hr.	Daily	Bare Costs	Incl. O&P
1 Labor Foreman (outside)	$43.05	$344.40	$65.10	$520.80	$44.04	$66.31
5 Laborers	41.05	1642.00	62.10	2484.00		
1 Cement Finisher	48.90	391.20	72.20	577.60		
1 Equip. Oper. (medium)	55.10	440.80	82.70	661.60		
2 Gas Engine Vibrators		53.10		58.41		
1 Concrete Conveyer		186.45		205.10	3.74	4.12
64 L.H., Daily Totals		$3057.95		$4507.51	$47.78	$70.43

Crew No.	Bare Costs		Incl. Subs O&P		Cost Per Labor-Hour	
Crew C-22	Hr.	Daily	Hr.	Daily	Bare Costs	Incl. O&P
1 Rodman Foreman (outside)	$56.85	$454.80	$86.05	$688.40	$55.16	$83.44
4 Rodmen (reinf.)	54.85	1755.20	83.00	2656.00		
.125 Equip. Oper. (crane)	57.45	57.45	86.25	86.25		
.125 Equip. Oper. (oiler)	49.20	49.20	73.85	73.85		
.125 Hyd. Crane, 25 Ton		72.71		79.98	1.73	1.90
42 L.H., Daily Totals		$2389.36		$3584.48	$56.89	$85.34

Crew No.	Bare Costs		Incl. Subs O&P		Cost Per Labor-Hour	
Crew C-23	Hr.	Daily	Hr.	Daily	Bare Costs	Incl. O&P
2 Skilled Worker Foremen (out)	$55.40	$886.40	$84.25	$1348.00	$53.78	$81.61
6 Skilled Workers	53.40	2563.20	81.25	3900.00		
1 Equip. Oper. (crane)	57.45	459.60	86.25	690.00		
1 Equip. Oper. (oiler)	49.20	393.60	73.85	590.80		
1 Lattice Boom Crane, 90 Ton		1698.00		1867.80	21.23	23.35
80 L.H., Daily Totals		$6000.80		$8396.60	$75.01	$104.96

Crew No.	Bare Costs		Incl. Subs O&P		Cost Per Labor-Hour	
Crew C-23A	Hr.	Daily	Hr.	Daily	Bare Costs	Incl. O&P
1 Labor Foreman (outside)	$43.05	$344.40	$65.10	$520.80	$46.36	$69.88
2 Laborers	41.05	656.80	62.10	993.60		
1 Equip. Oper. (crane)	57.45	459.60	86.25	690.00		
1 Equip. Oper. (oiler)	49.20	393.60	73.85	590.80		
1 Crawler Crane, 100 Ton		1879.00		2066.90		
3 Conc. Buckets, 8 C.Y.		632.70		695.97	62.79	69.07
40 L.H., Daily Totals		$4366.10		$5558.07	$109.15	$138.95

Crew No.	Bare Costs		Incl. Subs O&P		Cost Per Labor-Hour	
Crew C-24	Hr.	Daily	Hr.	Daily	Bare Costs	Incl. O&P
2 Skilled Worker Foremen (out)	$55.40	$886.40	$84.25	$1348.00	$53.78	$81.61
6 Skilled Workers	53.40	2563.20	81.25	3900.00		
1 Equip. Oper. (crane)	57.45	459.60	86.25	690.00		
1 Equip. Oper. (oiler)	49.20	393.60	73.85	590.80		
1 Lattice Boom Crane, 150 Ton		2021.00		2223.10	25.26	27.79
80 L.H., Daily Totals		$6323.80		$8751.90	$79.05	$109.40

Crew No.	Bare Costs		Incl. Subs O&P		Cost Per Labor-Hour	
Crew C-25	Hr.	Daily	Hr.	Daily	Bare Costs	Incl. O&P
2 Rodmen (reinf.)	$54.85	$877.60	$83.00	$1328.00	$44.35	$69.70
2 Rodmen Helpers	33.85	541.60	56.40	902.40		
32 L.H., Daily Totals		$1419.20		$2230.40	$44.35	$69.70

Crew No.	Bare Costs		Incl. Subs O&P		Cost Per Labor-Hour	
Crew C-27	Hr.	Daily	Hr.	Daily	Bare Costs	Incl. O&P
2 Cement Finishers	$48.90	$782.40	$72.20	$1155.20	$48.90	$72.20
1 Concrete Saw		110.15		121.17	6.88	7.57
16 L.H., Daily Totals		$892.55		$1276.37	$55.78	$79.77

Crew No.	Bare Costs		Incl. Subs O&P		Cost Per Labor-Hour	
Crew C-28	Hr.	Daily	Hr.	Daily	Bare Costs	Incl. O&P
1 Cement Finisher	$48.90	$391.20	$72.20	$577.60	$48.90	$72.20
1 Portable Air Compressor, Gas		16.05		17.66	2.01	2.21
8 L.H., Daily Totals		$407.25		$595.26	$50.91	$74.41

Crew No.	Bare Costs		Incl. Subs O&P		Cost Per Labor-Hour	
Crew C-29	Hr.	Daily	Hr.	Daily	Bare Costs	Incl. O&P
1 Laborer	$41.05	$328.40	$62.10	$496.80	$41.05	$62.10
1 Pressure Washer		63.05		69.36	7.88	8.67
8 L.H., Daily Totals		$391.45		$566.15	$48.93	$70.77

Crew No.	Bare Costs		Incl. Subs O&P		Cost Per Labor-Hour	
Crew C-30	Hr.	Daily	Hr.	Daily	Bare Costs	Incl. O&P
1 Laborer	$41.05	$328.40	$62.10	$496.80	$41.05	$62.10
1 Concrete Mixer, 10 C.F.		162.00		178.20	20.25	22.27
8 L.H., Daily Totals		$490.40		$675.00	$61.30	$84.38

Crew No.	Bare Costs		Incl. Subs O&P		Cost Per Labor-Hour	
Crew C-31	Hr.	Daily	Hr.	Daily	Bare Costs	Incl. O&P
1 Cement Finisher	$48.90	$391.20	$72.20	$577.60	$48.90	$72.20
1 Grout Pump		313.00		344.30	39.13	43.04
8 L.H., Daily Totals		$704.20		$921.90	$88.03	$115.24

Crew No.	Bare Costs		Incl. Subs O&P		Cost Per Labor-Hour	
Crew C-32	Hr.	Daily	Hr.	Daily	Bare Costs	Incl. O&P
1 Cement Finisher	$48.90	$391.20	$72.20	$577.60	$44.98	$67.15
1 Laborer	41.05	328.40	62.10	496.80		
1 Crack Chaser Saw, Gas, 6 H.P.		24.50		26.95		
1 Vacuum Pick-Up System		72.15		79.36	6.04	6.64
16 L.H., Daily Totals		$816.25		$1180.71	$51.02	$73.79

Crew No.	Bare Costs		Incl. Subs O&P		Cost Per Labor-Hour	
Crew D-1	Hr.	Daily	Hr.	Daily	Bare Costs	Incl. O&P
1 Bricklayer	$51.05	$408.40	$77.95	$623.60	$45.73	$69.83
1 Bricklayer Helper	40.40	323.20	61.70	493.60		
16 L.H., Daily Totals		$731.60		$1117.20	$45.73	$69.83

Crew No.	Bare Costs		Incl. Subs O&P		Cost Per Labor-Hour	
Crew D-2	Hr.	Daily	Hr.	Daily	Bare Costs	Incl. O&P
3 Bricklayers	$51.05	$1225.20	$77.95	$1870.80	$47.23	$72.06
2 Bricklayer Helpers	40.40	646.40	61.70	987.20		
.5 Carpenter	51.65	206.60	78.15	312.60		
44 L.H., Daily Totals		$2078.20		$3170.60	$47.23	$72.06

Crew No.	Bare Costs		Incl. Subs O&P		Cost Per Labor-Hour	
Crew D-3	Hr.	Daily	Hr.	Daily	Bare Costs	Incl. O&P
3 Bricklayers	$51.05	$1225.20	$77.95	$1870.80	$47.02	$71.77
2 Bricklayer Helpers	40.40	646.40	61.70	987.20		
.25 Carpenter	51.65	103.30	78.15	156.30		
42 L.H., Daily Totals		$1974.90		$3014.30	$47.02	$71.77

Crew No.	Bare Costs		Incl. Subs O&P		Cost Per Labor-Hour	
Crew D-4	Hr.	Daily	Hr.	Daily	Bare Costs	Incl. O&P
1 Bricklayer	$51.05	$408.40	$77.95	$623.60	$45.88	$69.72
2 Bricklayer Helpers	40.40	646.40	61.70	987.20		
1 Equip. Oper. (light)	51.65	413.20	77.55	620.40		
1 Grout Pump, 50 C.F./hr.		125.60		138.16	3.92	4.32
32 L.H., Daily Totals		$1593.60		$2369.36	$49.80	$74.04

Crew No.	Bare Costs		Incl. Subs O&P		Cost Per Labor-Hour	
Crew D-5	Hr.	Daily	Hr.	Daily	Bare Costs	Incl. O&P
1 Bricklayer	51.05	408.40	77.95	623.60	51.05	77.95
8 L.H., Daily Totals		$408.40		$623.60	$51.05	$77.95

Crew No.	Bare Costs		Incl. Subs O&P		Cost Per Labor-Hour	
Crew D-6	Hr.	Daily	Hr.	Daily	Bare Costs	Incl. O&P
3 Bricklayers	$51.05	$1225.20	$77.95	$1870.80	$45.96	$70.16
3 Bricklayer Helpers	40.40	969.60	61.70	1480.80		
.25 Carpenter	51.65	103.30	78.15	156.30		
50 L.H., Daily Totals		$2298.10		$3507.90	$45.96	$70.16

Crew No.	Bare Costs		Incl. Subs O&P		Cost Per Labor-Hour	
Crew D-7	Hr.	Daily	Hr.	Daily	Bare Costs	Incl. O&P
1 Tile Layer	$47.85	$382.80	$70.55	$564.40	$42.80	$63.10
1 Tile Layer Helper	37.75	302.00	55.65	445.20		
16 L.H., Daily Totals		$684.80		$1009.60	$42.80	$63.10

Crew No.	Bare Costs		Incl. Subs O&P		Cost Per Labor-Hour	
Crew D-8	Hr.	Daily	Hr.	Daily	Bare Costs	Incl. O&P
3 Bricklayers	$51.05	$1225.20	$77.95	$1870.80	$46.79	$71.45
2 Bricklayer Helpers	40.40	646.40	61.70	987.20		
40 L.H., Daily Totals		$1871.60		$2858.00	$46.79	$71.45

Crew No.	Bare Costs Hr.	Daily	Incl. Subs O&P Hr.	Daily	Cost Per Labor-Hour Bare Costs	Incl. O&P
Crew D-9					Bare Costs	Incl. O&P
3 Bricklayers	$51.05	$1225.20	$77.95	$1870.80	$45.73	$69.83
3 Bricklayer Helpers	40.40	969.60	61.70	1480.80		
48 L.H., Daily Totals		$2194.80		$3351.60	$45.73	$69.83
Crew D-10	Hr.	Daily	Hr.	Daily	Bare Costs	Incl. O&P
1 Bricklayer Foreman (outside)	$53.05	$424.40	$81.00	$648.00	$50.49	$76.72
1 Bricklayer	51.05	408.40	77.95	623.60		
1 Bricklayer Helper	40.40	323.20	61.70	493.60		
1 Equip. Oper. (crane)	57.45	459.60	86.25	690.00		
1 S.P. Crane, 4x4, 12 Ton		371.85		409.04	11.62	12.78
32 L.H., Daily Totals		$1987.45		$2864.24	$62.11	$89.51
Crew D-11	Hr.	Daily	Hr.	Daily	Bare Costs	Incl. O&P
1 Bricklayer Foreman (outside)	$53.05	$424.40	$81.00	$648.00	$48.17	$73.55
1 Bricklayer	51.05	408.40	77.95	623.60		
1 Bricklayer Helper	40.40	323.20	61.70	493.60		
24 L.H., Daily Totals		$1156.00		$1765.20	$48.17	$73.55
Crew D-12	Hr.	Daily	Hr.	Daily	Bare Costs	Incl. O&P
1 Bricklayer Foreman (outside)	$53.05	$424.40	$81.00	$648.00	$46.23	$70.59
1 Bricklayer	51.05	408.40	77.95	623.60		
2 Bricklayer Helpers	40.40	646.40	61.70	987.20		
32 L.H., Daily Totals		$1479.20		$2258.80	$46.23	$70.59
Crew D-13	Hr.	Daily	Hr.	Daily	Bare Costs	Incl. O&P
1 Bricklayer Foreman (outside)	$53.05	$424.40	$81.00	$648.00	$49.00	$74.46
1 Bricklayer	51.05	408.40	77.95	623.60		
2 Bricklayer Helpers	40.40	646.40	61.70	987.20		
1 Carpenter	51.65	413.20	78.15	625.20		
1 Equip. Oper. (crane)	57.45	459.60	86.25	690.00		
1 S.P. Crane, 4x4, 12 Ton		371.85		409.04	7.75	8.52
48 L.H., Daily Totals		$2723.85		$3983.03	$56.75	$82.98
Crew D-14	Hr.	Daily	Hr.	Daily	Bare Costs	Incl. O&P
3 Bricklayers	$51.05	$1225.20	$77.95	$1870.80	$48.39	$73.89
1 Bricklayer Helper	40.40	323.20	61.70	493.60		
32 L.H., Daily Totals		$1548.40		$2364.40	$48.39	$73.89
Crew E-1	Hr.	Daily	Hr.	Daily	Bare Costs	Incl. O&P
1 Welder Foreman (outside)	$58.00	$464.00	$92.50	$740.00	$55.22	$86.47
1 Welder	56.00	448.00	89.35	714.80		
1 Equip. Oper. (light)	51.65	413.20	77.55	620.40		
1 Welder, Gas Engine, 300 amp		95.65		105.22	3.99	4.38
24 L.H., Daily Totals		$1420.85		$2180.42	$59.20	$90.85
Crew E-2	Hr.	Daily	Hr.	Daily	Bare Costs	Incl. O&P
1 Struc. Steel Foreman (outside)	$58.00	$464.00	$92.50	$740.00	$55.52	$87.14
4 Struc. Steel Workers	56.00	1792.00	89.35	2859.20		
1 Equip. Oper. (crane)	57.45	459.60	86.25	690.00		
1 Equip. Oper. (oiler)	49.20	393.60	73.85	590.80		
1 Lattice Boom Crane, 90 Ton		1698.00		1867.80	30.32	33.35
56 L.H., Daily Totals		$4807.20		$6747.80	$85.84	$120.50
Crew E-3	Hr.	Daily	Hr.	Daily	Bare Costs	Incl. O&P
1 Struc. Steel Foreman (outside)	$58.00	$464.00	$92.50	$740.00	$56.67	$90.40
1 Struc. Steel Worker	56.00	448.00	89.35	714.80		
1 Welder	56.00	448.00	89.35	714.80		
1 Welder, Gas Engine, 300 amp		95.65		105.22	3.99	4.38
24 L.H., Daily Totals		$1455.65		$2274.82	$60.65	$94.78

Crew No.	Bare Costs Hr.	Daily	Incl. Subs O&P Hr.	Daily	Cost Per Labor-Hour Bare Costs	Incl. O&P
Crew E-3A					Bare Costs	Incl. O&P
1 Struc. Steel Foreman (outside)	$58.00	$464.00	$92.50	$740.00	$56.67	$90.40
1 Struc. Steel Worker	56.00	448.00	89.35	714.80		
1 Welder	56.00	448.00	89.35	714.80		
1 Welder, Gas Engine, 300 amp		95.65		105.22		
1 Telescoping Boom Lift, to 40'		283.15		311.46	15.78	17.36
24 L.H., Daily Totals		$1738.80		$2586.28	$72.45	$107.76
Crew E-4	Hr.	Daily	Hr.	Daily	Bare Costs	Incl. O&P
1 Struc. Steel Foreman (outside)	$58.00	$464.00	$92.50	$740.00	$56.50	$90.14
3 Struc. Steel Workers	56.00	1344.00	89.35	2144.40		
1 Welder, Gas Engine, 300 amp		95.65		105.22	2.99	3.29
32 L.H., Daily Totals		$1903.65		$2989.61	$59.49	$93.43
Crew E-5	Hr.	Daily	Hr.	Daily	Bare Costs	Incl. O&P
2 Struc. Steel Foremen (outside)	$58.00	$928.00	$92.50	$1480.00	$55.87	$88.12
5 Struc. Steel Workers	56.00	2240.00	89.35	3574.00		
1 Equip. Oper. (crane)	57.45	459.60	86.25	690.00		
1 Welder	56.00	448.00	89.35	714.80		
1 Equip. Oper. (oiler)	49.20	393.60	73.85	590.80		
1 Lattice Boom Crane, 90 Ton		1698.00		1867.80		
1 Welder, Gas Engine, 300 amp		95.65		105.22	22.42	24.66
80 L.H., Daily Totals		$6262.85		$9022.61	$78.29	$112.78
Crew E-6	Hr.	Daily	Hr.	Daily	Bare Costs	Incl. O&P
3 Struc. Steel Foremen (outside)	$58.00	$1392.00	$92.50	$2220.00	$55.77	$88.04
9 Struc. Steel Workers	56.00	4032.00	89.35	6433.20		
1 Equip. Oper. (crane)	57.45	459.60	86.25	690.00		
1 Welder	56.00	448.00	89.35	714.80		
1 Equip. Oper. (oiler)	49.20	393.60	73.85	590.80		
1 Equip. Oper. (light)	51.65	413.20	77.55	620.40		
1 Lattice Boom Crane, 90 Ton		1698.00		1867.80		
1 Welder, Gas Engine, 300 amp		95.65		105.22		
1 Air Compressor, 160 cfm		115.85		127.44		
2 Impact Wrenches		37.00		40.70	15.21	16.73
128 L.H., Daily Totals		$9084.90		$13410.35	$70.98	$104.77
Crew E-7	Hr.	Daily	Hr.	Daily	Bare Costs	Incl. O&P
1 Struc. Steel Foreman (outside)	$58.00	$464.00	$92.50	$740.00	$55.87	$88.12
4 Struc. Steel Workers	56.00	1792.00	89.35	2859.20		
1 Equip. Oper. (crane)	57.45	459.60	86.25	690.00		
1 Equip. Oper. (oiler)	49.20	393.60	73.85	590.80		
1 Welder Foreman (outside)	58.00	464.00	92.50	740.00		
2 Welders	56.00	896.00	89.35	1429.60		
1 Lattice Boom Crane, 90 Ton		1698.00		1867.80		
2 Welders, Gas Engine, 300 amp		191.30		210.43	23.62	25.98
80 L.H., Daily Totals		$6358.50		$9127.83	$79.48	$114.10
Crew E-8	Hr.	Daily	Hr.	Daily	Bare Costs	Incl. O&P
1 Struc. Steel Foreman (outside)	$58.00	$464.00	$92.50	$740.00	$55.56	$87.50
4 Struc. Steel Workers	56.00	1792.00	89.35	2859.20		
1 Welder Foreman (outside)	58.00	464.00	92.50	740.00		
4 Welders	56.00	1792.00	89.35	2859.20		
1 Equip. Oper. (crane)	57.45	459.60	86.25	690.00		
1 Equip. Oper. (oiler)	49.20	393.60	73.85	590.80		
1 Equip. Oper. (light)	51.65	413.20	77.55	620.40		
1 Lattice Boom Crane, 90 Ton		1698.00		1867.80		
4 Welders, Gas Engine, 300 amp		382.60		420.86	20.01	22.01
104 L.H., Daily Totals		$7859.00		$11388.26	$75.57	$109.50

Crews - Standard

Crew No.	Bare Costs		Incl. Subs O&P		Cost Per Labor-Hour	

Crew E-9	Hr.	Daily	Hr.	Daily	Bare Costs	Incl. O&P
2 Struc. Steel Foremen (outside)	$58.00	$928.00	$92.50	$1480.00	$55.77	$88.04
5 Struc. Steel Workers	56.00	2240.00	89.35	3574.00		
1 Welder Foreman (outside)	58.00	464.00	92.50	740.00		
5 Welders	56.00	2240.00	89.35	3574.00		
1 Equip. Oper. (crane)	57.45	459.60	86.25	690.00		
1 Equip. Oper. (oiler)	49.20	393.60	73.85	590.80		
1 Equip. Oper. (light)	51.65	413.20	77.55	620.40		
1 Lattice Boom Crane, 90 Ton		1698.00		1867.80		
5 Welders, Gas Engine, 300 amp		478.25		526.08	17.00	18.70
128 L.H., Daily Totals		$9314.65		$13663.08	$72.77	$106.74

Crew E-10	Hr.	Daily	Hr.	Daily	Bare Costs	Incl. O&P
1 Welder Foreman (outside)	$58.00	$464.00	$92.50	$740.00	$57.00	$90.92
1 Welder	56.00	448.00	89.35	714.80		
1 Welder, Gas Engine, 300 amp		95.65		105.22		
1 Flatbed Truck, Gas, 3 Ton		245.95		270.55	21.35	23.48
16 L.H., Daily Totals		$1253.60		$1830.56	$78.35	$114.41

Crew E-11	Hr.	Daily	Hr.	Daily	Bare Costs	Incl. O&P
2 Painters, Struc. Steel	$44.65	$714.40	$73.00	$1168.00	$45.50	$71.41
1 Building Laborer	41.05	328.40	62.10	496.80		
1 Equip. Oper. (light)	51.65	413.20	77.55	620.40		
1 Air Compressor, 250 cfm		167.95		184.75		
1 Sandblaster, Portable, 3 C.F.		20.70		22.77		
1 Set Sand Blasting Accessories		14.90		16.39	6.36	7.00
32 L.H., Daily Totals		$1659.55		$2509.11	$51.86	$78.41

Crew E-11A	Hr.	Daily	Hr.	Daily	Bare Costs	Incl. O&P
2 Painters, Struc. Steel	$44.65	$714.40	$73.00	$1168.00	$45.50	$71.41
1 Building Laborer	41.05	328.40	62.10	496.80		
1 Equip. Oper. (light)	51.65	413.20	77.55	620.40		
1 Air Compressor, 250 cfm		167.95		184.75		
1 Sandblaster, Portable, 3 C.F.		20.70		22.77		
1 Set Sand Blasting Accessories		14.90		16.39		
1 Telescoping Boom Lift, to 60'		451.25		496.38	20.46	22.51
32 L.H., Daily Totals		$2110.80		$3005.48	$65.96	$93.92

Crew E-11B	Hr.	Daily	Hr.	Daily	Bare Costs	Incl. O&P
2 Painters, Struc. Steel	$44.65	$714.40	$73.00	$1168.00	$43.45	$69.37
1 Building Laborer	41.05	328.40	62.10	496.80		
2 Paint Sprayer, 8 C.F.M.		85.20		93.72		
1 Telescoping Boom Lift, to 60'		451.25		496.38	22.35	24.59
24 L.H., Daily Totals		$1579.25		$2254.90	$65.80	$93.95

Crew E-12	Hr.	Daily	Hr.	Daily	Bare Costs	Incl. O&P
1 Welder Foreman (outside)	$58.00	$464.00	$92.50	$740.00	$54.83	$85.03
1 Equip. Oper. (light)	51.65	413.20	77.55	620.40		
1 Welder, Gas Engine, 300 amp		95.65		105.22	5.98	6.58
16 L.H., Daily Totals		$972.85		$1465.62	$60.80	$91.60

Crew E-13	Hr.	Daily	Hr.	Daily	Bare Costs	Incl. O&P
1 Welder Foreman (outside)	$58.00	$464.00	$92.50	$740.00	$55.88	$87.52
.5 Equip. Oper. (light)	51.65	206.60	77.55	310.20		
1 Welder, Gas Engine, 300 amp		95.65		105.22	7.97	8.77
12 L.H., Daily Totals		$766.25		$1155.42	$63.85	$96.28

Crew E-14	Hr.	Daily	Hr.	Daily	Bare Costs	Incl. O&P
1 Welder Foreman (outside)	$58.00	$464.00	$92.50	$740.00	$58.00	$92.50
1 Welder, Gas Engine, 300 amp		95.65		105.22	11.96	13.15
8 L.H., Daily Totals		$559.65		$845.22	$69.96	$105.65

Crew E-16	Hr.	Daily	Hr.	Daily	Bare Costs	Incl. O&P
1 Welder Foreman (outside)	$58.00	$464.00	$92.50	$740.00	$57.00	$90.92
1 Welder	56.00	448.00	89.35	714.80		
1 Welder, Gas Engine, 300 amp		95.65		105.22	5.98	6.58
16 L.H., Daily Totals		$1007.65		$1560.02	$62.98	$97.50

Crew E-17	Hr.	Daily	Hr.	Daily	Bare Costs	Incl. O&P
1 Struc. Steel Foreman (outside)	$58.00	$464.00	$92.50	$740.00	$57.00	$90.92
1 Structural Steel Worker	56.00	448.00	89.35	714.80		
16 L.H., Daily Totals		$912.00		$1454.80	$57.00	$90.92

Crew E-18	Hr.	Daily	Hr.	Daily	Bare Costs	Incl. O&P
1 Struc. Steel Foreman (outside)	$58.00	$464.00	$92.50	$740.00	$56.22	$88.65
3 Structural Steel Workers	56.00	1344.00	89.35	2144.40		
1 Equipment Operator (med.)	55.10	440.80	82.70	661.60		
1 Lattice Boom Crane, 20 Ton		869.00		955.90	21.73	23.90
40 L.H., Daily Totals		$3117.80		$4501.90	$77.94	$112.55

Crew E-19	Hr.	Daily	Hr.	Daily	Bare Costs	Incl. O&P
1 Struc. Steel Foreman (outside)	$58.00	$464.00	$92.50	$740.00	$55.22	$86.47
1 Structural Steel Worker	56.00	448.00	89.35	714.80		
1 Equip. Oper. (light)	51.65	413.20	77.55	620.40		
1 Lattice Boom Crane, 20 Ton		869.00		955.90	36.21	39.83
24 L.H., Daily Totals		$2194.20		$3031.10	$91.42	$126.30

Crew E-20	Hr.	Daily	Hr.	Daily	Bare Costs	Incl. O&P
1 Struc. Steel Foreman (outside)	$58.00	$464.00	$92.50	$740.00	$55.58	$87.42
5 Structural Steel Workers	56.00	2240.00	89.35	3574.00		
1 Equip. Oper. (crane)	57.45	459.60	86.25	690.00		
1 Equip. Oper. (oiler)	49.20	393.60	73.85	590.80		
1 Lattice Boom Crane, 40 Ton		1315.00		1446.50	20.55	22.60
64 L.H., Daily Totals		$4872.20		$7041.30	$76.13	$110.02

Crew E-22	Hr.	Daily	Hr.	Daily	Bare Costs	Incl. O&P
1 Skilled Worker Foreman (out)	$55.40	$443.20	$84.25	$674.00	$54.07	$82.25
2 Skilled Workers	53.40	854.40	81.25	1300.00		
24 L.H., Daily Totals		$1297.60		$1974.00	$54.07	$82.25

Crew E-24	Hr.	Daily	Hr.	Daily	Bare Costs	Incl. O&P
3 Structural Steel Workers	$56.00	$1344.00	$89.35	$2144.40	$55.77	$87.69
1 Equipment Operator (med.)	55.10	440.80	82.70	661.60		
1 Hyd. Crane, 25 Ton		581.70		639.87	18.18	20.00
32 L.H., Daily Totals		$2366.50		$3445.87	$73.95	$107.68

Crew E-25	Hr.	Daily	Hr.	Daily	Bare Costs	Incl. O&P
1 Welder Foreman (outside)	$58.00	$464.00	$92.50	$740.00	$58.00	$92.50
1 Cutting Torch		12.50		13.75	1.56	1.72
8 L.H., Daily Totals		$476.50		$753.75	$59.56	$94.22

Crew E-26	Hr.	Daily	Hr.	Daily	Bare Costs	Incl. O&P
1 Struc. Steel Foreman (outside)	$58.00	$464.00	$92.50	$740.00	$57.37	$90.65
1 Struc. Steel Worker	56.00	448.00	89.35	714.80		
1 Welder	56.00	448.00	89.35	714.80		
.25 Electrician	60.05	120.10	89.60	179.20		
.25 Plumber	63.15	126.30	94.75	189.50		
1 Welder, Gas Engine, 300 amp		95.65		105.22	3.42	3.76
28 L.H., Daily Totals		$1702.05		$2643.51	$60.79	$94.41

For customer support on your Electrical Costs with RSMeans data, call 800.448.8182.

505

Crew No.	Bare Costs		Incl. Subs O&P		Cost Per Labor-Hour	
Crew E-27	**Hr.**	**Daily**	**Hr.**	**Daily**	**Bare Costs**	**Incl. O&P**
1 Struc. Steel Foreman (outside)	$58.00	$464.00	$92.50	$740.00	$55.58	$87.42
5 Struc. Steel Workers	56.00	2240.00	89.35	3574.00		
1 Equip. Oper. (crane)	57.45	459.60	86.25	690.00		
1 Equip. Oper. (oiler)	49.20	393.60	73.85	590.80		
1 Hyd. Crane, 12 Ton		471.75		518.92		
1 Hyd. Crane, 80 Ton		1487.00		1635.70	30.61	33.67
64 L.H., Daily Totals		$5515.95		$7749.43	$86.19	$121.08
Crew F-3	**Hr.**	**Daily**	**Hr.**	**Daily**	**Bare Costs**	**Incl. O&P**
4 Carpenters	$51.65	$1652.80	$78.15	$2500.80	$52.81	$79.77
1 Equip. Oper. (crane)	57.45	459.60	86.25	690.00		
1 Hyd. Crane, 12 Ton		471.75		518.92	11.79	12.97
40 L.H., Daily Totals		$2584.15		$3709.72	$64.60	$92.74
Crew F-4	**Hr.**	**Daily**	**Hr.**	**Daily**	**Bare Costs**	**Incl. O&P**
4 Carpenters	$51.65	$1652.80	$78.15	$2500.80	$52.21	$78.78
1 Equip. Oper. (crane)	57.45	459.60	86.25	690.00		
1 Equip. Oper. (oiler)	49.20	393.60	73.85	590.80		
1 Hyd. Crane, 55 Ton		981.50		1079.65	20.45	22.49
48 L.H., Daily Totals		$3487.50		$4861.25	$72.66	$101.28
Crew F-5	**Hr.**	**Daily**	**Hr.**	**Daily**	**Bare Costs**	**Incl. O&P**
1 Carpenter Foreman (outside)	$53.65	$429.20	$81.15	$649.20	$52.15	$78.90
3 Carpenters	51.65	1239.60	78.15	1875.60		
32 L.H., Daily Totals		$1668.80		$2524.80	$52.15	$78.90
Crew F-6	**Hr.**	**Daily**	**Hr.**	**Daily**	**Bare Costs**	**Incl. O&P**
2 Carpenters	$51.65	$826.40	$78.15	$1250.40	$48.57	$73.35
2 Building Laborers	41.05	656.80	62.10	993.60		
1 Equip. Oper. (crane)	57.45	459.60	86.25	690.00		
1 Hyd. Crane, 12 Ton		471.75		518.92	11.79	12.97
40 L.H., Daily Totals		$2414.55		$3452.93	$60.36	$86.32
Crew F-7	**Hr.**	**Daily**	**Hr.**	**Daily**	**Bare Costs**	**Incl. O&P**
2 Carpenters	$51.65	$826.40	$78.15	$1250.40	$46.35	$70.13
2 Building Laborers	41.05	656.80	62.10	993.60		
32 L.H., Daily Totals		$1483.20		$2244.00	$46.35	$70.13
Crew G-1	**Hr.**	**Daily**	**Hr.**	**Daily**	**Bare Costs**	**Incl. O&P**
1 Roofer Foreman (outside)	$47.05	$376.40	$78.40	$627.20	$42.14	$70.20
4 Roofers Composition	45.05	1441.60	75.05	2401.60		
2 Roofer Helpers	33.85	541.60	56.40	902.40		
1 Application Equipment		181.15		199.26		
1 Tar Kettle/Pot		178.20		196.02		
1 Crew Truck		154.35		169.79	9.17	10.09
56 L.H., Daily Totals		$2873.30		$4496.27	$51.31	$80.29
Crew G-2	**Hr.**	**Daily**	**Hr.**	**Daily**	**Bare Costs**	**Incl. O&P**
1 Plasterer	$47.75	$382.00	$71.40	$571.20	$43.32	$65.02
1 Plasterer Helper	41.15	329.20	61.55	492.40		
1 Building Laborer	41.05	328.40	62.10	496.80		
1 Grout Pump, 50 C.F./hr.		125.60		138.16	5.23	5.76
24 L.H., Daily Totals		$1165.20		$1698.56	$48.55	$70.77

Crew No.	Bare Costs		Incl. Subs O&P		Cost Per Labor-Hour	
Crew G-2A	**Hr.**	**Daily**	**Hr.**	**Daily**	**Bare Costs**	**Incl. O&P**
1 Roofer Composition	$45.05	$360.40	$75.05	$600.40	$39.98	$64.52
1 Roofer Helper	33.85	270.80	56.40	451.20		
1 Building Laborer	41.05	328.40	62.10	496.80		
1 Foam Spray Rig, Trailer-Mtd.		514.30		565.73		
1 Pickup Truck, 3/4 Ton		109.90		120.89	26.01	28.61
24 L.H., Daily Totals		$1583.80		$2235.02	$65.99	$93.13
Crew G-3	**Hr.**	**Daily**	**Hr.**	**Daily**	**Bare Costs**	**Incl. O&P**
2 Sheet Metal Workers	$60.95	$975.20	$92.55	$1480.80	$51.00	$77.33
2 Building Laborers	41.05	656.80	62.10	993.60		
32 L.H., Daily Totals		$1632.00		$2474.40	$51.00	$77.33
Crew G-4	**Hr.**	**Daily**	**Hr.**	**Daily**	**Bare Costs**	**Incl. O&P**
1 Labor Foreman (outside)	$43.05	$344.40	$65.10	$520.80	$41.72	$63.10
2 Building Laborers	41.05	656.80	62.10	993.60		
1 Flatbed Truck, Gas, 1.5 Ton		195.20		214.72		
1 Air Compressor, 160 cfm		115.85		127.44	12.96	14.26
24 L.H., Daily Totals		$1312.25		$1856.56	$54.68	$77.36
Crew G-5	**Hr.**	**Daily**	**Hr.**	**Daily**	**Bare Costs**	**Incl. O&P**
1 Roofer Foreman (outside)	$47.05	$376.40	$78.40	$627.20	$40.97	$68.26
2 Roofers Composition	45.05	720.80	75.05	1200.80		
2 Roofer Helpers	33.85	541.60	56.40	902.40		
1 Application Equipment		181.15		199.26	4.53	4.98
40 L.H., Daily Totals		$1819.95		$2929.67	$45.50	$73.24
Crew G-6A	**Hr.**	**Daily**	**Hr.**	**Daily**	**Bare Costs**	**Incl. O&P**
2 Roofers Composition	$45.05	$720.80	$75.05	$1200.80	$45.05	$75.05
1 Small Compressor, Electric		14.85		16.34		
2 Pneumatic Nailers		44.70		49.17	3.72	4.09
16 L.H., Daily Totals		$780.35		$1266.31	$48.77	$79.14
Crew G-7	**Hr.**	**Daily**	**Hr.**	**Daily**	**Bare Costs**	**Incl. O&P**
1 Carpenter	$51.65	$413.20	$78.15	$625.20	$51.65	$78.15
1 Small Compressor, Electric		14.85		16.34		
1 Pneumatic Nailer		22.35		24.59	4.65	5.12
8 L.H., Daily Totals		$450.40		$666.12	$56.30	$83.27
Crew H-1	**Hr.**	**Daily**	**Hr.**	**Daily**	**Bare Costs**	**Incl. O&P**
2 Glaziers	$49.65	$794.40	$74.75	$1196.00	$52.83	$82.05
2 Struc. Steel Workers	56.00	896.00	89.35	1429.60		
32 L.H., Daily Totals		$1690.40		$2625.60	$52.83	$82.05
Crew H-2	**Hr.**	**Daily**	**Hr.**	**Daily**	**Bare Costs**	**Incl. O&P**
2 Glaziers	$49.65	$794.40	$74.75	$1196.00	$46.78	$70.53
1 Building Laborer	41.05	328.40	62.10	496.80		
24 L.H., Daily Totals		$1122.80		$1692.80	$46.78	$70.53
Crew H-3	**Hr.**	**Daily**	**Hr.**	**Daily**	**Bare Costs**	**Incl. O&P**
1 Glazier	$49.65	$397.20	$74.75	$598.00	$44.25	$67.20
1 Helper	38.85	310.80	59.65	477.20		
16 L.H., Daily Totals		$708.00		$1075.20	$44.25	$67.20
Crew H-4	**Hr.**	**Daily**	**Hr.**	**Daily**	**Bare Costs**	**Incl. O&P**
1 Carpenter	$51.65	$413.20	$78.15	$625.20	$48.21	$73.04
1 Carpenter Helper	38.85	310.80	59.65	477.20		
.5 Electrician	60.05	240.20	89.60	358.40		
20 L.H., Daily Totals		$964.20		$1460.80	$48.21	$73.04

Crew No.	Bare Costs Hr.	Daily	Incl. Subs O&P Hr.	Daily	Cost Per Labor-Hour Bare Costs	Incl. O&P
Crew J-1	Hr.	Daily	Hr.	Daily	Bare Costs	Incl. O&P
3 Plasterers	$47.75	$1146.00	$71.40	$1713.60	$45.11	$67.46
2 Plasterer Helpers	41.15	658.40	61.55	984.80		
1 Mixing Machine, 6 C.F.		135.35		148.88	3.38	3.72
40 L.H., Daily Totals		$1939.75		$2847.28	$48.49	$71.18
Crew J-2	Hr.	Daily	Hr.	Daily	Bare Costs	Incl. O&P
3 Plasterers	$47.75	$1146.00	$71.40	$1713.60	$45.97	$68.52
2 Plasterer Helpers	41.15	658.40	61.55	984.80		
1 Lather	50.25	402.00	73.80	590.40		
1 Mixing Machine, 6 C.F.		135.35		148.88	2.82	3.10
48 L.H., Daily Totals		$2341.75		$3437.68	$48.79	$71.62
Crew J-3	Hr.	Daily	Hr.	Daily	Bare Costs	Incl. O&P
1 Terrazzo Worker	$47.85	$382.80	$70.55	$564.40	$44.10	$65.03
1 Terrazzo Helper	40.35	322.80	59.50	476.00		
1 Floor Grinder, 22" Path		138.30		152.13		
1 Terrazzo Mixer		178.85		196.74	19.82	21.80
16 L.H., Daily Totals		$1022.75		$1389.27	$63.92	$86.83
Crew J-4	Hr.	Daily	Hr.	Daily	Bare Costs	Incl. O&P
2 Cement Finishers	$48.90	$782.40	$72.20	$1155.20	$46.28	$68.83
1 Laborer	41.05	328.40	62.10	496.80		
1 Floor Grinder, 22" Path		138.30		152.13		
1 Floor Edger, 7" Path		43.45		47.80		
1 Vacuum Pick-Up System		72.15		79.36	10.58	11.64
24 L.H., Daily Totals		$1364.70		$1931.29	$56.86	$80.47
Crew J-4A	Hr.	Daily	Hr.	Daily	Bare Costs	Incl. O&P
2 Cement Finishers	$48.90	$782.40	$72.20	$1155.20	$44.98	$67.15
2 Laborers	41.05	656.80	62.10	993.60		
1 Floor Grinder, 22" Path		138.30		152.13		
1 Floor Edger, 7" Path		43.45		47.80		
1 Vacuum Pick-Up System		72.15		79.36		
1 Floor Auto Scrubber		265.15		291.67	16.22	17.84
32 L.H., Daily Totals		$1958.25		$2719.76	$61.20	$84.99
Crew J-4B	Hr.	Daily	Hr.	Daily	Bare Costs	Incl. O&P
1 Laborer	$41.05	$328.40	$62.10	$496.80	$41.05	$62.10
1 Floor Auto Scrubber		265.15		291.67	33.14	36.46
8 L.H., Daily Totals		$593.55		$788.47	$74.19	$98.56
Crew J-6	Hr.	Daily	Hr.	Daily	Bare Costs	Incl. O&P
2 Painters	$43.25	$692.00	$64.75	$1036.00	$44.80	$67.29
1 Building Laborer	41.05	328.40	62.10	496.80		
1 Equip. Oper. (light)	51.65	413.20	77.55	620.40		
1 Air Compressor, 250 cfm		167.95		184.75		
1 Sandblaster, Portable, 3 C.F.		20.70		22.77		
1 Set Sand Blasting Accessories		14.90		16.39	6.36	7.00
32 L.H., Daily Totals		$1637.15		$2377.11	$51.16	$74.28
Crew J-7	Hr.	Daily	Hr.	Daily	Bare Costs	Incl. O&P
2 Painters	$43.25	$692.00	$64.75	$1036.00	$43.25	$64.75
1 Floor Belt Sander		17.75		19.52		
1 Floor Sanding Edger		13.65		15.02	1.96	2.16
16 L.H., Daily Totals		$723.40		$1070.54	$45.21	$66.91

Crew No.	Bare Costs Hr.	Daily	Incl. Subs O&P Hr.	Daily	Cost Per Labor-Hour Bare Costs	Incl. O&P
Crew K-1	Hr.	Daily	Hr.	Daily	Bare Costs	Incl. O&P
1 Carpenter	$51.65	$413.20	$78.15	$625.20	$48.58	$73.33
1 Truck Driver (light)	45.50	364.00	68.50	548.00		
1 Flatbed Truck, Gas, 3 Ton		245.95		270.55	15.37	16.91
16 L.H., Daily Totals		$1023.15		$1443.74	$63.95	$90.23
Crew K-2	Hr.	Daily	Hr.	Daily	Bare Costs	Incl. O&P
1 Struc. Steel Foreman (outside)	$58.00	$464.00	$92.50	$740.00	$53.17	$83.45
1 Struc. Steel Worker	56.00	448.00	89.35	714.80		
1 Truck Driver (light)	45.50	364.00	68.50	548.00		
1 Flatbed Truck, Gas, 3 Ton		245.95		270.55	10.25	11.27
24 L.H., Daily Totals		$1521.95		$2273.34	$63.41	$94.72
Crew L-1	Hr.	Daily	Hr.	Daily	Bare Costs	Incl. O&P
1 Electrician	$60.05	$480.40	$89.60	$716.80	$61.60	$92.17
1 Plumber	63.15	505.20	94.75	758.00		
16 L.H., Daily Totals		$985.60		$1474.80	$61.60	$92.17
Crew L-2	Hr.	Daily	Hr.	Daily	Bare Costs	Incl. O&P
1 Carpenter	$51.65	$413.20	$78.15	$625.20	$45.25	$68.90
1 Carpenter Helper	38.85	310.80	59.65	477.20		
16 L.H., Daily Totals		$724.00		$1102.40	$45.25	$68.90
Crew L-3	Hr.	Daily	Hr.	Daily	Bare Costs	Incl. O&P
1 Carpenter	$51.65	$413.20	$78.15	$625.20	$56.08	$84.61
.5 Electrician	60.05	240.20	89.60	358.40		
.5 Sheet Metal Worker	60.95	243.80	92.55	370.20		
16 L.H., Daily Totals		$897.20		$1353.80	$56.08	$84.61
Crew L-3A	Hr.	Daily	Hr.	Daily	Bare Costs	Incl. O&P
1 Carpenter Foreman (outside)	$53.65	$429.20	$81.15	$649.20	$56.08	$84.95
.5 Sheet Metal Worker	60.95	243.80	92.55	370.20		
12 L.H., Daily Totals		$673.00		$1019.40	$56.08	$84.95
Crew L-4	Hr.	Daily	Hr.	Daily	Bare Costs	Incl. O&P
2 Skilled Workers	$53.40	$854.40	$81.25	$1300.00	$48.55	$74.05
1 Helper	38.85	310.80	59.65	477.20		
24 L.H., Daily Totals		$1165.20		$1777.20	$48.55	$74.05
Crew L-5	Hr.	Daily	Hr.	Daily	Bare Costs	Incl. O&P
1 Struc. Steel Foreman (outside)	$58.00	$464.00	$92.50	$740.00	$56.49	$89.36
5 Struc. Steel Workers	56.00	2240.00	89.35	3574.00		
1 Equip. Oper. (crane)	57.45	459.60	86.25	690.00		
1 Hyd. Crane, 25 Ton		581.70		639.87	10.39	11.43
56 L.H., Daily Totals		$3745.30		$5643.87	$66.88	$100.78
Crew L-5A	Hr.	Daily	Hr.	Daily	Bare Costs	Incl. O&P
1 Struc. Steel Foreman (outside)	$58.00	$464.00	$92.50	$740.00	$56.86	$89.36
2 Structural Steel Workers	56.00	896.00	89.35	1429.60		
1 Equip. Oper. (crane)	57.45	459.60	86.25	690.00		
1 S.P. Crane, 4x4, 25 Ton		664.40		730.84	20.76	22.84
32 L.H., Daily Totals		$2484.00		$3590.44	$77.63	$112.20

507

Crew No.	Bare Costs		Incl. Subs O&P		Cost Per Labor-Hour	
	Hr.	Daily	Hr.	Daily	Bare Costs	Incl. O&P
Crew L-5B					$58.29	$89.16
1 Struc. Steel Foreman (outside)	$58.00	$464.00	$92.50	$740.00		
2 Structural Steel Workers	56.00	896.00	89.35	1429.60		
2 Electricians	60.05	960.80	89.60	1433.60		
2 Steamfitters/Pipefitters	63.95	1023.20	95.95	1535.20		
1 Equip. Oper. (crane)	57.45	459.60	86.25	690.00		
1 Equip. Oper. (oiler)	49.20	393.60	73.85	590.80		
1 Hyd. Crane, 80 Ton		1487.00		1635.70	20.65	22.72
72 L.H., Daily Totals		$5684.20		$8054.90	$78.95	$111.87

Crew L-6	Hr.	Daily	Hr.	Daily	Bare Costs	Incl. O&P
1 Plumber	$63.15	$505.20	$94.75	$758.00	$62.12	$93.03
.5 Electrician	60.05	240.20	89.60	358.40		
12 L.H., Daily Totals		$745.40		$1116.40	$62.12	$93.03

Crew L-7	Hr.	Daily	Hr.	Daily	Bare Costs	Incl. O&P
2 Carpenters	$51.65	$826.40	$78.15	$1250.40	$49.82	$75.20
1 Building Laborer	41.05	328.40	62.10	496.80		
.5 Electrician	60.05	240.20	89.60	358.40		
28 L.H., Daily Totals		$1395.00		$2105.60	$49.82	$75.20

Crew L-8	Hr.	Daily	Hr.	Daily	Bare Costs	Incl. O&P
2 Carpenters	$51.65	$826.40	$78.15	$1250.40	$53.95	$81.47
.5 Plumber	63.15	252.60	94.75	379.00		
20 L.H., Daily Totals		$1079.00		$1629.40	$53.95	$81.47

Crew L-9	Hr.	Daily	Hr.	Daily	Bare Costs	Incl. O&P
1 Labor Foreman (inside)	$41.55	$332.40	$62.85	$502.80	$46.59	$71.38
2 Building Laborers	41.05	656.80	62.10	993.60		
1 Struc. Steel Worker	56.00	448.00	89.35	714.80		
.5 Electrician	60.05	240.20	89.60	358.40		
36 L.H., Daily Totals		$1677.40		$2569.60	$46.59	$71.38

Crew L-10	Hr.	Daily	Hr.	Daily	Bare Costs	Incl. O&P
1 Struc. Steel Foreman (outside)	$58.00	$464.00	$92.50	$740.00	$57.15	$89.37
1 Structural Steel Worker	56.00	448.00	89.35	714.80		
1 Equip. Oper. (crane)	57.45	459.60	86.25	690.00		
1 Hyd. Crane, 12 Ton		471.75		518.92	19.66	21.62
24 L.H., Daily Totals		$1843.35		$2663.72	$76.81	$110.99

Crew L-11	Hr.	Daily	Hr.	Daily	Bare Costs	Incl. O&P
2 Wreckers	$41.05	$656.80	$63.55	$1016.80	$47.80	$72.72
1 Equip. Oper. (crane)	57.45	459.60	86.25	690.00		
1 Equip. Oper. (light)	51.65	413.20	77.55	620.40		
1 Hyd. Excavator, 2.5 C.Y.		1441.00		1585.10		
1 Loader, Skid Steer, 78 H.P.		392.50		431.75	57.30	63.03
32 L.H., Daily Totals		$3363.10		$4344.05	$105.10	$135.75

Crew M-1	Hr.	Daily	Hr.	Daily	Bare Costs	Incl. O&P
3 Elevator Constructors	$84.00	$2016.00	$124.90	$2997.60	$79.80	$118.65
1 Elevator Apprentice	67.20	537.60	99.90	799.20		
5 Hand Tools		49.50		54.45	1.55	1.70
32 L.H., Daily Totals		$2603.10		$3851.25	$81.35	$120.35

Crew M-3	Hr.	Daily	Hr.	Daily	Bare Costs	Incl. O&P
1 Electrician Foreman (outside)	$62.05	$496.40	$92.60	$740.80	$63.08	$94.16
1 Common Laborer	41.05	328.40	62.10	496.80		
.25 Equipment Operator (med.)	55.10	110.20	82.70	165.40		
1 Elevator Constructor	84.00	672.00	124.90	999.20		
1 Elevator Apprentice	67.20	537.60	99.90	799.20		
.25 S.P. Crane, 4x4, 20 Ton		146.94		161.63	4.32	4.75
34 L.H., Daily Totals		$2291.54		$3363.03	$67.40	$98.91

Crew M-4	Hr.	Daily	Hr.	Daily	Bare Costs	Incl. O&P
1 Electrician Foreman (outside)	$62.05	$496.40	$92.60	$740.80	$62.44	$93.23
1 Common Laborer	41.05	328.40	62.10	496.80		
.25 Equipment Operator, Crane	57.45	114.90	86.25	172.50		
.25 Equip. Oper. (oiler)	49.20	98.40	73.85	147.70		
1 Elevator Constructor	84.00	672.00	124.90	999.20		
1 Elevator Apprentice	67.20	537.60	99.90	799.20		
.25 S.P. Crane, 4x4, 40 Ton		187.56		206.32	5.21	5.73
36 L.H., Daily Totals		$2435.26		$3562.52	$67.65	$98.96

Crew Q-1	Hr.	Daily	Hr.	Daily	Bare Costs	Incl. O&P
1 Plumber	$63.15	$505.20	$94.75	$758.00	$56.83	$85.28
1 Plumber Apprentice	50.50	404.00	75.80	606.40		
16 L.H., Daily Totals		$909.20		$1364.40	$56.83	$85.28

Crew Q-1A	Hr.	Daily	Hr.	Daily	Bare Costs	Incl. O&P
.25 Plumber Foreman (outside)	$65.15	$130.30	$97.75	$195.50	$63.55	$95.35
1 Plumber	63.15	505.20	94.75	758.00		
10 L.H., Daily Totals		$635.50		$953.50	$63.55	$95.35

Crew Q-1C	Hr.	Daily	Hr.	Daily	Bare Costs	Incl. O&P
1 Plumber	$63.15	$505.20	$94.75	$758.00	$56.25	$84.42
1 Plumber Apprentice	50.50	404.00	75.80	606.40		
1 Equip. Oper. (medium)	55.10	440.80	82.70	661.60		
1 Trencher, Chain Type, 8' D		1837.00		2020.70	76.54	84.20
24 L.H., Daily Totals		$3187.00		$4046.70	$132.79	$168.61

Crew Q-2	Hr.	Daily	Hr.	Daily	Bare Costs	Incl. O&P
2 Plumbers	$63.15	$1010.40	$94.75	$1516.00	$58.93	$88.43
1 Plumber Apprentice	50.50	404.00	75.80	606.40		
24 L.H., Daily Totals		$1414.40		$2122.40	$58.93	$88.43

Crew Q-3	Hr.	Daily	Hr.	Daily	Bare Costs	Incl. O&P
1 Plumber Foreman (inside)	$63.65	$509.20	$95.50	$764.00	$60.11	$90.20
2 Plumbers	63.15	1010.40	94.75	1516.00		
1 Plumber Apprentice	50.50	404.00	75.80	606.40		
32 L.H., Daily Totals		$1923.60		$2886.40	$60.11	$90.20

Crew Q-4	Hr.	Daily	Hr.	Daily	Bare Costs	Incl. O&P
1 Plumber Foreman (inside)	$63.65	$509.20	$95.50	$764.00	$60.11	$90.20
1 Plumber	63.15	505.20	94.75	758.00		
1 Welder (plumber)	63.15	505.20	94.75	758.00		
1 Plumber Apprentice	50.50	404.00	75.80	606.40		
1 Welder, Electric, 300 amp		56.10		61.71	1.75	1.93
32 L.H., Daily Totals		$1979.70		$2948.11	$61.87	$92.13

Crew Q-5	Hr.	Daily	Hr.	Daily	Bare Costs	Incl. O&P
1 Steamfitter	$63.95	$511.60	$95.95	$767.60	$57.55	$86.35
1 Steamfitter Apprentice	51.15	409.20	76.75	614.00		
16 L.H., Daily Totals		$920.80		$1381.60	$57.55	$86.35

Crew No.	Bare Costs		Incl. Subs O&P		Cost Per Labor-Hour	
	Hr.	Daily	Hr.	Daily	Bare Costs	Incl. O&P
Crew Q-6					$59.68	$89.55
2 Steamfitters	$63.95	$1023.20	$95.95	$1535.20		
1 Steamfitter Apprentice	51.15	409.20	76.75	614.00		
24 L.H., Daily Totals		$1432.40		$2149.20	$59.68	$89.55
Crew Q-7	Hr.	Daily	Hr.	Daily	Bare Costs	Incl. O&P
1 Steamfitter Foreman (inside)	$64.45	$515.60	$96.70	$773.60	$60.88	$91.34
2 Steamfitters	63.95	1023.20	95.95	1535.20		
1 Steamfitter Apprentice	51.15	409.20	76.75	614.00		
32 L.H., Daily Totals		$1948.00		$2922.80	$60.88	$91.34
Crew Q-8	Hr.	Daily	Hr.	Daily	Bare Costs	Incl. O&P
1 Steamfitter Foreman (inside)	$64.45	$515.60	$96.70	$773.60	$60.88	$91.34
1 Steamfitter	63.95	511.60	95.95	767.60		
1 Welder (steamfitter)	63.95	511.60	95.95	767.60		
1 Steamfitter Apprentice	51.15	409.20	76.75	614.00		
1 Welder, Electric, 300 amp		56.10		61.71	1.75	1.93
32 L.H., Daily Totals		$2004.10		$2984.51	$62.63	$93.27
Crew Q-9	Hr.	Daily	Hr.	Daily	Bare Costs	Incl. O&P
1 Sheet Metal Worker	$60.95	$487.60	$92.55	$740.40	$54.85	$83.30
1 Sheet Metal Apprentice	48.75	390.00	74.05	592.40		
16 L.H., Daily Totals		$877.60		$1332.80	$54.85	$83.30
Crew Q-10	Hr.	Daily	Hr.	Daily	Bare Costs	Incl. O&P
2 Sheet Metal Workers	$60.95	$975.20	$92.55	$1480.80	$56.88	$86.38
1 Sheet Metal Apprentice	48.75	390.00	74.05	592.40		
24 L.H., Daily Totals		$1365.20		$2073.20	$56.88	$86.38
Crew Q-11	Hr.	Daily	Hr.	Daily	Bare Costs	Incl. O&P
1 Sheet Metal Foreman (inside)	$61.45	$491.60	$93.30	$746.40	$58.02	$88.11
2 Sheet Metal Workers	60.95	975.20	92.55	1480.80		
1 Sheet Metal Apprentice	48.75	390.00	74.05	592.40		
32 L.H., Daily Totals		$1856.80		$2819.60	$58.02	$88.11
Crew Q-12	Hr.	Daily	Hr.	Daily	Bare Costs	Incl. O&P
1 Sprinkler Installer	$61.50	$492.00	$92.50	$740.00	$55.35	$83.25
1 Sprinkler Apprentice	49.20	393.60	74.00	592.00		
16 L.H., Daily Totals		$885.60		$1332.00	$55.35	$83.25
Crew Q-13	Hr.	Daily	Hr.	Daily	Bare Costs	Incl. O&P
1 Sprinkler Foreman (inside)	$62.00	$496.00	$93.25	$746.00	$58.55	$88.06
2 Sprinkler Installers	61.50	984.00	92.50	1480.00		
1 Sprinkler Apprentice	49.20	393.60	74.00	592.00		
32 L.H., Daily Totals		$1873.60		$2818.00	$58.55	$88.06
Crew Q-14	Hr.	Daily	Hr.	Daily	Bare Costs	Incl. O&P
1 Asbestos Worker	$57.35	$458.80	$88.55	$708.40	$51.63	$79.70
1 Asbestos Apprentice	45.90	367.20	70.85	566.80		
16 L.H., Daily Totals		$826.00		$1275.20	$51.63	$79.70
Crew Q-15	Hr.	Daily	Hr.	Daily	Bare Costs	Incl. O&P
1 Plumber	$63.15	$505.20	$94.75	$758.00	$56.83	$85.28
1 Plumber Apprentice	50.50	404.00	75.80	606.40		
1 Welder, Electric, 300 amp		56.10		61.71	3.51	3.86
16 L.H., Daily Totals		$965.30		$1426.11	$60.33	$89.13

Crew No.	Bare Costs		Incl. Subs O&P		Cost Per Labor-Hour	
	Hr.	Daily	Hr.	Daily	Bare Costs	Incl. O&P
Crew Q-16					$58.93	$88.43
2 Plumbers	$63.15	$1010.40	$94.75	$1516.00		
1 Plumber Apprentice	50.50	404.00	75.80	606.40		
1 Welder, Electric, 300 amp		56.10		61.71	2.34	2.57
24 L.H., Daily Totals		$1470.50		$2184.11	$61.27	$91.00
Crew Q-17	Hr.	Daily	Hr.	Daily	Bare Costs	Incl. O&P
1 Steamfitter	$63.95	$511.60	$95.95	$767.60	$57.55	$86.35
1 Steamfitter Apprentice	51.15	409.20	76.75	614.00		
1 Welder, Electric, 300 amp		56.10		61.71	3.51	3.86
16 L.H., Daily Totals		$976.90		$1443.31	$61.06	$90.21
Crew Q-17A	Hr.	Daily	Hr.	Daily	Bare Costs	Incl. O&P
1 Steamfitter	$63.95	$511.60	$95.95	$767.60	$57.52	$86.32
1 Steamfitter Apprentice	51.15	409.20	76.75	614.00		
1 Equip. Oper. (crane)	57.45	459.60	86.25	690.00		
1 Hyd. Crane, 12 Ton		471.75		518.92		
1 Welder, Electric, 300 amp		56.10		61.71	21.99	24.19
24 L.H., Daily Totals		$1908.25		$2652.24	$79.51	$110.51
Crew Q-18	Hr.	Daily	Hr.	Daily	Bare Costs	Incl. O&P
2 Steamfitters	$63.95	$1023.20	$95.95	$1535.20	$59.68	$89.55
1 Steamfitter Apprentice	51.15	409.20	76.75	614.00		
1 Welder, Electric, 300 amp		56.10		61.71	2.34	2.57
24 L.H., Daily Totals		$1488.50		$2210.91	$62.02	$92.12
Crew Q-19	Hr.	Daily	Hr.	Daily	Bare Costs	Incl. O&P
1 Steamfitter	$63.95	$511.60	$95.95	$767.60	$58.38	$87.43
1 Steamfitter Apprentice	51.15	409.20	76.75	614.00		
1 Electrician	60.05	480.40	89.60	716.80		
24 L.H., Daily Totals		$1401.20		$2098.40	$58.38	$87.43
Crew Q-20	Hr.	Daily	Hr.	Daily	Bare Costs	Incl. O&P
1 Sheet Metal Worker	$60.95	$487.60	$92.55	$740.40	$55.89	$84.56
1 Sheet Metal Apprentice	48.75	390.00	74.05	592.40		
.5 Electrician	60.05	240.20	89.60	358.40		
20 L.H., Daily Totals		$1117.80		$1691.20	$55.89	$84.56
Crew Q-21	Hr.	Daily	Hr.	Daily	Bare Costs	Incl. O&P
2 Steamfitters	$63.95	$1023.20	$95.95	$1535.20	$59.77	$89.56
1 Steamfitter Apprentice	51.15	409.20	76.75	614.00		
1 Electrician	60.05	480.40	89.60	716.80		
32 L.H., Daily Totals		$1912.80		$2866.00	$59.77	$89.56
Crew Q-22	Hr.	Daily	Hr.	Daily	Bare Costs	Incl. O&P
1 Plumber	$63.15	$505.20	$94.75	$758.00	$56.83	$85.28
1 Plumber Apprentice	50.50	404.00	75.80	606.40		
1 Hyd. Crane, 12 Ton		471.75		518.92	29.48	32.43
16 L.H., Daily Totals		$1380.95		$1883.33	$86.31	$117.71
Crew Q-22A	Hr.	Daily	Hr.	Daily	Bare Costs	Incl. O&P
1 Plumber	$63.15	$505.20	$94.75	$758.00	$53.04	$79.72
1 Plumber Apprentice	50.50	404.00	75.80	606.40		
1 Laborer	41.05	328.40	62.10	496.80		
1 Equip. Oper. (crane)	57.45	459.60	86.25	690.00		
1 Hyd. Crane, 12 Ton		471.75		518.92	14.74	16.22
32 L.H., Daily Totals		$2168.95		$3070.13	$67.78	$95.94

509

Crew No.	Bare Costs		Incl. Subs O&P		Cost Per Labor-Hour	

Crew Q-23	Hr.	Daily	Hr.	Daily	Bare Costs	Incl. O&P
1 Plumber Foreman (outside)	$65.15	$521.20	$97.75	$782.00	$61.13	$91.73
1 Plumber	63.15	505.20	94.75	758.00		
1 Equip. Oper. (medium)	55.10	440.80	82.70	661.60		
1 Lattice Boom Crane, 20 Ton		869.00		955.90	36.21	39.83
24 L.H., Daily Totals		$2336.20		$3157.50	$97.34	$131.56

Crew R-1	Hr.	Daily	Hr.	Daily	Bare Costs	Incl. O&P
1 Electrician Foreman	$60.55	$484.40	$90.35	$722.80	$56.13	$83.76
3 Electricians	60.05	1441.20	89.60	2150.40		
2 Electrician Apprentices	48.05	768.80	71.70	1147.20		
48 L.H., Daily Totals		$2694.40		$4020.40	$56.13	$83.76

Crew R-1A	Hr.	Daily	Hr.	Daily	Bare Costs	Incl. O&P
1 Electrician	$60.05	$480.40	$89.60	$716.80	$54.05	$80.65
1 Electrician Apprentice	48.05	384.40	71.70	573.60		
16 L.H., Daily Totals		$864.80		$1290.40	$54.05	$80.65

Crew R-1B	Hr.	Daily	Hr.	Daily	Bare Costs	Incl. O&P
1 Electrician	$60.05	$480.40	$89.60	$716.80	$52.05	$77.67
2 Electrician Apprentices	48.05	768.80	71.70	1147.20		
24 L.H., Daily Totals		$1249.20		$1864.00	$52.05	$77.67

Crew R-1C	Hr.	Daily	Hr.	Daily	Bare Costs	Incl. O&P
2 Electricians	$60.05	$960.80	$89.60	$1433.60	$54.05	$80.65
2 Electrician Apprentices	48.05	768.80	71.70	1147.20		
1 Portable Cable Puller, 8000 lb.		152.70		167.97	4.77	5.25
32 L.H., Daily Totals		$1882.30		$2748.77	$58.82	$85.90

Crew R-2	Hr.	Daily	Hr.	Daily	Bare Costs	Incl. O&P
1 Electrician Foreman	$60.55	$484.40	$90.35	$722.80	$56.32	$84.11
3 Electricians	60.05	1441.20	89.60	2150.40		
2 Electrician Apprentices	48.05	768.80	71.70	1147.20		
1 Equip. Oper. (crane)	57.45	459.60	86.25	690.00		
1 S.P. Crane, 4x4, 5 Ton		259.45		285.39	4.63	5.10
56 L.H., Daily Totals		$3413.45		$4995.80	$60.95	$89.21

Crew R-3	Hr.	Daily	Hr.	Daily	Bare Costs	Incl. O&P
1 Electrician Foreman	$60.55	$484.40	$90.35	$722.80	$59.73	$89.23
1 Electrician	60.05	480.40	89.60	716.80		
.5 Equip. Oper. (crane)	57.45	229.80	86.25	345.00		
.5 S.P. Crane, 4x4, 5 Ton		129.72		142.70	6.49	7.13
20 L.H., Daily Totals		$1324.33		$1927.30	$66.22	$96.36

Crew R-4	Hr.	Daily	Hr.	Daily	Bare Costs	Incl. O&P
1 Struc. Steel Foreman (outside)	$58.00	$464.00	$92.50	$740.00	$57.21	$90.03
3 Struc. Steel Workers	56.00	1344.00	89.35	2144.40		
1 Electrician	60.05	480.40	89.60	716.80		
1 Welder, Gas Engine, 300 amp		95.65		105.22	2.39	2.63
40 L.H., Daily Totals		$2384.05		$3706.42	$59.60	$92.66

Crew R-5	Hr.	Daily	Hr.	Daily	Bare Costs	Incl. O&P
1 Electrician Foreman	$60.55	$484.40	$90.35	$722.80	$52.39	$78.78
4 Electrician Linemen	60.05	1921.60	89.60	2867.20		
2 Electrician Operators	60.05	960.80	89.60	1433.60		
4 Electrician Groundmen	38.85	1243.20	59.65	1908.80		
1 Crew Truck		154.35		169.79		
1 Flatbed Truck, 20,000 GVW		199.45		219.40		
1 Pickup Truck, 3/4 Ton		109.90		120.89		
.2 Hyd. Crane, 55 Ton		196.30		215.93		
.2 Hyd. Crane, 12 Ton		94.35		103.79		
.2 Earth Auger, Truck-Mtd.		77.17		84.89		
1 Tractor w/Winch		368.50		405.35	13.64	15.00
88 L.H., Daily Totals		$5810.02		$8252.42	$66.02	$93.78

Crew R-6	Hr.	Daily	Hr.	Daily	Bare Costs	Incl. O&P
1 Electrician Foreman	$60.55	$484.40	$90.35	$722.80	$52.39	$78.78
4 Electrician Linemen	60.05	1921.60	89.60	2867.20		
2 Electrician Operators	60.05	960.80	89.60	1433.60		
4 Electrician Groundmen	38.85	1243.20	59.65	1908.80		
1 Crew Truck		154.35		169.79		
1 Flatbed Truck, 20,000 GVW		199.45		219.40		
1 Pickup Truck, 3/4 Ton		109.90		120.89		
.2 Hyd. Crane, 55 Ton		196.30		215.93		
.2 Hyd. Crane, 12 Ton		94.35		103.79		
.2 Earth Auger, Truck-Mtd.		77.17		84.89		
1 Tractor w/Winch		368.50		405.35		
3 Cable Trailers		648.60		713.46		
.5 Tensioning Rig		214.32		235.76		
.5 Cable Pulling Rig		1233.50		1356.85	37.46	41.21
88 L.H., Daily Totals		$7906.44		$10558.49	$89.85	$119.98

Crew R-7	Hr.	Daily	Hr.	Daily	Bare Costs	Incl. O&P
1 Electrician Foreman	$60.55	$484.40	$90.35	$722.80	$42.47	$64.77
5 Electrician Groundmen	38.85	1554.00	59.65	2386.00		
1 Crew Truck		154.35		169.79	3.22	3.54
48 L.H., Daily Totals		$2192.75		$3278.59	$45.68	$68.30

Crew R-8	Hr.	Daily	Hr.	Daily	Bare Costs	Incl. O&P
1 Electrician Foreman	$60.55	$484.40	$90.35	$722.80	$53.07	$79.74
3 Electrician Linemen	60.05	1441.20	89.60	2150.40		
2 Electrician Groundmen	38.85	621.60	59.65	954.40		
1 Pickup Truck, 3/4 Ton		109.90		120.89		
1 Crew Truck		154.35		169.79	5.51	6.06
48 L.H., Daily Totals		$2811.45		$4118.27	$58.57	$85.80

Crew R-9	Hr.	Daily	Hr.	Daily	Bare Costs	Incl. O&P
1 Electrician Foreman	$60.55	$484.40	$90.35	$722.80	$49.51	$74.72
1 Electrician Lineman	60.05	480.40	89.60	716.80		
2 Electrician Operators	60.05	960.80	89.60	1433.60		
4 Electrician Groundmen	38.85	1243.20	59.65	1908.80		
1 Pickup Truck, 3/4 Ton		109.90		120.89		
1 Crew Truck		154.35		169.79	4.13	4.54
64 L.H., Daily Totals		$3433.05		$5072.68	$53.64	$79.26

Crew R-10	Hr.	Daily	Hr.	Daily	Bare Costs	Incl. O&P
1 Electrician Foreman	$60.55	$484.40	$90.35	$722.80	$56.60	$84.73
4 Electrician Linemen	60.05	1921.60	89.60	2867.20		
1 Electrician Groundman	38.85	310.80	59.65	477.20		
1 Crew Truck		154.35		169.79		
3 Tram Cars		438.60		482.46	12.35	13.59
48 L.H., Daily Totals		$3309.75		$4719.44	$68.95	$98.32

Crew No.	Bare Costs		Incl. Subs O&P		Cost Per Labor-Hour	

Crew R-11	Hr.	Daily	Hr.	Daily	Bare Costs	Incl. O&P
1 Electrician Foreman	$60.55	$484.40	$90.35	$722.80	$57.04	$85.30
4 Electricians	60.05	1921.60	89.60	2867.20		
1 Equip. Oper. (crane)	57.45	459.60	86.25	690.00		
1 Common Laborer	41.05	328.40	62.10	496.80		
1 Crew Truck		154.35		169.79		
1 Hyd. Crane, 12 Ton		471.75		518.92	11.18	12.30
56 L.H., Daily Totals		$3820.10		$5465.51	$68.22	$97.60

Crew R-12	Hr.	Daily	Hr.	Daily	Bare Costs	Incl. O&P
1 Carpenter Foreman (inside)	$52.15	$417.20	$78.90	$631.20	$48.55	$73.81
4 Carpenters	51.65	1652.80	78.15	2500.80		
4 Common Laborers	41.05	1313.60	62.10	1987.20		
1 Equip. Oper. (medium)	55.10	440.80	82.70	661.60		
1 Steel Worker	56.00	448.00	89.35	714.80		
1 Dozer, 200 H.P.		1290.00		1419.00		
1 Pickup Truck, 3/4 Ton		109.90		120.89	15.91	17.50
88 L.H., Daily Totals		$5672.30		$8035.49	$64.46	$91.31

Crew R-13	Hr.	Daily	Hr.	Daily	Bare Costs	Incl. O&P
1 Electrician Foreman	$60.55	$484.40	$90.35	$722.80	$57.95	$86.58
3 Electricians	60.05	1441.20	89.60	2150.40		
.25 Equip. Oper. (crane)	57.45	114.90	86.25	172.50		
1 Equipment Oiler	49.20	393.60	73.85	590.80		
.25 Hydraulic Crane, 33 Ton		236.65		260.32	5.63	6.20
42 L.H., Daily Totals		$2670.75		$3896.82	$63.59	$92.78

Crew R-15	Hr.	Daily	Hr.	Daily	Bare Costs	Incl. O&P
1 Electrician Foreman	$60.55	$484.40	$90.35	$722.80	$58.73	$87.72
4 Electricians	60.05	1921.60	89.60	2867.20		
1 Equipment Oper. (light)	51.65	413.20	77.55	620.40		
1 Telescoping Boom Lift, to 40'		283.15		311.46	5.90	6.49
48 L.H., Daily Totals		$3102.35		$4521.86	$64.63	$94.21

Crew R-15A	Hr.	Daily	Hr.	Daily	Bare Costs	Incl. O&P
1 Electrician Foreman	$60.55	$484.40	$90.35	$722.80	$52.40	$78.55
2 Electricians	60.05	960.80	89.60	1433.60		
2 Common Laborers	41.05	656.80	62.10	993.60		
1 Equip. Oper. (light)	51.65	413.20	77.55	620.40		
1 Telescoping Boom Lift, to 40'		283.15		311.46	5.90	6.49
48 L.H., Daily Totals		$2798.35		$4081.86	$58.30	$85.04

Crew R-18	Hr.	Daily	Hr.	Daily	Bare Costs	Incl. O&P
.25 Electrician Foreman	$60.55	$121.10	$90.35	$180.70	$52.70	$78.64
1 Electrician	60.05	480.40	89.60	716.80		
2 Electrician Apprentices	48.05	768.80	71.70	1147.20		
26 L.H., Daily Totals		$1370.30		$2044.70	$52.70	$78.64

Crew R-19	Hr.	Daily	Hr.	Daily	Bare Costs	Incl. O&P
.5 Electrician Foreman	$60.55	$242.20	$90.35	$361.40	$60.15	$89.75
2 Electricians	60.05	960.80	89.60	1433.60		
20 L.H., Daily Totals		$1203.00		$1795.00	$60.15	$89.75

Crew R-21	Hr.	Daily	Hr.	Daily	Bare Costs	Incl. O&P
1 Electrician Foreman	$60.55	$484.40	$90.35	$722.80	$60.05	$89.61
3 Electricians	60.05	1441.20	89.60	2150.40		
.1 Equip. Oper. (medium)	55.10	44.08	82.70	66.16		
.1 S.P. Crane, 4x4, 25 Ton		66.44		73.08	2.03	2.23
32.8 L.H., Daily Totals		$2036.12		$3012.44	$62.08	$91.84

Crew R-22	Hr.	Daily	Hr.	Daily	Bare Costs	Incl. O&P
.66 Electrician Foreman	$60.55	$319.70	$90.35	$477.05	$54.97	$82.02
2 Electricians	60.05	960.80	89.60	1433.60		
2 Electrician Apprentices	48.05	768.80	71.70	1147.20		
37.28 L.H., Daily Totals		$2049.30		$3057.85	$54.97	$82.02

Crew R-30	Hr.	Daily	Hr.	Daily	Bare Costs	Incl. O&P
.25 Electrician Foreman (outside)	$62.05	$124.10	$92.60	$185.20	$48.51	$72.91
1 Electrician	60.05	480.40	89.60	716.80		
2 Laborers (Semi-Skilled)	41.05	656.80	62.10	993.60		
26 L.H., Daily Totals		$1261.30		$1895.60	$48.51	$72.91

Crew R-31	Hr.	Daily	Hr.	Daily	Bare Costs	Incl. O&P
1 Electrician	$60.05	$480.40	$89.60	$716.80	$60.05	$89.60
1 Core Drill, Electric, 2.5 H.P.		46.55		51.20	5.82	6.40
8 L.H., Daily Totals		$526.95		$768.01	$65.87	$96.00

Crew W-41E	Hr.	Daily	Hr.	Daily	Bare Costs	Incl. O&P
.5 Plumber Foreman (outside)	$65.15	$260.60	$97.75	$391.00	$54.71	$82.29
1 Plumber	63.15	505.20	94.75	758.00		
1 Laborer	41.05	328.40	62.10	496.80		
20 L.H., Daily Totals		$1094.20		$1645.80	$54.71	$82.29

For customer support on your Electrical Costs with RSMeans data, call 800.448.8182.

Historical Cost Indexes

The table below lists both the RSMeans® historical cost index based on Jan. 1, 1993 = 100 as well as the computed value of an index based on Jan. 1, 2019 costs. Since the Jan. 1, 2019 figure is estimated, space is left to write in the actual index figures as they become available through the quarterly *RSMeans Construction Cost Indexes.*

To compute the actual index based on Jan. 1, 2019 = 100, divide the historical cost index for a particular year by the actual Jan. 1, 2019 construction cost index. Space has been left to advance the index figures as the year progresses.

Year	Historical Cost Index Jan. 1, 1993 = 100		Current Index Based on Jan. 1, 2019 = 100		Year	Historical Cost Index Jan. 1, 1993 = 100	Current Index Based on Jan. 1, 2019 = 100		Year	Historical Cost Index Jan. 1, 1993 = 100	Current Index Based on Jan. 1, 2019 = 100	
	Est.	Actual	Est.	Actual		Actual	Est.	Actual		Actual	Est.	Actual
Oct 2019*					July 2004	143.7	63.2		July 1986	84.2	37.1	
July 2019*					2003	132.0	58.1		1985	82.6	36.3	
April 2019*					2002	128.7	56.6		1984	82.0	36.1	
Jan 2019*	227.3		100.0	100.0	2001	125.1	55.0		1983	80.2	35.3	
July 2018		222.9	98.1		2000	120.9	53.2		1982	76.1	33.5	
2017		213.6	94.0		1999	117.6	51.7		1981	70.0	30.8	
2016		207.3	91.2		1998	115.1	50.6		1980	62.9	27.7	
2015		206.2	90.7		1997	112.8	49.6		1979	57.8	25.4	
2014		204.9	90.1		1996	110.2	48.5		1978	53.5	23.5	
2013		201.2	88.5		1995	107.6	47.3		1977	49.5	21.8	
2012		194.6	85.6		1994	104.4	45.9		1976	46.9	20.6	
2011		191.2	84.1		1993	101.7	44.7		1975	44.8	19.7	
2010		183.5	80.7		1992	99.4	43.7		1974	41.4	18.2	
2009		180.1	79.2		1991	96.8	42.6		1973	37.7	16.6	
2008		180.4	79.4		1990	94.3	41.5		1972	34.8	15.3	
2007		169.4	74.5		1989	92.1	40.5		1971	32.1	14.1	
2006		162.0	71.3		1988	89.9	39.5		1970	28.7	12.6	
2005		151.6	66.7		1987	87.7	38.6		1969	26.9	11.8	

Adjustments to Costs

The "Historical Cost Index" can be used to convert national average building costs at a particular time to the approximate building costs for some other time.

Example:

Estimate and compare construction costs for different years in the same city.

To estimate the national average construction cost of a building in 1970, knowing that it cost $900,000 in 2019:

INDEX in 1970 = 28.7

INDEX in 2019 = 227.3

Note: The city cost indexes for Canada can be used to convert U.S. national averages to local costs in Canadian dollars.

Time Adjustment Using the Historical Cost Indexes:

$$\frac{\text{Index for Year A}}{\text{Index for Year B}} \times \text{Cost in Year B} = \text{Cost in Year A}$$

$$\frac{\text{INDEX 1970}}{\text{INDEX 2019}} \times \text{Cost 2019} = \text{Cost 1970}$$

$$\frac{28.7}{227.3} \times \$900,000 = .126 \times \$900,000 = \$113,400$$

The construction cost of the building in 1970 was $113,400.

Example:

To estimate and compare the cost of a building in Toronto, ON in 2019 with the known cost of $600,000 (US$) in New York, NY in 2019:

INDEX Toronto = 110.1

INDEX New York = 132.1

$$\frac{\text{INDEX Toronto}}{\text{INDEX New York}} \times \text{Cost New York} = \text{Cost Toronto}$$

$$\frac{110.1}{132.1} \times \$600,000 = .834 \times \$600,000 = \$500,076$$

The construction cost of the building in Toronto is $500,076 (CN$).

*Historical Cost Index updates and other resources are provided on the following website:
http://info.thegordiangroup.com/RSMeans.html

How to Use the City Cost Indexes

What you should know before you begin

RSMeans City Cost Indexes (CCI) are an extremely useful tool for when you want to compare costs from city to city and region to region.

This publication contains average construction cost indexes for 731 U.S. and Canadian cities covering over 930 three-digit zip code locations, as listed directly under each city.

Keep in mind that a City Cost Index number is a percentage ratio of a specific city's cost to the national average cost of the same item at a stated time period.

In other words, these index figures represent relative construction factors (or, if you prefer, multipliers) for material and installation costs, as well as the weighted average for Total In Place costs for each CSI MasterFormat division. Installation costs include both labor and equipment rental costs. When estimating equipment rental rates only for a specific location, use 01 54 33 EQUIPMENT RENTAL COSTS in the Reference Section.

The 30 City Average Index is the average of 30 major U.S. cities and serves as a national average.

Index figures for both material and installation are based on the 30 major city average of 100 and represent the cost relationship as of July 1, 2018. The index for each division is computed from representative material and labor quantities for that division. The weighted average for each city is a weighted total of the components listed above it. It does not include relative productivity between trades or cities.

As changes occur in local material prices, labor rates, and equipment rental rates (including fuel costs), the impact of these changes should be accurately measured by the change in the City Cost Index for each particular city (as compared to the 30 city average).

Therefore, if you know (or have estimated) building costs in one city today, you can easily convert those costs to expected building costs in another city.

In addition, by using the Historical Cost Index, you can easily convert national average building costs at a particular time to the approximate building costs for some other time. The City Cost Indexes can then be applied to calculate the costs for a particular city.

Quick calculations

Location Adjustment Using the City Cost Indexes:

$$\frac{\text{Index for City A}}{\text{Index for City B}} \times \text{Cost in City B} = \text{Cost in City A}$$

Time Adjustment for the National Average Using the Historical Cost Index:

$$\frac{\text{Index for Year A}}{\text{Index for Year B}} \times \text{Cost in Year B} = \text{Cost in Year A}$$

Adjustment from the National Average:

$$\frac{\text{Index for City A}}{100} \times \text{National Average Cost} = \text{Cost in City A}$$

Since each of the other RSMeans data sets contains many different items, any *one* item multiplied by the particular city index may give incorrect results. However, the larger the number of items compiled, the closer the results should be to actual costs for that particular city.

The City Cost Indexes for Canadian cities are calculated using Canadian material and equipment prices and labor rates in Canadian dollars. Therefore, indexes for Canadian cities can be used to convert U.S. national average prices to local costs in Canadian dollars.

How to use this section

1. Compare costs from city to city.

In using the RSMeans Indexes, remember that an index number is not a fixed number but a ratio: It's a percentage ratio of a building component's cost at any stated time to the national average cost of that same component at the same time period. Put in the form of an equation:

$$\frac{\text{Specific City Cost}}{\text{National Average Cost}} \times 100 = \text{City Index Number}$$

Therefore, when making cost comparisons between cities, do not subtract one city's index number from the index number of another city and read the result as a percentage difference. Instead, divide one city's index number by that of the other city. The resulting number may then be used as a multiplier to calculate cost differences from city to city.

The formula used to find cost differences between cities for the purpose of comparison is as follows:

$$\frac{\text{City A Index}}{\text{City B Index}} \times \text{City B Cost (Known)} = \text{City A Cost (Unknown)}$$

In addition, you can use RSMeans CCI to calculate and compare costs division by division between cities using the same basic formula. (Just be sure that you're comparing similar divisions.)

2. Compare a specific city's construction costs with the national average.

When you're studying construction location feasibility, it's advisable to compare a prospective project's cost index with an index of the national average cost.

For example, divide the weighted average index of construction costs of a specific city by that of the 30 City Average, which = 100.

$$\frac{\text{City Index}}{100} = \% \text{ of National Average}$$

As a result, you get a ratio that indicates the relative cost of construction in that city in comparison with the national average.

3. Convert U.S. national average to actual costs in Canadian City.

$$\frac{\text{Index for Canadian City}}{100} \times \text{National Average Cost} = \text{Cost in Canadian City in \$ CAN}$$

4. Adjust construction cost data based on a national average.

When you use a source of construction cost data which is based on a national average (such as RSMeans cost data), it is necessary to adjust those costs to a specific location.

$$\frac{\text{City Index}}{100} \times \frac{\text{Cost Based on}}{\text{National Average Costs}} = \frac{\text{City Cost}}{\text{(Unknown)}}$$

5. When applying the City Cost Indexes to demolition projects, use the appropriate division installation index. For example, for removal of existing doors and windows, use the Division 8 (Openings) index.

What you might like to know about how we developed the Indexes

The information presented in the CCI is organized according to the Construction Specifications Institute (CSI) MasterFormat 2016 classification system.

To create a reliable index, RSMeans researched the building type most often constructed in the United States and Canada. Because it was concluded that no one type of building completely represented the building construction industry, nine different types of buildings were combined to create a composite model.

The exact material, labor, and equipment quantities are based on detailed analyses of these nine building types, and then each quantity is weighted in proportion to expected usage. These various material items, labor hours, and equipment rental rates are thus combined to form a composite building representing as closely as possible the actual usage of materials, labor, and equipment in the North American building construction industry.

The following structures were chosen to make up that composite model:

1. Factory, 1 story
2. Office, 2–4 stories
3. Store, Retail
4. Town Hall, 2–3 stories
5. High School, 2–3 stories
6. Hospital, 4–8 stories
7. Garage, Parking
8. Apartment, 1–3 stories
9. Hotel/Motel, 2–3 stories

For the purposes of ensuring the timeliness of the data, the components of the index for the composite model have been streamlined. They currently consist of:

- specific quantities of 66 commonly used construction materials;
- specific labor-hours for 21 building construction trades; and
- specific days of equipment rental for 6 types of construction equipment (normally used to install the 66 material items by the 21 trades.) Fuel costs and routine maintenance costs are included in the equipment cost.

Material and equipment price quotations are gathered quarterly from cities in the United States and Canada. These prices and the latest negotiated labor wage rates for 21 different building trades are used to compile the quarterly update of the City Cost Index.

The 30 major U.S. cities used to calculate the national average are:

Atlanta, GA	Memphis, TN
Baltimore, MD	Milwaukee, WI
Boston, MA	Minneapolis, MN
Buffalo, NY	Nashville, TN
Chicago, IL	New Orleans, LA
Cincinnati, OH	New York, NY
Cleveland, OH	Philadelphia, PA
Columbus, OH	Phoenix, AZ
Dallas, TX	Pittsburgh, PA
Denver, CO	St. Louis, MO
Detroit, MI	San Antonio, TX
Houston, TX	San Diego, CA
Indianapolis, IN	San Francisco, CA
Kansas City, MO	Seattle, WA
Los Angeles, CA	Washington, DC

What the CCI does not indicate

The weighted average for each city is a total of the divisional components weighted to reflect typical usage. It does not include the productivity variations between trades or cities.

In addition, the CCI does not take into consideration factors such as the following:

- managerial efficiency
- competitive conditions
- automation
- restrictive union practices
- unique local requirements
- regional variations due to specific building codes

City Cost Indexes

Table 1

DIVISION		UNITED STATES 30 CITY AVERAGE			ANNISTON 362			BIRMINGHAM 350 - 352			BUTLER 369			DECATUR 356			DOTHAN 363		
		MAT.	INST.	TOTAL	MAT.	INST.	TOTAL	MAT.	INST.	TOTAL	MAT.	INST.	TOTAL	MAT.	INST.	TOTAL	MAT.	INST.	TOTAL
015433	CONTRACTOR EQUIPMENT		100.0	100.0		104.5	104.5		104.7	104.7		102.1	102.1		104.5	104.5		102.1	102.1
0241, 31 - 34	SITE & INFRASTRUCTURE, DEMOLITION	100.0	100.0	100.0	90.9	92.7	92.1	92.1	93.1	92.8	105.4	88.1	93.3	86.0	92.0	90.2	103.0	88.7	93.1
0310	Concrete Forming & Accessories	100.0	100.0	100.0	87.3	65.3	68.6	90.7	66.5	70.1	83.8	65.9	68.6	91.5	61.4	65.9	92.2	68.2	71.8
0320	Concrete Reinforcing	100.0	100.0	100.0	86.9	72.3	79.5	96.0	72.3	83.9	91.9	72.9	82.2	89.8	68.7	79.1	91.9	72.3	82.0
0330	Cast-in-Place Concrete	100.0	100.0	100.0	87.8	66.3	79.8	102.6	68.4	89.8	85.6	66.6	78.5	100.7	66.1	87.8	85.6	68.0	79.0
03	CONCRETE	100.0	100.0	100.0	88.4	68.6	79.6	90.6	69.9	81.4	89.3	69.1	80.3	90.4	66.1	79.6	88.6	70.6	80.6
04	MASONRY	100.0	100.0	100.0	85.1	69.1	75.2	86.0	67.7	74.7	89.3	69.1	76.8	84.1	68.6	74.5	90.6	60.2	71.8
05	METALS	100.0	100.0	100.0	99.8	94.5	98.2	99.6	95.0	98.2	98.7	95.5	97.7	101.9	93.2	99.3	98.7	95.4	97.7
06	WOOD, PLASTICS & COMPOSITES	100.0	100.0	100.0	84.9	65.9	74.6	94.0	65.9	78.7	79.4	65.9	72.1	98.1	60.4	77.6	91.4	68.7	79.1
07	THERMAL & MOISTURE PROTECTION	100.0	100.0	100.0	94.3	63.9	81.1	95.9	67.3	83.4	94.4	67.3	82.6	95.1	66.4	82.6	94.4	65.0	81.6
08	OPENINGS	100.0	100.0	100.0	96.8	67.7	90.0	103.3	67.7	95.1	96.8	67.8	90.1	109.4	63.7	98.8	96.9	69.3	90.5
0920	Plaster & Gypsum Board	100.0	100.0	100.0	90.5	65.5	74.0	92.0	65.5	74.5	87.1	65.5	72.9	100.4	59.8	73.6	97.8	68.3	78.4
0950, 0980	Ceilings & Acoustic Treatment	100.0	100.0	100.0	80.8	65.5	70.4	82.1	65.5	70.8	80.8	65.5	70.4	82.4	59.8	67.1	80.8	68.3	72.3
0960	Flooring	100.0	100.0	100.0	83.8	80.6	82.9	94.8	76.3	89.4	88.6	80.6	86.3	85.6	80.6	84.2	94.0	82.2	90.6
0970, 0990	Wall Finishes & Painting/Coating	100.0	100.0	100.0	90.3	58.9	71.7	88.5	60.1	71.7	90.3	48.0	65.3	80.4	61.4	69.2	90.3	80.9	84.8
09	FINISHES	100.0	100.0	100.0	80.0	67.8	73.4	86.5	67.4	76.1	83.0	66.7	74.1	82.3	64.7	72.7	85.9	72.3	78.4
COVERS	DIVS. 10 - 14, 25, 28, 41, 43, 44, 46	100.0	100.0	100.0	100.0	70.2	93.4	100.0	83.4	96.3	100.0	77.0	94.9	100.0	69.5	93.3	100.0	72.0	93.8
21, 22, 23	FIRE SUPPRESSION, PLUMBING & HVAC	100.0	100.0	100.0	100.7	50.4	80.4	100.0	66.4	86.4	98.1	65.7	85.0	100.1	64.7	85.8	98.1	65.4	84.9
26, 27, 3370	ELECTRICAL, COMMUNICATIONS & UTIL.	100.0	100.0	100.0	96.6	58.4	77.2	95.3	61.2	77.9	98.7	65.4	81.8	91.2	64.0	77.4	97.5	75.3	86.2
MF2016	WEIGHTED AVERAGE	100.0	100.0	100.0	95.2	67.4	83.3	96.5	71.7	85.9	95.5	71.6	85.3	96.6	70.1	85.2	95.6	72.9	85.9

Table 2 — ALABAMA

DIVISION		EVERGREEN 364			GADSDEN 359			HUNTSVILLE 357 - 358			JASPER 355			MOBILE 365 - 366			MONTGOMERY 360 - 361		
		MAT.	INST.	TOTAL	MAT.	INST.	TOTAL	MAT.	INST.	TOTAL	MAT.	INST.	TOTAL	MAT.	INST.	TOTAL	MAT.	INST.	TOTAL
015433	CONTRACTOR EQUIPMENT		102.1	102.1		104.5	104.5		104.5	104.5		104.5	104.5		102.1	102.1		102.1	102.1
0241, 31 - 34	SITE & INFRASTRUCTURE, DEMOLITION	105.9	88.0	93.4	91.3	92.7	92.3	85.7	92.6	90.5	91.4	92.8	92.4	98.1	89.1	91.8	96.0	88.5	90.8
0310	Concrete Forming & Accessories	80.7	66.7	68.8	84.0	65.8	68.5	91.5	65.2	69.1	88.8	66.1	69.5	91.3	66.7	70.4	94.5	65.9	70.1
0320	Concrete Reinforcing	92.0	72.8	82.2	95.4	72.3	83.6	89.8	73.7	81.6	89.8	72.4	81.0	89.6	72.2	80.8	97.6	71.8	84.5
0330	Cast-in-Place Concrete	85.6	66.3	78.4	100.7	67.3	88.3	98.0	67.0	86.5	111.6	67.1	95.1	89.7	66.3	81.0	86.8	67.6	79.7
03	CONCRETE	89.6	69.3	80.6	94.6	69.2	83.3	89.2	69.0	80.3	98.3	69.2	85.4	85.3	69.2	78.2	83.7	69.2	77.3
04	MASONRY	89.3	68.7	76.5	82.6	65.8	72.3	85.2	65.4	73.0	80.4	70.3	74.1	87.9	58.8	69.9	85.2	59.2	69.2
05	METALS	98.7	95.4	97.7	99.8	95.0	98.4	101.9	95.3	99.9	99.8	94.9	98.3	100.7	94.7	98.9	99.6	94.4	98.0
06	WOOD, PLASTICS & COMPOSITES	76.0	67.4	71.3	88.5	65.9	76.3	98.1	65.3	80.3	95.2	65.9	79.3	89.9	67.4	77.7	93.3	65.9	78.5
07	THERMAL & MOISTURE PROTECTION	94.4	66.5	82.2	95.3	66.8	82.9	95.0	66.6	82.6	95.3	62.4	81.0	93.9	64.9	81.3	92.5	65.1	80.5
08	OPENINGS	96.8	68.6	90.3	105.9	67.7	97.0	109.1	67.7	99.5	105.9	67.7	97.0	99.5	68.6	92.3	98.8	67.7	91.6
0920	Plaster & Gypsum Board	86.5	67.0	73.7	92.6	65.5	74.7	100.4	64.8	77.0	97.2	65.5	76.3	94.2	67.0	76.3	89.9	65.5	73.8
0950, 0980	Ceilings & Acoustic Treatment	80.8	67.0	71.4	79.0	65.5	69.8	84.3	64.8	71.1	79.0	65.5	69.8	86.1	67.0	73.1	84.8	65.5	71.7
0960	Flooring	86.4	80.6	84.7	82.2	77.3	80.8	85.6	77.3	83.2	84.1	80.6	83.1	93.5	63.7	84.8	94.0	63.7	85.2
0970, 0990	Wall Finishes & Painting/Coating	90.3	48.0	65.3	80.4	58.9	67.7	80.4	67.3	72.6	80.4	58.9	67.7	93.6	49.9	67.7	91.3	58.9	72.1
09	FINISHES	82.3	67.4	74.2	80.0	67.1	72.9	82.7	67.5	74.4	81.1	68.1	74.0	85.8	64.2	74.0	86.1	64.4	74.2
COVERS	DIVS. 10 - 14, 25, 28, 41, 43, 44, 46	100.0	70.1	93.4	100.0	82.8	96.2	100.0	82.6	96.2	100.0	71.4	93.7	100.0	83.7	96.4	100.0	83.0	96.2
21, 22, 23	FIRE SUPPRESSION, PLUMBING & HVAC	98.1	65.5	84.9	102.0	62.8	86.2	100.1	63.5	85.3	102.0	66.2	87.5	100.0	66.2	84.9	100.0	65.0	85.8
26, 27, 3370	ELECTRICAL, COMMUNICATIONS & UTIL.	95.9	65.4	80.4	91.2	61.2	75.9	92.1	65.7	78.7	90.9	58.4	74.4	99.4	58.1	78.4	99.9	73.9	86.7
MF2016	WEIGHTED AVERAGE	95.2	71.4	85.0	96.6	70.6	85.5	96.5	71.4	85.7	97.1	71.0	85.9	96.1	68.8	84.4	95.6	71.4	85.2

Table 3 — ALABAMA / ALASKA

DIVISION		PHENIX CITY 368			SELMA 367			TUSCALOOSA 354			ANCHORAGE 995 - 996			FAIRBANKS 997			JUNEAU 998		
		MAT.	INST.	TOTAL	MAT.	INST.	TOTAL	MAT.	INST.	TOTAL	MAT.	INST.	TOTAL	MAT.	INST.	TOTAL	MAT.	INST.	TOTAL
015433	CONTRACTOR EQUIPMENT		102.1	102.1		102.1	102.1		104.5	104.5		115.9	115.9		115.9	115.9		115.9	115.9
0241, 31 - 34	SITE & INFRASTRUCTURE, DEMOLITION	109.8	88.5	95.0	102.9	88.5	92.9	86.3	92.7	90.8	121.2	130.7	127.8	120.6	130.7	127.7	139.5	130.7	133.3
0310	Concrete Forming & Accessories	87.2	66.1	69.2	85.0	65.9	68.8	91.4	66.0	69.8	116.4	117.7	117.5	120.5	117.9	118.3	123.6	117.7	118.5
0320	Concrete Reinforcing	91.9	69.2	80.4	91.9	72.4	82.0	89.8	72.3	80.9	152.4	121.3	136.6	141.8	121.3	131.4	127.6	121.3	124.4
0330	Cast-in-Place Concrete	85.6	67.7	78.9	85.6	66.6	78.5	102.1	67.6	89.3	97.6	117.2	104.9	108.1	117.6	111.7	120.4	117.2	117.7
03	CONCRETE	92.3	68.9	81.9	88.1	69.0	79.6	91.1	69.4	81.5	108.3	117.3	112.3	101.3	117.6	108.5	115.2	117.3	116.1
04	MASONRY	89.2	69.1	76.8	92.5	69.1	78.1	84.3	66.3	73.2	183.5	119.5	144.0	185.0	119.8	144.7	157.6	119.5	134.1
05	METALS	98.6	94.2	97.3	98.6	95.1	97.6	101.1	95.0	99.3	119.7	104.8	115.2	120.8	104.9	116.0	117.1	104.8	113.4
06	WOOD, PLASTICS & COMPOSITES	84.6	65.8	74.4	81.6	65.9	73.1	98.1	65.9	80.6	114.0	115.8	115.0	125.8	115.8	120.4	123.4	115.8	119.3
07	THERMAL & MOISTURE PROTECTION	94.8	67.8	83.0	94.2	67.3	82.5	95.1	67.0	82.9	168.7	117.6	146.4	179.9	118.7	153.2	184.7	117.6	155.5
08	OPENINGS	96.8	66.8	89.8	96.8	67.7	90.0	109.1	67.7	99.5	130.6	116.5	127.3	135.8	116.5	131.4	130.4	116.5	127.2
0920	Plaster & Gypsum Board	91.7	65.3	74.3	89.9	65.5	73.8	100.4	65.5	77.4	142.8	116.1	125.2	172.8	116.1	135.4	154.4	116.1	129.5
0950, 0980	Ceilings & Acoustic Treatment	80.8	65.3	70.3	80.8	65.5	70.4	85.6	65.5	71.5	132.6	116.1	121.4	127.9	116.1	119.9	138.9	116.1	123.4
0960	Flooring	90.6	80.6	87.7	89.1	80.6	86.6	85.6	77.3	83.2	125.1	122.6	124.4	122.1	122.6	122.2	129.4	122.6	127.4
0970, 0990	Wall Finishes & Painting/Coating	90.3	81.0	84.8	90.3	58.9	71.7	80.4	58.9	67.7	108.7	113.2	111.3	113.7	117.7	116.1	110.8	113.2	112.2
09	FINISHES	84.7	70.2	76.8	83.2	67.8	74.8	82.6	67.2	74.2	125.7	118.3	121.7	128.6	118.9	123.3	129.3	118.3	123.3
COVERS	DIVS. 10 - 14, 25, 28, 41, 43, 44, 46	100.0	82.9	96.2	100.0	70.7	93.6	100.0	82.9	96.2	100.0	109.3	102.0	100.0	109.4	102.1	100.0	109.3	102.0
21, 22, 23	FIRE SUPPRESSION, PLUMBING & HVAC	98.1	65.0	84.7	98.1	64.3	84.5	100.1	66.5	86.5	100.6	106.1	102.8	100.4	110.3	104.4	100.6	106.1	102.8
26, 27, 3370	ELECTRICAL, COMMUNICATIONS & UTIL.	98.1	65.7	81.6	97.1	73.9	85.3	91.7	61.2	76.2	113.8	108.8	111.3	116.7	108.8	112.7	100.4	108.8	104.7
MF2016	WEIGHTED AVERAGE	96.1	72.0	85.8	95.3	72.4	85.5	96.6	71.5	85.8	117.5	113.6	115.8	118.2	114.7	116.7	116.5	113.6	115.3

ALASKA / ARIZONA

DIVISION		ALASKA KETCHIKAN 999 MAT.	INST.	TOTAL	CHAMBERS 865 MAT.	INST.	TOTAL	FLAGSTAFF 860 MAT.	INST.	TOTAL	GLOBE 855 MAT.	INST.	TOTAL	KINGMAN 864 MAT.	INST.	TOTAL	MESA/TEMPE 852 MAT.	INST.	TOTAL
015433	CONTRACTOR EQUIPMENT		115.9	115.9		91.1	91.1		91.1	91.1		92.2	92.2		91.1	91.1		92.2	92.2
0241, 31 - 34	SITE & INFRASTRUCTURE, DEMOLITION	175.1	130.7	144.1	71.9	95.0	88.0	90.9	95.0	93.8	102.3	95.9	97.8	71.8	95.0	88.0	92.4	95.9	94.9
0310	Concrete Forming & Accessories	112.3	117.7	116.9	97.1	68.0	72.3	102.5	69.2	74.1	93.4	68.0	71.8	95.3	67.9	72.0	96.5	68.1	72.3
0320	Concrete Reinforcing	108.4	121.3	114.9	103.2	79.1	91.0	103.0	79.0	90.8	109.0	79.1	93.8	103.4	79.1	91.0	109.7	79.1	94.2
0330	Cast-in-Place Concrete	218.8	117.2	181.0	89.7	69.4	82.2	89.8	69.4	82.2	84.1	69.0	78.5	89.5	69.4	82.0	84.8	69.1	79.0
03	CONCRETE	173.5	117.3	148.6	89.8	70.3	81.2	108.8	70.9	92.0	104.4	70.3	89.3	89.5	70.3	81.0	94.9	70.4	84.0
04	MASONRY	192.7	119.5	147.5	95.2	60.7	73.9	95.3	59.0	72.9	100.4	60.6	75.8	95.2	60.7	73.9	100.6	60.6	75.9
05	METALS	121.0	104.8	116.1	98.1	72.4	90.3	98.6	72.3	90.7	101.8	73.2	93.1	98.8	72.3	90.8	102.1	73.3	93.4
06	WOOD, PLASTICS & COMPOSITES	116.0	115.8	115.9	102.8	69.3	84.6	109.3	70.9	88.4	94.2	69.4	80.8	97.5	69.3	82.2	98.1	69.4	82.5
07	THERMAL & MOISTURE PROTECTION	184.9	117.6	155.6	96.8	70.1	85.1	98.8	69.6	86.1	98.7	68.7	85.7	96.7	69.9	85.0	98.5	68.4	85.4
08	OPENINGS	131.5	116.5	128.0	108.2	66.4	98.5	108.3	72.2	99.9	97.2	66.7	90.2	108.4	68.7	99.2	97.3	68.8	90.7
0920	Plaster & Gypsum Board	158.3	116.1	130.5	92.6	68.6	76.8	96.2	70.2	79.1	88.0	68.6	75.2	84.6	68.6	74.1	90.8	68.6	76.2
0950, 0980	Ceilings & Acoustic Treatment	121.3	116.1	117.8	107.2	68.6	81.1	108.1	70.2	82.4	93.5	68.6	76.6	108.1	68.6	81.3	93.5	68.6	76.6
0960	Flooring	122.1	122.6	122.2	89.9	58.8	80.9	92.1	58.8	82.4	104.0	58.8	90.9	88.7	58.8	80.0	105.6	58.8	92.0
0970, 0990	Wall Finishes & Painting/Coating	113.7	113.2	113.4	88.5	68.3	76.6	88.5	68.3	76.6	92.8	68.3	78.3	88.5	68.3	76.6	92.8	68.3	78.3
09	FINISHES	129.1	118.3	123.2	90.1	66.1	77.0	93.2	67.0	78.9	96.5	66.2	80.0	88.7	66.1	76.4	96.3	66.2	79.9
COVERS	DIVS. 10 - 14, 25, 28, 41, 43, 44, 46	100.0	109.3	102.0	100.0	83.0	96.3	100.0	83.1	96.3	100.0	83.2	96.3	100.0	82.9	96.2	100.0	83.2	96.3
21, 22, 23	FIRE SUPPRESSION, PLUMBING & HVAC	98.6	106.1	101.6	97.7	77.0	89.3	100.2	77.2	90.9	96.6	77.0	88.7	97.7	77.0	89.3	100.0	77.2	90.8
26, 27, 3370	ELECTRICAL, COMMUNICATIONS & UTIL.	116.6	108.8	112.7	102.7	83.6	93.0	101.7	62.2	81.6	100.5	64.4	82.1	102.7	64.4	83.2	97.2	64.4	80.5
MF2016	WEIGHTED AVERAGE	128.4	113.6	122.0	97.2	74.4	87.4	101.0	71.7	88.4	99.5	71.8	87.6	97.2	71.8	86.3	98.6	72.0	87.2

ARIZONA / ARKANSAS

DIVISION		PHOENIX 850, 853 MAT.	INST.	TOTAL	PRESCOTT 863 MAT.	INST.	TOTAL	SHOW LOW 859 MAT.	INST.	TOTAL	TUCSON 856 - 857 MAT.	INST.	TOTAL	BATESVILLE 725 MAT.	INST.	TOTAL	CAMDEN 717 MAT.	INST.	TOTAL
015433	CONTRACTOR EQUIPMENT		93.2	93.2		91.1	91.1		92.2	92.2		92.2	92.2		91.9	91.9		91.9	91.9
0241, 31 - 34	SITE & INFRASTRUCTURE, DEMOLITION	92.8	94.5	94.0	78.6	95.0	90.0	104.5	95.9	98.5	87.9	95.9	93.5	76.7	89.7	85.8	80.5	89.7	86.9
0310	Concrete Forming & Accessories	101.2	71.6	76.0	98.7	68.0	72.5	100.6	68.1	72.9	96.9	69.3	73.4	80.1	56.4	59.9	79.6	56.3	59.8
0320	Concrete Reinforcing	107.8	79.2	93.2	103.0	79.1	90.9	109.7	79.1	94.2	91.0	79.0	84.9	93.3	65.3	79.1	98.6	65.0	81.5
0330	Cast-in-Place Concrete	84.5	69.3	78.8	89.8	69.4	82.2	84.2	69.1	78.5	87.3	69.1	80.5	74.8	76.4	75.4	78.1	76.4	77.5
03	CONCRETE	97.6	72.1	86.3	95.0	70.3	84.1	106.8	70.3	90.7	93.7	70.9	83.6	74.0	65.7	70.3	76.7	65.7	71.8
04	MASONRY	93.5	62.7	74.5	95.3	60.7	73.9	100.4	60.6	75.8	87.5	58.9	69.8	92.1	42.3	61.4	104.9	36.9	62.9
05	METALS	103.5	75.2	94.9	98.7	72.3	90.7	101.6	73.2	93.0	102.8	73.2	93.8	98.6	72.8	90.8	103.4	72.4	94.0
06	WOOD, PLASTICS & COMPOSITES	101.1	72.9	85.7	104.4	69.5	85.4	102.7	69.4	84.6	98.3	71.0	83.5	92.9	57.2	73.5	91.0	57.2	72.6
07	THERMAL & MOISTURE PROTECTION	99.8	69.2	86.5	97.4	70.2	85.5	99.0	69.0	85.9	99.5	68.1	85.9	100.3	57.4	81.6	97.0	55.8	79.1
08	OPENINGS	98.6	72.8	92.6	108.3	66.7	98.7	96.5	68.8	90.1	93.7	72.3	88.7	102.0	53.1	90.7	105.6	54.9	93.8
0920	Plaster & Gypsum Board	98.0	71.9	80.8	93.0	68.8	77.1	92.6	68.6	76.8	95.3	70.2	78.8	78.6	56.4	64.0	87.9	56.4	67.1
0950, 0980	Ceilings & Acoustic Treatment	104.1	71.9	82.3	106.4	68.8	80.9	93.5	68.6	76.6	94.3	70.2	78.0	85.7	56.4	65.8	80.0	56.4	64.0
0960	Flooring	107.7	58.8	93.5	90.7	58.8	81.5	107.3	58.8	93.2	96.1	58.8	85.3	86.3	53.0	76.6	92.4	35.5	75.9
0970, 0990	Wall Finishes & Painting/Coating	97.5	65.0	78.3	88.5	68.3	76.6	92.8	68.3	78.3	93.3	68.3	78.5	90.7	53.5	68.7	94.5	53.5	70.2
09	FINISHES	100.9	68.5	83.2	90.7	66.2	77.3	98.3	66.2	80.8	94.3	67.2	79.5	76.9	54.9	64.9	80.4	51.7	64.7
COVERS	DIVS. 10 - 14, 25, 28, 41, 43, 44, 46	100.0	84.7	96.6	100.0	82.9	96.2	100.0	83.2	96.3	100.0	83.4	96.3	100.0	65.9	92.5	100.0	65.9	92.5
21, 22, 23	FIRE SUPPRESSION, PLUMBING & HVAC	99.9	78.6	91.2	100.2	77.0	90.8	96.6	77.0	88.7	100.0	77.1	90.7	96.6	52.2	78.6	96.5	56.9	80.5
26, 27, 3370	ELECTRICAL, COMMUNICATIONS & UTIL.	103.0	62.2	82.2	101.4	64.4	82.6	97.5	64.4	80.7	99.5	59.5	79.2	95.8	60.7	78.0	93.9	57.2	75.3
MF2016	WEIGHTED AVERAGE	100.0	73.0	88.4	98.7	71.7	87.1	99.7	71.9	87.8	97.6	71.4	86.4	92.6	60.0	78.7	94.8	59.6	79.7

ARKANSAS

DIVISION		FAYETTEVILLE 727 MAT.	INST.	TOTAL	FORT SMITH 729 MAT.	INST.	TOTAL	HARRISON 726 MAT.	INST.	TOTAL	HOT SPRINGS 719 MAT.	INST.	TOTAL	JONESBORO 724 MAT.	INST.	TOTAL	LITTLE ROCK 720 - 722 MAT.	INST.	TOTAL
015433	CONTRACTOR EQUIPMENT		91.9	91.9		91.9	91.9		91.9	91.9		91.9	91.9		115.7	115.7		91.9	91.9
0241, 31 - 34	SITE & INFRASTRUCTURE, DEMOLITION	76.1	89.7	85.6	81.3	89.8	87.2	81.5	89.7	87.2	83.8	89.7	87.9	100.6	107.5	105.4	89.2	90.1	89.8
0310	Concrete Forming & Accessories	75.9	56.5	59.4	94.2	61.7	66.5	84.3	56.3	60.5	77.3	56.5	59.6	83.2	56.5	60.5	90.9	62.0	66.3
0320	Concrete Reinforcing	93.4	62.6	77.7	94.5	65.9	80.0	92.9	65.4	78.9	96.7	65.3	80.8	90.1	65.5	77.6	100.0	67.9	83.7
0330	Cast-in-Place Concrete	74.8	76.5	75.4	85.6	76.7	82.2	83.0	76.4	80.6	79.9	76.5	78.6	81.5	77.7	80.1	82.9	76.8	80.6
03	CONCRETE	73.7	65.3	70.0	81.1	68.3	75.4	80.6	65.7	74.0	79.9	65.8	73.6	78.2	67.3	73.4	81.9	68.9	76.1
04	MASONRY	83.5	41.8	57.8	90.1	62.1	72.8	92.6	43.3	62.1	78.8	42.0	56.1	84.8	39.9	57.1	86.0	62.1	71.2
05	METALS	98.6	72.0	90.5	100.8	74.6	92.8	99.7	72.8	91.6	103.3	72.9	94.1	95.0	87.8	92.8	101.2	75.6	93.4
06	WOOD, PLASTICS & COMPOSITES	89.2	57.2	71.8	110.0	63.6	84.8	98.9	57.2	76.2	88.4	57.2	71.4	97.1	57.5	75.6	96.5	63.6	78.6
07	THERMAL & MOISTURE PROTECTION	101.1	57.2	82.0	101.5	63.6	85.0	100.6	57.7	81.9	97.3	57.3	79.9	105.8	56.6	84.4	95.9	63.9	81.9
08	OPENINGS	102.0	54.9	91.1	104.1	60.1	93.8	102.8	52.0	91.0	105.5	54.8	93.8	107.8	56.0	95.8	97.2	59.7	88.5
0920	Plaster & Gypsum Board	78.0	56.4	63.8	84.8	63.0	70.4	83.9	56.4	65.8	86.6	56.4	66.7	89.9	56.4	67.8	94.5	63.0	73.8
0950, 0980	Ceilings & Acoustic Treatment	85.7	56.4	65.8	87.4	63.0	70.9	87.4	56.4	66.4	80.0	56.4	64.0	90.7	56.4	67.4	89.5	63.0	71.5
0960	Flooring	83.4	51.9	74.3	92.9	73.3	87.2	88.6	53.0	78.3	91.3	51.9	79.8	62.6	47.0	58.1	94.0	82.1	90.5
0970, 0990	Wall Finishes & Painting/Coating	90.7	53.5	68.7	90.7	53.5	68.7	90.7	53.5	68.7	94.5	53.5	70.2	80.6	53.5	64.6	91.2	53.5	68.9
09	FINISHES	76.1	54.5	64.3	80.3	63.1	70.9	79.0	54.7	65.8	80.2	54.7	66.3	74.9	53.8	63.4	86.3	64.9	74.6
COVERS	DIVS. 10 - 14, 25, 28, 41, 43, 44, 46	100.0	65.9	92.5	100.0	78.4	95.2	100.0	67.0	92.7	100.0	65.9	92.5	100.0	64.9	92.3	100.0	78.4	95.3
21, 22, 23	FIRE SUPPRESSION, PLUMBING & HVAC	96.6	61.2	82.3	100.1	49.1	79.5	96.6	48.5	77.1	96.5	52.5	78.7	100.4	52.2	80.9	99.8	52.6	80.7
26, 27, 3370	ELECTRICAL, COMMUNICATIONS & UTIL.	90.1	53.1	71.3	93.3	58.4	75.6	94.5	58.9	76.4	95.8	66.8	81.1	99.4	60.9	79.9	100.4	66.8	83.3
MF2016	WEIGHTED AVERAGE	91.5	60.7	78.3	95.1	63.6	81.6	94.0	59.1	79.0	94.2	61.0	79.9	94.7	62.7	81.0	95.4	65.9	82.8

City Cost Indexes

ARKANSAS / CALIFORNIA

DIVISION		PINE BLUFF 716 MAT.	INST.	TOTAL	RUSSELLVILLE 728 MAT.	INST.	TOTAL	TEXARKANA 718 MAT.	INST.	TOTAL	WEST MEMPHIS 723 MAT.	INST.	TOTAL	ALHAMBRA 917-918 MAT.	INST.	TOTAL	ANAHEIM 928 MAT.	INST.	TOTAL
015433	CONTRACTOR EQUIPMENT		91.9	91.9		91.9	91.9		93.2	93.2		115.7	115.7		97.7	97.7		101.8	101.8
0241, 31 - 34	SITE & INFRASTRUCTURE, DEMOLITION	85.8	89.8	88.6	77.9	89.7	86.1	96.6	91.9	93.3	107.4	107.7	107.6	99.9	110.3	107.1	100.5	110.6	107.5
0310	Concrete Forming & Accessories	77.0	61.9	64.1	80.8	56.3	59.9	82.6	56.4	60.3	88.5	56.8	61.5	115.3	140.0	136.3	101.2	140.1	134.3
0320	Concrete Reinforcing	98.4	67.9	82.9	94.0	65.2	79.4	98.0	65.3	81.4	90.1	63.7	76.7	102.5	130.1	116.5	95.4	129.9	112.9
0330	Cast-in-Place Concrete	79.9	76.7	78.7	78.4	76.4	77.6	87.1	76.4	83.2	85.5	77.8	82.6	83.4	129.8	100.7	85.2	133.2	103.1
03	CONCRETE	80.6	68.8	75.3	76.9	65.6	71.9	79.0	65.7	73.1	84.7	67.1	76.9	94.0	133.3	111.4	94.2	134.6	112.1
04	MASONRY	110.4	62.1	80.5	89.4	40.9	59.4	92.9	41.3	61.0	73.6	39.8	52.7	111.7	141.9	130.4	78.5	140.2	116.6
05	METALS	104.2	75.4	95.5	98.6	72.6	90.7	96.0	72.8	89.0	94.1	87.4	92.0	82.3	110.5	90.8	106.1	111.7	107.8
06	WOOD, PLASTICS & COMPOSITES	88.0	63.6	74.8	94.2	57.2	74.1	95.6	57.2	74.7	103.0	57.5	78.3	102.8	137.1	121.4	97.9	137.4	119.3
07	THERMAL & MOISTURE PROTECTION	97.4	64.0	82.9	101.4	57.0	82.0	98.0	57.1	80.2	106.3	56.6	84.7	97.2	133.2	112.9	106.8	136.3	119.7
08	OPENINGS	106.8	60.5	96.0	102.0	54.1	90.9	111.8	54.1	98.4	105.3	54.8	93.5	89.8	135.7	100.4	106.2	135.9	113.1
0920	Plaster & Gypsum Board	86.3	63.0	71.0	78.6	56.4	64.0	89.2	56.4	67.6	92.1	56.4	68.6	96.3	138.3	124.0	111.2	138.3	129.1
0950, 0980	Ceilings & Acoustic Treatment	80.0	63.0	68.5	85.7	56.4	65.8	83.4	56.4	65.1	88.8	56.4	66.8	112.6	138.3	130.0	112.6	138.3	130.0
0960	Flooring	91.1	73.3	85.9	85.9	53.0	76.3	93.2	44.8	79.1	64.9	47.0	59.7	100.0	120.4	105.9	97.3	120.4	104.0
0970, 0990	Wall Finishes & Painting/Coating	94.5	53.5	70.2	90.7	53.5	68.7	94.5	53.5	70.2	80.6	53.5	64.6	105.4	121.1	114.7	90.7	121.1	108.7
09	FINISHES	80.2	63.1	70.9	77.1	54.8	64.9	82.4	53.4	66.6	76.2	54.0	64.1	100.8	134.3	119.1	98.1	134.5	118.0
COVERS	DIVS. 10 - 14, 25, 28, 41, 43, 44, 46	100.0	78.4	95.2	100.0	65.9	92.5	100.0	65.9	92.5	100.0	66.7	92.7	100.0	119.6	104.3	100.0	120.1	104.4
21, 22, 23	FIRE SUPPRESSION, PLUMBING & HVAC	100.0	52.6	80.8	96.6	51.9	78.5	100.0	55.6	82.1	96.9	64.2	83.7	96.5	131.7	110.8	100.0	131.8	112.8
26, 27, 3370	ELECTRICAL, COMMUNICATIONS & UTIL.	94.0	58.9	76.2	93.3	58.9	75.8	95.7	58.9	77.0	100.9	62.7	81.5	117.4	130.5	124.1	90.2	111.0	100.8
MF2016	WEIGHTED AVERAGE	96.7	64.6	82.9	92.7	59.6	78.5	95.6	60.4	80.5	94.3	65.5	81.9	96.8	129.4	110.8	98.9	126.9	110.9

CALIFORNIA

DIVISION		BAKERSFIELD 932-933 MAT.	INST.	TOTAL	BERKELEY 947 MAT.	INST.	TOTAL	EUREKA 955 MAT.	INST.	TOTAL	FRESNO 936-938 MAT.	INST.	TOTAL	INGLEWOOD 903-905 MAT.	INST.	TOTAL	LONG BEACH 906-908 MAT.	INST.	TOTAL
015433	CONTRACTOR EQUIPMENT		99.8	99.8		102.8	102.8		99.6	99.6		99.8	99.8		99.0	99.0		99.0	99.0
0241, 31 - 34	SITE & INFRASTRUCTURE, DEMOLITION	97.6	108.4	105.1	111.6	110.7	110.9	111.8	107.4	108.8	101.6	107.8	106.0	90.5	106.7	101.8	97.6	106.7	104.0
0310	Concrete Forming & Accessories	101.0	138.9	133.3	112.6	167.2	159.1	109.4	152.9	146.4	98.3	152.1	144.1	104.1	140.3	134.9	99.2	140.3	134.2
0320	Concrete Reinforcing	103.5	129.9	116.9	89.1	131.8	110.8	103.9	131.3	117.8	86.5	130.8	109.0	103.0	130.1	116.8	102.1	130.1	116.4
0330	Cast-in-Place Concrete	89.3	132.3	105.3	111.5	133.6	119.7	92.5	129.8	106.3	96.8	129.4	108.9	79.5	132.3	99.2	90.5	132.3	106.1
03	CONCRETE	89.6	133.7	109.1	102.8	147.1	122.4	104.6	139.3	120.0	93.5	138.7	113.6	89.2	134.4	109.2	98.7	134.4	114.5
04	MASONRY	92.5	139.6	121.6	124.9	156.3	144.3	103.7	152.4	133.8	97.2	146.3	127.5	75.6	142.1	116.6	84.7	142.1	120.1
05	METALS	98.8	111.1	102.5	105.7	114.0	108.2	105.8	113.6	108.2	99.1	112.1	103.0	91.7	112.1	97.9	91.6	112.1	97.8
06	WOOD, PLASTICS & COMPOSITES	92.8	136.2	116.4	106.7	172.6	142.5	111.8	156.0	135.8	98.8	156.0	129.9	96.9	137.5	119.0	90.7	137.5	116.1
07	THERMAL & MOISTURE PROTECTION	103.9	128.2	114.5	107.4	157.2	129.1	110.6	156.0	128.0	95.0	134.4	112.2	100.6	134.9	115.5	100.9	134.9	115.7
08	OPENINGS	96.1	133.1	104.7	92.7	159.5	108.2	105.5	141.6	113.9	97.0	144.0	108.0	89.4	135.9	100.2	89.4	135.9	100.2
0920	Plaster & Gypsum Board	97.4	137.0	123.5	113.1	174.1	153.3	116.2	157.4	143.3	94.3	157.4	135.8	105.2	138.3	127.0	101.3	138.3	125.7
0950, 0980	Ceilings & Acoustic Treatment	104.7	137.0	126.6	98.5	174.1	149.8	116.0	157.4	144.0	103.9	157.4	140.1	110.8	138.3	129.5	110.8	138.3	129.5
0960	Flooring	102.9	120.4	108.0	115.9	142.0	123.5	101.1	142.0	113.0	102.1	118.0	106.7	101.8	120.4	107.2	98.9	120.4	105.1
0970, 0990	Wall Finishes & Painting/Coating	92.1	109.3	102.3	113.3	162.7	142.5	92.1	137.6	119.0	104.1	123.9	115.8	105.2	121.1	114.6	105.2	121.1	114.6
09	FINISHES	94.0	132.5	115.0	107.6	163.7	138.3	103.1	150.9	129.2	93.5	145.3	121.8	104.5	134.6	120.9	103.5	134.6	120.4
COVERS	DIVS. 10 - 14, 25, 28, 41, 43, 44, 46	100.0	120.2	104.4	100.0	129.9	106.6	100.0	126.7	105.9	100.0	126.7	105.9	100.0	120.4	104.5	100.0	120.4	104.5
21, 22, 23	FIRE SUPPRESSION, PLUMBING & HVAC	100.1	129.0	111.8	96.7	168.0	125.6	96.5	129.1	109.7	100.2	130.3	112.3	96.1	131.8	110.5	96.1	131.8	110.5
26, 27, 3370	ELECTRICAL, COMMUNICATIONS & UTIL.	107.1	110.7	109.0	92.3	161.0	132.1	96.8	124.3	110.8	96.8	108.0	102.5	100.3	130.5	115.7	100.0	130.5	115.5
MF2016	WEIGHTED AVERAGE	98.0	125.3	109.7	102.3	151.2	123.3	102.1	133.2	115.4	97.7	129.3	111.2	94.2	129.6	109.4	95.9	129.6	110.3

CALIFORNIA

DIVISION		LOS ANGELES 900-902 MAT.	INST.	TOTAL	MARYSVILLE 959 MAT.	INST.	TOTAL	MODESTO 953 MAT.	INST.	TOTAL	MOJAVE 935 MAT.	INST.	TOTAL	OAKLAND 946 MAT.	INST.	TOTAL	OXNARD 930 MAT.	INST.	TOTAL
015433	CONTRACTOR EQUIPMENT		105.2	105.2		99.6	99.6		99.6	99.6		99.8	99.8		102.8	102.8		98.7	98.7
0241, 31 - 34	SITE & INFRASTRUCTURE, DEMOLITION	96.1	112.4	107.4	108.3	107.3	107.6	103.0	107.3	106.0	94.7	108.4	104.3	116.5	110.7	112.4	101.9	106.5	105.1
0310	Concrete Forming & Accessories	102.2	140.6	134.9	99.8	152.5	144.6	96.2	152.3	143.9	108.3	138.8	134.3	101.9	167.2	157.5	100.5	140.2	134.3
0320	Concrete Reinforcing	103.8	130.2	117.2	103.9	130.6	117.5	107.6	130.5	119.3	104.1	129.9	117.2	91.1	131.8	111.8	102.3	129.8	116.3
0330	Cast-in-Place Concrete	82.1	133.4	101.2	103.3	129.5	113.1	95.3	129.4	106.2	83.7	132.3	101.8	105.8	133.6	116.2	97.5	132.5	110.6
03	CONCRETE	96.5	135.0	113.6	105.0	138.9	120.0	96.1	138.8	115.1	85.4	133.6	106.8	103.5	147.1	122.8	93.5	134.3	111.6
04	MASONRY	91.0	142.1	122.5	104.7	141.7	127.5	102.1	141.7	126.6	95.4	139.6	122.7	132.8	156.3	147.3	98.3	138.0	122.8
05	METALS	97.8	114.5	102.9	105.3	112.3	107.4	102.2	112.1	105.2	96.7	111.0	101.0	101.8	114.0	105.5	94.6	111.3	99.6
06	WOOD, PLASTICS & COMPOSITES	99.9	137.7	120.4	98.8	156.0	129.9	94.3	156.0	127.8	98.4	136.2	118.9	94.6	172.6	137.0	93.3	137.4	117.3
07	THERMAL & MOISTURE PROTECTION	96.9	134.8	113.4	110.0	139.1	122.7	109.5	135.5	120.8	100.8	125.6	111.6	105.9	157.2	128.3	104.2	134.3	117.3
08	OPENINGS	98.9	136.1	107.5	104.7	144.9	114.1	103.5	144.9	113.1	92.0	133.1	101.6	92.8	159.5	108.3	94.6	135.9	104.2
0920	Plaster & Gypsum Board	100.5	138.3	125.4	108.7	157.4	140.8	111.2	157.4	141.6	103.8	137.0	125.7	107.2	174.1	151.3	97.8	138.3	124.5
0950, 0980	Ceilings & Acoustic Treatment	117.7	138.3	131.7	115.1	157.4	143.8	112.6	157.4	143.0	106.1	137.0	127.1	101.3	174.1	150.6	107.7	138.3	128.4
0960	Flooring	100.0	120.4	105.9	96.8	123.5	104.6	97.2	129.9	106.7	102.3	120.4	107.5	109.6	142.0	119.0	94.9	120.4	102.3
0970, 0990	Wall Finishes & Painting/Coating	101.9	121.1	113.2	92.1	137.6	119.0	92.1	137.6	119.0	88.9	109.3	100.9	113.3	162.7	142.5	88.9	115.6	104.7
09	FINISHES	105.5	134.8	121.5	100.1	147.7	126.1	99.6	148.8	126.5	93.8	132.5	114.9	105.9	163.7	137.5	91.3	133.9	114.6
COVERS	DIVS. 10 - 14, 25, 28, 41, 43, 44, 46	100.0	121.1	104.6	100.0	126.7	105.9	100.0	126.7	105.9	100.0	120.2	104.4	100.0	129.9	106.6	100.0	120.4	104.5
21, 22, 23	FIRE SUPPRESSION, PLUMBING & HVAC	99.9	131.9	112.8	96.5	130.3	110.2	100.0	130.3	112.2	96.6	129.0	109.7	100.2	168.0	127.6	100.1	131.8	112.9
26, 27, 3370	ELECTRICAL, COMMUNICATIONS & UTIL.	98.6	130.5	114.8	93.6	122.4	108.3	96.0	108.4	102.3	95.3	110.7	103.1	101.5	161.0	131.8	101.0	117.5	109.4
MF2016	WEIGHTED AVERAGE	98.7	130.4	112.3	101.2	131.3	114.1	100.2	129.4	112.7	94.7	125.2	107.7	102.7	151.2	123.5	97.2	127.1	110.0

For customer support on your Electrical Costs with RSMeans data, call 800.448.8182.

517

CALIFORNIA

DIVISION		PALM SPRINGS 922 MAT.	INST.	TOTAL	PALO ALTO 943 MAT.	INST.	TOTAL	PASADENA 910-912 MAT.	INST.	TOTAL	REDDING 960 MAT.	INST.	TOTAL	RICHMOND 948 MAT.	INST.	TOTAL	RIVERSIDE 925 MAT.	INST.	TOTAL
015433	CONTRACTOR EQUIPMENT		100.7	100.7		102.8	102.8		97.7	97.7		99.6	99.6		102.8	102.8		100.7	100.7
0241, 31 - 34	SITE & INFRASTRUCTURE, DEMOLITION	92.1	108.6	103.6	107.8	110.7	109.8	96.5	110.3	106.1	129.0	107.3	113.9	115.7	110.6	112.2	99.0	108.6	105.7
0310	Concrete Forming & Accessories	98.0	140.1	133.8	100.0	167.2	157.2	103.7	140.0	134.6	102.8	152.7	145.3	115.4	166.8	159.2	101.8	140.1	134.4
0320	Concrete Reinforcing	109.3	129.9	119.8	89.1	131.1	110.5	103.4	130.1	116.9	132.1	130.6	131.3	89.1	131.3	110.5	106.2	129.9	118.2
0330	Cast-in-Place Concrete	81.2	133.2	100.6	94.4	133.6	109.0	79.1	129.8	98.0	101.6	129.5	112.0	108.5	133.5	117.8	88.4	133.2	105.0
03	CONCRETE	88.5	134.6	108.9	92.8	147.0	116.9	89.7	133.3	109.0	108.0	139.0	121.8	105.4	146.8	123.8	94.4	134.6	112.2
04	MASONRY	76.7	139.8	115.7	105.9	146.3	130.8	97.9	141.9	125.1	132.2	141.7	138.1	124.6	150.4	140.6	77.7	139.8	116.1
05	METALS	106.7	111.7	108.2	99.4	113.3	103.6	82.3	110.5	90.9	112.5	112.3	112.5	99.4	113.2	103.6	106.2	111.7	107.8
06	WOOD, PLASTICS & COMPOSITES	92.6	137.4	116.9	92.2	172.6	135.9	88.1	137.1	114.7	110.1	156.3	135.2	110.0	172.6	144.0	97.9	137.4	119.3
07	THERMAL & MOISTURE PROTECTION	106.1	136.2	119.2	105.1	154.4	126.6	96.8	133.2	112.6	131.9	139.2	135.1	105.8	154.5	127.0	107.0	136.2	119.7
08	OPENINGS	102.1	135.9	109.9	92.8	159.5	108.3	89.8	135.7	100.4	118.1	143.8	124.0	92.8	159.5	108.3	104.8	135.9	112.0
0920	Plaster & Gypsum Board	106.2	138.3	127.4	105.7	174.1	150.7	90.3	138.3	121.9	108.9	157.8	141.1	114.2	174.1	153.7	110.5	138.3	128.8
0950, 0980	Ceilings & Acoustic Treatment	109.2	138.3	128.9	99.4	174.1	150.0	112.6	138.3	130.0	148.5	157.8	154.8	99.4	174.1	150.0	118.7	138.3	132.0
0960	Flooring	99.4	120.4	105.5	108.2	142.0	118.0	94.2	120.4	101.8	93.5	132.9	105.0	118.2	142.0	125.1	100.9	120.4	106.6
0970, 0990	Wall Finishes & Painting/Coating	89.1	121.1	108.1	113.3	162.7	142.5	105.4	121.1	114.7	100.0	137.6	122.3	113.3	162.7	142.5	89.1	121.1	108.1
09	FINISHES	96.6	134.5	117.3	104.3	163.7	136.8	98.0	134.3	117.8	106.4	149.4	129.8	109.0	163.7	138.9	99.8	134.5	118.7
COVERS	DIVS. 10 - 14, 25, 28, 41, 43, 44, 46	100.0	120.2	104.4	100.0	129.9	106.6	100.0	119.6	104.3	100.0	126.8	105.9	100.0	129.8	106.6	100.0	120.2	104.4
21, 22, 23	FIRE SUPPRESSION, PLUMBING & HVAC	96.5	131.8	110.8	96.7	168.7	125.8	96.5	131.7	110.8	100.4	130.3	112.5	96.7	159.7	122.2	100.0	131.8	112.8
26, 27, 3370	ELECTRICAL, COMMUNICATIONS & UTIL.	93.2	113.5	103.5	101.4	169.9	136.2	114.1	130.5	122.4	103.9	122.4	113.3	102.0	136.5	119.5	90.0	113.5	101.9
MF2016	WEIGHTED AVERAGE	96.9	127.1	109.8	98.5	151.4	121.2	94.8	129.4	109.7	109.1	131.5	118.7	101.8	145.2	120.4	98.8	127.1	110.9

CALIFORNIA

DIVISION		SACRAMENTO 942, 956-958 MAT.	INST.	TOTAL	SALINAS 939 MAT.	INST.	TOTAL	SAN BERNARDINO 923-924 MAT.	INST.	TOTAL	SAN DIEGO 919-921 MAT.	INST.	TOTAL	SAN FRANCISCO 940-941 MAT.	INST.	TOTAL	SAN JOSE 951 MAT.	INST.	TOTAL
015433	CONTRACTOR EQUIPMENT		101.6	101.6		99.8	99.8		100.7	100.7		102.5	102.5		110.9	110.9		101.3	101.3
0241, 31 - 34	SITE & INFRASTRUCTURE, DEMOLITION	97.8	116.3	110.7	115.4	107.9	110.2	78.4	108.6	99.5	104.7	107.8	106.8	117.5	114.2	115.2	134.2	102.3	112.0
0310	Concrete Forming & Accessories	100.2	155.0	146.9	103.8	155.3	147.7	105.4	140.1	134.9	104.4	128.0	124.5	105.7	167.7	158.5	102.0	166.9	157.3
0320	Concrete Reinforcing	84.6	130.9	108.2	102.8	131.0	117.2	106.2	129.9	118.2	100.1	129.8	115.2	103.7	132.6	118.4	94.9	131.4	113.5
0330	Cast-in-Place Concrete	89.5	130.5	104.7	96.3	129.8	108.8	61.0	133.2	87.9	90.6	123.5	102.9	118.1	133.8	123.9	107.5	132.7	116.9
03	CONCRETE	95.1	140.1	115.0	103.0	140.4	119.6	69.5	134.6	98.3	100.1	125.8	111.5	114.7	148.1	129.5	102.0	147.0	121.9
04	MASONRY	106.8	147.2	131.8	95.4	147.3	127.5	84.0	139.8	118.5	89.8	132.3	116.0	142.9	153.6	149.5	132.3	150.4	143.5
05	METALS	97.6	107.5	100.6	99.2	113.2	103.4	106.1	111.7	107.8	97.8	113.7	102.6	106.7	122.6	111.5	100.3	119.0	106.0
06	WOOD, PLASTICS & COMPOSITES	88.9	159.3	127.1	98.1	159.0	131.2	101.6	137.4	121.0	99.4	124.9	113.2	97.0	172.6	138.1	106.5	172.4	142.3
07	THERMAL & MOISTURE PROTECTION	115.1	141.0	126.4	101.5	145.2	120.5	105.1	136.2	118.7	101.2	120.3	109.5	111.4	155.7	130.7	105.9	155.6	127.6
08	OPENINGS	105.8	146.7	115.3	95.7	152.0	108.8	102.1	135.9	110.0	98.3	126.1	104.7	98.5	159.8	112.7	94.6	159.4	109.7
0920	Plaster & Gypsum Board	103.2	160.5	140.9	98.9	160.5	139.5	111.8	138.3	129.3	96.3	125.3	115.4	105.3	174.1	150.6	107.3	174.1	151.3
0950, 0980	Ceilings & Acoustic Treatment	99.4	160.5	140.8	106.1	160.5	143.0	112.6	138.3	130.0	123.1	125.3	124.6	108.6	174.1	153.0	113.6	174.1	154.6
0960	Flooring	108.2	127.6	113.9	96.9	140.6	109.6	102.8	120.8	108.0	103.6	120.4	108.4	110.0	142.0	119.3	90.8	142.0	105.7
0970, 0990	Wall Finishes & Painting/Coating	109.9	137.6	126.3	89.7	162.7	132.9	89.1	121.1	108.1	103.0	121.1	113.7	111.2	179.4	151.6	92.4	162.7	134.0
09	FINISHES	103.2	150.4	128.9	93.4	155.2	127.1	98.0	134.5	117.9	105.5	125.5	116.4	107.5	165.6	139.2	98.9	163.6	134.2
COVERS	DIVS. 10 - 14, 25, 28, 41, 43, 44, 46	100.0	127.6	106.1	100.0	127.1	106.0	100.0	120.2	104.4	100.0	117.0	103.7	100.0	130.0	106.6	100.0	129.4	106.5
21, 22, 23	FIRE SUPPRESSION, PLUMBING & HVAC	100.1	129.9	112.2	96.6	136.3	112.6	96.5	131.8	110.8	99.9	128.4	111.4	100.1	185.2	134.5	100.0	168.6	127.7
26, 27, 3370	ELECTRICAL, COMMUNICATIONS & UTIL.	97.0	122.4	109.9	96.3	127.6	112.2	93.2	111.8	102.7	100.6	102.7	101.7	101.6	176.3	139.6	98.9	169.9	135.0
MF2016	WEIGHTED AVERAGE	100.1	132.7	114.1	98.3	135.7	114.3	94.5	126.8	108.4	99.6	120.8	108.7	106.3	158.1	128.5	102.1	151.7	123.4

CALIFORNIA

DIVISION		SAN LUIS OBISPO 934 MAT.	INST.	TOTAL	SAN MATEO 944 MAT.	INST.	TOTAL	SAN RAFAEL 949 MAT.	INST.	TOTAL	SANTA ANA 926-927 MAT.	INST.	TOTAL	SANTA BARBARA 931 MAT.	INST.	TOTAL	SANTA CRUZ 950 MAT.	INST.	TOTAL
015433	CONTRACTOR EQUIPMENT		99.8	99.8		102.8	102.8		102.0	102.0		100.7	100.7		99.8	99.8		101.3	101.3
0241, 31 - 34	SITE & INFRASTRUCTURE, DEMOLITION	107.5	108.4	108.1	113.6	110.7	111.5	110.5	116.2	114.5	90.5	108.6	103.1	101.9	108.4	106.4	133.8	102.2	111.7
0310	Concrete Forming & Accessories	110.0	140.2	135.7	105.9	167.1	158.0	111.1	167.0	158.7	105.7	140.1	135.0	101.1	140.1	134.3	102.0	155.6	147.6
0320	Concrete Reinforcing	104.1	129.8	117.2	89.1	131.6	110.7	89.7	131.6	111.0	109.8	129.9	120.1	102.3	129.8	116.3	117.4	131.1	124.3
0330	Cast-in-Place Concrete	103.6	132.4	114.3	105.0	133.6	115.7	122.2	132.6	126.1	77.9	133.2	98.5	93.4	134.3	111.5	104.3	141.4	120.7
03	CONCRETE	101.0	134.3	115.8	102.0	147.0	122.0	124.7	146.4	134.3	86.1	134.6	107.6	93.4	134.3	111.5	104.3	141.4	120.7
04	MASONRY	97.1	137.3	121.9	124.3	153.6	142.4	101.5	153.6	133.7	73.6	140.2	114.7	95.7	137.3	121.4	136.1	147.5	143.1
05	METALS	97.4	111.2	101.6	99.3	113.7	103.7	100.4	110.0	103.3	106.2	111.7	107.9	95.1	111.2	100.0	107.8	117.6	110.8
06	WOOD, PLASTICS & COMPOSITES	100.5	137.4	120.6	99.7	172.6	139.3	97.4	172.4	138.1	103.5	137.4	121.9	93.3	137.4	117.3	106.5	159.2	135.1
07	THERMAL & MOISTURE PROTECTION	101.6	133.7	115.6	105.5	155.9	127.5	109.9	156.2	130.0	106.5	136.3	119.5	101.1	133.7	115.3	105.5	147.5	123.8
08	OPENINGS	93.9	133.8	103.2	92.8	159.5	108.3	103.4	159.4	116.4	101.4	135.9	109.4	95.5	135.9	104.9	95.9	152.1	108.9
0920	Plaster & Gypsum Board	104.7	138.3	126.8	110.8	174.1	152.5	112.7	174.1	153.2	113.3	138.3	129.8	97.8	138.3	124.5	115.3	160.5	145.1
0950, 0980	Ceilings & Acoustic Treatment	106.1	138.3	127.9	99.4	174.1	150.0	106.2	174.1	152.2	112.6	138.3	130.0	107.7	138.3	128.4	113.4	160.5	145.3
0960	Flooring	103.1	120.4	108.1	111.8	142.0	120.6	123.4	140.6	128.4	103.4	120.4	108.3	96.0	120.4	103.1	94.8	140.6	108.1
0970, 0990	Wall Finishes & Painting/Coating	88.9	115.6	104.7	113.3	162.7	142.5	109.3	169.9	145.2	89.1	121.1	108.1	88.9	115.6	104.7	92.5	162.7	134.0
09	FINISHES	95.1	133.9	116.3	106.5	163.7	137.8	109.1	164.1	139.2	99.5	134.5	118.6	91.8	133.9	114.8	101.2	155.2	130.7
COVERS	DIVS. 10 - 14, 25, 28, 41, 43, 44, 46	100.0	126.4	105.8	100.0	129.9	106.6	100.0	129.2	106.4	100.0	120.2	104.4	100.0	119.5	104.3	100.0	127.4	106.0
21, 22, 23	FIRE SUPPRESSION, PLUMBING & HVAC	96.6	131.8	110.8	96.7	163.6	123.8	96.7	185.1	132.5	96.5	131.8	110.8	100.1	131.8	112.9	100.0	136.7	114.8
26, 27, 3370	ELECTRICAL, COMMUNICATIONS & UTIL.	95.3	113.1	104.3	101.4	163.0	132.7	98.3	124.3	111.5	93.2	112.8	103.2	94.2	114.2	104.3	98.1	127.6	113.1
MF2016	WEIGHTED AVERAGE	97.5	126.6	110.0	100.9	150.2	122.1	103.9	149.3	123.4	96.5	127.0	109.6	96.5	126.7	109.4	104.0	136.0	117.7

City Cost Indexes

CALIFORNIA / COLORADO

DIVISION		SANTA ROSA 954 MAT.	INST.	TOTAL	STOCKTON 952 MAT.	INST.	TOTAL	SUSANVILLE 961 MAT.	INST.	TOTAL	VALLEJO 945 MAT.	INST.	TOTAL	VAN NUYS 913-916 MAT.	INST.	TOTAL	ALAMOSA 811 MAT.	INST.	TOTAL
015433	CONTRACTOR EQUIPMENT		100.1	100.1		99.6	99.6		99.6	99.6		102.0	102.0		97.7	97.7		93.1	93.1
0241, 31 - 34	SITE & INFRASTRUCTURE, DEMOLITION	104.5	107.3	106.5	102.8	107.3	105.9	137.1	107.3	116.4	99.1	116.1	111.0	114.9	110.3	111.7	141.1	88.5	104.5
0310	Concrete Forming & Accessories	99.0	166.2	156.2	100.1	154.6	146.5	104.0	152.5	145.3	101.7	165.9	156.4	110.6	140.0	135.6	101.3	62.7	68.4
0320	Concrete Reinforcing	104.8	131.5	118.4	107.6	130.5	119.3	132.1	130.4	131.2	90.8	131.4	111.5	103.4	130.1	116.9	111.5	69.8	90.3
0330	Cast-in-Place Concrete	101.5	131.1	112.5	90.1	129.5	104.8	92.5	129.5	106.3	97.4	131.7	110.2	83.5	129.8	100.7	100.1	73.5	90.2
03	CONCRETE	105.0	145.9	123.1	95.3	139.9	115.0	110.9	138.9	123.3	99.6	145.6	120.0	104.4	133.3	117.2	107.8	68.4	90.4
04	MASONRY	102.3	152.5	133.3	102.0	141.7	126.5	130.7	139.2	136.0	77.4	152.5	123.8	111.7	141.9	130.4	129.7	60.6	87.0
05	METALS	106.5	115.1	109.1	102.4	112.2	105.4	111.4	112.1	111.6	100.3	109.3	103.1	81.5	110.5	90.3	100.3	78.4	93.7
06	WOOD, PLASTICS & COMPOSITES	94.1	172.2	136.5	99.9	159.0	132.0	111.9	156.0	135.9	86.6	172.4	133.2	97.2	137.1	118.9	99.6	63.0	79.7
07	THERMAL & MOISTURE PROTECTION	106.9	152.9	126.9	110.0	138.1	122.2	133.8	138.4	135.8	107.9	153.6	127.8	97.9	133.2	113.3	107.7	68.2	90.5
08	OPENINGS	103.0	159.3	116.1	103.5	146.6	113.5	119.0	144.9	125.0	105.1	159.4	117.7	89.6	135.7	100.3	98.5	66.4	91.0
0920	Plaster & Gypsum Board	108.2	174.1	151.6	111.2	160.5	143.7	109.6	157.4	141.1	106.9	174.1	151.2	94.4	138.3	123.3	81.5	61.9	68.6
0950, 0980	Ceilings & Acoustic Treatment	112.6	174.1	154.3	122.1	160.5	148.1	140.0	157.4	151.8	108.1	174.1	152.8	109.2	138.3	128.9	103.8	61.9	75.4
0960	Flooring	99.9	133.5	109.7	97.2	129.9	106.7	94.0	132.9	105.3	117.5	142.0	124.6	97.3	120.4	104.0	108.4	71.3	97.6
0970, 0990	Wall Finishes & Painting/Coating	89.1	169.9	136.9	92.1	137.6	119.0	100.0	137.6	122.3	110.3	169.9	145.6	105.4	121.1	114.7	102.7	60.8	77.9
09	FINISHES	98.6	162.5	133.5	101.5	150.6	128.3	106.0	149.3	129.6	105.8	164.1	137.6	100.1	134.3	118.8	98.4	63.5	79.4
COVERS	DIVS. 10 - 14, 25, 28, 41, 43, 44, 46	100.0	128.3	106.2	100.0	127.0	106.0	100.0	126.7	105.9	100.0	128.9	106.4	100.0	119.6	104.3	100.0	84.8	96.6
21, 22, 23	FIRE SUPPRESSION, PLUMBING & HVAC	96.5	184.5	132.1	100.0	130.3	112.2	96.9	130.3	110.4	100.2	147.9	119.5	96.5	131.7	110.8	96.5	66.0	84.2
26, 27, 3370	ELECTRICAL, COMMUNICATIONS & UTIL.	93.5	124.3	109.2	96.0	114.3	105.3	104.3	122.4	113.5	94.5	127.9	111.4	114.1	130.5	122.4	97.0	64.8	80.6
MF2016	WEIGHTED AVERAGE	100.8	148.5	121.2	100.3	130.8	113.4	108.8	131.2	118.4	99.5	141.5	117.5	97.9	129.4	111.4	102.3	68.8	87.9

COLORADO

DIVISION		BOULDER 803 MAT.	INST.	TOTAL	COLORADO SPRINGS 808-809 MAT.	INST.	TOTAL	DENVER 800-802 MAT.	INST.	TOTAL	DURANGO 813 MAT.	INST.	TOTAL	FORT COLLINS 805 MAT.	INST.	TOTAL	FORT MORGAN 807 MAT.	INST.	TOTAL
015433	CONTRACTOR EQUIPMENT		95.2	95.2		93.1	93.1		99.4	99.4		93.1	93.1		95.2	95.2		95.2	95.2
0241, 31 - 34	SITE & INFRASTRUCTURE, DEMOLITION	99.4	95.8	96.9	101.3	91.3	94.3	106.3	102.3	103.5	134.6	88.5	102.5	112.4	95.7	100.7	101.9	95.5	97.4
0310	Concrete Forming & Accessories	102.1	73.6	77.8	92.8	65.1	69.2	101.4	64.8	70.3	107.4	62.5	69.2	99.8	68.6	73.3	102.6	62.4	68.4
0320	Concrete Reinforcing	98.3	69.8	83.8	97.5	69.9	83.5	97.5	69.9	83.5	111.5	69.7	90.3	98.3	69.7	83.8	98.5	69.7	83.9
0330	Cast-in-Place Concrete	112.5	72.8	97.7	115.5	73.9	100.0	120.5	71.3	102.2	115.1	73.4	99.6	127.0	71.4	106.3	110.3	72.7	96.3
03	CONCRETE	101.1	73.0	88.7	104.5	69.6	89.1	107.2	68.6	90.1	110.1	68.2	91.5	112.0	70.3	93.5	99.6	67.9	85.6
04	MASONRY	101.3	63.3	77.8	103.2	63.6	78.7	104.4	63.8	79.3	116.7	60.6	82.1	119.2	60.9	83.2	116.7	60.9	82.2
05	METALS	90.8	77.4	86.7	93.8	78.1	89.0	96.1	78.6	90.8	100.3	78.2	93.6	92.0	77.3	87.6	90.5	77.3	86.5
06	WOOD, PLASTICS & COMPOSITES	106.7	76.3	90.2	96.1	65.5	79.5	104.3	65.3	83.1	109.5	63.0	84.2	104.2	70.9	86.1	106.7	62.5	82.7
07	THERMAL & MOISTURE PROTECTION	106.0	72.7	91.5	106.7	71.4	91.4	104.4	70.9	89.8	107.7	68.2	90.5	106.3	70.4	90.7	105.9	69.6	90.1
08	OPENINGS	97.5	73.4	91.9	101.7	67.8	93.8	104.5	67.7	96.0	105.4	66.4	96.4	97.4	70.4	91.2	97.4	66.1	90.1
0920	Plaster & Gypsum Board	116.5	76.0	89.8	100.0	64.8	76.8	111.7	64.8	80.8	95.3	61.9	73.3	110.7	70.5	84.2	116.5	61.9	80.5
0950, 0980	Ceilings & Acoustic Treatment	86.2	76.0	79.3	93.0	64.8	73.9	97.5	64.8	75.3	103.8	61.9	75.4	86.2	70.5	75.5	86.2	61.9	69.7
0960	Flooring	108.3	71.3	97.6	99.7	71.3	91.5	105.9	71.3	95.9	113.4	71.3	101.2	104.6	71.3	95.0	108.7	71.3	97.9
0970, 0990	Wall Finishes & Painting/Coating	99.4	63.0	77.8	99.1	63.0	77.7	106.2	63.0	80.6	102.7	60.8	77.9	99.4	60.8	76.5	99.4	60.8	76.5
09	FINISHES	100.5	72.2	85.0	97.0	65.5	79.8	102.3	65.3	82.1	101.0	63.5	80.6	99.1	68.2	82.2	100.6	63.2	80.2
COVERS	DIVS. 10 - 14, 25, 28, 41, 43, 44, 46	100.0	86.1	96.9	100.0	84.7	96.6	100.0	84.2	96.5	100.0	84.7	96.6	100.0	84.7	96.6	100.0	83.8	96.4
21, 22, 23	FIRE SUPPRESSION, PLUMBING & HVAC	96.6	74.6	87.7	100.2	73.8	89.5	100.0	73.6	89.4	96.5	57.2	80.6	100.1	73.3	89.2	96.6	73.3	87.2
26, 27, 3370	ELECTRICAL, COMMUNICATIONS & UTIL.	98.0	80.0	88.9	101.4	74.9	87.9	103.2	80.2	91.5	96.4	51.5	73.6	98.0	77.2	87.4	98.3	80.0	89.0
MF2016	WEIGHTED AVERAGE	97.7	75.9	88.3	100.0	73.0	88.4	101.8	74.4	90.0	102.7	65.0	86.6	101.1	73.7	89.4	98.3	72.9	87.4

COLORADO

DIVISION		GLENWOOD SPRINGS 816 MAT.	INST.	TOTAL	GOLDEN 804 MAT.	INST.	TOTAL	GRAND JUNCTION 815 MAT.	INST.	TOTAL	GREELEY 806 MAT.	INST.	TOTAL	MONTROSE 814 MAT.	INST.	TOTAL	PUEBLO 810 MAT.	INST.	TOTAL
015433	CONTRACTOR EQUIPMENT		96.3	96.3		95.2	95.2		96.3	96.3		95.2	95.2		94.7	94.7		93.1	93.1
0241, 31 - 34	SITE & INFRASTRUCTURE, DEMOLITION	150.4	96.4	112.8	112.9	95.8	101.0	134.0	96.6	107.9	98.0	95.7	96.4	143.8	92.5	108.0	125.5	88.9	100.0
0310	Concrete Forming & Accessories	98.2	52.4	59.2	95.2	63.4	68.1	106.3	73.0	78.0	97.7	72.6	76.3	97.7	62.7	67.9	103.6	65.4	71.1
0320	Concrete Reinforcing	110.3	69.7	89.7	98.5	69.7	83.9	110.7	69.8	89.9	98.3	69.7	83.8	110.2	69.8	89.6	106.6	69.8	87.9
0330	Cast-in-Place Concrete	100.1	72.3	89.7	110.4	74.1	96.9	110.8	73.7	97.0	106.1	72.7	93.7	100.1	72.9	90.0	99.4	74.2	90.0
03	CONCRETE	112.8	63.3	90.8	109.5	68.9	91.5	106.7	73.2	91.8	96.3	72.6	85.8	104.1	68.2	88.2	97.1	69.9	85.0
04	MASONRY	101.6	60.8	76.4	119.7	63.3	84.9	136.8	64.0	91.8	112.5	60.9	80.7	109.5	60.6	79.3	97.8	61.6	75.4
05	METALS	99.9	77.9	93.3	90.7	77.3	86.7	101.5	78.2	94.5	92.0	77.4	87.6	99.1	78.3	92.8	103.2	79.0	95.9
06	WOOD, PLASTICS & COMPOSITES	94.6	49.2	69.9	98.5	62.5	79.0	107.2	75.7	90.1	101.3	76.2	87.7	95.6	62.8	77.8	102.3	65.8	82.5
07	THERMAL & MOISTURE PROTECTION	107.6	66.7	89.8	107.1	71.3	91.6	106.6	71.6	91.3	105.5	71.0	90.5	107.7	68.2	90.5	106.0	69.5	90.1
08	OPENINGS	104.4	58.8	93.8	97.5	66.1	90.2	105.1	73.1	97.7	97.4	73.7	91.9	105.6	66.3	96.5	100.3	67.9	92.8
0920	Plaster & Gypsum Board	122.5	48.0	73.4	108.0	61.9	77.6	135.9	75.2	96.0	109.2	76.0	87.3	80.6	61.9	68.3	84.9	64.8	71.7
0950, 0980	Ceilings & Acoustic Treatment	103.0	48.0	65.7	86.2	61.9	69.7	103.0	75.2	84.2	86.2	76.0	79.3	103.8	61.9	75.4	109.8	64.8	79.3
0960	Flooring	107.5	71.3	97.0	102.3	71.3	93.3	112.8	71.3	100.7	103.5	71.3	94.2	110.8	71.3	99.4	102.7	75.5	99.7
0970, 0990	Wall Finishes & Painting/Coating	102.7	60.8	77.9	99.4	63.0	77.8	102.7	63.0	79.2	99.4	60.8	76.5	102.7	60.8	77.9	102.7	60.8	77.9
09	FINISHES	104.3	55.4	77.6	98.9	64.1	79.9	106.0	71.8	87.3	97.8	71.3	83.3	99.0	63.4	79.6	98.6	66.3	80.9
COVERS	DIVS. 10 - 14, 25, 28, 41, 43, 44, 46	100.0	82.6	96.2	100.0	84.6	96.6	100.0	86.3	97.0	100.0	85.3	96.8	100.0	84.4	96.6	100.0	85.4	96.8
21, 22, 23	FIRE SUPPRESSION, PLUMBING & HVAC	96.5	67.0	84.6	96.6	74.6	87.7	99.9	75.8	90.1	100.1	73.3	89.2	96.5	75.8	88.1	99.9	73.8	89.4
26, 27, 3370	ELECTRICAL, COMMUNICATIONS & UTIL.	93.9	51.5	72.3	98.4	80.0	89.0	96.1	51.5	73.4	98.0	77.2	87.4	96.1	51.5	73.4	97.0	64.8	80.6
MF2016	WEIGHTED AVERAGE	102.5	65.4	86.6	99.8	73.7	88.6	104.5	72.2	90.7	98.3	74.7	88.2	101.4	69.3	87.6	100.5	71.4	88.0

For customer support on your Electrical Costs with RSMeans data, call 800.448.8182.

519

Table 1

DIVISION		COLORADO SALIDA 812 MAT.	INST.	TOTAL	CONNECTICUT BRIDGEPORT 066 MAT.	INST.	TOTAL	BRISTOL 060 MAT.	INST.	TOTAL	HARTFORD 061 MAT.	INST.	TOTAL	MERIDEN 064 MAT.	INST.	TOTAL	NEW BRITAIN 060 MAT.	INST.	TOTAL
015433	CONTRACTOR EQUIPMENT		94.7	94.7		97.9	97.9		97.9	97.9		97.9	97.9		98.3	98.3		97.9	97.9
0241, 31 - 34	SITE & INFRASTRUCTURE, DEMOLITION	134.3	92.4	105.1	103.3	102.3	102.6	102.4	102.3	102.3	97.8	102.3	100.9	100.2	103.1	102.2	102.6	102.3	102.4
0310	Concrete Forming & Accessories	106.2	52.5	60.5	98.1	119.5	116.3	98.1	119.3	116.2	99.2	119.3	116.3	97.9	119.3	116.1	98.6	119.3	116.2
0320	Concrete Reinforcing	109.9	69.7	89.5	125.5	145.1	135.4	125.5	145.1	135.4	120.3	145.1	132.9	125.5	145.1	135.4	125.5	145.1	135.4
0330	Cast-in-Place Concrete	114.7	72.8	99.1	102.1	128.8	112.1	95.7	128.7	108.0	100.2	128.7	110.8	92.1	128.7	105.7	97.2	128.7	108.9
03	CONCRETE	105.3	63.5	86.8	100.1	126.2	111.7	97.1	126.1	109.9	97.3	126.1	110.1	95.4	126.1	109.0	97.9	126.1	110.4
04	MASONRY	138.1	60.6	90.2	116.6	130.9	125.4	107.4	130.9	121.9	106.8	130.9	121.7	107.0	130.9	121.8	109.7	130.9	122.8
05	METALS	98.7	78.1	92.5	87.9	115.7	96.4	87.9	115.6	96.3	92.3	115.6	99.4	85.5	115.6	94.6	84.6	115.6	94.0
06	WOOD, PLASTICS & COMPOSITES	103.6	49.3	74.1	100.8	117.5	109.9	100.8	117.5	109.9	91.0	117.5	105.4	100.8	117.5	109.9	100.8	117.5	109.9
07	THERMAL & MOISTURE PROTECTION	106.5	66.7	89.2	97.8	125.8	110.0	97.9	121.9	108.4	102.6	121.9	111.0	97.9	121.9	108.4	97.9	122.2	108.5
08	OPENINGS	98.6	58.8	89.3	98.4	122.7	104.0	98.4	122.7	104.0	96.5	122.7	102.5	100.6	122.7	105.7	98.4	122.7	104.0
0920	Plaster & Gypsum Board	80.9	48.0	59.2	116.7	117.7	117.4	116.7	117.7	117.4	104.7	117.7	113.3	117.8	117.7	117.7	116.7	117.7	117.4
0950, 0980	Ceilings & Acoustic Treatment	103.8	48.0	66.0	97.9	117.7	111.3	97.9	117.7	111.3	94.1	117.7	110.1	101.1	117.7	112.4	97.9	117.7	111.3
0960	Flooring	116.1	71.3	103.1	90.6	131.2	102.4	90.6	128.7	101.6	94.9	131.2	105.4	90.6	128.7	101.6	90.6	128.7	101.6
0970, 0990	Wall Finishes & Painting/Coating	102.7	60.8	77.9	89.8	128.8	112.9	89.8	128.8	112.9	94.3	128.8	114.8	89.8	128.8	112.9	89.8	128.8	112.9
09	FINISHES	99.5	55.5	75.5	90.9	122.2	108.0	90.9	121.8	107.8	92.0	122.2	108.5	91.7	121.8	108.1	90.9	121.8	107.8
COVERS	DIVS. 10 - 14, 25, 28, 41, 43, 44, 46	100.0	83.0	96.3	100.0	113.4	102.9	100.0	113.4	102.9	100.0	113.4	102.9	100.0	113.4	102.9	100.0	113.4	102.9
21, 22, 23	FIRE SUPPRESSION, PLUMBING & HVAC	96.5	65.0	83.8	100.1	118.5	107.5	100.1	118.5	107.5	100.0	118.5	107.5	96.6	118.5	105.5	100.1	118.5	107.5
26, 27, 3370	ELECTRICAL, COMMUNICATIONS & UTIL.	96.3	64.8	80.3	96.0	107.6	101.9	96.0	106.2	101.2	95.7	109.3	102.6	96.0	108.4	102.3	96.1	106.2	101.3
MF2016	WEIGHTED AVERAGE	102.0	66.6	86.8	97.6	118.4	106.5	96.8	118.1	105.9	97.2	118.5	106.4	95.5	118.4	105.4	96.4	118.1	105.7

Table 2 — CONNECTICUT

DIVISION		NEW HAVEN 065 MAT.	INST.	TOTAL	NEW LONDON 063 MAT.	INST.	TOTAL	NORWALK 068 MAT.	INST.	TOTAL	STAMFORD 069 MAT.	INST.	TOTAL	WATERBURY 067 MAT.	INST.	TOTAL	WILLIMANTIC 062 MAT.	INST.	TOTAL
015433	CONTRACTOR EQUIPMENT		98.3	98.3		98.3	98.3		97.9	97.9		97.9	97.9		97.9	97.9		97.9	97.9
0241, 31 - 34	SITE & INFRASTRUCTURE, DEMOLITION	102.5	103.1	102.9	94.9	102.8	100.4	103.1	102.3	102.5	103.7	102.3	102.8	103.1	102.3	102.6	103.0	102.1	102.3
0310	Concrete Forming & Accessories	97.9	119.4	116.2	97.9	119.3	116.1	98.1	119.5	116.3	98.1	119.8	116.6	98.1	119.4	116.2	98.1	119.2	116.0
0320	Concrete Reinforcing	125.5	145.1	135.4	98.3	145.1	122.1	125.5	145.2	135.5	125.5	145.3	135.5	125.5	145.1	135.4	125.5	145.1	135.4
0330	Cast-in-Place Concrete	98.9	128.7	110.0	84.3	128.7	100.8	100.5	130.0	111.5	102.1	130.1	112.6	102.1	128.7	112.0	95.4	127.0	107.1
03	CONCRETE	111.9	126.1	118.2	85.8	126.0	103.7	99.4	126.6	111.4	100.1	126.8	112.0	100.1	126.1	111.6	97.0	125.4	109.6
04	MASONRY	107.7	130.9	122.0	106.0	130.9	121.4	107.1	132.2	122.6	108.0	132.2	122.9	108.0	130.9	122.1	107.3	130.9	121.9
05	METALS	84.8	115.6	94.1	84.5	115.6	93.9	87.9	115.9	96.4	87.9	116.2	96.5	87.9	115.6	96.3	87.7	115.4	96.1
06	WOOD, PLASTICS & COMPOSITES	100.8	117.5	109.9	100.8	117.5	109.9	100.8	117.5	109.9	100.8	117.5	109.9	100.8	117.5	109.9	100.8	117.5	109.9
07	THERMAL & MOISTURE PROTECTION	98.0	122.5	108.7	97.8	121.9	108.3	98.0	126.4	110.3	97.9	126.4	110.3	97.9	122.5	108.6	98.1	121.3	108.2
08	OPENINGS	98.4	122.7	104.0	98.4	122.7	104.0	98.4	122.7	104.0	98.4	122.7	104.0	98.4	122.7	104.0	100.8	122.7	105.9
0920	Plaster & Gypsum Board	116.7	117.7	117.4	116.7	117.7	117.4	116.7	117.7	117.4	116.7	117.7	117.4	116.7	117.7	117.4	116.7	117.7	117.4
0950, 0980	Ceilings & Acoustic Treatment	97.9	117.7	111.3	96.0	117.7	110.7	97.9	117.7	111.3	97.9	117.7	111.3	97.9	117.7	111.3	96.0	117.7	110.7
0960	Flooring	90.6	131.2	102.4	90.6	128.7	101.6	90.6	128.7	101.6	90.6	131.2	102.4	90.6	131.2	102.4	90.6	125.2	100.6
0970, 0990	Wall Finishes & Painting/Coating	89.8	128.8	112.9	89.8	128.8	112.9	89.8	128.8	112.9	89.8	128.8	112.9	89.8	128.8	112.9	89.8	128.8	112.9
09	FINISHES	91.0	122.2	108.0	90.1	121.8	107.4	90.9	121.8	107.8	91.0	122.2	108.0	90.8	122.2	107.9	90.6	121.2	107.3
COVERS	DIVS. 10 - 14, 25, 28, 41, 43, 44, 46	100.0	113.4	102.9	100.0	113.4	102.9	100.0	113.4	102.9	100.0	113.6	103.0	100.0	113.4	103.0	100.0	113.4	102.9
21, 22, 23	FIRE SUPPRESSION, PLUMBING & HVAC	100.1	118.5	107.5	96.6	118.5	105.5	100.1	118.5	107.5	100.1	118.6	107.5	100.1	118.5	107.5	100.1	118.4	107.5
26, 27, 3370	ELECTRICAL, COMMUNICATIONS & UTIL.	96.0	108.6	102.4	93.3	108.4	101.0	96.0	107.4	101.8	96.0	153.8	125.4	95.6	109.7	102.8	96.0	109.3	102.8
MF2016	WEIGHTED AVERAGE	98.1	118.6	106.9	93.6	118.4	104.2	97.0	118.6	106.3	97.2	125.2	109.2	97.1	118.6	106.3	97.0	118.2	106.1

Table 3

DIVISION		D.C. WASHINGTON 200 - 205 MAT.	INST.	TOTAL	DELAWARE DOVER 199 MAT.	INST.	TOTAL	NEWARK 197 MAT.	INST.	TOTAL	WILMINGTON 198 MAT.	INST.	TOTAL	FLORIDA DAYTONA BEACH 321 MAT.	INST.	TOTAL	FORT LAUDERDALE 333 MAT.	INST.	TOTAL
015433	CONTRACTOR EQUIPMENT		105.6	105.6		121.3	121.3		121.3	121.3		121.5	121.5		102.1	102.1		95.2	95.2
0241, 31 - 34	SITE & INFRASTRUCTURE, DEMOLITION	103.8	95.9	98.3	100.7	113.6	109.7	102.3	113.6	110.2	104.9	114.0	111.2	113.4	88.9	96.3	92.5	76.8	81.5
0310	Concrete Forming & Accessories	96.7	74.0	77.3	97.3	101.9	101.2	96.7	101.9	101.2	95.3	101.9	101.0	98.4	62.7	68.0	92.4	63.3	67.6
0320	Concrete Reinforcing	100.9	88.7	94.7	109.1	118.4	113.8	103.3	118.4	111.0	107.4	118.4	113.0	93.3	62.6	77.7	94.5	61.4	77.7
0330	Cast-in-Place Concrete	112.6	80.6	100.7	100.6	107.5	103.1	90.0	107.5	96.5	102.3	107.5	104.3	88.1	65.9	79.8	90.1	62.0	79.6
03	CONCRETE	106.8	79.8	94.8	99.0	107.7	102.9	93.4	107.7	99.8	99.5	107.7	103.2	83.7	65.5	75.6	83.9	64.2	75.2
04	MASONRY	100.4	80.9	88.3	97.5	100.1	99.1	101.3	100.1	100.5	97.1	100.1	98.9	88.4	59.5	70.6	93.3	62.3	74.1
05	METALS	101.1	95.8	99.5	103.0	123.7	109.2	103.9	123.7	109.9	103.0	123.7	109.2	102.1	89.0	98.1	98.9	89.2	96.0
06	WOOD, PLASTICS & COMPOSITES	95.0	72.7	82.9	88.7	100.2	95.0	91.1	100.2	96.0	81.7	100.2	91.8	96.4	61.8	77.6	80.2	65.2	72.1
07	THERMAL & MOISTURE PROTECTION	102.7	84.9	95.0	104.7	112.7	108.2	108.2	112.7	110.2	103.5	112.7	107.5	100.8	65.7	85.5	101.1	62.5	84.3
08	OPENINGS	99.3	75.9	93.9	90.4	110.4	95.1	93.4	110.4	97.4	88.8	110.4	93.9	95.1	60.9	87.2	96.4	62.4	88.5
0920	Plaster & Gypsum Board	105.7	71.9	83.5	97.8	100.2	99.4	95.9	100.2	98.7	96.8	100.2	99.1	95.0	61.2	72.8	109.2	64.7	79.9
0950, 0980	Ceilings & Acoustic Treatment	105.6	71.9	82.8	99.1	100.2	99.9	96.6	100.2	99.1	90.3	100.2	97.0	83.7	61.2	68.5	86.1	64.7	71.6
0960	Flooring	96.4	78.1	91.1	90.4	109.0	95.8	87.7	109.0	93.9	91.5	109.0	96.6	101.0	56.7	88.2	96.8	76.7	91.0
0970, 0990	Wall Finishes & Painting/Coating	99.1	75.0	84.8	88.6	122.2	108.5	84.6	122.2	106.8	86.8	122.2	107.7	97.1	63.1	77.0	88.4	58.6	70.8
09	FINISHES	97.9	74.0	84.9	91.5	104.5	98.6	87.7	104.5	96.9	90.1	104.5	97.9	90.4	61.0	74.4	90.5	65.1	76.7
COVERS	DIVS. 10 - 14, 25, 28, 41, 43, 44, 46	100.0	96.3	99.2	100.0	105.4	101.2	100.0	105.4	101.2	100.0	105.4	101.2	100.0	82.7	96.2	100.0	82.7	96.2
21, 22, 23	FIRE SUPPRESSION, PLUMBING & HVAC	100.1	90.3	96.1	100.0	122.0	108.9	100.0	122.0	109.0	100.1	122.0	108.9	99.9	76.7	90.6	100.0	68.2	87.1
26, 27, 3370	ELECTRICAL, COMMUNICATIONS & UTIL.	100.6	102.4	101.5	96.0	112.4	104.3	98.5	112.4	105.6	97.3	112.4	105.0	95.5	60.3	77.6	92.7	69.1	80.7
MF2016	WEIGHTED AVERAGE	101.0	87.7	95.3	98.2	112.3	104.2	98.3	112.3	104.3	98.2	112.3	104.2	96.3	70.3	85.2	95.3	69.5	84.2

		FLORIDA																
	DIVISION	FORT MYERS			GAINESVILLE			JACKSONVILLE			LAKELAND			MELBOURNE			MIAMI		
		339, 341			326, 344			320, 322			338			329			330 - 332, 340		
		MAT.	INST.	TOTAL	MAT.	INST.	TOTAL	MAT.	INST.	TOTAL	MAT.	INST.	TOTAL	MAT.	INST.	TOTAL	MAT.	INST.	TOTAL
015433	CONTRACTOR EQUIPMENT		102.1	102.1		102.1	102.1		102.1	102.1		102.1	102.1		102.1	102.1		95.2	95.2
0241, 31 - 34	SITE & INFRASTRUCTURE, DEMOLITION	104.3	89.0	93.7	122.6	88.7	99.0	113.3	88.9	96.3	106.2	88.9	94.2	121.2	89.1	98.8	93.9	77.1	82.2
0310	Concrete Forming & Accessories	88.5	62.1	66.0	93.5	56.9	62.4	98.2	60.1	65.8	85.3	61.5	65.0	94.7	62.9	67.6	97.7	63.3	68.4
0320	Concrete Reinforcing	95.5	78.9	87.1	99.0	60.7	79.5	93.3	60.8	76.8	97.9	78.3	87.9	94.4	67.4	80.7	101.4	60.9	80.8
0330	Cast-in-Place Concrete	93.9	63.4	82.6	101.2	63.0	87.0	89.0	64.0	79.7	96.1	65.2	84.6	106.2	66.0	91.3	87.0	63.2	78.2
03	CONCRETE	84.6	67.1	76.9	93.9	61.6	79.6	84.1	63.4	74.9	86.2	67.4	77.8	94.5	66.5	82.1	82.5	64.6	74.5
04	MASONRY	87.1	64.8	73.3	100.7	55.9	73.0	88.1	56.5	68.6	101.9	58.3	74.9	85.0	59.6	69.3	92.6	54.7	69.2
05	METALS	101.3	95.2	99.4	101.0	87.9	97.0	100.5	88.1	96.8	101.2	94.2	99.1	111.1	90.8	104.9	99.1	88.3	95.9
06	WOOD, PLASTICS & COMPOSITES	77.1	63.2	69.6	90.0	56.2	71.6	96.4	60.3	76.8	72.4	61.0	66.2	91.8	61.8	75.5	90.3	65.2	76.7
07	THERMAL & MOISTURE PROTECTION	100.8	64.4	85.0	101.2	60.7	83.6	101.1	61.5	83.9	100.8	63.0	84.3	101.3	64.3	85.2	101.0	61.3	83.7
08	OPENINGS	97.7	65.2	90.2	94.8	57.4	86.1	95.1	59.7	86.9	97.7	64.0	89.8	94.4	62.0	86.9	99.9	62.4	91.2
0920	Plaster & Gypsum Board	105.3	62.7	77.2	91.8	55.4	67.8	95.0	59.7	71.8	101.6	60.4	74.5	91.8	61.2	71.7	95.5	64.7	75.2
0950, 0980	Ceilings & Acoustic Treatment	81.6	62.7	68.8	78.1	55.4	62.7	83.7	59.7	67.4	81.6	60.4	67.3	82.9	61.2	68.2	84.9	64.7	71.2
0960	Flooring	93.7	79.8	89.7	98.4	56.7	86.3	101.0	58.5	88.7	91.6	58.4	82.0	98.6	57.8	86.8	98.8	58.4	87.1
0970, 0990	Wall Finishes & Painting/Coating	92.6	63.6	75.4	97.1	63.6	77.3	97.1	63.6	77.3	92.6	63.6	75.4	97.1	83.0	88.8	88.9	58.6	70.9
09	FINISHES	90.3	65.3	76.6	89.1	56.9	71.6	90.5	59.7	73.7	89.2	60.6	73.6	89.8	63.4	75.4	89.1	61.6	74.1
COVERS	DIVS. 10 - 14, 25, 28, 41, 43, 44, 46	100.0	79.4	95.5	100.0	81.1	95.8	100.0	78.2	95.2	100.0	80.4	95.7	100.0	82.8	96.2	100.0	83.0	96.2
21, 22, 23	FIRE SUPPRESSION, PLUMBING & HVAC	98.1	57.5	81.7	98.8	62.3	84.0	99.9	62.3	84.7	98.1	59.4	82.5	99.9	75.3	90.0	100.0	62.5	84.8
26, 27, 3370	ELECTRICAL, COMMUNICATIONS & UTIL.	94.7	61.5	77.8	95.8	57.2	76.2	95.2	63.4	79.1	93.0	60.0	76.2	96.6	64.3	80.1	96.5	72.3	84.2
MF2016	WEIGHTED AVERAGE	95.6	68.3	83.9	97.8	64.9	83.7	96.1	66.5	83.4	96.2	67.1	83.8	99.2	71.2	87.2	95.8	67.4	83.7

		FLORIDA																
	DIVISION	ORLANDO			PANAMA CITY			PENSACOLA			SARASOTA			ST. PETERSBURG			TALLAHASSEE		
		327 - 328, 347			324			325			342			337			323		
		MAT.	INST.	TOTAL	MAT.	INST.	TOTAL	MAT.	INST.	TOTAL	MAT.	INST.	TOTAL	MAT.	INST.	TOTAL	MAT.	INST.	TOTAL
015433	CONTRACTOR EQUIPMENT		102.1	102.1		102.1	102.1		102.1	102.1		102.1	102.1		102.1	102.1		102.1	102.1
0241, 31 - 34	SITE & INFRASTRUCTURE, DEMOLITION	113.8	89.2	96.6	126.6	88.6	99.8	126.6	88.6	100.1	117.2	89.0	97.5	107.8	88.5	94.4	106.8	88.7	94.2
0310	Concrete Forming & Accessories	102.4	62.5	68.4	97.5	60.3	65.8	95.4	63.6	68.3	94.3	61.6	66.5	91.8	59.3	64.1	100.6	56.9	63.4
0320	Concrete Reinforcing	98.5	67.4	82.7	97.5	69.8	83.4	100.0	69.2	84.3	93.7	78.3	85.8	97.9	78.2	87.9	99.1	60.7	79.6
0330	Cast-in-Place Concrete	102.6	65.8	88.9	93.5	58.8	80.6	115.6	63.9	96.4	103.7	65.4	89.5	97.1	61.0	83.7	93.1	62.8	81.9
03	CONCRETE	89.7	66.2	79.3	92.0	63.2	79.2	101.5	66.4	85.9	93.1	67.5	81.8	87.6	64.9	77.5	87.2	61.5	75.8
04	MASONRY	92.3	59.5	72.1	92.4	57.4	70.8	113.1	57.4	78.7	90.1	58.8	70.7	138.6	52.1	85.2	90.1	55.4	68.7
05	METALS	99.1	90.4	96.5	101.8	91.4	98.6	102.8	90.4	99.2	103.0	93.6	100.2	102.1	93.9	99.6	99.6	88.0	96.1
06	WOOD, PLASTICS & COMPOSITES	93.5	61.8	76.3	95.2	64.2	78.4	93.1	64.2	77.4	94.4	61.0	76.3	81.5	61.0	70.4	95.5	56.2	74.1
07	THERMAL & MOISTURE PROTECTION	105.8	65.7	88.3	101.5	58.8	82.9	101.4	61.8	84.1	98.7	63.2	83.3	101.0	58.8	82.6	98.7	61.2	82.4
08	OPENINGS	99.4	62.0	90.7	93.2	63.7	86.4	93.2	63.7	86.4	100.0	63.6	91.5	96.5	64.0	88.9	100.3	57.4	90.3
0920	Plaster & Gypsum Board	96.0	61.2	73.1	94.3	63.7	74.1	102.8	63.7	77.0	100.9	60.4	74.2	107.7	60.4	76.6	99.3	55.4	70.4
0950, 0980	Ceilings & Acoustic Treatment	90.0	61.2	70.5	82.9	63.7	69.9	82.9	63.7	69.9	88.7	60.4	69.6	83.5	60.4	67.9	88.3	55.4	66.0
0960	Flooring	97.2	58.4	85.9	100.6	75.0	93.2	96.4	56.7	84.9	103.5	56.7	89.9	95.7	56.7	84.4	99.0	56.7	86.7
0970, 0990	Wall Finishes & Painting/Coating	93.5	61.9	74.8	97.1	63.6	77.3	97.1	63.6	77.3	98.2	63.6	77.7	92.6	63.6	75.4	92.3	63.6	75.3
09	FINISHES	92.1	61.2	75.2	91.5	63.5	76.2	91.1	62.1	75.3	95.8	60.4	76.5	91.7	58.9	73.8	91.2	56.8	72.4
COVERS	DIVS. 10 - 14, 25, 28, 41, 43, 44, 46	100.0	82.7	96.2	100.0	76.5	94.8	100.0	81.0	95.8	100.0	80.6	95.7	100.0	77.1	95.0	100.0	78.7	95.3
21, 22, 23	FIRE SUPPRESSION, PLUMBING & HVAC	100.1	57.1	82.7	99.9	51.5	80.4	99.9	63.2	85.1	99.9	58.8	83.3	100.0	55.7	82.1	100.0	66.6	86.5
26, 27, 3370	ELECTRICAL, COMMUNICATIONS & UTIL.	99.8	64.7	81.9	94.3	57.2	75.4	97.9	51.9	74.5	95.8	60.0	77.6	93.0	62.1	77.3	99.9	57.2	78.2
MF2016	WEIGHTED AVERAGE	97.9	67.1	84.7	97.6	64.2	83.3	100.3	66.4	85.8	98.7	67.0	85.1	99.0	65.2	84.5	97.3	65.7	83.7

		FLORIDA						GEORGIA											
	DIVISION	TAMPA			WEST PALM BEACH			ALBANY			ATHENS			ATLANTA			AUGUSTA		
		335 - 336, 346			334, 349			317, 398			306			300 - 303, 399			308 - 309		
		MAT.	INST.	TOTAL	MAT.	INST.	TOTAL	MAT.	INST.	TOTAL	MAT.	INST.	TOTAL	MAT.	INST.	TOTAL	MAT.	INST.	TOTAL
015433	CONTRACTOR EQUIPMENT		102.1	102.1		95.2	95.2		96.1	96.1		94.6	94.6		96.5	96.5		94.6	94.6
0241, 31 - 34	SITE & INFRASTRUCTURE, DEMOLITION	108.3	89.0	94.8	89.3	76.8	80.6	102.3	79.4	86.3	101.7	94.9	97.0	98.7	95.3	96.3	95.0	95.3	95.2
0310	Concrete Forming & Accessories	94.5	61.9	66.7	95.5	62.9	67.7	88.5	67.7	70.8	91.7	44.3	51.4	96.3	71.9	75.5	92.7	72.7	75.7
0320	Concrete Reinforcing	94.5	78.3	86.2	97.1	59.2	77.8	90.4	73.3	81.7	98.9	65.4	81.9	98.2	73.4	85.6	99.3	68.0	83.4
0330	Cast-in-Place Concrete	94.9	63.8	84.1	85.6	61.8	76.8	85.7	67.1	79.0	109.6	68.5	94.3	113.1	71.6	97.6	103.5	70.1	91.1
03	CONCRETE	86.3	67.8	78.1	81.0	63.6	73.3	81.2	70.4	76.4	102.1	57.7	82.4	104.3	72.7	90.3	94.6	71.6	84.4
04	MASONRY	94.2	59.3	72.6	92.8	53.3	68.4	91.9	67.7	77.0	76.6	77.8	77.3	89.9	69.7	77.4	90.1	69.2	77.2
05	METALS	101.1	94.3	99.1	98.0	87.7	94.9	102.9	97.5	101.2	97.6	79.5	92.1	98.4	84.6	94.2	97.2	80.5	92.2
06	WOOD, PLASTICS & COMPOSITES	85.3	61.0	72.1	85.0	65.2	74.3	79.5	68.4	73.5	95.0	36.8	63.4	97.8	73.1	84.4	96.5	75.4	85.0
07	THERMAL & MOISTURE PROTECTION	101.3	63.5	84.8	100.8	61.2	83.5	98.6	68.7	85.6	100.2	69.5	86.9	101.7	72.4	89.0	100.0	71.3	87.5
08	OPENINGS	97.7	64.0	89.8	96.0	62.0	88.1	87.6	70.5	83.6	93.0	50.9	83.3	97.6	73.4	92.0	93.1	73.3	88.5
0920	Plaster & Gypsum Board	110.1	60.4	77.4	113.8	64.7	81.5	103.8	68.0	80.2	97.4	35.5	56.6	99.6	72.6	81.8	98.7	75.1	83.1
0950, 0980	Ceilings & Acoustic Treatment	86.1	60.4	68.7	81.6	64.7	70.2	81.9	68.0	72.5	101.2	35.5	56.7	94.0	72.6	79.5	102.2	75.1	83.8
0960	Flooring	96.8	58.4	85.6	98.6	56.7	86.5	102.3	65.9	91.8	94.0	84.7	91.3	97.0	68.0	88.6	94.1	65.9	85.9
0970, 0990	Wall Finishes & Painting/Coating	92.6	63.6	75.4	88.4	58.6	70.8	94.5	96.5	95.6	101.8	96.5	98.7	106.2	96.5	100.5	101.8	80.6	89.2
09	FINISHES	93.0	60.8	75.4	90.6	61.1	74.5	92.4	70.0	80.2	97.7	55.2	74.5	97.8	73.5	84.5	97.4	72.5	83.8
COVERS	DIVS. 10 - 14, 25, 28, 41, 43, 44, 46	100.0	80.8	95.8	100.0	82.7	96.2	100.0	83.3	96.3	100.0	79.6	95.5	100.0	86.4	97.0	100.0	83.9	96.5
21, 22, 23	FIRE SUPPRESSION, PLUMBING & HVAC	100.0	60.0	83.8	98.1	59.9	82.7	99.9	69.9	87.8	96.6	67.3	84.8	100.0	72.0	88.7	100.1	63.6	85.4
26, 27, 3370	ELECTRICAL, COMMUNICATIONS & UTIL.	92.7	64.7	78.5	93.8	69.1	81.2	95.3	63.4	79.1	97.0	65.4	81.0	96.6	72.6	84.4	97.6	69.8	83.4
MF2016	WEIGHTED AVERAGE	96.8	68.1	84.5	94.3	66.0	82.2	95.2	72.6	85.5	96.8	67.9	84.4	99.0	75.6	89.0	97.1	72.7	86.6

GEORGIA

DIVISION		COLUMBUS 318 - 319			DALTON 307			GAINESVILLE 305			MACON 310 - 312			SAVANNAH 313 - 314			STATESBORO 304		
		MAT.	INST.	TOTAL	MAT.	INST.	TOTAL	MAT.	INST.	TOTAL	MAT.	INST.	TOTAL	MAT.	INST.	TOTAL	MAT.	INST.	TOTAL
015433	CONTRACTOR EQUIPMENT		96.1	96.1		110.1	110.1		94.6	94.6		105.9	105.9		97.0	97.0		97.7	97.7
0241, 31 - 34	SITE & INFRASTRUCTURE, DEMOLITION	102.2	79.5	86.4	101.5	100.3	100.7	101.7	94.9	97.0	103.9	94.6	97.4	102.1	80.9	87.3	102.6	81.4	87.8
0310	Concrete Forming & Accessories	88.5	67.8	70.8	84.6	65.5	68.3	95.2	41.8	49.7	88.3	67.8	70.9	92.0	70.4	73.6	79.5	52.1	56.2
0320	Concrete Reinforcing	90.3	73.3	81.7	98.4	62.5	80.2	98.7	65.3	81.7	91.5	73.3	82.3	97.3	68.0	82.4	98.0	36.6	66.8
0330	Cast-in-Place Concrete	85.4	67.8	78.8	106.5	67.2	91.8	115.2	68.3	97.7	84.2	69.7	78.8	89.0	67.6	81.0	109.4	67.2	93.7
03	CONCRETE	81.0	70.5	76.4	101.4	67.4	86.3	103.9	56.4	82.9	80.6	71.2	76.5	82.6	70.7	77.3	101.7	57.0	81.9
04	MASONRY	91.9	67.8	77.0	77.4	77.1	77.2	85.4	77.8	80.7	104.2	67.9	81.8	86.8	67.7	75.0	79.4	77.8	78.4
05	METALS	102.5	97.9	101.1	98.9	93.8	97.3	96.9	78.9	91.4	97.6	98.0	97.8	98.6	95.3	97.6	102.5	84.0	96.9
06	WOOD, PLASTICS & COMPOSITES	79.5	68.4	73.5	78.2	67.1	72.1	98.9	34.1	63.7	85.0	68.4	76.0	88.2	72.3	79.6	71.5	48.5	59.0
07	THERMAL & MOISTURE PROTECTION	98.5	69.4	85.8	102.4	71.7	89.0	100.2	69.2	86.7	97.0	71.5	85.9	96.4	68.9	84.5	100.8	68.1	86.6
08	OPENINGS	87.6	70.5	83.6	94.2	66.9	87.9	93.0	49.4	82.9	87.4	70.5	83.5	95.1	71.3	89.6	95.1	49.5	84.5
0920	Plaster & Gypsum Board	103.8	68.0	80.2	85.0	66.6	72.9	99.9	32.7	55.6	108.4	68.0	81.8	102.7	72.0	82.5	85.8	47.5	60.6
0950, 0980	Ceilings & Acoustic Treatment	81.9	68.0	72.5	117.4	66.6	83.0	101.2	32.7	54.8	77.1	68.0	70.9	88.3	72.0	77.3	111.8	47.5	68.2
0960	Flooring	102.3	65.9	91.8	94.5	84.7	91.6	95.5	84.7	92.3	81.0	65.9	76.6	98.3	65.9	88.9	111.2	84.7	103.5
0970, 0990	Wall Finishes & Painting/Coating	94.5	96.5	95.6	91.6	71.9	80.0	101.8	96.5	98.7	96.5	96.5	96.5	91.5	26.3	51.2	100.0	71.9	83.4
09	FINISHES	92.3	70.0	80.1	107.5	70.2	87.2	98.3	53.6	73.9	82.3	70.0	75.6	92.3	71.2	80.8	110.2	59.4	82.5
COVERS	DIVS. 10 - 14, 25, 28, 41, 43, 44, 46	100.0	83.3	96.3	100.0	28.4	84.2	100.0	36.2	85.9	100.0	83.4	96.3	100.0	83.3	96.3	100.0	42.2	87.3
21, 22, 23	FIRE SUPPRESSION, PLUMBING & HVAC	100.0	66.2	86.3	96.7	62.0	82.7	96.6	67.1	84.7	100.0	69.3	87.6	100.1	67.1	86.7	97.1	67.8	85.3
26, 27, 3370	ELECTRICAL, COMMUNICATIONS & UTIL.	95.5	69.4	82.2	106.2	63.7	84.6	97.0	69.3	82.9	94.0	64.2	78.9	99.1	70.2	84.4	97.3	58.6	77.6
MF2016	WEIGHTED AVERAGE	95.1	72.7	85.5	98.7	71.0	86.8	97.4	66.5	84.2	93.9	74.0	85.4	95.6	73.1	86.0	98.9	65.8	84.7

DIVISION		GEORGIA VALDOSTA 316			WAYCROSS 315			HAWAII HILO 967			HONOLULU 968			STATES & POSS., GUAM 969			IDAHO BOISE 836 - 837		
		MAT.	INST.	TOTAL	MAT.	INST.	TOTAL	MAT.	INST.	TOTAL	MAT.	INST.	TOTAL	MAT.	INST.	TOTAL	MAT.	INST.	TOTAL
015433	CONTRACTOR EQUIPMENT		96.1	96.1		96.1	96.1		100.4	100.4		100.4	100.4		165.0	165.0		97.6	97.6
0241, 31 - 34	SITE & INFRASTRUCTURE, DEMOLITION	111.5	79.4	89.1	108.5	80.6	89.0	149.7	107.5	120.3	158.4	107.5	122.9	194.0	103.5	130.9	87.5	97.0	94.1
0310	Concrete Forming & Accessories	79.7	42.0	47.6	81.5	64.7	67.2	105.8	123.7	121.0	118.7	123.7	122.9	108.3	53.0	61.2	98.0	79.8	82.5
0320	Concrete Reinforcing	92.4	61.2	76.5	92.4	61.4	76.7	137.5	126.6	131.9	159.9	126.6	142.9	237.2	28.2	131.0	105.1	78.8	91.8
0330	Cast-in-Place Concrete	83.9	67.5	77.8	94.8	67.4	84.6	170.9	125.6	154.1	137.5	125.6	133.1	148.2	98.5	129.7	90.6	83.1	87.8
03	CONCRETE	85.4	56.7	72.7	88.4	66.7	78.8	137.6	123.7	131.5	130.8	123.7	127.7	138.8	65.7	106.4	93.8	81.0	88.1
04	MASONRY	97.2	77.8	85.2	98.0	77.8	85.5	148.6	120.0	130.9	135.5	120.0	125.9	205.8	35.5	100.6	125.2	87.2	101.7
05	METALS	102.1	93.1	99.3	101.1	88.8	97.4	119.9	106.6	115.9	134.3	106.6	125.9	155.0	76.6	131.3	104.5	82.1	97.7
06	WOOD, PLASTICS & COMPOSITES	68.1	33.9	49.5	69.6	65.2	67.2	113.9	124.5	119.6	131.3	124.5	127.6	126.6	55.4	87.9	94.6	78.5	85.9
07	THERMAL & MOISTURE PROTECTION	98.8	67.5	85.2	98.5	70.3	86.2	132.2	120.2	127.0	150.1	120.2	137.1	155.5	59.8	113.9	97.5	85.5	92.3
08	OPENINGS	84.2	48.2	75.9	84.6	62.9	79.5	116.3	123.0	117.9	130.9	123.0	129.1	120.7	45.0	103.1	99.2	73.1	93.1
0920	Plaster & Gypsum Board	95.8	32.5	54.1	95.8	64.7	75.4	112.1	124.9	120.6	155.7	124.9	135.5	228.9	43.7	106.9	94.2	77.9	83.5
0950, 0980	Ceilings & Acoustic Treatment	80.2	32.5	47.9	78.3	64.7	69.1	137.5	124.9	129.0	144.3	124.9	131.2	254.9	43.7	111.8	101.2	77.9	85.4
0960	Flooring	96.1	86.3	93.2	97.3	84.7	93.6	110.8	140.7	119.5	130.2	140.7	133.3	130.1	40.8	104.2	93.2	79.6	89.3
0970, 0990	Wall Finishes & Painting/Coating	94.5	96.5	95.6	94.5	71.9	81.1	94.7	140.6	121.9	105.1	140.6	126.1	100.0	32.4	60.0	93.4	40.5	62.1
09	FINISHES	90.0	53.8	70.2	89.6	69.3	78.5	110.5	129.0	120.6	125.7	129.0	127.5	183.1	49.4	110.1	91.4	75.7	82.8
COVERS	DIVS. 10 - 14, 25, 28, 41, 43, 44, 46	100.0	79.1	95.4	100.0	49.8	89.0	100.0	112.6	102.8	100.0	112.6	102.8	100.0	67.3	92.8	100.0	88.6	97.5
21, 22, 23	FIRE SUPPRESSION, PLUMBING & HVAC	100.0	69.3	87.6	97.8	64.2	84.2	100.4	112.0	105.1	100.4	112.0	105.1	103.1	34.3	75.3	100.0	74.8	89.8
26, 27, 3370	ELECTRICAL, COMMUNICATIONS & UTIL.	93.8	52.9	73.0	97.5	54.0	75.4	111.0	124.3	117.8	112.8	124.3	118.7	162.9	36.8	98.8	97.3	69.8	83.3
MF2016	WEIGHTED AVERAGE	95.3	66.2	82.8	95.4	68.4	83.8	116.3	118.3	117.1	120.9	118.3	119.8	138.8	52.6	101.9	99.7	79.4	91.0

DIVISION		IDAHO COEUR D'ALENE 838			IDAHO FALLS 834			LEWISTON 835			POCATELLO 832			TWIN FALLS 833			ILLINOIS BLOOMINGTON 617		
		MAT.	INST.	TOTAL	MAT.	INST.	TOTAL	MAT.	INST.	TOTAL	MAT.	INST.	TOTAL	MAT.	INST.	TOTAL	MAT.	INST.	TOTAL
015433	CONTRACTOR EQUIPMENT		92.5	92.5		97.6	97.6		92.5	92.5		97.6	97.6		97.6	97.6		104.7	104.7
0241, 31 - 34	SITE & INFRASTRUCTURE, DEMOLITION	86.5	91.0	89.6	85.4	96.8	93.4	93.3	91.9	92.3	88.7	97.0	94.5	95.6	96.7	96.4	97.0	100.8	99.6
0310	Concrete Forming & Accessories	105.8	82.0	85.5	91.8	78.6	80.6	110.9	83.4	87.5	98.1	79.5	82.2	99.2	78.5	81.5	84.1	119.2	114.0
0320	Concrete Reinforcing	113.2	99.0	106.0	107.2	77.4	92.0	113.2	99.0	106.0	105.6	78.8	92.0	107.6	77.3	92.2	96.1	105.3	100.8
0330	Cast-in-Place Concrete	98.0	84.1	92.8	86.2	83.3	85.2	101.8	84.5	95.4	93.1	83.0	89.4	100.4	80.2	91.4	91.9	116.7	102.0
03	CONCRETE	99.8	85.7	93.6	86.1	80.2	83.5	103.3	86.5	95.8	93.0	80.8	87.6	100.4	80.2	91.4	91.9	116.7	102.9
04	MASONRY	126.8	85.0	100.9	120.2	84.6	98.2	127.2	86.5	102.0	122.7	87.2	100.8	125.4	84.6	100.2	110.9	124.2	119.2
05	METALS	98.2	89.0	95.4	112.6	81.4	103.1	97.6	89.2	95.0	112.7	81.7	103.3	112.7	81.2	103.2	94.9	123.8	103.6
06	WOOD, PLASTICS & COMPOSITES	97.6	82.1	89.2	88.1	78.5	82.9	103.7	82.1	92.0	94.6	78.5	85.9	95.7	78.5	86.3	83.4	116.3	101.3
07	THERMAL & MOISTURE PROTECTION	151.8	84.1	122.3	96.6	75.4	87.4	152.2	85.5	123.2	97.2	76.4	88.2	98.1	82.7	91.4	98.1	115.2	105.5
08	OPENINGS	116.4	78.1	107.5	102.1	72.8	95.3	110.1	82.0	103.6	99.9	68.1	92.5	102.9	63.6	93.8	93.3	122.1	100.0
0920	Plaster & Gypsum Board	170.4	81.8	112.0	80.8	77.9	78.9	172.2	81.8	112.7	82.2	77.9	79.4	84.4	77.9	80.1	88.8	116.9	107.3
0950, 0980	Ceilings & Acoustic Treatment	136.3	81.8	99.4	104.7	77.9	86.6	136.3	81.8	99.4	109.8	77.9	88.2	102.7	77.9	87.4	89.7	116.9	108.1
0960	Flooring	131.9	79.6	116.7	93.1	79.6	89.2	135.1	79.6	119.0	96.5	79.6	91.6	97.7	79.6	92.5	87.3	124.9	98.2
0970, 0990	Wall Finishes & Painting/Coating	112.5	74.5	90.0	93.4	40.5	62.1	112.5	73.2	89.3	93.3	40.5	62.1	93.4	40.5	62.1	88.7	138.3	118.0
09	FINISHES	155.0	80.7	114.4	89.7	75.0	81.7	156.3	81.2	115.3	92.3	75.7	83.2	93.1	75.0	83.2	86.9	122.5	106.3
COVERS	DIVS. 10 - 14, 25, 28, 41, 43, 44, 46	100.0	93.6	98.6	100.0	87.7	97.3	100.0	94.6	98.8	100.0	88.6	97.5	100.0	87.7	97.3	100.0	106.7	101.5
21, 22, 23	FIRE SUPPRESSION, PLUMBING & HVAC	99.5	81.7	92.3	100.9	73.4	89.7	100.7	86.9	95.1	99.9	74.8	89.7	99.9	69.6	87.6	96.5	107.7	101.0
26, 27, 3370	ELECTRICAL, COMMUNICATIONS & UTIL.	89.3	83.5	86.3	88.9	69.4	79.0	87.4	80.1	83.7	94.7	67.6	80.9	90.3	69.4	79.7	94.2	92.5	93.3
MF2016	WEIGHTED AVERAGE	106.8	84.4	97.2	99.2	78.2	90.2	106.9	85.7	97.8	100.7	78.6	91.2	101.9	77.2	91.3	95.2	112.2	102.5

ILLINOIS

DIVISION		CARBONDALE 629			CENTRALIA 628			CHAMPAIGN 618 - 619			CHICAGO 606 - 608			DECATUR 625			EAST ST. LOUIS 620 - 622		
		MAT.	INST.	TOTAL	MAT.	INST.	TOTAL	MAT.	INST.	TOTAL	MAT.	INST.	TOTAL	MAT.	INST.	TOTAL	MAT.	INST.	TOTAL
015433	CONTRACTOR EQUIPMENT		113.3	113.3		113.3	113.3		105.6	105.6		98.9	98.9		105.6	105.6		113.3	113.3
0241, 31 - 34	SITE & INFRASTRUCTURE, DEMOLITION	98.5	101.5	100.6	98.8	103.4	102.0	106.2	101.6	103.0	103.6	103.7	103.7	93.5	101.2	98.9	100.9	102.0	101.7
0310	Concrete Forming & Accessories	89.6	108.2	105.4	91.1	114.7	111.2	90.6	116.7	112.8	99.5	162.0	152.7	91.8	115.9	112.3	87.0	111.5	107.9
0320	Concrete Reinforcing	90.2	106.3	98.4	90.2	107.0	98.8	96.1	100.6	98.4	100.2	154.7	128.0	88.5	100.9	94.8	90.1	107.0	98.7
0330	Cast-in-Place Concrete	88.9	102.3	93.8	89.3	121.6	101.3	113.8	111.9	113.1	126.4	154.6	136.9	97.0	113.2	103.0	80.5	121.5	102.3
03	CONCRETE	79.1	107.3	91.6	79.5	117.1	96.2	104.2	112.8	108.0	109.4	157.3	130.6	89.3	113.1	99.9	80.5	115.7	96.1
04	MASONRY	76.0	112.4	98.5	76.0	125.5	106.6	133.9	121.8	126.4	104.5	166.2	142.6	71.8	120.2	101.7	76.3	125.5	106.7
05	METALS	103.2	130.7	111.6	103.3	133.8	112.5	94.9	119.7	102.4	102.1	146.6	115.6	107.2	119.2	110.9	104.4	133.5	113.2
06	WOOD, PLASTICS & COMPOSITES	88.7	105.4	97.8	91.2	110.9	101.9	90.4	114.9	103.7	101.1	161.7	134.0	89.8	114.9	103.4	85.8	106.9	97.3
07	THERMAL & MOISTURE PROTECTION	95.0	102.3	98.2	95.1	113.2	102.9	98.9	114.7	105.8	93.8	152.0	119.1	100.7	113.0	106.1	95.1	112.5	102.6
08	OPENINGS	89.2	116.9	95.6	89.2	120.0	96.3	93.9	116.8	99.2	104.4	170.9	119.9	100.3	116.9	104.2	89.3	117.7	95.9
0920	Plaster & Gypsum Board	95.3	105.7	102.2	96.6	111.3	106.3	90.6	115.4	107.0	94.8	163.5	140.1	97.8	115.4	109.4	93.8	107.2	102.6
0950, 0980	Ceilings & Acoustic Treatment	84.6	105.7	98.9	84.6	111.3	102.7	89.7	115.4	107.1	99.8	163.5	143.0	90.5	115.4	107.4	84.6	107.2	99.9
0960	Flooring	111.2	113.8	112.0	112.0	127.7	116.5	90.4	114.7	97.4	98.7	164.8	117.9	99.5	115.2	104.1	110.2	127.7	115.3
0970, 0990	Wall Finishes & Painting/Coating	109.1	107.7	108.3	109.1	114.6	112.3	88.7	115.4	104.5	104.9	162.3	138.9	100.8	114.4	108.9	109.1	114.6	112.3
09	FINISHES	92.8	108.1	101.1	93.2	117.1	106.3	88.8	116.6	104.0	98.4	163.5	133.9	92.8	116.2	105.6	92.4	114.4	104.4
COVERS	DIVS. 10 - 14, 25, 28, 41, 43, 44, 46	100.0	102.3	100.5	100.0	104.3	100.9	100.0	105.4	101.2	100.0	128.1	106.2	100.0	104.8	101.1	100.0	103.8	100.8
21, 22, 23	FIRE SUPPRESSION, PLUMBING & HVAC	96.4	106.5	100.5	96.4	98.8	97.4	96.5	106.3	100.4	99.8	138.0	115.3	99.9	96.4	98.5	99.9	99.3	99.7
26, 27, 3370	ELECTRICAL, COMMUNICATIONS & UTIL.	91.2	108.8	100.2	92.5	108.9	100.8	97.4	92.6	94.9	97.4	134.1	116.1	94.9	89.8	92.3	92.1	102.1	97.2
MF2016	WEIGHTED AVERAGE	93.1	109.8	100.2	93.3	113.0	101.8	98.7	109.8	103.5	101.7	146.1	120.7	97.2	107.0	101.4	94.4	111.3	101.6

ILLINOIS

DIVISION		EFFINGHAM 624			GALESBURG 614			JOLIET 604			KANKAKEE 609			LA SALLE 613			NORTH SUBURBAN 600 - 603		
		MAT.	INST.	TOTAL	MAT.	INST.	TOTAL	MAT.	INST.	TOTAL	MAT.	INST.	TOTAL	MAT.	INST.	TOTAL	MAT.	INST.	TOTAL
015433	CONTRACTOR EQUIPMENT		105.6	105.6		104.7	104.7		95.4	95.4		95.4	95.4		104.7	104.7		95.4	95.4
0241, 31 - 34	SITE & INFRASTRUCTURE, DEMOLITION	97.8	100.4	99.6	99.5	99.8	99.7	99.3	101.7	101.0	93.1	101.3	98.8	98.9	101.3	100.6	98.6	102.0	101.0
0310	Concrete Forming & Accessories	96.3	116.6	113.6	90.5	119.5	115.2	98.8	161.1	151.8	92.1	146.0	138.0	104.5	126.0	122.8	98.1	158.8	149.8
0320	Concrete Reinforcing	91.4	96.2	93.9	95.5	108.6	102.1	100.2	140.6	120.7	101.0	138.9	120.3	95.7	137.5	116.9	100.2	153.3	127.2
0330	Cast-in-Place Concrete	96.7	109.2	101.3	101.0	111.5	104.9	115.0	148.1	127.3	107.1	138.1	118.6	100.9	125.0	109.9	115.0	154.5	129.7
03	CONCRETE	90.0	111.0	99.3	94.7	115.4	103.9	102.2	151.9	124.3	96.0	141.3	116.1	95.6	128.2	110.0	102.2	155.4	125.8
04	MASONRY	80.2	114.1	101.2	111.1	123.7	118.9	103.7	160.0	138.5	100.0	148.4	129.9	111.1	145.6	132.4	100.7	162.1	138.6
05	METALS	104.2	114.5	107.3	94.9	124.1	103.7	100.0	135.6	110.8	100.0	133.8	110.2	95.0	142.8	109.4	101.1	142.0	113.5
06	WOOD, PLASTICS & COMPOSITES	92.1	117.3	105.8	90.3	118.7	105.7	99.2	162.3	133.5	92.1	144.6	120.6	106.5	123.3	115.6	98.1	157.7	130.5
07	THERMAL & MOISTURE PROTECTION	100.1	109.9	104.4	98.3	112.1	104.3	98.1	148.2	119.9	97.1	141.1	116.2	98.5	128.9	111.7	98.4	148.9	120.3
08	OPENINGS	95.3	116.2	100.2	93.3	118.6	99.2	101.9	166.5	116.9	94.8	156.8	109.2	93.3	135.2	103.0	102.0	168.4	117.4
0920	Plaster & Gypsum Board	97.8	117.9	111.0	90.6	119.3	109.5	87.3	164.1	137.9	85.5	146.0	125.3	98.3	124.1	115.2	90.9	159.5	136.1
0950, 0980	Ceilings & Acoustic Treatment	84.6	117.9	107.1	89.7	119.3	109.8	99.4	164.1	143.3	99.4	146.0	131.0	89.7	124.1	113.0	99.4	159.5	140.1
0960	Flooring	100.5	116.9	105.3	90.2	124.9	100.3	96.7	154.8	113.5	93.7	155.7	111.7	96.7	122.9	104.3	97.2	164.8	116.8
0970, 0990	Wall Finishes & Painting/Coating	100.8	112.2	107.6	88.7	101.1	96.0	97.0	167.3	138.6	97.0	138.3	121.4	88.7	136.5	117.0	99.0	168.2	139.9
09	FINISHES	91.9	117.4	105.8	88.1	119.4	105.2	93.7	162.0	131.0	92.2	147.5	122.4	91.1	125.3	109.8	94.5	161.8	131.2
COVERS	DIVS. 10 - 14, 25, 28, 41, 43, 44, 46	100.0	77.2	95.0	100.0	101.4	100.3	100.0	125.0	105.5	100.0	117.8	103.9	100.0	102.7	100.6	100.0	121.3	104.7
21, 22, 23	FIRE SUPPRESSION, PLUMBING & HVAC	96.5	104.7	99.8	96.5	105.7	100.2	99.9	133.0	113.3	96.4	132.4	111.0	96.5	127.0	108.8	99.8	136.4	114.6
26, 27, 3370	ELECTRICAL, COMMUNICATIONS & UTIL.	92.9	108.8	101.0	95.1	83.8	89.3	96.5	136.5	116.9	91.8	133.8	113.2	92.2	133.9	113.4	96.4	135.7	116.4
MF2016	WEIGHTED AVERAGE	95.7	109.2	101.5	95.9	109.5	101.8	99.7	142.2	117.9	96.4	136.1	113.4	96.1	128.9	110.2	99.8	144.1	118.8

ILLINOIS

DIVISION		PEORIA 615 - 616			QUINCY 623			ROCK ISLAND 612			ROCKFORD 610 - 611			SOUTH SUBURBAN 605			SPRINGFIELD 626 - 627		
		MAT.	INST.	TOTAL	MAT.	INST.	TOTAL	MAT.	INST.	TOTAL	MAT.	INST.	TOTAL	MAT.	INST.	TOTAL	MAT.	INST.	TOTAL
015433	CONTRACTOR EQUIPMENT		104.7	104.7		105.6	105.6		104.7	104.7		104.7	104.7		95.4	95.4		105.6	105.6
0241, 31 - 34	SITE & INFRASTRUCTURE, DEMOLITION	99.9	100.7	100.5	96.7	101.2	99.9	97.6	99.6	99.0	99.3	101.8	101.0	98.6	101.7	100.7	98.4	101.6	100.6
0310	Concrete Forming & Accessories	93.6	119.2	115.4	94.1	116.3	113.0	92.1	101.9	100.4	97.9	131.3	126.4	98.1	158.8	149.8	86.6	113.0	113.0
0320	Concrete Reinforcing	93.1	107.7	100.5	91.0	88.5	89.7	95.5	101.8	98.7	88.1	138.0	113.5	100.2	153.3	127.2	91.3	100.9	96.2
0330	Cast-in-Place Concrete	98.0	119.0	105.8	96.8	109.6	101.6	98.8	101.3	99.8	100.3	130.5	111.5	115.0	154.4	129.6	92.1	111.1	99.2
03	CONCRETE	92.1	117.7	103.4	89.6	109.7	98.5	92.7	102.7	97.2	92.8	132.6	110.5	102.2	155.3	125.7	87.5	112.5	98.6
04	MASONRY	110.8	124.0	119.0	100.2	114.1	108.8	110.9	102.4	105.6	86.3	145.3	122.8	100.7	162.1	138.6	81.7	121.6	106.4
05	METALS	97.5	125.0	105.8	104.3	113.0	106.9	94.9	119.6	102.4	97.5	142.8	111.2	101.1	142.0	113.5	104.6	119.3	109.1
06	WOOD, PLASTICS & COMPOSITES	98.8	116.5	108.4	89.6	117.3	104.6	92.0	100.4	96.5	98.8	127.2	114.2	98.1	157.7	130.5	91.0	114.9	104.0
07	THERMAL & MOISTURE PROTECTION	99.0	115.6	106.2	100.1	108.5	103.8	98.2	101.2	99.5	101.4	132.6	115.0	98.4	148.9	120.3	100.8	114.6	108.1
08	OPENINGS	98.8	123.0	104.4	96.0	114.9	100.4	93.3	106.3	96.3	98.8	140.5	108.5	102.0	168.4	117.4	100.8	117.7	104.6
0920	Plaster & Gypsum Board	95.1	117.0	109.5	96.6	117.9	110.6	90.6	100.5	97.1	95.1	128.0	116.8	90.9	159.5	136.1	99.5	115.4	110.0
0950, 0980	Ceilings & Acoustic Treatment	94.8	117.0	109.8	84.6	117.9	107.1	89.7	100.5	97.0	94.8	128.0	117.3	99.4	159.5	140.1	91.0	115.4	107.5
0960	Flooring	93.7	123.5	102.3	99.5	111.5	103.0	91.3	97.8	93.2	93.7	129.1	103.9	97.2	164.8	116.8	102.0	115.2	105.8
0970, 0990	Wall Finishes & Painting/Coating	88.7	138.3	118.0	100.8	114.4	108.9	88.7	98.0	94.2	88.7	146.5	122.9	99.0	168.2	139.9	102.9	113.1	108.9
09	FINISHES	90.8	122.2	107.9	91.4	116.5	105.1	88.4	100.4	94.9	90.8	132.2	113.4	94.5	161.8	131.2	97.3	116.4	107.7
COVERS	DIVS. 10 - 14, 25, 28, 41, 43, 44, 46	100.0	106.6	101.4	100.0	79.8	95.5	100.0	99.5	99.9	100.0	116.5	103.6	100.0	121.3	104.7	100.0	105.3	101.2
21, 22, 23	FIRE SUPPRESSION, PLUMBING & HVAC	99.9	102.4	100.9	96.5	103.5	99.3	96.5	99.7	97.8	100.0	118.7	107.6	99.8	136.4	114.6	99.9	102.4	101.0
26, 27, 3370	ELECTRICAL, COMMUNICATIONS & UTIL.	96.1	93.7	94.9	90.6	84.2	87.3	87.4	95.4	91.5	96.4	128.7	112.8	96.4	135.7	116.4	97.4	84.2	90.7
MF2016	WEIGHTED AVERAGE	97.8	111.5	103.7	96.3	105.1	100.1	94.9	102.1	98.0	96.8	128.7	110.5	99.8	144.0	118.7	97.9	107.7	102.1

523

INDIANA

DIVISION		ANDERSON 460 MAT.	INST.	TOTAL	BLOOMINGTON 474 MAT.	INST.	TOTAL	COLUMBUS 472 MAT.	INST.	TOTAL	EVANSVILLE 476-477 MAT.	INST.	TOTAL	FORT WAYNE 467-468 MAT.	INST.	TOTAL	GARY 463-464 MAT.	INST.	TOTAL
015433	CONTRACTOR EQUIPMENT		97.4	97.4		83.9	83.9		83.9	83.9		113.7	113.7		97.4	97.4		97.4	97.4
0241, 31 - 34	SITE & INFRASTRUCTURE, DEMOLITION	99.4	93.7	95.5	87.3	92.1	90.6	84.0	91.9	89.5	92.7	121.2	112.6	100.5	93.6	95.7	100.1	97.3	98.2
0310	Concrete Forming & Accessories	96.5	79.6	82.1	101.2	80.1	83.2	95.2	79.9	82.2	94.4	80.8	82.8	94.4	75.7	78.5	96.6	112.6	110.2
0320	Concrete Reinforcing	95.6	85.4	90.4	89.4	85.1	87.2	89.8	87.2	88.5	97.9	83.4	90.5	95.6	80.2	87.8	95.6	112.7	104.3
0330	Cast-in-Place Concrete	103.7	77.1	93.8	99.3	76.2	90.7	98.8	74.1	89.6	94.9	85.7	91.5	110.2	76.2	97.6	108.4	113.7	110.4
03	CONCRETE	91.8	80.3	86.7	97.1	79.2	89.2	96.4	78.7	88.6	97.3	83.1	91.0	94.7	77.4	87.0	94.0	112.7	102.3
04	MASONRY	87.3	75.6	80.1	88.1	72.0	78.1	87.9	73.1	78.7	83.7	78.9	80.7	90.4	72.9	79.6	88.7	109.2	101.3
05	METALS	98.8	89.1	95.9	99.5	74.2	91.9	99.6	74.4	91.9	92.5	84.0	90.0	98.8	87.2	95.3	98.8	105.1	100.7
06	WOOD, PLASTICS & COMPOSITES	97.5	79.8	87.9	112.3	80.6	95.1	106.6	80.4	92.4	93.2	79.9	86.0	97.2	76.0	85.6	94.7	111.0	103.5
07	THERMAL & MOISTURE PROTECTION	109.6	77.2	95.5	96.7	77.7	88.5	96.1	78.1	88.2	100.9	84.0	93.6	109.4	78.9	96.1	108.1	106.1	107.2
08	OPENINGS	95.3	78.0	91.3	99.7	78.6	94.8	95.9	79.0	92.0	93.8	78.2	90.2	95.3	73.2	90.1	95.3	114.2	99.7
0920	Plaster & Gypsum Board	104.6	79.5	88.1	98.6	80.7	86.8	95.5	80.6	85.7	94.0	79.0	84.1	103.7	75.6	85.2	97.3	111.6	106.7
0950, 0980	Ceilings & Acoustic Treatment	91.9	79.5	83.5	79.6	80.7	80.4	79.6	80.6	80.3	83.4	79.0	80.4	91.9	75.6	80.8	91.9	111.6	105.2
0960	Flooring	95.1	77.5	90.0	99.1	83.1	94.4	94.0	83.1	90.9	94.0	73.8	88.1	95.1	73.5	88.9	95.1	114.3	100.7
0970, 0990	Wall Finishes & Painting/Coating	94.0	67.6	78.4	85.4	79.5	81.9	85.4	79.5	81.9	91.0	86.6	88.4	94.0	72.1	81.0	94.0	121.4	110.2
09	FINISHES	91.8	77.9	84.2	90.1	80.7	85.0	88.2	80.6	84.0	88.8	79.9	84.0	91.5	75.1	82.5	90.6	113.9	103.3
COVERS	DIVS. 10 - 14, 25, 28, 41, 43, 44, 46	100.0	89.2	97.6	100.0	85.8	96.9	100.0	85.8	96.9	100.0	93.1	98.5	100.0	89.1	97.6	100.0	105.1	101.1
21, 22, 23	FIRE SUPPRESSION, PLUMBING & HVAC	99.9	78.7	91.4	99.8	78.9	91.3	96.3	78.6	89.2	100.0	80.4	92.0	99.9	73.8	89.4	99.9	105.1	102.0
26, 27, 3370	ELECTRICAL, COMMUNICATIONS & UTIL.	86.5	85.2	85.8	99.2	86.6	92.8	98.4	87.0	92.6	95.2	83.4	89.2	87.2	76.7	81.8	98.1	110.7	104.5
MF2016	WEIGHTED AVERAGE	95.9	81.8	89.8	97.7	80.3	90.3	96.1	80.4	89.3	95.4	84.8	90.9	96.4	78.2	88.6	97.2	108.3	102.0

INDIANA

DIVISION		INDIANAPOLIS 461-462 MAT.	INST.	TOTAL	KOKOMO 469 MAT.	INST.	TOTAL	LAFAYETTE 479 MAT.	INST.	TOTAL	LAWRENCEBURG 470 MAT.	INST.	TOTAL	MUNCIE 473 MAT.	INST.	TOTAL	NEW ALBANY 471 MAT.	INST.	TOTAL
015433	CONTRACTOR EQUIPMENT		85.3	85.3		97.4	97.4		83.9	83.9		103.6	103.6		95.5	95.5		93.3	93.3
0241, 31 - 34	SITE & INFRASTRUCTURE, DEMOLITION	100.2	90.9	93.7	95.8	93.6	94.3	84.8	92.0	89.9	82.3	107.3	99.8	87.3	92.6	91.0	79.5	94.5	89.9
0310	Concrete Forming & Accessories	101.2	85.1	87.5	99.7	78.0	81.2	92.6	80.9	82.6	91.6	76.6	78.9	92.3	79.1	81.0	90.4	77.4	79.3
0320	Concrete Reinforcing	95.6	87.6	91.5	86.4	87.3	86.9	89.4	85.3	87.3	88.7	77.2	82.9	98.9	85.3	92.0	90.0	83.0	86.4
0330	Cast-in-Place Concrete	100.6	84.7	94.7	102.7	80.9	94.6	99.4	80.0	92.2	92.9	74.4	86.0	104.3	76.2	93.9	95.9	74.3	87.9
03	CONCRETE	97.5	84.7	91.8	88.9	81.2	85.5	96.7	80.9	89.7	89.8	76.6	84.0	95.3	79.7	88.4	95.2	77.5	87.4
04	MASONRY	88.0	79.1	82.5	87.0	73.6	78.7	93.3	75.5	82.3	73.2	71.8	72.3	89.8	75.7	81.1	79.8	68.1	72.6
05	METALS	96.1	75.7	89.9	95.2	89.6	93.5	97.9	74.4	90.8	94.5	84.4	91.4	101.3	88.9	97.6	96.5	82.0	92.1
06	WOOD, PLASTICS & COMPOSITES	98.6	86.0	91.8	100.8	77.4	88.1	103.6	81.5	91.6	91.3	76.5	83.3	104.9	79.4	91.0	93.7	78.6	85.5
07	THERMAL & MOISTURE PROTECTION	99.8	81.4	91.8	108.5	76.5	94.6	96.1	79.4	88.8	101.7	76.2	90.6	99.2	78.5	90.2	88.2	72.2	81.3
08	OPENINGS	104.1	82.2	99.0	90.3	77.3	87.3	94.4	78.9	90.8	95.8	73.8	90.7	93.0	77.8	89.5	93.4	77.5	89.7
0920	Plaster & Gypsum Board	96.4	85.8	89.4	109.5	77.1	88.1	93.3	81.7	85.7	71.9	76.5	74.9	94.0	79.5	84.5	91.9	78.5	83.0
0950, 0980	Ceilings & Acoustic Treatment	94.9	85.8	88.7	91.9	77.1	81.9	76.2	81.7	80.0	86.0	76.5	79.5	79.6	79.5	79.6	83.4	78.5	80.1
0960	Flooring	96.9	81.8	92.5	99.0	91.3	96.8	93.0	81.8	89.8	69.0	82.9	73.1	93.4	77.5	88.8	91.4	55.8	81.1
0970, 0990	Wall Finishes & Painting/Coating	99.3	79.5	87.6	94.0	69.7	79.6	85.4	83.5	84.3	86.2	74.5	79.3	85.4	67.6	74.8	91.0	68.0	77.4
09	FINISHES	96.0	84.2	89.6	93.4	79.6	85.8	86.9	81.5	83.9	78.6	77.7	78.1	87.4	77.6	82.0	88.0	72.3	79.5
COVERS	DIVS. 10 - 14, 25, 28, 41, 43, 44, 46	100.0	90.6	97.9	100.0	86.4	97.0	100.0	88.4	97.5	100.0	84.8	96.6	100.0	88.1	97.4	100.0	83.5	96.4
21, 22, 23	FIRE SUPPRESSION, PLUMBING & HVAC	99.9	79.7	91.7	96.5	78.8	89.3	96.3	77.9	88.9	97.2	74.5	88.0	99.8	78.6	91.2	96.5	76.0	88.2
26, 27, 3370	ELECTRICAL, COMMUNICATIONS & UTIL.	101.9	87.0	94.3	90.9	79.1	84.9	97.8	80.5	89.0	93.1	72.9	82.9	91.0	75.9	83.3	93.7	76.4	84.9
MF2016	WEIGHTED AVERAGE	98.7	83.0	92.0	94.1	81.0	88.4	95.8	80.1	89.1	92.6	78.5	86.5	96.1	80.3	89.3	93.9	77.2	86.7

INDIANA / IOWA

DIVISION		SOUTH BEND 465-466 MAT.	INST.	TOTAL	TERRE HAUTE 478 MAT.	INST.	TOTAL	WASHINGTON 475 MAT.	INST.	TOTAL	BURLINGTON 526 MAT.	INST.	TOTAL	CARROLL 514 MAT.	INST.	TOTAL	CEDAR RAPIDS 522-524 MAT.	INST.	TOTAL
015433	CONTRACTOR EQUIPMENT		108.9	108.9		113.7	113.7		113.7	113.7		101.4	101.4		101.4	101.4		98.2	98.2
0241, 31 - 34	SITE & INFRASTRUCTURE, DEMOLITION	99.3	94.3	95.8	94.4	121.6	113.4	94.4	121.8	113.5	98.5	96.5	97.1	87.8	97.3	94.4	100.1	96.0	97.3
0310	Concrete Forming & Accessories	95.5	78.1	80.7	95.2	79.2	81.6	96.2	82.1	84.2	94.4	96.8	96.4	81.9	79.8	80.1	100.5	86.3	88.4
0320	Concrete Reinforcing	95.8	86.0	90.8	97.9	85.4	91.6	90.5	84.9	87.7	94.4	100.6	97.6	95.1	88.3	91.6	95.1	82.5	88.7
0330	Cast-in-Place Concrete	103.6	79.9	94.8	91.8	80.6	87.7	99.9	86.0	94.7	107.9	56.4	88.7	107.9	84.4	99.2	108.2	85.9	99.9
03	CONCRETE	91.0	81.7	86.9	100.2	81.0	91.7	106.0	84.1	96.3	93.6	83.7	89.2	92.4	83.7	88.5	93.7	86.1	90.3
04	MASONRY	92.9	75.6	82.2	91.0	74.8	81.0	83.9	78.6	80.7	99.7	73.4	83.5	101.3	74.0	84.4	105.3	81.7	90.7
05	METALS	102.3	104.8	103.0	93.2	85.2	90.8	87.7	85.2	87.0	85.5	99.9	89.8	85.5	95.8	88.6	87.9	93.9	89.7
06	WOOD, PLASTICS & COMPOSITES	95.8	77.2	85.7	95.5	78.9	86.5	95.9	81.5	88.1	93.0	101.0	97.4	78.7	82.9	81.0	100.4	86.1	92.6
07	THERMAL & MOISTURE PROTECTION	102.4	80.4	92.8	101.0	81.4	92.5	100.9	84.2	93.6	103.9	79.2	93.1	104.2	76.3	92.1	105.0	82.3	95.1
08	OPENINGS	95.8	77.7	91.6	94.3	77.5	90.4	91.2	79.4	88.4	96.5	98.6	97.0	95.1	82.2	92.0	101.5	83.6	97.3
0920	Plaster & Gypsum Board	94.5	76.9	82.9	94.0	78.0	83.4	94.3	80.6	85.3	105.0	101.2	102.5	100.5	82.6	88.7	110.1	86.0	94.3
0950, 0980	Ceilings & Acoustic Treatment	90.3	76.9	81.2	83.4	78.0	79.7	79.2	80.6	80.1	95.0	101.2	99.2	95.0	82.6	86.6	97.5	86.0	89.7
0960	Flooring	93.6	89.1	92.3	94.0	78.2	89.4	94.9	79.3	90.3	95.3	71.4	88.4	89.6	82.2	87.5	109.6	87.1	103.0
0970, 0990	Wall Finishes & Painting/Coating	88.9	84.7	86.4	91.0	81.9	85.6	91.0	86.6	88.4	92.3	87.5	89.5	92.3	85.8	88.4	93.9	72.6	81.3
09	FINISHES	90.0	80.5	84.9	88.8	79.2	83.6	88.6	81.9	85.0	92.4	92.4	92.4	88.7	81.2	84.6	97.8	85.1	90.9
COVERS	DIVS. 10 - 14, 25, 28, 41, 43, 44, 46	100.0	90.3	97.9	100.0	90.5	97.9	100.0	93.3	98.5	100.0	92.1	98.3	100.0	85.8	96.9	100.0	94.0	98.7
21, 22, 23	FIRE SUPPRESSION, PLUMBING & HVAC	99.9	76.7	90.5	100.0	78.2	91.2	96.5	80.4	90.0	96.7	85.7	92.2	96.7	72.6	86.9	100.1	83.0	93.2
26, 27, 3370	ELECTRICAL, COMMUNICATIONS & UTIL.	97.7	86.0	91.8	93.5	85.5	89.4	94.0	83.5	88.6	100.3	74.0	86.9	101.0	78.8	89.7	97.9	82.0	89.8
MF2016	WEIGHTED AVERAGE	97.4	83.6	91.5	96.2	83.8	90.9	94.6	85.4	90.7	95.1	86.3	91.3	94.9	81.3	89.1	97.4	85.8	92.4

City Cost Indexes

IOWA

DIVISION		COUNCIL BLUFFS 515 MAT.	INST.	TOTAL	CRESTON 508 MAT.	INST.	TOTAL	DAVENPORT 527-528 MAT.	INST.	TOTAL	DECORAH 521 MAT.	INST.	TOTAL	DES MOINES 500-503, 509 MAT.	INST.	TOTAL	DUBUQUE 520 MAT.	INST.	TOTAL
015433	CONTRACTOR EQUIPMENT		97.5	97.5		101.4	101.4		101.4	101.4		101.4	101.4		103.2	103.2		97.0	97.0
0241, 31 - 34	SITE & INFRASTRUCTURE, DEMOLITION	103.9	93.0	96.3	95.1	98.4	97.4	98.7	99.4	99.2	97.1	96.3	96.5	101.1	99.9	100.3	98.0	93.2	94.7
0310	Concrete Forming & Accessories	81.2	75.6	76.4	76.5	82.6	81.7	100.0	98.6	98.8	91.4	73.7	76.4	93.4	86.1	87.2	82.7	83.8	83.6
0320	Concrete Reinforcing	97.0	82.1	89.4	94.5	88.3	91.3	95.1	101.9	98.6	94.4	88.0	91.1	99.0	85.7	92.3	93.7	82.3	87.9
0330	Cast-in-Place Concrete	112.5	81.4	100.9	114.7	62.5	95.3	104.2	98.0	101.9	105.0	78.2	95.0	97.6	90.5	95.0	105.9	85.4	98.3
03	CONCRETE	95.8	79.7	88.6	94.0	77.3	86.6	91.8	99.4	95.2	91.6	78.7	85.9	87.3	88.3	87.7	90.6	84.8	88.0
04	MASONRY	106.7	77.8	88.8	99.3	81.6	88.3	102.1	94.1	97.2	120.9	76.2	93.3	87.3	85.5	86.2	106.3	70.3	84.1
05	METALS	92.6	93.5	92.9	87.0	96.3	89.8	87.9	106.4	93.5	85.6	95.4	88.6	92.6	97.2	94.0	86.5	93.1	88.5
06	WOOD, PLASTICS & COMPOSITES	77.5	74.2	75.7	70.6	82.9	77.3	100.4	98.1	99.1	89.8	72.9	80.7	87.7	85.3	86.4	79.3	84.6	82.2
07	THERMAL & MOISTURE PROTECTION	104.3	76.9	92.4	105.0	81.5	94.8	104.4	94.2	99.9	104.1	73.4	90.8	96.6	85.9	92.0	104.6	78.6	93.3
08	OPENINGS	100.5	77.6	95.2	111.0	84.0	104.7	101.5	99.5	101.0	99.5	80.2	95.0	98.9	88.2	96.4	100.5	84.8	96.9
0920	Plaster & Gypsum Board	100.5	73.9	82.9	96.1	82.6	87.2	110.1	98.2	102.3	103.8	72.3	83.1	93.5	85.0	87.9	100.5	84.5	90.0
0950, 0980	Ceilings & Acoustic Treatment	95.0	73.9	80.7	88.5	82.6	84.5	97.5	98.2	98.0	90.6	72.3	79.6	91.9	85.0	87.3	95.0	84.5	87.9
0960	Flooring	88.5	87.1	88.1	83.9	71.4	80.3	97.7	92.4	96.2	94.9	71.4	88.1	97.9	94.8	96.9	100.8	71.4	92.3
0970, 0990	Wall Finishes & Painting/Coating	89.1	72.6	79.3	83.2	76.2	79.1	92.3	92.9	92.6	92.3	85.8	88.4	88.8	85.8	87.0	93.1	80.5	85.7
09	FINISHES	89.4	77.1	82.7	82.5	79.9	81.1	94.3	96.8	95.7	92.0	74.4	82.4	90.2	87.7	88.8	93.1	81.2	86.6
COVERS	DIVS. 10 - 14, 25, 28, 41, 43, 44, 46	100.0	89.4	97.7	100.0	88.4	97.4	100.0	97.6	99.5	100.0	86.2	97.0	100.0	94.4	98.8	100.0	92.9	98.4
21, 22, 23	FIRE SUPPRESSION, PLUMBING & HVAC	100.1	74.4	89.7	96.6	80.2	90.0	100.1	96.0	98.5	96.7	75.3	88.0	99.8	84.7	93.7	100.1	77.3	90.9
26, 27, 3370	ELECTRICAL, COMMUNICATIONS & UTIL.	103.2	82.3	92.6	93.8	78.8	86.2	95.9	89.7	92.8	97.9	48.4	72.7	105.4	84.4	94.7	101.7	78.8	90.1
MF2016	WEIGHTED AVERAGE	98.2	80.8	90.7	95.2	83.1	90.0	96.5	96.9	96.7	95.9	75.9	87.3	96.0	88.4	92.8	96.5	81.9	90.3

IOWA

DIVISION		FORT DODGE 505 MAT.	INST.	TOTAL	MASON CITY 504 MAT.	INST.	TOTAL	OTTUMWA 525 MAT.	INST.	TOTAL	SHENANDOAH 516 MAT.	INST.	TOTAL	SIBLEY 512 MAT.	INST.	TOTAL	SIOUX CITY 510-511 MAT.	INST.	TOTAL
015433	CONTRACTOR EQUIPMENT		101.4	101.4		101.4	101.4		97.0	97.0		97.5	97.5		101.4	101.4		101.4	101.4
0241, 31 - 34	SITE & INFRASTRUCTURE, DEMOLITION	103.5	95.1	97.6	103.6	96.2	98.4	98.4	91.0	93.3	102.2	93.1	95.9	108.2	96.2	99.8	109.9	97.3	101.1
0310	Concrete Forming & Accessories	77.1	74.1	74.5	81.0	73.3	74.4	89.8	73.7	76.1	82.9	79.2	79.8	83.5	38.7	45.4	100.5	74.4	78.3
0320	Concrete Reinforcing	94.5	88.0	91.2	94.3	88.0	91.1	94.4	100.6	97.5	97.0	88.3	92.6	97.0	87.9	92.4	95.1	79.3	87.1
0330	Cast-in-Place Concrete	107.6	43.0	83.6	107.6	73.3	94.9	108.6	65.6	92.6	108.8	85.0	99.9	106.6	57.1	88.1	107.2	71.9	94.1
03	CONCRETE	89.5	66.7	79.4	89.7	76.8	84.0	93.1	76.6	85.8	93.4	83.6	89.1	92.4	55.5	76.0	93.1	75.3	85.2
04	MASONRY	98.1	58.6	73.7	110.9	75.0	88.7	102.6	54.0	72.6	106.2	76.0	87.6	124.9	58.8	84.1	99.4	68.4	80.2
05	METALS	87.1	95.0	89.5	87.1	95.4	89.6	85.5	100.1	89.9	91.6	96.3	93.0	85.7	94.3	88.3	87.9	91.4	89.0
06	WOOD, PLASTICS & COMPOSITES	71.1	82.9	77.5	75.0	72.9	73.9	86.9	79.7	83.0	79.3	80.8	80.1	80.1	35.1	55.7	100.4	74.3	86.2
07	THERMAL & MOISTURE PROTECTION	104.3	67.9	88.4	103.7	74.0	90.8	104.8	68.7	89.1	103.5	74.4	90.8	103.8	55.8	82.9	104.4	72.6	90.6
08	OPENINGS	104.8	73.7	97.6	96.8	80.2	93.0	101.0	83.9	97.0	92.1	78.7	89.0	97.7	47.3	86.0	101.5	75.5	95.4
0920	Plaster & Gypsum Board	96.1	82.6	87.2	96.1	72.3	80.5	101.4	79.5	87.0	100.5	80.6	87.4	100.5	33.5	56.3	110.1	73.8	86.2
0950, 0980	Ceilings & Acoustic Treatment	88.5	82.6	84.5	88.5	72.3	77.6	95.0	79.5	84.5	95.0	80.6	85.2	95.0	33.5	53.3	97.5	73.8	81.4
0960	Flooring	85.2	71.4	81.2	87.2	71.4	82.7	103.8	71.4	94.4	89.2	76.4	85.5	90.5	71.4	85.0	97.7	74.0	90.9
0970, 0990	Wall Finishes & Painting/Coating	83.2	83.4	83.3	83.2	85.8	84.7	93.1	85.8	88.8	89.1	85.8	87.1	92.3	85.8	88.4	92.3	68.2	78.0
09	FINISHES	84.3	75.5	79.5	84.9	74.1	79.0	94.3	74.8	83.6	89.5	79.3	83.9	91.7	47.7	67.7	95.7	73.5	83.6
COVERS	DIVS. 10 - 14, 25, 28, 41, 43, 44, 46	100.0	82.4	96.1	100.0	85.8	96.9	100.0	82.8	96.2	100.0	85.8	96.9	100.0	76.3	94.8	100.0	91.1	98.0
21, 22, 23	FIRE SUPPRESSION, PLUMBING & HVAC	96.6	69.5	85.6	96.6	78.6	89.3	96.7	72.6	86.9	96.7	85.7	92.3	96.7	67.7	84.9	100.1	78.1	91.2
26, 27, 3370	ELECTRICAL, COMMUNICATIONS & UTIL.	100.2	71.1	85.4	99.3	48.4	73.4	100.1	72.7	86.2	97.9	80.4	89.0	97.9	48.4	72.7	97.9	75.8	86.7
MF2016	WEIGHTED AVERAGE	94.9	74.1	86.0	94.7	76.2	86.8	95.7	76.3	87.4	95.4	83.8	90.4	96.1	63.3	82.1	97.1	78.7	89.2

IOWA / KANSAS

DIVISION		SPENCER 513 (IOWA) MAT.	INST.	TOTAL	WATERLOO 506-507 (IOWA) MAT.	INST.	TOTAL	BELLEVILLE 669 (KANSAS) MAT.	INST.	TOTAL	COLBY 677 MAT.	INST.	TOTAL	DODGE CITY 678 MAT.	INST.	TOTAL	EMPORIA 668 MAT.	INST.	TOTAL
015433	CONTRACTOR EQUIPMENT		101.4	101.4		101.4	101.4		106.5	106.5		106.5	106.5		106.5	106.5		104.7	104.7
0241, 31 - 34	SITE & INFRASTRUCTURE, DEMOLITION	108.2	94.9	99.0	108.8	97.4	100.9	109.9	96.7	100.7	107.2	97.4	100.3	109.6	96.4	100.4	102.3	94.2	96.7
0310	Concrete Forming & Accessories	89.7	38.5	46.1	92.1	70.6	73.8	96.9	55.7	61.8	99.3	62.9	68.3	92.6	62.7	67.2	87.5	67.4	70.4
0320	Concrete Reinforcing	97.0	87.9	92.3	95.1	82.6	88.7	105.6	103.6	104.6	105.2	103.6	104.4	102.5	103.6	103.1	104.1	103.8	103.9
0330	Cast-in-Place Concrete	106.6	67.4	92.0	115.3	86.5	104.6	119.6	86.0	107.1	116.6	89.8	106.6	118.7	89.5	107.8	115.7	89.9	106.1
03	CONCRETE	92.8	59.0	77.8	95.3	79.3	88.2	110.0	75.9	94.9	107.3	80.6	95.5	108.7	80.3	96.1	102.6	82.6	93.7
04	MASONRY	124.9	58.8	84.1	98.9	76.7	85.2	90.4	59.7	71.5	99.5	66.0	78.8	108.0	65.9	82.0	96.0	67.4	78.3
05	METALS	85.7	94.1	88.2	89.4	93.8	90.7	88.9	99.3	92.0	89.2	100.2	92.5	90.7	99.1	93.3	88.6	100.3	92.1
06	WOOD, PLASTICS & COMPOSITES	86.8	35.1	58.7	88.7	66.2	76.5	97.0	52.6	72.9	101.1	59.0	78.2	92.5	59.0	74.3	87.3	64.8	75.1
07	THERMAL & MOISTURE PROTECTION	104.8	58.9	84.8	104.0	79.3	93.2	93.3	63.9	80.5	99.6	67.2	85.5	99.6	69.9	86.6	91.4	76.8	85.0
08	OPENINGS	109.0	47.3	94.7	97.3	75.1	92.1	96.2	63.7	88.7	100.2	71.0	93.4	100.1	67.2	92.5	94.2	72.8	89.2
0920	Plaster & Gypsum Board	101.4	33.5	56.7	104.9	65.4	78.9	89.2	51.4	64.3	99.3	58.0	72.1	92.9	58.0	69.9	86.2	63.9	71.5
0950, 0980	Ceilings & Acoustic Treatment	95.0	33.5	53.3	91.1	65.4	73.7	85.4	51.4	62.4	82.9	58.0	66.0	82.9	58.0	66.0	85.4	63.9	70.8
0960	Flooring	93.3	71.4	86.9	92.4	81.6	89.3	90.5	69.3	84.4	86.2	69.3	81.3	82.4	69.3	78.6	86.1	69.3	81.2
0970, 0990	Wall Finishes & Painting/Coating	92.3	58.7	72.4	83.2	83.4	83.3	92.0	56.5	71.0	92.8	56.5	71.3	92.8	56.5	71.3	92.0	56.5	71.0
09	FINISHES	92.6	43.5	65.8	88.4	73.0	80.0	85.9	56.7	70.0	86.0	62.0	72.9	84.1	62.0	72.0	83.2	65.4	73.5
COVERS	DIVS. 10 - 14, 25, 28, 41, 43, 44, 46	100.0	76.2	94.8	100.0	91.5	98.1	100.0	83.6	96.4	100.0	86.4	97.0	100.0	86.4	97.0	100.0	87.0	97.1
21, 22, 23	FIRE SUPPRESSION, PLUMBING & HVAC	96.7	69.7	85.7	100.0	81.9	92.7	96.4	71.3	86.3	96.6	72.1	86.7	100.1	72.1	88.8	96.4	74.9	87.7
26, 27, 3370	ELECTRICAL, COMMUNICATIONS & UTIL.	99.6	48.4	73.6	95.7	63.1	79.1	99.2	64.2	81.4	96.7	69.5	82.9	93.8	69.5	81.5	96.4	68.5	82.2
MF2016	WEIGHTED AVERAGE	97.7	63.6	83.1	96.2	79.3	89.0	96.6	72.1	86.1	97.1	75.7	87.9	98.3	75.4	88.5	94.9	77.2	87.3

For customer support on your Electrical Costs with RSMeans data, call 800.448.8182.

525

KANSAS

DIVISION		FORT SCOTT 667			HAYS 676			HUTCHINSON 675			INDEPENDENCE 673			KANSAS CITY 660 - 662			LIBERAL 679		
		MAT.	INST.	TOTAL	MAT.	INST.	TOTAL	MAT.	INST.	TOTAL	MAT.	INST.	TOTAL	MAT.	INST.	TOTAL	MAT.	INST.	TOTAL
015433	CONTRACTOR EQUIPMENT		105.6	105.6		106.5	106.5		106.5	106.5		106.5	106.5		103.1	103.1		106.5	106.5
0241, 31 - 34	SITE & INFRASTRUCTURE, DEMOLITION	99.0	94.2	95.7	111.5	96.9	101.3	91.1	97.4	95.5	110.5	97.4	101.4	93.8	94.5	94.3	111.4	97.0	101.3
0310	Concrete Forming & Accessories	105.5	82.1	85.6	96.8	60.5	65.9	87.3	58.1	62.4	107.9	69.6	75.3	101.7	97.2	97.8	93.0	60.4	65.3
0320	Concrete Reinforcing	103.4	103.7	103.6	102.5	103.6	103.1	102.5	103.6	103.1	101.9	103.7	102.8	100.3	105.3	102.9	104.0	103.6	103.8
0330	Cast-in-Place Concrete	107.2	84.8	98.9	92.0	86.2	89.8	85.2	89.4	86.8	119.2	89.6	108.2	91.8	98.6	94.4	92.0	86.0	89.7
03	CONCRETE	98.1	87.5	93.4	98.9	78.2	89.7	82.3	78.2	80.5	109.8	83.5	98.2	90.3	99.5	94.4	100.7	78.1	90.7
04	MASONRY	97.0	57.3	72.5	107.5	59.8	78.0	99.5	65.9	78.7	96.7	65.9	77.7	98.1	98.0	98.1	105.6	59.7	77.3
05	METALS	88.6	99.7	91.9	88.8	99.9	92.2	88.6	99.0	91.8	88.6	99.6	91.9	95.6	105.2	98.5	89.1	99.2	92.1
06	WOOD, PLASTICS & COMPOSITES	108.2	88.9	97.7	97.8	59.0	76.7	87.2	52.9	68.5	112.3	67.8	88.1	103.8	96.8	100.0	93.2	59.0	74.6
07	THERMAL & MOISTURE PROTECTION	92.5	74.4	84.6	100.0	64.6	84.6	98.3	66.2	84.3	99.7	77.0	89.8	92.0	99.8	95.4	100.1	64.0	84.4
08	OPENINGS	94.2	86.1	92.3	100.1	67.2	92.4	100.1	63.8	91.6	98.1	71.9	92.0	95.4	96.4	95.6	100.1	67.2	92.5
0920	Plaster & Gypsum Board	91.6	88.7	89.7	96.5	58.0	71.1	91.6	51.6	65.3	107.8	67.0	80.9	85.5	96.8	92.9	93.8	58.0	70.2
0950, 0980	Ceilings & Acoustic Treatment	85.4	88.7	87.7	82.9	58.0	66.0	82.9	51.6	61.7	82.9	67.0	72.1	85.4	96.8	93.1	82.9	58.0	66.0
0960	Flooring	99.7	69.1	90.9	85.0	69.3	80.4	79.7	69.3	76.7	90.4	69.1	84.2	80.9	97.5	85.7	82.7	68.1	78.5
0970, 0990	Wall Finishes & Painting/Coating	93.6	79.5	85.3	92.8	56.5	71.3	92.8	56.5	71.3	92.8	56.5	71.3	99.6	100.9	100.3	92.8	56.5	71.3
09	FINISHES	88.2	80.5	84.0	85.8	60.5	72.0	81.6	58.4	68.9	88.6	67.7	77.2	83.7	97.5	91.3	85.0	60.2	71.5
COVERS	DIVS. 10 - 14, 25, 28, 41, 43, 44, 46	100.0	86.7	97.1	100.0	84.3	96.5	100.0	85.7	96.8	100.0	87.4	97.2	100.0	94.5	98.8	100.0	84.3	96.5
21, 22, 23	FIRE SUPPRESSION, PLUMBING & HVAC	96.4	68.3	85.1	96.6	68.9	85.4	96.6	70.7	86.2	96.6	72.0	86.7	99.8	98.8	99.4	96.6	70.3	86.0
26, 27, 3370	ELECTRICAL, COMMUNICATIONS & UTIL.	95.6	69.5	82.4	95.8	69.5	82.4	91.1	62.4	76.5	93.1	74.6	83.7	101.4	97.9	99.6	93.8	69.5	81.5
MF2016	WEIGHTED AVERAGE	94.8	78.2	87.7	96.3	73.5	86.5	92.4	73.1	84.1	97.0	77.9	88.8	96.0	98.6	97.1	96.2	73.7	86.5

DIVISION		KANSAS									KENTUCKY								
		SALINA 674			TOPEKA 664 - 666			WICHITA 670 - 672			ASHLAND 411 - 412			BOWLING GREEN 421 - 422			CAMPTON 413 - 414		
		MAT.	INST.	TOTAL	MAT.	INST.	TOTAL	MAT.	INST.	TOTAL	MAT.	INST.	TOTAL	MAT.	INST.	TOTAL	MAT.	INST.	TOTAL
015433	CONTRACTOR EQUIPMENT		106.5	106.5		104.7	104.7		106.5	106.5		100.4	100.4		93.3	93.3		99.7	99.7
0241, 31 - 34	SITE & INFRASTRUCTURE, DEMOLITION	100.1	96.9	97.9	96.9	94.0	94.9	96.2	98.1	97.5	115.0	83.0	92.7	79.5	94.2	89.7	88.5	95.5	93.4
0310	Concrete Forming & Accessories	89.1	55.8	60.7	98.8	67.8	72.4	94.8	55.0	60.9	87.0	92.2	91.4	86.6	77.8	79.1	89.2	83.2	84.1
0320	Concrete Reinforcing	101.9	103.4	102.7	99.9	102.7	101.3	100.0	101.0	100.5	91.4	96.9	94.2	88.7	79.0	83.8	89.6	95.7	92.7
0330	Cast-in-Place Concrete	103.0	86.2	96.8	96.5	87.9	93.3	96.3	74.6	88.2	87.3	97.5	91.1	86.6	71.5	81.0	96.6	69.7	86.6
03	CONCRETE	95.9	76.0	87.1	92.0	82.0	87.5	91.6	71.2	82.6	91.0	95.9	93.2	89.5	76.2	83.6	93.0	80.9	87.6
04	MASONRY	122.3	56.0	81.4	91.8	65.6	75.6	96.6	50.5	68.1	89.6	94.0	92.3	91.6	70.6	78.6	89.0	55.9	68.5
05	METALS	90.5	99.6	93.3	92.5	100.4	94.9	92.5	97.8	94.1	95.6	108.8	99.6	97.2	83.8	93.1	96.5	90.7	94.7
06	WOOD, PLASTICS & COMPOSITES	88.7	52.6	69.1	102.0	66.9	82.9	98.6	52.9	73.8	75.3	90.2	83.4	87.9	78.4	82.7	83.6	90.9	88.8
07	THERMAL & MOISTURE PROTECTION	99.0	62.8	83.2	96.3	76.1	87.6	97.8	59.4	81.1	92.1	90.9	91.5	88.2	79.6	84.5	101.1	69.7	87.4
08	OPENINGS	100.1	63.7	91.6	103.6	75.2	97.0	104.5	63.3	94.9	92.6	91.6	92.4	93.4	80.5	90.4	94.6	89.9	93.5
0920	Plaster & Gypsum Board	91.6	51.4	65.1	97.9	66.0	76.9	94.8	51.6	66.4	60.0	90.2	79.9	87.3	78.2	81.3	87.3	90.2	89.2
0950, 0980	Ceilings & Acoustic Treatment	82.9	51.4	61.6	91.7	66.0	74.3	86.7	51.6	63.0	78.7	90.2	86.5	83.4	78.2	79.9	83.4	90.2	88.0
0960	Flooring	81.0	69.3	77.6	95.7	69.3	88.0	92.8	69.3	86.0	74.3	84.0	77.1	89.4	64.0	82.0	91.5	66.5	84.3
0970, 0990	Wall Finishes & Painting/Coating	92.8	56.5	71.3	96.1	69.5	80.4	91.5	56.5	70.8	92.9	92.8	92.8	91.0	90.4	90.7	91.0	55.3	69.9
09	FINISHES	82.8	56.7	68.6	92.5	67.2	78.7	89.6	56.3	71.5	75.9	90.7	84.0	86.7	76.3	81.0	87.4	78.1	82.4
COVERS	DIVS. 10 - 14, 25, 28, 41, 43, 44, 46	100.0	84.3	96.5	100.0	81.6	95.9	100.0	83.7	96.4	100.0	90.0	97.8	100.0	85.6	96.8	100.0	48.8	88.7
21, 22, 23	FIRE SUPPRESSION, PLUMBING & HVAC	100.1	70.9	88.3	100.0	74.7	89.8	99.8	68.5	87.2	96.3	86.9	92.5	100.0	78.8	91.4	96.6	78.1	89.1
26, 27, 3370	ELECTRICAL, COMMUNICATIONS & UTIL.	93.6	71.9	82.5	100.1	73.6	86.6	97.5	71.8	84.5	91.4	90.4	90.9	94.0	76.5	85.1	91.4	90.3	90.9
MF2016	WEIGHTED AVERAGE	96.9	72.8	86.6	97.0	77.7	88.7	96.7	70.8	85.6	93.2	92.0	92.7	94.5	78.9	87.8	94.4	79.9	88.2

DIVISION		KENTUCKY																	
		CORBIN 407 - 409			COVINGTON 410			ELIZABETHTOWN 427			FRANKFORT 406			HAZARD 417 - 418			HENDERSON 424		
		MAT.	INST.	TOTAL	MAT.	INST.	TOTAL	MAT.	INST.	TOTAL	MAT.	INST.	TOTAL	MAT.	INST.	TOTAL	MAT.	INST.	TOTAL
015433	CONTRACTOR EQUIPMENT		99.7	99.7		103.6	103.6		93.3	93.3		99.7	99.7		99.7	99.7		113.7	113.7
0241, 31 - 34	SITE & INFRASTRUCTURE, DEMOLITION	91.3	96.0	94.6	83.8	107.3	100.2	73.9	93.9	87.8	89.4	96.5	94.4	86.3	96.6	93.5	82.5	120.8	109.2
0310	Concrete Forming & Accessories	85.5	78.1	79.2	84.7	72.7	74.5	81.3	74.0	75.1	102.4	79.3	82.8	85.6	84.1	84.4	92.3	79.1	81.1
0320	Concrete Reinforcing	92.1	92.3	92.2	88.3	80.9	84.5	89.2	84.3	86.7	96.9	82.2	89.3	90.0	92.5	91.3	88.8	81.1	84.9
0330	Cast-in-Place Concrete	92.8	74.3	85.9	92.4	82.1	88.6	78.3	69.0	74.8	94.1	76.4	87.6	92.9	71.7	85.0	76.5	87.0	80.4
03	CONCRETE	84.1	79.6	82.1	91.4	78.2	85.6	81.6	74.5	78.5	87.5	79.2	83.8	89.8	81.5	86.1	86.7	82.5	84.9
04	MASONRY	81.0	61.3	68.8	103.0	73.5	84.8	76.5	63.0	68.2	79.5	71.8	74.7	87.5	58.0	69.3	94.8	80.4	85.9
05	METALS	92.2	89.3	91.3	94.4	88.9	92.8	96.4	84.9	92.9	95.0	86.0	92.3	96.5	89.7	94.4	87.5	85.8	87.0
06	WOOD, PLASTICS & COMPOSITES	75.0	79.8	77.6	83.7	70.2	76.4	82.5	76.1	79.0	102.1	79.8	90.0	82.9	90.9	87.2	90.8	78.0	83.8
07	THERMAL & MOISTURE PROTECTION	103.6	71.5	89.6	101.9	73.7	89.7	87.6	70.5	80.2	102.5	75.7	90.8	101.0	71.5	88.1	100.2	85.1	93.7
08	OPENINGS	89.2	71.9	85.2	96.6	73.9	91.4	93.4	74.7	89.0	96.7	79.6	92.7	95.0	89.1	93.6	91.6	79.4	88.7
0920	Plaster & Gypsum Board	93.1	78.9	83.7	68.6	70.0	69.5	86.1	75.8	79.3	95.2	78.9	84.4	86.1	90.2	88.8	91.0	77.0	81.8
0950, 0980	Ceilings & Acoustic Treatment	79.8	78.9	79.2	86.0	70.0	75.2	83.4	75.8	78.3	90.1	78.9	82.5	83.4	90.2	88.0	79.2	77.0	77.7
0960	Flooring	86.0	66.5	80.3	66.6	86.3	72.3	86.8	76.3	83.8	95.1	66.5	86.8	89.8	66.5	83.0	93.0	78.6	88.9
0970, 0990	Wall Finishes & Painting/Coating	90.7	63.6	74.7	86.2	73.0	78.4	91.0	69.9	78.6	97.5	94.1	95.5	91.0	55.3	69.9	91.0	89.0	89.9
09	FINISHES	81.3	74.6	77.6	77.6	74.8	76.1	85.3	73.7	79.0	89.1	78.9	83.5	86.6	78.7	82.3	86.8	80.1	83.2
COVERS	DIVS. 10 - 14, 25, 28, 41, 43, 44, 46	100.0	91.7	98.2	100.0	87.7	97.3	100.0	71.5	93.7	100.0	58.6	90.9	100.0	49.6	88.9	100.0	56.9	90.5
21, 22, 23	FIRE SUPPRESSION, PLUMBING & HVAC	96.7	74.9	87.9	97.2	77.7	89.3	96.7	77.8	89.0	100.0	78.7	91.4	96.6	79.2	89.6	96.7	78.1	89.2
26, 27, 3370	ELECTRICAL, COMMUNICATIONS & UTIL.	90.3	90.3	90.3	95.2	74.2	84.5	91.3	78.2	84.6	102.2	73.2	87.5	91.4	90.4	90.9	93.5	75.5	84.3
MF2016	WEIGHTED AVERAGE	91.1	79.6	86.2	94.4	79.8	88.1	91.3	76.7	85.1	95.5	78.7	88.3	93.8	80.6	88.1	92.2	82.4	88.0

KENTUCKY

DIVISION		LEXINGTON 403 - 405			LOUISVILLE 400 - 402			OWENSBORO 423			PADUCAH 420			PIKEVILLE 415 - 416			SOMERSET 425 - 426		
		MAT.	INST.	TOTAL	MAT.	INST.	TOTAL	MAT.	INST.	TOTAL	MAT.	INST.	TOTAL	MAT.	INST.	TOTAL	MAT.	INST.	TOTAL
015433	CONTRACTOR EQUIPMENT		99.7	99.7		93.3	93.3		113.7	113.7		113.7	113.7		100.4	100.4		99.7	99.7
0241, 31 - 34	SITE & INFRASTRUCTURE, DEMOLITION	93.3	97.8	96.4	88.6	94.1	92.4	92.6	121.4	112.7	85.3	120.9	110.1	126.4	82.1	95.5	78.8	96.0	90.8
0310	Concrete Forming & Accessories	97.8	75.6	78.9	94.4	76.9	79.5	90.7	79.0	80.7	88.6	80.8	82.0	96.2	88.0	89.3	86.7	73.4	75.4
0320	Concrete Reinforcing	101.2	82.8	91.8	96.9	82.6	89.6	88.8	79.5	84.1	89.4	82.9	86.1	91.9	96.8	94.4	89.2	62.1	75.4
0330	Cast-in-Place Concrete	95.0	86.9	92.0	89.9	70.6	82.7	89.3	86.4	88.2	81.6	83.1	82.1	96.0	92.1	94.5	76.5	90.1	81.6
03	CONCRETE	86.9	81.2	84.4	85.1	76.1	81.1	98.5	82.0	91.2	91.2	82.3	87.2	104.5	92.2	99.0	76.8	77.8	77.2
04	MASONRY	79.6	72.6	75.3	75.7	68.0	70.9	87.8	80.3	83.1	90.6	80.1	84.1	87.9	85.0	86.1	82.4	64.0	71.1
05	METALS	94.6	86.1	92.0	95.9	84.6	92.5	89.0	85.6	87.9	86.0	87.6	86.5	95.5	108.6	99.5	96.4	78.3	90.9
06	WOOD, PLASTICS & COMPOSITES	91.3	73.6	81.7	87.9	78.4	82.7	88.5	78.0	82.8	86.0	80.7	83.1	85.7	90.2	88.2	83.4	73.6	78.1
07	THERMAL & MOISTURE PROTECTION	103.9	75.5	91.6	100.0	72.8	88.2	101.0	78.8	91.3	100.3	84.3	93.3	92.9	82.8	88.5	100.2	71.8	87.8
08	OPENINGS	89.4	76.0	86.3	87.4	75.5	84.6	91.6	79.0	88.7	90.9	81.3	88.7	93.2	88.2	92.1	94.1	68.0	88.0
0920	Plaster & Gypsum Board	102.7	72.5	82.8	95.9	78.2	84.3	89.5	77.0	81.2	88.5	79.8	82.8	63.7	90.2	81.2	86.1	72.5	77.1
0950, 0980	Ceilings & Acoustic Treatment	83.2	72.5	76.0	88.3	78.2	81.5	79.2	77.0	77.7	79.2	79.8	79.6	78.7	90.2	86.5	83.4	72.5	76.0
0960	Flooring	90.8	66.5	83.7	93.5	64.0	84.9	92.5	64.0	84.2	91.4	78.6	87.7	78.4	66.5	74.9	90.1	66.5	83.2
0970, 0990	Wall Finishes & Painting/Coating	90.7	81.8	85.4	96.9	69.9	80.9	91.0	89.0	89.9	91.0	75.1	81.6	92.9	94.1	93.6	91.0	69.9	78.6
09	FINISHES	84.7	74.1	78.9	88.2	73.8	80.4	87.0	76.9	81.5	86.2	79.7	82.7	78.4	85.0	82.0	86.0	71.6	78.1
COVERS	DIVS. 10 - 14, 25, 28, 41, 43, 44, 46	100.0	90.3	97.9	100.0	87.5	97.2	100.0	94.9	98.9	100.0	86.3	97.0	100.0	49.2	88.8	100.0	88.9	97.6
21, 22, 23	FIRE SUPPRESSION, PLUMBING & HVAC	100.1	78.0	91.2	100.0	77.2	90.8	100.0	78.8	91.4	96.7	78.4	89.3	96.3	84.1	91.4	96.7	74.5	87.7
26, 27, 3370	ELECTRICAL, COMMUNICATIONS & UTIL.	92.8	76.5	84.5	95.5	76.5	85.9	93.5	75.5	84.4	95.8	77.5	86.5	94.4	90.4	92.4	91.9	90.3	91.1
MF2016	WEIGHTED AVERAGE	93.4	79.6	87.5	93.2	77.7	86.6	94.6	83.1	89.7	92.5	83.8	88.8	95.8	87.6	92.3	91.7	77.8	85.7

LOUISIANA

DIVISION		ALEXANDRIA 713 - 714			BATON ROUGE 707 - 708			HAMMOND 704			LAFAYETTE 705			LAKE CHARLES 706			MONROE 712		
		MAT.	INST.	TOTAL	MAT.	INST.	TOTAL	MAT.	INST.	TOTAL	MAT.	INST.	TOTAL	MAT.	INST.	TOTAL	MAT.	INST.	TOTAL
015433	CONTRACTOR EQUIPMENT		93.2	93.2		90.4	90.4		90.9	90.9		90.9	90.9		90.4	90.4		93.2	93.2
0241, 31 - 34	SITE & INFRASTRUCTURE, DEMOLITION	102.9	92.1	95.4	103.4	90.4	94.4	100.1	89.6	92.8	101.1	91.8	94.7	101.9	91.0	94.3	102.9	92.0	95.3
0310	Concrete Forming & Accessories	78.8	61.6	64.1	94.4	73.4	76.5	77.8	56.2	59.4	94.8	69.5	73.3	95.5	69.6	73.4	78.4	60.9	63.5
0320	Concrete Reinforcing	99.9	55.6	77.4	90.5	55.6	72.8	91.7	55.3	73.2	93.0	55.6	74.0	93.0	55.9	74.1	98.8	55.6	76.8
0330	Cast-in-Place Concrete	91.1	68.1	82.5	97.0	75.6	89.1	92.8	64.2	82.1	92.3	67.8	83.2	97.2	67.9	86.3	91.1	65.9	81.7
03	CONCRETE	83.9	63.5	74.9	88.1	71.7	80.8	87.7	59.6	75.2	88.7	66.9	79.0	91.0	67.0	80.4	83.8	62.5	74.3
04	MASONRY	109.4	61.6	79.9	91.6	64.0	74.5	94.4	53.6	69.2	94.4	64.0	75.6	93.9	68.2	78.0	104.1	60.3	77.1
05	METALS	94.7	71.0	87.5	98.1	74.3	90.9	89.7	68.7	83.3	89.0	69.3	83.0	89.0	69.7	83.1	94.6	70.8	87.4
06	WOOD, PLASTICS & COMPOSITES	91.0	60.5	74.5	99.7	75.1	86.3	83.4	55.8	68.4	104.2	70.0	85.6	102.4	70.0	84.8	90.4	60.5	74.2
07	THERMAL & MOISTURE PROTECTION	98.5	66.3	84.5	97.0	68.7	84.6	96.2	62.4	81.5	96.8	68.1	84.3	96.6	69.1	84.6	98.5	65.5	84.2
08	OPENINGS	113.4	60.3	101.1	101.2	72.3	94.5	98.2	56.5	88.5	101.9	63.9	93.1	101.9	63.9	93.1	113.4	58.7	100.7
0920	Plaster & Gypsum Board	86.8	59.8	69.0	97.1	74.9	82.4	97.9	55.0	69.6	106.1	69.6	82.0	106.1	69.6	82.0	86.5	59.8	68.9
0950, 0980	Ceilings & Acoustic Treatment	80.9	59.8	66.6	95.3	74.9	81.5	98.8	55.0	69.1	97.1	69.6	78.4	98.0	69.6	78.7	80.9	59.8	66.6
0960	Flooring	91.3	63.5	83.2	92.1	59.0	82.5	88.7	57.1	79.6	97.2	62.9	87.3	97.2	65.9	88.1	90.9	56.9	81.0
0970, 0990	Wall Finishes & Painting/Coating	94.5	61.3	74.9	91.5	61.3	73.7	95.5	61.4	75.3	95.5	61.3	75.3	95.5	61.3	75.3	94.5	61.3	74.9
09	FINISHES	81.5	61.6	70.6	87.8	69.9	78.0	89.0	56.4	71.2	92.3	67.6	78.8	92.5	68.2	79.2	81.4	59.9	69.7
COVERS	DIVS. 10 - 14, 25, 28, 41, 43, 44, 46	100.0	80.7	95.8	100.0	85.5	96.8	100.0	80.1	95.6	100.0	85.0	96.7	100.0	85.0	96.7	100.0	80.6	95.7
21, 22, 23	FIRE SUPPRESSION, PLUMBING & HVAC	100.0	64.7	85.7	100.0	65.5	86.1	96.7	62.3	82.8	100.1	65.6	86.2	100.1	65.9	86.3	100.0	63.3	85.2
26, 27, 3370	ELECTRICAL, COMMUNICATIONS & UTIL.	92.6	63.0	77.5	96.5	59.5	77.7	93.9	71.3	82.4	95.0	65.3	79.9	94.6	67.6	80.8	94.3	58.7	76.2
MF2016	WEIGHTED AVERAGE	96.7	66.6	83.8	96.6	69.7	85.1	93.8	64.5	81.2	95.6	68.8	84.1	95.8	69.8	84.6	96.6	65.1	83.1

LOUISIANA / MAINE

DIVISION		LOUISIANA NEW ORLEANS 700 - 701			SHREVEPORT 710 - 711			THIBODAUX 703			MAINE AUGUSTA 043			BANGOR 044			BATH 045		
		MAT.	INST.	TOTAL	MAT.	INST.	TOTAL	MAT.	INST.	TOTAL	MAT.	INST.	TOTAL	MAT.	INST.	TOTAL	MAT.	INST.	TOTAL
015433	CONTRACTOR EQUIPMENT		86.7	86.7		93.2	93.2		90.9	90.9		97.9	97.9		97.9	97.9		97.9	97.9
0241, 31 - 34	SITE & INFRASTRUCTURE, DEMOLITION	101.2	92.9	95.4	105.1	92.0	96.0	102.4	91.6	94.9	87.6	98.5	95.2	91.2	99.8	97.2	89.1	98.5	95.7
0310	Concrete Forming & Accessories	97.9	70.6	74.6	94.0	61.6	66.5	89.0	65.8	69.3	98.4	78.7	81.6	92.9	78.0	80.2	88.7	78.7	80.2
0320	Concrete Reinforcing	94.7	56.0	75.0	100.1	55.4	77.4	91.7	55.1	73.1	105.3	82.9	93.9	94.5	83.1	88.7	93.5	82.6	88.0
0330	Cast-in-Place Concrete	93.1	70.7	84.7	94.0	66.1	83.6	99.7	65.8	87.1	90.5	110.3	97.8	73.2	111.1	87.3	73.2	110.2	87.0
03	CONCRETE	95.7	68.0	83.4	87.3	62.9	76.5	92.6	64.5	80.2	91.8	90.9	91.4	85.8	90.9	88.0	85.8	90.8	88.0
04	MASONRY	98.4	65.6	78.2	94.7	57.7	71.8	118.4	52.8	77.9	102.9	90.3	95.2	115.5	93.5	101.9	122.6	85.9	99.9
05	METALS	100.4	63.1	89.1	98.9	70.7	90.3	89.7	68.4	83.2	92.3	91.1	91.9	83.1	92.2	85.9	81.6	90.8	84.4
06	WOOD, PLASTICS & COMPOSITES	99.4	73.0	85.0	98.0	61.3	78.0	90.9	67.4	78.1	95.3	77.6	85.7	94.4	75.1	83.9	88.4	77.6	82.6
07	THERMAL & MOISTURE PROTECTION	95.2	69.7	84.1	96.7	64.9	82.9	96.4	63.8	82.2	106.2	76.0	93.1	102.9	82.9	94.2	102.8	75.4	90.9
08	OPENINGS	100.8	66.0	92.7	107.2	58.2	95.8	102.8	58.6	92.5	100.3	80.7	95.8	98.8	79.3	94.3	98.8	80.7	94.6
0920	Plaster & Gypsum Board	96.9	72.6	80.9	95.4	60.6	72.5	99.1	66.9	77.9	114.0	76.6	89.4	117.1	74.0	88.7	112.3	76.6	88.8
0950, 0980	Ceilings & Acoustic Treatment	102.1	72.6	82.1	87.6	60.6	69.3	98.8	66.9	77.2	105.6	76.6	86.0	90.9	74.0	79.5	90.0	76.6	80.9
0960	Flooring	102.3	65.7	91.7	97.6	56.9	85.8	94.6	39.4	78.6	88.2	47.3	76.3	80.9	102.1	87.0	79.4	47.3	70.0
0970, 0990	Wall Finishes & Painting/Coating	100.2	62.2	77.7	89.8	61.3	72.9	96.7	61.4	75.8	95.0	91.3	92.8	90.0	91.3	90.7	90.0	91.3	90.7
09	FINISHES	98.7	69.0	82.5	88.2	60.4	73.1	91.1	60.4	74.4	92.1	72.9	81.6	86.4	83.1	84.6	85.1	72.9	78.4
COVERS	DIVS. 10 - 14, 25, 28, 41, 43, 44, 46	100.0	85.8	96.9	100.0	84.0	96.5	100.0	83.4	96.4	100.0	99.3	99.9	100.0	101.8	100.4	100.0	99.3	99.9
21, 22, 23	FIRE SUPPRESSION, PLUMBING & HVAC	100.1	63.8	85.4	99.9	64.1	85.4	96.7	63.8	83.4	100.0	74.1	89.5	100.1	71.6	88.6	96.7	74.1	87.5
26, 27, 3370	ELECTRICAL, COMMUNICATIONS & UTIL.	100.6	71.3	85.7	99.9	67.1	83.2	92.7	71.3	81.8	96.7	78.6	87.5	94.5	71.1	82.6	92.7	78.6	85.6
MF2016	WEIGHTED AVERAGE	99.4	69.5	86.6	97.8	66.4	84.3	96.2	66.4	83.4	96.7	83.1	90.9	94.3	83.5	89.7	93.2	82.6	88.6

For customer support on your Electrical Costs with RSMeans data, call 800.448.8182.

527

City Cost Indexes

MAINE

DIVISION		HOULTON 047 MAT.	INST.	TOTAL	KITTERY 039 MAT.	INST.	TOTAL	LEWISTON 042 MAT.	INST.	TOTAL	MACHIAS 046 MAT.	INST.	TOTAL	PORTLAND 040-041 MAT.	INST.	TOTAL	ROCKLAND 048 MAT.	INST.	TOTAL
015433	CONTRACTOR EQUIPMENT		97.9	97.9		97.9	97.9		97.9	97.9		97.9	97.9		97.9	97.9		97.9	97.9
0241, 31 - 34	SITE & INFRASTRUCTURE, DEMOLITION	90.9	98.5	96.2	81.7	98.5	93.4	88.9	99.8	96.5	90.3	98.5	96.0	89.7	99.8	96.7	86.9	98.5	95.0
0310	Concrete Forming & Accessories	96.8	78.5	81.2	87.9	79.0	80.3	98.6	78.0	81.1	93.8	78.4	80.7	99.9	78.0	81.3	94.9	78.7	81.1
0320	Concrete Reinforcing	94.5	81.1	87.7	89.3	82.6	85.9	116.1	83.1	99.3	94.5	81.1	87.7	104.7	83.1	93.7	94.5	82.9	88.6
0330	Cast-in-Place Concrete	73.2	110.2	86.9	71.9	110.4	86.2	74.7	111.1	88.2	73.2	110.1	86.9	90.8	111.1	98.4	74.7	110.3	87.9
03	CONCRETE	86.8	90.4	88.4	80.8	91.0	85.3	85.6	90.9	88.0	86.2	90.4	88.1	91.2	90.9	91.0	83.6	90.8	86.8
04	MASONRY	98.2	61.9	75.8	113.2	85.9	96.4	98.8	93.5	95.5	98.2	61.9	75.8	102.5	93.5	96.9	92.3	90.3	91.1
05	METALS	81.8	88.8	83.9	78.5	91.1	82.3	86.2	92.2	88.0	81.8	88.8	83.9	92.3	92.2	92.3	81.7	91.1	84.5
06	WOOD, PLASTICS & COMPOSITES	98.6	77.6	87.2	91.1	77.6	83.8	100.5	75.1	86.7	95.3	77.6	85.7	97.5	75.1	85.3	96.2	77.6	86.1
07	THERMAL & MOISTURE PROTECTION	103.0	69.1	88.3	103.7	74.5	91.0	102.7	82.9	94.1	102.9	68.4	87.9	106.0	82.9	95.9	102.6	76.4	91.2
08	OPENINGS	98.8	80.7	94.6	98.4	84.3	95.2	101.8	79.3	96.6	98.8	80.7	94.6	100.9	79.3	95.9	98.8	80.7	94.6
0920	Plaster & Gypsum Board	119.6	76.6	91.3	108.0	76.6	87.4	122.2	74.0	90.5	117.8	76.6	90.7	111.0	74.0	86.6	117.8	76.6	90.7
0950, 0980	Ceilings & Acoustic Treatment	90.0	76.6	80.9	101.6	76.6	84.7	100.3	74.0	82.5	90.0	76.6	80.9	104.7	74.0	83.9	90.0	76.6	80.8
0960	Flooring	82.1	44.2	71.1	85.2	49.3	74.8	83.5	102.1	88.9	81.3	44.2	70.6	87.7	102.1	91.9	81.7	44.2	70.8
0970, 0990	Wall Finishes & Painting/Coating	90.0	91.3	90.7	83.3	104.8	96.1	90.0	91.3	90.7	90.0	91.3	90.7	93.6	91.3	92.2	90.0	91.3	90.7
09	FINISHES	87.0	72.4	79.0	87.9	74.7	80.7	89.4	83.1	85.9	86.4	72.4	78.8	91.8	83.1	87.0	86.2	72.4	78.7
COVERS	DIVS. 10 - 14, 25, 28, 41, 43, 44, 46	100.0	99.3	99.8	100.0	99.3	99.9	100.0	101.8	100.4	100.0	99.3	99.9	100.0	101.8	100.4	100.0	99.3	99.9
21, 22, 23	FIRE SUPPRESSION, PLUMBING & HVAC	96.7	74.0	87.5	96.6	81.6	90.5	100.1	71.6	88.6	96.7	74.0	87.5	100.0	71.6	88.5	96.7	74.1	87.5
26, 27, 3370	ELECTRICAL, COMMUNICATIONS & UTIL.	96.2	78.6	87.3	90.1	78.6	84.2	96.2	75.3	85.6	96.2	78.6	87.3	97.9	75.3	86.4	96.1	78.6	87.2
MF2016	WEIGHTED AVERAGE	92.8	79.7	87.2	91.3	84.6	88.5	94.7	84.1	90.1	92.6	79.7	87.1	96.9	84.1	91.4	91.9	83.1	88.1

MAINE / MARYLAND

DIVISION		WATERVILLE 049 MAT.	INST.	TOTAL	ANNAPOLIS 214 MAT.	INST.	TOTAL	BALTIMORE 210-212 MAT.	INST.	TOTAL	COLLEGE PARK 207-208 MAT.	INST.	TOTAL	CUMBERLAND 215 MAT.	INST.	TOTAL	EASTON 216 MAT.	INST.	TOTAL
015433	CONTRACTOR EQUIPMENT		97.9	97.9		104.4	104.4		104.4	104.4		110.1	110.1		104.4	104.4		104.4	104.4
0241, 31 - 34	SITE & INFRASTRUCTURE, DEMOLITION	90.8	98.5	96.2	103.0	93.4	96.3	100.3	97.5	98.4	101.3	98.0	99.0	93.6	94.3	94.1	100.8	91.5	94.3
0310	Concrete Forming & Accessories	88.2	78.7	80.1	98.8	74.3	77.9	99.7	78.4	81.5	83.4	77.1	78.0	91.7	84.2	85.3	89.8	74.9	77.1
0320	Concrete Reinforcing	94.5	82.9	88.6	113.0	89.7	101.2	100.7	89.8	95.2	100.9	96.7	98.7	95.4	90.0	92.6	94.6	89.6	92.0
0330	Cast-in-Place Concrete	73.2	110.3	87.0	124.7	73.7	105.7	114.1	79.0	101.0	119.0	77.4	103.5	102.7	85.2	96.2	114.1	66.5	96.4
03	CONCRETE	87.3	90.8	88.9	107.9	78.2	94.7	106.6	81.5	95.5	106.4	82.1	95.6	93.5	86.7	90.5	102.0	75.8	90.4
04	MASONRY	109.3	90.3	97.6	96.6	69.1	79.6	99.4	74.6	84.1	113.9	75.1	89.9	97.9	87.3	91.3	111.7	59.8	79.7
05	METALS	81.8	91.1	84.6	103.1	103.1	103.1	101.3	98.3	100.4	87.8	108.8	94.1	98.3	104.7	100.3	98.6	99.9	99.0
06	WOOD, PLASTICS & COMPOSITES	87.8	77.6	82.3	96.2	75.9	85.2	102.9	79.0	89.9	78.1	76.3	77.2	88.4	83.3	85.6	86.2	82.3	84.1
07	THERMAL & MOISTURE PROTECTION	103.0	76.0	91.2	104.6	78.5	93.3	103.1	82.0	93.9	103.9	83.0	94.8	101.7	84.1	94.1	101.9	75.0	90.2
08	OPENINGS	98.8	80.7	94.6	100.1	81.4	95.7	98.8	83.1	95.1	94.7	83.4	92.1	99.7	87.1	96.8	98.1	84.9	95.0
0920	Plaster & Gypsum Board	112.3	76.6	88.8	103.0	75.6	84.9	104.8	78.4	87.4	96.7	75.8	83.0	105.0	83.1	90.6	105.0	82.2	89.9
0950, 0980	Ceilings & Acoustic Treatment	90.0	76.6	80.9	87.9	75.6	79.6	105.5	78.4	87.2	109.0	75.8	86.5	103.9	83.1	89.9	103.9	82.2	89.2
0960	Flooring	79.0	47.3	69.8	89.3	78.2	86.1	101.2	78.2	94.5	88.6	79.1	85.8	89.1	96.2	91.2	88.2	78.2	85.3
0970, 0990	Wall Finishes & Painting/Coating	90.0	91.3	90.7	91.7	76.6	82.7	104.1	77.3	88.2	100.0	74.4	84.9	96.9	86.9	91.0	96.9	74.4	83.6
09	FINISHES	85.1	72.9	78.5	86.5	74.8	80.1	101.7	77.9	88.7	94.6	76.5	84.8	94.9	86.7	90.4	95.1	76.1	84.7
COVERS	DIVS. 10 - 14, 25, 28, 41, 43, 44, 46	100.0	99.2	99.8	100.0	86.1	96.9	100.0	88.6	97.5	100.0	89.0	97.6	100.0	90.5	97.9	100.0	69.8	93.3
21, 22, 23	FIRE SUPPRESSION, PLUMBING & HVAC	96.7	74.1	87.5	100.0	82.7	93.0	100.0	83.8	93.5	96.7	88.1	93.2	96.4	73.6	87.2	96.4	74.4	87.5
26, 27, 3370	ELECTRICAL, COMMUNICATIONS & UTIL.	96.2	78.6	87.2	100.2	88.2	94.1	100.4	89.0	94.6	100.1	101.0	100.5	97.9	81.0	89.3	97.4	63.8	80.3
MF2016	WEIGHTED AVERAGE	93.1	83.1	88.8	100.5	83.1	93.0	101.2	85.0	94.2	97.7	88.6	93.8	97.1	85.5	92.1	98.8	76.0	89.0

MARYLAND / MASSACHUSETTS

DIVISION		ELKTON 219 MAT.	INST.	TOTAL	HAGERSTOWN 217 MAT.	INST.	TOTAL	SALISBURY 218 MAT.	INST.	TOTAL	SILVER SPRING 209 MAT.	INST.	TOTAL	WALDORF 206 MAT.	INST.	TOTAL	BOSTON 020-022, 024 MAT.	INST.	TOTAL
015433	CONTRACTOR EQUIPMENT		104.4	104.4		104.4	104.4		104.4	104.4		101.8	101.8		101.8	101.8		102.0	102.0
0241, 31 - 34	SITE & INFRASTRUCTURE, DEMOLITION	87.6	93.0	91.3	91.8	94.4	93.6	100.7	91.5	94.3	89.4	89.6	89.5	95.9	89.6	91.5	95.1	101.3	99.4
0310	Concrete Forming & Accessories	95.7	94.0	94.2	90.8	80.9	82.4	104.1	51.6	59.4	91.4	76.3	78.6	98.3	76.3	79.6	103.0	137.0	131.9
0320	Concrete Reinforcing	94.4	115.3	105.1	95.4	90.0	92.6	94.6	67.0	80.6	99.7	96.5	98.1	100.3	96.5	98.4	105.8	152.0	129.3
0330	Cast-in-Place Concrete	92.3	73.1	85.2	97.8	85.2	93.1	114.1	64.6	95.7	121.8	77.8	105.4	136.5	77.8	114.6	91.1	141.0	109.7
03	CONCRETE	85.9	91.2	88.2	89.7	85.2	87.7	102.9	60.7	84.2	104.4	81.6	94.3	115.3	81.7	100.4	97.1	140.3	116.2
04	MASONRY	96.6	68.4	79.2	103.4	87.3	93.4	111.4	56.3	77.4	113.4	75.4	89.9	97.4	75.4	83.8	108.3	148.1	132.9
05	METALS	98.6	112.4	102.8	98.5	104.7	100.4	98.6	91.3	96.4	92.2	104.6	96.0	92.2	104.6	96.0	94.3	133.0	106.0
06	WOOD, PLASTICS & COMPOSITES	94.1	103.0	98.9	87.5	78.9	82.8	105.2	52.8	76.8	84.9	75.6	79.9	92.1	75.6	83.2	103.2	136.5	121.3
07	THERMAL & MOISTURE PROTECTION	101.4	81.1	92.5	102.0	85.2	94.7	102.2	70.1	88.2	106.8	88.0	98.6	107.3	88.0	98.9	113.5	138.6	124.4
08	OPENINGS	98.1	103.0	99.2	98.0	83.1	94.6	98.3	62.5	90.0	86.7	83.0	85.9	87.4	83.0	86.4	98.9	143.1	108.4
0920	Plaster & Gypsum Board	108.9	103.5	105.3	105.0	78.6	87.6	116.2	51.8	73.8	103.1	75.8	85.1	106.4	75.8	86.3	107.2	137.4	127.1
0950, 0980	Ceilings & Acoustic Treatment	103.8	103.5	103.6	104.9	78.6	87.1	103.9	51.8	68.6	117.5	75.8	89.3	117.5	75.8	89.3	106.3	137.4	127.4
0960	Flooring	90.5	78.2	87.0	88.7	96.2	90.9	94.1	78.2	89.5	94.6	79.1	90.1	98.2	79.1	92.6	93.4	169.1	115.4
0970, 0990	Wall Finishes & Painting/Coating	96.9	74.4	83.6	96.9	74.4	83.6	96.9	74.4	83.6	107.2	74.4	87.8	107.2	74.4	87.8	97.4	153.2	130.4
09	FINISHES	95.5	90.4	92.7	94.8	82.7	88.2	98.5	57.8	76.2	95.1	75.9	84.6	96.9	76.0	85.5	101.6	145.2	125.4
COVERS	DIVS. 10 - 14, 25, 28, 41, 43, 44, 46	100.0	57.7	90.7	100.0	90.0	97.8	100.0	78.0	95.2	100.0	84.6	96.6	100.0	84.6	96.6	100.0	117.3	103.8
21, 22, 23	FIRE SUPPRESSION, PLUMBING & HVAC	96.4	80.4	90.0	99.9	84.9	93.8	96.4	72.3	86.7	96.7	87.9	93.1	96.7	87.9	93.1	100.1	127.5	111.2
26, 27, 3370	ELECTRICAL, COMMUNICATIONS & UTIL.	99.1	89.0	94.0	97.7	81.0	89.2	96.2	61.6	78.6	97.2	101.0	99.1	94.8	101.0	97.9	100.8	131.1	116.2
MF2016	WEIGHTED AVERAGE	96.0	87.6	92.4	97.5	87.0	93.0	99.3	68.5	86.1	96.9	87.4	92.8	97.7	87.4	93.3	99.4	133.3	113.9

MASSACHUSETTS

DIVISION		BROCKTON 023 MAT.	INST.	TOTAL	BUZZARDS BAY 025 MAT.	INST.	TOTAL	FALL RIVER 027 MAT.	INST.	TOTAL	FITCHBURG 014 MAT.	INST.	TOTAL	FRAMINGHAM 017 MAT.	INST.	TOTAL	GREENFIELD 013 MAT.	INST.	TOTAL
015433	CONTRACTOR EQUIPMENT		100.3	100.3		100.3	100.3		101.2	101.2		97.9	97.9		99.7	99.7		97.9	97.9
0241, 31 - 34	SITE & INFRASTRUCTURE, DEMOLITION	91.9	102.7	99.4	82.5	102.9	96.7	90.9	102.8	99.2	84.5	102.6	97.1	81.4	102.5	96.1	88.1	101.9	97.7
0310	Concrete Forming & Accessories	99.6	124.3	120.6	97.3	123.9	119.9	99.6	124.0	120.4	93.0	121.6	117.3	100.5	124.5	120.9	91.3	120.8	116.4
0320	Concrete Reinforcing	104.9	146.8	126.2	84.1	125.8	105.3	104.9	125.8	115.5	88.3	146.6	118.0	88.3	146.9	118.1	92.0	123.8	108.1
0330	Cast-in-Place Concrete	84.6	133.5	102.8	70.3	133.3	93.7	81.8	133.7	101.2	78.7	133.3	99.0	78.7	135.3	99.7	80.9	119.4	95.2
03	CONCRETE	90.2	130.8	108.2	76.5	126.9	98.9	88.9	127.1	105.8	78.3	129.2	100.9	81.1	131.5	103.4	81.7	120.1	98.7
04	MASONRY	101.9	138.0	124.2	94.8	138.0	121.5	102.5	137.9	124.4	100.3	134.7	121.6	106.8	139.0	126.7	105.0	121.6	115.2
05	METALS	89.5	126.4	100.7	84.7	117.7	94.7	89.5	118.0	98.1	87.6	122.8	98.2	87.6	126.5	99.4	89.9	111.6	96.4
06	WOOD, PLASTICS & COMPOSITES	99.2	123.1	112.2	95.9	123.1	110.7	99.2	123.4	112.3	96.4	119.9	109.2	103.2	122.9	113.9	94.4	123.1	110.0
07	THERMAL & MOISTURE PROTECTION	104.2	129.8	115.3	103.0	127.0	113.5	104.0	126.2	113.7	101.3	122.7	110.6	101.4	129.9	113.8	101.4	112.1	106.1
08	OPENINGS	99.5	132.1	107.0	95.4	122.4	101.7	99.5	121.8	104.6	101.5	130.3	108.2	91.8	132.0	101.1	101.7	122.1	106.4
0920	Plaster & Gypsum Board	90.2	123.4	112.1	86.0	123.4	110.7	90.2	123.4	112.1	110.9	120.1	117.0	114.0	123.4	120.2	111.8	123.4	119.5
0950, 0980	Ceilings & Acoustic Treatment	106.5	123.4	118.0	90.2	123.4	112.7	106.5	123.4	118.0	94.5	120.1	111.9	94.5	123.4	114.1	103.0	123.4	116.9
0960	Flooring	88.9	167.4	111.6	86.7	167.4	110.1	87.9	167.4	110.9	85.8	167.4	109.5	87.2	167.4	110.5	85.0	139.6	100.9
0970, 0990	Wall Finishes & Painting/Coating	91.0	142.9	121.7	91.0	142.9	121.7	91.0	142.9	121.7	87.2	142.9	120.1	88.1	142.9	120.5	87.2	112.2	102.0
09	FINISHES	90.8	134.6	114.7	85.8	134.6	112.4	90.5	134.8	114.7	86.5	132.7	111.7	87.2	134.4	113.0	88.4	124.4	108.0
COVERS	DIVS. 10 - 14, 25, 28, 41, 43, 44, 46	100.0	111.9	102.6	100.0	112.0	102.6	100.0	112.5	102.7	100.0	106.4	101.4	100.0	111.6	102.5	100.0	105.0	101.1
21, 22, 23	FIRE SUPPRESSION, PLUMBING & HVAC	100.3	106.1	102.6	96.8	105.8	100.4	100.3	105.8	102.5	97.0	106.8	101.0	97.0	123.7	107.8	97.0	100.6	98.5
26, 27, 3370	ELECTRICAL, COMMUNICATIONS & UTIL.	98.0	100.1	99.1	95.3	102.6	99.0	97.9	100.1	99.0	97.7	106.6	102.3	94.6	127.4	111.3	97.7	98.5	98.1
MF2016	WEIGHTED AVERAGE	96.1	119.2	106.0	91.1	117.7	102.5	95.9	117.4	105.1	93.0	118.7	104.0	92.4	127.0	107.2	94.3	110.9	101.4

MASSACHUSETTS

DIVISION		HYANNIS 026 MAT.	INST.	TOTAL	LAWRENCE 019 MAT.	INST.	TOTAL	LOWELL 018 MAT.	INST.	TOTAL	NEW BEDFORD 027 MAT.	INST.	TOTAL	PITTSFIELD 012 MAT.	INST.	TOTAL	SPRINGFIELD 010 - 011 MAT.	INST.	TOTAL
015433	CONTRACTOR EQUIPMENT		100.3	100.3		100.3	100.3		97.9	97.9		101.2	101.2		97.9	97.9		97.9	97.9
0241, 31 - 34	SITE & INFRASTRUCTURE, DEMOLITION	88.4	102.9	98.5	93.3	102.7	99.8	92.2	102.8	99.6	89.4	102.8	98.7	93.3	101.3	98.9	92.7	101.7	98.9
0310	Concrete Forming & Accessories	91.4	123.9	119.0	101.7	124.6	121.1	98.2	125.7	121.6	99.6	124.1	120.4	98.2	105.7	104.6	98.4	120.7	117.4
0320	Concrete Reinforcing	84.1	125.8	105.3	109.3	138.5	124.2	110.2	138.3	124.5	104.9	125.9	115.6	91.3	117.4	104.6	110.2	123.8	117.1
0330	Cast-in-Place Concrete	77.2	133.3	98.1	91.0	133.6	106.9	82.8	135.3	102.3	72.1	133.8	95.1	90.3	116.6	100.1	86.2	119.4	98.5
03	CONCRETE	82.3	126.9	102.1	94.6	129.5	110.1	86.5	130.4	106.0	84.4	127.2	103.3	87.8	111.2	98.2	88.1	120.1	102.3
04	MASONRY	100.9	138.0	123.8	112.5	138.8	128.7	99.3	137.8	123.1	100.8	138.6	124.1	99.9	113.8	108.5	99.6	121.6	113.2
05	METALS	86.2	117.7	95.7	90.2	123.4	100.3	90.2	120.0	99.2	89.5	118.1	98.2	90.0	108.6	95.7	92.7	111.5	98.4
06	WOOD, PLASTICS & COMPOSITES	89.0	123.1	107.5	103.7	123.1	114.2	102.9	123.1	113.9	99.2	123.4	112.3	102.9	105.5	104.3	102.9	123.1	113.9
07	THERMAL & MOISTURE PROTECTION	103.5	127.0	113.8	102.1	130.0	114.2	101.8	130.3	114.2	104.0	126.4	113.7	101.9	107.2	104.2	101.8	112.1	106.3
08	OPENINGS	96.0	122.4	102.1	95.5	129.9	103.5	102.8	129.9	109.0	99.5	126.6	105.8	102.8	110.7	104.6	102.8	122.1	107.2
0920	Plaster & Gypsum Board	81.4	123.4	109.1	116.4	123.4	121.0	116.4	123.4	121.0	90.2	123.4	112.1	116.4	105.3	109.1	116.4	123.4	121.0
0950, 0980	Ceilings & Acoustic Treatment	99.7	123.4	115.8	104.9	123.4	117.5	104.9	123.4	117.5	106.5	123.4	118.0	104.9	105.3	105.2	104.9	123.4	117.5
0960	Flooring	84.2	167.4	108.3	87.8	167.4	110.9	87.8	167.4	110.9	87.9	167.4	110.9	88.2	134.2	101.5	87.2	139.6	102.4
0970, 0990	Wall Finishes & Painting/Coating	91.0	142.9	121.7	87.3	142.9	120.2	87.2	142.9	120.1	91.0	142.9	121.7	87.2	112.2	102.0	88.0	112.2	102.3
09	FINISHES	86.6	134.6	112.8	90.4	134.6	114.6	90.4	135.4	115.0	90.4	134.8	114.6	90.5	111.7	102.1	90.3	124.4	108.9
COVERS	DIVS. 10 - 14, 25, 28, 41, 43, 44, 46	100.0	112.0	102.6	100.0	112.1	102.7	100.0	113.1	102.9	100.0	112.5	102.7	100.0	101.5	100.3	100.0	105.0	101.1
21, 22, 23	FIRE SUPPRESSION, PLUMBING & HVAC	100.3	106.5	102.8	100.1	121.9	108.9	100.1	123.4	109.5	100.3	105.7	102.5	100.1	96.3	98.5	100.1	100.6	100.3
26, 27, 3370	ELECTRICAL, COMMUNICATIONS & UTIL.	95.7	100.1	97.9	96.9	125.9	111.6	97.3	122.2	110.0	98.8	100.1	99.4	97.3	98.5	97.9	97.3	95.2	96.3
MF2016	WEIGHTED AVERAGE	93.4	117.5	103.7	96.7	125.8	109.2	95.8	124.4	108.5	95.3	117.6	104.9	96.0	105.1	99.9	96.4	110.4	102.4

MASSACHUSETTS / MICHIGAN

DIVISION		WORCESTER 015 - 016 MAT.	INST.	TOTAL	ANN ARBOR 481 MAT.	INST.	TOTAL	BATTLE CREEK 490 MAT.	INST.	TOTAL	BAY CITY 487 MAT.	INST.	TOTAL	DEARBORN 481 MAT.	INST.	TOTAL	DETROIT 482 MAT.	INST.	TOTAL
015433	CONTRACTOR EQUIPMENT		97.9	97.9		111.9	111.9		99.0	99.0		111.9	111.9		111.9	111.9		96.4	96.4
0241, 31 - 34	SITE & INFRASTRUCTURE, DEMOLITION	92.6	102.6	99.6	79.5	96.7	91.5	93.7	86.0	88.3	71.2	96.2	88.7	79.3	96.8	91.5	92.9	100.7	98.4
0310	Concrete Forming & Accessories	98.7	121.5	118.1	97.3	104.3	103.2	96.4	79.5	82.0	97.3	83.7	85.8	97.1	104.9	103.7	101.7	108.2	107.2
0320	Concrete Reinforcing	110.2	146.6	128.7	102.3	106.5	104.5	94.8	83.3	88.9	102.3	105.7	104.0	102.3	106.6	104.5	103.1	108.3	105.7
0330	Cast-in-Place Concrete	85.7	133.3	103.4	81.0	99.2	87.8	82.0	93.8	86.4	77.5	85.9	80.6	79.2	100.0	87.0	86.1	101.5	91.8
03	CONCRETE	87.9	129.2	106.2	85.1	103.7	93.3	82.0	84.9	83.3	83.5	89.7	86.2	84.2	104.3	93.1	93.6	105.1	98.7
04	MASONRY	99.1	134.7	121.1	99.9	98.4	99.0	97.3	80.1	86.7	99.5	79.3	87.0	99.7	100.3	100.1	102.0	101.9	102.0
05	METALS	92.8	122.7	101.8	104.3	116.6	108.1	105.4	84.5	99.1	105.0	114.9	108.0	104.4	116.7	108.1	106.9	95.7	103.5
06	WOOD, PLASTICS & COMPOSITES	103.4	119.9	112.3	92.4	106.0	99.8	92.1	77.8	84.3	92.4	84.3	88.0	92.4	106.0	99.8	99.2	109.8	105.0
07	THERMAL & MOISTURE PROTECTION	101.8	122.7	110.9	109.7	100.6	105.7	101.5	81.3	92.7	106.9	83.2	96.6	107.9	103.1	105.8	106.1	106.2	106.1
08	OPENINGS	102.8	130.3	109.2	96.3	101.6	97.5	88.4	76.3	85.6	96.3	85.9	93.9	96.3	101.9	97.6	98.7	104.1	99.9
0920	Plaster & Gypsum Board	116.4	120.1	118.8	100.4	105.8	104.0	90.2	73.8	79.4	100.4	83.5	89.3	100.4	105.8	104.0	97.3	109.9	105.6
0950, 0980	Ceilings & Acoustic Treatment	104.9	120.1	115.2	85.1	105.8	99.1	77.8	73.8	75.1	86.1	83.5	84.3	85.1	105.8	99.1	95.9	109.9	105.4
0960	Flooring	87.8	160.1	108.8	89.4	110.8	95.7	89.7	72.8	84.8	89.4	79.5	86.6	89.0	105.0	93.6	93.7	106.0	97.3
0970, 0990	Wall Finishes & Painting/Coating	87.2	142.9	120.1	83.5	98.0	92.1	86.8	78.3	81.7	83.5	81.2	82.1	83.5	96.3	91.1	91.4	97.9	95.3
09	FINISHES	90.4	131.2	112.7	89.4	105.1	98.0	83.3	77.4	80.1	89.2	82.2	85.4	89.2	104.3	97.5	96.3	107.1	102.2
COVERS	DIVS. 10 - 14, 25, 28, 41, 43, 44, 46	100.0	106.4	101.4	100.0	95.2	98.9	100.0	95.7	99.0	100.0	89.8	97.8	100.0	95.6	99.0	100.0	102.9	100.6
21, 22, 23	FIRE SUPPRESSION, PLUMBING & HVAC	100.1	106.8	102.8	100.1	93.4	97.4	100.1	84.5	93.8	100.1	80.4	92.1	100.1	102.3	101.0	100.0	103.7	101.5
26, 27, 3370	ELECTRICAL, COMMUNICATIONS & UTIL.	97.3	106.6	102.1	98.5	105.5	102.1	94.9	79.6	87.1	97.6	84.5	90.9	98.5	98.4	98.5	102.5	101.3	101.9
MF2016	WEIGHTED AVERAGE	96.4	118.5	105.9	97.2	101.7	99.1	95.3	82.5	89.8	96.6	87.5	92.7	97.0	102.9	99.5	100.2	103.0	101.4

For customer support on your Electrical Costs with RSMeans data, call 800.448.8182.

City Cost Indexes

MICHIGAN

DIVISION		FLINT 484-485			GAYLORD 497			GRAND RAPIDS 493, 495			IRON MOUNTAIN 498-499			JACKSON 492			KALAMAZOO 491		
		MAT.	INST.	TOTAL	MAT.	INST.	TOTAL	MAT.	INST.	TOTAL	MAT.	INST.	TOTAL	MAT.	INST.	TOTAL	MAT.	INST.	TOTAL
015433	CONTRACTOR EQUIPMENT		111.9	111.9		106.8	106.8		99.0	99.0		93.8	93.8		106.8	106.8		99.0	99.0
0241, 31 - 34	SITE & INFRASTRUCTURE, DEMOLITION	69.3	96.3	88.1	88.3	82.8	84.5	92.5	85.9	87.9	96.9	93.2	94.3	111.2	85.5	93.3	94.0	86.0	88.4
0310	Concrete Forming & Accessories	100.2	84.1	86.5	95.1	74.3	77.4	95.3	78.0	80.5	87.5	79.7	80.9	91.8	84.6	85.7	96.4	79.3	81.9
0320	Concrete Reinforcing	102.3	106.1	104.2	88.4	89.7	89.0	99.5	83.0	91.1	88.1	87.3	87.7	85.8	106.1	96.1	94.8	78.4	86.5
0330	Cast-in-Place Concrete	81.6	88.3	84.1	81.8	80.6	81.3	86.1	92.0	88.3	96.8	69.7	86.7	81.6	92.5	85.7	83.6	93.7	87.4
03	CONCRETE	85.5	90.8	87.8	79.0	80.9	79.8	85.7	83.6	84.7	87.0	78.2	83.1	74.4	92.3	82.3	85.0	84.0	84.6
04	MASONRY	99.9	87.7	92.4	107.9	74.2	87.1	90.6	75.7	81.4	93.7	79.4	84.8	87.9	88.8	88.4	95.9	80.1	86.1
05	METALS	104.4	115.5	107.8	106.9	111.6	108.3	102.3	84.0	96.7	106.3	91.8	101.9	107.1	113.9	109.2	105.4	82.7	98.5
06	WOOD, PLASTICS & COMPOSITES	96.2	83.0	89.1	85.5	73.8	79.1	95.8	76.3	85.2	81.7	79.8	80.7	84.2	82.1	83.1	92.1	77.8	84.3
07	THERMAL & MOISTURE PROTECTION	107.2	86.9	98.4	100.0	70.6	87.2	103.4	72.7	90.0	104.0	76.2	91.9	99.3	90.8	95.6	101.5	81.3	92.7
08	OPENINGS	96.3	85.0	93.7	87.9	80.1	86.1	99.0	76.0	93.6	94.6	71.1	89.2	87.1	88.0	87.3	88.4	75.6	85.4
0920	Plaster & Gypsum Board	102.6	82.2	89.1	90.0	72.1	78.2	97.1	72.2	80.7	48.2	79.8	69.0	88.5	80.6	83.3	90.2	73.8	79.4
0950, 0980	Ceilings & Acoustic Treatment	85.1	82.2	83.1	77.0	72.1	73.6	89.6	72.2	77.8	74.0	79.8	77.9	77.0	80.6	79.4	77.8	73.8	75.1
0960	Flooring	89.4	88.9	89.3	83.0	88.5	84.6	93.6	78.7	89.3	101.9	90.3	98.5	81.7	81.3	81.6	89.7	72.8	84.8
0970, 0990	Wall Finishes & Painting/Coating	83.5	80.4	81.7	83.1	79.5	81.0	93.6	78.9	84.9	100.5	71.5	83.3	83.1	96.0	86.8	86.8	78.3	81.7
09	FINISHES	88.9	84.0	86.2	82.7	76.9	79.6	89.5	77.6	83.0	82.7	80.7	81.6	83.8	84.3	84.1	83.3	77.4	80.1
COVERS	DIVS. 10 - 14, 25, 28, 41, 43, 44, 46	100.0	90.6	97.9	100.0	79.2	95.4	100.0	95.4	99.0	100.0	87.5	97.3	100.0	93.5	98.6	100.0	95.7	99.0
21, 22, 23	FIRE SUPPRESSION, PLUMBING & HVAC	100.1	85.1	94.1	96.9	80.6	90.3	100.1	81.4	92.5	96.8	84.8	91.9	96.9	87.6	93.1	100.1	80.0	92.0
26, 27, 3370	ELECTRICAL, COMMUNICATIONS & UTIL.	98.5	91.3	94.8	92.7	74.3	83.4	100.3	85.0	92.5	99.2	82.6	90.8	96.7	105.5	101.2	94.7	75.9	85.2
MF2016	WEIGHTED AVERAGE	96.9	90.8	94.3	94.3	81.3	88.8	97.1	81.6	90.5	96.2	83.0	90.6	93.8	93.0	93.5	95.6	80.7	89.2

DIVISION		MICHIGAN LANSING 488-489			MUSKEGON 494			ROYAL OAK 480, 483			SAGINAW 486			TRAVERSE CITY 496			MINNESOTA BEMIDJI 566		
		MAT.	INST.	TOTAL	MAT.	INST.	TOTAL	MAT.	INST.	TOTAL	MAT.	INST.	TOTAL	MAT.	INST.	TOTAL	MAT.	INST.	TOTAL
015433	CONTRACTOR EQUIPMENT		111.9	111.9		99.0	99.0		91.9	91.9		111.9	111.9		93.8	93.8		100.3	100.3
0241, 31 - 34	SITE & INFRASTRUCTURE, DEMOLITION	91.9	96.5	95.1	91.7	85.9	87.7	83.2	97.2	92.9	72.2	96.2	88.9	82.2	91.1	88.4	95.5	99.0	98.5
0310	Concrete Forming & Accessories	94.2	77.0	79.6	96.8	79.2	81.8	92.8	105.1	103.3	97.3	81.9	84.2	87.5	72.3	74.6	85.8	84.3	84.5
0320	Concrete Reinforcing	105.3	105.9	105.6	95.4	83.0	89.1	92.7	89.8	91.2	102.3	105.7	104.0	89.5	82.5	85.9	97.1	101.4	99.3
0330	Cast-in-Place Concrete	99.5	87.2	95.0	81.7	91.9	85.5	70.9	100.3	81.8	80.0	85.8	82.2	75.5	76.9	76.0	102.4	105.7	103.6
03	CONCRETE	90.5	87.2	89.0	80.5	84.1	82.1	73.1	100.2	85.1	84.6	88.8	86.5	71.5	76.5	73.7	88.8	96.0	92.0
04	MASONRY	94.7	86.3	89.5	94.5	75.7	82.9	93.1	99.4	97.0	101.4	79.3	87.7	91.9	75.3	81.6	99.1	102.4	101.2
05	METALS	102.3	115.0	106.1	102.9	83.8	97.1	108.0	91.0	102.9	104.4	114.7	107.5	106.3	89.4	101.2	88.7	116.8	97.2
06	WOOD, PLASTICS & COMPOSITES	90.1	73.3	81.0	89.2	78.1	83.2	88.1	106.0	97.8	89.1	82.0	85.2	81.7	72.2	76.6	69.9	79.2	75.0
07	THERMAL & MOISTURE PROTECTION	106.3	85.5	97.3	100.4	73.2	88.5	105.4	102.7	104.2	108.2	83.1	97.3	102.6	67.9	87.7	105.8	92.2	99.9
08	OPENINGS	100.5	79.4	95.6	87.7	77.6	85.3	96.1	101.3	97.3	94.4	84.6	92.1	94.6	66.4	88.0	101.9	102.3	102.0
0920	Plaster & Gypsum Board	96.3	72.2	80.4	72.6	74.1	73.6	97.9	105.8	103.1	100.4	81.1	87.7	48.2	72.1	63.9	104.3	79.0	87.7
0950, 0980	Ceilings & Acoustic Treatment	84.9	72.2	76.3	77.8	74.1	75.3	84.4	105.8	98.9	85.1	81.1	82.4	74.0	72.1	72.7	121.3	79.0	92.6
0960	Flooring	96.2	81.3	91.9	88.4	74.7	84.4	86.9	105.0	92.1	89.4	79.5	86.6	101.9	88.5	98.0	92.3	88.4	91.2
0970, 0990	Wall Finishes & Painting/Coating	95.0	78.9	85.5	85.0	78.9	81.4	85.1	96.3	91.7	83.5	81.2	82.1	100.5	40.5	65.0	89.1	106.5	99.4
09	FINISHES	88.8	76.8	82.2	79.8	77.2	78.4	88.3	104.2	97.0	89.1	80.9	84.6	81.6	71.8	76.2	96.0	86.9	91.1
COVERS	DIVS. 10 - 14, 25, 28, 41, 43, 44, 46	100.0	90.7	97.9	100.0	95.5	99.0	100.0	100.6	100.1	100.0	89.5	97.7	100.0	85.6	96.8	100.0	93.9	98.6
21, 22, 23	FIRE SUPPRESSION, PLUMBING & HVAC	100.0	85.5	94.1	99.9	81.4	92.4	96.9	103.6	99.6	100.1	80.0	92.0	96.8	80.0	90.0	100.0	82.0	90.8
26, 27, 3370	ELECTRICAL, COMMUNICATIONS & UTIL.	99.1	89.8	94.4	95.2	74.7	84.8	100.7	101.3	101.0	96.6	86.6	91.5	94.4	75.8	84.9	104.1	101.1	102.6
MF2016	WEIGHTED AVERAGE	97.9	88.7	93.9	94.1	80.3	88.2	95.2	100.6	97.5	96.5	87.3	92.6	93.2	78.3	86.8	96.0	95.4	95.8

MINNESOTA

DIVISION		BRAINERD 564			DETROIT LAKES 565			DULUTH 556-558			MANKATO 560			MINNEAPOLIS 553-555			ROCHESTER 559		
		MAT.	INST.	TOTAL	MAT.	INST.	TOTAL	MAT.	INST.	TOTAL	MAT.	INST.	TOTAL	MAT.	INST.	TOTAL	MAT.	INST.	TOTAL
015433	CONTRACTOR EQUIPMENT		102.9	102.9		100.3	100.3		103.2	103.2		102.9	102.9		107.6	107.6		103.2	103.2
0241, 31 - 34	SITE & INFRASTRUCTURE, DEMOLITION	97.2	104.8	102.5	93.7	100.0	98.1	98.5	102.9	101.6	93.9	104.4	101.2	94.9	108.4	104.3	96.5	102.5	100.7
0310	Concrete Forming & Accessories	86.3	85.0	85.2	82.6	84.2	83.9	96.9	97.0	97.0	94.6	93.7	93.8	99.4	117.1	114.4	98.3	97.2	97.4
0320	Concrete Reinforcing	95.9	101.8	98.9	97.1	101.4	99.3	96.3	101.6	99.0	96.3	109.0	102.5	97.5	112.5	105.1	94.1	108.7	101.5
0330	Cast-in-Place Concrete	111.4	110.3	111.0	99.4	108.6	102.9	99.3	99.6	99.4	102.5	99.3	101.3	103.3	117.7	108.7	99.1	99.6	98.7
03	CONCRETE	92.7	98.0	95.0	86.5	96.9	91.1	89.6	99.8	94.1	88.4	99.4	93.3	98.7	117.1	106.8	89.6	100.6	94.5
04	MASONRY	123.4	115.3	118.4	123.3	108.0	113.8	101.6	107.9	105.5	111.9	109.3	110.3	112.2	122.0	118.3	103.3	105.7	104.8
05	METALS	89.7	117.1	98.0	88.6	116.2	97.0	91.6	119.0	99.9	89.5	120.6	98.9	91.6	126.4	102.2	91.4	122.2	100.7
06	WOOD, PLASTICS & COMPOSITES	86.0	76.4	80.8	66.8	76.6	72.1	93.8	94.4	94.1	95.9	91.6	93.6	96.7	113.9	106.0	97.8	95.6	96.6
07	THERMAL & MOISTURE PROTECTION	103.9	103.9	103.9	105.6	101.2	103.7	105.3	104.4	104.9	104.4	93.6	99.7	103.0	118.5	109.8	109.1	92.5	101.9
08	OPENINGS	88.3	100.7	91.2	101.8	100.8	101.6	106.3	106.5	106.3	92.9	110.8	97.1	101.6	125.0	107.0	100.6	113.8	103.6
0920	Plaster & Gypsum Board	91.4	76.3	81.5	103.4	76.3	85.5	94.4	94.8	94.6	96.0	92.0	93.3	97.4	114.4	108.6	103.5	95.9	98.5
0950, 0980	Ceilings & Acoustic Treatment	56.8	76.3	70.0	121.3	76.3	90.8	93.2	94.8	94.2	56.8	92.0	80.6	99.1	114.4	109.2	93.3	95.9	95.1
0960	Flooring	91.1	88.4	90.3	91.0	88.4	90.3	96.5	119.0	103.3	92.9	84.6	90.5	99.7	118.5	105.2	93.2	84.6	90.1
0970, 0990	Wall Finishes & Painting/Coating	83.6	106.5	97.1	89.1	85.6	87.0	89.5	107.2	100.0	94.4	102.5	99.2	97.4	127.1	115.0	85.3	100.8	94.5
09	FINISHES	80.8	87.0	84.2	95.4	84.5	89.4	91.5	101.6	97.0	82.5	93.0	88.2	100.5	118.2	110.2	91.2	95.8	93.7
COVERS	DIVS. 10 - 14, 25, 28, 41, 43, 44, 46	100.0	95.5	99.0	100.0	95.3	99.0	100.0	96.2	99.2	100.0	95.1	98.9	100.0	107.3	101.6	100.0	100.3	100.1
21, 22, 23	FIRE SUPPRESSION, PLUMBING & HVAC	96.1	85.7	91.9	96.8	84.7	91.9	99.8	96.5	98.4	96.1	84.8	91.5	99.9	113.0	105.2	99.9	93.5	97.3
26, 27, 3370	ELECTRICAL, COMMUNICATIONS & UTIL.	101.7	101.8	101.7	103.8	70.4	86.8	97.4	101.8	99.6	108.3	92.4	100.2	105.4	113.0	109.3	101.3	92.4	96.7
MF2016	WEIGHTED AVERAGE	95.0	98.6	96.5	96.7	92.2	94.8	97.1	102.7	99.5	95.1	98.0	96.4	99.7	116.5	106.9	97.1	100.3	98.4

DIVISION		MINNESOTA															MISSISSIPPI		
		SAINT PAUL			ST. CLOUD			THIEF RIVER FALLS			WILLMAR			WINDOM			BILOXI		
		550 - 551			563			567			562			561			395		
		MAT.	INST.	TOTAL	MAT.	INST.	TOTAL	MAT.	INST.	TOTAL	MAT.	INST.	TOTAL	MAT.	INST.	TOTAL	MAT.	INST.	TOTAL
015433	CONTRACTOR EQUIPMENT		103.2	103.2		102.9	102.9		100.3	100.3		102.9	102.9		102.9	102.9		102.7	102.7
0241, 31 - 34	SITE & INFRASTRUCTURE, DEMOLITION	93.3	103.8	100.7	92.5	105.8	101.8	94.5	99.6	98.0	91.9	104.7	100.8	85.8	103.8	98.4	107.9	89.9	95.4
0310	Concrete Forming & Accessories	98.5	116.8	114.1	83.7	112.9	108.6	86.5	83.4	83.8	83.5	89.1	88.2	87.8	84.3	84.8	94.1	65.7	69.9
0320	Concrete Reinforcing	95.5	109.2	102.5	95.9	109.2	102.7	97.5	101.2	99.4	95.5	109.0	102.4	95.5	107.9	101.8	86.0	50.0	67.7
0330	Cast-in-Place Concrete	100.5	116.4	106.4	98.2	115.1	104.5	101.5	82.7	94.5	99.8	81.7	93.0	86.3	85.6	86.0	120.0	66.8	100.2
03	CONCRETE	93.3	115.9	103.4	84.4	113.6	97.4	87.6	87.5	87.6	84.4	91.2	87.4	75.8	90.2	82.1	98.9	65.1	83.9
04	MASONRY	102.5	122.0	114.6	107.8	121.0	115.9	99.1	102.4	101.1	112.2	115.3	114.1	122.7	88.3	101.4	97.8	60.2	74.6
05	METALS	91.6	124.1	101.4	90.4	122.0	99.9	88.8	115.5	96.9	89.5	120.4	98.9	89.4	118.4	98.2	91.5	84.2	89.3
06	WOOD, PLASTICS & COMPOSITES	95.4	113.6	105.3	83.1	109.7	97.5	71.0	79.2	75.5	82.8	82.2	82.5	87.4	82.2	84.6	95.3	66.4	79.6
07	THERMAL & MOISTURE PROTECTION	103.8	119.3	110.5	104.1	113.1	108.0	106.7	90.3	99.6	103.9	103.4	103.7	103.9	84.5	95.4	97.1	63.2	82.3
08	OPENINGS	98.8	123.7	104.6	93.4	121.6	99.9	101.9	102.3	102.0	90.3	104.7	93.6	93.9	104.7	96.4	98.6	56.6	88.9
0920	Plaster & Gypsum Board	99.0	114.4	109.2	91.4	110.5	104.0	104.0	79.0	87.6	91.4	82.3	85.4	91.4	82.3	85.4	105.2	65.9	79.3
0950, 0980	Ceilings & Acoustic Treatment	96.2	114.4	108.5	56.8	110.5	93.2	121.3	79.0	92.6	56.8	82.3	74.0	56.8	82.3	74.0	94.2	65.9	75.0
0960	Flooring	93.1	124.6	102.2	87.9	118.5	96.8	91.9	88.4	90.9	89.3	88.4	89.1	91.8	88.4	90.8	93.5	56.2	82.7
0970, 0990	Wall Finishes & Painting/Coating	91.3	127.0	112.5	94.4	127.0	113.7	89.1	85.6	87.0	89.1	85.6	87.0	89.1	102.5	97.0	83.0	48.1	62.4
09	FINISHES	92.0	119.3	106.9	80.3	115.5	99.5	95.9	84.6	89.7	80.3	87.7	84.4	80.5	86.8	83.9	88.0	62.4	74.1
COVERS	DIVS. 10 - 14, 25, 28, 41, 43, 44, 46	100.0	106.6	101.5	100.0	101.1	100.2	100.0	93.7	98.6	100.0	96.1	99.1	100.0	92.4	98.3	100.0	73.4	94.1
21, 22, 23	FIRE SUPPRESSION, PLUMBING & HVAC	99.9	112.7	105.1	99.5	110.9	104.1	96.8	81.7	90.7	96.1	103.7	99.2	96.1	82.0	90.4	100.0	56.5	82.4
26, 27, 3370	ELECTRICAL, COMMUNICATIONS & UTIL.	102.1	113.0	107.6	101.7	113.0	107.4	101.1	70.4	85.5	101.7	85.8	93.6	108.3	92.3	100.2	99.7	55.0	77.0
MF2016	WEIGHTED AVERAGE	97.2	115.8	105.2	94.4	114.0	102.8	95.6	89.4	92.9	93.3	99.9	96.1	93.7	92.3	93.1	97.3	64.6	83.3

DIVISION		MISSISSIPPI																	
		CLARKSDALE			COLUMBUS			GREENVILLE			GREENWOOD			JACKSON			LAUREL		
		386			397			387			389			390 - 392			394		
		MAT.	INST.	TOTAL	MAT.	INST.	TOTAL	MAT.	INST.	TOTAL	MAT.	INST.	TOTAL	MAT.	INST.	TOTAL	MAT.	INST.	TOTAL
015433	CONTRACTOR EQUIPMENT		102.7	102.7		102.7	102.7		102.7	102.7		102.7	102.7		102.7	102.7		102.7	102.7
0241, 31 - 34	SITE & INFRASTRUCTURE, DEMOLITION	100.0	88.5	92.0	106.8	89.4	94.7	105.8	89.9	94.7	103.0	88.2	92.7	102.7	89.9	93.8	112.2	88.6	95.7
0310	Concrete Forming & Accessories	83.9	44.8	50.6	82.6	47.2	52.5	80.6	63.8	66.3	92.9	45.0	52.1	93.2	66.1	70.2	92.8	60.7	64.0
0320	Concrete Reinforcing	101.0	66.0	83.2	92.1	66.4	79.0	101.6	67.3	84.2	101.0	66.1	83.3	95.6	51.9	73.4	92.7	32.8	62.3
0330	Cast-in-Place Concrete	102.8	50.0	83.1	122.4	55.3	97.4	105.9	56.6	87.5	110.5	49.7	87.9	104.5	65.9	90.2	119.7	51.2	94.2
03	CONCRETE	91.6	52.4	74.2	100.2	55.5	80.4	96.8	63.6	82.1	97.8	52.4	77.7	93.0	65.3	80.7	102.6	54.4	81.2
04	MASONRY	90.6	40.2	59.4	123.7	46.2	75.8	134.1	68.2	93.4	91.3	40.0	59.6	103.4	59.1	76.1	119.8	42.9	72.3
05	METALS	92.9	86.8	91.1	88.6	89.6	88.9	93.9	90.8	93.0	92.9	86.5	91.0	98.1	85.0	94.2	88.7	75.5	84.7
06	WOOD, PLASTICS & COMPOSITES	80.8	45.7	61.7	80.6	46.5	62.1	77.4	64.6	70.5	93.8	45.7	67.7	94.9	67.8	80.2	81.8	66.4	73.4
07	THERMAL & MOISTURE PROTECTION	96.3	48.6	75.5	97.0	53.5	78.1	96.8	63.4	82.3	96.7	50.8	76.8	95.2	62.6	81.0	97.3	54.4	78.6
08	OPENINGS	97.2	49.2	86.0	98.2	49.6	87.0	96.9	60.0	88.3	97.2	49.2	86.0	99.9	58.2	90.2	95.4	53.3	85.6
0920	Plaster & Gypsum Board	93.4	44.6	61.3	96.6	45.5	63.0	93.1	64.1	74.0	104.6	44.6	65.1	90.0	67.3	75.1	96.6	65.9	76.4
0950, 0980	Ceilings & Acoustic Treatment	86.2	44.6	58.0	88.1	45.5	59.2	90.0	64.1	72.5	86.2	44.6	58.0	94.8	67.3	76.2	88.1	65.9	73.1
0960	Flooring	97.7	46.5	82.9	87.1	52.2	76.9	96.1	46.5	81.7	103.5	46.5	87.0	91.7	57.9	81.9	85.7	46.5	74.3
0970, 0990	Wall Finishes & Painting/Coating	93.8	48.1	66.8	83.0	48.1	62.4	93.8	49.1	67.3	93.8	48.1	66.8	85.0	49.1	63.8	83.0	48.1	62.4
09	FINISHES	89.9	45.2	65.5	83.8	47.5	63.9	90.7	59.0	73.4	93.5	45.2	67.1	87.0	63.2	74.0	83.9	58.1	69.8
COVERS	DIVS. 10 - 14, 25, 28, 41, 43, 44, 46	100.0	48.9	88.8	100.0	50.0	89.0	100.0	72.6	94.0	100.0	48.9	88.8	100.0	72.9	94.0	100.0	34.9	85.7
21, 22, 23	FIRE SUPPRESSION, PLUMBING & HVAC	98.5	52.4	79.8	98.1	54.2	80.4	100.0	58.2	83.1	98.5	52.8	80.0	100.0	59.0	83.4	98.2	48.0	77.9
26, 27, 3370	ELECTRICAL, COMMUNICATIONS & UTIL.	95.7	42.6	68.7	97.2	55.0	75.7	95.7	56.1	75.6	95.7	39.9	67.4	101.3	56.1	78.3	98.7	57.7	77.8
MF2016	WEIGHTED AVERAGE	95.2	54.4	77.7	97.0	57.4	80.5	98.6	66.0	84.6	96.5	54.2	78.3	98.0	64.3	84.0	97.2	56.9	79.9

DIVISION		MISSISSIPPI									MISSOURI								
		MCCOMB			MERIDIAN			TUPELO			BOWLING GREEN			CAPE GIRARDEAU			CHILLICOTHE		
		396			393			388			633			637			646		
		MAT.	INST.	TOTAL	MAT.	INST.	TOTAL	MAT.	INST.	TOTAL	MAT.	INST.	TOTAL	MAT.	INST.	TOTAL	MAT.	INST.	TOTAL
015433	CONTRACTOR EQUIPMENT		102.7	102.7		102.7	102.7		102.7	102.7		110.5	110.5		110.5	110.5		104.5	104.5
0241, 31 - 34	SITE & INFRASTRUCTURE, DEMOLITION	99.1	88.4	91.6	103.5	90.0	94.1	97.5	88.4	91.2	90.6	94.7	93.5	92.1	94.6	93.9	104.0	92.8	96.2
0310	Concrete Forming & Accessories	82.6	46.3	51.7	79.9	64.2	66.5	81.2	46.9	52.0	97.3	90.2	91.3	89.6	80.7	82.0	85.0	96.1	94.5
0320	Concrete Reinforcing	93.3	34.5	63.4	92.1	51.9	71.7	98.9	66.1	82.2	96.5	99.3	98.0	97.7	78.2	87.8	100.7	105.6	103.2
0330	Cast-in-Place Concrete	106.2	49.5	85.1	113.7	66.5	96.1	102.8	69.7	90.5	90.0	97.0	92.6	89.1	86.8	88.2	97.2	87.8	93.7
03	CONCRETE	89.5	47.5	70.9	94.1	64.6	81.0	91.2	60.2	77.4	92.1	95.7	93.7	91.2	84.3	88.1	99.4	95.6	97.7
04	MASONRY	125.1	39.6	72.3	97.3	59.6	74.0	123.0	42.7	73.3	110.9	97.4	102.5	107.5	80.2	90.6	99.2	92.9	95.3
05	METALS	88.8	74.8	84.6	89.7	84.9	88.3	92.8	86.8	91.0	97.3	118.9	103.8	98.4	109.0	101.6	84.8	111.2	92.8
06	WOOD, PLASTICS & COMPOSITES	80.6	48.1	62.9	77.8	64.6	70.6	78.0	46.5	60.9	101.3	89.8	95.1	93.1	78.7	85.3	93.9	97.5	95.9
07	THERMAL & MOISTURE PROTECTION	96.5	50.9	76.6	96.7	62.6	81.8	96.3	52.1	77.1	99.7	99.5	99.6	99.1	85.3	93.1	93.4	94.2	93.8
08	OPENINGS	98.3	42.5	85.3	98.0	56.1	88.3	97.2	52.5	86.8	100.7	97.0	99.8	100.7	75.4	94.8	87.9	97.7	90.2
0920	Plaster & Gypsum Board	96.6	47.1	64.0	96.6	64.1	75.2	93.1	45.5	61.7	97.7	89.9	92.6	97.1	78.4	84.8	98.8	97.3	97.8
0950, 0980	Ceilings & Acoustic Treatment	88.1	47.1	60.3	90.0	64.1	72.5	86.2	45.5	58.6	91.9	89.9	90.5	91.9	78.4	82.8	88.9	97.3	94.6
0960	Flooring	87.1	46.5	75.3	85.7	56.2	77.1	96.3	46.5	81.7	96.9	100.1	97.9	93.6	87.4	91.8	99.1	103.0	100.2
0970, 0990	Wall Finishes & Painting/Coating	83.0	48.1	62.4	83.0	48.1	62.4	93.8	46.4	65.8	97.0	109.1	104.1	97.0	68.8	80.3	93.8	105.9	101.0
09	FINISHES	83.1	46.5	63.1	83.4	61.2	71.3	89.4	46.1	65.8	97.4	93.5	95.3	96.3	79.5	87.1	98.1	98.8	98.5
COVERS	DIVS. 10 - 14, 25, 28, 41, 43, 44, 46	100.0	51.9	89.4	100.0	73.0	94.0	100.0	50.0	89.0	100.0	81.4	95.9	100.0	91.8	98.2	100.0	82.4	96.1
21, 22, 23	FIRE SUPPRESSION, PLUMBING & HVAC	98.1	51.8	79.4	100.0	58.7	83.3	98.6	53.8	80.5	96.6	100.4	98.1	100.0	99.8	99.9	96.7	101.5	98.6
26, 27, 3370	ELECTRICAL, COMMUNICATIONS & UTIL.	95.8	56.7	75.9	98.7	57.0	77.5	95.5	54.9	74.9	96.5	78.7	87.5	96.5	100.9	98.7	93.9	77.7	85.7
MF2016	WEIGHTED AVERAGE	95.4	54.5	77.9	95.6	65.1	82.5	96.5	58.2	80.1	97.5	96.0	96.9	98.1	91.8	95.4	94.5	95.4	94.9

MISSOURI

DIVISION		COLUMBIA 652			FLAT RIVER 636			HANNIBAL 634			HARRISONVILLE 647			JEFFERSON CITY 650 - 651			JOPLIN 648		
		MAT.	INST.	TOTAL	MAT.	INST.	TOTAL	MAT.	INST.	TOTAL	MAT.	INST.	TOTAL	MAT.	INST.	TOTAL	MAT.	INST.	TOTAL
015433	CONTRACTOR EQUIPMENT		113.3	113.3		110.5	110.5		110.5	110.5		104.5	104.5		113.3	113.3		108.0	108.0
0241, 31 - 34	SITE & INFRASTRUCTURE, DEMOLITION	103.1	98.4	99.8	93.3	94.5	94.1	88.3	94.6	92.7	95.5	94.1	94.5	102.5	98.4	99.6	105.0	97.6	99.8
0310	Concrete Forming & Accessories	83.1	80.1	80.5	104.0	89.2	91.4	95.4	84.8	86.4	82.3	101.7	98.8	95.1	80.0	82.2	96.3	76.3	79.3
0320	Concrete Reinforcing	90.7	96.0	93.4	97.7	105.9	101.9	96.0	99.3	97.7	100.3	111.1	105.8	98.0	96.0	97.0	103.9	86.8	95.2
0330	Cast-in-Place Concrete	87.0	84.6	86.1	93.0	93.3	93.1	85.2	96.3	89.4	99.6	102.4	100.6	92.2	84.6	89.4	105.3	77.8	95.0
03	CONCRETE	81.6	86.3	83.7	95.1	95.1	95.1	88.4	93.1	90.5	95.3	104.2	99.2	88.5	86.2	87.5	98.2	79.8	90.0
04	MASONRY	139.0	86.5	106.6	107.9	78.0	89.4	103.0	96.6	99.0	93.9	101.5	98.6	97.7	86.5	90.8	93.0	81.9	86.1
05	METALS	94.0	116.3	100.8	97.2	120.4	104.3	97.3	118.5	103.7	85.2	114.7	94.1	93.6	116.2	100.4	87.8	99.8	91.4
06	WOOD, PLASTICS & COMPOSITES	80.9	77.6	79.1	110.7	90.0	99.4	99.3	83.5	90.7	90.4	101.6	96.5	95.8	77.6	85.9	106.0	75.7	89.5
07	THERMAL & MOISTURE PROTECTION	96.6	86.1	92.0	99.9	93.3	97.1	99.5	96.1	98.0	92.6	103.2	97.2	104.0	86.1	96.2	92.6	83.5	88.6
08	OPENINGS	96.5	82.3	93.2	100.7	99.1	100.3	100.7	86.2	97.3	87.6	104.8	91.6	96.9	82.3	93.5	88.9	78.5	86.5
0920	Plaster & Gypsum Board	84.6	77.1	79.7	104.1	90.0	94.8	97.4	83.4	88.2	94.8	101.5	99.2	93.9	77.1	82.8	105.7	74.9	85.4
0950, 0980	Ceilings & Acoustic Treatment	97.8	77.1	83.8	91.9	90.0	90.6	91.9	83.4	86.1	88.9	101.5	97.5	97.3	77.1	83.6	89.8	74.9	79.7
0960	Flooring	90.6	96.3	92.2	100.2	87.4	96.5	96.3	100.1	97.4	94.3	104.2	97.2	98.0	73.9	91.0	125.7	74.6	110.8
0970, 0990	Wall Finishes & Painting/Coating	91.6	80.2	84.7	97.0	73.6	83.2	97.0	96.6	96.7	98.1	105.6	102.5	88.4	80.2	83.6	93.4	77.9	84.2
09	FINISHES	84.9	82.2	83.4	99.4	86.3	92.3	96.9	87.8	92.0	95.7	102.6	99.4	90.8	78.3	84.0	105.1	76.6	89.6
COVERS	DIVS. 10 - 14, 25, 28, 41, 43, 44, 46	100.0	94.6	98.8	100.0	92.3	98.3	100.0	80.5	95.7	100.0	84.5	96.6	100.0	94.6	98.8	100.0	81.2	95.9
21, 22, 23	FIRE SUPPRESSION, PLUMBING & HVAC	99.9	98.1	99.2	96.6	99.0	97.6	96.6	99.9	97.9	96.6	103.0	99.2	99.9	98.1	99.2	100.1	72.7	89.0
26, 27, 3370	ELECTRICAL, COMMUNICATIONS & UTIL.	93.7	81.3	87.4	100.8	100.9	100.8	95.4	78.7	86.9	100.7	101.0	100.9	98.9	81.3	90.0	91.9	67.7	79.6
MF2016	WEIGHTED AVERAGE	96.2	91.3	94.1	98.5	96.2	97.5	96.5	94.0	95.4	94.0	102.6	97.7	96.3	90.8	94.0	95.9	79.9	88.9

MISSOURI

DIVISION		KANSAS CITY 640 - 641			KIRKSVILLE 635			POPLAR BLUFF 639			ROLLA 654 - 655			SEDALIA 653			SIKESTON 638		
		MAT.	INST.	TOTAL	MAT.	INST.	TOTAL	MAT.	INST.	TOTAL	MAT.	INST.	TOTAL	MAT.	INST.	TOTAL	MAT.	INST.	TOTAL
015433	CONTRACTOR EQUIPMENT		105.5	105.5		100.4	100.4		102.7	102.7		113.3	113.3		103.1	103.1		102.7	102.7
0241, 31 - 34	SITE & INFRASTRUCTURE, DEMOLITION	96.8	98.2	97.8	93.4	94.9	94.6	79.9	93.3	89.2	102.1	98.9	99.9	100.5	93.6	95.7	83.5	93.9	90.8
0310	Concrete Forming & Accessories	96.0	102.2	101.2	87.6	80.6	81.6	87.8	78.8	80.1	90.4	96.6	95.7	88.4	80.4	81.6	88.8	78.9	80.4
0320	Concrete Reinforcing	98.7	111.2	105.0	96.7	85.7	91.1	99.7	78.2	88.7	91.2	96.1	93.7	89.9	110.5	100.3	99.0	78.2	88.4
0330	Cast-in-Place Concrete	103.2	103.9	103.5	92.9	84.8	89.9	71.5	86.0	76.9	89.1	97.5	92.2	93.0	83.5	89.5	76.4	86.0	80.0
03	CONCRETE	98.2	104.8	101.1	108.1	84.1	97.5	82.2	82.4	82.3	83.3	98.2	89.9	97.2	87.8	93.0	85.9	82.5	84.4
04	MASONRY	99.7	103.7	102.2	114.4	86.6	97.3	105.8	75.7	87.2	112.7	87.2	96.9	119.0	83.3	96.9	105.6	75.7	87.1
05	METALS	94.1	113.4	99.9	96.9	102.1	98.5	97.5	98.7	97.9	93.4	116.9	100.5	92.4	112.2	98.4	97.9	98.8	98.2
06	WOOD, PLASTICS & COMPOSITES	101.3	101.5	101.4	85.8	78.8	82.0	84.8	78.7	81.5	88.9	98.4	94.0	82.3	79.0	80.5	86.5	78.7	82.3
07	THERMAL & MOISTURE PROTECTION	92.7	104.9	98.0	106.0	92.9	100.3	104.2	84.1	95.4	96.8	94.5	95.8	102.9	89.7	97.1	104.3	84.1	95.5
08	OPENINGS	94.4	104.8	96.8	106.2	81.1	100.4	107.2	75.4	99.8	96.5	93.7	95.9	101.7	88.8	98.7	107.2	75.4	99.8
0920	Plaster & Gypsum Board	101.3	101.5	101.5	92.2	78.5	83.2	92.6	78.4	83.3	87.0	98.4	94.5	80.6	78.5	79.2	94.4	78.4	83.9
0950, 0980	Ceilings & Acoustic Treatment	92.1	101.5	98.5	90.2	78.5	82.3	91.9	78.4	82.8	97.8	98.4	98.2	97.8	78.5	84.7	91.9	78.4	82.8
0960	Flooring	100.3	104.2	101.4	75.2	99.7	82.3	88.5	87.4	88.1	94.1	99.7	95.7	72.9	99.7	80.7	89.0	87.4	88.6
0970, 0990	Wall Finishes & Painting/Coating	96.8	105.9	102.2	92.8	80.4	85.4	92.0	68.8	78.3	91.6	92.6	92.2	91.6	105.9	100.1	92.0	68.8	78.3
09	FINISHES	99.0	102.8	101.1	96.4	83.4	89.3	95.9	78.6	86.4	86.4	96.7	92.0	83.6	85.5	84.7	96.6	78.9	86.9
COVERS	DIVS. 10 - 14, 25, 28, 41, 43, 44, 46	100.0	99.0	99.8	100.0	80.0	95.6	100.0	90.1	97.8	100.0	98.7	99.7	100.0	89.7	97.7	100.0	90.3	97.9
21, 22, 23	FIRE SUPPRESSION, PLUMBING & HVAC	100.0	103.0	101.2	96.6	99.3	97.7	96.6	97.0	96.8	96.5	100.8	98.2	96.4	97.0	96.7	96.6	97.0	96.8
26, 27, 3370	ELECTRICAL, COMMUNICATIONS & UTIL.	102.3	101.0	101.7	95.5	78.6	86.9	95.7	100.8	98.3	92.3	81.3	86.7	93.3	101.0	97.2	94.9	100.8	97.9
MF2016	WEIGHTED AVERAGE	98.1	103.6	100.5	100.2	88.8	95.3	96.3	89.2	93.3	94.3	96.6	95.3	96.6	93.7	95.4	96.9	89.4	93.7

MISSOURI / MONTANA

DIVISION		SPRINGFIELD 656 - 658			ST. JOSEPH 644 - 645			ST. LOUIS 630 - 631			BILLINGS 590 - 591			BUTTE 597			GREAT FALLS 594		
		MAT.	INST.	TOTAL	MAT.	INST.	TOTAL	MAT.	INST.	TOTAL	MAT.	INST.	TOTAL	MAT.	INST.	TOTAL	MAT.	INST.	TOTAL
015433	CONTRACTOR EQUIPMENT		105.6	105.6		104.5	104.5		111.0	111.0		100.5	100.5		100.3	100.3		100.3	100.3
0241, 31 - 34	SITE & INFRASTRUCTURE, DEMOLITION	102.7	96.1	98.1	99.0	91.9	94.1	97.7	99.2	98.7	95.2	97.5	96.8	102.1	96.3	98.0	105.8	97.2	99.8
0310	Concrete Forming & Accessories	96.4	77.5	80.3	95.1	91.5	92.1	100.8	102.6	102.3	99.1	66.4	71.3	85.8	66.5	69.4	99.1	65.7	70.7
0320	Concrete Reinforcing	87.4	95.6	91.6	97.5	110.6	104.1	89.3	107.2	98.4	90.7	80.6	85.6	98.4	81.1	89.6	90.8	80.6	85.6
0330	Cast-in-Place Concrete	94.6	75.9	87.6	97.8	98.2	97.9	100.5	103.5	101.6	115.6	69.8	98.6	127.2	69.8	105.9	134.5	69.7	110.4
03	CONCRETE	93.4	81.2	88.0	94.4	98.0	96.0	98.0	104.8	101.1	92.9	71.0	83.2	96.5	71.1	85.3	101.5	70.6	87.8
04	MASONRY	90.6	81.6	85.0	95.7	90.5	92.5	92.8	109.7	103.2	121.7	73.8	92.1	117.6	82.2	95.7	121.8	73.8	92.1
05	METALS	98.5	102.5	99.8	90.7	113.4	97.6	103.2	121.0	108.5	109.1	87.1	102.4	103.2	87.8	98.6	106.4	87.1	100.5
06	WOOD, PLASTICS & COMPOSITES	90.3	77.2	83.2	106.0	91.2	98.0	102.7	100.8	101.6	92.6	63.3	76.7	78.5	63.3	70.3	93.9	62.5	76.8
07	THERMAL & MOISTURE PROTECTION	101.0	77.7	90.8	93.0	90.5	91.9	97.5	106.0	101.2	105.3	69.7	89.8	105.0	73.3	91.2	105.7	71.0	90.6
08	OPENINGS	104.0	86.7	100.0	91.9	98.8	93.5	99.3	105.8	100.8	100.8	65.0	92.5	99.0	65.0	91.1	101.9	64.6	93.3
0920	Plaster & Gypsum Board	88.0	76.6	80.5	107.1	90.9	96.4	102.7	101.1	101.6	114.5	62.7	80.4	112.1	62.7	79.6	122.5	61.8	82.5
0950, 0980	Ceilings & Acoustic Treatment	97.8	76.6	83.5	96.6	90.9	92.7	93.6	101.1	98.7	90.6	62.7	71.7	96.5	62.7	73.6	98.2	61.8	73.5
0960	Flooring	93.0	73.9	87.5	104.0	102.4	103.5	99.3	100.1	99.5	92.9	78.0	88.6	90.8	94.0	91.7	97.5	78.0	91.8
0970, 0990	Wall Finishes & Painting/Coating	86.1	102.0	95.5	93.8	105.6	100.8	98.0	109.1	104.5	88.8	69.5	77.4	88.0	69.5	77.1	88.0	69.5	77.1
09	FINISHES	88.5	79.9	83.8	101.4	95.0	97.9	99.9	102.4	101.3	90.5	68.3	78.4	90.4	71.5	80.1	94.5	67.7	79.9
COVERS	DIVS. 10 - 14, 25, 28, 41, 43, 44, 46	100.0	92.4	98.3	100.0	96.0	99.1	100.0	102.3	100.5	100.0	90.3	97.9	100.0	90.3	97.9	100.0	90.2	97.9
21, 22, 23	FIRE SUPPRESSION, PLUMBING & HVAC	100.0	71.0	88.2	100.1	89.4	95.8	100.0	106.1	102.5	100.0	73.2	89.2	100.1	71.5	88.5	100.1	69.6	87.8
26, 27, 3370	ELECTRICAL, COMMUNICATIONS & UTIL.	97.5	70.5	83.7	100.8	77.7	89.0	98.9	100.9	99.9	98.5	69.9	84.0	105.6	70.9	88.0	97.9	69.1	83.2
MF2016	WEIGHTED AVERAGE	97.7	81.0	90.5	96.8	92.8	95.1	99.6	105.8	102.3	100.8	75.0	89.7	100.7	76.2	90.2	102.1	74.0	90.1

City Cost Indexes

MONTANA

DIVISION		HAVRE 595 MAT.	INST.	TOTAL	HELENA 596 MAT.	INST.	TOTAL	KALISPELL 599 MAT.	INST.	TOTAL	MILES CITY 593 MAT.	INST.	TOTAL	MISSOULA 598 MAT.	INST.	TOTAL	WOLF POINT 592 MAT.	INST.	TOTAL
015433	CONTRACTOR EQUIPMENT		100.3	100.3		100.3	100.3		100.3	100.3		100.3	100.3		100.3	100.3		100.3	100.3
0241, 31 - 34	SITE & INFRASTRUCTURE, DEMOLITION	109.5	96.4	100.3	95.4	96.3	96.0	92.0	96.2	94.9	98.5	96.3	96.9	84.5	96.0	92.5	116.0	96.4	102.3
0310	Concrete Forming & Accessories	78.6	66.4	68.2	100.3	66.5	71.5	89.1	67.6	70.8	97.4	66.6	71.2	89.1	66.5	69.8	89.9	65.9	69.5
0320	Concrete Reinforcing	99.2	81.1	90.0	104.7	79.9	92.1	101.0	85.9	93.3	98.8	81.2	89.8	100.1	85.9	92.9	100.2	80.4	90.2
0330	Cast-in-Place Concrete	137.1	69.7	112.0	99.5	69.8	88.4	110.5	69.8	95.3	120.9	69.8	101.9	93.7	69.7	84.8	135.5	69.8	111.1
03	CONCRETE	103.9	71.0	89.3	89.0	70.9	81.0	86.7	72.4	80.4	93.6	71.1	83.6	76.1	71.9	74.2	107.0	70.7	90.9
04	MASONRY	118.6	82.2	96.1	110.3	82.2	92.9	116.4	82.2	95.2	123.5	82.2	98.0	140.6	82.2	104.5	124.7	82.2	98.4
05	METALS	99.2	87.7	95.7	105.2	87.4	99.8	99.1	89.4	96.1	98.3	87.9	95.2	99.6	89.3	96.5	98.4	87.6	95.2
06	WOOD, PLASTICS & COMPOSITES	69.7	63.3	66.2	93.3	63.3	77.0	82.1	64.7	72.6	90.8	63.3	75.9	82.1	63.3	71.9	81.5	62.5	71.2
07	THERMAL & MOISTURE PROTECTION	105.4	67.3	88.8	100.3	73.3	88.5	104.6	74.0	91.3	105.0	68.8	89.2	104.1	72.8	90.5	106.0	68.7	89.8
08	OPENINGS	99.5	65.0	91.5	99.8	64.8	91.7	99.5	66.9	91.9	99.0	65.0	91.1	99.0	66.1	91.4	99.0	64.4	91.0
0920	Plaster & Gypsum Board	107.5	62.7	78.0	108.5	62.7	78.3	112.1	64.0	80.5	121.8	62.7	82.9	112.1	62.7	79.6	115.7	61.8	80.2
0950, 0980	Ceilings & Acoustic Treatment	96.5	62.7	73.6	96.6	62.7	74.9	96.5	64.0	74.5	94.8	62.7	73.0	96.5	62.7	73.6	94.8	61.8	72.4
0960	Flooring	88.4	94.0	89.3	101.6	94.0	99.4	92.5	94.0	92.9	97.4	94.0	96.4	92.5	94.0	92.9	93.9	94.0	93.9
0970, 0990	Wall Finishes & Painting/Coating	88.0	69.5	77.1	93.5	69.5	79.3	88.0	69.5	77.1	88.0	69.5	77.1	88.0	55.0	68.5	88.0	69.5	77.1
09	FINISHES	89.7	71.5	79.8	97.5	71.4	83.2	90.3	72.3	80.5	93.2	71.5	81.4	89.8	69.9	78.9	92.7	71.0	80.8
COVERS	DIVS. 10 - 14, 25, 28, 41, 43, 44, 46	100.0	90.3	97.9	100.0	90.3	97.9	100.0	90.5	97.9	100.0	90.3	97.9	100.0	90.3	97.9	100.0	90.3	97.9
21, 22, 23	FIRE SUPPRESSION, PLUMBING & HVAC	96.6	70.4	86.0	100.1	71.5	88.5	96.6	69.7	85.8	96.6	76.9	88.6	100.1	69.7	87.8	96.6	76.9	88.6
26, 27, 3370	ELECTRICAL, COMMUNICATIONS & UTIL.	97.9	70.9	84.2	104.7	70.9	87.5	102.3	69.8	85.7	97.9	76.3	86.9	103.4	69.7	86.3	97.9	76.3	86.9
MF2016	WEIGHTED AVERAGE	99.5	75.7	89.3	100.1	76.1	89.8	97.4	76.2	88.3	98.4	77.9	89.6	97.9	75.7	88.4	100.5	77.8	90.8

NEBRASKA

DIVISION		ALLIANCE 693 MAT.	INST.	TOTAL	COLUMBUS 686 MAT.	INST.	TOTAL	GRAND ISLAND 688 MAT.	INST.	TOTAL	HASTINGS 689 MAT.	INST.	TOTAL	LINCOLN 683 - 685 MAT.	INST.	TOTAL	MCCOOK 690 MAT.	INST.	TOTAL
015433	CONTRACTOR EQUIPMENT		97.8	97.8		104.7	104.7		104.7	104.7		104.7	104.7		104.7	104.7		104.7	104.7
0241, 31 - 34	SITE & INFRASTRUCTURE, DEMOLITION	101.0	100.8	100.9	105.7	95.0	98.2	110.6	95.2	99.9	109.4	95.0	99.3	98.8	95.2	96.3	103.9	94.9	97.7
0310	Concrete Forming & Accessories	87.4	56.6	61.1	96.6	77.0	79.9	96.2	70.7	74.5	99.5	74.5	78.2	92.5	77.4	79.6	92.8	57.3	62.6
0320	Concrete Reinforcing	110.9	89.9	100.2	100.7	88.4	94.5	100.1	78.3	89.0	100.1	78.3	89.1	101.2	78.6	89.7	103.3	78.5	90.7
0330	Cast-in-Place Concrete	108.2	83.9	99.1	109.4	84.0	99.9	115.8	79.7	102.4	115.8	75.9	101.0	89.3	84.3	87.4	117.1	75.9	101.7
03	CONCRETE	114.1	72.5	95.7	97.3	82.5	90.7	101.9	76.4	90.6	102.1	76.8	90.9	85.1	81.1	83.3	103.7	69.1	88.4
04	MASONRY	107.1	79.0	89.7	112.2	78.9	91.7	105.0	75.3	86.7	113.4	75.4	89.9	92.5	79.8	84.6	102.9	78.9	88.1
05	METALS	98.8	85.1	94.6	89.8	98.2	92.4	91.5	94.1	92.3	92.2	93.9	92.7	93.4	94.6	93.8	93.6	94.0	93.7
06	WOOD, PLASTICS & COMPOSITES	84.6	50.1	65.9	96.1	77.0	85.7	95.2	68.2	80.5	99.1	74.2	85.6	96.3	77.0	85.8	92.8	51.1	70.1
07	THERMAL & MOISTURE PROTECTION	103.8	67.7	88.0	104.5	81.7	94.6	104.6	79.6	93.7	104.7	79.0	93.5	100.5	81.8	92.4	98.7	77.4	89.4
08	OPENINGS	92.8	59.8	85.1	93.2	75.9	89.2	93.2	68.7	87.5	93.2	71.8	88.2	102.5	70.8	95.1	93.3	57.4	84.9
0920	Plaster & Gypsum Board	80.1	48.9	59.5	94.5	76.5	82.6	93.6	67.4	76.3	95.4	73.5	81.0	103.7	76.5	85.8	90.6	49.8	63.8
0950, 0980	Ceilings & Acoustic Treatment	93.7	48.9	63.3	88.3	76.5	80.3	88.3	67.4	74.1	88.3	73.5	78.3	92.2	76.5	81.6	89.7	49.8	62.7
0960	Flooring	92.6	90.3	92.0	85.8	90.3	87.1	85.5	81.5	84.4	86.8	84.0	86.0	99.5	89.5	96.6	91.1	90.3	90.9
0970, 0990	Wall Finishes & Painting/Coating	157.3	53.0	95.6	75.6	61.7	67.3	75.6	64.1	68.8	75.6	61.7	67.3	97.7	81.4	88.1	88.7	46.2	63.5
09	FINISHES	91.7	60.9	74.9	84.8	77.8	81.0	84.9	71.4	77.5	85.5	74.8	79.7	94.5	80.1	86.6	89.1	60.8	73.6
COVERS	DIVS. 10 - 14, 25, 28, 41, 43, 44, 46	100.0	85.6	96.8	100.0	85.2	96.7	100.0	87.8	97.3	100.0	84.9	96.7	100.0	88.7	97.5	100.0	85.8	96.9
21, 22, 23	FIRE SUPPRESSION, PLUMBING & HVAC	96.6	76.7	88.5	96.6	77.0	88.7	100.0	81.3	92.5	96.6	76.2	88.3	99.9	81.3	92.4	96.5	76.9	88.6
26, 27, 3370	ELECTRICAL, COMMUNICATIONS & UTIL.	92.1	67.2	79.5	95.6	83.4	89.4	94.2	67.3	80.5	93.5	81.4	87.3	106.4	67.3	86.5	94.2	67.3	80.5
MF2016	WEIGHTED AVERAGE	98.9	74.7	88.5	95.6	82.7	90.1	97.0	78.5	89.1	96.7	80.0	89.5	97.1	81.1	90.2	96.6	74.8	87.3

NEBRASKA / NEVADA

DIVISION		NORFOLK 687 MAT.	INST.	TOTAL	NORTH PLATTE 691 MAT.	INST.	TOTAL	OMAHA 680 - 681 MAT.	INST.	TOTAL	VALENTINE 692 MAT.	INST.	TOTAL	CARSON CITY 897 MAT.	INST.	TOTAL	ELKO 898 MAT.	INST.	TOTAL
015433	CONTRACTOR EQUIPMENT		94.1	94.1		104.7	104.7		94.1	94.1		97.6	97.6		97.6	97.6		97.6	97.6
0241, 31 - 34	SITE & INFRASTRUCTURE, DEMOLITION	86.4	94.1	91.7	105.2	94.7	97.9	90.4	94.3	93.1	88.0	99.7	96.2	88.2	97.7	94.8	71.3	96.3	88.7
0310	Concrete Forming & Accessories	83.3	76.1	77.2	95.3	77.2	79.9	95.8	75.5	78.5	83.8	54.4	58.8	104.0	78.3	82.1	108.8	99.1	100.5
0320	Concrete Reinforcing	100.8	68.4	84.3	102.8	78.4	90.4	97.7	78.5	88.0	103.3	68.2	85.5	108.1	119.5	113.9	116.1	117.9	117.0
0330	Cast-in-Place Concrete	109.9	73.1	96.2	117.1	62.0	96.6	92.6	79.8	87.9	103.2	55.3	85.4	97.0	83.6	92.0	93.9	74.7	86.7
03	CONCRETE	95.9	74.2	86.3	103.8	73.2	90.2	86.5	78.0	82.7	101.1	58.2	82.1	95.1	87.3	91.7	92.8	93.4	93.0
04	MASONRY	117.8	78.9	93.8	91.1	78.4	83.2	95.6	80.2	86.1	102.5	78.3	87.6	118.0	70.2	88.5	125.0	69.1	90.5
05	METALS	93.1	80.4	89.2	92.8	93.2	92.9	93.4	84.6	90.8	104.7	79.2	97.0	104.9	95.7	102.2	108.2	94.7	104.1
06	WOOD, PLASTICS & COMPOSITES	80.0	76.6	78.1	94.8	79.1	86.3	94.5	74.5	83.6	79.4	48.8	62.8	92.3	76.4	83.7	105.4	105.3	105.3
07	THERMAL & MOISTURE PROTECTION	104.2	79.7	93.5	98.6	79.5	90.3	101.3	80.5	92.2	99.3	74.9	88.7	111.0	79.7	97.4	107.2	75.7	93.5
08	OPENINGS	94.6	70.7	89.1	92.6	74.2	88.3	100.6	75.3	94.7	94.8	55.0	85.6	99.4	79.7	94.8	104.5	95.3	102.4
0920	Plaster & Gypsum Board	94.3	76.5	82.6	90.6	78.6	82.7	101.6	74.3	83.6	91.9	47.9	62.9	103.2	75.8	85.2	109.5	105.5	106.9
0950, 0980	Ceilings & Acoustic Treatment	103.1	76.5	85.1	89.7	78.6	82.2	104.1	74.3	83.9	106.1	47.9	66.7	102.3	75.8	84.4	102.7	105.5	104.6
0960	Flooring	106.9	90.3	102.1	92.2	81.5	89.1	101.8	90.3	98.4	118.5	90.3	110.3	97.8	69.4	89.6	101.1	69.4	91.9
0970, 0990	Wall Finishes & Painting/Coating	132.0	61.7	90.4	88.7	60.4	72.0	101.1	74.0	85.1	157.4	62.6	101.3	96.7	81.4	87.7	93.9	81.4	86.5
09	FINISHES	101.4	77.4	88.3	89.4	76.9	82.6	98.0	77.8	87.0	109.7	61.0	83.1	95.2	76.1	84.8	93.2	93.1	93.1
COVERS	DIVS. 10 - 14, 25, 28, 41, 43, 44, 46	100.0	84.8	96.7	100.0	62.4	91.7	100.0	87.5	97.2	100.0	57.2	90.6	100.0	102.0	100.4	100.0	92.9	98.4
21, 22, 23	FIRE SUPPRESSION, PLUMBING & HVAC	96.3	76.0	88.1	99.9	75.3	90.0	99.9	76.8	90.6	96.1	75.2	87.6	100.1	78.7	91.4	98.3	78.7	90.4
26, 27, 3370	ELECTRICAL, COMMUNICATIONS & UTIL.	94.5	83.4	88.8	92.4	67.3	79.6	102.7	83.4	92.9	89.7	67.3	78.3	98.9	90.4	94.6	96.2	90.4	93.3
MF2016	WEIGHTED AVERAGE	96.9	79.2	89.3	96.6	77.3	88.3	96.9	80.8	90.0	98.8	70.8	86.8	100.4	84.2	93.5	100.3	87.4	94.8

City Cost Indexes

NEVADA / NEW HAMPSHIRE

DIVISION		ELY 893 MAT.	INST.	TOTAL	LAS VEGAS 889-891 MAT.	INST.	TOTAL	RENO 894-895 MAT.	INST.	TOTAL	CHARLESTON 036 MAT.	INST.	TOTAL	CLAREMONT 037 MAT.	INST.	TOTAL	CONCORD 032-033 MAT.	INST.	TOTAL
015433	CONTRACTOR EQUIPMENT		97.6	97.6		97.6	97.6		97.6	97.6		97.9	97.9		97.9	97.9		97.9	97.9
0241, 31-34	SITE & INFRASTRUCTURE, DEMOLITION	77.4	97.5	91.4	80.7	100.1	94.2	77.1	97.7	91.5	83.7	100.4	95.3	77.8	100.4	93.5	91.1	100.6	97.7
0310	Concrete Forming & Accessories	101.8	104.5	104.1	102.9	107.0	106.4	98.3	78.3	81.3	85.6	85.0	85.0	91.4	85.0	85.9	95.3	94.5	94.6
0320	Concrete Reinforcing	114.8	118.1	116.5	105.8	122.0	114.0	108.5	120.9	114.8	89.3	86.2	87.7	89.3	86.2	87.7	104.0	86.6	95.2
0330	Cast-in-Place Concrete	100.8	98.1	99.8	97.6	107.9	101.4	100.6	83.6	93.8	86.1	113.0	96.1	79.1	113.0	91.7	105.0	113.8	108.3
03	CONCRETE	101.0	104.1	102.4	96.4	109.3	102.1	100.6	87.6	94.8	89.2	95.0	91.8	81.8	95.0	87.6	97.4	99.7	98.4
04	MASONRY	130.4	76.4	97.0	117.2	97.1	104.8	124.1	70.2	90.8	93.8	94.8	94.4	94.2	94.8	94.6	101.1	101.3	101.2
05	METALS	108.2	97.6	105.0	116.3	101.3	111.8	109.7	96.2	105.6	84.8	89.5	86.2	84.8	89.5	86.2	90.8	90.4	90.7
06	WOOD, PLASTICS & COMPOSITES	95.2	107.5	101.9	93.0	105.3	99.7	88.9	76.4	82.1	89.4	81.7	85.2	95.8	81.7	88.1	93.8	93.6	93.7
07	THERMAL & MOISTURE PROTECTION	107.7	94.8	102.1	121.5	98.9	111.7	107.2	79.7	95.2	103.2	104.5	103.7	102.9	104.5	103.6	109.6	109.4	109.5
08	OPENINGS	104.5	96.5	102.6	103.4	110.0	105.0	102.2	80.0	97.1	98.4	81.8	94.5	99.5	81.8	95.4	94.4	88.4	93.0
0920	Plaster & Gypsum Board	104.1	107.7	106.5	97.1	105.5	102.6	92.6	75.8	81.5	108.0	80.8	90.1	109.2	80.8	90.5	107.7	93.1	98.1
0950, 0980	Ceilings & Acoustic Treatment	102.7	107.7	106.1	108.6	105.5	106.5	106.1	75.8	85.6	101.6	80.8	87.5	101.6	80.8	87.5	101.7	93.1	95.8
0960	Flooring	98.8	69.4	90.2	90.5	99.9	93.2	95.5	69.4	88.0	83.7	104.3	89.7	86.0	104.3	91.3	93.2	104.3	96.5
0970, 0990	Wall Finishes & Painting/Coating	93.9	115.0	106.4	96.4	115.0	107.4	93.9	81.4	86.5	83.3	92.4	88.7	83.3	92.4	88.7	91.1	92.4	91.8
09	FINISHES	92.2	99.8	96.4	90.1	106.5	99.1	90.1	76.1	82.5	86.6	88.5	87.6	87.0	88.5	87.8	90.1	95.8	93.2
COVERS	DIVS. 10-14, 25, 28, 41, 43, 44, 46	100.0	72.3	93.9	100.0	101.8	100.4	100.0	102.0	100.4	100.0	90.3	97.9	100.0	90.3	97.9	100.0	104.6	101.0
21, 22, 23	FIRE SUPPRESSION, PLUMBING & HVAC	98.3	98.3	98.3	100.1	101.3	100.6	100.0	78.7	91.4	96.6	83.9	91.5	96.6	83.9	91.5	100.0	86.3	94.4
26, 27, 3370	ELECTRICAL, COMMUNICATIONS & UTIL.	96.4	95.2	95.8	100.6	110.4	105.6	96.8	90.4	93.6	91.2	77.1	84.0	91.2	77.1	84.0	92.5	77.1	84.7
MF2016	WEIGHTED AVERAGE	101.6	95.6	99.0	102.7	104.1	103.3	101.5	84.2	94.1	92.5	88.6	90.8	91.7	88.6	90.3	96.1	92.4	94.5

NEW HAMPSHIRE / NEW JERSEY

DIVISION		KEENE 034 MAT.	INST.	TOTAL	LITTLETON 035 MAT.	INST.	TOTAL	MANCHESTER 031 MAT.	INST.	TOTAL	NASHUA 030 MAT.	INST.	TOTAL	PORTSMOUTH 038 MAT.	INST.	TOTAL	ATLANTIC CITY 082,084 MAT.	INST.	TOTAL
015433	CONTRACTOR EQUIPMENT		97.9	97.9		97.9	97.9		97.9	97.9		97.9	97.9		97.9	97.9		95.4	95.4
0241, 31-34	SITE & INFRASTRUCTURE, DEMOLITION	90.7	100.4	97.5	77.9	99.3	92.9	88.8	100.6	97.0	92.2	100.5	98.0	85.9	101.0	96.4	88.6	104.0	99.3
0310	Concrete Forming & Accessories	90.0	85.2	85.9	101.7	79.0	82.4	95.8	94.7	94.9	98.1	94.7	95.2	87.0	94.2	93.1	112.3	141.1	136.8
0320	Concrete Reinforcing	89.3	86.2	87.7	90.0	86.2	88.1	104.4	86.6	95.4	111.5	86.6	98.9	89.3	86.6	87.9	80.2	141.7	111.4
0330	Cast-in-Place Concrete	86.5	113.1	96.4	77.6	104.0	87.4	98.0	116.0	104.7	81.8	115.9	94.5	77.6	115.8	91.8	81.4	139.4	103.0
03	CONCRETE	88.8	95.1	91.6	81.2	89.2	84.7	94.2	100.5	97.0	88.7	100.4	93.9	81.4	100.3	89.8	87.3	139.0	110.2
04	MASONRY	96.4	94.8	95.4	106.4	79.0	89.5	99.9	101.3	100.8	98.9	101.3	100.4	94.4	100.5	98.1	105.2	137.8	125.3
05	METALS	85.5	89.7	86.8	85.5	89.5	86.7	92.3	90.7	91.8	90.7	90.5	90.7	87.0	91.4	88.3	97.7	114.5	102.8
06	WOOD, PLASTICS & COMPOSITES	94.0	81.7	87.3	105.9	81.7	92.8	93.9	93.6	93.7	104.1	93.6	98.4	90.7	93.6	92.2	122.6	141.9	133.1
07	THERMAL & MOISTURE PROTECTION	103.8	104.5	104.1	103.1	97.3	100.5	105.2	109.7	107.1	104.2	109.7	106.6	103.7	108.5	105.8	98.8	134.5	114.3
08	OPENINGS	97.0	85.4	94.3	100.4	81.8	96.1	99.3	92.0	97.6	101.4	91.3	99.1	102.0	81.7	97.3	98.7	137.3	107.7
0920	Plaster & Gypsum Board	108.3	80.8	90.2	123.2	80.8	95.3	105.3	93.1	97.2	117.7	93.1	101.5	108.0	93.1	98.2	117.5	142.7	134.1
0950, 0980	Ceilings & Acoustic Treatment	101.6	80.8	87.5	101.6	80.8	87.5	102.6	93.1	96.1	113.6	93.1	99.7	102.6	93.1	96.1	90.9	142.7	126.0
0960	Flooring	85.7	104.3	91.1	96.0	103.2	98.1	94.1	108.6	98.3	89.3	104.3	93.7	83.8	104.3	89.8	92.5	164.3	113.4
0970, 0990	Wall Finishes & Painting/Coating	83.3	106.1	96.8	83.3	92.4	88.7	94.3	106.1	101.3	83.3	104.9	96.1	83.3	104.9	96.1	81.1	145.5	119.2
09	FINISHES	88.4	90.0	89.3	91.9	84.3	87.8	91.9	98.2	95.3	93.1	97.1	95.3	87.5	96.9	92.6	87.7	147.1	120.1
COVERS	DIVS. 10-14, 25, 28, 41, 43, 44, 46	100.0	96.6	99.3	100.0	97.5	99.5	100.0	104.6	101.0	100.0	104.6	101.0	100.0	104.9	100.9	100.0	116.6	103.7
21, 22, 23	FIRE SUPPRESSION, PLUMBING & HVAC	96.6	84.0	91.5	96.6	75.7	88.2	100.0	86.3	94.4	100.1	86.3	94.5	100.1	85.7	94.3	99.7	129.2	111.6
26, 27, 3370	ELECTRICAL, COMMUNICATIONS & UTIL.	91.2	77.1	84.0	92.1	50.3	70.8	91.3	77.1	84.1	93.1	77.1	85.0	91.6	77.1	84.2	94.5	143.6	119.4
MF2016	WEIGHTED AVERAGE	93.0	89.2	91.3	93.0	80.1	87.4	96.2	93.0	94.8	95.9	92.8	94.6	93.3	92.2	92.8	96.4	132.6	111.9

NEW JERSEY

DIVISION		CAMDEN 081 MAT.	INST.	TOTAL	DOVER 078 MAT.	INST.	TOTAL	ELIZABETH 072 MAT.	INST.	TOTAL	HACKENSACK 076 MAT.	INST.	TOTAL	JERSEY CITY 073 MAT.	INST.	TOTAL	LONG BRANCH 077 MAT.	INST.	TOTAL
015433	CONTRACTOR EQUIPMENT		95.4	95.4		97.9	97.9		97.9	97.9		97.9	97.9		95.4	95.4		95.1	95.1
0241, 31-34	SITE & INFRASTRUCTURE, DEMOLITION	89.8	104.0	99.7	106.0	105.9	105.9	110.9	105.9	107.4	107.0	106.1	106.4	97.4	106.0	103.4	102.2	105.2	104.3
0310	Concrete Forming & Accessories	102.7	141.0	135.3	97.9	150.2	142.4	110.3	150.2	144.3	97.9	150.1	142.3	102.2	150.2	143.1	102.8	142.7	136.8
0320	Concrete Reinforcing	105.2	132.2	118.9	81.2	151.8	117.1	81.2	151.8	117.1	81.2	151.8	117.1	105.3	151.8	129.0	81.2	151.4	116.9
0330	Cast-in-Place Concrete	78.9	137.5	100.7	82.4	134.8	101.9	70.8	138.0	95.8	80.6	137.9	101.9	64.3	134.9	90.6	71.5	137.5	96.1
03	CONCRETE	87.4	136.6	109.2	86.0	143.5	111.4	83.0	144.6	110.3	84.4	144.5	111.0	80.6	143.3	108.4	84.7	140.7	109.6
04	MASONRY	96.5	137.8	122.0	90.5	148.2	126.1	107.5	148.2	132.6	94.4	148.2	127.6	86.7	148.2	124.7	100.0	138.3	123.7
05	METALS	103.2	111.1	105.6	95.4	123.3	103.8	96.9	123.3	104.9	95.4	123.2	103.8	101.0	120.3	106.9	95.5	119.2	102.7
06	WOOD, PLASTICS & COMPOSITES	109.9	141.9	127.2	98.6	150.2	126.6	114.4	150.2	133.9	98.6	150.2	126.6	99.5	150.2	127.1	100.6	143.3	123.8
07	THERMAL & MOISTURE PROTECTION	98.6	135.5	114.7	109.0	145.5	124.9	109.2	146.0	125.2	108.7	138.4	121.6	108.4	145.5	124.6	108.6	132.4	119.0
08	OPENINGS	100.8	135.1	108.8	104.1	145.7	113.8	102.5	145.7	112.5	101.9	145.7	112.1	102.0	145.7	111.0	96.5	141.2	106.9
0920	Plaster & Gypsum Board	112.5	142.7	132.4	108.5	151.3	136.7	116.1	151.3	139.3	108.5	151.3	136.7	111.3	151.3	137.6	110.3	144.3	132.7
0950, 0980	Ceilings & Acoustic Treatment	99.5	142.7	128.8	87.3	151.3	130.7	89.2	151.3	131.3	87.3	151.3	130.7	96.9	151.3	133.8	87.3	144.3	125.9
0960	Flooring	88.9	164.3	110.8	81.5	190.7	113.2	86.3	190.7	116.6	81.5	190.7	113.2	82.5	190.7	113.9	82.7	182.9	111.8
0970, 0990	Wall Finishes & Painting/Coating	81.1	145.5	119.2	83.0	145.5	120.0	83.0	145.5	120.0	83.0	145.5	120.0	83.1	145.5	120.1	83.1	145.5	120.1
09	FINISHES	87.4	147.1	120.0	84.4	157.8	124.4	87.8	157.8	126.0	84.2	157.8	124.4	86.7	158.1	125.7	85.3	151.3	121.3
COVERS	DIVS. 10-14, 25, 28, 41, 43, 44, 46	100.0	116.6	103.7	100.0	133.7	107.4	100.0	133.7	107.4	100.0	133.8	107.4	100.0	133.8	107.4	100.0	116.8	103.7
21, 22, 23	FIRE SUPPRESSION, PLUMBING & HVAC	99.9	129.1	111.7	99.8	138.3	115.4	100.1	137.9	115.4	99.8	138.3	115.4	100.1	138.3	115.5	99.8	134.0	113.6
26, 27, 3370	ELECTRICAL, COMMUNICATIONS & UTIL.	99.1	135.3	117.5	95.6	141.9	119.1	96.2	141.9	119.4	95.6	142.5	119.4	100.2	142.5	121.7	95.3	133.1	114.5
MF2016	WEIGHTED AVERAGE	97.5	130.7	111.8	96.1	139.6	114.7	97.3	139.7	115.4	95.9	139.6	114.6	96.3	139.4	114.8	95.6	133.6	111.9

City Cost Indexes

NEW JERSEY

DIVISION		NEW BRUNSWICK 088-089 MAT.	INST.	TOTAL	NEWARK 070-071 MAT.	INST.	TOTAL	PATERSON 074-075 MAT.	INST.	TOTAL	POINT PLEASANT 087 MAT.	INST.	TOTAL	SUMMIT 079 MAT.	INST.	TOTAL	TRENTON 085-086 MAT.	INST.	TOTAL
015433	CONTRACTOR EQUIPMENT		95.1	95.1		97.9	97.9		97.9	97.9		95.1	95.1		97.9	97.9		95.1	95.1
0241, 31 - 34	SITE & INFRASTRUCTURE, DEMOLITION	101.2	105.6	104.3	110.5	105.9	107.3	108.9	106.1	106.9	102.7	105.2	104.5	108.3	105.9	106.6	88.1	105.2	100.0
0310	Concrete Forming & Accessories	106.0	149.9	143.4	101.8	150.2	143.0	99.9	150.1	142.6	100.1	142.7	136.3	100.7	150.2	142.8	101.3	142.5	136.3
0320	Concrete Reinforcing	81.1	151.5	116.9	103.6	151.8	128.1	105.3	151.8	129.0	81.1	151.4	116.8	81.2	151.8	117.1	106.5	119.6	113.1
0330	Cast-in-Place Concrete	100.7	140.4	115.4	91.0	138.0	108.5	82.0	137.9	102.8	100.7	139.6	115.1	68.5	138.0	94.3	99.1	137.4	113.3
03	CONCRETE	103.9	145.0	122.1	93.0	144.6	115.8	88.7	144.5	113.4	103.5	141.4	120.3	80.3	144.6	108.8	97.0	135.2	113.9
04	MASONRY	104.4	143.1	128.3	93.1	148.2	127.2	90.9	148.2	126.3	93.9	138.3	121.3	93.5	148.2	127.3	97.1	138.3	122.6
05	METALS	97.7	119.6	104.4	103.0	123.3	109.1	96.2	123.2	104.4	97.7	119.2	104.2	95.4	123.3	103.8	102.9	108.1	104.5
06	WOOD, PLASTICS & COMPOSITES	115.3	150.2	134.3	99.4	150.2	127.0	101.3	150.2	127.9	107.0	143.3	126.7	102.5	150.2	128.4	99.0	143.3	123.1
07	THERMAL & MOISTURE PROTECTION	99.1	141.6	117.6	110.6	146.0	126.0	109.1	138.4	121.8	99.1	134.8	114.6	109.4	146.0	125.3	99.6	136.5	115.7
08	OPENINGS	93.8	145.7	105.9	100.3	145.7	110.8	107.2	145.7	116.1	95.6	141.2	106.2	108.3	145.7	117.0	100.0	133.8	107.8
0920	Plaster & Gypsum Board	114.8	151.3	138.8	104.9	151.3	135.4	111.3	151.3	137.6	109.6	144.3	132.4	110.3	151.3	137.3	103.6	144.3	130.4
0950, 0980	Ceilings & Acoustic Treatment	90.9	151.3	131.8	95.8	151.3	133.4	96.9	151.3	133.8	90.9	144.3	127.1	87.3	151.3	130.7	94.9	144.3	128.4
0960	Flooring	90.2	190.7	119.4	94.0	190.7	122.1	82.5	190.7	113.9	87.8	164.3	110.0	82.7	190.7	114.0	94.9	182.9	120.4
0970, 0990	Wall Finishes & Painting/Coating	81.1	145.5	119.2	88.0	145.5	122.0	83.0	145.5	120.0	81.1	145.5	119.2	83.0	145.5	120.0	88.5	145.5	122.2
09	FINISHES	87.7	157.7	125.9	90.4	157.8	127.2	86.8	157.8	125.5	86.2	148.1	120.0	85.3	157.8	124.9	89.7	151.3	123.3
COVERS	DIVS. 10 - 14, 25, 28, 41, 43, 44, 46	100.0	133.6	107.4	100.0	133.8	107.4	100.0	133.8	107.4	100.0	113.9	103.1	100.0	133.7	107.4	100.0	116.8	103.7
21, 22, 23	FIRE SUPPRESSION, PLUMBING & HVAC	99.7	136.5	114.6	100.1	137.9	115.4	100.1	138.3	115.5	99.7	134.0	113.5	99.8	137.9	115.2	100.1	133.7	113.7
26, 27, 3370	ELECTRICAL, COMMUNICATIONS & UTIL.	95.1	137.0	116.4	104.5	142.5	123.8	100.2	141.9	121.4	94.5	133.1	114.1	96.2	141.9	119.4	103.0	131.8	117.7
MF2016	WEIGHTED AVERAGE	98.3	137.7	115.2	99.6	139.7	116.8	97.7	139.5	115.6	97.7	133.3	113.0	96.2	139.7	114.8	99.2	131.4	113.0

NEW JERSEY / NEW MEXICO

DIVISION		VINELAND 080, 083 MAT.	INST.	TOTAL	ALBUQUERQUE 870-872 MAT.	INST.	TOTAL	CARRIZOZO 883 MAT.	INST.	TOTAL	CLOVIS 881 MAT.	INST.	TOTAL	FARMINGTON 874 MAT.	INST.	TOTAL	GALLUP 873 MAT.	INST.	TOTAL
015433	CONTRACTOR EQUIPMENT		95.4	95.4		110.9	110.9		110.9	110.9		110.9	110.9		110.9	110.9		110.9	110.9
0241, 31 - 34	SITE & INFRASTRUCTURE, DEMOLITION	92.9	104.0	100.6	91.1	102.3	98.9	110.0	102.3	104.6	97.3	102.3	100.8	97.5	102.3	100.9	106.6	102.3	103.6
0310	Concrete Forming & Accessories	97.1	141.2	134.7	99.2	66.9	71.7	96.5	66.9	71.3	96.5	66.8	71.2	99.3	66.9	71.7	99.3	66.9	71.7
0320	Concrete Reinforcing	80.2	135.0	108.0	108.9	71.9	90.1	113.5	71.9	92.3	114.7	71.9	92.9	119.1	71.9	95.1	113.9	71.9	92.6
0330	Cast-in-Place Concrete	87.8	137.6	106.3	91.4	69.1	83.1	94.8	69.1	85.2	94.7	69.1	85.2	92.3	69.1	83.7	86.8	69.1	80.2
03	CONCRETE	91.8	137.3	112.0	87.5	69.8	79.6	111.4	69.8	93.0	100.1	69.8	86.6	90.6	69.8	81.4	95.5	69.8	84.1
04	MASONRY	93.5	138.3	121.2	105.7	58.7	76.7	106.9	58.7	77.1	106.9	58.7	77.1	114.3	58.7	80.0	100.9	58.7	74.9
05	METALS	97.6	112.7	102.2	102.5	90.6	98.9	101.6	90.6	98.3	101.3	90.5	98.0	100.3	90.6	97.4	99.5	90.6	96.8
06	WOOD, PLASTICS & COMPOSITES	103.5	141.9	124.3	104.0	68.8	84.9	94.6	68.8	80.6	94.6	68.8	80.6	104.1	68.8	84.9	104.1	68.8	84.9
07	THERMAL & MOISTURE PROTECTION	98.5	134.7	114.3	101.8	70.0	88.0	102.8	70.0	88.5	101.5	70.0	87.8	102.1	70.0	88.1	103.4	70.0	88.8
08	OPENINGS	95.2	136.1	104.7	100.9	67.7	93.2	98.7	67.7	91.5	98.8	67.7	91.6	103.3	67.7	95.1	103.4	67.7	95.1
0920	Plaster & Gypsum Board	108.1	142.7	130.9	107.8	67.7	81.4	80.9	67.7	72.2	80.9	67.7	72.2	98.8	67.7	78.3	98.8	67.7	78.3
0950, 0980	Ceilings & Acoustic Treatment	90.9	142.7	126.0	103.0	67.7	79.1	103.8	67.7	79.4	103.8	67.7	79.4	102.6	67.7	79.0	102.6	67.7	79.0
0960	Flooring	86.9	164.3	109.4	87.8	68.0	82.0	97.5	68.0	88.9	97.5	68.0	88.9	89.2	68.0	83.1	89.2	68.0	83.1
0970, 0990	Wall Finishes & Painting/Coating	81.1	145.5	119.2	96.1	51.9	69.9	93.4	51.9	68.8	93.4	51.9	68.8	90.8	51.9	67.7	90.8	51.9	67.7
09	FINISHES	85.0	147.2	119.0	89.8	65.5	76.5	92.9	65.5	78.0	91.7	65.5	77.4	88.8	65.5	76.1	90.0	65.5	76.6
COVERS	DIVS. 10 - 14, 25, 28, 41, 43, 44, 46	100.0	116.8	103.7	100.0	85.1	96.7	100.0	85.1	96.7	100.0	85.1	96.7	100.0	85.1	96.7	100.0	85.1	96.7
21, 22, 23	FIRE SUPPRESSION, PLUMBING & HVAC	99.7	129.4	111.7	100.1	68.7	87.4	97.9	68.7	86.1	97.9	68.4	85.9	99.9	68.7	87.3	97.8	68.7	86.1
26, 27, 3370	ELECTRICAL, COMMUNICATIONS & UTIL.	94.5	143.6	119.4	88.9	85.8	87.3	90.8	85.8	88.3	88.5	85.8	87.1	86.9	85.8	86.4	86.2	85.8	86.0
MF2016	WEIGHTED AVERAGE	95.8	132.3	111.4	97.1	75.0	87.6	100.2	75.0	89.4	98.0	74.9	88.1	97.7	75.0	88.0	97.3	75.0	87.7

NEW MEXICO

DIVISION		LAS CRUCES 880 MAT.	INST.	TOTAL	LAS VEGAS 877 MAT.	INST.	TOTAL	ROSWELL 882 MAT.	INST.	TOTAL	SANTA FE 875 MAT.	INST.	TOTAL	SOCORRO 878 MAT.	INST.	TOTAL	TRUTH/CONSEQUENCES 879 MAT.	INST.	TOTAL
015433	CONTRACTOR EQUIPMENT		86.2	86.2		110.9	110.9		110.9	110.9		110.9	110.9		110.9	110.9		86.2	86.2
0241, 31 - 34	SITE & INFRASTRUCTURE, DEMOLITION	98.0	81.1	86.3	96.8	102.3	100.6	99.5	102.3	101.5	100.8	102.3	101.8	93.1	102.3	100.5	112.8	81.2	90.0
0310	Concrete Forming & Accessories	93.2	65.8	69.9	99.3	66.9	71.7	96.5	66.9	71.3	97.8	66.9	71.5	99.3	66.9	71.7	97.2	65.8	70.5
0320	Concrete Reinforcing	110.5	71.7	90.8	115.9	71.9	93.5	114.7	71.9	92.9	108.3	71.9	89.8	118.2	71.9	94.6	111.1	71.7	91.1
0330	Cast-in-Place Concrete	89.5	61.9	79.2	89.7	69.1	82.1	94.7	69.1	85.2	95.3	69.1	85.5	87.9	69.1	80.9	96.4	61.9	83.5
03	CONCRETE	79.5	66.4	73.7	88.4	69.8	80.2	100.9	69.8	87.1	87.9	69.8	79.9	87.4	69.8	79.6	82.6	66.4	75.4
04	MASONRY	102.3	58.4	75.2	101.2	58.7	75.0	118.1	58.7	81.4	94.3	58.7	72.3	101.1	58.7	74.9	98.2	58.4	73.6
05	METALS	100.1	82.8	94.9	99.2	90.6	96.6	102.5	90.6	98.9	96.3	90.6	94.6	99.5	90.6	96.8	99.0	82.8	94.1
06	WOOD, PLASTICS & COMPOSITES	83.9	67.8	75.1	104.1	68.8	84.9	94.6	68.8	80.6	99.1	68.8	82.7	104.1	68.8	84.9	95.4	67.8	80.4
07	THERMAL & MOISTURE PROTECTION	88.5	65.3	78.4	101.7	70.0	87.9	101.7	70.0	87.9	104.5	70.0	89.5	101.6	70.0	87.8	89.5	65.3	78.9
08	OPENINGS	94.1	67.1	87.9	99.8	67.7	92.3	98.6	67.7	91.5	100.2	67.7	92.6	99.6	67.7	92.2	92.9	67.1	86.9
0920	Plaster & Gypsum Board	79.0	67.7	71.6	98.8	67.7	78.3	80.9	67.7	72.2	111.2	67.7	82.6	98.8	67.7	78.3	99.7	67.7	78.7
0950, 0980	Ceilings & Acoustic Treatment	87.0	67.7	73.9	102.6	67.7	79.0	103.8	67.7	79.4	101.4	67.7	78.6	102.6	67.7	79.0	86.9	67.7	73.9
0960	Flooring	127.8	68.0	110.4	89.2	68.0	83.1	97.5	68.0	88.9	99.4	68.0	90.3	89.2	68.0	83.1	117.2	68.0	103.0
0970, 0990	Wall Finishes & Painting/Coating	82.6	51.9	64.4	90.8	51.9	67.7	93.4	51.9	68.8	98.2	51.9	70.8	90.8	51.9	67.7	83.3	51.9	64.7
09	FINISHES	100.5	64.7	81.0	88.6	65.5	76.0	91.8	65.5	77.4	96.1	65.5	79.4	88.6	65.5	76.0	98.7	64.7	80.2
COVERS	DIVS. 10 - 14, 25, 28, 41, 43, 44, 46	100.0	82.6	96.2	100.0	85.1	96.7	100.0	85.1	96.7	100.0	85.1	96.7	100.0	85.1	96.7	100.0	82.6	96.2
21, 22, 23	FIRE SUPPRESSION, PLUMBING & HVAC	100.3	68.5	87.4	97.8	68.7	86.1	99.9	68.7	87.3	100.0	68.7	87.4	97.8	68.7	86.1	97.7	68.5	85.9
26, 27, 3370	ELECTRICAL, COMMUNICATIONS & UTIL.	90.5	85.8	88.1	88.4	85.8	87.1	89.8	85.8	87.8	101.6	85.8	93.6	86.7	85.8	86.2	90.5	85.8	88.1
MF2016	WEIGHTED AVERAGE	95.6	71.7	85.4	95.8	75.0	86.9	99.5	75.0	89.0	97.6	75.0	87.9	95.5	75.0	86.7	95.2	71.7	85.2

For customer support on your Electrical Costs with RSMeans data, call 800.448.8182.

535

City Cost Indexes

NEW MEXICO / NEW YORK

DIVISION		TUCUMCARI 884 MAT.	INST.	TOTAL	ALBANY 120-122 MAT.	INST.	TOTAL	BINGHAMTON 137-139 MAT.	INST.	TOTAL	BRONX 104 MAT.	INST.	TOTAL	BROOKLYN 112 MAT.	INST.	TOTAL	BUFFALO 140-142 MAT.	INST.	TOTAL
015433	CONTRACTOR EQUIPMENT		110.9	110.9		115.2	115.2		120.3	120.3		106.8	106.8		111.8	111.8		99.5	99.5
0241, 31 - 34	SITE & INFRASTRUCTURE, DEMOLITION	96.9	102.3	100.6	82.2	104.6	97.8	95.4	93.4	94.0	96.2	114.2	108.8	119.7	125.6	123.8	105.2	101.8	102.8
0310	Concrete Forming & Accessories	96.5	66.8	71.2	98.5	107.8	106.4	100.0	94.0	94.9	99.4	193.1	179.2	106.0	193.1	180.1	101.1	118.9	116.3
0320	Concrete Reinforcing	112.5	71.9	91.8	106.2	116.1	111.3	102.8	109.4	106.1	99.1	179.6	140.2	104.1	246.0	176.2	109.1	115.0	112.1
0330	Cast-in-Place Concrete	94.7	69.1	85.2	80.0	116.1	93.4	104.1	108.3	105.7	77.3	175.4	113.8	106.1	173.8	131.3	115.0	124.3	118.5
03	CONCRETE	99.4	69.8	86.3	86.2	113.1	98.1	93.6	103.8	98.1	83.4	183.7	127.8	105.5	193.3	144.4	110.6	119.5	114.6
04	MASONRY	117.9	58.7	81.3	87.4	116.9	105.6	106.7	106.2	106.4	87.5	189.3	150.4	117.1	189.2	161.7	123.5	123.6	123.5
05	METALS	101.3	90.5	98.0	92.3	125.9	102.5	93.6	134.4	105.9	84.7	173.7	111.6	101.5	174.0	123.4	100.6	107.1	102.6
06	WOOD, PLASTICS & COMPOSITES	94.6	68.8	80.6	95.6	104.7	100.5	103.7	90.7	96.6	103.5	194.1	152.7	105.9	193.8	153.6	99.6	118.2	109.7
07	THERMAL & MOISTURE PROTECTION	101.4	70.0	87.8	107.0	111.5	108.9	108.9	96.2	103.4	108.8	171.4	136.0	109.1	170.6	135.9	106.0	113.6	109.3
08	OPENINGS	98.6	67.7	91.4	97.2	105.0	99.0	92.4	95.3	93.1	94.0	202.1	119.1	89.8	201.9	115.8	99.1	112.0	102.1
0920	Plaster & Gypsum Board	80.9	67.7	72.2	103.7	104.7	104.4	105.9	90.2	95.5	95.9	196.6	162.2	102.3	196.6	164.4	103.6	118.6	113.5
0950, 0980	Ceilings & Acoustic Treatment	103.8	67.7	79.4	99.5	104.7	103.1	96.5	90.2	92.2	94.3	196.6	163.6	92.3	196.6	162.9	93.3	118.6	110.4
0960	Flooring	97.5	68.0	88.9	93.5	111.3	98.7	101.8	103.9	102.4	96.1	188.7	123.0	108.8	188.7	132.0	98.3	120.6	104.8
0970, 0990	Wall Finishes & Painting/Coating	93.4	51.9	68.8	93.4	103.9	99.6	87.8	106.9	99.1	104.8	166.6	141.3	114.1	166.6	145.1	98.9	118.6	110.6
09	FINISHES	91.6	65.5	77.4	90.4	107.6	99.8	91.7	96.5	94.3	97.5	190.5	148.2	104.7	190.2	151.4	98.7	119.9	110.3
COVERS	DIVS. 10 - 14, 25, 28, 41, 43, 44, 46	100.0	85.1	96.7	100.0	102.5	100.6	100.0	98.0	99.6	100.0	138.0	108.4	100.0	137.2	108.2	100.0	106.4	101.4
21, 22, 23	FIRE SUPPRESSION, PLUMBING & HVAC	97.9	68.4	85.9	100.1	111.8	104.8	100.6	98.7	99.8	100.2	176.4	131.0	99.8	176.3	130.7	100.0	99.7	99.9
26, 27, 3370	ELECTRICAL, COMMUNICATIONS & UTIL.	90.8	85.8	88.3	99.6	107.6	103.6	100.3	100.9	100.6	91.4	186.7	139.9	100.2	186.7	144.2	103.7	102.8	103.3
MF2016	WEIGHTED AVERAGE	98.7	74.9	88.5	95.0	111.5	102.1	97.3	102.8	99.7	93.3	176.8	129.1	101.8	178.9	134.9	103.0	110.0	106.0

NEW YORK

DIVISION		ELMIRA 148-149 MAT.	INST.	TOTAL	FAR ROCKAWAY 116 MAT.	INST.	TOTAL	FLUSHING 113 MAT.	INST.	TOTAL	GLENS FALLS 128 MAT.	INST.	TOTAL	HICKSVILLE 115,117,118 MAT.	INST.	TOTAL	JAMAICA 114 MAT.	INST.	TOTAL
015433	CONTRACTOR EQUIPMENT		123.0	123.0		111.8	111.8		111.8	111.8		115.2	115.2		111.8	111.8		111.8	111.8
0241, 31 - 34	SITE & INFRASTRUCTURE, DEMOLITION	104.3	93.8	97.0	123.0	125.6	124.8	123.1	125.6	124.8	73.0	104.2	94.8	112.7	124.4	120.9	117.3	125.6	123.1
0310	Concrete Forming & Accessories	85.0	97.4	95.5	92.4	193.1	178.1	96.3	193.1	178.7	83.0	100.7	98.0	88.9	161.5	150.7	96.3	193.1	178.7
0320	Concrete Reinforcing	108.5	108.2	108.4	104.1	246.0	176.2	105.9	246.0	177.1	101.8	116.0	109.0	104.1	179.4	142.4	104.1	246.0	176.2
0330	Cast-in-Place Concrete	103.2	107.8	104.9	114.8	173.8	136.7	114.8	173.8	136.7	77.1	114.0	90.9	97.4	166.2	123.0	106.1	173.8	131.3
03	CONCRETE	93.0	105.2	98.4	111.8	193.3	147.9	112.2	193.3	148.1	79.9	109.1	92.9	97.3	166.2	127.9	104.9	193.3	144.1
04	MASONRY	110.3	106.8	108.2	121.8	189.2	163.4	115.8	189.2	161.2	93.3	113.6	105.8	111.6	176.9	152.0	119.9	189.2	162.7
05	METALS	96.8	136.7	108.9	101.5	174.0	123.5	101.5	174.0	123.5	86.4	125.3	98.2	103.0	173.6	124.4	101.5	174.0	123.5
06	WOOD, PLASTICS & COMPOSITES	84.8	95.6	90.7	88.8	193.8	145.8	93.6	193.8	148.0	85.0	97.0	91.5	85.3	158.4	125.0	93.6	193.8	148.0
07	THERMAL & MOISTURE PROTECTION	105.1	96.4	101.3	109.0	170.6	135.8	109.0	170.6	135.9	99.9	109.2	103.9	108.6	161.9	131.8	108.8	170.6	135.7
08	OPENINGS	98.6	97.7	98.4	88.5	201.9	114.8	88.5	201.9	114.8	92.0	100.9	94.0	88.9	183.2	110.8	88.5	201.9	114.8
0920	Plaster & Gypsum Board	98.0	95.4	96.3	91.1	196.6	160.6	93.6	196.6	161.4	94.8	96.8	96.1	90.8	160.2	136.5	93.6	196.6	161.4
0950, 0980	Ceilings & Acoustic Treatment	99.9	95.4	96.9	81.3	196.6	159.4	81.3	196.6	159.4	86.5	96.8	93.5	80.3	160.2	134.5	81.3	196.6	159.4
0960	Flooring	88.8	103.9	93.2	103.7	188.7	128.3	105.2	188.7	129.4	84.9	111.3	92.6	102.6	185.7	126.7	105.2	188.7	129.4
0970, 0990	Wall Finishes & Painting/Coating	98.0	94.0	95.6	114.1	166.6	145.1	114.1	166.6	145.1	86.9	102.0	95.9	114.1	166.6	145.1	114.1	166.6	145.1
09	FINISHES	91.7	98.0	95.1	99.9	190.2	149.2	100.7	190.2	149.6	83.0	102.0	93.4	98.5	165.8	135.2	100.3	190.2	149.4
COVERS	DIVS. 10 - 14, 25, 28, 41, 43, 44, 46	100.0	99.1	99.8	100.0	137.2	108.2	100.0	137.2	108.2	100.0	98.1	99.6	100.0	134.8	107.7	100.0	137.2	108.2
21, 22, 23	FIRE SUPPRESSION, PLUMBING & HVAC	96.7	97.2	96.9	96.3	176.3	128.6	96.3	176.3	128.6	96.7	108.9	101.6	99.8	163.2	125.4	96.3	176.3	128.6
26, 27, 3370	ELECTRICAL, COMMUNICATIONS & UTIL.	99.7	104.3	102.0	107.2	186.7	147.6	107.2	186.7	147.6	94.1	107.6	100.9	99.6	144.4	122.4	98.6	186.7	143.4
MF2016	WEIGHTED AVERAGE	97.6	103.9	100.3	102.2	178.9	135.1	102.0	178.9	135.0	90.6	108.7	98.4	99.8	160.5	125.8	100.2	178.9	134.0

NEW YORK

DIVISION		JAMESTOWN 147 MAT.	INST.	TOTAL	KINGSTON 124 MAT.	INST.	TOTAL	LONG ISLAND CITY 111 MAT.	INST.	TOTAL	MONTICELLO 127 MAT.	INST.	TOTAL	MOUNT VERNON 105 MAT.	INST.	TOTAL	NEW ROCHELLE 108 MAT.	INST.	TOTAL
015433	CONTRACTOR EQUIPMENT		92.7	92.7		111.8	111.8		111.8	111.8		111.8	111.8		106.8	106.8		106.8	106.8
0241, 31 - 34	SITE & INFRASTRUCTURE, DEMOLITION	105.8	94.2	99.7	143.9	121.2	128.1	120.9	125.6	124.2	139.1	121.1	126.6	101.1	110.1	107.3	100.8	110.0	107.2
0310	Concrete Forming & Accessories	85.0	92.2	91.1	84.3	133.6	126.2	100.7	193.1	179.4	91.3	133.6	127.3	89.4	141.8	134.0	104.8	138.2	133.2
0320	Concrete Reinforcing	108.7	114.4	111.6	102.3	164.9	134.1	104.1	246.0	176.2	101.4	164.9	133.7	98.1	182.1	140.8	98.3	181.9	140.8
0330	Cast-in-Place Concrete	107.0	106.6	106.8	104.1	147.5	120.2	109.5	173.8	133.4	97.5	147.5	116.1	86.3	151.4	110.5	86.3	151.1	110.4
03	CONCRETE	96.1	101.0	98.3	100.7	143.2	119.5	108.0	193.3	145.8	95.8	143.2	116.8	91.9	152.2	118.6	91.3	150.5	117.5
04	MASONRY	120.0	104.7	110.6	107.8	154.7	136.8	113.9	189.2	160.5	100.1	154.7	133.8	92.8	156.2	132.0	92.8	156.2	132.0
05	METALS	94.2	102.5	96.7	93.8	136.5	106.7	101.5	174.0	123.4	93.7	136.7	106.7	84.5	165.6	109.0	84.7	164.9	109.0
06	WOOD, PLASTICS & COMPOSITES	83.4	88.7	86.3	86.2	127.6	108.7	99.9	193.8	150.9	93.1	127.6	111.8	92.4	136.3	116.2	111.5	132.4	122.9
07	THERMAL & MOISTURE PROTECTION	104.5	96.7	101.1	123.5	146.8	133.6	109.0	170.6	135.8	123.1	146.8	133.4	109.8	148.9	126.8	109.9	146.3	125.7
08	OPENINGS	98.4	95.7	97.8	93.7	146.1	105.8	88.5	201.9	114.8	89.5	146.0	102.6	94.0	160.4	111.1	94.1	167.2	111.1
0920	Plaster & Gypsum Board	85.7	88.3	87.4	96.9	128.5	117.7	98.5	196.6	163.1	97.5	128.5	117.9	90.9	137.1	121.3	104.3	133.1	123.3
0950, 0980	Ceilings & Acoustic Treatment	96.5	88.3	90.9	77.1	128.5	112.0	81.3	196.6	159.4	77.1	128.5	111.9	92.6	137.1	122.7	92.6	133.1	120.0
0960	Flooring	91.4	103.9	95.0	102.2	166.0	120.7	106.8	188.7	130.6	104.5	166.0	122.4	87.9	185.7	116.3	95.3	169.5	116.9
0970, 0990	Wall Finishes & Painting/Coating	99.6	101.0	100.4	113.6	130.5	123.6	114.1	166.6	145.1	113.6	126.6	121.3	103.2	166.6	140.7	103.2	166.6	140.7
09	FINISHES	90.2	94.3	92.5	96.5	138.5	119.4	101.6	190.2	150.0	96.9	138.0	119.3	94.5	151.4	125.5	98.4	145.8	124.3
COVERS	DIVS. 10 - 14, 25, 28, 41, 43, 44, 46	100.0	98.5	99.7	100.0	119.7	104.3	100.0	137.2	108.2	100.0	120.6	104.5	100.0	131.4	106.9	100.0	114.8	103.3
21, 22, 23	FIRE SUPPRESSION, PLUMBING & HVAC	96.6	95.3	96.0	96.6	136.7	112.8	99.8	176.3	130.7	96.6	141.0	114.6	96.8	144.9	116.2	96.8	144.6	116.1
26, 27, 3370	ELECTRICAL, COMMUNICATIONS & UTIL.	98.6	97.2	97.9	95.5	119.2	107.5	99.1	186.7	143.7	95.5	119.2	107.5	89.6	170.1	130.5	89.6	145.2	117.8
MF2016	WEIGHTED AVERAGE	97.8	97.8	97.8	98.9	136.0	114.8	101.5	178.9	134.7	97.4	136.9	114.3	93.5	151.3	118.3	93.9	146.0	116.2

NEW YORK

DIVISION		NEW YORK 100-102 MAT.	INST.	TOTAL	NIAGARA FALLS 143 MAT.	INST.	TOTAL	PLATTSBURGH 129 MAT.	INST.	TOTAL	POUGHKEEPSIE 125-126 MAT.	INST.	TOTAL	QUEENS 110 MAT.	INST.	TOTAL	RIVERHEAD 119 MAT.	INST.	TOTAL
015433	CONTRACTOR EQUIPMENT		104.8	104.8		92.7	92.7		96.4	96.4		111.8	111.8		111.8	111.8		111.8	111.8
0241, 31 - 34	SITE & INFRASTRUCTURE, DEMOLITION	105.1	111.5	109.6	108.2	95.2	99.1	112.4	101.3	104.7	140.1	120.3	126.3	116.0	125.6	122.7	113.8	123.9	120.8
0310	Concrete Forming & Accessories	108.9	192.3	179.9	85.0	116.3	111.6	88.4	93.9	93.1	84.3	172.3	159.2	89.1	193.1	177.7	93.5	160.5	150.5
0320	Concrete Reinforcing	105.2	183.4	144.9	107.3	115.2	111.3	106.3	115.3	110.8	102.3	164.7	134.0	105.9	246.0	177.1	106.1	221.8	164.9
0330	Cast-in-Place Concrete	89.3	174.6	121.0	110.7	127.7	117.0	94.3	103.5	97.7	100.8	139.3	115.2	100.8	173.8	128.0	99.1	164.4	123.4
03	CONCRETE	96.2	183.5	134.9	98.5	119.3	107.7	92.8	100.6	96.3	98.0	157.8	124.5	100.6	193.3	141.7	97.9	171.0	130.3
04	MASONRY	101.2	187.1	154.2	128.3	127.8	128.0	87.7	99.4	94.9	100.1	142.5	126.3	108.1	189.2	158.2	117.3	175.8	153.5
05	METALS	95.6	173.1	119.1	96.9	103.1	98.7	90.2	100.0	93.2	93.8	136.3	106.7	101.5	174.0	123.4	103.4	161.6	121.1
06	WOOD, PLASTICS & COMPOSITES	109.2	194.2	155.4	83.4	111.4	98.6	91.7	90.7	91.2	86.2	184.5	139.6	85.4	193.8	144.3	90.5	158.4	127.4
07	THERMAL & MOISTURE PROTECTION	111.6	170.7	137.3	104.6	113.1	108.3	117.8	99.9	110.0	123.4	147.6	133.9	108.6	170.6	135.6	109.6	158.8	131.0
08	OPENINGS	97.1	203.0	121.7	98.5	107.7	100.6	99.4	97.4	98.9	93.7	175.5	112.7	88.5	201.9	114.8	88.9	176.6	109.3
0920	Plaster & Gypsum Board	101.2	196.6	164.0	85.7	111.6	102.8	114.7	90.0	98.4	96.9	187.0	156.2	90.8	196.6	160.5	91.9	160.2	136.9
0950, 0980	Ceilings & Acoustic Treatment	109.6	196.6	168.5	96.5	111.6	106.7	106.5	90.0	95.3	77.1	187.0	151.6	81.3	196.6	159.4	81.2	160.2	134.7
0960	Flooring	97.3	188.7	123.8	91.4	115.8	98.5	107.4	108.6	107.8	102.2	164.0	120.1	102.6	188.7	127.6	103.7	172.5	123.6
0970, 0990	Wall Finishes & Painting/Coating	101.4	166.6	140.0	99.6	114.4	108.4	108.3	97.6	102.0	113.6	126.6	121.3	114.1	166.6	145.1	114.1	166.6	145.1
09	FINISHES	102.3	190.0	150.2	90.4	116.2	104.5	94.9	96.3	95.7	96.3	169.0	136.0	98.9	190.2	148.7	99.1	162.7	133.8
COVERS	DIVS. 10 - 14, 25, 28, 41, 43, 44, 46	100.0	143.1	109.5	100.0	104.8	101.1	100.0	94.6	98.8	100.0	118.8	104.1	100.0	137.2	108.2	100.0	122.6	105.0
21, 22, 23	FIRE SUPPRESSION, PLUMBING & HVAC	100.1	175.3	130.5	96.6	107.5	101.0	96.6	102.6	99.0	96.6	123.0	107.3	99.8	176.3	130.7	99.9	157.9	123.4
26, 27, 3370	ELECTRICAL, COMMUNICATIONS & UTIL.	97.7	186.7	142.9	97.2	101.3	99.3	92.1	94.0	93.1	95.5	124.2	110.1	99.6	186.7	143.9	101.2	138.2	120.0
MF2016	WEIGHTED AVERAGE	99.0	176.2	132.1	98.8	110.2	103.7	95.6	99.0	97.0	98.1	140.2	116.1	99.8	178.9	133.8	100.5	156.8	124.6

NEW YORK

DIVISION		ROCHESTER 144-146 MAT.	INST.	TOTAL	SCHENECTADY 123 MAT.	INST.	TOTAL	STATEN ISLAND 103 MAT.	INST.	TOTAL	SUFFERN 109 MAT.	INST.	TOTAL	SYRACUSE 130-132 MAT.	INST.	TOTAL	UTICA 133-135 MAT.	INST.	TOTAL
015433	CONTRACTOR EQUIPMENT		119.9	119.9		115.2	115.2		106.8	106.8		106.8	106.8		115.2	115.2		115.2	115.2
0241, 31 - 34	SITE & INFRASTRUCTURE, DEMOLITION	92.2	108.7	103.7	83.0	104.6	98.0	105.1	114.2	111.5	98.0	108.8	105.5	94.0	103.2	100.4	72.5	102.7	93.5
0310	Concrete Forming & Accessories	107.6	101.9	102.8	98.9	107.7	106.4	88.8	193.3	177.8	98.1	144.7	137.8	99.1	91.6	92.7	100.1	88.7	90.4
0320	Concrete Reinforcing	110.9	108.0	109.4	100.8	116.1	108.6	99.4	222.2	161.8	98.3	156.4	127.8	103.8	109.2	106.5	103.8	103.2	103.5
0330	Cast-in-Place Concrete	101.1	105.3	102.7	93.1	116.0	101.7	86.3	175.6	119.5	83.7	143.0	105.8	96.6	105.3	99.9	88.4	103.9	94.2
03	CONCRETE	97.9	105.4	101.2	93.5	113.0	102.1	93.7	190.3	136.5	88.8	145.1	113.8	96.8	100.8	98.6	95.0	98.0	96.3
04	MASONRY	101.4	105.6	104.0	90.2	116.9	106.7	99.5	189.3	155.0	92.7	145.2	125.1	99.4	103.2	101.7	90.5	101.5	97.3
05	METALS	103.4	121.5	108.9	90.3	125.8	101.1	82.9	173.9	110.5	82.9	132.9	98.0	97.0	120.6	104.1	95.1	118.3	102.1
06	WOOD, PLASTICS & COMPOSITES	108.4	101.4	104.6	103.5	104.7	104.1	90.9	194.1	147.0	103.0	146.6	126.7	100.3	88.5	93.9	100.3	85.1	92.0
07	THERMAL & MOISTURE PROTECTION	105.3	101.8	103.8	101.5	111.5	105.8	109.2	171.4	136.3	109.7	144.3	124.8	103.6	96.8	100.6	92.1	96.7	94.1
08	OPENINGS	102.1	101.3	101.9	97.4	105.0	99.1	94.0	202.1	119.1	94.1	154.5	108.1	94.5	92.1	94.0	97.4	88.8	95.4
0920	Plaster & Gypsum Board	103.7	101.6	102.3	104.1	104.7	104.5	91.0	196.6	160.5	95.2	147.7	129.8	95.9	88.1	90.8	95.9	84.6	88.5
0950, 0980	Ceilings & Acoustic Treatment	102.4	101.6	101.8	93.2	104.7	101.0	94.3	196.6	163.6	92.6	147.7	129.9	96.5	88.1	90.8	96.5	84.6	88.4
0960	Flooring	90.0	108.5	95.4	92.1	111.3	97.7	91.5	188.7	119.7	91.4	177.3	116.3	90.7	95.4	92.0	88.5	95.5	90.6
0970, 0990	Wall Finishes & Painting/Coating	97.1	101.0	99.4	86.9	103.9	97.0	104.8	166.6	141.3	103.2	130.3	119.2	90.4	101.3	96.9	84.2	101.3	94.3
09	FINISHES	94.1	103.4	99.2	88.4	107.6	98.8	96.2	190.5	147.7	95.8	150.1	125.4	90.5	92.6	91.6	88.9	90.4	89.7
COVERS	DIVS. 10 - 14, 25, 28, 41, 43, 44, 46	100.0	100.4	100.1	100.0	102.5	100.6	100.0	138.0	108.4	100.0	117.0	103.8	100.0	97.0	99.3	100.0	91.6	98.1
21, 22, 23	FIRE SUPPRESSION, PLUMBING & HVAC	100.0	90.1	96.0	100.1	107.7	103.2	100.2	176.4	131.0	96.8	126.5	108.8	100.2	96.6	98.8	100.2	94.6	98.0
26, 27, 3370	ELECTRICAL, COMMUNICATIONS & UTIL.	102.0	92.1	97.0	98.6	107.6	103.2	91.4	186.7	139.9	95.8	117.4	106.8	100.3	103.5	101.9	98.2	103.5	100.9
MF2016	WEIGHTED AVERAGE	100.3	101.3	100.8	95.5	110.6	101.9	94.9	177.8	130.5	93.6	133.5	110.7	97.8	100.8	99.1	95.9	98.9	97.2

NEW YORK / NORTH CAROLINA

DIVISION		WATERTOWN 136 MAT.	INST.	TOTAL	WHITE PLAINS 106 MAT.	INST.	TOTAL	YONKERS 107 MAT.	INST.	TOTAL	ASHEVILLE 287-288 MAT.	INST.	TOTAL	CHARLOTTE 281-282 MAT.	INST.	TOTAL	DURHAM 277 MAT.	INST.	TOTAL
015433	CONTRACTOR EQUIPMENT		115.2	115.2		106.8	106.8		106.8	106.8		100.9	100.9		100.9	100.9		106.2	106.2
0241, 31 - 34	SITE & INFRASTRUCTURE, DEMOLITION	80.1	103.2	96.2	96.0	110.1	105.8	102.7	110.1	107.8	95.8	81.1	85.5	98.0	81.3	86.4	101.0	90.1	93.4
0310	Concrete Forming & Accessories	85.2	95.6	94.1	103.3	152.6	145.3	103.5	149.2	142.4	88.6	62.1	66.1	95.5	61.7	66.7	94.8	62.1	67.0
0320	Concrete Reinforcing	104.5	109.2	106.9	98.3	182.1	140.9	102.2	182.0	142.8	92.3	67.8	79.9	96.5	68.6	82.3	95.8	68.6	82.0
0330	Cast-in-Place Concrete	103.0	107.7	104.7	76.7	151.4	104.5	85.8	151.4	110.2	104.8	69.7	91.7	106.8	69.0	92.7	104.2	69.7	91.3
03	CONCRETE	107.8	103.4	105.9	83.1	157.1	115.9	91.4	155.6	119.9	91.6	67.4	80.9	91.5	67.1	80.7	92.5	67.5	81.4
04	MASONRY	91.7	107.0	101.2	92.2	156.2	131.7	96.3	156.2	133.3	86.5	62.4	71.6	90.2	61.0	72.2	85.6	62.4	71.3
05	METALS	95.2	120.6	102.9	84.4	165.6	109.0	92.3	165.5	114.4	99.0	90.1	96.3	99.7	90.5	96.9	115.1	90.4	107.6
06	WOOD, PLASTICS & COMPOSITES	81.7	92.5	87.6	109.1	151.0	131.9	109.0	146.6	129.4	87.4	60.7	72.9	92.3	60.7	75.1	90.3	60.7	74.2
07	THERMAL & MOISTURE PROTECTION	92.4	99.9	95.7	109.5	150.4	127.3	109.9	150.0	127.4	100.3	63.9	84.5	94.5	63.3	80.9	105.4	63.9	87.3
08	OPENINGS	97.4	96.5	97.2	94.1	177.5	113.5	97.8	173.0	115.2	90.9	61.6	84.1	97.8	61.7	89.4	97.4	61.7	89.1
0920	Plaster & Gypsum Board	86.7	92.2	90.4	98.2	152.2	133.7	101.4	147.7	131.9	107.3	59.5	75.8	104.4	59.5	74.8	94.1	59.5	71.3
0950, 0980	Ceilings & Acoustic Treatment	96.5	92.2	93.6	92.6	152.2	132.9	108.7	147.7	135.1	81.6	59.5	66.6	87.3	59.5	68.5	84.3	59.5	67.5
0960	Flooring	81.8	95.5	85.8	93.7	185.7	120.4	93.2	188.7	120.9	94.7	62.1	85.2	95.2	62.1	85.6	98.3	62.1	87.8
0970, 0990	Wall Finishes & Painting/Coating	84.2	95.4	90.9	103.2	166.6	140.7	103.2	166.6	140.7	103.7	58.6	77.0	97.4	58.6	74.4	98.9	58.6	75.1
09	FINISHES	86.2	95.1	91.1	96.6	160.0	131.2	100.5	158.0	131.9	88.6	61.5	73.8	88.7	61.1	73.6	87.0	61.5	73.1
COVERS	DIVS. 10 - 14, 25, 28, 41, 43, 44, 46	100.0	98.4	99.6	100.0	129.6	106.5	100.0	126.8	105.9	100.0	83.4	96.3	100.0	83.0	96.3	100.0	83.4	96.3
21, 22, 23	FIRE SUPPRESSION, PLUMBING & HVAC	100.2	89.8	96.0	100.3	144.9	118.3	100.3	144.8	118.3	100.5	61.5	84.7	99.9	61.9	84.6	100.6	61.5	84.8
26, 27, 3370	ELECTRICAL, COMMUNICATIONS & UTIL.	100.2	92.4	96.2	89.6	170.1	130.5	95.8	159.8	128.4	101.0	59.3	79.8	100.1	61.8	80.6	95.2	58.8	76.7
MF2016	WEIGHTED AVERAGE	97.7	99.2	98.3	93.3	153.5	119.1	97.3	151.2	120.5	96.3	67.0	83.8	97.0	67.3	84.3	99.3	67.7	85.7

For customer support on your Electrical Costs with RSMeans data, call 800.448.8182.

537

NORTH CAROLINA

DIVISION		ELIZABETH CITY 279			FAYETTEVILLE 283			GASTONIA 280			GREENSBORO 270, 272 - 274			HICKORY 286			KINSTON 285		
		MAT.	INST.	TOTAL	MAT.	INST.	TOTAL	MAT.	INST.	TOTAL	MAT.	INST.	TOTAL	MAT.	INST.	TOTAL	MAT.	INST.	TOTAL
015433	CONTRACTOR EQUIPMENT		110.3	110.3		106.2	106.2		100.9	100.9		106.2	106.2		106.2	106.2		106.2	106.2
0241, 31 - 34	SITE & INFRASTRUCTURE, DEMOLITION	105.5	91.8	96.0	95.1	90.0	91.5	95.6	81.4	85.7	100.9	90.1	93.4	94.7	88.8	90.6	93.6	88.7	90.2
0310	Concrete Forming & Accessories	81.0	62.5	65.2	88.2	61.6	65.6	94.6	62.4	67.2	94.6	62.1	66.9	85.2	62.1	65.6	81.8	61.6	64.6
0320	Concrete Reinforcing	93.9	71.7	82.6	95.9	68.6	82.0	92.7	68.6	80.5	94.7	68.6	81.4	92.3	68.6	80.3	91.9	68.6	80.0
0330	Cast-in-Place Concrete	104.4	71.4	92.1	109.9	68.9	94.6	102.4	69.8	90.3	103.4	69.7	90.8	104.7	69.7	91.7	101.1	68.9	89.1
03	CONCRETE	92.6	68.8	82.1	93.5	67.0	81.7	90.3	67.7	80.3	92.0	67.5	81.1	91.4	67.5	80.8	88.5	67.0	79.0
04	MASONRY	96.8	61.0	74.7	89.6	61.0	72.0	90.5	62.4	73.1	82.9	62.4	70.2	75.9	62.4	67.6	82.6	61.0	69.3
05	METALS	101.0	92.6	98.5	119.6	90.4	110.8	99.5	90.6	96.8	107.4	90.3	102.3	99.0	90.4	96.4	97.7	90.3	95.5
06	WOOD, PLASTICS & COMPOSITES	75.0	61.4	67.6	86.6	60.7	72.5	95.3	60.7	76.5	90.1	60.7	74.1	82.3	60.7	70.6	79.3	60.7	69.2
07	THERMAL & MOISTURE PROTECTION	104.6	63.2	86.5	99.8	63.3	83.9	100.5	63.9	84.6	105.2	63.9	87.2	100.7	63.9	84.7	100.5	63.3	84.3
08	OPENINGS	94.7	62.7	87.3	91.0	61.7	84.2	94.3	61.7	86.7	97.4	61.7	89.1	90.9	61.7	84.2	91.0	61.7	84.2
0920	Plaster & Gypsum Board	88.8	59.5	69.5	111.2	59.5	77.2	113.2	59.5	77.9	95.5	59.5	71.8	107.3	59.5	75.8	107.1	59.5	75.8
0950, 0980	Ceilings & Acoustic Treatment	84.3	59.5	67.5	83.3	59.5	67.2	85.0	59.5	67.7	84.3	59.5	67.5	81.6	59.5	66.6	85.0	59.5	67.7
0960	Flooring	90.4	79.8	87.3	94.8	62.1	85.3	98.0	79.8	92.7	98.3	62.1	87.8	94.6	79.8	90.3	91.7	79.8	88.3
0970, 0990	Wall Finishes & Painting/Coating	98.9	58.6	75.1	103.7	58.6	77.0	103.7	58.6	77.0	98.9	58.6	75.1	103.7	58.6	77.0	103.7	58.6	77.0
09	FINISHES	84.3	65.2	73.9	89.5	61.1	74.0	91.0	65.1	76.8	87.2	61.5	73.2	88.8	65.1	75.8	88.4	64.7	75.5
COVERS	DIVS. 10 - 14, 25, 28, 41, 43, 44, 46	100.0	81.7	96.0	100.0	83.0	96.2	100.0	83.4	96.4	100.0	80.7	95.7	100.0	83.4	96.3	100.0	82.9	96.2
21, 22, 23	FIRE SUPPRESSION, PLUMBING & HVAC	97.0	59.9	82.0	100.3	60.8	84.3	100.5	60.4	84.3	100.5	61.5	84.7	97.0	60.4	82.2	97.0	59.7	81.9
26, 27, 3370	ELECTRICAL, COMMUNICATIONS & UTIL.	94.9	66.2	80.3	100.6	58.8	79.3	100.5	61.9	80.9	94.3	59.3	76.5	98.7	61.9	80.0	98.5	57.2	77.5
MF2016	WEIGHTED AVERAGE	96.1	69.2	84.6	100.0	67.3	86.0	97.0	67.7	84.4	97.7	67.7	84.8	94.7	68.3	83.4	94.3	67.1	82.7

NORTH CAROLINA / NORTH DAKOTA

DIVISION		MURPHY 289			RALEIGH 275 - 276			ROCKY MOUNT 278			WILMINGTON 284			WINSTON-SALEM 271			BISMARCK 585		
		MAT.	INST.	TOTAL	MAT.	INST.	TOTAL	MAT.	INST.	TOTAL	MAT.	INST.	TOTAL	MAT.	INST.	TOTAL	MAT.	INST.	TOTAL
015433	CONTRACTOR EQUIPMENT		100.9	100.9		106.2	106.2		106.2	106.2		100.9	100.9		106.2	106.2		100.3	100.3
0241, 31 - 34	SITE & INFRASTRUCTURE, DEMOLITION	97.0	79.6	84.9	100.9	90.1	93.3	103.5	90.1	94.1	96.9	81.0	85.8	101.2	90.1	93.5	100.9	98.4	99.2
0310	Concrete Forming & Accessories	95.2	61.5	66.5	96.0	61.6	66.7	87.0	61.8	65.5	89.9	61.6	65.8	96.5	62.1	67.3	103.9	73.4	77.9
0320	Concrete Reinforcing	91.9	66.1	78.8	96.8	68.6	82.5	93.9	68.6	81.1	93.1	68.6	80.6	94.7	68.6	81.4	103.8	94.5	99.1
0330	Cast-in-Place Concrete	108.4	68.9	93.7	107.1	68.9	92.9	102.1	69.0	89.8	104.3	68.9	91.2	106.0	69.7	92.5	102.2	83.9	95.4
03	CONCRETE	94.4	66.5	82.0	91.2	67.0	80.5	93.2	67.1	81.6	91.5	67.0	80.7	93.3	67.5	81.9	90.6	81.4	86.5
04	MASONRY	78.9	61.0	67.9	81.6	61.0	68.9	75.3	61.0	66.5	76.5	61.0	66.9	83.2	62.4	70.3	98.4	82.4	88.5
05	METALS	96.8	89.5	94.6	99.7	90.4	96.9	100.2	90.5	97.3	98.5	90.4	96.0	104.5	90.4	100.2	95.2	92.6	94.4
06	WOOD, PLASTICS & COMPOSITES	96.0	60.6	76.7	90.4	60.7	74.3	81.8	60.7	70.4	89.2	60.7	73.7	90.1	60.7	74.1	102.1	69.5	84.4
07	THERMAL & MOISTURE PROTECTION	100.5	63.3	84.3	99.4	63.3	83.7	105.1	63.3	86.9	100.3	63.3	84.2	105.2	63.9	87.2	105.1	83.2	95.6
08	OPENINGS	90.9	61.1	84.0	98.7	61.7	90.1	93.9	61.7	86.5	91.0	61.7	84.2	97.4	61.7	89.1	106.3	79.3	100.0
0920	Plaster & Gypsum Board	112.5	59.3	77.5	91.1	59.5	70.3	90.3	59.5	70.0	109.2	59.5	76.5	95.5	59.5	71.8	100.0	69.0	79.6
0950, 0980	Ceilings & Acoustic Treatment	81.6	59.3	66.5	82.2	59.5	66.8	82.6	59.5	67.0	83.3	59.5	67.2	84.3	59.5	67.5	109.7	69.0	82.1
0960	Flooring	98.2	79.8	92.9	93.3	62.1	84.3	93.9	79.8	89.8	95.4	62.1	85.7	98.3	62.1	87.8	91.3	51.4	79.8
0970, 0990	Wall Finishes & Painting/Coating	103.7	58.6	77.0	92.1	58.6	72.3	98.9	58.6	75.1	103.7	58.6	77.0	98.9	58.6	75.1	93.7	59.0	73.2
09	FINISHES	90.6	64.6	76.4	85.6	61.1	72.3	85.3	64.7	74.1	89.4	61.1	74.0	87.2	61.5	73.2	95.6	68.3	80.7
COVERS	DIVS. 10 - 14, 25, 28, 41, 43, 44, 46	100.0	82.9	96.2	100.0	83.0	96.2	100.0	83.0	96.2	100.0	83.0	96.2	100.0	83.4	96.3	100.0	89.1	97.6
21, 22, 23	FIRE SUPPRESSION, PLUMBING & HVAC	97.0	59.6	81.9	100.0	60.8	84.1	97.0	59.7	81.9	100.5	60.8	84.5	100.5	61.5	84.7	99.9	75.5	90.1
26, 27, 3370	ELECTRICAL, COMMUNICATIONS & UTIL.	101.9	57.2	79.2	97.6	57.7	77.3	96.8	58.8	77.5	101.3	57.2	78.9	94.3	59.3	76.5	97.5	74.4	85.7
MF2016	WEIGHTED AVERAGE	95.5	66.2	82.9	96.4	67.1	83.8	95.3	67.5	83.4	95.9	66.3	83.2	97.4	67.8	84.7	98.2	80.0	90.4

NORTH DAKOTA

DIVISION		DEVILS LAKE 583			DICKINSON 586			FARGO 580 - 581			GRAND FORKS 582			JAMESTOWN 584			MINOT 587		
		MAT.	INST.	TOTAL	MAT.	INST.	TOTAL	MAT.	INST.	TOTAL	MAT.	INST.	TOTAL	MAT.	INST.	TOTAL	MAT.	INST.	TOTAL
015433	CONTRACTOR EQUIPMENT		100.3	100.3		100.3	100.3		100.3	100.3		100.3	100.3		100.3	100.3		100.3	100.3
0241, 31 - 34	SITE & INFRASTRUCTURE, DEMOLITION	106.3	98.5	100.8	113.8	98.4	103.1	97.6	98.4	98.2	109.8	98.4	101.9	105.3	98.5	100.5	107.4	98.4	101.1
0310	Concrete Forming & Accessories	102.4	73.5	77.8	91.5	73.4	76.1	97.7	73.5	77.1	95.6	73.4	76.7	93.2	73.4	76.4	91.1	73.4	76.0
0320	Concrete Reinforcing	100.2	94.4	97.3	101.1	94.5	97.8	94.6	94.8	94.7	98.7	94.4	96.5	100.7	94.8	97.7	102.1	94.5	98.2
0330	Cast-in-Place Concrete	122.1	84.0	107.9	110.4	83.9	100.5	99.9	83.9	93.9	110.4	83.9	100.5	120.5	84.0	106.9	110.4	83.9	100.5
03	CONCRETE	100.2	81.4	91.9	99.1	81.4	91.2	92.9	81.5	87.8	96.7	81.4	89.9	98.9	81.5	91.2	95.3	81.4	89.1
04	MASONRY	114.6	80.9	93.8	116.6	82.4	95.5	103.2	87.1	93.3	108.9	80.9	91.6	127.0	87.2	102.4	107.2	82.4	91.9
05	METALS	91.5	92.7	91.9	91.4	92.5	91.7	94.0	93.0	93.7	91.4	92.4	91.7	91.4	93.1	92.0	91.7	92.5	92.0
06	WOOD, PLASTICS & COMPOSITES	99.9	69.5	83.4	86.2	69.5	77.1	94.9	69.5	81.1	91.0	69.5	79.3	88.3	69.5	78.0	85.9	69.5	77.0
07	THERMAL & MOISTURE PROTECTION	106.3	82.8	96.1	106.9	83.2	96.6	106.1	85.0	96.9	106.5	82.8	96.2	106.1	85.0	96.9	106.2	83.2	96.2
08	OPENINGS	103.1	79.3	97.6	103.1	79.3	97.6	100.2	79.3	95.4	101.7	79.3	96.5	103.1	79.3	97.6	101.9	79.3	96.6
0920	Plaster & Gypsum Board	114.9	69.0	84.7	105.2	69.0	81.3	98.2	69.0	78.9	106.4	69.0	81.7	106.1	69.0	81.6	105.2	69.0	81.3
0950, 0980	Ceilings & Acoustic Treatment	101.8	69.0	79.6	101.8	69.0	79.6	95.2	69.0	77.4	101.8	69.0	79.6	101.8	69.0	79.6	101.8	69.0	79.6
0960	Flooring	95.3	51.4	82.6	88.8	51.4	77.9	100.0	51.4	85.9	90.6	51.4	79.3	89.5	51.4	78.4	88.5	51.4	77.8
0970, 0990	Wall Finishes & Painting/Coating	88.1	57.8	70.2	88.1	57.8	70.2	92.8	69.1	78.8	88.1	67.8	76.1	88.1	57.8	70.2	88.1	59.0	70.9
09	FINISHES	93.8	68.0	79.7	91.4	68.0	78.6	94.9	69.2	80.9	91.6	69.1	79.3	90.8	68.0	78.4	90.6	68.1	78.3
COVERS	DIVS. 10 - 14, 25, 28, 41, 43, 44, 46	100.0	89.1	97.6	100.0	89.1	97.6	100.0	89.1	97.6	100.0	89.1	97.6	100.0	89.1	97.6	100.0	89.1	97.6
21, 22, 23	FIRE SUPPRESSION, PLUMBING & HVAC	96.7	80.0	90.0	96.7	74.9	87.9	99.9	75.6	90.1	100.2	73.8	89.5	96.7	73.8	87.5	100.2	73.6	89.4
26, 27, 3370	ELECTRICAL, COMMUNICATIONS & UTIL.	95.5	69.9	82.5	103.8	69.5	86.4	102.1	71.1	86.3	98.9	69.9	84.2	95.5	69.9	82.5	101.9	74.4	87.9
MF2016	WEIGHTED AVERAGE	98.3	80.2	90.5	99.0	79.1	90.5	98.2	80.3	90.5	98.4	79.0	90.1	98.3	79.6	90.3	98.3	79.6	90.3

City Cost Indexes

NORTH DAKOTA / OHIO

DIVISION		WILLISTON 588 MAT.	INST.	TOTAL	AKRON 442-443 MAT.	INST.	TOTAL	ATHENS 457 MAT.	INST.	TOTAL	CANTON 446-447 MAT.	INST.	TOTAL	CHILLICOTHE 456 MAT.	INST.	TOTAL	CINCINNATI 451-452 MAT.	INST.	TOTAL
015433	CONTRACTOR EQUIPMENT		100.3	100.3		91.4	91.4		87.5	87.5		91.4	91.4		98.4	98.4		96.0	96.0
0241, 31 - 34	SITE & INFRASTRUCTURE, DEMOLITION	107.8	96.2	99.7	99.8	99.1	99.3	115.7	89.8	97.6	99.9	98.8	99.2	102.0	100.0	100.6	97.5	99.0	98.6
0310	Concrete Forming & Accessories	97.4	73.1	76.7	101.2	85.2	87.6	94.3	79.2	81.4	101.2	75.7	79.5	97.1	83.9	85.8	100.6	82.1	84.9
0320	Concrete Reinforcing	103.1	94.5	98.7	95.4	93.7	94.6	90.7	91.8	91.2	95.4	77.2	86.2	88.0	91.4	89.7	93.3	80.0	86.6
0330	Cast-in-Place Concrete	110.4	83.8	100.5	107.0	91.7	101.3	111.5	98.3	106.6	108.0	89.0	100.9	101.2	95.2	99.0	96.6	77.5	89.5
03	CONCRETE	96.6	81.2	89.8	100.7	88.4	95.2	102.5	87.7	95.9	101.2	80.3	91.9	97.1	89.4	93.7	96.1	80.3	89.1
04	MASONRY	102.1	82.4	89.9	94.2	91.3	92.4	72.3	98.5	88.5	94.9	80.9	86.2	78.7	92.4	87.2	81.3	81.0	81.1
05	METALS	91.6	92.2	91.8	93.6	81.1	89.8	100.1	81.4	94.4	93.6	74.4	87.8	92.3	90.8	91.8	94.5	82.4	90.8
06	WOOD, PLASTICS & COMPOSITES	92.7	69.5	80.1	108.4	83.5	94.8	87.7	73.9	80.2	108.7	73.9	89.8	100.7	80.3	89.6	100.7	82.3	90.7
07	THERMAL & MOISTURE PROTECTION	106.5	83.2	96.3	103.0	93.7	99.0	98.9	93.5	96.5	104.1	89.0	97.5	101.0	90.9	96.6	98.8	82.0	91.5
08	OPENINGS	103.2	79.3	97.6	109.5	85.1	103.8	99.1	76.8	93.9	103.1	71.3	95.8	90.3	80.3	88.0	98.4	78.1	93.7
0920	Plaster & Gypsum Board	106.4	69.0	81.7	101.8	82.9	89.3	96.9	73.0	81.2	102.7	73.0	83.2	100.3	80.1	87.0	99.9	82.2	88.2
0950, 0980	Ceilings & Acoustic Treatment	101.8	69.0	79.6	94.9	82.9	86.8	110.6	73.0	85.1	94.9	73.0	80.1	103.2	80.1	87.6	97.0	82.2	87.0
0960	Flooring	91.6	51.4	79.9	98.1	86.1	94.6	121.9	77.8	109.1	98.3	76.6	92.0	99.5	77.8	93.2	100.8	81.2	95.1
0970, 0990	Wall Finishes & Painting/Coating	88.1	57.8	70.2	95.1	93.3	94.0	103.6	91.5	96.4	95.1	70.9	80.8	100.8	91.5	95.3	100.4	73.2	84.3
09	FINISHES	91.8	68.0	78.8	99.0	86.3	92.1	101.9	79.4	89.7	99.2	74.6	85.7	99.3	82.7	90.3	98.5	81.2	89.1
COVERS	DIVS. 10 - 14, 25, 28, 41, 43, 44, 46	100.0	89.0	97.6	100.0	92.2	98.3	100.0	89.2	97.6	100.0	89.8	97.7	100.0	88.7	97.5	100.0	89.8	97.8
21, 22, 23	FIRE SUPPRESSION, PLUMBING & HVAC	96.7	74.8	87.9	100.0	89.8	95.9	96.6	82.6	91.0	100.0	79.6	91.8	97.2	94.5	96.1	100.0	77.2	90.8
26, 27, 3370	ELECTRICAL, COMMUNICATIONS & UTIL.	99.2	70.4	84.6	96.5	83.9	90.1	97.0	91.9	94.4	95.8	87.1	91.4	96.5	91.9	94.2	95.4	72.9	84.0
MF2016	WEIGHTED AVERAGE	97.4	79.0	89.6	99.5	88.3	94.7	98.2	86.4	93.2	98.9	81.4	91.4	95.4	90.8	93.4	96.9	80.7	89.9

OHIO

DIVISION		CLEVELAND 441 MAT.	INST.	TOTAL	COLUMBUS 430-432 MAT.	INST.	TOTAL	DAYTON 453-454 MAT.	INST.	TOTAL	HAMILTON 450 MAT.	INST.	TOTAL	LIMA 458 MAT.	INST.	TOTAL	LORAIN 440 MAT.	INST.	TOTAL
015433	CONTRACTOR EQUIPMENT		92.5	92.5		92.0	92.0		91.6	91.6		98.4	98.4		90.9	90.9		91.4	91.4
0241, 31 - 34	SITE & INFRASTRUCTURE, DEMOLITION	98.4	97.4	97.7	101.0	92.5	95.1	97.6	99.3	98.7	97.5	99.5	98.9	109.2	89.5	95.4	99.0	99.2	99.2
0310	Concrete Forming & Accessories	100.6	89.2	90.9	98.6	77.9	81.0	98.9	79.4	82.3	99.0	79.6	82.5	94.2	78.5	80.8	101.3	76.6	80.3
0320	Concrete Reinforcing	95.9	91.8	93.8	99.2	81.0	90.0	93.3	81.4	87.3	93.3	79.5	86.3	90.7	81.6	86.1	95.4	94.0	94.7
0330	Cast-in-Place Concrete	103.7	98.7	101.8	100.6	82.4	93.8	86.7	83.0	85.3	92.9	83.4	89.4	102.4	92.2	98.6	101.8	94.2	99.0
03	CONCRETE	100.9	92.4	97.1	98.6	80.0	90.3	88.3	80.9	85.0	91.2	81.4	86.8	95.3	83.8	90.2	98.3	85.4	92.6
04	MASONRY	100.7	97.9	99.0	84.3	86.9	85.9	77.6	78.0	77.9	78.0	82.9	81.0	97.4	79.8	86.5	90.8	95.0	93.4
05	METALS	95.2	83.7	91.7	94.5	80.8	90.3	93.8	77.6	88.9	93.8	86.5	91.6	100.1	81.0	94.3	94.2	81.8	90.5
06	WOOD, PLASTICS & COMPOSITES	102.3	86.7	93.8	97.8	76.7	86.3	104.6	79.4	90.9	103.4	79.4	90.3	87.6	77.3	82.0	108.4	72.1	88.7
07	THERMAL & MOISTURE PROTECTION	100.1	97.8	99.1	93.6	84.9	89.9	105.1	81.7	94.9	101.1	82.2	92.9	98.5	86.0	93.0	104.0	93.6	99.5
08	OPENINGS	97.7	86.3	95.0	99.3	74.4	93.5	97.5	76.8	92.7	95.2	77.0	91.0	99.1	75.9	93.7	103.1	78.9	97.5
0920	Plaster & Gypsum Board	101.4	86.1	91.3	97.3	76.1	83.4	102.0	79.2	87.0	102.0	79.2	87.0	96.9	76.5	83.5	101.8	71.2	81.7
0950, 0980	Ceilings & Acoustic Treatment	90.4	86.1	87.5	91.4	76.1	81.1	105.0	79.2	87.5	104.1	79.2	87.2	109.7	76.5	87.2	94.9	71.2	78.9
0960	Flooring	99.5	90.5	96.9	94.2	77.8	89.5	103.2	74.8	95.0	100.6	81.2	95.0	120.8	79.8	108.9	98.3	92.8	96.7
0970, 0990	Wall Finishes & Painting/Coating	97.7	93.3	95.1	95.8	81.7	87.5	100.8	72.6	84.1	100.8	73.2	84.5	103.7	78.5	88.8	95.1	93.3	94.0
09	FINISHES	99.3	89.6	94.0	94.7	77.9	85.5	100.7	77.6	88.1	99.7	79.2	88.5	101.0	78.2	88.5	99.0	80.3	88.8
COVERS	DIVS. 10 - 14, 25, 28, 41, 43, 44, 46	100.0	98.0	99.6	100.0	89.5	97.7	100.0	86.9	97.1	100.0	87.1	97.2	100.0	86.2	97.0	100.0	92.8	98.4
21, 22, 23	FIRE SUPPRESSION, PLUMBING & HVAC	100.0	92.1	96.8	100.0	85.9	94.3	100.8	83.0	93.6	100.6	76.6	90.9	96.6	93.8	95.5	100.0	90.4	96.1
26, 27, 3370	ELECTRICAL, COMMUNICATIONS & UTIL.	96.1	93.0	94.5	98.4	82.2	90.1	94.1	77.5	85.7	94.4	77.3	85.7	97.3	77.5	87.2	95.9	80.1	87.9
MF2016	WEIGHTED AVERAGE	98.7	92.2	95.9	97.4	83.2	91.3	95.9	81.3	89.6	95.9	81.5	89.7	98.3	83.8	92.1	98.4	86.8	93.4

OHIO

DIVISION		MANSFIELD 448-449 MAT.	INST.	TOTAL	MARION 433 MAT.	INST.	TOTAL	SPRINGFIELD 455 MAT.	INST.	TOTAL	STEUBENVILLE 439 MAT.	INST.	TOTAL	TOLEDO 434-436 MAT.	INST.	TOTAL	YOUNGSTOWN 444-445 MAT.	INST.	TOTAL
015433	CONTRACTOR EQUIPMENT		91.4	91.4		91.8	91.8		91.6	91.6		95.6	95.6		95.6	95.6		91.4	91.4
0241, 31 - 34	SITE & INFRASTRUCTURE, DEMOLITION	95.4	99.0	97.9	95.3	95.8	95.6	97.9	99.2	98.8	140.7	103.8	114.9	99.0	95.9	96.8	99.6	98.9	99.1
0310	Concrete Forming & Accessories	90.3	75.6	77.8	96.5	79.4	81.9	98.9	79.3	82.2	97.5	81.2	83.6	100.2	86.7	88.7	101.2	78.8	82.1
0320	Concrete Reinforcing	86.9	81.3	84.1	91.4	81.3	86.3	93.3	81.4	87.3	89.2	97.6	93.4	99.2	85.1	92.0	95.4	87.7	91.5
0330	Cast-in-Place Concrete	99.0	90.4	95.8	87.2	90.9	88.6	89.0	82.8	86.7	94.7	92.2	93.7	95.5	92.3	94.3	105.9	89.9	100.0
03	CONCRETE	92.9	81.5	87.8	85.3	83.6	84.5	89.4	80.7	85.5	89.4	87.3	88.5	93.1	88.4	91.0	100.2	83.8	92.9
04	MASONRY	93.3	90.8	91.7	86.5	91.8	89.7	77.8	77.5	77.6	76.8	91.4	85.8	91.7	90.2	90.8	94.5	88.5	90.8
05	METALS	94.5	76.7	89.1	93.6	79.8	89.4	93.8	77.6	88.9	90.2	82.7	87.9	94.3	86.2	91.8	93.7	78.6	89.1
06	WOOD, PLASTICS & COMPOSITES	94.8	72.1	82.5	95.0	76.8	85.1	106.0	79.4	91.5	90.2	78.6	83.9	99.3	86.4	92.3	108.4	76.8	91.2
07	THERMAL & MOISTURE PROTECTION	102.2	90.9	97.3	90.5	91.0	90.7	104.9	81.5	94.7	101.9	89.5	96.5	100.9	92.2	97.1	104.2	90.7	98.4
08	OPENINGS	104.2	73.1	97.0	93.1	75.6	89.0	95.6	76.8	91.3	93.3	80.7	90.4	95.7	83.7	92.9	103.1	77.4	97.2
0920	Plaster & Gypsum Board	94.6	71.2	79.2	98.5	76.1	83.8	102.0	79.2	87.0	96.7	77.6	84.1	100.3	86.0	90.9	101.8	76.1	84.8
0950, 0980	Ceilings & Acoustic Treatment	95.7	71.2	79.1	97.6	76.1	83.1	105.0	79.2	87.5	94.7	77.6	83.1	97.6	86.0	89.8	94.9	76.1	82.1
0960	Flooring	93.3	95.8	94.0	93.6	95.8	94.3	103.2	74.8	95.0	121.5	93.3	113.3	94.0	93.8	93.9	98.3	93.0	96.8
0970, 0990	Wall Finishes & Painting/Coating	95.1	79.9	86.1	100.2	79.9	88.2	100.8	72.6	84.1	113.6	89.4	99.3	100.2	89.3	93.8	95.1	82.1	87.4
09	FINISHES	96.4	78.8	86.8	96.1	81.8	88.3	100.7	77.5	88.0	113.9	83.5	97.3	96.8	88.0	92.0	99.1	81.1	89.3
COVERS	DIVS. 10 - 14, 25, 28, 41, 43, 44, 46	100.0	90.2	97.8	100.0	87.4	97.2	100.0	86.7	97.1	100.0	88.0	97.4	100.0	92.9	98.4	100.0	90.4	97.9
21, 22, 23	FIRE SUPPRESSION, PLUMBING & HVAC	96.6	89.4	93.7	96.6	95.2	96.0	100.8	82.9	93.6	97.0	93.4	95.5	100.1	95.1	98.1	100.0	85.8	94.3
26, 27, 3370	ELECTRICAL, COMMUNICATIONS & UTIL.	93.6	93.0	93.3	92.5	93.0	92.7	94.1	82.2	88.1	87.4	109.1	98.4	98.6	106.0	102.3	95.9	76.2	85.9
MF2016	WEIGHTED AVERAGE	96.5	86.4	92.1	93.4	88.5	91.3	95.9	81.8	89.8	95.3	92.1	94.0	96.7	92.8	95.0	98.7	84.0	92.4

For customer support on your Electrical Costs with RSMeans data, call 800.448.8182.

539

City Cost Indexes

Ohio / Oklahoma

DIVISION		OHIO ZANESVILLE 437 - 438 MAT.	INST.	TOTAL	OKLAHOMA ARDMORE 734 MAT.	INST.	TOTAL	CLINTON 736 MAT.	INST.	TOTAL	DURANT 747 MAT.	INST.	TOTAL	ENID 737 MAT.	INST.	TOTAL	GUYMON 739 MAT.	INST.	TOTAL
015433	CONTRACTOR EQUIPMENT		91.8	91.8		82.7	82.7		81.8	81.8		81.8	81.8		81.8	81.8		81.8	81.8
0241, 31 - 34	SITE & INFRASTRUCTURE, DEMOLITION	98.3	95.5	96.4	99.4	96.0	97.0	100.8	94.5	96.4	94.3	93.0	93.4	102.5	94.5	96.9	105.2	95.0	98.1
0310	Concrete Forming & Accessories	93.6	78.0	80.3	86.2	56.6	61.0	84.9	56.6	60.8	82.6	56.4	60.3	88.2	56.8	61.5	91.1	56.7	61.8
0320	Concrete Reinforcing	90.9	95.2	93.1	87.7	67.2	77.3	88.2	67.2	77.5	93.2	65.5	79.1	87.6	67.3	77.3	88.2	65.1	76.4
0330	Cast-in-Place Concrete	91.9	88.8	90.7	87.0	72.4	81.6	84.1	72.4	79.7	90.2	72.3	83.5	84.1	72.4	79.8	84.1	72.1	79.6
03	CONCRETE	89.1	84.6	87.1	80.8	63.8	73.2	80.3	63.8	73.0	84.2	63.4	75.0	80.8	63.9	73.3	83.3	63.3	74.4
04	MASONRY	83.5	86.2	85.2	94.6	57.5	71.7	118.1	57.5	80.6	87.3	63.4	72.5	100.2	57.5	73.8	97.1	56.6	72.1
05	METALS	94.9	85.1	92.0	92.9	61.2	83.3	92.9	61.2	83.3	93.6	60.7	83.7	94.4	61.3	84.4	93.4	59.7	83.2
06	WOOD, PLASTICS & COMPOSITES	90.3	76.8	82.9	99.6	56.0	75.9	98.7	56.0	75.5	89.3	56.0	71.2	102.0	56.0	77.0	105.4	56.0	78.5
07	THERMAL & MOISTURE PROTECTION	90.7	85.8	88.6	101.5	64.5	85.4	101.6	64.5	85.4	97.1	66.1	83.6	101.7	64.5	85.5	102.1	64.0	85.5
08	OPENINGS	93.1	79.7	90.0	105.9	56.0	94.3	105.9	56.0	94.3	98.1	55.5	88.2	107.1	56.7	95.4	106.0	55.5	94.3
0920	Plaster & Gypsum Board	94.5	76.1	82.4	93.7	55.3	68.4	93.4	55.3	68.3	79.6	55.3	63.6	94.3	55.3	68.6	94.3	55.3	68.6
0950, 0980	Ceilings & Acoustic Treatment	97.6	76.1	83.1	83.2	55.3	64.3	83.2	55.3	64.3	79.4	55.3	63.1	83.2	55.3	64.3	83.2	55.3	64.3
0960	Flooring	91.8	77.8	87.8	87.3	59.2	79.1	86.2	59.2	78.3	91.4	53.4	80.4	87.8	73.1	83.6	89.3	59.2	80.6
0970, 0990	Wall Finishes & Painting/Coating	100.2	91.5	95.0	84.6	45.1	61.2	84.6	45.1	61.2	92.9	45.1	64.6	84.6	45.1	61.2	84.6	43.4	60.2
09	FINISHES	95.2	78.5	86.1	81.0	54.9	66.8	80.8	54.9	66.7	81.4	53.8	66.3	81.5	57.8	68.6	82.4	56.1	68.0
COVERS	DIVS. 10 - 14, 25, 28, 41, 43, 44, 46	100.0	84.5	96.6	100.0	79.0	95.4	100.0	79.0	95.4	100.0	79.0	95.4	100.0	79.0	95.4	100.0	79.0	95.4
21, 22, 23	FIRE SUPPRESSION, PLUMBING & HVAC	96.6	91.5	94.5	96.7	67.8	85.0	96.7	67.8	85.0	96.8	67.7	85.0	100.2	67.8	87.1	96.7	65.0	83.9
26, 27, 3370	ELECTRICAL, COMMUNICATIONS & UTIL.	92.6	91.9	92.2	93.3	70.9	81.9	94.4	70.9	82.4	95.5	70.9	83.0	94.4	70.9	82.4	96.0	67.2	81.4
MF2016	WEIGHTED AVERAGE	94.0	87.1	91.0	93.8	66.2	82.0	95.0	66.1	82.6	93.2	66.3	81.7	95.5	66.5	83.1	94.9	64.8	82.0

Oklahoma

DIVISION		LAWTON 735 MAT.	INST.	TOTAL	MCALESTER 745 MAT.	INST.	TOTAL	MIAMI 743 MAT.	INST.	TOTAL	MUSKOGEE 744 MAT.	INST.	TOTAL	OKLAHOMA CITY 730 - 731 MAT.	INST.	TOTAL	PONCA CITY 746 MAT.	INST.	TOTAL
015433	CONTRACTOR EQUIPMENT		82.7	82.7		81.8	81.8		93.2	93.2		93.2	93.2		83.0	83.0		81.8	81.8
0241, 31 - 34	SITE & INFRASTRUCTURE, DEMOLITION	98.6	96.0	96.8	87.6	95.1	92.8	88.9	92.4	91.3	89.3	92.2	91.4	96.7	96.5	96.6	94.9	95.4	95.2
0310	Concrete Forming & Accessories	90.9	56.8	61.9	80.7	42.5	48.2	93.5	56.6	62.1	97.9	56.3	62.5	91.1	63.5	67.6	89.1	56.8	61.6
0320	Concrete Reinforcing	87.8	67.3	77.4	92.9	65.1	78.7	91.4	67.3	79.1	92.3	64.7	78.2	97.1	67.3	81.9	92.3	67.2	79.5
0330	Cast-in-Place Concrete	81.3	72.4	78.0	79.1	72.0	76.4	82.9	73.4	79.3	83.8	73.2	79.9	81.3	74.3	78.7	92.6	72.4	85.1
03	CONCRETE	77.5	63.9	71.5	75.3	56.9	67.2	79.8	65.2	73.3	81.3	64.5	73.9	80.5	67.6	74.8	86.3	63.9	76.4
04	MASONRY	97.0	57.5	72.5	104.4	57.4	75.3	90.1	57.5	70.0	106.1	50.2	71.6	99.6	57.5	73.5	82.9	57.5	67.2
05	METALS	97.9	61.3	86.8	93.6	59.6	83.3	93.6	76.4	88.4	95.1	74.4	88.8	91.5	61.4	82.4	93.6	61.3	83.8
06	WOOD, PLASTICS & COMPOSITES	104.5	56.0	78.1	86.8	37.2	59.9	102.1	56.1	77.1	106.6	56.1	79.2	92.9	65.0	77.7	97.5	56.0	75.0
07	THERMAL & MOISTURE PROTECTION	101.4	64.5	85.4	96.7	62.4	81.8	97.2	65.4	83.3	97.5	62.8	82.4	93.3	65.8	81.3	97.3	64.7	83.1
08	OPENINGS	108.9	56.7	96.8	98.1	45.2	85.8	98.1	56.1	88.3	99.3	55.6	89.1	103.4	61.6	93.7	98.1	56.0	88.3
0920	Plaster & Gypsum Board	95.8	55.3	69.1	78.7	36.0	50.6	85.7	55.3	65.7	87.5	55.3	66.3	100.2	64.5	76.7	84.5	55.3	65.3
0950, 0980	Ceilings & Acoustic Treatment	90.0	55.3	66.5	79.4	36.0	50.0	79.4	55.3	63.1	87.9	55.3	65.8	91.5	64.5	73.2	79.4	55.3	63.1
0960	Flooring	89.6	73.1	84.8	90.3	59.2	81.3	97.4	53.4	84.6	99.9	34.3	80.8	91.1	73.1	85.9	94.5	59.2	84.2
0970, 0990	Wall Finishes & Painting/Coating	84.6	45.1	61.2	92.9	43.4	63.6	92.9	43.4	63.6	92.9	43.4	63.6	88.6	45.1	62.9	92.9	45.1	64.6
09	FINISHES	83.1	57.8	69.3	80.4	43.7	60.3	83.3	53.7	67.1	86.0	50.3	66.5	86.5	63.1	73.7	83.1	55.9	68.2
COVERS	DIVS. 10 - 14, 25, 28, 41, 43, 44, 46	100.0	79.0	95.4	100.0	77.0	94.9	100.0	77.9	95.1	100.0	77.9	95.1	100.0	80.0	95.6	100.0	79.0	95.4
21, 22, 23	FIRE SUPPRESSION, PLUMBING & HVAC	100.2	67.8	87.1	96.8	65.0	83.9	96.8	65.0	83.9	100.2	62.7	85.1	100.1	67.8	87.0	96.8	65.0	83.9
26, 27, 3370	ELECTRICAL, COMMUNICATIONS & UTIL.	96.0	70.9	83.3	94.0	68.6	81.1	95.4	68.7	81.8	93.6	68.8	81.0	101.5	70.9	86.0	93.6	67.2	80.2
MF2016	WEIGHTED AVERAGE	95.9	66.6	83.3	92.4	61.9	79.3	92.8	66.5	81.5	95.1	64.4	81.9	95.2	68.2	83.6	93.2	65.2	81.2

Oklahoma / Oregon

DIVISION		POTEAU 749 MAT.	INST.	TOTAL	SHAWNEE 748 MAT.	INST.	TOTAL	TULSA 740 - 741 MAT.	INST.	TOTAL	WOODWARD 738 MAT.	INST.	TOTAL	BEND 977 MAT.	INST.	TOTAL	EUGENE 974 MAT.	INST.	TOTAL
015433	CONTRACTOR EQUIPMENT		91.9	91.9		81.8	81.8		93.2	93.2		81.8	81.8		99.8	99.8		99.8	99.8
0241, 31 - 34	SITE & INFRASTRUCTURE, DEMOLITION	76.1	87.9	84.3	97.9	95.4	96.1	95.5	91.4	92.7	101.2	94.5	96.5	114.4	103.2	106.6	103.9	103.2	103.4
0310	Concrete Forming & Accessories	87.1	56.4	60.9	82.5	56.7	60.6	97.9	57.5	63.5	85.0	44.6	50.6	106.4	96.8	98.2	102.9	96.8	97.7
0320	Concrete Reinforcing	93.3	67.2	80.1	92.3	67.2	79.5	92.5	67.3	79.7	87.6	67.2	77.3	100.4	111.7	106.1	104.8	111.7	108.3
0330	Cast-in-Place Concrete	82.9	73.3	79.3	95.6	72.4	87.0	91.3	75.3	85.4	84.1	72.3	79.7	115.1	99.9	109.4	111.4	99.9	107.1
03	CONCRETE	81.9	65.0	74.4	87.9	63.9	77.2	86.4	66.3	77.5	80.6	58.3	70.7	106.0	99.9	103.3	97.7	99.9	98.7
04	MASONRY	90.4	57.5	70.1	105.8	57.5	75.9	91.0	57.5	70.3	90.0	57.5	69.9	101.6	101.6	101.6	98.5	101.6	100.4
05	METALS	93.6	76.1	88.3	93.5	61.3	83.7	98.1	76.5	91.5	93.0	61.1	83.3	101.0	94.9	99.2	101.7	94.9	99.7
06	WOOD, PLASTICS & COMPOSITES	94.1	56.1	73.5	89.1	56.0	71.1	105.8	57.2	79.4	98.8	39.8	66.8	101.1	96.1	98.4	96.9	96.1	96.5
07	THERMAL & MOISTURE PROTECTION	97.3	65.4	83.4	97.3	64.0	82.8	97.5	65.6	83.6	101.6	62.9	84.8	116.5	98.7	108.8	115.6	101.7	109.5
08	OPENINGS	98.1	56.1	88.3	98.1	56.0	88.3	101.0	57.3	90.9	105.9	47.1	92.2	99.2	99.3	99.2	99.5	99.3	99.3
0920	Plaster & Gypsum Board	82.6	55.3	64.6	79.6	55.3	63.6	87.5	56.4	67.0	93.4	38.7	57.3	119.9	95.8	104.0	118.1	95.8	103.4
0950, 0980	Ceilings & Acoustic Treatment	79.4	55.3	63.1	79.4	55.3	63.1	87.9	56.4	66.6	83.2	38.7	53.0	95.9	95.8	95.9	96.9	95.8	96.2
0960	Flooring	93.9	53.4	82.1	91.4	59.2	82.0	98.7	63.7	88.5	86.2	56.5	77.5	97.8	104.2	99.7	96.3	104.2	98.6
0970, 0990	Wall Finishes & Painting/Coating	92.9	45.1	64.6	92.9	43.4	63.6	92.9	52.6	69.1	84.6	45.1	61.2	96.1	71.4	81.5	96.1	71.4	81.5
09	FINISHES	81.1	53.9	66.3	81.6	54.7	66.9	85.9	57.4	70.3	80.9	44.8	61.2	96.5	95.2	95.8	94.9	95.2	95.1
COVERS	DIVS. 10 - 14, 25, 28, 41, 43, 44, 46	100.0	79.3	95.4	100.0	79.0	95.4	100.0	78.0	95.2	100.0	77.3	95.0	100.0	98.3	99.6	100.0	98.3	99.6
21, 22, 23	FIRE SUPPRESSION, PLUMBING & HVAC	96.8	65.0	83.9	96.8	67.8	85.1	100.2	65.1	86.0	96.7	67.8	85.0	96.6	107.5	101.0	100.0	100.8	100.3
26, 27, 3370	ELECTRICAL, COMMUNICATIONS & UTIL.	93.7	68.7	81.0	95.6	70.9	83.0	95.5	68.7	81.9	95.9	70.9	83.2	100.1	93.7	96.8	98.7	93.7	96.2
MF2016	WEIGHTED AVERAGE	92.4	66.1	81.1	94.6	66.1	82.4	96.0	67.1	83.6	93.9	63.4	80.8	100.7	99.8	100.3	99.8	98.5	99.3

540

		OREGON																
	DIVISION	**KLAMATH FALLS**			**MEDFORD**			**PENDLETON**			**PORTLAND**			**SALEM**			**VALE**		
		976			975			978			970 - 972			973			979		
		MAT.	INST.	TOTAL	MAT.	INST.	TOTAL	MAT.	INST.	TOTAL	MAT.	INST.	TOTAL	MAT.	INST.	TOTAL	MAT.	INST.	TOTAL
015433	CONTRACTOR EQUIPMENT		99.8	99.8		99.8	99.8		97.3	97.3		99.8	99.8		99.8	99.8		97.3	97.3
0241, 31 - 34	SITE & INFRASTRUCTURE, DEMOLITION	118.7	103.2	107.9	112.0	103.2	105.9	112.1	96.3	101.1	106.9	103.2	104.3	98.0	103.2	101.6	98.7	96.2	97.0
0310	Concrete Forming & Accessories	99.4	96.6	97.0	98.4	96.6	96.9	100.1	97.0	97.5	104.1	96.8	97.9	105.0	96.8	98.0	106.6	95.7	97.3
0320	Concrete Reinforcing	100.4	111.6	106.1	102.1	111.6	106.9	99.5	111.7	105.7	105.5	111.7	108.7	113.5	111.7	112.6	97.1	111.5	104.4
0330	Cast-in-Place Concrete	115.1	99.8	109.4	115.0	99.8	109.4	115.9	100.3	110.1	114.5	99.9	109.0	105.2	99.9	103.2	91.2	100.6	94.7
03	CONCRETE	108.6	99.8	104.7	103.5	99.8	101.9	91.1	100.2	95.1	99.3	99.9	99.6	95.2	99.9	97.3	75.8	99.6	86.4
04	MASONRY	114.5	101.6	106.5	95.4	101.6	99.2	105.6	101.7	103.2	100.0	101.6	101.0	104.3	101.6	102.6	103.4	101.7	102.3
05	METALS	101.0	94.6	99.1	101.3	94.6	99.3	108.4	95.5	104.5	102.9	94.9	100.5	109.2	94.9	104.9	108.3	94.2	104.0
06	WOOD, PLASTICS & COMPOSITES	91.9	96.1	94.2	90.8	96.1	93.7	94.0	96.2	95.2	98.1	96.1	97.0	93.1	96.1	94.7	102.6	96.2	99.1
07	THERMAL & MOISTURE PROTECTION	116.8	95.3	107.4	116.4	95.3	107.2	109.0	97.0	103.8	115.6	98.8	108.3	112.0	99.1	106.4	108.4	92.8	101.6
08	OPENINGS	99.2	99.3	99.3	102.0	99.3	101.4	95.5	99.4	96.4	97.4	99.3	97.9	101.2	99.3	100.7	95.4	89.2	94.0
0920	Plaster & Gypsum Board	114.8	95.8	102.3	114.2	95.8	102.1	100.3	95.8	97.4	118.0	95.8	103.4	113.0	95.8	101.7	106.4	95.8	99.4
0950, 0980	Ceilings & Acoustic Treatment	103.6	95.8	98.3	110.2	95.8	100.5	68.4	95.8	87.0	98.8	95.8	96.8	103.9	95.8	98.4	68.4	95.8	87.0
0960	Flooring	95.0	104.2	97.6	94.4	104.2	97.2	65.5	104.2	76.7	94.2	104.2	97.1	99.1	104.2	100.6	67.5	104.2	78.1
0970, 0990	Wall Finishes & Painting/Coating	96.1	67.0	78.9	96.1	67.0	78.9	87.6	76.2	80.9	95.9	76.2	84.3	93.4	75.1	82.6	87.6	71.4	78.0
09	FINISHES	96.9	94.7	95.7	97.1	94.7	95.8	68.8	95.8	83.5	94.7	95.7	95.2	95.1	95.6	95.4	69.3	95.3	83.5
COVERS	DIVS. 10 - 14, 25, 28, 41, 43, 44, 46	100.0	98.2	99.6	100.0	98.2	99.6	100.0	98.2	99.6	100.0	98.3	99.6	100.0	98.3	99.6	100.0	98.5	99.7
21, 22, 23	FIRE SUPPRESSION, PLUMBING & HVAC	96.6	107.5	101.0	100.0	107.5	103.0	98.3	114.5	104.9	100.0	104.7	101.9	100.0	107.5	103.1	98.3	72.6	87.9
26, 27, 3370	ELECTRICAL, COMMUNICATIONS & UTIL.	98.8	80.5	89.5	102.2	80.5	91.2	91.1	99.2	95.2	98.9	101.5	100.2	106.6	93.7	100.0	91.1	69.7	80.2
MF2016	WEIGHTED AVERAGE	101.5	97.8	99.9	101.3	97.8	99.8	96.8	101.7	98.9	100.2	100.4	100.3	101.7	99.9	100.9	94.5	87.9	91.7

		PENNSYLVANIA																
	DIVISION	**ALLENTOWN**			**ALTOONA**			**BEDFORD**			**BRADFORD**			**BUTLER**			**CHAMBERSBURG**		
		181			166			155			167			160			172		
		MAT.	INST.	TOTAL	MAT.	INST.	TOTAL	MAT.	INST.	TOTAL	MAT.	INST.	TOTAL	MAT.	INST.	TOTAL	MAT.	INST.	TOTAL
015433	CONTRACTOR EQUIPMENT		115.2	115.2		115.2	115.2		113.2	113.2		115.2	115.2		115.2	115.2		114.4	114.4
0241, 31 - 34	SITE & INFRASTRUCTURE, DEMOLITION	92.7	101.8	99.0	95.7	101.6	99.8	100.6	99.3	99.7	91.7	100.5	97.9	87.0	102.4	97.7	88.0	99.8	96.2
0310	Concrete Forming & Accessories	98.5	111.2	109.3	83.5	84.9	84.7	82.3	83.5	83.3	85.8	99.2	97.2	85.0	96.7	95.0	85.0	78.9	79.8
0320	Concrete Reinforcing	103.8	116.6	110.3	100.6	108.8	104.8	99.5	108.8	104.2	102.8	108.8	105.9	101.3	122.8	112.2	103.4	115.5	109.5
0330	Cast-in-Place Concrete	87.6	102.8	93.2	97.6	89.3	94.5	94.0	88.3	91.9	93.2	93.2	93.2	86.2	98.9	90.9	95.1	96.5	95.6
03	CONCRETE	91.6	109.9	99.7	86.8	91.9	89.1	94.9	90.9	93.1	93.4	99.6	96.2	78.9	103.0	89.6	99.8	92.9	96.8
04	MASONRY	94.2	95.1	94.8	97.9	85.1	90.0	109.3	81.9	92.4	95.3	83.7	88.1	99.9	94.9	96.8	95.1	80.8	86.3
05	METALS	97.3	121.3	104.6	91.4	115.6	98.7	100.7	114.1	104.8	95.2	114.0	100.9	91.1	122.5	100.6	96.4	119.6	103.4
06	WOOD, PLASTICS & COMPOSITES	99.6	114.8	107.8	78.0	84.1	81.3	80.6	84.1	82.5	84.2	104.2	95.0	79.5	96.8	88.9	85.2	77.8	81.1
07	THERMAL & MOISTURE PROTECTION	103.6	110.8	106.7	102.4	92.5	98.1	100.6	89.8	95.9	103.4	91.0	98.0	101.9	97.1	99.8	93.0	85.4	89.7
08	OPENINGS	94.5	110.4	98.2	88.2	88.1	88.2	95.5	88.1	93.7	94.5	98.0	95.3	88.2	102.2	91.4	91.2	83.8	89.5
0920	Plaster & Gypsum Board	94.2	115.1	108.0	85.6	83.6	84.3	100.1	83.6	89.2	86.7	104.2	98.2	85.6	96.6	92.9	102.2	77.1	85.7
0950, 0980	Ceilings & Acoustic Treatment	88.8	115.1	106.6	91.4	83.6	86.1	101.3	83.6	89.3	91.5	104.2	100.1	92.3	96.6	95.2	92.1	77.1	81.9
0960	Flooring	90.7	98.3	92.9	84.2	102.3	89.4	96.7	107.2	99.7	85.0	107.2	91.5	85.1	108.9	92.0	88.1	80.6	85.9
0970, 0990	Wall Finishes & Painting/Coating	90.4	104.5	98.8	86.0	109.0	99.6	94.8	108.5	102.9	90.4	108.5	101.1	86.0	109.0	99.6	91.5	103.4	98.5
09	FINISHES	88.8	108.9	99.8	86.4	90.0	88.4	99.0	89.9	94.1	86.6	102.2	95.2	86.4	100.0	93.8	85.9	81.1	83.3
COVERS	DIVS. 10 - 14, 25, 28, 41, 43, 44, 46	100.0	101.6	100.4	100.0	96.4	99.2	100.0	95.1	98.9	100.0	98.1	99.6	100.0	99.3	99.8	100.0	94.6	98.8
21, 22, 23	FIRE SUPPRESSION, PLUMBING & HVAC	100.2	117.1	107.1	99.8	85.5	94.0	96.5	86.3	92.4	96.8	92.6	95.1	96.3	97.6	96.8	96.7	91.7	94.7
26, 27, 3370	ELECTRICAL, COMMUNICATIONS & UTIL.	99.6	98.1	98.8	90.0	110.3	100.3	97.8	110.3	104.2	93.6	110.3	102.1	90.6	110.4	100.6	91.0	87.9	89.4
MF2016	WEIGHTED AVERAGE	96.7	108.4	101.7	93.2	95.1	94.0	98.2	94.3	96.5	94.9	99.5	96.8	91.3	103.1	96.4	94.9	91.6	93.4

		PENNSYLVANIA																
	DIVISION	**DOYLESTOWN**			**DUBOIS**			**ERIE**			**GREENSBURG**			**HARRISBURG**			**HAZLETON**		
		189			158			164 - 165			156			170 - 171			182		
		MAT.	INST.	TOTAL	MAT.	INST.	TOTAL	MAT.	INST.	TOTAL	MAT.	INST.	TOTAL	MAT.	INST.	TOTAL	MAT.	INST.	TOTAL
015433	CONTRACTOR EQUIPMENT		93.4	93.4		113.2	113.2		115.2	115.2		113.2	113.2		114.4	114.4		115.2	115.2
0241, 31 - 34	SITE & INFRASTRUCTURE, DEMOLITION	106.7	89.8	94.9	105.4	99.7	101.4	92.6	101.9	99.1	96.8	101.2	99.9	88.2	99.9	96.3	86.1	101.6	96.9
0310	Concrete Forming & Accessories	82.7	130.2	123.2	81.8	86.0	85.4	97.8	88.2	89.6	88.4	91.0	90.6	96.3	87.9	89.1	80.4	91.3	89.7
0320	Concrete Reinforcing	100.4	154.7	128.0	98.8	122.8	111.0	102.8	110.4	106.7	98.8	122.5	110.9	113.1	114.2	113.7	100.8	116.3	108.7
0330	Cast-in-Place Concrete	82.7	133.4	101.6	90.6	96.8	92.9	95.9	91.2	94.2	87.2	98.4	91.4	93.7	98.7	95.5	82.7	98.3	88.5
03	CONCRETE	87.3	134.6	108.2	97.1	97.3	97.2	85.7	94.4	89.6	90.4	100.2	94.7	94.4	97.5	95.8	84.3	99.3	90.9
04	MASONRY	97.6	128.4	116.6	109.8	95.4	100.9	86.9	88.7	88.0	120.0	90.9	102.1	89.9	85.6	87.3	107.2	91.4	97.4
05	METALS	94.9	123.9	103.7	100.7	119.6	106.4	91.6	116.3	99.1	100.6	120.5	106.6	103.0	119.3	107.9	97.0	120.5	104.1
06	WOOD, PLASTICS & COMPOSITES	79.8	129.4	106.7	79.5	84.1	82.0	96.0	86.8	91.0	86.8	89.6	88.3	98.0	87.6	92.4	78.3	89.5	84.4
07	THERMAL & MOISTURE PROTECTION	100.8	132.7	114.7	100.9	95.5	98.6	102.9	91.0	97.7	100.5	94.8	98.0	95.8	102.6	98.8	102.9	103.5	103.2
08	OPENINGS	96.9	137.0	106.2	95.5	91.2	94.5	88.3	90.5	88.9	95.4	98.3	96.1	98.0	88.9	95.9	95.0	96.0	95.3
0920	Plaster & Gypsum Board	84.7	130.2	114.7	98.9	83.6	88.8	94.2	86.4	89.1	101.1	89.3	93.4	108.1	87.2	94.3	85.1	89.2	87.8
0950, 0980	Ceilings & Acoustic Treatment	88.1	130.2	116.6	101.3	83.6	89.3	88.8	86.4	87.2	100.5	89.3	92.9	99.7	87.2	91.2	89.8	89.2	89.4
0960	Flooring	75.2	144.0	95.2	96.4	107.2	99.5	91.1	102.3	94.3	100.3	80.6	94.6	92.0	93.5	92.4	82.3	92.3	85.2
0970, 0990	Wall Finishes & Painting/Coating	90.0	147.5	124.0	94.8	109.0	103.2	96.5	94.8	95.5	94.8	109.0	103.2	96.6	87.0	90.9	90.4	106.8	100.1
09	FINISHES	80.5	134.3	109.9	99.2	91.4	95.0	89.8	91.0	90.5	99.7	90.6	94.8	91.1	88.2	89.5	84.9	92.8	89.2
COVERS	DIVS. 10 - 14, 25, 28, 41, 43, 44, 46	100.0	115.3	103.4	100.0	97.1	99.4	100.0	97.9	99.5	100.0	98.4	99.6	100.0	96.9	99.3	100.0	97.4	99.4
21, 22, 23	FIRE SUPPRESSION, PLUMBING & HVAC	96.3	135.6	112.2	96.5	89.5	93.7	99.8	95.7	98.1	96.5	89.3	93.6	100.1	93.7	97.5	96.8	97.7	97.2
26, 27, 3370	ELECTRICAL, COMMUNICATIONS & UTIL.	93.0	135.6	114.7	98.4	110.3	104.5	91.7	95.4	93.6	98.4	110.3	104.5	98.2	87.9	93.0	94.4	92.2	93.3
MF2016	WEIGHTED AVERAGE	94.0	129.2	109.1	98.7	98.3	98.5	93.1	96.2	94.5	98.2	98.7	98.4	97.8	94.9	96.5	94.4	98.3	96.1

For customer support on your Electrical Costs with RSMeans data, call 800.448.8182.

541

PENNSYLVANIA

DIVISION		INDIANA 157			JOHNSTOWN 159			KITTANNING 162			LANCASTER 175 - 176			LEHIGH VALLEY 180			MONTROSE 188		
		MAT.	INST.	TOTAL	MAT.	INST.	TOTAL	MAT.	INST.	TOTAL	MAT.	INST.	TOTAL	MAT.	INST.	TOTAL	MAT.	INST.	TOTAL
015433	CONTRACTOR EQUIPMENT		113.2	113.2		113.2	113.2		115.2	115.2		114.4	114.4		115.2	115.2		115.2	115.2
0241, 31 - 34	SITE & INFRASTRUCTURE, DEMOLITION	95.0	100.3	98.7	101.1	100.5	100.7	89.5	102.0	98.2	80.0	100.1	94.0	90.0	102.0	98.4	88.7	101.7	97.8
0310	Concrete Forming & Accessories	82.8	95.5	93.6	81.8	84.8	84.4	85.0	92.2	91.1	87.0	88.8	88.5	92.0	111.6	108.7	81.3	91.8	90.2
0320	Concrete Reinforcing	98.1	122.9	110.7	99.5	122.3	111.1	101.3	122.6	112.2	103.0	114.2	108.7	100.8	111.1	106.0	105.6	111.0	112.7
0330	Cast-in-Place Concrete	85.5	97.1	89.8	94.8	89.1	92.7	89.6	97.4	92.5	80.8	99.5	87.8	89.6	102.0	94.2	87.8	95.7	90.7
03	CONCRETE	88.0	101.7	94.1	95.8	94.0	95.0	81.4	100.4	89.8	87.5	98.2	92.3	90.6	108.9	98.7	89.3	99.2	93.7
04	MASONRY	105.6	95.5	99.3	106.4	85.5	93.5	102.5	90.7	95.2	100.9	88.6	93.3	94.3	98.5	96.9	94.2	93.5	93.7
05	METALS	100.7	120.7	106.8	100.7	118.7	106.2	91.2	121.4	100.3	96.4	119.5	103.4	97.0	119.7	103.8	95.2	121.2	103.1
06	WOOD, PLASTICS & COMPOSITES	81.3	96.7	89.7	79.5	84.1	82.0	79.5	91.9	86.2	88.1	87.6	87.8	90.9	113.5	103.2	79.1	89.5	84.8
07	THERMAL & MOISTURE PROTECTION	100.4	96.9	98.8	100.6	91.0	96.4	102.0	95.0	99.0	92.4	103.8	97.4	103.3	112.8	107.4	102.9	93.5	98.8
08	OPENINGS	95.5	98.2	96.1	95.5	91.2	94.5	88.2	99.5	90.8	91.2	88.9	90.7	95.0	108.5	98.1	91.8	94.5	92.4
0920	Plaster & Gypsum Board	100.4	96.6	97.9	98.7	83.6	88.8	85.6	91.6	89.6	104.6	87.2	93.1	87.5	113.8	104.8	85.5	89.2	87.9
0950, 0980	Ceilings & Acoustic Treatment	101.3	96.6	98.1	100.5	83.6	89.1	92.3	91.6	91.8	92.1	87.2	88.8	89.8	113.8	106.1	91.5	89.2	89.9
0960	Flooring	97.2	107.2	100.1	96.4	80.6	91.8	85.1	107.2	91.5	89.1	98.9	91.9	87.7	97.3	90.5	83.0	107.2	90.0
0970, 0990	Wall Finishes & Painting/Coating	94.8	109.0	103.2	94.8	109.0	103.2	86.0	109.0	99.6	91.5	88.9	90.0	90.4	103.6	98.2	90.4	106.8	100.1
09	FINISHES	98.8	98.9	98.9	98.6	86.3	91.9	86.5	96.1	91.7	86.0	89.9	88.1	87.0	108.3	98.6	85.6	95.4	91.0
COVERS	DIVS. 10 - 14, 25, 28, 41, 43, 44, 46	100.0	98.4	99.7	100.0	96.3	99.2	100.0	98.1	99.6	100.0	97.6	99.5	100.0	104.3	101.0	100.0	98.1	99.6
21, 22, 23	FIRE SUPPRESSION, PLUMBING & HVAC	96.5	87.3	92.8	96.5	80.7	90.1	96.3	92.7	94.9	96.7	94.8	95.9	96.8	118.6	105.6	96.8	96.9	96.9
26, 27, 3370	ELECTRICAL, COMMUNICATIONS & UTIL.	98.4	110.3	104.5	98.4	110.3	104.5	90.0	110.3	100.3	92.3	95.6	94.0	94.4	132.5	113.8	93.6	97.6	95.6
MF2016	WEIGHTED AVERAGE	97.1	100.1	98.4	98.2	94.2	96.5	91.8	100.4	95.5	93.5	96.8	94.9	95.0	113.6	103.0	93.8	99.2	96.1

PENNSYLVANIA

DIVISION		NEW CASTLE 161			NORRISTOWN 194			OIL CITY 163			PHILADELPHIA 190 - 191			PITTSBURGH 150 - 152			POTTSVILLE 179		
		MAT.	INST.	TOTAL	MAT.	INST.	TOTAL	MAT.	INST.	TOTAL	MAT.	INST.	TOTAL	MAT.	INST.	TOTAL	MAT.	INST.	TOTAL
015433	CONTRACTOR EQUIPMENT		115.2	115.2		99.0	99.0		115.2	115.2		99.3	99.3		102.0	102.0		114.4	114.4
0241, 31 - 34	SITE & INFRASTRUCTURE, DEMOLITION	87.4	102.3	97.8	95.3	99.3	98.1	85.9	100.8	96.3	97.3	100.0	99.2	100.8	99.9	100.2	82.9	100.3	95.1
0310	Concrete Forming & Accessories	85.0	96.2	94.5	83.0	128.7	121.9	85.0	95.3	93.8	100.2	140.0	134.1	98.4	97.2	97.3	78.8	90.1	88.4
0320	Concrete Reinforcing	100.1	103.1	101.6	101.9	154.7	128.7	101.3	97.7	99.5	105.2	126.5	116.0	100.0	123.1	111.8	102.2	117.4	109.9
0330	Cast-in-Place Concrete	87.0	96.9	90.7	85.0	131.0	102.1	84.5	97.1	89.2	93.3	135.3	108.9	93.5	99.2	95.6	86.1	99.1	91.0
03	CONCRETE	79.2	98.7	87.9	87.9	133.0	107.9	77.7	97.5	86.5	98.8	134.8	114.8	96.5	102.3	99.1	91.2	99.3	94.8
04	MASONRY	99.6	94.6	96.5	108.7	124.6	118.5	98.8	90.6	93.7	98.9	131.3	118.9	101.3	99.6	100.2	94.6	89.4	91.4
05	METALS	91.2	115.0	98.4	99.6	123.8	106.9	91.2	113.9	98.0	101.3	113.5	105.0	102.2	107.4	103.8	96.7	121.2	104.1
06	WOOD, PLASTICS & COMPOSITES	79.5	96.8	88.9	76.1	129.3	105.0	79.5	96.8	88.9	96.8	141.8	121.2	99.7	97.2	98.3	77.8	87.6	83.1
07	THERMAL & MOISTURE PROTECTION	101.9	94.6	98.7	107.7	131.1	117.9	101.8	93.3	98.1	103.3	136.4	117.7	100.8	98.4	99.8	92.5	102.4	96.8
08	OPENINGS	88.2	94.1	89.5	88.7	137.0	99.9	88.2	97.7	90.4	101.3	137.3	109.6	97.8	102.5	98.9	91.3	95.7	92.3
0920	Plaster & Gypsum Board	85.6	96.6	92.9	83.8	130.2	114.3	85.6	96.6	92.9	97.8	142.8	127.4	98.7	96.9	97.5	99.1	87.2	91.3
0950, 0980	Ceilings & Acoustic Treatment	92.3	96.6	95.2	91.5	130.2	117.7	92.3	96.6	95.2	101.6	142.8	129.5	95.1	96.9	96.3	92.1	87.2	88.8
0960	Flooring	85.1	108.9	92.0	83.2	144.0	100.8	85.1	107.2	91.5	97.5	145.7	111.5	104.9	107.5	105.7	85.2	107.2	91.6
0970, 0990	Wall Finishes & Painting/Coating	86.0	109.0	99.6	85.8	147.5	122.3	86.0	109.0	99.6	91.8	152.1	127.5	97.5	111.4	105.7	91.5	106.8	99.6
09	FINISHES	86.4	99.8	93.7	83.3	133.3	110.6	86.3	98.9	93.2	98.4	142.8	122.7	100.4	100.2	100.3	84.3	93.9	89.6
COVERS	DIVS. 10 - 14, 25, 28, 41, 43, 44, 46	100.0	99.3	99.8	100.0	111.3	102.5	100.0	98.6	99.7	100.0	120.0	104.4	100.0	101.5	100.3	100.0	99.5	99.9
21, 22, 23	FIRE SUPPRESSION, PLUMBING & HVAC	96.3	99.0	97.4	96.6	133.6	111.6	96.3	96.3	96.3	100.2	140.0	116.3	99.9	98.3	99.2	96.7	97.9	97.2
26, 27, 3370	ELECTRICAL, COMMUNICATIONS & UTIL.	90.6	100.6	95.7	93.4	145.5	119.9	92.3	110.3	101.5	100.4	160.5	131.0	100.8	110.8	105.9	90.6	93.5	92.1
MF2016	WEIGHTED AVERAGE	91.4	100.3	95.2	94.8	130.0	109.9	91.3	100.3	95.1	100.1	135.3	115.2	99.9	102.2	100.9	93.4	98.4	95.6

PENNSYLVANIA

DIVISION		READING 195 - 196			SCRANTON 184 - 185			STATE COLLEGE 168			STROUDSBURG 183			SUNBURY 178			UNIONTOWN 154		
		MAT.	INST.	TOTAL	MAT.	INST.	TOTAL	MAT.	INST.	TOTAL	MAT.	INST.	TOTAL	MAT.	INST.	TOTAL	MAT.	INST.	TOTAL
015433	CONTRACTOR EQUIPMENT		121.3	121.3		115.2	115.2		114.4	114.4		115.2	115.2		115.2	115.2		113.2	113.2
0241, 31 - 34	SITE & INFRASTRUCTURE, DEMOLITION	99.6	112.5	108.6	93.2	101.8	99.2	83.8	100.2	95.3	87.7	101.9	97.6	94.7	101.2	99.3	95.5	101.1	99.4
0310	Concrete Forming & Accessories	98.5	88.5	90.0	98.6	89.0	90.5	84.5	85.1	85.0	86.6	93.1	92.1	89.5	87.1	87.5	76.4	96.5	93.5
0320	Concrete Reinforcing	103.3	152.9	128.5	103.8	119.7	111.9	102.0	108.9	105.5	104.1	119.8	112.1	105.0	115.5	110.4	98.8	122.9	111.1
0330	Cast-in-Place Concrete	76.2	99.7	84.9	91.4	95.8	93.1	88.2	89.4	88.6	86.2	97.2	90.3	94.2	97.6	95.4	85.5	98.4	90.3
03	CONCRETE	87.1	104.7	94.9	93.4	98.0	95.4	93.8	92.1	93.0	88.0	100.3	93.5	94.6	97.1	95.7	87.7	102.7	94.3
04	MASONRY	97.7	93.4	95.0	94.6	95.8	95.4	100.2	85.7	91.2	92.4	98.3	96.1	94.9	82.7	87.4	122.0	97.4	106.8
05	METALS	99.9	133.7	110.1	99.3	121.3	105.9	95.0	115.8	101.3	97.0	121.7	104.5	96.4	120.2	103.6	100.4	121.0	106.7
06	WOOD, PLASTICS & COMPOSITES	93.7	84.8	88.8	99.6	85.7	92.1	86.1	84.1	85.0	85.1	89.5	87.5	86.0	87.6	86.9	73.7	96.7	86.2
07	THERMAL & MOISTURE PROTECTION	108.1	106.2	107.3	103.5	93.9	99.3	102.5	100.4	101.6	103.1	93.9	99.1	93.5	95.3	94.3	100.2	97.8	99.2
08	OPENINGS	93.2	102.3	95.3	94.5	92.4	94.0	91.6	88.1	90.8	95.1	97.3	95.6	91.3	92.5	91.6	95.4	102.2	97.0
0920	Plaster & Gypsum Board	93.2	84.3	87.4	95.9	85.3	88.9	87.4	83.6	84.9	86.3	89.2	88.2	98.4	87.2	91.0	96.9	96.6	96.7
0950, 0980	Ceilings & Acoustic Treatment	82.9	84.3	83.9	96.5	85.3	88.9	88.9	83.6	85.3	88.1	89.2	88.8	88.7	87.2	87.7	100.5	96.6	97.9
0960	Flooring	87.7	101.2	91.6	90.7	96.0	92.2	88.1	101.2	91.9	85.5	97.2	88.9	86.0	107.2	92.1	93.5	107.2	97.5
0970, 0990	Wall Finishes & Painting/Coating	84.6	104.5	96.4	90.4	111.1	102.7	90.4	109.0	101.4	90.4	106.8	100.1	91.5	106.8	100.6	94.8	112.9	105.5
09	FINISHES	84.5	91.0	88.0	90.5	91.9	91.3	86.2	89.8	88.2	85.8	94.5	90.5	85.1	92.2	89.0	97.1	99.8	98.6
COVERS	DIVS. 10 - 14, 25, 28, 41, 43, 44, 46	100.0	97.6	99.5	100.0	96.5	99.2	100.0	96.5	99.2	100.0	99.2	99.8	100.0	94.9	98.9	100.0	99.1	99.8
21, 22, 23	FIRE SUPPRESSION, PLUMBING & HVAC	100.2	109.2	103.8	100.2	98.2	99.4	96.8	93.4	95.4	96.8	100.1	98.2	96.7	89.3	93.7	96.5	93.7	95.4
26, 27, 3370	ELECTRICAL, COMMUNICATIONS & UTIL.	99.5	93.6	96.5	99.6	97.6	98.6	92.7	110.3	101.7	94.4	145.2	120.2	91.0	94.1	92.6	95.4	110.3	103.0
MF2016	WEIGHTED AVERAGE	96.5	104.0	99.7	97.4	98.9	98.1	94.5	97.4	95.6	94.3	107.3	99.9	94.3	95.0	94.6	97.2	102.2	99.4

PENNSYLVANIA

DIVISION		WASHINGTON 153			WELLSBORO 169			WESTCHESTER 193			WILKES-BARRE 186 - 187			WILLIAMSPORT 177			YORK 173 - 174		
		MAT.	INST.	TOTAL	MAT.	INST.	TOTAL	MAT.	INST.	TOTAL	MAT.	INST.	TOTAL	MAT.	INST.	TOTAL	MAT.	INST.	TOTAL
015433	CONTRACTOR EQUIPMENT		113.2	113.2		115.2	115.2		99.0	99.0		115.2	115.2		115.2	115.2		114.4	114.4
0241, 31 - 34	SITE & INFRASTRUCTURE, DEMOLITION	95.6	101.4	99.6	95.1	101.2	99.3	101.4	100.3	100.7	85.7	101.8	96.9	85.7	101.4	96.6	83.8	100.1	95.1
0310	Concrete Forming & Accessories	82.9	96.6	94.6	85.2	86.5	86.3	89.5	130.0	124.0	89.3	87.6	87.9	86.3	88.0	87.7	82.0	88.7	87.7
0320	Concrete Reinforcing	98.8	123.0	111.1	102.0	119.5	110.9	101.0	154.6	128.3	102.8	119.7	111.4	104.2	114.1	109.3	105.0	114.2	109.7
0330	Cast-in-Place Concrete	85.5	98.6	90.4	92.4	90.7	91.8	94.1	133.1	108.7	82.7	95.6	87.5	79.7	92.5	84.5	86.5	99.5	91.3
03	CONCRETE	88.1	102.8	94.6	96.0	95.0	95.6	95.8	134.4	112.9	85.1	97.3	90.5	82.3	95.4	88.1	92.3	98.2	94.9
04	MASONRY	105.3	97.5	100.5	101.4	83.1	90.1	103.2	128.4	118.8	107.5	95.5	100.1	86.9	88.0	87.6	95.9	88.6	91.3
05	METALS	100.4	121.4	106.8	95.1	120.5	102.8	99.6	123.9	106.9	95.2	121.2	103.1	96.4	118.9	103.2	98.1	119.4	104.5
06	WOOD, PLASTICS & COMPOSITES	81.4	96.7	89.7	83.6	86.8	85.4	82.7	129.3	108.0	87.6	84.1	85.7	82.5	87.6	85.3	81.3	87.6	84.7
07	THERMAL & MOISTURE PROTECTION	100.4	97.8	99.2	103.7	88.4	97.0	108.2	128.0	116.8	102.9	93.4	98.8	92.8	90.7	91.9	92.6	103.8	97.5
08	OPENINGS	95.4	102.2	96.9	94.5	92.6	94.0	88.7	137.0	99.9	91.8	91.5	91.7	91.3	92.2	91.5	91.2	88.9	90.7
0920	Plaster & Gypsum Board	100.2	96.6	97.8	86.1	86.4	86.3	84.4	130.2	114.5	87.0	83.6	84.7	99.1	87.2	91.3	99.5	87.2	91.4
0950, 0980	Ceilings & Acoustic Treatment	100.5	96.6	97.9	88.9	86.4	87.2	91.5	130.2	117.7	91.5	83.6	86.1	92.1	87.2	88.8	91.1	87.2	88.5
0960	Flooring	97.3	107.2	100.2	84.7	107.2	91.2	86.0	144.0	102.8	86.4	96.0	89.2	84.9	89.7	86.3	86.5	98.9	90.1
0970, 0990	Wall Finishes & Painting/Coating	94.8	112.9	105.5	90.4	106.8	100.1	85.8	147.5	122.3	90.4	104.5	98.8	91.5	104.5	99.2	91.5	87.0	88.8
09	FINISHES	98.7	99.8	99.3	86.3	92.2	89.5	84.6	134.3	111.7	86.6	90.1	88.5	84.8	89.2	87.2	84.6	89.6	87.3
COVERS	DIVS. 10 - 14, 25, 28, 41, 43, 44, 46	100.0	99.1	99.8	100.0	98.2	99.6	100.0	117.6	103.9	100.0	96.2	99.2	100.0	97.4	99.4	100.0	97.6	99.5
21, 22, 23	FIRE SUPPRESSION, PLUMBING & HVAC	96.5	95.3	96.0	96.8	89.3	93.8	96.6	135.5	112.4	96.8	98.8	97.6	96.7	94.3	95.7	100.1	94.8	98.0
26, 27, 3370	ELECTRICAL, COMMUNICATIONS & UTIL.	97.8	110.4	104.2	93.6	97.6	95.6	93.2	135.6	114.8	94.4	92.2	93.3	91.4	85.4	88.4	92.3	84.9	88.5
MF2016	WEIGHTED AVERAGE	96.9	102.7	99.4	95.5	95.2	95.4	95.8	129.9	110.4	94.1	97.9	95.7	92.1	94.6	93.2	94.9	95.3	95.1

DIVISION		PUERTO RICO SAN JUAN 009			RHODE ISLAND NEWPORT 028			PROVIDENCE 029			SOUTH CAROLINA AIKEN 298			BEAUFORT 299			CHARLESTON 294		
		MAT.	INST.	TOTAL	MAT.	INST.	TOTAL	MAT.	INST.	TOTAL	MAT.	INST.	TOTAL	MAT.	INST.	TOTAL	MAT.	INST.	TOTAL
015433	CONTRACTOR EQUIPMENT		86.5	86.5		100.1	100.1		100.1	100.1		105.8	105.8		105.8	105.8		105.8	105.8
0241, 31 - 34	SITE & INFRASTRUCTURE, DEMOLITION	129.4	89.2	101.4	86.8	102.2	97.5	86.7	102.2	97.5	126.6	89.0	100.4	121.8	88.4	98.5	107.3	88.7	94.3
0310	Concrete Forming & Accessories	90.1	21.3	31.6	99.5	121.2	118.0	99.7	121.2	118.0	90.6	63.7	67.7	89.6	39.1	46.6	89.0	63.7	67.5
0320	Concrete Reinforcing	198.3	18.3	106.8	104.9	118.9	112.0	105.8	118.9	112.5	92.7	62.4	77.3	91.8	26.5	58.6	91.6	67.7	79.4
0330	Cast-in-Place Concrete	94.9	32.2	71.6	68.9	120.7	88.1	88.1	120.7	100.2	91.9	66.1	82.3	91.8	65.7	82.1	107.8	66.1	92.3
03	CONCRETE	96.6	25.6	65.1	82.8	120.1	99.3	90.0	120.1	103.4	99.2	66.0	84.5	96.7	48.8	75.5	93.3	66.9	81.6
04	MASONRY	87.7	20.4	46.1	96.0	124.0	113.3	100.4	124.0	115.0	78.2	65.3	70.3	92.4	65.3	75.7	93.8	65.3	76.2
05	METALS	108.8	38.4	87.5	89.5	112.3	96.4	92.3	112.3	98.3	100.5	89.4	97.1	100.5	79.7	94.2	102.4	91.1	98.9
06	WOOD, PLASTICS & COMPOSITES	97.6	20.7	55.8	99.2	120.3	110.7	97.7	120.3	110.0	87.9	65.6	75.8	86.9	33.3	57.8	85.7	65.6	74.8
07	THERMAL & MOISTURE PROTECTION	131.4	25.4	85.3	103.6	119.5	110.5	103.7	119.5	110.6	97.1	64.9	83.1	96.7	56.9	79.4	95.7	64.5	82.1
08	OPENINGS	155.2	19.0	123.6	99.5	120.2	104.3	100.8	120.2	105.3	95.8	62.6	88.1	95.8	38.8	82.6	99.4	63.8	91.1
0920	Plaster & Gypsum Board	170.9	18.6	70.6	89.5	120.6	109.9	99.7	120.6	113.4	92.0	64.6	73.9	95.6	31.4	53.3	96.7	64.6	75.5
0950, 0980	Ceilings & Acoustic Treatment	254.2	18.6	94.5	93.6	120.6	111.9	98.6	120.6	113.5	82.6	64.6	70.4	85.1	31.4	48.7	85.1	64.6	71.2
0960	Flooring	197.7	20.8	146.4	87.9	130.5	100.2	91.6	130.5	102.9	98.2	92.8	96.6	99.7	78.4	93.5	99.3	79.8	93.7
0970, 0990	Wall Finishes & Painting/Coating	186.1	22.2	89.1	91.0	118.7	107.4	95.3	118.7	109.1	97.0	67.8	79.7	97.0	58.4	74.1	97.0	67.8	79.7
09	FINISHES	191.9	21.4	98.8	87.9	123.1	107.1	89.2	123.1	107.7	89.6	69.7	78.7	90.6	47.1	66.9	88.7	67.5	77.1
COVERS	DIVS. 10 - 14, 25, 28, 41, 43, 44, 46	100.0	19.5	82.3	100.0	106.1	101.3	100.0	106.1	101.3	100.0	70.6	93.5	100.0	81.4	95.9	100.0	70.6	93.5
21, 22, 23	FIRE SUPPRESSION, PLUMBING & HVAC	103.5	17.9	68.9	100.3	113.2	105.5	100.1	113.2	105.4	97.0	56.0	80.5	97.0	56.0	80.4	100.5	58.3	83.4
26, 27, 3370	ELECTRICAL, COMMUNICATIONS & UTIL.	120.6	16.9	67.9	98.8	97.9	98.3	97.4	97.9	97.6	94.0	63.4	78.4	97.4	31.6	64.0	95.9	58.3	76.8
MF2016	WEIGHTED AVERAGE	117.8	27.2	79.0	94.6	113.7	102.8	96.2	113.7	103.7	96.9	67.9	84.4	97.5	56.0	79.7	98.0	67.6	84.9

DIVISION		SOUTH CAROLINA COLUMBIA 290 - 292			FLORENCE 295			GREENVILLE 296			ROCK HILL 297			SPARTANBURG 293			SOUTH DAKOTA ABERDEEN 574		
		MAT.	INST.	TOTAL	MAT.	INST.	TOTAL	MAT.	INST.	TOTAL	MAT.	INST.	TOTAL	MAT.	INST.	TOTAL	MAT.	INST.	TOTAL
015433	CONTRACTOR EQUIPMENT		105.8	105.8		105.8	105.8		105.8	105.8		105.8	105.8		105.8	105.8		100.3	100.3
0241, 31 - 34	SITE & INFRASTRUCTURE, DEMOLITION	106.5	88.7	94.1	116.1	88.5	96.8	111.2	88.8	95.6	109.0	88.0	94.4	111.0	88.8	95.5	100.5	98.2	98.9
0310	Concrete Forming & Accessories	90.9	63.7	67.7	78.4	63.6	65.8	88.5	63.7	67.4	86.8	63.4	66.9	91.0	63.7	67.8	95.5	77.8	80.4
0320	Concrete Reinforcing	96.7	67.7	81.9	91.2	67.7	79.3	91.1	66.0	78.4	91.9	67.7	79.6	91.1	67.7	79.2	105.3	72.4	88.5
0330	Cast-in-Place Concrete	109.0	66.1	93.1	91.8	65.9	82.2	91.8	66.1	82.2	91.8	66.0	82.2	91.8	66.1	82.2	112.8	81.2	101.1
03	CONCRETE	91.7	66.9	80.7	91.7	66.7	80.6	90.4	66.6	79.9	88.5	66.7	78.8	90.6	66.9	80.1	98.1	78.6	89.5
04	MASONRY	87.6	65.3	73.8	78.3	65.3	70.3	76.0	65.3	69.4	99.5	65.3	78.4	78.3	65.3	70.3	114.9	74.9	90.2
05	METALS	99.4	91.9	96.9	101.3	90.5	98.0	101.3	91.1	98.2	100.5	90.8	97.6	101.3	91.1	98.2	91.4	82.4	88.7
06	WOOD, PLASTICS & COMPOSITES	90.0	65.6	76.7	73.6	65.6	69.3	85.3	65.6	74.6	84.0	65.6	74.0	89.0	65.6	76.3	97.6	76.4	86.1
07	THERMAL & MOISTURE PROTECTION	91.8	64.5	79.9	96.1	64.5	82.4	96.0	64.5	82.3	95.8	60.2	80.3	96.0	64.5	82.3	102.9	79.4	92.7
08	OPENINGS	98.1	63.8	90.1	95.9	63.8	88.4	95.8	63.8	88.4	95.8	63.8	88.4	95.8	63.8	88.4	99.5	63.8	91.2
0920	Plaster & Gypsum Board	97.3	64.6	75.7	86.4	64.6	72.0	91.1	64.6	73.6	90.8	64.6	73.5	93.9	64.6	74.6	108.3	76.1	87.1
0950, 0980	Ceilings & Acoustic Treatment	87.6	64.6	72.0	83.4	64.6	70.6	82.6	64.6	70.4	82.6	64.6	70.4	82.6	64.6	70.4	93.9	76.1	81.8
0960	Flooring	91.4	78.5	87.6	90.7	78.4	87.1	97.0	78.4	91.6	95.9	78.4	90.8	98.4	78.4	92.6	91.9	45.3	78.4
0970, 0990	Wall Finishes & Painting/Coating	94.9	67.8	78.9	97.0	67.8	79.7	97.0	67.8	79.7	97.0	67.8	79.7	97.0	67.8	79.7	87.8	34.3	56.1
09	FINISHES	86.7	67.2	76.1	85.6	67.2	75.6	87.5	67.2	76.4	86.9	67.2	76.1	88.3	67.2	76.8	89.1	67.9	77.5
COVERS	DIVS. 10 - 14, 25, 28, 41, 43, 44, 46	100.0	70.6	93.5	100.0	70.6	93.5	100.0	70.6	93.5	100.0	70.6	93.5	100.0	70.6	93.5	100.0	86.5	97.0
21, 22, 23	FIRE SUPPRESSION, PLUMBING & HVAC	100.0	57.7	82.9	100.5	57.7	83.2	100.5	57.7	83.2	97.0	56.1	80.5	100.5	57.7	83.2	100.1	55.4	82.0
26, 27, 3370	ELECTRICAL, COMMUNICATIONS & UTIL.	96.2	61.1	78.4	94.0	61.1	77.3	96.0	61.3	78.3	96.0	61.4	78.4	96.0	61.3	78.3	100.2	65.0	82.3
MF2016	WEIGHTED AVERAGE	96.5	67.9	84.2	96.2	67.8	84.0	96.2	67.9	84.1	96.0	67.3	83.7	96.4	67.9	84.2	98.3	71.5	86.8

SOUTH DAKOTA

DIVISION		MITCHELL 573			MOBRIDGE 576			PIERRE 575			RAPID CITY 577			SIOUX FALLS 570-571			WATERTOWN 572		
		MAT.	INST.	TOTAL	MAT.	INST.	TOTAL	MAT.	INST.	TOTAL	MAT.	INST.	TOTAL	MAT.	INST.	TOTAL	MAT.	INST.	TOTAL
015433	CONTRACTOR EQUIPMENT		100.3	100.3		100.3	100.3		100.3	100.3		100.3	100.3		101.2	101.2		100.3	100.3
0241, 31 - 34	SITE & INFRASTRUCTURE, DEMOLITION	97.2	97.9	97.7	97.2	98.0	97.7	100.3	97.4	98.2	98.8	97.4	97.8	90.9	99.7	97.1	97.1	98.0	97.7
0310	Concrete Forming & Accessories	94.6	45.4	52.7	85.1	45.8	51.7	95.7	46.6	53.9	103.2	58.1	64.8	100.0	78.4	81.6	81.6	76.6	77.3
0320	Concrete Reinforcing	104.7	72.6	88.4	107.4	72.5	89.7	107.6	96.7	102.0	98.4	96.9	97.6	105.2	97.1	101.1	101.8	72.5	86.9
0330	Cast-in-Place Concrete	109.7	53.8	88.9	109.7	79.5	98.4	110.2	79.1	98.6	108.8	79.4	97.9	87.1	80.3	84.6	109.7	79.4	98.4
03	CONCRETE	95.8	54.5	77.5	95.5	63.6	81.4	94.8	67.8	82.9	95.2	73.1	85.4	84.1	82.8	83.5	94.7	77.5	87.1
04	MASONRY	102.5	77.1	86.8	111.2	74.4	88.4	113.0	74.1	89.0	110.9	75.6	89.1	93.2	77.1	83.2	138.5	75.3	99.4
05	METALS	90.4	82.2	87.9	90.5	82.7	88.1	92.7	88.9	91.5	93.1	89.8	92.1	94.2	92.0	93.5	90.4	82.6	88.0
06	WOOD, PLASTICS & COMPOSITES	96.5	34.6	62.9	84.5	34.4	57.3	102.2	34.9	65.6	102.1	49.3	73.4	98.8	76.4	86.6	80.4	76.4	78.2
07	THERMAL & MOISTURE PROTECTION	102.7	74.9	90.6	102.7	77.2	91.6	104.6	74.2	91.3	103.4	80.1	93.3	101.6	84.2	94.0	102.5	78.8	92.2
08	OPENINGS	98.6	40.8	85.2	101.2	40.1	87.0	100.7	58.1	90.8	103.4	66.1	94.7	102.3	81.0	97.3	98.6	63.3	90.4
0920	Plaster & Gypsum Board	106.8	33.2	58.3	100.6	32.9	56.0	102.8	33.4	57.1	107.5	48.3	68.5	98.5	76.1	83.7	98.5	76.1	83.7
0950, 0980	Ceilings & Acoustic Treatment	89.7	33.2	51.4	93.9	32.9	52.6	91.4	33.4	52.1	94.8	48.3	63.3	92.8	76.1	81.5	89.7	76.1	80.5
0960	Flooring	91.6	45.3	78.1	87.3	48.0	75.9	97.5	32.3	78.6	91.3	75.1	86.6	97.0	77.5	91.4	86.0	45.3	74.2
0970, 0990	Wall Finishes & Painting/Coating	87.8	37.5	58.0	87.8	38.5	58.6	95.5	96.2	95.9	87.8	96.2	92.8	99.9	96.2	97.7	87.8	34.3	56.1
09	FINISHES	87.8	42.3	63.0	86.6	42.8	62.7	92.3	46.2	67.2	89.0	63.4	75.0	93.6	79.8	86.1	85.0	66.6	74.9
COVERS	DIVS. 10 - 14, 25, 28, 41, 43, 44, 46	100.0	83.7	96.4	100.0	83.6	96.4	100.0	83.7	96.4	100.0	85.3	96.8	100.0	88.3	97.4	100.0	87.6	97.3
21, 22, 23	FIRE SUPPRESSION, PLUMBING & HVAC	96.6	50.1	77.8	96.6	70.6	86.1	100.0	78.4	91.3	100.1	78.5	91.4	100.0	71.6	88.5	96.6	53.2	79.1
26, 27, 3370	ELECTRICAL, COMMUNICATIONS & UTIL.	98.5	64.9	81.5	100.2	40.7	70.0	103.6	48.9	75.8	96.7	48.9	72.4	100.4	65.0	82.4	97.7	65.0	81.0
MF2016	WEIGHTED AVERAGE	96.0	62.4	81.6	96.6	64.4	82.8	98.7	69.4	86.2	98.0	73.3	87.4	96.3	79.2	89.0	97.1	70.8	85.8

TENNESSEE

DIVISION		CHATTANOOGA 373 - 374			COLUMBIA 384			COOKEVILLE 385			JACKSON 383			JOHNSON CITY 376			KNOXVILLE 377 - 379		
		MAT.	INST.	TOTAL	MAT.	INST.	TOTAL	MAT.	INST.	TOTAL	MAT.	INST.	TOTAL	MAT.	INST.	TOTAL	MAT.	INST.	TOTAL
015433	CONTRACTOR EQUIPMENT		107.6	107.6		102.1	102.1		102.1	102.1		108.3	108.3		101.4	101.4		101.4	101.4
0241, 31 - 34	SITE & INFRASTRUCTURE, DEMOLITION	104.2	98.6	100.3	90.1	88.8	89.2	95.8	85.6	88.7	99.3	98.7	98.9	110.3	85.2	92.8	90.8	88.0	88.8
0310	Concrete Forming & Accessories	95.7	58.9	64.3	80.8	64.0	66.5	81.0	32.5	39.7	87.6	41.9	48.7	82.1	59.0	62.4	94.1	63.4	67.9
0320	Concrete Reinforcing	95.2	67.0	80.9	88.7	66.8	77.6	88.7	66.4	77.4	88.7	67.3	77.8	95.9	63.6	79.5	95.2	63.6	79.1
0330	Cast-in-Place Concrete	98.5	63.3	85.4	91.6	62.0	80.6	103.9	40.5	80.3	101.4	67.8	88.9	79.2	58.7	71.6	92.5	64.8	82.2
03	CONCRETE	94.2	63.6	80.7	90.1	65.5	79.2	100.0	43.7	75.0	91.0	57.6	76.2	104.7	61.4	85.5	91.8	65.5	80.1
04	MASONRY	100.1	57.5	73.8	112.7	56.7	78.1	107.9	40.3	66.1	113.1	42.5	69.5	112.9	44.1	70.4	77.9	51.7	61.7
05	METALS	95.3	88.8	93.4	94.0	89.5	92.6	94.1	87.9	92.2	96.4	89.2	94.2	92.6	86.9	90.9	95.8	87.1	93.2
06	WOOD, PLASTICS & COMPOSITES	106.5	57.9	80.1	69.1	65.2	67.0	69.3	30.3	48.1	84.4	40.8	60.7	78.8	64.4	71.0	93.3	64.4	77.6
07	THERMAL & MOISTURE PROTECTION	95.7	62.1	81.1	94.3	62.9	80.6	94.8	50.4	75.5	96.7	55.3	78.7	91.9	54.6	75.6	89.5	61.2	77.2
08	OPENINGS	104.1	59.2	93.6	93.1	54.9	84.2	93.1	36.5	79.9	99.8	45.2	87.2	100.3	61.3	91.3	97.6	56.8	88.1
0920	Plaster & Gypsum Board	83.1	57.2	66.1	91.2	64.7	73.8	91.2	28.8	50.1	93.6	39.6	58.1	101.7	63.9	76.8	109.5	63.9	79.5
0950, 0980	Ceilings & Acoustic Treatment	107.2	57.2	73.3	80.2	64.7	69.7	80.2	28.8	45.4	89.6	39.6	55.8	102.6	63.9	76.4	103.4	63.9	76.7
0960	Flooring	98.0	57.1	86.1	82.0	55.8	74.4	82.0	51.7	73.2	81.3	56.1	74.0	93.3	45.4	79.4	98.3	51.3	84.7
0970, 0990	Wall Finishes & Painting/Coating	97.7	61.1	76.0	84.9	57.9	69.0	84.9	57.9	69.0	86.6	57.9	69.6	94.4	60.3	74.2	94.4	59.1	73.5
09	FINISHES	96.6	58.2	75.6	85.9	61.9	72.8	86.4	37.2	59.6	86.2	44.6	63.5	100.5	56.9	76.7	93.5	60.7	75.6
COVERS	DIVS. 10 - 14, 25, 28, 41, 43, 44, 46	100.0	69.2	93.2	100.0	68.7	93.1	100.0	36.6	86.0	100.0	62.5	91.7	100.0	77.3	95.0	100.0	81.0	95.8
21, 22, 23	FIRE SUPPRESSION, PLUMBING & HVAC	100.1	60.6	84.2	98.0	75.3	88.8	98.0	68.5	86.1	100.1	60.0	83.9	99.9	56.8	82.5	99.9	61.8	84.5
26, 27, 3370	ELECTRICAL, COMMUNICATIONS & UTIL.	101.2	84.9	92.9	93.0	51.8	72.0	94.5	62.8	78.4	99.4	52.8	75.7	91.8	41.8	66.4	97.4	56.3	76.5
MF2016	WEIGHTED AVERAGE	98.9	69.6	86.3	94.7	67.9	83.2	96.1	57.3	79.5	97.5	59.7	81.3	99.1	59.9	82.3	95.6	65.2	82.6

TENNESSEE / TEXAS

DIVISION		MCKENZIE 382			MEMPHIS 375, 380 - 381			NASHVILLE 370 - 372			ABILENE 795 - 796			AMARILLO 790 - 791			AUSTIN 786 - 787		
		MAT.	INST.	TOTAL	MAT.	INST.	TOTAL	MAT.	INST.	TOTAL	MAT.	INST.	TOTAL	MAT.	INST.	TOTAL	MAT.	INST.	TOTAL
015433	CONTRACTOR EQUIPMENT		102.1	102.1		101.5	101.5		105.2	105.2		93.2	93.2		93.2	93.2		92.7	92.7
0241, 31 - 34	SITE & INFRASTRUCTURE, DEMOLITION	95.4	85.8	88.7	87.8	94.9	92.8	101.9	97.5	98.8	94.1	91.8	92.5	93.1	91.1	91.7	96.9	90.8	92.7
0310	Concrete Forming & Accessories	88.6	34.8	42.8	94.5	66.3	70.5	95.6	64.9	69.4	95.2	60.8	65.9	97.2	53.0	59.6	95.1	54.4	60.5
0320	Concrete Reinforcing	88.8	67.4	77.9	102.8	67.9	85.1	101.3	67.0	83.9	95.6	52.7	73.8	106.3	51.0	78.2	90.0	48.5	68.8
0330	Cast-in-Place Concrete	101.6	53.2	83.6	95.5	78.0	89.0	91.1	65.6	81.6	84.8	66.8	78.1	81.6	66.7	76.1	94.6	66.9	84.3
03	CONCRETE	98.6	49.3	76.7	96.7	71.8	85.6	94.9	66.8	82.5	79.5	62.3	71.9	81.3	58.4	71.1	87.3	58.7	74.6
04	MASONRY	111.8	46.1	71.2	95.5	57.5	72.0	87.7	57.0	68.8	94.3	61.9	74.3	93.1	61.4	73.5	92.9	61.9	73.8
05	METALS	94.1	88.5	92.4	89.9	84.1	88.2	103.5	85.8	98.1	102.7	70.7	93.0	94.3	69.8	86.9	105.3	68.8	94.2
06	WOOD, PLASTICS & COMPOSITES	78.2	32.6	53.4	97.0	67.1	80.7	97.6	65.6	80.2	102.6	62.5	80.8	100.8	52.3	74.5	93.2	54.0	71.9
07	THERMAL & MOISTURE PROTECTION	94.8	52.6	76.4	88.6	68.6	79.9	94.4	63.9	81.1	99.3	65.2	84.5	97.9	63.4	82.9	93.7	65.0	81.2
08	OPENINGS	93.1	38.2	80.3	100.5	63.0	91.8	98.9	64.4	90.9	102.9	58.2	92.5	103.7	52.2	91.8	104.0	50.9	91.7
0920	Plaster & Gypsum Board	94.5	31.2	52.8	94.1	66.4	75.8	93.8	66.4	74.7	88.6	61.9	71.0	95.9	51.4	66.6	90.4	53.1	65.8
0950, 0980	Ceilings & Acoustic Treatment	80.2	31.2	47.0	98.9	66.4	76.9	98.7	64.7	75.7	90.4	61.9	71.1	97.5	51.4	66.3	89.4	53.1	64.8
0960	Flooring	84.9	55.8	76.4	97.2	57.1	85.6	97.4	57.1	85.7	92.5	69.9	85.9	96.5	65.9	87.6	92.3	65.9	84.6
0970, 0990	Wall Finishes & Painting/Coating	84.9	45.3	61.5	94.0	60.9	74.4	99.5	69.1	81.5	94.4	53.0	69.9	89.4	53.0	67.9	94.7	45.4	65.5
09	FINISHES	87.6	38.6	60.8	98.1	63.6	79.3	98.3	63.5	79.3	82.7	61.8	71.3	89.7	55.0	70.8	87.8	55.1	70.0
COVERS	DIVS. 10 - 14, 25, 28, 41, 43, 44, 46	100.0	24.5	83.4	100.0	81.9	96.0	100.0	81.7	96.0	100.0	80.5	95.7	100.0	73.8	94.2	100.0	77.8	95.1
21, 22, 23	FIRE SUPPRESSION, PLUMBING & HVAC	98.0	59.9	82.5	100.0	71.9	88.6	100.0	75.5	90.1	100.3	53.3	81.3	100.0	52.2	80.7	100.0	60.0	83.8
26, 27, 3370	ELECTRICAL, COMMUNICATIONS & UTIL.	94.3	56.7	75.2	103.2	65.1	83.9	94.1	62.8	78.2	96.1	55.3	75.4	99.8	60.3	79.7	93.7	58.7	75.9
MF2016	WEIGHTED AVERAGE	96.2	56.0	79.0	97.3	71.1	86.1	98.4	71.1	86.7	96.0	62.8	81.8	95.7	61.1	80.9	97.4	62.7	82.5

TEXAS

DIVISION		BEAUMONT 776-777			BROWNWOOD 768			BRYAN 778			CHILDRESS 792			CORPUS CHRISTI 783-784			DALLAS 752-753		
		MAT.	INST.	TOTAL	MAT.	INST.	TOTAL	MAT.	INST.	TOTAL	MAT.	INST.	TOTAL	MAT.	INST.	TOTAL	MAT.	INST.	TOTAL
015433	CONTRACTOR EQUIPMENT		97.3	97.3		93.2	93.2		97.3	97.3		93.2	93.2		103.7	103.7		106.9	106.9
0241, 31 - 34	SITE & INFRASTRUCTURE, DEMOLITION	87.9	96.0	93.6	102.7	91.7	95.0	79.1	96.2	91.0	104.7	90.1	94.5	141.6	87.3	103.7	106.8	96.8	99.9
0310	Concrete Forming & Accessories	100.2	56.1	62.7	99.8	60.4	66.2	80.2	59.8	62.8	93.7	60.2	65.2	99.2	53.6	60.4	96.5	62.9	67.9
0320	Concrete Reinforcing	90.3	63.8	76.8	86.5	52.6	69.3	92.5	52.6	72.2	95.8	52.5	73.8	82.4	52.4	67.1	92.0	53.8	72.6
0330	Cast-in-Place Concrete	97.8	67.0	86.3	93.0	60.3	80.8	78.3	61.2	71.9	87.0	60.2	77.0	119.7	69.1	100.9	93.0	72.5	85.3
03	CONCRETE	97.4	62.2	81.8	87.9	59.8	75.4	80.9	60.0	71.6	86.9	59.7	74.9	97.9	60.8	81.5	96.4	66.2	83.0
04	MASONRY	94.3	63.0	75.0	124.4	61.9	85.8	123.8	61.9	85.6	98.3	61.4	75.5	84.5	62.0	70.6	98.2	60.7	75.0
05	METALS	103.7	76.5	95.4	103.6	70.2	93.5	103.8	72.3	94.3	100.2	70.1	91.1	101.6	85.3	96.7	103.1	84.0	97.3
06	WOOD, PLASTICS & COMPOSITES	111.1	55.6	81.0	109.6	62.5	84.0	76.8	61.0	68.2	101.9	62.5	80.5	118.0	53.0	82.7	98.8	64.7	80.2
07	THERMAL & MOISTURE PROTECTION	93.7	66.1	81.7	93.0	63.4	80.1	86.2	65.6	77.2	100.0	61.9	83.4	100.2	63.9	84.4	89.1	67.1	79.5
08	OPENINGS	97.1	56.6	87.7	102.4	58.2	92.1	97.9	57.6	88.5	98.2	58.2	88.9	103.7	51.5	91.6	97.5	60.4	88.9
0920	Plaster & Gypsum Board	104.4	54.8	71.7	86.6	61.9	70.3	91.4	60.4	71.0	88.2	61.9	70.9	99.4	51.9	68.1	94.2	63.7	74.1
0950, 0980	Ceilings & Acoustic Treatment	99.7	54.8	69.3	79.3	61.9	67.5	92.7	60.4	70.8	88.7	61.9	70.5	92.4	51.9	65.0	89.6	63.7	72.1
0960	Flooring	116.6	77.2	105.2	83.6	56.4	75.7	87.4	69.2	82.1	90.9	56.4	80.9	104.8	70.0	94.7	98.1	70.4	90.0
0970, 0990	Wall Finishes & Painting/Coating	92.7	55.3	70.6	91.7	53.0	68.8	90.0	58.9	71.6	94.4	53.0	69.9	112.2	45.4	72.7	101.4	55.2	74.0
09	FINISHES	94.3	59.4	75.3	78.1	59.1	67.7	82.4	61.1	70.7	83.1	59.1	70.0	97.7	55.4	74.6	95.8	63.5	78.2
COVERS	DIVS. 10 - 14, 25, 28, 41, 43, 44, 46	100.0	79.3	95.4	100.0	74.9	94.5	100.0	78.5	95.3	100.0	74.9	94.5	100.0	79.4	95.5	100.0	81.5	95.9
21, 22, 23	FIRE SUPPRESSION, PLUMBING & HVAC	100.1	64.4	85.7	96.6	50.6	78.0	96.7	63.8	83.4	96.8	52.9	79.0	100.1	49.8	79.8	100.0	60.7	84.1
26, 27, 3370	ELECTRICAL, COMMUNICATIONS & UTIL.	98.9	64.9	81.6	95.3	47.3	70.9	96.9	64.9	80.6	96.1	57.1	76.3	90.5	65.0	77.5	98.5	61.3	79.6
MF2016	WEIGHTED AVERAGE	98.8	67.1	85.2	97.4	60.2	81.4	95.5	66.4	83.1	95.8	61.8	81.2	99.7	63.0	84.0	99.1	67.7	85.6

TEXAS

DIVISION		DEL RIO 788			DENTON 762			EASTLAND 764			EL PASO 798 - 799, 885			FORT WORTH 760 - 761			GALVESTON 775		
		MAT.	INST.	TOTAL	MAT.	INST.	TOTAL	MAT.	INST.	TOTAL	MAT.	INST.	TOTAL	MAT.	INST.	TOTAL	MAT.	INST.	TOTAL
015433	CONTRACTOR EQUIPMENT		92.7	92.7		102.8	102.8		93.2	93.2		93.2	93.2		93.2	93.2		110.3	110.3
0241, 31 - 34	SITE & INFRASTRUCTURE, DEMOLITION	121.8	90.1	99.7	102.1	83.7	89.3	105.6	89.8	94.6	92.7	92.5	92.6	98.4	92.5	94.3	103.9	94.8	97.6
0310	Concrete Forming & Accessories	95.7	52.5	58.9	107.9	60.5	67.5	100.6	60.2	66.2	95.3	58.2	63.7	99.5	61.0	66.7	89.3	57.3	62.1
0320	Concrete Reinforcing	82.9	48.9	65.6	87.9	52.5	69.9	86.7	52.5	69.4	98.3	52.5	75.1	93.1	52.3	72.4	92.0	61.4	76.5
0330	Cast-in-Place Concrete	128.9	60.2	103.3	71.9	61.4	68.0	98.3	60.2	84.1	75.9	69.0	73.3	85.7	66.9	78.7	104.0	62.2	88.4
03	CONCRETE	118.6	55.5	90.7	69.0	61.3	65.6	92.2	59.7	77.8	77.5	61.8	70.6	83.8	62.3	74.3	98.7	61.8	82.4
04	MASONRY	99.5	61.8	76.2	133.1	60.6	88.3	94.2	61.9	74.2	85.5	64.0	72.2	93.8	60.5	73.2	89.5	62.0	72.5
05	METALS	101.6	68.0	91.4	103.2	85.5	97.8	103.4	70.0	93.3	101.3	70.5	92.0	105.3	70.7	94.8	105.6	91.8	101.4
06	WOOD, PLASTICS & COMPOSITES	98.2	51.9	73.0	122.3	62.7	89.9	117.0	62.5	87.4	90.5	58.6	73.2	101.2	62.5	80.2	93.5	57.5	73.9
07	THERMAL & MOISTURE PROTECTION	97.0	63.8	82.5	90.9	63.9	79.2	93.4	63.4	80.3	95.1	65.6	82.2	100.0	65.5	79.4	85.5	66.2	77.1
08	OPENINGS	98.9	49.2	87.3	120.9	57.9	106.2	71.7	58.2	68.6	97.8	54.3	87.7	102.0	58.1	91.8	101.9	57.7	91.6
0920	Plaster & Gypsum Board	96.0	50.9	66.3	91.3	61.9	71.9	86.6	61.9	70.3	97.2	57.8	71.2	93.6	61.9	72.7	98.4	56.6	70.9
0950, 0980	Ceilings & Acoustic Treatment	88.8	50.9	63.2	84.4	61.9	69.1	79.3	61.9	67.5	91.1	57.8	68.5	89.3	61.9	70.7	97.0	56.6	69.6
0960	Flooring	89.9	56.4	80.2	79.2	56.4	72.6	106.4	56.4	91.9	93.1	72.1	87.0	104.6	66.6	93.6	102.7	69.2	93.0
0970, 0990	Wall Finishes & Painting/Coating	99.5	43.7	66.5	101.9	50.2	71.3	93.1	53.0	69.4	99.7	50.9	70.8	97.2	53.2	71.1	101.6	55.1	74.1
09	FINISHES	90.3	51.8	69.3	77.0	58.9	67.1	85.1	59.1	70.9	88.4	59.7	72.7	90.7	61.2	74.6	91.9	58.6	73.7
COVERS	DIVS. 10 - 14, 25, 28, 41, 43, 44, 46	100.0	75.2	94.5	100.0	80.8	95.8	100.0	73.2	94.1	100.0	77.9	95.1	100.0	80.5	95.7	100.0	79.8	95.6
21, 22, 23	FIRE SUPPRESSION, PLUMBING & HVAC	96.6	60.7	82.1	96.6	57.1	80.6	96.6	50.6	78.0	99.9	65.9	86.2	99.9	56.6	82.4	96.6	63.9	83.4
26, 27, 3370	ELECTRICAL, COMMUNICATIONS & UTIL.	92.2	61.3	76.5	97.5	60.1	78.5	95.2	60.0	77.3	94.8	51.7	72.9	97.0	60.0	78.2	98.6	65.0	81.5
MF2016	WEIGHTED AVERAGE	100.6	62.0	84.0	97.5	64.4	83.3	94.0	61.8	80.2	94.6	64.7	81.8	97.4	64.1	83.1	98.5	68.1	85.5

TEXAS

DIVISION		GIDDINGS 789			GREENVILLE 754			HOUSTON 770 - 772			HUNTSVILLE 773			LAREDO 780			LONGVIEW 756		
		MAT.	INST.	TOTAL	MAT.	INST.	TOTAL	MAT.	INST.	TOTAL	MAT.	INST.	TOTAL	MAT.	INST.	TOTAL	MAT.	INST.	TOTAL
015433	CONTRACTOR EQUIPMENT		92.7	92.7		103.6	103.6		102.2	102.2		97.3	97.3		92.7	92.7		94.6	94.6
0241, 31 - 34	SITE & INFRASTRUCTURE, DEMOLITION	106.7	90.4	95.3	97.7	86.8	90.1	102.8	94.6	97.0	94.4	95.9	95.4	100.0	90.9	93.6	96.1	93.3	94.2
0310	Concrete Forming & Accessories	93.2	52.6	58.6	88.0	58.7	63.1	94.5	58.6	64.0	87.2	55.6	60.3	95.8	53.6	59.9	83.9	60.4	63.9
0320	Concrete Reinforcing	83.4	49.8	66.4	92.4	52.6	72.1	91.8	52.8	72.0	92.7	52.6	72.3	82.9	52.4	67.4	91.5	52.2	71.5
0330	Cast-in-Place Concrete	109.1	60.5	91.0	92.5	61.4	80.9	96.5	63.3	84.2	107.7	61.1	90.4	92.1	67.0	82.7	107.6	60.3	90.0
03	CONCRETE	95.0	55.9	77.6	89.3	60.4	76.5	97.9	60.6	81.3	105.8	58.0	84.6	89.2	59.0	75.8	106.3	59.7	85.6
04	MASONRY	106.7	61.9	79.1	154.1	60.5	96.3	88.7	62.9	72.8	124.6	61.9	85.9	92.5	61.9	73.6	150.1	60.5	94.7
05	METALS	101.1	68.9	91.3	100.6	84.1	95.6	108.4	77.8	99.1	103.7	72.1	94.1	104.1	70.1	93.8	93.2	69.1	85.9
06	WOOD, PLASTICS & COMPOSITES	97.4	51.9	72.7	87.7	60.2	72.7	98.9	59.0	77.2	86.1	55.6	69.5	98.2	52.9	73.6	81.6	62.6	71.3
07	THERMAL & MOISTURE PROTECTION	97.4	64.5	83.1	89.1	63.0	77.7	87.2	67.3	78.6	87.3	65.0	77.6	96.2	64.8	82.6	90.6	62.5	78.3
08	OPENINGS	98.0	50.1	86.9	94.1	56.9	85.4	105.1	56.5	93.8	97.8	54.0	87.7	99.0	51.2	87.9	85.1	57.8	78.7
0920	Plaster & Gypsum Board	95.1	50.9	66.0	84.9	59.2	68.0	97.7	58.0	71.5	96.0	54.8	68.9	96.7	51.9	67.2	84.0	61.9	69.4
0950, 0980	Ceilings & Acoustic Treatment	88.8	50.9	63.2	85.0	59.2	67.5	93.9	58.0	69.6	92.7	54.8	67.0	92.2	51.9	64.9	80.8	61.9	68.0
0960	Flooring	89.5	56.4	79.9	91.8	56.4	81.5	103.8	72.1	94.6	91.7	56.4	81.5	89.8	65.9	82.8	96.9	56.4	85.1
0970, 0990	Wall Finishes & Painting/Coating	99.5	45.4	67.5	95.8	53.0	70.5	99.4	58.9	75.5	90.0	58.9	71.6	99.5	45.4	67.5	86.0	50.2	64.8
09	FINISHES	89.0	52.0	68.8	90.3	57.7	72.5	97.9	60.9	77.7	85.1	55.6	69.0	89.4	54.4	70.3	95.5	58.8	75.5
COVERS	DIVS. 10 - 14, 25, 28, 41, 43, 44, 46	100.0	71.8	93.8	100.0	78.1	95.2	100.0	81.9	96.0	100.0	71.2	93.7	100.0	77.4	95.0	100.0	80.7	95.7
21, 22, 23	FIRE SUPPRESSION, PLUMBING & HVAC	96.7	62.8	83.0	96.7	59.3	81.6	100.1	65.5	86.1	96.7	63.8	83.4	100.1	60.2	84.0	96.6	58.8	81.3
26, 27, 3370	ELECTRICAL, COMMUNICATIONS & UTIL.	89.6	57.0	73.0	95.2	60.1	77.4	100.7	68.1	84.1	96.9	60.6	78.4	92.3	58.5	75.1	95.6	52.1	73.5
MF2016	WEIGHTED AVERAGE	97.0	61.9	82.0	98.2	64.5	83.8	100.7	67.9	86.6	99.4	64.3	84.4	97.1	62.7	82.4	98.4	62.6	83.0

For customer support on your Electrical Costs with RSMeans data, call 800.448.8182.

545

City Cost Indexes

TEXAS

DIVISION		LUBBOCK 793-794 MAT.	INST.	TOTAL	LUFKIN 759 MAT.	INST.	TOTAL	MCALLEN 785 MAT.	INST.	TOTAL	MCKINNEY 750 MAT.	INST.	TOTAL	MIDLAND 797 MAT.	INST.	TOTAL	ODESSA 797 MAT.	INST.	TOTAL
015433	CONTRACTOR EQUIPMENT		105.6	105.6		94.6	94.6		103.8	103.8		103.6	103.6		105.6	105.6		93.2	93.2
0241, 31-34	SITE & INFRASTRUCTURE, DEMOLITION	118.6	91.0	99.4	91.0	94.8	93.7	146.4	87.3	105.2	94.1	86.8	89.0	121.6	90.0	99.6	94.3	92.5	93.0
0310	Concrete Forming & Accessories	94.4	54.5	60.4	87.1	55.7	60.3	99.7	52.6	59.6	87.2	58.7	63.0	98.2	60.7	66.3	95.1	60.8	65.9
0320	Concrete Reinforcing	96.9	52.8	74.4	93.1	68.7	80.7	82.5	52.3	67.2	92.4	52.6	72.2	97.9	52.7	74.9	95.6	52.6	73.7
0330	Cast-in-Place Concrete	85.0	70.2	79.5	96.2	60.1	82.8	130.0	61.6	104.6	86.7	61.4	77.3	90.3	67.8	81.9	84.8	66.8	78.1
03	CONCRETE	78.4	61.6	71.0	98.3	60.2	81.4	105.8	57.7	84.5	84.4	60.4	73.8	82.3	63.6	74.0	79.5	62.3	71.9
04	MASONRY	93.7	62.3	74.3	115.0	61.8	82.1	99.2	62.0	76.2	165.6	60.5	100.7	110.1	61.5	80.1	94.3	61.4	74.0
05	METALS	106.4	86.3	100.3	100.8	74.0	92.7	101.4	85.1	96.5	100.6	84.2	95.6	104.7	85.8	99.0	102.1	70.6	92.5
06	WOOD, PLASTICS & COMPOSITES	102.2	53.6	75.8	89.0	55.7	70.9	116.6	52.0	81.5	86.6	60.2	72.2	106.9	62.7	82.9	102.6	62.5	80.8
07	THERMAL & MOISTURE PROTECTION	88.6	65.3	78.5	90.3	61.6	77.8	100.1	64.7	84.7	88.9	63.0	77.6	88.9	65.1	78.5	99.3	64.2	84.1
08	OPENINGS	110.9	53.8	97.6	64.5	58.2	63.1	102.7	50.0	90.5	94.1	56.9	85.4	112.4	58.3	99.8	102.9	58.2	92.5
0920	Plaster & Gypsum Board	88.8	52.6	64.9	83.1	54.8	64.5	100.4	50.9	67.8	84.9	59.2	68.0	90.6	61.9	71.7	88.6	61.9	71.0
0950, 0980	Ceilings & Acoustic Treatment	91.3	52.6	65.0	76.5	54.8	61.8	92.2	50.9	64.3	85.0	59.2	67.5	89.6	61.9	70.8	90.4	61.9	71.1
0960	Flooring	86.4	71.3	82.0	129.5	56.4	108.3	104.4	75.2	95.9	91.4	56.4	81.2	87.6	65.9	81.3	92.5	65.9	84.8
0970, 0990	Wall Finishes & Painting/Coating	104.9	55.2	75.5	86.0	53.0	66.5	112.2	43.7	71.7	95.8	53.0	70.5	104.9	53.0	74.2	94.4	53.0	69.9
09	FINISHES	85.1	57.1	69.8	103.7	55.5	77.4	98.1	55.7	74.9	89.9	57.7	72.3	85.8	61.1	72.3	82.7	61.0	70.9
COVERS	DIVS. 10-14, 25, 28, 41, 43, 44, 46	100.0	79.9	95.6	100.0	71.5	93.7	100.0	77.5	95.0	100.0	80.8	95.8	100.0	75.3	94.6	100.0	75.0	94.5
21, 22, 23	FIRE SUPPRESSION, PLUMBING & HVAC	99.7	54.3	81.3	96.6	59.4	81.6	96.7	49.8	77.7	96.7	61.6	82.5	96.3	48.5	76.9	100.3	53.7	81.4
26, 27, 3370	ELECTRICAL, COMMUNICATIONS & UTIL.	94.9	61.6	77.9	96.9	62.9	79.6	90.3	32.4	60.8	95.3	60.1	77.4	94.9	60.4	77.3	96.2	60.3	78.0
MF2016	WEIGHTED AVERAGE	97.5	64.4	83.3	95.5	64.2	82.1	100.6	57.9	82.3	98.0	65.1	83.9	98.1	63.7	83.3	96.0	63.3	82.0

TEXAS

DIVISION		PALESTINE 758 MAT.	INST.	TOTAL	SAN ANGELO 769 MAT.	INST.	TOTAL	SAN ANTONIO 781-782 MAT.	INST.	TOTAL	TEMPLE 765 MAT.	INST.	TOTAL	TEXARKANA 755 MAT.	INST.	TOTAL	TYLER 757 MAT.	INST.	TOTAL
015433	CONTRACTOR EQUIPMENT		94.6	94.6		93.2	93.2		93.5	93.5		93.2	93.2		94.6	94.6		94.6	94.6
0241, 31-34	SITE & INFRASTRUCTURE, DEMOLITION	96.5	93.6	94.5	98.9	92.1	94.2	98.8	93.5	95.1	87.3	92.0	90.6	85.5	96.0	92.9	95.4	93.4	94.0
0310	Concrete Forming & Accessories	78.2	60.4	63.1	100.2	52.9	59.9	96.4	53.9	60.2	103.8	52.6	60.2	94.4	61.0	66.0	88.8	60.5	64.8
0320	Concrete Reinforcing	90.8	52.5	71.3	86.4	52.6	69.2	92.1	49.1	70.2	86.6	52.3	69.2	90.6	52.3	71.2	91.5	52.2	71.5
0330	Cast-in-Place Concrete	87.9	60.3	77.7	87.8	66.8	80.0	93.7	69.4	84.7	71.9	60.3	67.6	88.6	66.9	80.5	105.7	60.3	88.8
03	CONCRETE	100.7	59.8	82.6	83.9	58.7	72.7	93.9	59.3	78.6	71.6	56.2	64.8	90.8	62.3	78.2	105.8	59.8	85.4
04	MASONRY	109.4	60.5	79.2	121.0	61.9	84.5	95.5	62.0	74.8	132.2	61.9	88.8	169.9	60.5	102.3	159.7	60.5	98.4
05	METALS	100.5	69.2	91.1	103.8	70.6	93.7	106.5	66.7	94.4	103.5	69.7	93.2	93.1	69.7	86.0	100.4	69.2	91.0
06	WOOD, PLASTICS & COMPOSITES	79.5	62.6	70.3	109.9	51.9	78.4	99.5	53.0	74.3	119.9	51.9	83.0	94.1	62.6	77.0	90.8	62.6	75.5
07	THERMAL & MOISTURE PROTECTION	90.9	63.5	79.0	92.8	64.4	80.4	92.3	66.7	81.2	92.3	63.4	79.7	90.0	64.8	79.0	90.7	63.5	78.9
08	OPENINGS	64.5	58.3	63.0	102.4	52.3	90.8	101.7	50.8	89.9	68.3	52.3	64.5	85.0	58.2	78.8	64.4	58.2	63.0
0920	Plaster & Gypsum Board	81.3	61.9	68.5	86.6	50.9	63.1	98.5	51.9	67.9	86.6	50.9	63.1	88.6	61.9	71.0	83.1	61.9	69.1
0950, 0980	Ceilings & Acoustic Treatment	76.5	61.9	66.6	79.3	50.9	60.1	91.3	51.9	64.6	79.3	50.9	60.1	80.8	61.9	68.0	76.5	61.9	66.6
0960	Flooring	120.4	56.4	101.8	83.7	56.4	75.8	98.5	70.0	90.2	108.0	56.4	93.0	105.3	65.9	93.8	131.8	56.4	110.0
0970, 0990	Wall Finishes & Painting/Coating	86.0	53.0	66.5	91.7	53.0	68.8	99.4	45.4	67.5	93.1	45.4	64.9	86.0	53.0	66.5	86.0	53.0	66.5
09	FINISHES	101.4	59.2	78.3	77.8	52.8	64.2	99.0	55.4	75.2	84.3	52.0	66.6	98.0	61.1	77.8	104.7	59.2	79.8
COVERS	DIVS. 10-14, 25, 28, 41, 43, 44, 46	100.0	75.1	94.5	100.0	77.6	95.1	100.0	79.3	95.4	100.0	73.8	94.2	100.0	80.7	95.8	100.0	80.7	95.7
21, 22, 23	FIRE SUPPRESSION, PLUMBING & HVAC	96.6	58.0	81.0	96.6	53.4	79.1	100.1	61.1	84.3	96.6	54.4	79.5	96.6	56.8	80.5	96.6	59.5	81.6
26, 27, 3370	ELECTRICAL, COMMUNICATIONS & UTIL.	93.0	49.6	70.9	99.7	55.9	77.4	93.7	61.2	77.2	96.5	53.3	74.5	96.8	55.9	76.0	95.6	55.0	75.0
MF2016	WEIGHTED AVERAGE	95.1	62.0	80.9	97.1	60.8	81.6	99.3	63.5	83.9	92.5	60.0	78.5	97.5	63.6	83.0	98.6	63.2	83.5

TEXAS / UTAH

DIVISION		VICTORIA 779 MAT.	INST.	TOTAL	WACO 766-767 MAT.	INST.	TOTAL	WAXAHACHIE 751 MAT.	INST.	TOTAL	WHARTON 774 MAT.	INST.	TOTAL	WICHITA FALLS 763 MAT.	INST.	TOTAL	LOGAN 843 MAT.	INST.	TOTAL
015433	CONTRACTOR EQUIPMENT		109.0	109.0		93.2	93.2		103.6	103.6		110.3	110.3		93.2	93.2		96.9	96.9
0241, 31-34	SITE & INFRASTRUCTURE, DEMOLITION	108.9	92.0	97.1	95.8	92.4	93.4	95.9	87.0	89.7	113.9	94.2	100.2	96.6	92.5	93.7	99.7	94.1	95.8
0310	Concrete Forming & Accessories	88.5	52.7	58.1	102.2	60.9	67.0	87.2	60.7	64.6	84.0	54.5	58.9	102.2	60.9	67.0	101.1	67.5	72.5
0320	Concrete Reinforcing	88.4	52.4	70.1	86.3	48.6	67.1	92.4	52.6	72.2	91.9	52.4	71.8	86.3	51.1	68.4	105.6	86.4	95.8
0330	Cast-in-Place Concrete	116.8	62.1	96.5	77.8	66.8	73.7	91.5	61.6	80.4	120.0	61.1	98.1	83.1	66.8	77.1	86.4	75.2	82.2
03	CONCRETE	107.0	58.2	85.3	77.6	61.6	70.5	88.2	61.4	76.3	111.4	58.6	88.0	80.1	62.1	72.1	100.7	74.0	88.8
04	MASONRY	104.0	62.0	78.0	92.5	61.9	73.6	154.7	60.6	96.5	90.9	61.9	73.0	92.9	61.4	73.5	108.1	67.8	83.2
05	METALS	103.8	88.2	99.1	105.8	68.8	94.6	100.6	84.4	95.7	105.6	87.8	100.2	105.8	70.7	95.2	104.3	83.6	98.0
06	WOOD, PLASTICS & COMPOSITES	96.5	52.0	72.3	118.0	62.5	87.8	86.6	62.8	73.6	86.6	54.2	69.0	118.0	62.5	87.8	87.0	67.8	76.6
07	THERMAL & MOISTURE PROTECTION	88.9	62.8	77.6	93.2	65.5	81.2	89.0	63.2	77.8	85.8	65.5	77.0	93.2	64.5	80.7	99.7	71.7	87.5
08	OPENINGS	101.7	52.5	90.3	79.7	57.2	74.5	94.1	58.3	85.8	101.9	53.8	90.7	79.7	58.2	74.7	94.4	66.8	88.0
0920	Plaster & Gypsum Board	94.6	50.9	65.9	87.0	61.9	70.4	85.3	51.9	63.9	93.5	53.3	67.0	87.0	61.9	70.4	81.1	66.9	71.8
0950, 0980	Ceilings & Acoustic Treatment	97.8	50.9	66.1	81.0	61.9	68.0	86.7	51.9	63.9	97.0	53.3	67.4	81.0	61.9	68.0	104.7	66.9	79.1
0960	Flooring	101.4	56.4	88.3	107.3	65.9	95.3	91.4	56.4	81.2	100.0	68.8	90.9	108.0	74.3	98.3	97.5	59.7	86.5
0970, 0990	Wall Finishes & Painting/Coating	101.9	58.9	76.5	93.1	53.0	69.4	95.8	53.0	70.5	101.6	57.2	75.3	95.1	53.0	70.2	93.4	55.7	71.1
09	FINISHES	89.4	53.5	69.8	84.9	61.0	71.9	90.5	59.3	73.4	91.0	56.8	72.3	85.3	62.7	73.0	92.4	64.4	77.1
COVERS	DIVS. 10-14, 25, 28, 41, 43, 44, 46	100.0	71.1	93.6	100.0	80.5	95.7	100.0	81.0	95.8	100.0	71.3	93.7	100.0	74.9	94.5	100.0	83.7	96.4
21, 22, 23	FIRE SUPPRESSION, PLUMBING & HVAC	96.7	63.7	83.3	100.1	60.1	83.9	96.7	59.3	81.6	96.6	63.3	83.2	100.1	53.3	81.2	99.9	69.9	87.8
26, 27, 3370	ELECTRICAL, COMMUNICATIONS & UTIL.	103.7	55.3	79.1	99.8	55.4	77.2	95.2	60.1	77.4	102.6	60.6	81.2	101.5	55.3	78.0	95.7	70.5	82.9
MF2016	WEIGHTED AVERAGE	100.5	64.4	85.0	94.3	63.9	81.3	98.1	65.1	83.9	100.8	65.8	85.8	94.9	62.7	81.1	99.4	73.1	88.1

City Cost Indexes

DIVISION		UTAH												VERMONT					
		OGDEN 842, 844			PRICE 845			PROVO 846 - 847			SALT LAKE CITY 840 - 841			BELLOWS FALLS 051			BENNINGTON 052		
		MAT.	INST.	TOTAL	MAT.	INST.	TOTAL	MAT.	INST.	TOTAL	MAT.	INST.	TOTAL	MAT.	INST.	TOTAL	MAT.	INST.	TOTAL
015433	CONTRACTOR EQUIPMENT		96.9	96.9		95.9	95.9		95.9	95.9		96.8	96.8		97.9	97.9		97.9	97.9
0241, 31 - 34	SITE & INFRASTRUCTURE, DEMOLITION	87.6	94.1	92.1	97.1	92.2	93.7	95.9	92.5	93.5	87.4	94.0	92.0	90.1	100.7	97.5	89.5	100.7	97.3
0310	Concrete Forming & Accessories	101.1	67.5	72.5	103.5	66.0	71.6	102.8	67.5	72.8	103.4	67.5	72.9	96.5	104.2	103.1	93.9	104.1	102.6
0320	Concrete Reinforcing	105.2	86.4	95.7	113.2	86.4	99.6	114.2	86.4	100.1	107.5	86.4	96.8	85.4	84.3	84.9	85.4	84.3	84.8
0330	Cast-in-Place Concrete	87.7	75.2	83.1	86.5	73.8	81.8	86.5	75.2	82.3	95.8	75.2	88.2	91.3	112.8	99.3	91.3	112.7	99.3
03	CONCRETE	90.6	74.0	83.2	101.8	72.8	88.9	100.2	73.9	88.6	109.2	74.0	93.6	88.1	103.4	94.9	87.9	103.3	94.7
04	MASONRY	103.0	67.8	81.3	113.9	66.5	84.6	114.0	67.8	85.5	116.7	67.8	86.5	101.6	88.0	93.2	112.4	88.0	97.3
05	METALS	104.8	83.6	98.4	101.5	83.5	96.0	102.5	83.6	96.8	108.2	83.6	100.8	85.1	89.6	86.4	85.0	89.4	86.3
06	WOOD, PLASTICS & COMPOSITES	87.0	67.8	76.6	90.3	65.9	77.1	88.6	67.8	77.3	89.0	67.8	77.5	106.9	109.8	108.5	103.8	109.8	107.1
07	THERMAL & MOISTURE PROTECTION	98.5	71.7	86.8	101.6	70.2	87.9	101.6	71.7	88.6	106.2	71.7	91.2	98.7	90.7	95.2	98.7	90.7	95.2
08	OPENINGS	94.4	66.8	88.0	98.3	78.5	93.7	98.3	66.8	91.0	96.2	66.8	89.4	102.4	99.3	101.7	102.4	99.3	101.7
0920	Plaster & Gypsum Board	81.1	66.9	71.8	83.9	65.0	71.4	81.7	66.9	72.0	93.1	66.9	75.9	104.9	109.8	108.1	103.4	109.8	107.6
0950, 0980	Ceilings & Acoustic Treatment	104.7	66.9	79.1	104.7	65.0	77.8	104.7	66.9	79.1	96.1	66.9	76.3	97.0	109.8	105.7	97.0	109.8	105.7
0960	Flooring	95.4	59.7	85.0	98.6	58.0	86.8	98.4	59.7	87.1	99.3	59.7	87.8	91.2	102.2	94.4	90.4	102.2	93.8
0970, 0990	Wall Finishes & Painting/Coating	93.4	55.7	71.1	93.4	55.7	71.1	93.4	55.7	71.1	96.5	56.8	73.0	87.1	104.2	97.2	87.1	104.2	97.2
09	FINISHES	90.6	64.4	76.3	93.6	62.8	76.8	93.1	64.4	77.4	92.5	64.5	77.2	86.3	104.8	96.4	85.9	104.8	96.2
COVERS	DIVS. 10 - 14, 25, 28, 41, 43, 44, 46	100.0	83.7	96.4	100.0	83.4	96.3	100.0	83.7	96.4	100.0	83.7	96.4	100.0	93.7	98.6	100.0	93.6	98.6
21, 22, 23	FIRE SUPPRESSION, PLUMBING & HVAC	99.9	69.9	87.8	98.1	64.6	84.5	99.9	69.9	87.8	100.1	69.9	87.8	96.7	89.2	93.7	96.7	89.2	93.7
26, 27, 3370	ELECTRICAL, COMMUNICATIONS & UTIL.	96.0	70.5	83.0	101.0	70.5	85.5	96.3	70.3	83.1	98.6	70.5	84.3	98.9	81.5	90.0	98.9	55.8	77.0
MF2016	WEIGHTED AVERAGE	97.5	73.1	87.0	100.0	71.7	87.9	99.8	72.9	88.3	102.0	73.1	89.6	94.2	93.6	94.0	94.6	90.0	92.6

DIVISION		VERMONT																	
		BRATTLEBORO 053			BURLINGTON 054			GUILDHALL 059			MONTPELIER 056			RUTLAND 057			ST. JOHNSBURY 058		
		MAT.	INST.	TOTAL	MAT.	INST.	TOTAL	MAT.	INST.	TOTAL	MAT.	INST.	TOTAL	MAT.	INST.	TOTAL	MAT.	INST.	TOTAL
015433	CONTRACTOR EQUIPMENT		97.9	97.9		97.9	97.9		97.9	97.9		97.9	97.9		97.9	97.9		97.9	97.9
0241, 31 - 34	SITE & INFRASTRUCTURE, DEMOLITION	91.1	104.2	103.1	93.0	100.9	98.5	89.1	96.2	94.0	95.2	100.6	99.0	93.5	100.6	98.5	89.2	99.6	96.5
0310	Concrete Forming & Accessories	96.7	104.2	103.1	98.0	81.9	84.3	94.1	97.5	97.0	97.6	103.6	102.7	97.0	81.9	84.2	92.4	97.9	97.1
0320	Concrete Reinforcing	84.5	84.3	84.4	104.8	84.2	94.3	86.2	84.1	85.1	104.9	84.2	94.4	106.8	84.2	95.3	84.5	84.2	84.3
0330	Cast-in-Place Concrete	94.2	112.8	101.1	110.1	112.5	111.0	88.5	103.8	94.2	107.8	112.6	109.6	89.5	112.5	98.0	88.5	104.0	94.3
03	CONCRETE	90.2	103.4	96.0	100.0	93.2	97.0	85.6	97.1	90.7	99.0	103.0	100.7	90.6	93.1	91.7	85.3	97.4	90.7
04	MASONRY	112.0	88.0	97.2	113.8	88.1	97.9	112.3	73.0	88.0	104.3	88.1	94.3	92.5	88.1	89.8	140.6	73.0	98.8
05	METALS	85.0	89.6	86.4	92.3	88.8	91.2	85.1	88.5	86.1	90.2	88.9	89.8	90.5	88.7	90.0	85.1	88.9	86.2
06	WOOD, PLASTICS & COMPOSITES	107.2	109.8	108.6	100.8	80.6	89.8	102.8	109.8	106.6	96.5	109.8	103.8	107.4	80.6	92.8	97.5	109.8	104.2
07	THERMAL & MOISTURE PROTECTION	98.8	90.7	95.3	105.9	87.6	97.9	98.5	83.8	92.1	105.7	90.8	99.2	98.9	87.6	94.0	98.4	83.8	92.1
08	OPENINGS	102.4	99.6	101.8	104.1	79.6	98.4	102.4	95.7	100.9	106.0	95.7	103.6	105.6	79.6	99.5	102.4	95.7	100.9
0920	Plaster & Gypsum Board	104.9	109.8	108.1	106.4	79.7	88.8	110.7	109.8	110.1	106.7	109.8	108.7	105.3	79.7	88.4	112.2	109.8	110.6
0950, 0980	Ceilings & Acoustic Treatment	97.0	109.8	105.7	101.0	79.7	86.6	97.0	109.8	105.7	106.9	109.8	108.9	101.5	79.7	86.7	97.0	109.8	105.7
0960	Flooring	91.3	102.2	94.5	98.1	102.2	99.3	94.1	102.2	96.5	100.4	102.2	100.9	91.2	102.2	94.4	97.4	102.2	98.8
0970, 0990	Wall Finishes & Painting/Coating	87.1	104.2	97.2	96.4	90.7	93.0	87.1	90.7	89.2	95.3	90.7	92.5	87.1	90.7	89.2	87.1	90.7	89.2
09	FINISHES	86.5	104.8	96.5	92.5	86.1	89.0	87.9	99.5	94.3	94.2	103.4	99.2	87.4	86.1	86.7	89.1	99.5	94.8
COVERS	DIVS. 10 - 14, 25, 28, 41, 43, 44, 46	100.0	93.7	98.6	100.0	90.4	97.9	100.0	88.5	97.5	100.0	93.6	98.6	100.0	90.4	97.9	100.0	88.5	97.5
21, 22, 23	FIRE SUPPRESSION, PLUMBING & HVAC	96.7	89.2	93.7	100.1	69.4	87.7	96.7	61.5	82.5	96.6	69.4	85.6	100.2	69.4	87.8	96.7	61.5	82.5
26, 27, 3370	ELECTRICAL, COMMUNICATIONS & UTIL.	98.9	81.5	90.0	98.7	55.0	76.5	98.9	55.8	77.0	98.1	55.0	76.2	98.9	55.0	76.6	98.9	55.8	77.0
MF2016	WEIGHTED AVERAGE	95.0	93.6	94.4	99.1	80.6	91.2	94.5	80.2	88.3	97.7	85.3	92.4	96.3	80.5	89.5	95.8	80.5	89.3

DIVISION		VERMONT			VIRGINIA														
		WHITE RIVER JCT. 050			ALEXANDRIA 223			ARLINGTON 222			BRISTOL 242			CHARLOTTESVILLE 229			CULPEPER 227		
		MAT.	INST.	TOTAL	MAT.	INST.	TOTAL	MAT.	INST.	TOTAL	MAT.	INST.	TOTAL	MAT.	INST.	TOTAL	MAT.	INST.	TOTAL
015433	CONTRACTOR EQUIPMENT		97.9	97.9		107.5	107.5		106.2	106.2		106.2	106.2		110.4	110.4		106.2	106.2
0241, 31 - 34	SITE & INFRASTRUCTURE, DEMOLITION	93.5	99.7	97.8	114.3	92.2	98.9	124.3	89.9	100.4	108.9	89.2	95.2	113.4	91.7	98.3	112.0	89.9	96.6
0310	Concrete Forming & Accessories	91.3	98.4	97.4	90.2	70.2	73.1	89.0	70.1	73.0	85.1	65.6	68.5	83.5	47.6	52.9	80.8	72.9	74.0
0320	Concrete Reinforcing	85.4	84.2	84.8	83.9	87.4	85.7	94.7	87.4	91.0	94.7	69.8	82.0	94.1	70.1	81.9	94.7	87.4	91.0
0330	Cast-in-Place Concrete	94.2	104.8	98.2	110.1	76.5	97.6	107.1	76.5	95.7	106.7	45.4	83.9	111.0	76.9	98.3	109.7	76.4	97.3
03	CONCRETE	91.9	97.9	94.6	99.6	76.8	89.5	103.8	76.8	91.9	100.2	61.0	82.8	100.8	63.8	84.4	97.8	78.0	89.0
04	MASONRY	126.0	74.4	94.1	89.4	72.3	78.9	103.1	72.3	84.1	92.7	46.3	64.1	116.9	55.7	79.1	104.6	72.3	84.7
05	METALS	85.1	88.9	86.2	103.5	100.4	102.6	102.2	100.4	101.6	101.0	92.2	98.3	101.3	94.1	99.1	101.3	100.0	100.9
06	WOOD, PLASTICS & COMPOSITES	100.6	109.8	105.6	91.8	68.4	79.1	88.3	68.4	77.5	80.9	71.8	76.0	79.4	40.8	58.5	78.1	72.1	74.8
07	THERMAL & MOISTURE PROTECTION	98.9	84.5	92.6	102.7	79.8	92.7	104.7	79.8	93.9	104.0	59.5	84.6	103.6	68.0	88.1	103.8	79.7	93.3
08	OPENINGS	102.4	95.7	100.9	97.4	73.6	91.9	95.6	73.6	90.5	98.4	65.7	90.8	96.7	54.6	86.9	97.0	75.9	92.1
0920	Plaster & Gypsum Board	102.2	109.8	107.2	101.7	67.4	79.1	98.3	67.4	77.9	94.8	70.9	79.1	94.8	38.4	57.7	94.9	71.2	79.3
0950, 0980	Ceilings & Acoustic Treatment	97.0	109.8	105.7	93.4	67.4	75.8	91.7	67.4	75.2	90.8	70.9	77.4	90.8	38.4	55.3	91.7	71.2	77.8
0960	Flooring	89.3	102.2	93.0	97.0	76.9	91.2	95.5	76.9	90.1	92.1	57.6	82.1	90.6	57.6	81.0	90.6	76.9	86.7
0970, 0990	Wall Finishes & Painting/Coating	87.1	90.7	89.2	115.2	73.6	90.6	115.2	75.1	91.4	101.5	55.6	74.4	101.5	58.9	76.3	115.2	75.1	91.4
09	FINISHES	85.7	99.9	93.5	93.1	70.9	81.0	93.1	71.1	81.1	89.8	64.0	75.7	89.4	48.6	67.1	90.1	73.3	80.9
COVERS	DIVS. 10 - 14, 25, 28, 41, 43, 44, 46	100.0	89.0	97.6	100.0	87.3	97.2	100.0	84.9	96.7	100.0	76.0	94.7	100.0	80.4	95.7	100.0	85.3	96.8
21, 22, 23	FIRE SUPPRESSION, PLUMBING & HVAC	96.7	62.2	82.8	100.4	87.2	95.1	100.4	87.2	95.1	97.0	46.8	76.7	97.0	69.1	85.7	97.0	87.2	93.0
26, 27, 3370	ELECTRICAL, COMMUNICATIONS & UTIL.	98.9	55.7	76.9	96.2	96.7	96.4	93.9	99.3	96.6	95.8	37.2	66.0	95.8	70.0	82.7	98.2	99.3	98.7
MF2016	WEIGHTED AVERAGE	95.8	81.0	89.4	99.3	84.2	92.8	100.1	84.3	93.3	97.9	59.4	81.4	99.1	68.0	85.8	98.4	84.9	92.6

For customer support on your Electrical Costs with RSMeans data, call 800.448.8182.

547

City Cost Indexes

VIRGINIA

DIVISION		FAIRFAX 220-221			FARMVILLE 239			FREDERICKSBURG 224-225			GRUNDY 246			HARRISONBURG 228			LYNCHBURG 245		
		MAT.	INST.	TOTAL	MAT.	INST.	TOTAL	MAT.	INST.	TOTAL	MAT.	INST.	TOTAL	MAT.	INST.	TOTAL	MAT.	INST.	TOTAL
015433	CONTRACTOR EQUIPMENT		106.2	106.2		110.4	110.4		106.2	106.2		106.2	106.2		106.2	106.2		106.2	106.2
0241, 31 - 34	SITE & INFRASTRUCTURE, DEMOLITION	123.0	89.9	100.0	109.8	90.7	96.5	111.5	89.7	96.3	106.7	88.5	94.0	120.0	88.3	97.9	107.5	90.0	95.3
0310	Concrete Forming & Accessories	83.7	70.1	72.1	97.8	45.2	53.0	83.7	67.1	69.6	88.1	38.1	45.6	79.8	38.7	44.8	85.1	71.5	73.5
0320	Concrete Reinforcing	94.7	87.4	91.0	90.0	69.7	79.7	95.4	82.5	88.8	93.4	45.3	69.0	94.7	58.9	76.5	94.1	70.6	82.1
0330	Cast-in-Place Concrete	107.1	76.5	95.7	106.2	86.0	98.7	108.7	76.2	96.6	106.7	51.0	86.0	107.1	53.8	87.3	106.7	75.7	95.2
03	CONCRETE	103.5	76.7	91.6	99.3	65.7	84.4	97.6	74.5	87.3	98.8	46.0	75.4	101.3	49.8	78.5	98.7	74.3	87.9
04	MASONRY	103.0	72.3	84.0	101.3	51.0	70.2	103.7	68.2	81.7	94.2	54.6	69.7	101.5	59.1	75.3	109.0	64.4	81.5
05	METALS	101.4	100.3	101.1	98.9	90.6	96.4	101.4	98.4	100.5	101.0	75.1	93.2	101.3	87.2	97.0	101.2	94.2	99.1
06	WOOD, PLASTICS & COMPOSITES	80.9	68.4	74.1	94.1	39.5	64.4	80.9	66.5	73.1	83.8	31.4	55.3	77.1	34.9	54.2	80.9	73.2	76.7
07	THERMAL & MOISTURE PROTECTION	104.5	75.6	91.9	103.1	54.9	82.1	103.8	77.1	92.2	104.0	47.4	79.3	104.2	62.9	86.2	103.8	71.1	89.6
08	OPENINGS	95.6	73.9	90.6	96.2	44.4	84.2	96.7	70.5	90.6	98.4	30.9	82.7	97.0	44.3	84.8	97.0	66.4	89.9
0920	Plaster & Gypsum Board	94.9	67.4	76.8	103.8	37.0	59.8	94.9	65.4	75.5	94.8	29.4	51.7	94.8	33.0	54.1	94.8	72.3	80.0
0950, 0980	Ceilings & Acoustic Treatment	91.7	67.4	75.2	86.3	37.0	52.9	91.7	65.4	73.9	90.8	29.4	49.2	90.8	33.0	51.6	90.8	72.3	78.3
0960	Flooring	92.4	77.7	88.1	92.5	52.2	80.8	92.4	74.8	87.3	93.4	29.2	74.8	90.5	76.9	86.5	92.1	67.6	85.0
0970, 0990	Wall Finishes & Painting/Coating	115.2	78.0	93.2	101.8	58.7	76.3	115.2	58.5	81.6	101.5	32.1	60.4	115.2	49.6	76.4	101.5	55.6	74.4
09	FINISHES	91.7	71.5	80.7	88.7	46.3	65.6	90.6	66.9	77.7	90.3	34.4	59.6	90.6	45.3	65.8	89.6	69.2	78.5
COVERS	DIVS. 10 - 14, 25, 28, 41, 43, 44, 46	100.0	84.9	96.7	100.0	76.2	94.8	100.0	79.8	95.6	100.0	74.3	94.4	100.0	61.2	91.5	100.0	79.5	95.5
21, 22, 23	FIRE SUPPRESSION, PLUMBING & HVAC	97.0	87.2	93.0	97.0	53.1	79.3	97.0	81.7	90.8	97.0	66.1	84.5	97.0	66.8	84.8	97.0	69.4	85.8
26, 27, 3370	ELECTRICAL, COMMUNICATIONS & UTIL.	96.9	99.3	98.1	89.9	70.0	79.8	94.1	94.0	94.1	95.8	70.0	82.7	96.0	88.5	92.2	96.9	69.4	82.9
MF2016	WEIGHTED AVERAGE	99.2	84.2	92.8	97.1	62.8	82.4	97.9	80.6	90.5	97.8	59.2	81.2	98.7	66.0	84.7	98.4	73.7	87.8

VIRGINIA

DIVISION		NEWPORT NEWS 236			NORFOLK 233-235			PETERSBURG 238			PORTSMOUTH 237			PULASKI 243			RICHMOND 230-232		
		MAT.	INST.	TOTAL	MAT.	INST.	TOTAL	MAT.	INST.	TOTAL	MAT.	INST.	TOTAL	MAT.	INST.	TOTAL	MAT.	INST.	TOTAL
015433	CONTRACTOR EQUIPMENT		110.4	110.4		111.0	111.0		110.4	110.4		110.3	110.3		106.2	106.2		110.4	110.4
0241, 31 - 34	SITE & INFRASTRUCTURE, DEMOLITION	108.7	91.7	96.9	105.4	92.7	96.6	112.4	91.7	97.9	107.2	91.0	96.2	106.0	89.0	94.2	106.3	91.7	96.1
0310	Concrete Forming & Accessories	97.1	61.3	66.6	103.1	61.4	67.6	90.5	62.5	66.6	86.5	48.0	53.7	88.1	41.2	48.2	94.5	80.7	82.7
0320	Concrete Reinforcing	89.8	67.3	78.3	98.0	67.3	82.4	89.4	70.7	79.9	89.4	66.6	77.8	93.4	86.2	89.7	98.8	70.7	84.5
0330	Cast-in-Place Concrete	103.2	76.2	93.1	108.0	77.2	96.6	109.6	77.4	97.6	102.2	61.2	86.9	106.7	84.6	98.5	97.1	77.5	89.8
03	CONCRETE	96.4	69.3	84.4	98.6	69.7	85.8	101.9	70.8	88.1	95.3	58.0	78.7	98.8	66.2	84.3	93.1	79.1	86.9
04	MASONRY	96.2	63.2	75.8	98.4	63.2	76.6	109.6	63.2	80.9	101.9	48.4	68.9	88.1	55.4	67.9	90.5	63.2	73.6
05	METALS	101.1	93.5	98.8	103.0	93.6	100.1	98.9	95.1	97.8	100.1	91.9	97.6	101.1	96.2	99.6	103.0	95.2	100.6
06	WOOD, PLASTICS & COMPOSITES	92.8	60.3	75.1	103.1	60.3	79.8	84.0	61.2	71.6	80.2	46.1	61.7	83.8	32.9	56.2	93.1	85.7	89.1
07	THERMAL & MOISTURE PROTECTION	103.1	70.9	89.1	100.6	71.1	87.8	103.0	72.1	89.6	103.0	60.8	84.7	104.0	57.9	83.9	100.6	74.8	89.4
08	OPENINGS	96.6	59.0	87.9	98.3	60.3	89.5	96.0	65.8	89.0	96.7	48.8	85.6	98.4	44.1	85.8	100.1	79.4	95.3
0920	Plaster & Gypsum Board	104.7	58.4	74.2	99.9	58.4	72.5	98.2	59.4	72.6	98.6	43.8	62.5	94.8	30.9	52.7	101.7	84.6	90.4
0950, 0980	Ceilings & Acoustic Treatment	90.5	58.4	68.8	94.8	58.4	70.1	87.1	59.4	68.3	90.5	43.8	58.9	90.8	30.9	50.2	92.7	84.6	87.2
0960	Flooring	92.5	67.6	85.3	90.6	67.6	83.9	88.7	72.7	84.0	85.8	57.6	77.6	93.4	57.6	83.0	93.6	72.7	87.5
0970, 0990	Wall Finishes & Painting/Coating	101.8	58.9	76.4	101.5	58.9	76.3	101.8	58.9	76.4	101.8	58.9	76.4	101.5	49.6	70.8	99.6	58.9	75.5
09	FINISHES	89.5	61.7	74.3	89.0	61.7	74.1	87.1	63.3	74.1	86.7	49.5	66.4	90.0	42.0	63.8	91.6	77.8	84.1
COVERS	DIVS. 10 - 14, 25, 28, 41, 43, 44, 46	100.0	79.5	95.5	100.0	79.5	95.5	100.0	82.2	96.1	100.0	65.2	92.3	100.0	74.6	94.4	100.0	84.8	96.7
21, 22, 23	FIRE SUPPRESSION, PLUMBING & HVAC	100.5	63.5	85.5	100.1	66.0	86.3	97.0	68.5	85.5	100.5	63.2	85.4	97.0	67.5	85.1	100.0	68.5	87.2
26, 27, 3370	ELECTRICAL, COMMUNICATIONS & UTIL.	92.4	64.7	78.3	94.8	61.9	78.1	92.5	70.0	81.1	90.8	61.9	76.1	95.8	85.4	90.5	96.9	70.0	83.2
MF2016	WEIGHTED AVERAGE	97.9	69.7	85.8	98.8	70.0	86.5	97.9	72.4	87.0	97.4	63.1	82.7	97.5	68.3	85.0	98.3	76.4	88.9

VIRGINIA / WASHINGTON

DIVISION		ROANOKE 240-241			STAUNTON 244			WINCHESTER 226			CLARKSTON 994			EVERETT 982			OLYMPIA 985		
		MAT.	INST.	TOTAL	MAT.	INST.	TOTAL	MAT.	INST.	TOTAL	MAT.	INST.	TOTAL	MAT.	INST.	TOTAL	MAT.	INST.	TOTAL
015433	CONTRACTOR EQUIPMENT		106.2	106.2		110.4	110.4		106.2	106.2		92.5	92.5		102.5	102.5		102.5	102.5
0241, 31 - 34	SITE & INFRASTRUCTURE, DEMOLITION	106.8	89.9	95.1	110.0	90.6	96.5	118.7	89.8	98.6	101.5	90.4	93.7	93.2	110.8	105.5	93.0	110.8	105.4
0310	Concrete Forming & Accessories	94.3	71.6	75.0	87.8	49.4	55.1	82.1	68.1	70.2	103.1	66.0	71.5	110.9	101.2	102.7	100.2	101.1	100.9
0320	Concrete Reinforcing	94.5	70.6	82.3	94.1	86.1	90.0	94.1	82.5	88.2	104.8	98.7	101.7	111.1	111.6	111.4	118.0	111.6	114.8
0330	Cast-in-Place Concrete	121.2	85.7	108.0	111.0	86.5	101.9	107.1	65.6	91.7	81.1	85.3	82.7	99.4	108.9	103.0	87.2	108.8	95.2
03	CONCRETE	103.2	77.8	92.0	100.2	70.5	87.0	100.8	71.3	87.7	88.5	78.7	84.2	92.1	105.1	97.9	84.9	105.0	93.8
04	MASONRY	95.6	64.4	76.3	104.5	54.4	73.6	99.0	71.2	81.8	99.6	93.6	95.9	117.9	103.8	109.2	110.3	103.8	106.3
05	METALS	103.3	94.3	100.6	101.2	95.9	99.6	101.4	98.3	100.4	91.0	88.0	90.1	110.0	96.0	105.8	108.5	95.6	104.6
06	WOOD, PLASTICS & COMPOSITES	92.7	73.2	82.1	83.8	44.7	62.5	79.4	66.7	72.5	103.6	60.2	80.0	107.5	99.8	103.3	93.6	99.8	97.0
07	THERMAL & MOISTURE PROTECTION	103.8	74.5	91.0	103.6	56.3	83.0	104.4	76.7	92.3	156.1	85.2	125.2	113.6	106.8	110.6	113.1	104.3	109.3
08	OPENINGS	97.4	66.4	90.2	97.0	50.9	86.3	98.5	70.3	91.9	119.0	66.9	106.9	108.2	103.6	107.2	110.7	102.1	108.7
0920	Plaster & Gypsum Board	101.7	72.3	82.4	94.8	42.4	60.3	94.9	65.6	75.6	151.9	58.9	90.7	109.4	99.9	103.1	103.2	99.9	101.0
0950, 0980	Ceilings & Acoustic Treatment	93.4	72.3	79.1	90.8	42.4	58.0	91.7	65.6	74.0	106.8	58.9	74.4	106.8	99.9	102.1	103.9	99.9	101.1
0960	Flooring	97.0	67.6	88.5	93.0	34.2	75.9	91.9	76.9	87.5	88.3	80.1	85.9	105.2	99.3	103.5	94.0	99.3	95.6
0970, 0990	Wall Finishes & Painting/Coating	101.5	55.6	74.4	101.5	31.5	60.1	115.2	79.6	94.1	83.8	74.0	78.0	91.1	95.6	93.8	90.2	95.6	93.4
09	FINISHES	92.1	69.2	79.6	89.8	43.7	64.6	91.1	70.3	79.7	107.3	67.7	85.7	103.8	99.7	101.5	92.3	99.7	96.3
COVERS	DIVS. 10 - 14, 25, 28, 41, 43, 44, 46	100.0	79.4	95.5	100.0	77.2	95.0	100.0	84.3	96.6	100.0	91.2	98.1	100.0	100.4	100.1	100.0	100.8	100.2
21, 22, 23	FIRE SUPPRESSION, PLUMBING & HVAC	100.4	65.2	86.2	97.0	59.2	81.7	97.0	83.2	91.4	97.0	82.4	91.1	100.1	102.8	101.2	100.0	102.8	101.1
26, 27, 3370	ELECTRICAL, COMMUNICATIONS & UTIL.	95.8	56.1	75.6	94.8	70.1	82.2	94.5	94.0	94.3	84.8	97.9	91.5	104.5	101.5	102.9	102.8	100.1	101.4
MF2016	WEIGHTED AVERAGE	99.7	71.6	87.7	98.3	65.6	84.3	98.6	81.3	91.2	98.9	83.9	92.5	103.4	102.6	103.1	100.9	102.3	101.5

WASHINGTON

DIVISION		RICHLAND 993			SEATTLE 980-981, 987			SPOKANE 990-992			TACOMA 983-984			VANCOUVER 986			WENATCHEE 988		
		MAT.	INST.	TOTAL	MAT.	INST.	TOTAL	MAT.	INST.	TOTAL	MAT.	INST.	TOTAL	MAT.	INST.	TOTAL	MAT.	INST.	TOTAL
015433	CONTRACTOR EQUIPMENT		92.5	92.5		104.0	104.0		92.5	92.5		102.5	102.5		98.8	98.8		102.5	102.5
0241, 31 - 34	SITE & INFRASTRUCTURE, DEMOLITION	103.5	90.6	94.5	100.0	110.0	107.0	103.0	90.5	94.3	96.4	110.8	106.4	106.7	98.1	100.7	105.8	108.5	107.7
0310	Concrete Forming & Accessories	103.2	82.2	85.3	107.0	103.6	104.1	108.2	81.7	85.6	101.4	101.1	101.1	101.7	93.2	94.5	103.2	78.6	82.3
0320	Concrete Reinforcing	100.5	98.1	99.3	114.5	111.7	113.0	101.2	98.0	99.6	109.7	111.5	110.6	110.7	110.8	110.7	110.6	98.7	104.6
0330	Cast-in-Place Concrete	81.3	86.9	83.4	107.8	108.7	108.1	84.5	86.7	85.3	102.2	108.8	104.7	114.2	99.0	108.6	104.4	93.6	100.4
03	CONCRETE	88.2	86.5	87.4	105.0	106.3	105.6	90.1	86.1	88.4	93.9	105.0	98.8	103.8	98.1	101.2	101.8	87.3	95.4
04	MASONRY	100.6	85.5	91.3	125.2	103.8	112.0	101.2	85.5	91.5	116.6	102.4	107.8	116.6	90.6	100.6	120.0	96.1	105.2
05	METALS	91.3	87.8	90.3	111.7	99.1	107.9	93.5	87.3	91.6	111.8	95.5	106.9	109.2	96.0	105.2	109.2	88.2	102.9
06	WOOD, PLASTICS & COMPOSITES	103.8	80.2	91.0	101.3	102.7	102.1	112.7	80.2	95.0	97.0	99.8	98.5	89.5	92.2	91.0	98.8	75.5	86.1
07	THERMAL & MOISTURE PROTECTION	157.2	86.7	126.5	109.0	106.0	107.7	153.7	87.0	124.7	113.3	104.1	109.3	113.3	96.9	106.1	112.8	90.8	103.2
08	OPENINGS	117.0	76.9	107.7	108.4	105.2	107.7	117.6	76.8	108.1	109.0	102.1	107.4	105.0	96.6	103.0	108.5	75.1	100.7
0920	Plaster & Gypsum Board	151.9	79.5	104.2	107.8	102.9	104.5	144.5	79.5	101.7	109.2	99.9	103.0	107.4	92.4	97.5	112.5	74.9	87.7
0950, 0980	Ceilings & Acoustic Treatment	113.5	79.5	90.4	120.3	102.9	108.5	108.7	79.5	88.9	112.5	99.9	103.9	109.6	92.4	97.9	104.7	74.9	84.5
0960	Flooring	88.6	80.1	86.1	105.3	99.3	103.5	87.7	80.1	85.5	98.2	99.3	98.5	103.9	81.7	97.5	101.0	80.1	94.9
0970, 0990	Wall Finishes & Painting/Coating	83.8	77.0	79.8	101.9	95.6	98.2	83.9	78.4	80.7	91.1	95.6	93.8	93.8	73.8	82.0	91.1	72.8	80.3
09	FINISHES	108.9	80.4	93.3	109.6	101.4	105.1	106.8	80.5	92.4	102.5	99.7	101.0	100.5	88.5	93.9	103.2	77.1	89.0
COVERS	DIVS. 10 - 14, 25, 28, 41, 43, 44, 46	100.0	94.1	98.7	100.0	101.2	100.3	100.0	94.1	98.7	100.0	100.8	100.2	100.0	96.3	99.2	100.0	93.1	98.5
21, 22, 23	FIRE SUPPRESSION, PLUMBING & HVAC	100.6	110.9	104.8	100.1	112.5	105.1	100.5	85.5	94.4	100.1	102.8	101.2	100.2	100.0	100.1	96.7	94.1	95.6
26, 27, 3370	ELECTRICAL, COMMUNICATIONS & UTIL.	82.6	95.1	89.0	103.7	115.4	109.6	81.2	78.9	80.0	104.3	100.1	102.2	109.5	98.5	103.9	105.2	96.8	100.9
MF2016	WEIGHTED AVERAGE	99.6	92.2	96.4	106.0	107.4	106.6	99.9	84.5	93.3	103.8	102.1	103.1	104.8	96.2	101.1	104.1	91.0	98.5

DIVISION		WASHINGTON YAKIMA 989			WEST VIRGINIA BECKLEY 258 - 259			BLUEFIELD 247 - 248			BUCKHANNON 262			CHARLESTON 250 - 253			CLARKSBURG 263 - 264		
		MAT.	INST.	TOTAL	MAT.	INST.	TOTAL	MAT.	INST.	TOTAL	MAT.	INST.	TOTAL	MAT.	INST.	TOTAL	MAT.	INST.	TOTAL
015433	CONTRACTOR EQUIPMENT		102.5	102.5		106.2	106.2		106.2	106.2		106.2	106.2		106.2	106.2		106.2	106.2
0241, 31 - 34	SITE & INFRASTRUCTURE, DEMOLITION	98.9	109.3	106.2	99.1	91.3	93.6	100.6	91.3	94.1	106.8	91.2	95.9	98.6	92.3	94.2	107.6	91.2	96.2
0310	Concrete Forming & Accessories	102.0	96.1	97.0	83.6	89.7	88.8	85.2	89.5	88.8	84.6	85.9	85.7	94.9	89.6	90.4	82.1	86.0	85.4
0320	Concrete Reinforcing	110.2	98.2	104.1	88.8	89.4	89.1	93.0	84.9	88.9	93.7	84.7	89.1	96.0	89.4	92.7	93.7	98.0	95.9
0330	Cast-in-Place Concrete	109.2	85.6	100.4	103.1	95.1	100.1	104.4	95.0	100.9	104.0	94.4	100.5	99.6	96.1	98.3	114.0	90.0	105.1
03	CONCRETE	98.5	92.4	95.8	90.8	92.4	91.5	94.7	91.6	93.3	97.7	89.7	94.1	89.5	92.8	91.0	101.7	90.5	96.8
04	MASONRY	109.2	82.8	92.9	88.7	90.9	90.1	89.7	90.9	90.5	101.4	86.5	92.2	85.8	92.3	89.8	105.0	86.5	93.6
05	METALS	109.9	88.2	103.3	96.3	103.3	98.5	101.2	101.7	101.4	101.4	101.6	101.5	95.1	103.5	97.7	101.4	106.1	102.8
06	WOOD, PLASTICS & COMPOSITES	97.4	99.8	98.7	81.7	89.9	86.2	82.8	89.9	86.7	82.1	85.4	83.9	93.6	88.8	91.0	78.5	85.4	82.2
07	THERMAL & MOISTURE PROTECTION	113.5	87.7	102.2	106.1	88.4	98.4	103.7	88.4	97.1	104.1	88.6	97.4	102.3	88.9	96.5	104.0	88.3	97.2
08	OPENINGS	108.5	99.1	106.3	96.5	84.9	93.8	98.9	83.9	95.4	98.9	81.5	94.9	96.9	84.3	93.9	98.9	84.4	95.6
0920	Plaster & Gypsum Board	108.8	99.9	102.9	97.3	89.5	92.2	94.1	89.5	91.1	94.8	84.9	88.2	101.7	88.4	92.9	91.7	84.9	87.2
0950, 0980	Ceilings & Acoustic Treatment	106.6	99.9	102.0	81.9	89.5	87.1	89.1	89.5	89.4	90.8	84.9	86.8	93.8	88.4	90.1	90.8	84.9	86.8
0960	Flooring	99.0	80.1	93.5	91.3	101.6	94.3	89.9	101.6	93.3	89.6	96.4	91.6	97.3	101.6	98.5	88.6	96.4	90.9
0970, 0990	Wall Finishes & Painting/Coating	91.1	78.4	83.6	92.3	94.6	93.7	101.5	92.2	96.0	101.5	90.3	94.9	90.7	94.6	93.0	101.5	90.3	94.9
09	FINISHES	101.7	91.3	96.0	86.2	92.7	89.7	88.1	92.4	90.5	89.0	88.2	88.5	92.4	92.4	92.4	88.2	88.2	88.2
COVERS	DIVS. 10 - 14, 25, 28, 41, 43, 44, 46	100.0	97.3	99.4	100.0	92.1	98.3	100.0	92.1	98.3	100.0	90.8	98.0	100.0	92.1	98.3	100.0	90.8	98.0
21, 22, 23	FIRE SUPPRESSION, PLUMBING & HVAC	100.1	109.8	104.0	97.2	93.2	95.6	97.0	85.9	92.5	97.0	91.7	94.8	100.1	94.3	97.8	97.0	91.8	94.9
26, 27, 3370	ELECTRICAL, COMMUNICATIONS & UTIL.	107.4	95.1	101.2	93.1	86.2	89.6	94.8	86.2	90.4	96.2	91.5	93.8	98.1	86.2	92.0	96.2	91.5	93.8
MF2016	WEIGHTED AVERAGE	104.0	96.6	100.8	94.9	92.1	93.7	96.7	90.2	93.9	98.1	90.7	94.9	96.1	92.5	94.6	98.7	91.4	95.5

WEST VIRGINIA

DIVISION		GASSAWAY 266			HUNTINGTON 255 - 257			LEWISBURG 249			MARTINSBURG 254			MORGANTOWN 265			PARKERSBURG 261		
		MAT.	INST.	TOTAL	MAT.	INST.	TOTAL	MAT.	INST.	TOTAL	MAT.	INST.	TOTAL	MAT.	INST.	TOTAL	MAT.	INST.	TOTAL
015433	CONTRACTOR EQUIPMENT		106.2	106.2		106.2	106.2		106.2	106.2		106.2	106.2		106.2	106.2		106.2	106.2
0241, 31 - 34	SITE & INFRASTRUCTURE, DEMOLITION	104.3	91.2	95.2	103.7	92.3	95.8	116.2	91.2	98.8	102.9	92.0	95.3	101.6	92.0	94.9	109.8	92.2	97.5
0310	Concrete Forming & Accessories	84.0	89.2	88.4	94.7	91.7	92.2	82.4	89.2	88.1	83.6	81.7	82.0	82.4	86.1	85.6	86.5	88.3	88.0
0320	Concrete Reinforcing	93.7	94.2	94.0	90.1	95.8	93.0	93.7	84.9	89.2	88.8	97.9	93.4	93.7	98.0	95.9	93.0	84.7	88.8
0330	Cast-in-Place Concrete	108.9	94.4	103.5	112.4	96.5	106.5	104.4	94.9	100.9	108.0	90.8	101.6	104.0	94.5	100.5	106.3	91.4	100.8
03	CONCRETE	98.1	92.8	95.8	96.0	95.0	95.5	104.4	91.4	98.6	94.3	88.8	91.9	94.4	92.1	93.4	99.7	89.8	95.3
04	MASONRY	105.9	89.7	95.9	89.0	95.4	92.9	93.6	90.9	91.9	90.5	85.7	87.5	122.8	86.5	100.4	80.0	87.0	84.3
05	METALS	101.4	104.9	102.4	98.8	105.9	100.9	101.3	101.4	101.3	96.7	105.0	99.2	101.5	106.2	102.9	102.1	101.6	101.9
06	WOOD, PLASTICS & COMPOSITES	81.0	89.9	85.8	92.8	91.2	91.9	78.8	89.9	84.8	81.7	80.9	81.3	78.8	85.4	82.4	82.2	88.1	85.4
07	THERMAL & MOISTURE PROTECTION	103.7	90.0	97.8	106.3	90.4	99.4	104.7	88.4	97.6	106.4	83.0	96.2	103.8	88.6	97.2	103.8	89.1	97.4
08	OPENINGS	97.2	86.0	94.6	95.7	87.0	93.7	98.9	83.9	95.4	98.2	76.0	93.1	100.1	84.4	96.5	97.8	82.7	94.3
0920	Plaster & Gypsum Board	93.2	89.5	90.8	104.1	90.8	95.4	91.7	89.5	90.3	97.4	80.2	86.1	91.7	84.9	87.2	94.8	87.6	90.1
0950, 0980	Ceilings & Acoustic Treatment	90.8	89.5	89.9	82.8	90.8	88.2	90.8	89.5	89.9	82.8	80.2	81.0	90.8	84.9	86.8	90.8	87.6	88.7
0960	Flooring	89.4	101.6	93.0	98.6	106.7	100.9	88.7	101.6	92.4	91.3	102.3	94.5	88.7	96.4	90.9	92.5	95.0	93.2
0970, 0990	Wall Finishes & Painting/Coating	101.5	94.6	97.4	92.3	92.2	92.3	101.5	66.2	80.6	92.3	85.9	88.6	101.5	90.3	94.9	101.5	85.9	92.3
09	FINISHES	88.4	92.4	90.6	89.4	94.7	92.3	89.2	89.6	89.4	86.6	85.4	86.0	87.8	88.2	88.0	89.9	89.3	89.6
COVERS	DIVS. 10 - 14, 25, 28, 41, 43, 44, 46	100.0	91.3	98.1	100.0	92.5	98.3	100.0	66.7	92.7	100.0	83.0	96.3	100.0	85.2	96.7	100.0	91.5	98.1
21, 22, 23	FIRE SUPPRESSION, PLUMBING & HVAC	97.0	85.3	92.2	100.7	92.6	97.4	97.0	93.1	95.4	97.2	82.7	91.4	97.0	91.8	94.9	100.4	90.3	96.3
26, 27, 3370	ELECTRICAL, COMMUNICATIONS & UTIL.	96.2	86.2	91.1	96.8	91.8	94.3	92.4	86.2	89.2	98.9	78.7	88.7	96.3	91.5	93.9	96.3	90.4	93.3
MF2016	WEIGHTED AVERAGE	98.0	90.5	94.8	97.5	94.3	96.1	98.4	90.5	95.0	96.4	86.1	92.0	98.5	91.5	95.5	98.2	90.7	95.0

For customer support on your Electrical Costs with RSMeans data, call 800.448.8182.

549

DIVISION		WEST VIRGINIA									WISCONSIN								
		PETERSBURG			ROMNEY			WHEELING			BELOIT			EAU CLAIRE			GREEN BAY		
		268			267			260			535			547			541 - 543		
		MAT.	INST.	TOTAL	MAT.	INST.	TOTAL	MAT.	INST.	TOTAL	MAT.	INST.	TOTAL	MAT.	INST.	TOTAL	MAT.	INST.	TOTAL
015433	CONTRACTOR EQUIPMENT		106.2	106.2		106.2	106.2		106.2	106.2		101.8	101.8		102.6	102.6		100.3	100.3
0241, 31 - 34	SITE & INFRASTRUCTURE, DEMOLITION	100.9	92.2	94.8	103.8	92.2	95.7	110.5	92.0	97.6	95.8	107.0	103.6	96.1	104.0	101.7	99.7	100.2	100.0
0310	Concrete Forming & Accessories	85.5	88.8	88.3	81.7	89.1	88.0	88.2	86.1	86.4	97.6	98.9	98.7	96.3	97.4	97.3	106.4	100.3	101.2
0320	Concrete Reinforcing	93.0	94.0	93.5	93.7	98.2	96.0	92.5	98.0	95.3	96.4	141.1	119.2	93.9	117.5	105.9	91.9	112.6	102.4
0330	Cast-in-Place Concrete	104.0	94.3	100.4	108.9	85.7	100.3	106.3	94.5	101.9	103.9	103.2	103.6	100.5	97.6	99.4	103.9	102.9	103.5
03	CONCRETE	94.5	92.5	93.6	98.0	90.4	94.6	99.8	92.1	96.4	95.8	107.6	101.0	90.8	101.1	95.5	93.8	103.4	98.1
04	MASONRY	96.7	86.5	90.4	94.0	89.7	91.4	104.2	86.4	93.2	98.5	102.4	100.9	91.3	98.1	95.5	122.4	99.9	108.5
05	METALS	101.5	104.2	102.3	101.6	106.2	102.9	102.2	106.2	103.4	101.1	113.0	104.7	91.2	106.7	95.9	93.7	105.1	97.1
06	WOOD, PLASTICS & COMPOSITES	83.1	89.9	86.8	77.9	89.9	84.4	83.8	85.4	84.7	98.6	96.9	97.6	102.0	97.0	99.3	109.6	100.0	104.4
07	THERMAL & MOISTURE PROTECTION	103.8	84.5	95.4	103.9	86.1	96.2	104.3	88.8	97.5	99.8	96.1	98.2	103.5	97.2	100.8	105.7	100.2	103.3
08	OPENINGS	100.1	86.0	96.8	100.0	86.9	97.0	98.6	84.4	95.3	100.4	111.4	103.0	105.4	102.2	104.7	101.3	104.2	102.0
0920	Plaster & Gypsum Board	94.5	89.5	91.2	91.4	89.5	90.2	94.8	84.9	88.3	93.7	97.2	96.0	105.7	97.2	100.1	104.2	100.4	101.7
0950, 0980	Ceilings & Acoustic Treatment	90.8	89.5	89.9	90.8	89.5	89.9	90.8	84.9	86.8	82.5	97.2	92.5	87.3	97.2	94.0	80.7	100.4	94.0
0960	Flooring	90.5	96.4	92.2	88.5	102.3	92.5	93.4	96.4	94.3	94.1	122.4	102.3	81.9	110.1	90.1	97.6	118.7	103.7
0970, 0990	Wall Finishes & Painting/Coating	101.5	90.3	94.9	101.5	90.3	94.9	101.5	90.3	94.9	94.3	106.3	101.4	83.0	78.4	80.3	91.5	82.1	85.9
09	FINISHES	88.7	90.9	89.9	87.9	92.1	90.2	90.1	88.2	89.1	90.0	103.7	97.5	86.0	97.7	92.4	90.2	102.0	96.6
COVERS	DIVS. 10 - 14, 25, 28, 41, 43, 44, 46	100.0	56.0	90.3	100.0	91.2	98.1	100.0	86.0	96.9	100.0	100.7	100.2	100.0	96.1	99.1	100.0	97.7	99.5
21, 22, 23	FIRE SUPPRESSION, PLUMBING & HVAC	97.0	85.6	92.4	97.0	85.4	92.3	100.4	91.8	96.9	100.1	97.6	99.1	100.1	90.2	96.1	100.3	85.3	94.2
26, 27, 3370	ELECTRICAL, COMMUNICATIONS & UTIL.	99.4	78.8	88.9	98.6	78.8	88.5	93.6	91.5	92.5	99.7	86.3	92.9	103.6	86.5	94.9	98.0	83.2	90.5
MF2016	WEIGHTED AVERAGE	97.7	87.7	93.4	97.9	89.2	94.2	99.3	91.5	96.0	98.7	101.4	99.9	96.8	96.5	96.7	98.7	95.8	97.5

DIVISION		WISCONSIN																	
		KENOSHA			LA CROSSE			LANCASTER			MADISON			MILWAUKEE			NEW RICHMOND		
		531			546			538			537			530, 532			540		
		MAT.	INST.	TOTAL	MAT.	INST.	TOTAL	MAT.	INST.	TOTAL	MAT.	INST.	TOTAL	MAT.	INST.	TOTAL	MAT.	INST.	TOTAL
015433	CONTRACTOR EQUIPMENT		99.8	99.8		102.6	102.6		101.8	101.8		101.8	101.8		89.6	89.6		102.9	102.9
0241, 31 - 34	SITE & INFRASTRUCTURE, DEMOLITION	101.7	103.6	103.0	89.9	104.0	99.8	94.8	106.3	102.9	93.7	106.9	102.9	93.7	95.2	94.8	94.6	104.7	101.6
0310	Concrete Forming & Accessories	107.6	109.1	108.9	83.0	97.2	95.1	96.9	97.8	97.7	101.9	98.3	98.9	100.1	110.5	109.0	91.3	95.1	94.5
0320	Concrete Reinforcing	96.2	113.7	105.1	93.5	109.3	101.5	97.8	109.3	103.7	100.6	109.5	105.1	99.9	113.9	107.0	91.0	117.5	104.5
0330	Cast-in-Place Concrete	113.0	105.8	110.4	90.3	99.4	93.7	103.3	102.7	103.1	101.3	102.5	101.7	90.5	110.4	97.9	104.6	105.8	105.0
03	CONCRETE	100.6	108.5	104.1	82.4	100.2	90.3	95.5	101.5	98.1	93.7	101.7	97.3	95.4	110.1	101.9	88.8	102.9	95.0
04	MASONRY	96.2	110.3	104.9	90.4	98.1	95.2	98.6	102.3	100.9	97.1	101.2	99.6	100.6	113.0	108.3	117.5	100.5	107.0
05	METALS	102.1	104.2	102.7	91.1	103.7	95.0	98.5	101.2	99.3	104.0	102.1	103.5	101.5	96.5	100.0	91.5	106.3	96.0
06	WOOD, PLASTICS & COMPOSITES	106.0	108.8	107.5	86.2	97.0	92.1	97.7	96.9	97.3	98.4	96.9	97.5	101.3	109.2	105.6	90.8	92.8	91.9
07	THERMAL & MOISTURE PROTECTION	100.2	106.9	103.1	102.8	96.1	99.9	99.5	96.3	98.1	96.8	102.0	99.4	102.3	109.0	105.2	104.1	103.1	103.7
08	OPENINGS	94.7	108.5	97.9	105.4	94.7	102.9	96.2	94.6	95.8	103.5	101.7	103.0	101.4	109.1	103.3	90.6	93.0	91.1
0920	Plaster & Gypsum Board	82.7	109.5	100.4	100.7	97.2	98.4	93.0	97.2	95.8	105.1	97.2	99.9	98.0	109.5	105.6	92.1	93.2	92.8
0950, 0980	Ceilings & Acoustic Treatment	82.5	109.5	100.8	86.5	97.2	93.8	79.1	97.2	91.4	89.5	97.2	94.7	89.7	109.5	103.1	55.9	93.2	81.2
0960	Flooring	111.8	118.7	113.8	76.0	115.4	87.4	93.7	114.7	99.8	93.4	114.7	99.6	100.9	114.7	104.9	91.2	118.4	99.1
0970, 0990	Wall Finishes & Painting/Coating	104.4	121.9	114.8	83.0	83.9	83.5	94.3	100.3	97.8	93.6	106.3	101.1	106.1	122.7	115.9	94.4	83.9	88.1
09	FINISHES	94.4	112.6	104.3	83.0	99.3	91.9	89.1	101.3	95.7	90.2	102.1	96.7	99.4	112.8	106.7	81.4	98.0	90.5
COVERS	DIVS. 10 - 14, 25, 28, 41, 43, 44, 46	100.0	99.4	99.9	100.0	94.7	98.8	100.0	86.0	96.9	100.0	98.6	99.7	100.0	102.7	100.6	100.0	97.1	99.4
21, 22, 23	FIRE SUPPRESSION, PLUMBING & HVAC	100.3	98.1	99.4	100.1	90.1	96.0	96.6	91.2	94.4	100.0	97.4	98.9	100.0	105.6	102.3	96.1	91.9	94.0
26, 27, 3370	ELECTRICAL, COMMUNICATIONS & UTIL.	100.1	99.5	99.8	103.9	86.5	95.1	99.4	86.9	93.1	100.7	96.3	98.5	99.3	101.6	100.5	101.6	86.9	94.2
MF2016	WEIGHTED AVERAGE	99.4	104.6	101.6	95.2	95.9	95.5	96.9	96.7	96.8	99.1	100.3	99.6	99.6	105.9	102.3	94.7	97.0	95.7

DIVISION		WISCONSIN																	
		OSHKOSH			PORTAGE			RACINE			RHINELANDER			SUPERIOR			WAUSAU		
		549			539			534			545			548			544		
		MAT.	INST.	TOTAL	MAT.	INST.	TOTAL	MAT.	INST.	TOTAL	MAT.	INST.	TOTAL	MAT.	INST.	TOTAL	MAT.	INST.	TOTAL
015433	CONTRACTOR EQUIPMENT		100.3	100.3		101.8	101.8		101.8	101.8		100.3	100.3		102.9	102.9		100.3	100.3
0241, 31 - 34	SITE & INFRASTRUCTURE, DEMOLITION	91.5	100.1	97.5	85.8	106.2	100.0	95.6	107.4	103.8	103.5	100.1	101.1	91.4	104.4	100.4	87.3	100.7	96.6
0310	Concrete Forming & Accessories	88.8	97.4	96.1	88.9	98.4	97.0	97.9	109.1	107.4	86.4	97.3	95.7	89.3	91.0	90.8	88.3	100.1	98.3
0320	Concrete Reinforcing	92.1	112.5	102.4	97.9	109.5	103.8	96.4	113.7	105.2	92.3	109.5	101.0	91.0	109.9	100.6	91.3	109.9	100.9
0330	Cast-in-Place Concrete	96.3	100.9	98.0	88.7	101.3	93.4	101.9	106.5	103.6	109.3	100.9	106.1	98.5	101.2	99.5	89.8	95.2	91.8
03	CONCRETE	84.2	101.4	91.8	83.3	101.5	91.3	94.9	108.4	100.8	95.6	100.8	97.9	83.6	98.2	90.1	79.4	100.1	88.6
04	MASONRY	104.9	99.8	101.8	97.4	102.3	100.4	98.5	110.3	105.8	122.2	99.7	108.3	116.7	102.5	107.9	104.3	99.7	101.5
05	METALS	91.7	104.4	95.6	99.2	105.3	101.0	102.8	104.2	103.3	91.6	103.3	95.1	92.4	104.2	96.0	91.4	103.8	95.2
06	WOOD, PLASTICS & COMPOSITES	87.9	97.0	92.8	87.1	96.9	92.4	98.9	108.8	104.3	85.3	97.0	91.6	89.0	88.2	88.6	87.3	100.0	94.2
07	THERMAL & MOISTURE PROTECTION	104.6	85.6	96.4	98.9	103.0	100.7	99.8	105.6	102.3	105.6	85.4	96.8	103.8	98.6	101.5	104.4	97.8	101.5
08	OPENINGS	97.7	99.0	98.0	96.4	102.6	97.8	100.4	108.5	102.3	97.7	94.6	97.0	90.0	93.8	90.9	97.9	99.9	98.4
0920	Plaster & Gypsum Board	90.8	97.2	95.0	85.6	97.2	93.3	93.7	109.5	104.1	90.8	97.2	95.0	92.0	88.5	89.7	90.8	100.4	97.1
0950, 0980	Ceilings & Acoustic Treatment	80.7	97.2	91.9	81.6	97.2	92.2	82.5	109.5	100.8	80.7	97.2	91.9	56.8	88.5	78.3	80.7	100.4	94.0
0960	Flooring	89.3	118.7	97.8	90.0	118.4	98.3	94.1	118.7	101.2	88.7	118.4	97.3	92.3	127.2	102.4	89.2	118.4	97.7
0970, 0990	Wall Finishes & Painting/Coating	89.1	107.0	99.7	94.3	100.3	97.8	94.3	119.0	108.9	89.1	82.1	84.9	83.6	108.6	98.3	89.1	85.4	86.9
09	FINISHES	85.1	102.9	94.8	86.9	102.3	95.3	90.0	112.3	102.1	86.0	100.1	93.7	80.8	99.2	90.8	84.8	102.3	94.4
COVERS	DIVS. 10 - 14, 25, 28, 41, 43, 44, 46	100.0	87.3	97.2	100.0	87.4	97.2	100.0	99.4	99.9	100.0	87.6	97.3	100.0	95.8	99.1	100.0	97.7	99.5
21, 22, 23	FIRE SUPPRESSION, PLUMBING & HVAC	96.8	84.9	92.0	96.6	98.0	97.2	100.1	98.2	99.3	96.8	90.6	94.3	96.1	93.0	94.8	96.8	90.8	94.4
26, 27, 3370	ELECTRICAL, COMMUNICATIONS & UTIL.	102.2	81.1	91.5	103.2	96.3	99.7	99.5	100.4	100.0	101.6	80.0	90.6	106.6	101.0	103.7	103.4	80.0	91.5
MF2016	WEIGHTED AVERAGE	94.8	94.2	94.5	95.3	100.5	97.5	98.8	105.0	101.5	97.3	94.6	96.1	94.5	98.7	96.3	94.1	95.8	94.8

City Cost Indexes

WYOMING

DIVISION		CASPER 826 MAT.	INST.	TOTAL	CHEYENNE 820 MAT.	INST.	TOTAL	NEWCASTLE 827 MAT.	INST.	TOTAL	RAWLINS 823 MAT.	INST.	TOTAL	RIVERTON 825 MAT.	INST.	TOTAL	ROCK SPRINGS 829-831 MAT.	INST.	TOTAL
015433	CONTRACTOR EQUIPMENT		97.6	97.6		97.6	97.6		97.6	97.6		97.6	97.6		97.6	97.6		97.6	97.6
0241, 31 - 34	SITE & INFRASTRUCTURE, DEMOLITION	96.3	94.7	95.2	91.6	94.7	93.8	84.8	94.2	91.4	98.1	94.2	95.4	91.8	94.1	93.4	88.8	94.2	92.6
0310	Concrete Forming & Accessories	98.5	54.3	60.9	99.9	67.2	72.0	90.9	72.5	75.2	94.9	72.7	76.0	89.8	61.9	66.1	96.9	62.8	67.9
0320	Concrete Reinforcing	109.6	84.9	97.0	105.0	84.9	94.8	112.9	84.7	98.6	112.6	84.8	98.4	113.6	84.7	98.9	113.6	83.0	98.0
0330	Cast-in-Place Concrete	104.4	76.0	93.8	98.6	76.0	90.2	99.6	75.0	90.4	99.7	75.1	90.5	99.6	75.0	90.5	99.6	75.0	90.4
03	CONCRETE	97.7	67.9	84.5	95.3	73.7	85.7	95.5	75.7	86.7	108.2	75.9	93.9	103.1	71.0	88.9	95.9	71.1	84.9
04	MASONRY	99.3	64.7	77.9	102.2	63.8	78.5	98.9	60.5	75.2	98.9	60.5	75.2	98.9	60.5	75.2	159.3	56.3	95.7
05	METALS	98.1	80.9	92.9	100.4	80.9	94.5	96.7	80.5	91.8	96.7	80.7	91.9	96.8	80.5	91.9	97.5	79.7	92.1
06	WOOD, PLASTICS & COMPOSITES	94.5	49.0	69.8	93.4	66.5	78.8	83.7	74.8	78.9	87.7	74.8	80.7	82.6	60.4	70.6	92.9	61.7	75.9
07	THERMAL & MOISTURE PROTECTION	103.3	66.2	87.2	98.1	67.9	85.0	99.8	65.5	84.8	101.3	76.8	90.6	100.7	69.4	87.0	99.9	68.3	86.1
08	OPENINGS	106.0	60.0	95.3	109.6	69.7	100.3	114.0	74.3	104.7	113.6	74.3	104.5	113.8	66.3	102.8	114.4	66.7	103.3
0920	Plaster & Gypsum Board	103.3	47.6	66.6	92.9	65.6	74.9	89.9	74.1	79.5	90.2	74.1	79.6	89.9	59.4	69.8	102.4	60.7	74.9
0950, 0980	Ceilings & Acoustic Treatment	105.4	47.6	66.2	99.3	65.6	76.5	101.2	74.1	82.9	101.2	74.1	82.9	101.2	59.4	72.9	101.2	60.7	73.7
0960	Flooring	104.2	72.0	94.8	102.2	72.0	93.5	96.1	46.0	81.6	98.9	46.0	83.6	95.6	62.3	85.9	101.2	46.0	85.2
0970, 0990	Wall Finishes & Painting/Coating	94.7	59.9	74.1	98.9	59.9	75.8	95.5	62.1	75.7	95.5	62.1	75.7	95.5	62.1	75.7	95.5	62.1	75.7
09	FINISHES	96.7	56.6	74.8	93.7	67.0	79.1	89.2	66.5	76.8	91.3	66.5	77.8	89.9	61.3	74.3	92.5	58.8	74.1
COVERS	DIVS. 10 - 14, 25, 28, 41, 43, 44, 46	100.0	93.7	98.6	100.0	89.0	97.6	100.0	93.8	98.6	100.0	93.8	98.6	100.0	79.0	95.4	100.0	84.3	96.5
21, 22, 23	FIRE SUPPRESSION, PLUMBING & HVAC	100.0	71.3	88.4	99.9	71.3	88.4	98.1	70.5	86.9	98.1	70.5	86.9	98.1	70.5	86.9	99.9	70.5	88.0
26, 27, 3370	ELECTRICAL, COMMUNICATIONS & UTIL.	96.7	61.8	78.9	98.1	62.7	80.1	96.9	63.9	80.1	96.9	63.9	80.1	96.9	60.6	78.4	95.2	78.5	86.7
MF2016	WEIGHTED AVERAGE	99.3	69.5	86.5	99.6	72.1	87.8	98.1	72.3	87.1	100.3	72.7	88.5	99.4	69.7	86.6	101.9	71.6	88.9

WYOMING / CANADA

DIVISION		SHERIDAN 828 MAT.	INST.	TOTAL	WHEATLAND 822 MAT.	INST.	TOTAL	WORLAND 824 MAT.	INST.	TOTAL	YELLOWSTONE NAT'L PA 821 MAT.	INST.	TOTAL	BARRIE, ONTARIO MAT.	INST.	TOTAL	BATHURST, NEW BRUNSWICK MAT.	INST.	TOTAL
015433	CONTRACTOR EQUIPMENT		97.6	97.6		97.6	97.6		97.6	97.6		97.6	97.6		103.0	103.0		102.7	102.7
0241, 31 - 34	SITE & INFRASTRUCTURE, DEMOLITION	92.1	94.7	93.9	89.4	94.2	92.7	86.8	94.2	91.9	86.9	94.2	92.0	118.2	99.7	105.3	106.1	95.5	98.7
0310	Concrete Forming & Accessories	98.1	62.7	68.0	92.6	49.7	56.1	92.7	62.7	67.2	92.7	62.8	67.2	124.2	84.6	90.5	103.9	59.4	66.1
0320	Concrete Reinforcing	113.6	84.8	98.9	112.9	84.2	98.3	113.6	84.7	98.9	115.8	84.7	99.9	177.7	89.4	132.8	144.5	59.2	101.1
0330	Cast-in-Place Concrete	103.0	76.0	92.9	104.0	74.9	93.2	99.6	75.0	90.4	99.6	75.0	90.4	156.6	84.6	129.8	111.1	57.9	91.3
03	CONCRETE	103.4	71.6	89.3	100.3	65.3	84.8	95.7	71.3	84.9	95.9	71.3	85.0	136.7	85.8	114.1	107.5	59.9	86.4
04	MASONRY	99.2	62.0	76.2	99.2	52.2	70.2	98.9	60.5	75.2	98.9	60.5	75.2	165.9	91.2	119.7	160.0	58.7	97.4
05	METALS	100.2	80.7	94.3	96.6	79.6	91.5	96.8	80.4	91.8	97.4	79.9	92.1	110.2	92.4	104.8	118.4	74.3	105.0
06	WOOD, PLASTICS & COMPOSITES	95.0	60.4	76.2	85.5	44.2	63.1	85.6	61.7	72.6	85.6	61.7	72.6	118.5	83.2	99.3	99.7	59.6	77.9
07	THERMAL & MOISTURE PROTECTION	101.0	66.7	86.1	100.0	59.5	82.4	99.8	66.4	85.3	99.2	66.4	84.9	114.6	88.5	103.3	110.9	60.1	88.8
08	OPENINGS	114.5	66.3	103.3	112.5	57.4	99.7	114.2	67.0	103.2	106.8	66.7	97.5	92.6	83.2	90.4	85.8	52.7	78.1
0920	Plaster & Gypsum Board	112.7	59.4	77.6	89.9	42.7	58.8	89.9	60.7	70.7	90.1	60.7	70.7	151.9	82.7	106.3	121.5	58.4	79.9
0950, 0980	Ceilings & Acoustic Treatment	104.0	59.4	73.8	101.2	42.7	61.6	101.2	60.7	73.7	102.0	60.7	74.0	95.1	82.7	86.7	115.9	58.4	76.9
0960	Flooring	99.5	62.3	88.7	97.6	45.2	82.4	97.6	46.0	82.6	97.6	46.0	82.6	118.9	90.2	110.6	100.1	42.9	83.5
0970, 0990	Wall Finishes & Painting/Coating	97.7	59.9	75.3	95.5	59.9	74.5	95.5	62.1	75.7	95.5	62.1	75.7	105.0	86.0	93.7	110.7	49.2	74.3
09	FINISHES	96.5	61.7	77.5	89.9	48.1	67.1	89.7	58.8	72.8	89.8	58.9	72.9	109.7	85.9	96.7	105.7	55.4	78.2
COVERS	DIVS. 10 - 14, 25, 28, 41, 43, 44, 46	100.0	94.9	98.9	100.0	83.2	96.3	100.0	77.8	95.1	100.0	77.8	95.1	139.2	66.4	123.2	131.1	59.1	115.3
21, 22, 23	FIRE SUPPRESSION, PLUMBING & HVAC	98.1	71.3	87.3	98.1	70.5	86.9	98.1	70.5	86.9	98.1	70.5	86.9	103.2	95.6	100.1	103.1	66.2	88.2
26, 27, 3370	ELECTRICAL, COMMUNICATIONS & UTIL.	99.2	60.6	79.6	96.9	78.5	87.5	96.9	78.5	87.5	95.9	88.2	92.0	115.0	84.9	99.7	108.3	57.2	82.3
MF2016	WEIGHTED AVERAGE	100.9	70.6	87.9	98.8	68.1	85.6	98.3	71.8	87.0	97.6	73.2	87.1	116.0	89.3	104.6	110.4	63.9	90.5

CANADA

DIVISION		BRANDON, MANITOBA MAT.	INST.	TOTAL	BRANTFORD, ONTARIO MAT.	INST.	TOTAL	BRIDGEWATER, NOVA SCOTIA MAT.	INST.	TOTAL	CALGARY, ALBERTA MAT.	INST.	TOTAL	CAP-DE-LA-MADELEINE, QUEBEC MAT.	INST.	TOTAL	CHARLESBOURG, QUEBEC MAT.	INST.	TOTAL
015433	CONTRACTOR EQUIPMENT		104.7	104.7		103.1	103.1		102.7	102.7		132.6	132.6		103.5	103.5		103.5	103.5
0241, 31 - 34	SITE & INFRASTRUCTURE, DEMOLITION	128.6	98.0	107.3	117.9	100.1	105.5	102.3	97.4	98.9	127.2	125.7	126.2	98.1	98.9	98.7	98.1	98.9	98.7
0310	Concrete Forming & Accessories	142.7	68.0	79.1	124.4	91.7	96.5	97.0	69.8	73.9	119.4	97.4	100.6	130.4	81.4	88.7	130.4	81.4	88.7
0320	Concrete Reinforcing	184.3	56.1	119.1	171.2	88.1	128.9	145.8	49.0	96.6	136.9	84.2	110.1	145.8	74.6	109.6	145.8	74.6	109.6
0330	Cast-in-Place Concrete	104.8	72.4	92.7	127.3	105.0	119.0	130.8	69.0	107.8	137.0	107.0	125.8	102.7	90.2	98.1	102.7	90.2	98.1
03	CONCRETE	115.2	68.4	94.5	119.1	95.9	108.8	117.4	66.9	95.0	121.8	99.3	111.8	105.1	83.8	95.6	105.1	83.8	95.6
04	MASONRY	213.4	62.0	119.8	166.5	95.4	122.6	162.1	67.0	103.3	204.0	88.8	132.8	162.8	77.9	110.4	162.8	77.9	110.4
05	METALS	133.1	79.1	116.8	116.8	93.1	109.6	116.1	76.9	104.2	134.9	104.4	125.7	115.2	86.0	106.4	115.2	86.0	106.4
06	WOOD, PLASTICS & COMPOSITES	154.7	68.8	108.0	121.8	90.3	104.7	90.4	69.4	79.0	99.9	97.2	98.4	133.5	81.3	105.2	133.5	81.3	105.2
07	THERMAL & MOISTURE PROTECTION	128.6	70.6	103.4	121.0	94.0	109.3	115.4	69.6	95.5	128.7	96.5	114.6	114.2	86.5	102.2	114.2	86.5	102.2
08	OPENINGS	102.4	61.6	92.9	89.7	89.2	89.6	84.2	63.1	79.3	82.9	86.0	83.7	91.1	74.4	87.2	91.1	74.4	87.2
0920	Plaster & Gypsum Board	113.7	67.5	83.3	113.1	90.0	97.9	118.4	68.4	85.5	125.1	96.0	106.0	142.5	80.7	101.8	142.5	80.7	101.8
0950, 0980	Ceilings & Acoustic Treatment	122.4	67.5	85.2	104.7	90.0	94.7	104.7	68.4	80.1	150.3	96.0	113.5	104.7	80.7	88.4	104.7	80.7	88.4
0960	Flooring	131.3	64.0	111.8	113.6	90.2	106.8	96.1	60.9	85.9	121.3	86.7	111.2	113.6	88.8	106.5	113.6	88.8	106.5
0970, 0990	Wall Finishes & Painting/Coating	117.6	55.5	80.8	109.8	94.4	100.7	109.8	60.0	80.8	115.0	109.2	111.6	109.8	85.8	95.6	109.8	85.8	95.6
09	FINISHES	119.9	66.2	90.6	106.6	92.0	98.6	101.7	67.5	83.0	124.6	97.2	109.7	110.2	83.6	95.7	110.2	83.6	95.7
COVERS	DIVS. 10 - 14, 25, 28, 41, 43, 44, 46	131.1	61.4	115.8	131.1	68.4	117.3	131.1	62.1	115.9	131.1	94.0	123.0	131.1	77.2	119.3	131.1	77.2	119.3
21, 22, 23	FIRE SUPPRESSION, PLUMBING & HVAC	103.3	80.7	94.1	103.1	98.5	101.3	103.1	81.3	94.3	104.9	91.6	99.5	103.5	86.1	96.5	103.5	86.1	96.5
26, 27, 3370	ELECTRICAL, COMMUNICATIONS & UTIL.	110.5	64.7	87.2	107.3	84.5	95.7	111.7	60.1	85.5	102.0	96.2	99.1	106.3	66.8	86.2	106.3	66.8	86.2
MF2016	WEIGHTED AVERAGE	120.9	72.4	100.1	113.1	93.1	104.5	111.2	72.2	94.5	119.0	97.6	109.8	110.7	82.1	98.4	110.7	82.1	98.4

For customer support on your Electrical Costs with RSMeans data, call 800.448.8182.

City Cost Indexes

CANADA

DIVISION		CHARLOTTETOWN, PRINCE EDWARD ISLAND			CHICOUTIMI, QUEBEC			CORNER BROOK, NEWFOUNDLAND			CORNWALL, ONTARIO			DALHOUSIE, NEW BRUNSWICK			DARTMOUTH, NOVA SCOTIA		
		MAT.	INST.	TOTAL	MAT.	INST.	TOTAL	MAT.	INST.	TOTAL	MAT.	INST.	TOTAL	MAT.	INST.	TOTAL	MAT.	INST.	TOTAL
015433	CONTRACTOR EQUIPMENT		120.9	120.9		103.3	103.3		103.2	103.2		103.1	103.1		103.0	103.0		102.0	102.0
0241, 31 - 34	SITE & INFRASTRUCTURE, DEMOLITION	133.4	108.0	115.7	103.1	98.4	99.8	132.7	95.4	106.7	116.1	99.6	104.6	101.4	95.7	97.4	119.9	97.0	103.9
0310	Concrete Forming & Accessories	117.5	54.4	63.8	133.0	89.2	95.7	119.3	77.7	83.9	122.2	84.7	90.3	103.1	59.7	66.1	108.9	69.8	75.6
0320	Concrete Reinforcing	164.4	47.9	105.2	109.0	93.9	101.3	169.0	50.2	108.6	171.2	87.8	128.8	146.3	59.2	102.1	176.7	49.0	111.8
0330	Cast-in-Place Concrete	147.6	58.3	114.4	104.7	94.9	101.0	122.0	65.5	101.0	114.5	95.7	107.5	106.9	58.1	88.7	117.7	68.8	99.5
03	CONCRETE	130.6	56.5	97.8	99.3	92.1	96.1	145.1	69.4	111.6	113.0	89.5	102.6	109.7	60.0	87.7	128.4	66.8	101.1
04	MASONRY	176.6	56.0	102.1	162.3	88.4	116.7	209.3	75.9	126.9	165.5	87.4	117.2	163.7	58.7	98.8	223.5	67.0	126.8
05	METALS	141.8	81.5	123.6	118.1	91.2	110.0	133.6	76.0	116.1	116.6	91.8	109.1	109.7	74.6	99.1	134.2	76.6	116.7
06	WOOD, PLASTICS & COMPOSITES	103.7	53.8	76.6	135.4	89.8	110.0	131.2	83.8	105.5	120.2	84.0	100.5	97.6	59.6	76.9	118.4	69.3	91.7
07	THERMAL & MOISTURE PROTECTION	129.6	58.4	98.6	111.4	96.0	104.7	133.2	67.8	104.8	120.8	88.6	106.8	120.7	60.1	94.3	130.2	69.5	103.8
08	OPENINGS	85.3	47.0	76.4	89.7	76.1	86.6	108.6	68.5	99.3	91.1	82.9	89.2	87.3	52.7	79.3	93.2	63.0	86.2
0920	Plaster & Gypsum Board	118.6	52.0	74.8	141.4	89.3	107.1	145.6	83.2	104.6	168.5	83.5	112.5	126.5	58.4	81.6	140.7	68.4	93.1
0950, 0980	Ceilings & Acoustic Treatment	128.3	52.0	76.6	115.1	89.3	97.6	123.2	83.2	96.1	107.2	83.5	91.2	107.0	58.4	74.1	130.0	68.4	88.3
0960	Flooring	116.7	57.1	99.4	115.8	88.8	108.0	113.2	51.2	95.2	113.6	88.9	106.5	102.3	65.6	91.7	108.1	60.9	94.4
0970, 0990	Wall Finishes & Painting/Coating	116.4	40.9	71.7	110.7	103.2	106.3	117.4	57.9	82.2	109.8	87.9	96.8	113.4	49.2	75.4	117.4	60.9	84.0
09	FINISHES	118.8	53.8	83.3	112.6	91.6	101.1	119.8	72.3	93.8	115.1	86.0	99.2	106.5	60.0	81.1	117.6	67.5	90.2
COVERS	DIVS. 10 - 14, 25, 28, 41, 43, 44, 46	131.1	59.3	115.3	131.1	80.9	120.1	131.1	61.9	115.9	131.1	66.0	116.8	131.1	59.1	115.3	131.1	62.1	115.9
21, 22, 23	FIRE SUPPRESSION, PLUMBING & HVAC	103.4	60.3	86.0	103.1	80.0	93.7	103.3	67.4	88.8	103.5	96.4	100.7	103.2	66.2	88.2	103.3	81.3	94.4
26, 27, 3370	ELECTRICAL, COMMUNICATIONS & UTIL.	107.2	48.6	77.4	105.0	83.1	93.9	107.9	52.8	79.9	107.6	85.4	96.3	108.5	54.1	80.8	112.1	60.1	85.7
MF2016	WEIGHTED AVERAGE	120.1	61.8	95.1	110.3	87.3	100.4	125.0	70.2	101.5	113.1	89.6	103.0	109.8	64.1	90.2	121.8	72.1	100.5

CANADA

DIVISION		EDMONTON, ALBERTA			FORT MCMURRAY, ALBERTA			FREDERICTON, NEW BRUNSWICK			GATINEAU, QUEBEC			GRANBY, QUEBEC			HALIFAX, NOVA SCOTIA		
		MAT.	INST.	TOTAL	MAT.	INST.	TOTAL	MAT.	INST.	TOTAL	MAT.	INST.	TOTAL	MAT.	INST.	TOTAL	MAT.	INST.	TOTAL
015433	CONTRACTOR EQUIPMENT		132.2	132.2		105.3	105.3		120.6	120.6		103.5	103.5		103.5	103.5		119.0	119.0
0241, 31 - 34	SITE & INFRASTRUCTURE, DEMOLITION	124.4	125.5	125.1	122.9	100.7	107.4	112.2	109.3	110.2	97.9	98.9	98.6	98.4	98.9	98.8	104.1	109.4	107.8
0310	Concrete Forming & Accessories	121.1	97.4	100.9	122.1	89.8	94.6	122.4	60.2	69.5	130.4	81.3	88.6	130.4	81.3	88.6	117.8	83.3	88.5
0320	Concrete Reinforcing	137.3	84.2	110.3	158.5	84.1	120.6	161.2	59.5	109.5	154.2	74.5	113.7	154.2	74.5	113.7	155.1	72.1	112.9
0330	Cast-in-Place Concrete	131.0	106.9	122.0	169.0	101.1	143.7	108.4	58.7	89.9	101.3	90.2	97.2	104.7	90.2	99.3	102.5	84.6	95.8
03	CONCRETE	119.2	99.2	110.3	136.8	93.0	117.4	112.3	61.3	89.7	105.4	83.7	95.8	107.0	83.7	96.6	107.9	83.1	96.9
04	MASONRY	199.0	88.8	130.9	206.9	85.5	131.9	183.6	60.0	107.2	162.7	77.9	110.3	163.0	77.9	110.4	181.0	85.3	121.9
05	METALS	142.1	104.2	130.6	145.7	91.6	129.3	142.1	86.5	125.3	115.2	85.9	106.3	115.4	85.8	106.5	142.0	97.6	128.6
06	WOOD, PLASTICS & COMPOSITES	98.8	97.2	97.9	116.0	89.3	101.5	109.0	59.8	82.3	133.5	81.3	105.2	133.5	81.3	105.2	104.1	82.5	92.4
07	THERMAL & MOISTURE PROTECTION	140.5	96.5	121.3	129.5	92.0	113.2	131.1	60.5	100.4	114.2	86.5	102.2	114.2	85.0	101.5	134.2	84.9	112.7
08	OPENINGS	82.0	86.0	82.9	91.1	81.6	88.9	86.7	51.7	78.6	91.1	69.9	86.2	91.1	69.9	86.2	87.1	75.2	84.4
0920	Plaster & Gypsum Board	122.4	96.0	105.0	113.6	88.6	97.1	121.7	58.4	80.0	112.4	80.7	91.5	112.4	80.7	91.5	111.5	81.7	91.9
0950, 0980	Ceilings & Acoustic Treatment	145.4	96.0	112.0	112.3	88.6	96.3	132.9	58.4	82.4	104.7	80.7	88.4	104.7	80.7	88.4	142.1	81.7	102.1
0960	Flooring	122.0	86.7	111.7	113.6	86.7	105.8	118.0	68.5	103.6	113.6	88.8	106.5	113.6	88.8	106.5	111.2	81.9	102.7
0970, 0990	Wall Finishes & Painting/Coating	114.8	109.2	111.5	109.9	90.6	98.5	116.0	62.7	84.4	109.8	85.8	95.6	109.8	85.8	95.6	119.2	90.0	101.9
09	FINISHES	119.6	97.2	107.4	109.5	89.6	98.6	120.1	62.1	88.5	105.8	83.6	93.7	105.8	83.6	93.7	115.1	84.2	98.2
COVERS	DIVS. 10 - 14, 25, 28, 41, 43, 44, 46	131.1	94.0	123.0	131.1	90.4	122.2	131.1	59.7	115.4	131.1	77.2	119.3	131.1	77.2	119.3	131.1	67.4	117.1
21, 22, 23	FIRE SUPPRESSION, PLUMBING & HVAC	104.8	91.6	99.5	103.6	95.3	100.2	103.5	75.3	92.1	103.5	86.1	96.5	103.1	86.1	96.2	104.5	80.9	95.0
26, 27, 3370	ELECTRICAL, COMMUNICATIONS & UTIL.	105.7	96.2	100.9	101.7	79.4	90.4	106.5	71.4	88.6	106.3	66.8	86.2	106.9	66.8	86.5	108.2	84.4	96.1
MF2016	WEIGHTED AVERAGE	119.7	97.5	110.2	122.1	90.2	108.4	117.9	71.2	97.9	110.4	81.9	98.2	110.6	81.8	98.3	117.1	85.8	103.7

CANADA

DIVISION		HAMILTON, ONTARIO			HULL, QUEBEC			JOLIETTE, QUEBEC			KAMLOOPS, BRITISH COLUMBIA			KINGSTON, ONTARIO			KITCHENER, ONTARIO		
		MAT.	INST.	TOTAL	MAT.	INST.	TOTAL	MAT.	INST.	TOTAL	MAT.	INST.	TOTAL	MAT.	INST.	TOTAL	MAT.	INST.	TOTAL
015433	CONTRACTOR EQUIPMENT		120.9	120.9		103.5	103.5		103.5	103.5		106.7	106.7		105.3	105.3		105.5	105.5
0241, 31 - 34	SITE & INFRASTRUCTURE, DEMOLITION	109.6	114.9	113.3	97.9	98.9	98.6	98.5	98.9	98.8	120.7	102.6	108.1	116.1	103.4	107.3	96.2	104.9	102.3
0310	Concrete Forming & Accessories	122.2	94.0	98.2	130.4	81.3	88.6	130.4	81.4	88.7	121.0	84.6	90.0	122.3	84.8	90.4	113.8	86.8	90.9
0320	Concrete Reinforcing	151.4	98.6	124.5	154.2	74.5	113.7	145.8	74.6	109.6	114.3	78.6	96.2	171.2	87.8	128.8	105.9	98.4	102.1
0330	Cast-in-Place Concrete	113.9	102.2	109.5	101.3	90.2	97.2	105.5	90.2	99.9	91.4	94.7	92.6	114.5	95.7	107.5	111.0	97.3	105.9
03	CONCRETE	110.0	98.3	104.8	105.4	83.7	95.8	106.4	83.8	96.4	114.2	87.4	102.3	114.8	89.5	103.6	96.6	92.7	94.9
04	MASONRY	167.0	99.7	125.4	162.7	77.9	110.3	163.1	77.9	110.5	170.0	85.7	117.9	172.4	87.5	119.9	143.0	96.9	114.5
05	METALS	141.7	105.7	130.8	115.4	85.9	106.5	115.4	86.0	106.5	117.3	88.2	108.5	118.0	91.8	110.1	130.3	96.3	120.0
06	WOOD, PLASTICS & COMPOSITES	102.6	93.1	97.4	133.5	81.3	105.2	133.5	81.3	105.2	103.4	83.0	92.3	120.2	84.1	100.6	108.5	84.8	95.6
07	THERMAL & MOISTURE PROTECTION	124.7	100.0	114.0	114.2	86.5	102.1	114.2	86.5	102.1	130.7	84.1	110.4	120.8	89.8	107.3	115.3	96.8	107.2
08	OPENINGS	83.5	91.9	85.4	91.1	69.9	86.2	91.1	74.4	87.2	87.9	80.8	86.3	91.1	82.6	89.1	81.8	85.6	82.7
0920	Plaster & Gypsum Board	118.6	92.3	101.3	112.4	80.7	91.5	142.5	80.7	101.8	98.6	82.1	87.8	171.0	83.6	113.5	107.6	84.4	90.6
0950, 0980	Ceilings & Acoustic Treatment	134.8	92.3	106.0	104.7	80.7	88.4	104.7	80.7	88.4	104.7	82.1	89.4	119.1	83.6	95.1	107.6	84.4	91.8
0960	Flooring	116.4	95.6	110.4	113.6	88.8	106.5	113.6	88.8	106.5	112.8	50.8	94.8	113.6	88.9	106.5	101.4	95.6	99.7
0970, 0990	Wall Finishes & Painting/Coating	110.4	103.5	106.3	109.8	85.8	95.6	109.8	85.8	95.6	109.8	78.4	91.2	109.8	81.3	92.9	103.8	93.3	97.6
09	FINISHES	115.6	95.3	104.5	105.8	83.6	93.7	110.2	83.6	95.7	106.1	78.0	90.8	117.8	85.4	100.1	99.6	88.7	93.6
COVERS	DIVS. 10 - 14, 25, 28, 41, 43, 44, 46	131.1	91.8	122.5	131.1	77.2	119.3	131.1	77.2	119.3	131.1	86.2	121.2	131.1	66.0	116.8	131.1	89.0	121.9
21, 22, 23	FIRE SUPPRESSION, PLUMBING & HVAC	104.9	90.5	99.1	103.1	86.1	96.2	103.1	86.1	96.3	103.1	89.5	97.6	103.5	96.6	100.7	104.0	89.2	98.1
26, 27, 3370	ELECTRICAL, COMMUNICATIONS & UTIL.	106.3	99.6	102.9	108.2	66.8	87.2	106.9	66.8	86.5	110.2	76.0	92.8	107.6	84.2	95.7	107.7	97.1	102.3
MF2016	WEIGHTED AVERAGE	116.1	98.1	108.4	110.5	81.9	98.2	110.9	82.1	98.5	113.0	85.7	101.3	114.1	89.7	103.6	109.4	93.4	102.5

City Cost Indexes

CANADA

DIVISION		LAVAL, QUEBEC MAT.	INST.	TOTAL	LETHBRIDGE, ALBERTA MAT.	INST.	TOTAL	LLOYDMINSTER, ALBERTA MAT.	INST.	TOTAL	LONDON, ONTARIO MAT.	INST.	TOTAL	MEDICINE HAT, ALBERTA MAT.	INST.	TOTAL	MONCTON, NEW BRUNSWICK MAT.	INST.	TOTAL
015433	CONTRACTOR EQUIPMENT		103.5	103.5		105.3	105.3		105.3	105.3		122.4	122.4		105.3	105.3		102.7	102.7
0241, 31 - 34	SITE & INFRASTRUCTURE, DEMOLITION	98.4	98.9	98.8	115.5	101.3	105.6	115.4	100.7	105.2	104.1	114.7	111.5	114.2	100.8	104.8	105.5	95.8	98.8
0310	Concrete Forming & Accessories	130.6	81.3	88.6	123.6	89.9	94.9	121.7	80.3	86.5	122.4	88.4	93.4	123.6	80.3	86.7	103.9	62.0	68.2
0320	Concrete Reinforcing	154.2	74.5	113.7	158.5	84.1	120.6	158.5	84.1	120.6	151.4	97.3	123.9	158.5	84.0	120.6	144.5	70.8	107.0
0330	Cast-in-Place Concrete	104.7	90.2	99.3	126.8	101.1	117.2	117.6	97.5	110.1	120.4	100.1	112.8	117.6	97.5	110.1	107.0	68.2	92.6
03	CONCRETE	107.0	83.7	96.7	117.3	93.1	106.6	112.9	87.5	101.6	113.0	94.9	105.0	113.0	87.5	101.7	105.6	66.9	88.4
04	MASONRY	162.9	77.9	110.4	180.7	85.5	121.9	162.0	79.2	110.9	182.7	97.9	130.3	162.0	79.2	110.9	159.7	60.1	98.2
05	METALS	115.3	85.9	106.4	140.2	91.6	125.5	117.3	91.5	109.5	141.7	107.0	131.2	117.5	91.4	109.6	118.4	84.9	108.2
06	WOOD, PLASTICS & COMPOSITES	133.6	81.3	105.2	119.5	89.3	103.1	116.0	79.7	96.3	109.2	86.3	96.7	119.5	79.7	97.9	99.7	61.5	78.9
07	THERMAL & MOISTURE PROTECTION	114.7	86.5	102.4	126.7	92.0	111.6	123.0	87.5	107.6	124.5	97.6	112.8	130.0	87.5	111.5	115.6	64.2	93.2
08	OPENINGS	91.1	69.9	86.2	91.1	81.6	88.9	91.1	76.4	87.7	80.2	87.0	81.8	91.1	76.4	87.7	85.8	60.2	79.8
0920	Plaster & Gypsum Board	112.7	80.7	91.6	103.7	88.6	93.7	99.9	78.8	86.0	123.9	85.3	98.5	102.1	78.8	86.7	121.5	60.3	81.2
0950, 0980	Ceilings & Acoustic Treatment	104.7	80.7	88.4	112.3	88.6	96.3	104.7	78.8	87.1	142.7	85.3	103.8	104.7	78.8	87.1	115.9	60.3	78.2
0960	Flooring	113.6	88.8	106.5	113.6	86.7	105.8	113.6	86.7	105.8	112.8	95.6	107.8	113.6	86.7	105.8	100.1	67.2	90.5
0970, 0990	Wall Finishes & Painting/Coating	109.8	85.8	95.6	109.9	99.0	103.4	109.9	77.1	90.5	113.8	100.1	105.7	109.7	77.1	90.4	110.7	50.4	75.1
09	FINISHES	105.9	83.6	93.7	107.2	90.5	98.1	105.2	80.9	92.0	118.5	90.6	103.3	105.4	80.9	92.0	105.7	61.7	81.6
COVERS	DIVS. 10 - 14, 25, 28, 41, 43, 44, 46	131.1	77.2	119.3	131.1	90.4	122.2	131.1	87.3	121.5	131.1	90.7	122.2	131.1	87.3	121.5	131.1	60.7	115.6
21, 22, 23	FIRE SUPPRESSION, PLUMBING & HVAC	104.1	86.1	96.8	103.4	92.1	98.8	103.5	92.0	98.9	104.9	87.8	98.0	103.1	88.8	97.3	103.1	68.6	89.1
26, 27, 3370	ELECTRICAL, COMMUNICATIONS & UTIL.	105.4	66.8	85.8	103.1	79.4	91.1	100.9	79.4	90.0	103.5	97.1	100.3	100.9	79.4	90.0	112.0	59.0	85.1
MF2016	WEIGHTED AVERAGE	110.7	81.9	98.3	117.2	89.7	105.4	111.6	86.5	100.8	116.8	95.6	107.7	111.7	85.8	100.6	110.6	68.1	92.4

CANADA

DIVISION		MONTREAL, QUEBEC MAT.	INST.	TOTAL	MOOSE JAW, SASKATCHEWAN MAT.	INST.	TOTAL	NEW GLASGOW, NOVA SCOTIA MAT.	INST.	TOTAL	NEWCASTLE, NEW BRUNSWICK MAT.	INST.	TOTAL	NORTH BAY, ONTARIO MAT.	INST.	TOTAL	OSHAWA, ONTARIO MAT.	INST.	TOTAL
015433	CONTRACTOR EQUIPMENT		121.9	121.9		101.7	101.7		102.0	102.0		102.7	102.7		102.4	102.4		105.5	105.5
0241, 31 - 34	SITE & INFRASTRUCTURE, DEMOLITION	113.8	112.2	112.7	115.3	95.0	101.2	114.1	97.0	102.2	106.1	95.5	98.7	129.0	98.7	107.9	107.0	103.9	104.8
0310	Concrete Forming & Accessories	126.2	89.7	95.1	106.8	56.1	63.7	108.8	69.8	75.6	103.9	59.7	66.2	143.6	82.3	91.4	119.2	87.0	91.8
0320	Concrete Reinforcing	134.8	94.0	114.0	111.8	63.2	87.1	169.0	49.0	108.0	144.5	59.2	101.2	199.7	87.3	142.6	167.8	93.3	129.9
0330	Cast-in-Place Concrete	123.6	96.6	113.5	114.1	65.7	96.1	117.7	68.8	99.5	111.1	58.0	91.4	113.4	82.6	101.9	128.3	85.4	112.3
03	CONCRETE	112.8	93.6	104.3	99.2	61.7	82.6	127.5	66.8	100.6	107.5	60.0	86.5	129.4	83.7	109.2	117.7	87.9	104.5
04	MASONRY	168.0	88.4	118.9	160.7	57.6	97.0	208.9	67.0	121.2	160.0	58.7	97.4	215.0	83.6	133.8	146.0	89.9	111.4
05	METALS	149.0	102.6	134.9	114.2	75.9	102.6	131.6	76.6	114.9	118.4	74.5	105.1	132.4	91.2	119.9	119.9	95.0	112.4
06	WOOD, PLASTICS & COMPOSITES	108.8	90.2	98.7	100.7	54.7	75.7	118.4	69.3	91.7	99.7	59.6	77.9	157.8	82.6	117.0	115.5	85.9	99.4
07	THERMAL & MOISTURE PROTECTION	118.8	96.2	109.0	113.5	61.8	91.0	130.2	69.5	103.8	115.6	60.1	91.4	136.7	85.0	114.2	116.2	89.3	104.5
08	OPENINGS	85.7	78.2	83.9	87.1	52.9	79.2	93.2	63.0	86.2	85.8	52.7	78.1	100.8	80.5	96.1	86.9	87.1	86.9
0920	Plaster & Gypsum Board	117.7	89.3	99.0	96.1	53.4	68.0	139.1	68.4	92.6	121.5	58.4	79.9	136.4	82.1	100.6	105.7	85.4	92.4
0950, 0980	Ceilings & Acoustic Treatment	140.2	89.3	105.7	104.7	53.4	70.0	122.4	68.4	85.8	115.9	58.4	76.9	122.4	82.1	95.1	103.3	85.4	91.2
0960	Flooring	114.0	91.0	107.3	103.9	55.5	89.8	108.1	60.9	94.4	100.1	65.6	90.0	131.3	88.9	119.0	104.3	98.1	102.5
0970, 0990	Wall Finishes & Painting/Coating	112.2	103.2	106.9	109.8	62.6	81.9	117.4	60.9	84.0	110.7	49.2	74.3	117.4	87.3	99.6	113.8	107.5	106.0
09	FINISHES	116.5	92.2	103.2	101.7	56.4	76.9	115.9	67.5	89.4	105.7	59.9	80.7	122.9	84.3	101.9	100.7	91.0	95.4
COVERS	DIVS. 10 - 14, 25, 28, 41, 43, 44, 46	131.1	81.8	120.3	131.1	58.7	115.2	131.1	62.1	115.9	131.1	59.1	115.3	131.1	64.7	116.5	131.1	88.6	121.8
21, 22, 23	FIRE SUPPRESSION, PLUMBING & HVAC	105.1	80.1	95.0	103.5	71.9	90.7	103.3	81.3	94.4	103.1	66.2	88.2	103.3	94.5	99.7	104.0	100.0	102.4
26, 27, 3370	ELECTRICAL, COMMUNICATIONS & UTIL.	107.2	83.1	95.0	109.2	57.7	83.0	108.2	60.1	83.8	107.7	57.2	82.0	108.9	85.3	96.9	108.8	87.6	98.0
MF2016	WEIGHTED AVERAGE	118.1	89.9	106.0	109.0	65.6	90.4	119.9	72.1	99.4	110.5	64.5	90.8	122.8	87.4	107.7	111.6	92.9	103.6

CANADA

DIVISION		OTTAWA, ONTARIO MAT.	INST.	TOTAL	OWEN SOUND, ONTARIO MAT.	INST.	TOTAL	PETERBOROUGH, ONTARIO MAT.	INST.	TOTAL	PORTAGE LA PRAIRIE, MANITOBA MAT.	INST.	TOTAL	PRINCE ALBERT, SASKATCHEWAN MAT.	INST.	TOTAL	PRINCE GEORGE, BRITISH COLUMBIA MAT.	INST.	TOTAL
015433	CONTRACTOR EQUIPMENT		123.1	123.1		103.0	103.0		103.1	103.1		105.3	105.3		101.7	101.7		106.7	106.7
0241, 31 - 34	SITE & INFRASTRUCTURE, DEMOLITION	108.6	114.7	112.8	118.2	99.6	105.2	117.9	99.5	105.1	116.6	98.4	103.9	111.0	95.2	100.0	124.1	102.6	109.1
0310	Concrete Forming & Accessories	123.5	88.4	93.6	124.2	80.9	87.3	124.4	83.3	89.4	123.8	67.6	75.9	106.8	55.9	63.5	112.2	79.7	84.5
0320	Concrete Reinforcing	135.1	97.3	115.9	177.7	89.4	132.8	171.2	87.8	128.8	158.5	56.1	106.4	116.8	63.1	89.5	114.3	78.0	95.9
0330	Cast-in-Place Concrete	121.6	102.9	114.6	156.6	78.9	127.7	127.3	84.3	111.3	117.6	71.9	100.6	103.4	65.6	89.3	114.5	94.7	107.1
03	CONCRETE	111.8	96.0	104.8	136.7	82.2	112.5	119.1	84.9	103.9	107.1	68.1	89.8	94.8	61.6	80.1	124.4	85.2	107.0
04	MASONRY	164.0	100.4	124.7	165.9	88.9	118.3	166.5	89.9	119.2	165.2	61.0	100.8	159.9	57.6	96.7	172.0	85.7	118.7
05	METALS	142.5	108.4	132.2	110.2	92.2	104.7	116.8	91.9	109.3	117.5	79.3	105.9	114.2	75.7	102.6	117.3	88.2	108.5
06	WOOD, PLASTICS & COMPOSITES	107.8	85.2	95.5	118.5	79.5	97.3	121.8	81.4	99.8	119.5	68.8	91.9	100.7	54.7	75.7	103.4	76.3	88.7
07	THERMAL & MOISTURE PROTECTION	131.6	98.8	117.4	114.6	85.8	102.1	121.0	90.6	107.8	114.0	70.2	94.9	113.4	60.7	90.4	124.2	83.3	106.4
08	OPENINGS	88.5	86.3	88.0	92.6	79.9	89.6	89.7	82.3	88.0	91.1	61.7	84.2	86.1	52.9	78.4	87.9	77.1	85.4
0920	Plaster & Gypsum Board	120.1	84.1	96.4	151.9	78.9	103.8	113.1	80.8	91.8	101.7	67.5	79.2	96.1	53.4	68.0	98.6	75.2	83.2
0950, 0980	Ceilings & Acoustic Treatment	136.4	84.1	101.0	95.1	78.9	84.1	104.7	80.8	88.5	104.7	67.5	79.5	104.7	53.4	70.0	104.7	75.2	84.7
0960	Flooring	110.7	91.1	105.0	118.9	90.2	110.6	113.6	88.9	106.5	113.6	64.0	99.2	103.9	55.5	89.8	109.2	69.7	97.7
0970, 0990	Wall Finishes & Painting/Coating	110.6	94.9	101.3	105.0	86.0	93.7	109.8	89.4	97.7	109.9	55.5	77.7	109.8	53.4	76.4	109.8	78.4	91.2
09	FINISHES	117.0	89.1	101.8	109.7	83.2	95.2	106.6	85.0	94.8	105.3	66.0	83.8	101.7	55.4	76.4	105.0	77.3	89.9
COVERS	DIVS. 10 - 14, 25, 28, 41, 43, 44, 46	131.1	89.3	121.9	139.2	65.2	122.9	131.1	66.2	116.8	131.1	61.0	115.7	131.1	58.7	115.2	131.1	85.5	121.1
21, 22, 23	FIRE SUPPRESSION, PLUMBING & HVAC	104.9	89.5	98.7	103.2	94.4	99.6	103.1	97.9	101.0	103.1	80.1	93.8	103.5	64.8	87.9	103.1	89.5	97.6
26, 27, 3370	ELECTRICAL, COMMUNICATIONS & UTIL.	103.4	97.9	100.6	116.4	84.2	100.0	107.3	84.9	95.9	108.9	55.8	81.9	109.2	57.7	83.0	107.4	76.0	91.4
MF2016	WEIGHTED AVERAGE	116.9	96.4	108.1	116.1	87.6	103.9	113.1	89.4	102.9	111.6	70.9	94.1	108.2	63.9	89.2	113.9	85.0	101.5

For customer support on your Electrical Costs with RSMeans data, call 800.448.8182.

553

City Cost Indexes

CANADA

DIVISION		QUEBEC CITY, QUEBEC MAT.	INST.	TOTAL	RED DEER, ALBERTA MAT.	INST.	TOTAL	REGINA, SASKATCHEWAN MAT.	INST.	TOTAL	RIMOUSKI, QUEBEC MAT.	INST.	TOTAL	ROUYN-NORANDA, QUEBEC MAT.	INST.	TOTAL	SAINT HYACINTHE, QUEBEC MAT.	INST.	TOTAL
015433	CONTRACTOR EQUIPMENT		122.8	122.8		105.3	105.3		127.7	127.7		103.5	103.5		103.5	103.5		103.5	103.5
0241, 31 - 34	SITE & INFRASTRUCTURE, DEMOLITION	116.2	112.1	113.3	114.2	100.8	104.8	127.0	119.3	121.6	98.2	113.2	108.6	97.9	98.9	98.6	98.4	98.9	98.8
0310	Concrete Forming & Accessories	123.5	89.8	94.8	136.5	80.3	88.6	125.4	93.8	98.5	130.4	89.3	95.4	130.4	81.3	88.6	130.4	81.3	88.6
0320	Concrete Reinforcing	135.0	94.0	114.2	158.5	84.0	120.6	137.5	91.4	114.1	110.0	93.9	101.8	154.2	74.5	113.7	154.2	74.5	113.7
0330	Cast-in-Place Concrete	118.2	96.3	110.1	117.6	97.5	110.1	138.1	99.2	123.6	106.6	94.9	102.2	101.3	90.2	97.2	104.7	90.2	99.3
03	CONCRETE	110.2	93.7	102.9	113.8	87.5	102.2	121.2	96.1	110.1	102.7	92.1	98.0	105.4	83.7	95.8	107.0	83.7	96.7
04	MASONRY	162.6	88.4	116.8	162.0	79.2	110.9	192.6	89.2	128.7	162.5	88.4	116.7	162.7	77.9	110.3	162.9	77.9	110.4
05	METALS	142.3	104.4	130.8	117.5	91.4	109.6	148.8	103.1	134.9	115.0	91.3	107.8	115.4	85.9	106.5	115.4	85.9	106.5
06	WOOD, PLASTICS & COMPOSITES	113.7	90.1	100.9	119.5	79.7	97.9	110.3	94.6	101.8	133.5	89.8	109.7	133.5	81.3	105.2	133.5	81.3	105.2
07	THERMAL & MOISTURE PROTECTION	119.0	96.2	109.1	140.7	87.5	117.5	139.2	88.5	117.1	114.2	96.0	106.3	114.2	86.5	102.1	114.4	86.5	102.3
08	OPENINGS	87.4	85.6	87.0	91.1	76.4	87.7	86.3	81.8	85.3	90.7	76.1	87.3	91.1	69.9	86.2	91.1	69.9	86.2
0920	Plaster & Gypsum Board	121.7	89.3	100.4	102.1	78.8	86.7	131.2	93.8	106.6	142.3	89.3	107.4	112.2	80.7	91.4	112.2	80.7	91.4
0950, 0980	Ceilings & Acoustic Treatment	139.5	89.3	105.5	104.7	78.8	87.1	147.1	93.8	111.0	103.8	89.3	94.0	103.8	80.7	88.1	103.8	80.7	88.1
0960	Flooring	112.3	91.0	106.1	116.3	86.7	107.7	123.9	95.6	115.7	115.0	88.8	107.4	113.6	88.8	106.5	113.6	88.8	106.5
0970, 0990	Wall Finishes & Painting/Coating	122.0	103.2	110.9	109.7	77.1	90.4	115.3	91.4	101.2	113.0	103.2	107.2	109.8	85.8	95.6	109.8	85.8	95.6
09	FINISHES	116.5	92.2	103.2	106.1	80.9	92.4	128.0	94.9	109.9	110.7	91.6	100.2	105.7	83.6	93.6	105.7	83.6	93.6
COVERS	DIVS. 10 - 14, 25, 28, 41, 43, 44, 46	131.1	81.6	120.2	131.1	87.3	121.5	131.1	70.8	117.8	131.1	80.9	120.1	131.1	77.2	119.3	131.1	77.2	119.3
21, 22, 23	FIRE SUPPRESSION, PLUMBING & HVAC	104.8	80.1	94.8	103.1	88.8	97.3	104.4	88.8	98.1	103.1	80.0	93.8	103.1	86.1	96.2	99.7	86.1	94.2
26, 27, 3370	ELECTRICAL, COMMUNICATIONS & UTIL.	104.5	83.1	93.6	100.9	79.4	90.0	107.6	93.7	100.6	106.9	83.1	94.8	106.9	66.8	86.5	107.5	66.8	86.8
MF2016	WEIGHTED AVERAGE	116.4	90.3	105.2	112.2	85.8	100.8	122.1	94.2	110.1	110.3	87.3	100.4	110.4	81.9	98.1	109.8	81.9	97.8

CANADA

DIVISION		SAINT JOHN, NEW BRUNSWICK MAT.	INST.	TOTAL	SARNIA, ONTARIO MAT.	INST.	TOTAL	SASKATOON, SASKATCHEWAN MAT.	INST.	TOTAL	SAULT STE MARIE, ONTARIO MAT.	INST.	TOTAL	SHERBROOKE, QUEBEC MAT.	INST.	TOTAL	SOREL, QUEBEC MAT.	INST.	TOTAL
015433	CONTRACTOR EQUIPMENT		102.7	102.7		103.1	103.1		101.7	101.7		103.1	103.1		103.5	103.5		103.5	103.5
0241, 31 - 34	SITE & INFRASTRUCTURE, DEMOLITION	106.2	97.2	99.9	116.5	99.6	104.7	112.7	97.9	102.4	106.5	99.2	101.4	98.4	98.9	98.8	98.5	98.9	98.8
0310	Concrete Forming & Accessories	124.2	65.0	73.8	123.1	90.1	95.0	106.8	93.2	95.2	112.3	86.2	90.1	130.4	81.3	88.6	130.4	81.4	88.7
0320	Concrete Reinforcing	144.5	71.4	107.3	122.2	89.2	105.4	117.1	91.3	104.0	109.8	87.9	98.7	154.2	74.5	113.7	145.8	74.6	109.6
0330	Cast-in-Place Concrete	109.5	70.0	94.8	117.4	97.2	109.9	109.7	96.0	104.6	105.5	83.0	97.1	104.7	90.2	99.3	105.5	90.2	99.9
03	CONCRETE	108.2	68.9	90.8	108.7	92.6	101.6	98.5	93.8	96.4	95.3	85.8	91.1	107.0	83.7	96.7	106.4	83.8	96.4
04	MASONRY	181.3	70.0	112.6	178.2	92.5	125.2	172.6	89.2	121.1	163.4	91.9	119.2	163.0	77.9	110.4	163.1	77.9	110.5
05	METALS	118.3	86.2	108.6	116.8	92.3	109.4	110.6	89.6	104.3	116.0	94.0	109.3	115.2	85.9	106.3	115.4	86.0	106.5
06	WOOD, PLASTICS & COMPOSITES	123.8	63.2	90.9	121.0	89.3	103.8	98.0	94.0	95.8	107.8	87.1	96.6	133.5	81.3	105.2	133.5	81.3	105.2
07	THERMAL & MOISTURE PROTECTION	116.0	71.0	96.4	121.1	94.1	109.3	116.8	87.4	104.0	114.9	89.4	106.3	114.2	86.5	102.2	114.2	86.5	102.1
08	OPENINGS	85.7	60.0	79.7	92.4	86.0	90.9	87.1	81.4	85.8	84.2	85.3	84.4	91.1	69.9	86.2	91.1	74.4	87.2
0920	Plaster & Gypsum Board	135.3	62.1	87.1	137.6	89.0	105.6	113.0	93.8	100.4	104.4	86.8	92.8	112.2	80.7	91.4	142.3	80.7	101.7
0950, 0980	Ceilings & Acoustic Treatment	121.0	62.1	81.1	108.9	89.0	95.4	124.0	93.8	103.6	104.7	86.8	92.5	103.8	80.7	88.1	103.8	80.7	88.1
0960	Flooring	111.6	67.2	98.7	113.6	97.2	108.9	106.8	95.6	103.6	106.7	94.2	103.1	113.6	88.8	106.5	113.6	88.8	106.5
0970, 0990	Wall Finishes & Painting/Coating	110.7	82.7	94.1	109.8	101.1	104.7	113.4	91.4	100.4	109.8	93.7	100.3	109.8	85.8	95.6	109.8	85.8	95.6
09	FINISHES	112.1	66.8	87.4	111.0	92.9	101.1	109.9	94.4	101.4	102.7	88.7	95.0	105.7	83.6	93.6	110.0	83.6	95.6
COVERS	DIVS. 10 - 14, 25, 28, 41, 43, 44, 46	131.1	61.8	115.9	131.1	67.5	117.1	131.1	69.6	117.6	131.1	87.7	121.6	131.1	77.2	119.3	131.1	77.2	119.3
21, 22, 23	FIRE SUPPRESSION, PLUMBING & HVAC	103.1	76.5	92.3	103.1	103.8	103.4	104.3	88.7	98.0	103.1	92.3	98.7	103.5	86.1	96.5	103.1	86.1	96.3
26, 27, 3370	ELECTRICAL, COMMUNICATIONS & UTIL.	114.5	83.7	98.8	110.2	87.1	98.4	107.4	93.7	100.4	108.9	85.3	96.9	106.9	66.8	86.5	106.9	66.8	86.5
MF2016	WEIGHTED AVERAGE	113.0	75.6	96.9	113.2	93.7	104.8	109.6	90.8	101.5	108.6	90.0	100.6	110.6	81.9	98.3	110.9	82.1	98.5

CANADA

DIVISION		ST CATHARINES, ONTARIO MAT.	INST.	TOTAL	ST JEROME, QUEBEC MAT.	INST.	TOTAL	ST. JOHN'S, NEWFOUNDLAND MAT.	INST.	TOTAL	SUDBURY, ONTARIO MAT.	INST.	TOTAL	SUMMERSIDE, PRINCE EDWARD ISLAND MAT.	INST.	TOTAL	SYDNEY, NOVA SCOTIA MAT.	INST.	TOTAL
015433	CONTRACTOR EQUIPMENT		103.4	103.4		103.5	103.5		122.7	122.7		103.4	103.4		102.0	102.0		102.0	102.0
0241, 31 - 34	SITE & INFRASTRUCTURE, DEMOLITION	96.3	101.4	99.9	97.9	98.9	98.6	119.1	111.9	114.0	96.5	101.0	99.6	123.9	94.3	103.3	110.2	97.0	101.0
0310	Concrete Forming & Accessories	111.8	93.0	95.8	130.4	81.3	88.6	121.0	86.4	91.5	107.8	88.1	91.1	109.1	54.0	62.2	108.8	69.8	75.6
0320	Concrete Reinforcing	106.8	98.4	102.5	154.2	74.5	113.7	168.2	84.7	125.8	107.7	102.2	104.9	166.8	47.8	106.3	169.0	49.0	108.0
0330	Cast-in-Place Concrete	106.0	99.1	103.4	101.3	90.2	97.2	132.0	99.6	119.9	106.9	95.8	102.8	111.8	56.4	91.2	90.9	68.8	82.7
03	CONCRETE	94.2	96.2	95.1	105.4	83.7	95.8	124.6	91.7	110.0	94.5	93.4	94.0	135.3	54.9	99.7	115.0	66.8	93.7
04	MASONRY	142.6	99.2	115.8	162.7	77.9	110.3	187.8	91.7	128.5	142.7	95.6	113.6	208.2	56.0	114.2	206.6	67.0	120.3
05	METALS	119.8	96.3	112.7	115.4	85.9	106.5	144.7	100.2	131.2	119.2	95.8	112.2	131.6	70.0	112.9	131.6	76.6	114.9
06	WOOD, PLASTICS & COMPOSITES	106.0	92.5	98.7	133.5	81.3	105.2	105.1	84.3	93.8	101.9	87.2	93.9	119.0	53.4	83.3	118.4	69.3	91.7
07	THERMAL & MOISTURE PROTECTION	115.2	100.9	109.0	114.2	86.5	102.1	141.3	96.5	121.8	114.7	95.6	106.4	129.6	58.9	98.8	130.2	65.5	103.8
08	OPENINGS	81.3	90.2	83.4	91.1	69.9	86.2	85.0	75.4	82.8	82.0	85.7	82.8	104.7	46.8	91.2	93.2	63.0	86.2
0920	Plaster & Gypsum Board	94.2	92.3	92.9	112.2	80.7	91.4	138.3	83.2	102.0	95.6	86.8	89.8	140.0	52.0	82.1	139.1	68.4	92.6
0950, 0980	Ceilings & Acoustic Treatment	103.3	92.3	95.8	103.8	80.7	88.1	136.9	83.2	100.6	98.2	86.8	90.5	122.4	52.0	74.7	122.4	68.4	85.8
0960	Flooring	100.2	92.4	98.0	113.6	88.8	106.5	117.5	53.0	98.8	98.4	94.2	97.2	108.1	57.1	93.3	108.1	60.9	94.4
0970, 0990	Wall Finishes & Painting/Coating	103.8	93.5	103.6	109.8	85.8	95.6	118.0	101.2	108.0	103.8	94.4	98.2	117.4	40.9	72.1	117.4	60.9	84.0
09	FINISHES	97.2	94.1	95.5	105.7	83.6	93.6	123.2	81.7	100.5	95.9	89.7	92.5	117.0	53.4	82.3	115.9	67.5	89.4
COVERS	DIVS. 10 - 14, 25, 28, 41, 43, 44, 46	131.1	69.2	117.5	131.1	77.2	119.3	131.1	68.6	117.4	131.1	89.0	121.9	131.1	58.2	115.1	131.1	62.1	115.9
21, 22, 23	FIRE SUPPRESSION, PLUMBING & HVAC	104.0	89.1	98.0	103.1	86.1	96.2	104.5	84.8	96.5	103.0	88.6	97.2	103.3	60.2	85.9	103.3	81.3	94.4
26, 27, 3370	ELECTRICAL, COMMUNICATIONS & UTIL.	109.3	97.8	103.5	107.6	66.8	86.8	105.8	83.4	94.4	109.2	98.3	103.6	107.3	48.6	77.4	108.2	60.1	83.8
MF2016	WEIGHTED AVERAGE	107.3	94.3	101.7	110.4	81.9	98.2	120.8	88.8	107.0	106.9	93.1	101.0	122.3	59.4	95.3	118.1	72.1	98.4

For customer support on your Electrical Costs with RSMeans data, call 800.448.8182.

CANADA

DIVISION		THUNDER BAY, ONTARIO			TIMMINS, ONTARIO			TORONTO, ONTARIO			TROIS RIVIERES, QUEBEC			TRURO, NOVA SCOTIA			VANCOUVER, BRITISH COLUMBIA		
		MAT.	INST.	TOTAL	MAT.	INST.	TOTAL	MAT.	INST.	TOTAL	MAT.	INST.	TOTAL	MAT.	INST.	TOTAL	MAT.	INST.	TOTAL
015433	CONTRACTOR EQUIPMENT		103.4	103.4		103.1	103.1		122.5	122.5		102.9	102.9		102.7	102.7		142.3	142.3
0241, 31 - 34	SITE & INFRASTRUCTURE, DEMOLITION	100.9	101.3	101.2	117.9	99.1	104.8	116.0	114.5	114.9	110.5	98.4	102.1	102.5	97.4	99.0	114.8	130.8	125.9
0310	Concrete Forming & Accessories	119.2	91.3	95.5	124.4	82.3	88.6	121.0	99.1	102.4	151.5	81.4	91.8	97.0	69.8	73.9	122.2	92.8	97.1
0320	Concrete Reinforcing	95.5	97.9	96.7	171.2	87.3	128.5	151.4	100.3	125.4	169.0	74.5	121.0	145.8	49.0	96.6	122.5	93.4	107.7
0330	Cast-in-Place Concrete	116.7	98.4	109.9	127.3	82.8	110.7	113.9	111.9	113.1	94.1	90.1	92.6	132.2	69.0	108.7	110.9	92.3	104.0
03	CONCRETE	101.3	95.0	98.5	119.1	83.8	103.4	109.9	104.4	107.5	117.3	83.7	102.4	118.0	66.9	95.4	107.7	94.0	101.6
04	MASONRY	143.2	99.4	116.1	166.5	83.6	115.3	166.8	106.1	129.3	210.6	77.9	128.6	162.2	67.0	103.3	161.7	86.3	115.1
05	METALS	119.8	95.3	112.4	116.8	91.5	109.1	142.3	108.9	132.2	130.9	85.6	117.2	116.1	76.9	104.2	136.4	111.8	129.0
06	WOOD, PLASTICS & COMPOSITES	115.5	89.9	101.6	121.8	82.6	100.5	105.7	96.9	100.9	173.4	81.3	123.4	90.4	69.4	79.0	103.3	92.8	97.6
07	THERMAL & MOISTURE PROTECTION	115.5	98.2	108.0	121.0	85.0	105.4	132.3	106.4	121.0	128.9	86.5	110.4	115.4	69.6	95.5	137.7	88.0	116.1
08	OPENINGS	80.6	88.1	82.4	89.7	80.5	87.6	82.1	96.0	85.4	102.4	74.4	95.9	84.2	63.1	79.3	85.0	89.7	86.1
0920	Plaster & Gypsum Board	120.6	89.6	100.2	113.1	82.1	92.7	116.8	96.2	103.2	169.8	80.7	111.1	118.4	68.4	85.5	118.0	91.5	100.5
0950, 0980	Ceilings & Acoustic Treatment	98.2	89.6	92.4	104.7	82.1	89.4	133.9	96.2	108.4	120.7	80.7	93.6	104.7	68.4	80.1	143.7	91.5	108.3
0960	Flooring	104.3	99.0	102.8	113.6	88.9	106.5	109.1	101.4	106.9	131.3	88.8	119.0	96.1	60.9	85.9	127.9	90.7	117.1
0970, 0990	Wall Finishes & Painting/Coating	103.8	95.3	98.8	109.8	87.3	96.5	107.0	107.5	107.3	117.4	85.8	98.7	109.8	60.9	80.8	115.0	87.6	98.8
09	FINISHES	101.6	93.2	97.0	106.6	84.4	94.5	112.2	100.4	105.7	126.7	83.5	103.1	101.7	67.5	83.0	122.8	92.2	106.1
COVERS	DIVS. 10 - 14, 25, 28, 41, 43, 44, 46	131.1	69.3	117.5	131.1	64.8	116.5	131.1	93.9	122.9	131.1	77.2	119.3	131.1	62.1	115.9	131.1	91.7	122.5
21, 22, 23	FIRE SUPPRESSION, PLUMBING & HVAC	104.0	89.3	98.1	103.1	94.5	99.6	104.8	97.2	101.7	103.3	86.1	96.3	103.1	81.3	94.3	104.8	77.1	93.6
26, 27, 3370	ELECTRICAL, COMMUNICATIONS & UTIL.	107.7	96.5	102.0	108.9	85.3	96.9	103.2	100.0	101.6	108.5	66.8	87.3	106.6	60.1	83.0	101.7	80.5	90.9
MF2016	WEIGHTED AVERAGE	108.5	93.7	102.1	113.2	87.5	102.2	115.8	102.4	110.1	120.7	82.0	104.1	110.8	72.2	94.2	115.4	91.6	105.2

CANADA

DIVISION		VICTORIA, BRITISH COLUMBIA			WHITEHORSE, YUKON			WINDSOR, ONTARIO			WINNIPEG, MANITOBA			YARMOUTH, NOVA SCOTIA			YELLOWKNIFE, NWT		
		MAT.	INST.	TOTAL	MAT.	INST.	TOTAL	MAT.	INST.	TOTAL	MAT.	INST.	TOTAL	MAT.	INST.	TOTAL	MAT.	INST.	TOTAL
015433	CONTRACTOR EQUIPMENT		109.9	109.9		132.7	132.7		103.4	103.4		126.9	126.9		102.0	102.0		131.8	131.8
0241, 31 - 34	SITE & INFRASTRUCTURE, DEMOLITION	124.0	106.8	112.0	139.5	120.4	126.2	92.6	101.3	98.6	116.0	115.2	115.5	113.9	97.0	102.1	148.2	125.6	132.4
0310	Concrete Forming & Accessories	111.9	85.9	89.8	130.4	58.1	68.8	119.2	89.6	94.0	126.4	65.6	74.6	108.8	69.8	75.6	132.6	77.7	85.9
0320	Concrete Reinforcing	114.5	93.2	103.7	153.0	65.1	108.3	104.7	97.2	100.9	140.0	61.1	99.9	169.0	49.0	108.0	150.2	67.7	108.2
0330	Cast-in-Place Concrete	113.7	88.9	104.5	159.0	71.9	126.6	108.6	99.9	105.3	138.6	72.6	114.0	116.5	68.8	98.8	157.1	89.1	131.8
03	CONCRETE	125.6	88.5	109.1	139.7	66.0	107.0	95.6	94.7	95.2	121.8	68.8	98.3	126.9	66.8	100.3	138.6	81.3	113.2
04	MASONRY	175.6	86.2	120.4	240.3	59.2	128.4	142.7	98.4	115.3	188.9	64.9	112.3	208.8	67.0	121.2	235.0	70.3	133.2
05	METALS	111.7	91.4	105.6	149.2	91.8	131.8	119.4	95.8	112.5	148.8	89.3	130.8	131.6	76.6	114.9	147.3	94.7	131.4
06	WOOD, PLASTICS & COMPOSITES	101.8	84.6	92.5	123.9	56.7	87.4	115.5	88.2	100.7	110.1	66.2	86.3	118.4	69.3	91.7	132.9	79.2	103.7
07	THERMAL & MOISTURE PROTECTION	127.0	87.6	109.8	143.3	64.2	108.9	115.3	97.6	107.6	132.2	69.9	105.1	130.2	69.5	103.8	141.8	80.3	115.0
08	OPENINGS	88.6	80.9	86.8	102.8	54.4	91.5	80.4	87.6	82.1	85.6	59.8	79.6	93.2	63.0	86.2	97.1	67.4	90.2
0920	Plaster & Gypsum Board	106.4	83.8	91.5	172.9	54.6	95.0	106.6	87.8	94.2	124.1	64.2	84.6	139.1	68.4	92.6	179.7	77.7	112.5
0950, 0980	Ceilings & Acoustic Treatment	106.2	83.8	91.0	168.6	54.6	91.4	98.2	87.8	91.2	147.1	64.2	90.9	122.4	68.4	85.8	164.1	77.7	105.5
0960	Flooring	112.2	69.7	99.9	130.8	57.0	109.4	104.3	96.4	102.0	118.6	68.8	104.2	108.1	60.9	94.4	124.1	85.7	112.9
0970, 0990	Wall Finishes & Painting/Coating	113.4	87.6	98.1	121.7	54.5	82.0	103.8	96.3	99.4	111.5	53.4	77.1	117.4	60.9	84.0	124.4	77.9	96.9
09	FINISHES	108.4	83.7	94.9	145.3	57.2	97.2	99.2	91.5	95.0	123.1	65.2	91.5	115.9	67.5	89.4	142.1	78.7	107.5
COVERS	DIVS. 10 - 14, 25, 28, 41, 43, 44, 46	131.1	66.1	116.8	131.1	61.3	115.8	131.1	68.8	117.4	131.1	63.8	116.3	131.1	62.1	115.9	131.1	64.8	116.5
21, 22, 23	FIRE SUPPRESSION, PLUMBING & HVAC	103.2	75.2	91.9	103.6	73.2	91.3	104.0	89.2	98.0	104.8	64.3	88.5	103.3	81.3	94.4	104.1	90.4	98.6
26, 27, 3370	ELECTRICAL, COMMUNICATIONS & UTIL.	107.4	80.0	93.4	124.4	58.5	90.9	112.5	97.9	105.1	107.0	63.2	84.7	108.2	60.1	83.8	121.7	79.3	100.2
MF2016	WEIGHTED AVERAGE	113.7	84.2	101.0	131.9	70.5	105.6	107.8	93.5	101.7	121.1	71.2	99.7	119.8	72.1	99.4	130.4	85.0	110.9

For customer support on your Electrical Costs with RSMeans data, call 800.448.8182.

555

Location Factors - Commercial

Costs shown in RSMeans cost data publications are based on national averages for materials and installation. To adjust these costs to a specific location, simply multiply the base cost by the factor and divide by 100 for that city. The data is arranged alphabetically by state and postal zip code numbers. For a city not listed, use the factor for a nearby city with similar economic characteristics.

STATE/ZIP	CITY	MAT.	INST.	TOTAL
ALABAMA				
350-352	Birmingham	96.5	71.7	85.9
354	Tuscaloosa	96.6	71.5	85.8
355	Jasper	97.1	71.0	85.9
356	Decatur	96.6	70.1	85.2
357-358	Huntsville	96.5	71.4	85.7
359	Gadsden	96.6	70.6	85.5
360-361	Montgomery	95.6	71.4	85.2
362	Anniston	95.2	67.4	83.3
363	Dothan	95.6	72.9	85.9
364	Evergreen	95.2	71.4	85.0
365-366	Mobile	96.1	68.8	84.4
367	Selma	95.3	72.4	85.5
368	Phenix City	96.1	72.0	85.8
369	Butler	95.5	71.6	85.3
ALASKA				
995-996	Anchorage	117.5	113.6	115.8
997	Fairbanks	118.2	114.7	116.7
998	Juneau	116.5	113.6	115.3
999	Ketchikan	128.4	113.6	122.0
ARIZONA				
850,853	Phoenix	100.0	73.0	88.4
851,852	Mesa/Tempe	98.6	72.0	87.2
855	Globe	99.5	71.8	87.6
856-857	Tucson	97.6	71.4	86.4
859	Show Low	99.7	71.9	87.8
860	Flagstaff	101.0	71.7	88.4
863	Prescott	98.7	71.7	87.1
864	Kingman	97.2	71.8	86.3
865	Chambers	97.2	74.4	87.4
ARKANSAS				
716	Pine Bluff	96.7	64.6	82.9
717	Camden	94.8	59.6	79.7
718	Texarkana	95.6	60.4	80.5
719	Hot Springs	94.2	61.0	79.9
720-722	Little Rock	95.4	65.9	82.8
723	West Memphis	94.3	65.5	81.9
724	Jonesboro	94.7	62.7	81.0
725	Batesville	92.6	60.0	78.7
726	Harrison	94.0	59.1	79.0
727	Fayetteville	91.5	60.7	78.3
728	Russellville	92.7	59.6	78.5
729	Fort Smith	95.1	63.6	81.6
CALIFORNIA				
900-902	Los Angeles	98.7	130.4	112.3
903-905	Inglewood	94.2	129.6	109.4
906-908	Long Beach	95.9	129.6	110.3
910-912	Pasadena	94.8	129.4	109.7
913-916	Van Nuys	97.9	129.4	111.4
917-918	Alhambra	96.8	129.4	110.8
919-921	San Diego	99.6	120.8	108.7
922	Palm Springs	96.9	127.1	109.8
923-924	San Bernardino	94.5	126.8	108.4
925	Riverside	98.8	127.1	110.9
926-927	Santa Ana	96.5	127.0	109.6
928	Anaheim	98.9	126.9	110.9
930	Oxnard	97.2	127.1	110.0
931	Santa Barbara	96.5	126.7	109.4
932-933	Bakersfield	98.0	125.3	109.7
934	San Luis Obispo	97.5	126.6	110.0
935	Mojave	94.7	125.2	107.7
936-938	Fresno	97.7	129.3	111.2
939	Salinas	98.3	135.7	114.3
940-941	San Francisco	106.3	158.1	128.5
942,956-958	Sacramento	100.1	132.7	114.1
943	Palo Alto	98.5	151.4	121.2
944	San Mateo	100.9	150.2	122.1
945	Vallejo	99.5	141.5	117.5
946	Oakland	102.7	151.2	123.5
947	Berkeley	102.3	151.2	123.3
948	Richmond	101.8	145.2	120.4
949	San Rafael	103.9	149.3	123.4
950	Santa Cruz	104.0	136.0	117.7

STATE/ZIP	CITY	MAT.	INST.	TOTAL
CALIFORNIA (CONT'D)				
951	San Jose	102.1	151.7	123.4
952	Stockton	100.3	130.8	113.4
953	Modesto	100.2	129.4	112.7
954	Santa Rosa	100.8	148.5	121.2
955	Eureka	102.1	133.2	115.4
959	Marysville	101.2	131.3	114.1
960	Redding	109.1	131.5	118.7
961	Susanville	108.8	131.2	118.4
COLORADO				
800-802	Denver	101.8	74.4	90.0
803	Boulder	97.7	75.9	88.3
804	Golden	99.8	73.7	88.6
805	Fort Collins	101.1	73.7	89.4
806	Greeley	98.3	74.7	88.2
807	Fort Morgan	98.3	72.9	87.4
808-809	Colorado Springs	100.0	73.0	88.4
810	Pueblo	100.5	71.4	88.0
811	Alamosa	102.3	68.8	87.9
812	Salida	102.0	66.6	86.8
813	Durango	102.7	65.0	86.6
814	Montrose	101.4	69.3	87.6
815	Grand Junction	104.5	72.2	90.7
816	Glenwood Springs	102.5	65.4	86.6
CONNECTICUT				
060	New Britain	96.4	118.1	105.7
061	Hartford	97.2	118.5	106.4
062	Willimantic	97.0	118.2	106.1
063	New London	93.6	118.4	104.2
064	Meriden	95.5	118.4	105.4
065	New Haven	98.1	118.5	106.9
066	Bridgeport	97.6	118.4	106.5
067	Waterbury	97.1	118.6	106.3
068	Norwalk	97.0	118.6	106.3
069	Stamford	97.2	125.2	109.2
D.C.				
200-205	Washington	101.0	87.7	95.3
DELAWARE				
197	Newark	98.3	112.3	104.3
198	Wilmington	98.2	112.3	104.2
199	Dover	98.2	112.3	104.2
FLORIDA				
320,322	Jacksonville	96.1	66.5	83.4
321	Daytona Beach	96.3	70.3	85.2
323	Tallahassee	97.3	65.7	83.7
324	Panama City	97.6	64.2	83.3
325	Pensacola	100.3	66.4	85.8
326,344	Gainesville	97.8	64.9	83.7
327-328,347	Orlando	97.9	67.1	84.7
329	Melbourne	99.2	71.2	87.2
330-332,340	Miami	95.8	67.4	83.7
333	Fort Lauderdale	95.3	69.5	84.2
334,349	West Palm Beach	94.3	66.0	82.2
335-336,346	Tampa	96.8	68.1	84.5
337	St. Petersburg	99.0	65.2	84.5
338	Lakeland	96.2	67.1	83.8
339,341	Fort Myers	95.6	68.3	83.9
342	Sarasota	98.7	67.0	85.1
GEORGIA				
300-303,399	Atlanta	99.0	75.6	89.0
304	Statesboro	98.9	65.8	84.7
305	Gainesville	97.4	66.5	84.2
306	Athens	96.8	67.9	84.4
307	Dalton	98.7	71.0	86.8
308-309	Augusta	97.1	72.7	86.6
310-312	Macon	93.9	74.0	85.4
313-314	Savannah	95.6	73.1	86.0
315	Waycross	95.4	68.4	83.8
316	Valdosta	95.3	66.2	82.8
317,398	Albany	95.2	72.6	85.5
318-319	Columbus	95.1	72.7	85.5

556

Location Factors - Commercial

STATE/ZIP	CITY	MAT.	INST.	TOTAL	STATE/ZIP	CITY	MAT.	INST.	TOTAL
HAWAII					**KANSAS (CONT'D)**				
967	Hilo	116.3	118.3	117.1	678	Dodge City	98.3	75.4	88.5
968	Honolulu	120.9	118.3	119.8	679	Liberal	96.2	73.7	86.5
STATES & POSS.					**KENTUCKY**				
969	Guam	138.8	52.6	101.9	400-402	Louisville	93.2	77.7	86.6
					403-405	Lexington	93.4	79.6	87.5
IDAHO					406	Frankfort	95.5	78.7	88.3
832	Pocatello	100.7	78.6	91.2	407-409	Corbin	91.1	79.6	86.2
833	Twin Falls	101.9	77.2	91.3	410	Covington	94.4	79.8	88.1
834	Idaho Falls	99.2	78.2	90.2	411-412	Ashland	93.2	92.0	92.7
835	Lewiston	106.9	85.7	97.8	413-414	Campton	94.4	79.9	88.2
836-837	Boise	99.7	79.4	91.0	415-416	Pikeville	95.8	87.6	92.3
838	Coeur d'Alene	106.8	84.4	97.2	417-418	Hazard	93.8	80.6	88.1
					420	Paducah	92.5	83.8	88.8
ILLINOIS					421-422	Bowling Green	94.5	78.9	87.8
600-603	North Suburban	99.8	144.1	118.8	423	Owensboro	94.6	83.1	89.7
604	Joliet	99.7	142.2	117.9	424	Henderson	92.2	82.4	88.0
605	South Suburban	99.8	144.0	118.7	425-426	Somerset	91.7	77.8	85.7
606-608	Chicago	101.7	146.1	120.7	427	Elizabethtown	91.3	76.7	85.1
609	Kankakee	96.4	136.1	113.4					
610-611	Rockford	96.8	128.7	110.5	**LOUISIANA**				
612	Rock Island	94.9	102.1	98.0	700-701	New Orleans	99.4	69.5	86.6
613	La Salle	96.1	128.9	110.2	703	Thibodaux	96.2	66.4	83.4
614	Galesburg	95.9	109.5	101.8	704	Hammond	93.8	64.5	81.2
615-616	Peoria	97.8	111.5	103.7	705	Lafayette	95.6	68.9	84.1
617	Bloomington	95.2	112.2	102.5	706	Lake Charles	95.8	69.8	84.6
618-619	Champaign	98.7	109.8	103.5	707-708	Baton Rouge	96.6	69.7	85.1
620-622	East St. Louis	94.4	111.3	101.6	710-711	Shreveport	97.8	66.4	84.3
623	Quincy	96.3	105.1	100.1	712	Monroe	96.6	65.1	83.1
624	Effingham	95.7	109.2	101.5	713-714	Alexandria	96.7	66.6	83.8
625	Decatur	97.2	107.0	101.4					
626-627	Springfield	97.9	107.7	102.1	**MAINE**				
628	Centralia	93.3	113.0	101.8	039	Kittery	91.3	84.6	88.5
629	Carbondale	93.1	109.8	100.2	040-041	Portland	96.9	84.1	91.4
					042	Lewiston	94.7	84.1	90.1
INDIANA					043	Augusta	96.7	83.1	90.9
460	Anderson	95.9	81.8	89.8	044	Bangor	94.3	83.5	89.7
461-462	Indianapolis	98.7	83.0	92.0	045	Bath	93.2	82.6	88.6
463-464	Gary	97.2	108.3	102.0	046	Machias	92.6	79.7	87.1
465-466	South Bend	97.4	83.6	91.5	047	Houlton	92.8	79.7	87.2
467-468	Fort Wayne	96.4	78.2	88.6	048	Rockland	91.9	83.1	88.1
469	Kokomo	94.1	81.0	88.4	049	Waterville	93.1	83.1	88.8
470	Lawrenceburg	92.6	78.5	86.5					
471	New Albany	93.9	77.2	86.7	**MARYLAND**				
472	Columbus	96.1	80.4	89.3	206	Waldorf	97.7	87.4	93.3
473	Muncie	96.1	80.3	89.3	207-208	College Park	97.7	88.6	93.8
474	Bloomington	97.7	80.3	90.3	209	Silver Spring	96.9	87.4	92.8
475	Washington	94.6	85.4	90.7	210-212	Baltimore	101.2	85.0	94.2
476-477	Evansville	95.4	84.8	90.9	214	Annapolis	100.5	83.1	93.0
478	Terre Haute	96.2	83.8	90.9	215	Cumberland	97.1	85.5	92.1
479	Lafayette	95.8	80.1	89.1	216	Easton	98.8	76.0	89.0
					217	Hagerstown	97.5	87.0	93.0
IOWA					218	Salisbury	99.3	68.5	86.1
500-503,509	Des Moines	96.0	88.4	92.8	219	Elkton	96.0	87.6	92.4
504	Mason City	94.7	76.2	86.8					
505	Fort Dodge	94.9	74.1	86.0	**MASSACHUSETTS**				
506-507	Waterloo	96.2	79.3	89.0	010-011	Springfield	96.4	110.4	102.4
508	Creston	95.2	83.1	90.0	012	Pittsfield	96.0	105.1	99.9
510-511	Sioux City	97.1	78.7	89.2	013	Greenfield	94.3	110.9	101.4
512	Sibley	96.1	63.3	82.1	014	Fitchburg	93.0	118.7	104.0
513	Spencer	97.7	63.6	83.1	015-016	Worcester	96.4	118.5	105.9
514	Carroll	94.9	81.3	89.1	017	Framingham	92.4	127.0	107.2
515	Council Bluffs	98.2	80.8	90.7	018	Lowell	95.8	125.4	108.5
516	Shenandoah	95.4	83.8	90.4	019	Lawrence	96.7	125.8	109.2
520	Dubuque	96.5	81.9	90.3	020-022, 024	Boston	99.4	133.3	113.9
521	Decorah	95.9	75.9	87.3	023	Brockton	96.1	119.2	106.0
522-524	Cedar Rapids	97.4	85.8	92.4	025	Buzzards Bay	91.1	117.7	102.5
525	Ottumwa	95.7	76.3	87.4	026	Hyannis	93.4	117.5	103.7
526	Burlington	95.1	86.3	91.3	027	New Bedford	95.3	117.6	104.9
527-528	Davenport	96.5	96.9	96.7					
					MICHIGAN				
KANSAS					480,483	Royal Oak	95.2	100.6	97.5
660-662	Kansas City	96.0	98.6	97.1	481	Ann Arbor	97.2	101.7	99.1
664-666	Topeka	97.0	77.7	88.7	482	Detroit	100.2	103.0	101.4
667	Fort Scott	94.8	78.2	87.7	484-485	Flint	96.9	90.8	94.3
668	Emporia	94.9	77.2	87.3	486	Saginaw	96.5	87.3	92.6
669	Belleville	96.6	72.1	86.1	487	Bay City	96.6	87.5	92.7
670-672	Wichita	96.7	70.8	85.6	488-489	Lansing	97.9	88.7	93.9
673	Independence	97.0	77.9	88.8	490	Battle Creek	95.3	82.5	89.8
674	Salina	96.9	72.8	86.6	491	Kalamazoo	95.6	80.7	89.2
675	Hutchinson	92.4	73.1	84.1	492	Jackson	93.8	93.0	93.5
676	Hays	96.3	73.5	86.5	493,495	Grand Rapids	97.1	81.6	90.5
677	Colby	97.1	75.7	87.9	494	Muskegon	94.1	80.3	88.2

557

Location Factors - Commercial

STATE/ZIP	CITY	MAT.	INST.	TOTAL	STATE/ZIP	CITY	MAT.	INST.	TOTAL
MICHIGAN (CONT'D)					**NEW HAMPSHIRE (CONT'D)**				
496	Traverse City	93.2	78.3	86.8	032-033	Concord	96.1	92.4	94.5
497	Gaylord	94.3	81.3	88.8	034	Keene	93.0	89.2	91.3
498-499	Iron Mountain	96.2	83.0	90.6	035	Littleton	93.0	80.1	87.4
					036	Charleston	92.5	88.6	90.8
MINNESOTA					037	Claremont	91.7	88.6	90.3
550-551	Saint Paul	97.2	115.8	105.2	038	Portsmouth	93.3	92.2	92.8
553-555	Minneapolis	99.7	116.5	106.9					
556-558	Duluth	97.1	102.7	99.5	**NEW JERSEY**				
559	Rochester	97.1	100.3	98.4	070-071	Newark	99.6	139.7	116.8
560	Mankato	95.1	98.0	96.4	072	Elizabeth	97.3	139.7	115.4
561	Windom	93.7	92.3	93.1	073	Jersey City	96.3	139.4	114.8
562	Willmar	93.3	99.9	96.1	074-075	Paterson	97.7	139.5	115.6
563	St. Cloud	94.4	114.0	102.8	076	Hackensack	95.9	139.6	114.6
564	Brainerd	95.0	98.6	96.5	077	Long Branch	95.6	133.6	111.9
565	Detroit Lakes	96.7	92.2	94.8	078	Dover	96.1	139.6	114.7
566	Bemidji	96.0	95.4	95.8	079	Summit	96.2	139.7	114.8
567	Thief River Falls	95.6	89.4	92.9	080,083	Vineland	95.8	132.3	111.4
					081	Camden	97.5	130.7	111.8
MISSISSIPPI					082,084	Atlantic City	96.4	132.6	111.9
386	Clarksdale	95.2	54.4	77.7	085-086	Trenton	99.2	131.4	113.0
387	Greenville	98.6	66.0	84.6	087	Point Pleasant	97.7	133.3	113.0
388	Tupelo	96.5	58.2	80.1	088-089	New Brunswick	98.3	137.7	115.2
389	Greenwood	96.5	54.2	78.3					
390-392	Jackson	98.0	65.4	84.0	**NEW MEXICO**				
393	Meridian	95.6	65.1	82.5	870-872	Albuquerque	97.1	75.0	87.6
394	Laurel	97.2	56.9	79.9	873	Gallup	97.3	75.0	87.7
395	Biloxi	97.3	64.6	83.3	874	Farmington	97.7	75.0	88.0
396	McComb	95.4	54.5	77.9	875	Santa Fe	97.6	75.0	87.9
397	Columbus	97.0	58.4	80.5	877	Las Vegas	95.8	75.0	86.9
					878	Socorro	95.5	75.0	86.7
MISSOURI					879	Truth/Consequences	95.2	71.7	85.2
630-631	St. Louis	99.6	105.8	102.3	880	Las Cruces	95.6	71.7	85.4
633	Bowling Green	97.5	96.0	96.9	881	Clovis	98.0	74.9	88.1
634	Hannibal	96.5	94.0	95.4	882	Roswell	99.5	75.0	89.0
635	Kirksville	100.2	88.8	95.3	883	Carrizozo	100.2	75.0	89.4
636	Flat River	98.5	96.2	97.5	884	Tucumcari	98.7	74.9	88.5
637	Cape Girardeau	98.1	91.8	95.4					
638	Sikeston	96.9	89.4	93.7	**NEW YORK**				
639	Poplar Bluff	96.3	89.2	93.3	100-102	New York	99.0	176.2	132.1
640-641	Kansas City	98.1	103.6	100.5	103	Staten Island	94.9	177.8	130.5
644-645	St. Joseph	96.8	92.8	95.1	104	Bronx	93.3	176.8	129.1
646	Chillicothe	94.5	95.4	94.9	105	Mount Vernon	93.5	151.3	118.3
647	Harrisonville	94.0	102.6	97.7	106	White Plains	93.3	153.5	119.1
648	Joplin	95.9	79.7	88.9	107	Yonkers	97.3	151.2	120.5
650-651	Jefferson City	96.3	90.8	94.0	108	New Rochelle	93.9	146.0	116.2
652	Columbia	96.2	91.3	94.1	109	Suffern	93.6	133.5	110.7
653	Sedalia	96.6	93.7	95.4	110	Queens	99.8	178.9	133.8
654-655	Rolla	94.3	96.6	95.3	111	Long Island City	101.5	178.9	134.7
656-658	Springfield	97.7	81.0	90.5	112	Brooklyn	101.8	178.9	134.9
					113	Flushing	102.0	178.9	135.0
MONTANA					114	Jamaica	100.2	178.9	134.0
590-591	Billings	100.8	75.0	89.7	115,117,118	Hicksville	99.8	160.5	125.8
592	Wolf Point	100.5	77.8	90.8	116	Far Rockaway	102.2	178.9	135.1
593	Miles City	98.4	77.9	89.6	119	Riverhead	100.5	156.8	124.6
594	Great Falls	102.1	74.0	90.1	120-122	Albany	95.0	111.5	102.1
595	Havre	99.5	75.7	89.3	123	Schenectady	95.5	110.6	101.9
596	Helena	100.1	76.1	89.8	124	Kingston	98.9	136.0	114.8
597	Butte	100.7	76.2	90.2	125-126	Poughkeepsie	98.1	140.2	116.1
598	Missoula	97.9	75.7	88.4	127	Monticello	97.4	136.9	114.3
599	Kalispell	97.4	76.2	88.3	128	Glens Falls	90.6	108.7	98.4
					129	Plattsburgh	95.6	99.0	97.0
NEBRASKA					130-132	Syracuse	97.8	100.8	99.1
680-681	Omaha	96.9	80.8	90.0	133-135	Utica	95.9	98.9	97.2
683-685	Lincoln	97.1	81.1	90.2	136	Watertown	97.7	99.2	98.3
686	Columbus	95.6	82.7	90.1	137-139	Binghamton	97.3	102.8	99.7
687	Norfolk	96.9	79.2	89.3	140-142	Buffalo	103.0	110.0	106.0
688	Grand Island	97.0	78.5	89.1	143	Niagara Falls	98.8	110.2	103.7
689	Hastings	96.7	80.0	89.5	144-146	Rochester	100.3	101.3	100.8
690	McCook	96.6	74.8	87.3	147	Jamestown	97.8	97.8	97.8
691	North Platte	96.6	77.3	88.3	148-149	Elmira	97.6	103.9	100.3
692	Valentine	98.8	70.8	86.8					
693	Alliance	98.9	74.7	88.5	**NORTH CAROLINA**				
					270,272-274	Greensboro	97.7	67.7	84.8
NEVADA					271	Winston-Salem	97.4	67.8	84.7
889-891	Las Vegas	102.7	104.1	103.3	275-276	Raleigh	96.4	67.1	83.8
893	Ely	101.6	95.6	99.0	277	Durham	99.3	67.7	85.7
894-895	Reno	101.5	84.2	94.1	278	Rocky Mount	95.3	67.5	83.4
897	Carson City	100.4	84.2	93.5	279	Elizabeth City	96.1	69.2	84.6
898	Elko	100.3	87.4	94.8	280	Gastonia	97.0	67.7	84.4
					281-282	Charlotte	97.0	67.3	84.3
NEW HAMPSHIRE					283	Fayetteville	100.0	67.3	86.0
030	Nashua	95.9	92.8	94.6	284	Wilmington	95.9	66.3	83.2
031	Manchester	96.2	93.0	94.8	285	Kinston	94.3	67.1	82.7

STATE/ZIP	CITY	MAT.	INST.	TOTAL
NORTH CAROLINA (CONT'D)				
286	Hickory	94.7	68.3	83.4
287-288	Asheville	96.3	67.0	83.8
289	Murphy	95.5	66.2	82.9
NORTH DAKOTA				
580-581	Fargo	98.2	80.3	90.5
582	Grand Forks	98.4	79.0	90.1
583	Devils Lake	98.3	80.2	90.5
584	Jamestown	98.3	79.6	90.3
585	Bismarck	98.2	80.0	90.4
586	Dickinson	99.0	79.1	90.5
587	Minot	98.3	79.6	90.3
588	Williston	97.4	79.0	89.6
OHIO				
430-432	Columbus	97.4	83.2	91.3
433	Marion	93.4	88.5	91.3
434-436	Toledo	96.7	92.8	95.0
437-438	Zanesville	94.0	87.1	91.0
439	Steubenville	95.3	92.1	94.0
440	Lorain	98.4	86.8	93.4
441	Cleveland	98.7	92.2	95.9
442-443	Akron	99.5	88.3	94.7
444-445	Youngstown	98.7	84.0	92.4
446-447	Canton	98.9	81.4	91.4
448-449	Mansfield	96.5	86.4	92.1
450	Hamilton	95.9	81.5	89.7
451-452	Cincinnati	96.9	80.7	89.9
453-454	Dayton	95.9	81.3	89.6
455	Springfield	95.9	81.8	89.8
456	Chillicothe	95.4	90.8	93.4
457	Athens	98.2	86.4	93.2
458	Lima	98.3	83.8	92.1
OKLAHOMA				
730-731	Oklahoma City	95.2	68.2	83.6
734	Ardmore	93.8	66.2	82.0
735	Lawton	95.9	66.6	83.3
736	Clinton	95.0	66.1	82.6
737	Enid	95.5	66.5	83.1
738	Woodward	93.9	63.4	80.8
739	Guymon	94.9	64.8	82.0
740-741	Tulsa	96.0	67.1	83.6
743	Miami	92.8	66.5	81.5
744	Muskogee	95.1	64.4	81.9
745	McAlester	92.4	61.9	79.3
746	Ponca City	93.2	65.2	81.2
747	Durant	93.2	66.3	81.7
748	Shawnee	94.6	66.1	82.4
749	Poteau	92.4	66.1	81.1
OREGON				
970-972	Portland	100.2	100.4	100.3
973	Salem	101.7	99.9	100.9
974	Eugene	99.8	98.5	99.3
975	Medford	101.3	97.8	99.8
976	Klamath Falls	101.5	97.8	99.9
977	Bend	100.7	99.8	100.3
978	Pendleton	96.8	101.7	98.9
979	Vale	94.5	87.9	91.7
PENNSYLVANIA				
150-152	Pittsburgh	99.9	102.2	100.9
153	Washington	96.9	102.7	99.4
154	Uniontown	97.2	102.2	99.4
155	Bedford	98.2	94.3	96.5
156	Greensburg	98.2	98.7	98.4
157	Indiana	97.1	100.1	98.4
158	Dubois	98.7	98.3	98.5
159	Johnstown	98.2	94.2	96.5
160	Butler	91.3	103.1	96.4
161	New Castle	91.4	100.3	95.2
162	Kittanning	91.8	100.4	95.5
163	Oil City	91.3	100.3	95.1
164-165	Erie	93.1	96.2	94.5
166	Altoona	93.2	95.1	94.0
167	Bradford	94.9	99.5	96.8
168	State College	94.5	97.0	95.6
169	Wellsboro	95.5	95.2	95.4
170-171	Harrisburg	97.8	94.9	96.5
172	Chambersburg	94.9	91.6	93.4
173-174	York	94.9	95.3	95.1
175-176	Lancaster	93.5	96.8	94.9

STATE/ZIP	CITY	MAT.	INST.	TOTAL
PENNSYLVANIA (CONT'D)				
177	Williamsport	92.1	94.6	93.2
178	Sunbury	94.3	95.0	94.6
179	Pottsville	93.4	98.4	95.6
180	Lehigh Valley	95.0	113.6	103.0
181	Allentown	96.7	108.4	101.7
182	Hazleton	94.4	98.3	96.1
183	Stroudsburg	94.3	107.3	99.9
184-185	Scranton	97.4	98.9	98.1
186-187	Wilkes-Barre	94.1	97.9	95.7
188	Montrose	93.8	99.2	96.1
189	Doylestown	94.0	129.2	109.1
190-191	Philadelphia	100.1	135.3	115.2
193	Westchester	95.8	129.9	110.4
194	Norristown	94.8	130.0	109.9
195-196	Reading	96.5	104.0	99.7
PUERTO RICO				
009	San Juan	117.8	27.2	79.0
RHODE ISLAND				
028	Newport	94.6	113.7	102.8
029	Providence	96.2	113.7	103.7
SOUTH CAROLINA				
290-292	Columbia	96.5	67.9	84.2
293	Spartanburg	96.4	67.9	84.2
294	Charleston	98.0	67.6	84.9
295	Florence	96.2	67.8	84.0
296	Greenville	96.2	67.9	84.1
297	Rock Hill	96.0	67.3	83.7
298	Aiken	96.9	67.9	84.4
299	Beaufort	97.5	56.0	79.7
SOUTH DAKOTA				
570-571	Sioux Falls	96.3	79.2	89.0
572	Watertown	97.1	70.8	85.8
573	Mitchell	96.0	62.4	81.6
574	Aberdeen	98.3	71.5	86.8
575	Pierre	98.7	69.4	86.2
576	Mobridge	96.6	64.4	82.8
577	Rapid City	98.0	73.3	87.4
TENNESSEE				
370-372	Nashville	98.4	71.1	86.7
373-374	Chattanooga	98.9	69.6	86.3
375,380-381	Memphis	97.3	71.1	86.1
376	Johnson City	99.1	59.9	82.3
377-379	Knoxville	95.6	65.2	82.6
382	McKenzie	96.2	56.0	79.0
383	Jackson	97.5	59.7	81.3
384	Columbia	94.7	67.9	83.2
385	Cookeville	96.1	57.3	79.5
TEXAS				
750	McKinney	98.0	65.1	83.9
751	Waxahachie	98.1	65.1	83.9
752-753	Dallas	99.1	67.7	85.6
754	Greenville	98.2	64.5	83.8
755	Texarkana	97.5	63.6	83.0
756	Longview	98.4	62.6	83.0
757	Tyler	98.6	63.2	83.5
758	Palestine	95.1	62.0	80.9
759	Lufkin	95.5	64.2	82.1
760-761	Fort Worth	97.4	64.1	83.1
762	Denton	97.5	64.4	83.3
763	Wichita Falls	94.9	62.7	81.1
764	Eastland	94.0	61.8	80.2
765	Temple	92.5	60.0	78.5
766-767	Waco	94.3	63.9	81.3
768	Brownwood	97.4	60.2	81.4
769	San Angelo	97.1	60.8	81.6
770-772	Houston	100.7	67.9	86.6
773	Huntsville	99.4	64.3	84.4
774	Wharton	100.8	65.8	85.8
775	Galveston	98.5	68.1	85.5
776-777	Beaumont	98.8	67.1	85.2
778	Bryan	95.5	66.4	83.1
779	Victoria	100.5	64.4	85.0
780	Laredo	97.1	62.7	82.4
781-782	San Antonio	99.3	63.5	83.9
783-784	Corpus Christi	99.7	63.0	84.0
785	McAllen	100.6	57.9	82.3
786-787	Austin	97.4	62.7	82.5

559

STATE/ZIP	CITY	MAT.	INST.	TOTAL
TEXAS (CONT'D)				
788	Del Rio	100.6	62.0	84.0
789	Giddings	97.0	61.9	82.0
790-791	Amarillo	95.7	61.1	80.9
792	Childress	95.8	61.8	81.2
793-794	Lubbock	97.5	64.4	83.3
795-796	Abilene	96.0	62.8	81.8
797	Midland	98.1	63.7	83.3
798-799,885	El Paso	94.6	64.7	81.8
UTAH				
840-841	Salt Lake City	102.0	73.1	89.6
842,844	Ogden	97.5	73.1	87.0
843	Logan	99.4	73.1	88.1
845	Price	100.0	71.7	87.9
846-847	Provo	99.8	72.9	88.3
VERMONT				
050	White River Jct.	95.8	81.0	89.4
051	Bellows Falls	94.2	93.6	94.0
052	Bennington	94.6	90.0	92.6
053	Brattleboro	95.0	93.6	94.4
054	Burlington	99.1	80.6	91.2
056	Montpelier	97.7	85.3	92.4
057	Rutland	96.3	80.5	89.5
058	St. Johnsbury	95.8	80.5	89.3
059	Guildhall	94.5	80.2	88.3
VIRGINIA				
220-221	Fairfax	99.2	84.2	92.8
222	Arlington	100.1	84.3	93.3
223	Alexandria	99.3	84.2	92.8
224-225	Fredericksburg	97.9	80.6	90.5
226	Winchester	98.6	81.3	91.2
227	Culpeper	98.4	84.9	92.6
228	Harrisonburg	98.7	66.0	84.7
229	Charlottesville	99.1	68.0	85.8
230-232	Richmond	98.3	76.4	88.9
233-235	Norfolk	98.8	70.0	86.5
236	Newport News	97.9	69.7	85.8
237	Portsmouth	97.4	63.1	82.7
238	Petersburg	97.9	72.4	87.0
239	Farmville	97.1	62.8	82.4
240-241	Roanoke	99.7	71.6	87.7
242	Bristol	97.9	59.4	81.4
243	Pulaski	97.5	68.3	85.0
244	Staunton	98.3	65.6	84.3
245	Lynchburg	98.4	73.7	87.8
246	Grundy	97.8	59.2	81.2
WASHINGTON				
980-981,987	Seattle	106.0	107.4	106.6
982	Everett	103.4	102.6	103.1
983-984	Tacoma	103.8	102.1	103.1
985	Olympia	100.9	102.3	101.5
986	Vancouver	104.8	96.2	101.1
988	Wenatchee	104.1	91.0	98.5
989	Yakima	104.0	96.6	100.8
990-992	Spokane	99.9	84.5	93.3
993	Richland	99.6	92.2	96.4
994	Clarkston	98.9	83.9	92.5
WEST VIRGINIA				
247-248	Bluefield	96.7	90.2	93.9
249	Lewisburg	98.4	90.5	95.0
250-253	Charleston	96.1	92.5	94.6
254	Martinsburg	96.4	86.1	92.0
255-257	Huntington	97.5	94.3	96.1
258-259	Beckley	94.9	92.1	93.7
260	Wheeling	99.3	91.5	96.0
261	Parkersburg	98.2	90.7	95.0
262	Buckhannon	98.1	90.7	94.9
263-264	Clarksburg	98.7	91.4	95.5
265	Morgantown	98.5	91.5	95.5
266	Gassaway	98.0	90.5	94.8
267	Romney	97.9	89.2	94.2
268	Petersburg	97.7	87.7	93.4
WISCONSIN				
530,532	Milwaukee	99.6	105.9	102.3
531	Kenosha	99.4	104.6	101.6
534	Racine	98.8	105.0	101.5
535	Beloit	98.7	101.4	99.9
537	Madison	99.1	100.3	99.6

STATE/ZIP	CITY	MAT.	INST.	TOTAL
WISCONSIN (CONT'D)				
538	Lancaster	96.9	96.7	96.8
539	Portage	95.3	100.5	97.5
540	New Richmond	94.7	97.0	95.7
541-543	Green Bay	98.7	95.8	97.5
544	Wausau	94.1	95.8	94.8
545	Rhinelander	97.3	94.6	96.1
546	La Crosse	95.2	95.9	95.5
547	Eau Claire	96.8	96.5	96.7
548	Superior	94.5	98.7	96.3
549	Oshkosh	94.8	94.2	94.5
WYOMING				
820	Cheyenne	99.6	72.1	87.8
821	Yellowstone Nat'l Park	97.6	73.2	87.1
822	Wheatland	98.8	68.1	85.6
823	Rawlins	100.3	72.7	88.5
824	Worland	98.3	71.8	87.0
825	Riverton	99.4	69.7	86.6
826	Casper	99.3	69.5	86.5
827	Newcastle	98.1	72.3	87.1
828	Sheridan	100.9	70.6	87.9
829-831	Rock Springs	101.9	71.6	88.9
CANADIAN FACTORS (reflect Canadian currency)				
ALBERTA				
	Calgary	119.0	97.6	109.8
	Edmonton	119.7	97.5	110.2
	Fort McMurray	122.1	90.2	108.4
	Lethbridge	117.2	89.7	105.4
	Lloydminster	111.6	86.5	100.8
	Medicine Hat	111.7	85.8	100.6
	Red Deer	112.2	85.8	100.8
BRITISH COLUMBIA				
	Kamloops	113.0	85.7	101.3
	Prince George	113.9	85.0	101.5
	Vancouver	115.4	91.6	105.2
	Victoria	113.7	84.2	101.0
MANITOBA				
	Brandon	120.9	72.4	100.1
	Portage la Prairie	111.6	70.9	94.1
	Winnipeg	121.1	71.2	99.7
NEW BRUNSWICK				
	Bathurst	110.4	63.9	90.5
	Dalhousie	109.8	64.1	90.2
	Fredericton	117.9	71.2	97.9
	Moncton	110.6	68.1	92.4
	Newcastle	110.5	64.5	90.8
	Saint John	113.0	75.6	96.9
NEWFOUNDLAND				
	Corner Brook	125.0	70.2	101.5
	St. John's	120.8	88.8	107.0
NORTHWEST TERRITORIES				
	Yellowknife	130.4	85.0	110.9
NOVA SCOTIA				
	Bridgewater	111.2	72.2	94.5
	Dartmouth	121.8	72.1	100.5
	Halifax	117.1	85.8	103.7
	New Glasgow	119.9	72.1	99.4
	Sydney	118.1	72.1	98.4
	Truro	110.8	72.2	94.2
	Yarmouth	119.8	72.1	99.4
ONTARIO				
	Barrie	116.0	89.3	104.6
	Brantford	113.1	93.1	104.5
	Cornwall	113.1	89.6	103.0
	Hamilton	116.1	98.1	108.4
	Kingston	114.1	89.7	103.6
	Kitchener	109.4	93.4	102.5
	London	116.8	95.6	107.7
	North Bay	122.8	87.4	107.7
	Oshawa	111.6	92.9	103.6
	Ottawa	116.9	96.4	108.1
	Owen Sound	116.1	87.6	103.9
	Peterborough	113.1	89.4	102.9
	Sarnia	113.2	93.7	104.8

Location Factors - Commercial

STATE/ZIP	CITY	MAT.	INST.	TOTAL
ONTARIO (CONT'D)				
	Sault Ste. Marie	108.6	90.0	100.6
	St. Catharines	107.3	94.3	101.7
	Sudbury	106.9	93.1	101.0
	Thunder Bay	108.5	93.7	102.1
	Timmins	113.2	87.5	102.2
	Toronto	115.8	102.4	110.1
	Windsor	107.8	93.5	101.7
PRINCE EDWARD ISLAND				
	Charlottetown	120.1	61.8	95.1
	Summerside	122.3	59.4	95.3
QUEBEC				
	Cap-de-la-Madeleine	110.7	82.1	98.4
	Charlesbourg	110.7	82.1	98.4
	Chicoutimi	110.3	87.3	100.4
	Gatineau	110.4	81.9	98.2
	Granby	110.6	81.8	98.3
	Hull	110.5	81.9	98.2
	Joliette	110.9	82.1	98.5
	Laval	110.7	81.9	98.3
	Montreal	118.1	89.9	106.0
	Quebec City	116.4	90.3	105.2
	Rimouski	110.3	87.3	100.4
	Rouyn-Noranda	110.4	81.9	98.1
	Saint-Hyacinthe	109.8	81.9	97.8
	Sherbrooke	110.6	81.9	98.3
	Sorel	110.9	82.1	98.5
	Saint-Jerome	110.4	81.9	98.2
	Trois-Rivieres	120.7	82.0	104.1
SASKATCHEWAN				
	Moose Jaw	109.0	65.6	90.4
	Prince Albert	108.2	63.9	89.2
	Regina	122.1	94.2	110.1
	Saskatoon	109.6	90.8	101.5
YUKON				
	Whitehorse	131.9	70.5	105.6

R011105-05   Tips for Accurate Estimating

1. Use pre-printed or columnar forms for orderly sequence of dimensions and locations and for recording telephone quotations.

2. Use only the front side of each paper or form except for certain pre-printed summary forms.

3. Be consistent in listing dimensions: For example, length x width x height. This helps in rechecking to ensure that the total length of partitions is appropriate for the building area.

4. Use printed (rather than measured) dimensions where given.

5. Add up multiple printed dimensions for a single entry where possible.

6. Measure all other dimensions carefully.

7. Use each set of dimensions to calculate multiple related quantities.

8. Convert foot and inch measurements to decimal feet when listing. Memorize decimal equivalents to .01 parts of a foot (1/8″ equals approximately .01′).

9. Do not "round off" quantities until the final summary.

10. Mark drawings with different colors as items are taken off.

11. Keep similar items together, different items separate.

12. Identify location and drawing numbers to aid in future checking for completeness.

13. Measure or list everything on the drawings or mentioned in the specifications.

14. It may be necessary to list items not called for to make the job complete.

15. Be alert for: Notes on plans such as N.T.S. (not to scale); changes in scale throughout the drawings; reduced size drawings; discrepancies between the specifications and the drawings.

16. Develop a consistent pattern of performing an estimate. For example:
    a. Start the quantity takeoff at the lower floor and move to the next higher floor.
    b. Proceed from the main section of the building to the wings.
    c. Proceed from south to north or vice versa, clockwise or counterclockwise.
    d. Take off floor plan quantities first, elevations next, then detail drawings.

17. List all gross dimensions that can be either used again for different quantities, or used as a rough check of other quantities for verification (exterior perimeter, gross floor area, individual floor areas, etc.).

18. Utilize design symmetry or repetition (repetitive floors, repetitive wings, symmetrical design around a center line, similar room layouts, etc.). Note: Extreme caution is needed here so as not to omit or duplicate an area.

19. Do not convert units until the final total is obtained. For instance, when estimating concrete work, keep all units to the nearest cubic foot, then summarize and convert to cubic yards.

20. When figuring alternatives, it is best to total all items involved in the basic system, then total all items involved in the alternates. Therefore, you work with positive numbers in all cases. When adds and deducts are used, it is often confusing whether to add or subtract a portion of an item; especially on a complicated or involved alternate.

R011105-10   Unit Gross Area Requirements

The figures in the table below indicate typical ranges in square feet as a function of the "occupant" unit. This table is best used in the preliminary design stages to help determine the probable size requirement for the total project.

Building Type	Unit	Gross Area in S.F.		
		1/4	Median	3/4
Apartments	Unit	660	860	1,100
Auditorium & Play Theaters	Seat	18	25	38
Bowling Alleys	Lane		940	
Churches & Synagogues	Seat	20	28	39
Dormitories	Bed	200	230	275
Fraternity & Sorority Houses	Bed	220	315	370
Garages, Parking	Car	325	355	385
Hospitals	Bed	685	850	1,075
Hotels	Rental Unit	475	600	710
Housing for the elderly	Unit	515	635	755
Housing, Public	Unit	700	875	1,030
Ice Skating Rinks	Total	27,000	30,000	36,000
Motels	Rental Unit	360	465	620
Nursing Homes	Bed	290	350	450
Restaurants	Seat	23	29	39
Schools, Elementary	Pupil	65	77	90
Junior High & Middle		85	110	129
Senior High		102	130	145
Vocational		110	135	195
Shooting Ranges	Point		450	
Theaters & Movies	Seat		15	

562

R011105-50   Metric Conversion Factors

Description: This table is primarily for converting customary U.S. units in the left hand column to SI metric units in the right hand column. In addition, conversion factors for some commonly encountered Canadian and non-SI metric units are included.

	If You Know		Multiply By		To Find
Length	Inches	x	25.4[a]	=	Millimeters
	Feet	x	0.3048[a]	=	Meters
	Yards	x	0.9144[a]	=	Meters
	Miles (statute)	x	1.609	=	Kilometers
Area	Square inches	x	645.2	=	Square millimeters
	Square feet	x	0.0929	=	Square meters
	Square yards	x	0.8361	=	Square meters
Volume (Capacity)	Cubic inches	x	16,387	=	Cubic millimeters
	Cubic feet	x	0.02832	=	Cubic meters
	Cubic yards	x	0.7646	=	Cubic meters
	Gallons (U.S. liquids)[b]	x	0.003785	=	Cubic meters[c]
	Gallons (Canadian liquid)[b]	x	0.004546	=	Cubic meters[c]
	Ounces (U.S. liquid)[b]	x	29.57	=	Milliliters[c, d]
	Quarts (U.S. liquid)[b]	x	0.9464	=	Liters[c, d]
	Gallons (U.S. liquid)[b]	x	3.785	=	Liters[c, d]
Force	Kilograms force[d]	x	9.807	=	Newtons
	Pounds force	x	4.448	=	Newtons
	Pounds force	x	0.4536	=	Kilograms force[d]
	Kips	x	4448	=	Newtons
	Kips	x	453.6	=	Kilograms force[d]
Pressure, Stress, Strength (Force per unit area)	Kilograms force per square centimeter[d]	x	0.09807	=	Megapascals
	Pounds force per square inch (psi)	x	0.006895	=	Megapascals
	Kips per square inch	x	6.895	=	Megapascals
	Pounds force per square inch (psi)	x	0.07031	=	Kilograms force per square centimeter[d]
	Pounds force per square foot	x	47.88	=	Pascals
	Pounds force per square foot	x	4.882	=	Kilograms force per square meter[d]
Flow	Cubic feet per minute	x	0.4719	=	Liters per second
	Gallons per minute	x	0.0631	=	Liters per second
	Gallons per hour	x	1.05	=	Milliliters per second
Bending Moment Or Torque	Inch-pounds force	x	0.01152	=	Meter-kilograms force[d]
	Inch-pounds force	x	0.1130	=	Newton-meters
	Foot-pounds force	x	0.1383	=	Meter-kilograms force[d]
	Foot-pounds force	x	1.356	=	Newton-meters
	Meter-kilograms force[d]	x	9.807	=	Newton-meters
Mass	Ounces (avoirdupois)	x	28.35	=	Grams
	Pounds (avoirdupois)	x	0.4536	=	Kilograms
	Tons (metric)	x	1000	=	Kilograms
	Tons, short (2000 pounds)	x	907.2	=	Kilograms
	Tons, short (2000 pounds)	x	0.9072	=	Megagrams[e]
Mass per Unit Volume	Pounds mass per cubic foot	x	16.02	=	Kilograms per cubic meter
	Pounds mass per cubic yard	x	0.5933	=	Kilograms per cubic meter
	Pounds mass per gallon (U.S. liquid)[b]	x	119.8	=	Kilograms per cubic meter
	Pounds mass per gallon (Canadian liquid)[b]	x	99.78	=	Kilograms per cubic meter
Temperature	Degrees Fahrenheit	(F-32)/1.8		=	Degrees Celsius
	Degrees Fahrenheit	(F+459.67)/1.8		=	Degrees Kelvin
	Degrees Celsius	C+273.15		=	Degrees Kelvin

[a]The factor given is exact
[b]One U.S. gallon = 0.8327 Canadian gallon
[c]1 liter = 1000 milliliters = 1000 cubic centimeters
  1 cubic decimeter = 0.001 cubic meter

[d]Metric but not SI unit
[e]Called "tonne" in England and "metric ton" in other metric countries

563

R011105-60   Weights and Measures

Measures of Length
1 Mile = 1760 Yards = 5280 Feet
1 Yard = 3 Feet = 36 inches
1 Foot = 12 Inches
1 Mil = 0.001 Inch
1 Fathom = 2 Yards = 6 Feet
1 Rod = 5.5 Yards = 16.5 Feet
1 Hand = 4 Inches
1 Span = 9 Inches
1 Micro-inch = One Millionth Inch or 0.000001 Inch
1 Micron = One Millionth Meter + 0.00003937 Inch

Surveyor's Measure
1 Mile = 8 Furlongs = 80 Chains
1 Furlong = 10 Chains = 220 Yards
1 Chain = 4 Rods = 22 Yards = 66 Feet = 100 Links
1 Link = 7.92 Inches

Square Measure
1 Square Mile = 640 Acres = 6400 Square Chains
1 Acre = 10 Square Chains = 4840 Square Yards =
   43,560 Sq. Ft.
1 Square Chain = 16 Square Rods = 484 Square Yards =
   4356 Sq. Ft.
1 Square Rod = 30.25 Square Yards = 272.25 Square Feet = 625 Square
   Lines
1 Square Yard = 9 Square Feet
1 Square Foot = 144 Square Inches
An Acre equals a Square 208.7 Feet per Side

Cubic Measure
1 Cubic Yard = 27 Cubic Feet
1 Cubic Foot = 1728 Cubic Inches
1 Cord of Wood = 4 x 4 x 8 Feet = 128 Cubic Feet
1 Perch of Masonry = 16½ x 1½ x 1 Foot = 24.75 Cubic Feet

Avoirdupois or Commercial Weight
1 Gross or Long Ton = 2240 Pounds
1 Net or Short Ton = 2000 Pounds
1 Pound = 16 Ounces = 7000 Grains
1 Ounce = 16 Drachms = 437.5 Grains
1 Stone = 14 Pounds

Power
1 British Thermal Unit per Hour = 0.2931 Watts
1 Ton (Refrigeration) = 3.517 Kilowatts
1 Horsepower (Boiler) = 9.81 Kilowatts
1 Horsepower (550 ft-lb/s) = 0.746 Kilowatts

Shipping Measure
For Measuring Internal Capacity of a Vessel:
   1 Register Ton = 100 Cubic Feet

For Measurement of Cargo:
   Approximately 40 Cubic Feet of Merchandise is considered a Shipping
   Ton, unless that bulk would weigh more than 2000 Pounds, in which case
   Freight Charge may be based upon weight.

40 Cubic Feet = 32.143 U.S. Bushels = 31.16 Imp. Bushels

Liquid Measure
1 Imperial Gallon = 1.2009 U.S. Gallon = 277.42 Cu. In.
1 Cubic Foot = 7.48 U.S. Gallons

R011110-30   Engineering Fees

Typical **Structural Engineering Fees** based on type of construction and total project size. These fees are included in Architectural Fees.

Type of Construction	Total Project Size (in thousands of dollars)			
	$500	$500-$1,000	$1,000-$5,000	Over $5000
Industrial buildings, factories & warehouses	Technical payroll times 2.0 to 2.5	1.60%	1.25%	1.00%
Hotels, apartments, offices, dormitories, hospitals, public buildings, food stores		2.00%	1.70%	1.20%
Museums, banks, churches and cathedrals		2.00%	1.75%	1.25%
Thin shells, prestressed concrete, earthquake resistive		2.00%	1.75%	1.50%
Parking ramps, auditoriums, stadiums, convention halls, hangars & boiler houses		2.50%	2.00%	1.75%
Special buildings, major alterations, underpinning & future expansion		Add to above 0.5%	Add to above 0.5%	Add to above 0.5%

For complex reinforced concrete or unusually complicated structures, add 20% to 50%.

Typical **Mechanical and Electrical Engineering Fees** are based on the size of the subcontract. The fee structure for both is shown below. These fees are included in Architectural Fees.

Type of Construction	Subcontract Size							
	$25,000	$50,000	$100,000	$225,000	$350,000	$500,000	$750,000	$1,000,000
Simple structures	6.4%	5.7%	4.8%	4.5%	4.4%	4.3%	4.2%	4.1%
Intermediate structures	8.0	7.3	6.5	5.6	5.1	5.0	4.9	4.8
Complex structures	10.1	9.0	9.0	8.0	7.5	7.5	7.0	7.0

For renovations, add 15% to 25% to applicable fee.

R012153-10   Repair and Remodeling

Cost figures are based on new construction utilizing the most cost-effective combination of labor, equipment and material with the work scheduled in proper sequence to allow the various trades to accomplish their work in an efficient manner.

The costs for repair and remodeling work must be modified due to the following factors that may be present in any given repair and remodeling project.

1. Equipment usage curtailment due to the physical limitations of the project, with only hand-operated equipment being used.

2. Increased requirement for shoring and bracing to hold up the building while structural changes are being made and to allow for temporary storage of construction materials on above-grade floors.

3. Material handling becomes more costly due to having to move within the confines of an enclosed building. For multi-story construction, low capacity elevators and stairwells may be the only access to the upper floors.

4. Large amounts of cutting and patching and attempting to match the existing construction is required. It is often more economical to remove entire walls rather than create many new door and window openings. This sort of trade-off has to be carefully analyzed.

5. Cost of protection of completed work is increased since the usual sequence of construction usually cannot be accomplished.

6. Economies of scale usually associated with new construction may not be present. If small quantities of components must be custom fabricated due to job requirements, unit costs will naturally increase. Also, if only small work areas are available at a given time, job scheduling between trades becomes difficult and subcontractor quotations may reflect the excessive start-up and shut-down phases of the job.

7. Work may have to be done on other than normal shifts and may have to be done around an existing production facility which has to stay in production during the course of the repair and remodeling.

8. Dust and noise protection of adjoining non-construction areas can involve substantial special protection and alter usual construction methods.

9. Job may be delayed due to unexpected conditions discovered during demolition or removal. These delays ultimately increase construction costs.

10. Piping and ductwork runs may not be as simple as for new construction. Wiring may have to be snaked through walls and floors.

11. Matching "existing construction" may be impossible because materials may no longer be manufactured. Substitutions may be expensive.

12. Weather protection of existing structure requires additional temporary structures to protect building at openings.

13. On small projects, because of local conditions, it may be necessary to pay a tradesman for a minimum of four hours for a task that is completed in one hour.

All of the above areas can contribute to increased costs for a repair and remodeling project. Each of the above factors should be considered in the planning, bidding and construction stage in order to minimize the increased costs associated with repair and remodeling jobs.

R012153-60   Security Factors

Contractors entering, working in, and exiting secure facilities often lose productive time during a normal workday. The recommended allowances in this section are intended to provide for the loss of productivity by increasing labor costs. Note that different costs are associated with searches upon entry only and searches upon entry and exit. Time spent in a queue is unpredictable and not part of these allowances. Contractors should plan ahead for this situation.

Security checkpoints are designed to reflect the level of security required to gain access or egress. An extreme example is when contractors, along with any materials, tools, equipment, and vehicles, must be physically searched and have all materials, tools, equipment, and vehicles inventoried and documented prior to both entry and exit.

Physical searches without going through the documentation process represent the next level and take up less time.

Electronic searches—passing through a detector or x-ray machine with no documentation of materials, tools, equipment, and vehicles—take less time than physical searches.

Visual searches of materials, tools, equipment, and vehicles represent the next level of security.

Finally, access by means of an ID card or displayed sticker takes the least amount of time.

Another consideration is if the searches described above are performed each and every day, or if they are performed only on the first day with access granted by ID card or displayed sticker for the remainder of the project. The figures for this situation have been calculated to represent the initial check-in as described and subsequent entry by ID card or displayed sticker for up to 20 days on site. For the situation described above, where the time period is beyond 20 days, the impact on labor cost is negligible.

There are situations where tradespeople must be accompanied by an escort and observed during the work day. The loss of freedom of movement will slow down productivity for the tradesperson. Costs for the observer have not been included. Those costs are normally born by the owner.

R012909-80   Sales Tax by State

State sales tax on materials is tabulated below (5 states have no sales tax). Many states allow local jurisdictions, such as a county or city, to levy additional sales tax.

Some projects may be sales tax exempt, particularly those constructed with public funds.

State	Tax (%)	State	Tax (%)	State	Tax (%)	State	Tax (%)
Alabama	4	Illinois	6.25	Montana	0	Rhode Island	7
Alaska	0	Indiana	7	Nebraska	5.5	South Carolina	6
Arizona	5.6	Iowa	6	Nevada	6.85	South Dakota	4.5
Arkansas	6.5	Kansas	6.5	New Hampshire	0	Tennessee	7
California	7.25	Kentucky	6	New Jersey	6.625	Texas	6.25
Colorado	2.9	Louisiana	4.45	New Mexico	5.125	Utah	5.95
Connecticut	6.35	Maine	5.5	New York	4	Vermont	6
Delaware	0	Maryland	6	North Carolina	4.75	Virginia	5.3
District of Columbia	5.75	Massachusetts	6.25	North Dakota	5	Washington	6.5
Florida	6	Michigan	6	Ohio	5.75	West Virginia	6
Georgia	4	Minnesota	6.875	Oklahoma	4.5	Wisconsin	5
Hawaii	4	Mississippi	7	Oregon	0	Wyoming	4
Idaho	6	Missouri	4.225	Pennsylvania	6	Average	5.10%

Sales Tax by Province (Canada)

GST - a value-added tax, which the government imposes on most goods and services provided in or imported into Canada. PST - a retail sales tax, which five of the provinces impose on the prices of most goods and some services. QST - a value-added tax, similar to the federal GST, which Quebec imposes. HST - Three provinces have combined their retail sales taxes with the federal GST into one harmonized tax.

Province	PST (%)	QST (%)	GST(%)	HST(%)
Alberta	0	0	5	0
British Columbia	7	0	5	0
Manitoba	8	0	5	0
New Brunswick	0	0	0	15
Newfoundland	0	0	0	15
Northwest Territories	0	0	5	0
Nova Scotia	0	0	0	15
Ontario	0	0	0	13
Prince Edward Island	0	0	0	15
Quebec	0	9.975	5	0
Saskatchewan	6	0	5	0
Yukon	0	0	5	0

R012909-85 Unemployment Taxes and Social Security Taxes

State unemployment tax rates vary not only from state to state, but also with the experience rating of the contractor. The federal unemployment tax rate is 6.0% of the first $7,000 of wages. This is reduced by a credit of up to 5.4% for timely payment to the state. The minimum federal unemployment tax is 0.6% after all credits.

Social security (FICA) for 2019 is estimated at time of publication to be 7.65% of wages up to $128,400.

R012909-86 Unemployment Tax by State

Information is from the U.S. Department of Labor, state unemployment tax rates.

State	Tax (%)	State	Tax (%)	State	Tax (%)	State	Tax (%)
Alabama	6.74	Illinois	7.75	Montana	6.12	Rhode Island	9.79
Alaska	5.4	Indiana	7.474	Nebraska	5.4	South Carolina	5.46
Arizona	8.91	Iowa	8	Nevada	5.4	South Dakota	9.5
Arkansas	6.0	Kansas	7.6	New Hampshire	7.5	Tennessee	10.0
California	6.2	Kentucky	10.0	New Jersey	5.8	Texas	7.5
Colorado	8.9	Louisiana	6.2	New Mexico	5.4	Utah	7.2
Connecticut	6.8	Maine	5.4	New York	8.5	Vermont	8.4
Delaware	8.0	Maryland	7.50	North Carolina	5.76	Virginia	6.27
District of Columbia	7	Massachusetts	11.13	North Dakota	10.72	Washington	5.7
Florida	5.4	Michigan	10.3	Ohio	8.7	West Virginia	7.5
Georgia	5.4	Minnesota	9.0	Oklahoma	5.5	Wisconsin	12.0
Hawaii	5.6	Mississippi	5.4	Oregon	5.4	Wyoming	8.8
Idaho	5.4	Missouri	9.75	Pennsylvania	10.89	Median	7.47%

R012909-90 Overtime

One way to improve the completion date of a project or eliminate negative float from a schedule is to compress activity duration times. This can be achieved by increasing the crew size or working overtime with the proposed crew.

To determine the costs of working overtime to compress activity duration times, consider the following examples. Below is an overtime efficiency and

cost chart based on a five, six, or seven day week with an eight through twelve hour day. Payroll percentage increases for time and one half and double times are shown for the various working days.

Days per Week	Hours per Day	Production Efficiency					Payroll Cost Factors	
		1st Week	2nd Week	3rd Week	4th Week	Average 4 Weeks	@ 1-1/2 Times	@ 2 Times
5	8	100%	100%	100%	100%	100%	1.000	1.000
	9	100	100	95	90	96	1.056	1.111
	10	100	95	90	85	93	1.100	1.200
	11	95	90	75	65	81	1.136	1.273
	12	90	85	70	60	76	1.167	1.333
6	8	100	100	95	90	96	1.083	1.167
	9	100	95	90	85	93	1.130	1.259
	10	95	90	85	80	88	1.167	1.333
	11	95	85	70	65	79	1.197	1.394
	12	90	80	65	60	74	1.222	1.444
7	8	100	95	85	75	89	1.143	1.286
	9	95	90	80	70	84	1.183	1.365
	10	90	85	75	65	79	1.214	1.429
	11	85	80	65	60	73	1.240	1.481
	12	85	75	60	55	69	1.262	1.524

R013113-40   Builder's Risk Insurance

Builder's risk insurance is insurance on a building during construction. Premiums are paid by the owner or the contractor. Blasting, collapse and underground insurance would raise total insurance costs.

R013113-50   General Contractor's Overhead

There are two distinct types of overhead on a construction project: Project overhead and main office overhead. Project overhead includes those costs at a construction site not directly associated with the installation of construction materials. Examples of project overhead costs include the following:

1. Superintendent
2. Construction office and storage trailers
3. Temporary sanitary facilities
4. Temporary utilities
5. Security fencing
6. Photographs
7. Cleanup
8. Performance and payment bonds

The above project overhead items are also referred to as general requirements and therefore are estimated in Division 1. Division 1 is the first division listed in the CSI MasterFormat but it is usually the last division estimated. The sum of the costs in Divisions 1 through 49 is referred to as the sum of the direct costs.

All construction projects also include indirect costs. The primary components of indirect costs are the contractor's main office overhead and profit. The amount of the main office overhead expense varies depending on the following:

1. Owner's compensation
2. Project managers' and estimators' wages
3. Clerical support wages
4. Office rent and utilities
5. Corporate legal and accounting costs
6. Advertising
7. Automobile expenses
8. Association dues
9. Travel and entertainment expenses

These costs are usually calculated as a percentage of annual sales volume. This percentage can range from 35% for a small contractor doing less than $500,000 to 5% for a large contractor with sales in excess of $100 million.

R013113-55   Installing Contractor's Overhead

Installing contractors (subcontractors) also incur costs for general requirements and main office overhead.

Included within the total incl. overhead and profit costs is a percent mark-up for overhead that includes:

1. Compensation and benefits for office staff and project managers
2. Office rent, utilities, business equipment, and maintenance
3. Corporate legal and accounting costs

4. Advertising
5. Vehicle expenses (for office staff and project managers)
6. Association dues
7. Travel, entertainment
8. Insurance
9. Small tools and equipment

R013113-60 Workers' Compensation Insurance Rates by Trade

The table below tabulates the national averages for workers' compensation insurance rates by trade and type of building. The average "Insurance Rate" is multiplied by the "% of Building Cost" for each trade. This produces the "Workers' Compensation" cost by % of total labor cost, to be added for each trade by building type to determine the weighted average workers' compensation rate for the building types analyzed.

Trade	Insurance Rate (% Labor Cost) Range		Insurance Rate (% Labor Cost) Average	% of Building Cost Office Bldgs.	% of Building Cost Schools & Apts.	% of Building Cost Mfg.	Workers' Compensation Office Bldgs.	Workers' Compensation Schools & Apts.	Workers' Compensation Mfg.
Excavation, Grading, etc.	2.3 % to	19.2%	8.5%	4.8%	4.9%	4.5%	0.41%	0.42%	0.38%
Piles & Foundations	3.9 to	27.3	13.4	7.1	5.2	8.7	0.95	0.70	1.17
Concrete	3.5 to	25.6	11.8	5.0	14.8	3.7	0.59	1.75	0.44
Masonry	3.6 to	50.4	13.8	6.9	7.5	1.9	0.95	1.04	0.26
Structural Steel	4.9 to	43.8	21.2	10.7	3.9	17.6	2.27	0.83	3.73
Miscellaneous & Ornamental Metals	2.8 to	27.0	10.6	2.8	4.0	3.6	0.30	0.42	0.38
Carpentry & Millwork	3.7 to	31.9	13.0	3.7	4.0	0.5	0.48	0.52	0.07
Metal or Composition Siding	5.1 to	113.7	19.0	2.3	0.3	4.3	0.44	0.06	0.82
Roofing	5.7 to	103.7	29.0	2.3	2.6	3.1	0.67	0.75	0.90
Doors & Hardware	3.1 to	31.9	11.0	0.9	1.4	0.4	0.10	0.15	0.04
Sash & Glazing	3.9 to	24.3	12.1	3.5	4.0	1.0	0.42	0.48	0.12
Lath & Plaster	2.7 to	34.4	10.7	3.3	6.9	0.8	0.35	0.74	0.09
Tile, Marble & Floors	2.2 to	21.5	8.7	2.6	3.0	0.5	0.23	0.26	0.04
Acoustical Ceilings	1.9 to	29.7	8.5	2.4	0.2	0.3	0.20	0.02	0.03
Painting	3.7 to	36.7	11.2	1.5	1.6	1.6	0.17	0.18	0.18
Interior Partitions	3.7 to	31.9	13.0	3.9	4.3	4.4	0.51	0.56	0.57
Miscellaneous Items	2.1 to	97.7	11.2	5.2	3.7	9.7	0.58	0.42	1.09
Elevators	1.4 to	11.4	4.7	2.1	1.1	2.2	0.10	0.05	0.10
Sprinklers	2.0 to	16.4	6.7	0.5	—	2.0	0.03	—	0.13
Plumbing	1.5 to	14.6	6.3	4.9	7.2	5.2	0.31	0.45	0.33
Heat., Vent., Air Conditioning	3.0 to	17.0	8.3	13.5	11.0	12.9	1.12	0.91	1.07
Electrical	1.9 to	11.3	5.2	10.1	8.4	11.1	0.53	0.44	0.58
Total	1.4 % to	113.7%	—	100.0%	100.0%	100.0%	11.71%	11.15%	12.52%
	Overall Weighted Average				11.79%				

Workers' Compensation Insurance Rates by States

The table below lists the weighted average Workers' Compensation base rate for each state with a factor comparing this with the national average of 11.8%.

State	Weighted Average	Factor	State	Weighted Average	Factor	State	Weighted Average	Factor
Alabama	13.9%	129	Kentucky	11.3%	105	North Dakota	7.4%	69
Alaska	11.0	102	Louisiana	20.0	185	Ohio	6.2	57
Arizona	9.1	84	Maine	9.8	91	Oklahoma	8.7	81
Arkansas	5.7	53	Maryland	11.2	104	Oregon	9.0	83
California	23.6	219	Massachusetts	9.1	84	Pennsylvania	24.9	231
Colorado	8.1	75	Michigan	8.0	74	Rhode Island	10.6	98
Connecticut	17.0	157	Minnesota	16.2	150	South Carolina	20.3	188
Delaware	10.7	99	Mississippi	11.4	106	South Dakota	10.3	95
District of Columbia	9.1	84	Missouri	14.0	130	Tennessee	7.9	73
Florida	11.3	105	Montana	7.8	72	Texas	6.2	57
Georgia	32.9	305	Nebraska	13.4	124	Utah	7.4	69
Hawaii	9.2	85	Nevada	8.9	82	Vermont	10.8	100
Idaho	9.2	85	New Hampshire	11.3	105	Virginia	7.4	69
Illinois	20.1	186	New Jersey	15.4	143	Washington	8.9	82
Indiana	3.7	34	New Mexico	13.7	127	West Virginia	4.7	44
Iowa	12.1	112	New York	19.1	177	Wisconsin	13.3	123
Kansas	6.6	61	North Carolina	17.1	158	Wyoming	6.3	58
			Weighted Average for U.S. is	11.8% of payroll = 100%				

The weighted average skilled worker rate for 35 trades is 11.8%. For bidding purposes, apply the full value of Workers' Compensation directly to total labor costs, or if labor is 38%, materials 42% and overhead and profit 20% of total cost, carry 38/80 x 11.8% = 6.0% of cost (before overhead and profit)

into overhead. Rates vary not only from state to state but also with the experience rating of the contractor.

Rates are the most current available at the time of publication.

569

For customer support on your Electrical Costs with RSMeans data, call 800.448.8182.

R013113-80   Performance Bond

This table shows the cost of a Performance Bond for a construction job scheduled to be completed in 12 months. Add 1% of the premium cost per month for jobs requiring more than 12 months to complete. The rates are "standard" rates offered to contractors that the bonding company considers financially sound and capable of doing the work. Preferred rates are offered by some bonding companies based upon financial strength of the contractor. Actual rates vary from contractor to contractor and from bonding company to bonding company. Contractors should prequalify through a bonding agency before submitting a bid on a contract that requires a bond.

Contract Amount	Building Construction Class B Projects			Highways & Bridges					
				Class A New Construction			Class A-1 Highway Resurfacing		
First $ 100,000 bid	$25.00 per M			$15.00 per M			$9.40 per M		
Next 400,000 bid	$ 2,500	plus	$15.00 per M	$ 1,500	plus	$10.00 per M	$ 940	plus	$7.20 per M
Next 2,000,000 bid	8,500	plus	10.00 per M	5,500	plus	7.00 per M	3,820	plus	5.00 per M
Next 2,500,000 bid	28,500	plus	7.50 per M	19,500	plus	5.50 per M	15,820	plus	4.50 per M
Next 2,500,000 bid	47,250	plus	7.00 per M	33,250	plus	5.00 per M	28,320	plus	4.50 per M
Over 7,500,000 bid	64,750	plus	6.00 per M	45,750	plus	4.50 per M	39,570	plus	4.00 per M

R015113-65   Temporary Power Equipment

Cost data for the temporary equipment was developed utilizing the following information.

1) Re-usable material-services, transformers, equipment and cords are based on new purchase and prorated to three projects.
2) PVC feeder includes trench and backfill.
3) Connections include disconnects and fuses.
4) Labor units include an allowance for removal.
5) No utility company charges or fees are included.
6) Concrete pads or vaults are not included.
7) Utility company conduits not included.

R015423-10   Steel Tubular Scaffolding

On new construction, tubular scaffolding is efficient up to 60' high or five stories. Above this it is usually better to use hung scaffolding if construction permits. Swing scaffolding operations may interfere with tenants. In this case, the tubular is more practical at all heights.

In repairing or cleaning the front of an existing building the cost of tubular scaffolding per S.F. of building front goes up as the height increases above the first tier. The first tier cost is relatively high due to leveling and alignment.

The minimum efficient crew for erecting and dismantling is three workers. They can set up and remove 18 frame sections per day up to 5 stories high. For 6 to 12 stories high, a crew of four is most efficient. Use two or more on top and two on the bottom for handing up or hoisting. They can

also set up and remove 18 frame sections per day. At 7' horizontal spacing, this will run about 800 S.F. per day of erecting and dismantling. Time for placing and removing planks must be added to the above. A crew of three can place and remove 72 planks per day up to 5 stories. For over 5 stories, a crew of four can place and remove 80 planks per day.

The table below shows the number of pieces required to erect tubular steel scaffolding for 1000 S.F. of building frontage. This area is made up of a scaffolding system that is 12 frames (11 bays) long by 2 frames high.

For jobs under twenty-five frames, add 50% to the rental cost. Rental rates will be lower for jobs over three months duration. Large quantities for long periods can reduce rental rates by 20%.

Description of Component	Number of Pieces for 1000 S.F. of Building Front	Unit
5' Wide Standard Frame, 6'-4" High	24	Ea.
Leveling Jack & Plate	24	
Cross Brace	44	
Side Arm Bracket, 21"	12	
Guardrail Post	12	
Guardrail, 7' section	22	
Stairway Section	2	
Stairway Starter Bar	1	
Stairway Inside Handrail	2	
Stairway Outside Handrail	2	
Walk-Through Frame Guardrail	2	

Scaffolding is often used as falsework over 15' high during construction of cast-in-place concrete beams and slabs. Two foot wide scaffolding is generally used for heavy beam construction. The span between frames depends upon the load to be carried with a maximum span of 5'.

Heavy duty shoring frames with a capacity of 10,000#/leg can be spaced up to 10' O.C. depending upon form support design and loading.

Scaffolding used as horizontal shoring requires less than half the material required with conventional shoring.

On new construction, erection is done by carpenters.

Rolling towers supporting horizontal shores can reduce labor and speed the job. For maintenance work, catwalks with spans up to 70' can be supported by the rolling towers.

R015433-10   Contractor Equipment

Rental Rates shown elsewhere in the data set pertain to late model high quality machines in excellent working condition, rented from equipment dealers. Rental rates from contractors may be substantially lower than the rental rates from equipment dealers depending upon economic conditions; for older, less productive machines, reduce rates by a maximum of 15%. Any overtime must be added to the base rates. For shift work, rates are lower. Usual rule of thumb is 150% of one shift rate for two shifts; 200% for three shifts.

For periods of less than one week, operated equipment is usually more economical to rent than renting bare equipment and hiring an operator.

Costs to move equipment to a job site (mobilization) or from a job site (demobilization) are not included in rental rates, nor in any Equipment costs on any Unit Price line items or crew listings. These costs can be found elsewhere. If a piece of equipment is already at a job site, it is not appropriate to utilize mob/demob costs in an estimate again.

Rental rates vary throughout the country with larger cities generally having lower rates. Lease plans for new equipment are available for periods in excess of six months with a percentage of payments applying toward purchase.

Rental rates can also be treated as reimbursement costs for contractor-owned equipment. Owned equipment costs include depreciation, loan payments, interest, taxes, insurance, storage, and major repairs.

Monthly rental rates vary from 2% to 5% of the cost of the equipment depending on the anticipated life of the equipment and its wearing parts. Weekly rates are about 1/3 the monthly rates and daily rental rates are about 1/3 the weekly rates.

The hourly operating costs for each piece of equipment include costs to the user such as fuel, oil, lubrication, normal expendables for the equipment, and a percentage of the mechanic's wages chargeable to maintenance. The hourly operating costs listed do not include the operator's wages.

The daily cost for equipment used in the standard crews is figured by dividing the weekly rate by five, then adding eight times the hourly operating cost to give the total daily equipment cost, not including the operator. This figure is in the right hand column of the Equipment listings under Equipment Cost/Day.

Pile Driving rates shown for the pile hammer and extractor do not include leads, cranes, boilers or compressors. Vibratory pile driving requires an added field specialist during set-up and pile driving operation for the electric model. The hydraulic model requires a field specialist for set-up only. Up to 125 reuses of sheet piling are possible using vibratory drivers. For normal conditions, crane capacity for hammer type and size is as follows.

Crane Capacity	Hammer Type and Size		
	Air or Steam	Diesel	Vibratory
25 ton	to 8,750 ft.-lb.		70 H.P.
40 ton	15,000 ft.-lb.	to 32,000 ft.-lb.	170 H.P.
60 ton	25,000 ft.-lb.		300 H.P.
100 ton		112,000 ft.-lb.	

Cranes should be specified for the job by size, building and site characteristics, availability, performance characteristics, and duration of time required.

Backhoes & Shovels rent for about the same as equivalent size cranes but maintenance and operating expenses are higher. The crane operator's rate must be adjusted for high boom heights. Average adjustments: for 150' boom add 2% per hour; over 185', add 4% per hour; over 210', add 6% per hour; over 250', add 8% per hour and over 295', add 12% per hour.

Tower Cranes of the climbing or static type have jibs from 50' to 200' and capacities at maximum reach range from 4,000 to 14,000 pounds. Lifting capacities increase up to maximum load as the hook radius decreases.

Typical rental rates, based on purchase price, are about 2% to 3% per month.

Erection and dismantling run between 500 and 2000 labor hours. Climbing operation takes 10 labor hours per 20' climb. Crane dead time is about 5 hours per 40' climb. If crane is bolted to side of the building add cost of ties and extra mast sections. Climbing cranes have from 80' to 180' of mast while static cranes have 80' to 800' of mast.

Truck Cranes can be converted to tower cranes by using tower attachments. Mast heights over 400' have been used.

A single 100' high material **Hoist and Tower** can be erected and dismantled in about 400 labor hours; a double 100' high hoist and tower in about 600 labor hours. Erection times for additional heights are 3 and 4 labor hours per vertical foot respectively up to 150', and 4 to 5 labor hours per vertical foot over 150' high. A 40' high portable Buck hoist takes about 160 labor hours to erect and dismantle. Additional heights take 2 labor hours per vertical foot to 80' and 3 labor hours per vertical foot for the next 100'. Most material hoists do not meet local code requirements for carrying personnel.

A 150' high **Personnel Hoist** requires about 500 to 800 labor hours to erect and dismantle. Budget erection time at 5 labor hours per vertical foot for all trades. Local code requirements or labor scarcity requiring overtime can add up to 50% to any of the above erection costs.

Earthmoving Equipment: The selection of earthmoving equipment depends upon the type and quantity of material, moisture content, haul distance, haul road, time available, and equipment available. Short haul cut and fill operations may require dozers only, while another operation may require excavators, a fleet of trucks, and spreading and compaction equipment. Stockpiled material and granular material are easily excavated with front end loaders. Scrapers are most economically used with hauls between 300' and 1-1/2 miles if adequate haul roads can be maintained. Shovels are often used for blasted rock and any material where a vertical face of 8' or more can be excavated. Special conditions may dictate the use of draglines, clamshells, or backhoes. Spreading and compaction equipment must be matched to the soil characteristics, the compaction required and the rate the fill is being supplied.

R015433-15   Heavy Lifting

Hydraulic Climbing Jacks

The use of hydraulic heavy lift systems is an alternative to conventional type crane equipment. The lifting, lowering, pushing, or pulling mechanism is a hydraulic climbing jack moving on a square steel jackrod from 1-5/8" to 4" square, or a steel cable. The jackrod or cable can be vertical or horizontal, stationary or movable, depending on the individual application. When the jackrod is stationary, the climbing jack will climb the rod and push or pull the load along with itself. When the climbing jack is stationary, the jackrod is movable with the load attached to the end and the climbing jack will lift or lower the jackrod with the attached load. The heavy lift system is normally operated by a single control lever located at the hydraulic pump.

The system is flexible in that one or more climbing jacks can be applied wherever a load support point is required, and the rate of lift synchronized.

Economic benefits have been demonstrated on projects such as: erection of ground assembled roofs and floors, complete bridge spans, girders and trusses, towers, chimney liners and steel vessels, storage tanks, and heavy machinery. Other uses are raising and lowering offshore work platforms, caissons, tunnel sections and pipelines.

R015436-50 Mobilization

Costs to move rented construction equipment to a job site from an equipment dealer's or contractor's yard (mobilization) or off the job site (demobilization) are not included in the rental or operating rates, nor in the equipment cost on a unit price line or in a crew listing. These costs can be found consolidated in the Mobilization section of the data and elsewhere in particular site work sections. If a piece of equipment is already on the job site, it is not appropriate to include mob/demob costs in a new estimate that requires use of that equipment. The following table identifies approximate sizes of rented construction equipment that would be hauled on a towed trailer. Because this listing is not all-encompassing, the user can infer as to what size trailer might be required for a piece of equipment not listed.

3-ton Trailer	20-ton Trailer	40-ton Trailer	50-ton Trailer
20 H.P. Excavator	110 H.P. Excavator	200 H.P. Excavator	270 H.P. Excavator
50 H.P. Skid Steer	165 H.P. Dozer	300 H.P. Dozer	Small Crawler Crane
35 H.P. Roller	150 H.P. Roller	400 H.P. Scraper	500 H.P. Scraper
40 H.P. Trencher	Backhoe	450 H.P. Art. Dump Truck	500 H.P. Art. Dump Truck

R024119-10   Demolition Defined

Whole Building Demolition - Demolition of the whole building with no concern for any particular building element, component, or material type being demolished. This type of demolition is accomplished with large pieces of construction equipment that break up the structure, load it into trucks and haul it to a disposal site, but disposal or dump fees are not included. Demolition of below-grade foundation elements, such as footings, foundation walls, grade beams, slabs on grade, etc., is not included. Certain mechanical equipment containing flammable liquids or ozone-depleting refrigerants, electric lighting elements, communication equipment components, and other building elements may contain hazardous waste, and must be removed, either selectively or carefully, as hazardous waste before the building can be demolished.

Foundation Demolition - Demolition of below-grade foundation footings, foundation walls, grade beams, and slabs on grade. This type of demolition is accomplished by hand or pneumatic hand tools, and does not include saw cutting, or handling, loading, hauling, or disposal of the debris.

Gutting - Removal of building interior finishes and electrical/mechanical systems down to the load-bearing and sub-floor elements of the rough building frame, with no concern for any particular building element, component, or material type being demolished. This type of demolition is accomplished by hand or pneumatic hand tools, and includes loading into trucks, but not hauling, disposal or dump fees, scaffolding, or shoring. Certain mechanical equipment containing flammable liquids or ozone-depleting refrigerants, electric lighting elements, communication equipment components, and other building elements may contain hazardous waste, and must be removed, either selectively or carefully, as hazardous waste, before the building is gutted.

Selective Demolition - Demolition of a selected building element, component, or finish, with some concern for surrounding or adjacent elements, components, or finishes (see the first Subdivision (s) at the beginning of appropriate Divisions). This type of demolition is accomplished by hand or pneumatic hand tools, and does not include handling, loading, storing, hauling, or disposal of the debris, scaffolding, or shoring. "Gutting" methods may be used in order to save time, but damage that is caused to surrounding or adjacent elements, components, or finishes may have to be repaired at a later time.

Careful Removal - Removal of a piece of service equipment, building element or component, or material type, with great concern for both the removed item and surrounding or adjacent elements, components or finishes. The purpose of careful removal may be to protect the removed item for later re-use, preserve a higher salvage value of the removed item, or replace an item while taking care to protect surrounding or adjacent elements, components, connections, or finishes from cosmetic and/or structural damage. An approximation of the time required to perform this type of removal is 1/3 to 1/2 the time it would take to install a new item of like kind (see Reference Number R260105-30). This type of removal is accomplished by hand or pneumatic hand tools, and does not include loading, hauling, or storing the removed item, scaffolding, shoring, or lifting equipment.

Cutout Demolition - Demolition of a small quantity of floor, wall, roof, or other assembly, with concern for the appearance and structural integrity of the surrounding materials. This type of demolition is accomplished by hand or pneumatic hand tools, and does not include saw cutting, handling, loading, hauling, or disposal of debris, scaffolding, or shoring.

Rubbish Handling - Work activities that involve handling, loading or hauling of debris. Generally, the cost of rubbish handling must be added to the cost of all types of demolition, with the exception of whole building demolition.

Minor Site Demolition - Demolition of site elements outside the footprint of a building. This type of demolition is accomplished by hand or pneumatic hand tools, or with larger pieces of construction equipment, and may include loading a removed item onto a truck (check the Crew for equipment used). It does not include saw cutting, hauling or disposal of debris, and, sometimes, handling or loading.

R024119-20   Dumpsters

Dumpster rental costs on construction sites are presented in two ways.

The cost per week rental includes the delivery of the dumpster; its pulling or emptying once per week, and its final removal. The assumption is made that the dumpster contractor could choose to empty a dumpster by simply bringing in an empty unit and removing the full one. These costs also include the disposal of the materials in the dumpster.

The Alternate Pricing can be used when actual planned conditions are not approximated by the weekly numbers. For example, these lines can be used when a dumpster is needed for 4 weeks and will need to be emptied 2 or 3 times per week. Conversely the Alternate Pricing lines can be used when a dumpster will be rented for several weeks or months but needs to be emptied only a few times over this period.

R053100-10   Decking Descriptions

General - All Deck Products

A steel deck is made by cold forming structural grade sheet steel into a repeating pattern of parallel ribs. The strength and stiffness of the panels are the result of the ribs and the material properties of the steel. Deck lengths can be varied to suit job conditions, but because of shipping considerations, are usually less than 40 feet. Standard deck width varies with the product used but full sheets are usually 12″, 18″, 24″, 30″, or 36″. The deck is typically furnished in a standard width with the ends cut square. Any cutting for width, such as at openings or for angular fit, is done at the job site.

The deck is typically attached to the building frame with arc puddle welds, self-drilling screws, or powder or pneumatically driven pins. Sheet to sheet fastening is done with screws, button punching (crimping), or welds.

Composite Floor Deck

After installation and adequate fastening, a floor deck serves several purposes. It (a) acts as a working platform, (b) stabilizes the frame, (c) serves as a concrete form for the slab, and (d) reinforces the slab to carry the design loads applied during the life of the building. Composite decks are distinguished by the presence of shear connector devices as part of the deck. These devices are designed to mechanically lock the concrete and deck together so that the concrete and the deck work together to carry subsequent floor loads. These shear connector devices can be rolled-in embossments, lugs, holes, or wires welded to the panels. The deck profile can also be used to interlock concrete and steel.

Composite deck finishes are either galvanized (zinc coated) or phosphatized/painted. Galvanized deck has a zinc coating on both the top and bottom surfaces. The phosphatized/painted deck has a bare (phosphatized) top surface that will come into contact with the concrete. This bare top surface can be expected to develop rust before the concrete is placed. The bottom side of the deck has a primer coat of paint.

A composite floor deck is normally installed so the panel ends do not overlap on the supporting beams. Shear lugs or panel profile shapes often prevent a tight metal to metal fit if the panel ends overlap; the air gap caused by overlapping will prevent proper fusion with the structural steel supports when the panel end laps are shear stud welded.

Adequate end bearing of the deck must be obtained as shown on the drawings. If bearing is actually less in the field than shown on the drawings, further investigation is required.

Roof Deck

A roof deck is not designed to act compositely with other materials. A roof deck acts alone in transferring horizontal and vertical loads into the building frame. Roof deck rib openings are usually narrower than floor deck rib openings. This provides adequate support of the rigid thermal insulation board.

A roof deck is typically installed to endlap approximately 2″ over supports. However, it can be butted (or lapped more than 2″) to solve field fit problems. Since designers frequently use the installed deck system as part of the horizontal bracing system (the deck as a diaphragm), any fastening substitution or change should be approved by the designer. Continuous perimeter support of the deck is necessary to limit edge deflection in the finished roof and may be required for diaphragm shear transfer.

Standard roof deck finishes are galvanized or primer painted. The standard factory applied paint for roof decks is a primer paint and is not intended to weather for extended periods of time. Field painting or touching up of abrasions and deterioration of the primer coat or other protective finishes is the responsibility of the contractor.

Cellular Deck

A cellular deck is made by attaching a bottom steel sheet to a roof deck or composite floor deck panel. A cellular deck can be used in the same manner as a floor deck. Electrical, telephone, and data wires are easily run through the chase created between the deck panel and the bottom sheet.

When used as part of the electrical distribution system, the cellular deck must be installed so that the ribs line up and create a smooth cell transition at abutting ends. The joint that occurs at butting cell ends must be taped or otherwise sealed to prevent wet concrete from seeping into the cell. Cell interiors must be free of welding burrs, or other sharp intrusions, to prevent damage to wires.

When used as a roof deck, the bottom flat plate is usually left exposed to view. Care must be maintained during erection to keep good alignment and prevent damage.

A cellular deck is sometimes used with the flat plate on the top side to provide a flat working surface. Installation of the deck for this purpose requires special methods for attachment to the frame because the flat plate, now on the top, can prevent direct access to the deck material that is bearing on the structural steel. It may be advisable to treat the flat top surface to prevent slipping.

A cellular deck is always furnished galvanized or painted over galvanized.

Form Deck

A form deck can be any floor or roof deck product used as a concrete form. Connections to the frame are by the same methods used to anchor floor and roof decks. Welding washers are recommended when welding a deck that is less than 20 gauge thickness.

A form deck is furnished galvanized, prime painted, or uncoated. A galvanized deck must be used for those roof deck systems where a form deck is used to carry a lightweight insulating concrete fill.

R078413-30   Firestopping

Firestopping is the sealing of structural, mechanical, electrical and other penetrations through fire-rated assemblies. The basic components of firestop systems are safing insulation and firestop sealant on both sides of wall penetrations and the top side of floor penetrations.

Pipe penetrations are assumed to be through concrete, grout, or joint compound and can be sleeved or unsleeved. Costs for the penetrations and sleeves are not included. An annular space of 1″ is assumed. Escutcheons are not included.

A metallic pipe is assumed to be copper, aluminum, cast iron or similar metallic material. An insulated metallic pipe is assumed to be covered with a thermal insulating jacket of varying thickness and materials.

A non-metallic pipe is assumed to be PVC, CPVC, FR Polypropylene or similar plastic piping material. Intumescent firestop sealant or wrap strips are included. Collars on both sides of wall penetrations and a sheet metal plate on the underside of floor penetrations are included.

Ductwork is assumed to be sheet metal, stainless steel or similar metallic material. Duct penetrations are assumed to be through concrete, grout or joint compound. Costs for penetrations and sleeves are not included. An annular space of 1/2″ is assumed.

Multi-trade openings include costs for sheet metal forms, firestop mortar, wrap strips, collars and sealants as necessary.

Structural penetrations joints are assumed to be 1/2″ or less. CMU walls are assumed to be within 1-1/2″ of the metal deck. Drywall walls are assumed to be tight to the underside of metal decking.

Metal panel, glass or curtain wall systems include a spandrel area of 5′ filled with mineral wool foil-faced insulation. Fasteners and stiffeners are included.

R224000-10   Water Consumption Rates

Fixture Type	Water Supply Fixture Unit Value		
	Hot Water	Cold Water	Combined
Bathtub	1.0	1.0	1.4
Clothes Washer	1.0	1.0	1.4
Dishwasher	1.4	0.0	1.4
Kitchen Sink	1.0	1.0	1.4
Laundry Tub	1.0	1.0	1.4
Lavatory	0.5	0.5	0.7
Shower Stall	1.0	1.0	1.4
Water Closet (tank type)	0.0	2.2	2.2
Full Bath Group			
w/bathtub or shower stall	1.5	2.7	3.6
Half Bath Group			
w/W.C. and Lavatory	0.5	2.5	2.6
Kitchen Group			
w/Dishwasher and Sink	1.9	1.0	2.5
Laundry Group			
w/Clothes Washer and			
Laundry Tub	1.8	1.8	2.5
Hose bibb (sillcock)	0.0	2.5	2.5

Notes:
Typically, WSFU = 1GPM

Supply loads in the building water-distribution system shall be determined by total load on the pipe being sized, in terms of water supply fixture units (WSFU) and gallons per minute (GPM) flow rates. For fixtures not listed, choose a WSFU value of a fixture with similar flow characteristics. Water Fixture Supply Units determined the required water supply to fixtures and their service systems. Fixture units are equal to one (1) cubic foot of water drained in a 1-1/4″ pipe per minute. It is not a flow rate unit but a design factor.

Excerpted from the 2012 *International Plumbing Code,* Copyright 2011.
Washington, D.C.: International Code Council. Reproduced with permission.
All rights reserved. www.ICCSAFE.org

R224000-20   Fixture Demands in Gallons per Fixture per Hour

The table below is based on a 140°F final temperature except for dishwashers in public places where 180°F water is mandatory.

Supply Systems for Flush Tanks			Supply Systems for Flushometer Valves		
Load	Demand		Load	Demand	
WSFU	GPM	CU. FT.	WSFU	GPM	CU. FT.
1.0	3.0	0.041040	-	-	-
2.0	5.0	0.068400	-	-	-
3.0	6.5	0.868920	-	-	-
4.0	8.0	1.069440	-	-	-
5.0	9.4	1.256592	5.0	15	2.0052
6.0	10.7	1.430376	6.0	17.4	2.326032
7.0	11.8	1.577424	7.0	19.8	2.646364
8.0	12.8	1.711104	8.0	22.2	2.967696
9.0	13.7	1.831416	9.0	24.6	3.288528
10.0	14.6	1.951728	10.0	27	3.60936
11.0	15.4	2.058672	11.0	27.8	3.716304
12.0	16.0	2.138880	12.0	28.6	3.823248
13.0	16.5	2.205720	13.0	29.4	3.930192
14.0	17.0	2.272560	14.0	30.2	4.037136
15.0	17.5	2.339400	15.0	31	4.14408
16.0	18.0	2.906240	16.0	31.8	4.241024
17.0	18.4	2.459712	17.0	32.6	4.357968
18.0	18.8	2.513184	18.0	33.4	4.464912
19.0	19.2	2.566656	19.0	34.2	4.571856
20.0	19.6	2.620218	20.0	35.0	4.678800
25.0	21.5	2.874120	25.0	38.0	5.079840
30.0	23.3	3.114744	30.0	42.0	5.611356
35.0	24.9	3.328632	35.0	44.0	5.881920
40.0	26.3	3.515784	40.0	46.0	6.149280
45.0	27.7	3.702936	45.0	48.0	6.416640
50.0	29.1	3.890088	50.0	50.0	6.684000

Notes:

When designing a plumbing system that utilizes fixtures other than, or in addition to, water closets, use the data provided in the Supply Systems for Flush Tanks section of the above table.

To obtain the probable maximum demand, multiply the total demands for the fixtures (gal./fixture/hour) by the demand factor. The heater should have a heating capacity in gallons per hour equal to this maximum. The storage tank should have a capacity in gallons equal to the probable maximum demand multiplied by the storage capacity factor.

Excerpted from the 2012 *International Residential Code,* Copyright 2011. Washington, D.C.: International Code Council. Reproduced with permission. All rights reserved. www.ICCSAFE.org

Fig. R238313-11

R238313-10 Heat Trace Systems

Before you can determine the cost of a HEAT TRACE installation, the method of attachment must be established. There are four (4) common methods:

1. Cable is simply attached to the pipe with polyester tape every 12'.
2. Cable is attached with a continuous cover of 2" wide aluminum tape.
3. Cable is attached with factory extruded heat transfer cement and covered with metallic raceway with clips every 10'.
4. Cable is attached between layers of pipe insulation using either clips or polyester tape.

In all of the above methods each component of the system must be priced individually.

Example: Components for method 3 must include:

A. Heat trace cable by voltage and watts per linear foot.
B. Heat transfer cement, 1 gallon per 60 linear feet of cover.
C. Metallic raceway by size and type.
D. Raceway clips by size of pipe.

When taking off linear foot lengths of cable add the following for each valve in the system. (E)

SCREWED OR WELDED VALVE:

1/2"	=	6"
3/4"	=	9"
1"	=	1' -0"
1-1/2"	=	1' -6"
2"	=	2'
2-1/2"	=	2' -6"
3"	=	2' -6"
4"	=	4' -0"
6"	=	7' -0"
8"	=	9' -6"
10"	=	12' -6"
12"	=	15' -0"
14"	=	18' -0"
16"	=	21' -6"
18"	=	25' -6"
20"	=	28' -6"
24"	=	34' -0"
30"	=	40' -0"

FLANGED VALVE:

1/2"	=	1' -0"
3/4"	=	1' -6"
1"	=	2' -0"
1-1/2"	=	2' -6"
2"	=	2' -6"
2-1/2"	=	3' -0"
3"	=	3' -6"
4"	=	4' -0"
6"	=	8' -0"
8"	=	11' -0"
10"	=	14' -0"
12"	=	16' -6"
14"	=	19' -6"
16"	=	23' -0"
18"	=	27' -0"
20"	=	30' -0"
24"	=	36' -0"
30"	=	42' -0"

BUTTERFLY VALVES:

1/2"	=	0'
3/4"	=	0'
1"	=	1' -0"
1-1/2"	=	1' -6"
2"	=	2' -0"
2-1/2"	=	2' -6"
3"	=	2' -6"
4"	=	3' -0"
6"	=	3' -6"
8"	=	4' -0"
10"	=	4' -0"
12"	=	5' -0"
14"	=	5' -6"
16"	=	6' -0"
18"	=	6' -6"
20"	=	7' -0"
24"	=	8' -0"
30"	=	10' -0"

578

For customer support on your Electrical Costs with RSMeans data, call 800.448.8182.

R238313-10   Heat Trace Systems (cont.)

Add the following quantities of heat transfer cement to linear foot totals for each valve:

Nominal Valve Size	Gallons of Cement per Valve
1/2"	0.14
3/4"	0.21
1"	0.29
1-1/2"	0.36
2"	0.43
2-1/2"	0.70
3"	0.71
4"	1.00
6"	1.43
8"	1.48
10"	1.50
12"	1.60
14"	1.75
16"	2.00
18"	2.25
20"	2.50
24"	3.00
30"	3.75

The following must be added to the list of components to accurately price HEAT TRACE systems:

1. Expediter fitting and clamp fasteners (F)
2. Junction box and nipple connected to expediter fitting (G)
3. Field installed terminal blocks within junction box
4. Ground lugs
5. Piping from power source to expediter fitting
6. Controls
7. Thermostats
8. Branch wiring
9. Cable splices
10. End of cable terminations
11. Branch piping fittings and boxes

Deduct the following percentages from labor if cable lengths in the same area exceed:

150' to 250'	10%	351' to 500'	20%
251' to 350'	15%	Over 500'	25%

Add the following percentages to labor for elevated installations:

15' to 20' high	10%	31' to 35' high	40%
21' to 25' high	20%	36' to 40' high	50%
26' to 30' high	30%	Over 40' high	60%

R238313-20   Spiral-Wrapped Heat Trace Cable (Pitch Table)

In order to increase the amount of heat, occasionally heat trace cable is wrapped in a spiral fashion around a pipe, increasing the number of feet of heater cable per linear foot of pipe.

Engineers first determine the heat loss per foot of pipe (based on the insulating material, its thickness, and the temperature differential across it). A ratio is then calculated by the formula:

$$\text{Feet of Heat Trace per Foot of Pipe} = \frac{\text{Watts/Foot of Heat Loss}}{\text{Watts/Foot of the Cable}}$$

The linear distance between wraps (pitch) is then taken from a chart or table. Generally, the pitch is listed on a drawing leaving the estimator to calculate the total length of heat tape required. An approximation may be taken from this table.

Feet of Heat Trace Per Foot of Pipe

Pitch In Inches	\	\	\	\	Nominal Pipe Size in Inches	\	\	\	\	\	\	\	\	\	\	\
	1	1¼	1½	2	2½	3	4	6	8	10	12	14	16	18	20	24
3.5	1.80															
4	1.65															
5	1.46	1.60	1.80													
6	1.34	1.45	1.55	1.75												
7	1.25	1.35	1.43	1.57	1.75											
8	1.20	1.28	1.34	1.45	1.60	1.80										
9	1.16	1.23	1.28	1.37	1.51	1.68										
10	1.13	1.19	1.24	1.32	1.44	1.57	1.82									
15	1.06	1.08	1.10	1.15	1.21	1.29	1.42	1.78								
20	1.04	1.05	1.06	1.08	1.13	1.17	1.25	1.49	1.73							
25		1.04	1.04	1.06	1.08	1.11	1.17	1.33	1.51	1.72						
30				1.04	1.05	1.07	1.12	1.24	1.37	1.54	1.70	1.80				
35						1.06	1.09	1.17	1.28	1.42	1.54	1.64	1.78			
40						1.05	1.07	1.14	1.22	1.33	1.44	1.52	1.64	1.75		
50							1.05	1.09	1.15	1.22	1.29	1.35	1.44	1.53	1.64	1.83
60								1.06	1.11	1.16	1.21	1.25	1.31	1.39	1.46	1.62
70								1.05	1.08	1.12	1.17	1.19	1.24	1.30	1.35	1.47
80									1.06	1.09	1.13	1.15	1.19	1.24	1.28	1.38
90									1.04	1.06	1.10	1.13	1.16	1.19	1.23	1.32
100										1.05	1.08	1.10	1.13	1.15	1.19	1.23

Note: Common practice would normally limit the lower end of the table to 5% of additional heat. Above 80% an engineer would likely opt for two (2) parallel cables.

R260102-20   Labor Adjustment Factors

Labor Adjustment Factors are provided for Divisions 26, 27, and 28 to assist the mechanical estimator account for the various complexities and special conditions of any particular project. While a single percentage has been entered on each line of Division 26 01 02.20, it should be understood that these are just suggested midpoints of ranges of values commonly used by mechanical estimators. They may be increased or decreased depending on the severity of the special conditions.

The group for "existing occupied buildings" has been the subject of requests for explanation. Actually there are two stages to this group: buildings that are existing and "finished" but unoccupied, and those that also are occupied. Buildings that are "finished" may result in higher labor costs due to the workers having to be more careful not to damage finished walls, ceilings, floors, etc. and may necessitate special protective coverings and barriers. Also corridor bends and doorways may not accommodate long pieces of pipe or larger pieces of equipment. Work above an already hung ceiling can be very time consuming. The addition of occupants may force the work to be done on premium time (nights and/or weekends), eliminate the possible use of some preferred tools such as pneumatic drivers, powder charged drivers, etc. The estimator should evaluate the access to the work area and just how the work is going to be accomplished to arrive at an increase in labor costs over "normal" new construction productivity.

R260105-30   Electrical Demolition (Removal for Replacement)

The purpose of this reference number is to provide a guide to users for electrical "removal for replacement" by applying the rule of thumb: 1/3 of new installation time (typical range from 20% to 50%) for removal. Remember to use reasonable judgment when applying the suggested percentage factor. For example:

Contractors have been requested to remove an existing fluorescent lighting fixture and replace with a new fixture utilizing energy saver lamps and electronic ballast:

In order to fully understand the extent of the project, contractors should visit the job site and estimate the time to perform the renovation work in accordance with applicable national, state and local regulation and codes.

The contractor may need to add extra labor hours to his estimate if he discovers unknown concealed conditions such as: contaminated asbestos ceiling, broken acoustical ceiling tile and need to repair, patch and touch-up paint all the damaged or disturbed areas, tasks normally assigned to general contractors. In addition, the owner could request that the contractors salvage the materials removed and turn over the materials to the owner or dispose of the materials to a reclamation station. The normal removal item is 0.5 labor-hour for a lighting fixture and 1.5 labor-hours for new installation time. Revise the estimate times from 2 labor-hours work up to a minimum 4 labor-hours work for just a fluorescent lighting fixture.

For removal of large concentrations of lighting fixtures in the same area, apply an "economy of scale" to reduce estimating labor hours.

Figure R260110-10 Typical Overhead Service Entrance

Utility Pole

Service Entrance Cap 3"

Galvanized Conduit 3"

Two Hole Pipe Clip 3"

Coupling

Grade Line

Elbow

3" PVC Conduit

3" PVC To
Galv. Adapter

Adapter

3" Galv.
Conduit

4 - 350 kcmil XHHW
In Above Conduit

3" Galv. Conduit Used For Safety
On Pole And Through Foundation

Building
Wall

3" Locknut

Pullbox

3" Locknut
& Bushing Insulated
In Pullbox

For 600 AMP Service, Wiring
2 Parallel Runs of 3" Galv. Conduit to
600 AMP. Circuit Breaker Below

2 Locknuts And 1 Insulating Bushing
Required Where #4 Or Larger Wire Used

Figure R260110-11 Typical Commercial Electric System

Panel
First
Floor
100 AMP.

Floor
Box
In
Slab

Plate for
Concrete Ring   Concrete Ring

Meter

600 AMP.
Circuit Breaker

To Pull Box

CT Cabinet

200A
CB

Cutaway Box
Showing Wire
Room

Ground Rod

Ground Clamp

Panel
Basement
600 AMP.

NEMA-3R
200 AMP.

100
AMP.

Condenser
Unit

Elevator
Controller

A = 1" Conduit w/1-#1/0 Wire XHHW
B = 2-3" Conduits w/4-350kcmil XHHW in Each
C = 1¼" Conduit w/4 #3 THHN
D = 1¼" Conduit w/3 #1 XHHW
E = 1" Conduit w/4 #6 THHN
F = ½" Conduit w/2 #12 THHN

Figure R260110-12 Preliminary Procedure

1. Determine building size and use.
2. Develop total load in watts.
   a. Lighting
   b. Receptacle
   c. Air Conditioning
   d. Elevator
   e. Other power requirements
3. Determine best voltage available from utility company.
4. Determine cost from tables for loads (a) through (e) above.
5. Determine size of service from formulas (R260110-17).
6. Determine costs for service, panels, and feeders from tables.

Figure R260110-13 Office Building 90′ x 210′, 3 story, w/garage

Garage Area = 18,900 S.F.
Office Area = 56,700 S.F.
Elevator = 2 @ 125 FPM

Tables	Power Required		Watts
D5010-1151	Garage Lighting .5 Watt/S.F.		9,450
	Office Lighting 3 Watts/S.F.		170,100
D5010-1151	Office Receptacles 2 Watts/S.F.		113,400
D5010-1151, R262213-27	Low Rise Office A.C. 4.3 Watts/S.F.		243,810
D5010-1152, 1153	Elevators - 2 @ 20 HP = 2 @ 17,404 Watts/Ea.		34,808
D5010-1151	Misc. Motors + Power 1.2 Watts/S.F.		68,040
		Total	639,608　Watts

Voltage Available

277/480V, 3 Phase, 4 Wire

Formula

$$\text{D5010-116} \qquad \text{Amperes} = \frac{\text{Watts}}{\text{Volts x Power Factor x 1.73}} = \frac{639,608}{480V \text{ x } .8 \text{ x } 1.73} = 963 \text{ Amps}$$

Use 1200 Amp Service

System	Description	Unit Cost	Unit	Total
D5020 210 0200	Garage Lighting (Interpolated)	$2.82	S.F.	$　53,298
D5020 210 0540	Office Lighting (Using 32 W lamps)	8.57	S.F.	485,919
D5020 115 0880	Receptacle-Undercarpet	3.85	S.F.	218,295
D5020 140 0280	Air Conditioning	0.64	S.F.	36,288
D5020 135 0320	Misc. Pwr.	0.36	S.F.	20,412
D5020 145 2120	Elevators - 2 @ 20HP	2,825.00	Ea.	5,650
D5010 130 1250	Underground Electric Service - 1200 Amp	58,900.00	Ea.	58,900
D5010 230 0480	Feeder - Assume 200 feet	314.00	L.F.	62,800
D5010 240 0580	Panels - 277/480V, 1200 Amp	35,050.00	Ea.	35,050
D5030 910 0400	Fire Detection	36,900.00	Ea.	36,900
			Total	$1,013,512
			or	$1,013,500

Table R260110-14 Nominal Watts Per S.F. for Electric Systems for Various Building Types

Type Construction	1. Lighting	2. Devices	3. HVAC	4. Misc.	5. Elevator	Total Watts
Apartment, luxury high rise	2	2.2	3	1		
Apartment, low rise	2	2	3	1		
Auditorium	2.5	1	3.3	.8		
Bank, branch office	3	2.1	5.7	1.4		
Bank, main office	2.5	1.5	5.7	1.4		
Church	1.8	.8	3.3	.8		
College, science building	3	3	5.3	1.3		
College, library	2.5	.8	5.7	1.4		
College, physical education center	2	1	4.5	1.1		
Department store	2.5	.9	4	1		
Dormitory, college	1.5	1.2	4	1		
Drive-in doughnut shop	3	4	6.8	1.7		
Garage, commercial	.5	.5	0	.5		
Hospital, general	2	4.5	5	1.3		
Hospital, pediatric	3	3.8	5	1.3		
Hotel, airport	2	1	5	1.3		
Housing for the elderly	2	1.2	4	1		
Manufacturing, food processing	3	1	4.5	1.1		
Manufacturing, apparel	2	1	4.5	1.1		
Manufacturing, tools	4	1	4.5	1.1		
Medical clinic	2.5	1.5	3.2	1		
Nursing home	2	1.6	4	1		
Office building, high rise	3	2	4.7	1.2		
Office building, low rise	3	2	4.3	1.2		
Radio-TV studio	3.8	2.2	7.6	1.9		
Restaurant	2.5	2	6.8	1.7		
Retail store	2.5	.9	5.5	1.4		
School, elementary	3	1.9	5.3	1.3		
School, junior high	3	1.5	5.3	1.3		
School, senior high	2.3	1.7	5.3	1.3		
Supermarket	3	1	4	1		
Telephone exchange	1	.6	4.5	1.1		
Theater	2.5	1	3.3	.8		
Town Hall	2	1.9	5.3	1.3		
U.S. Post Office	3	2	5	1.3		
Warehouse, grocery	1	.6	0	.5		

Rule of Thumb: 1 KVA = 1 HP (Single Phase)

Three Phase:

Watts = 1.73 x Volts x Current x Power Factor x Efficiency

$$\text{Horsepower} = \frac{\text{Volts x Current x 1.73 x Power Factor}}{746 \text{ Watts}}$$

See R260120 for an additional reference

Table R260110-15 Horsepower Requirements for Elevators with 3 Phase Motors

Type	Maximum Travel Height in Ft.	Travel Speeds in FPM	Capacity of Cars in Lbs.		
			1200	1500	1800
Hydraulic	70	70	10	15	15
		85	15	15	15
		100	15	15	20
		110	20	20	20
		125	20	20	20
		150	25	25	25
		175	25	30	30
		200	30	30	40
Geared Traction	300	200			
		350			
			2000	2500	3000
Hydraulic	70	70	15	20	20
		85	20	20	25
		100	20	25	30
		110	20	25	30
		125	25	30	40
		150	30	40	50
		175	40	50	50
		200	40	50	60
Geared Traction	300	200	10	10	15
		350	15	15	23
			3500	4000	4500
Hydraulic	70	70	20	25	30
		85	25	30	30
		100	30	40	40
		110	40	40	50
		125	40	50	50
		150	50	50	60
		175	60		
		200	60		
Geared Traction	300	200	15		23
		350	23		35

The power factor of electric motors varies from 80% to 90% in larger size motors. The efficiency likewise varies from 80% on a small motor to 90% on a large motor.

Table R260110-16 Watts per Motor

90% Power Factor & Efficiency @ 200 or 460V			
HP	Watts	HP	Watts
10	9024	30	25784
15	13537	40	33519
20	17404	50	41899
25	21916	60	49634

Table R260110-17 Electrical Formulas

Ohm's Law

Ohm's Law is a method of explaining the relation between voltage, current, and resistance in an electrical circuit. It is practically the basis of all electrical calculations. The term "electromotive force" is often used to designate pressure in volts. This formula can be expressed in various forms.

To find the current in amperes:

$$\text{Current} = \frac{\text{Voltage}}{\text{Resistance}} \quad \text{or} \quad \text{Amperes} = \frac{\text{Volts}}{\text{Ohms}} \quad \text{or} \quad I = \frac{E}{R}$$

The flow of current in amperes through any circuit is equal to the voltage or electromotive force divided by the resistance of that circuit.

To find the pressure or voltage:

Voltage = Current x Resistance   or   Volts = Amperes x Ohms

or   E = I x R

The voltage required to force a current through a circuit is equal to the resistance of the circuit multiplied by the current.

To find the resistance:

$$\text{Resistance} = \frac{\text{Voltage}}{\text{Current}} \quad \text{or} \quad \text{Ohms} = \frac{\text{Volts}}{\text{Amperes}} \quad \text{or} \quad R = \frac{E}{I}$$

The resistance of a circuit is equal to the voltage divided by the current flowing through that circuit.

Power Formulas

One horsepower = 746 watts   One kilowatt = 1000 watts

The power factor of electric motors varies from 80% to 90% in the larger size motors.

Single-Phase Alternating Current Circuits

Power in Watts = Volts x Amperes x Power Factor

To find current in amperes:

$$\text{Current} = \frac{\text{Watts}}{\text{Volts x Power Factor}} \quad \text{or}$$

$$\text{Amperes} = \frac{\text{Watts}}{\text{Volts x Power Factor}} \quad \text{or} \quad I = \frac{W}{E \times PF}$$

To find current of a motor, single phase:

$$\text{Current} = \frac{\text{Horsepower x 746}}{\text{Volts x Power Factor x Efficiency}} \quad \text{or}$$

$$I = \frac{HP \times 746}{E \times PF \times Eff.}$$

To find horsepower of a motor, single phase:

$$\text{Horsepower} = \frac{\text{Volts x Current x Power Factor x Efficiency}}{746 \text{ Watts}}$$

$$HP = \frac{E \times I \times PF \times Eff.}{746}$$

To find power in watts of a motor, single phase:

Watts = Volts x Current x Power Factor x Efficiency   or

Watts = E x I x PF x Eff.

To find single phase kVA:

$$1 \text{ Phase kVA} = \frac{\text{Volts x Amps}}{1000}$$

Three-Phase Alternating Current Circuits

Power in Watts = Volts x Amperes x Power Factor x 1.73

To find current in amperes in each wire:

$$\text{Current} = \frac{\text{Watts}}{\text{Voltage x Power Factor x 1.73}} \quad \text{or}$$

$$\text{Amperes} = \frac{\text{Watts}}{\text{Volts x Power Factor x 1.73}} \quad \text{or} \quad I = \frac{W}{E \times PF \times 1.73}$$

To find current of a motor, 3 phase:

$$\text{Current} = \frac{\text{Horsepower x 746}}{\text{Volts x Power Factor x Efficiency x 1.73}} \quad \text{or}$$

$$I = \frac{HP \times 746}{E \times PF \times Eff. \times 1.73}$$

To find horsepower of a motor, 3 phase:

$$\text{Horsepower} = \frac{\text{Volts x Current x 1.73 x Power Factor}}{746 \text{ Watts}}$$

$$HP = \frac{E \times I \times 1.73 \times PF}{746}$$

To find power in watts of a motor, 3 phase:

Watts = Volts x Current x 1.73 x Power Factor x Efficiency   or

Watts = E x I x 1.73 x PF x Eff.

To find 3 phase kVA:

$$3 \text{ phase kVA} = \frac{\text{Volts x Amps x 1.73}}{1000} \quad \text{or}$$

$$kVA = \frac{V \times A \times 1.73}{1000}$$

Power Factor (PF) is the percentage ratio of the measured watts (effective power) to the volt-amperes (apparent watts)

$$\text{Power Factor} = \frac{\text{Watts}}{\text{Volts x Amperes}} \times 100\%$$

Table R260120-10 ASHRAE Climate Zone Map

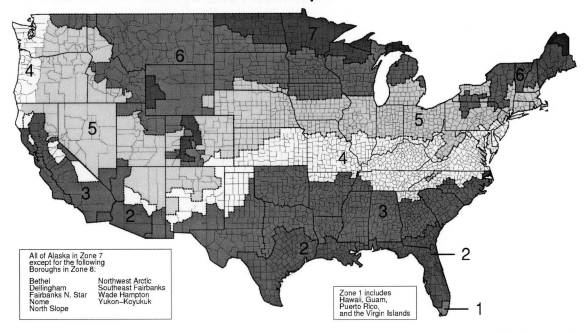

All of Alaska in Zone 7
except for the following
Boroughs in Zone 8:

Bethel	Northwest Arctic
Dellingham	Southeast Fairbanks
Fairbanks N. Star	Wade Hampton
Nome	Yukon–Koyukuk
North Slope	

Zone 1 includes
Hawaii, Guam,
Puerto Rico,
and the Virgin Islands

Table R260120-20 Watts per Square Foot for Lighting and Misc. Power by Building Type & Space

Building / Space	Percent	Lighting	Misc
Apartment, Low: Area = 20,000 SF; Floors = 2; Heating = DX			
Coils; Cooling = DX Coils			
Room	85%	0.50	0.10
Storage	8%	1.19	0.00
Laundry	7%	1.28	0.15
Weighted Average		**0.61**	**0.10**
Apartment, Mid: Area = 40,000 SF; Floors = 5; Heating = DX			
Coils; Cooling = DX Coils			
Room	71%	0.50	0.30
Corridor	16%	0.57	0.00
Storage	7%	1.19	0.00
Laundry	6%	1.28	0.15
Average		**0.61**	**0.22**
Apartment, High: Area = 72,000 SF; Floors = 8; Heating = DX			
Coils; Cooling = DX Coils			
Room	71%	0.50	0.30
Corridor	16%	0.57	0.00
Storage	7%	1.19	0.00
Laundry	6%	1.28	0.15
Weighted Average		**0.61**	**0.22**
Assisted Living: Area = 10,000 SF; Floors = 1; Heating = Furnace;			
Cooling = DX Coils			
Room	70%	1.46	0.25
Dining	15%	1.85	0.10
Food Prep	10%	1.19	1.00
Care	5%	1.34	1.00
Weighted Average		**1.49**	**0.34**

Building / Space	Percent	Lighting	Misc
Auditorium: Area = 70,000 SF; Floors = 1; Heating = Hot Water			
Coils; Cooling = Chilled Water Coils			
Theater	70%	0.90	
Lobby	15%	1.50	
Office	5%	1.30	
Corridor	5%	0.60	
MEP	5%	0.70	
Weighted Average		**0.99**	
Bank: Area = 25,000 SF; Floors = 2; Heating = Furnace; Cooling =			
DX Coils			
Bank	60%	2.33	0.45
Corridor	10%	0.57	0.00
Office	10%	1.49	0.75
Lobby	5%	1.52	0.25
Restrooms	5%	0.77	0.10
Conference	4%	0.92	0.10
MEP	4%	0.81	0.10
Production	2%	1.64	0.70
Weighted Average		**1.82**	**0.38**
Community Center: Area = 35,000 SF; Floors = 1; Heating =			
Furnace; Cooling = DX Coils			
Convention	50%	1.27	
Dining	15%	1.85	
Food Prep	15%	1.19	
Storage	10%	1.19	
Corridor	5%	0.57	
Restrooms	5%	0.77	
Weighted Average		**1.28**	
Convention Center: Area = 150,000 SF; Floors = 2; Heating = Hot			
Water Coils; Cooling = Chilled Water Coils			
Convention	65%	1.27	0.25
Conference	20%	0.92	0.10
Dining	5%	1.85	0.10
Corridor	4%	0.57	0.00
Food Prep	3%	1.19	1.00
Restrooms	3%	0.77	0.10
Weighted Average		**1.18**	**0.22**

Building / Space	Percent	Lighting	Misc
Day Care: Area = 2,500 SF; Floors = 1; Heating = Furnace;			
Cooling = DX Coils			
Classroom	45%	1.45	
Corridor	15%	0.57	
Library	15%	1.67	
Food Prep	10%	1.19	
Restrooms	7%	0.77	
Office	5%	1.49	
Storage	3%	1.19	
Weighted Average		**1.27**	
Fitness Center: Area = 20,000 SF; Floors = 1; Heating = Furnace;			
Cooling = DX Coils			
Exercise	80%	1.13	0.10
Corridor	5%	0.57	0.00
MEP	5%	0.81	0.10
Office	5%	1.24	0.75
Restrooms	7%	0.77	0.10
Weighted Average		**1.07**	**0.13**
Hospital: Area = 250,000 SF; Floors = 4; Heating = Hot Water			
Coils; Cooling = Chilled Water Coils			
Medical	60%	1.34	1.00
Laboratory	15%	3.19	1.00
Corridor	10%	0.57	0.00
Laundry	5%	1.28	0.15
MEP	5%	0.81	0.10
Restrooms	5%	0.77	0.10
Weighted Average		**1.48**	**0.77**
Hotel: Area = 180,000 SF; Floors = 10; Heating = Furnace;			
Cooling = DX Coils			
Room	75%	1.46	
Meeting Center	15%	1.27	
Corridor	8%	0.57	
MEP	2%	0.81	
Weighted Average		**1.35**	

Steps:
1. Determine climate zone using R260120-10
2. Use R260120-20 to determine lighting and miscellaneous power for appropriate building type
3. Use corresponding building type and climate zone in R260120-30 to determine other systems

*Values for R260120-20 & -30 were derived by performing an energy analysis using the eQUEST® software from James J. Hirsch & Associates. eQUEST is a trademark of James J. Hirsch
**Additional systems may be necessary

585

Table R260120-20 Watts per Square Foot for Lighting and Misc. Power by Building Type & Space (cont)

Manufacturing, General: Area = 100,000 SF; Floors = 10; Heating = Furnace; Cooling = DX Coils

Building / Space	Percent	Lighting	Misc
Work	60%	2.66	1.00
Storage	25%	1.19	0.00
Office	10%	1.24	0.75
Restrooms	3%	0.77	0.10
MEP	2%	0.81	0.10
Weighted Average		**2.06**	**0.68**

Manufacturing, High Tech: Area = 200,000 SF; Floors = 1; Heating = Furnace; Cooling = DX Coils

Building / Space	Percent	Lighting	Misc
Work	45%	3.19	1.00
Office	27%	1.24	0.75
Corridor	18%	0.57	0.00
Dining	3%	1.85	0.10
Conference	2%	0.92	0.10
Server	2%	1.24	5.00
MEP	2%	0.81	0.10
Food Prep	1%	1.19	1.00
Weighted Average		**2.00**	**0.77**

Medical Clinic: Area = 20,000 SF; Floors = 1; Heating = Furnace; Cooling = DX Coils

Building / Space	Percent	Lighting	Misc
Clinic	50%	1.40	
General Office	15%	1.30	
Private Office	10%	1.30	
Laboratory	10%	1.30	
Storage	5%	0.60	
Restrooms	5%	0.60	
MEP	5%	0.70	
Weighted Average		**1.25**	

Motel: Area = 10,000 SF; Floors = 1; Heating = DX Coils; Cooling = DX Coils

Building / Space	Percent	Lighting	Misc
Room	95%	1.46	0.25
Office	3%	1.24	0.75
MEP	2%	0.81	0.10
Weighted Average		**1.44**	**0.26**

Museum: Area = 20,000 SF; Floors = 1; Heating = Hot Water Coils; Cooling = Chilled Water Coils

Building / Space	Percent	Lighting	Misc
Exhibit Area	65%	1.03	0.25
Theater	15%	1.02	0.25
Lobby	5%	1.77	0.25
MEP	5%	0.81	0.10
Office	5%	1.49	0.75
Restrooms	5%	0.77	0.10
Weighted Average		**1.06**	**0.26**

Office, Low: Area = 25,000 SF; Floors = 2; Heating = Furnace; Cooling = DX Coils

Building / Space	Percent	Lighting	Misc
Office	70%	1.30	0.75
Corridor	10%	0.60	0.00
Lobby	5%	1.10	0.25
Restrooms	5%	0.60	0.10
Conference	4%	1.60	0.10
MEP	4%	0.70	0.10
Production	2%	1.50	0.70
Weighted Average		**1.18**	**0.56**

Office, Mid: Area = 125,000 SF; Floors = 4; Heating = Hot Water Coils; Cooling = Chilled Water Coils

Building / Space	Percent	Lighting	Misc
Office Open	40%	1.24	0.75
Office Private	30%	1.49	0.75
Corridor	10%	0.57	0.00
Lobby	5%	1.52	0.25
Restrooms	5%	0.77	0.10
Conference	4%	0.92	0.10
MEP	4%	0.81	0.10
Production	2%	1.64	0.70
Weighted Average		**1.22**	**0.56**

Office, High: Area = 250,000 SF; Floors = 8; Heating = Hot Water Coils; Cooling = Chilled Water Coils

Building / Space	Percent	Lighting	Misc
Office Open	45%	1.24	0.75
Office Private	25%	1.49	0.75
Corridor	10%	0.57	0.00
Lobby	5%	1.52	0.25
Conference	4%	0.92	0.10
Production	2%	1.64	0.70
Restrooms	5%	0.77	0.10
MEP	4%	0.81	0.10
Weighted Average		**1.20**	**0.56**

Restaurant, Bar/Lounge: Area = 4,000 SF; Floors = 1; Heating = Furnace; Cooling = DX Coils

Building / Space	Percent	Lighting	Misc
Lounge	35%	1.85	0.10
Dining	30%	1.85	0.10
Food Prep	25%	1.19	1.00
MEP	5%	0.81	0.10
Restrooms	5%	0.77	0.10
Weighted Average		**1.58**	**0.33**

Restaurant, Full: Area = 5,000 SF; Floors = 1; Heating = Furnace; Cooling = DX Coils

Building / Space	Percent	Lighting	Misc
Dining	60%	1.85	0.10
Food Prep	30%	1.19	1.00
Restrooms	5%	0.77	0.10
MEP	5%	0.81	0.10
Weighted Average		**1.55**	**0.37**

Restaurant, Quick: Area = 2,500 SF; Floors = 1; Heating = Furnace; Cooling = DX Coils

Building / Space	Percent	Lighting	Misc
Dining	50%	1.85	0.10
Food Prep	35%	1.19	1.00
Restrooms	10%	0.77	0.10
MEP	5%	0.81	0.10
Weighted Average		**1.46**	**0.42**

Store, Convenience: Area = 3,000 SF; Floors = 1; Heating = Furnace; Cooling = DX Coils

Building / Space	Percent	Lighting	Misc
Retail	80%	2.38	0.25
Office	10%	1.49	0.75
Restrooms	10%	0.77	0.10
Weighted Average		**2.13**	**0.29**

Store, Department: Area = 120,000 SF; Floors = 2; Heating = Hot Water Coils; Cooling = Chilled Water Coils

Building / Space	Percent	Lighting	Misc
Retail	55%	2.38	0.25
Exhibit Area	15%	1.03	0.25
Storage	16%	1.19	0.00
Office	5%	1.24	0.75
Smoking Lounge	2%	1.85	0.10
Restrooms	2%	0.77	0.10
MEP	5%	0.81	0.10
Weighted Average		**1.81**	**0.22**

Store, Large: Area = 70,000 SF; Floors = 1; Heating = Furnace; Cooling = DX Coils

Building / Space	Percent	Lighting	Misc
Retail	65%	2.38	0.25
Storage	20%	1.19	0.00
Office	5%	1.24	0.75
Dining	3%	1.85	0.10
Food Prep	1%	1.19	1.00
Lounge	2%	1.85	0.10
Restrooms	2%	0.77	0.10
MEP	2%	0.81	0.10
Weighted Average		**1.98**	**0.22**

Store, Small: Area = 5,000 SF; Floors = 1; Heating = Furnace; Cooling = DX Coils

Building / Space	Percent	Lighting	Misc
Retail	58%	2.38	0.25
Exhibit Area	15%	1.03	0.25
Storage	15%	1.19	0.00
Office	5%	1.24	0.75
Restrooms	5%	0.77	0.10
MEP	2%	0.81	0.10
Weighted Average		**1.83**	**0.23**

Store, Strip Mall: Area = 15,000 SF; Floors = 1; Heating = Furnace; Cooling = DX Coils

Building / Space	Percent	Lighting	Misc
Retail	38%	2.38	0.25
Atrium	20%	1.55	0.13
Exhibit Area	15%	1.03	0.25
Storage	15%	1.19	0.00
Office	5%	1.24	0.75
Restrooms	5%	0.77	0.10
MEP	2%	0.81	0.10
Weighted Average		**1.66**	**0.20**

Store, Warehouse: Area = 150,000 SF; Floors = 1; Heating = Furnace; Cooling = DX Coils

Building / Space	Percent	Lighting	Misc
Retail	75%	2.38	0.25
Storage	20%	1.19	0.00
Office	4%	1.24	0.75
Restrooms	1%	0.77	0.10
Weighted Average		**2.08**	**0.22**

School, Elementary: Area = 35,000 SF; Floors = 1; Heating = Furnace; Cooling = DX Coils

Building / Space	Percent	Lighting	Misc
Classroom	50%	1.45	0.50
Corridor	15%	0.57	0.00
Gym	10%	1.13	0.10
Dining	5%	1.85	0.10
Food Prep	5%	1.19	1.00
Library	5%	1.78	0.25
Auditorium	5%	1.62	0.25
Restrooms	5%	0.77	0.10
Weighted Average		**1.28**	**0.35**

School, Middle: Area = 75,000 SF; Floors = 1; Heating = Furnace; Cooling = DX Coils

Building / Space	Percent	Lighting	Misc
Classroom	50%	1.45	0.50
Corridor	14%	0.57	0.00
Gym	9%	1.13	0.10
Auditorium	7%	1.62	0.25
Dining	6%	1.85	0.10
Food Prep	5%	1.19	1.00
Library	5%	1.78	0.25
Restrooms	4%	0.77	0.10
Weighted Average		**1.31**	**0.35**

School, High: Area = 150,000 SF; Floors = 2; Heating = Furnace; Cooling = DX Coils

Building / Space	Percent	Lighting	Misc
Classroom	45%	1.45	0.50
Corridor	10%	0.57	0.00
Gym	10%	1.13	0.10
Dining	10%	1.85	0.10
Food Prep	5%	1.19	1.00
Library	5%	1.78	0.25
Auditorium	10%	1.62	0.25
Restrooms	5%	0.77	0.10
Weighted Average		**1.36**	**0.34**

School, College: Area = 250,000 SF; Floors = 4; Heating = Furnace; Cooling = DX Coils

Building / Space	Percent	Lighting	Misc
Classroom	50%	1.45	0.50
Office	15%	1.49	0.75
Corridor	12%	0.57	0.00
Gym	10%	1.13	0.10
Library	5%	1.78	0.25
Dining	3%	1.85	0.10
Restrooms	3%	0.77	0.10
Food Prep	2%	1.19	1.00
Weighted Average		**1.32**	**0.41**

Service Station: Area = 2,000 SF; Floors = 1; Heating = Furnace; Cooling = DX Coils

Building / Space	Percent	Lighting	Misc
Workshop	70%	1.31	0.50
Office	10%	1.49	0.75
Restrooms	10%	0.77	0.10
Retail	10%	2.38	0.25
Weighted Average		**1.38**	**0.46**

Storage, High Bay: Area = 200,000 SF; Floors = 1; Heating = Furnace; Cooling = DX Coils

Building / Space	Percent	Lighting	Misc
Storage	85%	1.19	0.00
Office	10%	1.24	0.75
Restrooms	3%	0.77	0.10
MEP	2%	0.81	0.10
Weighted Average		**1.17**	**0.08**

Storage, Low Bay: Area = 70,000 SF; Floors = 1; Heating = Furnace; Cooling = DX Coils

Building / Space	Percent	Lighting	Misc
Storage	85%	1.19	0.00
Office	10%	1.24	0.75
Restrooms	3%	0.77	0.10
MEP	2%	0.81	0.10
Weighted Average		**1.17**	**0.08**

Table R260120-30 Watts per S.F. for Various Systems by Building Type & Climate Zone

Building / Zone	Heating	Cooling	Heat rejection	Auxiliary (pumps)	Vent fan	Supp heat pump	Building / Zone	Heating	Cooling	Heat rejection	Auxiliary (pumps)	Vent fan	Supp heat pump
Apartment, Low							**Community Center**						
Zone 1	0.42	3.65	0.00	0.00	0.12	1.47	Zone 1	0.00	2.70	0.00	0.00	0.44	0.00
Zone 2	0.63	3.53	0.00	0.00	0.12	4.01	Zone 2	0.00	2.66	0.00	0.00	0.44	0.00
Zone 3	0.64	3.45	0.00	0.00	0.13	4.63	Zone 3	0.00	2.66	0.00	0.00	0.45	0.00
Zone 4	0.65	3.09	0.00	0.00	0.13	7.41	Zone 4	0.00	2.44	0.00	0.00	0.45	0.00
Zone 5	0.67	2.88	0.00	0.00	0.12	10.30	Zone 5	0.00	2.42	0.00	0.00	0.44	0.00
Zone 6	0.69	2.66	0.00	0.00	0.13	10.68	Zone 6	0.00	2.16	0.00	0.00	0.44	0.00
Zone 7	0.85	2.11	0.00	0.00	0.12	11.76	Zone 7	0.00	1.91	0.00	0.00	0.44	0.00
Average	**0.65**	**3.05**	**0.00**	**0.00**	**0.13**	**7.18**	**Average**	**0.00**	**2.42**	**0.00**	**0.00**	**0.45**	**0.00**
Apartment, Mid							**Convention Center**						
Zone 1	0.00	1.90	0.00	0.00	0.11	0.00	Zone 1	0.00	1.46	0.25	0.27	0.31	0.00
Zone 2	0.43	1.82	0.00	0.00	0.11	2.12	Zone 2	0.00	1.44	0.25	0.27	0.29	0.00
Zone 3	0.48	1.69	0.00	0.00	0.11	2.32	Zone 3	0.00	1.43	0.24	0.26	0.30	0.00
Zone 4	0.67	1.67	0.00	0.00	0.12	4.25	Zone 4	0.00	1.42	0.23	0.26	0.28	0.00
Zone 5	0.70	1.54	0.00	0.00	0.12	4.79	Zone 5	0.00	1.30	0.29	0.31	0.22	0.00
Zone 6	0.72	1.45	0.00	0.00	0.12	5.07	Zone 6	0.00	1.17	0.12	0.29	0.16	0.00
Zone 7	0.93	1.39	0.00	0.00	0.12	5.11	Zone 7	0.00	0.90	0.20	0.31	0.16	0.00
Average	**0.56**	**1.63**	**0.00**	**0.00**	**0.12**	**3.38**	**Average**	**0.00**	**1.30**	**0.22**	**0.28**	**0.25**	**0.00**
Apartment, High							**Day Care**						
Zone 1	0.00	2.11	0.00	0.00	0.14	0.00	Zone 1	0.00	3.47	0.00	0.02	0.29	0.00
Zone 2	0.00	2.03	0.00	0.00	0.13	0.00	Zone 2	0.00	3.42	0.00	0.02	0.28	0.00
Zone 3	0.48	1.88	0.00	0.00	0.14	1.75	Zone 3	0.00	3.40	0.00	0.02	0.29	0.00
Zone 4	0.57	1.77	0.00	0.00	0.14	4.22	Zone 4	0.00	3.29	0.00	0.02	0.29	0.00
Zone 5	0.66	1.64	0.00	0.00	0.14	5.40	Zone 5	0.00	3.08	0.00	0.02	0.29	0.00
Zone 6	0.71	1.63	0.00	0.00	0.15	5.64	Zone 6	0.00	2.29	0.00	0.02	0.29	0.00
Zone 7	0.91	1.18	0.00	0.00	0.14	5.75	Zone 7	0.00	2.12	0.00	0.02	0.24	0.00
Average	**0.48**	**1.75**	**0.00**	**0.00**	**0.14**	**3.25**	**Average**	**0.05**	**3.01**	**0.05**	**0.07**	**0.29**	**0.05**
Assisted Living							**Fitness Center**						
Zone 1	0.00	2.65	0.00	0.04	0.31	0.00	Zone 1	0.00	4.97	0.00	0.00	0.59	0.00
Zone 2	0.00	2.63	0.00	0.04	0.31	0.00	Zone 2	0.00	4.94	0.00	0.00	0.59	0.00
Zone 3	0.00	2.57	0.00	0.04	0.31	0.00	Zone 3	0.00	4.80	0.00	0.00	0.60	0.00
Zone 4	0.00	2.50	0.00	0.04	0.31	0.00	Zone 4	0.00	4.76	0.00	0.00	0.60	0.00
Zone 5	0.00	2.42	0.00	0.04	0.30	0.00	Zone 5	0.00	4.75	0.00	0.00	0.59	0.00
Zone 6	0.00	1.99	0.00	0.04	0.32	0.00	Zone 6	0.00	4.13	0.00	0.00	0.60	0.00
Zone 7	0.00	1.90	0.00	0.04	0.30	0.00	Zone 7	0.00	3.27	0.00	0.00	0.60	0.00
Average	**0.00**	**2.38**	**0.00**	**0.04**	**0.31**	**0.00**	**Average**	**0.00**	**4.52**	**0.00**	**0.00**	**0.59**	**0.00**
Auditorium							**Hospital**						
Zone 1	0.00	2.32	0.34	0.37	1.03	0.00	Zone 1	0.00	1.54	0.23	0.26	0.80	0.00
Zone 2	0.00	2.32	0.34	0.38	1.03	0.00	Zone 2	0.00	1.47	0.22	0.25	0.80	0.00
Zone 3	0.00	2.29	0.33	0.38	1.04	0.00	Zone 3	0.00	1.44	0.22	0.25	0.81	0.00
Zone 4	0.00	2.26	0.29	0.35	1.05	0.00	Zone 4	0.00	1.42	0.22	0.25	0.81	0.00
Zone 5	0.00	2.20	0.34	0.39	1.04	0.00	Zone 5	0.00	1.37	0.23	0.27	0.81	0.00
Zone 6	0.00	1.71	0.14	0.34	1.05	0.00	Zone 6	0.00	1.26	0.15	0.25	0.82	0.00
Zone 7	0.00	1.38	0.23	0.36	1.04	0.00	Zone 7	0.00	1.07	0.19	0.26	0.82	0.00
Average	**0.00**	**2.07**	**0.29**	**0.37**	**1.04**	**0.00**	**Average**	**0.00**	**1.36**	**0.21**	**0.26**	**0.81**	**0.00**
Bank							**Hotel**						
Zone 1	0.00	3.32	0.00	0.00	0.60	0.00	Zone 1	0.00	1.77	0.00	0.01	0.26	0.00
Zone 2	0.00	3.28	0.00	0.00	0.61	0.00	Zone 2	0.00	1.64	0.00	0.01	0.26	0.00
Zone 3	0.00	3.28	0.00	0.00	0.61	0.00	Zone 3	0.00	1.62	0.00	0.01	0.27	0.00
Zone 4	0.00	3.21	0.00	0.00	0.60	0.00	Zone 4	0.00	1.62	0.00	0.01	0.27	0.00
Zone 5	0.00	3.06	0.00	0.00	0.59	0.00	Zone 5	0.00	1.51	0.00	0.01	0.27	0.00
Zone 6	0.00	2.82	0.00	0.00	0.61	0.00	Zone 6	0.00	1.25	0.00	0.01	0.28	0.00
Zone 7	0.00	2.66	0.00	0.00	0.60	0.00	Zone 7	0.00	1.23	0.00	0.01	0.28	0.00
Average	**0.00**	**3.09**	**0.00**	**0.00**	**0.60**	**0.00**	**Average**	**0.00**	**1.52**	**0.00**	**0.01**	**0.27**	**0.00**

587

Table R260120-30 Watts per S.F. for Various Systems by Building Type & Climate Zone (cont.)

Building / Zone	Heating	Cooling	Heat rejection	Auxiliary (pumps)	Vent fan	Supp heat pump	Building / Zone	Heating	Cooling	Heat rejection	Auxiliary (pumps)	Vent fan	Supp heat pump
Manufacturing, General							**Office, Mid**						
Zone 1	0.00	1.46	0.00	0.00	0.13	0.00	Zone 1	0.00	1.61	0.31	0.33	0.45	0.00
Zone 2	0.00	1.45	0.00	0.00	0.13	0.00	Zone 2	0.00	1.60	0.29	0.32	0.46	0.00
Zone 3	0.00	1.43	0.00	0.00	0.13	0.00	Zone 3	0.00	1.59	0.28	0.34	0.42	0.00
Zone 4	0.00	1.39	0.00	0.00	0.13	0.00	Zone 4	0.00	1.56	0.27	0.33	0.43	0.00
Zone 5	0.00	1.39	0.00	0.00	0.13	0.00	Zone 5	0.00	1.55	0.34	0.36	0.37	0.00
Zone 6	0.00	1.21	0.00	0.00	0.13	0.00	Zone 6	0.00	1.29	0.16	0.34	0.31	0.00
Zone 7	0.00	1.00	0.00	0.00	0.13	0.00	Zone 7	0.00	1.10	0.22	0.35	0.32	0.00
Average	**0.00**	**1.33**	**0.00**	**0.00**	**0.13**	**0.00**	**Average**	**0.00**	**1.47**	**0.27**	**0.34**	**0.40**	**0.00**
Manufacturing, High Tech							**Office, High**						
Zone 1	0.00	2.15	0.00	0.00	0.17	0.00	Zone 1	0.00	1.70	0.30	0.33	0.44	0.00
Zone 2	0.00	2.12	0.00	0.00	0.17	0.00	Zone 2	0.00	1.59	0.29	0.32	0.41	0.00
Zone 3	0.00	2.04	0.00	0.00	0.17	0.00	Zone 3	0.00	1.59	0.32	0.34	0.42	0.00
Zone 4	0.00	2.01	0.00	0.00	0.17	0.00	Zone 4	0.00	1.58	0.28	0.35	0.42	0.00
Zone 5	0.00	1.88	0.00	0.00	0.17	0.00	Zone 5	0.00	1.58	0.32	0.36	0.41	0.00
Zone 6	0.00	1.63	0.00	0.00	0.17	0.00	Zone 6	0.00	1.11	0.16	0.29	0.33	0.00
Zone 7	0.00	1.50	0.00	0.00	0.17	0.00	Zone 7	0.00	0.86	0.23	0.31	0.37	0.00
Average	**0.00**	**1.90**	**0.00**	**0.00**	**0.17**	**0.00**	**Average**	**0.00**	**1.43**	**0.27**	**0.33**	**0.40**	**0.00**
Medical Clinic							**Restaurant, Bar/Lounge**						
Zone 1	0.00	3.15	0.00	0.00	0.54	0.00	Zone 1	0.00	3.07	0.00	0.06	0.24	0.00
Zone 2	0.00	3.09	0.00	0.00	0.53	0.00	Zone 2	0.00	2.87	0.00	0.06	0.25	0.00
Zone 3	0.00	3.07	0.00	0.00	0.54	0.00	Zone 3	0.00	2.85	0.00	0.06	0.24	0.00
Zone 4	0.00	3.02	0.00	0.00	0.54	0.00	Zone 4	0.00	2.83	0.00	0.06	0.24	0.00
Zone 5	0.00	2.99	0.00	0.00	0.54	0.00	Zone 5	0.00	2.80	0.00	0.06	0.24	0.00
Zone 6	0.00	2.58	0.00	0.00	0.55	0.00	Zone 6	0.00	1.88	0.00	0.06	0.24	0.00
Zone 7	0.00	2.45	0.00	0.00	0.52	0.00	Zone 7	0.00	1.87	0.00	0.06	0.23	0.00
Average	**0.00**	**2.91**	**0.00**	**0.00**	**0.54**	**0.00**	**Average**	**0.00**	**2.59**	**0.00**	**0.06**	**0.24**	**0.00**
Motel							**Restaurant, Full**						
Zone 1	0.97	2.68	0.00	0.00	0.19	2.15	Zone 1	0.00	3.37	0.00	0.05	0.38	0.00
Zone 2	1.19	2.52	0.00	0.00	0.18	4.43	Zone 2	0.00	3.30	0.00	0.05	0.39	0.00
Zone 3	1.19	2.52	0.00	0.00	0.19	4.81	Zone 3	0.00	3.25	0.00	0.05	0.39	0.00
Zone 4	1.19	2.42	0.00	0.00	0.21	7.39	Zone 4	0.00	3.20	0.00	0.05	0.39	0.00
Zone 5	1.28	2.26	0.00	0.00	0.21	7.82	Zone 5	0.00	3.03	0.00	0.05	0.38	0.00
Zone 6	1.29	2.07	0.00	0.00	0.23	8.89	Zone 6	0.00	2.35	0.00	0.05	0.39	0.00
Zone 7	1.39	1.97	0.00	0.00	0.23	9.51	Zone 7	0.00	2.33	0.00	0.05	0.38	0.00
Average	**1.21**	**2.35**	**0.00**	**0.00**	**0.21**	**6.43**	**Average**	**0.00**	**2.98**	**0.00**	**0.05**	**0.38**	**0.00**
Museum							**Restaurant, Quick**						
Zone 1	0.00	3.77	0.00	0.11	0.67	0.00	Zone 1	0.00	3.78	0.00	0.08	0.48	0.00
Zone 2	0.00	3.71	0.00	0.12	0.77	0.00	Zone 2	0.00	3.72	0.00	0.08	0.50	0.00
Zone 3	0.00	3.59	0.00	0.12	0.72	0.00	Zone 3	0.00	3.64	0.00	0.08	0.50	0.00
Zone 4	0.00	3.47	0.00	0.14	0.74	0.00	Zone 4	0.00	3.60	0.00	0.08	0.50	0.00
Zone 5	0.00	3.42	0.00	0.16	0.51	0.00	Zone 5	0.00	3.42	0.00	0.08	0.49	0.00
Zone 6	0.00	2.39	0.00	0.15	0.23	0.00	Zone 6	0.00	2.74	0.00	0.08	0.51	0.00
Zone 7	0.00	1.69	0.00	0.17	0.24	0.00	Zone 7	0.00	2.69	0.00	0.08	0.49	0.00
Average	**0.00**	**3.15**	**0.00**	**0.14**	**0.55**	**0.00**	**Average**	**0.00**	**3.37**	**0.00**	**0.08**	**0.50**	**0.00**
Office, Low							**Store, Convenience**						
Zone 1	0.00	2.90	0.00	0.02	0.29	0.00	Zone 1	0.00	4.08	0.00	0.02	0.29	0.00
Zone 2	0.00	2.90	0.00	0.02	0.30	0.00	Zone 2	0.00	3.66	0.00	0.02	0.30	0.00
Zone 3	0.00	2.81	0.00	0.02	0.30	0.00	Zone 3	0.00	3.60	0.00	0.02	0.30	0.00
Zone 4	0.00	2.70	0.00	0.02	0.29	0.00	Zone 4	0.00	3.46	0.00	0.02	0.29	0.00
Zone 5	0.00	2.66	0.00	0.02	0.29	0.00	Zone 5	0.00	3.28	0.00	0.02	0.29	0.00
Zone 6	0.00	2.04	0.00	0.02	0.30	0.00	Zone 6	0.00	2.66	0.00	0.02	0.27	0.00
Zone 7	0.00	1.82	0.00	0.02	0.29	0.00	Zone 7	0.00	2.50	0.00	0.02	0.24	0.00
Average	**0.00**	**2.55**	**0.00**	**0.02**	**0.29**	**0.00**	**Average**	**0.00**	**3.32**	**0.00**	**0.02**	**0.28**	**0.00**

Table R260120-30 Watts per S.F. for Various Systems by Building Type & Climate Zone (cont.)

Building / Zone	Heating	Cooling	Heat rejection	Auxiliary (pumps)	Vent fan	Supp heat pump	Building / Zone	Heating	Cooling	Heat rejection	Auxiliary (pumps)	Vent fan	Supp heat pump
Store, Department							**School, Middle**						
Zone 1	0.00	1.65	0.20	0.23	0.63	0.00	Zone 1	0.00	2.48	0.00	0.00	0.17	0.00
Zone 2	0.00	1.61	0.20	0.23	0.63	0.00	Zone 2	0.00	2.45	0.00	0.00	0.17	0.00
Zone 3	0.00	1.59	0.19	0.22	0.64	0.00	Zone 3	0.00	2.41	0.00	0.00	0.18	0.00
Zone 4	0.00	1.52	0.19	0.23	0.64	0.00	Zone 4	0.00	2.35	0.00	0.00	0.17	0.00
Zone 5	0.00	1.51	0.21	0.24	0.63	0.00	Zone 5	0.00	2.26	0.00	0.00	0.17	0.00
Zone 6	0.00	1.32	0.13	0.21	0.63	0.00	Zone 6	0.00	1.61	0.00	0.00	0.17	0.00
Zone 7	0.00	0.98	0.17	0.23	0.63	0.00	Zone 7	0.00	1.49	0.00	0.00	0.17	0.00
Average	**0.00**	**1.45**	**0.18**	**0.23**	**0.63**	**0.00**	**Average**	**0.00**	**2.15**	**0.00**	**0.00**	**0.17**	**0.00**
Store, Large							**School, High**						
Zone 1	0.00	2.68	0.00	0.00	0.16	0.00	Zone 1	0.00	3.24	0.00	0.01	0.22	0.00
Zone 2	0.00	2.61	0.00	0.00	0.17	0.00	Zone 2	0.00	3.08	0.00	0.01	0.22	0.00
Zone 3	0.00	2.61	0.00	0.00	0.17	0.00	Zone 3	0.00	3.05	0.00	0.01	0.23	0.00
Zone 4	0.00	2.56	0.00	0.00	0.16	0.00	Zone 4	0.00	3.00	0.00	0.01	0.23	0.00
Zone 5	0.00	2.53	0.00	0.00	0.16	0.00	Zone 5	0.00	2.94	0.00	0.01	0.23	0.00
Zone 6	0.00	1.84	0.00	0.00	0.16	0.00	Zone 6	0.00	1.98	0.00	0.01	0.23	0.00
Zone 7	0.00	1.50	0.00	0.00	0.16	0.00	Zone 7	0.00	1.81	0.00	0.01	0.23	0.00
Average	**0.00**	**2.33**	**0.00**	**0.00**	**0.16**	**0.00**	**Average**	**0.00**	**2.73**	**0.00**	**0.01**	**0.23**	**0.00**
Store, Small							**School, College**						
Zone 1	0.00	3.60	0.00	0.01	0.25	0.00	Zone 1	0.00	2.92	0.00	0.00	0.20	0.00
Zone 2	0.00	3.51	0.00	0.01	0.26	0.00	Zone 2	0.00	2.91	0.00	0.00	0.20	0.00
Zone 3	0.00	3.38	0.00	0.01	0.25	0.00	Zone 3	0.00	2.78	0.00	0.00	0.20	0.00
Zone 4	0.00	3.32	0.00	0.01	0.25	0.00	Zone 4	0.00	2.72	0.00	0.00	0.20	0.00
Zone 5	0.00	3.23	0.00	0.01	0.24	0.00	Zone 5	0.00	2.43	0.00	0.00	0.20	0.00
Zone 6	0.00	2.49	0.00	0.01	0.23	0.00	Zone 6	0.00	1.77	0.00	0.00	0.20	0.00
Zone 7	0.00	2.17	0.00	0.01	0.21	0.00	Zone 7	0.00	1.51	0.00	0.00	0.19	0.00
Average	**0.00**	**3.10**	**0.00**	**0.01**	**0.24**	**0.00**	**Average**	**0.00**	**2.43**	**0.00**	**0.00**	**0.20**	**0.00**
Store, Strip Mall							**Service Station**						
Zone 1	0.00	3.36	0.00	0.02	0.25	0.00	Zone 1	0.00	10.42	0.00	0.03	0.47	0.00
Zone 2	0.00	3.30	0.00	0.02	0.25	0.00	Zone 2	0.00	9.01	0.00	0.03	0.46	0.00
Zone 3	0.00	3.14	0.00	0.02	0.25	0.00	Zone 3	0.00	7.82	0.00	0.03	0.49	0.00
Zone 4	0.00	3.14	0.00	0.02	0.25	0.00	Zone 4	0.00	7.02	0.00	0.03	0.47	0.00
Zone 5	0.00	2.24	0.00	0.02	0.24	0.00	Zone 5	0.00	6.13	0.00	0.03	0.46	0.00
Zone 6	0.00	1.96	0.00	0.02	0.25	0.00	Zone 6	0.00	4.32	0.00	0.03	0.45	0.00
Zone 7	0.00	1.64	0.00	0.02	0.24	0.00	Zone 7	0.00	4.18	0.00	0.03	0.40	0.00
Average	**0.00**	**2.52**	**0.00**	**0.02**	**0.25**	**0.00**	**Average**	**0.00**	**6.99**	**0.00**	**0.03**	**0.46**	**0.00**
Store, Warehouse							**Storage, High Bay**						
Zone 1	0.00	2.99	0.00	0.00	0.21	0.00	Zone 1	0.00	1.98	0.00	0.00	0.17	0.00
Zone 2	0.00	2.97	0.00	0.00	0.21	0.00	Zone 2	0.00	1.97	0.00	0.00	0.18	0.00
Zone 3	0.00	2.94	0.00	0.00	0.21	0.00	Zone 3	0.00	1.90	0.00	0.00	0.18	0.00
Zone 4	0.00	2.83	0.00	0.00	0.21	0.00	Zone 4	0.00	1.89	0.00	0.00	0.18	0.00
Zone 5	0.00	2.70	0.00	0.00	0.21	0.00	Zone 5	0.00	1.85	0.00	0.00	0.17	0.00
Zone 6	0.00	2.10	0.00	0.00	0.21	0.00	Zone 6	0.00	1.29	0.00	0.00	0.17	0.00
Zone 7	0.00	1.78	0.00	0.00	0.21	0.00	Zone 7	0.00	1.10	0.00	0.00	0.17	0.00
Average	**0.00**	**2.61**	**0.00**	**0.00**	**0.21**	**0.00**	**Average**	**0.00**	**1.71**	**0.00**	**0.00**	**0.17**	**0.00**
School, Elementary							**Storage, Low Bay**						
Zone 1	0.00	2.73	0.00	0.01	0.21	0.00	Zone 1	0.00	1.87	0.00	0.00	0.16	0.00
Zone 2	0.00	2.69	0.00	0.01	0.21	0.00	Zone 2	0.00	1.85	0.00	0.00	0.16	0.00
Zone 3	0.00	2.63	0.00	0.01	0.21	0.00	Zone 3	0.00	1.82	0.00	0.00	0.16	0.00
Zone 4	0.00	2.61	0.00	0.01	0.20	0.00	Zone 4	0.00	1.77	0.00	0.00	0.16	0.00
Zone 5	0.00	2.51	0.00	0.01	0.20	0.00	Zone 5	0.00	1.76	0.00	0.00	0.16	0.00
Zone 6	0.00	1.71	0.00	0.01	0.20	0.00	Zone 6	0.00	1.20	0.00	0.00	0.16	0.00
Zone 7	0.00	1.64	0.00	0.01	0.20	0.00	Zone 7	0.00	1.00	0.00	0.00	0.16	0.00
Average	**0.00**	**2.36**	**0.00**	**0.01**	**0.20**	**0.00**	**Average**	**0.00**	**1.61**	**0.00**	**0.00**	**0.16**	**0.00**

R260519-20 Armored Cable

Armored Cable - Quantities are taken off in the same manner as wire.

Bx Type Cable - Productivities are based on an average run of 50' before terminating at a box fixture, etc. Each 50' section includes field preparation of (2) ends with hacksaw, identification and tagging of wire. Set up is open coil type without reels attaching cable to snake and pulling across a suspended ceiling or open face wood or steel studding, price does not include drilling of studs.

Cable in Tray - Productivities are based on an average run of 100 L.F. with set up of pulling equipment for (2) 90° bends, attaching cable to pull-in means, identification and tagging of wires, set up of reels. Wire termination to breakers equipment, etc., are not included.

Job Conditions - Productivities are based on new construction to a height of 10' in an unobstructed area. Material staging is assumed to be within 100' of work being performed.

R260519-80 Undercarpet Systems

Takeoff Procedure for Power Systems: List components for each fitting type, tap, splice, and bend on your quantity takeoff sheet. Each component must be priced separately. Start at the power supply transition fittings and survey each circuit for the components needed. List the quantities of each component under a specific circuit number. Use the floor plan layout scale to get cable footage.

Reading across the list, combine the totals of each component in each circuit and list the total quantity in the last column. Calculate approximately 5% for scrap for items such as cable, top shield, tape, and spray adhesive. Also provide for final variations that may occur on-site.

Suggested guidelines are:
1. Equal amounts of cable and top shield should be priced.
2. For each roll of cable, price a set of cable splices.
3. For every 1 ft. of cable, price 2-1/2 ft. of hold-down tape.
4. For every 3 rolls of hold-down tape, price 1 can of spray adhesive.

Adjust final figures wherever possible to accommodate standard packaging of the product. This information is available from the distributor.

Each transition fitting requires:
1. 1 base
2. 1 cover
3. 1 transition block

Each floor fitting requires:
1. 1 frame/base kit
2. 1 transition block
3. 2 covers (duplex/blank)

Each tap requires:
1. 1 tap connector for each conductor
2. 1 pair insulating patches
3. 2 top shield connectors

Each splice requires:
1. 1 splice connector for each conductor
2. 1 pair insulating patches
3. 3 top shield connectors

Each cable bend requires:
1. 2 top shield connectors

Each cable dead end (outside of transition block) requires:
1. 1 pair insulating patches

Labor does not include:
1. Patching or leveling uneven floors.
2. Filling in holes or removing projections from concrete slabs.
3. Sealing porous floors.
4. Sweeping and vacuuming floors.
5. Removal of existing carpeting.
6. Carpet square cut-outs.
7. Installation of carpet squares.

Takeoff Procedures for Telephone Systems: After reviewing floor plans identify each transition. Number or letter each cable run from that fitting.

Start at the transition fitting and survey each circuit for the components needed. List the cable type, terminations, cable length, and floor fitting type under the specific circuit number. Use the floor plan layout scale to get the cable footage. Add some extra length (next higher increment of 5 feet) to preconnectorized cable.

Transition fittings require:
1. 1 base plate
2. 1 cover
3. 1 transition block

Floor fittings require:
1. 1 frame/base kit
2. 2 covers
3. Modular jacks

Reading across the list, combine the list of components in each circuit and list the total quantity in the last column. Calculate the necessary scrap factors for such items as tape, bottom shield and spray adhesive. Also provide for final variations that may occur on-site.

Adjust final figures whenever possible to accommodate standard packaging. Check that items such as transition fittings, floor boxes, and floor fittings that are to utilize both power and telephone have been priced as combination fittings, so as to avoid duplication.

Make sure to include marking of floors and drilling of fasteners if fittings specified are not the adhesive type.

Labor does not include:
1. Conduit or raceways before transition of floor boxes
2. Telephone cable before transition boxes
3. Terminations before transition boxes
4. Floor preparation as described in power section

Be sure to include all cable folds when pricing labor.

Takeoff Procedure for Data Systems: Start at the transition fittings and take off quantities in the same manner as the telephone system, keeping in mind that data cable does not require top or bottom shields.

The data cable is simply cross-taped on the cable run to the floor fitting.

Data cable can be purchased in either bulk form in which case coaxial connector material and labor must be priced, or in preconnectorized cut lengths.

Data cable cannot be folded and must be notched at 1 inch intervals. A count of all turns must be added to the labor portion of the estimate. (Note: Some manufacturers have prenotched cable.)

Notching required:
1. 90 degree turn requires 8 notches per side.
2. 180 degree turn requires 16 notches per side.

Floor boxes, transition boxes, and fittings are the same as described in the power and telephone procedures.

Since undercarpet systems require special hand tools, be sure to include this cost in proportion to number of crews involved in the installation.

Job Conditions: Productivity is based on new construction in an unobstructed area. Staging area is assumed to be within 200' of work being performed.

R260519-90   Wire

Wire quantities are taken off by either measuring each cable run or by extending the conduit and raceway quantities times the number of conductors in the raceway. Ten percent should be added for waste and tie-ins. Keep in mind that the unit of measure of wire is C.L.F. not L.F. as in raceways so the formula would read:

$$\frac{\text{(L.F. Raceway x No. of Conductors) x 1.10}}{100} = \text{C.L.F.}$$

Price per C.L.F. of wire includes:
1. Setting up wire coils or spools on racks
2. Attaching wire to pull in means
3. Measuring and cutting wire
4. Pulling wire into a raceway
5. Identifying and tagging

Price does not include:
1. Connections to breakers, panelboards, or equipment
2. Splices

Job Conditions: Productivity is based on new construction to a height of 10′ in an unobstructed area. Material staging is assumed to be within 100′ of work being performed.

Economy of Scale: If more than three wires at a time are being pulled, deduct the following percentages from the labor of that grouping:

4-5 wires	25%
6-10 wires	30%
11-15 wires	35%
over 15	40%

If a wire pull is less than 100′ in length and is interrupted several times by boxes, lighting outlets, etc., it may be necessary to add the following lengths to each wire being pulled:

Junction box to junction box	2 L.F.
Lighting panel to junction box	6 L.F.
Distribution panel to sub panel	8 L.F.
Switchboard to distribution panel	12 L.F.
Switchboard to motor control center	20 L.F.
Switchboard to cable tray	40 L.F.

Measure of Drops and Riser: It is important when taking off wire quantities to include the wire for drops to electrical equipment. If heights of electrical equipment are not clearly stated, use the following guide:

	Bottom A.F.F.	Top A.F.F.	Inside Cabinet
Safety switch to 100A	5′	6′	2′
Safety switch 400 to 600A	4′	6′	3′
100A panel 12 to 30 circuit	4′	6′	3′
42 circuit panel	3′	6′	4′
Switch box	3′	3′6″	1′
Switchgear	0′	8′	8′
Motor control centers	0′	8′	8′
Transformers - wall mount	4′	8′	2′
Transformers - floor mount	0′	12′	4′

R260519-92 Minimum Copper and Aluminum Wire Size Allowed for Various Types of Insulation

| Amperes | Copper | | Aluminum | | Amperes | Copper | | Aluminum | |
	THW THWN or XHHW	THHN XHHW *	THW XHHW	THHN XHHW *		THW THWN or XHHW	THHN XHHW *	THW XHHW	THHN XHHW *
15A	#14	#14	#12	#12	195	3/0	2/0	250kcmil	4/0
20	#12	#12	#10	#10	200	3/0	3/0	250kcmil	4/0
25	#10	#10	#10	#10	205	4/0	3/0	250kcmil	4/0
30	#10	#10	#8	#8	225	4/0	3/0	300kcmil	250kcmil
40	#8	#8	#8	#8	230	4/0	4/0	300kcmil	250kcmil
45	#8	#8	#6	#8	250	250kcmil	4/0	350kcmil	300kcmil
50	#8	#8	#6	#6	255	250kcmil	4/0	400kcmil	300kcmil
55	#6	#8	#4	#6	260	300kcmil	4/0	400kcmil	350kcmil
60	#6	#6	#4	#6	270	300kcmil	250kcmil	400kcmil	350kcmil
65	#6	#6	#4	#4	280	300kcmil	250kcmil	500kcmil	350kcmil
75	#4	#6	#3	#4	285	300kcmil	250kcmil	500kcmil	400kcmil
85	#4	#4	#2	#3	290	350kcmil	250kcmil	500kcmil	400kcmil
90	#3	#4	#2	#2	305	350kcmil	300kcmil	500kcmil	400kcmil
95	#3	#4	#1	#2	310	350kcmil	300kcmil	500kcmil	500kcmil
100	#3	#3	#1	#2	320	400kcmil	300kcmil	600kcmil	500kcmil
110	#2	#3	1/0	#1	335	400kcmil	350kcmil	600kcmil	500kcmil
115	#2	#2	1/0	#1	340	500kcmil	350kcmil	600kcmil	500kcmil
120	#1	#2	1/0	1/0	350	500kcmil	350kcmil	700kcmil	500kcmil
130	#1	#2	2/0	1/0	375	500kcmil	400kcmil	700kcmil	600kcmil
135	1/0	#1	2/0	1/0	380	500kcmil	400kcmil	750kcmil	600kcmil
150	1/0	#1	3/0	2/0	385	600kcmil	500kcmil	750kcmil	600kcmil
155	2/0	1/0	3/0	3/0	420	600kcmil	500kcmil		700kcmil
170	2/0	1/0	4/0	3/0	430		500kcmil		750kcmil
175	2/0	2/0	4/0	3/0	435		600kcmil		750kcmil
180	3/0	2/0	4/0	4/0	475		600kcmil		

Minimum Wire Sizes

*Dry Locations Only

Notes:

1. Size #14 to 4/0 is in AWG units (American Wire Gauge).
2. Size 250 to 750 is in kcmil units (Thousand Circular Mils).
3. Use next higher ampere value if exact value is not listed in table.
4. For loads that operate continuously increase ampere value by 25% to obtain proper wire size.
5. Table R260519-92 has been written for estimating only. It is based on an ambient temperature of 30°C (86° F). For an ambient temperature other than 30°C (86° F), ampacity correction factors will be applied.

R260519-93 Metric Equivalent, Wire

United States		European	
Size AWG or kcmil	**Area Cir. Mils. (cmil) mm²**	**Size mm²**	**Area Cir. Mils.**
18	1620/.82	.75	1480
16	2580/1.30	1.0	1974
14	4110/2.08	1.5	2961
12	6530/3.30	2.5	4935
10	10,380/5.25	4	7896
8	16,510/8.36	6	11,844
6	26,240/13.29	10	19,740
4	41,740/21.14	16	31,584
3	52,620/26.65	25	49,350
2	66,360/33.61	–	–
1	83,690/42.39	35	69,090
1/0	105,600/53.49	50	98,700
2/0	133,100/67.42	–	–
3/0	167,800/85.00	70	138,180
4/0	211,600/107.19	95	187,530
250	250,000/126.64	120	236,880
300	300,000/151.97	150	296,100
350	350,000/177.30	–	–
400	400,000/202.63	185	365,190
500	500,000/253.29	240	473,760
600	600,000/303.95	300	592,200
700	700,000/354.60	–	–
750	750,000/379.93	–	–

Table title: U.S. vs. European Wire – Approximate Equivalents

R260519-94 Size Required and Weight (Lbs./1000 L.F.) of Aluminum and Copper THW Wire by Ampere Load

Amperes	Copper Size	Aluminum Size	Copper Weight	Aluminum Weight
15	14	12	24	11
20	12	10	33	17
30	10	8	48	39
45	8	6	77	52
65	6	4	112	72
85	4	2	167	101
100	3	1	205	136
115	2	1/0	252	162
130	1	2/0	324	194
150	1/0	3/0	397	233
175	2/0	4/0	491	282
200	3/0	250	608	347
230	4/0	300	753	403
255	250	400	899	512
285	300	500	1068	620
310	350	500	1233	620
335	400	600	1396	772
380	500	750	1732	951

R260526-80   Grounding

Grounding When taking off grounding systems, identify separately the type and size of wire.

Example:

Bare copper & size

Bare aluminum & size

Insulated copper & size

Insulated aluminum & size

Count the number of ground rods and their size.

Example:

1.   8′ grounding rod – 5/8″ dia. 20 Ea.
2.   10′ grounding rod – 5/8″ dia. 12 Ea.
3.   15′ grounding rod – 3/4″ dia.   4 Ea.

Count the number of connections; the size of the largest wire will determine the productivity.

Example:

Braze a #2 wire to a #4/0 cable

The 4/0 cable will determine the L.H. and cost to be used.

Include individual connections to:

1. Ground rods
2. Building steel
3. Equipment
4. Raceways

Price does not include:

1. Excavation
2. Backfill
3. Sleeves or raceways used to protect grounding wires
4. Wall penetrations
5. Floor cutting
6. Core drilling

Job Conditions: Productivity is based on a ground floor area, using cable reels in an unobstructed area. The material staging area is assumed to be within 100′ of work being performed.

R260533-20   Conduit To 10′ High

List conduit by quantity, size, and type. Do not deduct for lengths occupied by fittings, since this will be allowance for scrap. Example:

A. Aluminum — size
B. Rigid galvanized — size
C. Steel intermediate (IMC) — size
D. Rigid steel, plastic-coated 20 Mil. — size
E. Rigid steel, plastic-coated 40 Mil. — size
F. Electric metallic tubing (EMT) — size
G. PVC Schedule 40 — size

Types (A) through (E) listed above contain the following per 100 L.F.:

1. (11) Threaded couplings
2. (11) Beam-type hangers
3. (2) Factory sweeps
4. (2) Fiber bushings
5. (4) Locknuts
6. (2) Field threaded pipe terminations
7. (2) Removal of concentric knockouts

Type (F) contains per 100 L.F.:

1. (11) Set screw couplings
2. (11) Beam clamps
3. (2) Field bends on 1/2″ and 3/4″ diameter
4. (2) Factory sweeps for 1″ and above
5. (2) Set screw steel connectors
6. (2) Removal of concentric knockouts

Type (G) contains per 100 L.F.:

1. (11) Field cemented couplings
2. (34) Beam clamps

3. (2) Factory sweeps
4. (2) Adapters
5. (2) Locknuts
6. (2) Removal of concentric knockouts

Labor-hours for all conduit to 10′ include:

1. Unloading by hand
2. Hauling by hand to an area up to 200′ from loading dock
3. Installation of conduit and fittings as described in Conduit models (A) through (G)

Not included in the material and labor are:

1. Staging rental or purchase
2. Structural modifications
3. Wire
4. Junction boxes
5. Fittings in excess of those described in conduit models (A) through (G)
6. Painting of conduit

Fittings

Only those fittings listed above are included in the linear foot totals, although they should be listed separately from conduit lengths, without prices, to ensure proper quantities for material procurement.

If the fittings required exceed the quantities included in the model conduit runs, then material and labor costs must be added to the difference. If actual needs per 100 L.F. of conduit are: (2) sweeps, (4) LBs and (1) field bend, then, (4) LBs and (1) field bend must be priced additionally.

R260533-21 Hangers

It is sometimes desirable to substitute an alternate style of hanger if the support being used is not the type described in the conduit models.

One approach is the substitution method:
1. Find the cost of the type hanger described in the conduit model.
2. Calculate the cost of the desired type hanger (it may be necessary to calculate individual components such as drilling, expansion shields, etc.).
3. Calculate the cost difference (delta) between the two types of hangers.
4. Multiply the cost delta by the number of hangers in the model.
5. Divide the total delta cost for hangers in the model by the length of the model to find the delta cost per L.F. for that model.
6. Modify the given unit costs per L.F. for the model by the delta cost per L.F.

Another approach to hanger configurations would be to start with the conduit only and add all the supports and any other items as separate lines. This procedure is most useful if the project involves racking many runs of conduit on a single hanger; for instance, a trapeze type hanger.

Example: Five (5) 2″ RGS conduits, 50 L.F. each, are to be run on trapeze hangers from one pull box to another. The run includes one 90° bend.

1. List the hangers' components to create an assembly cost for each 2′-wide trapeze.
2. List the components for the 50′ conduit run, noting that 6 trapeze supports will be required.

Job Conditions: Productivities are based on new construction to 10′ high, using scaffolding, scissor lift, bucket lift in an unobstructed area. Material storage is assumed to be within 100′ of work being performed.

Add to labor for elevated installations:

10′ to 14.5′ High–10%	30′ to 35′ High–40%
15′ to 20′ High–20%	35′ to 40′ High–50%
20′ to 25′ High–25%	Over 40′ High–55%
25′ to 30′ High–35%	

Add these percentages to the L.F. labor cost, but not to fittings. Add these percentages only to quantities exceeding the different height levels, rather than the total conduit quantities.

Linear foot price for labor does not include penetrations in walls or floors and must be added to the estimate.

R260533-22   Conductors in Conduit

The table below lists the maximum number of conductors for various sized conduit using THW, TW or THWN insulations.

Copper Wire Size	1/2″ TW	1/2″ THW	1/2″ THWN	3/4″ TW	3/4″ THW	3/4″ THWN	1″ TW	1″ THW	1″ THWN	1-1/4″ TW	1-1/4″ THW	1-1/4″ THWN	1-1/2″ TW	1-1/2″ THW	1-1/2″ THWN	2″ TW	2″ THW	2″ THWN	2-1/2″ TW	2-1/2″ THW	2-1/2″ THWN	3″ THW	3″ THWN	3-1/2″ THW	3-1/2″ THWN	4″ THW	4″ THWN
#14	9	6	13	15	10	24	25	16	39	44	29	69	60	40	94	99	65	154	142	93		143		192			
#12	7	4	10	12	8	18	19	13	29	35	24	51	47	32	70	78	53	114	111	76	164	117		157			
#10	5	4	6	9	6	11	15	11	18	26	19	32	36	26	44	60	43	73	85	61	104	95	160	127		163	
#8	2	1	3	4	3	5	7	5	9	12	10	16	17	13	22	28	22	36	40	32	51	49	79	66	106	85	136
#6		1	1		2	4		4	6		7	11		10	15		16	26		23	37	36	57	48	76	62	98
#4		1	1		1	2		3	4		5	7		7	9		12	16		17	22	27	35	36	47	47	60
#3		1	1		1	1		2	3		4	6		6	8		10	13		15	19	23	29	31	39	40	51
#2		1	1		1	1		2	3		4	5		5	7		9	11		13	16	20	25	27	33	34	43
#1					1	1		1	1		3	3		4	5		6	8		9	12	14	18	19	25	25	32
1/0					1	1		1	1		2	3		3	4		5	7		8	10	12	15	16	21	21	27
2/0					1	1		1	1		1	2		3	3		5	6		7	8	10	13	14	17	18	22
3/0					1	1		1	1		1	1		2	3		4	5		6	7	9	11	12	14	15	18
4/0						1		1	1		1	1		1	2		3	4		5	6	7	9	10	12	13	15
250 kcmil								1	1		1	1		1	1		2	3		4	4	6	7	8	10	10	12
300								1	1		1	1		1	1		2	3		3	4	5	6	7	8	9	11
350									1		1	1		1	1		1	2		3	3	4	5	6	7	8	9
400											1	1		1	1		1	1		2	3	4	5	5	6	7	8
500												1		1	1		1	1		1	2	3	4	4	5	6	7
600												1		1	1		1	1		1	1	3	3	4	4	5	5
700														1	1		1	1		1	1	2	3	3	4	4	5
750														1	1		1	1		1	1	2	2	3	3	4	4

Reprinted with permission from NFPA 70-2014, *National Electrical Code®*, Copyright © 2013, National Fire Protection Association, Quincy, MA. This reprinted material is not the complete and official position of the NFPA on the referenced subject, which is represented solely by the standard in its entirety.

R260533-25 Conduit Weight Comparisons (Lbs. per 100 ft.) with Maximum Cable Fill*

Type	1/2"	3/4"	1"	1-1/4"	1-1/2"	2"	2-1/2"	3"	3-1/2"	4"	5"	6"
Rigid Galvanized Steel (RGS)	104	140	235	358	455	721	1022	1451	1749	2148	3083	4343
Intermediate Steel (IMC)	84	113	186	293	379	611	883	1263	1501	1830		
Electrical Metallic Tubing (EMT)	54	116	183	296	368	445	641	930	1215	1540		

*Conduit & Heaviest Conductor Combination

R260533-30 Labor for Couplings and Fittings

The labor included in the unit price lines for couplings is for their installation separate from a fitting. However, the labor included in the unit price lines for fittings covers their complete installation, which may include the labor installation of the coupling(s). If coupling(s) are required to complete a fitting installation, only the material for the required number of couplings needs to be added to the appropriate unit price lines.

R260533-60 Wireway

When "taking off" Wireway, list by size and type.

Example:
1. Screw cover, unflanged + size
2. Screw cover, flanged + size
3. Hinged cover, flanged + size
4. Hinged cover, unflanged + size

Each 10′ length on Wireway contains:
1. 10′ of cover either screw or hinged type
2. (1) Coupling or flange gasket
3. (1) Wall type mount

All fittings must be priced separately.

Substitution of hanger types is done the same as described in R260533-21, "HANGERS," keeping in mind that the wireway model is based on 10′ sections instead of a 100′ conduit run.

Labor-hours for wireway include:
1. Unloading by hand
2. Hauling by hand up to 100′ from loading dock
3. Measuring and marking
4. Mounting wall bracket using two (2) anchor type lead fasteners
5. Installing wireway on brackets, to 10′ high (for higher elevations use factors in R260533-20)

Job Conditions: Productivity is based on new construction, to a height of 10′ in an unobstructed area.

Material staging area is assumed to be within 100′ of work being performed.

R260533-65 Outlet Boxes

Outlet boxes should be included on the same takeoff sheet as branch piping or devices to better explain what is included in each circuit.

Each unit price in this section is a stand alone item and contains no other component unless specified. For example, to estimate a duplex outlet, components that must be added are:
1. 4″ square box
2. 4″ plaster ring
3. Duplex receptacle
4. Device cover

The method of mounting outlet boxes is two (2) plastic shield fasteners.

Outlet boxes plastic, labor-hours include:
1. Marking box location on wood studding
2. Mounting box

Economy of Scale – For large concentrations of plastic boxes in the same area deduct the following percentages from labor-hour totals:

1	to	10	0%
11	to	25	20%
26	to	50	25%
51	to	100	30%
	over	100	35%

Note: It is important to understand that these percentages are not used on the total job quantities, but only areas where concentrations exceed the levels specified.

R260533-70 Pull Boxes and Cabinets

List cabinets and pull boxes by NEMA type and size.

Example:	TYPE	SIZE
	NEMA 1	6″W x 6″H x 4″D
	NEMA 3R	6″W x 6″H x 4″D

Labor-hours for wall mount (indoor or outdoor) installations include:
1. Unloading and uncrating
2. Handling of enclosures up to 200′ from loading dock using a dolly or pipe rollers
3. Measuring and marking
4. Drilling four (4) anchor type lead fasteners using a hammer drill
5. Mounting and leveling boxes

Note: A plywood backboard is not included.

Labor-hours for ceiling mounting include:
1. Unloading and uncrating
2. Handling boxes up to 100′ from loading dock

3. Measuring and marking
4. Drilling four (4) anchor type lead fasteners using a hammer drill
5. Installing and leveling boxes to a height of 10′ using rolling staging

Labor-hours for free standing cabinets include:
1. Unloading and uncrating
2. Handling of cabinets up to 200′ from loading dock using a dolly or pipe rollers
3. Marking of floor
4. Drilling four (4) anchor type lead fasteners using a hammer drill
5. Leveling and shimming

Labor-hours for telephone cabinets include:
1. Unloading and uncrating
2. Handling cabinets up to 200′ using a dolly or pipe rollers
3. Measuring and marking
4. Mounting and leveling, using four (4) lead anchor type fasteners

Fig. R260536-11

R260536-10 Cable Tray

Cable Tray - When taking off cable tray it is important to identify separately the different types and sizes involved in the system being estimated. (Fig. R260536-11)

- A. – Ladder Type, galvanized or aluminum
- B. – Trough Type, galvanized or aluminum
- C. – Solid Bottom, galvanized or aluminum

The unit of measure is calculated in linear feet. Do not deduct from this footage any length occupied by fittings. This will be the only allowance for scrap. Be sure to include all vertical drops to panels, switch gear, etc.

Hangers - Included in the linear footage of cable tray is

- D. – 1 - Pair of connector plates per 12 L.F.
- E. – 1 - Pair clamp type hangers and 4' of 3/8" threaded rod per 12 L.F.

Not included are structural supports, which must be priced in addition to the hangers.

Fittings - Identify separately the different types of fittings

- A. – Ladder Type, galvanized or aluminum
- B. – Trough Type, galvanized or aluminum
- C. – Solid Bottom Type, galvanized or aluminum

The configuration, radius and rung spacing must also be listed. The unit of measure is "Ea."

- D. – Elbow, vertical Ea.
- E. – Elbow, horizontal Ea.
- F. – Tee, vertical Ea.
- G. – Cross, horizontal Ea.
- H. – Wye, horizontal Ea.
- I. – Tee, horizontal Ea.
- J. – Reducing fitting Ea.

Depending on the use of the system other examples of units which must be included are:

- K. – Divider strip L.F.
- L. – Drop-outs Ea.

- M. – End caps Ea.
- N. – Panel connectors Ea.

Wire and cable are not included and should be taken off separately. See Unit Price sections.

Job Conditions - Unit prices are based on a new installation to a work plane of 10' in an unobstructed area.

Add to labor for elevated installations:

10' to 14.5' High	10%
15' to 20' High	20%
20' to 25' High	25%
25' to 30' High	35%
30' to 35' High	40%
35' to 40' High	50%
Over 40' High	55%

Add these percentages for L.F. totals but not to fittings. Add percentages to only those quantities that fall in the different elevations. In other words, if the total quantity of cable tray is 200' but only 75' is above 15' then the 10% is added to the 75' only.

Linear foot costs do not include penetrations through walls and floors which must be added to the estimate.

Cable Tray Covers

Covers - Cable tray covers are taken off in the same manner as the tray itself, making distinctions as to the type of cover. (Fig. R260536-11)

- O. – Vented, galvanized or aluminum
- P. – Solid, galvanized or aluminum

Cover configurations are taken off separately noting type, specific radius and widths.

Note: Care should be taken to identify from plans and specifications exactly what is being covered. In many systems only vertical fittings are covered to retain wire and cable.

R260539-30 Conduit In Concrete Slab

List conduit by quantity, size and type.

Example:

  A. Rigid galvanized steel + size
  B. P.V.C. + size

Rigid galvanized steel (A) contains per 100 L.F.:

  1. (20) Ties to slab reinforcing
  2. (11) Threaded steel couplings
  3. (2) Factory sweeps
  4. (2) Field threaded conduit terminations

  5. (2) Fiber bushings + locknuts
  6. (2) Removal of concentric knockouts

P.V.C. (B) contains per 100 L.F.:

  1. (20) Ties to slab reinforcing
  2. (11) Field cemented couplings
  3. (2) Factory sweeps
  4. (2) Adapters
  5. (2) Removal of concentric knockouts

R260539-40 Conduit In Trench

Conduit in trench is galvanized steel and contains per 100 L.F.:

  1. (11) Threaded couplings
  2. (2) Factory sweeps
  3. (2) Fiber bushings + (4) locknuts
  4. (2) Field threaded conduit terminations
  5. (2) Removal of concentric knockouts

Note:

Conduit in Unit Price sections do not include:

  1. Floor cutting
  2. Excavation or backfill
  3. Grouting or patching

Conduit fittings in excess of those listed in the above Conduit model must be added. (Refer to R260533-20 for Procedure example.)

Fig. R260543-51

R260543-50 Underfloor Duct

When pricing Underfloor Duct it is important to identify and list each component, since costs vary significantly from one type of fitting to another. Do not deduct boxes or fittings from linear foot totals; this will be your allowance for scrap.

The first step is to identify the system as either:

FIG. R260543-51 Single Level

FIG. R260543-52 Dual Level

Single Level System

Include on your "takeoff sheet" the following unit price items, making sure to distinquish between Standard and Super duct:

A. Feeder duct (blank) in L.F.
B. Distribution duct (Inserts 2' on center) in L.F.
C. Elbows (Vertical) Ea.
D. Elbows (Horizontal) Ea.
E. Cabinet connector Ea.
F. Single duct junction box Ea.
G. Double duct junction box Ea.
H. Triple duct junction box Ea.
I. Support, single cell Ea.
J. Support, double cell Ea.
K. Support, triple cell Ea.
L. Carpet pan Ea.
M. Terrazzo pan Ea.
N. Insert to conduit adapter Ea.
O. Conduit adapter Ea.
P. Low tension outlet Ea.
Q. High tension outlet Ea.
R. Galvanized nipple Ea.
S. Wire per C.L.F.
T. Offset (Duct type) Ea.

Fig. R260543-52

R260543-50   Underfloor Duct (cont.)

Dual Level System + Labor

Dual Level

Include the following when "taking off" Dual Level systems:
Distinguish between Standard and Super duct.

- A. Feeder duct (blank) in L.F.
- B. Distribution duct (Inserts 2' on center) in L.F.
- C. Elbows (Vertical), Ea.
- D. Elbows (Horizontal), Ea.
- E. Cabinet connector, Ea.
- F. Single duct, 2 level, junction box, Ea.
- G. Double duct, 2 level, junction box, Ea.
- H. Support, single cell, Ea.
- I. Support, double cell, Ea.
- J. Support, triple cell, Ea.
- K. Carpet pan, Ea.
- L. Terrazzo pan, Ea.
- M. Insert to conduit adapter, Ea.
- N. Conduit adapter, Ea.
- O. Low tension outlet, Ea.
- P. High tension outlet, Ea.
- Q. Wire per C.L.F.

Note: Make sure to include risers in linear foot totals. High tension outlets include box, receptacle, covers and related mounting hardware.

Labor-hours for both Single and Dual Level systems include:
1. Unloading and uncrating
2. Hauling up to 200' from loading dock

3. Measuring and marking
4. Setting raceway and fittings in slab or on grade
5. Leveling raceway and fittings

Labor-hours do not include:
1. Floor cutting
2. Excavation or backfill
3. Concrete pour
4. Grouting or patching
5. Wire or wire pulls
6. Additional outlets after concrete is poured
7. Piping to or from Underfloor Duct

Note: Installation is based on installing up to 150' of duct. If quantities exceed this, deduct the following percentages:
1. 150' to 250' - 10%
2. 250' to 350' - 15%
3. 350' to 500' - 20%
4. over 500' - 25%

Deduct these percentages from labor only.

Deduct these percentages from straight sections only.

Do not deduct from fittings or junction boxes.

Job Conditions: Productivity is based on new construction.

Underfloor duct to be installed on first three floors.

The material staging area is within 100' of work being performed.

The area unobstructed and duct are not subject to physical damage.

R260580-75 Motor Connections

Motor connections should be listed by size and type of motor.
Included in the material and labor cost are:
1. (2) Flex connectors
2. 18″ of flexible metallic wireway
3. Wire identification and termination
4. Test for rotation
5. (2) or (3) Conductors

Price does not include:
1. Mounting of motor
2. Disconnect switch
3. Motor starter
4. Controls
5. Conduit or wire ahead of flex

Note: When "taking off" motor connections, it is advisable to list connections on the same quantity sheet as motors, motor starters and controls.

R260913-80 Switchboard Instruments

Switchboard instruments are added to the price of switchboards according to job specifications. This equipment is usually included when ordering "gear" from the manufacturer and will arrive factory installed.

Included in the labor cost are:
1. Internal wiring connections
2. Wire identification
3. Wire tagging

Transition sections include:
1. Uncrating
2. Hauling sections up to 100′ from loading dock
3. Positioning sections
4. Leveling sections
5. Bolting enclosures
6. Bolting vertical bus bars

Price does not include:
1. Equipment pads
2. Steel channels embedded or grouted in concrete
3. Special knockouts
4. Rigging

Job Conditions: Productivity is based on new construction and equipment to be installed on the first floor within 200′ of the loading dock.

For customer support on your Electrical Costs with RSMeans data, call 800.448.8182.

R262213-10   Electric Circuit Voltages

General: The following method provides the user with a simple non-technical means of obtaining comparative costs of wiring circuits. The circuits considered serve the electrical loads of motors, electric heating, lighting and transformers, for example, that require low voltage 60 Hertz alternating current.

The method used here is suitable only for obtaining estimated costs. It is **not** intended to be used as a substitute for electrical engineering design applications.

Conduit and wire circuits can represent from twenty to thirty percent of the total building electrical cost. By following the described steps and using the tables the user can translate the various types of electric circuits into estimated costs.

Wire Size: Wire size is a function of the electric load which is usually listed in one of the following units:

1. Amperes (A)
2. Watts (W)
3. Kilowatts (kW)
4. Volt amperes (VA)
5. Kilovolt amperes (kVA)
6. Horsepower (HP)

These units of electric load must be converted to amperes in order to obtain the size of wire necessary to carry the load. To convert electric load units to amperes one must have an understanding of the voltage classification of the power source and the voltage characteristics of the electrical equipment or load to be energized. The seven A.C. circuits commonly used are illustrated in Figures R262213-11 through R262213-17 showing the tranformer load voltage and the point of use voltage at the point on the circuit where the load is connected. The difference between the source and point of use voltage is attributed to the circuit voltage drop and is considered to be approximately 4%.

Motor Voltages: Motor voltages are listed by their point of use voltage and not the power source voltage.

For example: 460 volts instead of 480 volts
200 instead of 208 volts
115 volts instead of 120 volts

Lighting and Heating Voltages: Lighting and heating equipment voltages are listed by the power source voltage and not the point of wire voltage.

For example: 480, 277, 120 volt lighting
480 volt heating or air conditioning unit
208 volt heating unit

Transformer Voltages: Transformer primary (input) and secondary (output) voltages are listed by the power source voltage.

For example: Single phase 10 kVA
Primary 240/480 volts
Secondary 120/240 volts

In this case, the primary voltage may be 240 volts with a 120 volts secondary or may be 480 volts with either a 120V or a 240V secondary.

For example: Three phase 10 kVA
Primary 480 volts
Secondary 208Y/120 volts

In this case the transformer is suitable for connection to a circuit with a 3 phase, 3 wire or a 3 phase, 4 wire circuit with a 480 voltage. This application will provide a secondary circuit of 3 phase, 4 wire with 208 volts between phase wires and 120 volts between any phase wire and the neutral (white) wire.

R262213-11

3 Wire, 1 Phase, 120/240 Volt System

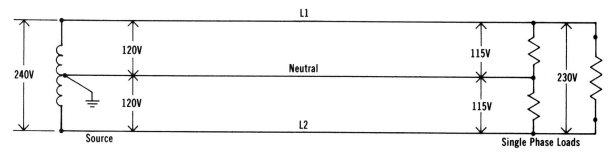

R262213-12

4 Wire, 3 Phase, 208Y/120 Volt System

R262213-13   Electric Circuit Voltages (cont.)

3 Wire, 3 Phase 240 Volt System

R262213-14

4 Wire, 3 Phase, 240/120 Volt System

R262213-15

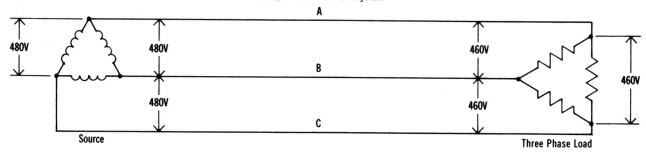

3 Wire, 3 Phase 480 Volt System

R262213-16

4 Wire, 3 Phase, 480Y/277 Volt System

R262213-17   Electric Circuit Voltages (cont.)

3 Wire, 3 Phase, 600 Volt System

R262213-20   kW Value/Cost Determination

General: Lighting and electric heating loads are expressed in watts and kilowatts.

Cost Determination:

The proper ampere values can be obtained as follows:

1. Convert watts to kilowatts
   (watts 1000 ÷ kilowatts)
2. Determine voltage rating of equipment.
3. Determine whether equipment is single phase or three phase.
4. Refer to Table R262213-21 to find ampere value from kW, Ton and BTU/hr. values.
5. Determine type of wire insulation – TW, THW, THWN.
6. Determine if wire is copper or aluminum.
7. Refer to Table R260519-92 to obtain copper or aluminum wire size from ampere values.
8. Next refer to Table R260533-22 for the proper conduit size to accommodate the number and size of wires in each particular case.
9. Next refer to unit cost data for the per linear foot cost of the conduit.
10. Next refer to unit cost data for the per linear foot cost of the wire. Multiply cost of wire per L.F. x number of wires in the circuits to obtain total wire cost per L.F.

11. Add values obtained in Step 9 and 10 for total cost per linear foot for conduit and wire x length of circuit = Total Cost.

Notes:

1. 1 Phase refers to single phase, 2 wire circuits.
2. 3 Phase refers to three phase, 3 wire circuits.
3. For circuits which operate continuously for 3 hours or more, multiply the ampere values by 1.25 for a given kW requirement.
4. For kW ratings not listed, add ampere values.

   For example: Find the ampere value of
   9 kW at 208 volt, single phase.

$$\begin{array}{r} 4\ kW = 19.2A \\ 5\ kW = 24.0A \\ \hline 9\ kW = 43.2A \end{array}$$

5. "Length of Circuit" refers to the one way distance of the run, not to the total sum of wire lengths.

R262213-21 Ampere Values as Determined by kW Requirements, BTU/HR or Ton, Voltage and Phase Values

			Ampere Values						
			120V	208V		240V		277V	480V
kW	Ton	BTU/HR	1 Phase	1 Phase	3 Phase	1 Phase	3 Phase	1 Phase	3 Phase
0.5	.1422	1,707	4.2A	2.4A	1.4A	2.1A	1.2A	1.8A	0.6A
0.75	.2133	2,560	6.2	3.6	2.1	3.1	1.9	2.7	.9
1.0	.2844	3,413	8.3	4.9	2.8	4.2	2.4	3.6	1.2
1.25	.3555	4,266	10.4	6.0	3.5	5.2	3.0	4.5	1.5
1.5	.4266	5,120	12.5	7.2	4.2	6.3	3.1	5.4	1.8
2.0	.5688	6,826	16.6	9.7	5.6	8.3	4.8	7.2	2.4
2.5	.7110	8,533	20.8	12.0	7.0	10.4	6.1	9.1	3.1
3.0	.8532	10,239	25.0	14.4	8.4	12.5	7.2	10.8	3.6
4.0	1.1376	13,652	33.4	19.2	11.1	16.7	9.6	14.4	4.8
5.0	1.4220	17,065	41.6	24.0	13.9	20.8	12.1	18.1	6.1
7.5	2.1331	25,598	62.4	36.0	20.8	31.2	18.8	27.0	9.0
10.0	2.8441	34,130	83.2	48.0	27.7	41.6	24.0	36.5	12.0
12.5	3.5552	42,663	104.2	60.1	35.0	52.1	30.0	45.1	15.0
15.0	4.2662	51,195	124.8	72.0	41.6	62.4	37.6	54.0	18.0
20.0	5.6883	68,260	166.4	96.0	55.4	83.2	48.0	73.0	24.0
25.0	7.1104	85,325	208.4	120.2	70.0	104.2	60.0	90.2	30.0
30.0	8.5325	102,390		144.0	83.2	124.8	75.2	108.0	36.0
35.0	9.9545	119,455		168.0	97.1	145.6	87.3	126.0	42.1
40.0	11.3766	136,520		192.0	110.8	166.4	96.0	146.0	48.0
45.0	12.7987	153,585			124.8	187.5	112.8	162.0	54.0
50.0	14.2208	170,650			140.0	208.4	120.0	180.4	60.0
60.0	17.0650	204,780			166.4		150.4	216.0	72.0
70.0	19.9091	238,910			194.2		174.6		84.2
80.0	22.7533	273,040			221.6		192.0		96.0
90.0	25.5975	307,170					225.6		108.0
100.0	28.4416	341,300							120.0

R262213-25 kVA Value/Cost Determination

General: Control transformers are listed in VA. Step-down and power transformers are listed in kVA.

Cost Determination:

1. Convert VA to kVA. Volt amperes (VA) ÷ 1000 = Kilovolt amperes (kVA).
2. Determine voltage rating of equipment.
3. Determine whether equipment is single phase or three phase.
4. Refer to Table R262213-26 to find ampere value from kVA value.
5. Determine type of wire insulation – TW, THW, THWN.
6. Determine if wire is copper or aluminum.
7. Refer to Table R260519-92 to obtain copper or aluminum wire size from ampere values.

Example: A transformer rated 10 kVA 480 volts primary, 240 volts secondary, 3 phase has the capacity to furnish the following:

1. Primary amperes = 10 kVA x 1.20 = 12 amperes
   (from Table R262213-26)
2. Secondary amperes = 10 kVA x 2.40 = 24 amperes
   (from Table R262213-26)

Note: Transformers can deliver generally 125% of their rated kVA. For instance, a 10 kVA rated transformer can safely deliver 12.5 kVA.

8. Next refer to Table R260533-22 for the proper conduit size to accommodate the number and size of wires in each particular case.
9. Next refer to unit price data for the per linear foot cost of the conduit.
10. Next refer to unit price data for the per linear foot cost of the wire. Multiply cost of wire per L.F. x number of wires in the circuits to obtain total wire cost.
11. Add values obtained in Step 9 and 10 for total cost per linear foot for conduit and wire x length of circuit = Total Cost.

R262213-26 Multiplier Values for kVA to Amperes Determined by Voltage and Phase Values

Volts	Multiplier for Circuits	
	2 Wire, 1 Phase	3 Wire, 3 Phase
115	8.70	
120	8.30	
230	4.30	2.51
240	4.16	2.40
200	5.00	2.89
208	4.80	2.77
265	3.77	2.18
277	3.60	2.08
460	2.17	1.26
480	2.08	1.20
575	1.74	1.00
600	1.66	0.96

R262213-60   Oil Filled Transformers

Transformers in this section include:
1. Rigging (as required)
2. Rental of crane and operator
3. Setting of oil filled transformer
4. (4) Anchor bolts, nuts and washers in concrete pad

Price does not include:
1. Primary and secondary terminations
2. Transformer pad
3. Equipment grounding
4. Cable
5. Conduit locknuts or bushings

Transformers in Unit Price sections for dry type, back-boost and isolating transformers include:
1. Unloading and uncrating
2. Hauling transformer to within 200' of loading dock

3. Setting in place
4. Wall mounting hardware
5. Testing

Price does not include:
1. Structural supports
2. Suspension systems
3. Welding or fabrication
4. Primary & secondary terminations

Add the following percentages to the labor for ceiling mounted transformers:

10' to 15'	= + 15%
15' to 25'	= + 30%
Over 25'	= + 35%

Job Conditions: Productivities are based on new construction. Installation is assumed to be on the first floor, in an obstructed area to a height of 10'. Material staging area is within 100' of final transformer location.

R262416-50   Load Centers and Panelboards

When pricing load centers list panels by size and type. List breakers in a separate column of the "quantity sheet," and define by phase and ampere rating.

Material and labor prices include breakers; for example: a 100A, 3-Wire, 102/240V, 18 circuit panel w/ main breaker, as described in the unit cost section, contains 18 single pole 20A breakers.

If you do not choose to include a full panel of single pole breakers, use the following method to adjust material and labor costs.

Example: In an 18 circuit panel only 16 single pole breakers are desired, requiring that the cost of 2 breakers be subtracted from the panel cost.

1. Go to the appropriate unit cost section of the data set to find the unit prices of the given circuit breaker type.
2. Modify those costs as follows:
    Bare material price x 0.50
    Bare labor cost x 0.60.
3. Multiply those modified bare costs by 2 (breakers in this example).
4. Subtract the modified costs for the 2 breakers from the given cost of the panel.

Labor-hours for load center installation includes:
1. Unloading, uncrating, and handling enclosures 200' from unloading area

2. Measuring and marking
3. Drilling (4) lead anchor type fasteners using a hammer drill
4. Mounting and leveling panel to a height of 6'
5. Preparation and termination of feeder cable to lugs or main breaker
6. Branch circuit identification
7. Lacing using tie wraps
8. Testing and load balancing
9. Marking panel directory

Not included in the material and labor are:
1. Modifications to enclosure
2. Structural supports
3. Additional lugs
4. Plywood backboards
5. Painting or lettering

Note: Knockouts are included in the price of terminating pipe runs and need not be added to the load center costs.

Job Conditions: Productivity is based on new construction to a height of 6', in an unobstructed area. Material staging area is assumed to be within 100' of work being performed.

609

Fig. R262419-61

R262419-60   Motor Control Centers

When taking off motor control centers, list the size, type and height of structures.

Example:
1. 600A, 22,000 RMS, 72" high
2. 600A, back to back, 72" high

Next take off individual starters; the number of structures can also be determined by adding the height in inches of starters divided by the height of the structure, and list on the same quantity sheet as the structures. Identify starters by type, horsepower rating, size and height in inches.

Example:
A. Class l, Type B, FVNR starter, 25 H.P., 18" high.
    Add to the list with starters, factory installed controls.

Example:

B. Pilot lights
C. Push buttons
D. Auxiliary contacts

Identify starters and structures as either copper or aluminum and by the NEMA type of enclosure.

When pricing starters and structures, be sure to add or deduct adjustments, using lines in the unit price section.

Included in the cost of motor control structures are:
1. Uncrating
2. Hauling to location within 100' of loading dock
3. Setting structures
4. Leveling
5. Aligning
6. Bolting together structure frames
7. Bolting horizontal bus bars

Labor-hours do not include:
1. Equipment pad
2. Steel channels embedded or grouted in concrete
3. Pull boxes
4. Special knockouts
5. Main switchboard section
6. Transition section
7. Instrumentation
8. External control wiring
9. Conduit or wire

Material for starters include:
1. Circuit breaker or fused disconnect
2. Magnetic motor starter
3. Control transformer
4. Control fuse and fuse block

Labor-hours for starters include:
1. Handling
2. Installing starter within structure
3. Internal wiring connections
4. Lacing within enclosure
5. Testing
6. Phasing

Job Conditions: Productivity is based on new construction. The motor control location is assumed to be on first floor in an unobstructed area. The material staging area is within 100' of the final location.

Note: Additional labor-hours must be added if M.C.C. is to be installed on floors other than the first, or if rigging is required.

R262419-65   Motor Starters and Controls

Motor starters should be listed on the same "quantity sheet" as motors and motor connections. Identify each starter by:

1. Size
2. Voltage
3. Type

Example:

  A. FVNR, 480V, 2HP, size 00
  B. FVNR, 480V, 5HP, size 0, combination type
  C. FVR, 480V, size 2

The NEMA type of enclosure should also be identified.

Included in the labor-hours are:

1. Unloading, uncrating and handling of starters up to 200' from loading area
2. Measuring and marking
3. Drilling (4) anchor type lead fasteners using a hammer drill
4. Mounting and leveling starter
5. Connecting wire or cable to line and load sides of starter (when already lugged)
6. Installation of (3) thermal type heaters
7. Testing

The following are not included unless specified in the unit price description.

1. Control transformer
2. Controls, either factory or field installed
3. Conduit and wire to or from starter
4. Plywood backboard
5. Cable terminations

The following material and labor were included for combination type starters:

1. Unloading, uncrating and handling of starter up to 200' of loading dock
2. Measuring and marking
3. Drilling (4) anchor type lead fasteners using a hammer drill
4. Mounting and leveling
5. Connecting prepared cable conductors
6. Installation of (3) dual element cartridge type fuses
7. Installation of (3) thermal type heaters
8. Test for rotation

MOTOR CONTROLS

When pricing motor controls make sure you consider the type of control system being utilized. If the controls are factory installed and located in the enclosure itself, then you would add to your starter price the items 5200 through 5800 in the Unit Price section. If control voltage is different from line voltage, add the material and labor cost of a control transformer.

For external control of starters include the following items:

1. Raceways
2. Wire & terminations
3. Control enclosures
4. Fittings
5. Push button stations
6. Indicators

Job Conditions: Productivity is based on new construction to a height of 10'.

Material staging area is assumed to be within 100' of work being performed.

Fig. R262419-81

R262419-80   Distribution Section

After "taking off" the switchboard section, include on the same "quantity sheet" the distribution section; identify by:

1. Voltage
2. Ampere rating
3. Type

Example:

(Fig. R262419-81) (B) Distribution Section

Included in the labor costs of the distribution section are:

1. Uncrating
2. Hauling to 200' of loading dock
3. Setting of distribution panel
4. Leveling & shimming
5. Anchoring of equipment to pad or floor
6. Bolting of horizontal bus bars between
7. Testing of equipment

Not included in the Distribution section are:

1. Breakers (C)
2. Equipment pads
3. Steel channels embedded or grouted in concrete
4. Pull boxes
5. Special knockouts
6. Transition section
7. Conduit or wire

R262419-82   Feeder Section

List quantities on the same sheet as the distribution section. Identify breakers by: (Fig. R262419-81)

1. Frame type
2. Number of poles
3. Ampere rating

Installation includes:

1. Handling
2. Placing breakers in distribution panel
3. Preparing wire or cable
4. Lacing wire
5. Marking each phase with colored tape
6. Marking panel legend
7. Testing
8. Balancing

R262419-84 Switchgear

It is recommended that "switchgear" or those items contained in the following sections be quoted from equipment manufacturers as a package price.

Switchboard instruments

Switchboard distribution sections

Switchboard feeder sections

Switchboard Service Disc.

Switchboard In-plant Dist.

Included in these sections are the most common types and sizes of factory assembled equipment.

The recommended procedure for low voltage switchgear would be to price (Fig. R262419-81) (A) Main Switchboard

Identify by:
1. Voltage
2. Amperage
3. Type

Example:
1. 120/208V, 4-wire, 600A, nonfused
2. 277/480V, 4-wire, 600A, nonfused
3. 120/208V, 4-wire, 400A w/fused switch & CT compartment
4. 277/480V, 4-wire, 400A, w/fused switch & CT compartment
5. 120/208V, 4-wire, 800A, w/pressure switch & CT compartment
6. 277/480V, 4-wire, 800A, w/molded CB & CT compartment

Included in the labor costs for switchboards are:
1. Uncrating
2. Hauling to 200' of loading dock
3. Setting equipment
4. Leveling and shimming
5. Anchoring
6. Cable identification
7. Testing of equipment

Not included in the switchboard price are:
1. Rigging
2. Equipment pads
3. Steel channels embedded or grouted in concrete
4. Special knockouts
5. Transition or Auxiliary sections
6. Instrumentation
7. External control
8. Conduit and wire
9. Conductor terminations

Fig. R16450-101

R262513-10 Aluminum Bus Duct

When taking off bus duct identify the system as either:
1. Aluminum
2. Copper

List straight lengths by type and size

A. Feeder — 800 A

B. Plug in — 800 A

Do not measure through fittings as you would on conduit, since there is no allowance for scrap in bus duct systems.

If upon taking off linear foot quantities of bus duct you find your quantities are not divisible evenly by 10 ft., then the remainder must be priced as a special item and quoted from the manufacturer. Do not use the bare material cost per L.F. for these special items. You can, however, safely use the bare labor cost per L.F. for the entire length.

Identify fittings by type and ampere rating.

Example:

C. Switchboard stub, 800 A

D. Elbows, 800 A

E. End box, 800 A

F. Cable tap box, 800 A

G. Tee Fittings, 800 A

H. Hangers

Plug-in Units – List separately plug-in units and identify by type and ampere rating

I. Plug-in switches 600 Volt 3 phase 60 A

J. Plug-in molded case C.B. 60 A

K. Combination starter FVNR NEMA 1

L. Combination contactor & fused switch NEMA 1

M. Combination fusible switch & lighting control 60 A

Labor-hours for feeder and plug-in sections include:
1. Unloading and uncrating
2. Hauling up to 200 ft. from loading dock
3. Measuring and marking
4. Installing hangers
5. Hanging and bolting sections
6. Aligning and leveling
7. Testing

Labor-hours do not include:
1. Modifications to existing structure for hanger supports
2. Threaded rod in excess of 2 ft.
3. Welding
4. Penetrations through walls
5. Staging rental

Deduct the following percentages from labor only:

| 150 ft. to 250 ft. — 10% |
| 251 ft. to 350 ft. — 15% |
| 351 ft. to 500 ft. — 20% |
| Over 500 ft. — 25% |

Deduct percentage only if runs are contained in the same area.

Example: If the job entails running 100 ft. in 5 different locations do not deduct 20%, but if the duct is being run in 1 area and the quantity is 500 ft. then you would deduct 20%.

Deduct only from straight lengths, not fittings or plug-in units.

R262513-10   Aluminum Bus Duct (cont.)

Add to labor for elevated installations:

10 ft. to 14.5 ft. high	10%
15 ft. to 20 ft. high	20%
21 ft. to 25 ft. high	25%
26 ft. to 30 ft. high	35%
31 ft. to 35 ft. high	40%
36 ft. to 40 ft. high	50%
Over 40 ft. high	55%

Bus Duct Fittings:

Labor-hours for fittings include:
1. Unloading and uncrating
2. Hauling up to 200 ft. from loading dock
3. Installing, fitting, and bolting all ends to in-place sections

Plug-in units include:
1. Unloading and uncrating
2. Hauling up to 200 ft. from loading dock
3. Installing plug-in into in-place duct
4. Connecting load wire to lugs
5. Marking wire
6. Checking phase rotation

Labor-hours for plug-ins do not include:
1. Conduit runs from plug-in
2. Wire from plug-in
3. Conduit termination

Economy of Scale – For large concentrations of plug-in units in the same area deduct the following percentages:

11	to	25		15%
26	to	50		20%
51	to	75		25%
76	to	100		30%
100	and	over		35%

Job Conditions: Productivities are based on new construction in an unobstructed first floor area to a height of 10' in an unobstructed area.

Material staging area is within 100 ft. of work being performed.
Add to the duct fittings and hangers:
  Plug-in switches (fused)
  Plug-in breakers
  Combination starters
  Combination contactors
  Combination fusible switch and lighting control

Labor-hours for duct and fittings include:
1. Unloading and uncrating
2. Hauling up to 200 ft. from loading dock
3. Measuring and marking
4. Installing duct runs
5. Leveling
6. Sound testing

Labor-hours for plug-ins include:
1. Unloading and uncrating
2. Hauling up to 200 ft. from loading dock
3. Installing plug-ins
4. Preparing wire
5. Wire connections and marking

615

R262716-40 Standard Electrical Enclosure Types

NEMA Enclosures

Electrical enclosures serve two basic purposes: they protect people from accidental contact with enclosed electrical devices and connections, and they protect the enclosed devices and connections from specified external conditions. The National Electrical Manufacturers Association (NEMA) has established the following standards. Because these descriptions are not intended to be complete representations of NEMA listings, consultation of NEMA literature is advised for detailed information.

The following definitions and descriptions pertain to NONHAZARDOUS locations.

NEMA   Type 1: General purpose enclosures intended for use indoors, primarily to prevent accidental contact of personnel with the enclosed equipment in areas that do not involve unusual conditions.

NEMA   Type 2: Dripproof indoor enclosures intended to protect the enclosed equipment against dripping noncorrosive liquids and falling dirt.

NEMA   Type 3: Dustproof, raintight and sleet-resistant (ice-resistant) enclosures intended for use outdoors to protect the enclosed equipment against wind-blown dust, rain, sleet, and external ice formation.

NEMA   Type 3R: Rainproof and sleet-resistant (ice-resistant) enclosures intended for use outdoors to protect the enclosed equipment against rain. These enclosures are constructed so that the accumulation and melting of sleet (ice) will not damage the enclosure and its internal mechanisms.

NEMA   Type 3S: Enclosures intended for outdoor use to provide limited protection against wind-blown dust, rain, and sleet (ice) and to allow operation of external mechanisms when ice-laden.

NEMA   Type 4: Watertight and dust-tight enclosures intended for use indoors and out – to protect the enclosed equipment against splashing water, seepage of water, falling or hose-directed water, and severe external condensation.

NEMA   Type 4X: Watertight, dust-tight, and corrosion-resistant indoor and outdoor enclosures featuring the same provisions as Type 4 enclosures, plus corrosion resistance.

NEMA   Type 5: Indoor enclosures intended primarily to provide limited protection against dust and falling dirt.

NEMA   Type 6: Enclosures intended for indoor and outdoor use – primarily to provide limited protection against the entry of water during occasional temporary submersion at a limited depth.

NEMA   Type 6R: Enclosures intended for indoor and outdoor use – primarily to provide limited protection against the entry of water during prolonged submersion at a limited depth.

NEMA   Type 11: Enclosures intended for indoor use – primarily to provide, by means of oil immersion, limited protection to enclosed equipment against the corrosive effects of liquids and gases.

NEMA   Type 12: Dust-tight and driptight indoor enclosures intended for use indoors in industrial locations to protect the enclosed equipment against fibers, flyings, lint, dust, and dirt, as well as light splashing, seepage, dripping, and external condensation of noncorrosive liquids.

NEMA   Type 13: Oil-tight and dust-tight indoor enclosures intended primarily to house pilot devices, such as limit switches, foot switches, push buttons, selector switches, and pilot lights, and to protect these devices against lint and dust, seepage, external condensation, and sprayed water, oil, and noncorrosive coolant.

The following definitions and descriptions pertain to HAZARDOUS, or CLASSIFIED, locations:

NEMA   Type 7: Enclosures intended to use in indoor locations classified as Class 1, Groups A, B, C, or D, as defined in the National Electrical Code.

NEMA   Type 9: Enclosures intended for use in indoor locations classified as Class 2, Groups E, F, or G, as defined in the National Electrical Code.

R262726-90 Wiring Devices

Wiring devices should be priced on a separate takeoff form which includes boxes, covers, conduit and wire.

Labor-hours for devices include:
1. Stripping of wire
2. Attaching wire to device using terminators on the device itself, lugs, set screws etc.
3. Mounting of device in box

Labor-hours do not include:
1. Conduit
2. Wire
3. Boxes
4. Plates

Economy of Scale – for large concentrations of devices in the same area deduct the following percentages from labor-hours:

1	to	10	0%
11	to	25	20%
26	to	50	25%
51	to	100	30%
	over	100	35%

R262726-90   Wiring Devices (cont.)

NEMA No.	15 R	20 R	30 R	50 R	60 R
1 125V 2 Pole, 2 Wire	⊙				
2 250V 2 Pole, 2 Wire		⊙	⊙		
5 125V 2 Pole, 3 Wire	⊙	⊙	⊙	⊙	
6 250V 2 Pole, 3 Wire	⊙	⊙	⊙	⊙	
7 277V, AC 2 Pole, 3 Wire	⊙	⊙	⊙	⊙	
10 125/250V 3 Pole, 3 Wire		⊙	⊙	⊙	
11 3 Phase 250V 3 Pole, 3 Wire	⊙	⊙	⊙	⊙	
14 125/250V 3 Pole, 4 Wire	⊙	⊙	⊙	⊙	⊙
15 3 Phase 250V 3 Pole, 4 Wire	⊙	⊙	⊙	⊙	⊙
18 3 Phase 208Y/120V 4 Pole, 4 Wire	⊙	⊙	⊙	⊙	⊙

R262726-90 Wiring Devices (cont.)

NEMA No.	15 R	20 R	30 R	NEMA No.	15 R	20 R	30 R
L 1 125V 2 Pole, 2 Wire	◯			**L 13** 3 Phase 600V 3 Pole, 3 Wire			◯
L 2 250V 2 Pole, 2 Wire		◯ 15A		**L 14** 125/250V 3 Pole, 4 Wire		◯	◯
L 5 125 V 2 Pole, 3 Wire	◯	◯	◯	**L 15** 3 Phase 250 V 3 Pole, 4 Wire		◯	◯
L 6 250 V 2 Pole, 3 Wire	◯	◯	◯	**L 16** 3 Phase 480V 3 Pole, 4 Wire		◯	◯
L 7 227 V, AC 2 Pole, 3 Wire	◯	◯	◯	**L 17** 3 Phase 600 V 3 Pole, 4 Wire			◯
L 8 480 V 2 Pole, 3 Wire		◯	◯	**L 18** 3 Phase 208Y/120V 4 Pole, 4 Wire		◯	◯
L 9 600 V 2 Pole, 3 Wire		◯	◯	**L 19** 3 Phase 480Y/277V 4 Pole, 4 Wire		◯	◯
L 10 125/250V 3 Pole, 3 Wire		◯	◯	**L 20** 3 Phase 600Y/347V 4 Pole, 4 Wire		◯	◯
L 11 3 Phase 250 V 3 Pole, 3 Wire		◯	◯	**L 21** 3 Phase 208Y/120V 4 Pole, 5 Wire		◯	◯
L 12 3 Phase 480 V 3 Pole, 3 Wire		◯	◯	**L 22** 3 Phase 480Y/277V 4 Pole, 5 Wire		◯	◯
				L 23 3 Phase 600Y/347V 4 Pole, 5 Wire		◯	◯

R262816-80 Safety Switches

List each safety switch by type, ampere rating, voltage, single or three phase, fused or nonfused.

Example:

A. General duty, 240V, 3-pole, fused
B. Heavy Duty, 600V, 3-pole, nonfused
C. Heavy Duty, 240V, 2-pole, fused
D. Heavy Duty, 600V, 3-pole, fused

Also include NEMA enclosure type and identify as:

1. Indoor
2. Weatherproof
3. Explosionproof

Installation of safety switches includes:

1. Unloading and uncrating
2. Handling disconnects up to 200′ from loading dock
3. Measuring and marking location
4. Drilling (4) anchor type lead fasteners using a hammer drill
5. Mounting and leveling safety switch
6. Installing (3) fuses
7. Phasing and tagging line and load wires

Price does not include:

1. Modifications to enclosure
2. Plywood backboard
3. Conduit or wire
4. Fuses
5. Termination of wires

Job Conditions: Productivities are based on new construction to an installed height of 6′ above finished floor.

Material staging area is assumed to be within 100′ of work in progress.

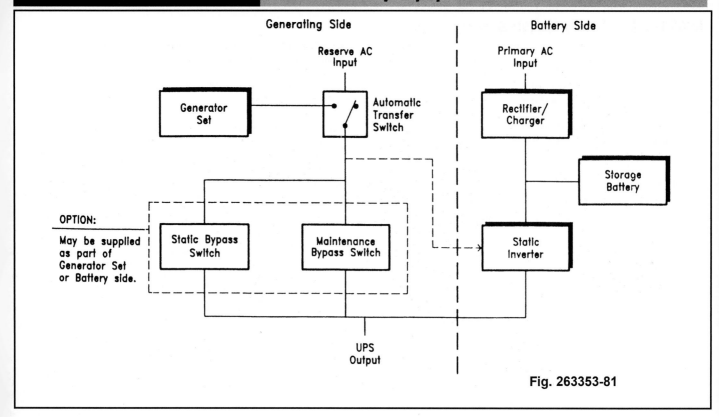

Generating Side   **Battery Side**

Reserve AC Input   Primary AC Input

Fig. 263353-81

R263353-80   Uninterruptible Power Supply Systems

General: Uninterruptible Power Supply (UPS) Systems are used to provide power for legally required standby systems. They are also designed to protect computers and provide optional additional coverage for any or all loads in a facility. Figure R263353-81 shows a typical configuration for a UPS System with a generator set.

Cost Determination:

It is recommended that UPS System material costs be obtained from equipment manufacturers as a package. Installation costs should be obtained from a vendor or contractor.

The recommended procedure for pricing UPS Systems would be to identify:

1. Frequency – by Hertz (Hz.)
   For example: 50, 60, and 400/415 Hz.
2. Apparent power and real power
   For example: 200 kVA/170 kW
3. Input/Output voltage
   For example: 120V, 208V or 240V

4. Phase – either single or three-phase
5. Options and accessories – For extended run time more batteries would be required. Other accessories include battery cabinets, battery racks, remote control panel, power distribution unit (PDU), and warranty enhancement plans.

Note:

1. Larger systems can be configured by paralleling two or more standard small size single modules.
   For example: two 15 kVA modules combined together and configured to 30 kVA.
2. Maximum input current during battery recharge is typically 15% higher than normal current.
3. UPS Systems weights vary depending on options purchased and power rating in kVA.

R263413-30   HP Value/Cost Determination

General: Motors can be powered by any of the seven systems shown in Figure R262213-11 through Figure R262213-17 provided the motor voltage characteristics are compatible with the power system characteristics.

Cost Determination:

Motor Amperes for the various size H.P. and voltage are listed in Table R263413-31. To find the amperes, locate the required H.P. rating and locate the amperes under the appropriate circuit characteristics.

For example:

   A. 100 H.P., 3 phase, 460 volt motor = 124 amperes (Table R263413-31)

   B. 10 H.P., 3 phase, 200 volt motor = 32.2 amperes (Table R263413-31)

Motor Wire Size: After the amperes are found in Table R263413-31 the amperes must be increased 25% to compensate for power losses. Next refer to Table R260519-92. Find the appropriate insulation column for copper or aluminum wire to determine the proper wire size.

For example:

   A. 100 H.P., 3 phase, 460 volt motor has an ampere value of 124 amperes from Table R263413-31

   B. 124A x 1.25 = 155 amperes

   C. Refer to Table R260519-92 for THW or THWN wire insulations to find the proper wire size. For a 155 ampere load using copper wire a size 2/0 wire is needed.

   D. For the 3 phase motor three wires of 2/0 size are required.

Conduit Size: To obtain the proper conduit size for the wires and type of insulation used, refer to Table R260533-22.

For example: For the 100 H.P., 460V, 3 phase motor, it was determined that three 2/0 wires are required. Assuming THWN insulated copper wire, use Table R260533-22 to determine that three 2/0 wires require 1-1/2″ conduit.

Material Cost of the conduit and wire system depends on:

  1. Wire size required

  2. Copper or aluminum wire

  3. Wire insulation type selected

  4. Steel or plastic conduit

  5. Type of conduit raceway selected.

Labor Cost of the conduit and wire system depends on:

  1. Type and size of conduit

  2. Type and size of wires installed

  3. Location and height of installation in building or depth of trench

  4. Support system for conduit.

Magnetic starters, switches, and motor connection:

To complete the cost picture from H.P. to Costs additional items must be added to the cost of the conduit and wire system to arrive at a total cost.

  1. Assembly Table D5020-160 Magnetic Starters Installed Cost lists the various size starters for single phase and three phase motors.

  2. Assembly Table D5020-165 Heavy Duty Safety Switches Installed Cost lists safety switches required at the beginning of a motor circuit and also one required in the vicinity of the motor location.

  3. Assembly Table D5020-170 Motor Connection lists the various costs for single and three phase motors.

Worksheet to obtain total motor wiring costs:

It is assumed that the motors or motor driven equipment are furnished and installed under other sections for this estimate and the following work is done under this section:

  1. Conduit

  2. Wire (add 10% for additional wire beyond conduit ends for connections to switches, boxes, starters, etc.)

  3. Starters

  4. Safety switches

  5. Motor connections

621

R263413-31 Ampere Values Determined by Horsepower, Voltage and Phase Values

H.P.	Amperes							
	Single Phase			Three Phase				
	115V	208V	230V	200V	208V	230V	460V	575V
1/6	4.4A	2.4A	2.2A					
1/4	5.8	3.2	2.9					
1/3	7.2	4.0	3.6					
1/2	9.8	5.4	4.9	2.5A	2.4A	2.2A	1.1A	0.9A
3/4	13.8	7.6	6.9	3.7	3.5	3.2	1.6	1.3
1	16	8.8	8	4.8	4.6	4.2	2.1	1.7
1-1/2	20	11	10	6.9	6.6	6.0	3.0	2.4
2	24	13.2	12	7.8	7.5	6.8	3.4	2.7
3	34	18.7	17	11.0	10.6	9.6	4.8	3.9
5	56	30.8	28	17.5	16.7	15.2	7.6	6.1
7-1/2	80	44	40	25.3	24.2	22	11	9
10	100	55	50	32.2	30.8	28	14	11
15				48.3	46.2	42	21	17
20				62.1	59.4	54	27	22
25				78.2	74.8	68	34	27
30				92	88	80	40	32
40				120	114	104	52	41
50				150	143	130	65	52
60				177	169	154	77	62
75				221	211	192	96	77
100				285	273	248	124	99
125				359	343	312	156	125
150				414	396	360	180	144
200				552	528	480	240	192
250							302	242
300							361	289
350							414	336
400							477	382

Figure   R263413-32

Worksheet for Motor Circuits					
				Cost	
Item	Type	Size	Quantity	Unit	Total
Wire					
Conduit					
Switch					
Starter					
Switch					
Motor Connection					
Other					
Total Cost					

R263413-33   Maximum Horsepower for Starter Size by Voltage

Starter	Maximum HP (3φ)			
Size	208V	240V	480V	600V
00	1½	1½	2	2
0	3	3	5	5
1	7½	7½	10	10
2	10	15	25	25
3	25	30	50	50
4	40	50	100	100
5		100	200	200
6		200	300	300
7		300	600	600
8		450	900	900
8L		700	1500	1500

R263623-60   Automatic Transfer Switches

When taking off automatic transfer switches identify by voltage, amperage and number of poles.

Example: Automatic transfer switch, 480V, 3 phase, 30A, NEMA 1 enclosure

Labor-hours for transfer switches include:
1. Unloading, uncrating and handling switches up to 200' from loading dock
2. Measuring and marking
3. Drilling (4) lead type anchors using a hammer drill
4. Mounting and leveling
5. Circuit identification
6. Testing and load balancing

Labor-hours do not include:
1. Modifications in enclosure
2. Structural supports
3. Additional lugs
4. Plywood backboard
5. Painting or lettering
6. Conduit runs to or from transfer switch
7. Wire
8. Termination of wires

R265113-40   Interior Lighting Fixtures

When taking off interior lighting fixtures, it is advisable to set up your quantity worksheet to conform to the lighting schedule as it appears on the print. Include the alpha-numeric code plus the symbol on your worksheet.

Take off a particular section or floor of the building and count each type of fixture before going on to another type. It would also be advantageous to include on the same worksheet the pipe, wire, fittings and circuit number associated with each type of lighting fixture. This will help you identify the costs associated with any particular lighting system and in turn make material purchases more specific as to when and how much to order under the classification of lighting.

By taking off lighting first you can get a complete "walk through" of the job. This will become helpful when doing other phases of the project.

Materials for a recessed fixture include:
1. Fixture
2. Lamps
3. 6' of jack chain
4. (2) S hooks
5. (2) Wire nuts

Labor for interior recessed fixtures include:
1. Unloading by hand
2. Hauling by hand to an area up to 200' from loading dock
3. Uncrating
4. Layout
5. Installing fixture
6. Attaching jack chain & S hooks
7. Connecting circuit power
8. Reassembling fixture
9. Installing lamps
10. Testing

Material for surface mounted fixtures include:
1. Fixture
2. Lamps
3. Either (4) lead type anchors, (4) toggle bolts, or (4) ceiling grid clips
4. (2) Wire nuts

Material for pendent mounted fixtures include:
1. Fixture
2. Lamps
3. (2) Wire nuts
4. Rigid pendents as required by type of fixtures

5. Canopies as required by type of fixture

Labor hours include the following for both surface and pendent fixtures:
1. Unloading by hand
2. Hauling by hand to an area up to 200' from loading dock
3. Uncrating
4. Layout and marking
5. Drilling (4) holes for either lead anchors or toggle bolts using a hammer drill
6. Installing fixture
7. Leveling fixture
8. Connecting circuit power
9. Installing lamps
10. Testing

Labor for surface or pendent fixtures does not include:
1. Conduit
2. Boxes or covers
3. Connectors
4. Fixture whips
5. Special support
6. Switching
7. Wire

Economy of Scale: For large concentrations of lighting fixtures in the same area deduct the following percentages from labor:

25	to	50	fixtures 15%
51	to	75	fixtures 20%
76	to	100	fixtures 25%
101 and over			30%

Job Conditions: Productivity is based on new construction in an unobstructed first floor location, to 10' high.

Material staging is assumed to be within 100' of work being performed.

Add the following percentages to labor for elevated installations:

10'	to	14.5'	high	10%
15'	to	20'	high	20%
21'	to	25'	high	25%
26'	to	30'	high	35%
31'	to	35'	high	40%
36'	to	40'	high	50%
41' and over				55%

R265723-10　For Other than Regular Cool White (CW) Lamps

	Multiply Material Costs as Follows:				
Regular Lamps	Cool white deluxe (CWX)	x 1.35	Energy Saving Lamps	Cool white (CW/ES)	x 1.35
	Warm white deluxe (WWX)	x 1.35		Cool white deluxe (CWX/ES)	x 1.65
	Warm white (WW)	x 1.30		Warm white (WW/ES)	x 1.55
	Natural (N)	x 2.05		Warm white deluxe (WWX/ES)	x 1.65

R271323-40 Fiber Optics

Fiber optic systems use optical fiber such as plastic, glass, or fused silica, a transparent material, to transmit radiant power (i.e., light) for control, communication, and signaling applications. The types of fiber optic cables can be nonconductive, conductive, or composite. The composite cables contain fiber optics and current-carrying electrical conductors. The configuration for one of the fiber optic systems is as follows:

The transceiver module acts as transmitting and receiving equipment in a common house, which converts electrical energy to light energy or vice versa.

Pricing the fiber optic system is not an easy task. The performance of the whole system will affect the cost significantly. New specialized tools and techniques decrease the installing cost tremendously. In the fiber optic section of Means Electrical Cost Data, a benchmark for labor-hours and material costs is set up so that users can adjust their costs according to unique project conditions.

Units for Measure: Fiber optic cable is measured in hundred linear feet (C.L.F.) or industry units of measure - meter (m) or kilometer (km). The connectors are counted as units (EA.)

Material Units: Generally, the material costs include only the cable. All the accessories shall be priced separately.

Labor Units: The following procedures are generally included for the installation of fiber optic cables:

- Receiving
- Material handling
- Setting up pulling equipment
- Measuring and cutting cable
- Pulling cable

These additional items are listed and extended: Terminations

Takeoff Procedure: Cable should be taken off by type, size, number of fibers, and number of terminations required. List the lengths of each type of cable on the takeoff sheets. Total and add 10% for waste. Transfer the figures to a cost analysis sheet and extend.

R271513-75 High Performance Cable

There are several categories used to describe high performance cable. The following information includes a description of categories CAT 3, 5, 5e, 6, and 7, and details classifications of frequency and specific standards. The category standards have evolved under the sponsorship of organizations such as the Telecommunication Industry Association (TIA), the Electronic Industries Alliance (EIA), the American National Standards Institute (ANSI), the International Organization for Standardization (ISO), and the International Electrotechnical Commission (IEC), all of which have catered to the increasing complexities of modern network technology. For network cabling, users must comply with national or international standards. A breakdown of these categories is as follows:

Category 3: Designed to handle frequencies up to 16 MHz.

Category 5: (TIA/EIA 568A) Designed to handle frequencies up to 100 MHz.

Category 5e: Additional transmission performance to exceed Category 5.

Category 6: Development by TIA and other international groups to handle frequencies of 250 MHz.

Category 7: Development to handle a frequency range up to 600 MHz.

R337116-60 Average Transmission Line Material Requirements (Per Mile)

Conductor: 397,500 – Cir. Mil., 26/7-ACSR

Terrain:		Flat				Rolling				Mountain			
Item		69kV	161kV	161kV	500kV	69kV	161kV	161kV	500kV	69kV	161kV	161kV	500kV
Pole Type	Unit	Wood	Wood	Steel	Steel	Wood	Wood	Steel	Steel	Wood	Wood	Steel	Steel
Structures	Ea.	12[1]				9[2]				7[2]			
Poles	Ea.	12				18				14			
Crossarms	Ea.	24[3]				9[4]				7[4]			
Conductor	Ft.	15,990				15,990				15,990			
Insulators[5]	Ea.	180				135				105			
Ground Wire	Ft.	5,330				10,660				10,660			

Conductor: 636,000 – Cir. Mil., 26/7-ACSR

Item	Unit	69kV Wood	161kV Wood	161kV Steel	500kV Steel	69kV Wood	161kV Wood	161kV Steel	500kV Steel	69kV Wood	161kV Wood	161kV Steel	500kV Steel
Structures	Ea.	13[1]	11[6]			10[2]	9[2]	6[9]		8[2]	8[2]	6[9]	
Excavation	C.Y.	–	–			–	–	120		–	–	120	
Concrete	C.Y.	–	–			–	–	10		–	–	10	
Steel Towers	Tons	–	–			–	–	32		–	–	32	
Poles	Ea.	13	11			20	18	–		16	16	–	
Crossarms	Ea.	26[3]	33[7]			10[4]	9[8]	–		8[4]	8[8]	–	
Conductor	Ft.	15,990	15,990			15,990	15,990	15,990		15,990	15,990	15,990	
Insulators[5]	Ea.	195	165			150	297	297		120	264	297	
Ground Wire	Ft.	5330	5330			10,660	10,660	10,660		10,660	10,660	10,660	

Conductor: 954,000 – Cir. Mil., 45/7-ACSR

Item	Unit	69kV Wood	161kV Wood	161kV Steel	500kV Steel	69kV Wood	161kV Wood	161kV Steel	500kV Steel	69kV Wood	161kV Wood	161kV Steel	500kV Steel
Structures	Ea.	14[1]	12[6]		4[11]	10[2]	9[2]	6[9]	4[11]	8[2]	8[2]	6[9]	4[13]
Excavation	C.Y.	–	–		200	–	–	125	214	–	–	125	233
Concrete	C.Y.	–	–		20	–	–	10	21	–	–	10	21
Steel Towers	Tons	–	–		57	–	–	33	57	–	–	33	63
Poles	Ea.	14	12		–	20	18	–	–	16	16	–	–
Crossarms	Ea.	28[10]	36[7]		–	10[4]	9[8]	–	–	8[4]	8[8]	–	–
Conductor	Ft.	15,990	15,990		47,970[12]	15,990	15,990	15,990	47,970[12]	15,990	15,990	15,990	47,970[12]
Insulators[5]	Ea.	210	180		288	150	297	297	288	120	264	297	576
Ground Wire	Ft.	5330	5330		10,660	10,660	10,660	10,660	10,660	10,660	10,660	10,660	10,660

Conductor: 1,351,500 – Cir. Mil., 45/7-ACSR

Item	Unit	Flat	Rolling
Structures	Ea.	8[14]	8[14]
Excavation	C.Y.	220	220
Concrete	C.Y.	28	28
Steel Towers	Ton	46	46
Conductor	Ft.	31,680[15]	31,680[15]
Insulators	Ea.	528	528
Ground Wire	Ft.	10,660	10,660

1. Single pole two-arm suspension type construction
2. Two-pole wood H-frame construction
3. 4¾" x 5¾" x 8' and 4¾" x 5¾" x 10' wood crossarms
4. 6" x 8" x 26'-0" wood crossarm
5. 5¾" x 10" disc insulator
6. Single pole construction with 3 fiberglass crossarms (5 fog-type insulators per phase)
7. 7'-0" fiberglass crossarms
8. 6" x 10" x 35'-0" wood crossarm
9. Laced steel tower, single circuit construction
10. 5" x 7" x 8' and 5" x 7" x 10' wood crossarms
11. Laced steel tower, single circuit 500-kV construction
12. Bundled conductor (3 sub-conductors per phase)
13. Laced steel tower, single circuit restrained phases (500-kV)
14. Laced steel tower, double circuit construction
15. Both sides of double circuit strung

Note: To allow for sagging, a mile (5280 Ft.) of transmission line uses 5330 Ft. of conductor per wire (called a wire mile).

R337119-30 Concrete for Conduit Encasement

The table below lists C.Y. of concrete for 100 L.F. of trench. Conduits separation center to center should meet 7.5″ (N.E.C.).

Number of Conduits	1	2	3	4	6	8	9	Number of Conduits
Trench Dimension	11.5″ x 11.5″	11.5″ x 19″	11.5″ x 27″	19″ x 19″	19″ x 27″	19″ x 38″	27″ x 27″	Trench Dimension
Conduit Diameter 2.0″	3.29	5.39	7.64	8.83	12.51	17.66	17.72	Conduit Diameter 2.0″
2.5″	3.23	5.29	7.49	8.62	12.19	17.23	17.25	2.5″
3.0″	3.15	5.13	7.24	8.29	11.71	16.59	16.52	3.0″
3.5″	3.08	4.97	7.02	7.99	11.26	15.98	15.84	3.5″
4.0″	2.99	4.80	6.76	7.65	10.74	15.30	15.07	4.0″
5.0″	2.78	4.37	6.11	6.78	9.44	13.57	13.12	5.0″
6.0″	2.52	3.84	5.33	5.74	7.87	11.48	10.77	6.0″

For customer support on your Electrical Costs with RSMeans data, call 800.448.8182.

Change Orders

Change Order Considerations

A change order is a written document usually prepared by the design professional and signed by the owner, the architect/engineer, and the contractor. A change order states the agreement of the parties to: an addition, deletion, or revision in the work; an adjustment in the contract sum, if any; or an adjustment in the contract time, if any. Change orders, or "extras", in the construction process occur after execution of the construction contract and impact architects/ engineers, contractors, and owners.

Change orders that are properly recognized and managed can ensure orderly, professional, and profitable progress for everyone involved in the project. There are many causes for change orders and change order requests. In all cases, change orders or change order requests should be addressed promptly and in a precise and prescribed manner. The following paragraphs include information regarding change order pricing and procedures.

The Causes of Change Orders

Reasons for issuing change orders include:

- Unforeseen field conditions that require a change in the work
- Correction of design discrepancies, errors, or omissions in the contract documents
- Owner-requested changes, either by design criteria, scope of work, or project objectives
- Completion date changes for reasons unrelated to the construction process
- Changes in building code interpretations, or other public authority requirements that require a change in the work
- Changes in availability of existing or new materials and products

Procedures

Properly written contract documents must include the correct change order procedures for all parties—owners, design professionals, and contractors—to follow in order to avoid costly delays and litigation.

Being "in the right" is not always a sufficient or acceptable defense. The contract provisions requiring notification and documentation must be adhered to within a defined or reasonable time frame.

The appropriate method of handling change orders is by a written proposal and acceptance by all parties involved. Prior to starting work on a project, all parties should identify their authorized agents who may sign and accept change orders, as well as any limits placed on their authority.

Time may be a critical factor when the need for a change arises. For such cases, the contractor might be directed to proceed on a "time and materials" basis, rather than wait for all paperwork to be processed—a delay that could impede progress. In this situation, the contractor must still follow the prescribed change order procedures including, but not limited to, notification and documentation.

Lack of documentation can be very costly, especially if legal judgments are to be made, and if certain field personnel are no longer available. For time and material change orders, the contractor should keep accurate daily records of all labor and material allocated to the change.

Owners or awarding authorities who do considerable and continual building construction (such as the federal government) realize the inevitability of change orders for numerous reasons, both predictable and unpredictable. As a result, the federal government, the American Institute of Architects (AIA), the Engineers Joint Contract Documents Committee (EJCDC), and other contractor, legal, and technical organizations have developed standards and procedures to be followed by all parties to achieve contract continuance and timely completion, while being financially fair to all concerned.

Pricing Change Orders

When pricing change orders, regardless of their cause, the most significant factor is when the change occurs. The need for a change may be perceived in the field or requested by the architect/engineer *before* any of the actual installation has begun, or may evolve or appear *during* construction when the item of work in question is partially installed. In the latter cases, the original sequence of construction is disrupted, along with all contiguous and supporting systems. Change orders cause the greatest impact when they occur *after* the installation has been completed and must be uncovered, or even replaced. Post-completion changes may be caused by necessary design changes, product failure, or changes in the owner's requirements that are not discovered until the building or the systems begin to function.

Specified procedures of notification and record keeping must be adhered to and enforced regardless of the stage of construction: *before, during,* or *after* installation. Some bidding documents anticipate change orders by requiring that unit prices including overhead and profit percentages—for additional as well as deductible changes—be listed. Generally these unit prices do not fully take into account the ripple effect, or impact on other trades, and should be used for general guidance only.

When pricing change orders, it is important to classify the time frame in which the change occurs. There are two basic time frames for change orders: *pre-installation change orders,* which occur before the start of construction, and *post-installation change orders,* which involve reworking after the original installation. Change orders that occur between these stages may be priced according to the extent of work completed using a combination of techniques developed for pricing *pre-* and *post-installation* changes.

Factors To Consider When Pricing Change Orders

As an estimator begins to prepare a change order, the following questions should be reviewed to determine their impact on the final price.

General

- *Is the change order work* pre-installation *or* post-installation?

  Change order work costs vary according to how much of the installation has been completed. Once workers have the project scoped in their minds, even though they have not started, it can be difficult to refocus. Consequently they may spend more than the normal amount of time understanding the change. Also, modifications to work in place, such as trimming or refitting, usually take more time than was initially estimated. The greater the amount of work in place, the more reluctant workers are to change it. Psychologically they may resent the change and as a result the rework takes longer than normal. Post-installation change order estimates must include demolition of existing work as required to accomplish the change. If the work is performed at a later time, additional obstacles, such as building finishes, may be present which must be protected. Regardless of whether the change occurs

For customer support on your Electrical Costs with RSMeans data, call 800.448.8182.

629

pre-installation or post-installation, attempt to isolate the identifiable factors and price them separately. For example, add shipping costs that may be required pre-installation or any demolition required post-installation. Then analyze the potential impact on productivity of psychological and/or learning curve factors and adjust the output rates accordingly. One approach is to break down the typical workday into segments and quantify the impact on each segment.

Change Order Installation Efficiency

The labor-hours expressed (for new construction) are based on average installation time, using an efficiency level. For change order situations, adjustments to this efficiency level should reflect the daily labor-hour allocation for that particular occurrence.

- *Will the change substantially delay the original completion date?*

A significant change in the project may cause the original completion date to be extended. The extended schedule may subject the contractor to new wage rates dictated by relevant labor contracts. Project supervision and other project overhead must also be extended beyond the original completion date. The schedule extension may also put installation into a new weather season. For example, underground piping scheduled for October installation was delayed until January. As a result, frost penetrated the trench area, thereby changing the degree of difficulty of the task. Changes and delays may have a ripple effect throughout the project. This effect must be analyzed and negotiated with the owner.

- *What is the net effect of a deduct change order?*

In most cases, change orders resulting in a deduction or credit reflect only bare costs. The contractor may retain the overhead and profit based on the original bid.

Materials

- *Will you have to pay more or less for the new material, required by the change order, than you paid for the original purchase?*

The same material prices or discounts will usually apply to materials purchased for change orders as new construction. In some instances, however, the contractor may forfeit the advantages of competitive pricing for change orders. Consider the following example:

A contractor purchased over $20,000 worth of fan coil units for an installation and obtained the maximum discount. Some time later it was determined the project required an additional matching unit. The contractor has to purchase this unit from the original supplier to ensure a match. The supplier at this time may not discount the unit because of the small quantity, and he is no longer in a competitive situation. The impact of quantity on purchase can add between 0% and 25% to material prices and/or subcontractor quotes.

- *If materials have been ordered or delivered to the job site, will they be subject to a cancellation charge or restocking fee?*

Check with the supplier to determine if ordered materials are subject to a cancellation charge. Delivered materials not used as a result of a change order may be subject to a restocking fee if returned to the supplier. Common restocking charges run between 20% and 40%. Also, delivery charges to return the goods to the supplier must be added.

Labor

- *How efficient is the existing crew at the actual installation?*

Is the same crew that performed the initial work going to do the change order? Possibly the change consists of the installation of a unit identical to one already installed; therefore, the change should take less time. Be sure to consider this potential productivity increase and modify the productivity rates accordingly.

- *If the crew size is increased, what impact will that have on supervision requirements?*

Under most bargaining agreements or management practices, there is a point at which a working foreman is replaced by a nonworking foreman. This replacement increases project overhead by adding a nonproductive worker. If additional workers are added to accelerate the project or to perform changes while maintaining the schedule, be sure to add additional supervision time if warranted. Calculate the hours involved and the additional cost directly if possible.

- *What are the other impacts of increased crew size?*

The larger the crew, the greater the potential for productivity to decrease. Some of the factors that cause this productivity loss are: overcrowding (producing restrictive conditions in the working space) and possibly a shortage of any special tools and equipment required. Such factors affect not only the crew working on the elements directly involved in the change order, but other crews whose movements may also be hampered. As the crew increases, check its basic composition for changes by the addition or deletion of apprentices or nonworking foreman, and quantify the potential effects of equipment shortages or other logistical factors.

- *As new crews, unfamiliar with the project, are brought onto the site, how long will it take them to become oriented to the project requirements?*

The orientation time for a new crew to become 100% effective varies with the site and type of project. Orientation is easiest at a new construction site and most difficult at existing, very restrictive renovation sites. The type of work also affects orientation time. When all elements of the work are exposed, such as concrete or masonry work, orientation is decreased. When the work is concealed or less visible, such as existing electrical systems, orientation takes longer. Usually orientation can be accomplished in one day or less. Costs for added orientation should be itemized and added to the total estimated cost.

- *How much actual production can be gained by working overtime?*

Short term overtime can be used effectively to accomplish more work in a day. However, as overtime is scheduled to run beyond several weeks, studies have shown marked decreases in output. The following chart shows the effect of long term overtime on worker efficiency. If the anticipated change requires extended overtime to keep the job on schedule, these factors can be used as a guide to predict the impact on time and cost. Add project overhead, particularly supervision, that may also be incurred.

For customer support on your Electrical Costs with RSMeans data, call 800.448.8182.

Days per Week	Hours per Day	Production Efficiency					Payroll Cost Factors	
		1st Week	2nd Week	3rd Week	4th Week	Average 4 Weeks	@ 1-1/2 Times	@ 2 Times
5	8	100%	100%	100%	100%	100%	100%	100%
	9	100	100	95	90	96.25	105.6	111.1
	10	100	95	90	85	92.50	110.0	120.0
	11	95	90	75	65	81.25	113.6	127.3
	12	90	85	70	60	76.25	116.7	133.3
6	8	100	100	95	90	96.25	108.3	116.7
	9	100	95	90	85	92.50	113.0	125.9
	10	95	90	85	80	87.50	116.7	133.3
	11	95	85	70	65	78.75	119.7	139.4
	12	90	80	65	60	73.75	122.2	144.4
7	8	100	95	85	75	88.75	114.3	128.6
	9	95	90	80	70	83.75	118.3	136.5
	10	90	85	75	65	78.75	121.4	142.9
	11	85	80	65	60	72.50	124.0	148.1
	12	85	75	60	55	68.75	126.2	152.4

Effects of Overtime

Caution: Under many labor agreements, Sundays and holidays are paid at a higher premium than the normal overtime rate.

The use of long-term overtime is counterproductive on almost any construction job; that is, the longer the period of overtime, the lower the actual production rate. Numerous studies have been conducted, and while they have resulted in slightly different numbers, all reach the same conclusion. The figure above tabulates the effects of overtime work on efficiency.

As illustrated, there can be a difference between the *actual* payroll cost per hour and the *effective* cost per hour for overtime work. This is due to the reduced production efficiency with the increase in weekly hours beyond 40. This difference between actual and effective cost results from overtime work over a prolonged period. Short-term overtime work does not result in as great a reduction in efficiency and, in such cases, effective cost may not vary significantly from the actual payroll cost. As the total hours per week are increased on a regular basis, more time is lost due to fatigue, lowered morale, and an increased accident rate.

As an example, assume a project where workers are working 6 days a week, 10 hours per day. From the figure above (based on productivity studies), the average effective productive hours over a 4-week period are:

$$0.875 \times 60 = 52.5$$

Depending upon the locale and day of week, overtime hours may be paid at time and a half or double time. For time and a half, the overall (average) *actual* payroll cost (including regular and overtime hours) is determined as follows:

$$\frac{40 \text{ reg. hrs.} + (20 \text{ overtime hrs.} \times 1.5)}{60 \text{ hrs.}} = 1.167$$

Based on 60 hours, the payroll cost per hour will be 116.7% of the normal rate at 40 hours per week. However, because the effective production (efficiency) for 60 hours is reduced to the equivalent of 52.5 hours, the effective cost of overtime is calculated as follows:

For time and a half:

$$\frac{40 \text{ reg. hrs.} + (20 \text{ overtime hrs.} \times 1.5)}{52.5 \text{ hrs.}} = 1.33$$

The installed cost will be 133% of the normal rate (for labor).

Thus, when figuring overtime, the actual cost per unit of work will be higher than the apparent overtime payroll dollar increase, due to the reduced productivity of the longer work week. These efficiency calculations are true only for those cost factors determined by hours worked. Costs that are applied weekly or monthly, such as equipment rentals, will not be similarly affected.

Equipment

- *What equipment is required to complete the change order?*

Change orders may require extending the rental period of equipment already on the job site, or the addition of special equipment brought in to accomplish the change work. In either case, the additional rental charges and operator labor charges must be added.

Summary

The preceding considerations and others you deem appropriate should be analyzed and applied to a change order estimate. The impact of each should be quantified and listed on the estimate to form an audit trail.

Change orders that are properly identified, documented, and managed help to ensure the orderly, professional, and profitable progress of the work. They also minimize potential claims or disputes at the end of the project.

Back by customer demand!

You asked and we listened. For customer convenience and estimating ease, we have made the 2019 Project Costs available for download at **RSMeans.com/2019books**. You will also find sample estimates, an RSMeans data overview video, and a book registration form to receive quarterly data updates throughout 2019.

Estimating Tips

- The cost figures available in the download were derived from hundreds of projects contained in the RSMeans database of completed construction projects. They include the contractor's overhead and profit. The figures have been adjusted to January of the current year.

- These projects were located throughout the U.S. and reflect a tremendous variation in square foot (S.F.) costs. This is due to differences, not only in labor and material costs, but also in individual owners' requirements. For instance, a bank in a large city would have different features than one in a rural area. This is true of all the different types of buildings analyzed. Therefore, caution should be exercised when using these Project Costs. For example, for courthouses, costs in the database are local courthouse costs and will not apply to the larger, more elaborate federal courthouses.

- None of the figures "go with" any others. All individual cost items were computed and tabulated separately. Thus, the sum of the median figures for plumbing, HVAC, and electrical will not normally total up to the total mechanical and electrical costs arrived at by separate analysis and tabulation of the projects.

- Each building was analyzed as to total and component costs and percentages. The figures were arranged in ascending order with the results tabulated as shown. The 1/4 column shows that 25% of the projects had lower costs and 75% had higher. The 3/4 column shows that 75% of the projects had lower costs and 25% had higher. The median column shows that 50% of the projects had lower costs and 50% had higher.

- Project Costs are useful in the conceptual stage when no details are available. As soon as details become available in the project design, the square foot approach should be discontinued and the project should be priced as to its particular components. When more precision is required, or for estimating the replacement cost of specific buildings, the current edition of *Square Foot Costs with RSMeans data* should be used.

- In using the figures in this section, it is recommended that the median column be used for preliminary figures if no additional information is available. The median figures, when multiplied by the total city construction cost index figures (see City Cost Indexes) and then multiplied by the project size modifier at the end of this section, should present a fairly accurate base figure, which would then have to be adjusted in view of the estimator's experience, local economic conditions, code requirements, and the owner's particular requirements. There is no need to factor in the percentage figures, as these should remain constant from city to city.

- The editors of this data would greatly appreciate receiving cost figures on one or more of your recent projects, which would then be included in the averages for next year. All cost figures received will be kept confidential, except that they will be averaged with other similar projects to arrive at square foot cost figures for next year.

See the website above for details and the discount available for submitting one or more of your projects.

Did you know?

RSMeans data is available through our online application:

- Search for costs by keyword
- Leverage the most up-to-date data
- Build and export estimates

Try it free
rsmeans.com/2019freetrial

---

50 17 00 | Project Costs

			UNIT	UNIT COSTS			% OF TOTAL			
				1/4	MEDIAN	3/4	1/4	MEDIAN	3/4	
01	0000	**Auto Sales with Repair**	S.F.							**01**
	0100	Architectural		104	116	126	59.8%	59.2%	63.7%	
	0200	Plumbing		8.70	9.10	12.15	5%	5%	5%	
	0300	Mechanical		11.65	15.60	17.20	6.7%	6.7%	8.6%	
	0400	Electrical		17.90	22	27.50	10.3%	10.3%	12.1%	
	0500	Total Project Costs	↓	174	182	187				
02	0000	**Banking Institutions**	S.F.							**02**
	0100	Architectural		157	192	234	60.4%	60.4%	65.5%	
	0200	Plumbing		6.30	8.80	12.25	2.4%	2.4%	3%	
	0300	Mechanical		12.55	17.35	20.50	4.8%	4.8%	5.9%	
	0400	Electrical		30.50	37	57	11.7%	11.7%	12.6%	
	0500	Total Project Costs	↓	260	293	360				
03	0000	**Court House**	S.F.							**03**
	0100	Architectural		82.50	162	162	54.6%	54.5%	58.3%	
	0200	Plumbing		3.12	3.12	3.12	2.1%	2.1%	1.1%	
	0300	Mechanical		19.50	19.50	19.50	12.9%	12.9%	7%	
	0400	Electrical		25	25	25	16.6%	16.6%	9%	
	0500	Total Project Costs	↓	151	278	278				
04	0000	**Data Centers**	S.F.							**04**
	0100	Architectural		187	187	187	68%	67.9%	68%	
	0200	Plumbing		10.20	10.20	10.20	3.7%	3.7%	3.7%	
	0300	Mechanical		26	26	26	9.5%	9.4%	9.5%	
	0400	Electrical		24.50	24.50	24.50	8.9%	9%	8.9%	
	0500	Total Project Costs	↓	275	275	275				
05	0000	**Detention Centers**	S.F.							**05**
	0100	Architectural		173	183	194	59.2%	59.2%	59%	
	0200	Plumbing		18.25	22	27	6.3%	6.2%	7.1%	
	0300	Mechanical		23	33	39.50	7.9%	7.9%	10.6%	
	0400	Electrical		38	45	58.50	13%	13%	14.5%	
	0500	Total Project Costs	↓	292	310	365				
06	0000	**Fire Stations**	S.F.							**06**
	0100	Architectural		95	121	171	47%	50%	52.4%	
	0200	Plumbing		9.95	13.50	15.65	4.9%	4.9%	5.8%	
	0300	Mechanical		13.45	18.40	25.50	6.7%	6.6%	8%	
	0400	Electrical		22.50	28.50	32.50	11.1%	11%	12.3%	
	0500	Total Project Costs	↓	202	231	300				
07	0000	**Gymnasium**	S.F.							**07**
	0100	Architectural		86.50	114	114	64.6%	64.4%	57.3%	
	0200	Plumbing		2.12	6.95	6.95	1.6%	1.6%	3.5%	
	0300	Mechanical		3.25	29	29	2.4%	2.4%	14.6%	
	0400	Electrical		10.65	20.50	20.50	7.9%	7.9%	10.3%	
	0500	Total Project Costs	↓	134	199	199				
08	0000	**Hospitals**	S.F.							**08**
	0100	Architectural		105	172	187	42.9%	43%	47.1%	
	0200	Plumbing		7.70	14.70	32	3.1%	3.1%	4%	
	0300	Mechanical		51	57.50	74.50	20.8%	20.8%	15.8%	
	0400	Electrical		23	46.50	60.50	9.4%	9.5%	12.7%	
	0500	Total Project Costs	↓	245	365	395				
09	0000	**Industrial Buildings**	S.F.							**09**
	0100	Architectural		44	70	227	56.4%	56.3%	68.6%	
	0200	Plumbing		1.70	6.40	12.95	2.2%	2.2%	6.3%	
	0300	Mechanical		4.71	8.95	42.50	6%	6%	8.8%	
	0400	Electrical		7.20	8.20	68.50	9.2%	9.2%	8%	
	0500	Total Project Costs	↓	78	102	425				
10	0000	**Medical Clinics & Offices**	S.F.							**10**
	0100	Architectural		87.50	120	158	53.7%	51%	56.3%	
	0200	Plumbing		8.50	12.85	20.50	5.2%	5.2%	6%	
	0300	Mechanical		14.20	22	45	8.7%	8.8%	10.3%	
	0400	Electrical		19.55	26	36.50	12%	11.9%	12.2%	
	0500	Total Project Costs	↓	163	213	286				

50 17 00 | Project Costs

			UNIT	UNIT COSTS			% OF TOTAL			
				1/4	MEDIAN	3/4	1/4	MEDIAN	3/4	
11	0000	**Mixed Use**	S.F.							11
	0100	Architectural		86.50	126	207	47%	47.5%	59.4%	
	0200	Plumbing		6	9.15	11.55	3.3%	3.2%	4.3%	
	0300	Mechanical		14.80	24	46.50	8%	7.8%	11.3%	
	0400	Electrical		15.25	24	40.50	8.3%	8.3%	11.3%	
	0500	Total Project Costs		184	212	335				
12	0000	**Multi-Family Housing**	S.F.							12
	0100	Architectural		75	125	167	60.5%	67.6%	56.6%	
	0200	Plumbing		6.40	12.75	15.10	5.2%	5.6%	5.8%	
	0300	Mechanical		6.95	11.70	37.50	5.6%	6.3%	5.3%	
	0400	Electrical		10.10	18	22.50	8.1%	8.8%	8.1%	
	0500	Total Project Costs		124	221	282				
13	0000	**Nursing Home & Assisted Living**	S.F.							13
	0100	Architectural		70	92	116	58.3%	58.4%	59%	
	0200	Plumbing		7.55	11.35	12.50	6.3%	5.9%	7.3%	
	0300	Mechanical		6.20	9.15	17.95	5.2%	5.2%	5.9%	
	0400	Electrical		10.25	16.20	22.50	8.5%	8.6%	10.4%	
	0500	Total Project Costs		120	156	188				
14	0000	**Office Buildings**	S.F.							14
	0100	Architectural		92.50	126	177	60.1%	60%	64.6%	
	0200	Plumbing		4.98	7.85	15.40	3.2%	3.1%	4%	
	0300	Mechanical		10.75	17.65	25.50	7%	6.8%	9.1%	
	0400	Electrical		12.35	21	34	8%	7.9%	10.8%	
	0500	Total Project Costs		154	195	285				
15	0000	**Parking Garage**	S.F.							15
	0100	Architectural		31	38	39.50	82.7%	82.1%	82.6%	
	0200	Plumbing		1.02	1.07	2	2.7%	2.7%	2.3%	
	0300	Mechanical		.79	1.22	4.62	2.1%	2.1%	2.7%	
	0400	Electrical		2.72	2.98	6.25	7.3%	7.1%	6.5%	
	0500	Total Project Costs		37.50	46	49.50				
16	0000	**Parking Garage/Mixed Use**	S.F.							16
	0100	Architectural		100	110	112	61%	61.2%	64.3%	
	0200	Plumbing		3.22	4.22	6.45	2%	2%	2.5%	
	0300	Mechanical		13.80	15.50	22.50	8.4%	8.4%	9.1%	
	0400	Electrical		14.45	21	21.50	8.8%	8.8%	12.3%	
	0500	Total Project Costs		164	171	177				
17	0000	**Police Stations**	S.F.							17
	0100	Architectural		113	127	160	53.3%	54%	48.5%	
	0200	Plumbing		15	18	18.10	7.1%	7%	6.9%	
	0300	Mechanical		34	47.50	49	16%	16.1%	18.1%	
	0400	Electrical		25.50	28	29.50	12%	12.1%	10.7%	
	0500	Total Project Costs		212	262	297				
18	0000	**Police/Fire**	S.F.							18
	0100	Architectural		110	110	340	67.9%	68.2%	65.9%	
	0200	Plumbing		8.65	9.15	34	5.3%	5.5%	5.5%	
	0300	Mechanical		13.55	21.50	77.50	8.4%	8.4%	12.9%	
	0400	Electrical		15.40	19.70	88.50	9.5%	9.6%	11.8%	
	0500	Total Project Costs		162	167	610				
19	0000	**Public Assembly Buildings**	S.F.							19
	0100	Architectural		115	156	218	62.5%	63%	61.7%	
	0200	Plumbing		5.95	8.75	12.90	3.2%	3%	3.5%	
	0300	Mechanical		13.60	22.50	34.50	7.4%	8%	8.9%	
	0400	Electrical		18.60	25.50	40.50	10.1%	10.5%	10.1%	
	0500	Total Project Costs		184	253	360				
20	0000	**Recreational**	S.F.							20
	0100	Architectural		108	170	231	56.3%	55.7%	59.2%	
	0200	Plumbing		8.35	15.35	24.50	4.3%	4.6%	5.3%	
	0300	Mechanical		12.90	19.60	31	6.7%	6.9%	6.8%	
	0400	Electrical		15.80	28	39	8.2%	7.7%	9.8%	
	0500	Total Project Costs		192	287	435				

50 17 00 | Project Costs

			UNIT	UNIT COSTS			% OF TOTAL			
				1/4	MEDIAN	3/4	1/4	MEDIAN	3/4	
21	0000	**Restaurants**	S.F.							21
	0100	Architectural		123	198	245	60.6%	77.9%	59.1%	
	0200	Plumbing		9.95	31	39	4.9%	14.6%	9.3%	
	0300	Mechanical		14.55	19.30	47	7.2%	11.2%	5.8%	
	0400	Electrical		14.45	30.50	51	7.1%	17.9%	9.1%	
	0500	Total Project Costs		203	335	420				
22	0000	**Retail**	S.F.							22
	0100	Architectural		54	60.50	109	73%	59.9%	64.4%	
	0200	Plumbing		5.65	7.85	9.95	7.6%	6.2%	8.4%	
	0300	Mechanical		5.15	7.40	9.05	7%	5.6%	7.9%	
	0400	Electrical		7.20	11.55	18.50	9.7%	7.9%	12.3%	
	0500	Total Project Costs		74	94	148				
23	0000	**Schools**	S.F.							23
	0100	Architectural		94	120	160	58%	58.8%	55.6%	
	0200	Plumbing		7.50	10.40	15.15	4.6%	4.7%	4.8%	
	0300	Mechanical		17.85	24.50	36.50	11%	11.1%	11.3%	
	0400	Electrical		17.45	24	30.50	10.8%	11%	11.1%	
	0500	Total Project Costs		162	216	286				
24	0000	**University, College & Private School Classroom & Admin Buildings**	S.F.							24
	0100	Architectural		121	150	188	60.2%	61%	54%	
	0200	Plumbing		6.90	10.70	15.10	3.4%	3.4%	3.8%	
	0300	Mechanical		26	37.50	45	12.9%	12.9%	13.5%	
	0400	Electrical		19.50	27.50	33.50	9.7%	9.8%	9.9%	
	0500	Total Project Costs		201	278	370				
25	0000	**University, College & Private School Dormitories**	S.F.							25
	0100	Architectural		79	139	147	67.5%	67.1%	62.6%	
	0200	Plumbing		10.45	14.80	22	8.9%	8.9%	6.7%	
	0300	Mechanical		4.69	19.95	31.50	4%	4%	9%	
	0400	Electrical		5.55	19.35	29.50	4.7%	4.8%	8.7%	
	0500	Total Project Costs		117	222	263				
26	0000	**University, College & Private School Science, Eng. & Lab Buildings**	S.F.							26
	0100	Architectural		136	144	188	48.7%	49.1%	50.5%	
	0200	Plumbing		9.35	14.20	26	3.4%	3.4%	5%	
	0300	Mechanical		42.50	67	68.50	15.2%	15.3%	23.5%	
	0400	Electrical		27.50	32	37.50	9.9%	9.8%	11.2%	
	0500	Total Project Costs		279	285	320				
27	0000	**University, College & Private School Student Union Buildings**	S.F.							27
	0100	Architectural		108	283	283	50.9%	60%	54.4%	
	0200	Plumbing		16.25	16.25	24	7.7%	4.3%	3.1%	
	0300	Mechanical		31	50	50	14.6%	9.7%	9.6%	
	0400	Electrical		27	47	47	12.7%	13.3%	9%	
	0500	Total Project Costs		212	520	520				
28	0000	**Warehouses**	S.F.							28
	0100	Architectural		46	72	171	66.7%	67.4%	58.5%	
	0200	Plumbing		2.40	5.15	9.90	3.5%	3.5%	4.2%	
	0300	Mechanical		2.84	16.20	25.50	4.1%	4.1%	13.2%	
	0400	Electrical	S.F.	5.15	19.40	32.50	7.5%	7.5%	15.8%	
	0500	Total Project Costs	S.F.	69	123	238				

Square Foot Project Size Modifier

One factor that affects the S.F. cost of a particular building is the size. In general, for buildings built to the same specifications in the same locality, the larger building will have the lower S.F. cost. This is due mainly to the decreasing contribution of the exterior walls plus the economy of scale usually achievable in larger buildings. The Area Conversion Scale shown below will give a factor to convert costs for the typical size building to an adjusted cost for the particular project.

Example: Determine the cost per S.F. for a 152,600 S.F. Multi-family housing.

$$\frac{\text{Proposed building area} = 152{,}600 \text{ S.F.}}{\text{Typical size from below} = 76{,}300 \text{ S.F.}} = 2.00$$

Enter Area Conversion scale at 2.0, intersect curve, read horizontally the appropriate cost multiplier of .94. Size adjusted cost becomes .94 x $194.00 = $182.36 based on national average costs.

Note: For Size Factors less than .50, the Cost Multiplier is 1.1
For Size Factors greater than 3.5, the Cost Multiplier is .90

The Square Foot Base Size lists the median costs, most typical project size in our accumulated data, and the range in size of the projects.

The Size Factor for your project is determined by dividing your project area in S.F. by the typical project size for the particular Building Type. With this factor, enter the Area Conversion Scale at the appropriate Size Factor and determine the appropriate cost multiplier for your building size.

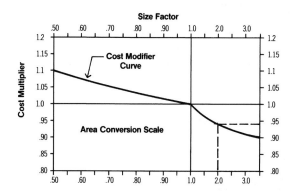

System	Median Cost (Total Project Costs)	Typical Size Gross S.F. (Median of Projects)	Typical Range (Low – High) (Projects)
Auto Sales with Repair	$182.00	24,900	4,700 – 29,300
Banking Institutions	293.00	9,300	3,300 – 38,100
Detention Centers	310.00	37,800	12,300 – 183,300
Fire Stations	231.00	12,300	6,300 – 29,600
Hospitals	365.00	87,100	22,400 – 410,300
Industrial Buildings	$102.00	22,100	5,100 – 200,600
Medical Clinics & Offices	213.00	22,500	2,300 – 327,000
Mixed Use	212.00	27,200	7,200 – 109,800
Multi-Family Housing	221.00	54,700	2,500 – 1,161,500
Nursing Home & Assisted Living	156.00	38,200	1,500 – 242,600
Office Buildings	195.00	20,600	1,100 – 930,000
Parking Garage	46.00	151,800	99,900 – 287,000
Parking Garage/Mixed Use	171.00	254,200	5,300 – 318,000
Police Stations	262.00	28,500	15,400 – 88,600
Public Assembly Buildings	253.00	22,600	2,200 – 235,300
Recreational	287.00	19,900	1,000 – 223,800
Restaurants	335.00	6,100	5,500 – 42,000
Retail	94.00	28,700	5,200 – 84,300
Schools	216.00	73,500	1,300 – 410,800
University, College & Private School Classroom & Admin Buildings	278.00	48,300	9,400 – 196,200
University, College & Private School Dormitories	222.00	28,900	1,500 – 126,900
University, College & Private School Science, Eng. & Lab Buildings	285.00	73,400	25,700 – 117,600
Warehouses	123.00	10,400	600 – 303,800

Abbreviation	Meaning
A	Area Square Feet; Ampere
AAFES	Army and Air Force Exchange Service
ABS	Acrylonitrile Butadiene Stryrene; Asbestos Bonded Steel
A.C., AC	Alternating Current; Air-Conditioning; Asbestos Cement; Plywood Grade A & C
ACI	American Concrete Institute
ACR	Air Conditioning Refrigeration
ADA	Americans with Disabilities Act
AD	Plywood, Grade A & D
Addit.	Additional
Adh.	Adhesive
Adj.	Adjustable
af	Audio-frequency
AFFF	Aqueous Film Forming Foam
AFUE	Annual Fuel Utilization Efficiency
AGA	American Gas Association
Agg.	Aggregate
A.H., Ah	Ampere Hours
A hr.	Ampere-hour
A.H.U., AHU	Air Handling Unit
A.I.A.	American Institute of Architects
AIC	Ampere Interrupting Capacity
Allow.	Allowance
alt., alt	Alternate
Alum.	Aluminum
a.m.	Ante Meridiem
Amp.	Ampere
Anod.	Anodized
ANSI	American National Standards Institute
APA	American Plywood Association
Approx.	Approximate
Apt.	Apartment
Asb.	Asbestos
A.S.B.C.	American Standard Building Code
Asbe.	Asbestos Worker
ASCE	American Society of Civil Engineers
A.S.H.R.A.E.	American Society of Heating, Refrig. & AC Engineers
ASME	American Society of Mechanical Engineers
ASTM	American Society for Testing and Materials
Attchmt.	Attachment
Avg., Ave.	Average
AWG	American Wire Gauge
AWWA	American Water Works Assoc.
Bbl.	Barrel
B&B, BB	Grade B and Better; Balled & Burlapped
B&S.	Bell and Spigot
B.&W.	Black and White
b.c.c.	Body-centered Cubic
B.C.Y.	Bank Cubic Yards
BE	Bevel End
B.F.	Board Feet
Bg. cem.	Bag of Cement
BHP	Boiler Horsepower; Brake Horsepower
B.I.	Black Iron
bidir.	bidirectional
Bit., Bitum.	Bituminous
Bit., Conc.	Bituminous Concrete
Bk.	Backed
Bkrs.	Breakers
Bldg., bldg	Building
Blk.	Block
Bm.	Beam
Boil.	Boilermaker
bpm	Blows per Minute
BR	Bedroom
Brg., brng.	Bearing
Brhe.	Bricklayer Helper
Bric.	Bricklayer
Brk., brk	Brick
brkt	Bracket
Brs.	Brass
Brz.	Bronze
Bsn.	Basin
Btr.	Better
BTU	British Thermal Unit
BTUH	BTU per Hour
Bu.	Bushels
BUR	Built-up Roofing
BX	Interlocked Armored Cable
°C	Degree Centigrade
c	Conductivity, Copper Sweat
C	Hundred; Centigrade
C/C	Center to Center, Cedar on Cedar
C-C	Center to Center
Cab	Cabinet
Cair.	Air Tool Laborer
Cal.	Caliper
Calc	Calculated
Cap.	Capacity
Carp.	Carpenter
C.B.	Circuit Breaker
C.C.A.	Chromate Copper Arsenate
C.C.F.	Hundred Cubic Feet
cd	Candela
cd/sf	Candela per Square Foot
CD	Grade of Plywood Face & Back
CDX	Plywood, Grade C & D, exterior glue
Cefi.	Cement Finisher
Cem.	Cement
CF	Hundred Feet
C.F.	Cubic Feet
CFM	Cubic Feet per Minute
CFRP	Carbon Fiber Reinforced Plastic
c.g.	Center of Gravity
CHW	Chilled Water; Commercial Hot Water
C.I., CI	Cast Iron
C.I.P., CIP	Cast in Place
Circ.	Circuit
C.L.	Carload Lot
CL	Chain Link
Clab.	Common Laborer
Clam	Common Maintenance Laborer
C.L.F.	Hundred Linear Feet
CLF	Current Limiting Fuse
CLP	Cross Linked Polyethylene
cm	Centimeter
CMP	Corr. Metal Pipe
CMU	Concrete Masonry Unit
CN	Change Notice
Col.	Column
CO2	Carbon Dioxide
Comb.	Combination
comm.	Commercial, Communication
Compr.	Compressor
Conc.	Concrete
Cont., cont	Continuous; Continued, Container
Corkbd.	Cork Board
Corr.	Corrugated
Cos	Cosine
Cot	Cotangent
Cov.	Cover
C/P	Cedar on Paneling
CPA	Control Point Adjustment
Cplg.	Coupling
CPM	Critical Path Method
CPVC	Chlorinated Polyvinyl Chloride
C.Pr.	Hundred Pair
CRC	Cold Rolled Channel
Creos.	Creosote
Crpt.	Carpet & Linoleum Layer
CRT	Cathode-ray Tube
CS	Carbon Steel, Constant Shear Bar Joist
Csc	Cosecant
C.S.F.	Hundred Square Feet
CSI	Construction Specifications Institute
CT	Current Transformer
CTS	Copper Tube Size
Cu	Copper, Cubic
Cu. Ft.	Cubic Foot
cw	Continuous Wave
C.W.	Cool White; Cold Water
Cwt.	100 Pounds
C.W.X.	Cool White Deluxe
C.Y.	Cubic Yard (27 cubic feet)
C.Y./Hr.	Cubic Yard per Hour
Cyl.	Cylinder
d	Penny (nail size)
D	Deep; Depth; Discharge
Dis., Disch.	Discharge
Db	Decibel
Dbl.	Double
DC	Direct Current
DDC	Direct Digital Control
Demob.	Demobilization
d.f.t.	Dry Film Thickness
d.f.u.	Drainage Fixture Units
D.H.	Double Hung
DHW	Domestic Hot Water
DI	Ductile Iron
Diag.	Diagonal
Diam., Dia	Diameter
Distrib.	Distribution
Div.	Division
Dk.	Deck
D.L.	Dead Load; Diesel
DLH	Deep Long Span Bar Joist
dlx	Deluxe
Do.	Ditto
DOP	Dioctyl Phthalate Penetration Test (Air Filters)
Dp., dp	Depth
D.P.S.T.	Double Pole, Single Throw
Dr.	Drive
DR	Dimension Ratio
Drink.	Drinking
D.S.	Double Strength
D.S.A.	Double Strength A Grade
D.S.B.	Double Strength B Grade
Dty.	Duty
DWV	Drain Waste Vent
DX	Deluxe White, Direct Expansion
dyn	Dyne
e	Eccentricity
E	Equipment Only; East; Emissivity
Ea.	Each
EB	Encased Burial
Econ.	Economy
E.C.Y	Embankment Cubic Yards
EDP	Electronic Data Processing
EIFS	Exterior Insulation Finish System
E.D.R.	Equiv. Direct Radiation
Eq.	Equation
EL	Elevation
Elec.	Electrician; Electrical
Elev.	Elevator; Elevating
EMT	Electrical Metallic Conduit; Thin Wall Conduit
Eng.	Engine, Engineered
EPDM	Ethylene Propylene Diene Monomer
EPS	Expanded Polystyrene
Eqhv.	Equip. Oper., Heavy
Eqlt.	Equip. Oper., Light
Eqmd.	Equip. Oper., Medium
Eqmm.	Equip. Oper., Master Mechanic
Eqol.	Equip. Oper., Oilers
Equip.	Equipment
ERW	Electric Resistance Welded

E.S.	Energy Saver	H	High Henry	Lath.	Lather
Est.	Estimated	HC	High Capacity	Lav.	Lavatory
esu	Electrostatic Units	H.D., HD	Heavy Duty; High Density	lb.; #	Pound
E.W.	Each Way	H.D.O.	High Density Overlaid	L.B., LB	Load Bearing; L Conduit Body
EWT	Entering Water Temperature	HDPE	High Density Polyethylene Plastic	L. & E.	Labor & Equipment
Excav.	Excavation	Hdr.	Header	lb./hr.	Pounds per Hour
excl	Excluding	Hdwe.	Hardware	lb./L.F.	Pounds per Linear Foot
Exp., exp	Expansion, Exposure	H.I.D., HID	High Intensity Discharge	lbf/sq.in.	Pound-force per Square Inch
Ext., ext	Exterior; Extension	Help.	Helper Average	L.C.L.	Less than Carload Lot
Extru.	Extrusion	HEPA	High Efficiency Particulate Air	L.C.Y.	Loose Cubic Yard
f.	Fiber Stress		Filter	Ld.	Load
F	Fahrenheit; Female; Fill	Hg	Mercury	LE	Lead Equivalent
Fab., fab	Fabricated; Fabric	HIC	High Interrupting Capacity	LED	Light Emitting Diode
FBGS	Fiberglass	HM	Hollow Metal	L.F.	Linear Foot
F.C.	Footcandles	HMWPE	High Molecular Weight	L.F. Hdr	Linear Feet of Header
f.c.c.	Face-centered Cubic		Polyethylene	L.F. Nose	Linear Foot of Stair Nosing
f'c.	Compressive Stress in Concrete;	HO	High Output	L.F. Rsr	Linear Foot of Stair Riser
	Extreme Compressive Stress	Horiz.	Horizontal	Lg.	Long; Length; Large
F.E.	Front End	H.P., HP	Horsepower; High Pressure	L & H	Light and Heat
FEP	Fluorinated Ethylene Propylene	H.P.F.	High Power Factor	LH	Long Span Bar Joist
	(Teflon)	Hr.	Hour	L.H.	Labor Hours
F.G.	Flat Grain	Hrs./Day	Hours per Day	L.L., LL	Live Load
F.H.A.	Federal Housing Administration	HSC	High Short Circuit	L.L.D.	Lamp Lumen Depreciation
Fig.	Figure	Ht.	Height	lm	Lumen
Fin.	Finished	Htg.	Heating	lm/sf	Lumen per Square Foot
FIPS	Female Iron Pipe Size	Htrs.	Heaters	lm/W	Lumen per Watt
Fixt.	Fixture	HVAC	Heating, Ventilation & Air-	LOA	Length Over All
FJP	Finger jointed and primed		Conditioning	log	Logarithm
Fl. Oz.	Fluid Ounces	Hvy.	Heavy	L-O-L	Lateralolet
Flr.	Floor	HW	Hot Water	long.	Longitude
Flrs.	Floors	Hyd.; Hydr.	Hydraulic	L.P., LP	Liquefied Petroleum; Low Pressure
FM	Frequency Modulation;	Hz	Hertz (cycles)	L.P.F.	Low Power Factor
	Factory Mutual	I.	Moment of Inertia	LR	Long Radius
Fmg.	Framing	IBC	International Building Code	L.S.	Lump Sum
FM/UL	Factory Mutual/Underwriters Labs	I.C.	Interrupting Capacity	Lt.	Light
Fdn.	Foundation	ID	Inside Diameter	Lt. Ga.	Light Gauge
FNPT	Female National Pipe Thread	I.D.	Inside Dimension; Identification	L.T.L.	Less than Truckload Lot
Fori.	Foreman, Inside	I.F.	Inside Frosted	Lt. Wt.	Lightweight
Foro.	Foreman, Outside	I.M.C.	Intermediate Metal Conduit	L.V.	Low Voltage
Fount.	Fountain	In.	Inch	M	Thousand; Material; Male;
fpm	Feet per Minute	Incan.	Incandescent		Light Wall Copper Tubing
FPT	Female Pipe Thread	Incl.	Included; Including	M²CA	Meters Squared Contact Area
Fr	Frame	Int.	Interior	m/hr.; M.H.	Man-hour
F.R.	Fire Rating	Inst.	Installation	mA	Milliampere
FRK	Foil Reinforced Kraft	Insul., insul	Insulation/Insulated	Mach.	Machine
FSK	Foil/Scrim/Kraft	I.P.	Iron Pipe	Mag. Str.	Magnetic Starter
FRP	Fiberglass Reinforced Plastic	I.P.S., IPS	Iron Pipe Size	Maint.	Maintenance
FS	Forged Steel	IPT	Iron Pipe Threaded	Marb.	Marble Setter
FSC	Cast Body; Cast Switch Box	I.W.	Indirect Waste	Mat; Mat'l.	Material
Ft., ft	Foot; Feet	J	Joule	Max.	Maximum
Ftng.	Fitting	J.I.C.	Joint Industrial Council	MBF	Thousand Board Feet
Ftg.	Footing	K	Thousand; Thousand Pounds;	MBH	Thousand BTU's per hr.
Ft lb.	Foot Pound		Heavy Wall Copper Tubing, Kelvin	MC	Metal Clad Cable
Furn.	Furniture	K.A.H.	Thousand Amp. Hours	MCC	Motor Control Center
FVNR	Full Voltage Non-Reversing	kcmil	Thousand Circular Mils	M.C.F.	Thousand Cubic Feet
FVR	Full Voltage Reversing	KD	Knock Down	MCFM	Thousand Cubic Feet per Minute
FXM	Female by Male	K.D.A.T.	Kiln Dried After Treatment	M.C.M.	Thousand Circular Mils
Fy.	Minimum Yield Stress of Steel	kg	Kilogram	MCP	Motor Circuit Protector
g	Gram	kG	Kilogauss	MD	Medium Duty
G	Gauss	kgf	Kilogram Force	MDF	Medium-density fibreboard
Ga.	Gauge	kHz	Kilohertz	M.D.O.	Medium Density Overlaid
Gal., gal.	Gallon	Kip	1000 Pounds	Med.	Medium
Galv., galv	Galvanized	KJ	Kilojoule	MF	Thousand Feet
GC/MS	Gas Chromatograph/Mass	K.L.	Effective Length Factor	M.F.B.M.	Thousand Feet Board Measure
	Spectrometer	K.L.F.	Kips per Linear Foot	Mfg.	Manufacturing
Gen.	General	Km	Kilometer	Mfrs.	Manufacturers
GFI	Ground Fault Interrupter	KO	Knock Out	mg	Milligram
GFRC	Glass Fiber Reinforced Concrete	K.S.F.	Kips per Square Foot	MGD	Million Gallons per Day
Glaz.	Glazier	K.S.I.	Kips per Square Inch	MGPH	Million Gallons per Hour
GPD	Gallons per Day	kV	Kilovolt	MH, M.H.	Manhole; Metal Halide; Man-Hour
gpf	Gallon per Flush	kVA	Kilovolt Ampere	MHz	Megahertz
GPH	Gallons per Hour	kVAR	Kilovar (Reactance)	Mi.	Mile
gpm, GPM	Gallons per Minute	KW	Kilowatt	MI	Malleable Iron; Mineral Insulated
GR	Grade	KWh	Kilowatt-hour	MIPS	Male Iron Pipe Size
Gran.	Granular	L	Labor Only; Length; Long;	mj	Mechanical Joint
Grnd.	Ground		Medium Wall Copper Tubing	m	Meter
GVW	Gross Vehicle Weight	Lab.	Labor	mm	Millimeter
GWB	Gypsum Wall Board	lat	Latitude	Mill.	Millwright
				Min., min.	Minimum, Minute

For customer support on your Electrical Costs with RSMeans data, call 800.448.8182.

639

Misc.	Miscellaneous	PCM	Phase Contrast Microscopy	SBS	Styrene Butadiere Styrene
ml	Milliliter, Mainline	PDCA	Painting and Decorating	SC	Screw Cover
M.L.F.	Thousand Linear Feet		Contractors of America	SCFM	Standard Cubic Feet per Minute
Mo.	Month	P.E., PE	Professional Engineer;	Scaf.	Scaffold
Mobil.	Mobilization		Porcelain Enamel;	Sch., Sched.	Schedule
Mog.	Mogul Base		Polyethylene; Plain End	S.C.R.	Modular Brick
MPH	Miles per Hour	P.E.C.I.	Porcelain Enamel on Cast Iron	S.D.	Sound Deadening
MPT	Male Pipe Thread	Perf.	Perforated	SDR	Standard Dimension Ratio
MRGWB	Moisture Resistant Gypsum	PEX	Cross Linked Polyethylene	S.E.	Surfaced Edge
	Wallboard	Ph.	Phase	Sel.	Select
MRT	Mile Round Trip	P.I.	Pressure Injected	SER, SEU	Service Entrance Cable
ms	Millisecond	Pile.	Pile Driver	S.F.	Square Foot
M.S.F.	Thousand Square Feet	Pkg.	Package	S.F.C.A.	Square Foot Contact Area
Mstz.	Mosaic & Terrazzo Worker	Pl.	Plate	S.F. Flr.	Square Foot of Floor
M.S.Y.	Thousand Square Yards	Plah.	Plasterer Helper	S.F.G.	Square Foot of Ground
Mtd., mtd., mtd	Mounted	Plas.	Plasterer	S.F. Hor.	Square Foot Horizontal
Mthe.	Mosaic & Terrazzo Helper	plf	Pounds Per Linear Foot	SFR	Square Feet of Radiation
Mtng.	Mounting	Pluh.	Plumber Helper	S.F. Shlf.	Square Foot of Shelf
Mult.	Multi; Multiply	Plum.	Plumber	S4S	Surface 4 Sides
MUTCD	Manual on Uniform Traffic Control	Ply.	Plywood	Shee.	Sheet Metal Worker
	Devices	p.m.	Post Meridiem	Sin.	Sine
M.V.A.	Million Volt Amperes	Pntd.	Painted	Skwk.	Skilled Worker
M.V.A.R.	Million Volt Amperes Reactance	Pord.	Painter, Ordinary	SL	Saran Lined
MV	Megavolt	pp	Pages	S.L.	Slimline
MW	Megawatt	PP, PPL	Polypropylene	Sldr.	Solder
MXM	Male by Male	P.P.M.	Parts per Million	SLH	Super Long Span Bar Joist
MYD	Thousand Yards	Pr.	Pair	S.N.	Solid Neutral
N	Natural; North	P.E.S.B.	Pre-engineered Steel Building	SO	Stranded with oil resistant inside
nA	Nanoampere	Prefab.	Prefabricated		insulation
NA	Not Available; Not Applicable	Prefin.	Prefinished	S-O-L	Socketolet
N.B.C.	National Building Code	Prop.	Propelled	sp	Standpipe
NC	Normally Closed	PSF, psf	Pounds per Square Foot	S.P.	Static Pressure; Single Pole; Self-
NEMA	National Electrical Manufacturers	PSI, psi	Pounds per Square Inch		Propelled
	Assoc.	PSIG	Pounds per Square Inch Gauge	Spri.	Sprinkler Installer
NEHB	Bolted Circuit Breaker to 600V.	PSP	Plastic Sewer Pipe	spwg	Static Pressure Water Gauge
NFPA	National Fire Protection Association	Pspr.	Painter, Spray	S.P.D.T.	Single Pole, Double Throw
NLB	Non-Load-Bearing	Psst.	Painter, Structural Steel	SPF	Spruce Pine Fir; Sprayed
NM	Non-Metallic Cable	P.T.	Potential Transformer		Polyurethane Foam
nm	Nanometer	P. & T.	Pressure & Temperature	S.P.S.T.	Single Pole, Single Throw
No.	Number	Ptd.	Painted	SPT	Standard Pipe Thread
NO	Normally Open	Ptns.	Partitions	Sq.	Square; 100 Square Feet
N.O.C.	Not Otherwise Classified	Pu	Ultimate Load	Sq. Hd.	Square Head
Nose.	Nosing	PVC	Polyvinyl Chloride	Sq. In.	Square Inch
NPT	National Pipe Thread	Pvmt.	Pavement	S.S.	Single Strength; Stainless Steel
NQOD	Combination Plug-on/Bolt on	PRV	Pressure Relief Valve	S.S.B.	Single Strength B Grade
	Circuit Breaker to 240V.	Pwr.	Power	sst, ss	Stainless Steel
N.R.C., NRC	Noise Reduction Coefficient/	Q	Quantity Heat Flow	Sswk.	Structural Steel Worker
	Nuclear Regulator Commission	Qt.	Quart	Sswl.	Structural Steel Welder
N.R.S.	Non Rising Stem	Quan., Qty.	Quantity	St.; Stl.	Steel
ns	Nanosecond	Q.C.	Quick Coupling	STC	Sound Transmission Coefficient
NTP	Notice to Proceed	r	Radius of Gyration	Std.	Standard
nW	Nanowatt	R	Resistance	Stg.	Staging
OB	Opposing Blade	R.C.P.	Reinforced Concrete Pipe	STK	Select Tight Knot
OC	On Center	Rect.	Rectangle	STP	Standard Temperature & Pressure
OD	Outside Diameter	recpt.	Receptacle	Stpi.	Steamfitter, Pipefitter
O.D.	Outside Dimension	Reg.	Regular	Str.	Strength; Starter; Straight
ODS	Overhead Distribution System	Reinf.	Reinforced	Strd.	Stranded
O.G.	Ogee	Req'd.	Required	Struct.	Structural
O.H.	Overhead	Res.	Resistant	Sty.	Story
O&P	Overhead and Profit	Resi.	Residential	Subj.	Subject
Oper.	Operator	RF	Radio Frequency	Subs.	Subcontractors
Opng.	Opening	RFID	Radio-frequency Identification	Surf.	Surface
Orna.	Ornamental	Rgh.	Rough	Sw.	Switch
OSB	Oriented Strand Board	RGS	Rigid Galvanized Steel	Swbd.	Switchboard
OS&Y	Outside Screw and Yoke	RHW	Rubber, Heat & Water Resistant;	S.Y.	Square Yard
OSHA	Occupational Safety and Health		Residential Hot Water	Syn.	Synthetic
	Act	rms	Root Mean Square	S.Y.P.	Southern Yellow Pine
Ovhd.	Overhead	Rnd.	Round	Sys.	System
OWG	Oil, Water or Gas	Rodm.	Rodman	t.	Thickness
Oz.	Ounce	Rofc.	Roofer, Composition	T	Temperature; Ton
P.	Pole; Applied Load; Projection	Rofp.	Roofer, Precast	Tan	Tangent
p.	Page	Rohe.	Roofer Helpers (Composition)	T.C.	Terra Cotta
Pape.	Paperhanger	Rots.	Roofer, Tile & Slate	T & C	Threaded and Coupled
P.A.P.R.	Powered Air Purifying Respirator	R.O.W.	Right of Way	T.D.	Temperature Difference
PAR	Parabolic Reflector	RPM	Revolutions per Minute	TDD	Telecommunications Device for
P.B., PB	Push Button	R.S.	Rapid Start		the Deaf
Pc., Pcs.	Piece, Pieces	Rsr	Riser	T.E.M.	Transmission Electron Microscopy
P.C.	Portland Cement; Power Connector	RT	Round Trip	temp	Temperature, Tempered, Temporary
P.C.F.	Pounds per Cubic Foot	S.	Suction; Single Entrance; South	TFFN	Nylon Jacketed Wire

Abbreviations

TFE	Tetrafluoroethylene (Teflon)	U.L., UL	Underwriters Laboratory	w/	With
T. & G.	Tongue & Groove;	Uld.	Unloading	W.C., WC	Water Column; Water Closet
	Tar & Gravel	Unfin.	Unfinished	W.F.	Wide Flange
Th., Thk.	Thick	UPS	Uninterruptible Power Supply	W.G.	Water Gauge
Thn.	Thin	URD	Underground Residential	Wldg.	Welding
Thrded	Threaded		Distribution	W. Mile	Wire Mile
Tilf.	Tile Layer, Floor	US	United States	W-O-L	Weldolet
Tilh.	Tile Layer, Helper	USGBC	U.S. Green Building Council	W.R.	Water Resistant
THHN	Nylon Jacketed Wire	USP	United States Primed	Wrck.	Wrecker
THW.	Insulated Strand Wire	UTMCD	Uniform Traffic Manual For Control	WSFU	Water Supply Fixture Unit
THWN	Nylon Jacketed Wire		Devices	W.S.P.	Water, Steam, Petroleum
T.L., TL	Truckload	UTP	Unshielded Twisted Pair	WT., Wt.	Weight
T.M.	Track Mounted	V	Volt	WWF	Welded Wire Fabric
Tot.	Total	VA	Volt Amperes	XFER	Transfer
T-O-L	Threadolet	VAT	Vinyl Asbestos Tile	XFMR	Transformer
tmpd	Tempered	V.C.T.	Vinyl Composition Tile	XHD	Extra Heavy Duty
TPO	Thermoplastic Polyolefin	VAV	Variable Air Volume	XHHW	Cross-Linked Polyethylene Wire
T.S.	Trigger Start	VC	Veneer Core	XLPE	Insulation
Tr.	Trade	VDC	Volts Direct Current	XLP	Cross-linked Polyethylene
Transf.	Transformer	Vent.	Ventilation	Xport	Transport
Trhv.	Truck Driver, Heavy	Vert.	Vertical	Y	Wye
Trlr	Trailer	V.F.	Vinyl Faced	yd	Yard
Trlt.	Truck Driver, Light	V.G.	Vertical Grain	yr	Year
TTY	Teletypewriter	VHF	Very High Frequency	Δ	Delta
TV	Television	VHO	Very High Output	%	Percent
T.W.	Thermoplastic Water Resistant	Vib.	Vibrating	~	Approximately
	Wire	VLF	Vertical Linear Foot	Ø	Phase; diameter
UCI	Uniform Construction Index	VOC	Volatile Organic Compound	@	At
UF	Underground Feeder	Vol.	Volume	#	Pound; Number
UGND	Underground Feeder	VRP	Vinyl Reinforced Polyester	<	Less Than
UHF	Ultra High Frequency	W	Wire; Watt; Wide; West	>	Greater Than
U.I.	United Inch			Z	Zone

For customer support on your Electrical Costs with RSMeans data, call 800.448.8182.

Index

649

For customer support on your Electrical Costs with RSMeans data, call 800.448.8182.

Index

For customer support on your Electrical Costs with RSMeans data, call 800.448.8182.

Index

For customer support on your Electrical Costs with RSMeans data, call 800.448.8182.

Index

ost Data Selection Guide

e following table provides definitive information on the content of each cost data publication. The number of lines of data provided in each unit price or emblies division, as well as the number of crews, is listed for each data set. The presence of other elements such as reference tables, square foot models, ipment rental costs, historical cost indexes, and city cost indexes, is also indicated. You can use the table to help select the RSMeans data set that has quantity and type of information you most need in your work.

Unit Cost Divisions	Building Construction	Mechanical	Electrical	Commercial Renovation	Square Foot	Site Work Landsc.	Green Building	Interior	Concrete Masonry	Open Shop	Heavy Construction	Light Commercial	Facilities Construction	Plumbing	Residential
1	584	406	427	531	0	516	200	326	467	583	521	273	1056	416	178
2	779	278	86	735	0	995	207	397	218	778	737	479	1222	285	274
3	1744	340	230	1265	0	1536	1041	354	2273	1744	1929	537	2027	316	444
4	961	21	0	921	0	726	180	615	1159	929	616	534	1176	0	448
5	1889	158	155	1093	0	852	1787	1106	729	1889	1025	979	1906	204	746
6	2453	18	18	2111	0	110	589	1528	281	2449	123	2141	2125	22	2661
7	1596	215	128	1634	0	580	763	532	523	1593	26	1329	1697	227	1049
8	2140	80	3	2733	0	255	1140	1813	105	2142	0	2328	2966	0	1552
9	2107	86	45	1931	0	309	455	2193	412	2048	15	1756	2356	54	1521
10	1089	17	10	685	0	232	32	899	136	1089	34	589	1180	237	224
11	1097	201	166	541	0	135	56	925	29	1064	0	231	1117	164	110
12	548	0	2	298	0	219	147	1551	14	515	0	273	1574	23	217
13	744	149	158	253	0	366	125	254	78	720	267	109	760	115	104
14	273	36	0	223	0	0	0	257	0	273	0	12	293	16	6
21	130	0	41	37	0	0	0	296	0	130	0	121	668	688	259
22	1165	7559	160	1226	0	1573	1063	849	20	1154	1682	875	7506	9416	719
23	1198	7001	581	940	0	157	901	789	38	1181	110	890	5240	1918	485
25	0	0	0	14	0	14	0	0	0	0	0	0	0	0	0
26	1512	491	10455	1293	0	811	644	1159	55	1438	600	1360	10236	399	636
27	94	0	447	101	0	0	0	71	0	94	39	67	388	0	56
28	143	79	223	124	0	0	28	97	0	127	0	70	209	57	41
31	1511	733	610	807	0	3266	289	7	1218	1456	3282	605	1570	660	614
32	838	49	8	905	0	4475	355	406	315	809	1891	440	1752	142	487
33	1248	1080	534	252	0	3040	38	0	239	525	3090	128	1707	2089	154
34	107	0	47	4	0	190	0	0	31	62	221	0	136	0	0
35	18	0	0	0	0	327	0	0	0	18	442	0	84	0	0
41	62	0	0	33	0	8	0	22	0	61	31	0	68	14	0
44	75	79	0	0	0	0	0	0	0	0	0	0	75	75	0
46	23	16	0	0	0	274	261	0	0	23	264	0	33	33	0
48	8	0	36	2	0	0	21	0	0	8	15	8	21	0	8
Totals	26136	19092	14584	20692	0	20952	10322	16446	8340	24902	16960	16134	51148	17570	12993

Assem Div	Building Construction	Mechanical	Electrical	Commercial Renovation	Square Foot	Site Work Landscape	Assemblies	Green Building	Interior	Concrete Masonry	Heavy Construction	Light Commercial	Facilities Construction	Plumbing	Asm Div	Residential
A		15	0	188	164	577	598	0	0	536	571	154	24	0	1	378
B		0	0	848	2554	0	5661	56	329	1976	368	2094	174	0	2	211
C		0	0	647	954	0	1334	0	1641	146	0	844	251	0	3	588
D		1057	941	712	1858	72	2538	330	824	0	0	1345	1104	1088	4	851
E		0	0	86	261	0	301	0	5	0	0	258	5	0	5	391
F		0	0	0	114	0	143	0	0	0	0	114	0	0	6	357
G		527	447	318	312	3378	792	0	0	535	1349	205	293	677	7	307
															8	760
															9	80
															10	0
															11	0
															12	0
Totals		1599	1388	2799	6217	4027	11367	386	2799	3193	2288	5014	1851	1765		3923

Reference Section	Building Construction Costs	Mechanical	Electrical	Commercial Renovation	Square Foot	Site Work Landscape	Assem.	Green Building	Interior	Concrete Masonry	Open Shop	Heavy Construction	Light Commercial	Facilities Construction	Plumbing	Resi.
Reference Tables	yes	yes	yes	yes	no	yes	yes	yes	yes	yes	yes	yes	yes	yes	yes	yes
Models					111			25					50			28
Crews	582	582	582	561		582		582	582	582	560	582	560	561	582	560
Equipment Rental Costs	yes	yes	yes	yes		yes		yes	yes	yes	yes	yes	yes	yes	yes	yes
Historical Cost Indexes	yes	yes	yes	yes	yes	yes	yes	yes	yes	yes	yes	yes	yes	yes	yes	no
City Cost Indexes	yes	yes	yes	yes	yes	yes	yes	yes	yes	yes	yes	yes	yes	yes	yes	yes

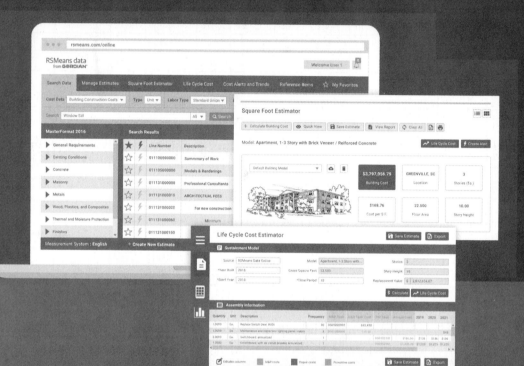

RSMeans Data Online

Cloud-based access to North America's
leading construction cost database.

Plan your next budget with
accurate prices and insight.

Estimate in great detail or
at a conceptual level.

Validate unfamiliar costs
or scopes of work.

Maintain existing buildings
with schedules and costs.

RSMeans data
from **GORDIAN**®

Learn more at rsmeans.com/online

RSMeans Data Online Core

The Core tier of RSMeans Data Online provides reliable construction cost data along with the tools necessary to quickly access costs at the material or task level. Users can create unit line estimates with RSMeans data from Gordian and share them with ease from the web-based application.

rsmeans.com/core

Key Features:

- Search
- Estimate
- Share

Data Available:

16 datasets and packages curated by project type

RSMeans Data Online Complete

The Complete tier of RSMeans Data Online is designed to provide the latest in construction costs with comprehensive tools for projects of varying scopes. Unit, assembly or square foot price data is available 24/7 from the web-based application. Harness the power of RSMeans data from Gordian with expanded features and powerful tools like the square foot model estimator and cost trends analysis.

rsmeans.com/complete

Key Features:

- Alerts
- Square Foot Estimator
- Trends

Data Available:

20 datasets and packages curated by project type

RSMeans Data Online Complete Plus

The Complete Plus tier of RSMeans Data Online was developed to take project planning and estimating to the next level with predictive cost data. Users gain access to an exclusive set of tools and features and the most comprehensive database in the industry. Leverage everything RSMeans Data Online has to offer with the all-inclusive Complete Plus tier.

rsmeans.com/completeplus

Key Features:

- Predictive Cost Data
- Life Cycle Costing
- Full History

Data Available:

Full library and renovation models

2019 Seminar Schedule

☎ 877-620-6245

Note: call for exact dates, locations, and details as some cities are subject to change.

Location	Dates	Location	Dates
Seattle, WA	January and August	San Francisco, CA	June
Dallas/Ft. Worth, TX	January	Bethesda, MD	June
Austin, TX	February	Dallas, TX	September
Jacksonville, FL	February	Raleigh, NC	October
Anchorage, AK	March and September	Baltimore, MD	November
Las Vegas, NV	March	Orlando, FL	November
Washington, DC	April and September	San Diego, CA	December
Charleston, SC	April	San Antonio, TX	December
Toronto	May		
Denver, CO	May		

Gordian also offers a suite of online RSMeans data self-paced offerings.
Check our website RSMeans.com/products/training.aspx for more information.

Facilities Construction Estimating

In this two-day course, professionals working in facilities management can get help with their daily challenges to establish budgets for all phases of a project.

Some of what you'll learn:
- Determining the full scope of a project
- Identifying the scope of risks and opportunities
- Creative solutions to estimating issues
- Organizing estimates for presentation and discussion
- Special techniques for repair/remodel and maintenance projects
- Negotiating project change orders

Who should attend: facility managers, engineers, contractors, facility tradespeople, planners, and project managers.

Mechanical & Electrical Estimating

This two-day course teaches attendees how to prepare more accurate and complete mechanical/electrical estimates, avoid the pitfalls of omission and double-counting, and understand the composition and rationale within the RSMeans mechanical/electrical database.

Some of what you'll learn:
- The unique way mechanical and electrical systems are interrelated
- M&E estimates—conceptual, planning, budgeting, and bidding stages
- Order of magnitude, square foot, assemblies, and unit price estimating
- Comparative cost analysis of equipment and design alternatives

Who should attend: architects, engineers, facilities managers, mechanical and electrical contractors, and others who need a highly reliable method for developing, understanding, and evaluating mechanical and electrical contracts.

Construction Cost Estimating: Concepts and Practice

This one or two day introductory course to improve estimating skills and effectiveness starts with the details of interpreting bid documents and ends with the summary of the estimate and bid submission.

Some of what you'll learn:
- Using the plans and specifications to create estimates
- The takeoff process—deriving all tasks with correct quantities
- Developing pricing using various sources; how subcontractor pricing fits in
- Summarizing the estimate to arrive at the final number
- Formulas for area and cubic measure, adding waste and adjusting productivity to specific projects
- Evaluating subcontractors' proposals and prices
- Adding insurance and bonds
- Understanding how labor costs are calculated
- Submitting bids and proposals

Who should attend: project managers, architects, engineers, owners' representatives, contractors, and anyone who's responsible for budgeting or estimating construction projects.

Assessing Scope of Work for Facilities Construction Estimating

This two-day practical training program addresses the vital importance of understanding the scope of projects in order to produce accurate cost estimates for facility repair and remodeling.

Some of what you'll learn:

• Discussions of site visits, plans/specs, record drawings of facilities, and site-specific lists
• Review of CSI divisions, including means, methods, materials, and the challenges of scoping each topic
• Exercises in scope identification and scope writing for accurate estimating of projects
• Hands-on exercises that require scope, take-off, and pricing

Who should attend: corporate and government estimators, planners, facility managers, and others who need to produce accurate project estimates.

Maintenance & Repair Estimating for Facilities

This two-day course teaches attendees how to plan, budget, and estimate the cost of ongoing and preventive maintenance and repair for existing buildings and grounds.

Some of what you'll learn:

• The most financially favorable maintenance, repair, and replacement scheduling and estimating
• Auditing and value engineering facilities
• Preventive planning and facilities upgrading
• Determining both in-house and contract-out service costs
• Annual, asset-protecting M&R plan

Who should attend: facility managers, maintenance supervisors, buildings and grounds superintendents, plant managers, planners, estimators, and others involved in facilities planning and budgeting.

Practical Project Management for Construction Professionals

In this two-day course, acquire the essential knowledge and develop the skills to effectively and efficiently execute the day-to-day responsibilities of the construction project manager.

Some of what you'll learn:

• General conditions of the construction contract
• Contract modifications: change orders and construction change directives
• Negotiations with subcontractors and vendors
• Effective writing: notification and communications
• Dispute resolution: claims and liens

Who should attend: architects, engineers, owners' representatives, and project managers.

Life Cycle Cost Estimating for Facility Asset Managers

Life Cycle Cost Estimating will take the attendee through choosing the correct RSMeans database to use and then correctly applying RSMeans data to their specific life cycle application. Conceptual estimating through RSMeans new building models, conceptual estimating of major existing building projects through RSMeans renovation models, pricing specific renovation elements, estimating repair, replacement and preventive maintenance costs today and forward up to 30 years will be covered.

Some of what you'll learn:
• Cost implications of managing assets
• Planning projects and initial & life cycle costs
• How to use RSMeans data online

Who should attend: facilities owners and managers and anyone involved in the financial side of the decision making process in the planning, design, procurement, and operation of facility real assets.

Please bring a laptop with ability to access the internet.

Building Systems and the Construction Process

This one-day course was written to assist novices and those outside the industry in obtaining a solid understanding of the construction process - from both a building systems and construction administration approach.

Some of what you'll learn:

• Various systems used and how components come together to create a building
• Start with foundation and end with the physical systems of the structure such as HVAC and Electrical
• Focus on the process from start of design through project closeout

This training session requires you to bring a laptop computer to class.

Who should attend: building professionals or novices to help make the crossover to the construction industry; suited for anyone responsible for providing high level oversight on construction projects.

RSMeans data Training

RSMeans data Training

Training for our Online Estimating Solution

Construction estimating is vital to the decision-making process at each state of every project. Our online solution works the way you do. It's systematic, flexible and intuitive. In this one-day class you will see how you can estimate any phase of any project faster and better.

Some of what you'll learn:
- Customizing our online estimating solution
- Making the most of RSMeans "Circle Reference" numbers
- How to integrate your cost data
- Generating reports, exporting estimates to MS Excel, sharing, collaborating and more

Also offered as a self-paced or on-site training program!

Training for our CD Estimating Solution

This one-day course helps users become more familiar with the functionality of the CD. Each menu, icon, screen, and function found in the program is explained in depth. Time is devoted to hands-on estimating exercises.

Some of what you'll learn:
- Searching the database using all navigation methods
- Exporting RSMeans data to your preferred spreadsheet format
- Viewing crews, assembly components, and much more
- Automatically regionalizing the database

This training session requires you to bring a laptop computer to class.

When you register for this course you will receive an outline for your laptop requirements.

Also offered as a self-paced or on-site training program!

Site Work Estimating with RSMeans data

This one-day program focuses directly on site work costs. Accurately scoping, quantifying, and pricing site preparation, underground utility work, and improvements to exterior site elements are often the most difficult estimating tasks on any project. Some of what you'll learn:
- Evaluation of site work and understanding site scope including: site clearing, grading, excavation, disposal and trucking of materials, backfill and compaction, underground utilities, paving, sidewalks, and seeding & planting.
- Unit price site work estimates—Correct use of RSMeans site work cost data to develop a cost estimate.
- Using and modifying assemblies—Save valuable time when estimating site work activities using custom assemblies.

Who should attend: Engineers, contractors, estimators, project managers, owner's representatives, and others who are concerned with the proper preparation and/or evaluation of site work estimates.

Please bring a laptop with ability to access the internet.

Facilities Estimating Using the CD

This two-day class combines hands-on skill-building with best estimating practices and real-life problems. You will learn key concepts, tips, pointers, and guidelines to save time and avoid cost oversights and errors.

Some of what you'll learn:
- Estimating process concepts
- Customizing and adapting RSMeans cost data
- Establishing scope of work to account for all known variables
- Budget estimating: when, why, and how
- Site visits: what to look for and what you can't afford to overlook
- How to estimate repair and remodeling variables

This training session requires you to bring a laptop computer to class.

Who should attend: facility managers, architects, engineers, contractors, facility tradespeople, planners, project managers, and anyone involved with JOC, SABRE, or IDIQ.

Registration Information

Register early to save up to $100!!!

Register 45+ days before date of a class and save $50 off each class. This savings cannot be combined with any other promotional or discounting of the regular price of classes!

How to register

By Phone
Register by phone at 877-620-6245

Online
Register online at
RSMeans.com/products/seminars.aspx

Note: Purchase Orders or Credits Cards are required to register.

Two-day seminar registration fee - $1,200.

One-Day Construction Cost Estimating or Building Systems and the Construction Process - $765.

Government pricing

All federal government employees save off the regular seminar price. Other promotional discounts cannot be combined with the government discount. Call 781-422-5115 for government pricing.

CANCELLATION POLICY:

If you are unable to attend a seminar, substitutions may be made at any time before the session starts by notifying the seminar registrar at 1-781-422-5115 or your sales representative.

If you cancel twenty-one (21) days or more prior to the seminar, there will be no penalty and your registration fees will be refunded. These cancellations must be received by the seminar registrar or your sales representative and will be confirmed to be eligible for cancellation.

If you cancel fewer than twenty-one (21) days prior to the seminar, you will forfeit the registration fee.

In the unfortunate event of an RSMeans cancellation, RSMeans will work with you to reschedule your attendance in the same seminar at a later date or will fully refund your registration fee. RSMeans cannot be responsible for any non-refundable travel expenses incurred by you or another as a result of your registration, attendance at, or cancellation of an RSMeans seminar.

ny on-demand training modules are not eligible for cancellation, substitution, nsfer, return or refund.

AACE approved courses

Many seminars described and offered here have been approved for 14 hours (1.4 recertification credits) of credit by the AACE International Certification Board toward meeting the continuing education requirements for recertification as a Certified Cost Engineer/Certified Cost Consultant.

AIA Continuing Education

We are registered with the AIA Continuing Education System (AIA/CES) and are committed to developing quality learning activities in accordance with the CES criteria. Many seminars meet the AIA/CES criteria for Quality Level 2. AIA members may receive 14 learning units (LUs) for each two-day RSMeans course.

Daily course schedule

The first day of each seminar session begins at 8:30 a.m. and ends at 4:30 p.m. The second day begins at 8:00 a.m. and ends at 4:00 p.m. Participants are urged to bring a hand-held calculator since many actual problems will be worked out in each session.

Continental breakfast

Your registration includes the cost of a continental breakfast and a morning and afternoon refreshment break. These informal segments allow you to discuss topics of mutual interest with other seminar attendees. (You are free to make your own lunch and dinner arrangements.)

Hotel/transportation arrangements

We arrange to hold a block of rooms at most host hotels. To take advantage of special group rates when making your reservation, be sure to mention that you are attending the RSMeans Institute data seminar. You are, of course, free to stay at the lodging place of your choice. (Hotel reservations and transportation arrangements should be made directly by seminar attendees.)

Important

Class sizes are limited, so please register as soon as possible.

Note: Pricing subject to change.